REVIEWS IN MINERALOGY
AND GEOCHEMISTRY

Volume 70 2009

Thermodynamics and Kinetics of Water-Rock Interaction

EDITORS

Eric H. Oelkers and Jacques Schott

CNRS/Université Paul Sabatier
Toulouse, France

ON THE COVER: Gullfoss waterfall in Iceland. This waterfall carries dissolved and suspended material from the land towards the ocean. Riverine transport coupled to water-rock interaction plays an essential role in the global cycling of the elements and long-term climate moderation (c.f. Gislason et al. 2006, *Geology* 34:49-52; Gislason et al. 2009, *Earth Planet Sci Lett* 277:213-222). Image by Pascale Bénézeth.

Series Editor: ***Jodi J. Rosso***

MINERALOGICAL SOCIETY OF AMERICA
GEOCHEMICAL SOCIETY

SHORT COURSE SERIES DEDICATION

Dr. William C. Luth has had a long and distinguished career in research, education and in the government. He was a leader in experimental petrology and in training graduate students at Stanford University. His efforts at Sandia National Laboratory and at the Department of Energy's headquarters resulted in the initiation and long-term support of many of the cutting edge research projects whose results form the foundations of these short courses. Bill's broad interest in understanding fundamental geochemical processes and their applications to national problems is a continuous thread through both his university and government career. He retired in 1996, but his efforts to foster excellent basic research, and to promote the development of advanced analytical capabilities gave a unique focus to the basic research portfolio in Geosciences at the Department of Energy. He has been, and continues to be, a friend and mentor to many of us. It is appropriate to celebrate his career in education and government service with this series of courses.

Reviews in Mineralogy and Geochemistry, Volume 70

Thermodynamics and Kinetics of Water-Rock Interaction

ISSN 1529-6466
ISBN 978-0-939950-84-3

The MINERALOGICAL SOCIETY of AMERICA
3635 Concorde Parkway, Suite 500
Chantilly, Virginia, 20151-1125, U.S.A.
www.minsocam.org

Thermodynamics and Kinetics of Water-Rock Interaction

70 *Reviews in Mineralogy and Geochemistry* **70**

FROM THE SERIES EDITOR

The chapters in this volume represent an extensive review of the material presented by the invited speakers at a short course on *Thermodynamics and Kinetics of Water-Rock Interaction* held prior to the 19th annual V. M. Goldschmidt Conference in Davos, Switzerland (June 19-21, 2009). This meeting was sponsored by the Mineralogical Society of America and the Geochemical Society.

Any supplemental material and errata (if any) can be found at the MSA website *www.minsocam.org*.

Jodi J. Rosso, Series Editor
West Richland, Washington
May 2009

PREFACE

This volume stems from a convergence of a number of factors. First, there is a compelling societal need to resuscitate the field of the thermodynamics and kinetics of natural processes. This field is essential to quantify and predict the response of the Earth's surface and crust to the disequilibria caused by the various natural and anthropic inputs of energy to our planet. As such, it serves as the basis for sustainable development and assuring the quality of life on the Earth; it serves as the key to understanding the long term future of radioactive waste storage, toxic metal mobility in the environment, the fate of CO_2 injected into the subsurface as part of carbon sequestration efforts, quantifying the quality of petroleum reservoirs and generating novel methods of petroleum extraction, and the identification of new ore deposits. The recent interest in the weathering of continental surfaces and its impact on global elemental cycles and climate evolution has also brought new attention to the thermodynamics and kinetics of water-rock interactions as it has become evident that only a true mechanistic approach based on robust thermodynamic and kinetic laws and parameters can accurately model these processes. Yet, this field has, in many ways, atrophied over the past two decades. Relatively few students have pursued graduate research in this field; many of the great contributors to this field have retired or otherwise moved on. No doubt some of this atrophy was caused by economic factors. For roughly two decades from the mid-1980's to the mid-2000's the price of base metals and petroleum, when adjusted for inflation, were at lows not seen for over a generation. Some of this atrophy was also caused by past successes in this field; the development and success of computer generated thermodynamic databases, for example, giving the illusion that the work of scientists in this field was complete.

A second factor motivating the creation of this volume was that it was requested by our graduate students. We currently coordinate two European Research Networks: MIR and MIN-

1529-6466/09/0070-0000$05.00 DOI: 10.2138/rmg.2009.70.0

GRO, and participate in two others GRASP and DELTA-MIN. As part of these networks we ran summer schools on the thermodynamics and kinetics of water-rock interaction in La Palma, Spain and in Anglet, France. In total theses classes were attended by roughly 100 students. By the end of these schools, we received numerous demands from our students requesting a book to help them follow the subject, as they, like most when introduced to thermodynamics and kinetics, got rapidly lost among the equations, symbols, and conventions, and standard states. This volume is an attempt to help these and others through these formalities towards applying the many advances available in thermodynamics and kinetics towards solving academic and societal problems.

A third factor is that we felt this volume would be a great way of getting many of our friends to write up that review paper that we have been hoping they would write for years. The chapters in this volume represent our effort to do just this. We recall Dave Sherman first explaining to us how to perform first principle thermodynamics calculations at an European Research Conference in Crete, Greece during 1999. We recall that his explanations were so clear that we wished to have recorded it. Manolo Prieto gave in La Palma, Spain a lecture summarizing decades of research on the thermodynamics of solid solutions. This lecture opened up our eyes to how little we know about the chemistry of minor and trace elements, and how they can drastically alter the pathways of reactions in nature. He also made us aware of the thermodynamic formalism available for advancing our ability to quantify the behavior of these elements in complex natural systems. Another lecture we left knowing that we needed a permanent record of was that of Dmitrii Kulik on the thermodynamics of sorption in Jena, Germany. After leaving Dmitrii's talk, we felt that we finally understood the differences between the various models used to describe sorption. Yet another chapter we felt essential to see published is a summary of the latest advances in mineral precipitation kinetics. We have followed the work of Bertrand Fritz for years as he developed a new formalism for quantifying mineral nucleation and growth, and in particular practical approaches to apply this formalism to complex systems. We are very pleased we were able to convince him to contribute his chapter to this volume.

Other chapters we believed were essential to include was that of Andrew Putnis, who has gathered extensive evidence for the existence of mineral transformation reactions, a novel and widespread mechanism in nature. Through this volume we were able to get Andrew to bring all this evidence together in a single place, where we can see clearly the significance and pervasiveness of these reactions. Similarly Jichwar Ganor has, over the past two decades, gathered a variety of evidence showing how organic compounds affect both thermodynamics and kinetics. Jichwar's chapter brings all this evidence together in one place for the first time.

This volume is completed with the future of this field, the application of thermodynamics and kinetics to natural phenomena. Two of the leaders in the development and application of reactive transport modeling are Carl Steefel and Chen Zhu. Carl, who has written what may be the most advanced reactive transport modeling code currently available, together with Kate Malher has written an informative summary of recent advances in reactive transport modeling. Chen then shows how the use of these models provides insight into the relative role of dissolution and precipitation kinetics in natural processes. This volume finishes with insightful applications of reactive transport modeling together with field observations to understand chemical weathering from the centimeter to the regional scale by Susan Brantley, Art White and Yves Goddéris.

A final motivating factor in creating this volume is as a way of honoring several great scientists who contributed immeasurably to this field and to us personally: Harold Helgeson, Robert Berner, and David Crerar.

EHO first met Harold Helgeson shortly after graduating from MIT, and shortly after Hal completed his landmark and seemingly endless American Journal of Science papers on the thermodynamic properties of minerals and aqueous fluids. A photo of Hal, with his research group at this time is shown in Figure 1. Mastering thermodynamics with Helgeson was challenging. Despite having taken two years of thermodynamics and kinetics as an undergraduate it took years of reading and classes to fully understand "Helgeson thermodynamics." Helgeson thermodynamics was a different language; Hal seemed to add four or five extra subscripts to every variable that appeared on the blackboard or in any of his publications and for good measure threw in virtually all of the Greek alphabet. Yet, through these equations, symbols, plots, and computer codes, Hal succeeded in transforming geochemistry into a quantified tool to interpret the evolution and consequences of natural processes through the development of the first widely used thermodynamic database of water-rock interaction and developing the basis for reactive transport modeling.

Figure 1. Prediction Central in 1986. Included from left to right are Harold Helgeson, Eric Oelkers, Peter Lichtner (currently at Sandia Laboratory), Chinh Nguyen-Trung (currently at the Université de Nancy), Ron Rossman, Mike Salas, Everett Shock (currently at Arizona State University), Barbara Ransom (currently at the National Science Foundation) and William Murphy (currently at Chico State University).

We sorely miss Hal, who passed in 2007. Helgeson lived the classic work hard-play hard life. Among his favorite expressions was "there's no tomorrow…" and he lived by this credo, often leading his friends and students onto all night adventures in Berkeley or at conferences. Everyone who spent an evening (or an unplanned weekend) with him left changed and with a story that would often last a lifetime. "Helgeson stories" were and continue to be a part of geochemistry folklore. Though we have been on many field trips, Hal's was the only one we can recall where there was a local farmer and donkey hired to walk along with us to continuously serve us cold beer, and where there were three cruise ships positioned at the end to bring us back to civilization.

After spending several years measuring chemical and isotopic fractionation in aqueous solutions and molten salts by thermal diffusion, JS moved to New Haven in 1979 at a time when Bob Berner (see Fig. 2) was fervently characterizing the mechanisms and rates of silicate mineral dissolution, and involved in fierce debates – on the existence, the extent and the formation mechanisms of leached layers at the mineral surface. At the same time Harold

Figure 2. Betty and Bob in December 1984 on Albi Pont-Vieux across the Tarn river.

Helgeson became interested in silicate dissolution kinetics and it is only after several days of discussion on pyroxene crystal chemistry and energetics with Bob and Hal in Spring 1980 in New Haven that JS started to feel he might be able to make a plausible story from his solution and XPS data on diopside and enstatite dissolution kinetics. By working with Bob JS benefited not only from his scientific brilliance and enthusiasm but also from his and Betty's wonderful hospitality. JS spent many Friday happy hours ending in Bob and Betty's house or back yard with pizzas or barbecue; students and post-docs were regular guests at their house for French wine, food and music. Bob is a true gourmet, and as a French visitor to New Haven JS had to practice and work hard before receiving Bob's approval of his soufflé au fromage and tarte Tatin. For the last twenty years most of Bob's efforts have concentrated on modeling over geologic times the global cycles of carbon, sulfur and oxygen; we would feel particularly honored if he could find in this volume some further inspiration to continue this work in the future.

It was during the memorable field trip organized in 1985 by Hal Helgeson in the Cyclades Islands that Greg Anderson proposed that he and David Crerar spend a sabbatical year with the "36 Ponts" Geochemistry group in Toulouse. The discovery of David and Scotia by our small group and reciprocally of the Toulouse lifestyle by our visitors was a mutual enchantment. All their friends on both sides of the Atlantic ocean can testify that David and Scotia fell in love with France; after their 1986-1987 sabbatical they came back every year to Toulouse culminating, in 1989, with trip of the entire Princeton's geochemistry group. David came to Toulouse at the time he was developing with Patricia Dove hydrothermal mixed flow reactors to investigate quartz dissolution kinetics in sodium chloride solutions to 300 °C (Dove and Crerar 1990, Geochim Cosmochim Acta 54:955-970). As a result, and thanks to the efforts of Patricia Dove and Gilles Berger, the Toulouse geochemistry group became conversant in hydrothermal mixed flow reactors and used them to investigate silicate mineral dissolution kinetics. We had only the time to publish one common study with David - on the dissolution kinetics of strained calcite (Schott et al. 1989, Geochim Cosmochim Acta 53:373-382) – yet this paper perfectly reflects the dense exchange of ideas between the two groups at that time. We have missed deeply David as a friend and a scientist for the past fifteen years; JS still has great memories of traveling with him, Scotia and Aurora to Strasbourg 1987 EUG meeting along the departmental roads of Rouergue, Berry, Burgundy, Morvan, discovering together places like, Conques, Bourges, Fontenay or the Vézelay "colline inspirée" which house the soul of France (see Fig. 3).

We would like to offer a special thanks to our close friend Sigurdur Gislason, who responded immediately to our request for potential cover photos, with more than 20 examples from around Iceland. Curiously, however, the cover photo on this volume is one supplied by Pascale Bénézeth, who had some spectacular photos of water-rock interaction in Iceland on her camera. Finally, and perhaps most of all we thank Jodi Rosso, who made the Herculean effort to complete this volume in record time, making up for our delays.

Eric H. Oelkers & Jacques Schott
Toulouse, France
May, 2009

Figure 3. David, Scotia, Aurora, and Jacques in Conques in the Spring of 1987.

Thermodynamics and Kinetics of Water-Rock Interaction

70 *Reviews in Mineralogy and Geochemistry* **70**

TABLE OF CONTENTS

1 Thermodynamic Databases for Water-Rock Interaction

Eric H. Oelkers, Pascale Bénézeth and Gleb S. Pokrovski

2 Thermodynamics of Solid Solution-Aqueous Solution Systems

Manuel Prieto

3 Mineral Replacement Reactions

Andrew Putnis

4 Thermodynamic Concepts in Modeling Sorption at the Mineral-Water Interface

Dmitrii A. Kulik

5 Surface Complexation Modeling: Mineral Fluid Equilbria at the Molecular Scale

David M. Sherman

6 The Link Between Mineral Dissolution/Precipitation Kinetics and Solution Chemistry

Jacques Schott, Oleg S. Pokrovsky and Eric H. Oelkers

7 Organics in Water-Rock Interactions

Jiwchar Ganor, Itay J. Reznik and Yoav O. Rosenberg

8 Mineral Precipitation Kinetics

Bertrand Fritz and Claudine Noguera

9 Towards an Integrated Model of Weathering, Climate, and Biospheric Processes

Yves Goddéris, Caroline Roelandt, Jacques Schott, Marie-Claire Pierret and Louis M. François

10 Approaches to Modeling Weathered Regolith

Susan L. Brantley and Art F. White

11 Fluid-Rock Interaction: A Reactive Transport Approach

Carl I. Steefel and Kate Maher

12 Geochemical Modeling of Reaction Paths and Geochemical Reaction Networks

Chen Zhu

Reviews in Mineralogy & Geochemistry
Vol. 70 pp. 1-46, 2009

Thermodynamic Databases for Water-Rock Interaction

Eric H. Oelkers, Pascale Bénézeth, and Gleb S. Pokrovski

Géochimie et Biogeochime Experimentale
LMTG, CNRS-UPS-OMP UMR5563
14 Avenue Edouard Belin
31400 Toulouse, FRANCE

oelkers@lmtg.obs-mip.fr

INTRODUCTION

The creation of thermodynamic databases may be one of the greatest advances in the field of geochemistry of the past century. These databases facilitate creation of phase diagrams describing which mineral phases are stable as a function of temperature and pressure, enabling detailed interpretation of metamorphic systems (e.g., Essene 1982; Spear and Cheney 1989; Zaho et al. 2000). The versatility of these databases provide insight into the fate and consequences of subsurface storage of radioactive waste (e.g., van der Lee and De Windt 2001; Lichtner et al. 2004; Zhang et al. 2008), toxic waste (e.g., Glynn and Brown 1996; Steefel et al. 2005), and CO_2 (e.g., Knauss et al. 2005; Oelkers and Schott 2005; Oelkers and Cole 2008; Oelkers et al. 2008).

The impressive utility of thermodynamic databases has lead to their incorporation into 'user-friendly' chemical speciation, reactive path, and reactive transport computer codes including EQ3 (Wolery 1983), PHREEQC (Parkhurst and Appelo 1999), and CHESS (van der Lee et al. 2002) allowing rapid calculation of mineral solubility and solute speciation in a variety of geochemical systems. A selected list of chemical speciation codes is provided in Table 1. These codes differ is ease of use, but all accurately solve for the equilibrium assemblages of minerals and aqueous species, and mineral solubilities within the limits of their thermodynamics databases. The quality of the results of each of these codes is directly related to the quality of these databases.

Geochemical modeling codes such as those listed in Table 1 have advantages and disadvantages. On one hand these computer algorithms allow calculation or prediction of the equilibrium state and/or the evolution of geochemical systems as a function of reaction progress with the press of a few buttons. Such calculations appear to have an amazing accuracy; results of these computer codes are commonly reported to 4 or more significant digits. Therein lines the disadvantage of these computer codes as they give the appearance that the thermodynamics databases on which they are based are perfect and accurate beyond all imagination. This, however, is a misunderstanding. The raw data on which common thermodynamic databases are based are sparse; for many species or minerals few data exist, and in many other cases, no experimental data exist at all. In such cases the thermodynamic 'data' were 'created' using correlations. The intent of such 'guesses' was not malicious, but rather to provide reasonable approximations until sufficient new experimental data became available to replace these estimates. Unfortunately, the success and widespread use of chemical speciation, reactive path, and reactive transport computer codes has obscured the provisional nature of existing thermodynamic databases. We are left with limited computational models that relatively few are currently improving and updating. This current state of affairs is bewildering considering that predictions made using currently available chemical speciation, reactive path, and reactive

1529-6466/09/0070-0001$05.00 DOI: 10.2138/rmg.2009.70.1

Table 1. Selected computer codes enabling calculation of equilibrium constants, chemical speciation, and/or mineral solubility.

Code	Web address	Cost
SUPCRT92 + SLOP98 database	*http://pdukonline.co.uk/download* *http://geopig.asu.edu/supcrt_data.html*	Free
EQ3/6	*https://ipo.llnl.gov/technology/software/softwaretitles/eq36.php*	$300. US Academic/Non Commercial $500. for non-US Academic/Noncommercial
PHREEQC	*http://wwwbrr.cr.usgs.gov/projects/GWC_coupled/phreeqc/index.html*	Free
MINTEQA2	*http://www.epa.gov/ceampubl/mmedia/minteq/index.htm*	Free
Visual MIN REQ	*http://www.lwr.kth.se/english/OurSoftware/Vminteq/*	Free
MINTEQA2 for Windows	*http://www.allisongeoscience.com/MINTEQ.htm*	$495/$396 Academic
CHESS	*http://chess.ensmp.fr/*	2900 Euros
Geochemists Workbench	*http://www.geology.uiuc.edu/Hydrogeology/hydro_gwb.htm*	'GWB Essentials': $799/$699 Academic 'GWB Standard' $3499/$1749 Academic
GEM-Selektor	*http://gems.web.psi.ch/*	**free**
THERMOCALC	*http://rock.esc.cam.ac.uk/astaff/holland/thermocalc.html*	free

transport computer codes are being used to address any number of critical societal problems such as where and how to store of nuclear waste, and how to best remediate toxic waste plumes.

A reason why it is difficult for the casual geochemist or environmental scientists to understand the limitations of existing thermodynamics data is that it is extremely difficult to determine exactly what data is being used. For example, many of us use the PHREEQC computer code. This code is currently delivered with three database options: the llnl, the phreeqc, and the miteq databases. These databases are not the original database sources. For example, much of the llnl database originates from SUPRCT92 (Johnson et al. 1992), a computer enabled thermodynamics database. SUPCRT92 is however, not an original source either. It is based on the regression of a variety of thermodynamics data, extrapolations, and estimates reported by Helgeson et al. (1978), Shock and Helgeson (1988, 1989, 1990), Shock et al. (1989, 1992), Sassani and Shock (1990), Shock and Koretsky (1993), Schulte and Shock (1993), Pokrovskii and Helgeson (1995, 1997), and Sverjensky et al. (1997). Moreover, the sources of the thermodynamic data regressed and interpreted in these studies often were taken from thermodynamics data compilations such as Wagman et al. (1982), rather than the original sources. The original sources of these data are commonly omitted from these data compilations, making it difficult to find and evaluate the quality of the original data. This chapter is aimed at reviewing how thermodynamic databases in geochemistry have been developed, and highlighting where and how they can be improved in the future.

FUNDAMENTAL THERMODYNAMIC RELATIONSHIPS

The key to performing geochemical solubility and speciation calculations stems from the second law of thermodynamics: each system attempts to attain the state of lowest energy. In most systems of geologic interest, the property to be minimized is the Gibbs free energy. The Gibbs free energy (G) of any substance or phase in a system depends on a variety of factors, notably temperature, pressure, and composition. A simple expression of this relationship is given by

$$dG = VdP - SdT + \sum_{i=1}^{k} \mu_i dN_i \tag{1}$$

where V refers to the volume, P denotes the pressure, S represents the entropy, T corresponds to the temperature, μ_i designates the chemical potential of the subscripted chemical component and N_i is the concentration of this component. In many natural systems the temperature and pressure are fixed, and chemical reactions primarily change the identity and composition of their component phases. To simplify calculations of reactions and equilibria at constant temperatures and pressures, one often adopts standard states. A standard state for a phase or a component is a merely a reference state. It allows us, for example, to calculate and tabulate the Gibbs free energy of a phase or a component at one specific composition for use in thermodynamic calculations. The Gibbs free energy, and other thermodynamic properties, when in this standard state, are referred to as standard properties and will be designated in this chapter using the superscript o. Many standard states are non-physical or "hypothetical states." Nevertheless, their thermodynamic properties are well-defined, usually by an extrapolation from some limiting condition, such as zero concentration, using an ideal extrapolating function, such as ideal solution or ideal gas behavior. Typical standard states adopted in geochemistry are given below.

The standard state for gases most commonly adopted in geochemistry is the hypothetical state it would have as a pure substance obeying the ideal gas equation at 1 bar. No real gas has perfectly ideal behavior, but this definition of the standard state allows corrections for non-ideality to be made consistently for all the different gases. The advantage of choosing this standard state is that the fugacity of a gas, which itself is a dimensionless quantity, have a value close to the partial pressure of this gas in units of bars.

For dissolved solutes in solution, the standard state is the hypothetical state it would have at the standard state molality but exhibiting infinite-dilution behavior. The reason for this unusual definition is that the behavior of a solute at the limit of infinite dilution is described by equations which are very similar to the equations for ideal gases. Hence taking infinite-dilution behavior to be the standard state allows corrections for non-ideality to be made consistently for all the different solutes. The most common standard state molality adopted in the Earth Sciences is 1 mol kg^{-1}. This standard state has the advantage that the activity of an aqueous species, which is dimensionless, will have a value close to the concentration of the species in mol kg^{-1}.

The typical standard state for liquids and solids adopted in geochemistry is taken as the pure substance. This choice has the advantage that the activity of a pure mineral or liquid (e.g., water) will be one, facilitating use of the law of mass action.

Geochemical modeling codes can perform speciation/solubility calculations by using directly the standard molal Gibbs free energy ($\Delta\bar{G}^{\circ}$) of aqueous species, minerals, and gases or alternatively such calculations can be performed using equilibrium constants (K) for reactions calculated from Gibbs free energies using

$$K = e^{-\Delta\bar{G}_r^{\circ}/RT} \tag{2}$$

where R refers to the gas constant (1.9872 cal mol^{-1} K^{-1}), T again stands for temperature in K, and $\Delta\bar{G}_r^{\circ}$ denotes the standard molal Gibbs free energy of reaction computed from

$$\Delta\bar{G}_r^o = \sum_i n_{i,r} \Delta\bar{G}_i^o \tag{3}$$

where $n_{i,r}$ represents the stoichiometric reaction coefficient of the ith species in the rth reaction, which is positive for products and negative for reactants, and $\Delta\bar{G}_i^o$ designates the standard partial molal Gibbs free energy of the subscripted reactant or product.

As discussed above, a critical step to determining the stability and equilibrium state of a natural system is the development of data bases for minerals, fluids, and gases containing standard molal Gibbs free energies at the temperatures and pressures of the natural systems that we are interested in. These Gibbs free energies, however, cannot be directly measured, but are determined through a combination of observations and measurements including heat capacity and volume measurements, and mineral solubility and transformation experiments. Moreover due to experimental limitations, it is often not possible or at least difficult to measure thermodynamics properties at temperatures of interest. For example, some aqueous species, like $NaCl^0$, the NaCl neutral ion pair, is negligibly concentrated at low temperatures making them difficult to measure at 25 °C. This species, however, dominates NaCl solutions at temperatures greater than 400 °C. Another example is measurement of univariant curves among minerals. These curves define the temperature and pressure where two minerals are at equilibrium, and thus $\Delta\bar{G}_r^o$ for the reaction among these minerals equal zero on univariant curves. Such equilibrium occur at only one temperature at a given pressure, so thermodynamic information obtained from univariant curves must be extrapolated to other temperature and pressure conditions for their information to be generally useful in geochemical modeling.

The variation of standard Gibbs free energy with temperature and pressure

The variation of standard Gibbs free energy with temperature can be computed by integrating the first two terms of Equation (1) leading to (c.f. Helgeson et al. 1978; Anderson and Crerar 1993)

$$\Delta\bar{G}_{i,T,P}^o = \Delta\bar{G}_{i,T_r,P_r}^o + \bar{S}_{i,T_r,P_r}^o + \int_{T_r}^{T} \bar{C}_{P_i}^o \, dT + \int_{T_r}^{T} \bar{C}_{P_i}^o \, d\ln T + \int_{P_r}^{P} \bar{V}_i^o dP \tag{4}$$

where $\Delta\bar{G}_{i,T,P}^o$ and $\Delta\bar{G}_{i,T_r,P_r}^o$ refer to the standard molar Gibbs free energy of formation from the elements of the subscripted substance at the temperature T and the pressure P and the reference temperature and pressure T_r and P_r, respectively. S_{i,T_r,P_r}^o denotes the standard molar entropy of the subscripted substance at the reference temperature and pressure. The reference temperature and pressure in such calculations is often taken to be 25 °C and 1 bar respectively. $\bar{C}_{P_i}^o$ and $\bar{V}_i^o$ in Equation (4) stand for the standard molar isobaric heat capacity and volume, respectively of the substance of interest. As the integration of Equation (4) is path independent, the integration of the heat capacity terms can be performed at a constant pressure equal to the reference pressure, and the volume integration can be performed at elevated temperature. It follows that the calculation of standard Gibbs free energy at any temperature requires knowledge of the entropy at the reference temperature and pressure, the heat capacity as a function of temperature at the reference pressure and the volume as a function of temperature and pressure. Equations describing the temperature and pressure variation of $\bar{C}_{P_i}^o$ and $\bar{V}_i^o$ are often referred to as Equations of State (EOS). Thermodynamic database development in geochemistry are focused to a great extent on defining these EOS for use in calculating standard Gibbs free energies and equilibrium constants, as will be discussed in detail below.

THERMODYNAMIC PROPERTIES OF MINERALS

Much of the thermodynamic data for minerals we use stems from regression of univariant equilibria curves among mineral. Examples of several univariant curves are shown in Figure 1.

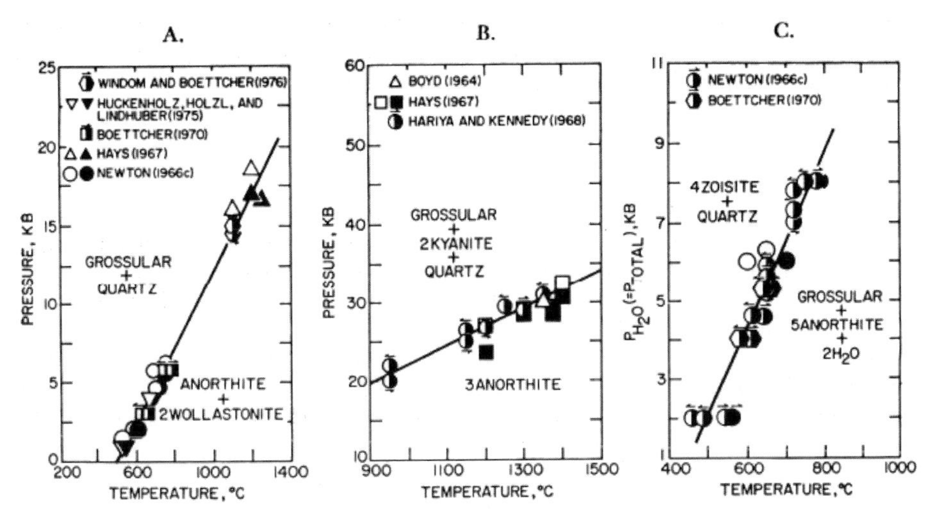

Figure 1. Univarient equilibrium curves and experimental observations of phase relations in the system CaO-Al_2O_3-SiO_2-H_2O at high temperatures and pressures [Figure used by permission of the American Journal of Science, from Helgeson et al. (1978), *American Journal of Science*, Vol. 278A, Fig. 81, p. 157.]

The reactions shown in this figure are at equilibrium along the indicated univariant curves. Thus, $\Delta\bar{G}_r^o$ for the reaction among these minerals equal zero anywhere on these curves. For example, the reaction

$$\text{Grossular} + \text{quartz} = \text{anorthite} + 2 \text{ wollastonite} \tag{5}$$

is at equilibrium at any temperature and pressure point on the univariant curve shown in Figure 1a. In accord with Equation (3) it follows that

$$\Delta\bar{G}^o_{anorthite} + 2\Delta\bar{G}^o_{wollastonite} - \Delta\bar{G}^o_{grossular} - \Delta\bar{G}^o_{quartz} = 0 \tag{6}$$

also holds at any temperature and pressure on this curve. As extensive sets of univariant curves have been measured experimentally, relations such as Equation (6) can be used for database generation so long as it's possible to accurately define Gibbs free energy variations with temperature and pressure. Special care needs to be taken to insure that such extrapolations are sufficiently accurate because the Gibbs free energy differences between minerals tend to be small compared with their absolute value. For example, the Gibbs free energy of calcite and aragonite, two forms of $CaCO_3$, are 123,456 and 124,356 cal/mol, respectively, at 25 °C and 1 bar. The difference in these Gibbs free energies is 900 cal/mol, which is just 0.7 percent of the Gibbs free energies of these minerals. Thus an uncertainty of just one percent in the Gibbs free energy of one or the other could lead to incorrect calculations of the relative stability of these $CaCO_3$ polymorphs. As such large efforts have been made to accurately measure and/or estimate the standard entropies, heat capacities and volumes of minerals required to quantify the Gibbs free energy variation with temperature and pressure as summarized below.

Mineral entropies

Entropies for minerals can be determined directly from their lattice vibration spectrums (e.g., Salje and Werneke 1982) and from calorimetry measurements (Robie et al. 1979). Such methods tend to be labor intensive, so directly measured entropies are available for only a limited number of minerals. To overcome this limitation a number of estimation schemes have been developed to estimate mineral entropies. Latimer (1951, 1952) and Fyfe et al. (1958)

summed the entropies of the component elements and of the component oxides to estimate the entropies of complex solids. Fyfe et al. (1958) also proposed that an improved estimate of entropies can be obtained from applying a volume correction in accord with

$$\bar{S}^{o}_{i,P_r,T_r} = \sum_j \nu_{j,i}\bar{S}^{o}_{j,P_r,T_r} + k\left(\bar{V}^{o}_{i,P_r T_r} + \sum_j \nu_{j,i}\bar{V}^{o}_{j,P_r,T_r}\right) \tag{7}$$

where $\nu_{i,j}$ stands for the number of moles of the jth oxide in one mole of the ith mineral. k in Equation (7) is a constant, which has a value of 0.6 for entropy in units of cal/(mol K) and volume in cm^3/mol (Fyfe et al. 1958). In most cases Equation (7) is able to estimate mineral entropies to within a few percent. Helgeson et al. (1978) improved on this estimation scheme by replacing structurally analogous mineral phases for oxides in the Fyfe et al. (1958) formalism and used this method to estimated the entropies of over 100 minerals that are generally within 0.6 cal/(mol K) of their experimental counterpart where such comparison were possible. Holland (1989) proposed a similar approach for estimating mineral entropies by first rearranging Equation (7) to obtain

$$\bar{S}^{o}_{i,P_r,T_r} = k\bar{V}^{o}_{i,P_r,T_r} + \sum_j \nu_{j,i}\left(\bar{S}^{o}_{j,P_r,T_r} - k\bar{V}^{o}_{j,P_r,T_r}\right) \tag{8}$$

The second term on the right is a constant for each oxide component. Holland (1989) then used Equation (8) to regress existing entropies for 60 minerals to obtain k and $(\bar{S}^{\circ}_{j,P_r,T_r} - k\bar{V}^{\circ}_{j,P_r,T_r})$ for 11 distinct oxide components. These oxide components were used to estimate the entropies of 40 other mineral end members. Holland (1989) estimated that the entropies for most minerals could be estimated to within ±2 cal/ (mol K) using this approach. The degree to which this level of uncertainty is acceptable depends on the application. A ±2 cal/(mol K) uncertainly in standard partial molal entropies leads to a ±200 cal/mol uncertainty in calculated Gibbs Free energy per 100 °C temperature variation, or a 1 kcal/mol uncertainty for a Gibbs free energy extrapolation of 500 °C. A 1 kcal uncertainty in the Gibbs free energy of a reaction at 25 °C translates into a ±0.73 uncertainty in log K values calculated from Equation (2).

Mineral heat capacities as a function of temperature

Heat capacity measurements are a major source of thermodynamic data. Knowledge of heat capacities as a function of temperature yield directly corresponding enthalpies and entropies. Due to their usefulness a large number of studies have been aimed at their measurement (c.f. Berman and Brown 1985). Summaries of methods to measure heat capacities are provided by McCullouch and Scott (1968) and Anderson and Crerar (1993).

Perhaps the simplest widely used equation to describe the temperature variation of isobaric heat capacities of minerals is the Maier-Kelley power function (Kelley 1960). This power function can be written:

$$\bar{C}_{P_i} = a_i + b_i T - c_i / T^2 \tag{9}$$

where a_i, b_i, and c_i stand for temperature-independent coefficients characteristic of the ith mineral. This equation was used extensively by Helgeson et al. (1978) in the creation of their mineral thermodynamic database. An example of mineral heat capacity variation with temperature is provided in Figure 2. Other similar power functions, however, have been used to describe isobaric heat capacities for minerals. For example, Haas and Fisher (1976) advocated the following five parameter fit to describe heat capacity data

$$\bar{C}_{P_i} = a_i + b_i T - c_i / T^2 + f_i T^2 + g_i / T^{1/2} \tag{10}$$

Berman and Brown (1985) concluded that the following four parameter fit provided an improved fit of existing heat capacity data:

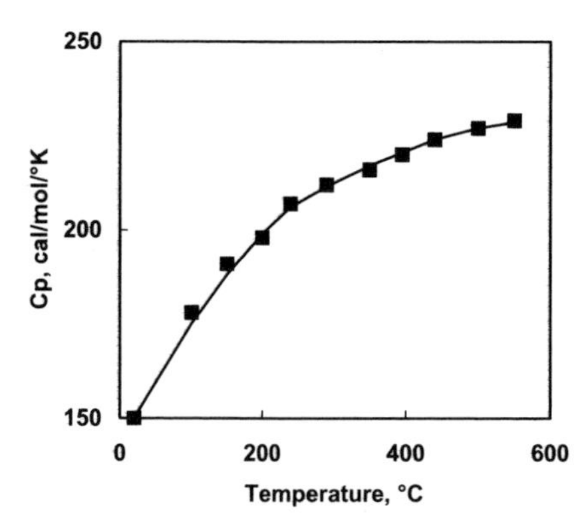

Figure 2. Molal isobaric heat capacity of tremolite. The symbols correspond to values measured by Krupka et al. (1977) whereas the curve illustrates a fit of these data in accord with the Maier-Kelley equation (Eqn. 9).

$$\bar{C}_{P_i} = k_{0,i} + k_{1,i}T^{-0.5} + k_{2,i}T^{-2} + k_{3,i}T^{-3} \qquad (11)$$

where $k_{0,i}$, $k_{1,i}$, $k_{2,i}$, and $k_{3,i}$ stand for fit coefficients. Berman and Brown (1985) used Equation (11) to regress the heat capacity data for 99 minerals of geologic interest as a function of temperature. Average absolute difference between calculated and measured heat capacities was found to be less than one percent for most minerals.

As heat capacities do not exist for all mineral of interest, several estimation approaches have been proposed. The simplest approach, analogous to that taken for mineral entropies, involve the weighted summation of the heat capacities of the mineral's constituent oxides. This summation can be performed directly with the constants on the heat capacity fit equations (Eqns. 9-11). For example, Helgeson et al. (1978) suggested that a_i, b_i, and c_i in Equation (9) can be estimated using:

$$a_i = \sum_j \nu_{i,j} a_j \qquad (12a)$$

$$b_i = \sum_j \nu_{i,j} b_j \qquad (12b)$$

and

$$c_i = \sum_j \nu_{i,j} c_j \qquad (12c)$$

where $\nu_{i,j}$ again stands for the number of moles of the jth oxide in one mole of the ith mineral. Helgeson et al. (1978) found that this method generally yielded heat capacities to within 1.2 cal/(mol K) of their experimental counterparts. These authors also found that improved heat capacity measurements could be obtained by assuming that $\Delta\bar{C}_{P_i} = 0$ for exchange reactions among structurally similar compounds. Comparison of experimentally measured and calculated heat capacities using the approximations of Helgeson et al. (1978) are shown in Figure 3. Robinson and Haas (1983) and Berman and Brown (1985) proposed similar linear summation schemes for estimating heat capacities for minerals for which data are unavailable.

Uncertainties on heat capacity estimates likely affect calculated standard Gibbs Free energies less than similar uncertainties in entropies. A 1 cal/(mol K) uncertainty in standard molal heat capacities leads to only a ~15 cal/mol uncertainty in Gibbs free energy extrapolations over the 100 °C temperature range from 125 to 25 °C, and only a ~50 cal/mol uncertainty for the 500 °C temperature extrapolation from 525 to 25 °C. This level of uncertainty may be negligible for most geochemical applications.

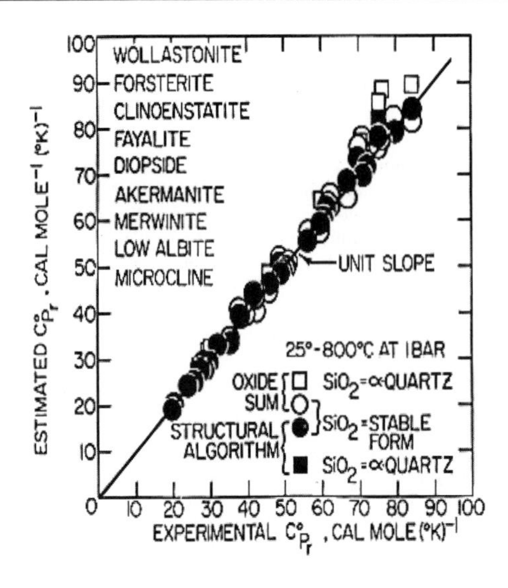

Figure 3. Comparison of estimated with corresponding standard molal heat capacities for various minerals at 25, 200, 500, and 800 °C and 1 bar. [Figure used by permission of the American Journal of Science, from Helgeson et al. (1978), *American Journal of Science*, Vol. 278A, Fig. 11, p. 63.]

Mineral volumes as a function of temperature and pressure

The simplest widely used description of mineral volumes is to assume they are independent of temperature and pressure. Helgeson et al. (1978) observed that for the most part minerals expanded less than 3% with increasing temperature from 25 to 800° C at 1 bar, and mineral compress by less than 4 percent with increasing pressure from 1 to 40 kbars. Some examples of mineral expansibilities are provided in Figure 4. Based on these observations Helgeson et al. (1978) argued that mineral expansibilities and compressibilities in the crust of the Earth are small and have an opposing effect on the standard partial molal volumes of the minerals with increasing temperature and pressure. Consequently, Helgeson et al. (1978) assumed that $\bar{V}^{\circ}_{P,T}$ for minerals can be taken to be equal to $\bar{V}^{\circ}_{P_r,T_r}$ without introducing undue uncertainty in calculated values of $(\bar{G}^{\circ}_{P,T} - \bar{G}^{\circ}_{P_r,T_r})$ at pressures 10 kb.

In contrast, Pawley et al. (1996) and Holland et al. (1996) argued that mineral expansibilities and compressibilities can have significant affect on calculated Gibbs Free energies as a function

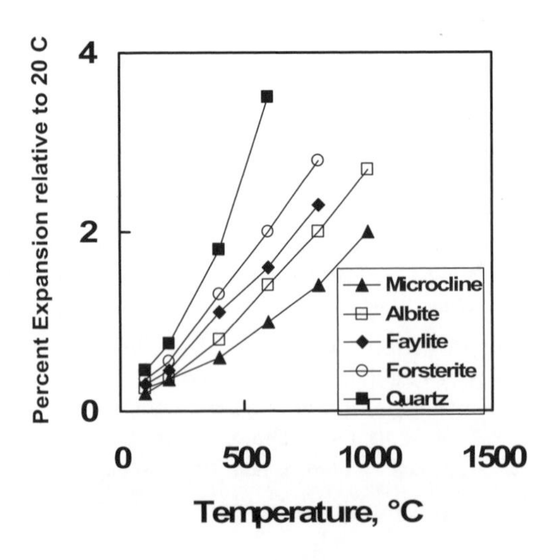

Figure 4. Percent expansion of selected minerals at 1 bar of pressure relative to their volume at 20 °C. Symbols correspond to data reported by Skinner (1996).

of temperature and pressure. Holland et al. (1996) adopted the Murnagham equation:

$$\bar{V}^{o}_{P,T} = \bar{V}^{o}_{P_r,T}\left(1+\frac{\kappa'}{\kappa}P\right)^{-\frac{1}{\kappa'}} \tag{13}$$

to describe the mineral compressibility, where κ stands for the bulk modulus of the minerals and κ′ was set to 4. Mineral expansibilities at 1 bar were calculated assuming thermal expansibility (α) is given by:

$$\alpha = \alpha_0\left(1-\frac{10}{\sqrt{T}}\right) \tag{14}$$

where α_0 refers to a mineral dependent thermal expansion parameter, which leads to the following dependence of mineral volume on temperature:

$$\bar{V}^{o}_{P,T} = \bar{V}^{o}_{P,T_r}\left(1+\alpha_0\left(T-T_r\right)-20\alpha_0\left(\sqrt{T}-\sqrt{T_r}\right)\right) \tag{15}$$

values of κ and α_0 for 154 mineral end members enabling description of the variation with temperature and pressure of mineral volumes were reported by Holland and Powell (1998) though many of these were estimated.

Mineral database compilations

Application of equations relating Gibbs free energy as a function of temperature and pressure has lead in part to the generation of "internally consistent" databases of minerals (Table 2). Among the earliest work on generating thermodynamic datasets for the rock forming are complementation of calorimetrically derived data such as Robie et al. (1978) and Robie and Hemmingway (1995).

A different approach was adopted by Helgeson et al. (1978). They argued that the calorimetric measurements performed on minerals were insufficient to afford reliable prediction of equilibrium constants at elevated temperature and pressure. Helgeson et al. (1978) concluded that the most reliable method to obtain internally consistent thermodynamic data was through

Table 2. Summary of selected mineral thermodynamic databases.

Reference	Number of phases	*P*/*T* limits	Method
Berman (1988)	67 minerals		Mathematical programming
Gottschalk (1997)	94 components		Linear Programming
Helgeson et al. (1978)	167 minerals	800 °C and 5 kBar	Manual Regression of mineral phase equilibria.
Holland and Powell (1985b)	43 phases		Linear Programming
Holland and Powell (1990)	123 minerals and fluids		Linear Programming
Holland and Powell (1998)	154 minerals, 13 silicate end-members and 22 aqueous fluid species	Up to 1000 kbar	Linear Programming
Robbie and Hemmingway (1985)		Up to 1400 °C at 1 bar	Complementation of calorimetric data

the regression of experimentally measured equilibrium pressures and temperatures for a series of equilibria among minerals and/or a co-existing fluid phase (e.g., univariant curves, see Fig. 1). To perform these regressions, Helgeson et al. (1978) assumed that mineral volumes, entropies, and heat capacities were known via methods described above. Assuming mineral volumes are independent of temperature and pressure, and heat capacities can be described using Equation (9) the integration of Equation (4) yields

$$\bar{G}^{\circ}_{P,T} - \bar{G}^{\circ}_{P_r,T_r} = -\bar{S}^{\circ}_{P_r,T_r}(T - T_r) + a\left(T - T_r - T\ln\left(\frac{T}{T_r}\right)\right) - \left(\frac{(c + bT_r^2T)(T - T_r)^2}{2T_r^2T}\right) + \bar{V}^{\circ}_{P_r,T_r}(P - P_r) \quad (16)$$

Versions of Equation (16) written for each mineral at equilibrium on a univariant temperature and pressure curve could be equated knowing that the Gibbs free energy of a reaction is zero at equilibrium and thus at any pressure and temperature on a univariant curve. Two examples of univariant curves in the system MgO-H_2O-CO_2 for the magnesite decarboxylation reaction and the brucite dehydration reaction are shown in Figure 5. The thermodynamic properties of periclase can be adopted from calorimetric data. Estimated entropies and heat capacity power function parameters and volumes for periclase, magnesite, and brucite were used together with Equations (3) and (16), and the data shown in Figure 5 to retrieve the 25 °C and 1 bar Gibbs Free energy of brucite and magnesite.

The success of the Helgeson et al. (1978) approach motivated others to improve this method using various numerical approaches to optimize the regression procedure. For example, Berman (1988) reported an internally consistent standard state thermodynamic data set for 67 minerals in the system Na_2O-K_2O-CaO-MgO-FeO-Fe_2O_3-Al_2O_3-SiO_2-TiO_2-H_2O-CO_2. Mathematical programming methods were used to achieve consistency of derived properties with phase equilibrium, calorimetric, and volumetric data, using equations that account for the thermodynamic consequences of first and second order phase transitions, and temperature-dependent disorder. Resulting properties were found to be in good agreement with the bulk of phase equilibrium data obtained in solubility studies, weight change experiments, and reversals involving both sin-

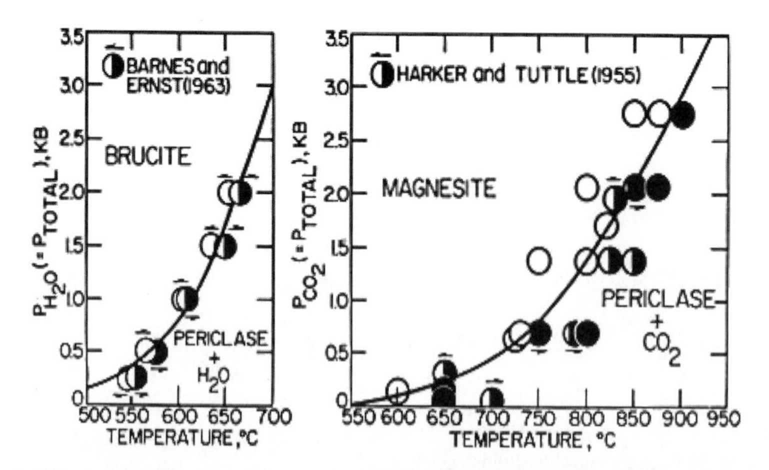

Figure 5. Univarient equilibrium curves and experimental observations of phase relations in the system MgO-CO_2-H_2O at high temperatures and pressures. [Figure used by permission of the American Journal of Science, from Helgeson et al. (1978), *American Journal of Science*, Vol. 278A, Fig. 20, p. 87.]

gle and mixed volatile species. The reliability of the thermodynamic data set was documented by these authors by comparisons between computed equilibria and phase equilibrium data.

Holland and Powell (1985a) outlined a linear programming method to simultaneously regress mineral phase relation data to generate internally consistent databases for mineral systems. This approach was applied by Holland and Powell (1985b) to general an internally consistent thermodynamic dataset for 43 phases, in the system K_2O–Na_2O–CaO–MgO–Al_2O_3–SiO_2–H_2O–CO_2.

Continuing this effort, Holland and Powell (1990) reported a revised thermodynamic dataset for 123 mineral and fluid end-members made consistent by the analysis of over 200 P–T–X_{CO_2}–f_{O_2} phase equilibrium experiments. Several improvements and advances were made to the original database. In addition to inclusion of numerous additional mineral phases these authors ranked their thermodynamic data into three groups according to reliability.

Holland and Powell (1998) further improved on their thermodynamic database through the use of a temperature-dependent thermal expansion and bulk modulus as described in Equations (13) to (15), and the use of high-pressure equations of state for solids and fluids, allowing calculation of mineral-fluid equilibria to 100 kbar or higher pressure. In addition, Holland and Powell (1998) added a simple aqueous species density model based on the work of Anderson et al. (1991) to enable simple speciation and mineral solubility calculations. The data set has also been improved over previous studies of these authors by consideration of many new phase equilibrium constraints, calorimetric studies and new measurements of molar volume, thermal expansion and compressibility. This led to improvements in the level of agreement with the available experimental phase equilibria and calorimetric data. Much of these advances were incorporated into the computer code THERMOCALC (Powell et al. 1998) allowing one to readily calculate phase relations among minerals as a function of temperature and pressure (see Table 1).

Gottschalk (1997) generated an internally consistent thermodynamic dataset by taking account of reactions among 94 components in the SiO_2-TiO_2-Al_2O_3-Fe_2O_3-CaO-MgO-Fe-O-K_2O-Na_2O-H_2O-CO_2 system. It was reported that the extracted data set reproduces 92% of the available phase equilibria experiments reported in the literature to that time.

There has tended to be a convergence in these mineral thermodynamic datasets with time as they are by and large based on the regression of the same experimental and calorimetric data. The major improvements include the mathematical data regression and uncertainty propagation methods allowing estimates of uncertainties associated with thermodynamic calculations. Uncertainties themselves are challenging to assess. Many of these mineral databases are based on measured temperature and pressures of mineral transformations. Ambiguities and uncertainties in these original data may exist and can stem from inadequate analysis of starting materials and reaction products, inaccurate data reduction or failure to demonstrate reversibility. Other uncertainties can stem from short-term experimental runs, failure to achieve stable equilibrium due to sluggish reaction kinetics and metastability and changes induced during quenching. As such even the best regression of existing experimental observations can lead to uncertainties in resulting databases.

There is little doubt that these mineral thermodynamic databases, based largely on the simultaneous regression of mineral phase equilibrium data as a function of temperature and pressure has lead to a revolution in our understanding of natural systems. These databases are unfortunately limited to those phases that form stable minerals at elevated temperature and generally major element silicate, carbonate, and oxide/hydroxide minerals. These phases are part of the system SiO_2-TiO_2-Al_2O_3-Fe_2O_3-CaO-MgO-Fe-O-K_2O-Na_2O-H_2O-CO_2. In numerous databases, these 'internally consistent' mineral thermodynamic data are supplemented with data for other minerals to improve their applicability to a larger variety of systems and geochemical problems. In many cases the additional data is of inferior quality or less consistent than those generated from the efforts described above. Some groups of minerals that are commonly added to databases, but are in critical need of additional research to improve their quality include:

- low-temperature clay minerals
- minerals of minor or trace elements
- phosphate minerals
- sulfate minerals

A brief description of these "problem systems" is presented below.

Clay minerals. Many low temperature phases such as clay minerals are not present in thermodynamics databases, or if they are present their quality is questionable. There have been numerous efforts to address these limitations and generate accurate thermodynamic data for these minerals (e.g., Giggenbach 1985; Aja 1991; Vieillard 2000). Many of the thermodynamic properties of 'low temperature' minerals such as clays are based on solubility experiments. Some of the difficulty in retrieving accurate thermodynamic properties for clay minerals is that they tend to be solid solutions whose compositions can vary dramatically with solution composition. As emphasized by Prieto (2009) it is very difficult to reverse the equilibrium of phases that form substantial solid solutions. Additional challenges arise due to difficulties in obtaining adequate crystallinity of precipitated phases, and due to the formation of secondary phases.

Minor and trace element bearing minerals. This limitation stems from two major sources: 1) lack of thermodynamic data for such phases, and 2) many minor and trace elements are incorporated into minerals of the major elements rather than in end-member minerals. As such robust solid solution models are generally required for their accurate quantification. The state of the art and current challenges of the thermodynamic treatment of trace elements is reviewed by Prieto (2009). This limitation is frustrating as recent analytical advances have enabled accurate measurement of minor and trace elements, but our ability to rigorously quantify these measurements is limited.

Phosphate minerals. Phosphate is the critical element limiting animal and plant productivity in most Earth surface environments. The solubility of phosphate minerals likely controls phosphate availability in most natural environments as well as the dissolved concentrations of Rare Earth Elements (Oelkers and Valsami-Jones 2008). Despite the need for an accurate thermodynamic database for phosphate minerals, the thermodynamic data in this system remains poorly quantified. Much of this stems from uncertainties in experimentally measured thermodynamics properties. As an example, reported values of the equilibrium constants for the apatite solubility reaction:

$$Ca_5(PO_4)_3F + 3H^+ = 5\,Ca^{2+} + 3HPO_4^{2-} + F^- \tag{17}$$

at 25 °C and 1 bar are reported in Table 3. Note that apatite is the most common phosphate mineral, present in most igneous, metamorphic, and sedimentary rocks. Equilibrium constants in Table 3 range over near 10 orders of magnitude! This example illustrates some of the challenges involved in choosing the most consistent value of an equilibrium constant for "low temperature" minerals. At least some of the variation shown among equilibrium constants listed in Table 3 stems from the difficulty in reversing the apatite solubility reaction due to the large number of metastable phases that can form (c.f. Chaïrat et al. 2007).

Sulfide minerals. Thermodynamic properties of sulfide minerals are challenging to quantify in part due to difficulties in controlling the oxygen fugacity during experiments and measuring/controlling the fugacity of sulfur. As a result the thermodynamic properties of numerous sulfide minerals are inconsistent and controversial. For example, a widely cited value of the standard molal Gibbs free energy of formation of arsenopyrite at 25 °C and 1 bar proposed by Barton (1969) is −26.2 kcal/mol, which is ~15 kcal/mol more negative than the corresponding value given by Wagman et al. (1982), who reported a value of −12.0 kcal/mol. A more recent experimentally measured value for this Gibbs free energy was reported by Pokrovski et al. (2002) of −33.8 kcal/mol! Again such experimental ambiguities make it difficult to choose the best value to use for accurate thermodynamic calculations.

Table 3. Summary of selected 25 °C equilibrium constants for the apatite dissolution reaction: $Ca_5(PO_4)_3F + 3H^+ = 5\ Ca^{2+} + 3HPO_4^{2-} + F^-$

Log *K*	Reference
−21.84	Lindsay 1979
−24.99	Robie et al. 1979
−23.10	Amjad et al. 1981
−23.09	Vieillard and Tardy 1984
−23.00	Chin and Nancollas 1991
−25.45	Elliot 1994
−21.95	Stumm and Morgan 1996
−32.95	Valsami-Jones et al. 1998
−21.13	Jaynes et al. 1998
−30.94	Stefansson 2001
−29.40	Harouiya et al. 2007

In part due to an attempt to overcome the limitations and uncertainties such as described for the four groups of minerals above, attempts have been made to develop methods to predict the Gibbs free energy of minerals (e.g., Tardy and Gartner 1977; Tardy and Vieillard 1977; Tardy and Fritz 1981; Chermak and Rimstidt 1989; Laiglesa and Felix 1994; Vieillard 2000, 2002; Gaboreau and Vieillard 2004). Sverjensky and Molling (1994) argued that such models are subject to too high uncertainties to afford accurate geochemical modeling results, and proposed an alternative predictive scheme for the Gibbs free energies of crystalline solids. Such predicted Gibbs free energies have not been in general embraced by widely used thermodynamic databases in geochemistry.

THERMODYNAMIC PROPERTIES OF AQUEOUS SPECIES

The determination of Gibbs Free energies of aqueous species poses different challenges from those of minerals. For the case of minerals, it is widely believed that the most accurate way to determine internally consistent properties of minerals is to extrapolate high temperature phase relations using estimated or measured entropies, heat capacities and volumes to low temperatures and pressures avoiding problems associated with slow kinetics and poor crystallinity. In contrast, due to rapid kinetics and experimental techniques, the properties of many aqueous species are readily obtained at ambient temperatures. Several experimental methods yield directly, through the law of mass action, the differences in the Gibbs free energy among aqueous species. Traditional methods include conductivity measurements (c.f. Quist and Marshall 1968; Oelkers and Helgeson 1988) electromotive force measurements (c.f. Ragnarsottir et al. 2001), and solubility measurements (e.g., Wesolowski 1992; Bourcier et al. 1993; Castet et al. 1993; Benezeth et al. 1994; Pokrovski et al. 1996, 2002). A review of some of these methods is provided by Nordstrom and Munoz (1994).

The point of departure for most aqueous species databases are critically assessed data compilations such as Stull and Prophet (1971), Wagman et al. (1982), Cox et al. (1989), and Chase (1998). These data compilations generally provide tabulations of standard molal Gibbs free energy, enthalpy, entropies, volumes, and heat capacities at a reference temperature and pressure, e.g., 25 °C and 1 bar. They were compiled by critical assessment of data from the literature. It is commonly difficult to track down these original sources used in these critical assessments as they may not be cited by in these compilations. It is likely that data generated and presented in different compilations for a given species rely on the same original sources although the fit obtained one the critical assessment may be different from another and therefore could lead to slight differences between the reported data.

The extent to which these data compilations are complete and therefore contain all the possible species that can be formed for a given set of components cannot be assessed without a user performing a literature review of their own. Such efforts are often essential as performing calculations with missing compounds can lead to totally misleading results. There are two reasons why the substances may be missing. The first is that there are no data for those substances and the second is that the formation of a particular species is unknown.

Like for the case of the thermodynamic properties of minerals, those for aqueous species also have internal consistence issues. Some of these issues stem from how thermodynamic properties for aqueous species are generated. In many cases regression of experimental data requires accepting the thermodynamic properties of other species present in the experimental system. For example, the Gibbs free energy of the aqueous sodium acetate neutral species, $NaCH_3COO^0$ has been determined from both measurement of pH and by the measurement of the Raman spectra in solutions containing NaCl and acetic acid (Fournier et al. 1998). Aqueous species that could be present in this solution include: CH_3COOH^0, CH_3COO^-, $NaCH_3COO^0$, Na^+, Cl^-, $NaCl^0$, OH^-, H^+ and H_2O. Retrieval of Gibbs free energies of $NaCH_3COO^0$ from either pH or spectroscopic measurements requires, therefore either knowledge of the Gibbs free energies of CH_3COOH^0, CH_3COO^-, Na^+, Cl^-, $NaCl^0$, OH^-, H^+ and H_2O. Alternatively, if one seeks only the equilibrium constant of the reaction

$$NaCH_3COO^0 = Na^+ + CH_3COO^- \tag{18}$$

one needs to accept from some different source corresponding equilibrium constants for the following reactions (c.f. Fournier et al. 1998)

$$CH_3COOH^0 = CH_3COO^- + H^+ \tag{19}$$

$$NaCl^0 = Na^+ + Cl^- \tag{20}$$

$$H_2O = OH^- + H^+ \tag{21}$$

The choice of thermodynamic data accepted for this calculation thus influences the Gibbs free energy or equilibrium constant retrieved from the interpretation of experimental data. This observation suggests that published equilibrium constants or Gibbs free energies stemming from the interpretation of experimental data are not necessarily consistent with other data available in the literature or available thermodynamic databases. The generation of internally consistent properties for aqueous species should likely follow the same integrated approach adopted for minerals, but it is unclear if such an approach has been pursued in generating the thermodynamic databases used for geochemical calculations.

To overcome missing data, creators of thermodynamic databases often estimate thermodynamic data; the use of estimated data is relatively common for substances where no data are available. For example, Shock and Koretsky (1993) proposed linear Gibbs free energy correlations to estimate the Gibbs free energies of aqueous metal acetate complexes. Haas et al. (1995) proposed linear Gibbs free energy relationships to estimate the Gibbs free energies of aqueous Rare-Earth Element – anion complexes. Gibbs free energies estimated from these linear correlations are incorporated both into current versions of the SUPCRT92 datafiles (Johnson et al. 1992; see Table 1) and the llnl database of PHREEQC (Parkhurst and Appelo 1999). Examples of some of the Gibbs free energy correlations proposed by Shock and Koretsky (1993) are shown in Figure 6. No doubt such estimates are useful, but such estimates are less certain than experimental measurements and should be considered to be provisional until experimental data are available. The quality of the estimated 'data' relies on the expertise and experience of the estimator. In some thermodynamic databases estimated data are clearly

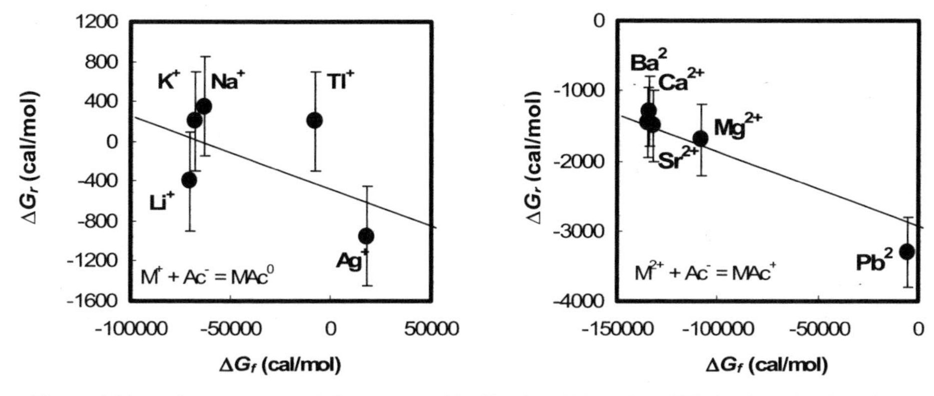

Figure 6. Linear free energy correlations proposed by Shock and Koretsky (1993) for the estimation of the standard molal Gibbs free energy of reaction for aqueous metal acetate complexes from the molal Gibbs free energy of formation of the corresponding aqueous metal. The reactions considered are provided in the respective plots, where Ac refers to aqueous acetate (CH_3COO^-).

labeled. However, as is often the case for databases supplied and used in conjunction with computer codes, the user will not be aware of whether the data are measured or estimated. This could be problematical as it is likely that estimated data have a large degree of uncertainty associated with them.

Once the thermodynamic properties are established at a reference temperature and pressure (e.g., 25 °C and 1 bar) they can be extrapolated to other temperature conditions using one of the methods described below.

Variation of the Gibbs free energy of aqueous species as a function of temperature and pressure

The Van't Hoff equation. The simplest method to extrapolate Gibbs free energies or equilibrium constants with temperature is to assume that the standard molal heat capacity for a reaction is zero. Although this assumption is false, this assumption allows for useful extrapolations in numerous cases over small temperature ranges from

$$\ln K_T = \ln K_{T_r} - \frac{\Delta \bar{H}_{r,T_r}}{R}\left(\frac{1}{T} - \frac{1}{T_r}\right) \tag{22}$$

Equation (22) implies that the logarithms of equilibrium constants are linear functions of reciprocal temperature. Numerous examples show that this assumption is extremely limited in most instances, though is somewhat more accurate for isocolumbic reactions, reactions that have the same change on both sides of the reaction. Several examples of the use of this extrapolation are provided by Anderson and Crerar (1993). Two examples of Van't Hoff plots are shown in Figure 7. It can be seen that the isocolumbic reaction in Figure 7A plots as a linear function of 1000/T, consistent with the behavior prescribed by Equation (22). In contrast, the reaction plotted in Figure 7B exhibits a non-linear behavior.

Density models. Frank (1956, 1961) noted that the ionization constant of water and many aqueous ions were linearly related to density over wide ranges of temperature and pressure. Anderson et al. (1991) suggested the variation of this relationship given by

$$\ln K = p_1 + \frac{p_2}{T} + \frac{p_3 \ln \rho}{T} \tag{23}$$

where p_1, p_2, and p_3 represent temperature and pressure independent constants and ρ stands for

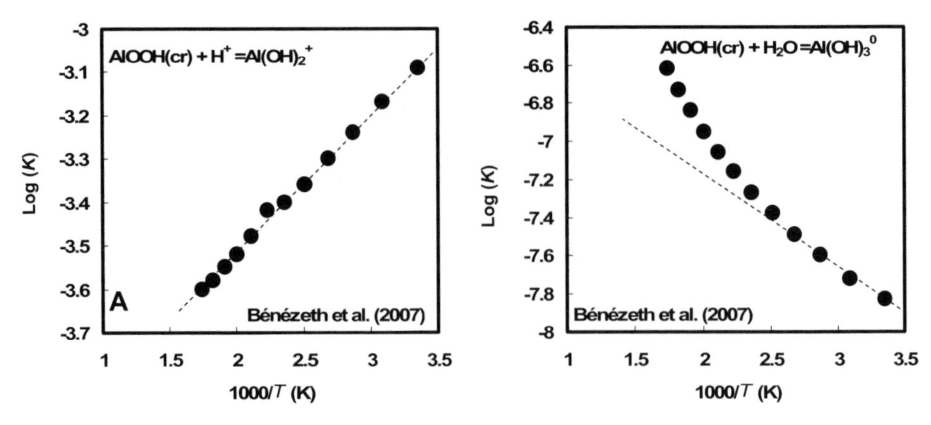

Figure 7. Examples of measured variations of log K for boehmite solubility in accord with the reactions listed in the plots as functions of reciprocal temperatures. Log K reported by Bénézeth et al. (2001). Note that the isocolumbic reaction in A plots as a linear function of $1000 \cdot T$, whereas the reaction plotted in B exhibits a non-linear behavior.

the density of H_2O. Taking into account of relations among the Gibbs free energy, enthalpy, and heat capacity, a relation consistent with Equation (23) is given by

$$\bar{G}^{o}_{P,T} - \bar{G}^{o}_{P_r,T_r} = -\bar{H}^{o}_{P_r,T_r}\left(1 - \frac{T}{T_r}\right) + \frac{\bar{C}_{P_{P_r,T_r}}}{RT_r\left(\partial\alpha/\partial T\right)_{P_r}}\left(\alpha_{P_r,T_r}\left(T - T_r\right) + \ln\frac{\rho}{\rho_r}\right) \tag{24}$$

where α refers to the thermal expansibility of H_2O. The density model thus allows for extrapolation of Gibbs free energies from the knowledge of enthalpies, heat capacities of the aqueous species of interest together with the density and thermal expansibility of H_2O. These latter values are known to fairly high accuracy to at least 800 °C and 5 kb (c.f. Helgeson and Kirkham 1974a). An example of a successful application of the density model to estimate the ionization constant of H_2O is shown in Figure 8.

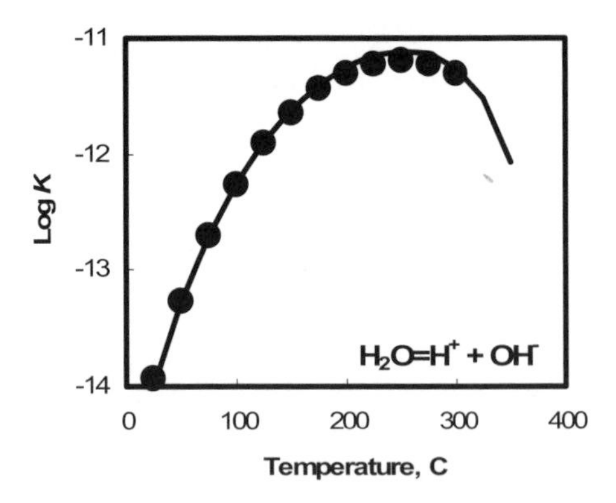

Figure 8. Comparison of the ionization constant of H_2O with corresponding values calculated using the density model. Symbols correspond to experimental data reported by Sweeton et al. (1994) – after Anderson et al. (1991).

The HKF model. The Helgeson-Kirkham-Flowers (1981) model was developed to describe the Gibbs free energy of an aqueous species at temperatures up to 1000 °C and at pressures to 5000 bars (Helgeson and Kirkham 1974a,b, 1976; Helgeson et al. 1981; Helgeson 1982, 1985). The original HKF model was subsequently revised by Tanger and Helgeson (1988) and Shock et al. (1991). This model links closely the thermodynamic properties of aqueous species with those of water. Phase diagrams of H_2O are shown in Figure 9. Liquid-vapor equilibrium is shown as a solid curve in Figure 9a. Water boils at 100 °C at 1 bar of pressure. This temperature increases systematically with increasing pressure until the critical point of H_2O at 221 bars and 374 °C. The critical point is relatively unstable, and a small addition of solute can drastically change the properties

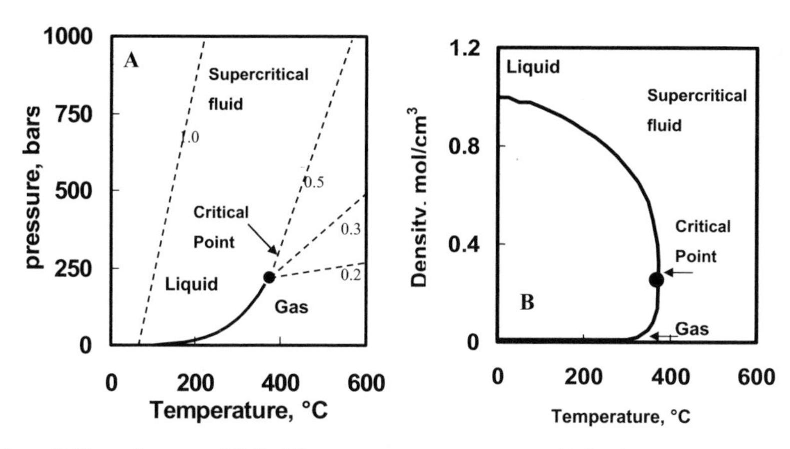

Figure 9. Phase diagrams of H_2O. (A) pressure versus temperature; A) density versus temperature.

of an aqueous solution near the critical pressure and temperature. As such partial molal heat capacities and volumes of aqueous species can vary dramatically near the critical point of H_2O. At high pressures and temperatures H_2O is a supercritical fluid, where 1) there is no phase change between liquid and vapor, and 2) this fluid has properties that vary continuously from vapor-like to liquid like with increasing temperature at constant pressure.

As was the case with minerals, the key to quantifying the thermodynamic properties of an aqueous species as a function of temperature lies in the accurate description of the corresponding volumes and heat capacities. Within the HKF models, heat capacities and volumes are assumed to stem from two distinct contributions: the solvation and non-solvation contribution such that

$$\bar{V}_i^{\mathrm{o}} = \bar{V}_{s.i}^{\mathrm{o}} + \bar{V}_{n,i}^{\mathrm{o}} \tag{25}$$

and

$$\bar{C}_{P_i}^{\mathrm{o}} = \bar{C}_{P_{s,i}}^{\mathrm{o}} + \bar{C}_{P_{n,i}}^{\mathrm{o}} \tag{26}$$

where the subscript s and n refer to the solvation and non-solvation contribution to the property. The solvation contribution to the thermodynamic properties of an aqueous species stems from the work associated from transferring a species from a vacuum to a solvent with a dielectric constant ε. Values of the dielectric constant of H_2O as a function of temperature and pressure are shown in Figure 10. Born (1920) showed that the Gibbs free energy of solvation was related to the solvent dielectric constant according to

$$\Delta\bar{G}_{s,i}^{\mathrm{o}} = \left(\frac{N_a (Z_i e)^2}{2r_{i,e}}\right)\left(\frac{1}{\varepsilon} - 1\right) \tag{27}$$

where N_a refers to Avogadro's number, e designates the charge of an electron, Z_j and r_j represent the charge and effective electrostatic radius of the subscripted species. Because

$$\Delta\bar{V}_{s,i}^{\mathrm{o}} = \frac{\partial \Delta\bar{G}_{s,i}^{\mathrm{o}}}{\partial P} \tag{28}$$

it follows that

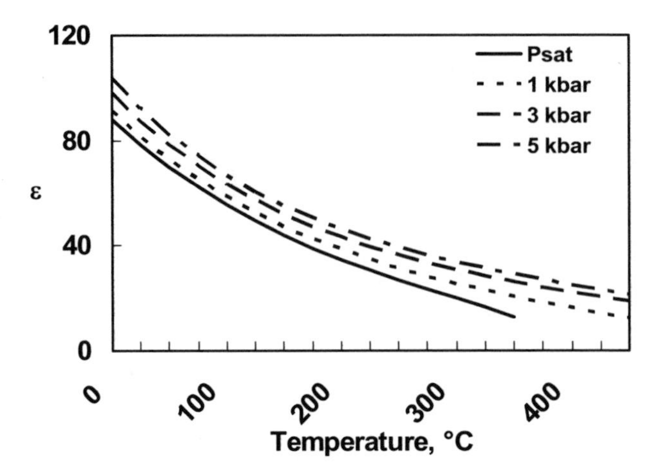

Figure 10. The dielectric constant of H_2O as a function of temperature at various pressures. The curves in this figure were calculated with the aid of equations reported by Helgeson and Kirkham (1974b).

$$\Delta \bar{V}^{o}_{s,i} = \left(\frac{N_a (Z_i e)^2}{2 r_{i,e}} \right) \frac{\partial}{\partial P} \left(\frac{1}{\varepsilon} - 1 \right) = \left(\frac{N_a (Z_i e)^2}{2 r_{i,e}} \right) Q \tag{29}$$

where Q stands for the Born function

$$Q = \frac{\partial}{\partial P} \left(\frac{1}{\varepsilon} - 1 \right) \tag{30}$$

Correspondingly for isobaric heat capacity because

$$\Delta \bar{C}^{o}_{P_{s,i}} = -T \frac{\partial^2 \Delta \bar{G}^{o}_{s,i}}{\partial P^2} \tag{31}$$

it follows that

$$\Delta \bar{C}^{o}_{P_{s,i}} = -T \left(\frac{N_a (Z_i e)^2}{2 r_{i,e}} \right) \frac{\partial^2}{\partial T^2} \left(\frac{1}{\varepsilon} - 1 \right) = T \left(\frac{N_a (Z_i e)^2}{2 r_{i,e}} \right) X \tag{32}$$

where X stands for the Born function

$$X = -\frac{\partial^2}{\partial T^2} \left(\frac{1}{\varepsilon} - 1 \right) \tag{33}$$

Comparisons between the measured partial molal volumes and heat capacities with the Born functions are shown in Figures 11 and 12.

The non-solvation contribution to the thermodynamics of an aqueous species originates from the intrinsic volume of the species and the collapse of the water structure around the solute. Empirical equations describing these contributions are

$$\bar{V}^{o}_{n,i} = a_1 + \frac{a_2}{(\Psi + P)} + \frac{a_3}{(T - \Theta)} + \frac{a_4}{(\Psi + P)(T - \Theta)} \tag{34}$$

and

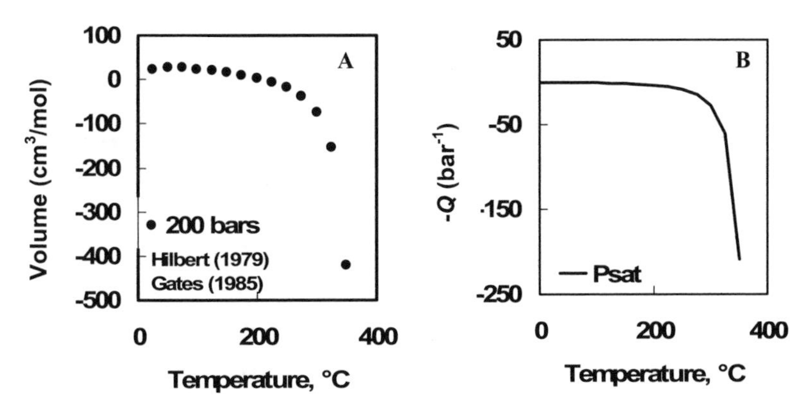

Figure 11. (A) Apparent partial molal volume of NaCl as a function of temperature at 200 bar reported by the indicated references; (B) Variation of the Q Born function as a function of temperature calculated using equations described by Tanger and Helgeson (1988). Note the similarity in form of the functions shown in A and B.

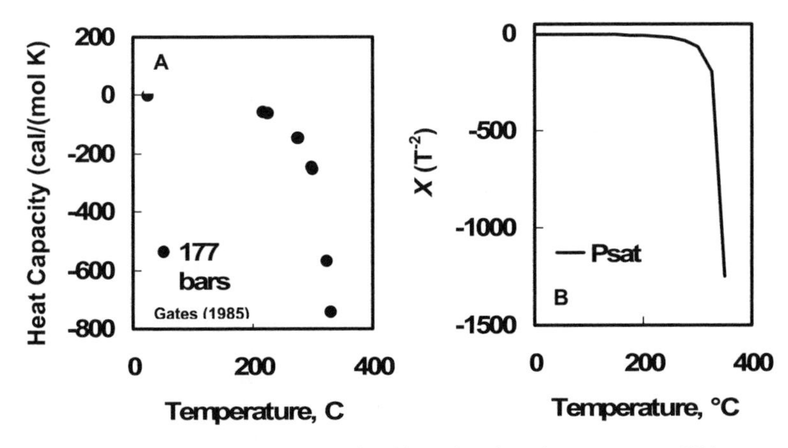

Figure 12. A) Apparent partial molal volume of NaCl as a function of temperature at 177 bar reported by the indicated references; B) Variation of the X Born function as a function of temperature calculated using equations described by Tanger and Helgeson (1988). Note the similarity in form of the functions shown in A and B.

$$\bar{C}_{P_{n,i}}^{o} = c_1 + \frac{c_2}{(T-\Theta)^2} \tag{35}$$

for volume and heat capacity, where a_1, a_2, a_3, a_4, c_1, and c_2 refer to temperature/pressure independent equation of state parameters for the species. Ψ and Θ in Equation (34) and (35) denote solvent constants equal to 2600 bars and 228 K, respectively. Combining Equations (25), (26), (29), (32), (34), and (35) yields the following equations describing the partial molal volumes and isobaric heat capacities of aqueous solutes:

$$\bar{V}_i^{o} = a_1 + \frac{a_2}{(\Psi+P)} + \frac{a_3}{(T-\Theta)} + \frac{a_4}{(\Psi+P)(T-\Theta)} + \left(\frac{N_a(Z_i e)^2}{2r_{i,e}}\right)Q \tag{36}$$

and

$$\bar{C}_{P_i}^{o} = c_1 + \frac{c_2}{(T-\Theta)^2} + T\left(\frac{N_a (Z_i e)^2}{2r_{i,e}}\right)X \tag{37}$$

Combining Equation (36) and (37) with Equation (4) yields the following expression describing the standard partial molal Gibbs free energy of an aqueous species as a function of temperature and pressure (Tanger and Helgeson 1988; Shock et al. 1992)

$$\begin{aligned}
\bar{G}_{P,T}^{o} - \bar{G}_{P_r,T_r}^{o} = & -\bar{S}_{P_r,T_r}^{o}(T-T_r) - c_1\left(T\ln\left(\frac{T}{T_r}\right) - T + T_r\right) + a_1(P-P_r) + a_2\ln\left(\frac{\Psi+P}{\Psi+P_r}\right) \\
& - c_2\left(\left(\left(\frac{1}{T-\Theta}\right) - \left(\frac{1}{T_r-\Theta}\right)\right)\left(\frac{\Theta-T}{\Theta}\right) - \left(\frac{T}{\Theta^2}\right)\ln\left(\frac{T_r(T-\Theta)}{T(T_r-\Theta)}\right)\right) \\
& + \left(\frac{1}{T-\Theta}\right)\left[a_3(P-P_r) + a_4\ln\left(\frac{\Psi+P}{\Psi+P_r}\right)\right] \\
& + \omega\left(\frac{1}{\varepsilon} - 1\right) - \omega_{P_r,T_r}\left(\frac{1}{\varepsilon_{P_r,T_r}} - 1\right) + \omega_{P_r,T_r} Y_{P_r,T_r}(T-T_r)
\end{aligned} \tag{38}$$

where $\bar{S}_{P_r,T_r}^{o}$ stands for the standard partial molal entropy of the species at the reference pressure and temperature, ε and $\varepsilon_{P_r T_r}$ designate the dielectric constant of H_2O at the temperature and pressure of interest and P_r, T_r, respectively, $Y_{P_r T_r}$ is given by

$$Y_{P_r,T_r} = \frac{1}{\varepsilon_{P_r,T_r}}\left(\left(\frac{\partial \ln\varepsilon}{\partial T}\right)_{P_r,T_r}\right) = \frac{1}{\varepsilon_{P_r,T_r}^2}\left(\left(\frac{\partial \varepsilon}{\partial T}\right)_{P_r,T_r}\right) \tag{39}$$

where ω represents the conventional Born coefficient of the species, which can be expressed as

$$\omega = \eta\left(Z^2\left(r_{e,P_r,T_r} + |Z|g\right)^{-1} - Z\left((3.082) + g\right)^{-1}\right) \tag{40}$$

where g designates a solvent function of temperature and density given by Shock et al. (1992), $\eta = 1.66027\times10^5$ Å cal mol^{-1}, Z again stands for the formal charge on the species, r_{e,P_r,T_r} refers to the effective electrostatic radius of the species at the reference pressure and temperature, which for monatomic ions is given by (Helgeson and Kirkham 1976)

$$r_{e,P_r,T_r} = r_x + k_Z|Z| \tag{41}$$

and for charged aqueous species by (Shock and Helgeson 1988)

$$r_{e,P_r,T_r} = \frac{Z^2\left(\eta Y_{P_r,T_r} - \kappa_Z\right)}{\bar{S}_{P_r,T_r}^{o} - \alpha_Z} \tag{42}$$

where r_x denotes the crystallographic radius of the ion, k_Z designates a constant equal to 0.94 Å for cations and 0.0 Å for anions, κ_Z represents a correlation parameter equal to 100 Å cal mol^{-1} K^{-1}, $\bar{S}_{P_r,T_r}^{o}$ again stands for the standard partial molal entropy of the species at the subscripted pressure and temperature, and α_Z is defined by

$$\alpha_Z \equiv 71.5\,|Z|\ \text{cal mol}^{-1}\ \text{K}^{-1} \tag{43}$$

The conventional Born correlation parameters of neutral aqueous species are taken to be independent of pressure and temperature. Hence, for these species

$$\omega = (-1514.4\ K)\,\bar{S}_{P_r,T_r}^{o} + \beta_Z \tag{44}$$

where β_Z refers to a correlation parameter equal to 0.0 cal mol^{-1} for noble or diatomic gases and 34,000 cal mol^{-1} for polyatomic and neutral aqueous species (Shock and Helgeson 1990).

This model has been demonstrated to be very versatile, being able to reproduce experimental thermodynamic data of aqueous species at temperatures from 0 to 1000 °C and pressures at or above the liquid-vapor saturation curve to 5000 bars (e.g., Sverjensky et al. 1997). This model, however, requires a large number of parameters for the successful calculation of Gibbs free energy over this pressure and temperature range. Specifically, the parameters required for this calculation include parameters describing the non-solvation volume and heat capacity: a_1, a_2, a_3, a_4, c_1, and c_2; the standard partial molal entropy of the species at the reference pressure and temperature: $\bar{S}^{\text{o}}_{P_r,T_r}$; the effective electrostatic radius of the species at the reference pressure and temperature: r_{e,P_r,T_r}; and the Gibbs free energy of an aqueous species at the reference pressure and temperature: $\bar{G}^{\text{o}}_{P_r,T_r}$. As sufficient thermodynamic data are available for the successful determination of all of these parameters for just a few aqueous species, Shock and Helgeson (1988) generated correlations to enable estimation of most of these parameters. Some of these correlations are shown in Figure 13. Care of these equations Shock and Helgeson (1988) suggested that the following linear equations could be used to obtain the parameters a_1, a_2, a_3, a_4, c_1, and c_2 from corresponding describing the non-solvation volume and heat capacity

$$\begin{aligned}
a_{1,i} &= 0.0136\bar{V}^{\text{o}}_{n,i} + 0.1765 \\
a_{2,i} &= 33.423\bar{V}^{\text{o}}_{n,i} - 347.23 \\
a_{3,i} &= 0.1435\bar{V}^{\text{o}}_{n,i} + 7.0274 \\
a_{4,i} &= -138.17\bar{V}^{\text{o}}_{n,i} - 26355 \\
c_{1,i} &= 0.6087\bar{C}p^{\text{o}}_{n,i} + 5.85 \\
c_{2,i} &= 2037\bar{C}p^{\text{o}}_{n,i} - 30460
\end{aligned} \tag{45}$$

where the parameters $a_{1,i}$, $a_{2,i}$, $a_{3,i}$, $a_{4,i}$, $c_{1,i}$, and $c_{2,i}$ are given in units of cal/(mol bar), cal/mol, (cal K)/(mol bar), (cal K)/mol, cal/(mol K), and (cal K)/mol, respectively and $\bar{V}^{\text{o}}_{n.i}$ and $\bar{C}p^{\text{o}}_{n.i}$ are in units of cm^3/mol and cal/(mol K). These empirical correlations, based on the regression of the few quantitative data available at that time, have been used to estimate equation of state parameters for several hundred other aqueous species for which experimental data are lacking. On one hand use of such empirical correlations, based on relatively few data points could enhance uncertainties in estimated thermodynamic properties. On the other hand they apparently provide useful estimates for the many cases where experimental data are lacking. Taking account of

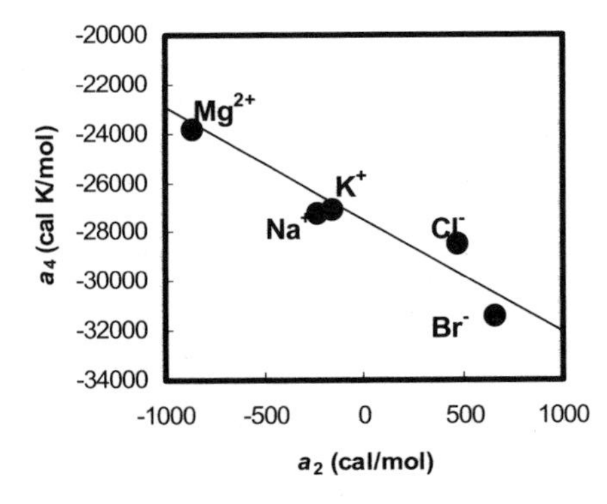

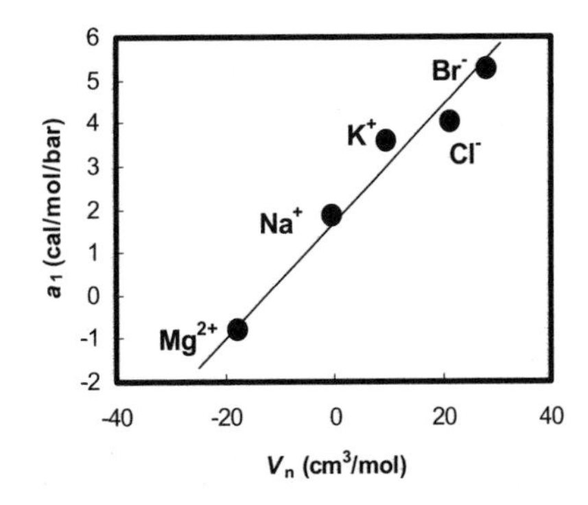

Figure 13. Correlations among equation of state parameters and non-solvation volumes: a) a_4 versus a_2, b) a_1 versus $\bar{V}_n$. After Shock and Helgeson (1988).

extensive recent measurements of the heat capacities and volumes of non-electrolytes Plyasunov and Shock (2001a) proposed revised correlations of similar form to that shown in Figure 13, to allow improved estimation of HKF equation of state parameters for aqueous non-electrolyte species.

Shock and Helgeson (1988) made further correlations between heat capacity and entropy, volumes and entropies and entropies and effective electrostatic radii of aqueous species such as shown in Figures 14 and 15. Taken together, these correlations permit determination of the parameters required to determine the standard partial molal Gibbs free energy of an aqueous species as a function of temperature and pressure (a_1, a_2, a_3, a_4, c_1, and c_2; and $\bar{S}^{\circ}_{P_r,T_r}$) from knowledge of $\bar{G}^{\circ}_{P_r,T_r}$ which are available for numerous aqueous species at a temperature of 25 °C and a pressure of 1 bar (see above) and r_{e,P_r,T_r}, which can be determined from Equations (41) to (44). Further similar correlations for other series of aqueous species have since been proposed (e.g., Sassani and Shock 1992). Once equations describing the temperature dependence of the standard partial molal Gibbs free energy were available, it was possible to regress high temperature equilibrium constants available in the literature to retrieve corresponding $\bar{G}^{\circ}_{P_r,T_r}$ for use in thermodynamic databases in a manner analogous to that used for generating internally consistent thermodynamic properties for minerals. This approach was used by Sverjensky et al. (1997) to retrieve the thermodynamic properties for many aqueous species that form in too low concentrations to allow their experimental characterization at ambient temperatures and pressures.

It should be emphasized, however that the correlations proposed by Shock and Helgeson (1988) and subsequent correlations to generate equation of state parameters for the HKF model (e.g., Sassani and Shock 1992; Shock and Koretsky 1993; Plyasunov and Shock 2001a) are empirical and based in many cases on relatively few data points. Nevertheless is remarkable how successful this approach appears to be. Several examples comparing estimates of

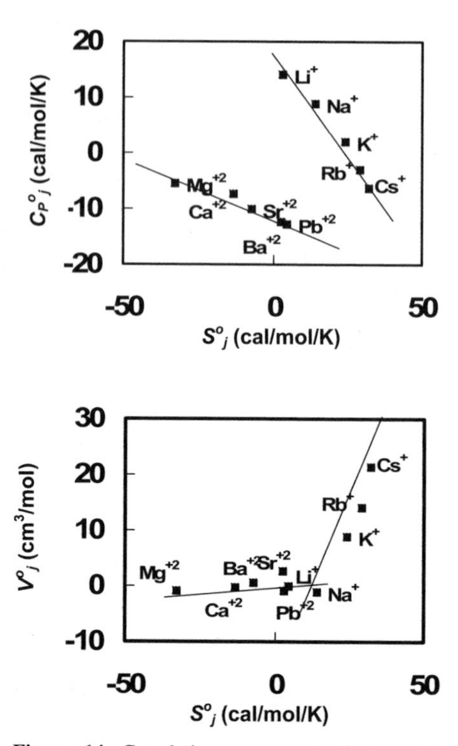

Figure 14. Correlations among standard partial molal heat capacities, volumes and entropies of aqueous species (after Shock and Helgeson 1988).

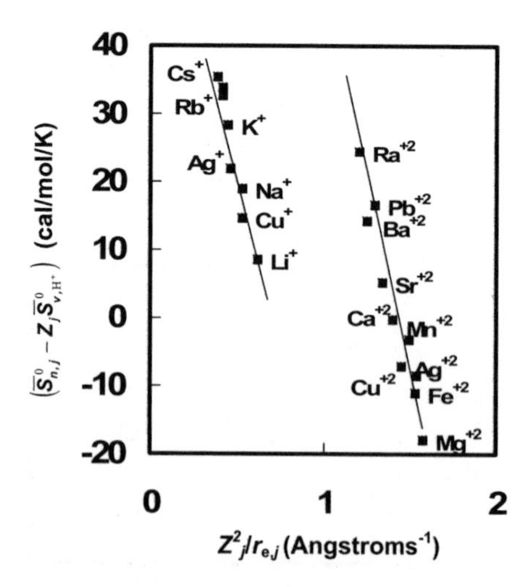

Figure 15. Correlations among standard partial molal entropies and effective electrostatic radii of aqueous species (after Shock and Helgeson 1988).

equilibrium constants obtained after the correlation schemes described above with estimates made using the HKF model are shown in Figures 16 to 18. Figure 16 compares equilibrium constants measured experimentally with corresponding estimates made using the HKF together with estimated equations of state parameters. For the case of $CsCH_3COO$ the Gibbs free energy at 25 °C and 1 bar was estimated as well, using linear correlations as reported by Shock and Koretsky (1993). The close correspondence between the experimental data and the thermodynamic estimates are remarkable. So too for the coherence shown between measured and estimated equilibrium constants shown in Figure 17. In this case not only were the thermodynamic properties of aqueous species extrapolated in temperature using the HKF model, but the thermodynamic properties of magnesite were generated from extrapolation of thermodynamic data collected at 600 to 900 °C using the equations and estimated thermodynamic properties reported by Helgeson et al. (1978) (see Fig. 5). A similar comparison for the equilibrium constant for the siderite dissolution reaction is shown in Figure 18. Despite the coherence of

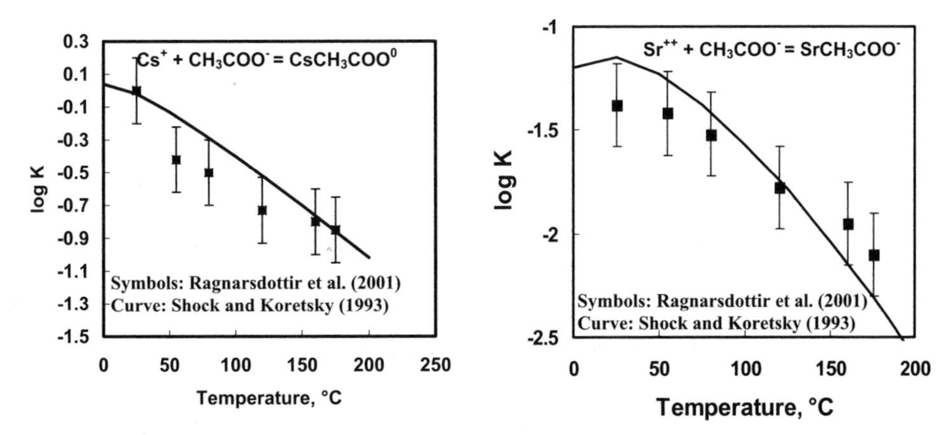

Figure 16. Comparison of experimentally measured equilibrium constants for metal acetate complexes indicated in the figure with those calculated using the HKF equations of state together with estimated equation of state parameters reported by Shock and Koretsky (1993).

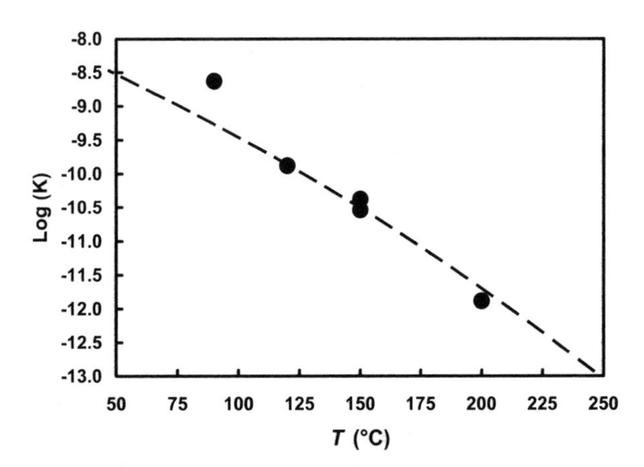

Figure 17. The solubility constant for the magnesite hydrolysis reaction: Magnesite = Mg^{2+} + CO_3^{-2}. The symbols correspond to values measured by Saldi (2009) but the curves corresponds to calculations performed using SUPCRT92 (Johnson et al. 1992) which generates aqueous species properties using the HKF equations of state and mineral properties from equations reported by Helegson et al. (1978).

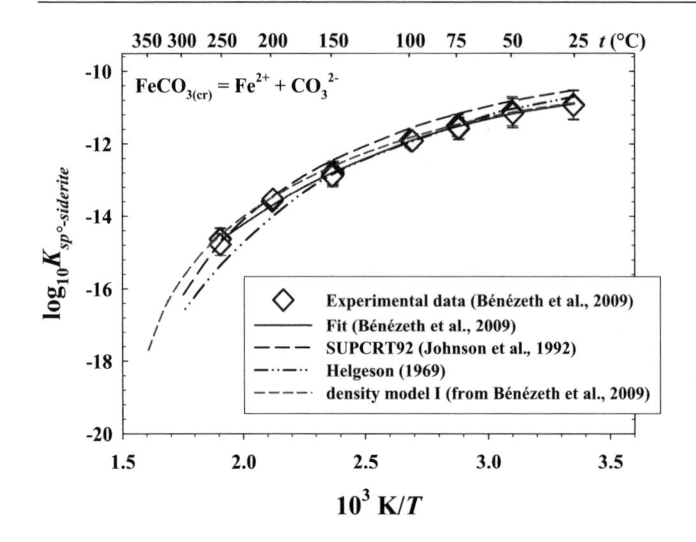

Figure 18. Comparison of experimental solubility products of siderite (Bénézeth et al. 2009) with SUPCRT (Johnson et al. 1992), Helgeson et al. (1969) and the density model (Anderson et al. 1991, Bénézeth et al. 2009) as a function of the reciprocal temperature.

the data shown in Figures 16 to 18, one must be aware that these correlations provide only estimates that will only be improved by upgrading as additional data become available. As these correlations are empirical, thermodynamic estimates made using these correlations should be checked, wherever possible, against experimental measurements to validate the quality of these estimates.

ACTIVITY COEFFICIENTS FOR AQUEOUS SPECIES

The stability of aqueous species depends strongly on the composition of the solution. The variation of this stability with solution composition can be quantified through the use of activity coefficients which are related to chemical potential in Equation (1) by

$$\mu_i = \mu_i^o + RT \ln a_i = \mu_i^o + RT \ln\left(\gamma_i c_i\right) \tag{46}$$

where μ_i and $\mu_i°$ stand for the chemical potential of the subscripted species at any concentration and at the standard state, and a_i, γ_i, and c_i in Equation (46) refer to the activity, activity coefficient and concentration of the subscripted aqueous species.

There are two general approaches used in geochemical modeling algorithms to calculate the activity of aqueous species. These two approaches stem from the Debye-Hückel equation and the Pitzer equations. These two approaches are summarized briefly below.

Debye-Hückel based methods

The Debye-Hückel equation is given by

$$\log\gamma_i = \frac{-Az_i^2\sqrt{I}}{1 + B\mathring{a}\sqrt{I}} \tag{47}$$

where A and B refer to solvent parameters that depend on the dielectric constant and density of water, å stands for an ion size parameter, and I refers to the ionic strength of the solution. Values of A and B for H_2O over wide ranges of temperature and pressure have been reported by Helgeson and Kirkham (1974b) and Archer and Wang (1991). This equation can provide activity coefficients to reasonable accuracy at ionic strengths to roughly 0.1 mol/kg or less. Efforts have been made to add additional terms to this equation to improve its ability to calculate activity coefficients at higher ionic strengths. Perhaps the simplest extension of this equation is given by

$$\log\gamma_i = \frac{-Az_i^2\sqrt{I}}{1+B\mathring{a}\sqrt{I}} + bI \tag{48}$$

where b designates an extended term parameter. Helgeson et al. (1981) observed that Equation (48) could provide accurate estimates of activity coefficient for a number of electrolytes to concentrations up to ~10 molal. Helgeson (1969) argued that at most natural solutions are dominated by NaCl, a single value of b could be adopted. This simplification has been widely used in geochemical calculations (Bethke 1996). Other approximations for NaCl rich fluids have been reported by Helgeson et al. (1981) and Oelkers and Helgeson (1991). Others have advocated adding a number of additional terms to the Debye-Hückel equation leading to (e.g., Staples and Nuttall 1997)

$$\log\gamma_i = \frac{-Az_i^2\sqrt{I}}{1+B\mathring{a}\sqrt{I}} + CI + \mathrm{D}I^2 + \mathrm{E}I^3 + ... \tag{49}$$

where C, D, and E represent empirically determined constants. Nevertheless, these Debye-Hückel equation based methods typically contain empirically determined parameters to allow estimated beyond an ionic strength of 0.1 mol/kg, so may not prove sufficiently accurate for some calculations.

Pitzer equations

The Pitzer equations (Harvie and Weare 1980; Pitzer 1987) were developed by combining the Debye-Hückel equation with additional terms forming a viral equation. These equations provide an impressively accurate description of the thermodynamic properties of aqueous systems to extremely high ionic strengths and are particularly applicable to describing the thermodynamic properties in seawater and evaporate systems (c.f. Monnin et al. 1999; Hacini et al. 2008). Equations describing the activity coefficients of cations and anions, denoted by M and X, within the Pitzer formalism are given by

$$\begin{aligned}\ln\gamma_M = {} & -z_M{}^2 A^{\Phi}\left(\frac{I^{0.5}}{1+bI^{0.5}} + \frac{2}{b}\ln\left(1+bI^{0.5}\right)\right) + \sum_a m_a\left(2B_{Ma} + ZC_{Ma}\right) \\ & + \sum_c m_c\left(2\Phi_{Mc} + \sum_a m_a \Psi_{Mca}\right) + \sum_{a<a'}\sum \mathrm{m_a} m_{a'}\Psi_{Maa'} \\ & + |\mathrm{z_M}|\sum_c\sum_a m_c m_a C_{ca} + ...\end{aligned} \tag{50}$$

and

$$\begin{aligned}\ln\gamma_X = {} & -z_X{}^2 A^{\Phi}\left(\frac{I^{0.5}}{1+bI^{0.5}} + \frac{2}{b}\ln\left(1+bI^{0.5}\right)\right) + \sum_c m_c\left(2B_{cX} + ZC_{cX}\right) \\ & + \sum_a m_a\left(2\Phi_{Xa} + \sum_c m_c \Psi_{cXa}\right) + \sum_{c<c'}\sum m_c m_{c'}\Psi_{cc'X} \\ & + |\mathrm{z}_X|\sum_c\sum_a m_c m_a C_{ca} + ...\end{aligned} \tag{51}$$

In these equations B_i, C_i, Φ_i, Ψ_I refer to parameters specific to the interaction of the subscripted cation and anion obtained from regression of activity coefficients and other thermodynamic properties of dissolved aqueous species. A^{Φ} and b in Equations (50) and (51) designate modified Debye-Hückel functions and z_i and m_i refer to the charge and molality of the subscripted species, respectively.

The impressive accuracy of the Pitzer equations stems from the large number of parameters describing interactions among each of the cations and anions present in solution. The large number of parameters necessary for the successful application of the Pitzer equations is also its major limitation. No general model is currently available to estimate interaction parameters for the Pitzer equations for aqueous species for which there are insufficient data to obtain these interaction parameters from data regression, nor are there general methods to extrapolate these parameters over wide ranges of temperature and pressure. As such the current application of these parameters is limited to where there is sufficient experimental data for the retrieval of the interaction parameters, which includes describing a limited number of major elements at 25 °C and 1 bar, and over limited temperature ranges (e.g., Monnin 1990; Clegg and Whitfield 1995).

THERMODYNAMIC PROPERTIES OF WATER

As water is pervasive at the surface and in the crust of the Earth its thermodynamic properties are critical to quantifying natural processes. As such a large number of experimental and computational studies have been performed to quantify these properties. Nordstrom and Munoz (1994) reported 52 sources for thermodynamic properties of H_2O. Among these, Hill (1990) provides a basis for the precise determination of the Gibbs free energy of H_2O based on a fit of experimental data obtained at temperatures from 0 to 900° C and pressures from 0 to 10 kb. Hill's (1990) fit provides equations describing the Helmholtz function with temperature and pressure. These can be converted to Gibbs Free energy for the stable phase of H_2O at a given pressure and temperature using (Johnson and Norton 1991)

$$(\bar{G}^{o}_{\mathrm{H_2O},P,T} - \bar{G}^{o}_{\mathrm{H_2O},P_r,T_r}) = \bar{A}^{o}_{P,T} + \rho_{P,T}(\frac{\partial \bar{A}^{o}}{\partial \rho})_T - (\bar{G}^{o}_{\mathrm{H_2O},P_r,T_r} - \bar{G}^{o}_{\mathrm{H_2O},tr}) \tag{52}$$

where $\rho_{P,T}$ and $\bar{A}^{o}_{P,T}$ represent the density and standard partial molal Helmholtz free energy of H_2O, respectively, at the subscripted pressure and temperature, and $(\bar{G}^{o}_{H_2O,P_r,T_r} - \bar{G}^{o}_{H_2O,tr})$ stands for the difference between the standard partial molal Gibbs free energy of H_2O at P_r, T_r and its value at the triple point of H_2O (0.0061173 bars and 273.16 K); this difference is equal to −398 cal mol^{-1}. The density of H_2O in Equation (52) can be expressed as (Haar et al. 1984; Johnson and Norton 1991)

$$\rho_{P,T} = (P\hat{C}M_w)^{1/2}\left(\frac{\partial \bar{A}^{o}}{\partial \rho}\right)_T^{-1/2} \tag{53}$$

where $\hat{C}$ in this equation stands for a conversion factor (0.02390054 cal bar^{-1} cm^{-3}), M_w signifies the molecular weight of H_2O (18.0152 g mol^{-1}) and P again represents the pressure of interest. Equations (52) and (53) can be used together with values of $\bar{A}^{o}_{P,T}$ and $(\partial \bar{A}^{o} / \partial \rho)_T$ computed from equations reported by Hill (1990) to calculate values of $\bar{G}^{o}_{\mathrm{H_2O}}$. Equation (52) was demonstrated to permit regular extrapolations of thermodynamic properties to a maximum temperature of 2000 °C and a maximum pressure of 250 kb.

SUMMARY OF THE CURRENT STATE OF THERMODYNAMIC DATABASES

From the summary presented above it is clear that although currently used thermodynamic databases provide excellent predictions in many cases, there still exist many significant limitations and uncertainties. The thermodynamic properties of minerals contained in these databases are at best extrapolated down to ambient temperatures using, for many minerals, estimated heat capacity power functions. In many other cases, the mineral properties contained in our databases are based on unreversed solubility experiments, which may or may not correspond to the actual thermodynamic properties of this mineral. The effects of solid solutions are also largely unaccounted for. Moreover the minerals present in the databases are mostly limited to those containing only the major elements; minerals that dominate minor and trace elements are often lacking. Such limitations make it difficult to interpret the measured concentrations of minor and trace elements using the results of geochemical modeling calculations. For example, measured fluid compositions from near the Mt. Hekla volcano, Iceland are compared to the results of PHREEQC calculations performed using its llnl database in Figure 19. It can be seen that the concentrations of major elements in the natural water samples are close to those calculated assuming the fluid was in equilibrium with the minerals present in the database, but minor and trace element concentrations in the natural waters are in many instances orders of magnitude different from those calculated by assuming equilibrium with minerals in this database. Due to the lack of accurate thermodynamic data, it is not possible to determine the reason for the poor coherence between the natural fluid compositions and the calculated results using the geochemical modeling.

The situation for the aqueous species is also provisional. Thermodynamics properties for many aqueous species are still missing, and many are based on empirical correlations. The ability of these correlations to accurately predict thermodynamic properties is untested for many aqueous species. The variation of these properties on temperature and pressure is based on correlations that contain relatively few data points.

As such it is best to consider the current state of thermodynamic databases as a work in progress rather than a completed task. This state of affairs was summarized roughly 30 years ago in the Preface of Helgeson et al. (1978) in a description that still rings true: "*The research reported in the following pages was carried out in recognition of the need for a well documented, comprehensive, and critical compilation of thermodynamic data for minerals that can be used with confidence to characterize chemical equilibrium in geologic systems. The credibility of such a compilation is largely a function of the extent to which the authors demonstrate that the thermodynamic data adopted for a given mineral are consistent with those derived from experimental data for all other minerals, and that none contravenes reliable geological observations. We make no pretense at having achieved these goals in the present contribution, which represents little more than a beginning step in this direction.*"

It is clear that additional experimental efforts could improve considerably the quality of geochemical databases. Moreover, there have been huge advances in our ability to experimentally characterize equilibrium in complex geochemical systems. A short summary of some of these advances is provided below.

RECENT ADVANCES IN EXPERIMENTAL TECHNIQUES

Some new and extremely powerful experimental techniques have recently become available which allow the acquisition of thermodynamic properties of minerals and co-existing aqueous fluids more rapidly and of a higher quality than ever before possible. In an attempt to motivate future research in this area, a brief summary of three of these methods is presented below.

In situ X-ray absorption fine structure spectroscopy

One of the major advances in experimental techniques over the past two decades has been the development and application of X-ray absorption fine structure spectroscopy (XAFS) to mineral-fluid systems. XAFS spectroscopy is element-selective and provides unique information, compared to other types of spectroscopy (e.g., UV-Vis, Raman, NMR), about oxidation state, site symmetry, ligand's identity and number and interatomic distances of an element in complex chemical and geological matrices at relatively low concentrations (see Brown and Sturchio 2002 for review of XAFS studies on environmental samples). The power of

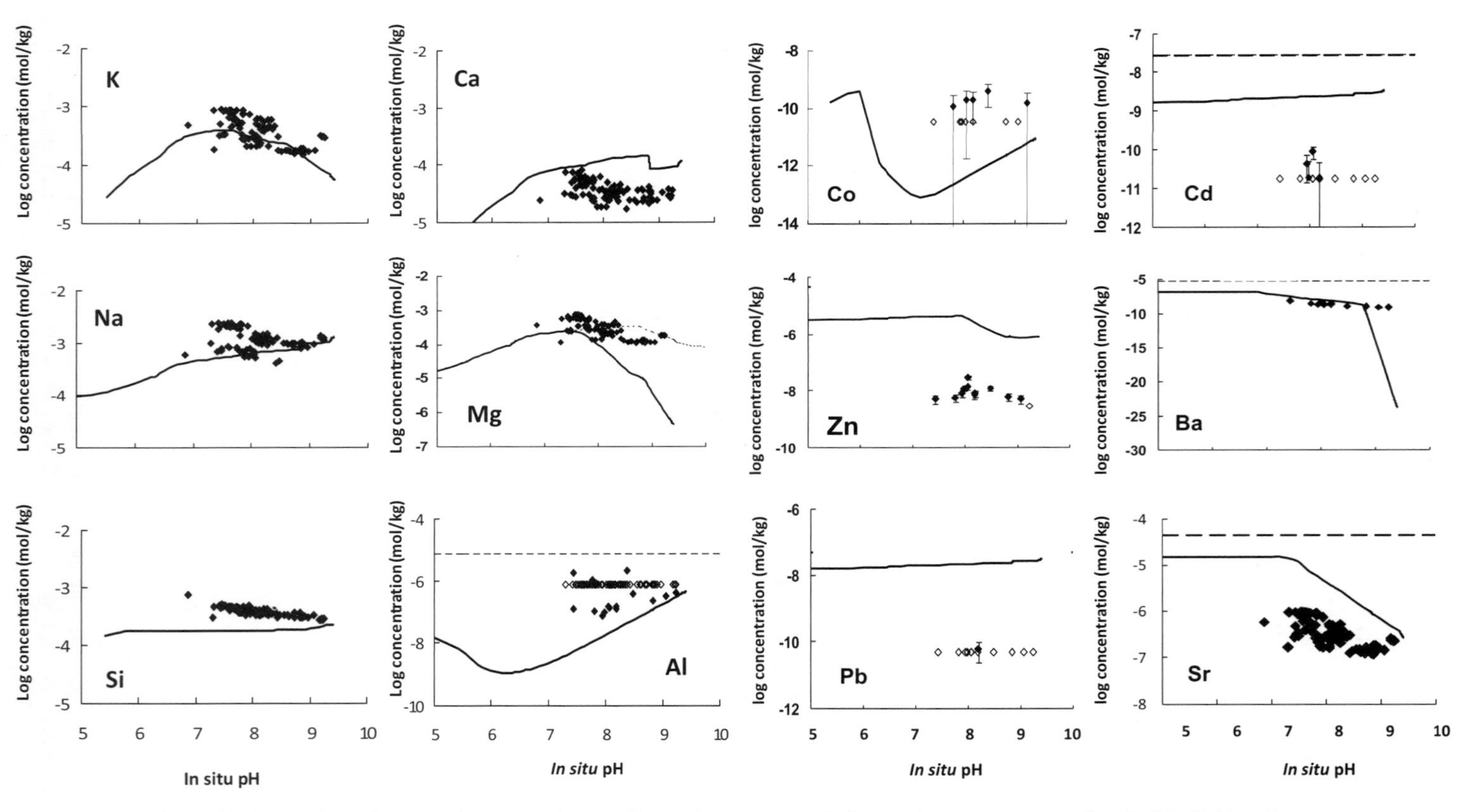

Figure 19. Comparison of measured concentrations of dissolved aqueous metals from spring waters surrounding the Mt. Hekla volcano, represented by symbols and corresponding values calculated assuming equilibrium with the most stable minerals phases using the PHREEQC computer code (Parkhurst and Appelo 1999) together with its llnl database. Filled and open symbols correspond to measured concentrations and the analytical detection limit for those concentrations lower than this limit. After Flaathen et al. (2009).

XAFS spectroscopy together with 1) the improvement of synchrotron sources and 2) progress in high *T-P* spectroscopic cell design have provoked an "explosion" of XAFS studies of high *T-P* systems (e.g., see Seward and Driesner 2004 for a recent review).

One advantage of coupling XAFS spectroscopy with newly designed cells is that they allow direct, *in situ* measurement of the state of aqueous fluids at elevated temperatures and pressures. These *in situ* data, complemented by *ab initio* and molecular dynamics modeling (e.g., Sherman et al. 2000a, b; Sherman 2001), provide unprecedented improvement of our knowledge of metal complex stoichiometries and structures in aqueous fluids, which were accessible until recently mostly using indirect bulk solubility methods. However, because XAFS spectroscopy probes an *average* atomic environment around the absorbing atom, in many cases ambiguity remains as for the true speciation, particularly in the case of a mixture of several complexes with different ligand numbers or different ligands simultaneously present in the system. Moreover, owing to its very weak sensitivity to *light atoms* like protons or distant atomic shells, XAFS spectroscopy is not capable of detecting directly the charge of species formed through deprotonation/hydrolysis or complexation reactions. Thus, to derive unambiguously species identities and distribution from XAFS data in complex high-temperature aqueous systems, information about solubility of solid phases as a function of ligand concentration and/or pH is often required.

Progress in designing spectroscopic cells in the last years (e.g., Testemale et al. 2005) has made it possible to measure simultaneously mineral solubilities and local atomic structures around a dissolved metal in two-phase mineral-fluid and three-phase mineral-brine-vapor systems (e.g., GeO_2(s)-H_2O, Pokrovski et al. 2005a; Sb_2O_3(s)-H_2O-NaCl-HCl, Pokrovski et al. 2006, 2008b; Au(s)-H_2S-S-NaOH, Pokrovski et al. 2006; Schott et al. 2006; Au(s)-H_2O-NaCl-HCl, Pokrovski et al. 2009; CuCl(s)-H_2O-HCl, Liu et al. 2008). Below we briefly discuss the essentials of solubility/partitioning measurements from X-ray absorption spectra and give a couple of representative examples of the derivation of stoichiometry and thermodynamic properties of aqueous metal species from XAFS data obtained using a recent designed X-ray cell (Testemale et al. 2005).

An important property of the X-ray absorption process is that the amount of the absorbed radiation, measured by the amplitude of the absorption edge height in transmission mode (Fig. 20A), is a direct function of the absorber concentration, according to the classical X-ray absorption relation (see Pokrovski et al. 2005a, 2006, 2008b; Testemale et al. 2005 for details):

$$C_i = \Delta\mu / (\Delta\sigma_i \times M_i \times l_{path} \times d_{fluid}) \qquad (54)$$

where C_i stands for the absorber aqueous concentration (mol kg^{-1} of fluid), $\Delta\sigma_i$ represents the change of the total absorption cross-section of the absorbing element over its absorption edge, l_{path} designates the optical path length inside the cell which remains constant through the experiment, M_i refers to the absorber atomic weight, and d_{fluid} signifies the density of the fluid at given T and P. The absorption cross-sections for most elements are known and typically agree within 5% amongst existing databases (e.g., as compiled in the Hephaestus software, Ravel and Newville 2005). For most fluid and vapor systems of typical natural compositions, the fluid density may be reasonably estimated using the Pressure-Volume-Temperature-Composition (*PVTX*) properties of pure water or the NaCl-H_2O system. The uncertainty in $\Delta\mu$ determinations is typically < 5%. With a relatively large cell having an optical path of several mm (Fig. 21A), this corresponds to a detection limit between 10^{-3}-10^{-5} mol/kg of solute, depending on the element. Thus this method that requires no *a priori* calibration is comparable in precision and reproducibility with typical solubility experiments using batch or flow-through reactors on relatively soluble minerals (≥ 10-100 ppm of metal).

An example of this approach is illustrated in Figure 20 for argutite solubility measurements in pure water to 500 °C and 400 bar (Pokrovski et al. 2005a). It can be seen that steady-state values of the absorption edge height as a function of T and time are attained in an hour at

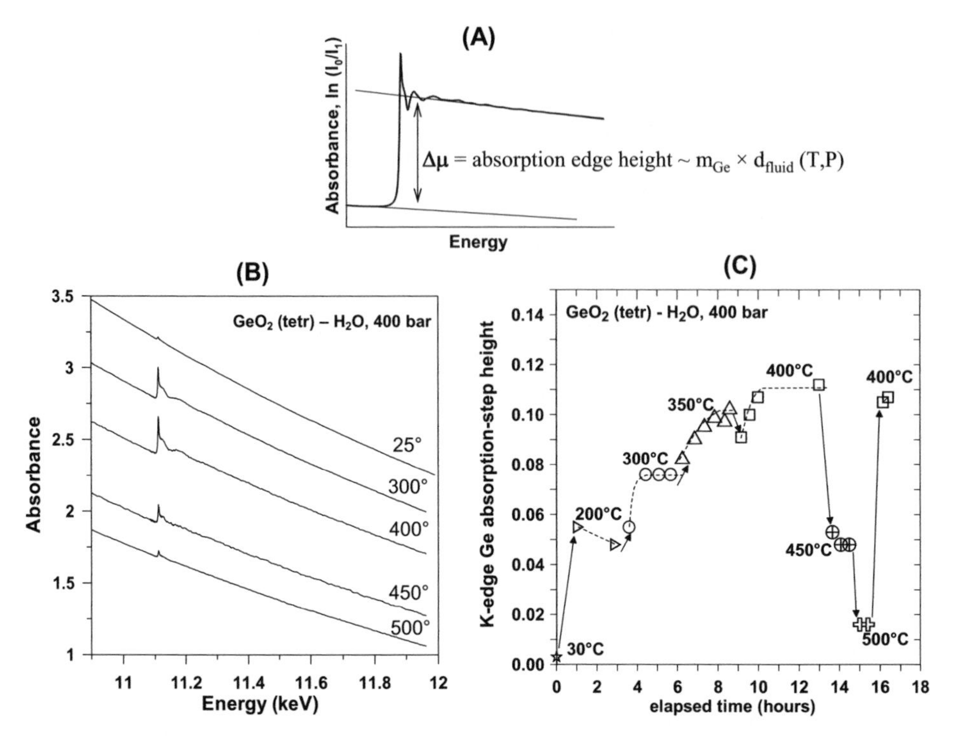

Figure 20. (A) Determination of the concentration of an absorber elements in the fluid phase can be obtained from the amplitude of the absorption edge height. (B) Raw XAFS transmission spectra at the Ge K-edge of aqueous solutions in contact with crystalline argutite (GeO_2, tetr) at 400 bar and the indicated temperatures. The decrease of the before-edge absorption with increasing temperature follows the corresponding decrease of water density, whereas the absorption-edge height provides the means to the aqueous Ge concentration. (C) Evolution of the absorption-edge height of aqueous Ge as a function of time and temperature during an argutite dissolution experiment. Each symbol corresponds to a XAFS scan. Arrows indicate temperature changes during the experiment; dashed lines linking the symbols at a given temperature are drawn to guide the eye. Error bars correspond to the symbol size.

temperatures from 300 to 500 °C and are reversible on cooling (see Fig. 20C). The GeO_2 solubilities derived from these values are in excellent agreement with previous batch-reactor measurements further demonstrating the attainment of equilibrium in the XAFS experiments. The analysis of fluorescence XANES (not shown) and EXAFS spectra (see Fig. 22A) recorded from the fluid phase simultaneously with solubility measurement provides structural parameters of Ge atomic environment in solution. These show that Ge is surrounded by 4±0.4 O atoms at 1.75±0.01 A in a tetrahedral symmetry. This is in agreement with the dominant formation of $Ge(OH)_4^0$ whose structure is analogous to that of aqueous silica, $Si(OH)_4^0$. The derived stoichiometry of the dominant Ge aqueous complexes and GeO_2 solubility allows refinement of the thermodynamic properties of $Ge(OH)_4(aq)$ as a function of temperature and pressure. Resulting GeO_2 solubilities are close to those of $Si(OH)_4^0$ and yield very similar solubility trends as quartz and argutite (Fig. 22B). The S-shaped solubility pattern of both solids at a function of T at P near to the critical point of water reflects the evolution of the water density and dielectric constant and is in agreement with the negative divergence of the partial molal heat capacity and volume of neutral Ge and Si hydroxides similarly to their As and B analogues as revealed by recent calorimetric measurements (see above and Perfetti et al. 2008).

A complementary XAFS approach for deriving thermodynamic properties of aqueous

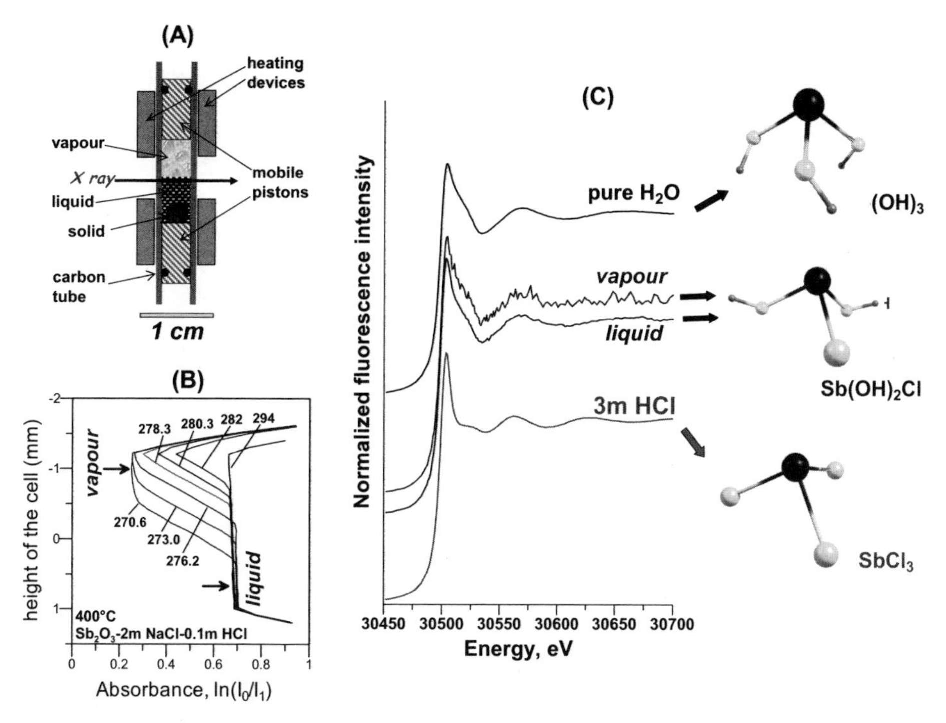

Figure 21. An immiscibility experiment in the system Sb_2O_3-H_2O-NaCl-HCl at 400 °C using a recently designed optical cell (A). The cell absorption was measured at ~30.6 keV (above Sb K-edge) at indicated pressures by moving the cell vertically relative to the X-ray beam position (B). The appearance and growth of the vapor phase with decreasing P due to vapor-liquid separation is manifested by the sharp decrease of the absorbance in the upper part of the cell. The absorbances provide direct measure of Sb concentration in each coexisting phase and thus its vapor-liquid partitioning coefficient ($K_{Sb} = m_{vapor}/m_{liquid} = 0.087 \pm 0.030$, where m is Sb molality in the corresponding phase) was derived from the measured Sb concentrations at 400 °C and 271 bar. Graph (C) shows XANES spectra of the vapor and liquid phases recorded at the cell positions indicated by horizontal arrows in graph (B). Despite a high noise of the vapor-phase spectrum due to low Sb concentrations, its shape is identical to that of the coexisting liquid but is different from spectra in pure water ($Sb(OH)_3$) and a concentrated HCl solution ($SbCl_3$). This demonstrates the formation of new mixed hydroxy-chloride species responsible for Sb volatility in the system H_2O-HCl-NaCl. Their optimized geometries were generated using the quantum chemical modeling code GAUSSIAN. They allow simulations of XANES spectra using new *ab initio* approaches like FDMNES (Pokrovski et al. 2008).

species at elevated T-P is largely inspired by UV-Vis spectroscopic data analysis techniques. It uses principal component (PCA) and/or linear combination (LCA) analyses of multiple XANES spectra from solution as a function of ligand concentration to derive the complex stoichiometry and distribution. For a mixture of species in solution, the challenge is to identify and extract spectra of each individual species and use an absorption relationship similar to Equation (54) to calculate their relative concentrations. This may appear to be an irresolvable task in cases where two or more species have quite similar XANES spectra. Support for these approaches is now provided by independent *ab initio* calculations of XANES spectra for aqueous clusters of a given geometry (e.g., Testemale et al. 2004; Pokrovski et al. 2009). Such a combination has been successful in a few recent studies, allowing, for example, a substantial improvement in our understanding of Cu(I) chloride complex stability constants to 400 °C (e.g., Brugger et al. 2007). Other studies using this approach are in progress. Thus, XAFS spectroscopy provides a valuable contribution to the acquisition of thermodynamic properties of aqueous species in the T-P range of the upper crust.

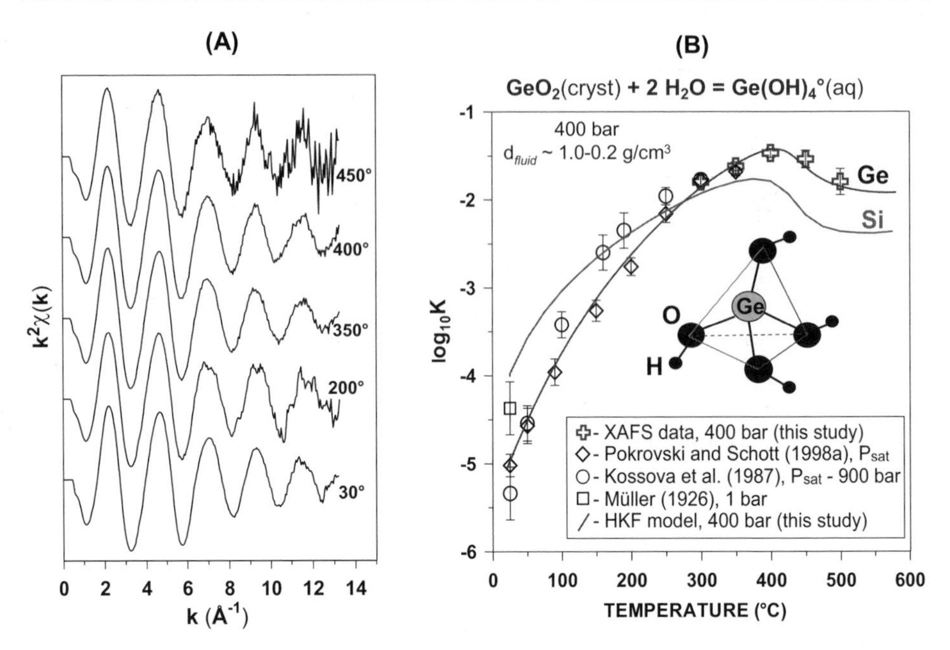

Figure 22. (A) Normalized k^2-weighted fluorescence EXAFS spectra of germanium aqueous solutions in equilibrium with argutite (GeO_2, tetr.) at 400 bar and the indicated temperatures. The spectra are consistent with a single atomic shell around Ge composed of four oxygens in a tetrahedral arrangement at 1.75±0.01 A. (B) Logarithm of GeO_2 (tetr) dissolution constant indicated in the figure as a function of temperature at 400 bar. The symbols were generated from measurements reported by Pokrovski et al. (2005a) and previous studies. The curve labeled Ge represents a regression of all experimental points with the revised HKF equation of state. Curve labeled Si shows the analogous reaction constant for aqueous silica according to Manning (1994). Note a large similarity between Ge and Si solubility patterns with temperature.

Another challenging topic that has been addressed by *in situ* XAFS spectroscopy is simultaneous measurement of vapor-liquid partition coefficients and the identities of metal species in the coexisting phases. Such information allows thermodynamic characterization of metals dissolved in water vapor and may prove to provide new insight into the key processes responsible for the formation and distribution of a wide range of noble and base metal ore deposits, from magmatic porphyry to epithermal (e.g., Hedenquist and Lowenstern 1994; Williams-Jones and Heinrich 2005). Prior to the development of XAFS spectroscopy, data on metal stabilities in water vapor were severely limited, particularly because of methodological and analytical difficulties inherent to experimental methods used to study vapor-brine systems (e.g., bulk solubility methods or synthetic fluid inclusion techniques).

An example of how XAFS spectroscopy can be used to define the thermodynamics of metals in water vapor for the system Sb_2O_3-H_2O-NaCl-HCl was reported by Pokrovski et al. (2008b). In this study, vapor-liquid immiscibility was created at 400 °C by slow decompression of the cell until the pressure crosses the two-phase vapor-liquid boundary of the water-salt system (see Fig. 21). Immiscibility was monitored by measuring the total cell absorbance by moving the cell vertically relative to the beam position (Fig. 21A,B). The spectra recorded by positioning the cell at the beam passage through the brine and vapor phases yield Sb concentrations in the two phases according to Equation (54) and thus bulk molal vapor-liquid partition coefficient for Sb^{III}, which is in good agreement with batch-reactor solubility measurements (Pokrovski et al. 2005b). The analysis of XANES spectra from both phases shows the dominant presence of oxy-chloride Sb species, likely $Sb(OH)_2Cl^0$ (Fig. 20c), similar to those forming in acidic

NaCl solutions at higher P (Pokrovski et al. 2006). The species geometry may be additionally constrained using quantum-chemical calculations with modern computer codes (e.g., Gaussian) of species structures, which may serve as input information for *ab initio* modeling of XANES spectra. Such *in situ* experiments using synchrotron radiation open new possibilities for studying *in situ* vapor-liquid and fluid-melt immiscibility phenomena thus avoiding certain artifacts common in less direct techniques and contributing to the acquisition of thermodynamic properties of components in multi-phase fluid systems ubiquitous in the Earth's crust.

In situ* calorimetric and volumetric measurements on aqueous species at elevated *T-P

Because the majority of mineral solubility experiments designed to obtain the thermodynamic properties of aqueous complexes over the last forty years have been performed over narrow temperature ranges (ΔT~50-100 °C) largely imposed by experimental and technical constraints, they allow fairly accurate derivation of ΔG, ΔH, and entropy (S) functions, but not partial molal heat capacities (C_p) and volume (V) functions required to estimate thermodynamics properties for aqueous species at elevated temperatures and pressures. Because the C_p is the second temperature derivative of the Gibbs free energy, the limited T range and the smooth character of $G(T)$ functions for most species below 300 °C, leads to large errors when deriving C_p from such data. This results in errors when using these heat capacity coefficients to predict Gibbs free energies of species at T-P outside the range covered by measurements.

The *direct* experimental determination of the derivative properties such as partial molal volumes and heat capacities of soluble aqueous species is thus both an alternative and complementary way for obtaining the Gibbs free energies at high T-P by integration using the data at 25 °C and 1 bar as integration constants. Recent advances in designing vibrating-tube densimeters and flow-through calorimeters allow acquisition of direct C_p and V data for a variety of relatively soluble (> 0.1 mol/kg H_2O) aqueous species at temperatures to 450 °C and pressures to 300 bar (for representative examples and details on these techniques see Majer et al. 1991, 1999, 2000; Majer and Wood 1994; Hnedkovsky et al. 1995, 1996; Hnedkovsky and Wood 1997; Obsil et al. 1996; Tremaine et al. 1997; Xiao et al. 1997; Clarke and Tremaine 1999; Clarke et al. 2000; Xie et al. 2004; Bulemela and Tremaine 2005; Censky et al. 2005a,b; Ballerat-Busseroles et al. 2007; Slavik et al. 2007; Perfetti et al. 2008). Details on these techniques are presented in the papers above, and only the methods principles will be resumed here.

The vibrating-tube technique is based on the comparison of vibration frequencies of a Pt-Rh tube placed in a magnetic field and through which pure water and the experimental solution pass alternatively. The density difference ($\Delta\rho$) between the two fluids is obtained from

$$\Delta\rho = \rho - \rho_w = K(\tau^2 - \tau_w^2) \qquad (55)$$

where ρ and ρ_w stand for the densities of the experimental solution and of pure water, respectively, τ and τ_w represent the oscillation periods of the vibrating tube filled with the experimental solution and pure water, respectively, and K denotes a calibration constant determined at each T-P by measurements with a NaCl solution up to 150 °C and with nitrogen gas at higher T (where the NaCl data are not of sufficient accuracy). Values of apparent molar volumes V_φ^{exp} are calculated from the experimental densities using

$$V_\varphi^{exp} = \frac{M_s}{\rho_w + \Delta\rho} - \frac{\Delta\rho}{m(\rho_w + \Delta\rho)\rho_w} \qquad (55)$$

where M_s represents the molar mass of the solute, m refers to the solution molality and ρ_w and $\Delta\rho$ again stands for the densities of the experimental solution and of pure water, respectively.

The high T-P flow-through calorimetric technique is based on the comparison of the electrical power required to increase temperature by a fixed value (typically 2 °C) of a cell consisting of a Pt-Rh tube, through which flow alternatively pure water and the experimental solution. The ratio of specific heat capacities of the solution and water, $C_{p,s}/C_{p,w}$, is related to the power required to heat the cell by

$$C_P / C_{p,w} = (\rho_w / \rho)_{SL} \{1 + f(P - P_w)/P_w\} \quad (57)$$

where $(\rho_w/\rho)_{SL}$ designate the ratio of water and solution densities at the temperature of the sample loop (at ~ 25 °C) and experimental pressure, and *f* represents a correction factor for heat loss, and P and P_w signify the electric power required to maintain the identical temperature rise when either the sample solution or water are flowing through the measuring cell. A second cell connected in series with the measuring cell is used to facilitate equilibration of the temperature bridge and to compensate for the flow rate and temperature fluctuations in the calorimeter. The heat loss correction factor *f* is commonly determined by changing the water flow rate, F_w, in the sample cell, to mimic a change in heat capacity, and by measuring the corresponding change in heat input. The resulting values of $f = \Delta F P_w/(\Delta P F_w)$ typically range from 1.02 (at 50 °C) to 1.10 (at 350 °C). The values of apparent molar heat capacities $C_{p,\varphi}^{exp}$ are calculated from the experimental data using

$$C_{p,\varphi}^{exp} = M_s c_p + (C_p - C_{p,w})/m = C_{p,w}\{(M_s + 1/m)C_p / C_{p,w} - 1/m)\} \quad (58)$$

where C_p and $C_{p,w}$ refer to the specific heat capacities (per unit of mass) of the solution and pure water, respectively, M_s stands for the molar mass of the solute, and *m* denotes the molality of the solution. Volumetric and calorimetric instruments and measurement procedures have been described in detail by Hynek et al. (1997) and Hnedkovsky et al. (2002), respectively, and a recent brief summary is given by Perfetti et al. (2008). The experimental values are used to obtain, by extrapolation to infinite dilution and correction for partial ionization, the standard molal volumes and heat capacities from the apparent molal properties Y_Φ^{exp} defined as

$$Y_\Phi^{exp} = (Y - 55.5 \times Y_w)/m \qquad Y = V, C_p \quad (59)$$

where *Y* relates to a 1 kg H_2O solution, *m* denotes the molality of the species, and Y_w corresponds to the molar property (volume or heat capacity) of pure water.

The C_p and *V* data obtained using these techniques for many aqueous nonelectrolyte (i.e., neutral) species over the last 15-10 years have allowed significant improvement of the HKF equation of state for non-electrolyte solutions (e.g., Plyasunov et al. 2000a,b; Plyasunov and Shock 2001a; Schulte et al. 2001). These data also motivated the development of alternative thermodynamic models for aqueous species (e.g., Sedlbauer et al. 2000; Sedlbauer and Majer 2000; Yezdimer et al. 2000; Majer et al. 2004; Akinfiev and Diamond 2003). Such models, however, are either limited for particular groups of species (e.g., volatiles, hydrocarbons) or require parameters not easily available for many solutes and aqueous complexes (e.g., Henry coefficients); they thus have received more limited use in the geochemical community than the HKF model.

An example of recent C_p and *V* measurements to 350 °C and 300 bar is shown in Figure 23. This figure shows results for the major AsIII and AsV aqueous complexes, $As(OH)_3(aq)$ and $AsO(OH)_3(aq)$. The negative divergence of $C_p°$ and $V°$ values near the critical point of water for both species is apparent in this figure, in marked contrast to the original HKF model predictions based on Gibbs free energy data below 300 °C (Pokrovski et al. 1996) and correlations amongst the HKF parameters established on the basis of aqueous electrolytes (Shock et al. 1997). The discrepancy between measured and predicted values of C_p and *V* yield differences ≥ 2 kcal/mol in the Gibbs free energy values for As-bearing species above 400-450 °C, obtained by integration of the two sets of data. This difference corresponds to an As dissolved concentration difference up to 1 order when calculating the solubilities of As minerals in high-temperature hydrothermal fluids typical for porphyry Cu-Mo-Sb-Au deposits. This example thus again illustrates the need to compare and upgrade our thermodynamic databases to insure accurate model results.

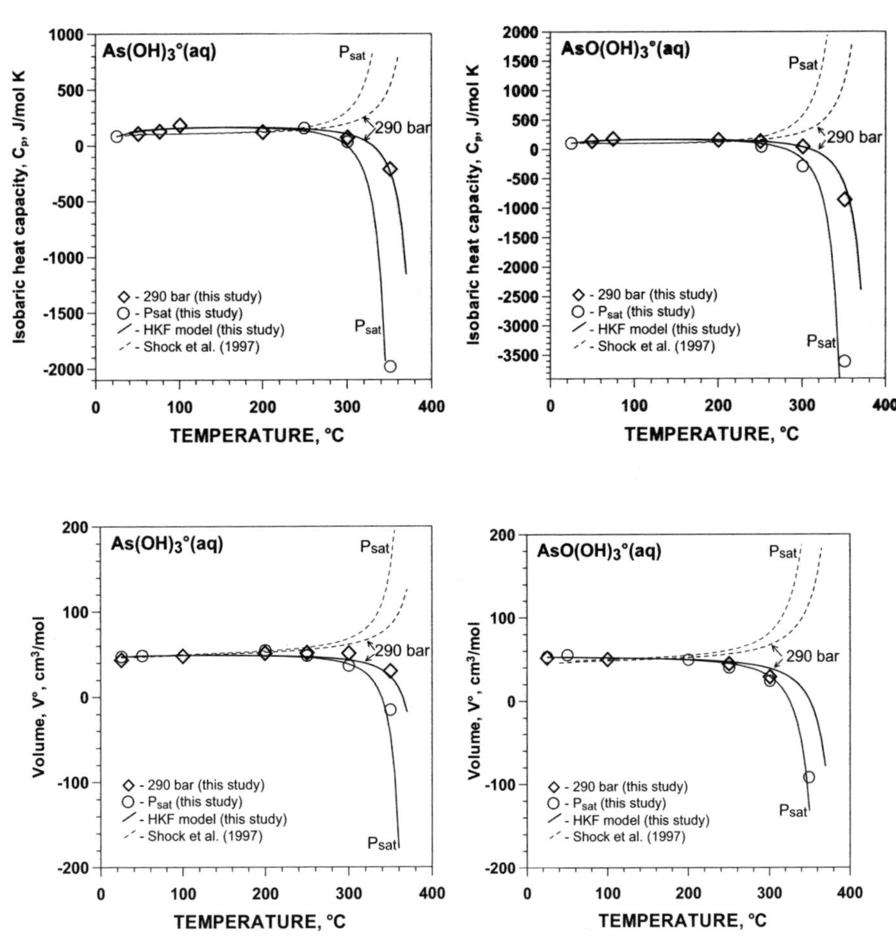

Figure 23. Standard partial molal heat capacities and volumes of aqueous arsenious and arsenic acids as a function of temperature at indicated pressures. The symbols represent experimental results reported by Perfetti et al. (2008). Whereas the solid lines correspond to a fit of these data by Perfetti et al. (2008) and estimates of these values made using equations and parameters reported by (Shock et al. 1997).

The tendency for the negative divergence of C_p and V values similar to that of aqueous ions, revealed by recent calorimetric measurements, is typical for many analogous neutral species like $B(OH)_3$ and H_3PO_4 (Hnedkovsky et al. 1995; Ballerat-Busseroles et al. 2007) and certain organic compounds potentially important in hydrothermal fluids (e.g., amino-acids and hydroxy derivatives of benzene, Clarke and Tremaine 1999; Clarke et al. 2000; Censky et al. 2005b). In contrast, other types of aqueous non-electrolytes, like carboxylic acids, alcohols, ketones and ethers (Schulte et al. 1997; Majer et al. 2000; Slavik et al. 2007) behave as dissolved gases (e.g., CH_4, CO_2, H_2S, NH_3, Hnedkovsky et al. 1995, 1996; Hnedkovsky and Wood 1997) showing negative ω values and a positive divergence of $\bar{V}_s^\circ$ and $\bar{C}_{P_s}^\circ$ (see also Plyasunov and Shock 2001a,b for global discussion of these phenomena). The reasons for this difference in extreme behavior lie in the solute-solvent interaction (hydration) which is the dominant contribution to the solute Gibbs free energy at near-critical conditions and is directly related, in the framework of the HKF model, to the Born electrostatic parameter ω (see Eqn. 40; Tanger and Helgeson 1988; and references therein). Thus, positive ω values reflect strong hydration due to attractive forces between solute and water typical for systems exhibiting negative near-critical divergence of $\bar{V}_s^\circ$ and $\bar{C}_{P_s}^\circ$. On the other hand, the negative ω values are typical for volatile non-electrolytes which are characterized by weak or repulsive interactions with the solvent water molecules and show a positive near-critical divergence of $\bar{V}_s^\circ$ and $\bar{C}_{P_s}^\circ$. While Helgeson and his collaborators expected that all nonelectrolyte solutes would exhibit positive divergence of their standard derivative properties (volume, enthalpy, heat capacity) near the critical point of water, the precise high-temperature volumetric and calorimetric measurements performed in the last 15 years demonstrate that the situation is much more complex. The derivative properties of many polar nonelectrolytes, and by inference of most neutral metal hydroxide and chloride complexes dominant in hydrothermal fluids, tend to have an "electrolyte-like" near-critical behavior. Accurate experimental volumetric and calorimetric data for aqueous non-electrolytes at high temperatures are thus needed; they allow quantitative constraining the type and strength of solute-solvent interactions and thus significantly improving thermodynamic models at high temperatures where the character of the near-critical behavior is difficult to predict *a priori*.

The hydrogen-electrode concentration cell

Another advance allowing efficient and accurate measurements of thermodynamic properties in mineral-aqueous solution systems is the hydrogen electrode concentration cell (HECC). The use of hydrogen electrodes in a concentration cell configuration, which provide highly accurate *in situ* pH measurement from 0 to 300 °C, was pioneered at Oak Ridge 40 years ago (Mesmer et al. 1970). The design and function of the hydrogen-electrode concentration cell has been described in numerous publications (e.g., Mesmer et al. 1970; Bénézeth et al. 1997a, 2007, 2009; Palmer et al. 2001) and has been used in a large number of studies of homogeneous reactions such as acid-base ionization (e.g., see the compilation in Tremaine et al. 2004); acid dissociation constants of organic acids (e.g., Bell et al. 1993; Mesmer et al. 1989; Kettler et al. 1991, 1992, 1995a, 1995b, 1998; Bénézeth et al. 1997b); the dissociation constants of tris(hydroxymethyl) aminometane, bis-tris″, 2,2-bis(hydroxymethyl)-2,2′,2″-nitrilotriethanol, EDTA, NTA, (Palmer and Wesolowski 1987; Wesolowski and Palmer 1989; Palmer and Nguyen-Trung 1995), and amines such as morpholine, dimethylamine and ethanolamine (Ridley et al. 2000; Bénézeth et al. 2001a, 2003; Tremaine et al. 2004); metal ion hydrolysis and complexation of Fe, Zn, Al and Cd by acetate (Palmer and Drummond 1988; Palmer and Hyde 1993; Giordano and Drummond 1991; Palmer and Bell 1994; Bénézeth and Palmer 2000), sulfate (Ridley et al. 1997) and malonate (Ridley et al. 1998a); Cd^{2+} by malonate (Ridley et al 1998b) and chloride (Palmer et al. 2000), and Nd^{3+} by acetate (Wood et al. 2000). Heterogeneous reactions have also been studied with the HECC including solubility measurements of aluminium hydroxide (Palmer et al. 2001; Bénézeth et al. 1997a, 2001b), zinc oxide (Wesolowski et al. 1998; Bénézeth et al. 1999, 2002) and neodymium hydroxide (Wood et al. 2002). Moreover, the HECC has been used to measure surface protonation of 1) bacteria by direct pH titration up to 75 °C (Wightman et al. 2001), 2) metal oxides at temperatures above 95 °C (Machesky et al. 1994, 1998; Wesolowski et al. 2000) as well as the sorption of Ca^{2+} (Ridley et al. 1999) and Nd^{3+} (Ridley et al. 2005) on rutile surfaces to 250 °C in NaCl aqueous solutions. A compilation of data on solubility and surface complexation of metal oxides is given by Wesolowski et al. (2004). The ability of such cell to perturb pH isothermally by addition of acidic or basic titrant also allows for precise measurement of dissolution/precipitation, rates at near to equilibrium conditions (Bénézeth et al. 2008; see Fig. 21 of Schott et al. 2009).

CONCLUSIONS

The development of thermodynamic databases for the modeling of geochemical processes is an ongoing process. The current available experimentally measured dataset only covers a small fraction of the conditions and compositions of natural solids and aqueous solutions. As such much of the 'data' currently found in our thermodynamic databases are values estimated using empirical correlations or estimations, having a varying uncertainties. In other cases where there are experimental measurements, such measurements can be conflicting and/or of questionable quality. It follows that an extensive effort to acquire new and high quality experimental measurements are essential to enable accurate predictions of the extent and consequences of geochemical processes in natural environments.

No doubt a complete understanding of the thermodynamic relations required to accurately model mineral solubility and solute speciation in water-rock systems is aided by reading several sources and several explanations of the essential concepts. We recommend that those that are interested in learning more about these relations and how they can be applied to the interpretation of natural processes consider reading one or more of these excellent reference books: Anderson and Crerar (1993), Nordstrom and Munoz 1994, Bethke (1996), Zhu and Anderson (2002).

ACKNOWLEDGMENTS

The summary presented above was motivated by the urging of the students and post-docs in the MIR and MIN-GRO Marie Curie Training Networks (MEST-2005-021120 and MRTN-2006-035112), and the CARB-FIX network. These students regularly use geochemical modeling codes to interpret their experimental and field observations but have difficulties determining on what the calculations were based, in particular in cases where calculated results did not match observations. EHO was introduced into the wonders of thermodynamic databases by Harold Helgeson, who spent decades regressing thermodynamic data to generate the basis of the thermodynamic calculations that many of us use routinely today. All of us owe a debt of gratitude to Jacques Schott who mentored us in the art of experimental geochemistry and thermodynamics. Our interest in this field has been encouraged by many of our friends and colleagues including Everett Shock, Dimitri Sverjensky, Barbara L. Ransom, William M. Murphy, Peter C. Lictner, Jan Amend, Christophe Monnin, Oleg Pokrovsky, Jean-Louis Dandurand, and Robert Gout.

REFERENCES

Akinfiev NN, Diamond LW (2003) Thermodynamic description of aqueous nonelectrolytes at infinite dilution over a wide range of state parameters. Geochim Cosmochim Acta 67:613-627

Amjad Z, Koutsoukos PG, Nancollas GH (1981) The crystallization of fluorapatite: A constant composition study. J Colloid Interface Sci 82:394-400

Anderson GM, Castet S, Schott J, Mesmer RE (1991) The density model for estimation of thermodynamic parameters of reactions at high temperature and pressures. Geochim Cosmochim Acta 55:1769-1799

Anderson GM, Crerar DA (1993) Thermodynamics in Geochemistry: The Equilibrium Model. Oxford University Press, New York

Aja S (1991) Illite equilibria in solutions: III A reinterpretation of the data of Sass et al. (1987). Geochim Cosmochim Acta 55:3431-3435

Archer DG, Wang PJ (1991) The dielectric constant of water and Debye-Hückel limiting law slopes. J Phys Chem Ref Data 19:371-411

Ballerat-Busserolles K, Sedlbauer J, Majer V (2007) Standard thermodynamic properties of H_3PO_4(aq) over a wide range of temperatures and pressures. J Phys Chem B 111:181-190

Barnes HL, Ernst WG (1963) Ideality and ionization in hydrothermal fluids: The system MgO-H_2O-NaOH. Am J Sci 261:129-150

Barton PB (1969) Thermodynamic study of the system Fe-As-S. Geochim Cosmochim Acta 33:841-857

Bell JLS, Wesolowski DJ, Palmer DA (1993) The dissociation of formic acid in sodium chloride solutions to 200°C. J Solution Chem 22:125-136

Benezeth P, Castet S, Dandurand JL, Gout R, Schott J (1994) Experimental study of aluminum-acetate complexing between 60 and 200 °C. Geochim Cosmochim Acta 58:4561-4571

Bénézeth P, Palmer DA, Wesolowski DJ (1997a) The aqueous chemistry of aluminium. A new approach to high temperature solubility measurements. Geothermics 26:465-481

Bénézeth P, Palmer DA, Wesolowski DJ (1997b) Dissociation constants for citric acid in aqueous sodium chloride media to 150 °C. J Solution Chem 26:63-84

Bénézeth P, Palmer DA, Wesolowski DJ (1999) The solubility of zinc oxide at 0.03m NaTr as a function of temperature, with *in situ* pH measurement. Geochim Cosmochim Acta 63:1571-1586

Bénézeth P, Palmer DA (2000) Potentiometric determination of cadmium acetate complexation in aqueous solutions to 250 °C. Chem Geol 167:11-24

Bénézeth P, Palmer DA, Wesolowski DJ (2001a) Potentiometric study of the dissociation quotients of aqueous dimethylammonium ion as a function of temperature and ionic strength. J Chem Eng Data 46:202-207

Bénézeth P, Palmer DA, Wesolowski DJ (2001b) Aqueous high temperature solubility studies. II. The solubility of boehmite at 0.03M ionic strength as a function of temperature and pH as determined by "*in situ*" measurements. Geochim Cosmochim Acta 65:2097-2111

Bénézeth P, Palmer DA, Wesolowski DJ, Xiao C (2002) New measurements of the solubility of zinc oxide at high temperatures. J Solution Chem 31:947-973

Bénézeth P, Wesolowski DJ, Palmer DA (2003) Potentiometric study of the dissociation quotients of aqueous ethanolammonium ion as a function of temperature and ionic strength. J Chem Eng Data 48:171-175

Bénézeth P, Palmer DA, Anovitz LM, Horita J (2007) Experimental dawsonite synthesis and reevaluation of its thermodynamic properties from solubility measurements: Implications for mineral trapping of CO_2. Geochim Cosmochim Acta 71:4438-4455

Bénézeth P, Palmer DA, Wesolowski DJ (2008) Dissolution/Precipitation kinetics of boehmite and gibbsite: Application of a pH-Relaxation technique to study near-equilibrium rates. Geochim Cosmochim Acta 72:2429-2453

Bénézeth P, Dandurand JL, Harrichoury JC (2009) The solubility product of siderite ($FeCO_3$) as a function of temperature. Chem Geol (in press)

Berman RG (1988) Internally-consistent thermodynamic data for minerals in the system Na_2O-K_2O-CaO-MgO-FeO-Fe_2O3-Al_2O3-SiO_2-TiO_2-H_2O-CO_2. J Petrology 29:445-522

Berman RG, Brown TH (1985) Heat Capacities of minerals in the system Na_2O-K_2O-CaO-MgO-FeO-Fe_2O3-Al_2O_3-SiO_2-TiO_2-H_2O-CO_2: Representation, estimation, and high temperature extrapolation. Contrib Mineral Petrol 89:168-183

Bethke CM (1996) Geochemical Reaction Modeling: Concepts and Applications. Oxford University Press, New York

Boettcher AL (1970) The system CaO-Al_2O_3-SiO_2-H_2O at high pressures and temperatures J Petrology 11:337-379

Born M (1920) Volumen und Hydratationswarme der Ionen. Zeitschr Physik 1:45-48

Bourcier WL, Knauss K, Jackson KJ (1993) Aluminum hydrolysis constants to 250 °C from bohemite solubilite measurements. Geochim Cosmochim Acta 57:747-762

Boyd FR (1964) Geological aspects of high pressure research. Science 145:13-20

Brown GE, Sturchio NC (2002) An overview of synchrotron radiation applications to low temperature geochemistry and environmental science. Rev Mineral Geochem 49:1-115

Brugger J, Etschmann B, Liu W, Testmale D, Hazemann JL, Emerich H, van Beek W, Proux O (2007) An XAS study of the structure and thermodynamics of Cu(I) chloride complexes in brines up to high temperature (400 degrees C, 600 bar). Geochim Cosmochim Acta 71:4920-494

Bulemela E, Tremaine PR (2005) Standard partial molar volumes of aqueous glycolic acid and tartaric acid from 25 to 350 °C: Evidence of a negative Krichevskii parameter for a neutral organic solute. J Phys Chem B 109:20539-20545

Castet S, Dandurand J-L, Schott J, Gout R (1993) Boehmite solubility and aqueous aluminum speciation in hydrothermal solutions (90-350 °C): Experimental study and modeling. Geochim Cosmochim Acta 57:4869-4884

Censky M, Hnedkovsky L, Majer V (2005a) Heat capacities of aqueous polar aromatic compounds over wide range of conditions. Part I: phenol, cresols, aniline, and toluidines. J Chem Thermo 37:205-219

Censky M, Hnedkovsky L, Majer V (2005b) Heat capacities of aqueous polar aromatic compounds over a wide range of conditions. Part II: dihydroxybenzenes, aminophenols, diaminobenzenes. J Chem Thermo 37:221-232

Chaïrat C, Oelkers EH, Schott J, Lartigue J-E, Harouiya N (2007) A combined potentiometric, electrokinetic and spectroscopic study of fluorapatite surface composition in aqueous solutions and consequences on solubility properties. Geochim Cosmochim Acta 71:5888-5900

Chase M (1998) NIST-JANAF Thermochemical Tables 4th Edition. J Phys Chem Ref Data Monograph 9:1-1952

Chermak JA, Rimstidt JD (1989) Estimating the thermodynamic properties (ΔG_f° and ΔH_f°) of silicate minerals at 298 K from the sum of polyhedral contributions. Am Mineral 74:1023-1031

Chin K, Nancollas GH (1991) Dissolution of fluorapatite – A constant composition kinetics study. Langmuir **7**: 175-2179

Clarke RG, Tremaine P (1999) Amino acids under hydrothermal conditions: Apparent molar volumes of aqueous alpha-alanine, beta-alanine, and proline at temperatures from 298 to 523 K and pressures up to 20.0 MPa. J Phys Chem B 103:5131-5144

Clarke RG, Hnedkovsky L, Tremaine PR, Majer V (2000) Amino acids under hydrothermal conditions: Apparent molar heat capacities of aqueous α-alanine, β-alanine, glycine, and proline at temperatures from 298 to 500 K and pressures up to 30.0 MPa. J Phys Chem B 104:11781-11793

Cleg SL, Whitfield M (1995) A chemical model of seawater including dissolved ammonia and the stoichiometric dissociation-constant of ammonia in estuarine water and seawater from 2 to 40 °C. Geochim Cosmochim Acta 59:2403-2421

Cox JD, Wagmann DD, Medvedev VA (1989) CODATA Key Values for Thermodynamics. John Benjamins Publishing Co

Elliott JC (1984) Structure and Chemistry of the Apatites and other Calcium Orthophosphates. Elsevier, Amsterdam

Essene EJ (1982) Geologic thermometry and barometry. Rev Mineral 10:153-206

Flaathen TK, Gislason SR, Oelkers EH, Sveinbjornsdottir AE (2009) Chemical evolution of the Mt. Hekla, Iceland, groundwaters. A natural analogue for CO_2 sequestration in basaltic rocks. App Geochem 24:463-474

Fournier P, Oelkers EH, Gout R, Pokrovski GS (1998) Experimental determination of aqueous sodium-acetate dissociation constants at temperatures from 20 to 240 °C. Chem Geol 151:69-84

Frank EU (1956) Hochverdichteter Wasserdampf II. Ionendissociation von KCl in H_2O bis 750 °C. Z Phys Chem 8:107-126

Frank EU (1961) Uberkritisches Wasser als electrolytisches Losungsmittel. Angew Chem 73:309-322

Fyfe WS, Turner FJ, Verhoogen J (1958) Metamorphic reactions and metamorphic facies. Geol Soc Am Mem 73, 259 p.

Gaboreau S, Vieillard P (2004) Prediction of Gibbs free energies of formation of minerals of the alumite supergroup. Geochim Cosmochim Acta 68:3307-3316

Gates JA (1985) The apparent volumes and heat capacities of electrolytes in H_2O at elevated temperatures. PhD thesis, University of Delaware, USA

Giggenback WF (1985) Construction of thermodynamic stability diagrams involving dioctahedral potassium clay minerals. Chem Geol 49:231-242

Giordano TH, Drummond SE (1991) The potentiometric determination of stability constants for zinc acetate complexes in aqueous solutions to 295 °C. Geochim Cosmochim Acta 55:2401-2415

Glynn P, Brown J (1996) Reactive transport modeling of acidic metal-contaminated groundwater at a site with sparse spatial information. Rev Mineral 34:377-438.

Gottschalk M (1997) Internally consistent thermodynamic data for rock-forming minerals in the system SiO_2-TiO_2-Al_2O_3-Fe_2O_3-CaO-MgO-Fe-O-K_2O-Na_2O-H_2O-CO_2. Eur J Mineral 9:175-223

Haar L, Gallagher JS, Kell GS (1984) *NBS/NRC Steam Tables.* Hemisphere Publishing Corporation

Haas JL Jr, Fisher JR (1976) Simultaneous evaluation and correlation of thermodynamic data. Am J Sci 276:525-545

Haas JR, Shock EL, Sassani DC (1995) Rare earth elements in hydrothermal systems – Estimates of standard partial molal thermodynamic properties of aqueous complexes of rare-earth elements at high pressures and temperatures. Geochim Cosmochim Acta 59:4329-4350

Hacini M, Oelkers EH, Kherici N (2008) Mineral precipitation rates during the complete evaporation of the Merouane Chott ephemeral lake. Geochim Cosmochim Acta 72:1583-1597

Harker RI Tuttle OF (1955) Studies on the system CaO-MgO-CO_2. Pt. 1. The thermal dissociation of calcite, dolomite and magnesite. Am J Sci 253:209-224

Harouiya N, Chairat C, Kohler SJ, Gout R, Oelkers EH (2007) the dissolution kinetics and apparent solubility of natural apatite in close reactors at temperatures from 5 to 50 °C. Chem Geol 244:554-568

Hariya Y Kennedy GC (1968) Equilibrium study of anorthite under high pressure and temperature. Am J Sci 266:193-203

Harvie CE, Weare JH (1980) The prediction of mineral solubilities in natural waters: The Na-K-Mg-Ca-Cl-SO_4-H_2O system from zero to high concentrations at 25 °C. Geochim Cosmochim Acta 44:981-987

Hays JF (1967) Lime-alumina-silica Carnegie Inst Washington Yearbook 65: 234-239

Hedenquist JW, Lowenstern JB (1994) The role of magmas in the formation of hydrothermal ore deposits. Nature 370:519-527

Helgeson HC (1969) Thermodynamics of hydrothermal systems at elevated temperatures and pressures. Am J Sci 267:729-804

Helgeson HC (1982) Errata: Thermodynamics of minerals, reactions, and aqueous solutions at high pressures and temperatures. Am J Sci 282:1143-1149

Helgeson HC (1985) Errata II: Thermodynamics of minerals, reactions, and aqueous solutions at high pressures and temperatures. Am J Sci 285:845-855

Helgeson HC, Kirkham DH (1974a) Theoretical prediction of the thermodynamic behavior of aqueous electrolytes at high pressures and temperatures: I. Summary of the thermodynamic/electrostatic properties of the solvent. Am J Sci 274:1089-1198

Helgeson HC, Kirkham DH (1974b) Theoretical prediction of the thermodynamic behavior of aqueous electrolytes at high pressures and temperatures: II. Debye-Hückel parameters for activity coefficients and relative partial molal properties. Am J Sci 274:1199-1261

Helgeson HC, Kirkham DH (1976) Theoretical prediction of the thermodynamic behavior of aqueous electrolytes at high pressures and temperatures: III. Equation of state for aqueous species at infinite dilution. Am J Sci 276:97-240

Helgeson HC, Delaney JM, Nesbitt HW, Bird DK (1978)Summary and critique of the thermodynamic properties of rock forming minerals. Am J Sci 278A:1-229

Helgeson HC, Kirkham DH, Flowers GC (1981) Theoretical prediction of the thermodynamic behavior of aqueous electrolytes at high pressures and temperatures: IV. Calculation of activity coefficients, osmotic coefficients and apparent molal and relative partial molal properties to 600° C and 5 kb. Am J Sci 281:1249 -1516

Hilbert R (1979) pVT-Daten von wasser und wassrigen Natriu,chloridLosungen bis 873 K, 4000 Bar und 25 Gewichsprozent NaCl. PhD Thesis, University of Karlsruhe, West Germany

Hill PG (1990) A unified fundamental equation for the thermodynamic properties of H_2O. J Phys Chem Ref Data 19:1233-1274

Hnedkovsky L, Majer V, Wood RH (1995) Volumes and heat capacities of H_3BO_3(aq) at temperatures from 298.15 K to 705 K and at pressures to 35 MPa. J Chem Thermo 27:801-814

Hnedkovsky L, Majer V, Wood RH (1996) Volumes of aqueous solutions of CH_4, CO_2, H_2S and NH_3 from 298.15 to 705 K and pressures up to 35 MPa. J Chem Thermo 28:125-142

Hnedkovsky L, Wood RH (1997) Apparent molar heat capacities of aqueous solutions of CH_4, CO_2, H_2S, and NH_3 at temperatures from 304 K to 704 K at a pressure of 28 MPa. J Chem Thermo 29:731-747

Hnedkovsky L, Hynek V, Majer V, Wood RH (2002) A new version of differential flow heat capacity calorimeter: Heat capacities of aqueous NaCl from 303 to 623 K. J Chem Thermo 34:759-785

Holland TJB (1989) Dependence of entropy on volume for silicate and oxide minerals: A review and a predictive tool. Am Mineral 74:5-13

Holland TJB, Powell R (1985a) An internally consistent thermodynamic dataset with uncertainties and correlations: 1. Data and Results. J Metamorph Petrol 3: 327-342

Holland TJB, Powell R (1985b) An internally consistent thermodynamic dataset with uncertainties and correlations: 2. Methods and worked examples. J Metamorph Petrol 3:343-370

Holland TJB, Powell R (1990) An enlarged and updated internally consistent thermodynamic dataset with uncertainties and correlations: the system K_2O-Na_2O-CaO-MgO-FeO-Fe_2O_3-Al_2O_3-TiO_2-SiO_2-C-H_2-O_2. J Metamorph Petrol 8:89-124

Holland TJB, Powell R (1998) An internally consistent thermodynamics dataset for phases of petrologic interest. J Metamorph Petrol 16:309-343

Holland TJB, Redfern SAT, Pawley AR (1996) Volume behavior of hydrous minerals at high pressure and temperature. 1. Compressibilities of lawsonite, zoisite, clinosoisite, and diaspore. Am Mineral 81:341-348

Huckenholz HG, Hölzl E, Lindhuber W (1956) Grossularite, its solidus and liquidus relations in the CaO-Al_2O_3-SiO_2-H_2O system up to 10 kbar. Neues Jahrb Mineralogie Abh 124:1-46

Hynek V, Hnedkovsky L, Cibulka I (1997) A new design of a vibrating-tube densimeter and partial molar volumes of phenol(aq) at temperatures from 298 to 573 K. J Chem Thermodyn 29:1237-1252

Jaynes WF, Moore PA, Miller DM (1999) Solubility and ion activity products of calcium phosphate minerals. J Environ Qual 28: 530-536

Johnson JW, Norton D (1991) Critical phenomena in hydrothermal systems: State, thermodynamic, electrostatic, and transport properties of H_2O in the critical region. Am J Sci 291:541-648

Johnson JW, Oelkers EH, Helgeson HC (1992) SUPCRT92: A software package for calculating the standard molal thermodynamic properties of minerals, gases, aqueous species, and reactions from 1 to 5000 bars and 0 to 1000 °C. Comput Geosci 18:889-947

Kelley KK (1960) Contributions to the data on theoretical metallurgy. XIII. High temperature heat content, heat capacity and entropy data for elements and inorganic compounds US Bur Mines Bull 548: 232 p

Kettler RM, Palmer DA, Wesolowski DJ (1991) Dissociation quotients of oxalic acid in aqueous sodium chloride media to 175 °C. J Solution Chem 20:905-927

Kettler RM, Wesolowski DJ, Palmer DA (1992). Dissociation quotients of malonic acid in aqueous sodium chloride media to 100 °C. J Solution Chem 21:883-900

Kettler RM, Wesolowski DJ, Palmer DA (1995a). Dissociation quotient of benzoic acid in aqueous sodium chloride media to 250 °C. J Solution Chem 24:385-407

Kettler RM, Palmer DA Wesolowski DJ (1995b). Dissociation quotients of succinic acid in aqueous sodium chloride media to 225 °C. J Solution Chem 24:65-87
Kettler RM, Wesolowski DJ, Palmer DA (1998). Dissociation quotients of oxalic acid in aqueous sodium chloride and sodium trifluoromethanesulfonate media to 175 °C. J Chem Eng Data 43:337-350
Knauss KG, Johnson JW, Steefel CI (2005) Evaluation of the impact of CO_2, co-contaminant gas, aqueous fluid and reservoir-rock interactions on the geological sequestration of CO_2. Chem Geol 217:339-350
Krupa KM, Robie RA, Hemmingway BS (1977) High temperature heat capacities of dolomite, talc, and tremolite, and implications to equilibrium in the siliceous dolomite system. Geol Soc Am Abs Prog 9:1060
Laiglesis A, Felix JF (1994) Estimation of thermodynamic properties of mineral carbonates at high and low temperatures from the sum of polyhedral contributions. Geochim Cosmochim Acta 58:3983-3991
Latimer WM (1951) Method of estimating the entropy of solid compounds. J Am Chem Soc 73:1480-1482
Latimer WM (1952) The Oxidation States of the Elements and their Potentials in Aqueous Solution. Prentice Hall., Englewood Cliffs, New Jersey
Lindsay WL (1979) Chemical Equilibria in Soils. Wiley, New York, 449 p.
Lichtner PC, Yahusaki S, Pruess K, Steefel CI (2004) Role of competitive cation exchange on chromatographic displacement of cesium in the valdose zone beneath the Hanford S/SX tank farm. Valdose Zone J 3:203-219
Liu W, Brugger J, Etchmann B, Testemale D, Hazemann J-L (2008) The solubility of nantokite (CuCl(s)) and Cu speciation in low-density fluids near the critical isochore: An *in situ* XAS study. Geochim Cosmochim Acta 72:4094-4106
Machesky ML, Palmer DA, Wesolowski DJ (1994) Hydrogen ion adsorption at the rutile-water interface to 250 °C. Geochim Cosmochim Acta 58:5627-5632
Machesky ML, Wesolowski DJ, Palmer DA Hayashy KI (1998) Potentiometric titrations of rutile suspensions to 250 °C. J Colloid Interface Sci 200:298-309
Majer V, Lu Hui V, Crovetto R, Wood RH (1991) Volumetric properties of aqueous 1-1 electrolyte solutions near and above the critical temperature of water II. Densities and apparent molar volumes of LiCl(aq) and NaBr(aq) at molalities from 0.0025 $mol.kg^{-1}$ to 3.0 $mol.kg^{-1}$, temperatures from 604.4 K to 725.5 K, and pressures from 18.5 MPa to 38.0 MPa. J Chem Thermo 23:365-378
Majer V, Wood RH (1994) Volumetric properties of aqueous 1-1 electrolyte solutions near and above the critical temperature of water III. Experimental densities and apparent molar volumes of CsBr(aq) up to the temperature 725.5 K and the pressure 38.0 MPa, comparison with other 1-1 electrolytes, and extrapolations to infinite dilution for NaCl(aq). J Chem Thermo 26:1143-1166
Majer V, Degrange S, Seldbauer J (1999) Temperature correlation of partial molar volumes of aqueous hydrocarbons at infinite dilution: test of equations. Fluid Phase Equilibria 158:419-428
Majer V, Sedlbauer J, Hnedkovsky L, Wood RH (2000) Thermodynamics of aqueous acetic and propionic acids and their anions over a wide range of temperatures and pressures. Phys Chem Chem Phys 2:2907-2917
Majer V, Sedlbauer J, Wood RH (2004) Calculation of standard thermodynamic properties of aqueous electrolytes and nonelectrolytes. *In*: Aqueous Systems at Elevated Temperatures and Pressures. Palmer DA, Fernandez-Prini R, Harvey AH (eds). Elsevier, Oxford, p 99-149
Manning CE (1994) The solubility of quartz in H_2O in the lower crust and upper mantle. Geochim Cosmochim Acta 58:4831-4839
McCullough JP, Scott DW (eds) (1968) Experimental Thermodynamics, v1. Calorimetry of non-reacting systems. Butterworths, London
Mesmer RE, Baes CF Jr, Sweeton H (1970) Acidity measurements at elevated temperatures. IV. Apparent dissociation of water in 1m potassium chloride up to 292°C. J Phys Chem 74:1937-1954
Mesmer RE, Patterson CS, Busey RH, Holmes HF (1989) Ionization of acetic acid in NaCl (aq) media: A potentiometric study to 572K and 130 bar. J Phys Chem 93:7483-7490
Monnin C (1990) the influence of pressure on the activity coefficients of the solutes and on the solubility of minerals in the system Na-Ca-Cl-SO_4-H_2O to 200 °C and 1 kbar, and to high NaCl concentration. Geochim Cosmochim Acta 54:3265-3282
Monnin C, Jeandel C, Cattaldo T, Dehairs F (1999) The marine barite saturation state of the worlds oceans. Mar Chem 65:253-261
Newton RC (1966c) Some calc-silicate equilibrium relations. Am J Sci 264:204-222
Nordstrom DK, Munoz JL (1994) Geochemical Thermodynamics. Blackwell Scientific Publications, Cambridge
Obsil M, Majer V, Grolier J-PE, Hefter GT (1996) Volumetric properties of and ion-pairing in aqueous solutions of alkali-metal sulfates at superambient conditions. J Chem Soc Faraday Trans 92:4445-4451
Oelkers EH, Cole DR (2008) Carbon dioxide sequestration: A solution to a global problem. Elements 4:305-310
Oelkers EH, Gislason SR, Matter J (2008) Mineral carbonation of CO_2. Elements 4:333-337

Oelkers EH, Helgeson HC (1988) Calculation of the thermodynamic and transport properties of aqueous species at high temperatures and pressures. Dissociation constants for supercritical alkali metal halides at temperatures from 400° to 800 °C and pressures from 500 to 4000 bars. J Phys Chem 92:1631-1639

Oelkers EH, Helgeson HC (1991) Calculation of the activity coefficients and degrees of formation of neutral ion pairs in supercritical electrolyte solutions. Geochim Cosmochim Acta 55:1235-1251

Oelkers EH, Schott J (2005) Geochemical aspects of CO_2 sequestration. Chem Geol 217:183-185

Oelkers EH, Valsami-Jones E (2008) Phosphate mineral reactivity and global sustainability. Elements 4:83-88

Palmer DA, Wesolowski DJ (1987) Acid association quotients of tris(hydroxymethyl)aminomethane in Aqueous NaCl Media to 200 °C. J Solution Chem 16:571-581

Palmer DA, Drummond SE (1988) Potentiometric determination of the molal formation constants of ferrous acetate complexes in aqueous solutions to high temperatures. J Phys Chem 92:6795-6800

Palmer DA, Hyde KE (1993) Ferrous chloride and acetate complexation in aqueous solutions at high temperatures. Geochim Cosmochim Acta 57:1393-1408

Palmer DA, Bell J (1994) Aluminium speciation and equilibria in aqueous Solution. IV. A potentiometric study of aluminium acetate complexation in acidic NaCl brines to 150 °C. Geochim Cosmochim Acta 58: 651-659

Palmer DA, Nguyen-Trung C (1995) A potentiometric study of the hydrolysis of ethylenediaminetetraacetic acid to 150 °C. *In:* Physical Chemistry of Aqueous Systems: Meeting the Needs of Industry: Proc. 12th Int Conf Properties Water Steam. Begell House, New York, 809-815

Palmer DA, Corti HR, Grotewold A, Hyde KE (2000) Potentiometric measurements of the thermodynamics of cadmium(ii) chloride complexation to high temperatures. *In:* Steam, Water, and Hydrothermal Systems Physics and Chemistry Meeting the Needs of Industry, Proc 13th Int Conf Properties Water Steam. Tremaine PR, Hill PG, Irish DE, Balakrishnan PV (eds), NRC Press, Ottawa, 736-743

Palmer DA, Bénézeth P, Wesolowski DJ (2001) Aqueous high temperature solubility studies. I. The solubility of boehmite at 150 °C as a function of ionic strength and pH as determined by "*in situ*" measurements. Geochim Cosmochim Acta 65:2081-2095

Parkhurst DL, Appelo CAJ (1999) User's guide to PHREEQC (Version 2)- A computer program for speciation, batch-reaction, one-dimensional transport, and inverse geochemical calculations. US Geol Surv Water Res Inv Report 99-4259

Pawley AR, Redfern SAT, Holland TJB (1996) Volume behavior of hydrous minerals at high pressure and temperature. 1. Thermal expansion of lawsonite, zoisite, clinosoisite, and diaspore. Am Mineral 81:335-340

Perfetti E, Pokrovski GS, Ballerat-Busseroles K, Majer V, Gilbert F (2008) Densitites and heat capacities of aqueous arsenious and arsenic acid solutions to 350 °C and 300 bar, and revised thermodynamics properties of $As(OH)_3$ (aq), $AsO(OH)_3$ (aq) and iron sulfarsenide minerals. Geochim Cosmochim Acta 72:713-731

Pitzer KS (1978) A thermodynamic model for aqueous solutions of liquid like density. Rev Mineral 17:97-142

Plyasunov AV, O'Connell JP, Wood RH (2000a) Infinite dilution partial molar properties of aqueous solutions of nonelectrolytes. I. Equations for partial molar volumes at infinite dilution and standard thermodynamic functions of hydration of volatile nonelectrolytes over wide ranges of conditions. Geochim Cosmochim Acta 64:495-512

Plyasunov AV, O'Connell JP, Wood RH, Shock EL (2000b) Infinite dilution partial molar properties of aqueous solutions of nonelectrolytes. II. Equations for the standard thermodynamic functions of hydration of volatile nonelectrolytes over wide ranges of conditions including subcritical temperatures. Geochim Cosmochim Acta 64:2779-2795

Plyasunov AV, Shock EL (2001a) Correlation strategy for determining the parameters of the revised Helgeson-Kirkham-Flowers model for aqueous nonelectrolytes. Geochim Cosmochim Acta 65:3879-3900

Plyasunov AV, Shock EL (2001b) Estimation of the Krichevskii parameter for aqueous nonelectrolytes. J Supercrit Fluids 20:91-103

Pokrovski GS, Gout R, Zotov A, Schott J, Harrichoury JC (1996) Thermodynamic properties and stoichiometry of the arsenic(III) hydroxide complexes at hydrothermal conditions. Geochim Cosmochim Acta 60:737-749

Pokrovski GS, Kara S, Roux J (2002). Stability and solubility of arsenopyrite, FeAsS, in crustal fluids. Geochim Cosmochim Acta 66:2361-2378.

Pokrovski GS, Roux J, Hazemann JL, Testemale D (2005a) An X-ray Absorption spectroscopy study of argutite solubility and germanium aqueous speciation in hydrothermal fluids to 500 °C and 400 bar. Chem Geol 217:127-145

Pokrovski GS, Roux J, Harrichoury JC (2005b) Fluid density control on vapor-liquid partitioning of metals in hydrothermal systems. Geology 33:657-660

Pokrovski GS, Borisova AYu, Roux J, Hazemann J-L, Petdang A, Tella M, Testemale D (2006). Antimony speciation in saline hydrothermal fluids: A combined X-ray absorption fine structure and solubility study. Geochim Cosmochim Acta 70:4196-4214

Pokrovski GS, Borisova AYu, Harrichoury J-C (2008a). The effect of sulfur on vapor-liquid fractionation of metals in hydrothermal systems. Earth Planet Sci Let 266:345-362

Pokrovski GS, Roux J, Hazemann J-L, Borisova AYu, Gonchar AA Lemeshko MP (2008b) In situ X-ray absorption spectroscopy measurement of vapor-brine fractionation of antimony at hydrothermal conditions. Mineral Mag 72:667-681

Pokrovski GS, Tagirov BR, Schott J, Bazarkina EF, Hazemann J-L, Proux O (2009) An *in situ* X-ray absorption spectroscopy study of gold-chloride complexing in hydrothermal fluids. Chem Geol 259:17-29

Pokrovskii VA, Helgeson HC (1995) Thermodynamic properties of aqueous species and the solubilities of minerals at high pressures and temperatures: The system Al_2O_3-H_2O-NaCl. Am J Sci 295:1255-1342

Pokrovskii VA, Helgeson HC, (1997) Thermodynamic properties of aqueous species and the solubilities of minerals at high pressures and temperatures: The system Al_2O_3-H_2O-KOH. Chem Geol 137: 221-242

Powell R, Holland T, Worley B (1998) Calculating diagrams involving solid solutions via non-linear equations with examples using THERMOCALC. J Metamorph Petrol 16: 577-588

Prieto M (2009) Thermodynamics of solid solution-aqueous solution systems. Rev Mineral Geochem 70:47–85

Quist AS, Marshall WL (1968) Electrical conductances of aqueous sodium chloride solutions from 0-800 °C and at pressures to 4000 bars. J Phys Chem 72:684-703

Ragnarsdottir KV, Fournier P, Oelkers EH, Harrichoury JC (2001) Experimental determination of the complexation of strontium and cesium with acetate in high temperature aqueous solutions. Geochim Cosmochim Acta 65:3955-3964

Ravel B, Newville M (2005) ATHENA, ARTEMIS, HEPHAESTUS: data analysis for X-ray absorption spectroscopy using IFEFFIT. J Synchrotron Rad 12:537-541

Ridley MK, Wesolowski DJ, Palmer DA, Benezeth P, Kettler RM (1997). The effect of sulphate on the release rate of Al^3+ from gibbsite in low temperature acidic water. Environ Sci Technol 31:1922-1925

Ridley MK, Palmer DA, Wesolowski DJ, Kettler RM (1998) Cadmium malonate complexation in aqueous sodium trifluoromethanesulfonate media to 75 °C; including dissociation quotients of malonic Acid. J Solution Chem 27:195-216

Ridley MK, Machesky ML, Wesolowski DJ, Palmer DA (1998b) Potentiometric and solubility studies of association quotients of aluminum malonate complexation in NaCl media to 75 °C. Geochim Cosmochim Acta 62:2279-2291

Ridley MK, Machesky ML, Wesolowski DJ, Palmer DA (1999) Calcium adsorption at the rutile-water interface: A potentiometric study in NaCl media to 250 °C. Geochim Cosmochim Acta 63:3087-3096

Ridley MK, Xiao C, Palmer DA, Wesolowski DJ (2000) Thermodynamic properties of the hydrolysis of morpholine as a function of temperature and ionic strength. J Chem Eng Data 45:502-507

Ridley MK, Machesky ML, Wesolowski DJ, Palmer DA (2005) Surface complexation of neodymium at the rutile-water interface: A potentiometric and modeling study in NaCl media to 250 °C. Geochim Cosmochim Acta 69:63-81

Robie RA, Hemmingway BS (1995) Thermodynamic properties of minerals and related substances at 298.15 K and 1 bar (10^5 Pascals) pressure and at higher temperatures. USGS Geol Survey Bull No. 2131

Robie RA, Hemmingway BS, Fisher JR (1979) Thermodynamic properties of minerals and related substances at 298.15 K and 1 bar (10^5 Pascals) pressure and at higher temperatures. USGS Geol Survey Bull No. 1452

Robinson GR Jr, Haas JL Jr (1983) Heat capacity relative enthalpy and calorimetric entropy of silicate minerals. An empirical method of prediction. Am Mineral 68:541-553

Salje E, Werneke C (1982) The phase-equilibrium between sillimanite and andalusite as determined from lattice vibrations. Contrib Mineral Petrol 79:56-67

Saldi G (2009) The thermodynamics and kinetics of magnesite-water-CO_2 interaction. PhD Thesis, Université Paul Sabatier, Toulouse, France.

Sassani DC, Shock EL (1990) Speciation and solubility of palladium in aqueous magmatic-hydrothermal solutions. Geology 18:925-928

Sassani DC, Shock EL (1992) Estimation of standard partial molal entropies of aqueous ions at 25 °C and 1 bar. Geochim Cosmochim Acta 56:3895-3908

Schott J, Pokrovski GS, Tagirov BR, Hazemann JL, Proux O (2006) First *in situ* XAFS determination of gold solubility and speciation in sulfur-rich hydrothermal solutions. Geochim Cosmochim Acta 70:A564

Schott J, Pokrovsky OS, Oelkers EH (2009) The link between mineral dissolution/precipitation kinetics and solution chemistry. Rev Mineral Geochem 70:207–258

Schulte MD, Shock EL (1993) Aldehydes in hydrothermal solution: Standard molal properties and relative stabilities at high temperatures and pressures. Geochim Cosmochim Acta 57:3835-3846

Schulte MD, Shock EL, Wood RH (2001) The temperature dependence of the standard-state thermodynamic properties of aqueous nonelectrolytes. Geochim Cosmochim Acta 65:3919-3930

Schulte MD, Shock EL, Obšil M, Majer V (1997) Volumes of aqueous alcohols, ethers and ketones up to 523 K and 28 MPa. J Chem Thermo 31:1195-1229

Sedlbauer J, Majer V (2000) Data and models for calculating the standard thermodynamic properties of aqueous nonelectrolyte solutes under hydrothermal conditions. Eur J Mineral 12:1109-1122

Sedlbauer J, O'Connell JP, Wood RH (2000) A new equation of state for correlation and prediction of standard molal thermodynamic properties of aqueous electrolytes and nonelectrolytes at high temperatures and pressures. Chem Geol 163:43-63

Seward TM, Driesner T (2004). Hydrothermal solution structure: experiments and computer simulations. *In:* Aqueous Systems at Elevated Temperatures and Pressures. Palmer DA, Fernandez-Prini R, Harvey AH (eds), Elsevier, p 149-182

Sherman DM (2001) Quantum chemistry and classical simulations of metal complexes in aqueous solutions. Rev Mineral Geochem 42:273-317

Sherman DM, Ragnarsdottir KV, Oelkers EH (2000a) Speciation of tin (Sn^{2+} and Sn^{4+}) in aqueous Cl solutions from 25 to 300 °C: An *in situ* EXAFS study. Chem Geol 167:169-176

Sherman DM, Ragnarsdottir KV, Oelkers EH (2000b) Antimony transport in hydrothermal solutions: An EXAFS study of antimony (V) complexation with sulphide and hydroxide ligands at temperatures from 25 to 300 °C at Psat. Chem Geol 167:161-168

Shvarov YV (1999) Algorithmization of the numerical equilibrium modeling of dynamic geochemical processes. Geochem Int 37:571–576

Shvarov Y, Bastrakov E (1999). A software package for geochemical equilibrium modeling. User's Guide. Australian Geological Survey Organization, Department of Industry, Science and Resources.

Shock EL, Helgeson HC (1988) Calculation of the thermodynamic and transport properties of aqueous species at high pressures and temperatures: Correlation algorithms for ionic species and equation of state predictions to 5 kb and 1000° C. Geochim Cosmochim Acta 52:2009-2036

Shock EL, Helgeson HC (1989) Corrections to: Calculation of the thermodynamic and transport properties of aqueous species at high pressures and temperatures: Correlation algorithms for ionic species and equation of state predictions to 5 kb and 1000° C. Geochim Cosmochim Acta 53:215-215

Shock EL, Helgeson HC (1990) Calculation of the thermodynamic and transport properties of aqueous species at high pressures and temperatures: Standard partial molal properties of organic aqueous species. Geochim Cosmochim Acta 54:915-945

Shock EL, Koretsky CM (1993) Metal-organic complexes in geochemical processes: Calculation of standard partial molal thermodynamic properties of aqueous acetate complexes at high pressures and temperatures. Geochim Cosmochim Acta 57:4899-4922

Shock EL, Helgeson HC Sverjensky DA (1989) Calculation of the thermodynamic and transport properties of aqueous species at high pressures and temperatures: Standard partial molal properties of inorganic neutral species. Geochim Cosmochim Acta 53:2157-2183

Shock EL, Oelkers EH, Johnson JW, Sverjensky DA, Helgeson HC (1992) Calculation of the thermodynamic and transport properties of aqueous species at high pressures and temperatures: Effective electrostatic radii, dissociation constants and the standard molal properties to 1000 °C and 5 kbar. J Chem Soc Faraday Trans 88:803-826

Shock EL, Sassani DC, Willis M, Sverjensky DA (1997) Inorganic species in geological fluids: Correlations among standard molal thermodynamic properties of aqueous ions and hydroxide complexes. Geochim Cosmochim Acta 61:907-950

Skinner BJ (1966) Thermal expansion. Geol Soc Am Mem 97:75-96

Slavik M, Sedlbauer J, Ballerat-Busserolles K, Majer V (2007) Heat capacities of aqueous solutions of acetone; 2,5-hexanedione; diethyl ether; 1,2-dimethoxyethane; benzyl alcohol; and cyclohexanol at temperatures to 523 K. J Solution Chem 36:107-134

Spear FS, Cheney JT (1989) A petrogenetic grid for pelitic schists in the system SiO_2-Al_2O_3-FeO-MgO-K_2O-H_2O. Contrib Mineral Petrol 101:149-164

Staples BR, Nuttall RL (1997) The activity and osmotic coefficients of aqueous calcium chloride at 298.15 K. J Phys Chem Ref Data 6:385-407

Stefansson A (2001) Dissolution of primary minerals in natural waters. I. Calculation of mineral solubilities from 0 to 350 °C. Chem Geol 172:225-250

Steefel CI, DePaolo DJ, Lichtner PC (2005) Reactive transport modeling: An essential tool and a new research approach for the Earth sciences. Earth Planet Sci Lett 240:539-558

Stull DR, Prophet H (1971) JANAF Thermochemical Tables 2nd Edition. NSRDS-NBS

Stumm W, Morgan JJ (1996) Aquatic Chemistry: Chemical Equilibria and Rates. Wiley-Interscience, New York

Sverjensky DA (1987) Calculation of the thermodynamic properties of aqueous species and the solubilities of minerals in supercritical electrolyte solutions. Rev Mineral 7:177-209

Sverjensky DA, Molling PA (1992) A linear free energy relationship for crystalline solids and aqueous ions. Nature 356:231-234

Sverjensky DA, Shock EL, Helgeson HC (1997) Prediction of the thermodynamic properties of aqueous metal complexes to 1000 °C and 5 kb. Geochim Cosmochim Acta 61:1359-1412

Sweeton FH, Mesmer RE, Baes CF (1974) Acidity measurements at elevated temperatures. VII. Dissociation of water. J Solution Chem 3:191-214

Tanger JC, Helgeson HC (1988) Calculation of the thermodynamic and transport properties of aqueous species at high pressures and temperatures: Revised equation of state for the standard partial molal properties of ions and electrolytes. Am J Sci 288:19-98

Tardy Y, Gartner L (1977) Relations among Gibbs free energies and enthalpies of formation of sulfates, nitrates, carbonates, oxides and aqueous ions. Contrib Mineral Petrol 63:89-102

Tardy Y, Vieillard P (1977) Relations among Gibbs free energies and enthalpies of formation of phosphates, oxides, and aqueous ions. Contrib Mineral Petrol 63:75-88

Tardy Y, Fritz B (1981) An ideal solid solution model for calculating solubilites of clay minerals. Clay Mineral 16:361-373

Testemale D, Hazemann J-L, Pokrovski GS, Joly Y, Roux J, Argoud R, Geaymond O (2004) Structural and electronic evolution of the $As(OH)_3$ molecule in high temperature aqueous solutions: An x-ray absorption investigation. J Chem Phys 121:8973-8982

Testemale D, Argoud R, Geaymond O, Hazemann J-L (2005). High pressure/high temperature cell for x-ray absorption and scattering techniques. Rev Sci Instrum 76:043905-043909

Tremaine PR, Shvedov D, Xiao C (1997) Thermodynamic properties of aqueous morpholine and morpholinium chloride at temperatures from 10 to 300°C: Apparent molar volumes, heat capacities, and temperature dependence of ionization. J Phys Chem B 101:409-419

Tremaine P, Zhang K, Bénézeth P, Xiao C (2004) Dissociation equilibria of weak acids and bases. *In:* Aqueous Systems at Elevated Temperatures and Pressure. Physical Chemistry in Water Steam and Hydrothermal Solutions. Palmer DA, Fernández-Prini R, Harvey AH (eds), Elsevier/Academic Press, p 441-492

Van der Lee J, De Windt L (2001) Present state and future directions of modeling of geochemistry on hydrogeological systems. J Contam Hydrol 47:265-282

Van der Lee J, De Windt L, Lagneau V, Goblet P. (2002). Presentation and application of the reactive transport code HYTEC. Dev Water Sci 47:599-606

Valsami-Jones E, Ragnarsdottir KV, Putnis A, Bosbach D, Kemp AJ, Cressey G (1998) The dissolution of apatite in the presence of aqueous metal cations at pH 2-7. Chem Geol 151:215-233

Vieillard P (2000) A new method for the prediction of Gibbs free energies of formation of hydrates clay minerals based on the electronegativity scale. Clays Clay Mineral 48:459-473

Vieillard P (2002) A new method for the prediction of Gibbs free energies of formation of phyllosilicates based on the electronegativity scale. Clays Clay Mineral 50:352-363.

Vieillard P, Tardy Y (1984) Thermochemical properties of phosphates. *In*: Phosphate Minerals. Niragu, JO, Moore PB (Eds), Springer-Verlag, New York, p 171–198

Wagman DD, Evans WH, Parker VB, Schumm RH, Hallow I, Bailey SS, Churney KL, Nuttall, RL (1982) The NBS tables of chemical thermodynamic properties. Selected values for inorganic and C_1 and C_2 organic substances in SI units. J Phys Chem Ref Data 11 suppl. 2:1-392

Wesolowski DJ (1992) Aluminum speciation and equilibria in aqueous solution: I. The solubility of gibbsite in the system Na-K-Cl-OH-$Al(OH)_4$ from 1 to 100 °C. Geochim Cosmochim Acta 61:1065-1091

Wesolowski DJ, Palmer DA (1989) Acid Association of "Bis-tris", 2,2-Bis(hydroxymethyl)-2,2',2"-nitrilotriethanol, in Aqueous Sodium Chloride Media to 125 °C. J Solution Chem 18:545-559

Wesolowski DJ, Bénézeth P, Palmer DA (1998) ZnO solubility and Zn^{2+} complexation by chloride and sulfate in acidic solutions to 290 °C with *in situ* pH measurement. Geochim Cosmochim Acta 62:971-984

Wesolowski DJ, Machesky ML, Palmer DA, Anovitz LM (2000) Magnetite surface charge studies to 290 °C from *in situ* pH titrations. Chem Geol 167:193-229

Wesolowski DJ, Ziemniak SE, Anovitz LM, Machesky ML, Bénézeth P, Palmer DA (2004) Solubility and surface adsorption characteristics of metal oxides. *In*: Aqueous Systems at Elevated Temperatures and Pressure. Physical Chemistry in Water Steam and Hydrothermal Solutions. Palmer DA, Fernández-Prini R, Harvey AH (eds), Elsevier/Academic Press, p 493-595

Williams-Jones AE, Heinrich CA (2005) 100th anniversary special paper. Vapor transport of metals and the formation of magmatic-hydrothermal ore deposits. Econ Geol 100:1287-1312

Wightman PG, Fein JB, Wesolowski DJ, Phelps TJ, Bénézeth P, Palmer D.A. (2001) Measurement of bacterial surface protonation constants for two species at elevated temperatures. Geochim Cosmochim Acta 65:3657-3669

Windom KE, Boettcher AL (1976) The effect of reduced activity of anorthite on the reaction grossular + quartz = anorthite + wollastonite: a model for plagioclase in the Earth's lower crust and upper mantle. Am Mineral 61:889-896

Wolery TJ (1983) EQ3NR, A computer program for geochemical aqueous speciation-solubility calculations: Users guide and documentation. UCRL-53414. Lawrence Livermore National Laboratory, Livermore, CA

Wood SA, Wesolowski DJ, Palmer DA (2002) The aqueous geochemistry of rare earth element IX. A potentiometric study of Nd^{3+} complexation with acetate in 0.1 molal NaCl solution from 25 to 225 °C. Chem Geol 167:231-253

Xiao C, Bianchi H, Tremaine PR (1997) Excess molar volumes and densities of (methanol+water) at temperatures between 323 K and 573 K and pressures of 7.0 MPa and 13.5 MPa. J Chem Thermo 29:261-286

Xie W, Trevani L, Tremaine PR (2004) Apparent and standard partial molar heat capacities and volumes of aqueous tartaric acid and its sodium salts at elevated temperature and pressure. J Chem Thermo 36:127-140

Yezdimer EM, Sedlbauer J, Wood RH (2000) Predictions of thermodynamic properties at infinite dilution of aqueous organic species at high temperatures via functional group additivity. Chem Geol 164:259-280

Zaho GC, Cawood PA, Wilde SA, Min S, Lu LZ (2000) Metamorphism of basement rock in the Central Zone of the North China Craton: Implications for Paleoproterozic tectonic evolution. Precamb Res 103:55-88

Zhang GX, Spycher N, Sonnenthal E, Steefel CI, Xu T (2008) Modeling reactive multiphase flow and transport of concentrated solutions. Nucl Technol 164:180-195

Zhu C, Anderson GM (2002) Environmental Applications of Geochemical Modeling. Cambridge University Press, New York

Reviews in Mineralogy & Geochemistry
Vol. 70 pp. 47-85, 2009

2

Thermodynamics of Solid Solution-Aqueous Solution Systems

Manuel Prieto

Department of Geology
University of Oviedo
Oviedo, Spain

mprieto@geol.uniovi.es

INTRODUCTION

When two solutes crystallize simultaneously from an aqueous phase and have similar crystal structures, a solid solution is likely to form. Indeed, although often disregarded, solid solution-aqueous solution (SS-AS) effects are ubiquitous in both natural and industrial crystallization processes. In nature, and particularly on the Earth's surface environments, the crystallization of minerals from multicomponent aqueous solutions provokes in most cases the formation of solids with more or less wide compositional ranges, i.e., solid solutions. Moreover, the interaction between existing minerals and water frequently leads to surface precipitation and dissolution-recrystallization processes, in which a number of substituting ions (major, minor, or trace) redistribute to adapt to the new conditions.

The study of SS-AS relationships between minerals and solutions can provide very valuable information about natural waters, contamination of soils and aquifers, and global element cycles. For instance, in diagenetic studies of sedimentary rocks, the minor and trace element concentrations are commonly used to infer the composition of the crystallizing fluids and have been demonstrated to play a major role in the dissolution-recrystallization processes that lead to the formation of authigenic minerals (e.g., Böttcher 1997b; Rimstidt et al. 1998; Kulik et al. 2000; Mucci 2004). Likewise, the concentrations of specific minor and trace elements in calcite and aragonite precipitated by marine organisms have been shown to correlate with various parameters of the growth environment, including temperature, salinity, nutrient levels, carbonate concentration, and water chemistry. These findings suggest that compositional signatures recorded in biogenic carbonates can be powerful tools in reconstruction of the past from the fossil record (e.g., Rosenthal et al. 1997; Stoll et al. 2002; Rickaby et al. 2002). Obviously, for these proxies to be effective, it is essential to rigorously investigate the underlying physical and chemical mechanisms both in the absence and in the presence of vital effects (e.g., Wasylenki et al. 2005; Stephenson et al. 2008).

Since the early eighties, the study of SS-AS systems has garnered much attention from environmental geochemists because they have been demonstrated to play major roles in the mobility of dissolved metals in contaminated environments. Particularly, the incorporation of divalent metals into the structure of rhombohedral carbonates has been one of the most studied SS-AS processes due to its significant potential for the removal of dissolved toxic-metals from water (e.g., Kornicker et al. 1985; Davis et al. 1987; Zachara et al. 1991; Stipp et al. 1992; Tesoriero and Pankow 1996; Chiarello et al. 1997; Freij et al. 2003; Godelistsas et al. 2003; Katsikopoulos et al. 2008a). More recently, solid solutions involving other mineral groups, anionic substitutions, coupled substitutions, etc., have also been widely studied from an environmental perspective (e.g., Prieto et al. 2002; Andara et al. 2005; Curti et al. 2005; Heberling et al. 2008).

1529-6466/09/0070-0002$05.00 DOI: 10.2138/rmg.2009.70.2

Unfortunately, despite the numerous studies and implications, solid-solution effects are often disregarded in water-rock interaction geochemistry, probably because the physicochemical reactions between aqueous solutions and solid solutions are complex and difficult to implement in computer codes for geochemical modeling. Moreover, although the most recent versions of programs like PHREEQC (Parkhurst and Appelo 1999) or GEMS (Kulik 2002, 2006) incorporate improved solid-solution tools, there are few available data about the extent and degree of ideality of many geochemically-relevant solid solutions. A goal of this chapter is to review the theoretical background developed in the last decades on this topic, and to alert the scientific community for the need to accurately account for solid solution formation.

First, this chapter reviews the key concepts describing equilibrium in binary SS-AS systems, including references to the most relevant contributions (e.g., Thorstenson and Plummer 1977; Lippmann 1980; Plummer and Busenberg 1987; Glynn and Reardon 1990; Gamsjäger et al. 2000; Glynn 2000). These works constitute the basis for understanding SS-AS effects, but in most cases do not consider the concept of supersaturation, a crucial parameter from the point of view of geochemical modeling. In reality, whereas the equilibrium behavior is well established, the concept of supersaturation in SS-AS systems is still contentious (Prieto et al. 2007). Here, the interrelation and thermodynamic meaning of the most common supersaturation expressions available in the geochemical literature (Prieto et al. 1993; Astilleros et al. 2003a; Monnin and Cividini 2006; Shtukenberg et al. 2006) are discussed in depth in a specific section. Finally, this chapter includes an introduction to kinetic controls on ion partitioning during nucleation and growth. Factors such as supersaturation, solution chemistry, and overall growth rates (e.g., Lorens 1981; Mucci and Morse 1983; Dromgoole and Walter 1990; Prieto et al. 1997; Pina and Putnis 2002) are known to have a significant influence on the effective distribution coefficients. Further, the correlation of these macroscopic effects with the study by surface-sensitive techniques of the mechanisms that operate at a molecular scale is without a doubt the key to developing consistent kinetic models (e.g., Paquete and Reeder 1995; Xu et al. 1996; Reeder et al. 1999; Stipp et al. 1992, 2003; Lea et al. 2003; Astilleros et al. 2000, 2003a). All these aspects are briefly introduced as essential complements to the previous thermodynamic background.

EQUILIBRIUM RELATIONSHIPS

Consider a hypothetical compound, *BA*. Equations for the dissolution of minerals are usually written as dissociation reactions. Thus, for a pure, stoichiometric mineral *BA*, the dissolution reaction may be written:

$$BA_{(s)} \leftrightarrow B^{n+}_{(aq)} + A^{n-}_{(aq)} \quad (1)$$

where the subscripts *s* and *aq* stand for solid and aqueous species, respectively. Now, according to the law of mass action, the equilibrium distribution between the species in Equation (1) is given by:

$$K_{BA} = \frac{[B^{n+}][A^{n-}]}{a_{BA}} \quad (2)$$

where the bracketed quantities denote the activities of the aqueous ions and a_{BA} represents the activity of *BA* in the solid phase. Thus, the equilibrium constant K_{BA} is in fact the solubility product of *BA*, since the activity of a pure solid is equal to one by definition (the standard state condition). Of course, the expression in Equation (2) refers to a very simple compound. For pure solids with a more complex stoichiometry, the solubility product is given by the generalized expression $\Pi[i]^{vi}$, where vi stands for the stoichiometric number of ion *i* in the solute formula. Here, we are going to consider simple *BA*-type solids in order to simplify the expressions, but the generalization to more complex stoichiometries is straightforward.

Whereas the concept of solubility is quite simple when applied to pure solids, the equilibrium behavior in SS-AS systems is a complex issue that has generated a number of controversies during the last decades (Thortenson and Plummer 1977; Lafont 1978; Garrels and Wollast 1978; Thortenson and Plummer 1978; Stoessell 1992; Glynn et al. 1992; Königsberger and Gamsjäger 1992; Glynn and Reardon 1992). At present, most of these controversies can be considered as being resolved, but from time to time new controversies arise (Shtukenberg et al. 2006; Prieto et al. 2007). Both now and in the past, the question has been how we can express in a simple way the solubility of a solid solution. Apparently the answer is quite simple in light of Equation (2). In the case of a binary solid solution $(B,C)A$, in which B^{n+} and C^{n+} ions substitute for each other, the solid phase consists of two components, BA and CA, and we have to write a second reaction in addition to Equation (1) to describe the dissolution, namely:

$$CA_{(s)} \leftrightarrow C^{n+}_{(aq)} + A^{n-}_{(aq)} \tag{3}$$

Now, the equilibrium condition for this second reaction is given by:

$$K_{CA} = \frac{[C^{n+}][A^{n-}]}{a_{CA}} \tag{4}$$

where K_{CA} is the thermodynamic solubility product of a pure CA solid and a_{CA} is the activity of the component CA in the solid phase. It is worth noting that Equations (2) and (4) cannot be used independently to describe equilibrium in the $(B,C)A$-H_2O system. Equilibrium requires both conditions to be fulfilled simultaneously and this is the first problem in dealing with these kinds of systems. The question is how to handle these two equations to find simple expressions that define the solubility of a given solid solution member and the distribution of the substituting ions between the solid and the aqueous phases.

A second difficulty arises from the fact that, in dealing with solid solutions, the activities of the solid-phase components are not equal to one, but depend in a complex way on the composition of the solid solution. As can be seen in Equations (2) and (4), describing thermodynamic equilibrium in SS-AS systems also requires the determination of the activities of the aqueous ions, but the Debye-Hückel-based methods offer a straightforward model to relate concentrations to activities in water for a wide range of ionic strengths. Moreover, a number of computer applications and databases compile the critical stability constants of the most relevant aqueous species (see e.g., Appelo and Postma 2005). Unfortunately, for the activities of the components of a solid solution, the situation is more complicated. In such a case, the substituting ions randomly occupy specific sites in a particular crystal structure with a specific chemical nature, and this prevents the use of a general empirical formula (in a way similar to the Debye-Hückel method) for the determination of activity coefficients in solids. Moreover, the available information about the thermodynamic properties of solid solutions is quite limited. For these reasons, and previous to any consideration of the equilibrium behavior in SS-AS systems, we are going to consider the solid-phase problem, understanding that this is the main difficulty in the case of non-ideal solid solutions.

Ideal and non-ideal solid solutions

In mineral thermodynamics, the standard state of a solid is defined as the pure phase and the activities of the components in the solid are related to their concentrations (in mole fractions) with respect to the standard state. Thus, for the BA component in the binary $(B,C)A$ solid solution, the activity is dimensionless and is given by:

$$a_{BA} = \gamma_{BA}\frac{X_{BA}}{X^{\circ}_{BA}} = \gamma_{BA}X_{BA} \tag{5}$$

where X_{BA} is the mole fraction of the component BA in the solid phase and $X^{\circ}_{BA} = 1$ is the mole fraction in the standard state. As for solutes, a dimensionless activity coefficient γ_{BA} is used to correct for non-ideal behavior. An analogous expression defines a_{CA} as a function of X_{CA} and γ_{CA}. Obviously, for the case of a binary solid solution, $X_{CA} = 1 - X_{BA}$ and both mole fractions stand for the same solid composition.

In ideal solid solutions, γ_{BA} and γ_{CA} are equal to unity, which means that $a_{BA} = X_{BA}$ and $a_{CA} = X_{CA}$ for the whole compositional range. However, in non-ideal solid solutions, both γ_{BA} and γ_{CA} vary with composition and their determination is crucial in describing the SS-AS thermodynamics. The ideal or non-ideal character of a solid solution depends on the so-called mixing properties, which express the difference between the thermodynamic properties of the solid solution and those of a compositionally-equivalent mechanical mixture of the two end-members. An extended review on classical solid-solution thermodynamics is available in a number of books (e.g., Saxena 1973; Putnis 1992), so we only present here an outline of the concepts relevant for the estimation of solid phase activity coefficients.

Whereas a solid solution is a single phase with non-stoichiometric composition and substitutional disorder, a mechanical mixture is an aggregation of two pure phases. Therefore, the free energy per mole of a mechanical mixture G_{MM} is a linear combination of the free energies of the pure end-members, according to:

$$G_{MM} = X_{BA}G_{BA} + X_{CA}G_{CA} \tag{6}$$

where G_{BA} and G_{CA} are the free energies per mole of *BA* and *CA*. Now, the free energy of mixing ΔG_M of the solid solution represents the change in the free energy during the formation of the solid solution from a mechanical mixture of the pure end-members, i.e.:

$$\Delta G_M = G_{SS} - G_{MM} \tag{7}$$

where G_{SS} is the free energy (per mole) of the solid solution. The free energy of mixing controls the stability of a solid solution and can be expressed as the sum of an enthalpy term (ΔH_M) and an entropy term (ΔS_M) at a given temperature, that is:

$$\Delta G_M = \Delta H_M - T\Delta S_M \tag{8}$$

Since the formation of solid solutions implies an increase in disorder due to the random substitution of one ion for another, the entropy of mixing is always positive. Consequently, the contribution of ΔS_M reduces the free energy of mixing and favors the formation of the solid solution. The only term in Equation (8) that can restrict the degree of solid solution, therefore, is ΔH_M. This parameter describes the change in enthalpy when a solid solution forms from the pure end-members and arises from the interactions between dissimilar atoms originated by atomic substitution, when *B* and *C* mix randomly in the crystal structure. When the presence of interactions *B-C* between dissimilar atoms produces no difference with respect to a mechanical mixture in which there are only interactions *B-B* and *C-C*, the solid solution is said to be ideal. In such a case $\Delta H_{M,id} = 0$, the enthalpy is independent of the distribution of *B* and *C*, and one can assume prefect random mixing because there is no trend for the dissimilar atoms to be near or distant, that is:

$$\Delta G_{M,id} = -T\Delta S_{M,id} = \mathrm{R}T(X_{BA}\ln X_{BA} + X_{CA}\ln X_{CA}) \tag{9}$$

where the label "*id*" has been added to specify that we are dealing with an ideal solid solution. The ideal solid solution is stabilized over the whole compositional range by the negative entropy term. In Equation (9), the ideal entropy of mixing is given by:

$$\Delta S_{M,id} = -\mathrm{R}(X_{BA}\ln X_{BA} + X_{CA}\ln X_{CA}) \tag{10}$$

where R is the gas constant. This formulation implies perfect random mixing and the absence of non-configurational contributions to the entropy of mixing. Note that $\Delta S_{M,id}$ is positive for any composition since both X_{BA} and X_{CA} are smaller than unity and, consequently, their logarithms are negative. In reality, no solid solution is absolutely ideal, but in many cases the ideal model can be used to approach solid-solution thermodynamics in a reasonable way.

In contrast, in a non-ideal solid solution, $\Delta H_M \neq 0$ and the atomic substitution can deviate from perfect randomness. The non-ideality degree of a solid solution is usually evaluated in terms of the excess free energy of mixing (ΔG_E), which is given by the difference between the free energy of mixing of the current solid solution and that of an equivalent ideal solid solution at the same temperature, i.e.:

$$\Delta G_E = \Delta G_M - \Delta G_{M,id} \tag{11}$$

The excess free energy of mixing can be expressed as the sum of an excess enthalpy term (ΔH_E) and an excess entropy term (ΔS_E), which can also be defined by comparison with an equivalent ideal solid solution. Thus, the excess entropy of mixing accounts for both the deviation from perfect randomness and the non-configurational contributions to the entropy of mixing. In all cases, $\Delta H_E = \Delta H_M$, given that, by definition, $\Delta H_{M,id} = 0$. In an analogous way, we can define the excess molar volume of mixing $\Delta V_E = \Delta V_M$ with $\Delta V_{M,id} = 0$. Obviously, for ideal solid solutions, all of the excess parameters are equal to zero.

When the two end-members of the solid solution are isomorphic, the mixing properties can be represented by single curves across the whole composition range. In practice, when the value of ΔG_E is experimentally known for certain compositions, the excess free energy of mixing can be expressed as a function using a fitting model. Here we are going to use the so-called Guggenheim expansion series (Guggenheim 1937), but there are other suitable fitting functions available in the literature (see Glynn and Reardon 1990 and references therein). The Guggenheim function satisfies the required property that $\Delta G_E = 0$ at the end-member compositions and, applying the Redlich and Kister (1948) modification, it is given by:

$$\Delta G_E = X_{BA} X_{CA} \mathrm{R}T[a_0 + a_1(X_{BA} - X_{CA}) + a_2(X_{BA} - X_{CA})^2 + \cdots] \tag{12}$$

where a_0, a_1, a_2,... are dimensionless fitting parameters independent of composition, but may be functions of temperature and pressure. Obviously, for ideal solid solutions, all of the Guggenheim fitting parameters are equal to zero. In contrast, to quantify the non-ideal character of a solid solution, the determination of Guggenheim parameters is a good option. In general, no more than two or three parameters are needed to quantify the non-ideality (Glynn and Reardon 1990, 1992). The simplest kind of non-ideal solid solution is the one in which all but the first parameter in Equation (12) are zero. In such a case, ΔG_E becomes a function of composition that is symmetric around $X_{BA} = X_{CA} = 0.5$, and the solid solution is termed "regular" (Hildebrand 1936). The "strictly regular" solid-solutions (Navrotsky 1987) constitute a special class of these kinds of symmetric solutions, for which ΔS_E and ΔV_E are zero, such that a_0 and ΔG_E are independent of temperature and pressure. Moreover, this requirement implies that $\Delta G_E = \Delta H_M$, as can be deduced from Equations (8), (9), and (11).

Activities in non-ideal solid solutions

According to Equation (7), the total free energy of an ideal solid solution is given by:

$$G_{SS,id} = G_{MM} + \Delta G_{M,id} \tag{13}$$

Therefore, combining Equations (6), (9), and (13) the free energy of an ideal solid solution may be written as:

$$G_{SS,id} = X_{BA} G_{BA} + X_{CA} G_{CA} + \mathrm{R}T(X_{BA} \ln X_{BA} + X_{CA} \ln X_{CA}) \tag{14}$$

and, rearranging this expression,

$$G_{SS,id} = X_{BA}(G_{BA} + \mathrm{R}T \ln X_{BA}) + X_{CA}(G_{CA} + \mathrm{R}T \ln X_{CA}) \tag{15}$$

Now, the two terms in parenthesis represent the chemical potentials (μ_{BA} and μ_{CA}) or partial molar free energies of the *BA* and *CA* components in the ideal solid solution, i.e.:

$$\mu_{BA} = G_{BA} + RT \ln X_{BA} \tag{16}$$

$$\mu_{CA} = G_{CA} + RT \ln X_{CA} \tag{17}$$

Thus, by combining Equations (15), (16), and (17), we get:

$$G_{SS} = \mu_{BA} X_{BA} + \mu_{CA} X_{CA} \tag{18}$$

This last expression holds for non-ideal solid solutions as well (note that the additional subscript *id* has been suppressed). However, in the case of non-ideal solid solutions, the expressions defining the chemical potentials (Eqns. 16 and 17) must be modified. When the solid solution is non-ideal, its total free energy differs from that of an equivalent ideal solid solution since $\Delta G_M \neq \Delta G_{M,id}$ in Equation (13). This deviation is accounted for by the introduction of the activities of the *BA* and *CA* components in the solid solution, such that Equations (16) and (17) become:

$$\mu_{BA} = G_{BA} + RT \ln a_{BA} \tag{19}$$

$$\mu_{CA} = G_{CA} + RT \ln a_{CA} \tag{20}$$

Since at the standard state a_{BA} or a_{CA} =1, the standard-state chemical potentials (μ°_{BA} and μ°_{CA}) of *BA* and *CA* are exactly equal to the molar Gibbs free energies of the pure end-members, G_{BA} and G_{CA}, and will be a function of temperature and pressure only. Using the activities, the free energy of mixing of a non-ideal solid solution can be written in a form reminiscent of Equation (9), i.e.:

$$\Delta G_M = RT(X_{BA} \ln a_{BA} + X_{CA} \ln a_{CA}) \tag{21}$$

as can be easily found by combining Equations (7), (18), (19), and (20). Now, taking into account the definition of the excess free energy of mixing:

$$\Delta G_E = RT(X_{BA} \ln a_{BA} + X_{CA} \ln a_{CA}) - RT(X_{BA} \ln X_{BA} + X_{CA} \ln X_{CA}) \tag{22}$$

Finally, considering the relationship between activity (a), activity coefficient (γ), and mole fraction (X) given in Equation (5), Equation (22) reduces to:

$$\Delta G_E = RT(X_{BA} \ln \gamma_{BA} + X_{CA} \ln \gamma_{CA}) \tag{23}$$

This last expression allows for the derivation of the activity coefficients from the excess free energy of mixing, according to:

$$\ln \gamma_{BA} = \left(\Delta G_E + X_{CA} \frac{\partial \Delta G_E}{\partial X_{BA}} \right) / RT \tag{24}$$

$$\ln \gamma_{CA} = \left(\Delta G_E - X_{BA} \frac{\partial \Delta G_E}{\partial X_{BA}} \right) / RT \tag{25}$$

Now, if we express the ΔG_E function by means of a Guggenheim expansion series, a combination of Equations (12), (24), and (25) leads to:

$$\ln \gamma_{BA} = X_{CA}^2 [a_0 - a_1(3X_{BA} - X_{CA}) + a_2(X_{BA} - X_{CA})(5X_{BA} - X_{CA}) + ...] \tag{26}$$

$$\ln \gamma_{CA} = X_{BA}^2 [a_0 + a_1(3X_{CA} - X_{BA}) + a_2(X_{CA} - X_{BA})(5X_{CA} - X_{BA}) + ...] \tag{27}$$

Equations (26) and (27) are usually termed Redlich and Kister's expressions (Redlich and Kister 1948; Plummer and Busenberg 1982) and are the most common recipes for expressing the activity coefficients in the case of non-ideal solid solutions.

The aqueous solubility of binary mineral systems

Once the problem of the activities of the solid-phase components has been addressed, we can return to the problem of the equilibrium solubility of binary solid solutions. As previously stated, description of solubility of solid solutions requires two expressions relating the activities of the components in the solid phase to the activities of the substituting ions in the aqueous solution. Thus, we can recall the two laws of mass action in Equations (2) and (4) to write them in the form:

$$[B^{n+}][A^{n-}] = K_{BA}a_{BA} = K_{BA}\gamma_{BA}X_{BA} \tag{28}$$

$$[C^{n+}][A^{n-}] = K_{CA}a_{CA} = K_{CA}\gamma_{CA}X_{CA} \tag{29}$$

Lippmann (1977, 1980) handled these equations to obtain two functions; "solidus" and "solutus," which allow for the construction of phase diagrams that depict the aqueous solubility of binary solid solutions. After some key contributions by Glynn and co-workers (Glynn and Reardon 1990; Glynn et al. 1990), the Lippmann model has become one of the most widely used tools for the description of thermodynamic equilibrium in SS-AS systems (Königsberger et al. 1991; Böttcher 2000; Gamsjäger et al. 2000; Glynn 2000) and has been demonstrated to be particularly useful in interpreting crystallization processes (Prieto et al. 1993, 1997). In the Lippmann model, the key concept is the "total activity product" variable, $\Sigma\Pi$, defined as the sum of the partial activity products contributed by each component of the solid solution. For the binary *(B,C)A* solid solution, $\Sigma\Pi$ is given by:

$$\Sigma\Pi = ([B^{n+}] + [C^{n+}])[A^{n-}] \tag{30}$$

Since Equations (28) and (29) describe a thermodynamic equilibrium, so does any combination of these equations. Thus, by adding these two equations, one obtains the value of the total activity product at thermodynamic equilibrium, $\Sigma\Pi_{eq}$, expressed as a function of the solid-phase composition:

$$\Sigma\Pi_{eq} = K_{BA}a_{BA} + K_{CA}a_{CA} = K_{BA}\gamma_{BA}X_{BA} + K_{CA}\gamma_{CA}X_{CA} \tag{31}$$

In the Lippmann model, $\Sigma\Pi_{eq}$ is termed the "total solubility product" and the relationship in Equation (31) is known as "solidus". $\Sigma\Pi_{eq}$ can also be expressed as a function of the aqueous solution composition by the introduction of the aqueous activity fractions of the substituting ions at thermodynamic equilibrium. The aqueous activity fractions $X_{B,aq}$ and $X_{C,aq}$ of B^{n+} and C^{n+} ions are defined as:

$$X_{B,aq} = \frac{[B^{n+}]}{[B^{n+}] + [C^{n+}]} \tag{32}$$

$$X_{C,aq} = \frac{[C^{n+}]}{[B^{n+}] + [C^{n+}]} \tag{33}$$

Obviously, $0 \leq X_{B,aq} \leq 1$ and $X_{C,aq} = 1 - X_{B,aq}$. Now, solving Equations (32) and (33) for $[B^{n+}]$ and $[C^{n+}]$ and substituting the result in Equations (28) and (29), we arrive at the expressions:

$$X_{BA} = \frac{X_{B,aq}([B^{n+}] + [C^{n+}])[A^{n-}]}{K_{BA}\gamma_{BA}} \tag{34}$$

$$X_{CA} = \frac{X_{C,aq}([B^{n+}] + [C^{n+}])[A^{n-}]}{K_{CA}\gamma_{CA}} \tag{35}$$

Finally, adding these two equations and keeping in mind that $X_{BA} + X_{CA} = 1$, we obtain the equilibrium "total solubility product" as a function of the aqueous phase composition:

$$\Sigma\Pi_{eq} = \frac{1}{\dfrac{X_{B,aq}}{K_{BA}\gamma_{BA}} + \dfrac{X_{C,aq}}{K_{CA}\gamma_{CA}}} \tag{36}$$

This relationship was termed "solutus" by Lippmann and expresses $\Sigma\Pi_{eq}$ as a function of $X_{B,aq}$. Note that, like the mole fractions X_{BA} and X_{CA}, the aqueous activity fractions $X_{B,aq}$ and $X_{C,aq}$ are not independent and represent the same compositional variable.

Lippmann and Roozeboom diagrams

The graphical representation of solidus and solutus yields a type of phase diagram, usually known as a "Lippmann diagram." Because "solidus" is a function of the solid-phase mole fractions, whereas "solutus" depends on the aqueous activity fractions, to construct a Lippmann diagram these two functions must be plotted against two superimposed scales (X_{BA} and $X_{B,aq}$) on the abscissa. Disregarding this last peculiarity, the Lippmann diagram can be used in a similar way to binary solid-melt phase diagrams. Every horizontal tie-line that is drawn between the solutus and solidus curves represents an aqueous phase-solid phase compositional pair at thermodynamic equilibrium. Figure 1a shows the Lippmann diagram of a hypothetical $(B,C)A$ solid solution, in which the dashed horizontal line connects an aqueous and a solid phase that can coexist at thermodynamic equilibrium. Eventually, these coexisting equilibrium compositions can be used to construct an $X_{B,aq}$-X_{BA} plot, which is useful for visualizing the equilibrium distribution of the substituting ions between the solid and aqueous phases. The equilibrium $X_{B,aq}$-X_{BA} pairs satisfy the expression (Prieto et al. 2000)

$$X_{BA} = \frac{K_{CA}\gamma_{CA}X_{B,aq}}{(K_{CA}\gamma_{CA} - K_{BA}\gamma_{CA})X_{B,aq} + K_{BA}\gamma_{BA}} \tag{37}$$

which can be derived by combining Equations (31) and (36) (solidus and solutus) and taking $X_{CA} = 1 - X_{BA}$ and $X_{C,aq} = 1 - X_{B,aq}$. Unfortunately, while Equation (37) is correct, it cannot be used in a functional way to calculate X_{BA} for a given value of $X_{B,aq}$. The reason is that the second term contains the solid phase activity coefficients, which in turn depend on X_{BA}. Thus,

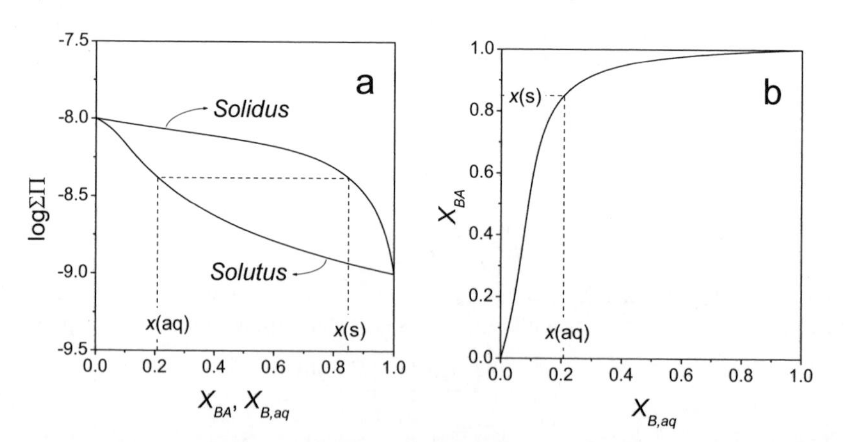

Figure 1. Lippmann diagram (a) and Roozeboom plot (b) for a hypothetical $(B,C)A$-H_2O system. In diagram (a), values for the solidus are represented on the ordinate (note the logarithmic scale for $\Sigma\Pi$) against X_{BA} on the abscissa, whereas values for the solutus are represented against $X_{B,aq}$ on the abscissa. In both plots, the dashed lines connect aqueous, x(aq), and solid phase, x(s), compositions at thermodynamic equilibrium.

any attempt at calculating X_{BA} from $X_{B,aq}$ using Equation (37) will lead to a vicious circle. In practice, to determine equilibrium pairs and construct a Lippmann diagram, the solidus relationship is calculated first using Equation (31) and, subsequently, the $\Sigma\Pi_{eq}$ values obtained for each given solid composition are substituted into the solutus equation. Finally, solving Equation (36) for $X_{B,aq}$ and taking $X_{C,aq} = 1 - X_{B,aq}$, we arrive at:

$$X_{B,aq} = \frac{\Sigma\Pi_{eq} K_{BA}\gamma_{BA} - K_{BA}\gamma_{BA}K_{CA}\gamma_{CA}}{\Sigma\Pi_{eq}(K_{BA}\gamma_{BA} - K_{CA}\gamma_{CA})} \tag{38}$$

Therefore, we need both solidus and solutus to construct an $X_{B,aq}$-X_{BA} plot. In other words, Equation (37) alone is a necessary but not sufficient condition for describing thermodynamic equilibrium in the $(B,C)A$-H_2O system. However, despite its apparent uselessness, Equation (37) has the virtue of illustrating that, in an $X_{B,aq}$-X_{BA} diagram, an important piece of information is missing, namely the solubility (represented by $\Sigma\Pi_{eq}$) of the solid solution as a function of its composition.

Figure 1b shows the $X_{B,aq}$-X_{BA} plot corresponding to the Lippmann diagram in Figure 1a. As can be observed, in this particular case there is a preferential partitioning of the B atoms towards the solid phase, as the solids are richer in B than the aqueous phase at thermodynamic equilibrium. These kinds of diagrams are usually known as Roozeboom diagrams (Roozeboom 1904) and have been widely used to represent the distribution of substituting atoms in binary solid-melt systems. In mineralogy, Roozeboom diagrams are commonly used to represent the distribution of substituting atoms between coexisting minerals (see e.g., Saxena 1968). However, typical Roozeboom diagrams utilize mole fractions to describe the composition of both phases, whereas the present $X_{B,aq}$-X_{BA} plots have the peculiarity of using aqueous activity fractions to describe the aqueous solution composition.

Equilibrium distribution coefficients

While Roozeboom diagrams display visual information about the distribution of the substitution ions between the aqueous and solid phases, we can also quantify this information by applying the law of mass action to the distribution reaction:

$$CA_{(s)} + B^{n+}_{(aq)} \leftrightarrow BA_{(s)} + C^{n+}_{(aq)} \tag{39}$$

for which the equilibrium condition can be written as:

$$K_D = \frac{a_{BA}[C^{n+}]}{a_{CA}[B^{n+}]} \tag{40}$$

where K_D is the equilibrium constant of this reaction. As we will see, this constant is in fact an equilibrium distribution coefficient, in some way comparable to that defined by Berthelot (1872) and Nerst (1891). As in the case of the Roozeboom diagrams, Equation (40) describes the equilibrium partitioning between the solid and the aqueous phases, but it does not completely describe equilibrium, because it does not include information about the aqueous activity of the non-substituting ion A^{n-}. Therefore, in Equation (40), the information about the solubility of the solid solution is missing.

In the geochemical literature, the partitioning in SS-AS systems is most frequently expressed in terms of the so-called distribution coefficient, which for the B atoms in the (B,C) A-H_2O system is defined by:

$$D_B = \frac{X_{BA}}{X_{CA}} \Big/ \frac{[B^{n+}]}{[C^{n+}]} \tag{41}$$

Values of D_B larger than unity indicate preferential partitioning of the B atoms towards the solid phase. On the contrary, for $D_B < 1$, the B atoms incorporate preferentially into the aqueous solution. Obviously, $D_C = 1/D_B$. As we will discuss later, the tendency of a given ion to incorporate preferentially into the solid phase depends not only on thermodynamics but also on kinetic factors. Parameters such as the supersaturation and the solid-phase growth rate have an effect on the effective distribution coefficients (e.g., Lorens 1981; Kornicker et al. 1991; Tesoriero and Pankow 1996; Prieto et al 1997; Rimstidt et al. 1998; Katsikopoulos et al. 2008b). For this reason, measured distribution coefficients frequently do not coincide with those for equilibrium. At thermodynamic equilibrium, combining Equation (41) with the two equilibrium conditions in Equations (28) and (29), gives an expression for $D_{B(\text{eq})}$ in terms of the end-member solubility products:

$$D_{B(\text{eq})} = \frac{K_{CA}\gamma_{CA}}{K_{BA}\gamma_{BA}} \tag{42}$$

It is worth noting that because both γ_{BA} and γ_{CA} depend on the solid solution composition, the equilibrium distribution coefficient as defined in Equation (42) is not constant, but rather a function that can change significantly with composition. This is an important point sometimes overlooked in the literature. In fact, distribution coefficients measured for different compositions can be used to estimate free energies of mixing, when there is certainty that equilibrium has been attained. In contrast, the equilibrium distribution parameter K_D in Equation (40) is a thermodynamic constant given by the quotient between the solubility products of the end-members, that is:

$$K_D = \frac{K_{CA}}{K_{BA}} \tag{43}$$

which can be easily deduced from the two equilibrium conditions in Equations (28) and (29). Distribution coefficients are sometimes expressed not using aqueous activities as in Equation (41), but rather using molal concentrations. This is equivalent to multiplying the values obtained using Equation (41) by $\gamma(B^{n+})/\gamma(C^{n+})$, where $\gamma(B^{n+})$ and $\gamma(C^{n+})$ are the activity coefficients of the corresponding aqueous ions. In general, distribution coefficients should be determined using activities, after a suitable speciation of the aqueous solution. Values calculated using total concentrations can only be correlated when they correspond to aqueous solutions with comparable ionic strengths and concentrations of major elements. Unfortunately, such is not the case with many correlations reported in the scientific literature.

Lippmann and Roozeboom diagrams for ideal solid solutions

For ideal solid solutions, the solid-phase activity coefficients are equal to unity, and all the previous expressions become simpler. Thus, for $\gamma_{BA} = \gamma_{CA} = 1$, the solidus equation reduces to:

$$\Sigma\Pi_{eq,id} = K_{BA}X_{BA} + K_{CA}X_{CA} \tag{44}$$

and the solutus expression becomes:

$$\Sigma\Pi_{eq,id} = \frac{1}{\dfrac{X_{B,aq}}{K_{BA}} + \dfrac{X_{C,aq}}{K_{CA}}} \tag{45}$$

Finally, Equation (37) simplifies to:

$$X_{BA} = \frac{K_{CA}X_{B,aq}}{(K_{\text{CA}} - K_{BA})X_{B,aq} + K_{BA}} \tag{46}$$

Now, different from Equation (37), the equilibrium $X_{B,aq}$-X_{BA} pairs can be obtained directly using Equation (46), which in fact expresses the Roozeboom curve in a functional $X_{BA}(X_{B,aq})$ form (Prieto et al. 1993). Finally, Equation (42) turns into:

$$D_{B(\text{eq},\,id)} = \frac{K_{CA}}{K_{BA}} \tag{47}$$

and, consequently, the equilibrium distribution coefficient becomes a constant equivalent to K_D. Inspection of Equations (44), (45), (46), and (47) indicates that the equilibrium behavior in ideal SS-AS systems only depends on the solubility products of the pure end-members. In all cases, the less soluble component tends to distribute preferentially towards the solid phase, which is evident considering Equation (47). Moreover, both the Lippmann and the Roozeboom diagrams develop a symmetric shape. Figure 2 shows both diagrams in the case of a hypothetical, ideal $(B,C)A$ solid solution with end-member solubility products (in cologarithms or p*K*s) of $pK_{BA} = 9.4$ and $pK_{CA} = 8.8$. For an ideal solid solution, the Lippmann diagram (Fig. 2a) is symmetric with respect to a 2-fold axis located at the center of the plot. In the same way, the Roozeboom curve (Fig. 2b) has the peculiarity of being symmetric with respect to the straight line joining the points (0,1) and (1,0), i.e., the line $X_{BA} = 1 - X_{B,aq}$.

In this particular case, the end-member solubility products differ only by 0.6 $\log\Sigma\Pi$ and the solutus and solidus curves in Figure 2a plot near each other. As a consequence, the solutus values are in equilibrium with an evenly distributed range of solid compositions. For example, values of X_{BA} between 0.25 and 0.75 correspond to equilibrium values of $X_{B,aq}$ between ≈ 0.08 and 0.92, i.e., a wide range of fluid compositions is at equilibrium with intermediate members of the solid solution. This feature becomes manifest in the Roozeboom curve, which shows a low degree of curvature due to the moderate ($D_{B(\text{eq})} = 3.98$) preferential partitioning of the less soluble component towards the solid phase.

The equilibrium behavior illustrated in Figure 2 is typical of ideal solid solutions with close end-member solubility products. Such is the case, for example, with $Ba(CrO_4,SO_4)$, a solid solution of anionic substitution that has been demonstrated to be close to ideal. The unit cell parameters within this series have been determined by X-ray diffraction to vary linearly with composition (Fernández-González et al. 1999a), i.e., $\Delta V_M = \Delta V_E = 0$. These experimental measurements agree with the data obtained by Becker et al. (2006) who, using molecular

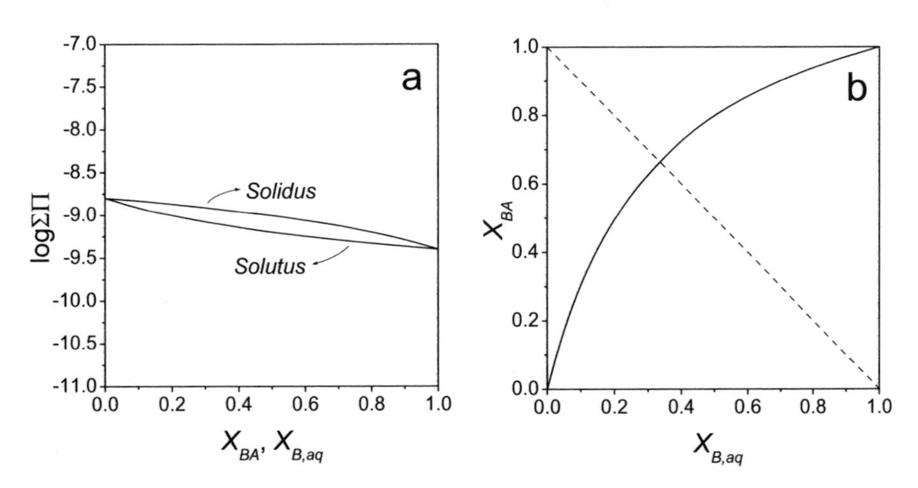

Figure 2. Lippmann diagram (a) and Roozeboom plot (b) for an ideal $(B,C)A$ solid solution with close end-member solubility products ($pK_{BA} = 9.4$ and $pK_{CA} = 8.8$). In diagram (b), the solid $X_{BA}(X_{B,aq})$ curve is symmetric with respect to the dashed line joining the points (0,1) and (1,0).

simulations, determined a linear variation of the unit cell parameters and a positive but significantly small enthalpy of mixing. In this case, the end-member solubility products differed even less (0.3 log ΣΠ) than in the case of Figure 2, which implies a very slight preferential partitioning ($D_{S(eq)} = 2.04$) of the sulfate ions (barite is less soluble than $BaCrO_4$) towards the solid phase.

When the end-member solubility products differ by several orders of magnitude, the solidus and solutus curves plot far from each other and there is a strong tendency of the less soluble component to partition into the solid phase. Figure 3 shows a hypothetical example in which the end-member p*K*s are 11 and 8 for *BA* and *CA*, respectively; that is, $D_{B(eq)} = 3000$. Due to the extremely low solubility of *BA* compared with *CA*, the Roozeboom curve approximates to two straight lines (one nearly horizontal and the other nearly vertical) that form a right angle. The horizontal branch parallels the abscissa axis from $X_{B,aq} \approx 0.1$ to 1, which means that extremely *BA*-rich ($X_{BA} > 0.99$) solid solutions are in equilibrium with *B*-poor aqueous solutions. Only a very narrow range ($0.0001 \leq X_{B,aq} \leq 0.009$) of aqueous-phase compositions can coexist in equilibrium with solid compositions in the range $0.1 \leq X_{BA} \leq 0.9$. Small changes in the fluid composition within this range imply dramatic changes in the solid equilibrium composition, which switch from *CA*-rich to *BA*-rich by modifying by some thousandths the value of $X_{B,aq}$. This behavior has been proposed to explain certain bimodal crystallization trends observed in some complete solid solutions (Prieto et al. 1993, 1997).

The otavite-calcite $(Cd,Ca)CO_3$ solid solution can be chosen as a real example with an equilibrium behavior similar to that shown in Figure 3. Although there is some disparity in the published solubility data for otavite (Stipp et al. 1993), the end-member solubility products differ by more than three orders of magnitude. Moreover, Königsberger et al. (1991) suggest that otavite and calcite form a virtually ideal solid solution (with a Guggenheim parameter, $a_0 = -0.038$, close to zero), which is also supported by the similar ionic radii of Cd^{2+} and Ca^{2+} and the linear shift in the lattice parameters (Reeder 1983). All these data point to a strong tendency of cadmium to partition into calcite ($D_{Cd(eq)} \approx 4200$) and explain the effectiveness of carbonate minerals in removing cadmium from the environment (Davis et al. 1987; Stipp et al. 1992; Tesoriero and Pankow 1996; Chiarello et al. 1997; Prieto et al. 2003).

Examples of Lippmann and Roozeboom diagrams for non-ideal solid solutions

The simplest model of a non-ideal solid solution is the regular solution, in which all but the first constant, a_0, in Equation (12) are zero. For regular solid solutions, ΔG_E is a symmetric function about the mid-point composition, $X_{BA} = 0.5$. This kind of behavior is quite frequent in solid solutions in which there is a small difference in the size of the substituting atoms, relative to the size of the non-substituting atom. In those cases, substituting *B* for *C* into the *CA* end-member is expected to have a similar energetic cost than substituting *C* for *B* into *BA*. As a result, the enthalpy of mixing tends to be symmetric, and the regular solid solution model becomes a good approach (Urusov 1974).

Figure 4 shows the free energy-composition curves of a hypothetical, regular solid solution with $a_0 = 2.2$. The two end-members are isomorphic and thus the free-energy functions are single curves across the whole composition range. Figure 4a displays ΔG_E, ΔG_M, and $\Delta G_{M,id}$. The three curves are related by Equation (11) and are symmetric about $X_{BA} = 0.5$. The fact that ΔG_E is positive for the whole range of compositions implies that the solid solution is non-ideal with $\Delta H_M > 0$ and tends toward unmixing. Moreover, because ΔG_M is negative across the whole range, the solid solution has a lower free energy than an equivalent mechanical mixture of the two end-members for all compositions. Finally, the two minima at $X_{g1} = 0.25$ and $X_{g2} = 0.75$ in the ΔG_M curve (Fig. 4b) mark the limits of a miscibility gap. A solid solution with a composition within the miscibility gap has a lower free energy than a mechanical mixture of the end-members, but it has a higher free energy than a compositionally-equivalent mechanical mixture of two solid phases with compositions of 0.25 and 0.75. Thus, to reach equilibrium, a

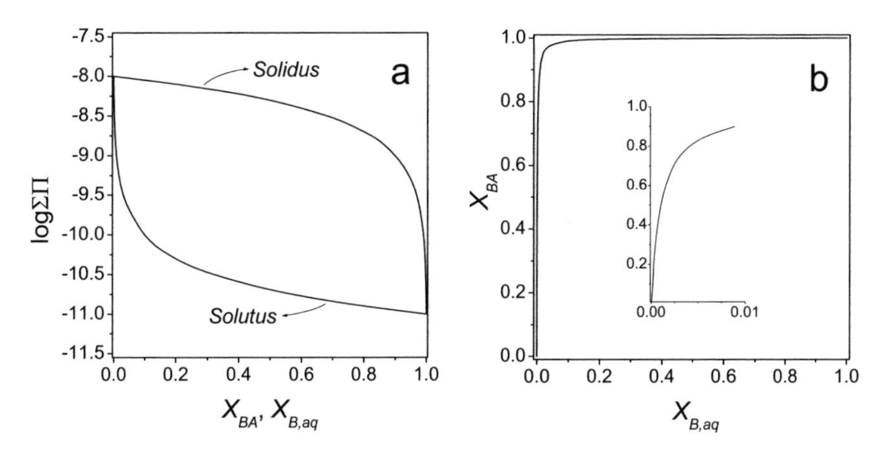

Figure 3. Lippmann diagram (a) and Roozeboom plot (b) for an ideal $(B,C)A$ solid solution with very different end-member solubility products (pK_{BA} = 11 and pK_{CA} = 8). The inset in (b) shows the vertical branch of the Roozeboom curve at a larger scale.

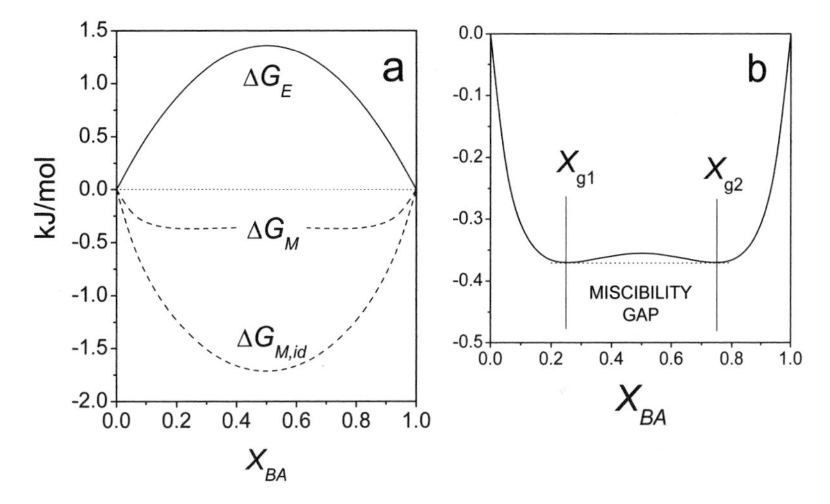

Figure 4. Free energy-composition curves for a regular solid solution with $a_0 = 2.2$. (a) The three curves are symmetric about $X_{BA} = 0.5$. In (b) the ΔG_M curve is plotted at a larger scale. The dashed line represents the miscibility gap determined by the common tangent points of the ΔG_M curve. Because the function ΔG_M is symmetric, the common tangent points and the two minima coincide at $X_{g1} = 0.25$ and $X_{g2} = 0.75$.

solid solution with a composition within the gap will tend to separate into two solid phases, with the compositions given by the gap boundaries.

Figure 5a displays the Lippmann diagram corresponding to the hypothetical solid solution shown in Figure 4. The end-member solubility products used in the calculation were pK_{BA} = 9.3 and pK_{CA} = 8. The main feature of this diagram is the presence of a "peritectic" point (its composition lies outside the gap limits) at $X_{B,aq}$ = 0.07. The "peritectic" aqueous solution is simultaneously at equilibrium with two solid compositions that represent the two extremes of the miscibility gap, X_{g1} = 0.25 and X_{g2} = 0.75. The dashed segment of the solidus curve corresponds to unstable compositions and is meaningless. The peritectic point and the miscibility limits are shown at a larger scale in Figure 5b. Because the solid solution is non-ideal, the Roozeboom

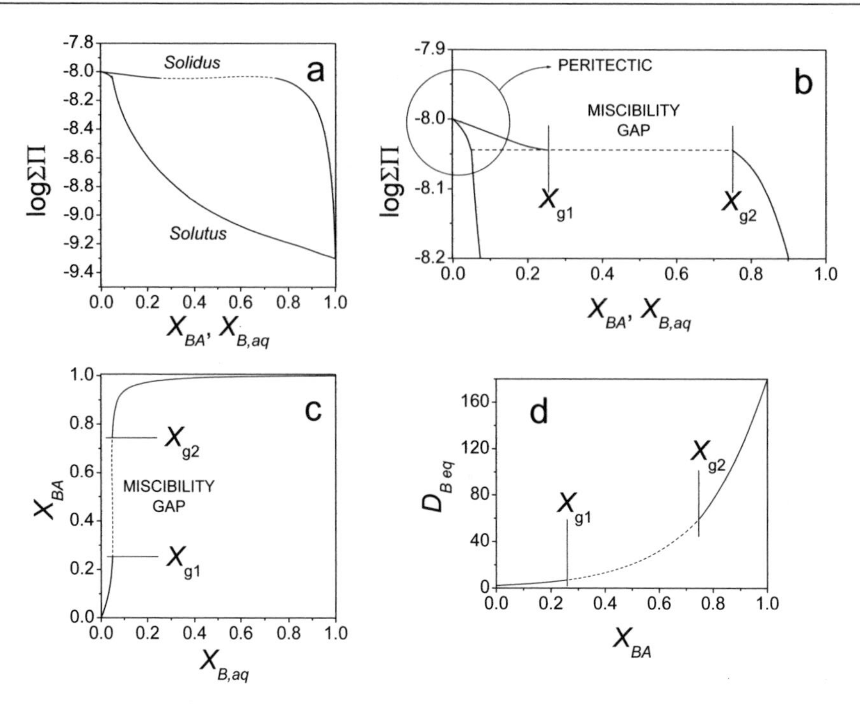

Figure 5. (a) Lippmann diagram for the regular solid solution shown in Figure 4 (a_0 = 2.2), computed using pK_{BA} = 9.3 and pK_{CA} = 8. In (b), the "peritectic" point and the miscibility limits are shown at a larger scale. (c) Roozeboom diagram. (d) Dependence of $D_{B(\text{eq})}$ on the solid composition. The dashed segments correspond to unstable compositions.

curve in Figure 5c is not symmetric with respect to the line $X_{BA} = 1 - X_{B,aq}$. Moreover, γ_{BA} and γ_{CA} are functions of the solid-phase composition and, as a consequence, $D_{B(\text{eq})}$ depends on X_{BA}. As shown in Figure 5d, the distribution coefficients are nearly two orders of magnitude smaller in the *C*-rich region than near the *BA* end-member. This means that there is a stronger tendency to uptake *C* by *BA* than vice versa and confirms that the dependence of $D_{B(\text{eq})}$ on composition cannot be overlooked, for example, in assessment of metal-ion behavior in the environment.

There are a number of examples of solid solutions with a symmetric miscibility gap that can be described using the regular solid solution model. Such is the case, for instance, with the $Ca(SeO_4,SO_4){\cdot}2H_2O$ solid solution, in which selenate substitutes for sulfate in the gypsum structure. This solid solution presents a symmetric miscibility gap between 0.23 and 0.77 and has been described using a regular model (a_0 = 2.24) by Fernández-González et al. (2006), who determine a peritectic point in the Lippmann diagram at $X_{S,aq}$ = 0.012. In a similar way, Glynn and Reardon (1990), on the basis of experimental distribution coefficients determined by Crocket and Winchester (1966), propose a regular model for the $(Zn,Ca)CO_3$ solid solution and calculate a symmetric miscibility gap (from 0.20 to 0.80) that is in reasonable agreement with the gap that can be estimated from natural occurrences.

While the regular model predicts miscibility gaps that are symmetric about X_{BA} = 0.5, most real gaps are more or less asymmetric. For example, it may be energetically less costly to substitute atoms of *C* into the *BA* end-member than vice versa. In general, substituting a smaller atom into a larger site may be expected to be easier than substituting a larger atom into a site commonly filled with a smaller one. Moreover, the "strictly regular" model implies completely random mixing ($\Delta S_E = 0$), but in practice there can be departures from ideal entropy. Figure 6a

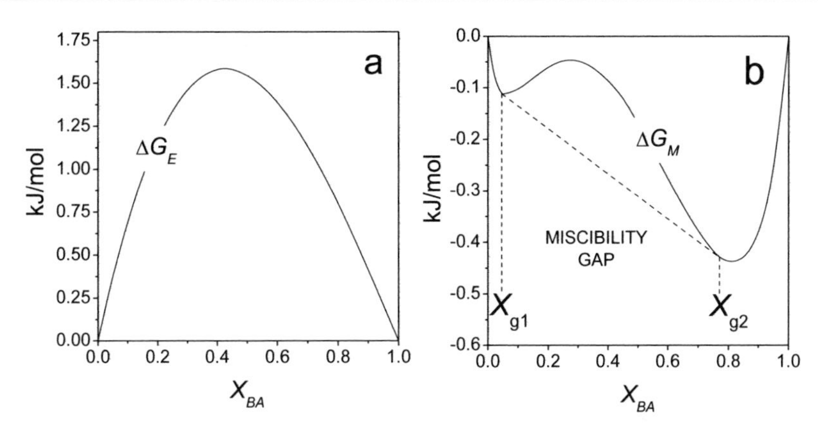

Figure 6. ΔG_E curve (a) for a hypothetical solid solution with a_0 = 2.5 and a_1 = –0.8. The curve is asymmetric, with a maximum at X_{BA} = 0.425. In (b), the corresponding ΔG_M curve is shown. The dashed line represents the common tangent to the convex downward segments of the ΔG_M curve. The common tangent points, X_{g1} = 0.046 and X_{g2} = 0.77, delimitate an asymmetric miscibility gap.

displays the ΔG_E curve of a hypothetical solid solution with a positive enthalpy of mixing (ΔH_M > 0), in which the asymmetry effect has been incorporated using two non-zero constants (a_0 = 2.5 and a_1 = –0.8) in Equation (12). As can be observed, the curve is asymmetric, with a maximum at X_{BA} = 0.425. The resultant ΔG_M curve (= $\Delta G_E + \Delta G_{M,id}$) is shown in Figure 6b. This curve has two minima, such that the common-tangent points X_{g1} and X_{g2} delimitate a miscibility gap. However, as the curve is asymmetric, the common-tangent points do not coincide with the minima. Thus, to determine the gap limits we must solve the set of equations (Saxena 1973):

$$\Delta G_{MBA}(X_{g1}) = \Delta G_{MBA}(X_{g2}) \tag{48}$$

$$\Delta G_{MCA}(X_{g1}) = \Delta G_{MCA}(X_{g2}) \tag{49}$$

where the functions ΔG_{MBA} and ΔG_{MCA} are given by:

$$\Delta G_{MBA} = \Delta G_M + X_{CA} \frac{\partial \Delta G_M}{\partial X_{BA}} \tag{50}$$

$$\Delta G_{MCA} = \Delta G_M - X_{BA} \frac{\partial \Delta G_M}{\partial X_{BA}} \tag{51}$$

In reality, the compositions defined by the common tangency must satisfy equilibrium conditions analogous to those in Equations (48) and (49) for any partial molar quantity. In the example, the gap boundaries occur at X_{g1} = 0.046 and X_{g2} = 0.77, which indicates that there is a greater tolerance for the substitution of *C* atoms into *BA*-rich solids than vice versa. In turn, Equations (48) and (49) can be used for determining the free energy of mixing (assuming a regular or sub-regular model) from experimentally measured miscibility gaps. As always the main problem with this kind of determination is to be sure that equilibrium has been attained and we are not dealing with metastable compositions.

Figure 7 displays the Lippmann diagram corresponding to the hypothetical solid solution shown in Figure 6. The end-member solubility products used in the calculation were pK_{BA} = 9.6 and pK_{CA} = 9. Now, the solutus curve shows a "eutectic" point at $X_{B,aq}$ = 0.17, whose composition lies between the gap limits. At the eutectic point, the aqueous solution is in equilibrium with the two solid compositions (0.046 and 0.77) that define the gap. This example is reminiscent of the strontianite-aragonite (Sr,Ca)CO_3 solid solution, which exhibits a miscibility gap similar

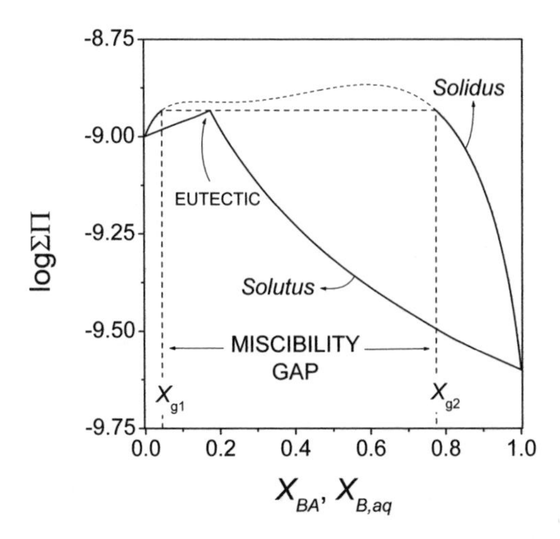

Figure 7. Lippmann diagram for the solid solution shown in Figure 6 (a_0 = 2.5 and a_1 = −0.8), computed using pK_{BA} = 9.6 and pK_{CA} = 9. The "eutectic" point is at equilibrium with the two extremes of the miscibility gap. The dashed part of the solidus curve corresponds to unstable compositions and is meaningless.

to that shown in Figure 6. Glynn and Reardon (1990), using congruent dissolution data from Plummer and Busenberg (1987) determined Guggenheim parameters for this solid solution at 25 °C (a_0 = 3.43 and a_1 = −1.82) and 76 °C (a_0 = 2.66 and a_1 = −1.15). In both cases, the Lippmann diagram exhibits one eutectic point and the miscibility gap is asymmetric, with a higher tolerance for the incorporation of the smaller calcium atoms into strontianite.

While in all the previous examples the enthalpy of mixing was positive, we are now going to consider a solid solution with $\Delta H_M < 0$. In the simplest case, a negative enthalpy of mixing means that the presence of interactions *B-C* between substituting atoms produces a decrease in enthalpy with respect to a mechanical mixture of the end-members. This, in turn, implies a tendency toward ordering, because an ordered BCA_2 phase of intermediate composition maximizes the number of *B-C* interactions between dissimilar atoms. Despite that, the existence of a negative enthalpy of mixing does not necessarily imply stability of the ordered phase, due to its lower entropy. Figure 8 displays the ΔG_E curve of a hypothetical solid solution with $\Delta H_M < 0$. As can be observed, ΔG_E is negative and has been chosen to be slightly asymmetric (a_0 = −5 and a_1 = −1) in order to exemplify that a tendency toward ordering may occur in conjunction with factors promoting asymmetry. In general, ΔG_E may be expected to differ from ΔH_M in these kinds of solid solutions, in which departures from perfect randomness ($\Delta S_E \neq 0$) are common.

Figure 9a displays the Lippmann diagram corresponding to the solid solution in Figure 8. The diagram has been computed using pK_{BA} = 9 and pK_{CA} = 8. The negative excess free energy of mixing produces a fall of the solutus curve with respect to the position of an equivalent ideal solutus. This means that the solubility of intermediate compositions is significantly smaller than that of an equivalent ideal solid solution. Indeed, for a certain range of solid compositions, the solutus values plot below the solubility product of the less soluble end-member. The

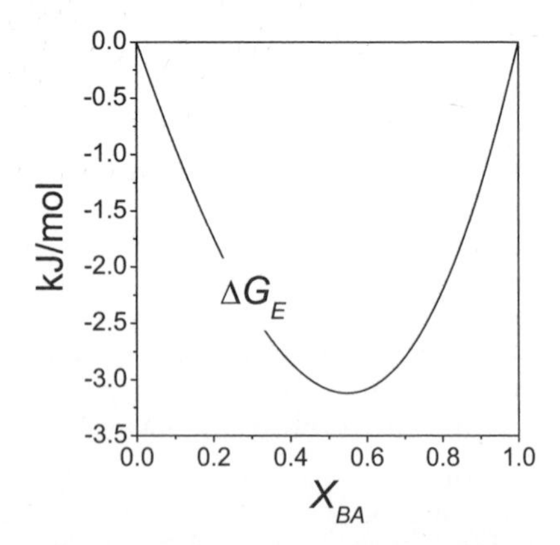

Figure 8. ΔG_E curve for a solid solution with a negative enthalpy of mixing (a_0 = −5 and a_1 = −1). The curve is slightly asymmetric, with a minimum at X_{BA} = 0.55.

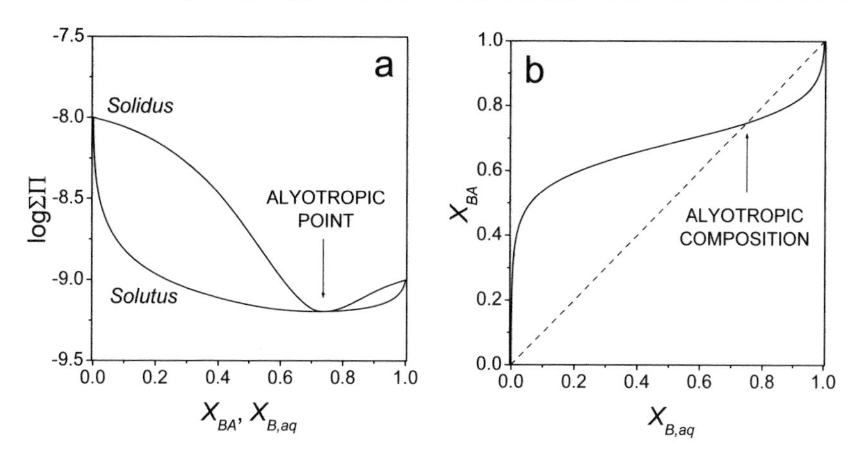

Figure 9. (a) Lippmann diagram for a hypothetical solid solution with a negative enthalpy of mixing (a_0 = −5 and a_1 = −1) and end-member p*Ks* of 9 and 8. The arrow indicates the "alyotropic" point. In (b), the corresponding Roozeboom diagram is shown. The line $X_{B,aq} = X_{BA}$ intersects the curve at the alyotropic composition.

system shows an "alyotropic" minimum at $X_{B,aq} = 0.71$ where the solutus and solidus curves meet. Such a point represents a thermodynamic equilibrium state in which the mole fraction of the substituting ions in the solid phase equals its activity fraction in the aqueous solution. Obviously, at the alyotropic point, the equilibrium distribution coefficient is equal to unity. This is shown in Figure 9b, in which the line $X_{B,aq} = X_{BA}$ intersects the Roozeboom curve at the alyotropic composition. At this composition, the Roozeboom curve has an inflection point at which the curvature changes from convex upward to downward. While for $0 < X_{B,aq} < 0.71$ the *B* atoms tend to incorporate preferentially towards the solid phase with $D_{B(\text{eq})} > 1$, for activity fractions in the range $0.71 < X_{B,aq} < 1$, the tendency switches and $D_{B(\text{eq})}$ becomes smaller than one.

The scheelite-powellite ($CaWO_4$-$CaMoO_4$) isomorphic series is an interesting example of a solid solution with a negative enthalpy of mixing. On one hand, the small difference in the cell parameters suggests the possibility of a wide range of miscibility. On the other hand, whereas the unit cell parameter *a* is larger in scheelite, *c* is larger in powellite and it has been observed that the *c*/*a* ratio is significantly larger for molybdates than for tungstates in sheelite-type compounds. Paradoxically, a number of structure determinations have shown that the tetrahedral units WO_4^{2-} and MoO_4^{2-} have essentially the same size, which means that the difference in the *c*/*a* ratio cannot be due to dissimilarities in the configuration of these anionic units. Sleight (1972) explains this systematic difference by assuming that MoO_4^{2-} is more covalent than WO_4^{2-}. These differences can explain the peculiar features of the scheelite-powellite series. Fernández-González et al. (2007) determined a negative molar volume of mixing with a significant minimum for intermediate compositions. Moreover, using low-temperature solution calorimetry, these authors determined a negative enthalpy of mixing at 25 °C. Their finding is in agreement with the data obtained by Kiseleva et al. (1980), who determined a negative enthalpy of mixing by high-temperature solution calorimetry. The negative character of both parameters points towards a non ideal solid solution (a_0 = −1.422, a_1 = 0.213, and a_2 = −0.130) with a tendency toward ordering, and some features reminiscent of those shown in Figures 8 and 9.

Lippmann diagrams for solid-solution systems involving ordered phases

Whereas in the previous example the tendency toward ordering did not result in the stability of an ordered phase, in solid-solution systems with $\Delta H_M < 0$ the formation of ordered phases is always a possibility, particularly at low temperatures. Now, as a final example, we are going to consider the case of binary rhombohedral carbonate systems with ordered dolomite-type phases involved. The example is especially complex since it deals with solid solutions that show tendencies to both order and unmix in the same system (Putnis 1992).

The fact that unmixing and ordering are not always mutually exclusive has been recognized in many alloys and solid solutions of silicates and sulfides (Carpenter 1980). In carbonates, Capobianco et al. (1987) studied the mixing properties in the $CdCO_3$-$MgCO_3$ system and distinguished between a disordered $(Cd,Mg)CO_3$ solid solution with positive enthalpy of mixing and an ordered $CdMg(CO_3)_2$ phase with a negative enthalpy of formation from $CdCO_3$ and $MgCO_3$. Due to a number of kinetic factors, it has been impossible to perform a similar in-depth study with both the $MgCO_3$-$CaCO_3$ archetype and $MnCO_3$-$CaCO_3$, but their order-disorder behaviors are estimated to be analogous to those of $CdCO_3$-$MgCO_3$. The disordering reaction in dolomite-type structures implies that alternating cation layers become crystallographically distinct when the cations are ordered, which leads to a change in the space group symmetry. Under these conditions, the fact that the enthalpy of formation of the ordered phase from the pure end-members is negative, whereas that of the disordered phase is positive, may be explained by assuming positive (repulsive, promoting segregation) interactions between neighboring dissimilar atoms within the same layer and negative interactions (attractive, promoting ordering) between dissimilar atoms of successive layers (Burton 1987).

Figure 10 shows a hypothetical ΔG_M scheme for these kinds of systems, following the examples by Carpenter (1980), Reeder (1983), and Goldsmith (1983). In the example, the free-energy curves of the disordered and ordered phases join with continuous first derivatives typical of non-first order transitions. For the sake of simplicity, the curves have been chosen to be symmetric about $X_{BA} = 0.5$. Here, we have three convex downward segments and the common-tangent points determine two miscibility gaps ($X_{g1} - X_{g2}$ and $X_{g3} - X_{g4}$), one on each side of the ordered-phase minimum.

The most practical way of constructing Lippmann diagrams for these kinds of systems is to choose the ordered phase as one of the end-members and to construct two diagrams, one for the *BA*-BCA_2 side and the other for BCA_2-*CA*. However, the application of the Lippmann relationships as written in Equations (31) and (36) is not possible since the BCA_2 end-member

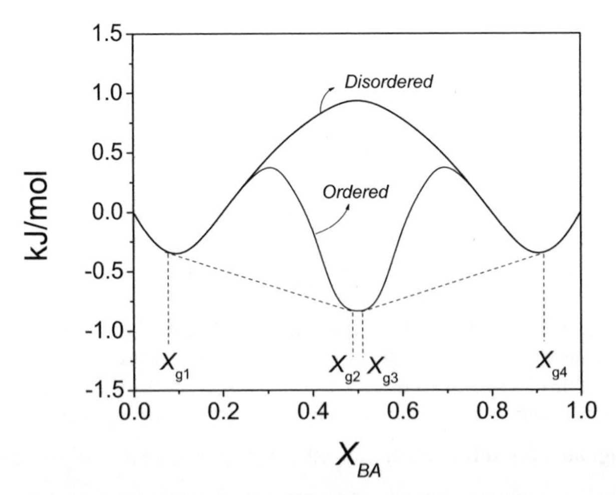

Figure 10. Order-disorder behavior of a hypothetical solid solution for which both ordering and unmixing tendencies are involved: the ordered phase is superimposed on what would be a broad miscibility gap for the disordered series. The common tangents of the dashed lines determine two miscibility gaps.

has a different stoichiometry (Lippmann 1991, Glynn 1991). If we keep the solubility product of BA in the form $K_{BA} = [B^{n+}][A^{n-}]$, the solubility product K_{ORD} of the ordered BCA_2 end-member must be written for the same number of ions, in order to make both solubility products dimensionally equivalent, that is:

$$K_{\mathrm{ORD}} = [B^{n+}]^{0.5}[C^{n+}]^{0.5}[A^{n-}] \tag{52}$$

Obviously, Equation (52) is equivalent to the square root of the typical solubility product expression used for this kind of stoichiometry: $[B^{n+}][C^{n+}][A^{n-}]^2$. In addition, the mole fraction X_{ORD} of the ordered component is defined as twice the number of C atoms in a mole of solid divided by the total number of C and B atoms. Obviously, for pure BCA_2, X_{ORD} is equal to one. Finally, the total activity product $\Sigma\Pi$ and the aqueous activity fraction $X_{\mathrm{ORD},aq}$ must be written as:

$$\Sigma\Pi = ([B^{n+}] + [B^{n+}]^{0.5}[C^{n+}]^{0.5})[A^{n-}] \tag{53}$$

and

$$X_{\mathrm{ORD},aq} = \frac{[B^{n+}]^{0.5}[C^{n+}]^{0.5}}{[B^{n+}] + [B^{n+}]^{0.5}[C^{n+}]^{0.5}} \tag{54}$$

Figure 11 shows the Lippmann diagram for the BA-BCA_2 side of this hypothetical system, in which the solvus can be observed to have a eutectic point at $X_{\mathrm{ORD},aq} \approx 0.42$. This eutectic solution is in equilibrium with a disordered solid solution of composition $X_{g1} \approx 0.15$ and an ordered phase of composition $X_{g2} \approx 0.98$, i.e. with the two limits of the "left-side" miscibility gap.

It is worth noting that, in order to calculate Lippmann diagrams for these kinds of systems, we need to know the solid-phase activity coefficients of both components in both the ordered and the disordered solid phase. Unfortunately, this is a main handicap in dealing with real systems. Glynn (1991), in order to overcome this problem for the $CaCO_3$-$MgCO_3$ system, assumed an isostructural solid-solution series between calcite and dolomite and used the Guggenheim parameters determined by Busenberg and Plummer (1989) for magnesian calcites. However, as Glynn (1991) already pointed out, calcite and dolomite have different structures, and the standard chemical potentials of the calcite component in the dolomite structure and the dolomite component in the calcite structure are not known.

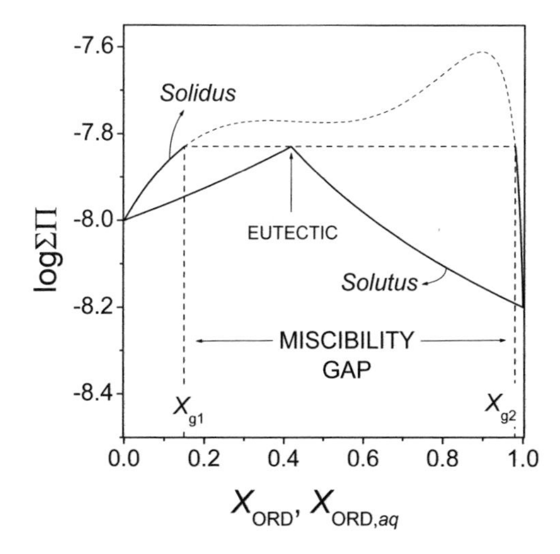

Figure 11. Lippmann diagram for the BA-BCA_2 side of the hypothetical system shown in Figure 10. The eutectic point is at equilibrium with the two phases of compositions X_{g1} and X_{g2} that delimit the miscibility gap, as delineated with the dashed line. The dashed part of the solidus curve corresponds to unstable compositions and is meaningless.

METASTABLE EQUILIBRIUM STATES

Stoichiometric saturation

Whereas the Lippmann diagram describes true equilibrium states, for practical purposes the so-called "stoichiometric" solubility product is frequently used to represent the solubility of solid solutions (Thorstenson and Plummer 1977). In reality, the stoichiometric solubility product does not represent a true equilibrium state, but rather a metastable equilibrium condition that considers the solid solution as a stoichiometric phase (i.e., a pure single-component solid) with a fixed formula, $B_xC_{1-x}A$, which dissolves congruently according to the reaction:

$$B_xC_{1-x}A_{(s)} \leftrightarrow xB^{n+}_{(aq)} + (1-x)C^{n+}_{(aq)} + A^{n-}_{(aq)} \tag{55}$$

If we assume that the solid behaves as if it were stoichiometric, the solid-phase activity becomes equal to unity (Glynn and Reardon 1990), since the activity of a single-component solid is equal to one by definition. Now, applying the mass action law, we obtain the expression:

$$K_{st} = [B^{n+}]^x[C^{n+}]^{1-x}[A^{n-}] \tag{56}$$

in which the "equilibrium" constant K_{st} is the stoichiometric solubility product. This stoichiometric solubility product can be expressed in terms of the solubility products of the end-members. Thus, raising the "true" equilibrium conditions in Equations (28) and (29) to the powers of x and $1 - x$, we obtain:

$$([B^{n+}][A^{n-}])^x = (K_{BA}a_{BA})^x = (K_{BA}\gamma_{BA}X_{BA})^x \tag{57}$$

$$([C^{n+}][A^{n-}])^{1-x} = (K_{CA}a_{CA})^{1-x} = (K_{CA}\gamma_{CA}X_{CA})^{1-x} \tag{58}$$

Finally, multiplying Equations (57) and (58), and taking into account that $x = X_{BA}$ and $1 - x = X_{CA}$, we arrive to:

$$K_{st} = (K_{BA}\gamma_{BA}X_{BA})^{X_{BA}}(K_{CA}\gamma_{CA}X_{CA})^{X_{CA}} \tag{59}$$

which expresses K_{st} in terms of K_{BA} and K_{CA}. The fact that Equation (59) has been obtained by handling the two equilibrium conditions has an essential implication: an aqueous solution that is at thermodynamic equilibrium with respect to a solid $B_xC_{1-x}A$ will always be at stoichiometric saturation with respect to that solid. However, as we have reduced two conditions to a single equation, the converse statement is not true, i.e., stoichiometric saturation does not necessarily imply true thermodynamic equilibrium. We will come back to this apparent paradox later. For now, we must note that the concept of stoichiometric saturation arises from the experimental evidence that solid solutions tend to dissolve congruently until an initial saturation is reached. However, an aqueous solution can remain metastable at stoichiometric saturation with respect to a given solid solution because, to reach true equilibrium, the solid solution would need to undergo a dissolution-recrystallization process whose kinetics can be very sluggish.

It is worth noting that the stoichiometric solubility product in Equation (59) is a function that depends on the solid composition under consideration. For an ideal solid solution, the solid-phase activity coefficients are equal to one and Equation (59) becomes:

$$K_{st} = (K_{BA}X_{BA})^{X_{BA}}(K_{CA}X_{CA})^{X_{CA}} \tag{60}$$

Obviously, in order to compute the K_{st} function for a non-ideal solid solution, we need to know the dependence of γ_{BA} and γ_{CA} on composition. Thus, taking into account the relation of γ_{BA} and γ_{CA} with the excess free energy of mixing in Equations (24) and (25), we can easily reach the expression:

$$K_{st} = (K_{BA}X_{BA})^{X_{BA}}(K_{CA}X_{CA})^{X_{CA}} \exp(\Delta G_E / RT) \tag{61}$$

which allows one to calculate K_{st} when the Guggenheim parameters are known. Conversely, solving Equation (61) for ΔG_E yields:

$$\Delta G_E = RT[\ln K_{st} - X_{BA}\ln(K_{BA}X_{BA}) - X_{CA}\ln(K_{CA}X_{CA})] \tag{62}$$

This equation was first derived by Glynn and Reardon (1990) and has been demonstrated to be an excellent tool for estimating non-ideality, since it allows a Guggenheim function to be fitted from experimental K_{st} values. However, in order to obtain reliable results, one must be sure that stoichiometric saturation has been reached and that the system has not undergone a later dissolution-recrystallization process towards equilibrium (Böttcher 1997a). Figure 12 shows the effect of non-ideality on the stoichiometric solubility of a hypothetical solid solution, with end-member pKs of 9 and 8. The dashed upper curve corresponds to a solid solution with a positive enthalpy of mixing ($a_0 = 2.5$) and the dash-dot lower one has been derived for an equivalent solid solution with a negative enthalpy of mixing ($a_0 = -2.5$). As can be observed, a solid solution with a negative enthalpy of mixing is less soluble than an equivalent ideal solid solution (the solid intermediate line), whereas a positive enthalpy of mixing produces the opposite effect.

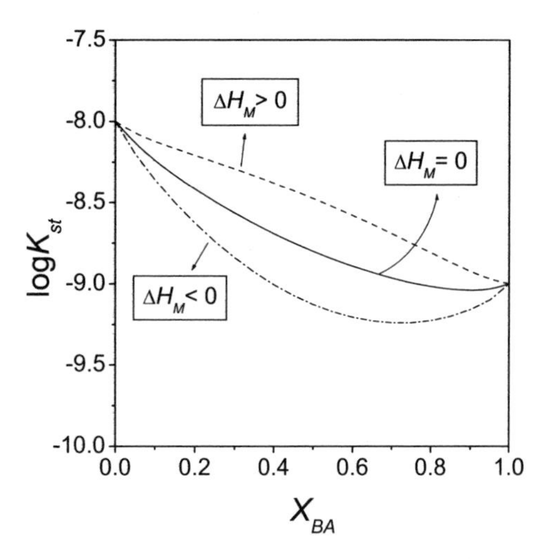

Figure 12. Stoichiometric solubility of a hypothetical solid solution (p$K_{BA} = 9$ and p$K_{BA} = 8$) as a function of composition. The dashed upper curve corresponds to a solid solution with a positive enthalpy of mixing ($a_0 = 2.5$). The dash-dot lower curve has been calculated for an equivalent solid solution with a negative enthalpy of mixing ($a_0 = -2.5$). The intermediate solid line represents an ideal solid solution.

Stoichiometric saturation and Lippmann diagrams

We may now consider the case where the solid solution undergoes congruent dissolution in contact with an aqueous solution with a given proportion of B^{n+}, C^{n+}, and A^{n-}, that is:

$$B_xC_{1-x}A_{(s)} + pB^{n+}_{(aq)} + qC^{n+}_{(aq)} + rA^{n-}_{(aq)} \rightarrow (p+x)B^{n+}_{(aq)} + (q+1-x)C^{n+}_{(aq)} + (r+1)A^{n-}_{(aq)} \tag{63}$$

In reality, the reaction in Equation (63) is exactly the same as that in Equation (55): the transfer of ions between the solid and the aqueous phases involves formula units, i.e., it occurs in a ratio $B^{n+}/C^{n+}/A^{n-}$ equal to that of the solid. The aim of writing this "useless" equation is simply to illustrate that we can imagine different scenarios of stoichiometric saturation for different proportions (aqueous activity fractions) of the substituting ions in the aqueous solution. Indeed, for a given solid-solution composition, we can represent all the different stoichiometric saturation conditions on the Lippmann diagram as a function of $X_{B,aq}$.

Consider a solid solution with a "fixed" (indicated by the additional subscript s to distinguish it from the variable X_{BA}) composition $X_{BAs} = 1 - X_{CAs}$. According to Equation (56), the stoichiometric solubility product for this particular solid solution is given by:

$$K_{st} = [B^{n+}]^{X_{BAs}}[C^{n+}]^{X_{CAs}}[A^{n-}] \tag{64}$$

Now, multiplying and dividing this expression by $[B^{n+}]+[C^{n+}]$ and rearranging the terms, we arrive at:

$$K_{st} = ([B^{n+}]+[C^{n+}])[A^{n-}]\left(\frac{[B^{n+}]}{[B^{n+}]+[C^{n+}]}\right)^{X_{BAs}}\left(\frac{[C^{n+}]}{[B^{n+}]+[C^{n+}]}\right)^{X_{CAs}} \tag{65}$$

and, considering the definitions of total activity product $\Sigma\Pi$ and aqueous activity fractions, $X_{B,aq}$ and $X_{C,aq}$, we finally get:

$$\Sigma\Pi_{st} = \frac{K_{st}}{X_{B,aq}^{X_{BAs}} X_{C,aq}^{X_{CAs}}} \tag{66}$$

where $\Sigma\Pi_{st}$ expresses the stoichiometric saturation on the $\Sigma\Pi$ (variable) scale. For a solid-solution with fixed composition X_{BAs}, $\Sigma\Pi_{st}$ describes a series of stoichiometric saturation scenarios as a function of the aqueous activity fraction $X_{B,aq}$ of the substituting ions in the aqueous solution. Figure 13 displays the stoichiometric saturation $\Sigma\Pi_{st}$ curve for a solid solution member with composition 0.7. The calculation has been performed for an ideal solid solution using $pK_{BA} = 9$ and $pK_{CA} = 8$. As can be observed, the stoichiometric saturation curve plots above the solutus curve for all aqueous activity fractions except for $X_{B,aq} = 0.19$. For this value, the stoichiometric saturation curve concurs with the solutus. The common point "EQ" corresponds to $\Sigma\Pi = 10^{-8.43}$ and is the only point on the $\Sigma\Pi_{st}$ curve that represents a true equilibrium situation. We can now go back to the paradox of the previous section and realize that stoichiometric saturation is a necessary but insufficient condition for equilibrium.

The fact that the stoichiometric saturation curve lies above the solutus curve is not surprising. In all metastable-equilibrium states the aqueous solution is supersaturated with respect to the stable phase, which is inevitably less soluble than the metastable one. To use a simple and well-known example, an aqueous solution saturated with respect to aragonite is supersaturated with respect to calcite, the stable phase at ambient conditions. Thus, in order to reach equilibrium, the metastable phase must undergo a dissolution-recrystallization process in which the stable and less soluble phase is formed.

Pure end-member saturation

A particular case of stoichiometric saturation occurs when we chose 0 or 1 as a "fixed" composition for the solid. In such a case, setting X_{BAs} equal to 0 or 1 in Equations (59) and (66), we arrive at:

$$\Sigma\Pi_{BA} = \frac{K_{BA}}{X_{B,aq}} \tag{67}$$

or

$$\Sigma\Pi_{CA} = \frac{K_{CA}}{X_{C,aq}} \tag{68}$$

respectively. These equations define the families of conditions for which a solution containing B^{n+}, C^{n+}, and A^{n-} ions can be in metastable equilibrium with respect to pure *BA* or *CA* solid phases. As with all the $\Sigma\Pi_{st}$ curves, the "pure end-member saturation" curves plot above the solutus for all aqueous activity fractions, concurring with the solutus only for $X_{B,aq}$ or $X_{C,aq}$ equal to one. This can be seen in Figure 14, in which the curves $\Sigma\Pi_{BA}$ and $\Sigma\Pi_{CA}$ have been represented for the same solid solution as in Figure 13. Note that the two curves intersect at the point "M". This point represents an aqueous solution at simultaneous stoichiometric saturation with respect to both end-members.

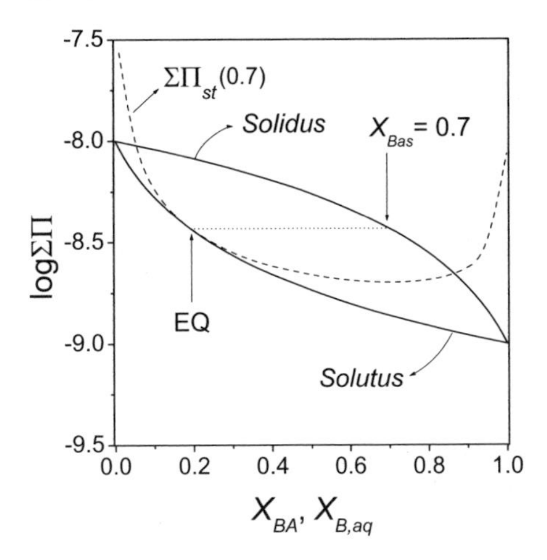

Figure 13. Stoichiometric saturation $\Sigma\Pi_{st}$ curve calculated for X_{BA} = 0.7, projected on the corresponding Lippmann diagram. The calculation has been performed for an ideal solid solution with pK_{BA} = 9 and pK_{CA} = 8. At the point "EQ", the $\Sigma\Pi_{st}$ curve concurs with the solutus. The dotted horizontal tie line connects "EQ" with the solidus curve at X_{BA} = 0.7. Note that both the solutus and the $\Sigma\Pi_{st}$ curve are plotted on the ordinate against $X_{B,aq}$ on the abscissa, whereas the solidus is plotted against X_{BA}.

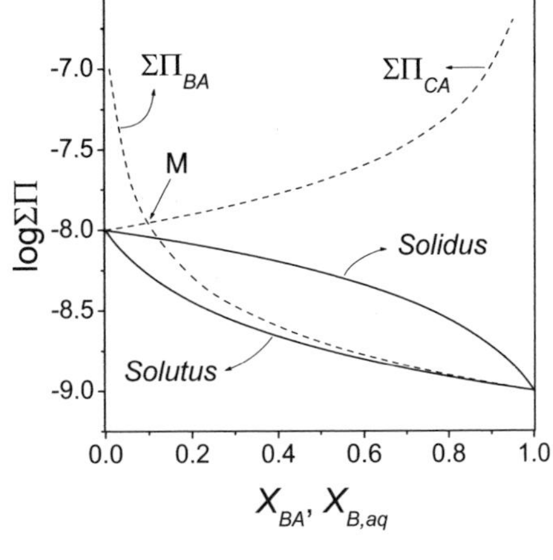

Figure 14. Pure end-member saturation $\Sigma\Pi_{BA}$ and $\Sigma\Pi_{CA}$ plotted as dashed curves projected on a Lippmann diagram. The calculation has been performed for an ideal solid solution with pK_{BA} = 9 and pK_{CA} = 8. At the point "M", the aqueous solution is simultaneously at stoichiometric saturation with respect to both end-members.

SUPERSATURATION IN BINARY SS-AS SYSTEMS

Law of mass action and thermodynamics

Whereas the equilibrium behavior in SS-AS systems is well established and non-controversial (Thortenson and Plummer 1977; Lafont 1978; Garrels and Wollast 1978; Thortenson and Plummer 1978; Stoessell 1992; Glynn et al. 1992; Königsberger and Gamsjäger 1992; Glynn and Reardon 1992), and the concepts of true-equilibrium and stoichiometric saturation may be considered resolved, the concept of supersaturation is still contentious. A number of expressions for supersaturation are available in the geochemical literature (Prieto et al. 1993; Astilleros et al. 2003a; Monnin and Cividini 2006; Shtukenberg et al. 2006), but their interrelation and their meaning in a thermodynamic framework are not always obvious (Prieto et al. 2007). Here, in order to make the fundamentals clear, we start from the basics, even at the risk of being too elementary.

We return to the dissolution reaction shown in Equation (1) for a pure mineral *BA*. The change, ΔG_{BA}, in the Gibbs free energy per mole for this reaction is given by:

$$\Delta G_{BA} = \Delta G^{\circ}_{BA} + \mathrm{R}T \ln \frac{[B^{n+}][A^{n-}]}{a_{BA}} \tag{69}$$

where ΔG°_{BA} is the standard Gibbs free energy of the reaction and a_{BA} = 1 for a pure solid. Obviously, when $\Delta G_{BA} > 0$, the reaction proceeds to the left and the solid grows, whereas for $\Delta G_{BA} < 0$, the solid dissolves. At equilibrium, $\Delta G_{BA} = 0$ and the ionic activity product becomes equal to the solubility product, K_{BA}. Therefore:

$$\Delta G^{\circ}_{BA} = -\mathrm{R}T \ln K_{BA} \tag{70}$$

Back substitution of Equation (70) into Equation (69), keeping $a_{BA} = 1$, results in

$$\Delta G_{BA} = RT \ln \frac{IAP}{K_{BA}} \tag{71}$$

where IAP stands for the ionic activity product, $[B^{n+}][A^{n-}]$, of the current solution and the quotient $\Omega = IAP/K_{BA}$ is the so-called saturation state. The term 'current solution' is used throughout this chapter to refer to the aqueous solution in the example that is currently being discussed in the text. Therefore, the saturation state (or the saturation index $SI = \log \Omega$) is the fundamental supersaturation when dealing with stoichiometric minerals. For a given composition of the aqueous solution, the saturation state takes a single value, and the aqueous solution is said to be saturated, undersaturated, or supersaturated for $\Omega = 1$, $\Omega < 1$, or $\Omega > 1$, respectively.

We can now introduce chemical potentials into the previous formulations. The chemical potential concept (see e.g., Langmuir 1997) provides a way to think about the tendency of a component to transfer from the solid to the aqueous phase and will be particularly useful in the next sections. Using chemical potentials, Equation (69) may be written:

$$\Delta G_{BA} = \mu_{B,aq} + \mu_{A,aq} - \mu_{BA} \tag{72}$$

where $\mu_{B,aq}$ and $\mu_{C,aq}$ are the chemical potentials of the ions B^{n+} and A^{n-} in the aqueous phase and μ_{BA} is the chemical potential of the component BA in the solid, as defined in Equation (19). In this particular case, as we are dealing with a pure solid, $a_{BA} = 1$ and μ_{BA} is equal to G_{BA}, i.e., to the Gibbs free energy of a mole of pure BA. For a given phase, the chemical potential of "i" is related to the activity of i through the expression:

$$\mu_i = \mu_i^\circ + RT \ln a_i \tag{73}$$

where μ_i° is the chemical potential of i at the standard state. Inspection of Equations (19) and (73) shows that

$$\mu_{BA}^\circ = G_{BA} \tag{74}$$

since the standard state for a component in the solid phase is the pure solid. In the case of the aqueous B^{n+} ions, Equation (73) yields:

$$\mu_{B,aq} = \mu_{B,aq}^\circ + RT \ln[B^{n+}] \tag{75}$$

in which the standard state for the aqueous B^{n+} ions is an ideal unimolal solution. Obviously, we can write expressions analogous to those in Equations (69) to (75) for CA, C^{n+}, and A^{n-}.

The previous fundamentals allow us to introduce the concept of supersaturation in SS-AS systems. The simplest case that we can consider is the one in which the solid solution behaves as a stoichiometric (single component) solid that dissolves congruently according to Equation (55). In such a case, by analogy with Equation (71), we can write:

$$\Delta G_{st} = RT \ln \frac{IAP_{st}}{K_{st}} \tag{76}$$

where K_{st} is the stoichiometric solubility product as defined in Equation (59), and IAP_{st} is the stoichiometric activity product in the current solution. Finally, ΔG_{st} is the change in the Gibbs free energy (per mole) for the congruent dissolution reaction in Equation (55). Similar to Equation (72), for a solid solution with a "fixed" composition $X_{BAs} = 1 - X_{CAs}$, the free energy change can also be written as (Glynn and Reardon 1990; Prieto et al. 1993):

$$\Delta G_{st} = X_{BAs}\mu_{B,aq} + X_{CAs}\mu_{C,aq} + \mu_{A,aq} - \mu_{st} \tag{77}$$

where μ_{st} is the chemical potential of the solid solution of composition X_{BAs} considered as a stoichiometric solid. It is worth noting that μ_{st} is quantitatively equal to the value that takes G_{SS} for $X_{BA} = X_{BAs}$, although its fundamental meaning is different. Equation (76) allows us to define the so-called "stoichiometric supersaturation" or stoichiometric saturation state (Prieto et al. 2003), which for a solid solution of composition X_{BAs} is given by:

$$\Omega_{st} = \frac{IAP_{st}}{K_{st}} = \frac{[B^{n+}]^{X_{BAs}}[C^{n+}]^{X_{CAs}}[A^{n-}]}{(K_{BA}\gamma_{BA}X_{BAs})^{X_{BAs}}(K_{CA}\gamma_{CA}X_{CAs})^{X_{CAs}}} \tag{78}$$

Thus, when Ω_{st} is equal to one, ΔG_{st} is zero and the aqueous solution is at stoichiometric saturation with respect to a solid solution of composition X_{BAs}. For $\Omega_{st} < 1$, if the solid solution dissolves congruently, the free energy decreases since $\Delta G_{st} < 0$. Finally, for $\Omega_{st} > 1$ and $\Delta G_{st} > 0$, the solid solution may grow congruently. Of course, as the stoichiometric saturation does not represent true equilibrium, but rather a metastable equilibrium situation, Ω_{st} does not represent the departure of the system from stability. For $\Omega_{st} < 1$, if the solid dissolves congruently until reaching stoichiometric saturation, the free energy of the system decreases but does not reach the minimum free energy possible for the given conditions of pressure, temperature, and composition. This fact does not invalidate the concept of stoichiometric supersaturation, but we must be aware of its true meaning. In reality, we use the stoichiometric supersaturation concept in daily practice when we calculate the saturation state of an aqueous solution containing, for example, Sr^{2+} or Mg^{2+} with respect to a pure mineral like calcite. In such a case, we are calculating the stoichiometric supersaturation with respect to a pure end-member (calcite) of a solid solution, even if we do not realize it. Something similar also occurs when we calculate the saturation state of an aqueous solution with respect to aragonite at ambient conditions. As when using Ω_{st}, in this last case we are not determining the departure of the system from true stability, but rather from a metastable equilibrium state.

The stoichiometric-supersaturation function

In practice, for a given aqueous solution, we may want to know the saturation state with respect to all possible solid phases. This is, in fact, what typical speciation programs like PHREEQC (Parkhurst and Appelo 1999) do for stoichiometric solids. Dealing with binary SS-AS systems, we know that all aqueous solutions that plot below the solutus are undersaturated with respect to any solid phase, including pure end-members. In contrast, aqueous solutions plotting above the solutus are supersaturated with respect to a series of solid compositions and all members of this series are susceptible to precipitation. This means that, in dealing with SS-AS systems, the saturation state of a particular aqueous solution cannot be expressed by a single value. To make predictions, we should know the range of solid solution compositions for which such a specific aqueous solution is supersaturated (or undersaturated) and the magnitude of the supersaturation (or undersaturation) with respect to each possible solid composition. The best way to solve this problem is to express the supersaturation in functional form, i.e., as a function of the solid composition, X_{BA}. With this aim, the stoichiometric supersaturation in Equation (78) can be expressed as a function of X_{BA} (variable), that is:

$$\Omega_{st}(X_{BA}) = \frac{[B^{n+}]^{X_{BA}}[C^{n+}]^{X_{CA}}[A^{n-}]}{(K_{BA}\gamma_{BA}X_{BA})^{X_{BA}}(K_{CA}\gamma_{CA}X_{CA})^{X_{CA}}} \tag{79}$$

For a particular aqueous solution, i.e., for given values of $[B^{n+}]$, $[C^{n+}]$, and $[A^{n-}]$, Equation (79) provides a description of the departure of stoichiometric saturation with respect to all the possible compositions of the solid solution. An example of a typical $\Omega_{st}(X_{BA})$ function is shown in Figure 15a. The calculation has been carried out for an ideal solid solution with $pK_{BA} = 9$ and $pK_{CA} = 8$ and for an aqueous solution with $X_{B,aq} = 0.25$ (the aqueous activities are given in the caption). As can be observed, the curve has a maximum at X^{max}, which represents the solid composition for which the aqueous solution is most supersaturated. The composition X^{max} is

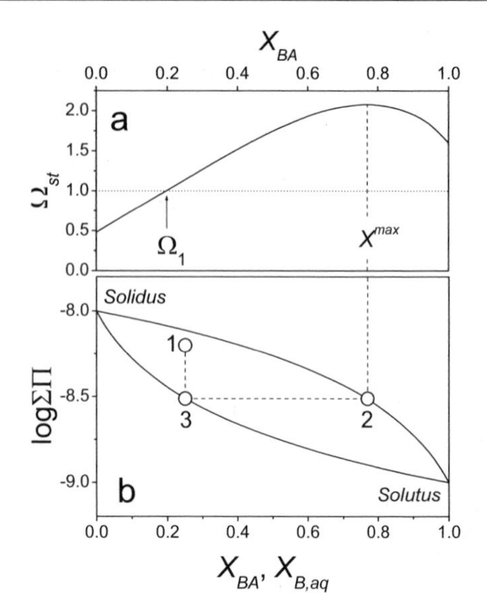

Figure 15. Stoichiometric-supersaturation function and Lippmann diagram. The calculation has been carried out for an ideal solid solution with pK_{BA} = 9 and pK_{CA} = 8 and an aqueous solution with activities $[B^{n+}] = 2\times10^{-5}$, $[C^{n+}] = 6\times10^{-5}$, and $[A^{n-}] = 8\times10^{-5}$. (a) Stoichiometric supersaturation $\Omega_{st}(X_{BA})$ function. In this particular case, the maximum $\Omega_{st,}(X^{max})$ occurs at 0.77. The value Ω_1 (≈0.2) delimitates the supersaturated range. (b) The current solution (point 1) corresponds to $X_{B,aq} = 0.25$ and $\Sigma\Pi = 10^{-8.2}$. On the solutus, point 3 represents an aqueous solution with the same activity fraction ($X_{B,aq} = 0.25$) as the current solution. This last solution is at equilibrium ($\Sigma\Pi_{eq} = 10^{-8.5}$), with a solid solution of composition 0.77 (point 2 on the solidus), i.e., with the composition corresponding to X^{max}.

not arbitrary, but rather has a very specific meaning that is illustrated on the Lippamnn diagram in Figure 15b. The current solution composition (point 1) plots at $X_{B,aq} = 0.25$ and $\Sigma\Pi = 10^{-8.2}$. On the solutus, point 3 represents an aqueous solution with the same aqueous activity fraction ($X_{B,aq} = 0.25$) as the current solution. An aqueous solution plotting at point 3 ($\Sigma\Pi_{eq} = 10^{-8.5}$) is in equilibrium with a solid solution that has a composition exactly equal to X^{max} (point 2 on the solidus). Therefore, the maximum $\Omega_{st}(X^{max})$ supersaturation represents a departure from a "true" equilibrium situation. In other words, for $X_{BA} = X^{max}$, the denominator in Equation (79) represents a stoichiometric saturation condition that is also a "true" equilibrium condition, like point "EQ" in Figure 13.

It is worth noting that, if a solid of composition X^{max} crystallizes, the effective distribution coefficient will be exactly equal to the equilibrium coefficient ($D_{B(\mathrm{eq})}$) as defined in Equation (42). That is, the maximum supersaturation corresponds to equilibrium partitioning. Although this property is obvious after inspection of Figure 15, it can also be demonstrated by equating to zero the derivative of Equation (79) and rearranging the resulting expression (Andara et al. 2005).

At this point, one might ask why we do not use the single value $\Omega_{st}(X^{max})$, which corresponds to the maximum as a "strict" supersaturation parameter. In fact, that is what Monnin and Cividini (2006) and Shtukenberg et al. (2006) did when they defined the saturation state of a particular aqueous solution as:

$$\xi = \frac{\Sigma\Pi}{\Sigma\Pi_{eq}} \tag{80}$$

where $\Sigma\Pi$ is the total activity product of the current solution (point 1 in Fig. 15b) and $\Sigma\Pi_{eq}$ is the total (equilibrium) solubility product ($\Sigma\Pi_{eq} = 10^{-8.5}$ in the example) corresponding to an aqueous solution (point 3 in Fig. 15b) with the same aqueous activity fraction ($X_{B,aq} = 0.25$ in the example) as the current solution. This last value plots on the Lippmann solutus curve on the same vertical line as the current solution (point 3 in Fig. 15b). In reality, Equation (80) defines the supersaturation of the aqueous solution with respect to the specific member X^{max} of the solid solution (point 2), i.e., ξ has exactly the same meaning as $\Omega_{st}(X^{max})$. Moreover, ξ is also quantitatively equal to $\Omega_{st}(X^{max})$, as can be demonstrated by handling Equations (79) and (80) in a similar way to that used to obtain Equation (66).

Unfortunately, despite being the maximum supersaturation, $\xi = \Omega_{st}(X^{max})$ has the limitation of giving information only for a single composition of the solid. In a first approximation, one could think that the composition of the solid solution that is most likely to precipitate is the one for which the aqueous solution is most supersaturated, i.e., X^{max}. However, there is ample evidence that demonstrates that nuclei compositions are frequently not determined by maximum supersaturation values (Prieto et al. 1997; Putnis et al. 2003), particularly at high supersaturations; this is simply an intrinsic feature of non-equilibrium partitioning. As Pina and Putnis (2002) have pointed out, in many cases "more soluble" solid-solution compositions are kinetically favored and tend to precipitate, even though the aqueous solution is less supersaturated for these compositions than for less soluble members. As a result, the precipitate tends to be enriched in the most soluble component, differing in this way from the composition expected from equilibrium partitioning considerations. In this framework, delineating the range of solid compositions that are thermodynamically susceptible to precipitate becomes crucial.

Such is the main advantage of using the $\Omega_{st}(X_{BA})$ function shown in Equation (79). It allows not only an immediate evaluation of $\xi = \Omega_{st}(X^{max})$, but also an estimation of the compositional range for which the aqueous solution is supersaturated. As can be observed in Figure 15, the aqueous solution is stoichiometrically supersaturated ($\Omega_{st} > 1$) with respect to the pure *BA* end-member and with respect to all solid compositions in the range $\Omega_1 < X_{BA} \leq 1$. This means that if a solid with a composition within this range grows congruently, the free energy of the system will decrease. On the contrary, for compositions in the range $0 \leq X_{BA} < \Omega_1$ the solid will tend to dissolve congruently. The only difficulty is that, while congruent dissolution is a real dissolution mechanism and Ω_{st} is a suitable estimator for the driving force of this process, congruent growth is not a common process in either nature or the laboratory (Lafon 1978). Initial or earlier-produced phases are not likely to maintain a fixed composition during growth and, therefore, the stoichiometric supersaturation should not be used as a driving force in modeling crystal growth or recrystallization reactions. In spite of this, the "Ω_{st}-supersaturated" range is a good qualitative tool to interpret the composition of the initial nuclei (e.g., Andara et al. 2005) and the growth mechanisms (Pina et al. 2000) in crystallization experiments.

The "δ-functions"

Astilleros et al. (2003a), in an attempt to find a more "all-purpose" expression for supersaturation, defined the functions:

$$\delta_{BA}(X_{BA}) = \frac{[B^{n+}][A^{n-}]}{K_{BA}\gamma_{BA}X_{BA}} \tag{81}$$

$$\delta_{BA}(X_{CA}) = \frac{[C^{n+}][A^{n-}]}{K_{BA}\gamma_{CA}X_{CA}} \tag{82}$$

According to these authors, for a given composition of an aqueous solution, Equation (81) defines the supersaturation for all the solid compositions with $X_{BA} \geq X^{max}$, whereas Equation (82) defines the supersaturation for all the solid compositions with $X_{BA} \leq X^{max}$.

In the original paper, the derivation of these equations is in some way "geometrical", and the underlying thermodynamic meaning is not straightforward. We will return to this point in the following section. For the moment, we can extract some conclusions by comparing the functions in Equations (81) and (82) with the stoichiometric-supersaturation function in Equation (79). Figure 16 shows such a comparison for the same case as in Figure 15. As can be observed, $\Omega_{st}(X_{BA})$ and the δ-functions have three common points, $\delta(0) = \Omega_{st}(0)$, $\delta(1) = \Omega_{st}(1)$,

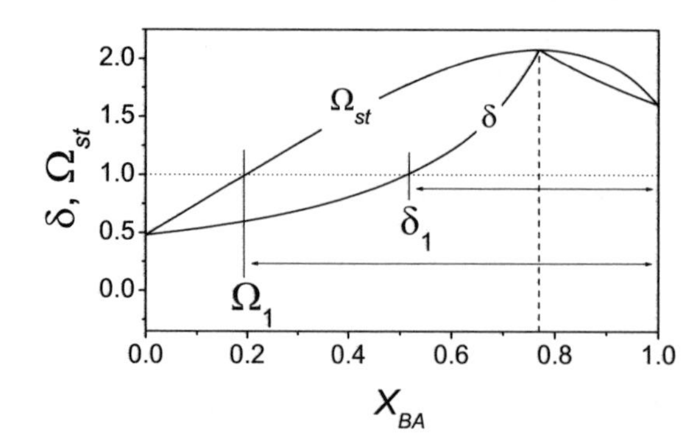

Figure 16. Correlation between supersaturation models. The calculation has been carried out for the same case as in Figure 15. Note the three common points at $\delta(0) = \Omega_{st}(0)$, $\delta(1) = \Omega_{st}(1)$, and $\delta(X^{max}) = \Omega_{st}(X^{max})$. The figure also shows the ranges, $\delta_1 - 1$ ($\delta_1 \approx 0.52$) and $\Omega_1 - 1$ ($\Omega_1 \approx 0.2$), of solid-solution compositions for which the aqueous solution is δ-supersaturated ($\delta > 1$) and stoichiometrically supersaturated ($\Omega_{st} > 1$).

and $\delta(X^{max}) = \Omega_{st}(X^{max})$. Again, the common maximum is the only point that represents a departure from a "true" equilibrium situation. For all other compositions the functions diverge, with δ always smaller than Ω_{st}. Moreover, the range $\delta_1 < X_{BA} \leq 1$ of compositions for which the aqueous solution is supersaturated is considerably narrower. Both observations seem to indicate that the δ-functions are more "restrictive", but the meaning of such differences is a priori difficult to explain.

Thermodynamic meaning of the supersaturated ranges

In order to clarify the meaning of the ranges previously defined, we can start from the two equilibrium conditions in Equations (28) and (29). Expressing both conditions in terms of chemical potentials (see Eqn. 72) we can write:

$$\Delta G_{BA} = \mu_{B,aq} + \mu_{A,aq} - \mu_{BA} = 0 \tag{83}$$

$$\Delta G_{CA} = \mu_{C,aq} + \mu_{A,aq} - \mu_{CA} = 0 \tag{84}$$

Equilibrium requires these two conditions to be satisfied simultaneously, whereas when one or both of them are not fulfilled (ΔG_{BA} and/or $\Delta G_{CA} \neq 0$), the system is out of equilibrium. Therefore, ΔG_{BA} and ΔG_{CA} can be used to establish the tendency of the two components *BA* and *CA* to transfer from the aqueous to the solid phase and vice versa. Note that, according to Equation (72) for $\Delta G_{BA} < 0$, the component *BA* will tend to transfer to the aqueous phase, whereas for $\Delta G_{BA} > 0$, the aqueous ions will tend to incorporate into the solid phase.

Shtukenberg et al. (2006) explored the conditions in which one of the two chemical potential differences, ΔG_{BA} or ΔG_{CA}, is equal to zero but the other is not. To do that, they fixed the composition of the solid solution at $X_{BAs} = 1 - X_{CAs}$ (indicated by the additional subscript *s*) and expressed $\Sigma\Pi$ as a function of $X_{B,aq} = 1 - X_{C,aq}$, that is:

$$\Sigma\Pi_B(X_{B,aq}) = \frac{K_{BA}\gamma_{BAs}X_{BAs}}{X_{B,aq}} \tag{85}$$

$$\Sigma\Pi_C(X_{C,aq}) = \frac{K_{CA}\gamma_{CAs}X_{CAs}}{X_{C,aq}} \tag{86}$$

These two functions can be derived directly from Equations (34) and (35) and cannot be used independently to describe equilibrium (Prieto et al. 2007). True equilibrium occurs for the specific value of $X_{B,aq}$ at which both equations give the same value of $\Sigma\Pi$, that is:

$$\Sigma\Pi_{eq} = \frac{K_{BA}\gamma_{BAs}X_{BAs}}{X_{B,aq}} = \frac{K_{CA}\gamma_{CAs}X_{CAs}}{X_{C,aq}} \tag{87}$$

This value corresponds to a point on the Lippmann solutus curve, and is the only one in the series that describes a true equilibrium situation. Figure 17 shows the projection of these functions (Eqns. 85 and 86) on the Lippmann diagram. The calculation has been carried out for the same case as in Figures 15 and 16 and for a fixed composition X_{BAs} of 0.8. As can be observed both functions are hyperbolas that intersect at the point "EQ" on the solutus. This point corresponds to the true equilibrium condition of Equation (87) and represents an aqueous solution that is at equilibrium with the solid solution of composition 0.8, for which both $\Sigma\Pi_B(X_{B,aq})$ and $\Sigma\Pi_C(X_{C,aq})$ have been calculated. The two hyperbolas divide the phase space into four regions, marked in Figure 17 by roman numerals. Keeping in mind the meaning of Equations (85) and (86), these four regions are useful for exploring the reaction trajectories that are thermodynamically possible.

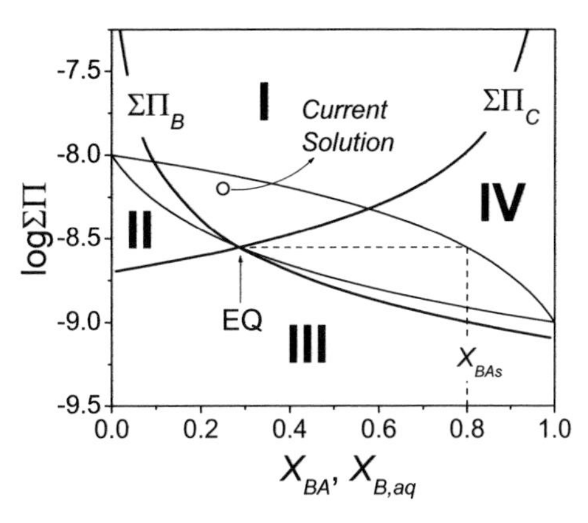

Figure 17. Projection of $\Sigma\Pi_B(X_{B,aq})$ and $\Sigma\Pi_C(X_{C,aq})$ (hyperbolas in bold solid lines) on the Lippmann diagram. The diagram corresponds to the same case as in Figure 15 and the hyperbolas have been calculated for a fixed (X_{BAs}) solid composition of 0.8. The intersection point "EQ" on the solutus is at equilibrium with the solid solution of composition X_{BA} = 0.8. The meaning of regions I, II, III, and IV is discussed in the text. Note that the current solution (point 1, $X_{B,aq}$ = 0.25 and $\Sigma\Pi = 10^{-8.2}$) plots in region I and is strictly supersaturated with respect to a solid of composition X_{BA} = 0.8.

For example, a solid solution with the fixed composition X_{BAs} (0.8 in Fig. 17), when maintained in contact with an aqueous solution plotted in region I, will tend to undergo a net growth. During this process, the total amount of both *BA* and *CA* in the solid phase will increase. This does not mean that the solid solution will grow without undergoing compositional changes. In practice, the actual reaction paths will depend on a number of kinetic factors. However, the thermodynamically-relevant argument is that a net increase in the amount of solid solution with this composition will lead to a decrease in the free energy of the SS-AS system. This is because both ΔG_{BA} and ΔG_{CA} are positive, i.e., $\mu_{BA} < \mu_{B,aq} + \mu_{A,aq} - \mu_{BA}$ and $\mu_{CA} < \mu_{C,aq} + \mu_{A,aq}$.

Net growth can also occur in contact with an aqueous solution that plots above the solutus in region II. However, in this case the solid will tend to get richer in *BA* and poorer in *CA* because $\Delta G_{BA} > 0$ and $\Delta G_{CA} < 0$. Obviously, such a compositional evolution can only occur by a dissolution-recrystallization process. The opposite will occur in contact with an aqueous solution that plots above the solutus in region IV because, in this case, $\Delta G_{BA} < 0$ and $\Delta G_{CA} > 0$. Finally, all the aqueous solutions that plot below the solutus will tend to dissolve.

In light of the previous graphs, we can now discuss the meaning of the range of solid compositions, $\delta_1 - 1$, for which the aqueous solution is δ-supersaturated. To do this, we can calculate the $\Sigma\Pi_B(X_{B,aq})$ and $\Sigma\Pi_C(X_{C,aq})$ functions for all compositions within this range and explore in each case the position of the current solution (point 1, $X_{B,aq}$ = 0.25 and $\Sigma\Pi = 10^{-8.2}$) with respect to the obtained hyperbolas. The result is unequivocal. In all cases, the aqueous solution plots in region I. This means that, when in contact with this particular aqueous solution, any solid solution with a composition within this range will tend to undergo a net growth process, since for such a reaction both ΔG_{BA} and ΔG_{CA} are positive.

This finding is not surprising. In reality, the two δ-functions explore the disequilibrium conditions in a way that is reminiscent of the one used by Shtukenberg and co-workers with Equations (85) and (86). This becomes clear if we write ΔG_{BA} and ΔG_{CA} in the following form:

$$\Delta G_{BA} = \mathrm{R}T \ln \frac{[B^{n+}][A^{n-}]}{K_{BA}\gamma_{BA}X_{BA}} \tag{88}$$

$$\Delta G_{CA} = \mathrm{R}T \ln \frac{[C^{n+}][A^{n-}]}{K_{CA}\gamma_{CA}X_{CA}} \tag{89}$$

Note that Equation (88) is equivalent to Equation (71), except that $a_{BA} = \gamma_{BA}X_{BA} \neq 1$ because we are not dealing with a pure solid. Considering Equations (81) and (82), we can write:

$$\Delta G_{BA}(X_{BA}) = \mathrm{R}T \ln \delta_{BA}(X_{BA}) \tag{90}$$

$$\Delta G_{CA}(X_{CA}) = \mathrm{R}T \ln \delta_{CA}(X_{CA}) \tag{91}$$

In practice, for a given composition of the aqueous solution, what Astilleros and co-workers did was to explore separately these two functions, which meet at X^{max}, i.e.:

$$\Delta G_{BA}(X^{\max}) = \Delta G_{CA}(X^{\max}) \tag{92}$$

As previously discussed, this point corresponds to the "true" or maximum supersaturation condition. Figure 18 represents Equations (90) and (91) for the example that we have been considering throughout this section. As can be observed, the intersection point occurs at X^{max}. Moreover, the ΔG_{CA} function intersect the line $y = 0$ at δ_1, i.e., when the δ-function is equal to one. For $X_{BA} < \delta_1$, the δ-function values are smaller than one and $\Delta G_{CA} < 0$. In reality, the δ-functions (Astilleros et al. 2003a) represent a species of "partial" supersaturation and should not be used separately as estimators of the driving force in modeling crystal growth processes. Such a task requires both Equations (90) and (91) to be used simultaneously. In spite of this, the δ-functions have the virtue of defining in a simple way the range of solid solution compositions that fulfill simultaneously the conditions $\Delta G_{BA} > 0$ and $\Delta G_{CA} > 0$ when grown in contact with the given aqueous solution.

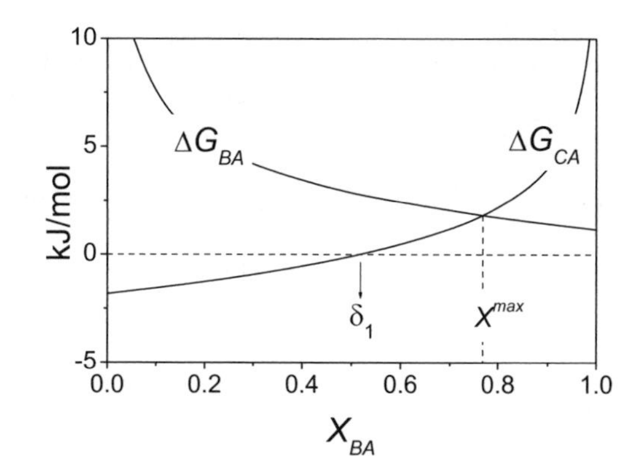

Figure 18. $\Delta G_{BA}(X_{BA})$ and $\Delta G_{CA}(X_{CA})$ functions. The calculation has been carried out for the same case as in Figure 15. Note that the intersection point occurs for $X_{BA} = 0.77$, i.e., at X^{max}. Note that $\Delta G_{CA} = 0$ for $X_{CA} = \delta_1 \approx 0.52$.

Finally, it is worth noting that, for the solid solutions in the range $\Omega_1 < X_{BA} < \delta_1$ (see Fig. 16), the current solution plots in region IV. The solid will tend to get richer in *CA* and poorer in *BA* because, in such a case, $\Delta G_{BA} < 0$ and $\Delta G_{CA} > 0$. Obviously, if the solid solution grows congruently, the free energy of the system will decrease according to Equation (77).

FROM THERMODYNAMICS TO CRYSTALLIZATION BEHAVIOR

Kinetic and mechanistic effects on partitioning

The thermodynamic relationships presented throughout the previous sections constitute an indispensable background for interpreting SS-AS processes, but they are insufficient to account for the actual behavior in these kinds of systems. Kinetic and mechanistic factors usually play a major role on the nucleation and growth of solid solutions. A detailed review of these factors is beyond the scope of this chapter, but identifying the main concerns and the work that has to be done is the first step to progress in any scientific field.

The first concern relates to non-equilibrium partitioning. Whereas the influence of factors such as the overall growth rate (Lorens 1981; Tesoriero and Pankow 1996) and the supersaturation rate (Prieto et al. 1997; Katsikopoulos et al. 2008b) has been experimentally demonstrated (see Fig. 19), the available predictive models still need to be very much improved. For example, Pina and Putnis (2002) applied classical nucleation theory to solid solutions in order to explain the experimental evidence, which indicated that the higher the supersaturation, the higher the deviation of the effective distribution coefficients from the equilibrium values. This deviation always leads to a reduction in the degree of preferential partitioning, which implies that the solid phase becomes richer in the more soluble component. The model by Pina and Putnis (2002) predicted this behavior well, but should be developed in order to fit quantitatively the experimental results. In classical nucleation theory, besides supersaturation, one of the key parameters is interfacial tension. Interfacial tension is directly related to the solubility of a substance and to its "ability" to form supersaturated solutions (e.g. Sangwal 1989; Kashchiev and Van Rosmalen 2003). The relationship between interfacial tension and solubility has a fundamental basis, since a high solubility implies strong crystal-solution interactions and a low interfacial tension. However, one of the main difficulties in modeling nucleation behavior in SS-AS systems is the absence of interfacial tension data for solid solutions. Moreover, there are other factors like the cation-anion ratio or the presence of background electrolytes in the aqueous solution that have recently been demonstrated to have a significant influence on the nucleation behavior of stoichiometric solids (Kowack et al. 2007; Kowack and Putnis 2008). These factors should also be considered when dealing with solid solutions.

Kinetic effects introduce complexities in attempting to apply equilibrium criteria to laboratory or field observations. For instance, thermodynamic measurements frequently predict much more narrow ranges of miscibility than observed in compositions in natural or synthetic samples (e.g., Casey et al. 1996). In dealing with incomplete solid solutions, the precipitation of intermediate, metastable compositions is always a possibility, and the kinetic models have to be open to it. Obviously, such metastable solid-phases will tend to undergo dissolution-recrystallization to reach equilibrium, which can be considered as a "solvent-mediated" unmixing process.

If predicting nucleation behavior in SS-AS systems is a complex task, modeling crystal growth is by far more complicated, since classical growth theory recognizes different growth

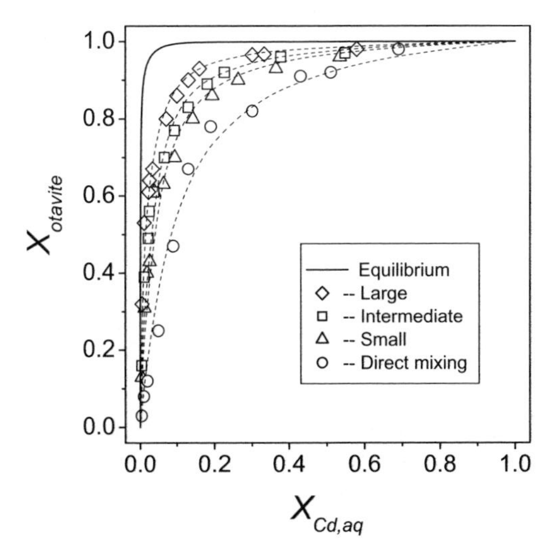

Figure 19. Effective $X_{Cd,aq}$-$X_{otavite}$ pairs obtained crystallizing $(Cd,Ca)CO_3$ solid solutions at different supersaturation rates. The data correspond to experiments carried out by counter-diffusion using diffusion columns with different lengths (direct mixing, short, intermediate, and large). The solid line represents the equilibrium Roozeboom diagram, whereas the dashed curves fit the different datasets. Due to the extremely low solubility of otavite compared with calcite, the equilibrium curve approximates to two straight lines forming a right angle. All the experimental values plot below the equilibrium curve, indicating that the effective distribution coefficients decrease as the supersaturation increases: the shorter the column length the higher the supersaturation rate and the deviation from equilibrium partitioning. (Modified after Katsikopoulos et al. 2008b).

laws and mechanisms even in stoichiometric solids. So far, the supersaturation functions Ω_{st} and δ have been demonstrated to be good tools for explaining the transitions between growth mechanisms in a qualitative way (Pina et al. 2000; Putnis et al 2003), but the available quantitative models are still unsatisfactory. Moreover, classical growth theory disregards surface speciation and a number of mechanistic phenomena that occur at the molecular scale on the surface of minerals and which may be expected to have a kinetic effect. For example, the importance of multiple surface sites for trace element incorporation and growth step kinetics on calcite was demonstrated in some pioneering papers by Reeder and co-workers (Paquette and Reeder 1995; Reeder 1996; Reeder et al. 1999). These studies showed a direct correlation between the preference for trace element incorporation and growth step orientation on the calcite surface and have been the basis for interpreting a number of nanoscale growth phenomena (e.g., Astilleros et al. 2000; Lea et al. 2004; Wasylensky et al. 2005; Pérez-Garrido et al. 2007). The necessity of similar studies on other mineral substrates is evident and reveals the key role of near-surface sensitive techniques (e.g., Xu et al. 1996; Reeder et al. 1999; Stipp et al. 1992, 2003) in the development of improved models for crystal growth.

Compositional zoning and "partial" equilibrium end-points

Equilibrium thermodynamics deals with bulk properties that do not depend on specific scenarios. However, in dealing with SS-AS systems, the reaction paths and the final equilibrium points depend on the specific starting amounts of solid and aqueous solutions. If the system is finite, during the growth process, both crystal and aqueous phase compositions tend to vary. This occurs because the substituting ions do not incorporate into the solid phase in the same stoichiometric proportion as in the aqueous phase. As a consequence, the individual crystals grow in a way such that the layers growing on the surface block the inner zones from contact with solution. This results in a compositional evolution from core to rim in the bulk crystal, i.e., in compositional zoning. In order to model this evolution, one can imagine a situation is which chemical equilibrium is maintained between each incremental crystal layer and the solution composition at the time of its formation. In such a case, we can write the typical differential equation for a Rayleigh process (e.g. Appelo and Posma 2005):

$$\frac{dm_{BA}}{dm_{CA}} = \frac{-dm_{B,aq}}{-dm_{C,aq}} = \lambda_0 \frac{m_{B,aq}}{m_{C,aq}} \tag{93}$$

where m_{iA} and $m_{i,aq}$ are the molar concentrations of the "i" component in the solid and the aqueous solution, respectively. The fractionation factor, λ_0, is usually called logarithmic distribution coefficient because, by integrating Equation (93) from an initial (*ini*) to a final (*fin*) composition, we obtain:

$$\ln\left(\frac{m_{B,aq}(ini)}{m_{B,aq}(fin)}\right) = \lambda_0 \ln\left(\frac{m_{C,aq}(ini)}{m_{C,aq}(fin)}\right) \tag{94}$$

This expression is usually known as Doerner-Hoskins equation or logarithmic distribution law (Doerner and Hoskins 1925). For $\lambda_0 > 1$, the component *BA* concentrates in the central zones of the crystal and decreases towards the periphery while, for $\lambda_0 < 1$, the sequence is the reverse. Unfortunately, this kind of Doerner-Hoskins behavior may only be expected to occur when growth rates and supersaturation are extremely small. In actual scenarios, the distribution coefficients depend on kinetic factors and will change during the process, as the crystal grows and the system approaches equilibrium.

In practice, the compositional gradients in the solid phase depend on the nature of the components and also on supersaturation and supersaturation rates. When the end-member solubility products are similar, the substituting ions tend to incorporate into the solid nearly in the same proportion as in the aqueous phase, particularly at high supersaturations. As a consequence, the compositional gradients in the solid are very small and the crystals grow virtually homogeneous. Such is the case, for example, with the $Ba(CrO_4,SO_4)$ solid solution (Fernández-González et al. 1999a), a nearly ideal solid solution in which the end-member solubility products differ only by $0.31 \log\Sigma\Pi$.

The situation is very different in the case of solid solutions with end-member solubility products differing by several orders of magnitude. In such a case, one of the substituting ions incorporates very preferentially into the solid. As a result, the aqueous phase depletes drastically in this ion as growth proceeds, giving rise to a growth period rich in the other component and generating in this way a sharp compositional zoning. Such is the behavior of the (Cd,Ca) CO_3 solid solution (Fernández-González et al. 1999b), in which the end-member solubility products differ by more than three orders of magnitude. Figure 20a shows the backscattered electron image of the central section of a $(Cd,Ca)CO_3$ crystal with a Cd-rich core and a Ca-rich rim. The crystal begins to grow with a Cd-rich nucleus but, as growth proceeds, the strong preferential depletion of Cd in the fluid leads to Ca-rich growth period. The transition occurs over a small distance, giving rise to a sharp concentric zoning. At the end, the Cd-rich core becomes "encapsulated" by a rim of nearly pure calcite. Obviously, this phenomenon is extremely interesting from the point of view of the removal of cadmium from contaminated environments.

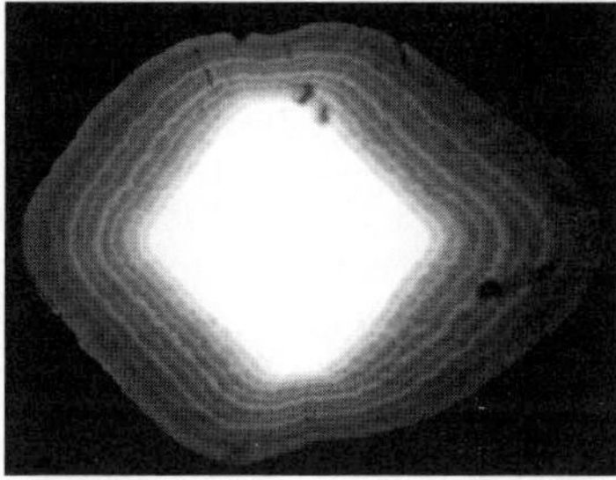

Figure 20. (a) Backscattered electron image of the central section of a (Cd,Ca) CO_3 crystal in which a Cd-rich core is encapsulated by a rim of nearly pure calcite. (b) Oscillatory zoning in a (Cd,Ca) CO_3 solid-solution crystal. (SEM pictures by A. Fernández-González).

A phenomenon closely related to concentric zoning is oscillatory zoning (Fig. 20b). Oscillatory zoning has usually been observed in crystallizing solid solutions with very different end-member solubility products (Putnis et al. 1992; Prieto et al. 1997). Inspection of the inset in Figure 3b indicates that, in these kinds of systems, there is a narrow range of aqueous solution compositions in equilibrium with intermediate solid-solution members. In that range, any cause that provokes small oscillations in the composition of the aqueous solution will be strongly amplified in the solid composition recording, resulting in oscillatory zoning. Oscillatory zoning is usually considered as a result of chemical self-organization during the growth process. Self-organization is the spontaneous patterning of a system without the intervention of an external template, which requires feedback and disequilibrium. Different feedback mechanisms have been proposed in the literature to account for oscillatory zoning in crystals (e.g., Reeder at al. 1990; Wang and Merino 1992; L'Heureux 2006), but this aspect is beyond the scope of the present review. Here, we only wish to point out that oscillatory zoning usually becomes apparent in solid solutions with a large difference between end-member solubility products. This does not mean that oscillatory behavior is exclusive of these kinds of solid solutions. Feedback and disequilibrium are obviously possible in crystallizing solid solutions with close end-member solubility products. However, the sharpness of the oscillations is favored in solid solutions with a strong preferential partitioning for one of the substituting ions.

The previous paragraphs clearly reveal the enormous difficulties in modeling actual growth processes of solid solutions. Whereas SS-AS thermodynamics provides qualitative explanations, the development of quantitative kinetic models is still poor. In the end, the growth process leads, in most cases, to a "partial" equilibrium situation, and the models should be able to predict quantitatively such a condition. Partial equilibrium (Helgeson 1968) is an ample expression intended to indicate that a multicomponent, heterogeneous system is in equilibrium with respect to at least one process or reaction, but out of equilibrium with respect to others. For instance, the reaction between a solid and an aqueous solution may produce a second solid that is compatible with the aqueous solution but not with the original solid phase. Equilibrium is then attained between the second solid phase and the aqueous phase, even though the original solid continues to react with the solution. In SS-AS systems, a typical case of partial equilibrium occurs when the initial reacting solid solution becomes isolated from the aqueous solution through coatings or outer layers of solid solution that have a different composition (Glynn et al. 1990). While the initial solid may become isolated from the aqueous solution, the reactive parts forming the outer layers of the solid may maintain equilibrium with the aqueous phase. This occurs because the mobility of the substituting ions in the aqueous phase is very high, while solid state diffusion is extremely slow.

Partial equilibrium effects are important in mineral-water interactions. For example, surface co-precipitation has been demonstrated to be an efficient mechanism of uptake of dissolved metal ions by carbonate minerals. The process involves the release of solute via dissolution of the mineral substrate and the subsequent reaction between the released solute and the metal ions to precipitate a metal-bearing solid solution on the mineral surface. Under these conditions, the precipitate can form a layer that "protects" the substrate from further dissolution. As a consequence, the removal stops because the system approaches a "partial" equilibrium situation. This phenomenon is particularly efficient when both the substrate and overgrowth are isostructural and have close lattice parameters (Prieto et al. 2003; Pérez-Garrido et al. 2007). In such a case the process can stop when a layer has formed that is only some few nanometers thick (Fig. 21). The consequences of these kinds of coatings on the kinetics of water-rock interactions are significant (Cubillas et al. 2005), although frequently disregarded in kinetic modeling.

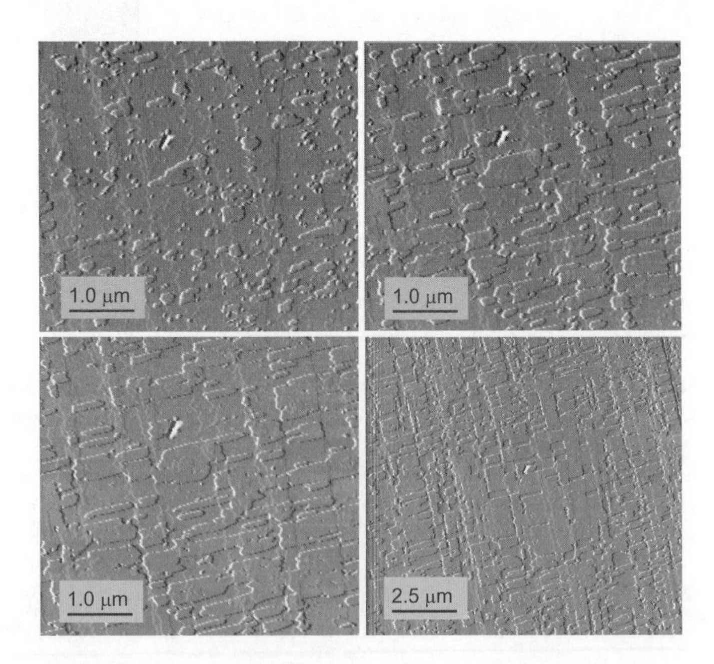

Figure 21. Sequence of AFM deflection images showing the spread of islands to form a layer (2.6 nm thick) of $(Cd,Ca)CO_3$ on the calcite surface. Eventually, the layer covers completely the surface. (AFM pictures by C. Pérez-Garrido)

CONCLUDING REMARKS

SS-AS systems are geochemically important because natural minerals always contain a certain proportion of minor or trace elements, which is a record of their crystallization environment. Solid solution effects are important in quantifying element budgets in different Earth reservoirs and biogeochemical cycles. SS-AS systems are also environmentally important. The interaction of dissolved toxic metals with minerals frequently results in the precipitation of metal-bearing solid solutions on the mineral surfaces or in rock or sediment pores. As a result, metals can be removed from contaminated waters and the thermodynamic properties of these solid solutions have a crucial influence on the transport and fate of toxic metals in the environment. Despite this, solid-solution effects are often disregarded in water-rock interaction geochemistry and geochemical modeling. The reason probably lies in the complexity of the physicochemical processes involved in these kinds of systems. Whereas the thermodynamics of SS-AS systems is firmly established, the kinetic models of SS-AS processes need to be very much improved. However, above all, the main difficulty is the lack of low-temperature thermodynamic data for non-ideal solid solutions. The experimental determination of mixing properties is very time consuming and has many sources of uncertainty. For example, stoichiometric solubility measurements may yield inaccuracies because, in many cases, there may be potential problems in achieving stoichiometric saturation. The same occurs with the experimental determination of distribution coefficients or miscibility gaps, in which formation of metastable compositions by kinetic effects is always a possibility. Fortunately, the use of semi-empirical expressions (e.g., Zhu 2004) and molecular simulation methods to obtain mixing parameters of solid solutions has been getting increasing attention in the last few years (e.g., Becker et al. 2000; Vinograd et al. 2006, 2008; Benny et al. 2009). Combining different simulation methods with experimental measurements (e.g., enthalpies of mixing, compositional evolution of lattice parameters, etc.) is probably the best way to improve calculation methods. This progress, together with the development of models describing the thermodynamics of multicomponent SS-AS systems, are the main goals to be attained in the next few years.

ACKNOWLEDGMENTS

I would like to thank the editors Eric Oelkers and Jacques Schott for the invitation to contribute to this volume and their editorial comments and suggestions. Many of the concepts and findings illustrated in this chapter have been matured over time by a number of researchers and collaborators. Initial works with Andrew Putnis, Lurdes Fernández-Díaz and a first generation of students, including Ángeles Fernández-González, Carlos Pina, and José Manuel Astilleros, were then followed by numerous collaborations with researchers from different countries. I would like to thank all these colleagues for discussions concerning solid-solutions and related subjects, particularly Udo Becker, Dirk Bosbach, Michael Böttcher, Mario Gonçalves, Athanasios Godelitsas, Stephan Köhler, Dimitrii Kulik, Christine Putnis, Susan Stipp, and Eric Oelkers. I would also like to express my gratitude to students past and present, too numerous to list, for their insightful contributions. Discussions with partners and students of the European Networks MIR (Mineral-Fluid Interface Reactivity) and MIN-GRO (Mineral Nucleation and Growth Kinetics) helped me to decide the structure of this chapter.

REFERENCES

Andara AJ, Heasman DM, Fernández-González A, Prieto M (2005) Characterization and crystallization of the $Ba(SO_4,SeO_4)$ solid solution. Cryst Growth Des 5:1371-1378

Appelo CAJ, Postma D (2005) Geochemistry, Groundwater and Pollution (2nd edition). A.A. Balkema Publishers, Leiden

Astilleros JM, Pina CM, Fernández-Díaz L, Putnis A (2000) Incorporation of barium on calcite $(10\bar{1}4)$ surfaces during growth. Geochim Cosmochim Acta 64:2965-2972

Astilleros JM, Pina CM, Fernández-Díaz L, Putnis A (2003a) Supersaturation functions in binary solid solution-aqueous solution systems. Geochim Cosmochim Acta 67:1601-1608

Astilleros JM, Pina CM, Fernández-Díaz L, Putnis A (2003b) Metastable phenomena on calcite $\{10\bar{1}4\}$ surfaces growing form Sr^{2+}-Ca^{2+}-CO_3^{2-} aqueous solutions. Chem Geol 193:93-107

Becker U, Fernández-González A, Prieto M, Harrison R, Putnis A (2000) Direct calculation of thermodynamic properties of the barite-celestite solid solution from molecular principles. Phys Chem Mineral 27:291-300

Becker U, Risthaus P, Brandt F, Bosbach D (2006) Thermodynamic properties and crystal growth behavior of the Hashemite ($BaSO_4$-$BaCrO_4$) solid solution. Chem Geol 225:244-255

Benny S, Grau-Crespo R, de Leeuw NH (2009) A theoretical investigation of α-Fe_2O_3–Cr_2O_3 solid solutions. Phys Chem Chem Phys 11:808-815

Berthelot M (1872) On the law which governs the distribution of a substance between two solvents. Ann Chim Phys (4th series) 26:408-417

Böttcher ME (1997a) Experimental dissolution of $CaCO_3$-$MnCO_3$ solid solutions in CO_2-H_2O solutions at 20 °C, I. Synthetic low-temperature carbonates. Solid State Ionics 101-103:1263-1266

Böttcher ME (1997b) The transformation of aragonite to $Mn_xCa_{1-x}CO_3$ solid-solutions at 20 °C: An experimental study. Mar Chem 57:97-106

Böttcher ME (2000) Editorial Honoring Friedrich Lippmann. Aquatic Geochem 6:115-117

Burton BP (1987) Theoretical analysis of cation ordering in binary rhombohedral carbonate systems. Am Mineral 72:329-336

Capobianco C, Burton BP, Davidson PM, Navrotsky A (1987) Structural and calorimetric studies of order-disorder in $CdMg(CO_3)_2$. J Solid State Chem 71:214-223

Carpenter MA (1980) Mechanisms of exsolution in sodic pyroxenes. Contrib Mineral Petrol 71:289-300

Casey WH, Chai L, Navrostsky A, Rock PA (1996) Thermochemistry of mixing strontianite [$SrCO_3$ (s)] and aragonite [$CaCO_3$ (s)] to form $Ca_xSr_{1-x}CO3$ (s) solid solutions. Geochim Cosmochim Acta 60:933-940

Chiarello RP, Sturchio NC, Grace JD, Geissbuhler P, Sorensen LB, Cheng L, Xu S (1997) Otavite-calcite solid-solution formation at the calcite-water interface studied in situ by synchrotron X-ray scattering. Geochim Cosmochim Acta 61:1467-1474

Crocket JH, Winchester JW (1966) Coprecipitation of zinc with calcium carbonate. Geochim Cosmochim Acta 30:1903-1109

Cubillas P, Köhler S, Prieto M, Causserand C, Oelkers EH (2005) How do mineral coatings affect dissolution rates? An experimental study of coupled $CaCO_3$ dissolution - $CdCO_3$ precipitation. Geochim Cosmochim Acta 69:5459-5476

Curti E, Kulik DA, Tits J (2005) Solid solutions of trace Eu(III) in calcite: thermodynamic evaluation of experimental data over a wide range of pH and pCO_2. Geochim Cosmochim Acta 69:1721-1737

Davis JA, Fuller CC, Cook AD (1987) Mechanisms of trace metal sorption by calcite: Adsorption of Cd^{2+} and subsequent solid solution formation. Geochim Cosmochim Acta 51:1477-1490

Doerner HA, Hoskins WM (1925) Coprecipitation of radium and barium sulphates. J Am Chem Soc 46:662-675

Dromgoole EL, Walter LM (1990) Iron and manganese incorporation into calcite: Effects of growth kinetics, temperature, and solution chemistry. Chem Geol 81:311-336

Fernández-González A, Martín-Díaz R, Prieto M (1999a) Crystallisation of $Ba(SO_4,CrO_4)$ solid solutions form aqueous solutions. J Cryst Growth 200:227-235

Fernández-González A, Prieto M, Putnis A, López Andrés S (1999b) Concentric zoning patterns in crystallizing $(Cd,Ca)CO_3$ solid-solutions from aqueous solutions. Mineral Mag 63:331-343

Fernández-González A, Andara A, Alía JM, Prieto M (2006) Miscibility in the $CaSO_4{\cdot}2H_2O$–$CaSeO_4{\cdot}2H_2O$ system: Implications for the crystallisation and dehydration behaviour. Chem Geol 225:256-265

Fernández González A, Andara A, Prieto M (2007) Mixing properties and crystallization behaviour of the scheelite-powellite solid solution. Cryst Growth Des 7:545-552

Freij SJ, Godelitsas A, Putnis A (2003) Crystal growth and dissolution processes at the calcite-water interface in the presence of zinc ions. J Cryst Growth 273:535-545

Gamsjäger H, Königsberger E, Preis W (2000) Lippmann diagrams: Theory and application to carbonate systems. Aquatic Geochem 6:119-132

Garrels RM, Wollast R (1978) Discussion of: "Equilibrium criteria for two-component solids reacting with fixed composition in an aqueous phase. Example: The magnesian calcites". Am J Sci 278:1469-1474

Glynn PD (1991) MBSSAS: A code for the computation of Margules parameters and equilibrium relations in binary solid-solution aqueous-solution systems. Comp Geosci 17:907-966

Glynn PD (2000) Solid-Solution Solubilities and Thermodynamics: Sulfates, carbonates and halides. Rev Mineral 40:481-511

Glynn PD, Reardon EJ (1990) Solid-solution aqueous solution equilibria: thermodynamic theory and representation. Am J Sci 278:164-201

Glynn PD, Reardon EJ (1992) Reply to a comment by Königsberger E and Gamsjäger H on "Solid-solution aqueous-solution equilibria: Thermodynamic theory and representation". Am J Sci 292:215-225

Glynn PD, Reardon EJ, Plummer LN, Busenberg, E (1990) Reaction paths and equilibrium end-points in solid-solution aqueous solution systems. Geochim Cosmochim Acta 54:267-282

Glynn PD, Reardon EJ, Plummer LN, Busenberg E (1992) Reply to Dr. Stoessell's comment on "Reaction paths and equilibrium end-points in solid solution aqueous solution systems". Geochim Cosmochim Acta 56:2559-2572

Godelitsas A, Astilleros JM, Hallam K, Harissopoulos S, Putnis A (2003) Interaction of calcium carbonates with lead in aqueous solutions. Environ Sci Technol 37:3351-3360

Goldsmith JR (1983) Phase relations of rhombohedral carbonates. Rev Mineral 11:49-76

Guggenheim EA (1937) Theoretical basis of Raoult's law. Trans Faraday Soc 33:151-159

Heberling F., Denecke MA, Bosbach D (2008) Neptunium (V) coprecipitation with calcite. Environ Sci Technol 42:471-476

Helgeson HC (1968) Evaluation of irreversible reactions in geochemical processes involving minerals and aqueous solutions: I. Thermodynamic relations. Geochim Cosmochim Acta 32:853-877

Hildebrand JH (1936) Solubility of Non-Electrolites. Reinhold Publ Co, New York

Kashchiev D, van Rosmalen GM (2003) Review: Nucleation in solutions revisited. Cryst Res Tech 38:555-574.

Katsikopoulos D, Fernández-González A, Prieto AC, Prieto M (2008a) Co-crystallization of Co(II) with calcite: implications for the mobility of cobalt in aqueous environments. Chem Geol 254:87-100

Katsikopoulos D, Fernández-González A, Prieto M (2008b) Crystallization of the $(Cd,Ca)CO_3$ solid solution in double diffusion systems: The partitioning behaviour of Cd^{2+} in calcite at different supersaturation rates. Mineral Mag 72:433-436

Kiseleva IA, Ogorodova LP, Topor ND (1980) High-temperature microcalorimetry of scheelite-powellite solid solutions. Int Geochem 5:764-768

Kornicker WA, Morse JW, Damasceno RN (1985) The chemistry of Co^{2+} interaction with calcite and aragonite surfaces. Chem Geol 53:229-236

Kornicker WA, Presta PA, Paige CR, Johnson DM, Hileman OE, Snodgrass WJ (1991) The aqueous dissolution kinetics of the barium/lead sulfate solid solution series at 25 and 60 °C. Geochim Cosmochim Acta 55:3531-3541

Königsberger E, Gamsjäger H (1992) Comment on "Solid-solution aqueous-solution equilibria: thermodynamic theory and representation" by Glynn P.D. and Reardon W.J. Am J Sci 292:199-214

Königsberger E, Hausner R, Gamsjäger H (1991) Solid-solute equilibria in aqueous solution: V. The system $CdCO_3$-$CaCO_3$-CO_2-H_2O. Geochim Cosmochim Acta 55:3505-3514

Kowack M, Putnis A (2008) The effect of specific background electrolytes on water structure and solute hydration: consequences for crystal dissolution and growth. Geochim Cosmochim Acta 72:4476-4487

Kowack M, Putnis CV, Putnis A (2007) The effect of cation:anion concentration ratio in solution on the mechanism of barite growth at constant supersaturation: Role of the desolvation process on the growth kinetics. Geochim Cosmochim Acta 71:5168-5179

Kulik DA (2002) Gibbs energy minimization approach to model sorption equilibria at the mineral-water interface: Thermodynamic relations for multi-site surface complexation. Am J Sci 302:227-279

Kulik DA (2006) Dual-thermodynamic estimation of stoichiometry and stability of solid solution end members in aqueous - solid solution systems. Chem Geol 225:189-212

Kulik DA, Kersten M, Heiser U, Neumann T (2000) Application of Gibbs energy minimization to model early-diagenetic solid-solution aqueous-solution equilibria involving authigenic rhodochrosites in anoxic Baltic sea sediments. Aquatic Geochem 6:147-149

Langmuir D (1997) Aqueous environmental geochemistry. Prentice Hall, New Jersey

Lafon (1978) Discussion of "Equilibrium criteria for two-component solids reacting with fixed composition in aqueous phase. Example: the magnesian calcites". Am J Sci 274:1455-1468

Lea AS, Hurt TT, El-Azab A, Amonette JE, Baer DR (2003) Heteroepitaxial growth of a manganese carbonate secondary nano-phase on the $(10\bar{1}4)$ surface of calcite in solution. Surf Sci 524:63-77

L'Heureux I, Kastev S (2006) Oscillatory zoning in a $(Ba,Sr)SO_4$ solid solution: Macroscopic and cellular automata models. Chem Geol 225:230-243

Lippmann F (1977) The solubility product of complex minerals, mixed crystals and three-layer clay minerals. N Jahrb Mineral Abh 130:243-263

Lippmann F (1980) Phase diagrams depicting the aqueous solubility of binary mineral systems. N Jahrb Mineral Abh 139:1-25

Lippmann F (1991) Aqueous solubility of magnesian calcites with different endmembers. Acta Mineral Petrogr 32:5-19

Lorens RB (1981) Strontium, cadmium, manganese, and cobalt distribution coefficients in calcite as a function of calcite precipitation rate. Geochim Cosmochim Acta 45:553-561.

Monnin C, Cividini D (2006) The saturation state of the world's ocean with respect to (Ba,Sr)SO_4 solid solutions. Geochim Cosmochim Acta 70:3290-3298

Mucci A (2004) The behaviour of mixed Ca-Mn carbonates in water and seawater: Controls of manganese concentrations in marine pore waters. Aquatic Geochem 10:139-169

Mucci A, Morse JW (1983) The incorporation of Mg^{2+} and Sr^{2+} into calcite overgrowths: influences of growth rate and solution composition. Geochim Cosmochim Acta 47:217-233

Navrotsky A (1987) Models of crystalline solutions. Rev Mineral 17:35–69

Nerst W (1891) Distribution of a substance between two solvents and between solvent and vapor. Z Phys Chem 8:110-139

Paquette J, Reeder RJ (1995) Relationship between surface structure, growth mechanism, and trace element incorporation in calcite. Geochim Cosmochim Acta 59:735-749

Parkhurst DL, Appelo CAJ (1999) User's guide to PHREEQC (Version 2) - A computer program for speciation, batch-reaction, one-dimensional transport and inverse geochemical calculations. U. S. Geological Survey Water Resources Investigations Report 99-4259, U. S. Geological Survey, Denver

Pérez-Garrido C, Fernández-Díaz L, Pina CM, Prieto M (2007) In situ AFM observations of the interaction between calcite ($10\bar{1}4$) surfaces and Cd-bearing aqueous solutions. Surf Sci 601:5499-5509

Plummer LN, Busenberg E (1987) Thermodynamics of the aragonite-strontianite solid solutions: Results from stoichiometric solubility at 25 and 76 °C. Geochim. Cosmochim. Acta 51:1393-1411

Pina CM, Putnis A (2002) The kinetics of nucleation of solid solutions from aqueous solutions: a new model for calculating non-equilibrium distribution coefficients. Geochim Cosmochim Acta 66:185-192

Pina CM, Enders M, Putnis A (2000) The composition of solid solutions crystallizing from aqueous solutions: The influence of supersaturation and growth mechanisms. Chem Geol 168:195-210

Prieto M, Putnis A, Fernández-Díaz L (1993) Crystallization of solid solutions from aqueous solutions in a porous medium: Zoning in (Ba,Sr)SO_4. Geol Mag 130:289-299

Prieto M, Fernández-González A, Putnis A, Fernández-Díaz L (1997) Nucleation, growth, and zoning phenomena in crystallizing (Ba,Sr)CO_3, Ba(SO_4,CrO_4), (Ba,Sr)SO_4, and (Cd,Ca)CO_3 solid solutions from aqueous solutions. Geochim Cosmochim Acta 61:3383-3397

Prieto M, Fernández-González A, Becker U, Putnis A (2000) Computing Lippmann diagrams from direct calculation of mixing properties of solid solutions: Application to the barite-celestite system. Aquatic Geochem 6:133-146

Prieto M, Fernández-González A, Martín-Díaz R (2002) Sorption of chromate ions diffusing through barite-hydrogel composites: Implications for the fate and transport of chromium in the environment. Geochim Cosmochim Acta 66:783-795

Prieto M, Cubillas P, Fernández-Gonzalez A (2003) Uptake of dissolved Cd by biogenic and abiogenic aragonite: a comparison with sorption onto calcite. Geochim Cosmochim Acta 67:3859-3869

Prieto M, Astilleros JM, Pina CM, Fernández-Díaz L, Putnis A (2007) Comment: Supersaturation in binary solid solution-aqueous solution systems. Am J Sci 307:1034-1045

Putnis A (1992) Introduction to Mineral Sciences. Cambridge University Press, Cambridge

Putnis A, Fernández-Díaz L, Prieto M (1992) Experimentally produced oscillatory zoning in the (Ba,Sr)SO_4 solid solution. Nature 358:743-745

Putnis A, Pina CM, Astilleros JM, Fernández-Díaz L, Prieto M (2003) Nucleation of solid solutions crystallizing from aqueous solutions. Phil Trans R Soc London 361:615-632

Redlich O, Kister AT (1948) Algebraic representation of the thermodynamic properties and the classification of solutions. Ind Eng Chem 40:345-348

Reeder RJ (1983) Crystal chemistry of the rhombohedral carbonates. Rev Mineral 11:1-47

Reeder RJ, Fagioly RO, Meyers W (1990) Oscillatory zoning of Mn in solution grown calcite crystals. Earth Sci Rev 29:39-46

Reeder RJ (1996) Interaction of divalent cobalt, zinc, cadmium, and barium with the calcite surface during layer growth. Geochim Cosmochim Acta 60:1543-1552

Reeder RJ, Lamble GM, Northrup PA (1999) XAFS study of the coordination and local relaxation around Co^{2+}, Zn^{2+}, Pb^{2+}, and Ba^{2+} trace elements in calcite. Am Mineral 84:1049-1060

Rickaby REM, Schrag DP, Zondervan I, Riebesell U (2002) Growth rate dependence of Sr incorporation during calcification of Emiliania huxleyi. Global Biogeochem Cycles 16:61-68

Rimstidt JD, Balog A, Webb J (1998) Distribution of trace elements between carbonate minerals and aqueous solutions. Geochim Cosmochim Acta 62:1851-1863

Roozeboom HWB (1904) Die Heterogenen Gleichgewichte vom Standpunkte der Phasenlehre II. Friedrich Vieweg und Sohn, Braunschweig

Rosenthal Y, Boyle EA, Slowey N (1997) Temperature control on the incorporation of magnesium, strontium, fluorine, and cadmium into benthic foraminiferal shells from Little Bahama Bank: Prospects for thermoclyne paleoceanography. Geochim Cosmochim Acta 61:3633-3643

Sangwal K (1989) On the estimation of surface entropy factor, interfacial tension, dissolution enthalpy and metastable zone-width for substances crystallizing from solution. J Cryst Growth 97:393-405

Saxena SK (1968) Distribution of elements between coexisting minerals and the nature of solid solution in garnet. Am Mineral 53:994-1014

Saxena SK (1973) Thermodynamics of Rock-Forming Crystalline Solutions. Springer-Verlag, Berlin

Shtukenberg AG, Punin YO, Azimov P (2006) Crystallization kinetics in binary solid solution-aqueous solutions systems. Am J Sci 290:164-201

Sleight AW (1972) Accurate cell dimensions for ABO_4 molybdates and tungstates. Acta Crystallogr B 28:2899-2902

Stephenson AE, De Yoreo JJ, Wu L, Wu KJ, Hoyer J, Dove PM (2008) Peptides enhance magnesium signature in calcite: insights into origins of vital effects. Science 322:724-727

Stipp SL, Hochella MF, Parks GA, Leckie JO (1992) Cd^{2+} uptake by calcite, solid-state diffusion, and the formation of solid-solution: Interface processes observed with near-surface sensitive techniques (XPS, LEED, and AES). Geochim Cosmochim Acta 56:1941-1954

Stipp SL, Parks GA, Nordstrom DK, Leckie JO (1993) Solubility-product constant and thermodynamic properties for synthetic otavite, $CdCO_{3(s)}$, and aqueous association constants for the Cd(II)-CO_2-H_2O system. Geochim Cosmochim Acta 57: 2699-2713

Stipp SL, Lakshtanov LZ, Jensen JT, Baker JA (2003) Eu^{3+} uptake by calcite: preliminary results from co-precipitation experiments and observations with near-surface sensitive techniques, J Contam Hydrol 61:33-43

Stoessel RK (1992) Comment on "Reaction paths and equilibrium end-points in solid-solution aqueous-solution systems" by P. D. Glynn, E. J. Reardon, L. N. Plummer, and E. Busenberg. Geochim Cosmochim Acta 56:2555-2557

Stoll HM, Rosenthal Y, Falkowski P (2002) Climate proxies from Sr/Ca of coccolith lattice: calibrations from continuous culture of Emiliania Husleyi. Geochim Cosmochim Acta 64:927-936

Tesoriero AJ, Pankow JF (1996) Solid solution partitioning of Sr^{2+}, Ba^{2+}, and Cd^{2+} to calcite. Geochim Cosmochim Acta 60:1053-1063

Thorstenson DC, Plummer LN (1977) Equilibrium criteria for two-component solids reacting with fixed composition in an aqueous-phase. Example: The magnesian calcites. Am J Sci 277:1203-1223

Thorstenson DC, Plummer LN (1978) Equilibrium criteria for two-component solids reacting with fixed composition in an aqueous-phase. Example: The magnesian calcites. Reply. Am J Sci 278:1478-1488

Urusov VS (1974) Energetic criteria for solid solution miscibility gap calculations. Bull Soc Fr Minéral Cristallogr 91:217-222

Vinograd VL, Winkler B, Putnis A, Gale JD, Sluiter MHF (2006) Static lattice energy calculations of mixing and ordering enthalpies in binary carbonate solid solutions. Chem Geol 225:304-313

Vinograd VL, Bosbach D, Winkler B, Gale JD (2008) Subsolidus phase relations in $Ca_2Mo_2O_8$–$NaEuMo_2O_8$-powellite solid solution predicted from static lattice energy calculations and Monte Carlo simulations. Phys Chem Chem Phys 10:3509-3518

Wang Y, Merino E (1992) Dynamic model of oscillatory zoning of trace elements in calcite: Double layer, inhibition, and self-organization. Geochim Cosmochim Acta 56:587-596

Wasylenki L, Dove PM, Wilson DS, De Yoreo JJ (2005) Nanoscale effects of strontium on calcite growth: An in situ AFM study in the absence of vital effects. Geochim Cosmochim Acta 69:3017-3027

Xu N, Hochella MF, Brown GE, Parks GA (1996) Co(II) sorption at the calcite-water interface: I. X-ray photoelectron spectroscopic study. Geochim Cosmochim Acta 60:2801-2815

Zachara JM, Cowan CE, Resh CT (1991) Sorption of divalent metals on calcite. Geochim Cosmochim Acta 55:1549-1562

Zhu C (2004) Coprecipitation in the barite isostructural family: 1. Binary mixing properties. Geochim Cosmochim Acta 68:3327-3337

Reviews in Mineralogy & Geochemistry
Vol. 70 pp. 87-124, 2009

Mineral Replacement Reactions

Andrew Putnis

Institut für Mineralogie
University of Münster
Münster, 48149 Germany

putnis@uni-muenster.de

INTRODUCTION

Whenever a mineral or mineral assemblage comes into contact with a fluid with which it is out of equilibrium, reequilibration will tend to take place to reduce the free energy of the whole system (i.e., of the solid + fluid). Such fluid-solid interactions span a very wide range of possible reactions, and are responsible for most of the mineral assemblages we see in the Earth's crust. However, before discussing mineral replacement reactions, we will put them into the broader context of fluid-solid interactions by considering some examples of such reequilibration.

In the simplest case, we could consider the situation where a mineral, thermodynamically stable under some specific temperature and pressure conditions, comes into contact with pure water, such as quartz within its stability field (e.g., at T = 100 °C and 1 atmosphere pressure). Clearly, quartz will tend to dissolve until, at equilibrium, the aqueous silica solution, which at neutral pH is $H_4SiO_{4(aq)}$, becomes saturated with respect to quartz. The reaction for this equilibration can be written:

$$SiO_{2\,(qtz)} + H_2O \leftrightarrow H_4SiO_{4(aq)} \tag{1}$$

The equilibrium solubility constant for reaction (1) is given by

$$K_{sp}(qtz) = \frac{a(H_4SiO_4)_{aq}}{a(SiO_2)_{qtz} \cdot a(H_2O)} = a(H_4SiO_4)_{aq}$$

where $a(i)_{aq}$ stands for the activity of the parenthetical aqueous species. For the case of pure quartz and pure water where the activity is 1, $K_{sp}(qtz) \sim 1.2 \times 10^{-3}$ at 100 °C and 1 atmosphere pressure. However, if under these conditions the solid silica phase in contact with pure water was cristobalite (the high temperature polymorph of SiO_2), the resulting aqueous solution, saturated with respect to cristobalite, would be supersaturated with respect to quartz, since the less stable phase is more soluble. The value of K_{sp} for cristobalite at 100 °C and 1 atmosphere pressure is $\sim 5.1 \times 10^{-3}$. Thus the thermodynamics would indicate that quartz *should* precipitate from such a solution. On the other hand, as is well known, the kinetics of nucleation may preclude any precipitation, and the solution may remain supersaturated with respect to every silica phase more stable than cristobalite. If quartz did precipitate from a solution which had equilibrated with respect to cristobalite, this transformation from cristobalite to quartz would be an example of what has been termed a "solvent-mediated phase transformation" (Cardew and Davey 1985) whose kinetics are many orders of magnitude faster than the solid state transformation of cristobalite to quartz under similar temperature conditions. However, Cardew and Davey considered that the dissolution of the metastable phase and the nucleation and growth of the more stable phase took place *independently*. In this chapter, we will be concerned about how these two processes may be coupled.

1529-6466/09/0070-0003$05.00 DOI: 10.2138/rmg.2009.70.3

There are many examples of polymorphic phase transformations achieved by dissolution and reprecipitation. One of the most studied is the transformation from aragonite to calcite. Early studies (e.g., Brown et al. 1962) already confirmed that the transformation at low temperatures was up to 10 orders of magnitude faster in the presence of H_2O than in the dry system. The important implication for geosciences is that aragonite preserved in high pressure rocks (within the stability field of aragonite) which have been exhumed to pressures where calcite is the stable polymorph must have remained dry during uplift (Carlson and Rosenfeld 1981; Carlson 1983; Essene 1983). The aragonite to calcite transformation is complicated in many natural environments because of the known (but poorly understood) effect of the presence of background ions in the solution, some of which accelerate and others inhibit the transformation. However, the details of the aragonite to calcite transformation in complex aqueous solutions such as sea-water are beyond the scope of this chapter.

Dissolution and precipitation are the principal driving mechanisms for all reequilibration reactions in the presence of a fluid phase. Relatively small free energy reductions can drive large-scale processes. For example, in a sedimentary sandstone deposit the stresses developed at quartz grain-grain contacts due to the weight of the overburden are sufficient to increase the solubility of the stressed contact relative to the unstressed mineral phase. This results in *intergranular pressure solution* (Fig. 1) which causes the dissolution of material at grain contacts, transport of material in solution and reprecipitation at sites of low stress, ultimately causing the compaction and lithification of the sediment (Rutter 1983; Gratier and Guiget 1986; Renard et al. 2000; Revil 2001; Chester et al. 2004; Lang 2004). However, even in such an apparently simple case of a dissolution–precipitation reaction, the rate controlling processes are still poorly understood. Furthermore, the pore fluid chemistry is known to strongly influence the dissolution and precipitation rates of quartz and hence the compaction rate due to pressure solution may either be increased or decreased depending on the specific ions in solution (Dove and Rimstidt 1994). The presence of other solid phases which in turn change the chemistry of the fluid by dissolution will also therefore affect the dissolution–precipitation rates. For example, the presence of a small clay fraction in the quartz can significantly enhance the pressure solution and hence the compaction rate (Renard et al. 1997).

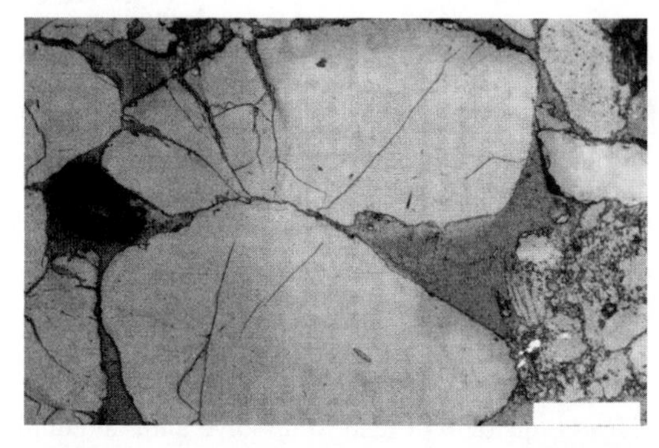

Figure 1. Indented quartz grains due to intergranular pressure solution in a sandstone. (Image: F. Renard)

A decrease in free energy of a solid + fluid system can also be achieved by grain coarsening. The larger surface:volume ratio of small crystals relative to large crystals translates into a higher solubility for the former, and hence the dissolution of small crystals and the precipitation onto larger crystals of the same phase is also mechanism of reducing the free energy of the system. That a reduction in interfacial free energy can drive a dissolution-precipitation reaction has been demonstrated by Nakamura and Watson (1981) who showed that dense pressed pellets of synthetic quartzite, when placed in contact with silica saturated fluid, recrystallize and coarsen by a dissolution-precipitation mechanism. Significantly, this process generates new porosity at the recrystallization front, as the fluid infiltrates the dense pellet. The porosity generation, which implies that some of the silica is transferred to the fluid phase, creates permeability in the recrystallized part of the pellet, and allows the fluid to infiltrate and remain in contact with the as yet unrecrystallized solid. Porosity generation is an important theme in this chapter.

Increasing the complexity of the fluid phase by the addition of other components dissolved in the solution, increases the complexity of the thermodynamics and the nature of the phases which can potentially precipitate. For example, if we consider the reaction between feldspar and an acidic fluid, the dissolution of the feldspar results in a fluid which becomes supersaturated with respect to kaolinite. The reaction with potassium feldspar may be written as:

$$2KAlSi_3O_8 + 2H^+ + 9H_2O \rightarrow Al_2Si_2O_5(OH)_4 + 2K^+ + 4H_4SiO_{4(aq)} \tag{2}$$

and the resultant solution becomes enriched in silica and potassium ions. A similar reaction can be written for the breakdown of other feldspar compositions, in which case the resultant solution would also be enriched in Na^+ and Ca^{2+}. In the presence of an applied stress, the dissolution reaction would be promoted by pressure solution, in other words the stress would drive the kaolinite-forming reaction.

If we consider reaction (2) in two sequential steps, with an acidic aqueous solution first dissolving the K-feldspar congruently to form a solution containing K, Al, and Si dissolved in some complexed form in the fluid, and then the subsequent precipitation of kaolinite when the supersaturation becomes sufficiently high for nucleation, we also need to ask how far this solution is transported from the dissolution site before kaolinite precipitates. We also might consider the role of the residual alkali and silica rich fluid in further reactions with other mineral phases. If we generalize such a dissolution–transport–precipitation scenario we can apply this to a conceptual model of how mineral-fluid reactions work in a polyphase rock. Disequilibrium between a fluid and a rock can arise by the infiltration of a fluid with composition such that it is out of equilibrium with the mineral or mineral assemblage, or by a change of temperature and/or pressure. In either case this will generally lead to a dissolution process, and the resulting fluid may become supersaturated with respect to a number of new solid phases. The application of stress in the same situation may enhance the dissolution reactions.

Such a situation has been described in the context of prograde metamorphic reaction mechanisms by Carmichael (1969) who demonstrated how in a closed system, in which the overall chemical composition of solid + fluid remains constant, a sequence of dissolution and precipitation sub-reactions can explain the textural development of a metamorphic rock on a thin-section scale. However, the sub-reactions involved in the individual steps of dissolution–transport–reaction–precipitation are metasomatic reactions i.e. with local changes in composition and redistribution of material. Thus on a small spatial scale the system is "open" while on a larger scale the system may be closed.

The linkage between deformation and induced chemical reactions in a rock has been explored by Wintsch (1985) and Wheeler (1987) and forms the basis for a unified conceptual framework in which fluid-driven dissolution-precipitation processes determine both the deformation mechanisms and the chemical reactions in the rock. Depending on the length-scale of transport of different components in the fluid, the redistribution of elements may result in very marked modification of textures during syntectonic mineral growth. Evidence for mass transfer by solution resulting in deformation is well documented in sedimentary and low grade metamorphic rocks (Passchier and Trouw 1996) and is increasingly recognized as a mechanism of deformation (dissolution and replacement creep) in higher grade mid-crustal rocks (Wintsch and Yi 2002) and high pressure metamorphic environments (Schwarz and Stöckert 1996; Stöckert et al. 1999).

In the case of kaolinite formation from granitic rocks, the development of a pure kaolinite deposit implies the long-range transport of the other elements removed from the granite. Under suitable conditions these may become concentrated and hence sufficiently supersaturated to precipitate metal-rich mineral deposits. In Cornwall, England a complex sequence of hydrothermal alteration processes eventually results in very pure kaolinite, mined as "china clay." Although most of the kaolinite-rich deposits preserve very little of the original rock texture, the formation of kaolinite *pseudomorphs* after Carlsbad-twinned feldspar is not uncommon. Figure 2 shows an example of such a pseudomorph, defined as a replacement of a parent mineral by a product phase or phase assemblage, while preserving the external shape and dimensions of the parent. Such a phenomenon is not merely a mineralogical curiosity, but raises quite fundamental issues about the nature of dissolution, transport, and precipitation during fluid-mineral interactions, including metaomatism, metamorphism, diagenesis, and chemical weathering. Clearly, in the case of pseudomorphism, the dissolution and precipitation are spatially coupled.

Figure 2. Pseudomorph of kaolinite after Carlsbad-twinned feldspar.

PSEUDOMORPHIC REPLACEMENT – VOLUMETRIC CONSIDERATIONS

Preservation of external shape and dimensions is a key feature of the pseudomorphic replacement of one mineral by another. There are many parent–product mineral pairs such as that shown in Figure 2 where the volume is apparently preserved. We now consider reaction (2) again, and determine the volume change involved in converting 2 moles of K-feldspar to one mole of kaolinite. Standard data bases give the molar volumes of K-feldspar and kaolinite as ~109 cm^3mol^{-1} and ~99 cm^3mol^{-1} respectively. Reaction (2) would therefore involve a volume decrease in the solids of ~50%. To preserve the external volume of the parent during a pseudomorphic replacement reaction, the reaction must be rewritten so that it is balanced on volume. The principle of balancing volumes in replacement reactions was pointed out almost 100 years ago by Lindgren (1912). To illustrate this, but approximating the molar volumes of K-feldspar and kaolinite as equal, reaction (3) is balanced so that one mole of K-feldspar is replaced by one mole of kaolinite:

$$KAlSi_3O_8 + Al^{3+} + 5H_2O \rightarrow Al_2Si_2O_5(OH)_4 + K^+ + 2H^+ + H_4SiO_{4(aq)} \tag{3}$$

There are a number of important principles which arise from considering a reaction of this kind, as pointed out by Merino et al. (1993) and Merino and Dewers (1998). First, the replacement requires input of Al and the loss of Si. The Al in solution may come from the dissolution of a more aluminous mineral elsewhere in the rock. In replacement reactions we must consider that every element involved may be mobile and that assuming immobility of an element to balance metamorphic reactions may not be valid. A second point to consider is that the product kaolinite in the reaction may be porous and hence a volume calculation would need to take into account any porosity generated by the reaction. As we shall see however, to understand porosity generation we will need to consider other factors as well as molar volume changes.

Another common replacement reaction in rocks is the serpentinization of olivine and orthopyroxene. If we assume olivine as forsterite, Mg_2SiO_4, the orthopyroxene as enstatite $MgSiO_3$ and serpentine as lizardite, $Mg_3[Si_2O_5](OH)_4$ and take the molar volumes as ~50 cm^3mol^{-1}, ~25 cm^3mol^{-1} and ~100 cm^3mol^{-1} respectively, reactions balanced on volume could be:

$$2Mg_2SiO_4 + 2H^+ + H_2O \rightarrow Mg_3[Si_2O_5](OH)_4 + Mg^{2+} \tag{4}$$

$$4MgSiO_3 + 2H^+ + 5H_2O \rightarrow Mg_3[Si_2O_5](OH)_4 + Mg^{2+} + 2H_4SiO_{4(aq)} \tag{5}$$

However, textural relations between serpentinites and their parent minerals indicate that constant volume replacement may not always apply. There are observations of pseudomorphic replacement (e.g., Dungan 1979), while in other cases expansion of 50-60% has been estimated for the formation of serpentine mesh textures in olivine (Shervais et al. 2005). Figure 3 is a back-scattered SEM image of serpentinized olivine, showing a typical mesh texture. The replacement appears to preserve the volume, although it would be difficult to claim that no expansion has taken place. Without some reference points it is not always obvious whether a replacement is truly pseudomorphic. Thus although serpentinization involves a dissolution and precipitation process it is not clear under what conditions a constant volume replacement takes place and under what conditions the reaction is isochemical for some elements. For example, if we assumed that there was no significant loss of Si in the reaction, the reaction for the serpentinization of orthopyroxene could be:

$$2MgSiO_3 + Mg^{2+} + 3H_2O \rightarrow Mg_3[Si_2O_5](OH)_4 + 2H^+ \quad (6)$$

which would imply a 50% increase in the volume. On the other hand, reaction (4) preserves both the volume and the silica composition. The release of Mg in reaction (4) is consistent with the observation in Figure 3 that the veins contain magnesite, $MgCO_3$, also indicating that the fluid contained dissolved carbonate ions.

The above examples suggest that the fluid chemistry and the stress generated by such reactions will play a role in determining whether a replacement takes place at constant volume.

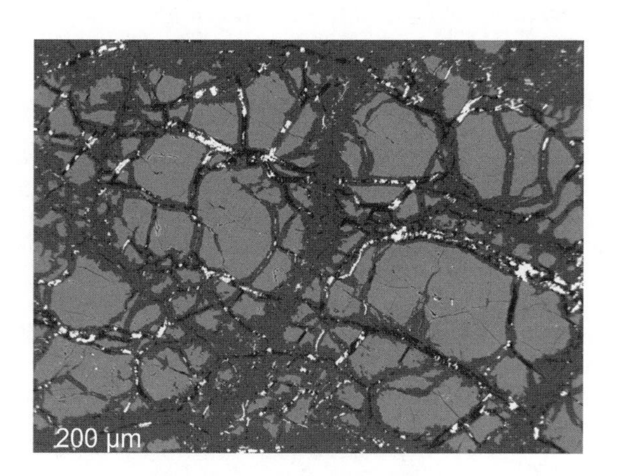

Figure 3. Back-scattered SEM image of a typical mesh texture in serpentinized olivine. The dark grey areas are serpentine, the lighter grey areas are olivine, and the black material in the veins is magnesite. The brightest phase in the veins is hematite. (Image: H. Austrheim)

STUDIES OF MINERAL REPLACEMENT REACTIONS: EXPERIMENT AND NATURE

Mineral replacement is the underlying process in the rock cycle of the Earth. The long time-scale of geological processes has frequently led to the assumption that the mechanism of mineral reequilibration is by slow reactions achieved by solid-state diffusion through crystal structures. However, the contention in this chapter is that in the presence of an aqueous fluid phase, solid state reactions compete kinetically with dissolution-precipitation reactions and that the latter will dominate over a wide range of crustal temperatures. The ubiquitous nature of mineral replacement in the crust, and the fast kinetics of dissolution-precipitation reactions means that unless reactions are incomplete and the parent and product phases can be texturally related, the mechanism of replacement and even the nature of the protolith may remain uncertain. Pseudomorphs provide a clue, but ultimately to understand the mechanism of pseudomorphism and replacement, we need to combine experimental studies with observations from nature.

The examples which follow are chosen to illustrate specific aspects of replacement processes which contribute to a better understanding of the mechanisms involved.

The replacement of calcite ($CaCO_3$) by fluorite (CaF_2)

It has long been known that carbonate fossils can be converted to fluorite by treatment in hydrofluoric acid to improve the appearance of fine anatomical details (e.g., Cookson and Singleton 1954; Grayson 1956). This somewhat surprising result was experimentally verified using cleavage rhombs of calcite immersed in hydrofluoric acid at room temperature by Glover and Sippel (1962). The principal conclusions were that (i) the replacement is pseudomorphic, (ii) the polycrystalline fluorite is composed of parallel oriented fibers which inherit their orientation from the calcite structure, (iii) the kinetics of the growth of the fluorite replacement rim is such that the thickness of the rim is linearly related to the square root of time, indicating that the rate of replacement is controlled by mass transfer through the reacted rim and (iv) the molar volume reduction results in a porous reaction rim which allows fluid infiltration to the replacement front.

The replacement of aragonite and calcite by hydroxyapatite $Ca_5(PO_4)_3(OH)$

When single crystals of aragonite or calcite are treated hydrothermally at 200 °C with an aqueous solution of diammonium phosphate, $(NH_4)_2HPO_4$, the carbonates are pseudomorphically replaced by polycrystalline hydroxyapatite, HAP (Kasioptas et al. 2008). Figure 4 shows the development of reaction rims in each case. As in the replacement by fluorite, the HAP crystals are fibrous and parallel to the reaction interface. If we write the reaction as:

$$5CaCO_3 + 3HPO_4^{2-} + H_2O \rightarrow Ca_5(PO_4)_3OH + 3CO_3^{2-} + 2H_2CO_3 \qquad (7)$$

the change in molar volume of the solid phases is −6% in the case of aragonite to HAP, and −12.7% in the case of calcite. For a pseudomorphic replacement, the molar volume change is compensated by generation of porosity in the product. This porosity allows fluid infiltration to the reaction front and also may explain the observation that the reaction is faster with calcite than with aragonite, even though aragonite is less stable than calcite at the temperature of the experiment.

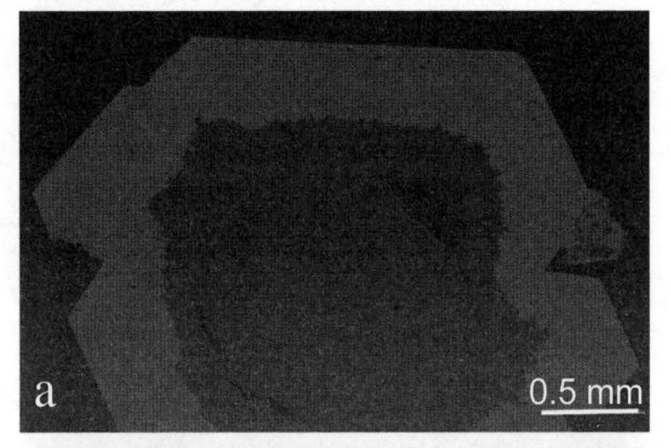

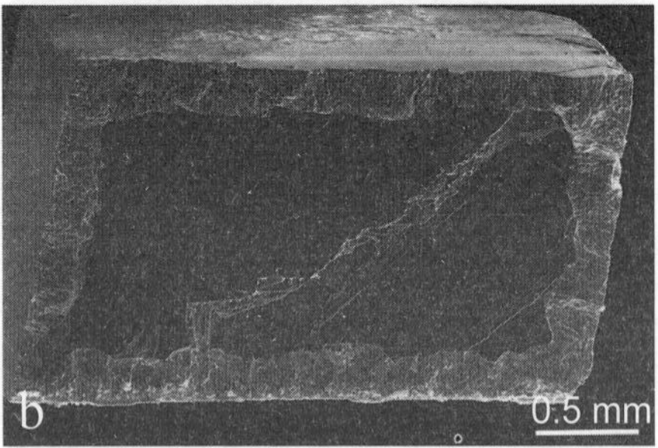

Figure 4. (a) Back-scattered SEM image of a cross-section through a partially pseudomorphed aragonite single crystal. The central darker core is aragonite, the rim is polycrystalline apatite. (b) Cleaved cross section through a partially pseudomorphed calcite crystal. The rim is polycrystalline apatite. (Image: A. Kasioptas)

A common factor in the above examples is that the carbonates are placed in a solution with which they are out of equilibrium, and will therefore begin to dissolve, introducing Ca ions into the solution. Secondly, both calcium fluoride and hydroxyapatite are significantly less soluble than the parent carbonate phases.

One problem in analyzing such experiments is that there is generally no available data for the solubilities of the parent and product phases in the specific solvent used, nor is there kinetic data to determine the rate controlling step in the overall replacement process. Some progress has been made in this direction by a study of replacement in a simple solid solution system KBr-KCl-H_2O in which the solubilities and solid-fluid equilibria are known.

The replacement of KBr by KCl

When a single crystal of KBr is placed in a saturated KCl solution at room temperature, an immediate replacement reaction begins from the original crystal surface with dissolution of KBr and the simultaneous reprecipitation of a K(Br,Cl) solid solution (Putnis and Mezger 2004; Putnis et al. 2005). As the replacement continues the original dimensions of the crystal remain unchanged, and, with excess KCl solution available for reaction, eventually the end result is a single crystal of almost pure KCl. This remarkable and simple experiment has given us many clues about what controls the mechanism of this process. KBr-KCl forms a solid solution at room temperature and to understand the replacement process in a more quantitative way, we must refer to the equilibrium solid solution – aqueous solution phase diagram, termed a Lippmann diagram (Lippmann 1980) (Fig. 5).

At the beginning of the experiment the KBr crystal is far out of equilibrium with the KCl solution and hence will begin to dissolve. As soon as some KBr is dissolved in the KCl solution, the Lippmann diagram shows that the solution will now be in equilibrium with a new composition of the solid solution. Experiments show that the first product phase formed on the surface of the dissolving KBr crystal is a solid solution rich in KBr and that as the replacement progresses the rim becomes increasingly KCl rich (points A-E) in Figure 5a. Figure 5b shows the development of the reaction rim during the replacement. The compositional evolution of the

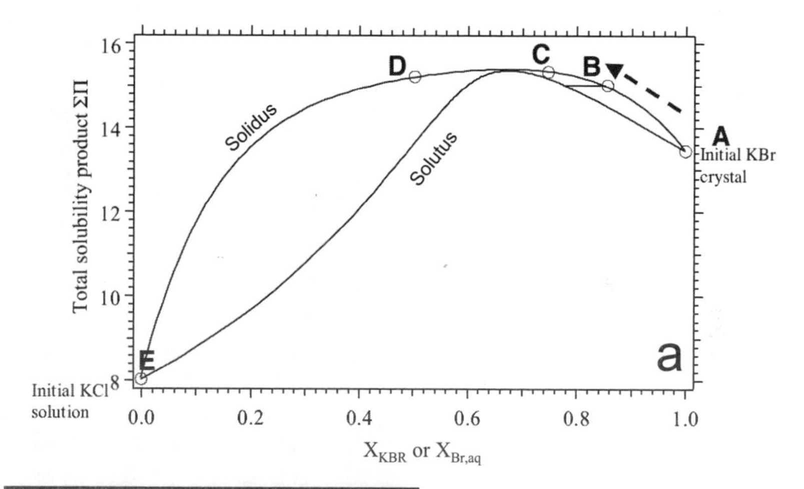

Figure 5. (a) Lippmann phase diagram for the system KBr-KCl-H_2O, with labels A-E showing the progressive change in the composition of the replaced rim shown in (b). The pale core in (b) is KBr.

reaction rim shows that it is controlled by the fluid composition at the interface with the solid. If the initial small quantity of dissolved KBr mixed with the larger volume of KCl solution, the Lippmann diagram would suggest equilibrium with a KCl-rich solid solution—i.e., the first precipitate would then be near point E—which was not found to be the case. The conclusion that the fluid composition at the interface, which also evolves with time, controls the replacement is a significant observation. Real time phase-shift interferometry (Putnis et al. 2005) also confirms that the dissolution and precipitation process is confined to a fluid boundary layer at the reaction interface, which has a different composition to the fluid in the bulk.

The kinetics of replacement is about 10 orders of magnitude faster than would be expected from a solid state ion-exchange process of Cl replacing Br. A dissolution–precipitation mechanism is further confirmed by isotope tracer experiments. When the KCl fluid is enriched in ^{40}K as a tracer, mass spectrometry confirms that the ^{40}K is incorporated in the product KCl (Putnis and Mezger 2004).

A further important observation is that the product K(Br,Cl) rim is porous and permeable to the infiltrating fluid (Fig. 6), and as the compositional equilibration is achieved, textural equilibration also evolves, with the coarsening of pores and eventual loss of permeability. When the crystal composition has equilibrated with the fluid, the driving force for textural equilibration is the reduction of interfacial surface area. This is a slower process, but if the crystal remains in the fluid, eventually the rim becomes clear and pore-free. The generation of porosity allows the replacement reaction to proceed, providing pathways for mass transport. During the replacement, the porosity generated, or in other words, the volume in the reaction rim occupied by the fluid phase, is higher than would be expected from the reduction of solid molar volume in replacing KBr by K(Br,Cl). This is another important observation which indicates that the porosity generation is a function of both the change in molar volume of the solids as well as the relative solubilities of the parent and product phases in the aqueous solution. A difference in solubility means that during a replacement process, more of the parent phase may be dissolved than the product precipitated (i.e., some solid material remains in the fluid phase). Further examples of this may be found in Putnis (2002) and more experiments on replacement processes in simple salt systems are described by Glikin (2008).

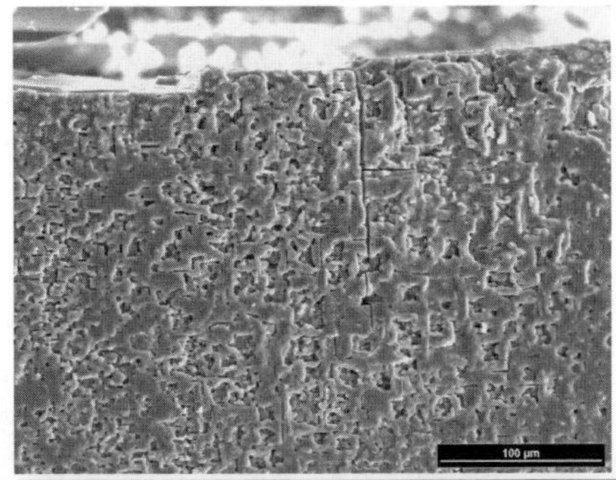

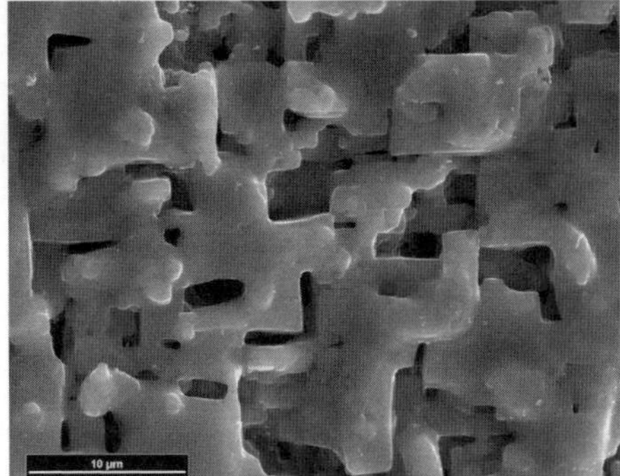

Figure 6. SEM image showing the porosity developed on the surface of a single crystal of KBr which has reacted with a saturated KCl solution. The surface composition is K(Br,Cl). [Used with permission of Elsevier from Putnis and Mezger (2004).]

The replacement of leucite $KAlSi_2O_6$ by analcime $NaAlSi_2O_6{\cdot}H_2O$

Given that both of these phases have very large structural channels (analcime is often classified as a zeolite) this pseudomorphic replacement would be considered as a typical ion exchange process where the alumino-silicate framework acts passively while the alkali ions and water molecules diffuse through the crystal structure. However, experiments by Putnis et al. (2007b), in which single crystals of leucite are reacted with a 3.5% NaCl solution at 150 °C, showed conclusively that the replacement proceeds by a dissolution-precipitation mechanism (Fig. 7a). There is no diffusion profile at the sharp reaction interface, and Raman spectroscopy shows that a tracer of ^{18}O in the aqueous solution is incorporated in the analcime aluminosilicate framework. A significant observation in this example is that although there is an increase in molar volume of 10%, porosity is generated in the analcime rim. Conventional wisdom would suggest that a volume increase should close any porosity, but in this case the porosity must be due to the fact that more leucite is dissolved than analcime precipitated, and this outweighs the effect of molar volume increase. More recent work by Xia et al. (2009a) has verified that the crystallographic orientation of the leucite is preserved after the replacement, and also that fine-scale twinning is therefore also preserved (Fig. 7b) They also noted that the rate of replacement is slowest at neutral pH and also that the porosity is dependent on the pH.

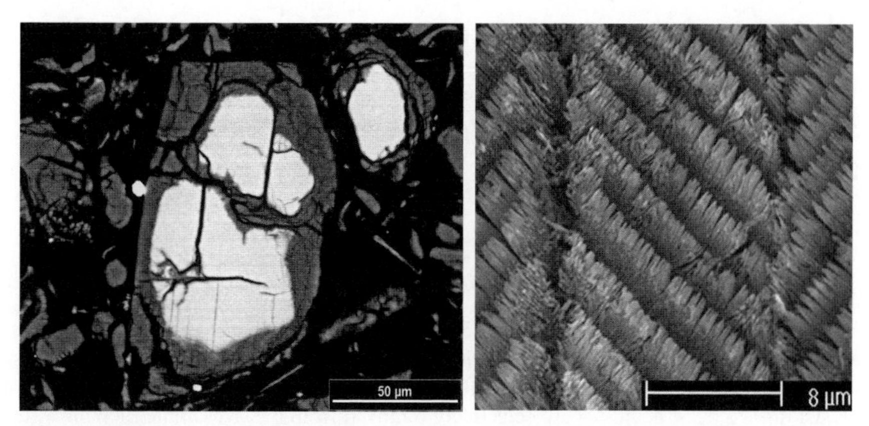

Figure 7. (a) Back-scattered SEM image of a cross section of a leucite crystal (white core), partially replaced by analcime (grey). (Image: C. V. Putnis) (b) SEM image of analcime which has replaced leucite, retaining the fine-scale twinning. (Image: F. Xia)

The replacement of pentlandite $(Fe,Ni)_9S_8$ by violarite $(Ni,Fe)_3S_4$

In a comprehensive study of the kinetics and mechanism of mineral replacement reactions, Tenailleau et al. (2006a) and Xia et al. (2009b) investigated the roles of temperature and fluid composition on the replacement of pentlandite by violarite at temperatures between 80 and 210 °C in fluids of varying pH and composition. Xia et al. (2009b) introduced the concept of the *scale of pseudomorphism* to describe the degree of spatial coupling between the dissolution and the precipitation process. When the rate controlling step (at $1 < pH < 6$) is pentlandite dissolution, nanometer scale coupling precisely preserves the morphology and internal details of the parent phase, as is seen texturally in natural pentlandite/violarite assemblages (Fig. 8). At higher pH values violarite precipitation appears to be rate limiting (and is then more "loosely coupled" to the dissolution), and the result is a less perfect pseudomorph, on a length scale of 10's of microns rather than nanometers. The rate of replacement increases with increasing oxidants, and decreases with Ni^{2+} or Fe^{2+} addition in the fluid. Furthermore, the rate increases with temperature up to 125 °C and then decreases at higher temperatures, again emphasizing that the equilibria are controlled by fluid-solid thermodynamics. Previous publications on sulfide-sulfide replacements favoured a solid state reaction in which Fe,Ni ions were exchanged by solid state diffusion, while the sulfur close-packing remained inert. Xia et al. (2009b) show conclusively that in their experiments, and in nature, by implication from textural comparisons, that this is not the case.

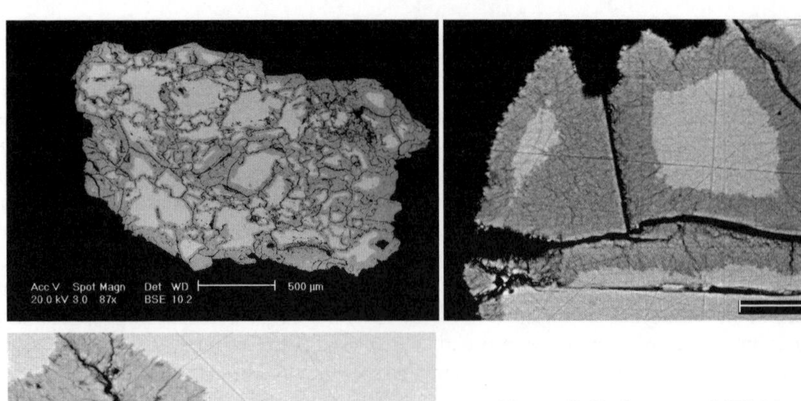

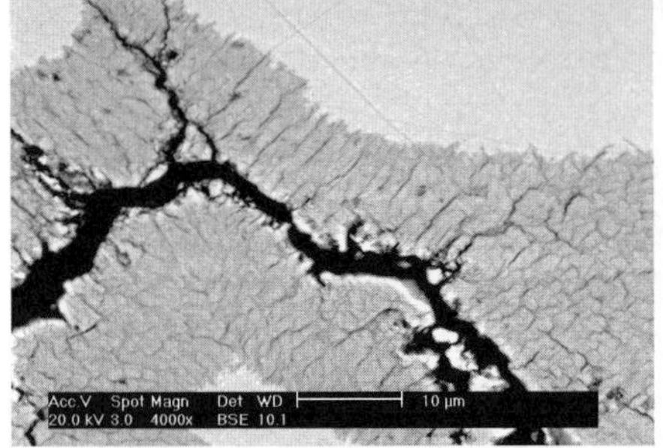

Figure 8. Back-scattered SEM images of a cross section of a crystal of pentlandite, partially replaced by violarite. [Images by F. Xia used with permission of the author and by the American Chemical Society for Xia et al. (2008) and by Elsevier for Xia et al. (2009b).]

Hydrothermal alteration of natural pyrochlore $(Ca_{1.23}Na_{0.75})Ta_{1.78}O_{6.28}F_{0.57}$

When a solid reacts with a fluid, the apparently selective removal of some components is often referred to as "leaching," a term which also carries with it the implication of a solid-state diffusion mechanism. When natural pyrochlore is treated in a solution containing 1M HCl and 1M $CaCl_2$ at 175 °C, Ca and Na are selectively removed from the pyrochlore, leaving a rim of depleted composition but with the crystal structure retained (see Fig. 9) (Geisler et al. 2005a). This may seem to support an "ion-exchange" reaction between the solid and the fluid. However, the rapid reaction rate at these moderate temperatures, the occurrence of a reaction interface that is sharp on the nanometer scale as observed by transmission electron microscopy (Geisler et al. 2005b), and the incorporation of ^{18}O from an enriched fluid into the pyrochlore structure are all inconsistent with a solid state diffusion mechanism. The data are consistent with a pseudomorphic reaction that involves the dissolution of the pyrochlore parent and the simultaneous re-precipitation of a defect pyrochlore at a moving reaction interface. Although the fluids used in the experiments are more aggressive than would be found in nature, it is remarkable that the experimental alteration features are very similar to those found in naturally altered pyrochlore samples (Geisler et al. 2004).

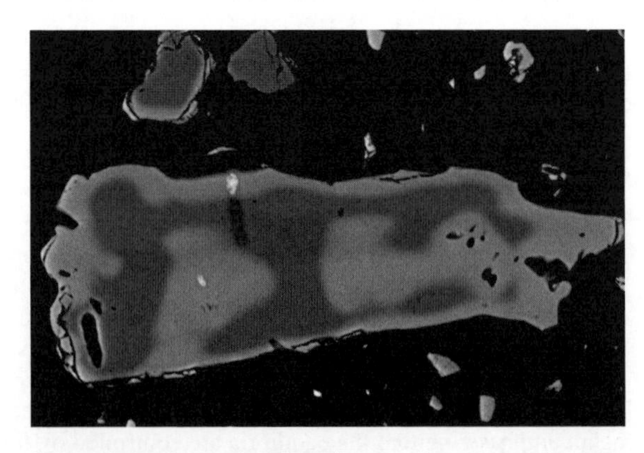

Figure 9. Back-scattered SEM images of a cross section of a crystal of pyrochlore. The darker core is the original pyrochlore composition, while the paler area is depleted in Ca and Na. (Image: T. Geisler)

The significance of pyrochlore is that ceramic materials based on pyrochlore are considered as good candidates for the immobilization of highly radioactive waste (Ewing 2005). One reason for such a choice is the low solubility of pyrochlore in aqueous fluids. However, because coupled dissolution-precipitation reactions take place at the parent-product interface, the relative solubility of the two solids is more important than their absolute solubility as only a small amount of material needs to be in solution at the interface at any one time. This same point comes up in the following example.

Reequilibration of zircon ($ZrSiO_4$) in aqueous fluids

Although ideal zircon is stable under crustal conditions, and is considered to be a very resistant mineral to alteration because of its low solubility (Tromans 2006), natural zircons formed at high temperatures can contain substituted thorium, uranium, scandium and hafnium substituting for Zr in solid solution. These solid solutions are non-ideal and are characterized by wide miscibility gaps at low temperatures. Thus the thermodynamics predicts that phases such as thorite ($ThSiO_4$), coffinite ($USiO_4$), thortveitite (Sc_2SiO_4) and hafnon ($HfSiO_4$) should exsolve at low temperatures. However, solid state exsolution textures have not been reported and are highly unlikely due to the slow diffusion of cations in zircon at crustal conditions (Cherniak and Watson 2003). However, natural zircons which have been exposed to hydrothermal fluids have regions with textures very similar to those expected from dissolution–precipitation reequilibration mechanisms. Figure 10 shows a typical example, in which the pristine zircon has a lighter back-scattered contrast and contains higher concentrations of substituted elements compared to the porous altered zircon (darker contrast) which contains inclusions of silicates such as thorite (Spandler et al. 2004; Soman et al. 2006; Geisler et al. 2007).

The interpretation of the pores and inclusions in altered zircons (Geisler et al. 2007) is that the metastable zircon solid solutions $(M,Zr)SiO_4$ have a higher solubility in a pure aqueous solution than the pure zircon end-member ($ZrSiO_4$). Therefore, even a very small amount of dissolution of a solid solution composition results in an interfacial fluid which is supersaturated with respect to a purer zircon composition, and eventually also with respect to the other end member of the solid solution ($MSiO_4$). The low temperature equilibrium expected in such a solid solution-aqueous solution (SS-AS) system (see Prieto 2009) is the coexistence of an M-poor zircon + a Zr-poor $MSiO_4$ end-member + some Zr and M ions and silicate in solution. The loss of some $(M,Zr)SiO_4$ to the solution gives rise to the porosity, which is also a necessary feature for the replacement to proceed, allowing access of the fluid to the reaction interface.

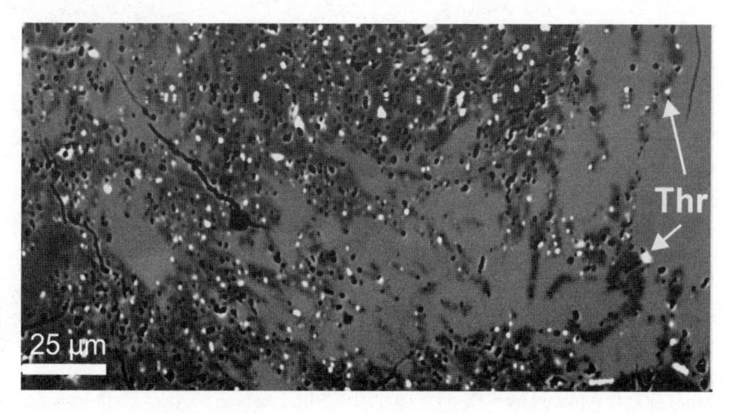

Figure 10. Back-scattered SEM images of a cross section of a zircon crystal. The pale grey area is the original Th-rich zircon, while the darker area are Th-depleted zircon, containing small bright inclusions of thorite (Thr) and pores (black). (Image: A. Soman)

This is an example of a phase separation of a solid solution driven by a fluid interaction. It could also be termed a fluid-driven "exsolution," but it would be better to restrict the term "exsolution" to a solid state process. One significant difference between solid state exsolution and a fluid-induced phase separation is that in the former, the bulk composition of the original solid solution and the total exsolved phases is unchanged, whereas in a fluid-induced phase separation components may be exchanged between the solid and the fluid, changing the bulk composition of both.

Even minor changes in chemical composition during a dissolution-reprecipitation process can have profound consequences when isotopes used in dating are involved (Martin et al. 2008). In a study of the reequilibration of magmatic zircons during metamorphism at low-temperature, high-pressure conditions, Rubatto et al. (2008) describe partial replacement textures in which zircon crystals preserve domains of primary zircon as well as domains of porous, recrystallized zircon containing inclusions of high-pressure pyroxene and epidote. The textural changes are coupled with chemical and isotopic modifications such that the primary zircon retains its U-Pb age of ~163 Ma, while the recrystallized zircon gives an age of ~46 Ma. The recrystallized zircon also loses Th which may explain the high-Th content of the epidote inclusions. The crystallographic orientation of the recrystallized zircon is inherited from the parent zircon, as expected when the dissolution and reprecipitation are closely coupled. From the compositions of the inclusions in the zircons, Rubatto et al. (2008) conclude that the recrystallization was induced by a relatively cold (<600 °C), alkaline, high pressure fluid.

Phase separation in monazite and xenotime by coupled dissolution-precipitation

Monazite [(Ce,LREE)PO_4] and xenotime [(Y,HREE)PO_4] are relatively common accessory minerals in igneous and metamorphic rocks and also often show fluid-induced reequilibration textures. Figure 11 shows a monazite sample in which the homogeneous Th-bearing monazite (Mnz1 with lighter contrast) is replaced by porous Th-depleted monazite (Mnz2 with darker contrast) + bright thorite inclusions (Seydoux-Guillaume et al. 2007). Similar textures in monazite and xenotime have been described from metasomatized pegmatites by Hetherington and Harlov (2008), who also interpret the breakdown of the solid solutions to a multiphase assemblage, and the resulting textures and compositions, by a dissolution-reprecipitation replacement process. From data on the relative solubilities of the phases involved, they estimate that the aqueous metasomatizing fluid was rich in Na^+ and K^+, and contained F^- and minor amounts of Cl^-.

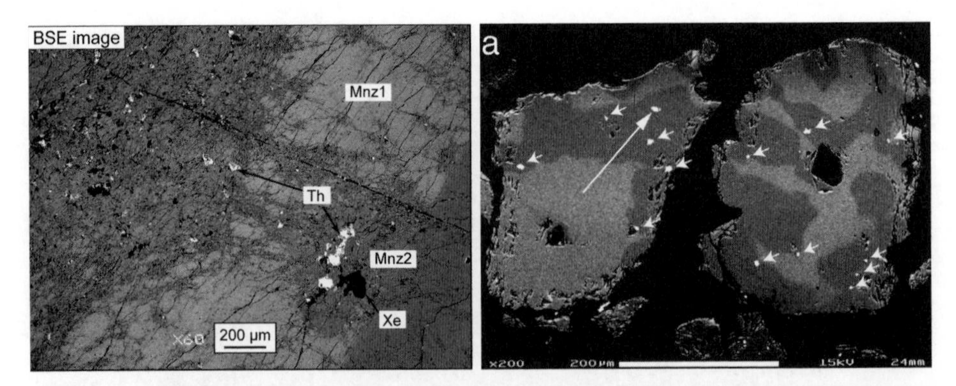

Figure 11. (a) Back-scattered SEM images of a cross section of monazite containing the parent Th-rich monazite (Mnz1) partly replaced by a Th-poor monazite (Mnz2) which contains bight inclusions of thorite (Th) (Image: A-M. Seydoux-Guillaume). (b) Experimentally metasomatized chlorapatite showing both metasomatized regions (dark) which contain monazite inclusions as well as the lighter parent monazite. (Image: D. Harlov)

Replacement processes in metasomatized apatites $Ca_5(PO_4)_3(OH,Cl,F)$

A series of papers by Harlov and co-workers (Harlov et al. 2002, 2003, 2005) demonstrate how the correspondence between natural observations and hydrothermal experiments can provide compelling evidence for the nature of aqueous fluids, and the mechanism of reactions. As in the above example, apatite minerals can contain Y + REE in solid solution, which can be released as REE-bearing phosphate phases occurring as inclusions when the apatite interacts with fluids. This is clearly seen from chemical analyses and microstructural observations of partially metasomatized apatites from the Bamble Sector, SE Norway, in which the metasomatized parts of the single crystals which are enriched in OH and F relative to the parent Cl-rich apatite, contain numerous inclusions (1-15 μm in size) of monazite and xenotime. A very similar texture can be reproduced (see Fig. 11b) by hydrothermally treating unaltered chlorapatite in a range of aqueous solutions (Harlov et al. 2002). The extent of metasomatism depended on temperature and the textures depended on the fluid composition. In pure H_2O the altered regions were porous, depleted in REE relative to the parent apatite and contained inclusions of monazite and xenotime. When a fluid containing CaF_2 was used, the metasomatized areas were enriched in F, but retained their Y + REE composition and hence had very few monazite or xenotime inclusions. Transmission electron microscopy (Engvik et al. 2009) of natural partly metasomatized chlorapatite from the Bamble Sector showed that the product OH-rich apatite is nanoporous, but shares the same crystallographic orientation with the parent apatite.

In a separate set of hydrothermal experiments using a wide range of fluid compositions at temperatures between 300 and 900 °C, Harlov et al. (2003, 2005) determined the conditions under which monazite inclusions could be induced to form from the LREE-enriched Durango fluorapatite. After reaction, only the metasomatized regions which were enriched in F and depleted in Y + REE contained monazite inclusions. Transmission electron microscopy showed that the reacted regions contained a pervasive nanoporosity, together with larger pore-spaces containing the monazite, while the crystallographic orientations of the reacted and unreacted regions were nearly identical.

The overall conclusions of these papers were that nucleation of monazite and xenotime inclusions in apatite is a function of composition of infiltrating fluid and to a much lesser extent, temperature and pressure. Even though the inclusions are often oriented they are not the product of a solid-state exsolution in fluid-absent conditions over millions of years but are metasomatically induced. This is another example of fluid-induced phase separation.

Fluoridation of bone apatite

Similar principles apply to the experimental fluoridation of bone apatite (essentially hydroxyapatite) by NaF solutions. In a study by Pasteris and Ding (2009) dentin and enamel from a modern horse-tooth were immersed in aqueous solutions with NaF concentrations from 0.1 M to 2 M, for periods up to 14 days at 19.5 °C. Dissolution of the bioapatite released Ca and P to the F-bearing solution which resulted in the nucleation and precipitation of essentially end-member fluorapatite "very near the place from which the calcium was released." The porosity and permeability in the product fluorapatite is increased due to the loss of constituents from the bioapatite to the fluid, and this permeability allows mass transport between the fluid and the reaction interface. The replacement process progresses due to the higher solubility of bioapatite relative to fluorapatite. The dissolution-precipitation mechanism, rather than a mechanism in which a diffusional front moves through the bioapatite, causing a solid state replacement of OH by F, is verified by *in situ* Raman spectroscopic analyses during the fluoridation reaction.

Replacement processes in feldspars – experiments

The initial fundamental experiments on feldspar reequilibration in alkali chloride solutions by Wyart and Sabatier (1958) and Orville (1962, 1963) were followed by O'Neil and Taylor (1967) who established the basic mechanisms which subsequent research has verified and elaborated. In experiments on the hydrothermal reaction between alkali feldspars and plagioclase and aqueous chloride solutions they found that complete oxygen isotope equilibration between the solutions and the feldspars accompanies the cation exchange which pseudomorphically transforms one feldspar composition to another. This refuted the previously accepted idea that the cation exchange was a simple solid state diffusion process in which the alumino-silicate framework remained inert. They concluded that the mechanism of cation and oxygen exchange involves a fine-scale solution and redeposition in a fluid film at the interface between exchanged and unexchanged feldspar. The oxygen isotope exchange is therefore an inevitable consequence of the replacement which is thermodynamically driven by the reequilibration of the feldspar in the aqueous fluid.

Similar experiments have been conducted using more sophisticated analytical techniques for characterizing the reaction interface in partly replaced feldspars. Labotka et al. (2004) experimentally replaced albite ($NaAlSi_3O_8$) by K-feldspar ($KAlSi_3O_8$) in an aqueous solution of KCl at 600 °C, and using a combination of electron microprobe and nano-SIMS (ion-probe) imaging showed the correspondence between Na-K and ^{18}O exchange. SIMS cannot specify whether the ^{18}O is in the feldspar structure, or in any other phases within the pore-spaces in the reaction rim, but Niedermeier et al. (2009) analyzing the products of similar hydrothermal experiments using Raman spectroscopy and imaging were able to conclusively show that the ^{18}O is in the aluminosilicate framework of the K-feldspar. Raman spectroscopy can identify the mass-dependent frequency shifts of the O-T-O vibrations in the tetrahedral rings of the feldspar structure and this can be mapped onto the Na-K chemical maps. Scanning and transmission electron microscopy (see Fig. 12) shows that the reaction interface is sharp on a nanoscale, often showing a small gap between the parent and product, as predicted 40 years earlier by O'Neil and Taylor (1967). The crystallographic orientation of the feldspar is preserved across the interface. The replaced rim of K-feldspar shows nano tube-like structures normal to the interface, which are assumed to be part of the porosity forming conduits for maintaining contact between the fluid reservoir and the reaction interface.

In the above experiments the Al:Si ratio in the feldspar remains unchanged during the replacement process, but this is not the usual case in nature. Plagioclase feldspar with compositions intermediate between albite ($NaAlSi_3O_8$) and anorthite ($CaAl_2Si_2O_8$) are replaced by albite in a wide variety of rock types and tectonic settings. This process known as "albitization" occurs at low temperatures during burial diagenesis (Perez and Boles 2005) and at higher temperatures during hydrothermal alteration of feldspar-bearing rocks (Lee and Parsons 1997; Engvik et al. 2008). On a regional scale, albitization is closely related to mineral and ore deposits (e.g., Oliver et al. 2004). Of interest therefore is the mobility of elements during albitization.

Hövelmann et al. (in press) have studied experimentally the albitization of oligoclase ($Ab_{73}An_{23}Or_4$) and labradorite ($Ab_{39}An_{60}Or_1$) in aqueous solutions of sodium silicate enriched in ^{18}O. The results confirmed that the replacement is pseudomorphic with a sharp chemical interface which progresses through the feldspar while preserving the crystallographic orientation (see Fig. 13), and that there is a direct correspondence between the replacement of the major elements and the oxygen isotopes in the feldspar structure. Transmission electron microscopy shows that the albite product is rich in defects and possibly nanopores, similar to naturally albitized plagioclase (see Engvik et al. 2009 below). The Ca from the plagioclase reacts with the fluid to form pectolite ($NaCa_2Si_3O_8OH$). Determining which elements are gained or lost to the solution requires making some assumptions. For example, balancing a replacement reaction can be achieved in a number of ways depending on which elements are mobile. For the reaction from oligoclase to albite we could make the assumption, which is usually made in metasomatic reactions, that Al is immobile. In this case we could write:

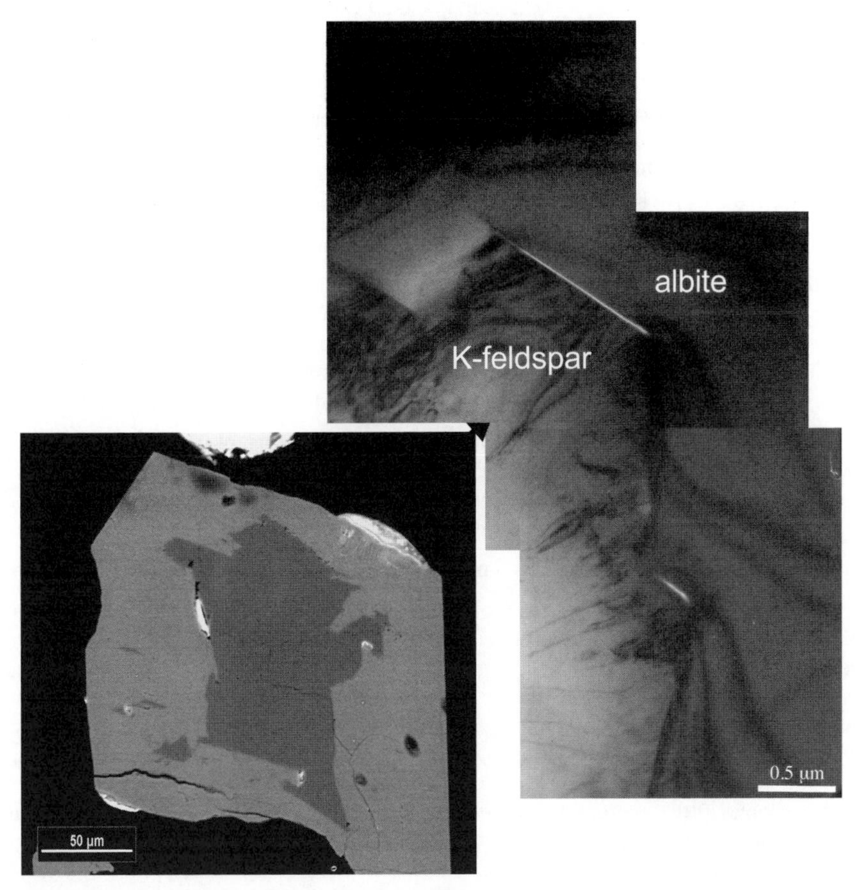

Figure 12. (a) Back-scattered SEM images of a cross section of an albite crystal (darker core) partially replaced by K-feldspar (pale grey rim). (b) TEM image of the interface between the albite parent and K-feldspar product. (Image: D. Niedermeier)

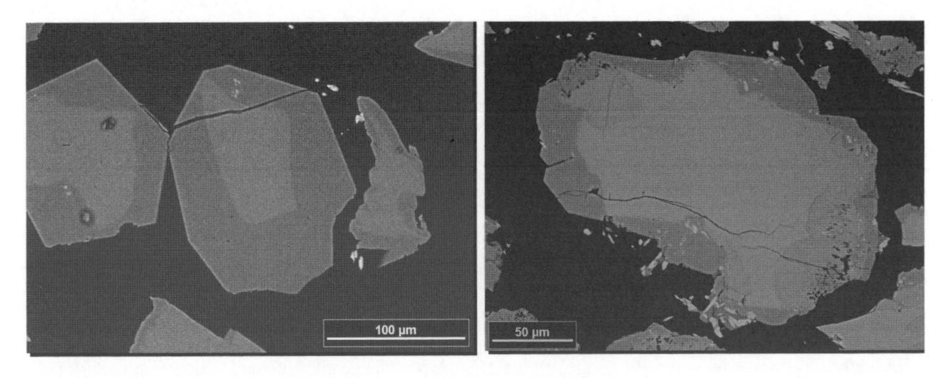

Figure 13. Back-scattered SEM images of a cross sections of (a) oligoclase crystals (ol) and (b) labradorite crystals (lab) partially replaced by albite (ab). (Image: J. Hövelmann)

$$Ca_{0.22}Na_{0.74}K_{0.04}Al_{1.22}Si_{2.78}O_8 + 0.59\,Na^+_{(aq)} + 0.59\,OH^- + 1.21SiO_2 \rightarrow 1.22NaAlSi_3O_8 + 0.11NaCa_2Si_3O_8OH + 0.04\,K^+_{(aq)} + 0.04\,OH^- + 0.22H_2O \quad (8)$$

However, this would involve a large volume increase of the feldspar crystal, which is not the case, as the replacement is pseudomorphic. If the reaction is balanced on volume, then neglecting the small molar volume difference between the feldspars and the porosity due to the fact that some of the material is lost to the fluid, we could write:

$$Ca_{0.22}Na_{0.74}K_{0.04}Al_{1.22}Si_{2.78}O_8 + 0.59\,Na^+_{(aq)} + 0.59\,OH^- + 0.55SiO_2 + 0.22H_2O \rightarrow NaAlSi_3O_8 + 0.11NaCa_2Si_3O_8OH + 0.22NaAl(OH)^0_4 + 0.04\,K^+_{(aq)} + 0.04\,OH^- \quad (9)$$

In reaction (9) aqueous Al is assumed to be complexed by Na, but in any case, the experiments confirm that Al as well as Ca are released by the reaction. Minor and trace element analyses record considerable losses of Ti, Fe, Mg, Sr, Ba, Y, K, Rb, Pb and the LREEs from the feldspar to the fluid. In a natural setting these "charged fluids" migrate away from the sites of albitization and may precipitate mineral deposits elsewhere, even hundreds of kilometers away.

A common feature of all the experimental work on feldspar replacement reactions is that they are fast on a laboratory time-scale (days or weeks at typical hydrothermal conditions) and therefore virtually instantaneous on a geological time scale. Thus a replacement may only be recognized when the reaction front is "fossilized" due to insufficient fluid to complete the reaction.

Replacement processes in feldspars – nature

Before describing some examples of replacement reactions which involve changes in major element composition it is instructive to consider the special, but very common case in which alkali feldspars with fine-scale coherent cryptoperthite intergrowths (cryptoperthite: a nanoscale intergrowth of albite and microcline) formed by true solid-state exsolution, react with aqueous fluids. The result of this interaction with fluids has been termed "mutual replacement" in which there is a pervasive recrystallization with no significant change in bulk composition, nor change in external morphology, nor of crystallographic orientation. However, there is a profound change in the internal microstructure of the crystal resulting in a range of textures, broadly referred to as "patch perthite" or "vein perthite" (see Fig. 14). In some cases the continued coarsening retains the initial spindle shaped intergrowths reminiscent of a solid state exsolution texture (see Fig. 14b). Mutual replacement has been studied in detail by Parsons and co-workers for many years and Parsons and Lee (2009) provide an excellent review of this literature, together with new observations.

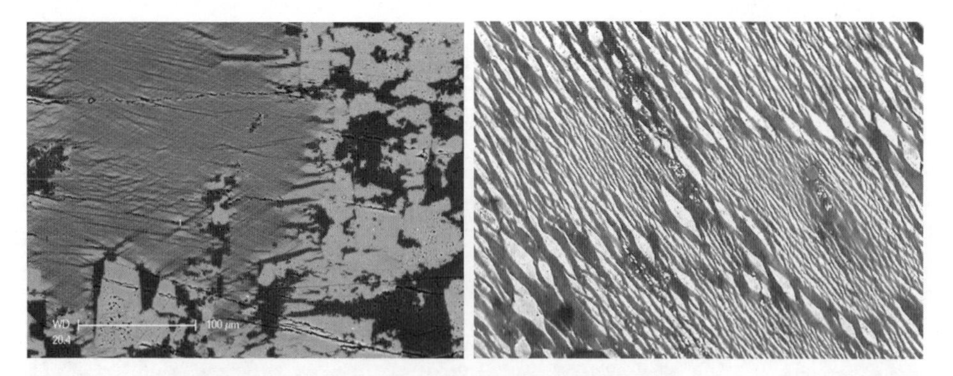

Figure 14. (a) Back-scattered SEM image of coarse patch-perthite (with black and white contrast – right) replacing the finely exsolved cryptoperthite (grey braid perthite – left). (Image: I. Parsons) (b) Optical image of a thin section of perthite showing the variable scale of phase separation due to fluid infiltration and replacement.

The thermodynamic driving force for such replacement is the reduction in the coherency strain energy generated by the exsolution. This strain energy translates into a higher solubility in an aqueous solution for cryptoperthite than for patch perthite and drives a dissolution-precipitation mechanism. In cases where there has been only partial replacement, the relationship between the strained cryptoperthite and the fluid-induced patch perthite is clearly defined. The observed development of micropores (~2% porosity reported by Parsons and Lee 2009) in the product (see Fig. 14) is a natural consequence of this mechanism since the solubility difference results in the removal of some feldspar material into the fluid phase. The development of porosity in feldspars leads to a turbidity or "milkiness" which is almost universally developed to varying extents and is evidence of pervasive fluid-rock interaction on a large scale. In contrast, feldspars which have not reacted with aqueous solutions appear dark or glassy in hand-specimen, and free of pores in electron microscope images (e.g., Fig. 15 below).

Although the bulk chemistry in this example does not significantly change, measurements of trace element compositions using LA ICPMS demonstrate that many trace elements, including REEs were lost from the cryptoperthite during the replacement (Parsons et al. 2009), further emphasizing the role of the fluid phase.

Feldspar replacement where the bulk chemistry of the feldspar is changed by interaction with alkali or alkali earth bearing fluids, are also very common, but usually not recognized unless the replacements are incomplete. In a study of a "fossil hydrothermal system" in the Rico dome, Colorado, Cole et al. (2004) describe plagioclase and K-feldspar partially replaced by albite, while retaining original grain shape and twinning. Oxygen isotope measurements show that while the plagioclase cores retain their igneous isotopic compositions, the replaced rims show low $\delta^{18}O$ values indicative of reequilibration of the feldspar with meteoric water.

In a similar study of an anorthosite rock which contains a bi-modal plagioclase assemblage of andesine (An_{34-48}) and bytownite-anorthite (An_{82-98}), the extreme heterogeneity of oxygen isotope compositions on a mm scale, and the textural relationships between the feldspars led Mora et al. (2009) to conclude that the anorthite-rich phase partially replaced the andesine during a high temperature hydrothermal alteration event. The replacement created new permeability allowing considerable volumes of meteoric-hydrothermal fluids to infiltrate the rock. In this case of "anorthitization" the fluid involved is assumed to be a Ca^{2+}-bearing aqueous solution.

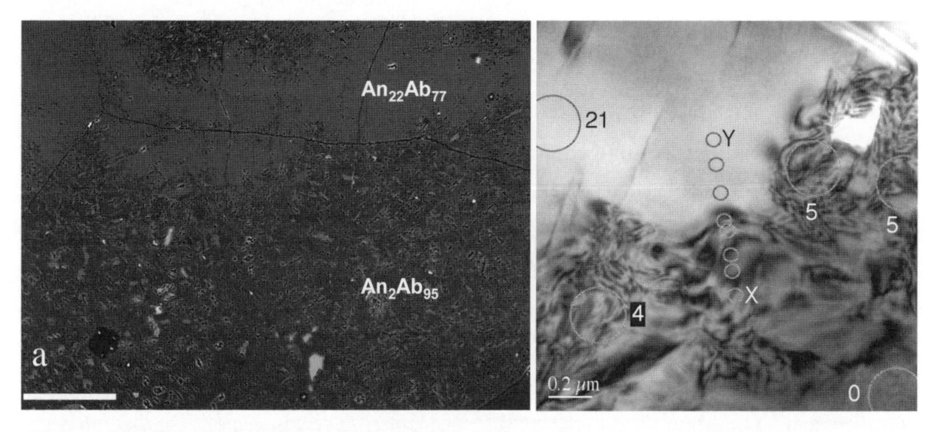

Figure 15. (a) Back-scattered SEM image of natural oligoclase (~An_{22}) partially replaced by albite (~An_2). The parent oligoclase has a smooth textured-surface, while the albite (darker grey) is full of small sericite inclusions (pale) and pores (black). (b) TEM image of an oligoclase-albite interface. The numbers refer to An-content, and the points between X and Y are positions of chemical analyses. (Engvik et al. 2008)

However to write a reaction for anorthitization of a natural rock poses the same questions as raised by reactions (8) and (9) above. A reaction which keeps the Al immobile results in a very large volume decrease due to loss of Si to the fluid. At the other extreme, if Al is present in a calcic fluid and is added to the feldspar it is possible to write a reaction in which volume is preserved. As the evidence for the compositions of fluids before and after reaction is lost in rocks, estimates of the porosity generated or the extent to which volume is preserved must be made to estimate the fluid compositions.

To illustrate the point in the simplest possible way, the anorthitization of albite could be written as:

$$2NaAlSi_3O_8 + Ca^{2+} + 8H_2O \rightarrow CaAl_2Si_2O_8 + 2Na^+ + 4H_4SiO_4 \quad (10)$$

would involve an approximately 50% loss of volume due to the amount of silica dissolved in the fluid, but preserves the Al as "immobile." This raises problems not only whether such a large volume loss is possible in a rock, but also given the relatively low solubility of silica, whether the amount of fluid required is realistic. On the other hand, a reaction which approximately conserves volume could be written as:

$$NaAlSi_3O_8 + Ca^{2+} + Al^{3+} \rightarrow CaAl_2Si_2O_8 + Na^+ + Si^{4+} \quad (11)$$

but involves the introduction of Al by the fluid and the loss of Si. While dissolved Al could be derived from albitization elsewhere in a hydrothermal system, the real situation is likely to lie somewhere between the extremes of reactions (10) and (11).

Regional scale albitization of granitic rocks from the Bamble Sector of SE Norway, can be recognized along reaction fronts where albitization occurs normal to fractures in essentially unaltered rock. Such reaction fronts represent the limits of fluid infiltration, and allow a detailed study of the "before" and "after" compositions and textures. In the field, these reaction fronts can be recognized by a reddening of the rock due to hematite precipitation associated with the albitization. Engvik et al. (2008) studied the replacement, on a nanometer scale, of plagioclase (oligoclase ~ $An_{21\text{-}23}$) in the unaltered rock by almost pure albite. Figure 15a shows a scanning electron micrograph of such a reaction front between oligoclase, which has a smooth and textureless contrast, and the product albite, which contains fine laths of sericite mica, occasional Fe-oxides, and numerous small pores. The replacement is pseudomorphic and there is no indication of any overall volume change i.e. the volume of oligoclase equals the volume of albite + sericite + pores. Figure 15b is a transmission electron micrograph of the reaction interface, showing both the sharpness of the interface and the complex diffraction contrast in the albite, indicative of a defect structure which includes dislocations and nanopores as well as the larger pores seen by scanning electron microscopy. As in the other cases described above, the crystallographic orientation is preserved across the interface.

Mechanisms of isotopic exchange in calcite marble

As a final example of replacement in nature, the recent study by Bowman et al. (2009) demonstrates how high spatial resolution measurements of $^{18}O/^{16}O$ ratios using modern ion microprobes with high precision measurements on ~10 μm spot sizes, together with textural observations can clarify the mechanism of isotopic exchange. In marble from a contact aureole of the Alta Stock, Utah, two isotopically and texturally distinct types of calcite exist on a mm scale. Clear calcite grains, with homogeneous $\delta^{18}O$ values are replaced by turbid calcite with more variable and lower $\delta^{18}O$ (see Fig. 16). The turbid calcite also contains small blebs of dolomite which are interpreted as phase separation due to the interaction of clear calcite with fluid, or precipitation from the fluid during the replacement. The conclusions are that the isotopic reequilibration and Mg exchange are the result of dissolution and reprecipitation of calcite during retrograde cooling.

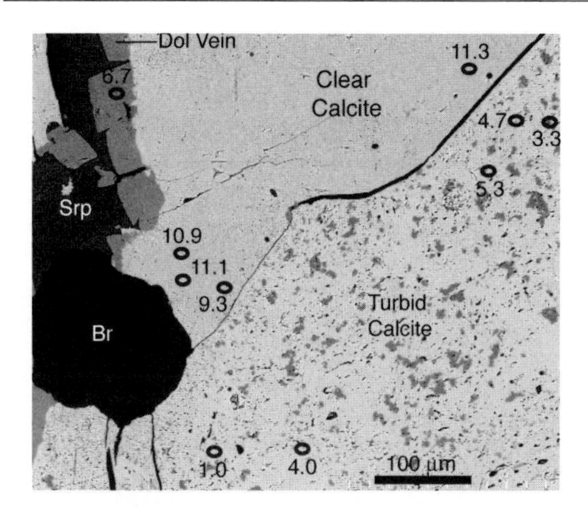

Figure 16. Back-scattered SEM image of calcite showing a parent clear calcite partially replaced by a later, turbid calcite. The numbers refer to $\delta^{18}O$ values. (Bowman et al. 2009)

Reaction induced fracturing during replacement processes

In all of the examples we have considered, we have focused on the chemical processes associated with mineral replacement. We have seen that, depending on the composition of the fluid, a reaction may be balanced to preserve the solid volume, or alternatively may involve a significant volume change. When replacement reactions involve volume changes between the parent and product, stresses are generated which may eventually cause fracturing of the mineral. On a large scale, similar processes operate during weathering, where the interplay of chemical and mechanical processes involve hydration reactions, and fracturing which allows further fluid infiltration and chemical reaction.

Experimental studies show that even when there are significant changes in the molar volume of the solid reactants, the replacement may be pseudomorphic. One example is the reaction of ilmenite ($FeTiO_3$) with acid solution, which results in a pseudomorph of the ilmenite made of polycrystalline rutile (TiO_2) (Janssen et al. 2008). In the experiments, polished cubes, prepared from a massive ilmenite sample were reacted in HCl solutions at 150 °C. Figure 17 shows a cross section through the cube after partial reaction. The dark rim is rutile, while the core is unreacted ilmenite. The most striking result is that although there is a decrease in molar volume in replacing ilmenite by rutile of ~40%, the reaction is pseudomorphic, and

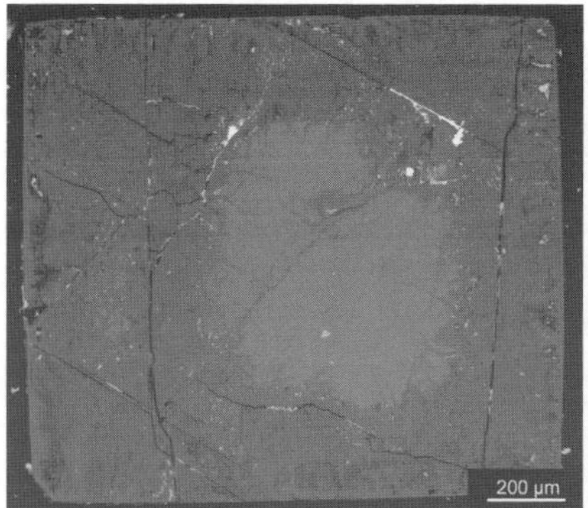

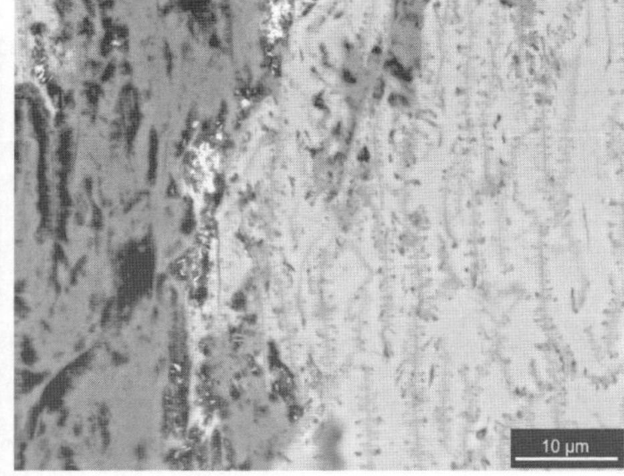

Figure 17. Back-scattered SEM images of ilmenite partially replaced by rutile. (a) Cross section of the entire sample. The lighter core is ilmenite. (b) High magnification image of the interface between rutile (dark) with black pores and ilmenite (light). Note the fine darker fracture network in the ilmenite. (Image: A. Janssen)

the difference in volume appears as a high porosity in the rutile product. Another interesting aspect of this dissolution-precipitation reaction is that the microtexture of the rutile shows three twin-related orientations, as would be expected from an epitaxial nucleation of tetragonal rutile on the surface of trigonal ilmenite. This illustrates the importance of epitaxy in controlling nucleation in a coupled dissolution-precipitation mechanism.

Ahead of the main reaction front, the ilmenite is extensively fractured throughout the whole apparently unreacted core, and the fracture pattern forms a fish-bone-like array. Within these fractures the ilmenite has been replaced by nanosized crystals of rutile. The pattern of replacement depends on the concentration of acid used. In more dilute acid the reaction is slower, but proceeds almost exclusively by the propagation of very fine fractures though the whole sample (see Fig. 18a). These fractures become sites for new lateral spreading of the reaction (see Fig. 18b), by allowing easier pathways for fluid access.

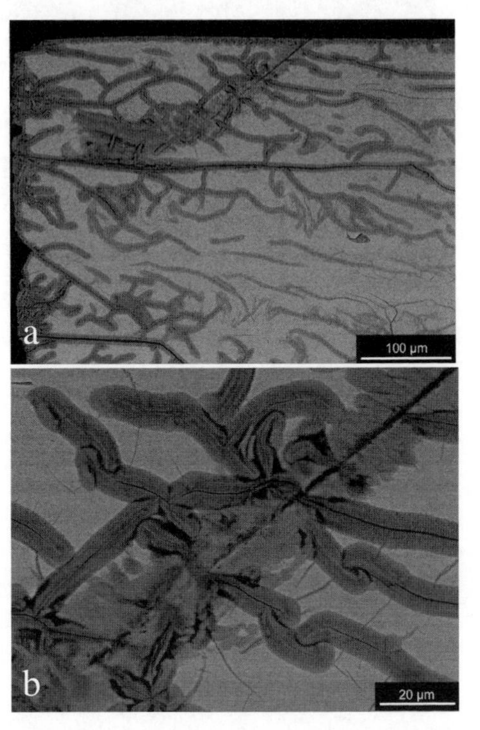

Figure 18. (a) Ilmenite, treated in dilute HCl, partially replaced by rutile along a fracture network which allows lateral spreading of the reaction. (Image: A. Janssen)

Fracturing is also a feature of pseudomorphic reactions where the molar volume of the reactants increases. One example is the replacement of leucite by analcime, described above. Despite the increase in molar volume the product phase is porous, but the strain build-up means that after a certain rim thickness of product is reached, in an unconstrained system, it tends to spall off the unreacted core. Fractures are also generated ahead of the reaction front, but not as extensively as in the ilmenite-rutile case. Another example discussed above is the serpentinization of olivine or orthopyroxene. The typical "mesh textures" of serpentinized olivine shown in Figure 3 is produced by the fracture generation associated with the replacement.

A general analysis and mechanical model of reaction induced fracturing has recently been presented by Jamtveit et al. (2009). The important result is that both porosity generation and fracturing provide pathways for fluid infiltration.

CHARACTERISTICS OF INTERFACE-COUPLED DISSOLUTION-PRECIPITATION REACTIONS

Although the above examples are not an exhaustive list, they are sufficient to identify the common characteristics of reequilibration processes in the presence of an aqueous phase. These can be summarized as follows:

(i) The dissolution and precipitation are closely spatially coupled at the interface between the parent and the product phases. This coupling preserves the external morphology of the parent.

(ii) The reaction front between parent and product is sharp with no significant diffusion profile in the parent.

(iii) The product phase develops intracrystalline porosity and permeability which allows the fluid to maintain contact with the reaction front.

(iv) In cases where there is a large increase or decrease in the solid molar volume, reaction induced fracturing produces a network of fractures ahead of the reaction front.

(v) There is an epitactic relationship between parent and product phases such that if they have the same crystal structure, this is preserved across the interface. In cases where the product phase has a different crystal structure to the parent, the product may be polycrystalline. Nevertheless in cases where the crystallographic relationships have been measured an epitactic relationship has been demonstrated.

"Porosity" as an integral microstructural feature of fluid-induced replacement

In solid state reactions, a specific transformation mechanism results in a microstructure which describes the spatial and crystallographic relationships between the parent and the product phases. This microstructure is controlled by both the thermodynamics and the kinetics of the reaction, and in a full analysis, interfaces generated by the transformation increase the free energy of the system. The microstructure reflects the competition between the thermodynamics and the kinetics of the process. It is also a *transient* feature of a reaction, and given sufficient time microstructures should coarsen to reduce the free energy. However, in solid state reactions, coarsening involves volume diffusion through the crystal, with a rate which is exponentially decreases with temperature. As the temperature falls, diffusion becomes negligible, and the microstructure becomes "frozen-in." This can be thought of in terms of a "closure temperature" for diffusion, which limits the length scale over which diffusion can occur at a particular temperature. As the temperature is decreased, this length becomes negligible in relation to our ability to measure or observe. The preservation of microstructure is dependent on this closure, and allows us to study the transformation mechanisms.

The same concepts of microstructural development apply to interface-coupled dissolution-precipitation reactions. The addition of a fluid phase, which both contributes to and removes components from the solids, has both a thermodynamic, kinetic and microstructural significance. The fluid occupies space, and in a dissolution-precipitation reaction, the generation of "porosity" can be considered as part of the microstructure associated with the mechanism. It is that part of the microstructure occupied by the fluid phase. In experiments such as the replacement of KBr by KCl (see above) the porosity also has a crystallographic orientation, minimizing the interfacial free energy. If the porous product phase remains in contact with the fluid, it would be expected that the microstructure will continue to evolve with time. This has been experimentally observed in the KBr-KCl system, where textural equilibration follows chemical equilibration with the fluid - the porosity coarsens and eventually clears (Putnis et al. 2005). Since the mechanism of textural equilibration is also by dissolution and reprecipitation, the "closure temperature" is very much lower, in many cases down to ambient temperatures. In contrast to solid state reactions, the low closure temperatures mean that microstructures associated with dissolution-precipitation may not be preserved and evidence for the mechanisms may be lost. The preservation of the reaction interface, as seen in many of the examples above, is assumed to be the limit of fluid infiltration.

What causes the coupling between dissolution and precipitation?

It is evident from the examples above that to produce a pseudomorph the dissolution and precipitation reactions must be closely coupled in space and in time. However, the complexity of the thermodynamics of such reactions, which must take into account the overall change in the free energy of solids + fluids, the stress generated by volume changes, as well as the stress imposed by the rigid rock if the replacement takes place in a confined space, has led to different conceptual models for the origin of the coupling.

Model 1. In the examples above it is evident that the formation of the product precipitate relies on the dissolution of the parent to provide the chemical ingredients, which when combined with the fluid chemistry, produce a solution which is supersaturated with respect to the product. However, this in itself does not imply pseudomorphism and it might be expected that the most likely situation is that the dissolution and precipitation would be spatially separated. Pseudomorphism demands that the rates of dissolution and nucleation are closely coupled at the parent-product interface. Such close coupling can be achieved when the controlling mechanism is the dissolution rate, and there is a low activation energy barrier for nucleation. On the other hand, if the nucleation is the rate controlling step, implying that dissolution is fast and nucleation is slower, the coupling may not be so closely related in space and time, leading to a less exact pseudomorph or complete loss of coupling. Such a situation has been explored by Xia et al. (2009b) where the rate controlling step in the replacement of pentlandite by violarite changes as the pH changes from acidic to neutral.

The nucleation rate is controlled by factors which include the degree of epitaxy between the product phase and the parent surface. If the crystal structures of the parent and product phases are closely related, then nucleation of the product on the parent surface, maintaining the crystallographic orientation is favoured, because the interfacial energy at this new interface is reduced (see Putnis 1992, pp. 338). Thus epitaxy transfers the crystallographic information from parent to product even though the parent dissolves. The experiments by Putnis and Mezger (2004) and Putnis et al. (2005) emphasize that the fluid composition *at the interface* is the most important factor, and that during the replacement this may be different from the fluid composition in the bulk, even in a small reaction vessel. This suggests that the rate of transport of the solution species to and from the interface is an important factor which needs to be considered in formulating a quantitative description of the replacement mechanism. The composition of this boundary layer of fluid at the interface is likely to be the most important factor which controls the coupling.

Thus even a monolayer of dissolved parent phase can supersaturate this boundary layer with respect to the product. When nucleation of the product phase takes place on the surface of the dissolving parent it lies within the normal diffusion profile which would be generated in the fluid by the dissolving crystal. This defines the external shape of the original crystal as the outside surface of the new phase. Such an argument has been used by Anderson et al. (1998a,b) to describe an autocatalytic feedback between the rates of dissolution and precipitation when a product of a solid + fluid reaction nucleates on the surface of the dissolving parent phase.

A second factor which controls the nucleation rate of the product phase is the degree of supersaturation of the fluid in the boundary layer. If there is a large difference between the solubility of the product and parent, even a small amount of dissolution will result in a highly supersaturated fluid, and even in the absence of obvious crystallographic relationships between the solid phases, nucleation will be rapid. This may well be the case when calcite is replaced by apatite (Kasioptas et al. 2008) or by fluorite (Glover and Sippel 1962), both of which are many of orders of magnitude less soluble than calcite. In some cases the chemistry of the fluid may not be too critical and the interfacial fluid becomes supersaturated with respect to the product over a wide range of fluid compositions. The nucleation rate may also not be particularly sensitive to fluid composition, only to the degree of supersaturation.

Porosity in the product phase is generated whenever there is a volume deficit reaction. This volume deficit does not only refer to the change in the molar volumes of the parent and product solids, but also to their relative solubilities, which will determine how much parent is dissolved and how much is precipitated. Thus it is possible that even when the molar volume of the parent is smaller than that of the product, porosity can still be generated, as seen in the case of the leucite to analcime replacement (Putnis et al. 2007a). Another important point is that the *absolute solubilities* of the phases are not the determining factors for a dissolution-precipitation reaction to take place. Because the reactions take place at the parent-product interface, the *relative* solubility is the most important factor, as at any one time only a small amount of material needs to be in solution.

The porosity allows the mass transfer of material from the solution reservoir to the reaction interface through the solution. In the case of the replacement of KBr by KCl (Putnis and Mezger 2004) the reaction rim increases in thickness as a linear function of $(\text{time})^{1/2}$ indicating that the rate control in this case may be the diffusion of the ions through the fluid-filled interconnected pores. This was also found by Glover and Sippel (1962) for the replacement of calcite by fluorite.

The evidence from the examples above suggests that interface-coupled dissolution-precipitation is controlled (i) by the chemistry of the fluid at the reaction front and (ii) by the role of the parent mineral as a substrate for nucleation of the product. The finding that the length scale of pseudomorphism can be manipulated by changing the fluid chemistry (Xia et al. 2009b) is a powerful argument in support of this model. Further details of this model for interface-coupled dissolution-precipitation can be found in Putnis (2002), Putnis et al. (2005) and Putnis and Putnis (2007).

Model 2. Maliva and Siever (1988) proposed a model for the diagenetic replacement of one phase by another based on the argument that a replacement does not require the undersaturation of the bulk fluid with respect to the rock (the "bulk host phase") undergoing diagenesis, nor its subsequent supersaturation with respect to a new precipitating phase. From the outset, Maliva and Siever emphasize that their description of a "bulk" phase refers to free surfaces of host phase grains in contact with free pore waters, "as opposed to intergranular or intercrystalline contacts." They propose that the replacement is controlled by the force of crystallization, where this is defined as a pressure which a crystal, growing in a supersaturated solution, can exert on its surroundings. The basic idea of this model (Nahon and Merino 1997) is that when a new crystal (A) begins to grow at some point in a rigid rock, it exerts a stress on its surroundings, leading to pressure induced dissolution of adjacent mineral or rock (B). The dissolution of B *decreases* the supersaturation of the fluid relative to A and hence decreases the growth rate of A. The rates of growth of A, and dissolution of B change in opposite directions until at some point in time they become equal.

Pressure solution is the basis of the coupling between dissolution and precipitation in this model and several criteria are given as textural indicators of force of crystallization controlled replacement (Maliva and Siever 1988).

(i) Dissolution of the host phase is restricted to the interface between the parent and product.

(ii) The preservation of microstructural features of the parent phase indicating that the rates of dissolution of the parent and precipitation of the product are equal "otherwise a gap would form between them."

(iii) The presence of a euhedral product crystal faces in planar contact with the unreplaced host. The argument here is that it would not be expected that the parent phase would dissolve in such a way as to accommodate the interfacial angles of the product, without the product crystal exerting a pressure on the host.

Merino and co-workers have taken up this model to explain replacement mechanisms, specifically in weathering phenomena (Merino et al. 1993; Nahon and Merino 1997; Merino and Banerjee 2008). The mathematical formulations of the Maliva and Siever model (e.g.,

Dewers and Ortoleva 1989; Merino and Dewers 1998 and Fletcher and Merino 2001) involve a coupling between the growth kinetics of crystals and the mechanical response of a viscously relaxing surrounding rock or medium.

However many mineral replacements, especially those in experimental studies, do not take place in a confined space, and in surface weathering, such as in the genesis of "terra rossa" (Merino and Banerjee 2008) it is questionable whether the "rigid rock" prerequisite applies. The generation of porosity in the secondary phase is also not a feature of the Maliva and Siever model.

Nevertheless, Merino and Banerjee (2008) argue that a force of crystallization will exist at the parent-product interface, even without an externally applied stress and in surficial environments. The examples used to argue a case against Model 1 for interface-coupled dissolution-precipitation are that there are observations of parent/product mineral pairs with no chemical components in common e.g., limestone replaced by silica, or sphalerite replaced by dolomite in which the botryoidal layering structure is inherited from the parent sphalerite. Such replacements are much easier to recognize than, for example, a feldspar–feldspar replacement, and have therefore been known for many years (e.g., Lindgren 1912; Bastin et al. 1931).

The question of whether the coupling between dissolution and precipitation in fluid-induced pseudomorphic replacement is driven by solution chemistry (Model 1) or local non-hydrostatic stress and pressure solution (Model 2) is still the subject of debate (e.g., Merino and Banerjee 2008; Xia et al. 2009b). The examples given in this chapter provide compelling arguments in favour of Model 1, but these are all cases in which parent and product phases share some common chemistry such that dissolution of the parent results in an interfacial fluid in which the supersaturation is *increased* with respect to the product phase. In Model 2, pressure dissolution of the parent results in a *decrease* in the supersaturation of the fluid with respect to the precipitating phase. While it is not immediately obvious how Model 1 can be applied when there is no apparent common chemistry between the parent and product phases, the demonstrated role of fluid chemistry is easier to appreciate with Model 1 than Model 2.

The question of the role of the fluid composition on dissolution and precipitation thermodynamics and kinetics is not straightforward, as it is known that components in the fluid which do not seem to be directly related to the chemistry of the dissolving or precipitating solids can have a major effect. For example, the role of pH or the redox state in the solution are well known factors which may indirectly influence reactions. The precipitation of dolomite is enhanced by sulfate species in solution (Brady et al. 1996), and may form a link between sphalerite dissolution and oxidation and dolomite precipitation from a supersaturated solution. Background electrolytes which do not enter into the precipitating phase, but affect the hydration/dehydration of the dissolving/precipitating ions also play an important role in both dissolution and precipitation (Kowacz and Putnis 2008). However, too little is known about the role of solution chemistry to explain the kind of replacement where a whole assemblage of fine grains in a schist for example, are replaced by a single pyrite crystal.

Although the two models have different explanations for apparently similar phenomena, the effects of chemistry and induced stress are both likely to play an important role in replacement processes in rocks. When discriminating between reactions such as reactions (8) and (9), the mechanism will depend on the fluid composition, the relative free energy changes for the reactions (including the fluid thermodynamics), the molar volume changes of the solid phases, and the presence or absence of an imposed stress, as well as the kinetics of all possible (i.e. $\Delta G < 0$) reactions. In such a situation it is likely that elements of both models will need to be included in a full explanation of mineral replacement in rocks.

IMPLICATIONS OF INTERFACE-COUPLED DISSOLUTION-PRECIPITATION AS A MECHANISM FOR MINERAL REPLACEMENT

Mineral replacement reactions take place primarily by dissolution-reprecipitation processes. At high temperatures, solid state reactions which involve volume diffusion may become kinetically significant on a local scale, but the wholesale replacement of one mineral assemblage by another, in the presence of any fluid, will kinetically favour dissolution and precipitation over a solid state reaction under most crustal conditions. Similarly, deformation by dissolution-precipitation creep, rather than by solid state creep, is also likely to be the dominant mechanism at metamorphic grades up to at least amphibolite facies (Wintsch and Yi 2002).

Given that processes such as cation exchange, chemical weathering, deuteric alteration, leaching, metasomatism, diagenesis and metamorphism are all linked by common features in which one mineral or mineral assemblage is replaced by a more stable assemblage, understanding mineral replacement has implications to virtually all aspects of the rock cycle on earth. In industrial processes, understanding crystal dissolution and growth is also fundamental to modelling the rates of chemical reactions and finding new routes for the synthesis of materials. In this section we will briefly review some of these applications.

The kinetics of dissolution-precipitation and metamorphic and metasomatic reactions

One of the important conclusions from this review of replacement processes is that they are fast and that they can be driven by small reductions in free energy. In the experiments on feldspars and apatites, significant replacement takes place under hydrothermal conditions even in a few days. In other examples replacement takes at room temperature. Although the solutions used in the experiments are initially significantly out of equilibrium with the parent phase, the rapid rates suggest that in a geological context, when a fluid interacts with a solid with which it is out of equilibrium, it is inevitable that dissolution will begin to take place. The contention that the departures from equilibrium need only be small is supported by the results by Nakamura and Watson (1981) who showed that a reduction in interfacial energy was sufficient to recrystallize quartz under hydrothermal conditions. The experimental results support the analysis by Wood and Walther (1983) and Walther and Wood (1984) that in the presence of a fluid phase metamorphic reaction rates are rapid, and that dissolution-precipitation reactions should proceed at very small departures from equilibrium ("rarely more than a few degrees").

A very well-studied example of the role of fluid in a metamorphic reaction is the eclogitization of Precambrian anorthositic granulites of the Bergen Arcs, western Norway. In a classic paper, Austrheim (1987) concluded, on the basis of detailed mapping of eclogite outcrops within the granulite facies rocks, that the eclogitization is a function of deformation and fluid access, rather than being controlled by temperature, pressure and rock composition alone. Figure 19a shows a rock outcrop in which a dark vein of eclogite cross cuts the paler granulite. The vein is associated with a shear plane and fluid infiltration along which the eclogitization has occurred. The eclogitization is limited by the extent of this fluid infiltration laterally from the shear plane and in the absence of fluid, the granulite remains essentially unaltered. If we assume that the whole rock outcrop experienced the same P-T conditions, then the conclusion that the fluid triggered the eclogitization reaction is inescapable.

The effect of eclogitization can be seen at every scale, from the rock outcrop, to interfaces within individual crystals. Figure 19b shows an image of a garnet crystal from the eclogite-granulite interface. The dark part of this partially pseudomorphed crystal has the original granulite garnet composition, while the lighter regions are the eclogite composition replacement product. Electron back-scattered diffraction (EBSD) analysis confirms that the crystallographic orientation of the garnet is preserved during the replacement, but with textural reequilibration of the original sub-grain microstructure of the original granulitic garnet (Pollok

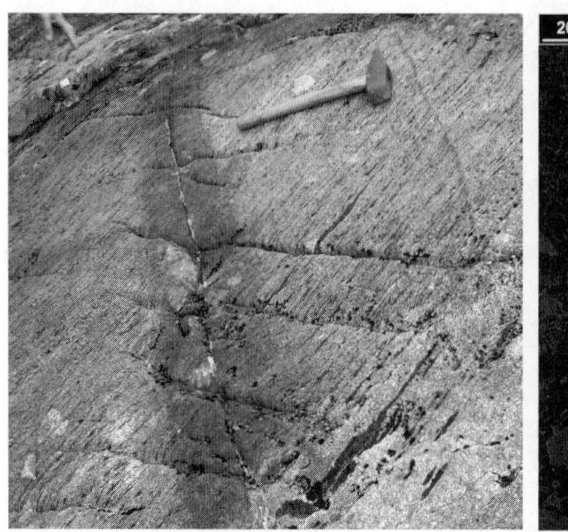

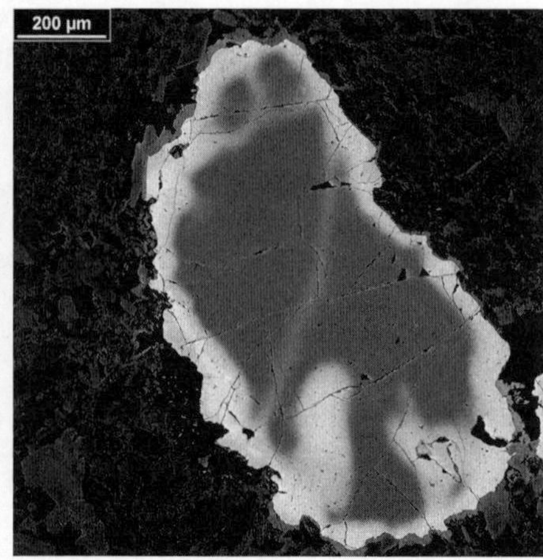

Figure 19. (a) Rock outcrop showing eclogitized rock (dark) along a fracture in the granulite. (b) A back-scattered SEM image of a single garnet crystal from the interface between granulite and eclogite. The dark parts of the partly pseudomorphed crystal have the original granulite composition, while the lighter parts of the same single crystal have a composition of eclogitic garnet. [Used with permission of Elsevier from Pollok et al. (2008).]

et al. 2008). The garnet replacement texture is interpreted as due to a coupled dissolution-precipitation mechanism.

The same principles apply to metasomatism, in which there is a significant change in the chemical composition of the rock, resulting from components added from the fluid. The distinction between metamorphism and metasomatism is a matter of degree and scale, rather than of principle. Whenever a fluid is involved in a replacement process some chemical components may be added to or subtracted from the solid phases. Even in an apparently isochemical recrystallization, trace elements may be lost to the fluid (Parsons et al. 2009), and as Carmichael (1969) has pointed out, metamorphic reactions are metasomatic if we consider a small enough sample volume.

Sub-solidus reequilibration of granitic rocks

Based on oxygen isotope studies of granites from many terrains, Taylor (1977) concluded that "gigantic meteoric-hydrothermal convective circulation systems were established in the epizonal portions of *all* batholiths, locally producing very low $\delta^{18}O$ values (particularly in feldspars) during sub-solidus exchange." The experimental work on mechanisms of cation and oxygen isotope exchange in feldspars, together with higher spatial resolution isotope data, show that the two processes are simultaneous, as would be expected from a dissolution-precipitation replacement process. It has also long been recognized that replaced feldspars are turbid due to porosity, and fluid and solid inclusions and may also be pink or red-colored (Boone 1969). More recently, Putnis et al. (2007a) showed by transmission electron microscopy that red-clouded potassium-rich feldspars from granites contain many pores up to several hundred nanometers in size, and that almost every pore contains rosettes or needles of crystalline hematite (see Fig. 20). This observation, together with the fact that the origin of the porosity is consistent with an interface-coupled dissolution-precipitation reaction, indicates that the hematite is a direct product of fluid-rock interaction. Hematite in pores is *not* consistent with solid state exsolution from a Fe-bearing feldspar. In examples where interfaces between a parent plagioclase and a product K-feldspar can be found, compositional data show that the plagioclase parent does not have enough Fe to account for the hematite by a fluid-induced phase separation mechanism,

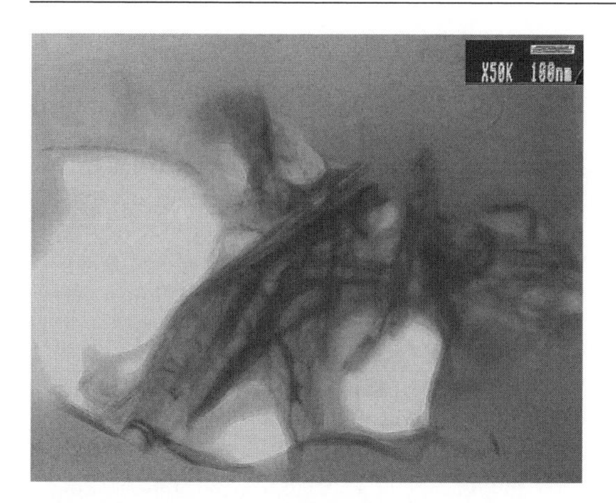

Figure 20. A single pore, containing hematite nano-platelets, in a pink feldspar from a hydrothermally altered granite. [Used with permisson of Elsevier from Putnis et al. (2007a).]

and that it was therefore introduced by the fluid.

The broader conclusions from these studies is that, as feldspars in granites are almost invariably turbid and are often red-clouded, many granites have been subject to regional alteration which have modified the feldspar compositions, and resulted in markedly discordant Rb-Sr isochron "ages" (Taylor 1977). More detailed high resolution studies by SEM and TEM of a grey granite from south-eastern Sweden, with pronounced red-staining associated with fractures (Plümper and Putnis 2009) suggest that virtually all of the feldspar in the grey *and* red parts of the rock is secondary and that the granite has a very complex sub-solidus history, such that it is not even possible to determine the nature of the original protolith. Depending on the composition of an infiltrating fluid, interaction with the rock may replace some minerals and not others, as noted by O'Neil and Taylor (1967). This leads to grain-scale isotope disequilibrium, a situation explored by Abart and Sperb (1997).

These studies suggest that the interpretation of geochemical and isotopic signatures of granites purely in terms of their magmatic origins, without considering the extent of sub-solidus reequilibration by externally derived fluids, is questionable.

Porosity and fracture generation and the mechanism of fluid transport through rocks

There is overwhelming evidence that fluids can flow through rocks. While tectonically-induced fracturing undoubtedly plays a major role in permeability development and introduction of fluid into rocks (Jamtveit and Yardley 1997), replacement reactions by dissolution-precipitation require that fluid infiltrates every part of a rock, moving *through* the minerals as they are replaced. Discussions about how fluids move through rocks are generally restricted to hydraulic fractures and grain boundaries (e.g., Kostenko et al. 2002), but porosity generation by a reactive fluid greatly increases the number of possible fluid pathways and the permeability of a rock. A replacement reaction creates this permeability behind the reaction front and as long as there is sufficient fluid and mass transport through the created porosity, rocks can be reequilibrated on a large scale. The fluid can literally react its way through a rock, by-passing some minerals with which there may be no reaction, or with which a reaction generates a non-porous product which effectively seals the mineral off from the fluid.

Fluid passing through rocks can result in large scale element mobilization in the earth (Oliver et al. 2004; Putnis et al. 2008). For example, albitization affects large volumes of rock in many parts of the crust, stripping the parent rock of major and trace elements, leaving behind pure porous albite. This "charged fluid" can migrate large distances before encountering conditions conducive to precipitation (Clark et al. 2005) and forming ore and mineral deposits.

How to recognize mineral replacement

In most of the examples above, a replacement interface could be recognized in the field, on the thin-section scale, and at the electron microscope scale in natural rocks, as well as in experimental samples. In these cases of partial replacement where the limit of fluid infiltration can be well defined, there is no doubt about the nature of the parent and product phases. However, in a rock which has been totally replaced, it is often not possible to determine the protolith, and there is a lot of scope for misunderstanding the petrogenesis of a mineral assemblage. At a thin section scale it is very important to study the textural relationships between minerals as well as internal microstructures. Porosity and the presence of mineral inclusions within pores is an indication that the host mineral is a replacement product. Porosity produces turbidity, especially in feldspars, and turbid feldspars are all secondary, whereas primary feldspars are clear and glassy. This has been recognized for many years (e.g., Folk 1955; O'Neil and Taylor 1967; Parsons and Lee 2009), but still deserves emphasis. Fluids can also preferentially move along cleavage planes and other defects leaving trails of fluid inclusions strung out along their length. It is a common observation that more calcic cores of zoned plagioclase crystals in igneous rocks are sericitized (i.e. forming an alteration product of fine mica + albite – e.g., Plümper and Putnis 2009) while more sodic rims are unaltered.

Replacement rims are often mistaken for and referred to as "overgrowths" representing a change in the chemical or physical conditions in the fluid from which the crystal is growing. While it is not always straightforward to distinguish between an overgrowth and a replacement, the presence of pores and inclusions in the rim is an indication for replacement.

An interesting example of recognizing the metasomatic origin of rocks has recently been described by Austrheim et al. (2008). Zircon rims around ilmenite occur in a wide variety of mafic igneous and granulite facies rocks (Bingen et al. 2001; Morisset and Scoates 2008). The rims may be continuous or may be trails of discrete single crystals along the oxide grain boundary. These rims and trails may serve as inert markers which remain in place after the ilmenite is replaced during a metasomatic or metamorphic event. In scapolitized gabbros from the Bamble and Kongsberg sectors in Norway, these markers can be seen in both the unaltered gabbro where the ilmenite is preserved, as well as in the scapolitized rock where the ilmenite is totally replaced by a new mineral assemblage. Only the unreacted zircon trails are evidence for the former ilmenite boundary and provide a physical reference frame which can be used to quantify the mobility of elements during replacement as well as to identify fluid involvement in the replacement mechanism (Austrheim et al. 2008). Again this study emphasizes the importance of studying the mineral and rock textures.

Replacement textures and "co-existing compositions"

An understanding of replacement textures aids in identifying when minerals are in equilibrium in a rock or when they represent a frozen-in parent-product relationship. The ease with which feldspars are partially replaced leads to many possibilities for very different feldspar compositions existing in a single hand specimen or thin section. If interpreted as "co-existing" from a thermodynamic point of view, this can lead to incorrectly proposed miscibility gaps.

"Leaching" and the formation of depleted surface layers

Many minerals and glasses, when exposed to aqueous solutions, do not dissolve congruently, so that the release of ions into solution is not in proportion to their composition in the solid. Such non-stoichiometric dissolution results in the development of chemically and structurally altered surface layers relative to the bulk. These are commonly referred to as "leached layers" and the prevailing view is that some components are selectively removed from the solid, and the depleted layer then may become structurally reorganized by a solid state mechanism.

Many of the examples above could be referred to as "leaching": the conversion of pentlandite $(Fe,Ni)_9S_8$ by violarite $(Ni,Fe)_3S_4$ involves removal of Fe and Ni; the reaction of natural pyrochlore $(Ca_{1.23}Na_{0.75})Ta_{1.78}O_{6.28}F_{0.57}$ with fluids results in the selective removal of Na and Ca; the reactions of REE-bearing zircon and monazite with fluid result in the removal of the REEs and the precipitation of secondary REE minerals within the replaced zones.

However, we have seen that using ^{18}O as a tracer in the fluid, this isotope has been *added* to the depleted layer, demonstrating an interaction between the whole fluid and the whole solid, and interpreted in terms of a coupled dissolution-reprecipitation mechanism.

Leached layers, depleted in metal ions, are commonly formed during incongruent dissolution of silicate minerals and glasses. Whether a silicate mineral dissolves congruently or incongruently depends on both the structure of the solid, and the composition of the fluid. Materials with the same chemistry but different crystal structure may dissolve differently, for example of the two forms of $CaSiO_3$, wollastonite dissolves incongruently while pseudo-wollastonite dissolves congruently (Casey et al. 1993). In feldspars, the pH determines whether dissolution is congruent (at high pH) or incongruent (at low pH) (Casey et al. 1989a,b; Petit et al. 1990).

If we consider the balance of free energies for such solid-fluid interactions, the overall reduction in free energy is achieved by considering the free energies of solids and fluids. Therefore it should not be surprising that as the structure and composition of either changes, a reaction Solid(1) + Fluid(1) $\rightarrow$ Solid(2) + Fluid(2) may be favoured over a reaction Solid(1) + Fluid(1) $\rightarrow$ Fluid(3). The thermodynamic argument is the same as that for congruent and incongruent melting – the phase diagram is merely a consequence of the relative positions of free energy curves for the phases involved, e.g., enstatite melts incongruently at atmospheric pressure but congruently at pressures above about 3 kbar. Small differences in the free energies of each phase can lead to very different phase equilibria.

However the mechanism of incongruent dissolution in silicate minerals has been the subject of much research and controversy. The generally accepted view is that H^+ ions from the solution diffuse into the structure and exchange for the metal cations which diffuse out of the crystal, leaving behind a silica-rich leached layer. This interdiffusion is accompanied by a spontaneous reconstruction of this layer to form amorphous silica (Casey et al. 1993). Weissbart and Rimstidt (2000) re-emphasized the role of the reconstruction of the leached layer in limiting the release of cations into solution, and summarize much of the earlier work on incongruent dissolution of wollastonite.

However, Hellmann et al. (2003) studied the interface between experimentally altered (i.e., leached), and non-altered plagioclase feldspar by high resolution transmission electron microscopy, and concluded that the interface was chemically and structurally sharp on an atomic scale, and did not show the compositional profiles that would be expected from a solid state interdiffusion mechanism. They concluded that the data were better explained by an interfacial dissolution-reprecipitation mechanism. In other words, the dissolution is initially stoichiometric, but is coupled with the precipitation of amorphous silica from a supersaturated boundary layer of fluid.

The mechanism of dissolution of glass is also a hotly debated topic, especially for glass which is used for the containment of radionuclides from spent nuclear fuel. The arguments are similar to those generally proposed for crystalline silicates: interdiffusion of H^+ and alkali cations in the glass to form a leached layer, accompanied by a structural reorganization of this layer to form glass with a different structure from the parent. The reaction with aqueous solution and production of a leached layer is generally referred to as "corrosion." While in contact with the fluid, the leached layer continues to get denser with time, leading to pore closure and a sharp drop in the migration of the reaction interface (Cailleteau et al. 2008). Geisler et al. (2008) have proposed a new mechanistic model to explain the same data, based on congruent dissolution of the glass which is spatially and temporally coupled to the precipitation of amorphous silica at an inward moving reaction interface.

The debate on the mechanism of silicate mineral and glass corrosion is reminiscent of the debate about cation exchange between feldspars and alkali and alkali-earth chloride solutions, a debate which was definitively quashed by O'Neil and Taylor (1967) by their concept of a dissolution-reprecipitation model. It is tempting to interpret all of the examples of fluid-mineral interaction given in this chapter in a similar way. The example of the replacement of KBr by KCl after contact with a saturated KCl solution at room temperature is one in which interface-coupled dissolution precipitation cannot be seriously questioned. A slightly undersaturated KCl solution in contact with a crystal of KBr immediately completely dissolves the crystal, while in a saturated KCl solution a virtually perfect pseudomorph is formed, preserving crystallographic orientations, and with a porosity which depends on both molar volume changes and relative solubilities of parent and product in the fluid. Subsequent coarsening and loss of connectivity of porosity by continued textural equilibration slows the reaction with time (Putnis and Mezger 2004; Putnis et al. 2005). These common features in the other examples given above outweigh the differences. When the parent and product crystal structures are different, the product is polycrystalline, but in the few cases where any crystallographic orientations have been measured, a degree of topotaxy has been found (Glover and Sippel 1962).

Many of the observations made on the incongruent dissolution of silicate minerals could also be interpreted in terms of a coupled dissolution-precipitation model as suggested by Hellmann (2003), with subsequent textural reequilibration of leached layers by continued dissolution and reprecipitation. However, at the present time, the most commonly accepted explanation of leaching is related to interdiffusion of protons which form Si-OH bonds in the silicate network, and release weakly bound cations which diffuse through the structure and into the solution. Exactly how this interdiffusion is accomplished at room temperature remains a matter of some conjecture.

Determining the mechanism of the incongruent dissolution which results in these depleted layers is not merely an academic issue. The main point of understanding mechanisms is to be able to model experimental results and extrapolate these to long time-scales appropriate to understanding weathering reactions, in the case of silicate minerals, or to the long-term aqueous durability of materials used to encapsulate nuclear waste.

Another interesting example of the formation of "depleted layers" is found in the gold-rich rims on electrum (Au-Ag alloy) nuggets in placer gold deposits (see Fig. 21). The Au-rich, Ag-depleted rims vary in thickness and may have a convoluted, lobate interface with a core of less-pure gold. The interface between the core and rim is chemically sharp and the rim is porous. It is also a well known observation that alluvial gold becomes richer in Au-content the further it is from the lode from which it was derived (Groen et al. 1990) and the general consensus is that the Ag-depletion takes place at low temperatures. The enhancement of gold content on the surfaces of impure gold objects ("depletion gilding"), has been known since the pre-Columbian cultures of Central and South America, who treated the surfaces with naturally-derived organic fluids (Lechtmann 1984). The literature on the origin of the Ag-depletion has generally used

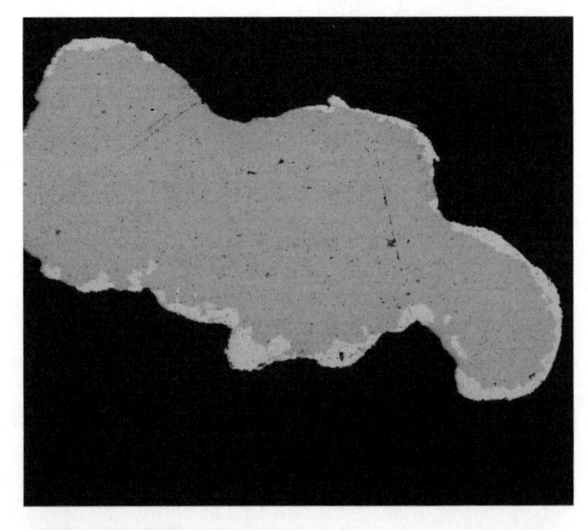

Figure 21. Reflected light image of a nugget of gold alloy (electrum Au-Ag), which has an Au-enriched rim (light grey). (Image: B. Grguric)

the terminology of "selective dissolution," whereby "the less noble element is preferentially removed from the alloy, leaving a gold-rich residue" (Fortey 1981). However, Groen et al. (1990) rejected a diffusion-based leaching mechanism to remove Ag as far too slow to account for the rim thicknesses (which would take between 10^{17} and 10^{18} years to form by solid state diffusion of Ag), and also inconsistent with the sharp compositional gradient at the interface between rim and core. Groen et al. (1990) invoke a mechanism termed self-electrorefining (Fontana 1986) (also used to explain "dezincification" of brass), where the electrum at the solid-fluid interface dissolves and the purer gold immediately precipitates back onto the surface of the grain. The porosity generation, due to the lower solubility of gold, allows continued contact between the fluid and the reaction interface. More recent crystallographic studies (Hough et al. 2007) have also shown that there is no variation in crystallographic orientation between the core and rim.

Solid state and dissolution-precipitation mechanisms: what's the difference?

This may seem a rather trivial question. In the extreme cases there is no ambiguity. As pointed out early in this Chapter, Cardew and Davey (1985) made the distinction between a polymorphic transformation in a dry system, in which all atomic motion is by volume or "solid state" diffusion and in a wet system where dissolution of the less stable polymorph is followed by precipitation of a more stable polymorph. Where there is no coupling between the dissolution and precipitation (i.e., the two processes are separate in both space and time) then the terminology of dissolution-precipitation has a well understood meaning. However, as these two processes become more closely coupled, a pseudomorphic replacement, which retains various degrees of crystallographic information, begins to have characteristics attributed to solid state reactions. Even in recent papers "structure inheriting solid state reactions under hydrothermal conditions" is a terminology in use (Eda et al. 2005, 2006) for reactions which preserve external morphology and aspects of the crystal structure. However, the degree of coupling between dissolution and precipitation can be manipulated by changing the fluid composition (e.g., Xia et al. 2009b) emphasizing that although the texture of the final product is very different, the fundamental mechanism remains the same.

There is an historical tendency to interpret the mechanisms of reactions in solid state terms even when fluids are present and other mechanisms are available. In the debate on the formation of leached layers in silicate minerals and glasses, a solid state terminology is also used, but modified to include terms such as "hydrolysis" of bonds and "condensation" of leached layers. On a nanoscale these terms may have similar meanings to "dissolution" and "precipitation." There is also, in general usage, the vague concept of "fluid-enhanced solid-state diffusion" (i.e., that "solid state" reactions are faster in the presence of fluid).

An example where a comparison between solid state diffusion and dissolution-precipitation has been made is in the reequilibration of cation distributions between coexisting minerals. In a perfectly dry system, when two minerals are in contact at high temperature, they may exchange cations across the contact by a volume interdiffusion process until the equilibrium cation distribution between the two phases is reached. For example, the interchange of Fe and Mg between two minerals, such as garnet and pyroxene, could be accomplished by a solid state diffusion mechanism which would initially produce a diffusion profile of Fe and Mg in each phase. However, if the two minerals were separated by a fluid, it is also possible that the same reequilibration could take place by a dissolution–precipitation process, retaining the morphology and crystallography but with a sharp interface between the parent and product compositions in both the garnet and the pyroxene. Analysis of experimental kinetic data and comparison with that expected from solid state diffusional exchange led Pattison and Newton (1989) to conclude that, in the presence of fluids, equilibration was by dissolution and precipitation. This has significant implications for the way thermodynamic data are extracted from experiments, as discussed by Pattison (1994).

It is well known that the presence of fluids enhances reactions which in a perfectly dry system would proceed at negligible rates. This suggests that the fluid acts as a catalyst, i.e. a substance that decreases the activation energy of a chemical reaction without itself being changed at the end of the chemical reaction. Thus in a reaction involving only anhydrous species

$$A + B \rightarrow X + Y \tag{12}$$

the thermodynamics of the solid phases will define the P,T slope of the univariant reaction curve. If water catalyses this reaction by providing a dissolution-precipitation mechanism, the P,T slope of the univariant reaction curve is not affected. This is the assumption made in petrology when determining P,T stability fields of a mineral assemblage, such as $X + Y$.

However, in an open system where a fluid with many dissolved components reacts with a rock, the fluid adds some components to the solids and removes others, such that the final assemblage $X + Y$ could have formed by a reaction such as

$$C + D + F_1 \rightarrow X + Y + F_2 \tag{13}$$

where F_1 and F_2 represent the compositions of ingoing and outgoing fluids. In such a case the fluid is not merely a catalyst but is a phase whose thermodynamics also needs to be considered, both before and after the reaction. It is therefore possible that an assemblage $X + Y$ could have formed at different P,T conditions by reactions (12) and (13).

A situation of this type has been discussed for the partial eclogitization of gabbros in Zambia, in which fluids not only enhance the reaction between anhydrous gabbros and anhydrous eclogites, but are also major contributors to the trace-element variations found in the reaction product (John and Schenk 2003; John et al. 2004).

Replacement processes in magma

In so far as a magma is a multicomponent solution from which minerals may precipitate or into which they may dissolve, the general principles of interface-coupled dissolution-precipitation are applicable. Tsuchiyama (1985) has described the reequilibration of plagioclase crystals of different compositions in melts in the system, diopside-albite-anorthite. Cross-sections of partially reequilibrated crystals show that the morphology is maintained while a sharp reaction interface moves into the crystal separating parent and product plagioclase compositions. The rim is porous, and is generally referred to as "dusty plagioclase," due to the many melt inclusions trapped in the pores. The melts in such inclusions are in equilibrium with the product plagioclase composition (Nakamura and Shimakita 1998). Johannes et al. (1994), described such processes in terms of a "crystallographically oriented solution-reprecipitation mechanism."

New strategies for materials synthesis

Synthesizing new materials with specific physical and chemical properties lies at the heart of materials science. Metal sulfide synthesis provides a good example of how coupled dissolution-precipitation replacement reactions can lead to new routes for synthesis of compounds. Many metal sulfides can be synthesized using traditional methods of reacting metal powders with molten sulfur at high temperatures (> 500 °C) in a vacuum. However, many sulfides are not stable at such high temperatures and their synthesis presents considerable problems. For example, violarite $(Ni,Fe)_3S_4$ is only stable below 373 °C, and below such temperatures the reaction rates are very slow and generate other unwanted impurities (Tenailleau et al. 2006b). However Xia et al. (2008) have demonstrated the synthesis of violarite and linnaeite (Co_3S_4) by replacement of pentlandite $(Fe,Ni)_9S_8$ and cobaltpentlandite (Co_9S_8) as precursors. The synthesis can be carried out at temperatures < 145 °C, and produces a much purer product in a fraction of the time.

Hydrothermal synthesis of materials is not new, but by using a suitable precursor phases and solution composition, the replacement route opens many new possibilities. Precursor phases in hydrothermal experiments have also been used by Eda et al. (2005, 2006) to synthesize alkali metal molybdenum oxides, but their interpretation of the mechanism is quite different. They define "structure-inheriting solid-state reactions under hydrothermal conditions" based on the fact that the product phase inherits the morphology of the parent (i.e., is a pseudomorph) as well as preservation of some aspects of the crystallographic structure. Structural models are then devised to explain how such a "solid state" reaction might be achieved. However, as we have seen above, preservation of morphology and structure is also a feature of the interface-coupled dissolution-precipitation mechanism.

The experimental conversion of sulfates to carbonates has also been interpreted in terms of a coupled dissolution-precipitation mechanism. Celestite ($SrSO_4$) can be readily converted to strontianite ($SrCO_3$) under hydrothermal conditions in carbonated solutions (Suárez-Orduna et al. 2004, 2007). The external morphology is preserved, the product is porous and there is a sharp reaction front between the parent and product phases. Since the crystal structures of the two phases are different, the product is polycrystalline. Similar experiments have been used to pseudomorphically replace celestite by $SrCrO_4$ (Rendón-Angeles et al. 2005), barite ($BaSO_4$) by $BaCO_3$ (Rendón-Angeles et al. 2008) as well as celestite by SrF_2 (Rendón-Angeles et al. 2006). In each case the features of the replacement are consistent with an interface-coupled dissolution-precipitation mechanism, although in earlier publications the pseudomorphism was thought to necessitate a solid state reaction.

In some cases the preservation of morphological and internal microstructural details of the parent becomes the main design rationale of materials synthesis. For example, in the conversion of leucite to analcime (see above), the very fine twinning common in leucite is preserved, and the reprecipitation reaction produces arrays of uniformly oriented analcime nanocrystals which inherit their crystallographic orientation from the leucite. There is considerable interest in developing methods of producing materials with self-assembled arrays of nanocrystals, and Xia et al. (2009a) have proposed that three-dimensional arrays can be produced by a suitable choice of parent phase and solution composition.

An early classic example of materials design using this principle of preservation of morphology during replacement was the conversion of natural coral to hydroxyapatite (Roy and Linnehan 1974) to produce porous material which would be compatible with bone for medical implants.

CONCLUSION

Understanding reaction mechanisms is a fundamental step for understanding processes in nature and in industry. In the examples described in this chapter we have seen that in the presence of a fluid phase, reequilibration processes in a wide range of materials are dominated by dissolution and reprecipitation mechanisms. The range of materials has included open framework structures where solid state ion exchange may have been expected, sulfides in which volume diffusion is relatively rapid, highly soluble salts and low solubility phosphates, oxides, silicates and metal alloys. The interaction of aqueous fluids with these materials results in a remarkable consistency of behavior, especially in the way that the dissolution and precipitation processes are coupled, and how the relative solubility and molar volume combine to create porosity which allows fluid to continuously migrate to the reaction interface. The recent discovery that the coupling can be manipulated by changing the fluid composition, not only emphasizes the role of the fluid in the coupling, but opens new avenues for materials syntheses.

The realization that fluids not only trigger metamorphic processes, but also contribute to the chemistry of the solids, introduces a new and largely unknown parameter into determining the thermodynamics of metamorphic and metasomatic reactions. The thermodynamics of the aqueous fluid phases, before and after the reaction must be included. However, before we can even begin on such a daunting task, we need to recognize the role of fluids in determining textures of rocks over a wide range of spatial scales.

ACKNOWLEDGMENTS

Many of the examples used in this chapter have been taken from the work of colleagues who have greatly influenced the increasing appreciation of the specific role of fluids in mineral replacement mechanisms. In particular I would like to acknowledge discussions with colleagues in the Institute für Mineralogie in Münster including Thorsten Geisler and Christine Putnis, at the Physics of Geological Processes Centre (PGP) in Oslo, including Håkon Austrheim and Bjørn Jamtveit, and Allan Pring at the South Australian Museum, all of whom have influenced the interpretations in this Chapter. However, any errors of fact or judgment are entirely the author's. I thank Eric Oelkers for a careful reading of an earlier version of this chapter and for constructive suggestions for its improvement. Financial support from the Marie Curie Actions of the European Union (MEST-2005-012210 and MRTN-2006-31482), the Deutsche Forschungsgemeinschaft and the Humboldt Foundation is also acknowledged.

REFERENCES

Abart R, Sperb R (1997) Grain-scale stable isotope disequilibrium during fluid-rock interaction. I: series approximations for advective-dispersive transport and first-order kinetic mineral-fluid exchange. Am J Sci 287:679-706

Anderson JG, Doraiswamy LK, Larson MA (1998a) Microphase assisted "autocatalysis" in a solid-liquid reaction with a precipitating product – I. Theory. Chem Eng Sci 53:2451-2458

Anderson JG, Larson MA, Doraiswamy LK (1998b) Microphase assisted "autocatalysis" in a solid-liquid reaction with a precipitating product – II. Experimental. Chem Eng Sci 53:2459-2468

Austrheim H (1987) Eclogitization of lower crystal granulites by fluid migration through shear zones. Earth Planet Sci Lett 81:221-232

Austrheim H, Putnis CV, Engvik AK, Putnis A (2008) Zircon coronas around Fe-Ti oxides: a physical reference frame for metamorphic and metasomatic reactions. Contrib Mineral Petrol 156:517-527

Bastin ES, Graton LC, Lindgren W, Newhouse WH, Schwartz GM, Short MN (1931) Criteria of age relations of minerals, with special reference to polished sections of ores. Econ Geol 26:561-610

Bingen B, Austrheim H, Whitehouse M (2001) Ilmenite as a source for zirconium during high-grade metamorphism? Textural evidence from the Caledonides of Western Norway and implications for zircon geochronology. J Petrol 42:355-375

Boone GM (1969) Origin of red-clouded feldspars: petrologic contrasts in a granitic porphyry intrusion. Am J Sci 267:633-668

Bowman JR, Valley JW, Kita NT (2009) Mechanisms of oxygen isotopic exchange and isotopic evolution of $^{18}O/^{16}O$- depleted periclase zone marbles in the Alta aureole Utah: insights from ion microprobe analysis of calcite. Contrib Mineral Petrol 157:77-93

Brady PV, Krumhansl JL, Papenguth HW (1996) Surface complexation clues to dolomite growth. Geochim Cosmochim Acta 60:727-731

Brown WH, Fyfe WS, Turner FJ (1962) Aragonite in California glaucophane schists, and the kinetics of the aragonite-calcite transformation. J Petrol 3:566-582

Cailleteau C, Angeli F, Devreux F, Gin S, Jestin J, Jollivet P, Spalla O (2008) Insight into silicate-glass corrosion mechanisms. Nat Mater 7:978-983

Cardew PT, Davey RJ (1985) The kinetics of solvent-mediated phase transformations. Proc R Soc A 398:415-428

Carlson WD (1983) Aragonite-calcite nucleation kinetics: An application and extension of Avrami transformation theory. J Geol 91:57-71

Carlson WD, Rosenfeld JL (1981) Optical determination of topotactic aragonite-calcite growth kinetics: Metamorphic implications. J Geol 89:615-638

Carmichael DM (1969) On the mechanism of prograde metamorphic reactions in quartz-bearing pelitic rocks. Contrib Mineral Petrol 20:244-267

Casey WH, Westrich HR, Banfield JF, Ferruzzi G, Arnold GW (1993) Leaching and reconstruction at the surfaces of dissolving chain-silicate minerals. Nature 366:253-256

Casey WH, Westrich HR, Arnold GW, Banfield JF (1989a) The surface chemistry of dissolving labradorite feldspar. Geochim Cosmochim Acta 53:821-832
Casey WH, Westrich HR, Massis T, Banfield JF, Arnold GW (1989b) The surface of labradorite feldspar after acid hydrolysis. Chem Geol 78:205-218
Chester JS, Lenz SC, Chester FM, Lang RA (2004) Mechanisms of compaction of quartz sand at diagenetic conditions. Earth Planetary Sci Lett 220:435-451
Cherniak DJ, Watson EB (2003) Discussion on zircon. Rev Mineral Geochem 53:469-500
Clark C, Schmidt-Mumm A, Faure K (2005) Timing and nature of fluid flow and alteration during mesoproterozoic shear zone formation, Olary domain, South Australia. J Metamorph Geol 23:147-164
Cole DR, Larson PB, Riciputi LR, Mora CI (2004) Oxygen isotope zoning profiles in hydrothermally altered feldspars: estimating the duration of water-rock interaction. Geology 32:29-32
Cookson IC, Singleton DP (1954) The preparation of translucent fossils by treatment with hydrofluoric acid. Geol Soc Australia News Bull 2:1-2
Dewers T, Ortoleva P (1989) Mechano-chemical coupling in stressed rocks. Geochim Cosmochim Acta 53:1243-1258
Dove PM, Rimstidt JD (1994) Silica-water interactions. Rev Mineral 29:259-308
Dungan MA (1979) Bastite pseudomorphs after orthopyroxene, clinopyroxene, and tremolite. *In:* Serpentine Mineralogy, Petrology, and Paragenesis. Proceedings of a symposium sponsored by the Mineralogical Association of Canada, held on October 25, 1978. Wicks FJ (ed), Can Mineral 17:729-740
Eda K, Chin K, Sotani N, Whittingham MS (2005) Hydrothermal synthesis of potassium molybdenum oxide bronzes: structure-inheriting solid-state route to blue bronze and dissolution/deposition route to red bronze. J Solid State Chem 178:158-165
Eda K, Uno Y, Nagai N, Sotani N, Chen C, Whittingham MS (2006) Structure-inheriting solid-state reactions under hydrothermal conditions. J Solid State Chem 179:1453-1458
Engvik AK, Putnis A, Fitz Gerald JD, Austrheim H (2008) Albitisation of granitic rocks: The mechanism of replacement of oligoclase by albite. Can Mineral 46:1401-1415
Engvik AK, Golla-Schindler U, Berndt-Gerdes J, Austrheim H, Putnis A (2009) Intragranular replacement of chlorapatite by hydroxy-fluor-apatite during metasomatism. Lithos doi:10.1016/j.lithos.2009.02.005
Essene EJ (1983) Solid solutions and solvi among metamorphic carbonates with applications to geologic thermobarometry. Rev Mineral Geochem 11:77-96
Ewing RC (2005) Plutonium and "minor" actinides: safe sequestration. Earth Planet Sci Lett 229:165-181
Fletcher R, Merino E (2001) Mineral growth in rocks: kinetic-rheological models of replacement, vein formation and syntectonic crystallization. Geochim Cosmochim Acta 65:3733-3748
Folk RL (1955) Note on the significance of "turbid" feldspars. Am Mineral 40:356
Fontana MG (1986) Corrosion Engineering (3rd ed). McGraw-Hill, New York
Fortey AJ (1981) Micromorphological studies of the corrosion of gold alloys. Gold Bull 14:25-35
Geisler T, Seydoux-Guillaume A-M, Wiedenbeck M, Wirth R, Berndt J, Zhang M, Mihailova B, Putnis A, Salje EKH, Schlüter J (2004) Periodic pattern formation in hydrothermally treated, metamict zircon. Am Mineral 89:1341-1347
Geisler T, Pöml P, Stephan T, Janssen A, Putnis A (2005a) Experimental observation of an interface-controlled pseudomorphic replacement reaction in natural crystalline pyrochlore. Am Mineral 90:1683-1687
Geisler T, Seydoux-Guillaume A-M, Poeml P, Golla-Schindler U, Berndt J, Wirth R, Pollok K, Janssen A, Putnis A (2005b) Experimental hydrothermal alteration of crystalline and radiation-damaged pyrochlore. J Nucl Mater 344:17-23
Geisler T, Schaltegger U, Tomaschek T (2007) Reequilibration of zircon in aqueous fluids and melts. Elements 3:43-50
Geisler T, Janssen A, Stephan T, Berndt J, Putnis A (2008) A new model for nuclear waste borosilicate glass alteration. Materials Research Society Fall Meeting, Boston, Scientific Basis for Nuclear Waste Management XXXII, Q3.1 (available online: *http://www.mrs.org/s_mrs/doc.asp?CID=16988&DID=217328*).
Glikin AE (2008) Polymineral-Metasomatic Crystallogenesis. Springer
Glover ED, Sippel RF (1962) Experimental pseudomorphs: replacement of calcite by fluorite. Am Mineral 47:1156-1165
Gratier JP, Guiguet R (1986) Experimental pressure solution-deposition on quartz grains: The crucial effect of the nature of the fluid. J Struct Geol 8:845-856
Grayson JF (1956) The conversion of calcite to fluorite. Micropaleo 2:71-78
Groen JC, Craig JR, Rimstidt JD (1990) Gold-rich rim formation on electrum grains in placers. Can Mineral 28:207-228
Hetherington CJ, Harlov DE (2008) Metasomatic thorite and uraninite inclusions in xenotime and monazite from granitic pegmatites, Hidra anorthosite massif, southwestern Norway: Mechanics and fluid chemistry. Am Mineral 93:806-820
Harlov DE, Förster H-J, Nijland TG (2002) Fluid-induced nucleation of (Y+REE)-phosphate minerals within apatite: Nature and experiment. Part I. Chlorapatite. Am Mineral 87:245-261

Harlov DE, Förster H-J (2003) Fluid-induced nucleation of (Y+REE)-phosphate minerals within apatite: Nature and experiment. Part II. Fluorapatite. Am Mineral 88:1209-1229

Harlov DE, Wirth R, Förster H-J (2005) An experimental study of dissolution-reprecipitation in fluorapatite: Fluid infiltration and the formation of monazite. Contrib Mineral Petrol 150:268-286

Hellmann R, Penisson J-M, Hervig RL, Thomassin J-H, Abrioux M-F (2003) An EFTEM/HRTEM high resolution study of the near surface of labradorite feldspar altered at acid pH: evidence for interfacial dissolution-reprecipitation. Phys Chem Mineral 30:192-197

Hough RM, Butt CRM, Reddy SM, Verrall M (2007) Gold nuggets: supergene or hypogene? Aust J Earth Sci 54:959-964

Hövelmann J, Putnis A, Geisler T, Schmidt BC, Golla-Schindler U (2009) The replacement of plagioclase feldspars by albite: observations from hydrothermal experiments. Contrib Mineral Petrol (in press)

Jamtveit B, Yardley BWD (1997) Fluid flow and transport in rocks: an overview *In:* Fluid Flow and Transport in Rocks: Mechanisms and Effects. Jamtveit B, Yardley BWD (eds) Chapman and Hall, London, p 1-14

Jamtveit B, Putnis CV, Malthe-Sørenssen A (2009) Reaction induced fracturing during replacement processes. Contrib Mineral Petrol 157:127-133

Janssen A, Putnis A, Geisler T, Putnis CV (2008) The mechanism of experimental oxidation and leaching of ilmenite in acid solution. Proc. Ninth International Congress for Applied Mineralogy. pp 503-506.

Johannes W, Koepke J, Behrens H (1994) Partial melting reactions of plagioclases and plagioclase-bearing systems. *In:* Feldspars and Their Reactions. Parsons I (ed) NATO ASI series, Kluwer Academic Publishers, Dordrecht, p 161-194

John T, Schenk V (2003) Partial eclogitisation of gabbroic rocks in a late Precambrian subduction zone (Zambia): prograde metamorphism triggered by fluid infiltration. Contrib Mineral Petrol 146:174-191

John T, Scherer EE, Haase K, Schenk V (2004) Trace element fractionation during fluid-induced eclogitisation in a subducting slab: trace element and Lu-Hf-Sm-Nd isotope systematics. Earth Planet Sci Lett 227:441-456

Kasioptas A, Perdikouri C, Putnis CV, Putnis A (2008) Pseudomorphic replacement of single calcium carbonate crystals by polycrystalline apatite. Mineral Mag 72:77-80

Kostenko O, Jamtveit B, Austrheim H, Pollok K, Putnis CV (2002) The mechanism of fluid infiltration in peridotites at Almklovdalen, western Norway. Geofluids 2:203-215

Kowacz M, Putnis A (2008) The effect of specific background electrolytes on water structure and solute hydration: Consequences for crystal dissolution and growth. Geochim Cosmochim Acta 72:4476-4487

Labotka TC, Cole DR, Fayek M, Riciputi LR, Staderman FJ (2004) Coupled cation and oxygen exchange between alkali feldspar and aqueous chloride solution. Am Mineral 89:1822-1825

Lang RA (2004) Mechanisms of compaction of quartz sand at diagenetic conditions. Earth Plan Sci Lett 220:435-451

Lechtmann H (1984) Pre-Columbian surface metallurgy. Sci Am 250:56-63

Lee MR, Parsons I (1997) Dislocation formation and albitization in alkali feldspars from the Shap granite. Am Mineral 82:557-570

Lindgren W (1912) The nature of replacement. Econ Geol 7:521-535

Lippmann F (1980) Phase diagrams depicting the aqueous solubility of mineral systems. Neues Jahrb Mineral Abh 139:1-25

Maliva RG, Siever R (1988) Diagenetic replacement controlled by force of crystallization. Geology 16:688-691

Martin LAJ, Duchene S, Deloule E, Vanderhaeghe O (2008) Mobility of trace elements and oxygen in zircon during metamorphsm: Consequences for geochemical tracing. Earth Planet Sci Lett 267:161-174

Merino E, Nahon D, Wang Y (1993) Kinetics and mass transfer in pseudomorphic replacement: Application to replacement of parent minerals and kaolinite by Al, Fe and Mn oxides during weathering. Am J Sci 293:135-155

Merino E, Dewers T (1998) Implications of replacement for reaction-transport modelling. J Hydrol 209:137-146

Merino E, Banerjee A (2008) Terra Rossa genesis, implications for karst, and eolian dust: a geodynamic thread. J Geol 116:62-75

Mora CI, Riciputi LR, Cole DR, Walker KD (2009) High-temperature hydrothermal alteration of the Boehls Butte anorthosite: origin of bimodal plagioclase assemblage. Contrib Mineral Petrol DOI 10.1007/s00410-008-0364-3

Morisset C-E, Scoates JS (2008) Origin of zircon rims around ilmenite in mafic plutonic rocks of proerozoic anorthosite suites. Can Mineral 46:289-304

Nahon D, Merino E (1997) Pseudomorphic replacement in tropical weathering: evidence, geochemical consequences, and kinetic-rheological origin. Am J Sci 297:393-417

Nakamura M, Shimakita S (1998) Dissolution origin and syn-entrapment compositional change of melt inclusions in plagioclase. Earth Planet Sci Lett 161:119-133

Nakamura M, Watson EB (1981) Experimental study of aqueous fluid infiltration into quartzite: implications for the kinetics of fluid redistribution and grain growth driven by interfacial energy reduction. Geofluids 1:73-89

Niedermeier DRD, Putnis A, Geisler T, Golla-Schindler U, Putnis CV (2009) The mechanism of cation and oxygen exchange in alkali feldspars under hydrothermal conditions. Contrib Mineral Petrol 157:65-76

Oliver NHS, Cleverley JS, Mark G, Pollard PJ, Fu B, Marshall LJ, Rubenach MJ, Williams PJ, Baker T (2004) Modeling the role of sodic alteration in the genesis of iron oxide-copper-gold deposits, Eastern Mount Isa block, Australia. Econ Geol 99:1145-1176

O'Neil JR, Taylor HP (1967) The oxygen isotope and cation exchange chemistry of feldspars. Am Mineral 52:1414-1437

Orville PM (1962) Alkali metasomatism and feldspars. Nor Geol Tidsskr 51:283-316

Orville PM (1963) Alkali ion exchange between vapor and feldspar phases. Am J Sci 261:201-237

Parsons I, Lee MR (2009) Mutual replacement reactions in alkali feldspars I: microtextures and mechanisms. Contrib Mineral Petrol DOI 10.1007/s00410-008-0355-4

Parsons I, Magee CW, Allen CM, Shelley JMG, Lee MR (2009) Mutual replacement reactions in alkali feldspars II: trace element partitioning and geothermometry. Contrib Mineral Petrol DOI 10.1007/s00410-008-0358-1

Passchier CW, Trouw RAJ (1996) Microtectonics. Springer, New-York

Pasteris JD, Ding DY (2009) Experimental fluoridation of nanocrystalline apatite. Am Mineral 94:53-63

Pattison DRM (1994) Are reversed Fe-Mg exchange and solid solution experiments really reversed? Am Mineral 79:938-950

Pattison DRM, Newton RC (1989) Reversed experimental calibration of the garnet-clinopyroxene Fe-Mg exchange thermometer. Contrib Mineral Petrol 101:87-103

Perez R, Boles AR (2005) An empirically derived kinetic model for albitization of detrital plagioclase. Am J Sci 305:312-343

Petit J-C, Della Mea G, Dran J-C, Magontheir M-C, Mando PA, Paccagnella A (1990) Hydrated layer formation during dissolution of complex silicate glasses and minerals. Geochim Cosmochim Acta 54:1941-1955

Plümper O, Putnis A (2009) The complex hydrothermal history of granitic rocks; multiple feldspar replacement reactions under sub-solidus conditions. J Petrol (in press)

Pollok K, Lloyd GE, Austrheim H, Putnis A (2008) Complex replacement patterns in garnets from Bergen Arcs eclogites: A combined EBSD and analytical TEM study. Chemie der Erde 68:177-191

Prieto M (2009) Thermodynamics of solid solution-aqueous solution systems. Rev Mineral Geochem 70:47–85

Putnis A (1992) Introduction to Mineral Sciences. Cambridge University Press

Putnis A (2002) Mineral replacement reactions: from macroscopic observations to microscopic mechanisms. Mineral Mag 66:689-708

Putnis A, Hinrichs R, Putnis CV, Golla-Schindler U, Collins LG (2007a) Hematite in porous red-clouded feldspars: evidence of large-scale crustal fluid-rock interaction. Lithos 95:10-18

Putnis A, Putnis CV (2007) The mechanism of reequilibration of solids in the presence of a fluid phase. J Solid State Chem 180:1783-1786

Putnis CV, Mezger K (2004) A mechanism of mineral replacement: isotope tracing in the model system KCl-KBr-H_2O. Geochim Cosmochim Acta 68:2839-2848

Putnis CV, Tsukamoto K, Nishimura Y (2005) Direct observations of pseudomorphism: compositional and textural evolution at a fluid-solid interface. Am Mineral 90:1909-1912

Putnis CV, Geisler T, Schmid-Beurmann P, Stephan T, Giampaolo C (2007b) An experimental study of the replacement of leucite by analcime. Am Mineral 92:19-26

Putnis CV, Austrheim H, Engvik AK, Putnis A (2008) A mechanism of fluid infiltration through minerals – implications for element mobilisation within the earth. Proc. Ninth International Congress for Applied Mineralogy. pp 625- 629

Renard F, Ortoleva P, Gratier JP (1997) Pressure solution in sandstones: Influence of clays and dependence on temperature and stress. Tectonophysics 280: 257-266

Renard F, Brosse É, Gratier JP (2000) The different processes involved in the mechanism of pressure solution in quartz rich rocks and their interactions. *In:* Quartz Cement in Oil Field Sandstones. Worden R, Morad S (eds) International Association of Sedimentologists Special Publication 29:67-78

Rendón-Angeles JC, Rangel-Hernández YM, López-Cuevas J, Pech-Canul MI, Matamoros-Veloza Z, Yanagisawa K (2005). Proc Joint 20th AIRAPT-43rd EHPRG Conf Science Technol High Pressure. Karlsruhe Germany July 2005, p 1-10

Rendón-Angeles JC, Pech-Canul MI, López-Cuevas J, Matamoros-Veloza Z, Yanagisawa K (2006) Differences on the conversion of celestite in solutions bearing monovalenet ions under hydrothermal conditions. J Solid State Chem 179:3645-3652

Rendón-Angeles JC, Matamoros-Veloza Z, López-Cuevas J, Pech-Canul MI, Yanagisawa K (2008) Stability and direct conversion of mineral barite crystals in carbonted hydrothermal fluids. J Mater Sci 42:2189-2197

Revil A (2001) Pervasive pressure solution transfer in a quartz sand. J Geophys Res 106(B5):8665-8686
Roy DM, Linnehan SK (1974) Hydroxyapatite formed from coral skeletal carbonate by hydrothermal exchange. Nature 247:220-222
Rubatto D, Müntener O, Barnhoorn A, Gregory C (2008) Dissolution-reprecipitation of zircon at low temperature, high pressure conditions (Lanzo Massif, Italy). Am Mineral 93:1519-1529
Rutter EH (1983) Pressure solution in nature, theory and experiment. Jour Geol Soc London 140:725-740
Schwarz S, Stöckert B (1996) Pressure solution in siliciclastic HP-LT metamorphic rocks – constraints on the state of stress in deep levels of accretionary complexes. Tectonophys 255:203-209
Seydoux-Guillaume A-M, Wirth R, Ingrin J (2007) Contrasting response of $ThSiO_4$ and monazite to natural irradiation. Eur J Mineral 19:7-14
Shervais JW, Kolesar P, Andreasen K (2005) A field and chemical study of serpentinization – Stonyford, California: Chemical Flux and Mass Balance. Int Geol Rev 47:1-23
Soman A, Tomaschek F, Berndt J, Geisler T, Scherer E (2006) Hydrothermal reequilibration of zircon from an alkali pegmatite of Malawi. Eur J Mineral Suppl 18:132
Spandler C, Hermann J, Rubatto D (2004) Exsolution of thortveitite, yttrialite and xenotime during low-temperature recrystallisation of zircon from New Caledonia, and their significance for trace element incorporation in zircon. Am Mineral 89:1795-1806
Stöckert B, Wachmann M, Kuster M, Bimmermann S (1999) Low effective viscosity during high pressure metamorphism due to dissolution precipitation creep: the record of HP ± LT metamorphic carbonates and siliciclastic rocks from Crete. Tectonophys 303:299-319
Suárez-Orduna R, Rendón-Angeles JC, López-Cuevas J, Yanagisawa K (2004) The conversion of mineral celestite to strontianite under alkaline hydrothermal conditions. J Phys Condens Matter 16:S1331-S1344
Suárez-Orduna R, Rendón-Angeles JC, Yanagisawa K (2007) Kinetic study of the conversion of mineral celestite to strontianite under alkaline hydrothermal conditions. Int J Mineral Process 83:12-18
Taylor HP (1977) Water/rock interactions and the origin of H_2O in granitic batholiths. J Geol Soc (Lond) 133:509-558
Tenailleau C, Pring A, Etschmann B, Brugger J, Grguric BA, Putnis A (2006a) Transformation of pentlandite to violarite under mild hydrothermal conditions. Am Mineral 91:706-709
Tenailleau C, Etschmann B, Ibberson RM, Pring A (2006b) A neutron powder diffraction study of Fe and Ni distributions in synthetic pentlandite and violarite using 60-Ni isotope. Am Mineral 91:1442-1447
Tromans D (2006) Solubility of crystalline and metamict zircon: A thermodynamic analysis. J Nucl Mater 357:221-233
Tsuchiyama A (1985) Dissolution kinetics of plagioclase in the melt of the system diopside-albite-anorthite, and the origin of dusty plagioclase in andesites. Contrib Mineral Petrol 89:1-16
Walther JV, Wood BJ (1984) Rate and mechanism in prograde metamorphism. Contrib Mineral Petrol 88:246-259
Weissbart EJ, Rimstidt JD (2000) Wollastonite: incongruent dissolution and leached layer formation. Geochim Cosmochim Acta 64:4007-4016
Wheeler J (1987) The significance of grain-scale stresses in the kinetics of metamorphism. Contrib Mineral Petrol 97:397-404
Wintsch RP (1985) The possible effects of deformation on chemical processes in metamorphic fault zones. *In:* Metamorphic Reactions. Kinetics, Textures and deformation. Thompson AB, Rubie DC (eds) Springer, Berlin, p 251-268
Wintsch RP, Yi K (2002) Dissolution and replacement creep: a significant deformation mechanism in crustal rocks. J Struct Geol 24:1179-1193
Wood BJ, Walther JV (1983) Rates of hydrothermal reactions. Science 222:413-415
Wyart J, Sabatier G (1958) The mobility of silicon and aluminium in feldspar crystals. Bull Soc Franc Mineral Cristallogr 82, 223-226 (In French)
Xia F, Zhou J, Brugger J, Ngothai Y, O'Neill B, Chen G, Pring A (2008) Novel route to synthesize complex metal sulfides: Hydrothermal coupled dissolution-reprecipitation replacement reactions. Chem Mater 20:2809-2817
Xia F, Brugger J, Ngothai Y, O'Neill B, Chen G, Pring A (2009a) Three dimensional ordered arrays of nanozeolites with uniform size and orientation by a pseudomorphic coupled dissolution-reprecipitation replacement route. Adv Mater (in press)
Xia F, Brugger J, Chen G, Ngothai Y, O'Neill B, Putnis A, Pring A (2009b) Mechanism and kinetics of pseudomorphic mineral replacement reactions: A case study of the replacement of pentlandite by violarite. Geochim Cosmochim Acta 73:1945-1969

Reviews in Mineralogy & Geochemistry
Vol. 70 pp. 125-180, 2009

4

Thermodynamic Concepts in Modeling Sorption at the Mineral-Water Interface

Dmitrii A. Kulik

Laboratory for Waste Management
Nuclear Energy and Safety Research Department
Paul Scherrer Institute
CH-5232 Villigen PSI
Switzerland

dmitrii.kulik@psi.ch

INTRODUCTION

Sorption phenomena on mineral-water interfaces (MWI) are known to control the retention of many trace elements and organic substances of environmental concern, as well as the kinetics of mineral dissolution/precipitation (Sposito 1984; Davis and Kent 1990; Dzombak and Morel 1990; Goldberg 1992; Stumm 1992; Morel and Hering 1993; Langmuir 1997; Zachara and Westall 1999; Benjamin 2002b; Ganor et al. 2009; and references therein). However, there is still no agreement about a uniform thermodynamic concept which can account for the relevant physicochemical interactions at MWI and can be implemented in popular geochemical modeling codes. This is a major obstacle for compiling a unified (ad)sorption thermodynamic database, also marred by disagreements in defining standard and reference states for (ad)sorbed species. In turn, confusion stems from the multiplicity of thermodynamic concepts used in the description of sorption phenomena on MWIs. The goal of this contribution is to summarize these concepts in a hope to illuminate ways along which they can eventually be made mutually consistent.

ADSORPTION AND SORPTION

A minor element (M) can be retained on mineral-water interfaces by a number of sorption mechanisms which also depend on time. On a kinetic basis, this can be summarized in the so-called *sorption continuum*. At short reaction times, aqueous complexation competes for M with *ion exchange* on permanently-charged surfaces (such as basal planes or interlayer of clay minerals), as well as the *outer-sphere surface complexation* on (hydr)oxides. These fast processes are (almost) reversible. At longer reaction times, *specific* (*multi-site*) *adsorption* on clay particle edges or (hydr)oxides takes place; this kind of binding into *inner-sphere surface species* often displays an energetic heterogeneity at low total M concentrations in the system and becomes progressively irreversible with time. At still longer reaction times, the incorporation of (ad)sorbed species deeper into mineral structure is expected, especially for carbonates and other host minerals prone to re-crystallization; this ultimately leads to formation of a *solid solution* layer. At rather high dissolved aqueous M concentrations, various *surface* (*co*)*precipitation* phenomena may be observed, in which epitaxial clusters or multiple atomic layers of a new phase are formed on surfaces of the host mineral.

These and other sorption phenomena are nicely summarized in Figure 1. Advanced methods of spectroscopy, microscopy and molecular modeling are usually needed to elucidate sorption mechanisms (Manceau et al. 2002), see also Sherman (2009). Part of this knowledge

1529-6466/09/0070-0004$10.00 DOI: 10.2138/rmg.2009.70.4

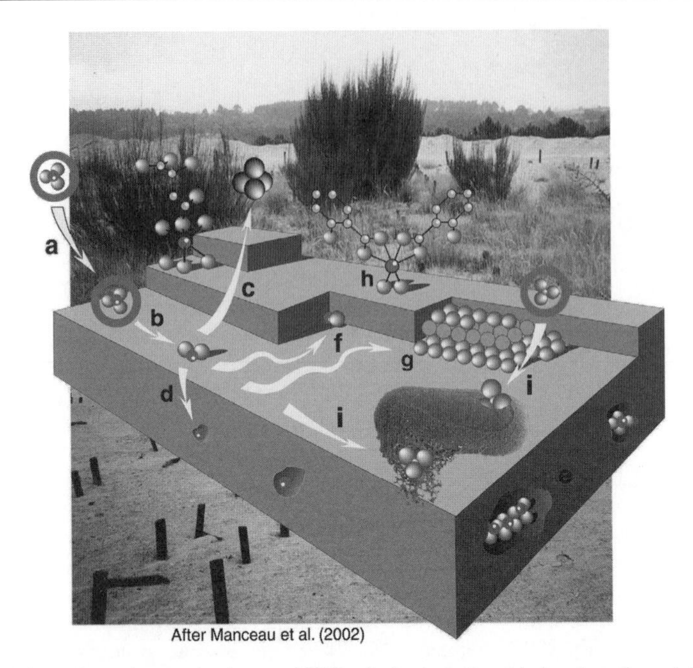

Figure 1. Variety of sorption mechanisms at MWIs: a) physisorption; b) chemisorption; c) detachment; d) absorption or inclusion (impurity ion that has a size and charge similar to those of one of the ions in the crystal); e) occlusion (pockets of impurity that are literally trapped inside the growing crystal); f) attachment; g) hetero-nucleation (epitaxial growth); h) organo-mineral complexation; i) complexation to bacterial exopolymer. After Manceau et al. (2002).

can be utilized in setting up *thermodynamic models* of sorption aimed at predicting the partitioning of M between the dissolved aqueous and the (ad)sorbed states. Such models are necessary in assessing *retention and mobility* of M in the environment or in (radio)toxic waste repositories. The chemical thermodynamic formalism can be applied to sorption phenomena in a uniform manner by defining a sorbed species stoichiometry, its standard and reference states, its standard chemical potential, ideal behavior, and the non-ideality model.

From the thermodynamic viewpoint, the equilibrium in a geochemical system at the Earth's surface is defined by temperature T, pressure P, bulk stoichiometry (composition) of the system b, and the potentially present phases with their stoichiometrically feasible components. In most cases, some solid phases are non-reactive or metastable, so *partial equilibria* must be considered.

To begin, the mineral sorbent will be denoted with S, and the sorbate with B (present in mole amounts n_S and n_B, respectively). The molar mass M_S (in kg·mol^{-1}) or volume V_S of the sorbent must be known together with the mass M_{aq} or volume V_{aq} of the aqueous solution. Often, only the sorbent/water (s/w) ratio is given as the sorbent concentration c_S (in mol·dm^3 or kg·dm^3) together with the aqueous concentration of sorbate c_B (in mol·dm^3) or its molality m_B (in mol·kg(H_2O)$^{-1}$).

The thermodynamic description of adsorption largely depends on what is known about the *reactive specific surface area* A_S of the sorbent. If A_S is known then the *adsorption* model can be developed using A_S as an additional variable of state. If only the sorption capacity q_C (in mol·kg^{-1} or eq·kg^{-1}) is known then *sorption* or *ion exchange* models can be developed, as done e.g., for clay minerals or zeolites. If neither A_S nor q_C is provided then only the chemical mass action can control the partitioning of B between the solid and the aqueous solution. This can be described using classic thermodynamic solubility concepts for solid solution phases (Gamsjäger and Königsberger 2003; Anderson 2005; Bruno et al. 2007), in which no internal metastability factors such as A_S or q_C are involved. These three cases are considered briefly below.

Adsorption: Specific surface is known

If there is evidence that aqueous sorbate B can be immobilized by coordination on the mineral surface of known area then *adsorption reactions* are considered. Usually, it is assumed that such reactions occur in a *surface monolayer*, and the amount of adsorbed B, expressed as a *number density* N_B (in molecules per nm^2, nm^{-2}), is limited by a *site number density* parameter N_C (in nm^{-2}).

For the mineral sorbent S, the specific surface area A_S (in m^2·kg^{-1}) is introduced as an additional variable of state. The total (reactive) surface of sorbent (in m^2) is then

$$A = A_S \cdot n_S \cdot M_S \tag{1}$$

It is convenient to express the adsorbed B concentration as a *surface density*:

$$\Gamma_B = n_{B,\mathrm{ads}} \cdot A^{-1} \text{ (in mol·m}^{-2}\text{)} \tag{2}$$

The number density N_B is connected to surface density of B via Avogadro's number as:

$$\Gamma_B = 1.66054 \times 10^{-6}\, N_B \tag{3}$$

At low total n_B in the system and large sorbent surface A, many surface sites remain unoccupied with B. The concentration of adsorbed B relative to total amount of available sites is called a *surface coverage fraction* θ_B:

$$\theta_B = \frac{\Gamma_B}{\Gamma_C} = \frac{N_B}{N_C} = \frac{n_{B,ads}}{\Gamma_C A} \tag{4}$$

where Γ_C is the site density parameter (in mol·m^{-2}). The θ_B parameter (sometimes called a *fractional coverage*) plays a key role in various equations for adsorption isotherms (see also Kulik 2006a,b).

Using Equations (1) and (2), an adsorbed concentration q_B of B per unit mass of the sorbent (in mol·kg^{-1}) can be defined as

$$q_B = \Gamma_B A_S \tag{5}$$

and in the same way, the *sorption capacity parameter* q_C (in mol kg^{-1}) can be defined as

$$q_C = \Gamma_C A_S \tag{6}$$

Both Equations (5) and (6) actually eliminate the specific surface area and normalize the adsorbed amount to unit mass of the sorbent.

Only sorption capacity is known

In many systems with internal porosity, such as soils, clays, and cements, it is difficult to measure reactive specific surfaces separately in pores, interlayers, and on particles. The experimental data, expressed as a distribution ratio Rd, are usually normalized by sorbent mass (or volume) defined at certain "standard" conditions (e.g., a homoionic form at certain relative humidity). Conversely, only the mass of the sorbent

$$M = n_S \cdot M_S \tag{7}$$

is known in the system, and the sorbed B concentration q_B (in mol·kg^{-1}) can be defined as

$$q_B = n_{B,sorb} \,/\, M$$

and connected to distribution ratio (in $dm^3 \cdot kg^{-1}$):

$$Rd_B = \frac{q_B}{c_B} \tag{8}$$

In most cases, the sorption is limited to *sorption capacity* q_C ($mol \cdot kg^{-1}$) which can be measured experimentally as e.g., cation exchange capacity (CEC) on clays, anion exchange capacity (AEC) on layered double hydroxides, or either of them on zeolites. These parameters are often expressed as the equivalent capacity q_E in charge equivalents per kg to emphasize the electrostatic nature of sorption by ion exchange (more in Helfferich 1962). The *sorbed capacity fraction* θ_B can be defined as

$$\theta_B = \frac{q_B}{q_C} = \frac{n_{B,ads}}{q_C M} \tag{9}$$

Similar to the quantity θ_B defined in Equation (4), the θ_B of Equation (9) can be regarded as the fraction of available sorption sites occupied by the sorbate *B*. Using the formula charge z_B of ionic *B*, the *sorbed equivalent fraction* $\theta_{E,B}$ can be defined as the fraction of available permanent charge compensated by *B* ions:

$$\theta_{E,B} = \frac{z_B q_B}{q_E} = \frac{z_B n_{B,ads}}{q_E M} \tag{10}$$

Sorption capacity is unknown

If sorption capacity is unknown or unlimited then the appropriate quantity to express the sorbed *B* concentration relative to the whole sorbent is the *mole fraction*

$$x_B = \frac{n_B}{\sum_j n_j} \tag{11}$$

where index *j* covers all components of the solid solution phase, including *B*. It is also possible to use q_B (Eqn. 8) with the drawback that the molar mass of the solid M_S will change at high fraction of sorbed component(s). Thermodynamic treatment of solid solutions is based on the mole fraction scale (Eqn. 11) for all end members (see Prieto 2009).

THERMODYNAMIC CONCEPTS

Four different chemical representations of (ad)sorption can be distinguished:

(1) transfer of *B* from aqueous solution to the MWI "monolayer phase";

(2) complexation of *B* with the (surface) site ligand;

(3) formation of a *B*-containing solid solution end member;

(4) transfer of *B* to surface layer on mineral (solid solution) carrier (sorbent).

These representations lead to different thermodynamic models, different representations of non-ideality, and different methods of calculating sorption equilibria. Variants (1) and (2) usually imply knowledge of the specific surface and, at least, of the limiting sorption capacity q_C (or number N_C of surface sites), utilized in equations for the Langmuir or other classic sorption isotherms.

Sorption phase with particle- or pore surface interface

As shown on Figure 2, the concept of surface "monolayer phase" implies that physically adsorbed molecules, in principle, can move within the surface layer. The limiting sorption

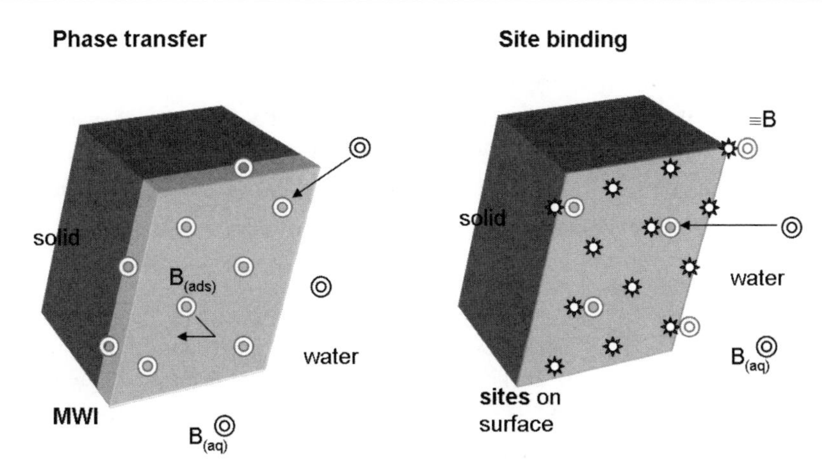

Figure 2. Schematic representations of the MWI "monolayer phase transfer" (left) and the "monolayer site binding" or "surface ligand" (right) concepts. B(aq) is the adsorbate molecule in aqueous phase, B(ads) is the molecule adsorbed in the interface monolayer, and ≡B is the adsorbate molecule bound to the surface ligand.

capacity is then defined by the area and the thickness (i.e., volume) of this "surface solution layer," and by volumes of adsorbed molecules. In contrast, the concept of "site binding" or "surface ligand" represents each surface site as a (immobile) ligand to which the sorbate molecule can be attached via chemical bonding. The limiting sorption capacity in the system is set by a total number of moles of reactive sites n_C, expressed either as site density parameter Γ_C or sorption capacity parameter q_C linked by Equation (6).

Phase transfer reactions in the MWI surface monolayer phase concept represent direct (ad)sorption or exchange between different sorbate species. For the (ad)sorption reaction

$$B(\mathrm{aq}) \leftrightarrow B(\mathrm{ads}) \text{ with equilibrium constant } K_B \tag{12}$$

the equilibrium is determined by the equality of chemical potentials

$$\mu_{B(\mathrm{aq})} = \mu_{B(\mathrm{ads})} \tag{13}$$

which leads to the mass-action expression

$$a_{B(\mathrm{ads})} = a_{B(\mathrm{aq})} \cdot K_B \tag{14}$$

By introducing concentrations and activity coefficients, Equation (14) takes the form

$$c_{\mathrm{B(ads)}} \cdot \gamma_{B(\mathrm{ads})} = c_{\mathrm{B(aq)}} \cdot \gamma_{B(\mathrm{aq})} \cdot K_B \tag{15}$$

where the equilibrium constant K_B

$$-RT \ln K_B = \Delta_r G = \mu^\circ{}_{B(\mathrm{ads})} - \mu^\circ{}_{B(\mathrm{aq})} \tag{16}$$

depends on the concentration scales, standard and reference states chosen for sorbate B in aqueous and adsorbed phases. In Equation (16), $R = 8.31451$ J K^{-1} mol^{-1} is the universal gas constant, T is temperature (K), $\Delta_r G$ is the standard Gibbs energy effect of reaction, and μ° stands for the standard-state chemical potential.

The *(ad)sorption isotherm equation* is the functional dependence of (measurable) B concentrations in the adsorbed and dissolved aqueous states:

$$c_{\mathrm{B(ads)}} = f(c_{\mathrm{B(aq)}}) \tag{17}$$

Using Equation (15), this can be written as

$$c_{\mathrm{B(ads)}} = K_B \cdot c_{\mathrm{B(aq)}} \cdot \gamma_{B\mathrm{(aq)}} / \gamma_{B\mathrm{(ads)}} \tag{18}$$

where, in general, the activity coefficient $\gamma_{B(\mathrm{aq})}$ is a function of composition of aqueous phase, and $\gamma_{B(\mathrm{ads})}$ is a function of composition of the adsorbed layer.

Experimental sorption data are often represented as distribution ratios (Eqn. 8). The thermodynamic equilibrium counterpart of *Rd* is the *distribution coefficient*:

$$Kd_B = \frac{c_{\mathrm{B(ads)}}}{c_{\mathrm{B(aq)}}} = \frac{K_B \gamma_{\mathrm{B(aq)}}}{\gamma_{\mathrm{B(ads)}}} \tag{19}$$

Mineral surfaces in pure water are usually covered with an adsorbed water layer (Parks 1990; Benjamin 2002b). Hence, the primary phase-transfer exchange reaction describes the displacement of *n* water molecules by one *B* molecule:

$$B(\mathrm{aq}) + n\mathrm{H_2O(ads)} \leftrightarrow B(\mathrm{ads}) + n\mathrm{H_2O(aq)} \qquad K_{B,n} \tag{20}$$

with the mass action expression

$$c_{\mathrm{B(ads)}} \cdot \gamma_{B\mathrm{(ads)}} \cdot [x_{\mathrm{H_2O(aq)}} \cdot \gamma_{\mathrm{H_2O(aq)}}]^n = c_{\mathrm{B(aq)}} \cdot \gamma_{B\mathrm{(aq)}} \cdot [c_{\mathrm{H_2O(ads)}} \cdot \gamma_{\mathrm{H_2O(ads)}}]^n \cdot K_{B,n} \tag{21}$$

In dilute aqueous solutions, the product $x_{\mathrm{H_2O(aq)}} \cdot \gamma_{\mathrm{H_2O(aq)}}$ is very close to unity, but it should not be ignored in concentrated solutions. Whether the product $c_{\mathrm{H_2O(ads)}} \cdot \gamma_{\mathrm{H_2O(ads)}}$ can be set constant or ignored, remains unclear.

Reaction (12) can be combined with a reaction for a different sorbate M

$$\mathrm{M(aq)} \leftrightarrow \mathrm{M(ads)} \qquad K_M \tag{22}$$

into a M-*B* exchange reaction (assuming that *m* of *M* molecules displace the same number of H_2O molecules from the surface layer as *n* of *B* molecules)

$$m\mathrm{M(aq)} + nB(\mathrm{ads}) \leftrightarrow m\mathrm{M(ads)} + nB(\mathrm{aq}) \qquad K_{M,B} \tag{23}$$

For this reaction, obviously, $\ln K_{M,B} = m \cdot \ln K_M - n \cdot \ln K_B$. Reactions of this type are often used to describe exchange of similar species on surface or in the volume of the sorbent, be it the ion exchange or the solid solution.

Surface complexation reactions are usually written (Benjamin 2002b) as reactions of a sorbate *B* with a "site ligand" ≡S or simply ≡:

$$\equiv + B(\mathrm{aq}) = \equiv B \qquad K_{\equiv B} \tag{24}$$

It is usually assumed that: (a) each adsorbed molecule *B* occupies one binding site (e.g., monodentate adsorption on a homogeneous surface); (b) adsorption is terminated upon the completion of a monomolecular layer (i.e., *B* does not bind to ≡*B*); and (c) there is no lateral movement of ≡*B* or lateral interactions between the site-bound *B* molecules.

The mass-action expression for Equation (24) is

$$a_{\equiv B} = a_{\equiv} \cdot a_{B(\mathrm{aq})} \cdot K_{\equiv B} \tag{25}$$

or, expanded to concentrations and activity coefficients,

$$c_{\equiv B} \cdot \gamma_{\equiv B} = c_{\equiv} \cdot \gamma_{\equiv} \cdot c_{B(\mathrm{aq})} \cdot \gamma_{B(\mathrm{aq})} \cdot K_{\equiv B} \tag{26}$$

where

$$-RT \ln K_{\equiv B} = \Delta_r G_{\equiv} = \mu^{\circ}_{\equiv B} - \mu^{\circ}_{B(\mathrm{aq})} - \mu^{\circ}_{\equiv} \tag{27}$$

Comparison with Equations (14)-(16) reveals the difference between the phase transfer and the surface complexation approach: in the latter, the activity $a_{\equiv}$, the concentration $c_{\equiv}$, the activity coefficient $\gamma_{\equiv}$, and the standard chemical potential $\mu^{\circ}_{\equiv}$ of the "site ligand" $\equiv$ are involved, whereas in the former, they are not. By the *ad hoc* setting of $a_{\equiv}$ to one and $\mu^{\circ}_{\equiv}$ to zero, $K_{\equiv B}$ can be made numerically equal to K_B. Conversely, serious inconsistencies between existing surface complexation model (SCM) approaches, intrinsic equilibrium constants, and methods of calculation of sorption equilibria can be explained mainly by the lack of conventions and by an insufficient understanding of the role and magnitudes of thermodynamic variables related to this "surface site ligand" and to the underlying concentration scales, standard and reference states (Kulik 2002a,b, 2006b; Sverjensky 2003).

For MWIs on oxides and hydroxides, it is common (Dzombak and Morel 1990; Benjamin 2002b; Mathur and Dzombak 2006) to write surface complexation reactions using the site ligand represented as an amphoteric *surface hydroxyl group* $\equiv$SOH or $\equiv$OH or a surface-coordinated water molecule $\equiv H_2O$. This provides a simple way to describe the development of variable pH-dependent surface charge which can be measured in potentiometric titrations.

In early models of surface complexation (e.g., constant capacitance CC, generalized double layer DL, triple layer TL models), the primary $\equiv$SOH group was assumed to undergo deprotonation and protonation. Accordingly, two reactions are considered in this so-called 2pK formalism:

$$\equiv OH_2^+ \leftrightarrow \equiv OH^0 + H^+(aq) \qquad K_{A1} \tag{28}$$

$$\equiv OH^0 \leftrightarrow \equiv O^- + H^+(aq) \qquad K_{A2} \tag{29}$$

Thus, the hydroxylated mineral surface immersed in pure water is modeled as being populated with "inner-sphere" $\equiv OH^0$, $\equiv OH_2^+$ and $\equiv O^-$ species which together occupy all reactive sites. This scheme reflects the relationships between water molecules, hydronium ion, and hydroxyl ion in bulk water. At zero charge, the surface is expected to be dominated with neutral $\equiv OH^0$ groups.

Introducing acid and base (e.g., HCl and NaOH) into the system allows titrations from neutral to low or high pH (i.e., high or low activity of H^+) in the aqueous solution. In classic CC and DL models, the response of surface speciation and charge to titration is directly accounted for by using Equations (28) and (29). In more advanced models like TL or basic Stern (BS), the formation of outer-sphere surface complexes (ion pairs) is considered in addition to "inner-sphere" species:

$$\equiv O^- + Na^+(aq) \leftrightarrow \equiv O^-Na^+ \qquad K_C \tag{30}$$

$$\equiv OH_2^+ + Cl^-(aq) \leftrightarrow \equiv OH_2^+Cl^- \qquad K_L \tag{31}$$

The distinction between "inner-sphere" and "outer-sphere" surface species is related to the model representation of the EIL (electrified interface layer), see below. Models which include the electrolyte adsorption reactions like (Eqns. 30, 31), in general, are capable of much better reproduction of ionic strength dependence of surface charge and of electrophoretic data (Lützenkirchen 2002a,b; Kallay et al. 2006).

In contrast to weakly (mainly electrostatically) adsorbed electrolyte ions, many cations such as transition metals or rare earth elements, and some polyvalent anions, show only little dependence of their adsorption on ionic strength. These ions are thought to be chemisorbed, i.e., to form chemical bonds to surface atoms of the mineral structure, and are treated as inner-sphere surface complexes forming via hydrolysis

$$\equiv OH^0 + M^{2+}(aq) \leftrightarrow \equiv OM^+ + H^+(aq) \qquad K_{M1} \tag{32}$$

$$\equiv OH^0 + M^{2+}(aq) + H_2O \leftrightarrow \equiv OMOH^0 + 2H^+(aq) \qquad K_{M2} \tag{33}$$

or ligand exchange mechanisms

$$\equiv OH^0 + A^{2-}(aq) \leftrightarrow \equiv A^- + OH^-(aq) \qquad K_{an} \tag{34}$$

In recent developments, such as the CD-MUSIC approach (cf. Hiemstra and van Riemsdijk 2002; van Riemsdijk and Hiemstra 2006) which takes the surface structure of minerals into detailed account, the surface deprotonation reaction has the following general form:

$$\equiv O^? + 2H^+(aq) \leftrightarrow \equiv OH^{?+1} + H^+(aq) \leftrightarrow \equiv OH_2^{?+2} \tag{35}$$

where the charge of surface oxygen (denoted with "?") depends on the mineral structure. Only one of two protonation steps actually occurs within the experimentally accessible pH window. The charge of surface oxygen is determined using the Pauling bond valence principle. For example, in gibbsite α-$Al(OH)_3$, the aluminum (valence 3) is coordinated with six oxygens, so each Al-O bond carries a valence 3/6 = +½. With the formal valence −2 of the oxygen, this gives for three possible surface species on gibbsite:

$$Al\equiv O^{-3/2} + 2H^+(aq) \leftrightarrow Al\equiv OH^{-1/2} + H^+(aq) \leftrightarrow Al\equiv OH_2^{+1/2}$$

In the pH range of 4 to 10, only the right part of the above reaction is necessary:

$$Al\equiv OH^{-1/2} + H^+(aq) \leftrightarrow Al\equiv OH_2^{+1/2} \qquad \log_{10}K_A = 10 \pm 0.5 = pH_{PPZC} \tag{36}$$

where pH_{PPZC} is the pH point of pristine zero charge, i.e., pH at which the surface is neutral in absence of adsorbed ions other than H^+ or OH^-. This picture of the oxide surface implies that even at zero total charge, the gibbsite surface is occupied with charged sites and no neutral inner-sphere surface $\equiv OH^0$ groups are present. The surface charge density is a rather small difference between the relatively large densities of protonated and deprotonated sites. In other words, the MUSIC representation of (hydr)oxide surface is completely different from that assumed in the classic 2pK formalism (Eqns. 28, 29).

However, there are connections between 2pK and 1pK formalisms. Addition of reactions (28) and (29) results in a new reaction

$$\equiv OH_2^+ + \equiv OH^0 \leftrightarrow \equiv OH^0 + \equiv O^- + 2H^+(aq) \qquad K'_A = K_{A1}\,K_{A2} \tag{37}$$

From its mass action expression, it is easy to see that $\log_{10}K'_A = -2pH_{PPZC}$. Reversing this reaction and re-defining reactive groups as $2(\equiv OH_{1/2})^- = \equiv O^- + \equiv OH^0$ and $2(\equiv OH_{3/2})^+ = \equiv OH_2^+ + \equiv OH^0$ one obtains

$$2(\equiv OH_{1/2})^- + 2H^+(aq) \leftrightarrow 2(\equiv OH_{3/2})^+ \qquad K'_A = K^2_A$$

By further dividing this reaction by 2 and representing $\equiv OH_{1/2}$ as $\equiv SO^{-1/2}$, one arrives at

$$\equiv SO^{-1/2} + H^+(aq) \leftrightarrow \equiv SOH^{+1/2} \qquad \log_{10}K_A = pH_{PPZC} \tag{38}$$

which is quite similar to Equation (36). From this exercise, one can conclude that for splitting the 1pK reaction (38) into two reactions as required in the 2pK formalism, an arbitrary parameter is necessary in addition to the experimentally measurable pH_{PPZC}. In the logarithmic scale, using Equation (37), $-\log_{10}K_{A2} - \log_{10}K_{A1} = 2\log_{10}K_A$ or

$$pK_{A2} + pK_{A1} = 2pH_{PPZC} \tag{39}$$

Denoting $\Delta pK_A = pK_{A1} - pK_{A2}$, it becomes clear that Equation (39) can be satisfied by any value of ΔpK_A because

$$pK_{A1} = pH_{PPZC} - ½\Delta pK_A \quad \text{and} \quad pK_{A2} = pH_{PPZC} + ½\Delta pK_A \tag{40}$$

For example, as shown by (Hayes et al. 1991) for the $2pK_A$ TL model, good fits to titration data are indeed possible at different values of ΔpK_A. This arbitrariness is often seen as a major drawback of 2pK formalism because there is no rigorous way to constrain ΔpK_A independently of the chosen EIL electrostatic model. Typical values for ΔpK_A lie between 1 and 9 (Sverjensky and Sahai 1996) (for ETL model these are equal to 5.6 to 8.4, Sahai and Sverjensky 1997). This is not consistent with MUSIC predictions (see Hiemstra and van Riemsdijk 2002) of the pK difference, which results in at least 12 units between the consecutive protonation steps in Equation (35).

In the 1pK framework, the electrolyte adsorption and inner-sphere chemisorption reactions are constructed in the same way as in the 2pK formalism. For instance, for outer-sphere complexes on gibbsite surface,

$$\equiv OH^{-1/2} + M^{z+}(aq) \leftrightarrow \equiv OH^{-1/2}M^{z+} \qquad K_C \tag{41}$$

$$\equiv OH_2^{+1/2} + L^{z-}(aq) \leftrightarrow \equiv OH_2^{+1/2}L^{z-} \qquad K_L \tag{42}$$

Following (Lützenkirchen et al. 1995), the usual assumption for 1:1 electrolytes such as NaCl is $K_C = K_L$ (symmetric electrolyte adsorption). Magnitudes of K_C and K_L are small (between 0.1 and 30) because of largely non-specific electrostatic nature of outer-sphere surface complexes.

A special case of surface complexation is ion exchange on permanently-charged surfaces of clay minerals. This is done in non-electrostatic models (e.g., Bradbury and Baeyens 2006) using an X^- ligand representing a permanent-charge site:

$$X^- + Na^+(aq) \leftrightarrow XNa \qquad K_{XNa} \tag{43}$$

$$XNa + M1^+(aq) \leftrightarrow XM1^0 + Na^+(aq) \qquad K_{Na\text{-}M1} \tag{44}$$

$$2XNa + M2^{2+}(aq) \leftrightarrow X_2M2^0 + 2Na^+(aq) \qquad K_{Na\text{-}M2} \tag{45}$$

The direct binding constant K_{XNa} is usually assigned a large arbitrary value (Fletcher and Sposito 1989) to make sure that no X^- sites remain free; in contrast, the exchange (selectivity) constants like $K_{Na\text{-}M1}$ are small. Several different conventions are in use to represent activities of sorbed components in heterovalent exchange reactions such as Equation (45), and one has to be careful with the magnitude of $K_{K\text{-}M2}$ (Helfferich 1962; Sposito 1984; Dzombak and Hudson 1995). The main effect of permanent charge is a competition of sorbed cations because of the limited sorption capacity q_C or equivalent capacity q_E, which is a property of the clay mineral (see below).

Solid solution reactions can provide an alternative thermodynamic description of sorption in a host mineral by the ion exchange mechanism in clays (Tardy and Fritz 1981), zeolites, or by the co-precipitation (Glynn 2000) when the sorption capacity is ill-defined, unlimited, or unknown. All end members in the solid solution phase are treated thermodynamically in the same way using the standard and reference states of a pure end member (Anderson 2005). The aqueous hydrolysis of a pure solid MLH can be expressed via the reaction

$$M_{\nu_M}L_{\nu_L} \cdot \nu_w H_2O(\text{solid}) = \nu_M M^{Z_M+} + \nu_L L^{Z_L-} + \nu_w H_2O \tag{46}$$

with a standard molar Gibbs energy and a thermodynamic solubility constant given by

$$\Delta G^\circ = \nu_M \mu^\circ_M + \nu_L \mu^\circ_L + \nu_w \mu^\circ_{H_2O} - \mu^\circ_{MLH} = -RT \ln \frac{a_M^{\nu_M} a_L^{\nu_L} a_{H_2O}^{\nu_w}}{a_{MLH}^{(solid)}} = -RT \ln K_{S,MLH}$$

Writing a reaction similar to (46) for a binary solid solution of BL and CL with a mole fraction x of the CL end member leads to a mass action law expression

$$\left(a_{B+}^{(aq)}\right)^{1-x} \left(a_{C+}^{(aq)}\right)^{x} a_{L-}^{(aq)} = K_{S,BCL}^{(x)} \tag{47}$$

where $K_{S,BCL}^{(x)}$ is an equilibrium solubility product of the solid solution. Equivalently one can write the law of mass action for each end member

$$a_{B+}^{(aq)} a_{L-}^{(aq)} = K_{S,BL} a_{BL} \quad \text{and} \quad a_{C+}^{(aq)} a_{L-}^{(aq)} = K_{S,CL} a_{CL} \tag{48}$$

where $a_{BL} = (1-x) \cdot f_{BL}$ and $a_{CL} = x \cdot f_{CL}$ stand for activities of end members. Equations (47) and (48) can be generalized to many end members (Gamsjäger and Königsberger 2003). In the simple ideal solid solution, activity coefficients f_{BL} and f_{CL} equal to unity.

Ion exchange reactions like Equation (44) can also be applied to solid solutions:

$$BL(\text{solid}) + C^+ = CL(\text{solid}) + B^+ \qquad K_{B\text{-}C}$$

Here $K_{B\text{-}C} = K_{S,BL}/K_{S,CL}$. The ligand L may also denote a charged structural core in zeolite or in clay mineral. The isotherm equation for C^+ "sorption" into a $(B,C)L$ solid solution follows from Equation (48):

$$x = \left\{ \frac{a_{B+}^{(aq)}}{a_{C+}^{(aq)}} \frac{f_{CL}}{K_{B-C} f_{BL}} + 1 \right\}^{-1} \tag{49}$$

This isotherm is linear with respect to a_{C+} when C is mixed into an ideal sparingly soluble solid solution, or when C is a trace component ($x << 1$) in a non-ideal solid solution. In the latter case, $1 - x \rightarrow 1$ and $f_{BL} \rightarrow 1$; $a_{B+}^{(aq)}$ is independent of $a_{C+}^{(aq)}$; and $f_{CL} \rightarrow const = f_{CL}^{\infty}$. For instance, in a regular mixing model, $\ln f_{CL} = (1-x)^2\alpha_0$ and $f_{CL}^{\infty} = \exp(\alpha_0)$. Defining a Henry constant $K_{H,CL} = (K_{B-C})/(a_{B+}^{(aq)} f_{CL}^{\infty})$ and re-arranging Equation (49) leads to a linear dependence

$$x = a_{C+} K_{H,CL} \tag{50}$$

More about thermodynamics of solid solutions can be found in textbooks (e.g., Anderson 2005) and in this book (Prieto 2009).

Classic adsorption isotherm equations

A thermodynamic (ad)sorption model of MWI must reproduce the experimentally measurable dependence of (ad)sorbed concentration $c_{B(\text{ads})}$ of component B on its aqueous dissolved concentration $c_{B(\text{aq})}$ (see Eqn. 17) – the *(ad)sorption isotherm.* In systems with dilute B and large reactive MWI area, the dependence of $c_{B(\text{ads})}$ on $c_{B(\text{aq})}$ is usually linear. Upon increasing total B or decreasing MWI area, the slope of isotherm gradually becomes less steep, finally approaching zero at the level set by the sorption capacity parameter q_C (or the limiting density parameter Γ_C).

Linear isotherm. Consider again a phase transfer reaction in Equation (12) on the basis of the molecular kinetic theory (e.g., De Boer 1968). Assume that the B molecule has a large residence time $\tau >> 0$ in the interface monolayer, represented for simplicity as a 2-D planar grid made of identical cells of total density Γ_C (in moles per unit area). Each cell of the grid can be empty or occupied with one B molecule. At low bulk B concentration, the incident flux F from the bulk solution to the interface is very small. In equilibrium, the density of cells (in moles per unit area) occupied by B

$$\Gamma_B = \tau \cdot F \cdot p_B \tag{50}$$

is also very small, and such cells are on average separated by very many empty cells (see Fig. 3, left). The sticking probability p_B for the next B molecule, defined as a ratio of number of nodes unoccupied with B to the total number of nodes, is very close to 1. This holds also when one B molecule sticks on n adjacent cells (multidentate binding), i.e., the "denticity" plays no role at such conditions. Note that p_B is practically independent of Γ_C as long as the latter is large compared to Γ_B.

Since the flux F is proportional to bulk concentration of B, and the residence time τ is proportional to the affinity of B molecules to sorbent surface, in the case of adsorption on MWI, Equation (50) can be replaced (at $p_B = 1$) with

$$\Gamma_B = K_B' \cdot m_B \tag{51}$$

where K_B' is the chemical affinity constant, and m_B is the molality of B in aqueous phase. This is a linear isotherm equation with the constant K_B' in units of (kg H_2O)$\cdot$m^{-2}. After defining the

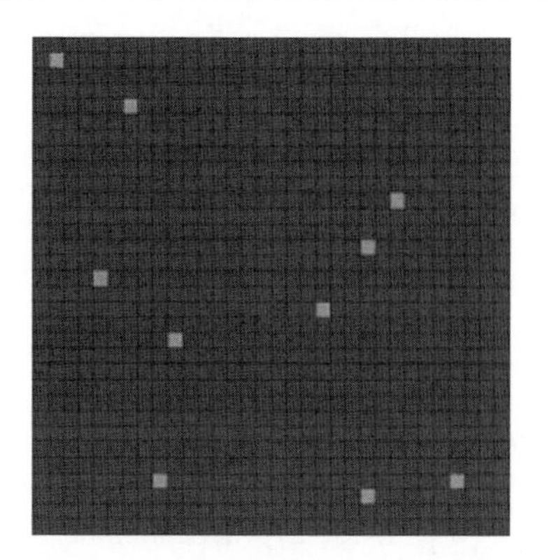
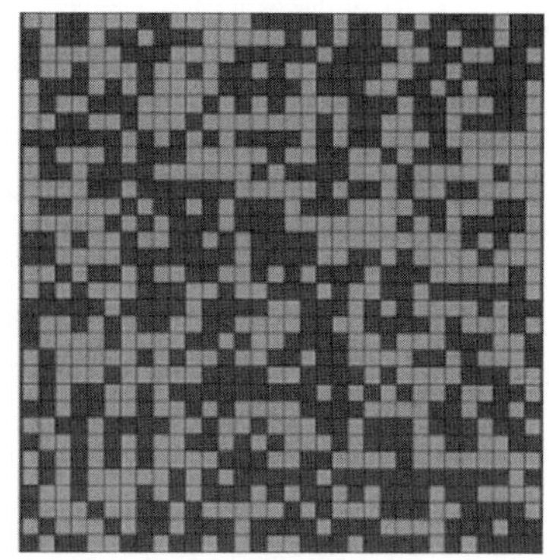

Figure 3. Sticking probability p_B in surface monolayer, illustrated with a 2D grid: at low surface coverage, $p_B = 0.99$ (1% of sites occupied), left; at high surface coverage, $p_B = 1 - \theta_B = 0.5$, right.

standard-state absorbed density as Γ_o (e.g., 1 mol·m^{-2}) and the standard state molality as m_o (1 mol·(kg H_2O)$^{-1}$), Equation (51) is converted into dimensionless concentrations

$$\frac{\Gamma_B}{\Gamma_o} = K_B \cdot \frac{m_B}{m_o} \tag{52}$$

where K_B is the equilibrium constant of Equation (12). Taking the infinite dilution reference states for B on the surface and in the bulk aqueous solution, both activity coefficients in the dilute system approach unity ($\gamma_{B(aq)} \rightarrow 1$; $\gamma_{B(ads)} \rightarrow 1$), and activities become

$$a_{B(ads)} = K_B^o \cdot a_{B(aq)} \qquad \text{where } K_B^o \rightarrow K_B,\ a_{B(ads)} \rightarrow \frac{\Gamma_B}{\Gamma_o},\ a_{B(aq)} \rightarrow \frac{m_B}{m_o} \tag{53}$$

Langmuir isotherm. Consider now the same system but at much higher bulk B concentration, when a large part of surface cells is occupied with B molecules. In addition to large residence time for B adsorbed on the surface $\tau_0 >> 0$ or a high stability of B(ads) species, assume that B does not stick to already adsorbed B(ads), i.e., the residence time $\tau_1 = 0$, or the B-B(ads) complex is absolutely unstable. Assume also that adsorbed B molecule occupies exactly one cell and cannot move on surface from one cell to another, unless B is detached back to the solution and then stuck in another empty surface cell.

Under such circumstances (e.g., Fig. 3, right), the density of unoccupied surface cells is $\Gamma_C - \Gamma_B$ and their fraction, i.e., sticking probability $p_B = (\Gamma_C - \Gamma_B)/\Gamma_C$ or (using Eqn. 4) $p_B = 1 - \theta_B$. The density of occupied cells is then

$$\Gamma_B = \tau_0 \cdot F \cdot (1 - \theta_B) \tag{54}$$

Replacing τ_0 with K'_B, and F with m_B, one obtains

$$\Gamma_B = K'_B \cdot m_B \cdot (1 - \theta_B) \tag{55}$$

Dividing both sides by Γ_C and substituting $K'_B = K_{L,B} \cdot \Gamma_C$ yields $\theta_B = K_{L,B} \cdot m_B \cdot (1 - \theta_B)$ or

$$K_{L,B} m_B = \frac{\theta_B}{1 - \theta_B} \tag{56}$$

This is the *Langmuir isotherm equation* which is often expressed as

$$\Gamma_B = \frac{\Gamma_C K_{L,B} m_B}{1 + K_{L,B} m_B} \tag{57}$$

Using the standard state density Γ_o and the standard state molality m_o, Equation (55) can be converted into thermodynamic relative concentrations:

$$\frac{\Gamma_B}{\Gamma_o} = K'_B \cdot \frac{m_B}{m_o} \cdot (1-\theta_B) \tag{58}$$

Comparing this with Equation (53) expanded into concentrations and activity coefficients

$$\frac{\Gamma_B}{\Gamma_o} y_{B(ads)} = K^o_B \cdot m_B \gamma_{B(aq)} \tag{59}$$

we see that the expression $(1 - \theta_B)^{-1}$ effectively replaces the activity coefficient $y_{B(ads)}$. Hence, the equation

$$\Gamma_B y_{B(ads)} = \Gamma_o K^o_B \cdot m_B \gamma_{B(aq)} \qquad \text{where } y_{B(ads)} = (1-\theta_B)^{-1} \tag{60}$$

is a general thermodynamic form of Langmuir isotherm for Equation (12) taking the standard state of B(ads) at surface density Γ_o and the reference state of B(ads) at infinite dilution (Kulik 2006a). Equation (60) reduces to Langmuir isotherm Equation (56) if $\Gamma_o = \Gamma_C$ and $K_{L,B} = K^o_B \cdot \gamma_{B(aq)} / m_o$, and to a linear isotherm when $m_B \rightarrow 0$ and $\Gamma_B << \Gamma_C$ (or $\theta_B \rightarrow 0$).

In the above derivation, it was assumed that each empty cell in the surface monolayer is reactive with respect to sorbate B. Site-binding reactions like Equation (24) are commonly considered for the specific inner-sphere adsorption on (hydr)oxides or for ion exchange on permanent-charge surfaces. In both cases, not all sites in the surface monolayer are reactive. For instance, the average density of permanent negative charge on basal surfaces of montmorillonite is close to one charge per nm^{-2} (1.66×10^{-6} $eq \cdot m^{-2}$) which corresponds to CEC $q_E \approx 1$ $eq \cdot kg^{-1}$ (Bradbury and Baeyens 2006). This is much less than the density of monolayer coverage with H_2O which is ca. 12 nm^{-2} (Davis and Kent 1990; Parks 1990). What changes in the derivation of Langmuir isotherm for MWI if the limiting site density parameter $\Gamma_{C,B}$ for B adsorption would be much less than the complete cell density Γ_C in the surface monolayer?

From the molecular-kinetic reasoning, $\Gamma_B = \tau \cdot F \cdot p_B$, but the sticking probability p_B at very low $\Gamma_B << \Gamma_{C,B}$ is now very close to the constant fraction of reactive sites to all sites $\Gamma_{C,B}/\Gamma_C$. Equation (51) takes the form

$$\Gamma_B = K'_B \cdot m_B \cdot \frac{\Gamma_{C,B}}{\Gamma_C} \tag{61}$$

The constant ratio $\Gamma_{C,B}/\Gamma_C$ can be included into equilibrium constant $K''_B = K'_B \cdot \Gamma_{C,B} / \Gamma_C$ which contains now contributions from both chemical affinity and site selectivity. Actually, the limiting site density $\Gamma_{C,B}$ cannot be determined solely from adsorption experimental data measured in dilute systems. In a more concentrated system,

$$p_B = \frac{\Gamma_{C,B} - \Gamma_B}{\Gamma_C} = \frac{\Gamma_{C,B}}{\Gamma_C} \cdot \frac{\Gamma_{C,B} - \Gamma_B}{\Gamma_{C,B}} \tag{62}$$

which leads to

$$\Gamma_B = K'_B \cdot m_B \cdot \frac{\Gamma_{C,B}}{\Gamma_C} \cdot \frac{\Gamma_{C,B} - \Gamma_B}{\Gamma_{C,B}} \tag{63}$$

Using $K''_B = K'_B \cdot \Gamma_{C,B} / \Gamma_C$ and defining $\theta_{C,B} = \Gamma_B / \Gamma_{C,B}$ as the fractional coverage of B on B-reactive sites, one comes back to the Langmuir isotherm:

$$\Gamma_B = K''_B \cdot m_B \cdot (1-\theta_{C,B}) \tag{64}$$

This time, because $\Gamma_{C,B} < \Gamma_C$, we expect the site saturation effect to occur at much lower bulk m_B concentration than in the case of Equation (55). Converting the Langmuir isotherm into the general thermodynamic form as done in deriving Equation (60) yields

$$\Gamma_B = (1-\theta_{C,B})\frac{\Gamma_{C,B}}{\Gamma_C}\Gamma_o K_B^o \cdot m_B \gamma_{B(aq)} \tag{65}$$

where $K_B^o \rightarrow K_B$ and K_B refers to K'_B. In principle, it is possible to regard the expression $\Gamma_C / (\Gamma_{C,B}(1-\theta_{C,B})) = \Gamma_C / (\Gamma_{C,B} - \Gamma_B)$ as the "activity coefficient" $y_{B(ads)}$. However, this is inconvenient for two reasons: (1) at the infinite dilution limit, $\gamma_{B(aq)} \rightarrow 1$, but $y_{B(ads)} \rightarrow \Gamma_C/\Gamma_{C,B}$ which is a constant that can be very different from 1; and (2) two parameters for $y_{B(ads)}$ instead of one are used. So, it is more convenient to take the factor $\Gamma_{C,B}/\Gamma_C$ into the equilibrium constant (i.e., $K_B^o \rightarrow K_B$ where K_B refers to K''_B) to have $y_{B(ads)} \rightarrow 1$ at infinite dilution, and Equation (65) taking the form identical to Equation (60). A practical consequence of this decision is that it makes no sense in the models to account for the concentration of sites that are non-reactive with respect to the sorbate *B*.

Langmuir isotherm and the site mole balance. Another (more popular) way to derive the Langmuir isotherm is to consider the site-binding reaction

$$\equiv S_{free} + B_{aq} \leftrightarrow \equiv SB \qquad K_{L,B} \tag{66}$$

where $\equiv S_{free}$ denotes the free *B*-reactive sites on the solid sorbent which is present in the system in a concentration c_S (in kg·dm^{-3}). The mass-action expression is

$$K_{L,B} \cdot c(\equiv S_{free}) \cdot c(B_{aq}) = c(\equiv SB) \tag{67}$$

(all concentrations in mol·dm^{-3}), and the balance of sites is

$$c(\equiv S_{free}) = c(\equiv S)_{tot} - c(\equiv SB) \tag{68}$$

Substituting Equation (68) into Equation (67) leads to $K_{L,B}[c(\equiv S)_{tot} - c(\equiv SB)] \cdot c(B) = c(\equiv SB)$ and, after re-arrangement,

$$c(\equiv SB) = \frac{K_{L,B}c(B) \cdot c(\equiv S)_{tot}}{1 + K_{L,B}c(B)} \tag{69}$$

Dividing Equation (69) by c_S and defining $q_B = c(\equiv SB)/c_S$ and $q_C = c(\equiv S)_{tot}/c_S$ yields

$$q_B = \frac{q_C K_{L,B} c(B_{aq})}{1 + K_{L,B} c(B_{aq})} \tag{70}$$

Because q_C is the limiting sorption capacity, the ratio q_B/q_C is, in fact, the fractional coverage θ_B. Equation (70) is the Langmuir isotherm, in a more familiar form written as

$$K_{L,B} c(B_{aq}) = \frac{\theta_B}{1-\theta_B} \tag{71}$$

The above derivation (see also Stumm 1992; Morel and Hering 1993; Benjamin 2002b) seems also to justify the "site mole balance" method for calculating surface speciation, widely used in FITEQL, MINTEQA2, PHREEQC, CHESS, and many similar speciation codes.

In some cases, the experimental data for specific cation adsorption on MWI can be closely described using the Langmuir isotherm equation. For example, the data for metastable uranyl sorption on goethite (Giammar 2001; Giammar and Hering 2001) obtained by titration nicely conform with Equation (70) with $K_{L,U} = 6.53 \times 10^{-7}$ M^{-1} and the sorption capacity $q_{C,U} = 114.4$ µmol·g^{-1} (density parameter $\Gamma_{C,U} = 2.71 \times 10^{-6}$ mol·m^{-2} or number density parameter $N_{C,U} = 1.632$ nm^{-2}), Figure 4.

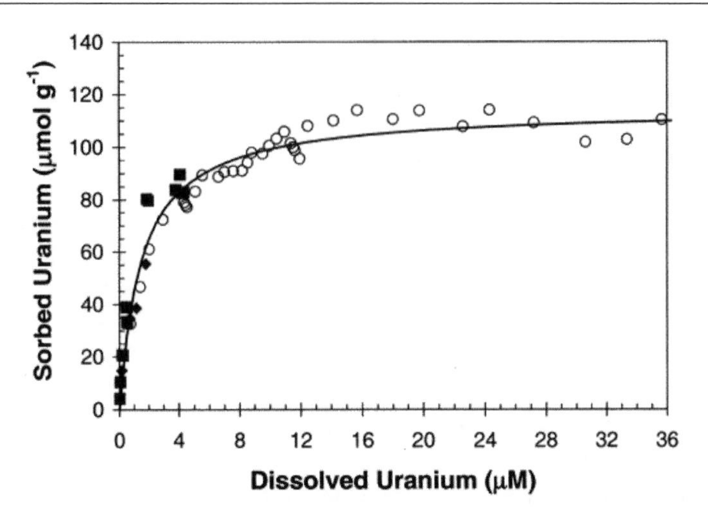

Figure 4. Metastable sorption of uranyl on goethite at pH 6 and I = 0.1 M. Symbols: data from (■) batch equilibration (0.1-2.5 g·dm³ goethite), (♦) endpoints of kinetics experiments (0.010-0.035 g·dm³ goethite), and (○) incremental loading experiments (0.10 g·dm³ goethite); (—) Langmuir isotherm. [Used with permission from the American Chemical Society, from Giammar and Hering (2001), *Environ Sci Technol*, Vol. 35, Fig. 1, p. 3334.]

Competitive Langmuir isotherm. If two or more sorbates (indexed with j) can bind monodentately, each with a residence time τ_j to surface sites of the same kind of limiting density Γ_C then, at the dilution limit, each sorbate will obey an own linear isotherm

$$\frac{\Gamma_j}{\Gamma_o} = K_j \cdot \frac{m_j}{m_o} \tag{72}$$

(see also Eqns. 50 to 53) as if there were no other competing species on surface sites. However, when a significant fraction of sites becomes occupied, the sticking probability for a next sorbate molecule of any j-th type will equal the fraction of unoccupied sites

$$1-\theta_\Sigma \quad \text{where} \quad \theta_\Sigma = \frac{\Gamma_\Sigma}{\Gamma_C} \quad \text{where} \quad \Gamma_\Sigma = \sum_j \Gamma_j \tag{73}$$

Similar to Equation (55), this leads to the adsorption isotherm

$$\Gamma_j = K'_j \cdot m_j \cdot (1-\theta_\Sigma) \tag{74}$$

Dividing both sides by Γ_C and substituting $K'_j = K_{L,j} \cdot \Gamma_C$ yields

$$\Gamma_j = K_{L,j} \cdot m_j \cdot (\Gamma_C - \Gamma_\Sigma) \quad \text{or} \quad K_{L,j} m_j = \frac{\theta_j}{1-\theta_\Sigma} \tag{75}$$

which is the competitive Langmuir isotherm. Transforming Equation (74) to relative concentrations,

$$\frac{\Gamma_j}{\Gamma_o} \cdot \frac{\Gamma_C}{\Gamma_C - \Gamma_\Sigma} = K_j^o \cdot \frac{m_j}{m_o} \cdot \gamma_{j(aq)} \tag{76}$$

Compare this with Equation (59) to see that all surface species competing for a specific site refer to the same standard state and are affected by the same "activity coefficient" $1 - \theta_\Sigma$. The generalized form of the competitive Langmuir isotherm follows from Equation (76):

$$\Gamma_j y_{\Sigma(ads)} = \Gamma_o K_j^o \cdot m_j \gamma_{j(aq)} \quad \text{where} \quad y_{\Sigma(ads)} = (1-\theta_\Sigma)^{-1} \tag{77}$$

Because the competitive Langmuir isotherm can be alternatively derived from the site mole balance in the same way as shown in Equations (66) to (71), it is implicit in surface complexation models implemented in FITEQL, CHESS, PHREEQC and similar speciation codes. This also includes non-electrostatic models of ion exchange based on the sorption capacity q_E (CEC) and permanent charge site density balance.

Generalized Langmuir isotherms for n-dentate adsorption. Binding of a sorbate molecule to two or more structural sites on the sorbent surface is usually described with a generalized reaction and the respective molar balance of sites as shown below:

$$n \equiv S_{free} + B_{aq} = \equiv S_n B \qquad K_{n,B} \tag{78}$$

$$c(\equiv S_{free}) = c(\equiv S)_{tot} - n \cdot c(\equiv S_n B) \tag{79}$$

It follows that the site-balance speciation codes cannot readily model bi- (tri-...) dentate adsorption because the latter obeys isotherm equations different than the competitive Langmuir isotherm (Benjamin 2002a,b). Moreover, the problem with constructing the mass-action expressions for Equation (78) requires answering the question: how the activity of free sites $a(\equiv S_{free})$ is related to the "denticity" n? As noted in (Morel and Hering 1993, p.518), one may be tempted to write the mass action law as

$$c(\equiv S_n B) = K_{n,B} \cdot c(B_{aq}) \cdot c(\equiv S_{free})^n$$

but the exponent n in this expression is doubtful because it covers also single free surface sites to which the n-dentate molecule cannot bind. As seen on Figure 5 (left), if one assumes that a bidentate molecule B can bind only to two closest adjacent free sites (e.g., shown by rectangles), there are only 5 such places (10 sites), and other 14 free sites to which B cannot bind. Hence, there are 24 free sites out of 100, but only 10 of sites are available to maximum 5 bidentate molecules. Hence, at high coverage, the sticking probablilty $p_{B,n}$ for n-dentate B on surface is much less than $(\Gamma_C - n\Gamma_{n,B})/\Gamma_C$.

The sticking probability $p_{B,n} \rightarrow (\Gamma_C - n\Gamma_{n,B})/\Gamma_C$ occurs only when $\equiv S_n B$ can move and rotate fast on the surface; this $p_{B,n}$ leads to an isotherm

$$K_{n,B} a(B_{aq}) = \frac{\theta_{n,B}}{1 - n\theta_{n,B}} \quad \text{where} \quad \theta_{n,B} = \frac{\Gamma_{n,B}}{\Gamma_C} \tag{80}$$

Such lateral movement is hard to imagine for inner-sphere adsorption on (hydr)oxides where

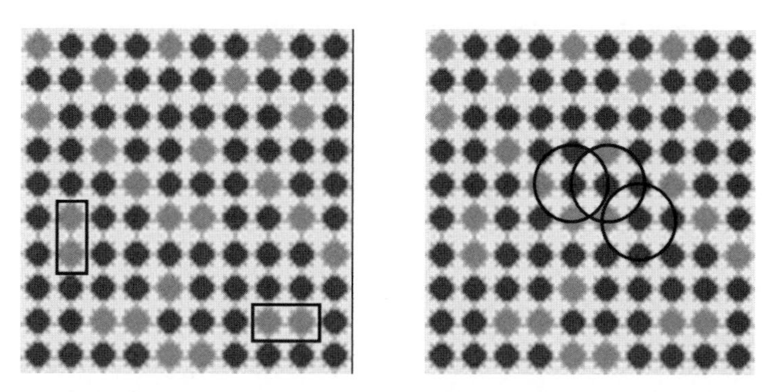

Figure 5. Reduced availability of free sites illustrated for bidentate adsorption (left); overlapping lateral interaction fields of adsorbed species, schematically shown with circles (right).

a strong chemical binding occurs on crystallographic surface sites. In this situation, a Quasi-chemical-approximation (QCA) equation (LaViolette and Redden 2002) seem to provide a better description of immobile n-dentate adsorption, which can be written as:

$$K_{n,B}a(\mathrm{B_{aq}}) = \delta_n \frac{\theta_{n,B}\left(1-\theta_{n,B}\right)^{n-1}}{\left(1-n\theta_{n,B}\right)^n} \tag{81}$$

where δ_n is the shape constant, and $1 \le n \le 4$. Both Equations (80) and (81) reduce to the Langmuir isotherm at $n = 1$. The QCA isotherm can also be converted into a generalized thermodynamic form (Kulik 2006b):

$$\Gamma_{n,B} = y_{L,n,B}^{-1} \cdot \Gamma_o \cdot K_B^o \cdot m_B \gamma_{B(aq)} \tag{82}$$

where the "surface activity correction term" is

$$y_{L,n,B} = \frac{\left(1-\theta_{n,B}\right)^{n-1}}{\left(1-n\theta_{n,B}\right)^n} = \frac{\Gamma_C(\Gamma_C - \Gamma_{n,B})^{n-1}}{(\Gamma_C - n\Gamma_{n,B})^n} \tag{83}$$

It is important that at the infinite dilution limit, $\gamma_{L,n,B} \rightarrow 1$ and Equation (82) reduces to the linear isotherm Equations (52) and (53). This means that the magnitude of equilibrium constant in Equation (82) is not influenced by "denticity" n of the adsorbed B species, so the whole impact of multidentate binding must be ascribed to configurational non-ideality (Eqn. 83). There is no easy way to obtain a mass balance equation for QCA isotherm (LaViolette and Redden 2002), i.e., there is no easy way of implementing it in site-balance-based adsorption models.

Langmuir isotherms for multi-site adsorption. (Ad)sorption experiments often provide evidence of more than one type of surface site when different sorbates do not compete, or when the shape of sorption isotherm at low loadings is not linear. This is usually rationalized using the *multi-site surface binding* concept, more details about which can be found in (Benjamin 2002b). The most popular multi-site models represent the MWI as having a small number of sites that bind the adsorbate strongly, and a much larger number of sites that bind it weakly (Dzombak and Morel 1990). In this two-site case, the adsorbate B is distributed between "weak" sites of Γ_{C1} density and "strong" sites of Γ_{C2} density ($\Gamma_{C1} >> \Gamma_{C2}$). The overall isotherm is a sum of Langmuir isotherms for each site type. Using Equations (63) and (64),

$$\Gamma_B = K''_{L1,B} \cdot m_B \cdot \frac{\Gamma_{C1} - \Gamma_{1,B}}{\Gamma_{C1}} + K''_{L2,B} \cdot m_B \cdot \frac{\Gamma_{C2} - \Gamma_{2.B}}{\Gamma_{C2}} \tag{84}$$

where $K''_{L1,B} << K''_{L2,B}$. This composite Langmuir isotherm can also be readily obtained using site-binding reactions and site molar balances in a way similar to Equations (66)-(71). Using Equation (65), Equation (84) can be recast into the generalized form

$$\Gamma_B = \Gamma_o \cdot [(1-\theta_{1,B})K_{1,B}^o + (1-\theta_{2,B})K_{2,B}^o] \cdot m_B \gamma_{B(aq)} \tag{85}$$

where $K_{1,B}^o < K_{2,B}^o$; both refer to the same standard state at Γ_o and to the infinite dilution. At very low surface coverage, Equation (85) reduces to the linear isotherm

$$\Gamma_B = \Gamma_o \cdot (K_{1,B}^o + K_{2,B}^o) \cdot m_B \gamma_{B(aq)} \tag{86}$$

Additive Equations (84) to (86) can be easily generalized to arbitrary number of surface site types.

The additive multi-site surface complexation model based on Equation (84) seems to provide the basis for popular frameworks, e.g., the CD-MUSIC approach to surface complexation on (hydr)oxides (more in Hiemstra and van Riemsdijk 2002; van Riemsdijk and Hiemstra 2006), or the Bradbury-Baeyens (B&B) quasi-mechanistic model of cation sorption on clay minerals montmorillonite and illite (Bradbury and Baeyens 2006). In the former, surface site types and limiting densities are determined from the crystallographic structure of MWI on dominant faces of (hydr)oxide particles. In the latter, site types are assigned to edges of clay platelets (one strong and two weak cation binding sites with limiting capacities adjusted from H^+ and cation sorption data) and to permanent charge basal surfaces (X- sites with capacity set to experimental CEC values).

In multi-site models, in principle, several sorbates are allowed to compete for at least one type of site (e.g., by including several surface species in the same site molar balance). Then Equations (84) and (85) must be replaced by

$$\Gamma_B = K''_{L1,B} \cdot m_B \cdot (1-\theta_{1,\Sigma}) + K''_{L2,B} \cdot m_B \cdot (1-\theta_{2,\Sigma}) + \ldots \tag{87}$$

$$\Gamma_B = \Gamma_o \cdot [(1-\theta_{1,\Sigma})K^o_{1,B} + (1-\theta_{2,\Sigma})K^o_{2,B} + \ldots] \cdot m_B \gamma_{B(aq)} \tag{88}$$

where $1-\theta_{1,\Sigma} = y^{-1}_{1,\Sigma(ads)} = (\Gamma_{C1} - \Gamma_{1,\Sigma}) / \Gamma_{C1}$ and so on, and $\Gamma_{1,\Sigma}$ is computed for all surface species (including Γ_B) assigned to the given site type. The real challenge here is to decide which sorbate competes for a site with other sorbates, and which site types compete for the same sorbate. This difficulty seems to induce a certain criticism against the complexity of multi-site competitive models.

Frumkin isotherm (see Stumm 1992) takes into account "soft" lateral interactions in the interlayer as opposed to "hard" effect of "denticity" when the adsorbed species blocks n sites from binding other sorbate molecules. "Soft" interactions may include electrostatic attraction/ repulsion, the overlap of hydration shells, etc., all more pronounced at high surface coverage (as shown very schematically on Fig. 5B). The Frumkin isotherm is derived in the same way as the monodentate single-site Langmuir isotherm, but using an additional exponential term containing θ_B and the interaction parameter α_F:

$$K_{F,B} m_B = \frac{\theta_B}{1-\theta_B} \exp(-2\alpha_F \theta_B) \tag{89}$$

Lateral repulsion between adsorbed B species is modeled when $\alpha_F < 0$; lateral attraction – when $\alpha_F > 0$; and at $\alpha_F = 0$, Equation (89) reduces to the Langmuir isotherm (Eqn. 56); both are reduced to the same linear isotherm at very low surface coverage. Equation (89) can be re-written into the generalized form (Kulik 2006a,b) illustrated on Figure 6:

$$\Gamma_B = y^{-1}_{F,B} \cdot \Gamma_o \cdot K^o_B \cdot m_B \gamma_{B(aq)} \quad \text{where} \quad y_{F,B} = \frac{\Gamma_C}{\Gamma_C - \Gamma_B} \exp\left(-2\alpha \frac{\Gamma_B}{\Gamma_C}\right) \tag{90}$$

This equation shows that the Frumkin isotherm is a kind of non-ideal correction for a real behaviour of adsorbed species, and it has no impact on the underlying thermodynamic equilibrium constant if the latter is referenced to the infinite dilution. There are interesting connections between the Frumkin isotherm on one side, the Coulombic EIL corrections on other side, and the non-ideal models in liquid or solid solutions on the third side.

Following Stumm (1992) for the constant capacitance (CC) EIL model, the apparent electrostatic potential on surface is $\phi_{CC} = \sigma_0 / C_0$ (in V) where C_0 is capacitance parameter ($F \cdot m^{-2}$), $\sigma_0 = F\theta_{\Sigma,z}\Gamma_C$ is the surface charge density (in $C \cdot m^{-2}$), $F = 96485.3$ $C \cdot mol^{-1}$ is the Faraday's constant, $\theta_{\Sigma,z} = \Gamma_{\Sigma,z} / \Gamma_C$ is the charge fractional coverage, and $\Gamma_{\Sigma,z}$ is the total density of charged surface species (in $eq \cdot m^{-2}$). Then the Frumkin interaction parameter can be expressed as $2\alpha_{F,CC} = (z_B F^2 \Gamma_C) / (RTC_0)$ and the Frumkin isotherm for the adsorbed species B with formula charge z_B (assuming that there are other sorbates binding to the same kind of surface sites) takes the form

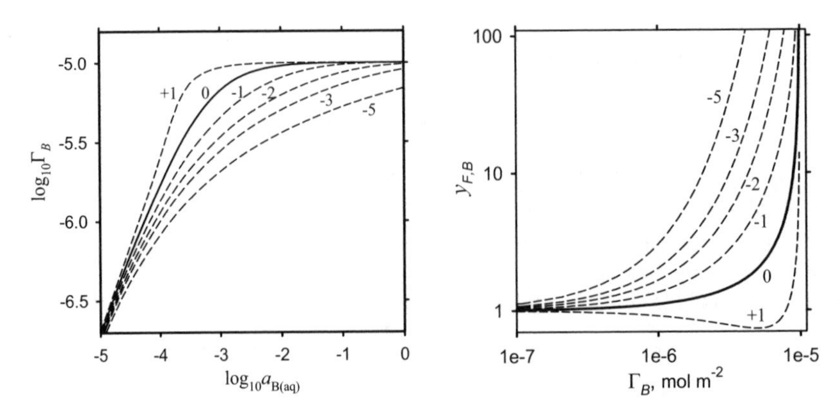

Figure 6. Plots of the Frumkin isotherm (Eqn. 90) assuming $\log_{10}K_B = 3$, $a_{B(aq)} = m_{B(aq)}\gamma_{B(aq)}$, $\Gamma_o = 2\times10^{-5}$, and $\Gamma_C = 1\times10^{-5}$ mol·m^{-2} (modified after Kulik 2006b). Numbers on the curves are Frumkin interaction parameter α_F values. The solid curve at $\alpha_F = 0$ is the Langmuir isotherm (Eqn. 60).

$$K_{F,B}a_B = \frac{\theta_B}{1-\theta_\Sigma}\exp\left(-2\alpha_{F,CC}\theta_{\Sigma,z}\right) \tag{91}$$

Comparison with Equation (89) shows that the electrostatic correction is scaled by the charge fractional coverage $\theta_{\Sigma,z}$ and there is also a non-electrostatic correction $1 - \theta_\Sigma$, the same as that in the competitive Langmuir isotherm (Eqn. 75). Note that electrostatic interactions are, in principle, long-range and extend over all kinds of sites on a surface patch. Hence, $\theta_{\Sigma,z}$ can even exceed θ_Σ, depending on the surface binding model.

Even though the CC EIL model is simple, other EIL models (see below) are expected to have qualitatively similar effects due to "soft long-range" interactions between charged surface species. Therefore, the Frumkin isotherm appears to be the most appropriate one for charged MWIs, even in its non-electrostatic form (Tamura 2004).

For a two-component $(B_{1-x}C_x)L$ solid solution, assuming the regular mixing model with interaction parameter α_0, Equation (49) can be re-arranged into

$$K_{B-C}\frac{a_{C+}^{(aq)}}{a_{B+}^{(aq)}} = \frac{x}{1-x}\exp[-2\alpha_0(x-0.5)] \tag{92}$$

It is easy to see that the structure of Equation (92) is similar to that of Equation (91). It also has a ratio of number of "sites" occupied by species $C+$ relative to the number of "sites" occupied by other species in the solid solution lattice, and an exponent (ratio of end-member activity coefficients) that accounts for "repulsive" or "attractive" interactions. When C is a trace component ($x \ll 1$), the activity coefficient f_{CL} reaches the infinite dilution limit $f_{CL}^{\infty} = e^{\alpha_0}$, and, Equation (92) becomes linear just as the Frumkin isotherm:

$$K_{B-C}\,a_{C+}^{(aq)} \big/ a_{B+}^{(aq)} = x\cdot\exp(\alpha_0)$$

BET isotherm (Brunauer-Emmett-Teller) equation (Brunauer et al. 1938) accounts for the condensation effects when the residence time τ_1 of B attached to an already adsorbed B molecule is comparable with the residence time τ_0 of B directly in the surface monolayer (or bound to a surface site). The BET isotherm has been mainly applied to adsorption of gases on solid surfaces at low to moderate temperatures (De Boer 1968; Somorjai 1994); it is widely used in measuring specific surface areas of particulate samples. The derivation of BET isotherm equation is rather lengthy (De Boer 1968); the isotherm it can be represented in the following form:

$$\theta_B = \frac{\Gamma_B}{\Gamma_C} = \frac{k_r p_r}{(q_r - p_r)\left[1 + (k_r - 1)^{p_r}/_{q_r}\right]} \qquad (93)$$

Here, the surface density Γ_B can exceed Γ_C due to formation of several layers of adsorbed B by increasing the relative partial pressure $p_r = p_B / p_{B,sat}$; $q_r = q / p_{B,sat}$ where q is an adjustable parameter; $p_{B.sat}$ is the saturated vapor pressure of condensed B; and $k_r = K_{Ads,B} / K_{Con,B}$ is the ratio of stability constants for adsorption and condensation reactions, proportional to the ratio of B residence times in surface- and in condensed layers.

A usual assumption is $q = p_{SAT}$; then $\theta_B = (k_r p_r)/((1 - p_r)(1 + (k_r - 1)p_r))$ and, as expected, $\theta_B \to \infty$ at $p_r \to 1$ leading to B surface condensation. However, in Equation (93), by setting q slightly greater than p_{SAT}, it is possible to limit the number of adsorbed layers forming at p_{SAT} and fix θ_B on this level upon further increase in pressure, eventually with formation of an independent condensed bulk phase of sorbate B.

The BET isotherm has already been used for describing the surface precipitation on MWIs. In a seminal paper, Rodda et al. (1996) compared the two-site Langmuir isotherm, the modified surface precipitation model (Farley et al. 1985), and the BET model, by applying them to the data on Pb(II) and Zn(II) sorption on goethite at constant pH, in a range of dissolved metal concentrations, and at temperatures between 10 and 70 °C. The following BET equation was used (see also Kulik 2006a):

$$\frac{\Gamma_M}{\Gamma_C} = \frac{k' c_M}{(c_{M,sat} - c_M)\left[1 + (k' - 1)^{c_M}/_{c_{M,sat}}\right]}; \qquad k' = \frac{K'_{Ads}}{K_{Ppt}} \qquad (94)$$

where Γ_M is the adsorbed density of cation M ($mol \cdot m^{-2}$); K'_{Ads} and K_{Ppt} are equilibrium constants for adsorption and precipitation of M, respectively; c_M is the dissolved aqueous molarity of M, and $c_{M,sat}$ is the molarity of dissolved M in equilibrium with precipitate at a given pH and temperature (estimated as $K_{sp} \cdot c(OH^-)^{-2}$). Rodda et al. (1996) have shown that Equation (94) with three adjustable parameters Γ_C, $c_{M,sat}$ and k' provides a good fit to Zn adsorption isotherm data on goethite.

The macroscopic BET model does not imply specific molecular mechanisms for adsorption and precipitation, hence the separation of fitted k' values into K'_{Ads} and K_{Ppt} remains somewhat arbitrary. $1/K_{Ppt}$ is not necessary the same value as the solubility product K_{ML} of the pure M (hydr)oxide phase.

Sparks et al. (1999) reviewed numerous papers about formation of surface precipitates and polynuclear surface complexes as important sorption mechanisms of hazardous elements on natural materials. In particular, nucleation products of Co, Cr^{III}, Cu, Ni and Pb on oxides and alumosilicates were observed at metal surface loadings far below the theoretical monolayer coverage, also below pH ranges expected for precipitation of metal hydroxides according to their thermodynamic solubility constants. Three different types of nucleation products were identified: formation or sorption of polymers (dimers, trimers, etc.) on the surface; a solid-solution or co-precipitate that involves co-ions from the adsorbent; and a surface precipitate composed of ions from the bulk solution or their hydrolysis products (Sparks et al. 1999, p.118).

Model description of such surface polymerization/precipitation phenomena remains a challenging task. However, using BET-like isotherms, this kind of sorption, which occurs before precipitation of bulk hydroxide phases, can be described with a single (monomeric) stoichiometry of the surface species, thus reducing the number of required model parameters. The BET equation can be represented as a sum of two isotherms (Tóth 1995): the Langmuir-type one for the monolayer adsorption ($\theta_{L,B}$), and another one for the multilayer adsorption ($\theta_{C,B}$). For instance, Equation (93) can be represented as

$$\theta'_B = \theta_{L,B} + \theta_{C,B} \quad \text{where} \quad \theta_{L,B} = \frac{k_r p_r}{q_r + (k_r - 1)p_r}; \quad \theta_{C,B} = \theta_{L,B}\frac{p_r}{q_r - p_r} \tag{95}$$

and q_r is slightly above one. At $p_r << ½$ and $k_r >> 1$, $\theta_{C,B} << \theta_{L,B}$ and $k_r - 1 \approx k_r$, so Equation (95) behaves as classic Langmuir isotherm (Eqn. 57). At $p_r > ½$, the fraction $p_r/(q_r - p_r)$ becomes large and causes $\theta'_B > 1$. Because $\theta_{L,B} < 1$, $\theta_{C,B}$ is limited by $1/(q_r - 1)$.

Let us recast the BET isotherm equation for surface adsorption/precipitation on MWIs into a general thermodynamic form using a method of (Kulik 2006a) for deriving an "activity coefficient" γ_{BET} which would provide a BET correction relative to the linear (ad)sorption isotherm found in Equations (52) and (53). Equation (94) is first re-formulated using aqueous molality m_B, activity a_B and a dimensionless parameter $q_B \geq 1$:

$$\frac{\Gamma_B}{\Gamma_C} = \frac{m_B k_r}{(qa_B^{(Ppt)} - a_B)\left[1 + (k_r - 1)^{a_B}\!\big/_{qa_B^{(Ppt)}}\right]}; \; k_r = \frac{K_{L,B}}{K^{\circ}_{Ppt}} \tag{95}$$

Here, K°_{Ppt} is the equilibrium constant of surface precipitation reaction B(aq) $\leftrightarrow$ B(Ppt); and $a_B^{(Ppt)} = (K^{\circ}_{Ppt})^{-1}$ is the activity of B(aq) in equilibrium with its surface precipitate, and $q \cdot a_B^{(Ppt)}$ slightly exceeds $a_B^{(Ppt)}$.

Now, substitute in the numerator the Langmuir-type adsorption equilibrium constant $K_{L,B}$ with the expression $K^{\circ}_B(\gamma_{B(aq)} / m_{\circ})(\Gamma_{\circ} / \Gamma_C)$ where K°_B refers to the reaction B(aq) $\leftrightarrow$ B(ads), standard density $\Gamma_{\circ}$, and infinite dilution. Re-arranging and re-scaling by q^{-1} to have $y_{BET,B} \rightarrow 1$ at $a_B \rightarrow 0$ yields:

$$\frac{\Gamma_B}{\Gamma_{\circ}} y_{BET,B} = K^{\circ}_B \cdot \frac{m_B}{q \cdot m_{\circ}} \gamma_{B(aq)} \;\; ; \;\; y_{BET,B} = (1 - \frac{a_B}{q} K^{\circ}_{Ppt}) \cdot \left\{1 + \frac{a_B}{q}\left(K_{L,B} - K^{\circ}_{Ppt}\right)\right\} \tag{96}$$

The product $a_{B(aq)}K^{\circ}_{Ppt}$ is the saturation index of B in "pure surface condensate" form. The product $a_{B(aq)}K_{L,B}$ is the activity of adsorbed B scaled to the complete monolayer coverage density Γ_C. By re-defining the parameters in Equation (96) as $\eta_B = q^{-1}$; $\beta_{P,B} = K^{\circ}_{Ppt}$; and $\beta_{M,B} = K_{L,B} \approx K^{\circ}_B(\Gamma_{\circ} / \Gamma_C)$, one arrives at a more elegant generalized BET isotherm:

$$\Gamma_B = \eta_B y^{-1}_{BET,B} \Gamma_{\circ} K^{\circ}_B \cdot a_{B(aq)} \tag{97}$$

where

$$y_{BET,B} = (1 - \eta_B a_{B(aq)} \beta_{P,B})\left[1 + \eta_B a_{B(aq)}\left(\beta_{M,B} - \beta_{P,B}\right)\right] \tag{98}$$

It is implicit that $\beta_{M,B} \geq \beta_{P,B}$ and $\gamma_{BET,B} > 0$. Precipitation of another bulk condensed phase of B in the same system must occur at $\hat{a}_{B(aq)}\beta_{P,B} = 1$. Substitution of the maximum possible aqueous activity $\hat{a}_{B(aq)}$ into Equation (98) brings it to the limit

$$\hat{y}_{BET,B} = (1 - \eta_B)\left[1 + \eta_B\left(\frac{\beta_{M,B}}{\beta_{P,B}} - 1\right)\right]; \quad \hat{\Gamma}_B = \eta_B \hat{y}^{-1}_{BET,B} \Gamma_{\circ} K^{\circ}_B \beta^{-1}_{P,B} \tag{99}$$

Behavior of the generalized BET isotherm for MWI is illustrated on Figure 7 by plotting Equation (97) over a range of B(aq) activities at certain parameters $\beta_{M,B}$, $\beta_{P,B}$ and η_B.

At $\beta_{M,B} >> \beta_{P,B}$, the function resembles the Langmuir isotherm, but raises above Γ_C upon further increase in $a_{B(aq)}$ up to $\hat{\Gamma}_B$ value between $\Gamma_C\eta_B(1 - \eta_B)^{-1}$ and $\Gamma_C[\eta_B(1 - \eta_B)^{-1} + 1]$. At $a_{B(aq)} = 10^{-3}$, an independent phase of B precipitates and the activities of B in all phases remain

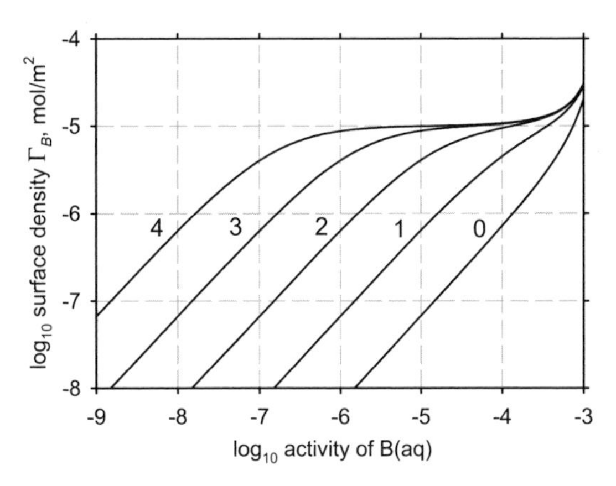

Figure 7. Plots of the generalized BET isotherm (Eqn. 97) at parameter values $\Gamma_o = 2\times10^{-5}$, $\Gamma_C = 1\times10^{-5}$ mol·m^{-2}; $\eta_B = 0.67$; $\log_{10}\beta_{P,B} = 3$; $\log_{10}\beta_{M,B}$ equals 3 (curve 0); 4 (curve 1); 5 (curve 2); 6 (curve 3) and 7 (curve 4).

fixed. Plots of Equation (98) on Figure 8A show that at $\beta_{M,B} >> \beta_{P,B}$, the surface activity correction term $y_{BET,B}$ increases from 1 to a maximum at $a_{B(aq)} \approx 0.5\beta_{P,B}^{-1}$ and then goes down to a constant value set by Equation (99) when the bulk condensed phase of *B* precipitates. From Equation (99), the maximum number of adsorbed layers is

$$\hat{\theta}_B = \frac{\hat{\Gamma}_B}{\Gamma_C} = \frac{\beta_{M,B}}{(1-\eta_B)\beta_{P,B}}\left[\frac{1}{\eta_B} + \frac{\beta_{M,B}}{\beta_{P,B}} - 1\right]^{-1} \tag{100}$$

BET isotherms expressed in units of monolayer coverage fraction θ_B are shown on Figure 8B where the chosen value of $\eta_B = 0.67$ results in the maximum adsorption between 2 monolayers (when $\beta_{M,B}$ parameter is weak) and 3 monolayers (when $\beta_{M,B}$ is strong). Conversely, realistic values of the η_B parameter should be confined between $\eta_B = 0.01$ (maximum ~1 monolayer) and $\eta_B = 0.95$ (up to ~20 monolayers), depending on the system. The variable $y_{BET,B}$ in Equation (97) is still not a "true" activity coefficient in the MWI "phase" because it is a function of composition of another (aqueous) phase. Another open issue is how to combine the generalized

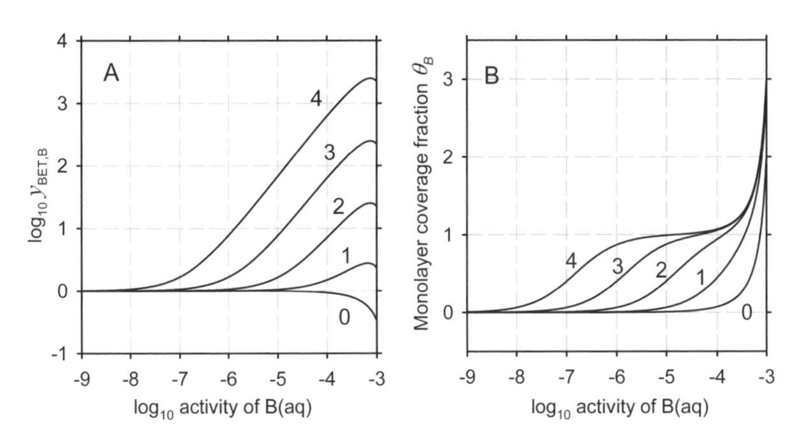

Figure 8. (A) Plots of BET "activity coefficient" (Eqn. 98), and (B) BET isotherms (Eqn. 97) expressed in the coverage fraction scale. For the parameter values, see caption to Figure 7.

BET isotherm with EIL corrections, and how to express it for several adsorbates in the same surface layer. More work on BET isotherm, a potentially very interesting model combining adsorption and surface precipitation on mineral surfaces, is still needed (Kulik, in preparation).

Freundlich isotherm equation is the simplest to describe the energetic site heterogeneity of naturally occurring materials (e.g., soils, mineral particulate or porous aggregates) which affects trace metal M (ad)sorption. The Freundlich isotherm can be derived assuming an exponential distribution of site energies, such that for all sites of the same energy, the Langmuir isotherm is applicable (Sposito 1984; Benjamin 2002b), see Figure 9. Integration over a continuum of Langmuir equations leads to

$$\Gamma_M = k_f \cdot c_M^n \quad \text{or} \quad q_M = k_f' \cdot c_M^n \tag{101}$$

where the parameters are $k_f (k'_f)$ describing the adsorption density Γ_M (q_M) under some "standard" conditions ($\Gamma_M = k_f$ when $c_M = 1$), and the power n indicating how strongly the binding strength changes with the Γ_M change. Plots of this isotherm are linear in logarithmic scale. Setting n = 1 reduces Equation (101) to linear (Henry) isotherm. Usually, $n < 1$ means that rare sites have higher binding energies.

In a seminal paper (Kinniburgh et al. 1983), the adsorption of Ca and Zn on ferrihydrite as function of c_M was described using a discrete-site Langmuir; a Generalized Freundlich; a Langmuir-Freundlich; and a Tóth isotherm. Such isotherm equations each have a characteristic SADF (site affinity distribution function); one has to keep in mind, however, that deriving SADFs from experimental titration data is not a mathematically trivial task (Borkovec et al. 1998).

Implementation of the Freundlich or other isotherms based on continuous SADF is desirable to reduce the number of sorption site types on an energetically heterogeneous MWI in the setup a computer-based surface complexation model. The difficulty with Equation (101) is that at $n \neq 1$, it is not easy to reference it to the infinite dilution limit (Kulik 2006b), which is needed for thermodynamic consistency and can easily be done with the Langmuir, Freundlich, or BET isotherm equations, as shown above.

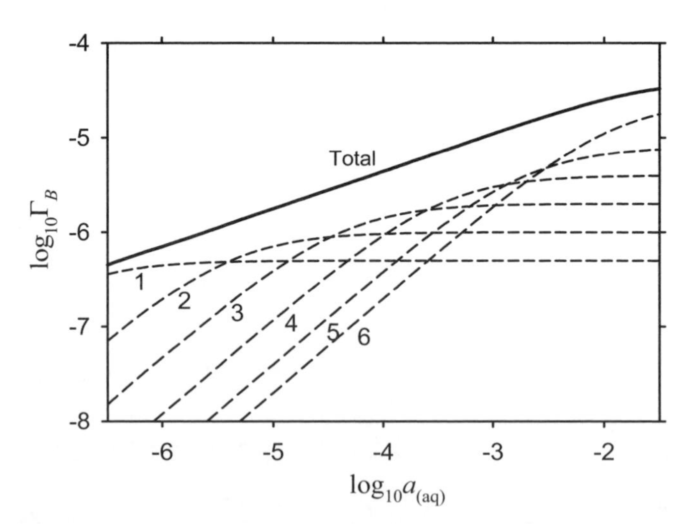

Figure 9. Hypothetical Freundlich isotherm approximated as a total of six generalized Langmuir isotherms (Eqn. 60) at $\Gamma_o = 2\times10^{-5}$ mol·m^{-2} and isotherm parameters: $\Gamma_{C1} = 5.0\times10^{-7}$, $K_{B1} = 2.0\times10^5$; $\Gamma_{C2} = 1.0\times10^{-6}$, $K_{B2} = 1.2\times10^4$; $\Gamma_{C3} = 2.0\times10^{-6}$, $K_{B3} = 2.4\times10^4$; $\Gamma_{C4} = 4.0\times10^{-6}$, $K_{B4} = 6.0\times10^2$; $\Gamma_{C5} = 8.0\times10^{-6}$, $K_{B5} = 2.0\times10^2$; $\Gamma_{C6} = 2.5\times10^{-5}$, $K_{B6} = 1.0\times10^2$ (density parameters in mol·m^{-2}).

Summary on adsorption isotherms. There are many other interesting isotherm equations, but most are used for describing the gas-solid or gas-liquid adsorption in simple systems (Tóth 1995; Aranovich and Donohue 1998, Hinz 2001; Limousin et al. 2007).

Except the Freundlich isotherm, isotherms considered above can be reduced in a uniform manner to the linear isotherm at the infinite dilution limit. Differences between isotherms due to e.g., the "*density*" of binding, lateral interactions, competition of different sorbates for sites, and surface condensation, can only be seen at relatively high total (ad)sorbed coverage.

Coulombic corrections in EIL surface complexation models are not very easy to combine with classic isotherms. However, simple EIL models such as the CC (constant capacitance) one can be converted into a form which is similar to the exponential part of the Frumkin isotherm equation. Further on, this part has something in common with (solid) solution mixing models based on excess enthalpy, such as the regular model.

The competitive multi-site Langmuir isotherm underlies many computer codes for modeling surface complexation and ion exchange on MWIs that use site mole balance(s) and LMA expressions for surface- and aqueous complexes. The extension to BET isotherm provides a promising way to represent surface condensation/precipitation phenomena in such models. However, the BET isotherm needs to be combined with competitive adsorption effects, and lateral interactions in surface monolayer, and re-formulated as function of activity/composition only in the sorption phase.

In many systems, (ad)sorption data for heterogeneous surfaces and sites, even with a continuous SADF, can be rationalized using a non-competitive multi-site Langmuir isotherm with two to four sites, optionally including the ion exchange (competitive Langmuir isotherm) for the weakest and most abundant site type (Westall 1995; Borkovec et al. 1998). However, no mechanistic conclusions can be drawn from the isotherm fits which are not supported with independent crystallographic, spectroscopic, or other atomistic-level data.

Standard and reference states for (ad)sorbed chemical species

In geochemical thermodynamics, the equilibrium is defined in terms of chemical potential

$$\mu_B = \mu_B^{\circ} + RT \ln a_B = \mu_B^{\circ} + RT \ln r_B + RT \ln \gamma_B \qquad (102)$$

where r_B is the relative amount (concentration) of the species B in the phase, and γ_B is the activity coefficient. To connect the standard chemical potential μ_B° to μ_B and to r_B in a convenient unequivocal way in all systems of interest, a *standard state* of the species B in the phase must be defined with respect to composition and ideality (Kallay et al. 2004). Three main approaches to define standard states for surface species exist (see a summary in Kulik 2006b):

(1) borrowed from those for aqueous electrolyte species, as used in the site-balance-based speciation codes (e.g., FITEQL, PHREEQC);

(2) expressed via the coverage fraction θ, usually in the context of adsorption isotherms;

(3) defined by combining physico-chemical states of the sorbent, its surface (sites), and the bulk solution phase.

Approach 1 has been criticized by Kulik (2002a) and Sverjensky (2003) because of its non-uniqueness. If a surface species is treated similar to an aqueous complex with 1.0 *m* or 1.0 M standard state then infinitely many systems can contain one mole of a surface species, but at different concentration c_S and/or specific surface A_S of the sorbent, i.e at different "standard" density $\Gamma_C^{(o)}$ or "standard" capacity $q_C^{(o)}$. Conversely, equilibrium constants for surface binding reactions will depend on the properties of sorbent surface, such as the complete-coverage density Γ_C and specific surface area A_S.

Approach 2 is straightforward when (ad)sorption obeys the Langmuir isotherm; then, setting $\gamma_{B(\text{ads})} = 1$, one can write $\mu_{B(ads)} = \mu^{o}_{B(ads)} + RT\ln(\theta/(1-\theta))$. As $\mu_{B(ads)} = \mu^{o}_{B(ads)}$ only at $\theta_B = 0.5$, it is tempting to define the standard state for B(ads) at $\theta_o = 0.5$. However, this will not work for n-dentate adsorbed species because of ambiguities between their isotherm Equations (80) and (81), and, in general, when $y_{B(\text{ads})} \neq 1$ (e.g., in presence of the Frumkin exponent or electrostatic correction term). Thus, for MWIs, it is more convenient to define standard states at the complete surface coverage: $\theta_o = 1$ for B(ads) and $\theta_o = 1$ for H_2O(ads). These standard states do not depend on "denticity" n and are similar to the "pure substance" standard states used for a component of solid or liquid solution. However, because the quantity $\theta_B = \Gamma_B/\Gamma_C$ is scaled to Γ_C—a property of the particular MWI—the "complete surface coverage" standard states are non-unique, unless the same Γ_C (or q_C) value is taken for all MWIs.

Approach 3 defines a standard state for adsorbed species together with a defined state of the solid sorbent relative to the bulk (gas or liquid) phase. (Kulik 2002a) proposed a standard state of a surface species at fixed density Γ_o (in mol·m^{-2}) and specific surface area (A^o_S in m^2·mol^{-1}) chosen to be common to all MWIs. For convenience, the amounts of the sorbent n^o_S; of surface species n_o; of water-solvent in the bulk solution n^o_w; and the molality of dissolved (ad)sorbate m_o are fixed so as to make the standard relative amount equal to unity. This standard state leads to μ^o and K values comparable between all MWIs and consistent with those of aqueous species. Instead of fixing n^o_S, n_o and n^o_w, (Sverjensky 2003) proposed to include the actual sorbent molarity c_S (in g· L^{-1}) into the standard state for surface complexes, and to use a different standard state (full coverage at actual Γ_C, A_S and c_S) for ≡SOH surface sites following (Dugger et al. 1964).

When discrepancies between the proposed definitions of standard states are sorted out, one of them will become a convention necessary to compile standard μ^o or K values into the unified chemical thermodynamic data base for adsorbed species on MWIs.

Reference states. The standard state determines the numerical value of the activity of surface species a_B at the actual state of interest (Eqn. 1). A *reference state* when the activity equals concentration (and the activity coefficient equals unity) is needed to scale the value of activity coefficient.

In computer-aided surface speciation models (Morel et al. 1981), the hypothetical "infinite dilution" reference state for B(ads) at electrostatic potential $\phi = 0$ was implicitly borrowed from that for aqueous species together with the 1.0 M (or 1.0 m) standard states and the "full coverage" or "pure surface" reference state for "sites" or ≡SOH groups. Upon the dilution of bulk aqueous electrolyte and/or upon increase of amount of sites, both reference states are approached simultaneously.

In the site-balance-based speciation calculation methods, the (competitive) Langmuir saturation effects are automatically reproduced via the site balance; the activity coefficients of surface complexes were traditionally ignored or taken only as electrostatic corrections (Dzombak and Morel 1990; Wingrave 1996; Kallay et al. 2004)

$$y_{El,M} = \exp\left(\frac{z_M F \phi_x}{RT}\right) \tag{103}$$

Here, ϕ_x stands for the relative electrostatic potential—a function of surface charge density, position, and other EIL parameters, as considered in EIL representations used in LMA surface complexation modeling (Lützenkirchen 2002a); see a summary below.

Non-electrostatic surface activity coefficients y_{Ne} acting on both the neutral and charged surface species comprise, perhaps, the most confusing topic, traditionally circumvented by setting all $y_{Ne} = 1$ (Wingrave 1996; Kallay et al. 2004) or assuming y_{Ne} to cancel out from the LMA expressions (Davis and Kent 1990; Dzombak and Morel 1990; Sverjensky 2003). However, it is hard to believe that no non-electrostatic interactions take place between the adsorbed species, especially looking at classic (ad)sorption isotherms at high coverage. Shortcomings of electrostatic corrections such as Equation (103) have been summarized by (Tamura 2004) who concluded that the non-electrostatic Frumkin isotherm (Eqn. 89) accounting for lateral interactions on surface is the most appropriate for ion-exchange reactions on metal (hydr)oxide surfaces.

When the specific surface area is known, the most reasonable interfacial concentration scale is the *surface density* Γ_B (Eqn. 2). In Equation (102), we can define $r_B = \Gamma_B/\Gamma_o$ and $a_{B(Ads)} = y_{B(Ads)}\Gamma_B / \Gamma_o$. A value of practical *standard surface density* $\Gamma_o = 2\times10^{-5}$ mol·m^{-2} (Kulik 2002a) was chosen close to the density of water molecules in the planar surface monolayer. Using this Γ_o scales the numeric values of molal equilibrium adsorption constants to the same order of magnitude as found in most literature sources. Another suggestion sets $\Gamma_o = 1.0$ mol·m^{-2} (Kallay et al. 2004) making values of these constants 5 orders of magnitude different. This is a matter of convention yet to be reached.

It is important to realize that the relative content $r_B = \Gamma_B/\Gamma_o$ is not the same as the coverage fraction $\theta_B = \Gamma_B/\Gamma_C$. The value r_B is just scaled to the standard surface density Γ_o taken conventionally the same for all MWIs, but not identical with the density of complete coverage Γ_C which is a sorbent-sample-surface property. By using Γ_o, it becomes possible (Kulik 2002b, 2006a,b) to construct surface complexation models without site balance constraints, based on a standard state where one mole interface species occupies at density $\Gamma_o = 2\times10^{-5}$ mol·m^{-2} all surface of one mole of the sorbent of specific surface area $A_o = 5\times10^5$ m^2·mol^{-1} immersed in 1 kg H_2O at 1 bar pressure and defined temperature. This state is referenced to a hypothetical "infinite surface dilution" at zero surface charge and potential for an adsorbed surface complex, and to a Γ_o coverage state for the adsorbed water or for the neutral ≡OH group. In this framework, the ideal behavior of a surface species is described with the linear isotherm (Eqns. 52, 53), deviations from which are attributed to an activity coefficient $y_{Ne,B} \cdot y_{El,B}$ split into an electrostatic part $y_{El,B}$ (defined in EIL models) and a non-electrostatic part $y_{Ne,B}$ (Kulik 2006a). The latter is derived from the appropriate isotherm (Eqns. 59, 65, 77, 83, 90, 98).

When the sorption capacity q_C but not the specific surface area of the sorbent A_S is known, the definition of standard state will depend on the concentration scale adopted for q_C. It may be difficult to determine the molar mass of such material as soil, clay, or humate, therefore q_C is usually expressed in mol·kg^{-1} or eq·kg^{-1} – per unit mass of the sorbent conditioned for composition, relative humidity, etc. Then the relative content in Equation (102) will be $r_B = q_B / q_o$ and a practical value of standard q_o will be e.g., 1 mol·kg^{-1}, close to the typical cation exchange capacity of clay mineral montmorillonite (Grim 1968; Meunier 2005). For such systems, a good choice seems to be a standard state with one mole of species sorbed at $q_o = 1$ mol·kg^{-1} on one kilogram of the particulate (porous) sorbent immersed in 1 kg H_2O at 1 bar pressure and defined temperature. This state can be referenced to a hypothetical "infinite dilution" at zero surface charge and potential for the sorbed species, and to a q_o capacity state for the sorbed water.

Concentrations of adsorbed species expressed in mol·m^{-2} scale, of course, can be reduced to those in mol·kg^{-1} scale by multiplying with A_S: $q_B = \Gamma_B \cdot A_S$; $q_C = \Gamma_C \cdot A_S$; θ_B values remain unaffected. To compare standard states and equilibrium constants defined at Γ_o and q_o, a fictive standard specific surface area $A_S^o = (M_S\Gamma_o)^{-1}$ (m^2·kg^{-1}) can be used, yielding a correction factor $f_{q_o \to \Gamma_o} = q_o / M_S$. This is only possible when the molar mass of the sorbent M_S is known.

SURFACE COMPLEXATION MODELS

Adsorption modeling with site balances

The framework of thermodynamic equilibrium speciation calculations involving site balances for the adsorption on (hydr)oxide surfaces, was originally implemented in the MINEQL/

FITEQL speciation codes (Morel et al. 1981; Dzombak and Morel 1990). Due to its simplicity, it became popular and did not evolve much during 30 years. A *surface complexation model* (SCM) of this kind combines an aqueous speciation model (with ions as master species and complexes as product species) with:

(1) Surface complexation reactions e.g., Equations (28)-(34) involving ≡SOH (≡OH) surface hydroxyl groups; the stability of a surface complex is expressed using an *intrinsic adsorption constant* K^{int} referenced to infinite dilution and zero charge/ potential;

(2) Mole (molar) balance of one or more "site ligands" ≡S (≡ or ≡SOH);

(3) EIL Coulombic corrections for the mass action of charged surface complexes.

Details on this SCM setup are available in numerous reviews (Davis and Kent 1990; Lützenkirchen 2002a; Turner et al. 2006) and textbooks (Sposito 1984; Stumm 1992; Morel and Hering 1993; Benjamin 2002a), thus, only a brief summary is provided below.

Variable surface charge, measurable in potentiometric titrations of mineral particle suspensions, is a distinct feature of (hydr)oxide and silicate surfaces, which also strongly influences the adsorption of anions and cations. Due to this fact, SCMs are more complex than single-species adsorption isotherms reviewed above. In the simplest case, the amphoteric MWI is represented by an absolutely stable "surface site ligand" ≡S of given total amount or total molar concentration. Because mineral surfaces are hydr(oxyl)ated in water (Parks 1990), SCMs usually do not include a "free" site ligand, but only hydrated, (de)protonated and hydrolyzed surface species. Their amounts are connected through an additional balance for each surface ligand, for example (in molarities []):

$$[\equiv S_wOH] + [\equiv S_wL^{-2}] + [\equiv S_wOH_2^+] + [\equiv S_wO^-] + [\equiv S_wOM^+] = \mathrm{Tot}[\equiv S_wOH]$$

$$[\equiv S_sOH] + [\equiv S_sOMOH] + [\equiv S_sOM^+] = \mathrm{Tot}[\equiv S_sOH]$$

In absence of electrostatic terms, each site balance acts as a competitive Langmuir isotherm (see Eqns 70, 75). Of course, each surface complex is also subject to another balance for the respective aqueous master species (Tot[M^{+2}], Tot[L^{-3}], Tot[H^+], …). In other words, surface complexes are treated as aqueous product species with one compulsory and one optional difference. The formation reaction for a surface complex *must* include one, two, … *n* surface site ligands. Optionally, its mass action law expression may contain the electrostatic term y_{El} (Eqn. 103) or the ratio of such terms, which makes the surface complex to obey a Frumkin-like isotherm (Stumm 1992) (see also Eqns. 89 and 91).

A common feature of site-balance-based SCMs is that no mineral sorbent is explicitly included into the chemical speciation model. Only total amount (or molar concentration) of reactive sites Tot[≡S] is used as the balance constraint. The sorbent properties, such as reactive specific surface area A_S, site density parameter Γ_C, concentration c_S (see Eqns. 1 to 6), or the solubility product K_S, stay behind the scenes, outside the SCM. In fact, only c_S (kg·dm^{-3}) and sorption capacity q_C (mol kg^{-1}) need to be known to set Tot[≡S] = $q_C \cdot c_S$. There are infinitely many systems with the same Tot[≡S] but different combinations of A_S, Γ_C and c_S. This seems to be one of reasons of the misunderstanding in defining standard states for intrinsic surface complexation constants (more about that in Kulik 2006b, pp.194-196).This is also a source of specific inconsistency in electrostatic SCMs. Most EIL models use the *surface charge density* σ_0 (σ_1, σ_β, σ_d …) expressed in C·m^{-2}, further converting it into the relative electrostatic potential ϕ_0 (ϕ_1, ϕ_β, ϕ_d) acting on the respective EIL plane. This implies that A_S must be known because $\sigma_0 = F \cdot \Sigma[\mathrm{Charge},0]/(c_S \cdot A_S)$ where

$$\Sigma[\mathrm{Charge},0] = [\equiv S_wOH_2^+] - [\equiv S_wO^-] + [\equiv S_wOM^+] - 2[\equiv S_wL^{-2}] + [\equiv S_sOM^+]$$

Conversely, although both A_S and Γ_C can be safely ignored in setting Tot[≡S] in the site balance, at least one of them must be known to evaluate the surface charge density. Such measurements are relatively easy for dilute laboratory suspensions, but may be difficult to impossible for natural soils or compacted clay systems. In the latter case, a more honest approach would be to use a (multi-site) non-electrostatic sorption model.

***EIL models* (*introduction*).** Electrostatic SCMs ignore non-electrostatic activity coefficients y_{Ne} and differ mainly in their treatment of electrostatic terms y_{El}, depending on the underlying EIL physicochemical picture. In early models, y_{El} were neglected or derived from a very simplified EIL concept such as a planar capacitor in the CC model. However, in the CD-MUSIC three-plane model (Hiemstra and van Riemsdijk 1996; van Riemsdijk and Hiemstra 2006), each adsorbed ion is placed at a different distance from the surface plane, and its charge is distributed between three reference planes separating the surface and the diffuse layer. Such a complex EIL picture requires several model parameters in addition to the set of K^{int} values for surface complexes, which makes its parameterization from experimental data a complex task, unless some parameters can be fixed on the basis of theoretical predictions or correlations.

Any electrostatic SCM is expected to describe potentiometic titrations of a dilute mineral suspension in a background electrolyte such as NaCl. In the 1pK formalism, at least two surface species must be involved in a site protonation reaction

$$\equiv SO^{z} + H^{+} \leftrightarrow \equiv SOH^{z+1} \qquad \log K_A^{int} = pH_{PPZC} \qquad (104)$$

where z is a negative charge of the surface oxygen (obtained e.g., from the bond valence). The site balance is

$$\mathrm{Tot}[\equiv SO] = [\equiv SO^{z}] + [\equiv SOH^{z+1}]$$

Together with the aqueous speciation model (involving H_2O, aqueous H^+, OH^-, Na^+, Cl^- with the respective balances, reactions, and activity coefficients), this is sufficient to model the response of mineral surface to pH changes in the aqueous solution. Because pH_{PZC} is an experimental quantity, the only adjustable parameter in the non-electrostatic case is Tot[≡SO], which at known c_S translates to the sorption capacity q_C, and at known A_S translates to the complete-coverage density Γ_C. Hence, both q_C and Γ_C parameters become purely empirical and model-dependent. The situation is not much better in 2pK formalism (Eqns. 28,29,39,40) because there is one more surface species $\equiv SOH^0$ in the balance, and one more parameter ΔpK_A. The latter becomes the only adjustable parameter if q_C (or Γ_C) is fixed on the basis of independent (e.g., crystallographic or tritium exchange) data. Surface species in such SCM will follow a competitive Langmuir isotherm, so the model will produce reasonable fits only in a limited pH region around pH_{PPZC}, departing strongly from the potentiometric data outside of that range.

Insufficiency of non-electrostatic SCMs in describing potentiometric data on oxides can be illustrated (Fig. 10) by comparing the surface H^+ charge curves of the rutile TiO_2 surface calculated using the same set of K^{int} values for (Eqns. 28-31) without electrostatic corrections, as well as according to the TL (triple layer) EIL model (see below).

Following (Machesky et al. 1998), the site density parameter was set to $\Gamma_C = 2.08\times10^{-5}$ mol·m^{-2} ($N_C = 12.5$ nm^{-2})—a value based on crystallographic considerations, tritium exchange, and weight loss experiments. It is seen on Figure 10 that already at pH = 3 or pH = 8 (ca. 2.5 units away from pH_{PPZC}), the non-electrostatic SCM tends to overestimate the surface charge by 8-9 times, and at pH = 8.3 would get close to -2×10^{-5} mol·m^{-2} or 15 times in excess of the measured value. This gives an impression about the magnitudes of electrostatic corrections through γ_{El} terms (Eqn. 103) and justifies why they cannot be avoided in modeling the surface charge and electrophoretic data away of pH_{PPZC}.

***Gouy-Chapman equation*.** Assuming that both surface complexes are located directly on mineral surface (0-plane), the charge density in 1pK formalism is obtained as

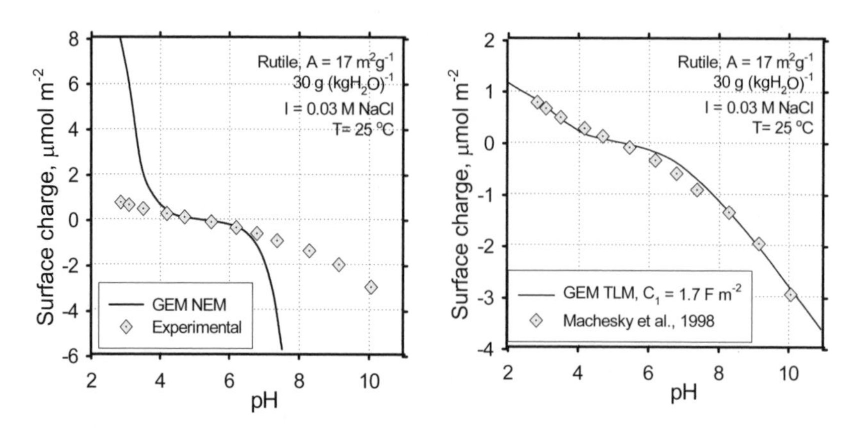

Figure 10. Surface proton charge on rutile modeled using a non-electrostatic (left) and the TL (triple layer model, TLM with $C_2 = 0.2$ F·m^{-2}) electrostatic (right) SCM with the same set of intrinsic constants $\log_{10}K_{A1} = -2.2$; $\log_{10}K_{A2} = -8.6$; $\log_{10}K_C = 1.9$; and $\log_{10}K_L = 1.7$ (for Eqns. 28-31, respectively). Experimental data from (Machesky et al. 1998); details of GEM SCM in (Kulik 2000).

$$\sigma_0 = ((z-1)[\equiv SO^{z-1}] + z[\equiv SOH^z])\cdot F/(c_S\cdot A_S) \quad (105)$$

In the 2pK formalism, the charge density is

$$\sigma_0 = (-[\equiv SO^-] + [\equiv SOH_2^+])\cdot F/(c_S\cdot A_S) \quad (106)$$

Within the *diffuse layer* on the aqueous electrolyte side, the relative electrostatic potential ϕ acting on the species charge Z drops exponentially from the value ϕ_d at the d-plane ("shear plane") to zero (conventional value) in the bulk aqueous solution. For planar interfaces, ϕ_d is connected to the charge density σ_d via the Gouy-Chapman equation

$$\sigma_d = -0.1174\sqrt{I}\sinh\frac{ZF\phi_d}{2RT} \quad (107)$$

(at 25 °C), where I is the molar ionic strength in the bulk electrolyte. Derivation of the Gouy-Chapman equation and methods of solving it for ϕ_d are readily available in textbooks, e.g., (Morel and Hering 1993). One popular solution is

$$\phi_d = \frac{2RT}{F}\ln\left[\sqrt{1+\left(\frac{\sigma_d}{2A\sqrt{I}}\right)^2} - \frac{\sigma_d}{2A\sqrt{I}}\right] \quad \text{where} \quad A = \sqrt{2000\varepsilon_0\varepsilon_d\rho_d RT} \quad (108)$$

ε_0 is the vacuum permittivity (8.854×10^{-12} C^2·J^{-1}·m^{-1}), ε_d is the dielectric constant, and ρ_d is the density of aqueous medium in the diffuse layer (at T,P of interest).

EIL models used in SCMs differ mainly in assigning charges of surface species to different EIL planes and restoring for each plane the respective electrostatic potential, applied then in the activity correction term (Eqn. 103) for each charged surface species.

Diffuse Layer (DL) model (Stumm et al. 1970; Dzombak and Morel 1990) assumes that the electrostatic potential drops only within the diffuse layer, i.e., $\phi_d = \phi_0$. The charge balance between EIL planes is then $\sigma_d + \sigma_0 = 0$, $\sigma_d = -\sigma_0$; ϕ_d is computed by solving the Gouy-Chapman equation. The mass-action expression for the 1pK reaction (Eqn. 104) then becomes

$$K_A^{\text{int}} = \frac{[\equiv SOH^z]}{[\equiv SO^{z-1}]\{H^+\}}\exp\left(\frac{+1\cdot F\phi_0}{RT}\right) \quad (109)$$

where braces denote the (molar) activity of aqueous H^+ ion. Now it becomes clear why the equilibrium constant K_A^{int} is called "intrinsic": in fact, it is a Langmuir isotherm constant related to activity of the aqueous ion and molar concentrations of surface species, but referenced to infinite dilution and zero surface potential. In the DL model, no adjustable parameters control the electrostatic correction term. This simplicity seems to be a reasons for its popularity, but also for its stiffness in 1pK setup (where $K_A^{int} = pH_{PPZC}$). In the 2pK case, there is one adjustable parameter ΔpK which can be used to fit the potentiometric titration data at certain electrolyte concentration and fixed sorption capacity or site density parameter. By doing this, Dzombak and Morel (1990) were able to produce an internally consistent data base of intrinsic adsorption constants of many cations and anions on hydrous ferric oxide (HFO), later extended also to goethite (Mathur and Dzombak 2006). The DL model is neither appropriate for describing the electrolyte adsorption nor the ionic strength dependences of cation/anion adsorption.

Constant Capacitance* (*CC*) *model was initially used for modelling adsorption in highly concentrated aqueous solutions, and later in comparing the surface chemistry of oxide minerals (Schindler and Stumm 1987). At high ionic strength, the compact Helmholtz layer can be approximated with a two-plane capacitor with the capacitance density C_1 (in $F{\cdot}m^{-2}$) leading to a simple linear charge-potential dependence

$$\phi_0 = \frac{\sigma_0}{C_1} \tag{110}$$

thus avoiding the need to solve the Gouy-Chapman equation. The C_1 parameter now controls the magnitude of EIL correction term y_{El} (Eqn. 103) and is usually treated as an adjustable parameter together with mass-action equations in 2pK or 1pK formalisms. The advantage of CC model is in its simplicity and also in applicability to 1pK formalism. However, realistic fits of potentiometric data can be obtained only at high ionic strength (> 1 M), and the electrolyte adsorption is not accounted for.

The Triple Layer* (*TL*) *model makes it possible for SCMs to account for the ionic strength dependence of surface proton charge and cation/anion adsorption. Early versions of the TL model (Yates et al. 1974; Davis et al. 1978) included only outer-sphere surface complexes of background electrolyte according to reactions such as Equations (30) and (31). More recent variants of TL model (Hayes and Leckie 1987) also included the formation of inner-sphere metal surface complexes according to hydrolysis reactions such as Equations (32) and (33), as well as inner- and outer-sphere adsorption of anions (Sahai and Sverjensky 1997; Sverjensky 2001, 2005; Sverjensky and Fukushi 2006). The predictive Extended TL model (ETLM) approach of Sverjensky resulted in a large internally consistent thermodynamic data set for adsorption of electrolyte ions, rare earth elements, and anions such as sulfate, selenate, As^{III} and arsenate on a number of mineral surfaces.

The TL model is based upon a distinction between dehydrated *inner-sphere* surface complexes located on mineral surface with charge assigned to zero EIL plane, and partially hydrated outer-sphere surface complexes with charges assigned to both 0-plane and to the so-called β-plane, separated from 0-plane by the inner Helmholtz layer with capacitance density C_1. The β-plane is separated from the d-plane (the onset of diffuse layer) by the outer Helmholtz layer with capacitance density C_2. Accordingly, three charge densities must be evaluated even in the simple 2pK case (see Eqns. 28 to 34):

$$\sigma_0 = (-[\equiv O^-] + [\equiv OH_2^+] - [\equiv O^-Na^+] + [\equiv OH_2^+Cl^-] + [\equiv OM^+] - [\equiv A^-]) \cdot F/(c_S \cdot A_S)$$

$$\sigma_\beta = ([\equiv O^-Na^+] - [\equiv OH_2^+Cl^-]) \cdot F/(c_S \cdot A_S)$$

$$\sigma_d = 0 - \sigma_0 - \sigma_\beta$$

From σ_d, the electrostatic potential ϕ_d is found by solving the Gouy-Chapman equation (Eqn.

108). The potential ϕ_β acting at the β-plane is then found as $\phi_\beta = \phi_d + \sigma_\beta/C_2$. Finally, the potential ϕ_0 acting on the 0-plane is found as $\phi_0 = \phi_\beta + \sigma_0/C_1$ (or, alternatively, as $\phi_0 = \phi_\beta - \sigma_\beta/C_1$; Kallay et al. 2006). Note that for an outer-sphere surface complex, as for any surface complex with charge distributed between different planes (see e.g., CD model below), Equation (103) takes the form

$$\gamma_{El,M} = \exp\left(F \cdot \frac{z_0\phi_0 + z_\beta\phi_\beta + ...}{RT} \right) \tag{111}$$

where z_0, z_β, ... is the species charge assigned to the respective EIL plane.

The TL model uses a more detailed EIL picture than CC or DL models, so it is not quite surprising that TL SCMs can describe the surface charge (even up to 250 °C, Kulik 2000), the ionic strength dependence of adsorption, and the electrophoretic data. The main disadvantage of TL SCMs lies in the ambiguity of assignment of surface complexes to inner- or outer-sphere type, and, especially, in their complexity - several ill-defined adjustable parameters such as ΔpK_A, K_C/K_L, C_1 and C_2 are required in addition to pH_{PPZC} and intrinsic cation/anion chemisorption constants.

As demonstrated by Hayes et al. (1991), good fits to experimental data can be obtained with quite different combinations of TL model parameters, although some combinations result in a failure of FITEQL or similar fitting codes. Recommendations to reduce on the number of TL model adjustable parameters include setting C_2 to a constant value (e.g., 0.2 $F \cdot m^{-2}$); symmetric electrolyte adsorption ($K_C = K_L$); as well as correlations of K_C, K_L and capacitance density parameters between different minerals and electrolytes.

Basic Stern (***BS***) model used in the early MUSIC approach (Hiemstra and van Riemsdijk 1991) together with the 1pK formalism, is a simplified variant of the TL EIL model. In BS model, the potential on d-plane ϕ_d is set equal to that ϕ_1 on the 1-plane (β-plane in the TL model) that separates the inner and outer Helmholtz layers. The condition $\phi_d = \phi_1$ means that $C_2 = \infty$. Conversely, charge densities and potentials are calculated for surface species e.g., from reactions (Eqns. 36,41,42) as follows:

$$\sigma_0 = (-0.5[\equiv OH^{-0.5}] + 0.5[\equiv OH_2^{+0.5}] - 0.5[\equiv OH^{-0.5}Na^+] + 0.5[\equiv OH_2^{+0.5}Cl^-]) \cdot F/(c_S \cdot A_S)$$

$$\sigma_1 = ([\equiv OH^{-0.5}Na^+] - [\equiv OH_2^{+0.5}Cl^-]) \cdot F/(c_S \cdot A_S)$$

$$\sigma_d = 0 - \sigma_0 - \sigma_1;\ \phi_d \text{ from Gouy-Chapman};\ \phi_1 = \phi_d;\ \phi_0 = \phi_1 + \sigma_0/C_1$$

Usually, the symmetric electrolyte adsorption is assumed ($K_C = K_L$). The chemically plausible BS model with one ill-defined adjustable parameter C_1 still provides good fits for titration data in a wide range of pH and ionic strength and combines inner- and outer-sphere complexes.

Charge Distribution (***CD***) ***three-plane*** model (Hiemstra and van Riemsdijk 1996) has been extensively used in combination with the 1pK multi-site (MUSIC) site-binding model (Hiemstra et al. 1989a,b; Hiemstra and van Riemsdijk 2002) to provide a detailed and accurate description of the potentiometric titration, adsorption, and electrophoretic data for goethite, gibbsite, rutile and other minerals, also in the hydrothermal region (Machesky et al. 2006).

As the TL or BS model, the CD model is based on the Stern picture of EIL with two layers having capacitance densities C_1 and C_2 (Fig. 11). However, the CD model recognizes the fact that surface complexes involve adsorbed ions in different hydration states. Depending on their size, shape, number and polarization of water molecules between them and the mineral surface, these ions are located at different distances from 0-plane, with the charge distributed between the 0, 1 and 2 planes. In this way, the CD model can account for "transitional" binding in Stern layers, where the charge of inner-sphere species is shared between 0 and 1 plane, and the charge of outer-sphere surface complexes is distributed between 1 and 2 plane (Rahnemaie et al. 2006).

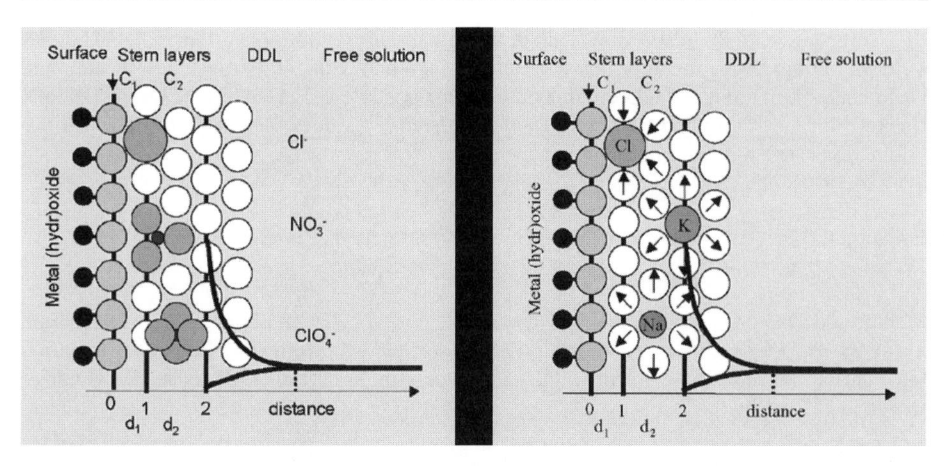

Figure 11. Schematic representation of the CD three-plane model with locations of various outer-sphere anionic and cationic surface complexes. [Used with permission of Elsevier, from Rahnemaie et al. (2006) *J Colloid Interface Sci,* Vol. 293, Figs. 5 and 6, p. 312–321.]

CD SCMs give a good fit to ion adsorption data on oxides in a wide range of ionic strength and pH, and account well for the observed effects in cation and oxoanion adsorption. A strong feature of CD-MUSIC approach is that most model parameters can be predicted from the mineral surface and EIL structure, properties of aqueous ions and water molecules, although some aspects are still disputed (Bickmore et al. 2006). A shortcoming of CD, as of other three-layer models, seems to consist in the usage of ill-defined capacitance density parameters.

EIL models (summary). The mean electrostatic field that develops on an amphoteric MWI due to ion adsorption/desorption may strongly influence the stability of a charged surface complex through Equations (103) and (111) and the underlying charge-potential relationships. The associated Frumkin-type isotherm corrections of adsorbed density may reach an order of magnitude or more. Hence, a non-electrostatic SCM where such corrections are neglected may be quite inaccurate in modeling the potentiometric and electrophoretic data measured away of pH_{PZC} in dilute suspensions.

However, the introduction of sophisticated EIL model corrections (TL, CD, BS) increases the number of SCM adjustable parameters by semi-empirical quantities such as capacitance densities which are difficult to constrain. Moreover, the C_1, C_2 parameters are interdependent with the site density parameter Γ_C and magnitudes of electrolyte adsorption intrinsic constants K_C, K_L. For this reason, almost equally good fits to titration or adsorption isotherm data can be obtained using many different parameter sets. Over the recent years, this problem has been made less acute by the development of predictive correlations for the surface acidity and electrolyte adsorption intrinsic constants (Hiemstra et al. 1989a, 1996; Sverjensky 1994, 2005; Sverjensky and Sahai 1996; Bourikas et al. 2001; Bickmore et al. 2004, 2006); uniform standard states for K^{int} (Sverjensky 2003); and correlations for capacitance parameters of the TL model (Sverjensky 2001, 2005).

EIL corrections appear to be much less important for adsorption of transitional metals, rare earths, actinide cations, as some oxoanions that all form relatively stable inner-sphere surface complexes on (hydr)oxide surfaces. Such chemical binding is usually described via hydrolysis-type (Eqns. 32, 33) or ligand exchange (Eqn. 34) reactions, in which the reaction stoichiometry is more crucial than the EIL correction. For neutral surface complexes—formed e.g., via Equation (33)—electrostatic corrections are not relevant.

Compared with the adsorption of alkali- and alkali-earth cations (e.g., Eqn. 30), the affinity of MWI to transitional metal or rare earth cations is much stronger. For instance, Mathur and Dzombak (2006) compiled adsorption constants for goethite α-FeOOH in the DL model 2pK framework. They presented many fitted and correlated K^{int} values for cations using Equations (32) and (33) and a direct binding reaction

$$\equiv OH^0 + M^{2+}(aq) \leftrightarrow \equiv OHM^{2+} \qquad K_{MO} \tag{112}$$

Equation (32) can be combined with Equation (29) into another binding reaction

$$\equiv O^- + M^{2+}(aq) \leftrightarrow \equiv OM^+ \qquad K_M = K_{M1}/K_{A2} \tag{113}$$

Mathur and Dzombak (2006) took for goethite $\Gamma_C = 3.32\times10^{-6}$ mol·m^{-2}, $pH_{PPZC} = 8.0$, and $\log K_{A2}^{int} = -9.65$ (corresponding to $\Delta pK_A = 3.3$). Now, the data from their Table 4 for Equation (32) can be converted into intrinsic $\log_{10}K_M$ values for Equation (112) by adding 9.65, as shown in Table 1 below.

Inspection of this table shows that binding constants K_M for Equation (113) have magnitudes similar to those of K_{MO} in Equation (112); the differences in $\log_{10}K$ values for these two reactions increase with the value of K_M from 0.6 to 2.2 but remain within 12% - 20% of $\log_{10}K_M$. Calcium has the smallest $\log_{10}K_M = 3.17$. Constants for Na, K, and other alkali metals in Equation (30) are not available in the DL model framework, but the $\log_{10}K_C$ value in the range 1.1 to 1.7 for the outer-sphere Na^+ surface complex can be extracted e.g., from the TL model constants for goethite in (Villalobos 2006). This value is at least 1.5 to 2.2 units weaker than $\log_{10}K_M$ for Ca^{2+} adsorption in Table 1.

In their Table 8.3, Morel and Hering (1993) provide the magnitudes of DL model Coulombic correction factor $\log_{10} y_{El} = \exp(-F\phi_0 / RT)$ for different pH and ionic strengths I on the surface of hydrous ferric oxide (HFO) with $pH_{PPZC} = 8.1$, rather similar to the goethite surface. At $I = 0.1$ M, these corrections range from −2.45 at pH = 4.0 to −0.09 at pH = 8 and up to 2.01 at pH = 11. Comparing these values with the data in Table 1 above, one can see that they exceed the intrinsic constant for Na^+ adsorption and are about 1 unit below $\log_{10}K_M$ for Ca^{2+}. Thus, the adsorption of alkali cations will be strongly, and that of alkali earth cations – moderately affected by variations in the surface charge. However, Co^{2+}, Ni^{2+}, Cd^{2+}, Zn^{2+} have $\log_{10}K_M$ values that exceed the Coulombic correction factor by 5-6 units; Cu^{2+} and Pb^{2+}—by about 8 units, hexavalent actinide oxocations—by 10 units, and Hg^{2+}, Sn^{2+} and Pd^{2+} by 11-12 logarithmic units. For such strongly adsorbed cations, Coulombic corrections play a negligible role compared with the affinities involved in hydrolysis or complexation reactions in the

Table 1. Intrinsic adsorption constants for divalent cations on goethite (DL model).

Ion	$\log_{10}K_{M1}$*	$\log_{10}K_M$	$\log_{10}K_{MO}$ *	Ion	$\log_{10}K_{M1}$ *	$\log_{10}K_M$	$\log_{10}K_{MO}$ *
Ba^{2+}	−5.62	4.03	3.43	Cu^{2+}	1.39	11.04	8.92
Sr^{2+}	−5.44	4.21	3.59	Pb^{2+}	0.44	10.09	8.25
Ca^{2+}	−6.48	3.17	3.98	UO_2^{2+}	2.28	11.93	10.26
Mg^{2+}	−3.02	6.63	5.24	PuO_2^{2+}	2.48	12.13	10.44
Ag^+	−4.11	5.54	4.74	NpO_2^{2+}	2.92	12.57	10.82
Mn^{2+}	−2.66	6.99	5.99	Be^{2+}	2.69	12.34	10.61
Co^{2+}	−0.79	8.86	7.28	Hg^{2+}	3.84	13.49	12.39
Ni^{2+}	−1.96	7.69	6.38	Sn^{2+}	4.75	14.40	12.39
Cd^{2+}	−1.96	7.69	6.28	Pd^{2+}	6.19	15.84	13.64
Zn^{2+}	−0.96	8.69	7.50	Na^+		1.4**	

* Data for K_1^{int} (K_{M1} in the table) and K_2^{int} (K_{MO}) from Table 4 of Mathur and Dzombak (2006).

** For comparison: mean value for the range 1.1 to 1.7 from TL model (Villalobos 2006).

bulk aqueous solution, as well as with the specific coordination, complexation, or hydrolysis reactions on the MWI.

The differences highlighted above may partially explain the success of non-electrostatic multi-site sorption models such as the Bradbury-Baeyens model for sorption of radionuclides and related cations on clay minerals ((Bradbury and Baeyens 2006) and references therein). This success has been further enhanced by the LFER (linear free energy relation) correlations between the intrinsic surface complexation constants of transitional, rare earth, and actinide cations, on one side, and their aqueous hydrolysis constants on the other side (Schindler and Stumm 1987; Dzombak and Morel 1990; Bradbury and Baeyens 2005; Mathur and Dzombak 2006). Such correlations provide an easy way to extend the database of intrinsic constants with the estimated values for cations and anions when there is no available experimental data.

The choice of SCM type is mainly dictated by research objectives. If the emphasis is set on reproducing the potentiometric titration and zeta-potential data measured in dilute suspensions of pre-conditioned (hydr)oxide with known specific surface area at various background electrolyte concentrations then any NE, DL, or CC SCM must be considered inadequate, and a BS, TL or CD model must be used. The final quality of results will greatly depend on the accuracy of experimental procedures and reduction of the raw titration data (Lützenkirchen 2002; Lefevre et al. 2006).

If the aim is to model pH adsorption edges or isotherms of strongly adsorbed trace ions measured in dilute suspensions of hydroxide such as goethite or gibbsite at relatively high background electrolyte concentration then a CC or DL SCM may be sufficient and a NE SCM may be satisfactory, especially for modeling adsorption isotherms in pH-buffered systems. In the latter case, a multi-site model may be adequate to account for site heterogeneity (Dzombak and Morel 1990); at high sorbate loadings, an extension of the model to surface precipitation may be needed (Zhu 2002).

In studies of the geochemistry of natural waters, the emphasis may be on sorption of trace elements and radionuclides on mineral particles present in river-, lake-, and seawater (see reviews by Honeyman and Santschi 1992 and Brown and Parks 2001). pH of natural waters is relatively limited because of buffering mineral associations or the atmosphere; the suspended particulate matter is very dilute and heterogeneous, containing detrital and secondary mineral particles, organic colloids, and bacteria, hence large difficulties in separating the particles and characterizing their surface areas and composition. These and other factors put direct applications of electrostatic SCMs under dispute in favor of simple multi-site non-electrostatic sorption models (Westall 1995) based on empirical binding constants or *Rd* values. The affinity of natural particle surfaces to transitional elements appears to be not less than that of synthetic goethite, HFO or gibbsite particles. Hence, a generally accepted conclusion (Krauskopf 1956; Brown and Parks 2001) is that sorption processes on natural particulate matter are likely responsible for the present trace element concentrations in seawater.

In geochemistry of waste disposal, the interest is shifted from dilute laboratory systems to compacted porous mineral aggregates such as soils, clay rocks, bentonites, cements, where pores are filled with aqueous electrolyte. Such systems pose specific difficulties in the application of electrostatic SCMs. The pH of pore solutions is usually strongly buffered by reactive minerals such as carbonates, sulfates, and clays present together in a high solid-to-liquid ratio. It is difficult to differentiate between "true" pore spaces where reactive amphoteric surfaces of minerals are exposed and the clay interlayer pore space limited by permanent-charge planes of structure packages of clay minerals montmorillonite, or illite. In other words, it is difficult to characterize reactive specific surfaces in such compact mineral aggregates. The "true" pores may be so small that the diffuse parts of EIL overlap, and the pore surfaces are so curved, that the classic Goyu-Chapman charge-potential relationship cannot be rigorously applied. In modeling such complex systems, the electrostatic SCMs in their present form seem to have little chances for success, and new approaches to sorption modeling may need to be developed.

There are two paradigms in modeling sorption on heterogeneous natural mineral mixtures, called the *Component Additivity* CA ("bottom-up") and the *Generalized Composite* GC ("top-down") approach (Davis et al. 1998; Payne et al. 2004, 2006; Turner et al. 2006). The CA approach represents the mixture by proportions of several reactive mineral phases, each with individual surface properties and appropriate (electrostatic or not) SCM parameters obtained from experiments on dilute single-mineral suspensions or predicted for this mineral. The CA approach combines then the sorption models for each phase within the chemical mass transfer in the whole system. In contrast, the GC approach assumes that the assemblage of mineral surfaces in the aggregate is too complex to be described by adding the contributions from individual mineral components. Instead, one ore more generic site types (with respective site capacities) are introduced, and sorption affinities to those sites are represented via surface binding reactions with empirical binding constants, commonly developed in the non-electrostatic framework because of large uncertainty in defining the appropriate charge-potential relationships. While extrapolation beyond experimental ranges used in fitting the model parameters is questionable, the GC approach may provide better fits with fewer adjustable parameters compared to the CA approach for the same system (Payne et al. 2006).

Multi-site (multi-surface) models

Structural heterogeneity of surface sites on oxides and silicates is the topic of intensive research during the last two decades, since seminal papers (Hiemstra et al. 1989a,b) introduced the MUSIC approach to surface complexation modeling. Studies of MWI surface structure and coordination of adsorbed cations and anions are supported by many advanced surface-sensitive microscopic, spectroscopic and atomistic modeling methods (more in Sherman 2009), also widely applied to elucidate sorption mechanisms of hazardous metals on mineral surfaces (Manceau et al. 2002). Results of these studies have some important implications for thermodynamic (ad)sorption modeling because they provide model-independent information such as:

- specific areas of structurally different mineral particle surfaces;
- charges and densities of reactive surface sites on particle faces, edges and kinks;
- constrains on "denticity" and stoichiometry of surface complexes;
- estimates or predictions of surface protonation and ion adsorption constants.

Here, some thermodynamic aspects of multi-site adsorption models (based on site mole balances) will be considered.

In the MUSIC surface structural approach as summarized in (Hiemstra and van Riemsdijk 2002; van Riemsdijk and Hiemstra 2006), Pauling bond valence principle and crystallography data are used in defining stoichiometry, charge and density of surface complexes, while crystal-chemical bond length/bond strength ratios are used in predicting the intrinsic surface protonation constants in 1pK framework. On this basis, reaction stoichiometry, Γ_C, K_A, and K_C (K_L) values, in principle, can be *a priori* fixed in the SCM setup, leaving capacitance densities as adjustable parameters. Consideration of mineral surface structure leads to different number of reactive site types, their charges and densities for different minerals and even different faces on the same mineral particle.

For instance, on gibbsite α-$Al(OH)_3$ platelets, mainly the edges (of effective specific surface area ca. 20 $m^2 \cdot g^{-1}$) are reactive with one type of singly-coordinated $Al\text{-}OH^{-1/2}$ proton-reactive sites of $N_C = 8.15$ nm^{-2} (Hiemstra et al. 1999). On SiO_2 minerals surfaces, a singly-coordinated $Si\text{-}O^{-1}$ group of density $N_C = 4.8$ nm^{-2} ($\Gamma_C \approx 8 \times 10^{-6}$ $mol \cdot m^{-2}$) is proton-reactive (Hiemstra et al. 1996). However, even the dominant (110) face of goethite α-FeOOH exposes four different types of sites – one singly, one doubly, and two triply-coodinated oxygens, of which a singly-coordinated $Fe\text{-}OH^{-1/2}$ (N_{C1} = 3 nm^{-2}) and a triply-coordinated $Fe{\equiv}O^{-1/2}$ (N_{C3} = 3 nm^{-2}) are proton-reactive (Hiemstra and van Riemsdijk 1996). On the (021) face of goethite, a doubly-coordinated $Fe{=}O^{-1}$ (N_{C1} = 3 nm^{-2}) is also reactive; all three have different K_A intrinsic constants (see also Table 4 in (Kulik 2006b)) that together produce $pH_{PZC} \approx 10$. Reactive HFO (ferrihydrite) surface can be represented similar to that of goethite, but with N_{C3} = 1.5 nm^{-2} ($Fe{\equiv}O^{-1/2}$) and N_{C1} = 5.25 nm^{-2} ($Fe\text{-}OH^{-1/2}$) (Stachowicz 2007). On the rutile α-TiO_2 particle <110> surface, singly-coordinated $Ti\text{-}OH^{-1/3}$ and doubly-coordinated $Ti{=}O^{-2/3}$ oxygens are proton-reactive, both with densities $N_{C1} = N_{C2}$ = 6 nm^{-2} and similar K_A values (Bourikas et al. 2001).

On carbonate minerals such as calcite, the pristine surface consists of the oxygens coordinated either to carbons in carbonate groups or to lattice metal ions. On the most stable $(10\bar{1}4)$ face, both postulated initial species (sites) $>Me^{1+}$ and $>CO_3^{1-}$ exposed on the cleavage plane are present in equal densities $\Gamma_{Me,C} = \Gamma_{CO_3,C} = 8.22\times10^{-6}$ mol·m^{-2}. Subsequent hydration and adsorption from the bulk solution leads to formation of the following surface species: $>MeO^-$, $>MeOH^0$, $>MeOH^{2+}$, $>MeHCO_3^0$, $>MeCO_3^-$, $>CO_3^-$, $>CO_3H^0$, and CO_3Me^+, treated in a CC SCM in 2pK_A fashion with rather high capacitance density parameter C_1 value of 17 or 31 F·m^{-2} (see Table 3 in Pokrovsky and Schott 2002). Furthermore, this SCM provides the surface speciation controls on dissolution kinetics of several carbonate minerals both in acidic and neutral to alkaline solutions. This can be described using a rate equation derived by combining the activated surface complex theory with the transitional state theory (Pokrovsky and Schott 2002)

$$R_T = (k_{CO_3}\Gamma^m_{>CO_3H^0} + k_{Me}\Gamma^n_{>MeOH_2^+})\cdot\left[1-\left(\frac{Q}{K_{SP}}\right)\right] \tag{114}$$

where k_{CO_3} and k_{Me} are the rate constants, m and n are the reaction orders with respect to surface protonated carbonate and hydrated metal sites, Q is the aqueous ion activity product, K_{SP} is the solubility product of metal carbonate, and units are mol·cm^{-2}·s^{-1} for RT and mol·cm^{-2} for Γ_j (more in Schott 2009). Note that Equation (114) directly refers to adsorbed densities Γ_j and not to coverage fractions θ_j and, hence, is not sensitive to the choice of $\Gamma_{Me,C}$ and $\Gamma_{CO_3,C}$ SCM parameters.

One thermodynamic aspect of this two-site model of carbonate-water interface (as of any other site-balance based multi-site SCM) consists in an independent competition of adsorbed species on each site type according to the competitive Langmuir isotherm (Eqn. 74). Both site types also compete with the aqueous solution (and other phases if present) through the overall chemical balance for a given chemical element in the system. However, because the sites of both kinds are tightly intermittent and close to each other on the mineral particle face, the (long-range electrostatic) EIL correction refers to charge density σ_0 and relative potential φ_0 *common to all* charged surface species on both site types. Recalling Equation (91) for the Frumkin isotherm implicit in the CC model, and combining it with Equations (84) and (74) yields for the sorbate B fixed on the carbonate surface:

$$\Gamma_B = a_B \exp(2\alpha_{F,CC}\theta_{\Sigma,z})\cdot\left[K''_{Me,B}(\Gamma_{Me,C}-\Gamma_{Me,\Sigma}) + K''_{CO_3,B}(\Gamma_{CO_3,C}-\Gamma_{CO_3,\Sigma})\right] \tag{115}$$

This expression can be regarded as a two-site competitive Frumkin isotherm in which the exponential term now is $(z_B F^2\Gamma_{\Sigma,z})/(RTC_0)$ where the density of surface charge $\Gamma_{\Sigma,z} = \sigma_0/F$ is limited as $|\Gamma_{\Sigma,z}| < (\Gamma_{Me,C} + \Gamma_{CO_3,C})$.

Common electrostatic corrections via the "multi-site competitive Frumkin isotherm" (but with more sophisticated charge-potential relationships) seem to act also in CD-MUSIC SCMs for goethite or rutile (see references above). Due to the existence of two or more kinds of proton-reactive surface oxygens, a sorbate (ion) can, in principle, be coordinated: to one such site; to two sites of the same kind; to two sites of different kind; and to four sites of two different kinds. This brings another thermodynamic aspect of "denticity" into SCMs for such mineral surfaces because of an ambiguity in treating surface species as either *multi-dentate* or *multiply-coordinated*. Problems behind the isotherms for multi-dentate species were addressed in the discussion around Equations (78)-(81).

Multi-site multi-dentate binding. According to the most comprehensive papers (Machesky et al. 2006; Ridley et al. 2009) and references therein, the recent X-ray spectroscopy and reflectivity data enhanced with atomistic simulations for the rutile <110> surface have shown that cations are principally adsorbed there in various inner-sphere configurations (Fig. 12), mainly tetradentate for Rb^+, Sr^{2+}, and Y^{3+} (by inference also for Na^+, Ca^{2+}, and Nd^{3+}), and combined bidentate and monodentate for Zn^{2+} (by inference, for Co^{2+} and Ni^{2+}). The rutile surface consists of intermittent rows of terminal and bridging oxygens, represented in the advanced CD-MUSIC BS SCM formulation (Machesky et al. 2008) as $Ti\text{-}OH^{-0.434}$ and $Ti{=}O^{-0.555}$ groups, respectively. For the above-mentioned cations and background anions (Cl^-, Br^-), this leads to a set of reactions (Table 2) for n-dentate surface species having charge $z0$ allocated to 0-plane and charge $z1$ to 1-plane according to CD BS EIL model. Fitted, predicted and optimized values of K^{int}, $z0$, $z1$, and Stern layer capacitance densities can be found in (Ridley et al. 2004, 2005, 2009; Zhang et al. 2004; Machesky et al. 2006).

Inspection of Table 2 and Figure 12 induces thoughts that there may be potential inconsistencies in site-balance-based SCM calculations involving bi- and tetradentate surface complexes because they do not follow the same (competitive Langmuir) isotherms as the monodentate surface species do. Moreover, QCA isotherms (Eqns. 81, 82) account for n-dentate adsorption only on sites of the same type; the exact form of isotherm for surface species of 1+1 or 2+2 "denticity" remains unknown. QCA isotherms have not yet been extended onto the competition of different adsorbates for the same kind of sites.

Nevertheless, Ridley et al. (2009) were able to obtain excellent fits to experimental titration data using the ECOSAT speciation code based on multiple site balances (i.e., reproducing multi-site competitive Langmuir isotherm augmented with CD BS electrostatic correction terms). This success may be to some extent fortuitous for two reasons. Firstly, the QCA isotherms for different n show different behavior only at high fractional coverage; at low coverage, they tend to be linear. Binding of an ion to two (or 2+2) sites of different types would bias the

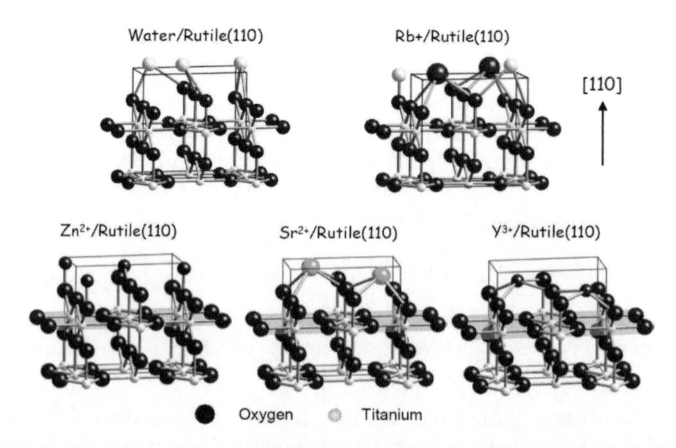

Figure 12. Water and cation bonding configurations on the rutile <110> surface as determined from in situ X-ray standing wave and X-ray reflectivity studies. [Used with permission of Elsevier, from Machesky et al. (2006) *Interface Science and Technology*, Vol. 11, p 324-358.]

Table 2. Summary of surface complexation reactions for rutile TiO_2 (see text for explanations).

Reaction	***n***	K^{int}	**Ions**
$\text{Ti-OH}^{-0.434} + H^+ \leftrightarrow \text{Ti-OH}_2^{+0.566}$	1	K_{H1}	H^+
$\text{Ti=O}^{-0.555} + H^+ \leftrightarrow \text{Ti=OH}^{+0.445}$	1	K_{H2}	H^+
$\text{Ti-OH}_2^{+0.566} + L^- \leftrightarrow [\text{Ti-OH}_2^{+0.566}]^{z0}\text{-}L^{z1}$	1	K_{L1}	Cl^-, Br^-
$\text{Ti=OH}^{+0.445} + L^- \leftrightarrow [\text{Ti=OH}^{+0.445}]^{z0}\text{-}L^{z1}$	1	K_{L2}	Cl^-, Br^-
$2\text{Ti-OH}^{-0.434} + 2\text{Ti=O}^{-0.555} + C^+ \leftrightarrow [(\text{Ti-OH}^{-0.434})_2(\text{Ti=O}^{-0.555})_2]^{z0}\text{-}C^{z1}$	2+2	$K_{C\text{-}22}$	K^+,Na^+,Rb^+,Cs^+
$2\text{Ti-OH}^{-0.434} + C^+ \leftrightarrow [(\text{Ti-OH}^{-0.434})_2]^{z0}\text{-}C^{z1}$	2	$K_{C1\text{-}2}$	K^+,Na^+,Rb^+,Cs^+
$2\text{Ti-OH}^{-0.434} + 2\text{Ti=O}^{-0.555} + M^{2+} \leftrightarrow [(\text{Ti-OH}^{-0.434})_2(\text{Ti=O}^{-0.555})_2]^{z0}\text{-}M^{z1}$	2+2	$K_{M\text{-}22}$	Ca^{2+},Sr^{2+}
$\text{Ti-OH}^{-0.434} + \text{Ti=O}^{-0.555} + M^{2+} \leftrightarrow [(\text{Ti-OH}^{-0.434})(\text{Ti=O}^{-0.555})]^{z0}\text{-}M^{z1}$	1+1	$K_{M\text{-}11}$	Ca^{2+},Sr^{2+}
$2\text{Ti-OH}^{-0.434} + M^{2+} \leftrightarrow [(\text{Ti-OH}^{-0.434})_2]^{z0}\text{-}M^{z1}$	2	$K_{M1\text{-}2}$	Zn^{2+},Ca^{2+},Sr^{2+}
$\text{Ti=O}^{-0.555} + M^{2+} \leftrightarrow [\text{Ti=O}^{-0.555}]^{z0}\text{-}M^{z1}$	1	K_{M1}	Zn^{2+},Co^{2+},Ni^{2+}
$\text{Ti=O}^{-0.555} + M^{2+} + H_2O \leftrightarrow [\text{Ti=O}^{-0.555}]^{z0}\text{-MOH}^{z1} + H^+$	1	K_{M1OH}	Zn^{2+},Co^{2+},Ni^{2+}
$2\text{Ti-OH}^{-0.434} + 2\text{Ti=O}^{-0.555} + Y^{3+} \leftrightarrow [(\text{Ti-OH}^{-0.434})_2(\text{Ti=O}^{-0.555})_2]\text{-}Y^{3+}$	2+2	$K_{Y\text{-}22}$	Y^{3+},Nd^{3+}
$2\text{Ti-OH}^{-0.434} + 2\text{Ti=O}^{-0.555} + Y^{3+} + H_2O \leftrightarrow [(\text{Ti-OH}^{-0.434})_2(\text{Ti=O}^{-0.555})_2]\text{-YOH}^{2+} + H^+$	2+2	K_{YOH}	Y^{3+},Nd^{3+}
$2\text{Ti-OH}^{-0.434} + 2\text{Ti=O}^{-0.555} + Y^{3+} + 2H_2O \leftrightarrow [(\text{Ti-OH}^{-0.434})_2(\text{Ti=O}^{-0.555})_2]\text{-Y(OH)}_2^+ + 2H^+$	2+2	K_{YOH2}	Y^{3+},Nd^{3+}
$2\text{Ti-OH}^{-0.434} + 2\text{Ti=O}^{-0.555} + Y^{3+} + 3H_2O \leftrightarrow [(\text{Ti-OH}^{-0.434})_2(\text{Ti=O}^{-0.555})_2]\text{-Y(OH)}_3^0 + 3H^+$	2+2	K_{YOH3}	Y^{3+},Nd^{3+}

n is the "denticity" of the surface-bound ion, given separately for bridging and terminal oxygen sites.

isotherm only if one site type has Γ_C very different from the other one. On rutile, both site types are ubiquitous and have (almost) equal Γ_C of about 1×10^{-5} mol·m^{-2}. Secondly, at moderate coverages, the impact of "denticity" and of the Frumkin exponent (or electrostatic term) produce similar effects on theoretical isotherm curves. Thus, the inconsistencies related to inadequacy of Langmuir isotherms dictated by site balances, might have been absorbed in the CD MUSIC SCM through the adjustment of electrostatic correction terms acting on all charged surface species regardless of the site type.

More about standard states for K^{int}. The rutile example is also instructive in the context that great care is needed to ensure the correct bookkeeping of multi-dentate surface species in site balances and in extracting consistent values of K^{int} with regard to the underlying standard states. In speciation codes such as MINEQL/FITEQL, the operational 1M standard state is used for the intrinsic adsorption constants, the drawback of which is that the standard 1M concentration of the surface species is not connected to the (standard) concentration of the solid surface (Kulik 2002a; Sverjensky 2003).

For monodentate adsorption reactions like $Ti{=}O^{-0.555} + M^{2+} \leftrightarrow [Ti{=}O^{-0.555}]^{z0} - M^{z1}$, the mass action expression looks like $K_B^{(1M)} = ([\equiv B])/([\equiv]\{B_{aq}\})$ where charges and electrostatic term are omitted for clarity, brackets denote molarity, and braces denote activity.

As equal total number of sites (either free or occupied with B) is implicit in the numerator and denominator, any related scaling factors are expected to cancel out. However, for n-dentate reaction $n(\equiv) + B_{aq} = \equiv_n B$, the mass action will take the form $K_{n,B}^{(1M)} = ([\equiv_n B])/([\equiv]\{B_{aq}\}[\equiv]^{(n-1)})$. A newly appearing factor $[\equiv]^{(1-n)}$ has the dimension M^{1-n}. At infinite dilution, $[\equiv] \rightarrow [\equiv]_{Tot} = c_S A_S \Gamma_C$ (c_S is the molar solid concentration, A_S is its specific surface area, and Γ_C is the site density parameter). At typical experimental values c_S = 1 to 10 g·L^{-1}, A_S = 10 m^{2}g^{-1}, and at $\Gamma_C = 10^{-6}$ to 10^{-5} mol m^{-2}, $[\equiv]_{Tot} = 10^{-5}$ to 10^{-3} M. The value of $K_{n,B}^{(1M)}$ becomes $10^{5(n-1)}$ to $10^{3(n-1)}$ times greater than the value of $K_B^{(1M)}$!

As an alternative, n adjacent sites can be represented as a single site $\times$ of density $\Gamma_{nC} = \Gamma_C/n$ to which the sorbate B binds according to reaction $\times + B_{aq} = \times B$ with the mass action $K_{\times B}^{(1M)} = ([\times B])/([\times]\{B_{aq}\})$. If fitted on the same titration data set as the monodentate binding constant, at infinite dilution $K_{\times B}^{(1M)} = n \cdot K_B^{(1M)}$ because $[\equiv]_{Tot} = n[\times]_{Tot}$ and $[\equiv B] = [\times B]$. However, as follows from Equation (82), the thermodynamic constant referenced to standard surface density and infinite dilution should not depend on the "denticity" at all.

So, which of the two fitted intrinsic constants—$K_{n,B}^{(1M)}$ and $K_{\times B}^{(1M)}$—is "more apparent," keeping in mind that the latter is only 2 (3, 4, …) times greater than the monodentate binding constant $K_B^{(1M)}$, but the former is $10^{3(n-1)}$ to $10^{5(n-1)}$ times greater? Apparent constants cannot be used in retrieval of such thermodynamic quantities as the standard entropy change in the reaction necessary to establish temperature corrections. This issue is discussed in detail in Kulik (2006b p. 229-233) on the basis of comparison of published mono-, bi- and tetradentate adsorption constants for ion adsorption on rutile (see Table 2) at elevated temperatures.

A more reasonable definition of standard state for surface species is the hypothetical full monolayer coverage $\theta^{\#} = 1$ (unity mole fraction) referenced to the infinite dilution for surface complexes, and to the "pure surface" (unity mole fraction) for >SOH sites, at standard density $N^{\#} = 10\times10^{18}$ m^{-2}, specific surface $A^{\#} = 10$ m^2·g^{-1}, and real molar sorbent concentration $c^{\#} = c_S$ in g·dm^{-3} (Sverjensky 2003). The suggested conversion of intrinsic constants at 1M standard state into the ones at $\theta^{\#} = 1$ standard state is

$$K_{n,B}^{\theta} = \frac{(c_S N_S A_S)^n}{c^{\#} N^{\#} A^{\#}} K_{n,B}^{1M} \tag{116}$$

where $N_S = 6.022\times10^{23}\,\Gamma_C$ (in m^{-2}). This means that $K_{n,M}^{\theta}$ has dimensionality of $M^{(n-1)}$ and for 2,3,.. dentate surface complexes is not a standard-state thermodynamic constant in the sense of IUPAC definition (Eqn. 102). This drawback of the "full site-occupancy" standard state might arise from an underlying fact that different amounts of adsorbate are involved in obtaining $\theta^{\#} = 1$ for 1, 2, … n-dentate surface species. At $c^{\#} = 10$ g·dm^{-3}, there will be 10^{21} surface sites per dm^3 (1.66×10^{-3} M sites) which defines standard states with 1.66 millimoles of monodentately adsorbed B in the system; 0.83 millimoles of bidentately adsorbed B; and 0.415 millimoles of tetradentately adsorbed B in otherwise the same system. Alas, the species of different "denticity" still have different standard states.

This inconsistency can be eliminated if the fractional coverage of surface sites $\theta^{\#}$ is removed from the definition of standard state and replaced with the standard adsorbed density Γ_o common to all surfaces at standard surface area $A_o = 1/\Gamma_o$ (in m^2·mol^{-1}) such that 1 mol of surface species exists on 1 mol of the sorbent immersed in 55.5084 mol (1 kg) H_2O (Kulik 2002a,b). The proposed value $\Gamma_o = 2\times10^{-5}$ mol·m^{-2} is close to the monolayer water density; for minerals like HFO, $A_o = 50000$ m^2·mol^{-1} or ca. 500-600 m^2·g^{-1}, the values realistic in many experiments. This 1m standard state is the same for n-dentate species at any n because it corresponds to the same molecular adsorbed density Γ_o. This is consistent with the expected infinite dilution behavior of Equation (82) and does not depend on the arbitrariness of selecting reactive sites on the surface. The choice of sites may still influence the non-ideality corrections like Equation (83), but it must not affect the value of thermodynamic equilibrium constant at infinite dilution.

Conversions between intrinsic adsorption constants. As discussed in detail in Kulik (2006b p.194-196), the values of K_B^{int} obtained from fitting the experimental data in site-balance-based codes like FITEQL are referenced to infinite dilution and to a 1M (1m) standard state scaled to the used site density parameter Γ_C. For the monodentate binding, the magnitudes of these constants are related via the underlying values of Γ_{C1}, Γ_{C2}, … (or q_{C1}, q_{C2}, …) and to the standard adsorbed density Γ_o (capacity q_o) as

$$K_B^{\text{int}} = K_B^{\circ}\frac{\Gamma_o}{\Gamma_C} \quad \text{or} \quad K_B^{\text{int},q} = K_B^{\circ}\frac{q_o}{q_C} \tag{117}$$

$$K_B^{\text{int},C2} = K_B^{\text{int},C1}\frac{\Gamma_{C1}}{\Gamma_{C2}} \quad \text{or} \quad K_B^{\text{int},q,C2} = K_B^{\text{int},q,C1}\frac{q_{C1}}{q_{C2}} \tag{118}$$

These conversions apply only to reactions where the adsorbed species is on the right-hand side and the "site" or ≡SOH group (or neither) on the left-hand side; otherwise the density ratios will be inverted. In 1pK or any exchange reaction, such corrections are not needed because the density ratios on both sides cancel out. This is a clear advantage of the 1pK formalism. Equations (117) and (118) permit to compare the magnitudes of apparent and intrinsic adsorption constants fitted at different site densities, and to reduce them to the same standard adsorbed density Γ_o, as shown schematically on Figure 13. This is a pre-requisite for compiling the unified thermodynamic adsorption data base with constants comparable between different mineral surfaces and adsorbed species (Kulik 2002b).

Many published adsorption constants are already presented as referenced to infinite dilution at the full site occupancy $\theta^{\#} = 1$ standard state (Sverjensky 2003). To convert such a constant K_B^{θ} into a Γ_o-based one K_B°, one can substitute Equation (116) into Equation (117) keeping in mind that $K_B^{\text{int}} = K_B^{(1M)}$ in the monodentate case, and $c^{\#} = c_S$:

$$K_B^{\theta} = K_B^{\circ}\frac{\Gamma_o}{\Gamma_C}\frac{c_S N_S A_S}{c^{\#} N^{\#} A^{\#}} = K_B^{\circ}\frac{\Gamma_o}{\Gamma_C^{\#}}\frac{A_S}{A^{\#}} \quad \text{where} \quad \Gamma_C^{\#} = \frac{N^{\#}}{6.022\times10^{23}} = 1.66\times10^{-5}\,\text{mol}\,\text{m}^{-2}$$

Thus, one correction factor is $\Gamma_o / \Gamma_C^{\#} = 1.205$. Another dimensionless factor $A_S / A^{\#} = 0.1A_S$ requires knowledge of the actual BET solid specific surface area in m^2·g^{-1}. Thus, the experimen-

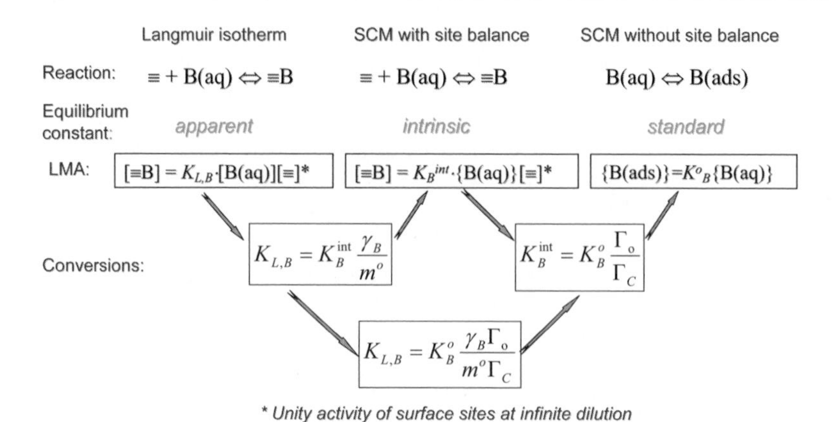

Figure 13. Diagram representing conversions between apparent, intrinsic, and standard-state equilibrium constants for simple monodentate surface binding reactions. See text for details.

tal range of A_S within 0.8 and 830 $m^2 \cdot g^{-1}$ would lead to correction factors from 0.1 to 100 for the whole conversion from K_B° to K_B^{θ}. Usually, these factors should not be so extreme because typical BET values of (hydr)oxides are within 10 and 150 $m^2 \cdot g^{-1}$.

Multi-site sorption models for clay minerals have received a lot of recent attention because of the prospective use of clay rocks and bentonites in nuclear waste disposal (Bradbury and Baeyens 2006, 2009), and references therein. Illite, montmorillonite, mixed layer illite/smectite, and kaolinite are ubiquitous in such systems and possess strong retention properties for radionuclide- and heavy metal cations.

In comparison with (hydr)oxides such as gibbsite and quartz, all clay minerals have a more complex layered structure with two different surfaces, namely amphoteric edges and permanent-charge basal siloxane planes of clay particles (Fig. 14). Thickness of smectite platelets may be as small as 1 nm corresponding to a single T-O-T layer where the T layer is made of SiO_4 tetrahedra, and the O layer – of the Al^{+3} ions in octahedral coordination. Negative permanent charge on montmorillonite basal planes arises mainly due to the isomorphic substitutions of Al^{3+} by Mg^{2+} or other divalent cations in the O layer and is responsible for the cation exchange capacity CEC of ca. q_{ch} = 0.9 eq·kg^{-1} (or charge density on basal planes Γ_{ch} about 1×10^{-6} mol·m^{-2}). This permanent charge is compensated by exchange cations such as Na^+, K^+, Ca^{2+} in the interlayer or sorbed on outer basal planes.

In illite, the permanent charge is mainly due to substitution of Si^{4+} by Al^{3+} in tetrahedral positions (in T layers), and this charge is mainly compensated with K^+ ions. Because of larger thickness of particles (10-50 T-O-T), only part of the interlayer surface in illite is accessible to ion exchange, which leads to lower CEC values about 0.2 eq kg^{-1}. Kaolinite has very low CEC and forms rather thick particles composed of T-O layers, thus having two different basal surfaces of low reactivity: the siloxane and the gibbsite-like plane. A related mineral, pyrophyllite, has the T-O-T structure, but no permanent charge (more about clays in (Grim 1968; Meunier 2005)).

Due to their structure, clay platelets have sorption properties of both amphoteric (hydr) oxides such as gibbsite and silica, and of ion exchangers (such as zeolites). This results in a strong dependence of M^{2+}/M^{+3} cation adsorption (such as Ni^{+2}) on the aqueous electrolyte concentration in the acidic pH region (due to the competition with major cations for the ion exchange sites), and in a weak dependence in the alkaline pH region (where mainly amphoteric edge sites play a role).

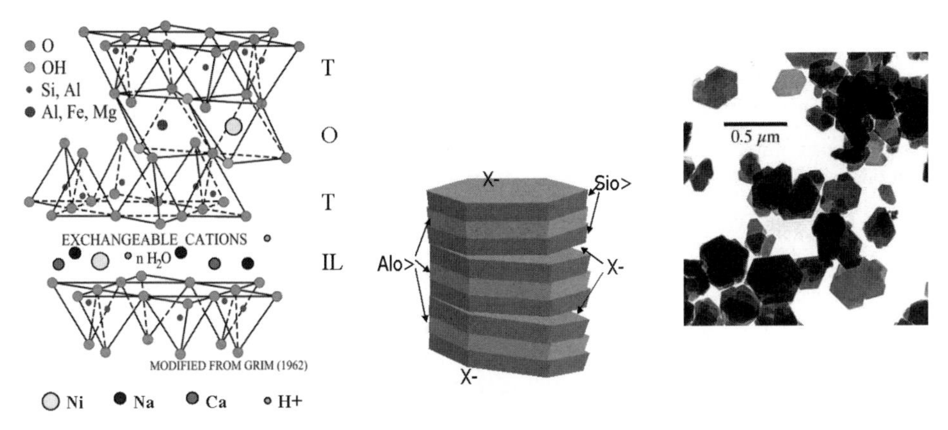

Figure 14. Structure of montmorillonite with Ni^{2+} sorbed on edge and in the interlayer (left, from R. Dähn, pers. comm.); schematic representation of adsorption sites Alo> on O-layer, Sio> on T-layers on edges, and X- on basal surfaces (middle); and TEM image of kaolinite particles (right, from *http://www.gly.uga.edu/schroeder/clay/clayfaculty.html*).

Unlike simple hydroxide surfaces, clay edges expose not only reactive M-OH^z (Alo>) or Si-O^- (Sio>) groups, but also doubly-coordinated M-O-Si oxygen sites. There are many ways to account for different reactive groups on clay particle edges in SCMs, up to applying the predictive MUSIC approach (Bickmore et al. 2003; Tournassat et al. 2004). In the latter case, 27 crystallographically different sites with their densities and protonation constants were determined on the edge surface of MX-80 smectite; only five (after adjustment of MUSIC-predicted K_H^{int} values within 0.3 - 0.9 pK units) were found sufficient to model potentiometric titration data with the multi-site NE model. As the multiplicity of adjustable parameters is considered a shortage of SCMs, a simpler approach has often been used in defining silanol Si≡O- and aluminol Al≡OH groups on clay edges (in 1:1 ratio for kaolinite and 2:1 ratio for smectite or illite), and applying a simple EIL model to describe the potentiometric data (Wanner et al. 1994; Kraepiel et al. 1998; Kulik et al. 2000; Tertre et al. 2006).

There is a rather large uncertainty in splitting the measured BET surface area of clay mineral samples (typically 20-90 $m^2 \cdot g^{-1}$) between the edge- and the basal surfaces because this directly influences the impact of site density parameters for edge-surface sites on modeled SCM curves. While the edge surface may be predominant for kaolinite particles, it is certainly minor for smectite platelets whose total reactive area reaches 800 $m^2 \cdot g^{-1}$. This issue has been addressed by (Tournassat et al. 2003) who used the dry and fluid-cell Atomic Force Microscopy (AFM) and Low-Pressure Gas Adsorption methods to determine the edge specific surface area A_{ES} of Na-smectite particles. With both techniques, a consistent value of $A_{ES} \approx$ 8 $m^2 \cdot g^{-1}$ has been obtained which also corresponds to the mean perimeter-to-area ratio that must be measured in order to get a reliable estimate of A_{ES} for a given clay sample. The large discrepancy between total and BET values of A_S for smectite is due to the stacking of platelets in dry conditions (ca. 20 T-O-T layers) compared to their dispersion in aqueous suspension (1-2 layers). These findings are important because they provide a consistent link between the bulk sorption capacity on edges q_{EC} and the geometric site densities per unit area which can be found from the surface structure data.

B&B sorption model for clays. To date, the most comprehensive experimental data set for sorption of di- and trivalent cations on smectites and illites, as well as the most extensive data base of intrinsic and ion-exchange sorption constants is due to Bradbury and Baeyens (Bradbury and Baeyens 2006, 2009) and references therein. Their (B&B) sorption model, called by its authors the "2 Site Protolysis Non Electrostatic Surface Complexation and Cation Exchange (2SPNE SC/CE) model", combines several site balances for the amphoteric groups on the clay particle edge with one balance for the basal-surface permanent charge sites on which the ion exchange occurs.

Surface complexation reactions on edge sites are represented in B&B model in the 2pK monodentate fashion for (de)protonation, and as the hydrolysis for metal sorption:

$$\equiv SOH^0 \leftrightarrow \equiv SO^- + H^+ \qquad K_{A2}$$

$$\equiv SOH^0 + H^+ \leftrightarrow \equiv SOH_2^+ \qquad 1/K_{A1}$$

$$\equiv SOH^0 + Me^{z+} + yH_2O \leftrightarrow \equiv SOMe(OH)_y^{z-y-1} + H^+ \qquad K_{y,Me} \tag{119}$$

Cation exchange reactions on basal permanent charge sites are given as

$$z_B A^{zA}\text{-clay} + z_A B^{zB} \leftrightarrow z_A B^{zB}\text{-clay} + z_B A^{zA} \qquad {}^B_AK_c \tag{120}$$

with the selectivity coefficients ${}^B_AK_c$ defined in the convention of (Gaines and Thomas 1953), i.e., in an equivalent (charge) fraction basis for sorbed- and in activities for aqueous ions.

The (de)protonation reactions were adjusted together with the sorption capacity parameters for two "weak" edge site types W1 and W2 separately against the potentiometric data for preconditioned Na-montmorillonite and Na-illite, and then kept constant (Table 3) in fitting all the metal sorption data. The B&B model was recently extended to account for the influence of carbonate on radionuclide sorption (Marques Fernandes et al. 2008), as illustrated on Figure 15.

The power of B&B sorption model consists not in fitting the potentiometric titration data (which is, of course, more accurate in electrostatic sorption models), but in describing pH edges and sorption isotherms of many polyvalent cations (Mn(II), Co(II), Cd(II), Ni(II), Zn(II), Eu(III), Am(III), Sn(IV), Th(IV), Np(V), U(VI)), see an example in Figure 15. Many of those show stronger sorption at trace concentrations, which is accounted for by using a small capacity of "strong" sites $\equiv^S OH$ with protolysis constants taken the same as for one of the "weak" sites ($\equiv^{W1}OH$). The fitted intrinsic constants for metal complexes on $\equiv^S OH$ sites on both montmorillonite and illite show good correlations with the respective aqueous hydrolysis constants, leading to the LFER correlations (Bradbury and Baeyens 2005, 2009) useful in predicting cation sorption constants in the absence of experimental data. These LFERs show that metal cations sorb as "inner-sphere" complexes on clay edges in a way similar to adsorption on (hydr)oxide surfaces where such correlations were established previously (Schindler and Stumm 1987; Dzombak and Morel 1990), and this is not much affected by the variable edge surface charge.

Table 3. Parameters fixed in B&B sorption models for montmorillonite and illite (Bradbury and Baeyens 2006, 2009).

Site type	**q_C(MO), mol·kg^{-1} (eq·kg^{-1})**	**$\log_{10}K_{A2}$ (MO)**	**$\log_{10}K_{A1}$ (MO)**	**ΔpK_A (MO)**	**q_C(IL), mol·kg^{-1} (eq·kg^{-1})**	**$\log_{10}K_{A2}$ (IL)**	**$\log_{10}K_{A1}$ (IL)**	**ΔpK_A (IL)**
$\equiv^{W1}OH$	4.0×10^{-2}	4.5	7.9	3.4	4.0×10^{-2}	4.0	6.2	2.2
$\equiv^{W2}OH$	4.0×10^{-2}	6.0	10.5	4.5	4.0×10^{-2}	8.5	10.5	2.0
$\equiv^{S}OH$	2.0×10^{-3}	4.5	7.9	3.4	2.0×10^{-3}	4.0	6.2	2.2
-clay	0.87				0.225			

MO is montmorillonite, IL is illite, -clay is the permanent ion exchange site (-MO and –IL, respectively). Specific surface areas of clay samples were measured but not used in the model.

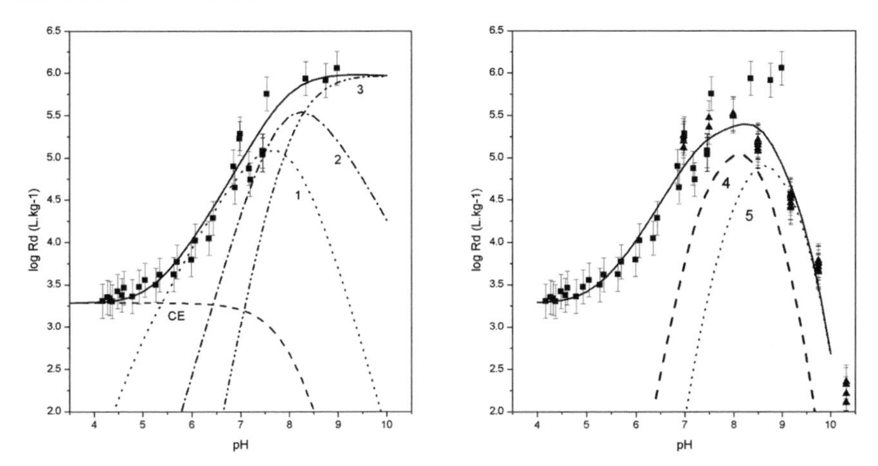

Figure 15. Sorption pH edges of trace Eu(III) (<2×10^{-9} M) on Na-montmorillonite in 0.1 M $NaClO_4$ in absence (left) and in presence (right) of atmospheric CO_2 ($pCO_2 = 10^{-3.5}$ bar). Scattered symbols: measured data in *Rd* scale; lines – calculated with the B&B sorption model. Surface species: (1) $\equiv^S OEu^{2+}$ (2) $\equiv^S OEuOH^+$ (3) $\equiv^S OEu(OH)_2^0$ (4) $\equiv^S OEuCO_3^0$ (5) $\equiv^S OEuOHCO_3^-$; (CE) Eu-clay cation exchange species. For sorption reactions and intrinsic constants see (Marques Fernandes et al. 2008). Lines for surface complexes CE, (1) to (3) are not shown on the right plot. Courtesy of M. Marques Fernandes (personal communication).

Direct comparison of intrinsic constants in the B&B model with those known for simple (hydr)oxides is not quite easy because only sorption capacity parameters, but not area densities, are used in the B&B model. In order to apply conversions given in Equations (118) and (119), the site capacities q_C from Table 3 must first be converted into site density parameters Γ_C. This can be done at least for montmorillonite using the Tournassat's estimate of edge specific surface $A_{ES} \approx 8$ m^2·g^{-1} (see above), yielding $\Gamma_{W1,C} = \Gamma_{W2,C} = 5\times10^{-6}$ mol·m^{-2} and $\Gamma_{S,C} = 2.5\times10^{-7}$ mol·m^{-2}. Thus obtained W1 and W2 site densities correspond to $N_S = 3.0$ nm^{-2} and appear realistic but somewhat less than MUSIC crystallographic densities of reactive sites on gibbsite (8.15 nm^{-2}) or SiO_2 (4.8 nm^{-2}). Using the estimate of basal and interlayer surface area A_X = 780 m^2·g^{-1} (Tournassat et al. 2004), the CEC value (-clay in Table 3) of 0.87×10^{-3} eq·g^{-1} can be converted into site density $\Gamma_{X,C} = 1.115\times10^{-6}$ mol·m^{-2} ($N_X = 0.67$ nm^{-2}) which is close to theoretical value $N_X = 1$ nm^{-2}.

There are two thermodynamic aspects of interest in the model representation of ion exchange on permanent-charge sites in clays, briefly discussed below.

The heterovalent ion exchange representation used in B&B model (Eqn. 120) is only one of several possible; the related Gaines-Thomas selectivity coefficient ${}^B_A K_c$ depends on composition and is actually not a true equilibrium constant. It may be not trivial to compare values of selectivity coefficients for the heterovalent ion exchange data without knowing exactly the conventions and concentration scales behind them. This important issue in thermodynamics of ion exchange has been discussed in soil chemistry (Sposito and Mattigod 1979), in chemical engineering (Grant and Fletcher 1993; Ioannidis et al. 2000), and in relation to ion exchange in clays (Townsend 1984) and zeolites (Fletcher et al. 1984; Pabalan and Bertetti 1994), to mention just a few references. Without going into this extensive theoretical issue, only definitions will be compared here following (Grant and Fletcher 1993) and (Townsend and Coker 2001).

The general ion exchange reaction is a combination of two phase transfer reactions

$$\nu_j M_j^{z_j}(ads) + \nu_i M_i^{z_i}(aq) \leftrightarrow \nu_j M_j^{z_j}(aq) + \nu_j M_j^{z_j}(ads)$$

where j and i are indexes of ions and ν are the reaction stoichiometry coefficients. However, traditionally, the permanent-charge solid phase interface is represented as a hypothetical exchanger ligand X- (for anion exchange X+), and the sorbed components (end members) are constructed in two alternative ways. A component of V-type (after Vanselow) has the composition $M_iX_{z_i}$, i.e., one mole of ion i combined with part of the sorbent carrying z_i moles of charge. A component of G-type (after Gapon) with the composition $(M_i)_{z_i^{-1}}X$ represents one mole of ion i charge combined with part of the sorbent carrying one mole of the opposite charge. Reactions of these components with ions in solution have different stoichiometries, which leads to different exchange equilibrium constants:

$$ {}_{\mathrm{j}}^{\mathrm{i}}K_{\mathrm{V}} = \frac{a_{\mathrm{j}(aq)}^{z_i}\, a_{\mathrm{i(V)}}^{z_j}}{a_{\mathrm{i}(aq)}^{z_j}\, a_{\mathrm{j(V)}}^{z_i}} \quad \text{and} \quad {}_{\mathrm{j}}^{\mathrm{i}}K_{\mathrm{G}} = \frac{a_{\mathrm{j}(aq)}^{1/z_j}\, a_{\mathrm{i(G)}}}{a_{\mathrm{i}(aq)}^{1/z_i}\, a_{\mathrm{j(G)}}} \quad \text{related by} \quad {}_{\mathrm{j}}^{\mathrm{i}}K_{\mathrm{V}} = {}_{\mathrm{j}}^{\mathrm{i}}K_{G}^{z_i z_j} $$

These equilibrium constants are connected to measurable selectivity coefficients by the ratios of activity coefficients of V-type and G-type end members, respectively:

$$ {}_{\mathrm{j}}^{\mathrm{i}}K_{\mathrm{V}}^{\mathrm{a}} = \frac{a_{\mathrm{j}(aq)}^{z_i}\, x_{\mathrm{i(V)}}^{z_j}}{a_{\mathrm{i}(aq)}^{z_j}\, x_{\mathrm{j(V)}}^{z_i}} \quad \text{and} \quad {}_{\mathrm{j}}^{\mathrm{i}}K_{\mathrm{V}} = {}_{\mathrm{j}}^{\mathrm{i}}K_{\mathrm{V}}^{\mathrm{a}} \frac{f_{\mathrm{i(V)}}^{z_j}}{f_{\mathrm{j(V)}}^{z_i}} \quad \text{(Vanselow selectivity coefficient)} \tag{121} $$

$$ {}_{\mathrm{j}}^{\mathrm{i}}K_{\mathrm{G}}^{\mathrm{a}} = \frac{a_{\mathrm{j}(aq)}^{1/z_j}\, x_{\mathrm{i(G)}}}{a_{\mathrm{i}(aq)}^{1/z_i}\, x_{\mathrm{j(G)}}} \quad \text{and} \quad {}_{\mathrm{j}}^{\mathrm{i}}K_{\mathrm{G}} = {}_{\mathrm{j}}^{\mathrm{i}}K_{\mathrm{G}}^{\mathrm{a}} \frac{\gamma_{\mathrm{i(G)}}}{\gamma_{\mathrm{j(G)}}} \quad \text{(Gapon selectivity coefficient)} \tag{122} $$

Here, x stays for the end member mole fraction, hence both selectivity coefficients (and respective activity coefficients) refer to the mole fraction concentration scale. For an ideal ion exchanger, all activity coefficients $f_{\mathrm{i(V)}}$ or $\gamma_{\mathrm{i(G)}}$ remain unity over the whole range if the ion exchange isotherm. Alternatively, Gaines and Thomas (1953) used V-type end members to define a selectivity coefficient (called ${}_{A}^{B}K_{c}$ in B&B model) in an ionic charge (equivalent) fraction scale:

$$ {}_{\mathrm{j}}^{\mathrm{i}}K_{\mathrm{GT}}^{\mathrm{a}} = \frac{a_{\mathrm{j}(aq)}^{z_i}\, \mathrm{E}_{\mathrm{i}}^{z_j}}{a_{\mathrm{i}(aq)}^{z_j}\, \mathrm{E}_{\mathrm{j}}^{z_i}} \quad \text{where } \mathrm{E}_{\mathrm{i}} = \frac{z_i x_{\mathrm{i(V)}}}{z_i x_{\mathrm{i(V)}} + z_j x_{\mathrm{j(V)}}} \quad \text{leading to} \quad {}_{\mathrm{j}}^{\mathrm{i}}K_{\mathrm{V}} = {}_{\mathrm{j}}^{\mathrm{i}}K_{\mathrm{GT}}^{\mathrm{a}} \frac{y_{\mathrm{i(V)}}^{z_j}}{y_{\mathrm{j(V)}}^{z_i}} \tag{123} $$

where the activity coefficients y are taken also in the equivalent fraction scale. Gaines-Thomas activity coefficients $y_{\mathrm{i(V)}}$ cannot be simultaneously equal to unity over the whole isotherm range even for the ideal exchanger (Grant and Fletcher 1993).

Definitions (Eqns. 121-123) show that for the same set of experimental heterovalent ion exchange data, values of selectivity- and activity coefficients will differ depending on the underlying conventions for ion-exchange components and reactions, and on the applied concentration scales. One needs to be careful in compiling selectivity coefficients from different sources into a data base for geochemical speciation codes.

SCM implementation of ion exchange on clays has been considered in (Fletcher and Sposito 1989) in the context of the permanent charge surface ligand X- site balance. In essence, their approach involves transforming classical ion exchange reactions and their equilibrium constants into sets of formation reactions and equilibrium constants for product species such as NaX, CaX_2 or CaClX. To do this, X- is considered as a hypothetical aqueous (master) species representing one mole of charged surface with unit valence. An ion exchange reaction $z_B A_{aq}^{z_A} + z_A BX_{z_B} \leftrightarrow z_B AX_{z_A} + z_A B_{aq}^{z_B}$ (cf. Eqn. 120) having the equilibrium constant and the Vanselow selectivity coefficient defined in Equation (121) is then split into two half-reactions for the hypothetical complex formation:

$$ A_{aq}^{z_A} + z_A X^- \leftrightarrow AX_{z_A} \quad \text{and} \quad B_{aq}^{z_B} + z_B X^- \leftrightarrow BX_{z_B} \tag{124} $$

These reactions are analogous to formation reactions of *n*-dentate surface complexes (see Eqn. 78). Accordingly, their equilibrium constants K_A and K_B are probably intrinsic constants at infinite dilution and 1 *m* standard states referred to site density Γ_C/n, and the isotherms will deviate from competitive Langmuir isotherms. The relation between the equilibrium constant for complete ion exchange reaction and intrinsic adsorption constants K_A and K_B is given by (Fletcher and Sposito 1989):

$$K_V^{\circ} = \frac{f_A^{z_B}}{f_B^{z_A}} \cdot \frac{K_A^{z_B}}{K_B^{z_A}} \left(m_{AX} + m_{BX} \right)^{(z_A - z_B)} \tag{125}$$

Both K_A and K_B are, in principle, unknown, but one of reactions can be arbitrarily chosen as reference with the constant value of K^*_B large enough to ensure that the concentration of "free" X- site ligand makes negligible contribution to mass balance. The value of another constant K_A (as all similar constants) can be obtained using equations like (125). The anion co-adsorption on X- sites is represented using intrinsic constants of reactions such as $\upsilon_A A_{aq}^{z_A} + \upsilon_Y Y_{aq}^{z_Y} + \upsilon_X X^- \leftrightarrow A_{\upsilon_A} Y_{\upsilon_Y} X_{\upsilon_X}$. For montmorillonite, Fletcher and Sposito set $K^*_{Na} = 10^{10}$ as a conventional value for the reference reaction $Na^+_{aq} + X^- \leftrightarrow NaX$ and provided K^{o}_V values for a number of mono- and divalent ions (all between 0.95 to 2.9) and species like CaClX (*K* values from 172 to 1000) (Fletcher and Sposito 1989).

Electrostatic SCM for ion exchange. In site-balance-based non-electrostatic B&B or Fletcher-Sposito SCM ion exchange models, the permanent charge must be completely compensated, which requires direct binding constants in reactions like Equations (124) to be very high (about 10^{10}). This is in contradiction with the physical picture of mainly electrostatic attachment of exchange ions, with intrinsic chemical affinities to the surface similar to those in electrolyte adsorption on amphoteric oxide surfaces, (10^0 to 10^2).

Perhaps because of the above-mentioned inconsistency, several electrostatic SCMs have been developed for ion exchange, summarized in (Dzombak and Hudson 1995). In principle, the electrostatic correction term can be included into the activity expression of X- free species (which is conventionally set to 1 in NE models) with the relative electrostatic potential derived from the Donnan volume model for the porous sorbent (Westall 1987). This approach seems to be potentially very interesting for future studies, especially in compacted clay systems where the typical SCMs fail because the planar EIL representation on particle surface in excess bulk electrolyte becomes inadequate.

(Ad)sorption modeling without site balances

It has been shown (Kulik 2006a) (see also Eqns. 53, 59, 76, 82, 90, 97), that classic adsorption isotherms used in site-balance-based SCMs can be represented as a product $\Gamma_B y_{B(Ads)}$ of the same linear isotherm $\Gamma_B = \Gamma_o K_B^{\circ} \cdot m_{B(aq)} \gamma_{B(aq)}$ and a *surface activity correction term* (SACT) $y_{B(Ads)}$ which is different for each isotherm type. SACT includes the actual complete site coverage density parameter Γ_C for the real MWI surface and obeys the infinite dilution limit $y_{B(Ads)} \to 1$. Close to this limit, any such isotherm is insensitive to Γ_C and, respectively, to the site balance.

Hence, instead of introducing the molar balances for sorption sites of different types, the same isotherms can be computed just by putting the respective Γ_C parameters (and other information about surface patches and site types) into the appropriate SACT expressions, whose magnitudes can be found iteratively in the process of speciation calculations. The EIL electrostatic model corrections come naturally in as part of the activity coefficient for the surface species, i.e., the full employed isotherm is

$$\Gamma_B y_{El,B} y_{B(Ads)} = \Gamma_o K_B^{\circ} \cdot m_{B(aq)} \gamma_{B(aq)} \tag{126}$$

where the electrostatic term $y_{El,B}$ is given by Equation (111) according to the EIL model of choice. The magnitude of equilibrium constant K_B° in Equation (126) depends on the reaction stoichiometry; on the chemical affinity of B to surface site; and on the standard adsorbed density Γ_o, but it does not depend on Γ_C, A_S, c_S, or other properties of the sorbent sample.

GEM implementation. This approach, used in the Gibbs energy minimization GEM-Selektor code package (*http://gems.web.psi.ch*), has been shown to describe well the potentiometric titration and adsorption data on some (hydr)oxides up to hydrothermal conditions (Kulik 2000, 2002a,b, 2006a), and even reproduce some CD-MUSIC adsorption models (Kulik 2006b). In this implementation framework, the mole amount of any j-th species adsorbed on t-th surface of S-th solid sorbent is

$$n_{j(\text{ads})} = \Gamma_j \cdot \phi_{t,S} n_S M_S A_S \tag{127}$$

where n_S is the amount of sorbent, M_S is its molar mass, A_S is its specific surface area, and $\phi_{t,S} = A_{t,S}/A_S$ is the fraction of the t-th surface type where the j-th surface species resides. Unlike in the site-balance-based method, the solid sorbent is included into overall elemental chemical mass balance through its formula (e.g., $Al(OH)_3$) and has the usual stability constant (standard molar Gibbs energy G°_S related to the solubility product K_S) which determines the equilibrium amount $n_{eq,S}$. As seen in Equation (127), A_S now plays the role of "adsorption variable of state" linked to the S-th sorbent, and $\phi_{t,S}$ is the additional such variable for the surface patch (e.g., particle crystal face). Hence, the sorbent together with its A_S, $\phi_{t,S}$ values, and all related surface species, comprises a *sorption phase*. In such phase, the sorbent can be a pure solid or a solid solution (stable or metastable) whose amount in the system depends very little on the surface speciation. Surface complexes on each surface patch form something like a "2D interface solution" which behaves ideal at infinite dilution and zero surface potential limits, but at higher coverage is affected by Langmuir or similar site-saturation, competition, or local interaction non-ideal effects on one side, and by Coulombic long-range non-ideal effects on the other side.

In the GEM-Selektor implementation, many sorption phases can be included into the system definition, each with its own A_S and up to six surface patch types, to which EIL model corrections apply separately. In other words, a sorbent like montmorillonite may have two surface types: one for amphoteric particle edges with silanol and aluminol groups and ion complexes on them acting under the BS electrostatic model; and another for the basal/interlayer siloxane planes, with a non-electrostatic model for ion exchange. For example, using the previously mentioned estimates for specific surfaces of montmorillonite (8 $m^2 \cdot g^{-1}$ for edges and 780 $m^2 \cdot g^{-1}$ for siloxane planes), one has $A_S = 788$ $m^2 \cdot g^{-1}$, $\phi_{1,S} = 0.01$ and $\phi_{2,S} = 0.99$. The sum of ϕ fractions may be less than one because part of the surface may be non-reactive and no sorption is expected to occur there (e.g., as on gibbsite-like planes on kaolinite particles).

Elemental stoichiometry and standard molal properties of surface species. Chemical model systems in GEM approach are normally set in the bulk elemental composition. Abandoning site balances creates a problem of defining the elemental stoichiometry of surface groups and complexes (details in Kulik 2006, p.196-217). This problem is solved in three steps. Firstly, a decision is made whether part of the mineral stoichiometry is included into surface species or not. Secondly, the molal Gibbs energy of the "primary functional group" is defined (this differs for the 2pK$_A$, 1pK$_A$ (MUSIC), and X^- (ion exchange) formalisms). Thirdly, the standard molal Gibbs energy values of surface complexes are obtained from the appropriate surface complexation or exchange reaction equilibrium constants. For the 2pK formulation, this sequence is exemplified in Table 4.

The (hydr)oxide MWI can be viewed as a mineral surface covered with a layer of adsorbed H_2O molecules, each producing two $\equiv OH$ groups, some of which can be deprotonated, and some others can be protonated. Thus, the full-coverage standard and reference states are convenient for $\equiv OH$ groups (adsorbed water) considered as a MWI "solvent". The choice is whether this "surface water species" includes the sorbent atoms or not (e.g., $[TiO_{1.5}]OH^0$ or simply OH^0).

Table 4. Example of the determination of G^o_{298} values for surface species on SiO_2. The constant K_{A2} is already corrected to Γ_o; other details in (Kulik 2006b).

Surface species	Formation reaction	Equilibrium constant	Elemental stoichiometry	Standard molal properties (at 25 °C)
$\equiv OH^o$ group	$0.5H_2O \leftrightarrow O_{0.5}H^o$	$K^o_n = 55.5084$	$O_{0.5}H^0$	G^o = **128.548** kJ·mol^{-1} S^o = 68.36 J·K^{-1}·mol^{-1}
$\equiv O^-$	$\equiv OH^o \leftrightarrow \equiv O^- + H^+$	$K^o_{A2} = 10^{-8.0}$; $\Delta_r G^o_{A2}$ = 45.664 kJ·mol^{-1}	$O_{0.5}^-$	G^o = 45.664 – 0 + (–128.548) = **–82.884** kJ·mol^{-1}
$\equiv O^-Na^+$	$\equiv O^-$ + Na+ $\leftrightarrow$ $\equiv O^-Na^+$	$K^o_{Na} = 10^{-1.5}$; $\Delta_r G^o_{A2}$ = 8.56 kJ·mol^{-1}	$O_{0.5}Na^0$	G^o = 8.56 + (–82.88) + (–261.88) = **–336.20** kJ·mol^{-1}

Choosing the former would make standard molal properties of surface species dependent on composition of the sorbent, rendering them not comparable between different mineral surfaces. But it is desirable to have standard molal properties of surface species comparable between different MWIs. Besides, for solid solutions and aluminosilicates it is difficult to identify a fixed part of stoichiometry that can be assigned to surface species. Hence, the choice *not to include* stoichiometry of the sorbent makes the molal properties of surface species comparable between different MWIs, and consistent with that of aqueous species, solids, and gases.

In CD MUSIC SCMs, the surface species have fractional charges; protonation reactions are expressed in 1pK notation; and "MWI water-solvent" ≡OH groups are not involved. This makes the determination of elemental stoichiometries not so straightforward. Yet, the Pauling bond valence concept helps in solving this task by implementing CD-MUSIC SCMs in the GEM framework without site balances (Kulik 2006b). In plain words, in this method, the surface oxygen has to be "divided" into two parts: one part compensating the bond valence is assigned back to the mineral, and remaining part is assigned to the surface complex. This eliminates the charge imbalance in solid stoichiometry which may affect its stability at large specific surface area and significant site density.

Taking as an example the primary gibbsite surface species (Eqn. 36), the bond valence of ½ must be compensated by ¼ oxygens to retain the bulk oxide stoichiometry $AlO_{1.5}$ which is electroneutral. The rest, 3/4 oxygens, can stay in place of the $\equiv AlO^{-1.5}$ moiety. This renders the $\equiv AlOH^{-1/2}$ surface species an elemental stoichiometry $O_{0.75}H^{-0.5}$, and the $\equiv AlOH_2^{+1/2}$ species - $O_{0.75}H_2^{0.5}$. The reaction (Eqn. 36) for the gibbsite edge surface protonation takes the form

$$O_{0.75}H^{-0.5} + H^+ = O_{0.75}H_2^{0.5} \qquad K^o_A \tag{128}$$

Definition of the elemental stoichiometry for bi- and tetradentate MUSIC surface complexes (e.g., on rutile) proceeds in the same way, by adding up the respective fractions of surface oxygens. For example, the Ca^{2+} 2+2 dentate surface complex on rutile (see Table 2) will have the elemental stoichiometry $2O_{0.717}H^{-0.434} + 2O_{0.278}^{-0.555} + Ca^{2+} = O_2H_2Ca^0$.

The determination of standard molal Gibbs energies g^o_{298} for Al-OH$^{-0.5}$ and Al-OH$_2^{+0.5}$ surface species in Equation (128) is not straightforward because that reaction fixes only the *difference* $-RT\ln K^o_A = \Delta_r G^o_A = g^o(O_{0.75}H^{-0.5}) - g^o(O_{0.75}H_2^{0.5})$. Keeping this difference constant, the standard molal Gibbs energies of both surface complexes can be made simultaneously more positive or more negative to any extent.

Of course, this fact plays no role in the site-balance-based SCMs, but is an important issue in GEM SCMs without site balances. In the latter case, the surface proton charge curve defined by the competition of both surface species in Equation (128) for the sites of Γ_C density always goes through the same pH_{PPZC} but will be the steeper the more negative *both* Gibbs energies $g^o(Al\text{-}OH_2^{+0.5})$ and $g^o(Al\text{-}OH^{-0.5})$ are. Hence, in addition to pH_{PPZC} (K^o_A), one more parameter defining the "strength" of proton binding to the particular oxide surface is needed, although it is hidden in site-balance-based MUSIC models (Kulik 2006b). This parameter (for gibbsite surface) can be represented via the reactions of formation of (fictive) MWI solvent $^3/_4H_2O = O_{0.75}H_{1.5}^0$ ($K_n = 55.5084$) and then its protonation:

$$O_{0.75}H_{1.5}^0 + \tfrac{1}{2}H^+ = O_{0.75}H_2^{0.5} \qquad K_{n2} \tag{129}$$

where the equilibrium constant can be fitted against the potentiometric titration data at the chosen set of EIL model parameters and electrolyte adsorption constants. To some extent, K_{n2} works as the ΔpK_A parameter in 2pK notation (Eqn. 40), so both formalisms are not as different as they seem. For the gibbsite edge surface, $\log_{10}K_{n2} = 5.3$ consistent with $C_1 = 0.9$ $F{\cdot}m^{-2}$; this K_{n2} is quite high and induces a strong non-ideality due to large SACT corrections of the kind shown in Equation (90). Weaker values of K_{n2} have been obtained for goethite surface sites ($\log_{10}K_{1,n2} = 2.7$ and $\log_{10}K_{3,n1} = 4.6$), and yet weaker for rutile surface sites ($\log_{10}K_{1,n2} = 2.0$, $\log_{10}K_{2,n1} = 2.0$) (Kulik 2006b).

Such reference stability constants determine the whole set of G^o_{298} values on oxide surfaces which will remain provisional until the robust values of constants such as $K_{1,n2}$ and $K_{2,n1}$ are clarified for all relevant mineral surfaces. Then it should be possible to compile a database of consistent G^o values for all MUSIC SCMs – a formidable task for the years to come.

CONCLUDING REMARKS

Main focus of this review has been on the thermodynamic concepts used in modeling sorption on MWIs, and neither on the detailed description of methods and codes for surface speciation modeling nor on the compilation of surface chemistry data necessary for development of internally consistent databases of thermodynamic constants for certain ions adsorbed on certain minerals. The reader interested in these aspects may consider fundamental monographs such as Dzombak and Morel (1990) and Kosmulski (2001). Several chemical thermodynamic issues were also left aside; some of these issues are mentioned below.

Temperature and pressure corrections

Development of surface charge and adsorption of cations and anions on sparingly soluble (hydr)oxides (TiO_2, SnO_2, ZrO_2, Fe_3O_4, Fe_2O_3, FeOOH) at hydrothermal conditions (up to 300 °C) have been experimentally studied over the past two decades, as summarized by (Machesky et al. 2006). Qualitative conclusions from these studies are that at increasing temperature, the pH_{PZC} generally decreases, cations are adsorbed stronger, anion adsorption becomes weaker, and the ion exchange on permanent-charge sites is rather insensitive. The extension of thermodynamic sorption models to hydrothermal region requires knowledge of enthalpy $\Delta_r H^o_{T_o}$ or entropy $\Delta_r S^o_{T_o}$ and heat capacity $\Delta_r Cp^o_{T_o}$ effects of surface protonation and complexation reactions at least at $T^o = 298.15$ K (Kulik 2006b; Machesky et al. 2006, and references therein). The temperature corrections in SCMs are usually based on the following set of equations:

$$\Delta_r G^o_T = \Delta_r G^o_{T_o} - (T - T_o)\Delta_r S^o_{T_o} + \left(T - T_o - T\ln\frac{T}{T_o}\right)\Delta_r Cp^o_{T_o}$$

where

$$\Delta_r G^o_T = -\ln(10)\cdot RT\log_{10}K_T \qquad \text{and} \qquad \log_{10}K_T = a + \frac{b}{T} + c\ln T$$

with

$$a = \frac{\Delta_r S^o_{To} - \Delta_r Cp^o_{T_o}(1 + \ln T_o)}{\ln 10 \cdot R}; \quad b = \frac{-\Delta_r H^o_{T_o} + T_o \Delta_r Cp^o_{T_o}}{\ln 10 \cdot R}; \quad c = \frac{\Delta_r Cp^o_{T_o}}{\ln 10 \cdot R}$$

The familiar Vant'Hoff temperature extrapolation is obtained by setting $\Delta_r Cp^o_{T_o} = 0$. To further calculate the $G^o{}_T$ values for individual surface species in GEM SCMs (without site balances), the respective values for aqueous ions are collected, for instance, from the SUPCRT98 data base (Shock et al. 1997) available in modeling packages such as GEMS-PSI. The $\Delta_r S^o_{T_o}$ and $\Delta_r Cp^o_{T_o}$ values for surface complexation reactions can be obtained by fitting K_T^{int} values for different temperatures, as well as estimated from correlation equations. For surface protonation reactions, the simplest yet accurate predictive equation (Kulik 2000) for the pH of pristine point of zero charge

$$\text{pH}_{PPZC,T} = -29.314 + \frac{T_o}{T}\left(\text{pH}_{PPZC,T_o} + 3.2385\right) + 4.545 \ln T \tag{130}$$

was derived by setting $\Delta S^o_{prot,T_o}$= 25 J·K^{-1}·mol^{-1} (single H^+ attachment) for all (hydr) oxides. Further extensions to electrolyte adsorption are facilitated using the symmetric assumption ($K_C = K_L$) with the Rudzinski's $\Delta_r H^o = 0$ hypothesis. In Stern-based EIL models, the caveat is, as expected, in defining temperature dependence for the capacitance density parameter(s), which still remains an issue for future studies.

Pressure corrections of adsorption constants can only be based on simple equations assuming $V^o{}_T = V^o{}_{T_o}$ because the compressibility of surface complexes is unknown:

$$\ln K_{P,T} = \ln K_{P_o,T} - \frac{\Delta_r V^o_T (P - P_o)}{RT} \quad \text{and} \quad G_{T,P} = G^o_T + \bar{V}^o_T (P - P_o)$$

where P_o = 1 bar. However, a few molar volume change $\Delta_r V^o{}_{T_o}$ values, obtained from dilatometry experiments for ion adsorption on HFO and goethite, were interpreted in a seminal paper (Kosmulski and Maczka 2004) as related to the loss of about ½ of ion hydration water shell upon adsorption of both cations and anions. Some $\Delta_r V^o{}_{T_o}$ estimates are 18 cm^3·mol^{-1} for SeO_4^{2-}, 21 cm^3·mol^{-1} for Cd^{2+}, 31 cm^3·mol^{-1} for Co^{2+}, Cu^{2+}, Ni^{2+}, Pb^{2+}, and 42 cm^3·mol^{-1} for Zn^{2+} adsorption. Within this concept, correlations between $\Delta_r V^o{}_{T_o}$ and $\Delta_r S^o{}_{T_o}$ should be possible for adsorption reactions. Pressure corrections do not seem to be dramatic ($K^{int}_{1kbar} = 0.48 K^{int}_{1bar}$ for selenate and $K^{int}_{1kbar} = 0.29 K^{int}_{1bar}$ for Ni^{2+}), and are expected to be insignificant for the outer-sphere electrolyte ion adsorption (see Kulik 2006b for details).

The impact of adsorption on solubility

This interesting topic has been discussed by (Parks 1990) in context of the interfacial free energy $\gamma = (\partial G / \partial A)_{P,T,n}$ where G is the Gibbs energy of the solid, and A is its surface area. The solubility of solid particles in pure water depends on γ according to the Freundlich-Ostwald equation, which for dissolution reactions like $SiO_{2,solid} \leftrightarrow SiO_{2,aq}$ (with the bulk equilibrium constant K) has the form (Parks 1990)

$$\ln K_{(s)} = \ln K + \frac{2\gamma \cdot A_S M_S}{3RT} \cdot \frac{A_\delta}{A_\delta + A_S} \tag{131}$$

where A_δ is the specific surface area (in m^2·g^{-1}) of spherical particles of this mineral having the Tolman reference diameter $\delta \approx 0.1$ nm. This equation can be used for determining the value of γ from solubility data for mineral samples with different nano-particle size.

The few known values of γ for minerals are usually between 0.1 and 1 J·m^{-2} (e.g., 0.35 J·m^{-2} for quartz at near-neutral pH), hence the solubility effect can be significant, up to 10-15 kJ·mol^{-1}. The smaller particles with higher A_S will be unstable with respect to the larger particles of the same mineral, rendering the driving force to slow recrystallization processes of Ostwald ripening. Unfortunately, too little is currently known about interfacial energies of MWIs in pure water and their temperature dependencies, especially for solid solutions or alumosilicates.

Upon adsorption, the interfacial free energy γ of MWI must change according to the Gibbs equation (e.g., Parks 1990)

$$d\gamma = -\sum_j \Gamma_j d\mu_j$$

where Γ_j is the Gibbs adsorbed density (in mol m^{-2}), and $\mu_j = RT \ln a_j$ is the chemical potential of j-th adsorbed component. Thus, a positive adsorption of anything in MWI must decrease γ and, hence, decrease the solubility of a (nano)particulate solid. This was indeed observed e.g., by (Fukushi and Sato 2005), but the body of such data is still fragmentary. Little progress has been made in accounting for the impact of surface area and adsorption on the solubility of particulate and porous mineral sorbents in the existing SCM frameworks. Because the sorbent is only imagined in site-balance-based SCMs, the expectations turn to GEM SCMs (Kulik 2002a) which explicitly include the sorbent with its stoichiometry, stability and surface area. Alternatively, one might attempt to integrate the specific surface and the interfacial energy into thermodynamic representation of sorption using the solid solution concepts. This seems to be an exciting area of future studies.

Sorption layer models (carbonates)

Trace element incorporation by co-precipitation with a ubiquitous mineral such as calcium carbonate is of great interest in geochemistry, representing the final stage of the sorption continuum (see Fig. 1). Trace cation uptake depends on the precipitation rate of the host mineral, resulting in a bias in incorporation of cations relative to their equilibrium aqueous-solid solution partitioning (Rimstidt et al. 1998; Curti 1999; Watson 2004; Gaetani and Cohen 2006, and references therein).

This bias is quantitatively described in the *surface entrapment model* (Watson 2004) which is based on the following postulate: a growing crystal assumes the composition of its surface unless diffusion in the near-surface region is effective during growth. The surface enrichment or depletion results from (metastable) adsorption equilibrium between the MWI and the bulk aqueous solution. When the crystal is growing, the surface layer is "buried" and trapped in the newly-formed lattice, resulting in lattice concentrations that deviate from those predicted by the aqueous-solid solution equilibrium.

The efficiency of this entrapment process depends on the interplay between the growth rate V of the crystal (i.e., the burial rate of the entrapped element M) and the diffusivity D_M in the near-surface layer (which determines how efficiently M can escape to the surface by diffusion). This competition is quantified by the Peclet number Pe = Vl/D_M where l is the half-thickness of the near-surface layer (Watson 1996). Growth entrapment is possible at Pe > 0.1, and is highly efficient at Pe > 10 when the newly-grown crystal preserves the surface composition within the bulk lattice. The M concentration in the near-surface layer is given by $C_M = C_{Mo} F^{\exp(x/l)}$ where x is the distance, and F is the surface entrapment factor, effectively a partition coefficient for the crystal surface relative to the crystal interior. Surface enrichment occurs at $F > 1$, and surface depletion at $F < 1$.

Recent studies (Watson 2004; Gaetani and Cohen 2006; Tang et al. 2008) corroborate the validity of the surface entrapment model which combines diffusion transport with thermodynamics of sorption and aqueous-solid solution partitioning. In this way, an amazing connection between adsorption in MWI and solid solution in the bulk solid is provided. A future research challenge might be to introduce a "real" SCM without site balances and, perhaps, based on an extended variant of the Frumkin or BET isotherm to provide an enhanced multi-species sorption thermodynamic model for the surface layer. In turn, this might eventually lead to a more sound understanding of trace elements behavior during the MWI evolution within the "sorption continuum" in space and time.

ACKNOWLEDGMENTS

The author is grateful to S. Aja, B. Baeyens, A. Bauer, U. Berner, M. Bradbury, V. Brendler, L. Charlet, S. Churakov, E. Curti, M. Fedoroff, O. Gaskova, T. Hiemstra, W. Hummel, N. Kallay, M. Kersten, L. Lakshtanov, J. Lützenkirchen, M. Machesky, E. Oelkers, T. Payne, W. Rudzinski, N. Sahai, V. Sinitsyn, G. Sposito, D. Sverjensky, J. Tits, W. van Riemsdijk, N. Vlasova, E. Wieland, and other colleagues for stimulating discussions. Partial financial support by Nagra, Wettingen, is gratefully appreciated.

REFERENCES

Anderson GM (2005) Thermodynamics of natural systems. Cambridge University Press, Cambridge

Aranovich G, Donohue M (1998) Analysis of adsorption isotherms: Lattice theory predictions, classification of isotherms for gas–solid equilibria, and similarities in gas and liquid adsorption behavior. J Colloid Interface Sci 200:273-290

Benjamin MM (2002a) Modeling the mass-action expression for bidentate adsorption. Environ Sci Technol 36:307-313

Benjamin MM (2002b) Water Chemistry. McGraw Hill, New York

Bickmore BR, Rosso KM, Mitchell SC (2006a) Is there hope for multi-site complexation (MUSIC) modeling? *In:* Surface Complexation Modelling. Interface Sciences and Technology, Vol. 11. Lützenkirchen J (ed) Elsevier, Amsterdam, p 269-283

Bickmore BR, Rosso KM, Nagy KL, Cygan RT, Tadanier CJ (2003) Ab initio determination of edge surface structures for dioctahedral 2:1 phyllosilicates: Implications for acid-base reactivity. Clays Clay Mineral 51:359–371

Bickmore BR, Rosso KM, Tadanier CJ, Bylaska EJ, Doud D (2006b) Bond-valence methods for pKa prediction. II. Bond-valence, electrostatic, molecular geometry, and solvation effects. Geochim Cosmochim Acta 70:4057–4071

Bickmore BR, Tadanier CJ, Rosso KM, Monn WD, Eggett DL (2004) Bond-valence methods for pKa prediction: critical reanalysis and a new approach. Geochim Cosmochim Acta 68:2025–2042

Borkovec M, Rusch U, Westall JC (1998) Modeling of competitive ion binding to heterogeneous materials with affinity distributions *In:* Adsorption of Metals by Geomedia. Variables, Mechamisms, and Model Applications. Jenne EA (ed) Academic Press, San Diego, p 467-482

Bourikas K, Hiemstra T, van Riemsdijk WH (2001) Ion pair formation and primary charging behavior of titanium oxide (anatase and rutile). Langmuir 17:749-756

Bradbury MH, Baeyens B (2005) Modelling the sorption of Mn(II), Co(II), Ni(II), Zn(II), Cd(II), Eu(III), Am(III), Sn(IV), Th(IV), Np(V) and U(VI) on montmorillonite: Linear free energy relationships and estimates of surface binding constants for some selected heavy metals and actinides. Geochim Cosmochim Acta 69:875–892

Bradbury MH, Baeyens B (2006) A quasi-mechanistic non-electrostatic modelling approach to metal sorption on clay minerals. *In:* Surface Complexation Modelling. Interface Sciences and Technology, Vol. 11. Lützenkirchen J (ed) Elsevier, Amsterdam, p 518-538

Bradbury MH, Baeyens B (2009a) Sorption modelling on illite. Part I: Titration measurements and the sorption of Ni, Co, Eu and Sn. Geochim Cosmochim Acta 73:990-1003

Bradbury MH, Baeyens B (2009b) Sorption modelling on illite. Part II: Actinide sorption and linear free energy relationships. Geochim Cosmochim Acta 73:1004-1013

Brown GEJ, Parks GA (2001) Sorption of trace elements on mineral surfaces: Modern perspectives from spectroscopic studies, and comments on sorption in the marine environment. Inter Geology Rev 43:963-1073

Brunauer S, Emmett PH, Teller E (1938) Adsorption of gases in multimolecular layers. J Phys Chem 60:309-319

Bruno J, Bosbach D, Kulik D, Navrotsky A (2007) Chemical thermodynamics of solid solutions of interest in radioactive waste management. A state-of-the-art report. OECD NEA, Paris

Curti E (1999) Coprecipitation of radionuclides with calcite: estimation of partition coefficients based on a review of laboratory investigations and geochemical data. Appl Geochem 14:433-445

Davis JA, Coston JA, Kent DB, Fuller CC (1998) Application of the surface complexation concept to complex mineral assemblages. Environ Sci Technol 32:2820-2828

Davis JA, James RO, Leckie JO (1978) Surface ionization and complexation at the oxide/water interface. I. Computation of electrical double layer properties in simple electrolytes. J Colloid Interface Sci 63:480-499

Davis JA, Kent DB (1990) Surface complexation modeling in aqueous geochemistry. Rev Mineral 23:177-260

De Boer JH (1968) The Dynamical Character of Adsorption. Clarendon Press, Oxford

Dugger DL, Stanton JH, Irby BN, McConnell BL, Cummings WW, Maatman RW (1964) The exchange of twenty metal ions with the weakly acidic silanol group of silica gel. J Phys Chem 68:757-760

Dzombak DA, Hudson RJM (1995) Ion exchange: The contributions of diffuse layer sorption and surface complexation *In:* Aquatic Chemistry: Interfacial and Interspecies Processes. Huang CP, O'Melia CR, Morgan JJ (eds) American Chemical Society, Washington, D.C., p 59-94

Dzombak DA, Morel FMM (1990) Surface Complexation Modeling. Hydrous ferric oxide. Wiley Interscience, New York

Farley KJ, Dzombak DA, Morel FMM (1985) A surface pretipitation model for sorption of cations on metal oxides. J Colloid Interface Sci 106:226-242

Fletcher P, Franklin KR, Townsend RP (1984) Thermodynamics of binary and ternary ion exchange in zeolites: The exchange of sodium, ammonium, and potassium ions in mordenite. Philos Trans R Soc London A 312:141-178

Fletcher P, Sposito G (1989) The chemical modelling of clay/electrolyte interactions for montmorillonite. Clay Minerals 24:375-391

Fukushi K, Sato T (2005) Using a surface complexation model to predict the nature and stability of nanoparticles. Environ Sci Technol 39:1250-1256

Gaetani GA, Cohen AL (2006) Element partitioning during precipitation of aragonite from seawater: A framework for understanding paleoproxies. Geochim Cosmochim Acta 70:4617–4634

Gaines GL, Thomas HC (1953) Adsorption studies on clay minerals. II. A formulation of the thermodynamics of exchange adsorption. J Chem Phys 21:714-718

Gamsjäger H, Königsberger E (2003) Solubility of sparingly soluble ionic solids in liquids. *In:* The Experimental Determination of Solubilities. Hefter GT, Tomkins RPT (eds) John Wiley and Sons, Ltd, p 315-358

Ganor J, Reznik IJ, Rosenberg YO (2009) Organics in water-rock interactions. Rev Mineral Geochem 70:259–369

Giammar D (2001) Geochemistry of uranium at mineral-water interfaces: Rates of sorption-desorption and dissolution-precipitation reactions. PhD Dissertation, California Institute of Technology, Pasadena, CA

Giammar DE, Hering JG (2001) Time scales for sorption-desorption and surface precipitation of uranyl on goethite. Environ Sci Technol 35:3332-3337

Glynn P (2000) Solid solution solubilities and thermodynamics: Sulfates, carbonates and halides. Rev Mineral Geochem 40:481-511

Goldberg S (1992) Use of surface complexation models in soil chemical systems. Adv Agronomy 47: 233-329

Grant SA, Fletcher P (1993) Chemical thermodynamics of cation exchange reactions: theoretical and practical considerations. Ion Exch Solvent Extr 11:1-108

Grim RE (1968) Clay Mineralogy. McGraw Hill, New York

Hayes KF, Leckie JO (1987) Modeling ionic strength effects on cation adsorption at hydrous oxide/solution interfaces. J Colloid Interface Sci 115:564-572

Hayes KF, Redden G, Ela W, Leckie JO (1991) Surface complexation models: An evaluation of model parameter estimation using FITEQL and oxide mineral titration data. J Colloid Interface Sci 142:448-469

Helfferich F (1962) Ion Exchange. McGraw-Hill, New York

Hiemstra T, van Riemsdijk WH (1991) Physical chemical interpretation of primary charging behaviour of metal (hydr) oxides. Colloids Surf 59:7-25

Hiemstra T, van Riemsdijk WH (1996) A surface structural approach to ion adsorption: The charge distribution (CD) model. J Colloid Interface Sci 179:488-508

Hiemstra T, van Riemsdijk WH (2002) On the relationship between surface structure and ion complexation of oxide- solution interfaces. *In:* Encyclopedia of Surface and Colloid Science. Hubbard A (ed) Marcel Dekker, New York, p 3773-3799

Hiemstra T, van Riemsdijk WH, Bolt GH (1989a) Multisite proton adsorption modeling at the solid/solution interface of (hydr)oxides: A new approach. I. Model description and evaluation of intrinsic reaction constants. J Colloid Interface Sci 133:91-104

Hiemstra T, van Riemsdijk WH, Bolt GH (1989b) Multisite proton adsorption modeling at the solid/solution interface of (hydr)oxides: A new approach. II. Application to various important (hydr)oxides. J Colloid Interface Sci 133:105-117

Hiemstra T, Venema P, van Riemsdijk WH (1996) Intrinsic proton affinity of reactive surface groups of metal (hydr)oxides: The bond valence principle. J Colloid Interface Sci 184:680-692

Hiemstra T, Yong H, van Riemsdijk WH (1999) Interfacial charging phenomena of aluminum (hydr)oxides. Langmuir 15:5942-5955

Hinz C (2001) Description of sorption data with isotherm equations. Geoderma 99:225–243

Honeyman BD, Santschi PH (1992) The role of particles and colloids in the transport of radionuclides and trace metals in the oceans. *In:* Environmental Particles. Buffle J, van Leeuwen H (eds) Lewis Publishers, Boca Raton, Ann Arbor, p 379-423

Ioannidis S, Anderko A, Sanders SJ (2000) Internally consistent representation of binary ion exchange equilibria. Chem Eng Sci 55:2687-2698

Kallay N, Kovačević D, Žalac S (2006) Thermodynamics of the solid/liquid interface - its application to adsorption and colloid stability *In:* Surface Complexation Modeling. Lützenkirchen J (ed) Elsevier, Amsterdam, p 133-170

Kallay N, Preočanin T, Žalac S (2004) Standard states and activity coefficients of interfacial species. Langmuir 20:2986-2988

Kinniburgh DG, Barker JA, Whitfield M (1983) A comparison of some simple adsorption isotherms for describing divalent cation adsorption by ferrihydrite. J Colloid Interface Sci 95:370-384

Kosmulski M (2001) Chemical Properties of Material Surfaces. Marcel Dekker, New York

Kosmulski M, Maczka E (2004) Dilatometric study of the adsorption of heavy-metal cations on goethite. Langmuir 20:2320-2323

Kraepiel AML, Keller K, Morel FMM (1998) On the acid-base chemistry of permanently charged minerals. Environ Sci Technol 32:2829-2838

Krauskopf KB (1956) Factors controlling the concentrations of thirteen rare metals in sea-water. Geochim Cosmochim Acta 10:1-26

Kulik DA (2000) Thermodynamic properties of surface species at the mineral-water interface under hydrothermal conditions: A Gibbs energy minimization triple layer model of rutile in NaCl electrolyte to 250 °C. Geochim Cosmochim Acta 64:3161-3179 (erratum 65:2027)

Kulik DA (2002a) Gibbs energy minimization approach to modelling sorption equilibria at the mineral-water interface: Thermodynamic relations for multi-site-surface complexation. Am J Sci 302:227-279

Kulik DA (2002b) Sorption modelling by Gibbs energy minimisation: Towards a uniform thermodynamic database for surface complexes of radionuclides. Radiochim Acta 90:815-832

Kulik DA (2006a) Classic adsorption isotherms incorporated in modern surface complexation models: Implications for sorption of actinides. Radiochim Acta 94:765–778

Kulik DA (2006b) Standard molar Gibbs energies and activity coefficients of surface complexes (Thermodynamic insights) *In:* Surface Complexation Modelling. Interface Sciences and Technology, Vol. 11. Lützenkirchen J (ed) Elsevier, Amsterdam, p 171-250

Kulik DA, Aja SU, Sinitsyn VA, Wood SA (2000) Acid-base surface chemistry and sorption of some lanthanides on K+ saturated Marblehead illite: II. A multisite-surface complexation modeling. Geochim Cosmochim Acta 64:195-213

Langmuir D (1997) Aqueous Environmental Geochemistry. Prentice Hall, NJ

LaViolette RA, Redden GD (2002) Comment on "Modeling the mass-action expression for bidentate adsorption". Environ Sci Technol 36:2279-2280

Lefèvre G, Duc M, Fédoroff M (2006) Accuracy in the determination of acid-base properties of metal oxides surfaces *In:* Surface Complexation Modelling. Interface Sciences and Technology, Vol. 11. Lützenkirchen J (ed) Elsevier, Amsterdam, p 35-66

Limousin G, Gaudet J-P, Charlet L, Szenknect S, Barthes V, Krimissa M (2007) Sorption isotherms: A review on physical bases, modeling and measurement. Appl Geochem 22:249–275

Lützenkirchen J (2002a) Surface complexation models of adsorption *In:* Encyclopedia of Surface and Colloid Science. Hubbard A (ed) Marcel Dekker, New York, p 5028-5046

Lützenkirchen J (2002b) Surface complexation models of adsorption: A critical survey in the context of experimental data *In:* Adsorption: Theory, Modeling, and Analysis. Tóth J (ed) Marcel Dekker, New York, p 631-710

Lützenkirchen J, Magnico P, Behra P (1995) Constraints upon electrolyte binding constants in triple-layer model calculations and consequences of the choice of the thermodynamic framework. J Colloid Interface Sci 170:326-334

Machesky ML, Předota M, Wesolowski DJ, Vlček L, Cummings PT, Rosenqvist J, Ridley MK, Kubicki JD, Bandura AV, Kumar N, Sofo JO (2008) Surface Protonation at the Rutile (110) Interface: Explicit Incorporation of Solvation Structure within the Refined MUSIC Model Framework. Langmuir 24:12331-12339

Machesky ML, Wesolowski DJ, Palmer DA, Ichiro-Hayashi K (1998) Potentiometric titrations of rutile suspensions to 250 C. J Colloid Interface Sci 200 298-309

Machesky ML, Wesolowski DJ, Palmer DA, Ridley MK, Bénézeth P, Lvov SN, Fedkin MV (2006) Ion adsorption into the hydrothermal regime: experimental and modeling approaches. *In:* Surface Complexation Modelling. Interface Sciences and Technology, Vol. 11. Lützenkirchen J (ed) Elsevier, Amsterdam, p 324-358

Manceau A, Marcus MA, Tamura N (2002) Quantitative speciation of heavy metals in soils and sediments by synchrotron X-ray techniques. Rev Mineral Geochem 49:341-428

Marques Fernandes M, Baeyens B, Bradbury MH (2008) The influence of carbonate complexation on lanthanide/actinide sorption on montmorillonite. Radiochim Acta 96:691-697

Mathur SS, Dzombak DA (2006) Surface complexation modeling: Goethite. *In:* Surface Complexation Modelling. Interface Sciences and Technology, Vol. 11. Lützenkirchen J (ed) Elsevier, Amsterdam, p 443-468

Meunier A (2005) Clays. Springer, Berlin

Morel FMM, Hering JG (1993) Principles and Applications of Aquatic Chemistry. Wiley Interscience, New York

Morel FMM, Yeasted JG, Westall JC (1981) Adsorption models: A mathematical analysis in the framework of general equilibrium calculations. *In:* Adsorption of inorganics at solid-liquid interfaces. Anderson MA, Rubin AJ (eds) Ann Arbor Sience, Ann Arbor, MI, p 263-294

Pabalan RT, Bertetti FP (1994) Thermodynamics of ion-exchange between Na^+/Sr^{2+} solutions and the zeolite mineral clinoptilolite. Mat Res Soc Symp Proc 333:731-738

Parks GA (1990) Surface energy and adsorption at mineral/water interfaces: An introduction. Rev Mineral 23:133-176

Payne TE, Davis JA, Ochs M, Olin M, Tweed CJ (2004) Uranium adsorption on weathered schist – intercomparison of modelling approaches. Radiochim Acta 92:651–661

Payne TE, Davis JA, Ochs M, Olin M, Tweed CJ, Altmann S, Askarieh MM (2006) Comparative evaluation of surface complexation models for radionuclide sorption by diverse geologic materials. *In:* Surface Complexation Modeling. Interface Sciences and Technology, Vol. 11. Lützenkirchen J (ed) Elsevier, Amsterdam, p 605-633

Pokrovsky OS, Schott J (2002) Surface chemistry and dissolution kinetics of divalent metal carbonates. Environ Sci Technol 36:426-432

Prieto M (2009) Thermodynamics of solid solution-aqueous solution systems. Rev Mineral Geochem 70:47–85

Rahnemaie R, Hiemstra T, van Riemsdijk WH (2006) A new surface structural approach to ion adsorption: Tracing the location of electrolyte ions. J Colloid Interface Sci 293:312–321

Ridley MK, Hiemstra T, van Riemsdijk WH, Machesky ML (2009) Inner-sphere complexation of cations at the rutile-water interface: A concise surface structural interpretation with the CD and MUSIC model. Geochim Cosmochim Acta 73:1841-1856

Ridley MK, Machesky ML, Wesolowski DJ, Palmer DA (2004) Modeling the surface complexation of calcium at the rutile-water interface to 250 °C. Geochim Cosmochim Acta 68:239-251

Ridley MK, Machesky ML, Wesolowski DJ, Palmer DA (2005) Surface complexation of neodymium at the rutile-water interface: A potentiometric and modeling study in NaCl media to 250 °C. Geochim Cosmochim Acta 69:63-81

Rimstidt JD, Balog A, Webb J (1998) Distribution of trace elements between carbonate minerals and aqueous solutions. Geochim Cosmochim Acta 62:1851-1863

Rodda DP, Johnson BB, Wells JD (1996) Modeling the effect of temperature on adsorption of lead(II) and zinc(II) onto goethite at constant pH. J Colloid Interface Sci 184:365-377

Sahai N, Sverjensky DA (1997a) Evaluation of internally consistent parameters for the triple-layer model by the systematic analysis of oxide surface titration data. Geochim Cosmochim Acta 61:2801-2826

Sahai N, Sverjensky DA (1997b) Solvation and electrostatic model for specific electrolyte adsorption. Geochim Cosmochim Acta 61:2827-2848

Schindler PW, Stumm W (1987) The surface chemistry of oxides, hydroxides, and oxide minerals. *In:* Aquatic Surface Chemistry. Stumm W (ed) Wiley-Interscience, New York, p 83-110

Schott J, Pokrovsky OS, Oelkers EH (2009) The link between mineral dissolution/precipitation kinetics and solution chemistry. Rev Mineral Geochem 70:207–258

Sherman DM (2009) Surface complexation modeling: mineral fluid equilbria at the molecular scale. Rev Mineral Geochem 70:181–205

Shock EL, Sassani DC, Willis M, Sverjensky DA (1997) Inorganic species in geologic fluids: Correlations among standard molal thermodynamic properties of aqueous ions and hydroxide complexes. Geochim Cosmochim Acta 61:907-950

Somorjai GA (1994) Thermodynamics of Surfaces. Introduction to Surface Chemistry and Catalysis. John Wiley & Sons, Hoboken NJ, p 271-318

Sparks DL, Scheidegger AM, Strawn DG, Scheckel KG (1999) Kinetics and mechanisms of metal sorption at the mineral-water interface. *In:* Mineral-Water Interfacial Reactions. Kinetics and mechanisms. Sparks DL, Grundl TJ (eds) American Chemical Society, Washington DC, p 108-135

Sposito G (1984) The Surface Chemistry of Soils. Oxford University Press, Oxford

Sposito G, Mattigod SV (1979) Ideal behavior in Na^+ - trace metal cation exchange on Camp Berteau montmorillonite. Clays Clay Mineral 27:125-128

Stachowicz M (2007) Solubility of arsenic in multi-component systems. From the misroscopic to the macroscopic scale Ph.D. Dissertation, University of Wageningen, p 207

Stumm W (1992) Chemistry of the solid-water interface. Wiley-Interscience, New York

Stumm W, Huang CP, Jenkins SR (1970) Specific chemical interactions affecting the stability of dispersed systems. Croatica Chemica Acta 42:223-244

Sverjensky DA (1994) Zero-point-of-charge prediction from crystal chemistry and solvation theory. Geochim Cosmochim Acta 58:3123-3129

Sverjensky DA (2001) Interpretation and prediction of triple-layer model capacitances and the structure of the oxide-electrolyte-water interface. Geochim Cosmochim Acta 65:3643-3655

Sverjensky DA (2003) Standard states for the activities of mineral surface sites and species. Geochim Cosmochim Acta 67:17-28

Sverjensky DA (2005) Prediction of surface charge on oxides in salt solutions: Revisions for 1:1 (M+L-) electrolytes. Geochim Cosmochim Acta 69:225–257

Sverjensky DA, Fukushi K (2006) A predictive model (ETLM) for As(III) adsorption and surface speciation on oxides consistent with spectroscopic data. Geochim Cosmochim Acta 70:3778–3802

Sverjensky DA, Sahai N (1996) Theoretical prediction of single-site surface-protonation equilibrium constants for oxides and silicates in water. Geochim Cosmochim Acta 60:3773-3797

Tamura H (2004) Theorization on ion-exchange equilibria: activity of species in 2-D phases. J Colloid Interface Sci 279:1–22

Tang J, Koehler SJ, Dietzel M (2008) Sr^{2+}/Ca^{2+} and $^{44}Ca/^{40}Ca$ fractionation during inorganic calcite formation: I. Sr incorporation. Geochim Cosmochim Acta 72:3718-3732

Tardy Y, Fritz B (1981) An ideal solid solution model for calculating solubility of clay minerals. Clays Clay Mineral 16:361-373

Tertre E, Castet S, Berger G, Loubet M, Giffaut E (2006) Surface chemistry of kaolinite and Na-montmorillonite in aqueous electrolyte solutions at 25 and 60 °C: Experimental and modeling study. Geochim Cosmochim Acta 70:4579–4599

Tóth J (1995) Uniform interpretation of gas/solid adsorption. Adv Colloid Interface Sci 55:1-239

Tournassat C, Ferrage E, Poinsignon C, Charlet L (2004) The titration of clay minerals. II. Structure-based model and implications for clay reactivity. J Colloid Interface Sci 273:234-246

Tournassat C, Neaman A, Villiéras F, Bosbach D, Charlet L (2003) Nanomorphology of montmorillonite particles: Estimation of the clay edge sorption site density by low-pressure gas adsorption and AFM observations. Am Mineral 88:1989–1995

Townsend RP (1984) Thermodynamics of ion exchange in clays. Philos Trans R Soc London A 311:301-314

Townsend RP, Coker EN (2001) Ion exchange in zeolites. Studies in surface science and catalysis. 137:467-524

Turner DR, Bertetti FP, Pabalan RT (2006) Applying surface complexation modeling to radionuclide sorption *In:* Surface Complexation Modelling. Interface Sciences and Technology, Vol. 11. Lützenkirchen J (ed) Elsevier, Amsterdam, p 553-604

van Riemsdijk WH, Hiemstra T (2006) The CD-MUSIC model as a framework for interpreting ion adsorption on metal (hydr)oxide surfaces. *In:* Surface Complexation Modelling. Interface Sciences and Technology, Vol. 11. Lützenkirchen J (ed) Elsevier, Amsterdam, p 251-268

Villalobos M (2006) Triple layer modelling of carbonate adsorption on goethites with variable adsorption capacities based on congruent site-occupancy. *In:* Surface Complexation Modelling. Interface Sciences and Technology, Vol. 11. Lützenkirchen J (ed) Elsevier, Amsterdam, p 417-442

Wanner H, Albinsson Y, Karnland O, Wieland E, Wersin P, Charlet L (1994) The acid/base chemistry of montmoryllonite. Radiochim Acta 66/67:157-162

Watson EB (1996) Surface enrichment and trace-element uptake during crystal growth. Geochim Cosmochim Acta 60:5013-5020

Watson EB (2004) A conceptual model for near-surface controls on the trace-element and stable isotope composition of abiogenic calcite crystals. Geochim Cosmochim Acta 68:1473-1488

Westall JC (1987) Adsorption mechanisms in aquatic surface chemistry. *In:* Aquatic Surface Chemistry. Stumm W (ed) Wiley, New York, p 3-32

Westall JC (1995) Modeling of the association of metal ions with heterogeneous environmental sorbents. Mater Res Soc Symp Proc 353:937-950

Wingrave JA (1996) Single cation adsorption equation for the solution - metal oxide interface. J Colloid Interface Sci 183:579-596

Yates DE, Levine S, Healy TW (1974) Site-binding model of the electrical double layer at the oxide/water interface. J Chem Soc Faraday Trans I,70:1807-1818

Zachara JM, Westall JC (1999) Chemical modeling of ion adsorption in soils. *In:* Soil Physical Chemistry, 2nd edition. Sparks DL (ed) CRC Press, Boca Raton FL, p 48-96

Zhang Z, Fenter P, Cheng L, Sturchio NC, Bedzyk MJ, Předota M, Bandura A, Kubicki JD, Lvov SN, Cummings PT, Chialvo AA, Ridley MK, Bénézeth P, Anovitz L, Palmer DA, Machesky ML, Wesolowski DJ (2004) Ion adsorption at the rutile-water Interface: Linking molecular and macroscopic properties. Langmuir 20: 4954-4969

Zhu C (2002) Estimation of surface precipitation constants for sorption of divalent metals onto hydrous ferric oxide and calcite. Chem Geol 188:23– 32

Reviews in Mineralogy & Geochemistry
Vol. 70 pp. 181-205, 2009

5

Surface Complexation Modeling: Mineral Fluid Equilbria at the Molecular Scale

David M. Sherman

Department of Earth Sciences
University of Bristol
Bristol BS8 1RJ, United Kingdom

dave.sherman@bristol.ac.uk

INTRODUCTION

Complexation of dissolved ions by mineral surfaces is one of the primary controls on the aqueous concentrations of trace micronutrients, toxic heavy metals and radionuclides. In many reactive transport simulations and geochemical box models, these sorption processes are modeled using a simple distribution coefficient

$$K_d = \frac{M_{solid}}{M_{solution}} \tag{1}$$

where M_{solid} is the mass concentration of component i in the solid and $M_{solution}$ is the mass concentration of i in the aqueous solution. However, this approach fails to give a correct description of reactive transport (Brady and Bethke 2000). Even in simple batch simulations (no transport), the K_d approach cannot predict how element partitioning will change with the chemistry (pH, ionic strength) of the system. To predict these changes, we need to consider the nature of the mineral water interface and the chemical equilibria involved in surface complexation.

The surfaces of oxides and silicates consist of coordinatively unsaturated oxygens which, in aqueous solutions, act as Bronsted bases:

$$\text{S-O}^- + \text{H}^+ = \text{S-OH} \tag{2a}$$

$$\text{S-OH} + \text{H}^+ = \text{S-OH}_2^+ \tag{2b}$$

Hence, the surface of an oxide mineral will have a pH-dependant charge. This charge may attract ions from solution to form outer-sphere complexes:

$$\text{S-O}^- + \text{M(H}_2\text{O)}_6^{2+} = \text{S-O}^- \cdots (\text{H}_2\text{O})_6\text{M}^{2+} \tag{3}$$

However, the surface oxygens themselves may act as Lewis basis and directly coordinate to cations to form inner-sphere complexes:

$$\text{S-OH} + \text{M}^{+2} = \text{S-O-M}^+ + \text{H}^+ \tag{4a}$$

$$2\text{S-OH} + \text{M}^{+2} = (\text{S-O})_2\text{M} + 2\text{H}^+ \tag{4b}$$

Anions may form inner-sphere surface complexes by exchanging with surface oxygens

$$\text{S-OH}_2^+ + \text{L}^- = \text{S-L} + \text{H}_2\text{O} \tag{5}$$

We can express the K_d for the complexation of M^{2+} by the monodentate complex (4a) in terms of the conditional equilibrium expressions for (2a), (2b) and (4a):

1529-6466/09/0070-0005$05.00 DOI: 10.2138/rmg.2009.70.5

$$K_{a1} = \frac{\{SOH_2^+\}}{\{SOH\}[H^+]}; \quad K_{a2} = \frac{\{SOH\}}{\{SO^-\}[H^+]}; \quad K_s = \frac{\{SOM^+\}}{\{SO^-\}[M^{2+}]} \tag{6}$$

Here, { } denotes the ideal activity of a surface species which is taken to be the mole-fraction of surface sites taken up by that species. Since

$$\{SOH_2^+\} + \{SOM^+\} + \{SOH\} + \{SO^-\} = 1 \tag{7}$$

we get

$$K_d = \frac{\{SOM^+\}S_{tot}}{[M^{+2}]} = \frac{S_{tot}K_s}{K_{a1}K_{a2}[H^+]^2 + K_{a2}[H^+] + K_s[M^{2+}] + 1} \tag{8}$$

where S_{tot} is the moles of surface sites/gram of sorbent. However, spectroscopic evidence is that most inner-sphere complexes are bidentate. For bidentate complexes, K_d is a function of the surface loading

$$K_d = \frac{\{(SO)_2M^+\}S_{tot}/2}{[M^{+2}]} = \frac{K_s\{SOH\}S_{tot}/2}{K_{a1}K_{a2}[H^+]^2 + K_{a2}[H^+] + K_s\{SOH\}[M^{2+}] + 1} \tag{9}$$

Hence, K_d's for surface complexes are complex functions of pH, surface loading and concentration.

A further problem is that the conditional equilibrium constants K_a, K_s, themselves, are a complex function of pH and surface loading. Protonation of surface oxygens and sorption of ions will give the surface a net charge. The electrostatic potential resulting from this charge makes a significant contribution to the energetics of sorption reactions. Sorption via outer-sphere complexation is mostly driven by the electrostatic potential. It is convenient to formulate the free-energy of sorption/protonation reactions as

$$\Delta G = \Delta G_{intrinsic} + \Delta G_{elect} \tag{10}$$

where $\Delta G_{intrinsic}$ is the free energy change at infinite dilution when the surface charge is neutral and ΔG_{elec} is that part of the free energy change associated with bringing a charged ion in the vicinity of a charged surface. The electrostatic contribution to the free energy is

$$\Delta G_{elect} = -\int_0^\infty \Delta\rho(x) F\psi(x)dx \tag{11}$$

where ψ is the electrostatic potential at the mineral-water interface, F is the Faraday constant, and $\Delta\rho(x)$ is the change in charge density associated with the sorption reaction. In nearly all surface complexation models, the activity coefficients of the surface species are given by the change in free energy associated with the electrostatic energy. Hence, the equilibrium constant for a sorption reaction

$$SO^- + M^{2+} = SOM^+ \tag{12}$$

is

$$K = \frac{\{SOM\}}{\{SO^-\}\gamma_B[M^{2+}]}\exp\left(\frac{-\Delta G_{elec}}{RT}\right) \tag{13}$$

where γ_B is the activity coefficient of B, R is the gas constant and T is the absolute temperature (Kelvin).

Modeling sorption reactions via the K_d approach was a practical necessity in older reactive transport simulations due to the computational demands of combining hydrologic transport and thermodynamic speciation. With the development of fast computers and new codes, full transport-speciation simulations are now feasible. However, before we can perform such simulations, we need thermodynamic models for surface complexation reactions. Setting up a surface complexation model requires that we know the reactive sites on a mineral surface, their proton affinities, the nature of the surface complex (e.g., outer-sphere, bidentate vs. monodentate inner-sphere) and how the electrostatic potential at the mineral-water interface (ψ) changes as a function of surface loading and pH. In short, we need an understanding of the mineral-water interface at a molecular scale. The goal of this review is to give an introduction to how a molecular scale picture of the mineral-water interface is being used to develop surface complexation models.

NATURE OF OXIDE SURFACES

Iron and manganese oxides provide, by far, the most reactive mineral surfaces in soils, sediments and aquatic environments. Of these minerals, the most studied is goethite (α-FeOOH) and this will be used as the type substrate in this review.

Goethite surface structure

TEM studies (e.g., Prelot et al. 2003) show that goethite crystallites are acicular needles dominated by the {101} and {001} surfaces. These are terminated by {210} and {010} surfaces. This is outlined in Figure 1. (Note: here, surfaces are indexed in terms of the standard space group setting *Pnma*; in many papers, surfaces are referenced in terms of the alternate setting *Pbnm* so that {101} becomes {110} etc.).

At the atomic level, each surface will be terminated by oxygens which may be protonated. We would not expect coordinatively unsaturated Fe atoms (e.g., five-fold coordinated Fe) on any of the goethite surfaces since, under ambient environmental conditions, such centers would be quickly hydrolyzed by liquid or vapor H_2O. The presence of terminal oxygens is the fundamental atomic-level description of oxide surfaces and is our starting point for developing thermodynamic models of surface complexation. There are several types of surface oxygens, differing in the number Fe atoms they are coordinated to: singly coordinated O (designated -OH or >FeOH), doubly coordinated O (designated μOH or >Fe_2OH) and triply coordinated O (designated as μ_3OH or >Fe_3OH). On the {210} surfaces there are 10.8 -OH sites/nm^2. On the {101} surfaces, there are 3.03 Fe-OH sites/nm^2 and 2.78 μ_3O sites/nm^2. Given typical surface areas of 35-85 m^2/g (Schwertmann and Cornell 1991), sorption capacities of goethite are,

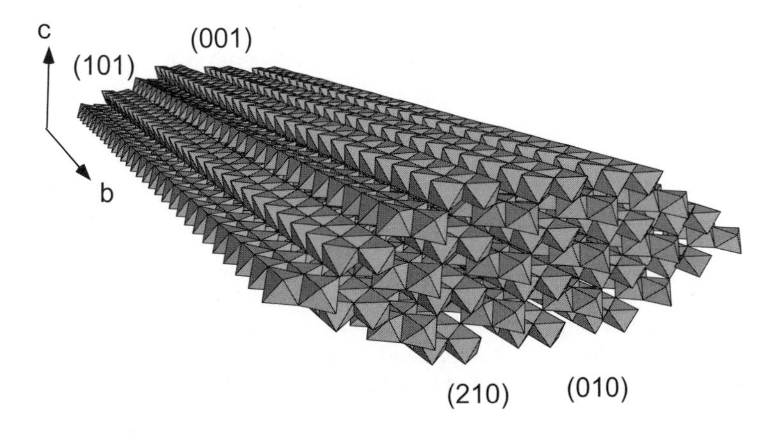

Figure 1. Structure of goethite and the dominant surfaces indexed in the *Pnma* setting.

therefore, on the order of 10^{-4} moles/gram. Higher apparent sorption capacities could result from surface precipitation or solid solution.

A starting point for predicting surface reactivity is provided by Pauling's second rule: in a stable structure, the charge on each anion will be neutralized by the sum of the "electrostatic bond strengths" donated by the surrounding cations (Pauling 1960). In Pauling's model, the electrostatic bond strength (EBS) donated by a cation equals its charge/coordination number. For example, the $>Fe_2OH$ oxygens on {101} and {001} (Fig. 2) receive a net bond strength of 2.0 (each Fe^{3+} center contributes a bond strength of 3/6 = 0.5; in the absence of hydrogen bonding to the solvent H_2O molecules, an H^+ contributes a bond strength of 1.0). Since Pauling's rule is satisfied, we predict that the $>Fe_2OH$ oxygens to be relatively unreactive. In contrast, the singly coordinated surface oxygens $>FeO$ receive an EBS of only 0.5. Hence we would predict these oxygens to be readily protonated in aqueous solution so that the resulting $>FeOH$ oxygen receives an EBS of 1.5. Since this is still < 2, we predict the $>FeOH$ sites should complex with cations or be protonated further to give $>FeOH_2^+$ with an EBS of 2.5. However, because of the excess EBS, the resulting $>FeOH_2^+$ oxygens could be displaced by another ligand (as in 5). The μ_3O or $>Fe_3O$ oxygens receive an EBS of 1.5; these could be singly protonated to give $>Fe_3OH$ with an EBS of 2.5.

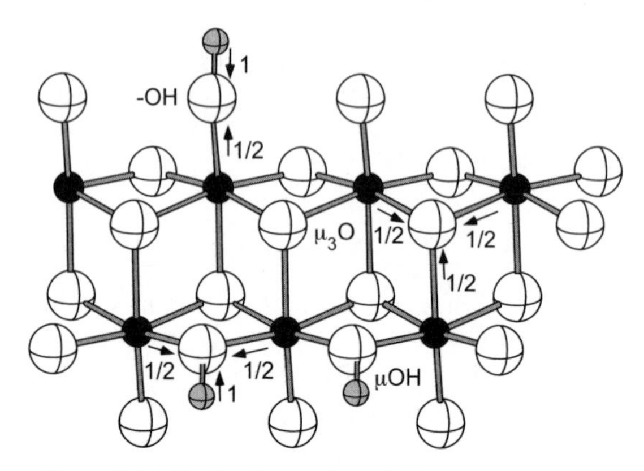

Figure 2. Pauling bond strength model for the three types of surface oxygens on {101} and {001}.

ACID-BASE CHEMISTRY OF OXIDE SURFACES

The surface oxygens are coordinatively unsaturated and in aqueous solutions will have variable degrees of protonation to yield a surface charge on the oxides that is pH dependent. Moreover, protonation of the surface oxygens will compete with surface complexation (or promote ligand exchange). Before we can develop surface complexation models, we need to know the proton affinities of the surface oxygens. Early models of surface protonation (e.g., Parks 1965; Dzombak and Morel 1990) assumed that surface oxygens were amphoteric:

$$K_{a1}: >Fe_nOH_2^+ = >Fe_nOH + H^+ \quad (14a)$$

$$K_{a2}: >Fe_nOH = >Fe_nO^- + H^+ \quad (14b)$$

For each type of surface oxygen, therefore, we would need both K's to model the surface charging (a "2pK model"). Potentiometric titrations of iron (hydr)oxide suspensions, however, are not able to resolve different surface sites and a practical strategy might be to model the surface in terms of one average surface site (e.g., Dzombak and Morel 1990). The surface protonation of goethite in terms of (14a,b) gives $pK_{a1} = 6.93$ and $pK_{a2} = 9.65$ (Mathur and Dzombak 2006); however, this difference between pK_{a1} and pK_{a2} is too small to be physically reasonable. For iron (hydr)oxides, the 2pK model is falling out of favor (Lützenkirchen 1998) and has been largely replaced by the multisite 1-pK model discussed below.

Bond valence theory and the MUSIC model

Bond valence theory (Brown 1981, 2002) extends Pauling's second rule to a more flexible description of charge distribution and geometry. The *bond valence* S_{ij} between two atoms i and j is defined by

$$S_{ij} = \exp\left(\frac{R_0 - R_{ij}}{B}\right) \tag{15}$$

where R_{ij} is the distance between the two atoms while R_0 and B are empirical constants.

As with Pauling's second rule, the net charge of a surface oxygen i will be

$$Z_i = -2 + \sum_{1}^{j} S_{ij} \tag{16}$$

The proton affinities of oxygen ligands correlate with the bond valence donated to those oxygens. Using this correlation, Hiemstra et al. (1989) developed the MUSIC* model of oxide surface reactivity by considering the bond valence sums donated to different surface oxygens. Hydrogen bonds to oxygens donate a Pauling bond strength of 0.2 while those hydrogens directly bonded to a surface oxygen donate a bond strength of 0.8. Estimates of the K_{a1} and K_{a2} (Eqn. 14) for different surface oxygens (Table 1) suggest that the overall protonation of goethite (and other iron(III) (hydr)oxides) can be modeled in terms of a 2-site 1 pK formalism:

$$>FeOH^{-0.5} + H^+ = FeOH_2^{+0.5}\ K_{(s)} \tag{17a}$$

$$>Fe_3O^{-0.5} + H^+ = Fe_3OH^{+0.5}\ K_{(t)} \tag{17b}$$

Determination of the bond valence sum, however, has several complications: first, the bond valence is function of bond lengths to the surrounding cations; moreover, we need to know the solvation structure to predict the bond valence donated by the hydrogen bonding.

Table 1. Predicted pK's for deprotonation reactions on the {101} surface of goethite.

	Calculated pK's				
	H89	**R96**	**V98**	**GH03**	**A08**
		S-OH₂⁺ = S-OH + H⁺			
>FeOH (-OH)	10.6	9.1	7.7	3.7,7.7, 11.7	12.1
>Fe₂OH (μ-OH)	−0.2	7.2	0.4	−3.6, 0.4	−4.0
>Fe₃OH (μ₃-OH)					
		S-OH = S-O⁻ + H⁺			
>FeOH (-OH)	24.4		19.6	15.6,19.6, 23.6	15.1
>Fe₂OH (μ-OH)	13.6	9.3	12.3	8.3,12.3	9.5
>Fe₃OH (μ₃-OH)	4.3	8.1	11.7	11.7	10.0

H89 = Hiemstra et al. (1989) using original MUSIC model; R96 = Rustad et al. (1996) calculated using interatomic potentials; V98 = Venema et al. (1998) calculated using bond valence version of MUSIC model; GH03 = Gaboriaud and Ehrhardt (20003) calculated using different H-bonding configurations. A08 = Aquino et al. (2008) from DFT in COSMO solvation field.

* MUSIC = MUltiSIte Complexation. The acronym is somewhat confusing since the concept of modeling surface complexation in terms of multiple types of surface sites is entirely rigorous and should not necessarily imply that one is invoking Bond Valence theory to predict site pK_a's. Accordingly, some authors distinguish the MUSIC method from the MUSIC model.

Gaboriaud and Ehrhardt (2003) give predictions of pK for different hydrogen bonding structures (Table 1). Several authors have used *ab initio* calculations (discussed below) on finite clusters representing surface sites to predict bond lengths to surface oxygens; the calculated bond lengths are then used to predict bond valence sums. However, this approach is unrealistic: calculations using even the most reliable exchange-correlation functionals and reasonably complete basis sets can still only predict bond lengths to within 2-3 %; such errors correspond to large uncertainties in pK values. Indeed, the actual empirical correlation between bond lengths, bond valence and proton affinity is very approximate and not accurate enough to enable quantitative predictions (for example, see Bickmore et al. 2006).

Classical atomistic simulations

A drastic, but very useful, approximation is to model the energetics of a chemical system as the sum of two- (or three-) body interactions described by an interatomic potential. For example, the structure and energetics of the goethite surface might be modeled in terms of Fe-O, O-O, O-H and H-O-H potential functions. From the potential functions, we can then calculate the forces on each atom and then minimize the total energy to find the optimized structure. We could then calculate the energetics of adding or removing protons from surface oxygens. Rustad et al. (1996a,b) developed a set of such potentials and used them to calculate model static ($T = 0$) internal energies of the gas-phase reactions

$$Fe(OH)_n(H_2O)_{6-n}{}^{3-n} = Fe(OH)_{n+1}(H_2O)_{5-n}{}^{2-n} + H^+ \quad (18)$$

(n = 1,2,3,4). These energies were found to correlate well to the experimental free energies (pK's) for protonation of the $Fe(OH)_n(H_2O)_m{}^{3-n}$ complexes in solution. Rustad et al. (1996b) then calculated the proton affinities of surface oxygens on model goethite surface slabs (including long-range electrostatics). Using the pK correlation obtained for the simple clusters, Rustad et al. (1996b) were then able to predict pK_a's of surface oxygens on goethite. The results differed substantially from those obtained using the MUSIC model (Table 1).

Quantum chemical studies of surface protonation

A much more rigorous approach than the MUSIC model and classical interatomic potentials would be to proton affinities of surface oxygens using quantum mechanical calculations. However, Fe and Mn (hydr)oxides provide a significant challenge for molecular level investigations as they have complex electronic structures with unpaired electrons, partially covalent bonding and, in some phases, metal atoms in more than one formal oxidation state. The theory and methodology of quantum mechanical calculations of chemical structures and energetics will not be given here. A previous review (Sherman 2001) in this series outlines the theory behind quantum chemical methods that are being used to address problems in aqueous geochemistry. Theoretical calculations on systems complex enough to be of geochemical interest must be done using density functional theory. The fundamental approximation such calculations involve is the exchange-correlation functional used to describe the interelectronic interactions. Moreover, most calculations need to approximate the problem by using a finite system such as a cluster of atoms to represent a surface site. The local aqueous solution may be approximated by several waters (for short-range interations) or a solvation field modeled as a dielectric continuum (e.g., the COSMO model of Klamt 1995).

Recently, Aquino et al. (2008) used a cluster taken from the {101} surface (Fig. 3) which also included 3 water molecules in the first hydration shell of each surface OH site and a COSMO solvation field to simulate the long-range solvation. Calculations on clusters such as this are often very difficult to converge if a correct spin-polarized (spin-unrestricted) charge density is used. To circumvent this, Aquino et al. (2008) used a spin-restricted calculation. This approach will most certainly will give an incorrect description of the Fe-O bonding since the spin-restricted calculation will mix in excited state configurations that involve Fe-O antibonding orbital's (Sherman 1985a,b). The consequence for proton affinities is unclear. Aquino et al.

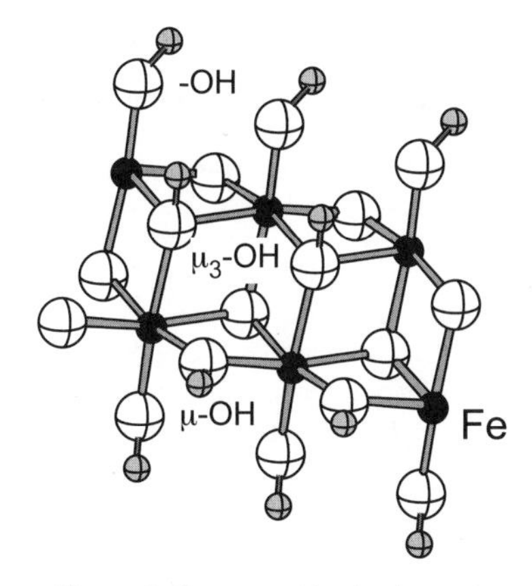

Figure 3. Cluster used by Aquino et al. (2008) to calculate proton affinities.

(2008) calculated protonation/deprotonation enthalpies of the >FeOH, >Fe_2OH and >Fe_3OH oxygens; the protonation/deprotonation free energies were obtained by subtracting the entropy ($T\Delta S$) terms obtained from calculated vibrational frequencies. Note, however, this would have been approximate since, with a cluster having some of the atomic positions fixed, many of the vibrational modes would have imaginary frequencies. The results (A08 in Table 1) differ substantially from previous MUSIC model predictions. Based on the predicted protonation/deprotonation energies, Aquino et al. (2008) assumed a 2pK model for the >FeOH and >Fe_2OH sites but a 1pK for the >Fe_3OH site. Given the ratio of site densities, they predict the pK_{pzc} of goethite to be 9.1 in close agreement with experiment. However, this close agreement is somewhat fortuitous given that the uncertainties in pK's are on the order of a log unit. The best quantum chemistry calculations have uncertainties of energies on the order of 4 kJ/mole. These would correspond to uncertainties in pK of 0.7 units.

Experimental measurements of surface protonation

Potentiometric titrations of goethite suspensions (Figure 4) are not able to resolve the protonation reactions of different surface oxygens (e.g., Boily et al. 2003). Such data can usually only be modeled in a 1pK formalism by assuming that there is one overall surface >Fe_nOH site with a protonation pK = pH_{pzc} (where pH_{pzc} is the pH at zero surface charge; for goethite which is free of adsorbed CO_3^{-2}, this is near 9.2). Moreover, potentiometric titrations are not able to measure the number of surface oxygens since, with increasing protonation, the increase surface charge prevents further protonation so that no real end-point can be reached (Lutzenkirchen 2005). How this surface potential is modelled is described in the next section.

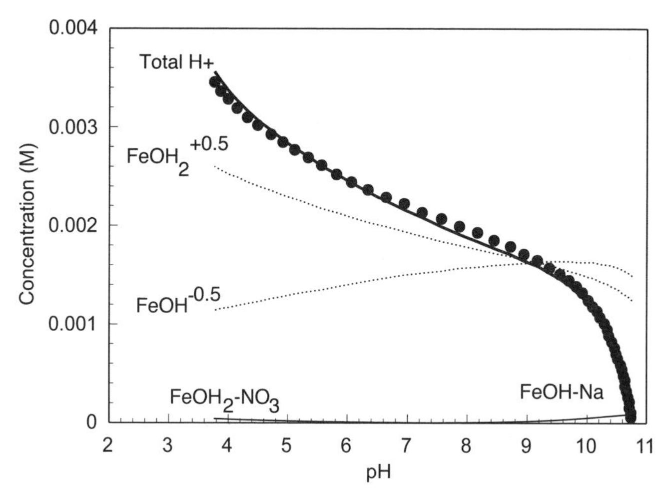

Figure 4. Potentiometric titration of goethite suspension in 0.1 m $NaNO_3$ fit to a 1pK basic Stern model. The charges of outer-sphere complexes of Na and NO_3^- are placed in the diffuse layer.

ELECTROSTATIC POTENTIAL AT THE MINERAL/WATER INTERFACE

The electrostatic work associated with sorption of an ion from the bulk solution to the charged surface is

$$\Delta G_{elect} = -\int_0^{\infty} \Delta\rho(x) F \psi(x) dx \tag{19}$$

where $\psi(x)$ is the electrostatic potential as a function of distance from the surface and $\Delta\rho(x)$ is the change in charge density. In nearly all the surface complexation models used, this is simplified by distributing the charge among a series of layers. Separating each layer is a plane, i, where the electrostatic potential is ψ_i. We make the abstraction that the charge in each layer is assigned to the boundary plane of that layer. Hence,

$$\Delta G_{elect} = \sum_i -\Delta z_i F \psi_i \tag{20}$$

where Δz_i is the change in charge of plane i.

In the diffuse layer model, the charged surface attracts a diffuse swarm (diffuse layer) of counter-ions from the bulk solution. The diffuse swarm of counter-ions neutralizes the surface charge (σ_0) so that $\sigma_o = -\sigma_d$. The excess charge of the diffuse layer (σ_d) corresponds to an electrostatic potential at the mineral surface (ψ_0) given by the Gouy-Chapman expression:

$$\sigma_d = \left(\frac{8\varepsilon\varepsilon_0 RTI}{10^3}\right)^{1/2} \sinh\left(\frac{zF\psi_0}{2RT}\right) \tag{21}$$

where z is the charge of the symmetrical electrolyte (e.g., 1 for $NaNO_3$), ε is the dielectric constant of the solution and ε_0 is the vacuum permittivity. Hence, if we know the surface charge (σ_o), we can determine the surface potential (ψ_0). Of course, we cannot know the surface charge without first knowing the surface potential; hence, we start with an initial guess and determine both charge and potential self-consistently as described in the next section.

A major shortcoming of the Gouy-Chapman model (21) is that it assumes that the electrolyte ions are point-charges in a dielectric continuum. However, the finite size of the real ions and their hydration shells means that they cannot approach the surface closer than their hydration radius. Moreover, as discussed below, the dielectric constant of water near the mineral surface will be very different from that in the bulk solution. Consequently, the diffuse layer model cannot give the correct potential very close to the mineral surface. The Stern model corrects for this by introducing a layer of constant capacitance (C_s) with thickness d between the mineral surface and the diffuse layer (Fig. 5). The electrostatic potential at the boundary between the Stern layer and the diffuse layer is ψ_d, and the surface electrostatic potential is now found from

$$\sigma_0 = C_s(\psi_0 - \psi_d) \tag{22}$$

The charge density in the diffuse layer is (21). With the constraint that $\sigma_o = -\sigma_d$ we can solve for ψ_0 and ψ_d. In principle, the capacitance of the Stern layer is $C_s = \varepsilon\, \varepsilon_0 / d$. Most studies treat the Stern layer capacitance as an empirical parameter that is optimized when fitting sorption and protonation data (e.g., Boily et al. 2001). As C_s increases, the basic Stern model approaches the diffuse layer model. In practice, we assign the charges of surface oxygens and inner-sphere complexes to the 0-plane. Outer-sphere complexes are assigned to the diffuse layer.

Extended stern models and charge distribution. We can extend the basic Stern model by introducing a second Stern layer so that we can distribute the charge of a sorbed species between the 0-plane, 1-plane and diffuse layer and an extended Stern Model. This approach,

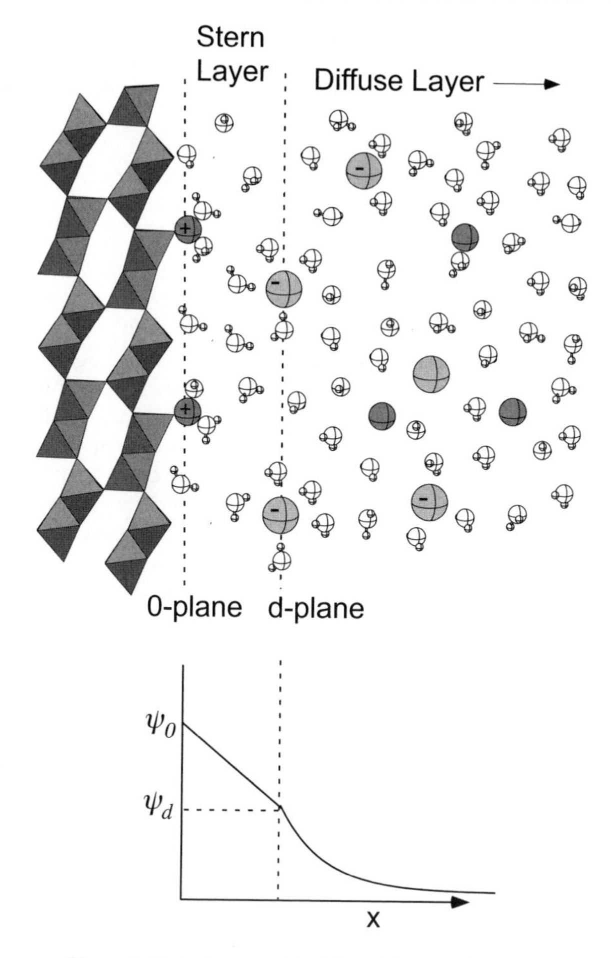

Figure 5. Basic Stern model of the oxide-water interface.

for example, might be used in models for the sorption of ternary complexes or oxyanions. Combining the charge distribution idea with the MUSIC approach to predict the charge distributions yields the CD-MUSIC model of Hiemstra and Van Riemsdijk (1996). This will be discussed in the examples given below.

Atomistic simulations

The Stern model still treats ions as point charges in a dielectric continuum. This approach, of course, fails for ions in bulk electrolytes and there is no reason it should work at the mineral-water interface. Moreover, in continuum model, we are assuming the dielectric constant is a bulk property that can be applied at the length scale of molecular interactions.

Atomistic simulations are needed to take us beyond the continuum model and determine the actual structure of the electrical double layer. Results of such simulations could then be used to develop more realistic models for the electrostatic potential at the mineral surface. The essential feature of such simulations is that they would model the structure and energetics of the solvent water at the mineral water interface. Static molecular orbital calculations (even on periodic systems) would not be very useful since they would not explore the many configurational degrees of freedom that define the problem. This is a problem for molecular dynamics where the equations of motion are solved and the time-averages of the atomic positions and velocities are evaluated. A large enough number of atoms, simulated over a long-enough time will allow us to evaluate thermodyamic quantities. However, simulations of the mineral-water interface that are large enough to be realistic (and geochemically interesting) are not really practical feasible using *ab initio* calculations. An alternative approach is to model the interatomic interactions using classical potentials so that the forces on atoms can be computed rapidly. Such approximate calculations can allow dynamic simulations with thousands of atoms over "long" (1 ns) time scales.

Keriset et al. (2006) used classical interatomic potentials in molecular dynamic simulations to determine the structure of NaCl solutions at the goethite {100} surface. At the interface, water forms a highly ordered structure that is different from the hydrogen bonded structure in the bulk solution (Fig. 6 -- bottom). The simulations also show the structure of the Na and Cl ions adopted at the interface (Fig. 6 -- top) with several counter-ion layers held together by the ordered solvent. There is little evidence of a diffuse layer as expected from Gouy-Chapman theory. The highly ordered water structure implies that the dielectric constant of water (ε) in the Stern layer will be much smaller than that in the bulk. We could use MD simulations to predict the capacitance of the mineral water interface; however, such simulations are computationally intensive since they are done by evaluating the time-fluctuations of the average dipole moment. In a rigid structure, long simulations would be needed to get reliable statistics.

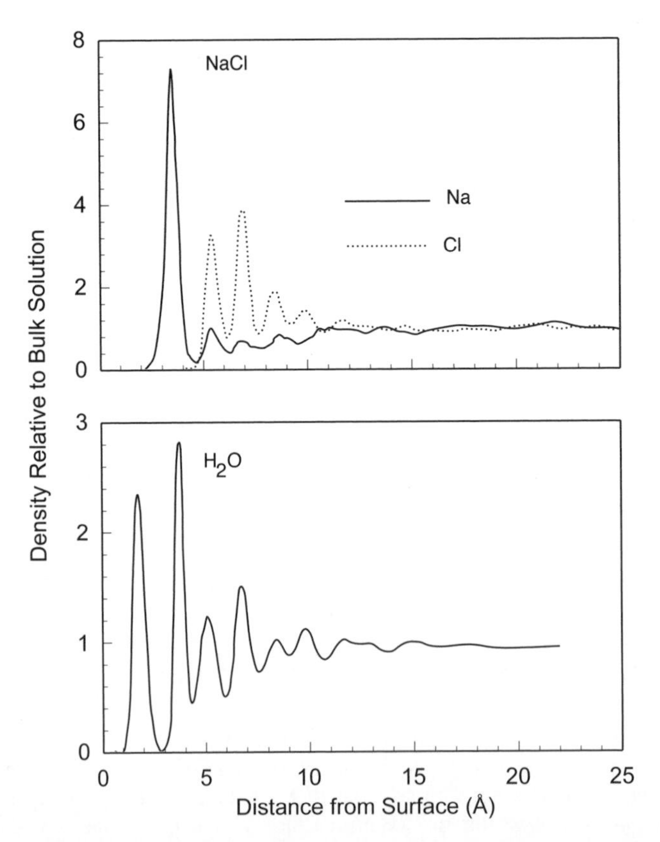

Figure 6. Structure and the goethite {001} water interface from MD simulations of Kerisit et al. (2006).

Given the layered structure adopted by the charged ions and solvent molecules, the electrostatic potential and near the goethite surface also shows an oscillatory structure (Fig. 7). However, much of this is due to the water structure and the dipole contribution to the electric field. For surface complexation modeling, we are interested in the electrostatic work required to bring a charged species to the surface. What are needed, therefore, are simulations which bring a charged test particle to the goethite surface. Such simulations would be easy to do: at each distance, the system is equilibrated and a time average of the total energy is determined. This is repeated for a number of distances and, using thermodynamic integration, the work associated with bringing the charged test particle is evaluated. From such simulations, we could develop model potentials that could be used in surface complexation models. Whether these would be a practical improvement over the simple continuum models (e.g., basic Stern), and how we would implement them, is unclear.

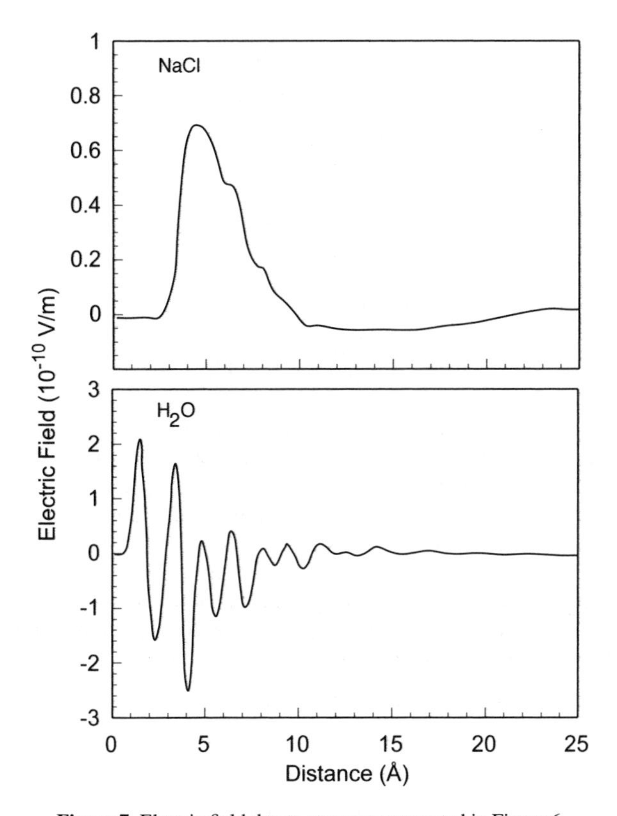

Figure 7. Electric field due to structure presented in Figure 6.

COMPUTATIONAL METHODS IN SURFACE COMPLEX MODELING

Given a molecular picture of the surface reactivity and the surface complexation, we can fit experimental sorption data to a surface complexation model that incorporates the oxygen site densities, the protonation reactions and a model for the surface electrostatic potential. The surface speciation calculated from a thermodynamic model must be solved iteratively since the surface electrostatic potential is a function of the surface speciation. There are two computational approaches that can be used: free energy minimization and the law of mass action. The chapter by Kulik (2009) describes modeling sorption equilibria using free energy minimization. In principle, free energy minimization should be a more flexible method since one can allow for changes in the sorbing surface. Calculations based on the law of mass action are computationally efficient, however, and are a convenient method when fitting experimental data to a set of equilibrium constants. The computational method is based the "tableau algorithm" that was coded as mineql (Morel and Morgan 1972).

Outline of the mass-action tableau method

A chemical system is modeled in terms of components and species. The species are linear combinations of components. Associated with each component is a species that is the "free component". This is the species that corresponds to the fraction of a component which has not combined with any other component. Hence, a system with components >Fe-OH, H+, Zn^{+2}, H_2O will have the actual species >Fe-OH, H^+, Zn^{+2}, H_2O that correspond to the free components. In addition, however, we will have species $>FeOH_2^+$, $>Fe(OH)_2Zn$ and OH^-

$$>FeOH_2^+ = >FeOH + H^+ \tag{23}$$

$$(>FeOH)_2Zn = 2 >FeOH + Zn^{+2}$$

$$OH^- = H_2O - H^+$$

with conditional equilibrium constants:

$$K_1 = \frac{[Zn^{+2}]\{>FeOH^{-1/2}\}^2}{\{>(FeOH)_2Zn^+\}}\exp\left(\frac{2\psi_0 F}{RT}\right) \tag{24}$$

$$K_w = [H^+][OH^-]$$

$$K_a = \frac{[H^+]\{>FeOH^{-1/2}\}}{\{>FeOH_2^{+1/2}\}}\exp\left(\frac{\psi_0 F}{RT}\right)$$

In addition to the mass action expressions, we have the constraints provided by conservation of mass and charge. To set these up, we need to recognize three types of components: type I components are those for which the total concentration is known; type II components are those for which neither the free concentration nor the total concentration is known; type III components are those for which only the free concentration is known. The usual type II components are the electrostatic potentials in each layer. In sorption experiments, the most common type III component is H^+ since usually the pH (and hence, $[H^+]$) is known at each data point. Another useful type III component is H_2CO_3 since, for a given partial pressure of CO_2, the concentration of the species H_2CO_3 is known and constant with pH.

For the Zn surface complexation example, we have two type I components: Zn and >FeOH.

$$\{FeOH^{-1/2}\} + \{FeOH_2^{+1/2}\} + \{(>FeOH)_2Zn^+\} = 1 \tag{25}$$

$$[Zn]_{total} = [Zn^{2+}] + \{(>FeOH)_2Zn^+\}\ S_{tot}/2$$

where S_{tot} is the total moles of surface sites/kg water. Finally, we have conservation of charge giving

$$C_s(\psi_0 - \psi_{dl}) - \left(\frac{8\varepsilon\varepsilon_0 RTI}{10^3}\right)^{1/2}\sinh\left(\frac{zF\psi_0}{2RT}\right) = 0 \tag{26}$$

$$(-\frac{1}{2})\{>FeOH^{-1/2}\}S_{tot} + (\frac{1}{2})\{>FeOH_2^{1/2}\}S_{tot} + \{(>FeOH)_2Zn^+\}S_{tot} = \frac{\psi_0 - \psi_d}{C_s}$$

Since we know $[H^+]$ (a type III component), we have seven Equations (24-26) in seven unknowns (i.e., {$>FeOH^{-1/2}$}, {$>FeOH_2^{+1/2}$}, {$(>FeOH)_2Zn^+$}, [Zn], [OH^-], ψ_0, and ψ_d). However, the simultaneous equations are non-linear and we must solve the equations iteratively with the Newton–Raphson method developed by Morel and Morgan (1972).

For a system with m-species, we have m mass-action expressions over the n components

$$\log c_i = \sum_{j=1}^{n} a_{ij} \log x_j + \log K_i \tag{27}$$

where c_i is the concentration of species i, x_j is the free-concentration of the component j and K_i is the equilibrium constant for the formation of species i using the reaction defined by the a_{ij} coefficients. For the basic Stern model of the sorption of Zn described above, we would need two type II components to describe the change in charge in the 0-plane and diffuse layer. The a_{ij} elements are set up as a tableau as in Table 2. Note that the a_{ij} elements for the ψ_0, and ψ_d components are simply the changes in charge of the Stern and diffuse layers associated with the formation of each species.

Table 2. A-Matrix tableau

	Type I		**Type II**		**Type III**
	>FeOH	**Zn**	**ΔZ (0-plane)**	**ΔZ (diffuse)**	**H+**
>FeOH	1	0	0	0	0
$>FeOH_2^+$	1	0	1	0	1
Zn^{+2}	0	1	0	0	0
H^+	0	0	0	0	1
OH^-	0	0	0	0	−1
$(>FeOH)_2Zn^{+2}$	2	1	2	0	0

We also have the mass (or charge) conservation constraints over the n components:

$$t_j = \sum_{i=1}^{m} b_{ji} c_i \tag{28}$$

where t_j is the total concentration of component j and b_{ji} are the stoichiometric coefficients for mass and charge balance. The b-matrix, b_{ij} is the same as the a-matrix except that the b_{ij} for surface species are the actual charges of the surface species and not the change in surface charge Δz associated with the formation of each species. Hence, the b_{ij} matrix is derived from the a-matrix by

$$b_{ij} = a_{ij} + \sum_{j'=1}^{nsurf} a_{ij'} z_{j'} \tag{29}$$

where the j' summation is over the n_{surf} surface sites and $z_{j'}$ is the reference charge of surface site j' (e.g., we assign a reference charge of −0.5 for a -OH site). The reference charges are entirely in the Stern layer. Equation (28) can be recast as

$$y_j = \sum_{i=1}^{m} b_{ji} c_i - t_j \quad (j = 1 \ldots n \text{ components}) \tag{30}$$

where y_j is the difference between the predicted and actual t_j. The Newton-Raphson method can be used to iteratively find the zeros of y_j

$$y_j = \sum_{k=1}^{n} \frac{\partial y_j}{\partial x_k} \Delta x_k \tag{31}$$

Where Δx_k is the correction to x_k. From (30) we have

$$\frac{\partial y_j}{\partial x_k} = z_{jk} = \sum_{i=1}^{m} b_{ij} \frac{\partial c_i}{\partial x_k} - \frac{\partial t_j}{\partial x_k} \tag{32}$$

z_{jk} is the Jacobian matrix. In matrix notation, we can write Equation (31) as $\mathbf{y} = \mathbf{z}\Delta\mathbf{x}$ and solve for $\Delta\mathbf{x}$ as $\Delta\mathbf{x} = \mathbf{z}^{-1}\mathbf{y}$.

For the non-electrostatic terms (e.g., the type I components) each t_j is constant and the $\partial t_j/\partial x_k$ terms in (32) are 0. From (29) we get

$$z_{jk} = \sum_{i=1}^{m} \frac{b_{ij} a_{ik} c_i}{x_k} \tag{33}$$

However, for the electrostatic components, the $\partial t_j/\partial x_k$ terms are more complex and depend on the surface model used (e.g., basic Stern, diffuse double layer, etc.) for the surface electrostatics. In the basic Stern model, we have the total charges

$$t_{0-plane} = \frac{\sigma_0 A}{F} \tag{34}$$

$$t_{d-plane} = \frac{\sigma_d A}{F}$$

where A is the total surface area/kg water. The x_k values corresponding to the 0-plane and d-plane are

$$x_{0-plane} = \log(e^{-\psi_0 \mathrm{F}/\mathrm{R}T}) = \frac{-\psi_0 \mathrm{F}}{\ln(10)\mathrm{R}T} \tag{35}$$

$$x_{d-plane} = \log(e^{-\psi_d \mathrm{F}/\mathrm{R}T}) = \frac{-\psi_d \mathrm{F}}{\ln(10)\mathrm{R}T}$$

The corrections to the Jacobian are then

$$\frac{\partial t_{0-plane}}{\partial x_{0-plane}} = \frac{C_s \mathrm{R}T}{\mathrm{F} x_{0-plane}} \frac{\mathrm{A}}{\mathrm{F}} \tag{36}$$

$$\frac{\partial t_{0-plane}}{\partial x_{d-plane}} = \frac{-C_s \mathrm{R}T}{\mathrm{F} x_{d-plane}} \frac{\mathrm{A}}{\mathrm{F}}$$

$$\frac{\partial t_{d-plane}}{\partial x_{0-plane}} = \frac{-C_s \mathrm{R}T}{\mathrm{F} x_{0-plane}} \frac{\mathrm{A}}{\mathrm{F}}$$

$$\frac{\partial t_{d-plane}}{\partial x_{d-plane}} = \left(\frac{C_s \mathrm{R}T}{\mathrm{F} x_{d-plane}} + \frac{1}{2 x_{d-plane}} \left(\frac{8\varepsilon\varepsilon_0 \mathrm{R}TI}{10^3} \right)^{1/2} \sinh\left(\frac{z\mathrm{F}\psi_0}{2\mathrm{R}T} \right) \right) \frac{\mathrm{A}}{\mathrm{F}}$$

An initial guess for x_j is provided as follows: For type I components, x_j is the total concentration of component i; for type II components, $x_j = 0$; for type III components, we already know x_j and, therefore, fix it to that value. Given the initial guess, we solve Equation (27) for c_j. From the c_j's we obtain initial y_j's from Equation (30). We then solve the simultaneous Equations (31) for Δx_k and evaluate the new x_k from

$$x_k^{(\text{new})} = x_k^{(\text{old})} - \Delta x \quad (37)$$

To insure that $x_k > 0$, we use the kludge proposed by Morel and Morgan (1972) and set

$$x_k^{(\text{new})} = (x_k^{(\text{old})} + \Delta x)/10 \text{ if } \Delta x > x_k^{(\text{old})} \quad (38)$$

Computer programs. The FITEQL program (Herbelin and Westall 1996) embedded the Morel and Morgan (1972) algorithm into a non-linear least squares routine so that fits of experimental data to a set of equilibria could be done. FITEQL also enabled one to model surface reactions by including models for the surface electrostatics (e.g., constant capacitance, diffuse double layer etc.). The original FITEQL required integer charges on surface species in accordance with the 2pK formalism. Subsequently, several modifications were made to allow fractional charges in order that the 1pK formalism and the CD model could be implemented (e.g., Tadanier and Eick 2002). The FITEQL code was written in BASIC and requires the DOS operating system for the GUI. Unfortunately, BASIC comes in a variety of flavors and the DOS operating system is (arguably) obsolete. It is not possible to port the code to other platforms such as UNIX/LINUX or Mac OS.

For convenience, the Morel and Morgan (1972) algorithms, along with the terms for surface electrostatics described above, were implemented into a Fortran 95 program called EQLFOR that is available upon request from the author. For fitting experimental sorption data, the speciation routine is embedded in a gradient search non-linear least squares method that is more forgiving than the Newton-Marquardt algorithm in FITEQL. Because the code is written in Fortran 95, it can be moved to other platforms and can, hopefully, be more easily modified by users attempting to develop new capabilities or correct errors. The code compiles using the free gfortran compiler. The input format is designed to be flexible so that data points with different surface areas, ionic strengths etc. can be modeled at the same time. Rather than using a GUI, the data and surface complexation model are read in from a file (produced by a text editor) and the fitting procedure is driven by a command-line interface. On Mac OSX, the program produces graphical output using the Aquaterm interface.

SURFACE COMPLEXATION OF METALS AND OXYANIONS

Using the models for oxide surface acidities and the computational methods for including the effect of the surface electrostatic potential, we can now illustrate some investigations on the molecular nature of surface complexes and how molecular information can be used to set up a surface complexation model. Sorption of ions on mineral surfaces can occur by several different structural mechanisms: inner-sphere complexes, outer-sphere complexes, surface precipitates or structural incorporation (solid solution). For this reason alone, molecular scale information is needed to even qualitatively understand sorption reactions. Here, however, we will focus on inner-sphere complexes since those are dominant sorption mechanism for the most biogeochemically significant trace metals and metalloids in aquatic environments.

In many of the earlier models of surface complexation (e.g., Dzombak and Morel 1990), sorption complexes were modeled as being monodentate complexes (e.g., Eqn. 4a) to some kind of average surface oxygen site. With the availability of synchrotron-based spectroscopic techniques, however, the actual nature of sorption complexes became directly observable. EXAFS spectroscopy, in particular, can show the structure (interatomic distances and coordination numbers) of the first few coordination shells surrounding an atom. Such information can readily identify sorption via inner-sphere complexes, surface precipitates and solid solutions. Distinguishing among different inner-sphere complexes is more difficult, however. The scattering by a neighboring shell is determined not only by the number of atoms in the shell but also by the Debye-Waller factor of the shell. Because coordination numbers and Debye-Waller factors are strongly correlated in fits to EXAFS spectra, coordination numbers are difficult to determine accurately. The primary constraints for distinguishing among different inner-sphere complexes are the apparent distances to next-nearest neighbor shells and some *a priori* prediction of the bond-lengths expected in different complexes. However, multiple scattering in EXAFS can often complicate spectra by masking the dominant complex and giving apparent features leading to incorrect structures. This problem is mostly associated with oxyions such as UO_2^{++} and $(AsO)_4^{-3}$. Finally, the most significant complexes may be at too low a concentration to be observed by EXAFS. Synchrotrons used in the past 15 years (e.g., the SRS facility at Daresbury Laboratory, UK) were only able to get reasonable spectra of when sorbed metals/metalloids concentrations were > 0.1 wt%. New synchrotron sources will allow EXAFS at much lower concentrations.

Because of the problems with modeling EXAFS spectra and the difficulty in observing surface complexes at low concentrations, it is often very helpful to supplement EXAFS spectra with quantum mechanical calculations of the structures and energetics of different possible surface complexes. In recent years, we have been attempting to develop surface complexation models that are constrained by structural information provided by EXAFS spectroscopy and first-principles calculations based on density functional theory.

Examples of surface complexation models developed from molecular information

Cu(II) on goethite. High quality EXAFS spectra of copper on iron oxides are difficult to obtain because the iron fluorescence interferes with that of copper. However, EXAFS spectra of Cu sorbed on various oxides do show evidence of a first shell with four oxygen atoms and a next nearest neighbor shells attributable to either Cu or Fe near 2.9 Å (Parkman et al. 1999; Peacock and Sherman 2004) and 3.1-3.3Å. The short 2.9 Å distance might be considered evidence for an E2 type complex forming on the {210} or {010} surfaces (Fig. 8). However, because those surfaces comprise only a small fraction of the total surface area, any Cu forming those complexes would not be observable in the EXAFS. (If Cu only sorbed via such complexes, then EXAFS spectra would be very difficult or impossible to obtain since the total loading of Cu would be < 0.1 wt%). Peacock and Sherman (2004) proposed that the dominant complex was a Cu-Cu dimer forming on the {101} and {001} surfaces (Fig. 9); this was at a surface loading of 0.38 and 0.75 wt% Cu. Using density functional calculations, they optimized the geometries of $Fe_2(OH)_2(H_2O)_8Cu(OH)_4$ and $Fe_3(OH)_4(H_2O)_7Cu_2O(OH)_6$ clusters

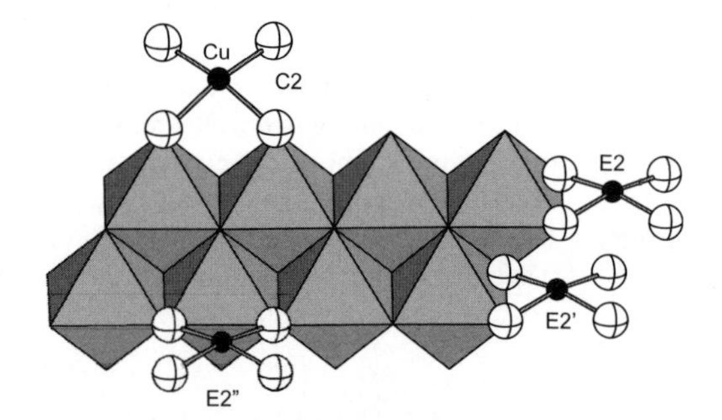

Figure 8. Possible complexes formed by Cu^{2+}.

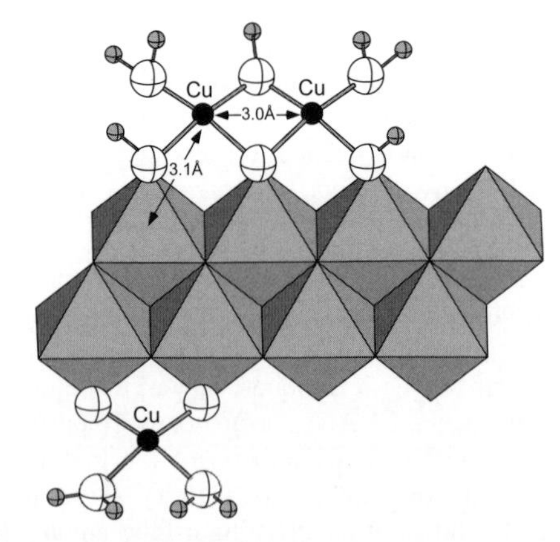

Figure 9. Proposed complexes formed by Cu on goethite. In a basic Stern model, all the charge of the surface complexes is assigned to the 0-plane.

corresponding to Cu and Cu-Cu complexes forming on the >Fe-OH sites on the {101} or {001} surface of goethite (Fig. 1). The optimized geometries agreed with the EXAFS, supporting the structural model. Given the structures of the Cu complexes, Peacock and Sherman (2004) modeled Cu sorption using the equilibria:

$$3\ {>}FeOH + 2Cu^{2+} + 3H_2O = ({>}FeOH)_2({>}FeO)Cu_2(OH)_3{}^0 + 4H^+ \tag{39}$$

$$2\ {>}FeOH + Cu^{2+} + 2H_2O = ({>}FeOH)_2Cu(OH)_2{}^0 + 2H^+$$

with a 2pK model for surface protonation and the diffuse layer model for electrostatics. Alternatively, the sorption of Cu can be modeled using a 1pK formalism ($pK_{a(t)} = pK_{a(s)} = 9.12$) with basic Stern electrostatics:

$$3\ {>}FeOH^{-1/2} + 2Cu^{2+} + H_2O = ({>}FeOH)_2({>}FeO)Cu_2(OH)^{+1/2} + 2H^+ \tag{40}$$

$$2\ {>}FeOH^{-1/2} + Cu^{2+} = ({>}FeOH)_2Cu^{+1/2}$$

For both reactions, the change in surface charge Δz is +2 which we assign to the 0-plane to give the fit shown in Figure 10. Note that the sorption edge at this loading cannot be fit very well with only a mononuclear $({>}FeOH)_2Cu^{+1/2}$ complex. The protonation state of the surface complexes is highly dependent upon the model used (1pK vs. 2pK) and cannot be easily determined from simple batch experiments. More recently, Ponthieu et al. (2006) modeled the sorption of Cu to goethite by invoking a bidentate edge-sharing (E2″) complex of Cu on the {101} surface (Fig. 8). This complex would involve the $>Fe_2OH$ sites which, as discussed above, were assumed unreactive. The argument was that EXAFS indicating the Cu-Cu distance could also be interpreted as a short Cu-Fe distance associated with an E2 complex. Pauling's rules, of course, would favor the C2 complex over the E2″. Perhaps for this reason, Peacock and Sherman (2004) never considered such a complex. Based on a DFT cluster optimization, the E2 complex (as in Fig. 8) was found to be less stable than the C2 complex and a cluster that would model the unlikely E2″ complex involving the two $>Fe_2OH$ sites on {101} was not considered. It is sobering that two wildly different structural models can fit the sorption data; moreover, spectroscopic data can sometimes confuse matters further! First-principles calculations can sometimes offer essential constraints of surface complexation models.

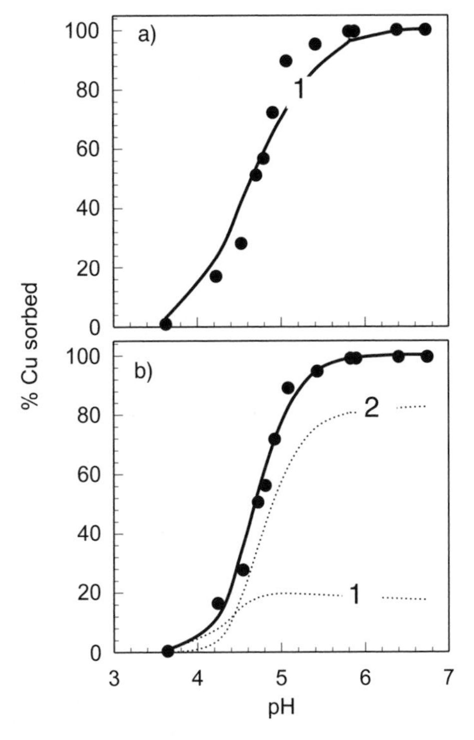

Figure 10. Sorption edge of Cu on goethite fit using a) a single $({>}FeOH)_2Cu^+$ complex and b) both the $({>}FeOH)_2Cu^+$ (dashed curve 1) and $({>}FeOH)_2{>}FeOCu_2OH$ (dashed curve 2) complexes in the 1pK basic Stern model.

U(VI) on goethite. The sorption of U(VI) as the UO_2^{++} ion poses a serious challenge to spectroscopy, quantum chemistry and surface complexation modeling. Based on EXAFS spectra, it has previously been proposed that U(VI), as the UO_2^{++} cation, sorbs to Fe oxide hydroxide phases by forming a bidentate edge-sharing (E2) surface complex, $>Fe(OH)_2UO_2(H_2O)n$

(e.g., Waite et al. 1994; Moyes et al. 2000). This is based on the observation of a strong scattering feature corresponding to a U-Fe distance near 3.5 Å. By analogy with the discussion given above for copper, if such a complex formed on goethite, it would most likely occur on the {210} and {010} surfaces making its observation with EXAFS impossible. Moreover, such {210} or {010} complexes alone could not account for the sorption capacity of goethite for U(VI) which is on the order of 2 wt%. A likely possibility is that a C2 complex forms on the {101} surfaces; however, there is no EXAFS evidence for such a complex. As with Cu, the formation of E2 complexes via the $>Fe_2OH$ sites on {101} seems at odds with our simple picture of reactivity based on Pauling's rules; the >FeOH sites should be far more reactive leading to C2 complexes. Sherman et al. (2008) optimized the geometries of clusters corresponding to possible C2 $(>FeOH)_2UO_2^{++}$ and E2 $>Fe(OH)_2UO_2^{++}$ complexes and found that the E2 cluster is only 12 kJ/mol more stable than the C2 cluster. Configurational entropy would therefore allow a substantial fraction of C2 complexes even at low surface loading. Sherman et al. (2008) also obtained EXAFS of U(VI) on goethite at several surface loadings and pH but modeled the spectra by including full multiple scattering. They found that the shell associated with an apparent U-Fe distance of 3.5 can be entirely attributed to multiple scattering; moreover, a further effect of multiple scattering is to "shadow" the paths corresponding to U-Fe distances near 4.2 Å that would be present if UO_2^{++} forms a C2 complex. Hence, the dominant surface complex in CO_2-free systems could be a bidentate corner-sharing (C2) complex, $(>FeOH)_2UO_2(H_2O)_3$ on the dominant {101} surface even though there is no direct evidence for such a complex from EXAFS.

A further complication is that UO_2^{++} forms a number of complexes with OH^- and CO_3^{-2}. At ambient partial pressures of CO_2, EXAFS spectra show a U-C scattering consistent with $(>FeOH)_2UO_2CO_3$ or $(>FeO)CO_2UO_2$ complexes.

Sherman et al. (2008) obtained sorption edges at several U concentrations (Fig. 11) and fit the data to the surface complexation equilibria

$$2{>}FeOH^{-1/2} + UO_2^{++} = ({>}FeOH)_2UO_2^{+} \tag{41}$$

$$2{>}FeOH^{-1/2} + UO_2^{++} + H_2CO_3 = ({>}FeOH)_2UO_2CO_3^{-1} + 2H^+$$

$${>}FeOH^{-1/2} + UO_2^{++} + H_2CO_3 = ({>}FeO)CO_2UO_2^{1/2} + H^+ + H_2O$$

with conditional equilibrium constants

$$K_1 = \frac{\{({>}\,FeOH)_2UO_2^{\ +}\}}{\{{>}\,FeOH^{-1/2}\}^2[UO_2^{+2}]}\exp\left(\frac{2\psi_0 F}{RT}\right) \tag{42a}$$

$$K_2 = \frac{\{({>}\,FeOH)_2UO_2CO_3^{\ -}\}[H^+]^2}{\{{>}\,FeOH^{-1/2}\}^2[UO_2^{+2}][H_2CO_3]}\exp\left(\frac{(2\psi_0 - 2\psi_1)F}{RT}\right) \tag{42b}$$

$$K_3 = \frac{\{({>}\,FeOCO_2UO_2^{\ -1/2}\}[H^+]}{\{{>}\,FeOH^{-1/2}\}[UO_2^{+2}][H_2CO_3]}\exp\left(\frac{(-\psi_0 + 2\psi_1)F}{RT}\right) \tag{42c}$$

using an extended Stern model for electrostatics (Fig. 12). The surface complexation model used the 2-site 1p*K* formalism and needed to include all the aqueous complexes of UO_2^{++} along with the complexation of the FeOOH surface by CO_3^{-2}. The model gives a reasonable fit to sorption experiments (Fig. 11) done in both ambient and reduced CO_2 environments at surface loadings of 0.02–0.2 wt% U. The possible contribution from an E2 complex cannot be resolved from the fits using only C2 complexes.

As(V) on goethite. Some of the first EXAFS spectra of AsO_4^{-3} on goethite (Fendorf et al. 1997; Farquar et al. 2002) were interpreted as showing an E2 complex together with a C2 complex (Fig. 13). Fendorf et al. (1997) also proposed the existence of a monodentate

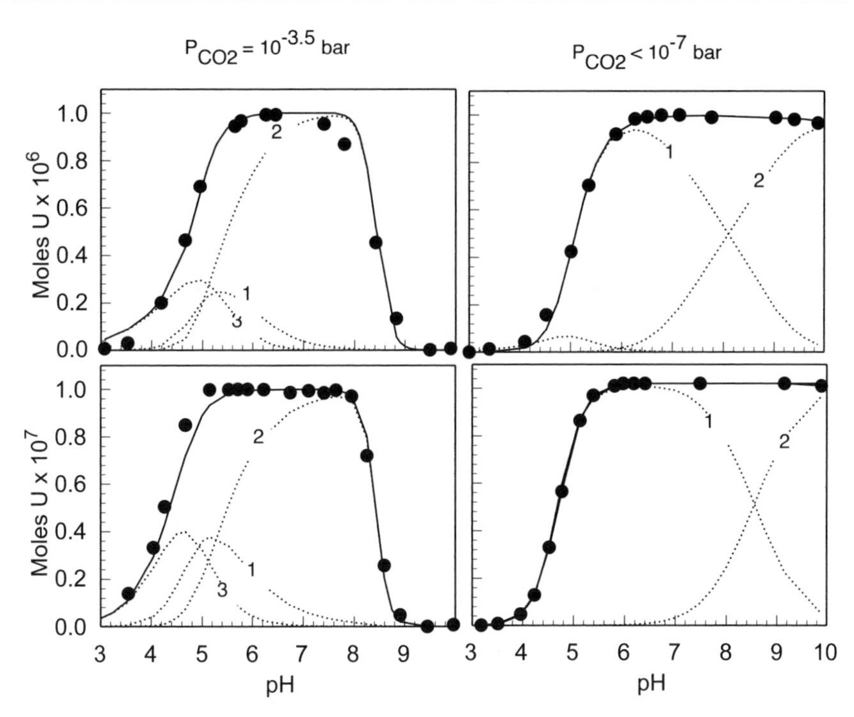

Figure 11. Sorption edges for U on goethite (Sherman et al. 2008) modeled using the equilibria given in Equation (41). Complex 1 is the $(>FeOH)_2UO_2^+$, complex 2 is the $(>FeOH)_2UO_2CO_3^-$ and complex 3 is the (tentative) $(>FeO)CO_2UO_2^{-1.5}$.

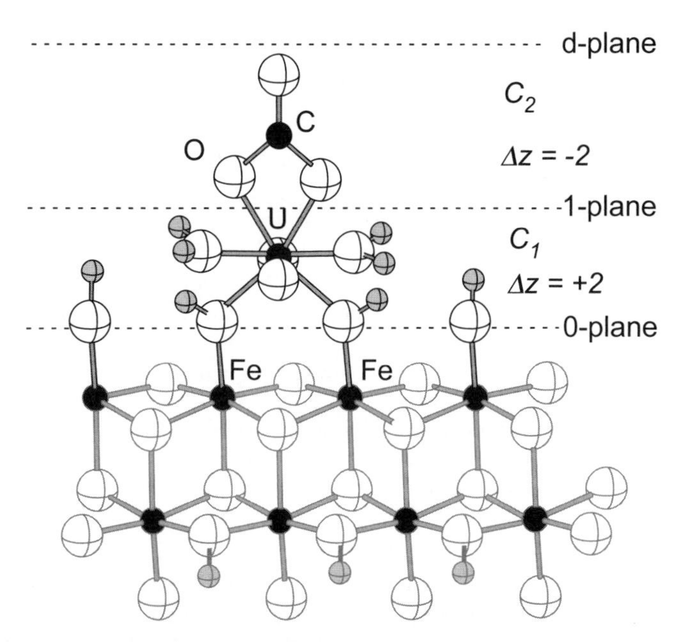

Figure 12. Extended Stern model showing charge distribution of a sorbed $UO_2(CO_3)$ complex. Dashed lines corresponding to the 0-plane, 1-plane and *d*-plane are shown. Here the charge of the CO_3^{-2} group is placed on the 1-plane while the charge of the UO_2^{+2} is placed on the 0-plane.

V1 complex. Based on Pauling's rules, we would only expect AsO_4^{-3} to exchange with singly coordinated >Fe-OH oxygens. The only surfaces on which AsO_4^{-3} could form an E2 complex with $>Fe(OH)_2$ oxygens is on the {210} and {010} surfaces. However, those surfaces make up only a small fraction of the total surface area. To reconcile the EXAFS and structural considerations, Sherman and Randall (2003) predicted the structures and energetics of clusters corresponding to the E2 and C2 complexes on goethite using density functional calculations. The E2 complex was found to be much higher in energy; moreover, it was shown that the apparent E2 feature in the EXAFS spectra is actually an artifact of multiple scattering within the AsO_4 tetrahedron.

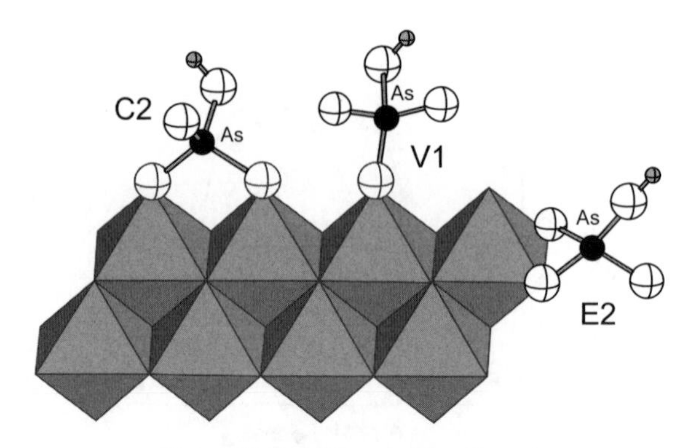

Figure 13. Possible complexes formed by tetrahedral oxyanions.

As(V) sorbs so strongly to FeOOH that sorption edges (showing anything to fit) can only be obtained at high surface loading (Fig. 14). Under such conditions, sorption edges of As(V) on FeOOH are difficult to model without invoking monodentate complexes in addition to the C2 complexes (e.g., Stachowicz et al. 2006). Such complexes were indicated in the EXAFS of Fendorf et al. (1997) but not in the EXAFS of Sherman and Randall (2003) where an As loading of only 0.3 wt% at pH 4 was used. In addition to having both C2 and V1 complexes, we need to consider the different degrees of protonation of the AsO_4^{-3} complex. Hence, we could model the sorption of As(V) using the bidentate C2 complexes

$$2{>}FeOH^{0.5-} + H_3AsO_4 = ({>}FeO)_2AsOOH^- + 2H_2O \qquad (43a)$$

$$2{>}FeOH^{0.5-} + H_3AsO_4 = ({>}FeO)_2AsO_2^{-2} + H^+ + 2H_2O \qquad (43b)$$

and the monodentate V1 complexes

$$>FeOH^{0.5-} + H_3AsO_4 = ({>}FeO)AsO(OH)_2^{0.5-} + H_2O \qquad (43c)$$

$$>FeOH^{0.5-} + H_3AsO_4 = ({>}FeO)AsO_2OH^{1.5-} + H_2O + H^+ \qquad (43d)$$

$$>FeOH^{0.5-} + H_3AsO_4 = ({>}FeO)AsO_3^{2.5-} + H_2O + 2H^+ \qquad (43e)$$

Even with all these complexes, however, it is difficult to fit the sorption edges using only a basic Stern model for the electrostatics. Using all five complexes with an extended Stern model for electrostatics yields a nearly perfect fit (not shown) of the data in Figure 14. However, the species that are present depend strongly on how we partition the charge between the two Stern layers. We could treat the charge distribution as an adjustable quantity used to optimize the fit (e.g., Tadanier and Eik 2002) but that suggests an overparameterization. Some kind of *a priori* model would be helpful. Using Pauling's second rule, we could partition the charge of a $({>}FeO)_2AsOOH$ complex as shown in Figure 15. This is the strategy in the CD-MUSIC model of Hiemstra and van Reimsdijk (1996); however, they modify the charge distribution to account for the protonation of the AsO_4^{-3} oxygens (values in parentheses in Fig. 15).

More recently, Hiemstra and coworkers (Hiemstra and van Riemsdijk 2006; Stachowicz et al. 2006) discussed "quantum chemically derived charge distributions" in their CD-MUSIC model. However, these charges are simply the atomic charges that are obtained from bond valence theory but using bond lengths optimized from density functional calculations on clusters approximating surface complexes. Indeed, the atomic charges in a complex are

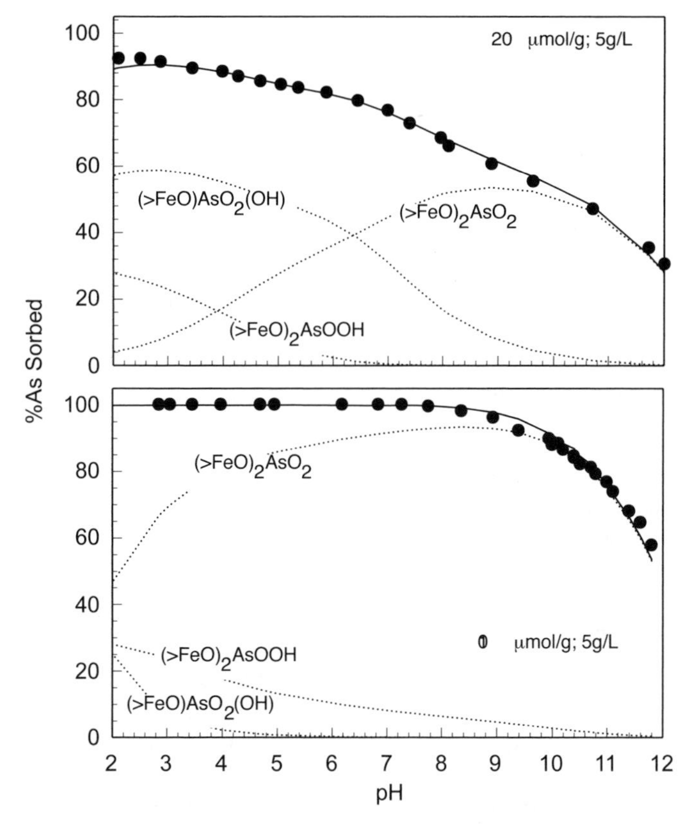

Figure 14. Sorption of As(V) at two surface loadings (Jonsson and Sherman, unpublished) in 0.1 m $NaNO_3$. Also shown is the fit to the data based on the CD-MUSIC model of Stachowicz et al. (2006).

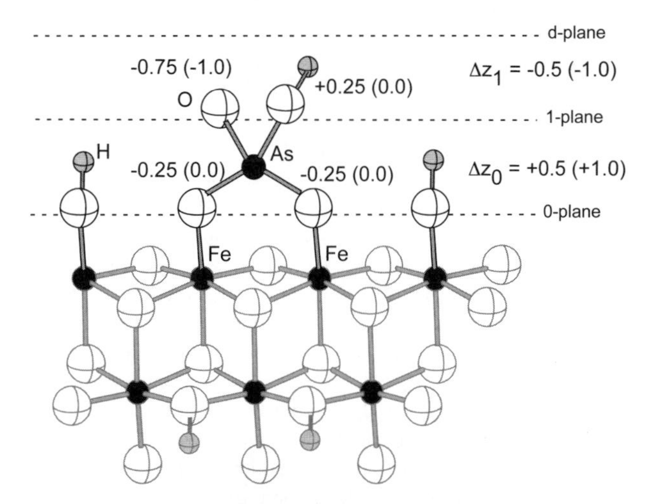

Figure 15. Extended Stern model and charge distribution based on Pauling's second rule and the charge-distribution proposed by Hiemstra and van Riemsdijk (1996). Dashed lines corresponding to the 0-plane, 1-plane and *d*-plane are shown.

not based on any quantum mechanical observable and there are no rigorous criteria to assign charges. The charges typically used in the CD-MUSIC model have little resemblance to those obtained from, for example, the Mulliken charges. The latter are based on the contributions of the atomic orbitals to each molecular orbital. For example, the Mulliken charge distribution in the $(>FeO)_2AsOOH$ complex is shown in Figure 16. Using the Mulliken scheme, the Δz_0 for the $(>FeO)_2AsOOH$ complex is −0.1. Of course the Mulliken scheme is not a unique or rigorous way to assign charge either and the absolute values are dependent upon the atomic basis set used in the calculation. However, it should provide a measure of the relative charges.

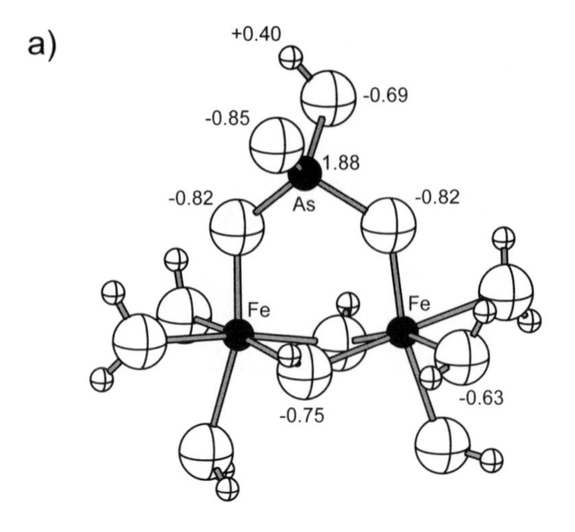

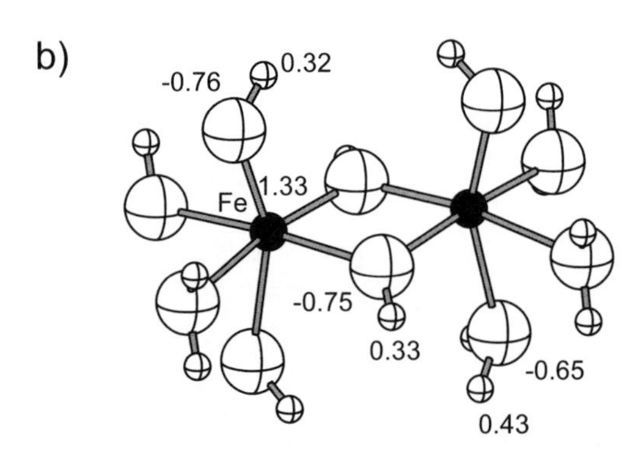

Figure 16. a) Mulliken charges for the $(>FeO)_2AsO(OH)^{-1}$ complex based on the triple zeta basis set used in the density functional calculations in Sherman and Randall (2003). b) Mulliken charges in a Fe-Fe dimer cluster.

Stachowicz et al. (2006) were able to fit their As(V) sorption data using the just the bidentate $(>FeO)_2AsO(OH)$, $(>FeO)_2AsO_2$ complexes and the monodentate $(>FeO)AsO_2OH$ complexes. The surface charges were obtained from the MUSIC model but based on bond lengths calculated from DFT. These gave Δz_0 values of 0.58, 0.47 and 0.3, respectively which are, as expected, close to those expected from Pauling's second rule (Fig. 15). We have obtained somewhat better sorption data at two surface loadings (Jonsson and Sherman 2007). Using the complexes and the charge distribution proposed by Stachowicz et al. (2006), we find that we get a reasonable fit to our data (Fig. 14). The monodentate complex is only important at higher surface loadings and at low pH, consistent with the EXAFS of Sherman and Randall (2003). Note, however, that an even better fit can be obtained at high surface loading if we replace the $(>FeO)AsO_2OH$ complex with a $(>FeO)AsO(OH)_2^{-0.5}$ complex at low pH and a $(>FeO)AsO_3^{-2.5}$ complex at high pH; however, the improved fit may simply reflect an over parameterization of the model. If, instead of the charges based on bond-valence theory, we used the change in charge of the surface oxygens based on the Mulliken charges, the fit is much worse. Why something as simple as Pauling's second rule (even with its extension in the bond-valence model) appears to give a good charge distribution is unclear. The $(>FeO)_nAsO_pH_q$ complexes are not even very ionic. Perhaps, as Rustad (2001) put it: "the MUSIC model gets the right answer but for the wrong reasons." More theoretical work is needed to clarify the physical meaning of the bond-valence charges and their relation to electrostatic work of surface complexation.

CONCLUSIONS AND FUTURE WORK

The first goal of this review is to give brief outline of surface complexation modeling. A further goal, however, is to try and illustrate how such models are dependent upon molecular-level information. Aside from spectroscopic measurements, results from computational quantum chemistry have great potential for resolving some of the current problems. In particular, quantum chemistry can be used to develop a better understanding of the reactivities of surface oxygens. For example, we can directly determine the proton affinities of different surface oxygens from quantum mechanical calculations on finite clusters that model sites on the mineral surface (e.g., Aquino et al. 2008). This approach should be much more reliable than semiclassical models such as bond-valence theory.

Understanding how the electrostatic potential at the mineral-water interface varies as a function of protonation and ion sorption is probably the most challenging problem for molecular theory. In the future, *ab initio* molecular dynamical simulations of surface complexation and protonation reactions may be practical; from such simulations, we would be able to determine actual free energies of surface reactions at a level of theory only limited by the exchange-correlation functional of density functional theory. In the meantime, we will need to use classical interatomic potentials. However, dynamical simulations of the mineral-water interface, even if based on classical potentials, are bound to give important insights on the structure of the electrical double layer and the nature of the surface electrostatic free energy for sorption.

Our understanding of how ions actually sorb to mineral surfaces is still incomplete and the structures of surface complexes are still controversial. Although EXAFS spectroscopy can be extremely useful, it has important limitations that must be recognized. Multiple scattering is a problem when interpreting the weak signals from next-nearest neighbor scattering. The most important complexes may occur only at concentrations too low to be observable with EXAFS. Many of these problems can be overcome using first-principles calculations on the geometries and energetics of surface complexes. Such calculations can help predict the relative stabilities of different possible surface complexes. The primary challenge to such calculations is that the model system (cluster of atoms) must be large enough to reliably define the problem. Moreover, the solvation must be treated using both explicit water molecules for the local field and a dielectric continuum model such as COSMO for the long-range solvation field. However, calculated energies are only relative and the finite cluster cannot give us the electrostatic component of the free energy. This would require a calculation on an extended (periodic) system. Moreover, the entropy of sorption can only be evaluated if we are able to sample the vibrational and configurational degrees of freedom using dynamical calculations. Nevertheless, even static calculations on small clusters have provided important constraints for models of surface complexation.

ACKNOWLEDGMENTS

This work was funded by NERC Grants NER/A/S/2003/00526 and NE/C513142/1. Insights on surface complexation modeling aided by discussions with Jorgen Jonsson, Caroline Peacock and David Moore.

REFERENCES

Aquino AJA, Tunega D, Haberhauer G, Gerzabek MH, Lischka H (2008) Acid-base properties of a goethite surface model: A theoretical view. Geochim Cosmochim Acta 72:3587-3602

Bethke CM, Brady PV (2000) How the K_d approach undermines ground water cleanup. Ground Water 38:435-443

Bickmore BR, Rosso KM, Mitchell SC (2006) Is there hope for multisite surface complexation (MUSIC) modeling? *In:* Surface Complexation Modeling. Lutzenkirchen J (ed) Interface Sci Technol 11:269-283

Boily JF, Lützenkirchen J, Balmès O, Beattie J, Sjöberg S (2001) Modeling proton binding at the goethite (α-FeOOH)-water interface. J Colloid Surf A 179:11-27

Brown ID (1981) The bond-valence method: an empirical approach to chemical structure and bonding. *In:* Structure and Bonding in Crystals. O'Keefe M, Navrotsky A (eds) Academic Press, New York, 2:1-30

Brown ID (2002) The Chemical Bond in Inorganic Chemistry. The Bond Valence Model. Oxford University Press

Dzombak DA, Morel FMM (1990) Surface Complexation Modeling. Hydrous Ferric Oxide. Wiley

Farquhar ML, Charnock JM, Livens FR, Vaughan DJ (2002) Mechanisms of arsenic uptake from aqueous solution by interaction with goethite, lepidocrocite, mackinawite, and pyrite: An X-ray absorption spectroscopy study. Environ Sci Technol 36:1757-1762

Fendorf S, Eick MJ, Grossl P, Sparks DJ (1997) Arsenate and chromate retention mechanisms on goethite. 1. Surface structure. Environ Sci Technol 31:315–320

Gaboriaud F, Ehrhardt JJ (2003) Effects of different crystal faces on the surface charge of colloidal goethite (α-FeOOH) particles: and experimental and modeling study. Geochim Cosmochim Acta 67:967-983

Herbelin A, Westall J (1996) A Computer Program for Determination of Chemical Equilibrium Constants from Experimental Data. Version 3.2, Department of Chemistry, Oregon State University, Corvallis, Oregon 97331

Hiemstra T, Vanriemsdijk WH, Bolt GH (1989a) Multisite proton adsorption modeling at the solid-solution interface of (hydr)oxides - a new approach. 1. Model description and evaluation of intrinsic reaction constants. J Colloid Interface Sci 133:91-104

Hiemstra T, Dewit JCM, Van Riemsdijk WH (1989b) Multisite proton adsorption modeling at the solid-solution interface of (hydr)oxides - a new approach. 2. Application to various important (hydr)oxides. J Colloid Interface Sci 133:105-117

Hiemstra T, Van Riemsdijk WH (1996) A surface structural approach to ion adsorption: The charge distribution (CD) model. J Colloid Interface Sci 179:488–508

Hiemstra T, Venema P, VanRiemsdijk WH (1996) Intrinsic proton affinity of reactive surface groups of metal (hydr)oxides: The bond valence principle. J Colloid Interface Sci 184:680-692

Hiemstra T, Van Riemsdijk WH (2006) On the relationship between charge distribution, surface hydration and structure of the interface of metal hydroxides. J Colloid Interface Sci 301:1-18

Jonsson J, Sherman DM (2007) Sorption of arsenic under oxic and anoxic conditions: Possible origins of elevated arsenic in groundwater. Geochim Cosmochim Acta 71:A450-A450

Kerisit S, Ilton ES, Parker SC (2006) Molecular dynamics simulations of electrolyte solutions at the (100) goethite surface. J Phys Chem B 110:20491-20501

Klamt A (1995) Conductor-like screening model for real solvents—a new approach to the quantitative calculation of solvation phenomena. J Phys Chem 99:2224–2235

Kulik DA (2009) Thermodynamic concepts in modeling sorption at the mineral-water interface. Rev Mineral Geochem 70:125–180

Lützenkirchen J (1998) Comparison of 1-p*K* and 2-p*K* versions of surface complexation theory by the goodness of fit in describing surface charge data of (hydr)oxides. Environ Sci Technol 32:3149-3154

Lützenkirchen J (2005) On derivatives of surface charge curves of minerals. J Colloid Interface Sci 290:489-497

Mathur SS, Dzombak DA (2006) Surface complexation modeling: goethite. *In:* Surface Complexation Modeling. Lützenkirchen J (ed.) Interface Sci Technol 11:442-468

Morel F, Morgan J (1972) Numerical method for computing equilibria in aqueous chemical systems. Environ Sci Technol 6:58-67

Moyes LN, Parkman RH, Charnock JM, Vaughan DJ, Livens FR, Hughes CR, Braithwaite A (2000) Uranium uptake from aqueous solution by interaction with goethite, lepidocrocite, muscovite and mackinawite: an X-ray absorption spectroscopy study. Environ Sci Technol 34:1062-1068

Pauling L (1960) The Nature of the Chemical Bond. Cornell University Press, Ithaca, New York

Parks GA (1965) The isoelectric points of solid oxides, solid hydroxides, and aqueous hydroxo complex systems. Chem Rev 65:177-198

Parkman RH, Charnock JM, Bryan ND, Livens FR, Vaughan DJ (1999) Reactions of copper and cadmium ions in aqueous solution with goethite, lepidocrocite, mackinawite, and pyrite. Am Mineral 84:407-419

Peacock CL, Sherman DM (2004a) Vanadium(V) adsorption onto goethite (α-FeOOH) at pH 1.5-12: a surface complexation model based on ab initio molecular geometries and EXAFS spectroscopy. Geochim Cosmochim Acta 68:1723-1733

Peacock CL, Sherman DM (2004b) Copper(II) sorption onto goethite, hematite and lepidocrocite: A surface complexation model base on *ab initio* molecular geometries and EXAFS spectroscopy. Geochim Cosmochim Acta 68:2623-2637

Peacock CL, Sherman DM (2007) Sorption of Ni by birnessite: equilibrium controls of Ni in seawater. Chem Geol 238:94-106

Ponthieu M, Juillot F, Hiemstr T, van Riemsdijk WH, Benedetti MF (2006) Metal binding to iron oxides. Geochim Cosmochim Acta 70:2679-2698

Prelot B, Villieras F, Pelletier M, Gerard G, Gaboriaud F, Ehrhardt JJ, Perrone J, Fedoroff M, Jeanjean J, Lefevre G, Mazerolles L, Pastol JL, Rouchaud JC, Lindecker C (2003) Morphology and surface heterogeneities in synthetic goethites. J Colloid Interface Sci 261:244-325

Rustad JR, Felmy AR, Hay BP (1996) Molecular statics calculations of proton binding to goethite surfaces: a new approach to estimation of stability constants for multisite complexation models. Geochim Cosmochim Acta 60:1563-1576

Rustad JR (2001) Molecular models of surface relaxation, hydroxylation and surface charging at oxide-water interfaces. Rev Mineral Geochem 42:169-198

Schwertmann U, Cornell, RM (1991) Iron Oxides in the Laboratory, VCH Verlag

Sherman DM (1985) Electronic structures of Fe^{3+} coordination sites in iron oxides; applications to spectra, bonding and magnetism. Phys Chem Mineral 12:161-175

Sherman DM (1985) SCF-Xα-SW study of Fe-O and Fe-OH chemical bonds: applications to the Mossbauer spectra and magnetochemistry of hydroxyl-bearing iron oxides and silicates. Phys Chem Mineral 12:311-314

Sherman DM (2001) Quantum chemistry and classical simulations of metal complexes in aqueous solutions. Rev Mineral Geochem 42:273-318

Sherman DM, Randall SR (2003) Surface complexation of arsenic (V) to iron (III) oxides and oxide hydroxides: structural mechanism from *ab initio* molecular geometries and EXAFS spectroscopy. Geochim Cosmochim Acta 67:4223-4230

Sherman DM, Peacock CL, Hubbard CG (2008) Surface complexation of U(VI) on goethite (α-FeOOH). Geochim Cosmochim Acta 72:298-310

Stachowicz M, Hiemstra T, van Riemsdijk WH (2006) Surface speciation of As(III) and As(V) in relation to charge distribution. J Colloid Interface Sci 302:62-75

Tadanier CJ, Eick MJ (2002) Formulating the charge-distribution multisite surface complexation model with FITEQL. Soil Sci Soc Am J 66:1505-1517

Venema P, Hiemstra T, Weidler PG, van Riemsdijk WH (1998) Intrinsic proton affinity of reactive surface groups of metal (hydr)oxides: Application to iron (hydr)oxides. J Colloid Interface Sci 198:282–295

Waite TD, Davis JA, Payne TE, Waychunas GA, Xu N (1994) Uranium(VI) adsorption to ferrihydrite—application of a surface complexation model. Geochim Cosmochim Acta 58:5465–5478

Reviews in Mineralogy & Geochemistry
Vol. 70 pp. 207-258, 2009

6

The Link Between Mineral Dissolution/Precipitation Kinetics and Solution Chemistry

Jacques Schott, Oleg S. Pokrovsky and Eric H. Oelkers

Laboratoire d'Etude des Mécanismes et Transferts en Géologie
Observatoire Midi-Pyrénées
CNRS-Université de Toulouse
Toulouse, France

schott@lmtg.obs-mip.fr

INTRODUCTION

Recent years have seen the growing development and application of reactive transport models to describe, at different spatial and temporal scales, natural and industrial processes involving water-rock interactions such as continental weathering and its impact on ocean chemistry and climate, cycling of metals and the formation of ore deposits, porosity formation/reduction and oil migration in sedimentary basins, transport and geological sequestration of CO_2, and geothermal power generation. The successful application of these models requires a comprehensive and robust kinetic data base of mineral-water interactions. To address this need significant efforts have been made over the past two decades to i) measure in the laboratory mineral dissolution/crystallization rates and ii) develop robust rate laws which could be incorporated in reactive transport algorithms. The aim of this chapter is to review the mechanisms which control the kinetics of mineral dissolution and precipitation and show that accurate knowledge of aqueous chemistry and thermodynamics is essential for quantifying the available kinetic data of mineral water interaction.

REACTIVE TRANSPORT AND THE CONTROL OF FLUID-MINERAL INTERACTIONS

The main processes involved in reactive transport in a porous and/or fractured media: advection, molecular diffusion, mechanical dispersion and fluid-solid reactions (dissolution and crystallization) are illustrated in Figure 1. Crystal dissolution and growth proceed via the transport of aqueous reactants and products to and from the surface coupled to chemical reactions occurring at the surface. The overall rate of dissolution or crystallization is controlled by the slowest of these coupled processes, either surface reaction or transport of aqueous species. When surface reactions are fast relative to molecular diffusion, dissolved species are depleted at the solid surface; the reaction is "transport" controlled[1]. If transport is fast relative to surface reactions, no depletion is observed; the overall reaction rate is "surface reaction" controlled. The rotating disk reactor method (Gregory and Riddiford 1956) allows discrimination between transport and surface-reaction control by changing the rotation speed of a solid disk subject to dissolution. It has been successfully used to measure the dissolution rate constants of fast reacting solids like calcite (Alkattan et al. 1998), halite (Alkattan et al.

---

1 As transport to and from the mineral surface is generally via diffusion, this condition is also often referred to as "diffusion control."

1529-6466/09/0070-0006$05.00 DOI: 10.2138/rmg.2009.70.6

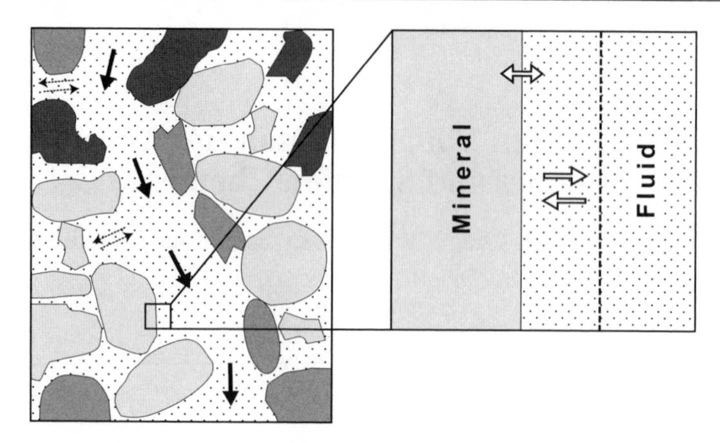

Figure 1. Schematic illustration of a mineral-fluid interface in a porous rock. Dissolved species can be transported to and from the surface of minerals by either advection or diffusion as shown by the solid and dashed arrows, respectively. The fluid adjacent to the mineral interface is often an immobile boundary layer fluid; transport of elements to and from the reactive surface requires diffusion through this boundary layer.

1997), and basaltic glass (Guy and Schott 1989). The impact of aqueous transport on solid surface reactivity and fluid chemistry is illustrated on Figure 2. This figure shows the fully coupled modeling at the pore scale of CO_2 reactive transport through a small porous calcite core. The 3D geometry of this core was generated from synchrotron micro-tomography spectra of natural samples (Flukiger and Bernard 2009). Stream lines showing the water flow, iso-pH-4 surface, and the surfaces exhibiting the highest dissolution rates are shown in Figures 2a, 2b, and 2c, respectively. One can see the strong impact of fluid velocity and pore geometry on solution chemistry. Note the distorted pH-4 surface and the distribution of the calcite surfaces displaying highest dissolution rates. This shows that, in the case of fast reacting solids like limestone, the micro-scale distribution of reactive surface area is tightly linked to pore-fluid velocity and pore geometry.

For most rock-forming minerals at ambient temperatures, chemical reactions at the solid surface are slow and thus rate limiting. As a result, the accurate modeling of fluid-rock interaction requires using a robust kinetic theory allowing a mechanistic description of chemical reactions occurring at the solid-solution interface.

DESCRIPTION OF SURFACE REACTIONS WITHIN THE FRAMEWORK OF TRANSITION STATE THEORY

Chemical reactions tend towards equilibrium. Such is certainly the case for reactions in natural systems. Many natural processes evolve over extremely long timeframes, in some instances thousands and millions of years. As such many natural reactions proceed at very near to equilibrium conditions. It is, however, experimentally challenging to perform laboratory experiments at very near to equilibrium conditions. One reason is that reactions slow as equilibrium is approached, stopping once equilibrium is attained. For this reason large efforts have been made to develop methods to extrapolate rates measured in the laboratory to the near to equilibrium conditions found in many natural environments. The most versatile method to extrapolate rates to near to equilibrium conditions stems from Transition State Theory (TST).

There are several different scales used to quantify the distance a system is from equilibrium. The one scale commonly used is the saturation index, Ω, defined by

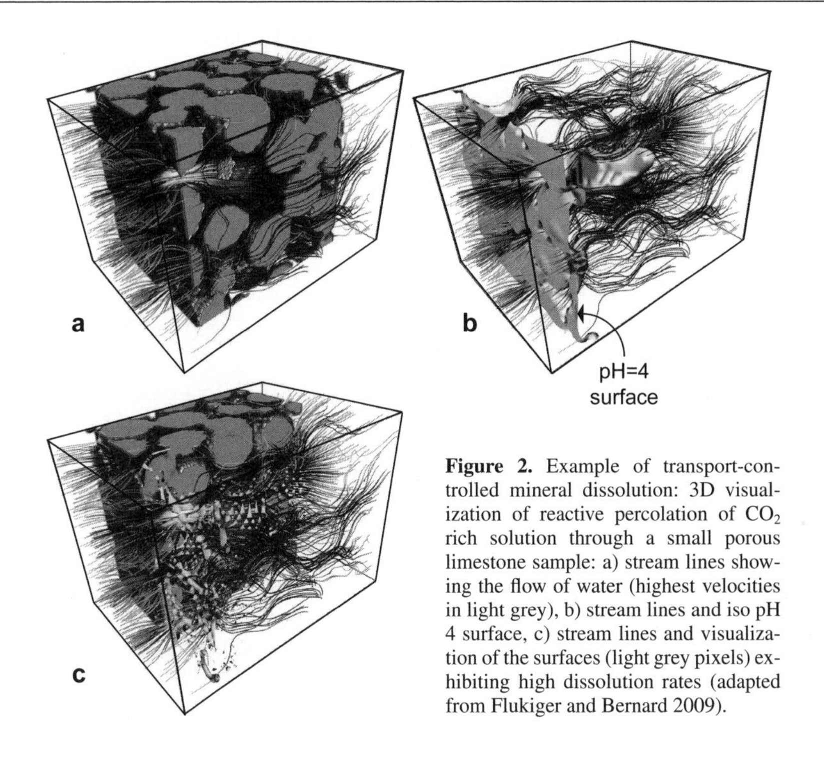

Figure 2. Example of transport-controlled mineral dissolution: 3D visualization of reactive percolation of CO_2 rich solution through a small porous limestone sample: a) stream lines showing the flow of water (highest velocities in light grey), b) stream lines and iso pH 4 surface, c) stream lines and visualization of the surfaces (light grey pixels) exhibiting high dissolution rates (adapted from Flukiger and Bernard 2009).

$$\Omega = Q/K \tag{1}$$

where Q stands for the activity quotient and K the equilibrium constant for the reaction of interest. Ω is dimensionless, it has a value of 1 at equilibrium, less than 1 when the reactants are undersaturated and the reaction will tend to go towards the products. When Ω is greater than 1, the reactants are supersaturated, and the reaction will tend towards the reactants. Another common scale describing the distance from equilibrium is the chemical affinity, A, which is related to the saturation index according to:

$$A = RT\ln(Q/K) = -RT\ln(\Omega) \tag{2}$$

where R designates the gas constant, and T represents absolute temperature. The chemical affinity describes the energy driving the reaction towards the products, and has units of energy per mole. Chemical affinity is positive when the reactants are undersaturated, zero at equilibrium and negative when the reactants are supersaturated. As the chemical affinity represents the energy difference between the reactants and the products, it is sometimes replaced by $\Delta G = -A$.

It is important to note that both the chemical affinity (A) and the energy difference between the reactant and products (ΔG) are both molar properties (e.g., in units of energy per mole). As such their value depends on what formula is used to represent one mole of reactant or product. For example, the chemical affinity for the enstatite dissolution reaction calculated assuming enstatite has the formula $MgSiO_3$ is half that calculated if one assumes enstatite has the formula $Mg_2Si_2O_6$. The scaling of chemical affinity to the stoichiometry of the reactants is critical to describing rates as a function of distance from equilibrium as will be shown below.

From the microscopic viewpoint, surface reactions consist of one or several successive elementary steps, each of which requires overcoming a potential energy barrier. Transition state theory developed in the 1930s (Eyring 1935) provides the means to characterize the rates of reactants over this energy barrier. TST focuses on the molecular configuration at the top of this barrier, the so-called "activated complex," and assumes that the reaction rate is equal to the product of two terms, the concentration of the activated complex and the frequency with which these complexes cross the energy barrier (see Fig. 3). This concept is consistent with:

$$r_+ = k\,[AB^*] \tag{3}$$

where r_+ stands for the forward reaction rate, k refers to a rate constant and $[AB^*]$ represents the concentration of the activated species (Fig. 3). TST treats the activated complex as a true chemical species and furthermore assumes that the reactants are in equilibrium with this activated complex. This makes TST particularly appealing because i) it provides a direct link between the thermodynamics and kinetics of elementary reactions and ii) it simplifies calculation of the activated complex concentration.

The reaction rate constant can be expressed in terms of the thermodynamic functions for the formation of the activated complex (see Fig. 3), and assuming the activated complex is the same for the forward and backward reactions yields the following classical expressions for the overall rate of an elementary reaction, r (e.g., Lasaga 1981; Aagaard and Helgeson 1982):

$$r = r_+(1 - e^{\Delta G/RT}) = r_+(1 - e^{-A/RT}) \tag{4}$$

where r_+ again stands for the forward reaction rate and its expression is given in Figure 3, ΔG represents the Gibbs free energy difference between reactants and products, A stands for the chemical affinity of the elementary reaction, and R and T denote the gas constant and absolute temperature, respectively. Equation (4) implies that the overall rate of an elementary reaction depends both on the chemical affinity of the reaction and on the concentrations of species involved in the formation of the activated complex through their effect on r_+. As there is much confusion in the literature, it needs to be reemphasized that Equation (4) is written for an elementary reaction. ΔG and A in this equation, therefore, correspond to energies normalized to one mole of the elementary reaction. The chemical affinity of an overall reaction is related to the chemical affinity of the elementary reaction by a stoichiometric coefficient, σ, equal to the number of moles of activated complex required to transform one mole of reactant. This is readily demonstrated numerically. For example, if the activated complex for the reaction

$$\mathrm{A + B = C + D} \tag{5}$$

is not AB* created by the reaction:

$$\mathrm{A + B = AB^*} \tag{6}$$

but $A_{05}B_{05}$* created by:

$$\mathrm{0.5\,A + 0.5\,B = A_{05}B_{05}^*} \tag{7}$$

the ΔG^* for the reaction (6) is exactly twice that of reaction (7). It follows that $\Delta G_{(7)}{}^* = \Delta G_{(6)}{}^*/2 = \Delta G_{(6)}{}^*/\sigma$. Equation (4) can be thus rewritten for the overall reaction

$$r = r_+(1 - e^{\Delta G/\sigma RT}) = r_+(1 - e^{-A/\sigma RT}) \tag{8}$$

σ in Equation (8) is known as the Temkin's stoichiometric number which is equal to the ratio of the rate of destruction of the activated complex relative to the overall reaction rate. The impact of σ value on the dependence of reaction rates on the ΔG of the overall reaction is illustrated in Figure 4. It can be seen in this figure that at far from equilibrium conditions ($A/\sigma RT > 3$) the $(1 - e^{-A/\sigma RT})$ term approaches 1 and $r \approx r_+$. As equilibrium is approached, rates decrease, reaching zero at equilibrium.

TST has proved to be a very efficient theoretical framework for describing reversible elementary reactions. Its application to solid-solution interactions—including sorption reactions, solid dissolution and crystal growth—requires identification of the activated complex and it's

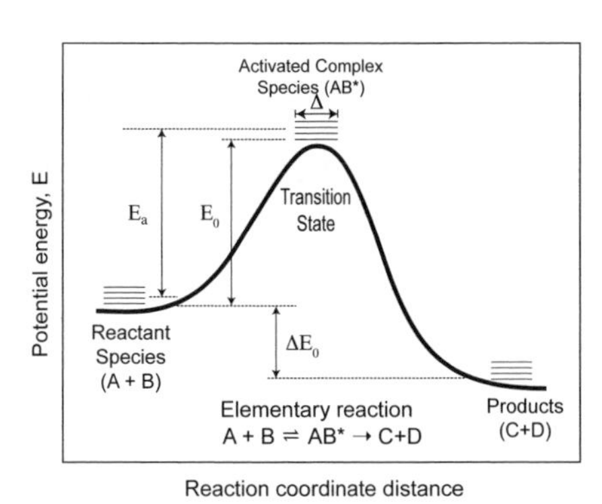

$$A + B \overset{K^*}{\leftrightarrow} AB^*$$

$$AB^* \rightarrow C + D$$

Activated complex (AB*) : high-energy ground state species formed from reactants and in equilibrium with them. AB* falls apart to yield products.

Rate of reaction $r_+ = \frac{k_B T}{h}[AB^*] = \frac{k_B T}{h} K^*[A][B]\frac{\gamma_A \gamma_B}{\gamma^*}$

Rate constant $k = \frac{k_B T}{h} K^* \frac{\gamma_A \gamma_B}{\gamma^*} = \frac{k_B T}{h}\frac{\gamma_A \gamma_B}{\gamma^*}\exp\left(-\frac{\Delta G^{*0}}{RT}\right)$

with ΔG^{*0} = standard free energy of activation

$$k = \frac{k_B T}{h}\frac{\gamma_A \gamma_B}{\gamma^*}\exp\left(\frac{\Delta S^{*0}}{R}\right)\exp\left(-\frac{\Delta H^{*0}}{RT}\right)$$

Reversible elementary reaction $r = r_+\left(1 - e^{\Delta G/RT}\right)$

ΔG=Gibbs free-energy difference between reactants and products

Figure 3. Slow reactions and transition state theory.

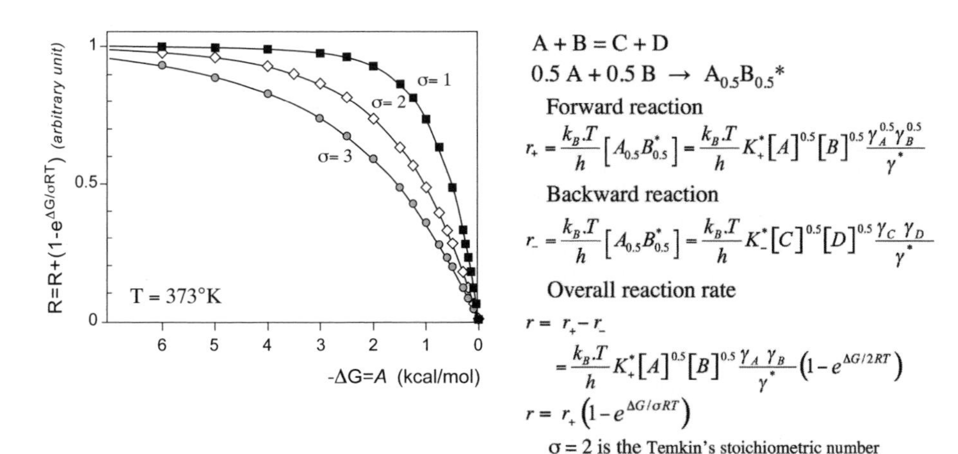

Figure 4. Temkin's average stoichiometric number and its impact on near-equilibrium reaction rates.

forming reactions. Moreover, it is believed that Equation (8) can be applied to an overall reaction if only a single step controls the rate (Boudart 1976) or if steady state conditions are met and the magnitude of ΔG for each elementary reaction is not much greater than RT (Nagy et al. 1991). One of the great advances in the use of TST to describe dissolution and precipitation rates stems from the understanding of the solid-solution interface, as quantified through the coordinative surface chemistry concepts developed by Schindler and Stumm (1987) as described below.

It is important to emphasize that rates as described in Equation (8) are the product of two contributions. The first term, r_+, the forward reaction rate, is proportional to the concentration of the activated complex, as described in Equation (3). Within TST the activated complex is assumed to be in equilibrium with the reactants that form this complex. The variation with the concentration of reactants of 1) activated complex concentration, and thus 2) the forward reaction rate can be calculated from the law of mass action for the activated complex forming reaction. This r_+ term, therefore, provides a link between the concentration of the reactants and *the overall reaction rate. Insight into the identity of the reactions forming the activated complex can be obtained from coordinative surface chemistry, as described below. The second term of Equation (8) describes how overall rates slow down, as the driving force for the reaction decreases, as one approaches equilibrium. Note that as both the chemical affinity term and the r_+ term may both depend on the concentration of the reactants, quantifying each of these terms independently requires studying the variation of rates as a function of solution concentration at either 1) far from equilibrium conditions ($A/\sigma RT > 3$ when $r \approx r_+$ or 2) at constant A.*

INSIGHTS ON THE ACTIVATED COMPLEX AND ITS PRECURSOR FROM COORDINATIVE SURFACE CHEMISTRY

The key to using Transition State Theory to describe rates as a function of solution composition is identification of the stoichiometry and reactions forming activated complexes from the reactants. Much research over the past several decades suggests that the activated complex for the dissolution and precipitation reactions have the same stoichiometry as surface complexes that form on mineral surfaces. Insight into these surface complexes can be obtained from coordinative surface chemistry as described below.

The surface of metal oxides in the presence of water

In the absence of water, the surface of a metal oxide is characterized by the presence of low-coordinated metals giving rise to Lewis acidity. In aqueous solution, water molecules coordinate to these metal centers and dissociative chemisorption occurs with formation, depending on solution pH, of positively, negatively charged and neutral hydroxyl groups that can be exchanged with the ligands present in solution (see Fig. 5). To facilitate representation of reactions and equilibrium, one usually considers the chemical reactions of the simple monodendate surface hydroxyl group, S-OH, although some different surface species may form (i.e., S_2OH, $SOH(OH_2)$ or $S(OH)_3$ groups, Schindler 1992; Kulik 2009) exhibiting distinct reactivities.

Ligand exchange and the reactivity of the solid surface

The reactivity of the oxide surface and its tendency to dissolve depends both on the 1) type and concentration of surface complexes present, and 2) their formation rate.

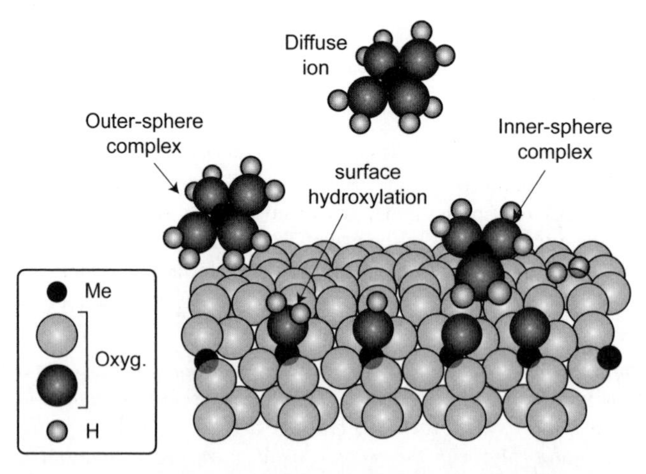

Figure 5. Cartoon of the surface of a metal oxide in the presence of water.

Stoichiometry and stability of precursor surface complex. As shown by Figure 5, when placed into an aqueous solution the coordinative environment of metals at the mineral-water interface can be changed by formation of inner or outer sphere complexes with a variety of aqueous ligands including fluoride, organic anions, etc. These ligands promote surface dissolution by polarizing and/or weakening interatomic bonds in the vicinity of the central metal, and, thus, facilitating its detachment into solution. As emphasized by Stumm and coworkers (Stumm et al. 1983, 1985; Stumm and Wieland 1990), the surface species present at the mineral-fluid interface can be viewed, within the framework of TST, as the precursor of the activated complex. The precursor 1) has the same stoichiometry and 2) its concentration is proportional to that of the activated complex (Wieland et al. 1988). Assuming that sorption of reactants is fast and thus surface species attain mutual equilibrium and that subsequent detachment of the metal species from the surface lattice is rate limiting, the following rate laws can be written to describe proton, hydroxyl and ligand promoted dissolution, by analogy with analogous reactions derived from TST (see Fig. 3)

$$r_{+,\mathrm{H}} = k_{\mathrm{H}}\{>\mathrm{MeOH_2}^+\}^{n_\mathrm{H}}$$
$$r_{+,\mathrm{OH}} = k_{\mathrm{OH}}\{>\mathrm{MeO}^-\}^{n_\mathrm{OH}} \qquad (9)$$
$$r_{+,L} = k_L\{<\mathrm{Me}L\}^{n_L}$$

where the k_i are the rate constants, {>MeX} stands for the concentrations of the various rate controlling surface precursor complexes, and n_i represents the reaction order with respect to the subscripted complex. It follows that characterization of the speciation of a simple oxide allows description of the dependence of its dissolution rate as a function of aqueous solution composition.

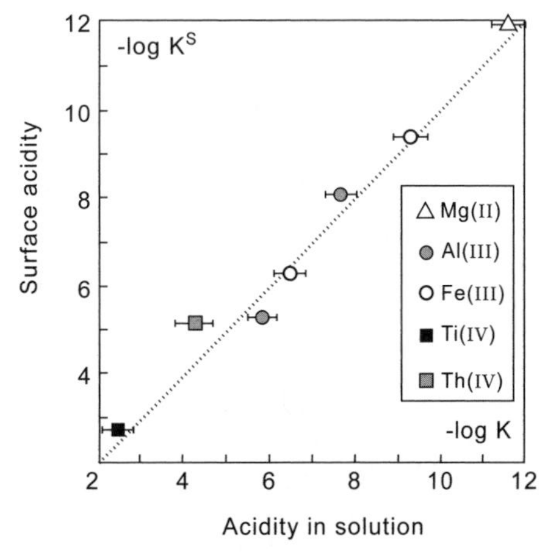

Figure 6. Correlation of metal oxide surface acidity and metal acidity in solution (adapted from Schindler 1987.)

Because the formation of surface and aqueous complexes is controlled by similar ligand exchange reactions, a number of authors suggested that the stability constants of surface metal-ligand complexes should correlate with the stability constants of corresponding aqueous metal-ligand complexes (see, for example, Schindler and Stumm 1980 and references therein). The data shown in Figure 6 illustrate this correlation. Similarly, as shown in Figure 7, the stability constants of metal surface complexes correlate with corresponding aqueous metal hydroxide complexes both for mono and binuclear complexes for the case of amorphous

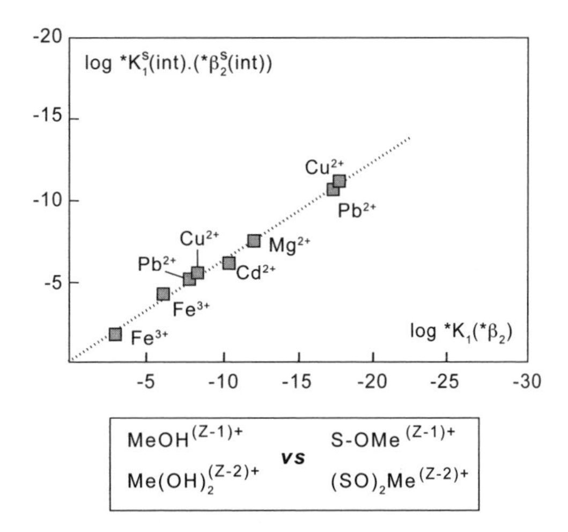

Figure 7. Stability constants of metal surface complexes formed on amorphous silica against stability constants of corresponding metal hydroxo complexes (adapted from Schindler 1987).

silica. This correlation occurs because metal adsorption to hydroxo surface groups of metal oxide surfaces acts like aqueous metal hydrolysis reactions (Wehrli 1990). Similar correlations are observed between ligand adsorption on mineral surface and ligand-metal complexation in solution as shown on Figure 8 for organic and inorganic ligands sorption at the brucite surface. Thus, in addition to surface titration, spectroscopy measurements and *ab initio* calculations, linear free-energy correlations between aqueous and surface complexes can be useful to estimate the stoichiometry and stability of surface complexes that control oxide minerals dissolution kinetics.

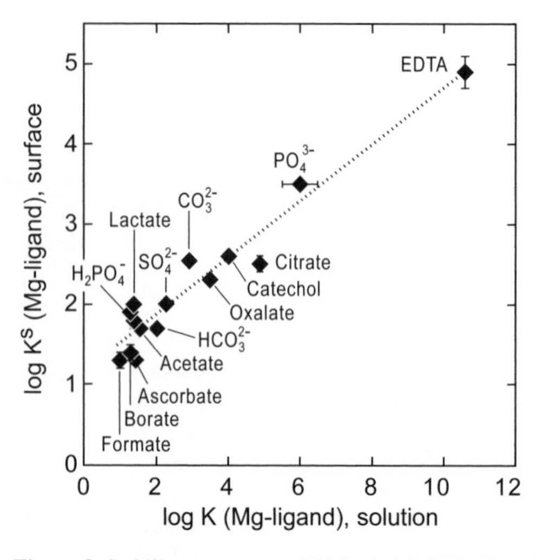

Figure 8. Stability constants at 25 °C of $>MgOH_2^+$-ligand complexes formed at brucite surface against the stability constants of corresponding aqueous Mg-ligand complexes (from Pokrovsky et al. 2005).

The rate of formation of surface complexes. Like for aqueous complexes, the rate of formation of surface complexes is proportional to the rate of exchange of water molecules from the metal coordination sphere. This constitutes an important link between surface and aqueous chemistry. This link stems from the similarity of these reactions. Metal ions in solution are complexes of water and the formation of an aqueous metal complex with a ligand proceeds via the exchange of incoming ligand with a water molecule. For example, the formation of the Al-acetate complex is shown in Figure 9a. Similarly, ligand adsorption at a surface site proceeds via the exchange of a water molecule with the incoming ligand. This process can obey the Eigen mechanism: formation of an outer-sphere complex be-

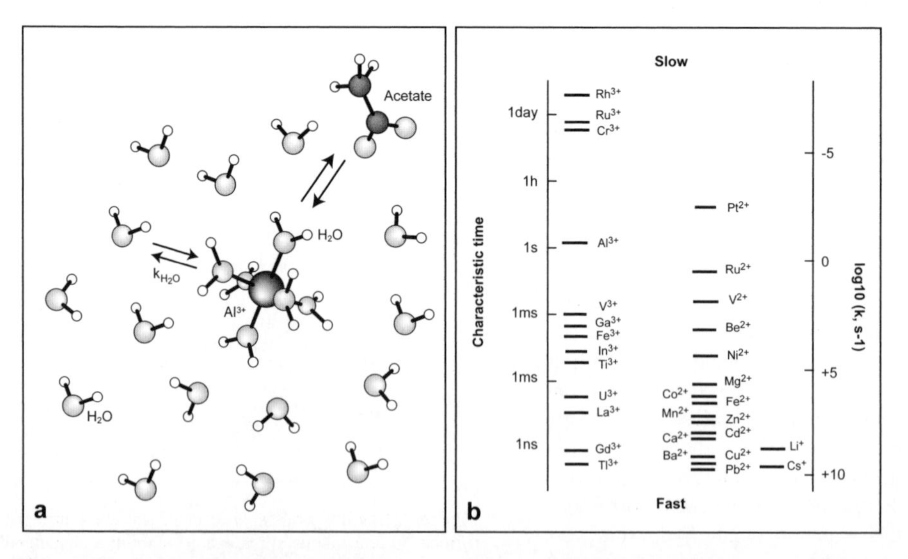

Figure 9. Mechanism of formation of a metal aqueous complexes ($AlAcO_2^+$) (a), and the rate of exchange and residence time of water molecules in metals inner coordination spheres from Burgess (1988) (b).

tween the metal and the ligand is followed by the loss of a water molecule from the metal inner coordination sphere (Margerun et al. 1978). The rate constants for water exchange, and mean residence times for water molecules in primary hydration shell can vary enormously between different cations (by more than 15 orders of magnitude at 25 °C). These differences depend on charge, radius, and crystal field effects (see Fig. 9b). Figure 10 shows that the rate constants for both the adsorption of metal ions on the γ-Al_2O_3 surface and the dissolution rates of simple metal oxide minerals are correlated to the rate constants for water exchange. These two correlations hold for reaction rates that vary by up to 8 orders of magnitude. It should be noted that substitution in the metal coordination sphere of a water molecule by an inorganic or organic ligand (OH^-, F^-, $C_2O_4^{2-}$, etc.) that reduces the metal complex formal charge and weakens metal bonds to remaining water molecules, enhances the rate of exchange of water molecules from inner-coordination sphere by one to two orders of magnitude (Phillips et al. 1997a). This labilizing effect explains why 1) ligand adsorption accelerates mineral dissolution, and 2) a transient increase in simple oxides dissolution rate is observed following solution pH changes (see below an example with brucite).

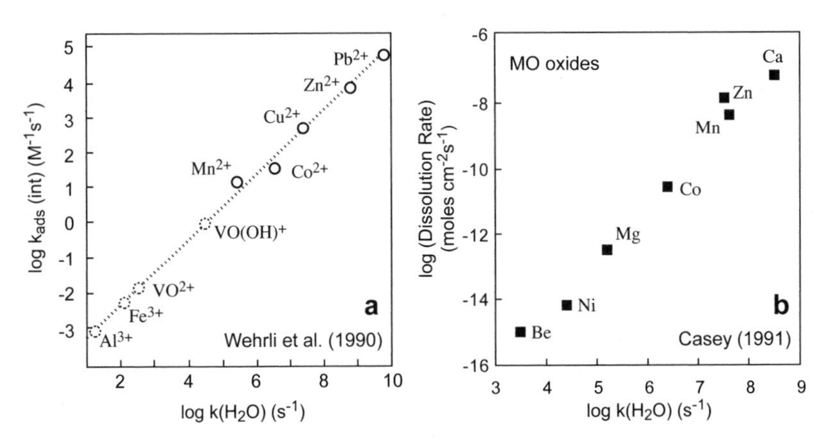

Figure 10. Rate of metal sorption on γ-Al_2O_3 (a) and rate of dissolution at 25 °C and pH = 2 of metal oxide minerals having a rock-salt structure (b) plotted against the rate constant for exchange of water molecules in the coordination sphere of corresponding metal in solution (adapted from Wehrli et al. 1990, and Casey 1991, respectively).

THE SCM/TST DESCRIPTION OF THE KINETICS OF DISSOLUTION/ PRECIPITATION OF OXIDE MINERALS

The dissolution of single oxide minerals, which requires breaking only one type of bond, provides an impressive illustration of the effectiveness of transition state theory and surface coordination concepts in describing the kinetics of mineral-water interactions. As first proposed by Furrer and Stumm (1986) and Zinder et al. (1986) for the case of BeO, δ-Al_2O_3, and Fe(III) oxides, the far from equilibrium dissolution rate for simple oxides can be described from the sum of parallel elementary reactions occurring at metal centers (Eqn. 9):

$$r_+ = k_H\{>MeOH_2^+\}^{n_H} + k_{OH}\{>MeO^-\}^{n_{OH}} + k_L\{>MeL\}^{n_L} + k_{H_2O} \quad (10)$$

where n_H and n_{OH}, the reaction order of proton and hydroxyl promoted dissolution, were proposed to correspond to the charge of the central metal ion (Stumm 1992). The rate of the ligand-promoted dissolution is generally assumed to be proportional to the ligand surface concentration ({>MeL}) so n_L is usually assumed to be equal to one. Note that most mineral dissolution experiments have been performed in dilute solutions in which water concentration is roughly constant, so the impact of water activity on mineral oxide dissolution kinetics, as incorporated in the final term of Equation (10), is poorly known.

The dissolution and precipitation kinetics of magnesium, silicon and aluminum oxide/hydroxides provide useful case studies in the application of surface coordination chemistry concepts coupled to transition state theory.

Brucite ($Mg(OH)_2$) possesses an isoelectic point (pH_{IEP}) around 11 and its dissolution rate uniformly decreases with increasing pH to at least pH 11.5 (Pokrovsky and Schott 2004; Fig. 11). The rate limiting step for brucite dissolution is breaking Mg-O bonds. Bond breaking is facilitated by surface protonation and the brucite dissolution rate found to be proportional to the square concentration of $MgOH_2^+$ surface complex, in accord with

$$r_+ = k_{MgOH_2^+} \{>MgOH_2^+\}^2 \tag{11}$$

which suggests that the surface precursor complex is composed of two adjacent protonated Mg sites (Fig. 11). There is also a strong correlation between brucite surface acidity and aqueous magnesium hydrolysis which explains the close correspondence between brucite dissolution rates and solubility as a function of pH (see Fig. 11).

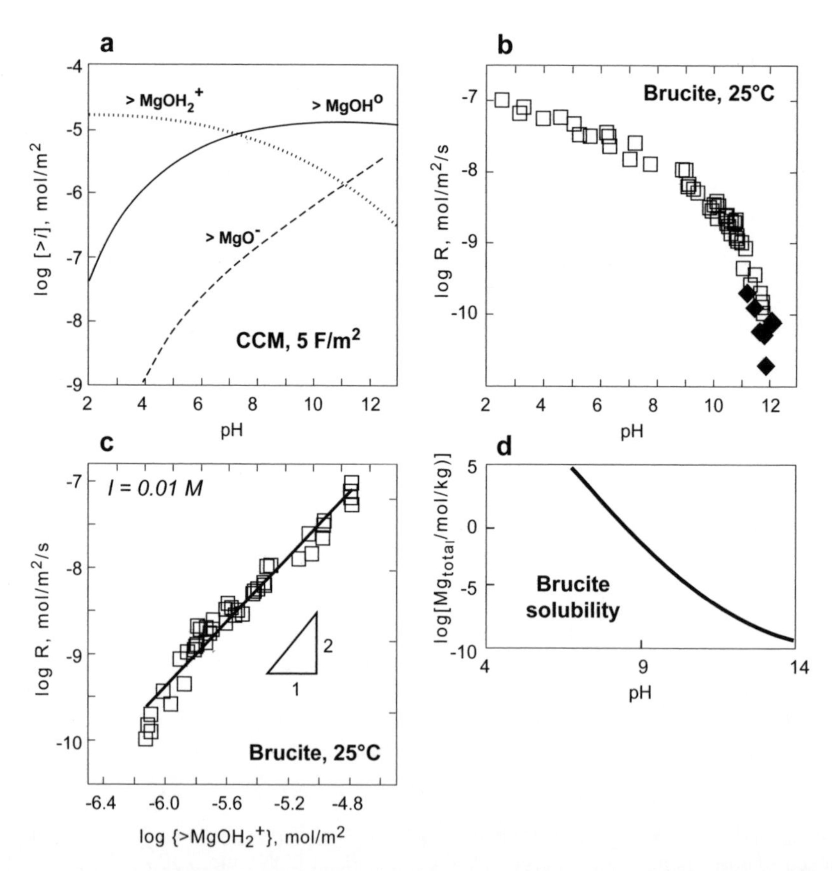

Figure 11. Brucite ($Mg(OH)_2$) surface speciation (a), dissolution rate at 25 °C as a function of solution pH (b) and concentration of $>MgOH_2^+$ (c), and brucite solubility at 25 °C as a function of pH (d); solid symbols in Figure 11b represent rates measured close to equilibrium ($\Omega > 0.5$); after Pokrovsky and Schott (2004).

The labilizing effect resulting from the substitution of a water molecule by a negatively charged oxygen donor compound in the Mg coordination sphere is demonstrated by non-steady-state brucite dissolution rates. One such non-steady-state effect is the change of reactor fluid pH. It can be seen in Figure 12a that a downward pH change from 12 to 3 results in a transient period of rapid brucite dissolution (i.e., ~ 3.5 higher) that lasts about one hour. As suggested by the correlation between the excess Mg release after a pH-jump and the concentration of $>MgO^-$ sites shown in Figure 12b, replacement of water molecules by oxygen donors in the Mg coordination sphere has a labilizing effect on the dynamics of the remaining water molecules, increasing dissolution rates. This transient dissolution rate enhancement occurs because proton sorption at brucite surface is faster than the re-equilibration of brucite surface speciation (transformation of $>MgO^-$ to $>MgOH_2^+$). The observation of this behavior implies that surface protonation occurs not only on the ligands present in the metal coordination (i.e., formation of H_2O at the expand of OH^-) but also on additional sites of a non-flat surface. Non-steady-state dissolution rates generated by pH-jumps were first observed for trivalent metal oxides by Samson and Eggleston (1998) and Samson et al. (2000).

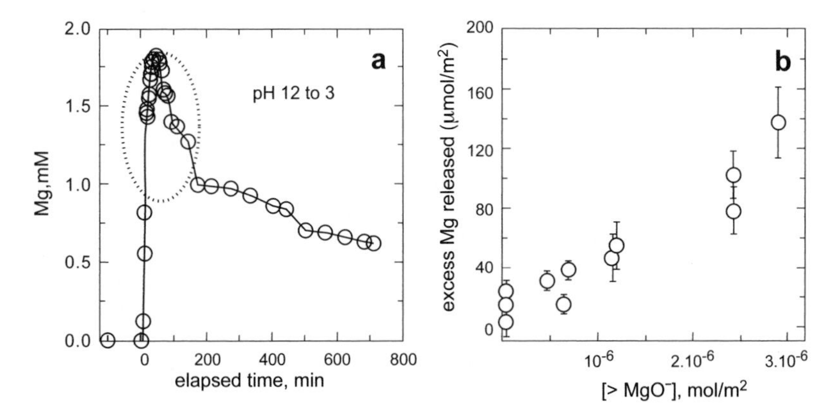

Figure 12. Brucite pH-jump dissolution experiment at 25 °C in 0.01 M NaCl using a mixed flow reactor: a) outlet Mg concentration as a function of time during pH jump from 12 to 3, b) excess of Mg released during the pH-jump experiments normalized to the total surface area of solid in reactor as a function of concentration of $>MgO^-$ sites (from Pokrovsky and Schott 2004).

An expression describing brucite dissolution and precipitation rates as a function of departure from equilibrium (e.g., saturation state) can be obtained by first invoking the law of detailed balancing. The law of detailed balancing assumes that the precursor complex is the same for dissolution and precipitation. Taking account of detailed balancing allows a straightforward derivation from Equation (8) of brucite dissolution and precipitation rates as a function of departure from equilibrium:

$$r = r_+(1-e^{-A/\sigma RT}) = k_{\mathrm{MgOH_2^+}}\{>\mathrm{MgOH_2^+}\}^2(1-\Omega^2) = k_{\mathrm{MgOH_2^+}}\{>\mathrm{MgOH_2^+}\}^2(1-e^{-A/\sigma RT}) \quad (12)$$

where Ω stands for the solution saturation index with respect to brucite. Brucite dissolution and precipitation rates, measured by Pokrovsky and Schott (2004) are plotted in Figure 13a as a function of departure from equilibrium at $10.6 < pH < 11.0$ and $11.3 < pH < 12.1$. In this figure, the dashed and solid lines labeled 1 and 2 correspond to a fit of these data by Equation (12) for pH of 10.8 and 11.5, respectively. A close agreement is observed between both measured and calculated dissolution and precipitation rates. For example, it can be seen on Figure 13b that, as predicted by Equation (12), brucite precipitation rates are proportional to the square of $>MgOH_2^+$ concentration for a constant solution saturation index $\Omega = 1.8$. Note if Equation (12) is expressed

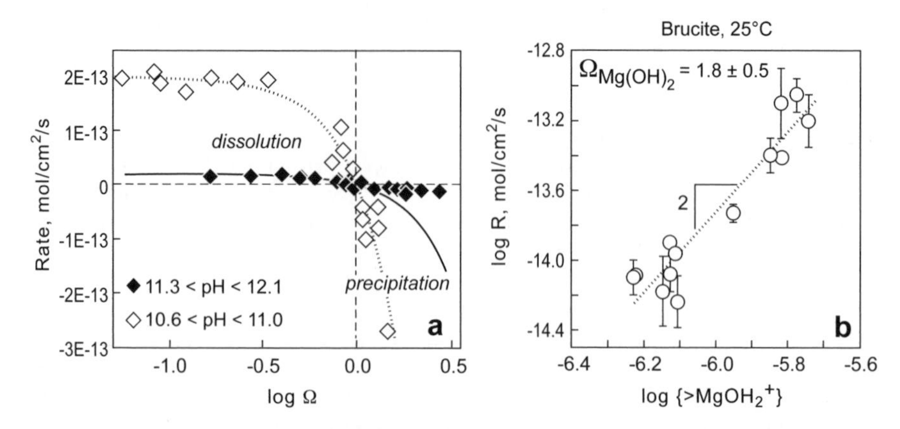

Figure 13. Brucite overall reaction rate at 25 °C for I = 0.01-0.001 mol/L and various pHs as a function of departure from equilibrium (a), and brucite precipitation rate at 25 °C, constant saturation index (Ω = 1.8 ± 0.5), pH of 10 to 12 and ionic strength of 0.01 to 0.001 mol/L as a function of $>MgOH_2^+$ concentration (b); after Pokrovsky and Schott 2004.

in terms of aqueous species concentrations rather than surface species concentrations, brucite precipitation rate is proportional at constant Ω to $((K{\cdot}a_{H^+})/(1 + K{\cdot}a_{H^+}))^2$, K being the formation constant of $>MgOH_2^+$ from MgOH surface sites. It should be noted that Equation (12) cannot be used to compute the rate of brucite crystallization in highly supersaturated solution ($\Omega > 3$). At such conditions rates become controlled by a different process: surface nucleation.

Mg-O bonds are labilized in the presence of organic ligands L^{n-}, like catechol or EDTA, which form bidendate surface complexes and bring electron density or negative charge into the surface Mg coordination sphere. As a result, brucite dissolution rates are found to be consistent with the sum of two parallel reactions: 1) proton promoted dissolution and 2) ligand promoted dissolution consistent with (Pokrovsky et al. 2005):

$$r_+ = k_{MgOH_2^+}\{>MgOH_2^+\}^2 + k_L.\ \{>MgL^{1-n}\} \tag{13}$$

Expressing concentrations of Mg surface complexes in terms of their stability constants and the concentration of aqueous species involved in their formation reaction, yields

$$r_+ = k^{\#}_{MgOH_2^+}\left(1 - \frac{K^*_{>Mg\text{-}L}\cdot[L^{n-}]}{1+K^*_{>Mg\text{-}L}\cdot[L^{n-}]}\right)^2 + k^{\#}_L.\frac{K^*_{>Mg\text{-}L}\cdot[L^{n-}]}{1+K^*_{>Mg\text{-}L}\cdot[L^{n-}]} \tag{14}$$

where $k_i^{\#} = k_i{\cdot}S_T$ with i = Mg, L and S_T = total number of surface sites, and $K^*_{>Mg\text{-}L}$ stands for the formation constant of $>MgL^{1-n}$. The close correspondence observed in Figure 14 between brucite dissolution rates measured as a function of EDTA and catechol concentrations, and dissolution rates computed with Equation (14) demonstrates the validity of Equations (13) and (14). Moreover this correspondence confirms that brucite ligand-promoted dissolution rate is proportional to ligand surface concentration.

Quartz. Far-from equilibrium quartz dissolution rates as a function of solution pH have been interpreted within the framework of surface coordination to be a function of the concentration of three surface species: a protonated, a neutral and a deprotonated surface species ($>SiOH_2^+$, $>SiOH$ and $>SiO^-$) consistent with (Wollast and Chou 1988; Hiemstra and Van Riemsdijk 1990; Berger et al. 1994)

$$r_+ = k_{H^+}\{>SiOH_2^+\}^{n_H} + k_{H_2O}\{>SiOH\} + k_{OH^-}\{>SiO^-\}^{n_{OH}} \tag{15}$$

where n_H and n_{OH} stand for reaction orders and k_{H+}, k_{H_2O} and k_{OH^-} refer to rate constants. The stoichiometry and concentration of these surface species (see Fig. 15a) have been characterized by surface titration (Schindler and Kamber 1968; Abendroth 1970) and X-ray photoelectron spectroscopy (Duval et al. 2002). Note this speciation scheme is a simplified view of the actual species present at quartz surface; *ab initio* investigation of the hydrolysis of protonated, neutral and deprotonated silica dimers (Xiao and Lasaga 1996; Nangia and Garrison 2008) show that

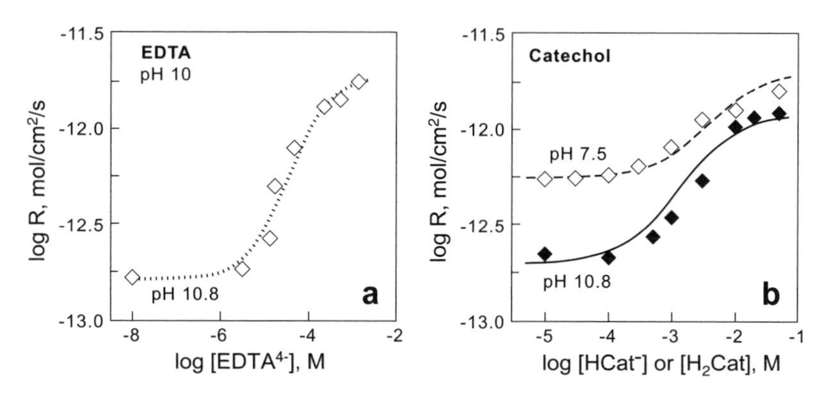

Figure 14. Brucite dissolution rate at 25 °C in presence of EDTA (a) and catechol (b). The symbols represent the experimental data but the curves were generated using Equation (14) (from Pokrovsky et al. (2005).

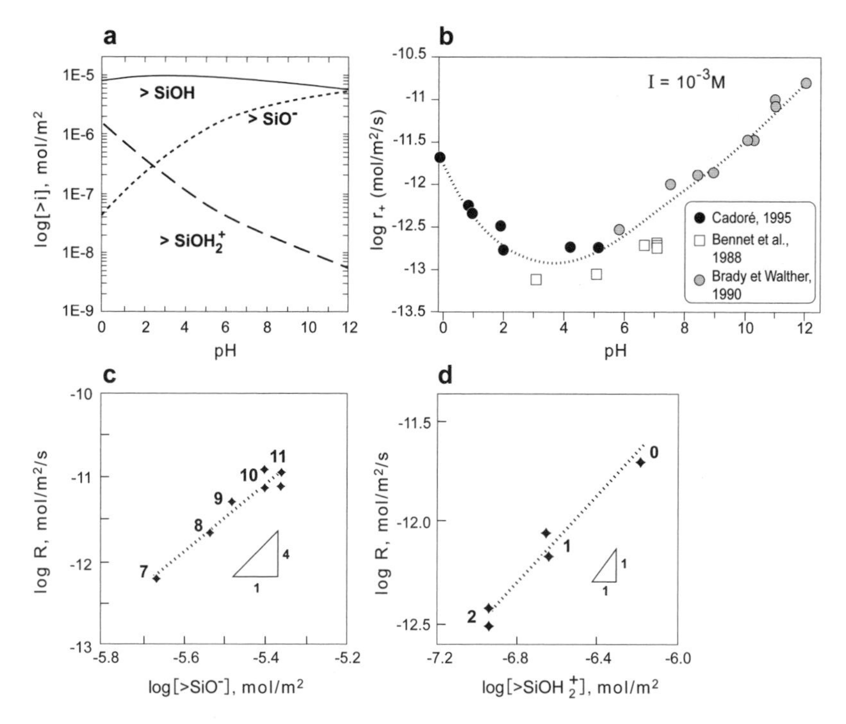

Figure 15. Quartz surface speciation (a) and dissolution rate (b) at 25 °C as a function of solution pH, and quartz dissolution rate at 25 °C as a function of the concentration of >SiO^- (c) and >$SiOH_2^+$ (d); adapted from Duval et al. 2002 (a, c, d) and Cadoré 1995 (b).

protonation of bridging oxygens is thermodynamically more favorable than that of a terminal OH group. Measured quartz dissolution rates as a function of pH is shown in Figure 15b. This figure shows that measured quartz far from equilibrium dissolution rates are in agreement with Equation (15). Rates minimize near the quartz pH_{IEP} which corresponds to the maximum concentration of >SiOH and increase both with decreasing pH below its pH_{IEP} and increasing pH above its pH_{IEP} consistent with the increase of $>SiOH_2^+$ and $>SiO^-$ concentrations, respectively. This observation is also consistent with calculations of TST rate constants reported by Nangia and Garrison (2008) which showed that dissolution rate constants of neutral surface species are several orders of magnitude lower than those for the protonated and deprotonated species. Adherence of experimental data to Equation (15) is illustrated in Figure 15c,d. A good coherence is observed, but it should be emphasized that derived values of reactions orders n_H and n_{OH} depend strongly on the surface speciation model selected for quartz. For example, XPS analysis of quartz surfaces yields, interpreted within the framework of electrical double layer (EDL) model, values of >SiOH deprotonation constant that are more than two orders of magnitude higher than those derived from amorphous silica surface titrations.

Insight into the quartz hydrolysis mechanism can be also gained from the dependence of quartz dissolution rate on water activity. Cadoré (1995) measured the rate of dissolution of quartz powder in binary mixtures of water with methanol, formamide or dioxane (3 organic compounds that do not react with aqueous or solid silica) at 25 °C and pH of 3 to 4. The results are depicted in Figure 16 where quartz dissolution rates normalized to that in pure water are plotted as a function of water activity. The linear data regression with a slope of 4 suggests that quartz far-from equilibrium dissolution rates are proportional to water activity to the fourth power. This observed dependence of quartz dissolution rates on water activity could be explained by a simple reaction mechanism involving the fast reversible adsorption of 4 water molecules at silica surface

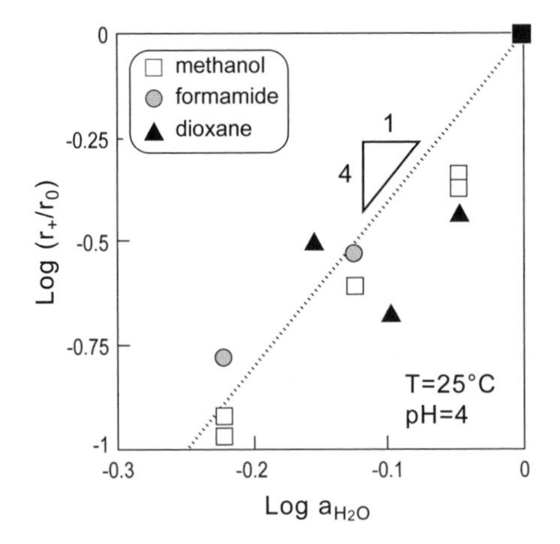

Figure 16. Quartz dissolution rate at 25 °C and pH = 4 as a function of water activity (adapted from Cadoré 1995).

$$>SiO + 4H_2O \leftrightarrow >Si\text{-}(OH)_4^{ads} \tag{16}$$

followed by a slow hydrolysis step

$$>Si\text{-}(OH)_4^{ads} \rightarrow H_4SiO_4(aq) \tag{17}$$

which, if $\{>Si\text{-}(OH)_4^{ads}\} << \{>Si\text{-}O\}$, yields

$$r_+ = k_{H_2O} \cdot a_{H_2O}^{\ 4} \tag{18}$$

This dependence of quartz dissolution rates on water activity is consistent with a) *ab initio* calculations showing that a single hydration step is sufficient to break silicon-bridging oxygen bonds in the silica dimer (Nangia and Garrison 2008; Fig. 17), and b) the need for successive coordination of 4 water molecules to weaken and break the 4 anchoring Si-O bonds for Si release as shown in Figure 18.

As was the case for brucite, an equation describing quartz dissolution rates as a function of degree of departure from equilibrium can be derived by invoking the law of detailed balancing.

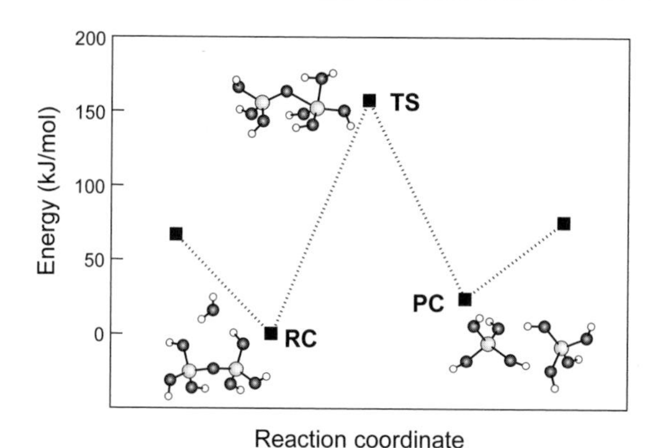

Figure 17. *Ab initio* calculation of the energy profile of the Si-O-Si hydrolysis reaction along the reaction coordinate for the neutral species. Geometries of the reactant complex (RC), transition state (TS) and product complex (PC) are shown along the path. The O, Si and H atoms are represented in dark grey, light grey and white, respectively; from Nangia and Garrison 2008.

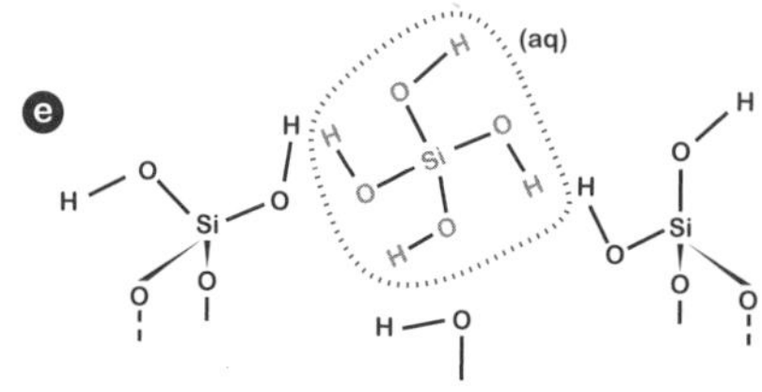

Figure 18. Successive steps involved in the hydrolysis of quartz by water molecules (adapted from Lasaga 1995).

For example in near neutral pH solutions, quartz forward rate of dissolution, as deduced from Equation (15), can be expressed as

$$r_+ = k_{H_2O}\{>SiOH\} \tag{19}$$

which can be combined with Equation (6) to yield the constant pH overall rate of quartz dissolution

$$r = k_{H_2O}\{>SiOH\}(1-\exp(-A/RT)) = k_{H_2O}'(1-\exp(-A/RT)) = k_{H_2O}'(1-\Omega) \tag{20}$$

where Ω again stands for the solution saturation index with respect to quartz ($\Omega = m_{SiO_2}/m_{SiO_2(eq)}$, where m_{SiO_2} and $m_{SiO_2(eq)}$ stand for aqueous silica concentration in solution and in equilibrium with quartz, respectively).

Quartz dissolution rates have been measured as a function of saturation state in mixed flow reactors. Mixed flow reactors allow regulation of the aqueous fluid composition and its saturation state with respect to the dissolving mineral by simply changing the flow rate or the composition of the input fluid (Dove and Crerar 1990; Berger et al. 1994). Dissolution rates measured at 200 °C (Gautier 1999) and 300 °C (Berger et al. 1994) and constant near-neutral pH (pH_{25} = 5.5 and 6.5 at 200 and 300 °C, respectively) are shown as a function of the chemical affinity of the quartz hydrolysis reaction in Figure 19 and compared to quartz dissolution rates computed using Equation (20). The close correspondence between experimental and calculated rates both at 200 and 300 °C confirms, as predicted by *ab initio* calculation and TST/SCM theory, that the rate controlling precursor complex is formed

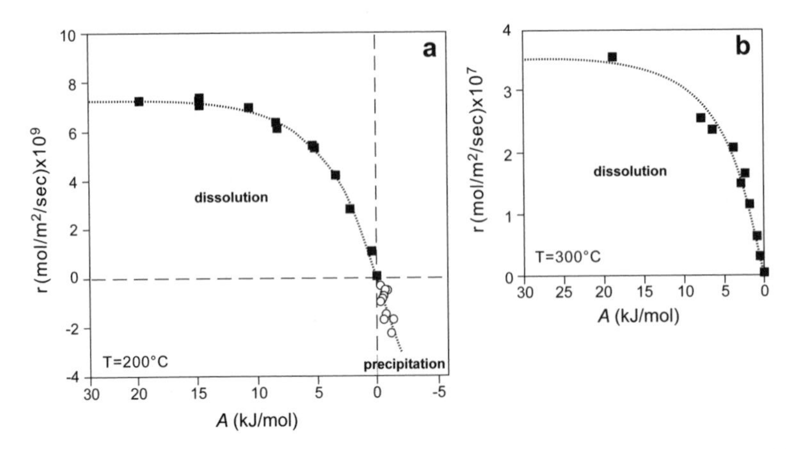

Figure 19. Quartz dissolution and precipitation rates at 200 °C and near-neutral pH from Gautier 1999 (a) and quartz dissolution rate at 300 °C from Berger et al. 1994 (b) as a function of reaction chemical affinity. The symbols correspond to experimental data but the curves were generated with Equation (20).

by the simple sorption of H_2O molecules on quartz surface. Note also that Equation (20) describes accurately quartz crystal growth rates measured by Gautier (1999) at 200 °C and close to equilibrium conditions ($A > -2$ kJ/mol, $\Omega < 1.7$) (see Fig. 19a) where quartz crystallization is controlled by sorption of aqueous silica on steps and kinks (linear rate law). For higher values of solution saturation index, quartz growth is controlled by a different mechanism: surface nucleation on F (flat) surfaces (Gautier 1999).

In contrast with the results obtained in pure water as shown in Figure 19, quartz dissolution rates measured by Berger et al. (1994) at 300 °C in 0.05 M $NaNO_3$ and 10^{-3} M $Pb(NO_3)_2$ solutions are not linear functions of the solution saturation index as might be expected from Equation (20) (see Fig. 20). A considerable increase of quartz dissolution rates is seen at far from equilibrium conditions which was attributed to the inner-sphere adsorption of Na^+ and Pb^{2+} on $>SiO^-$ sites. This adsorption polarizes and weakens bridging Si-O bonds (Dove and Elston 1992; Dove 1994; Berger et al 1994). The sharp decrease of dissolution rates observed for aqueous silica concentrations of 3×10^{-4} M (A = 16.7 kJ/mol) and 2×10^{-3} M (A = 8 kJ/mol) for sodium and lead, respectively, was attributed to the competitive adsorption between dissolved cations and aqueous silica at the quartz surface. It can be seen on Figure 20 that for aqueous silica concentrations > 2×10^{-3} M ($A < 8$ kJ/mol) and > 5×10^{-3} M ($A < 3$ kJ/mol) in sodium- and lead-bearing solutions, respectively, dissolution rates decrease to values similar to those obtained in cation-free solutions, and quartz dissolution rates as a function of solution saturation index again follow Equation (20). It should be emphasized that this sigmoidal behavior cannot be explained by a dislocation control of quartz dissolution at high chemical affinities since i) a linear dependence of dissolution on solution saturation index is observed in cation-free solutions and ii) the calculated critical Gibbs free energy for the opening of most common screw dislocation cores in quartz to form etch pits is −5 kJ/mol at 300 °C (Berger et al. 1994) which is significantly different from the Gibbs free energy at which a sharp drop of dissolution rate is observed in $NaNO_3$ ($\Delta G_r = -16.7$ kJ/mol) and $Pb(NO_3)_2$ solutions ($\Delta G_r = -8$ kJ/mol).

Boehmite and gibbsite. In accord with TST and SCM concepts, aluminum oxide/hydroxide dissolution in neutral to alkaline solutions can be described by (Furrer and Stumm 1986; Carroll-Webb and Walther 1988)

$$r_+ = k_{OH}\{>AlO^-\} \tag{21}$$

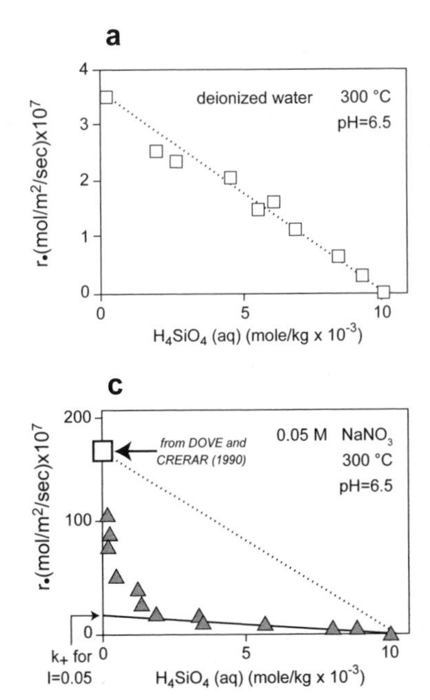

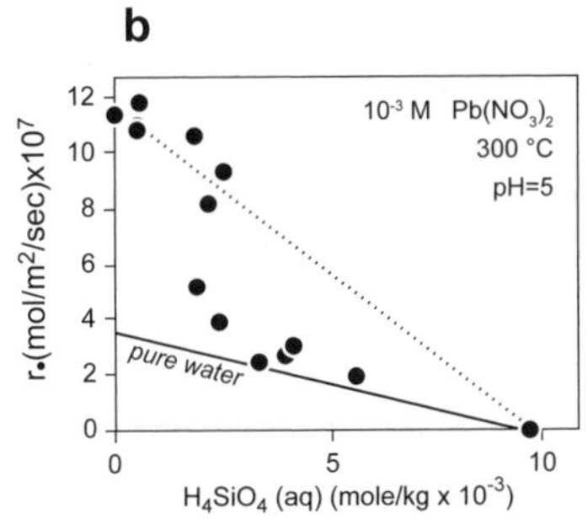

Figure 20. Quartz dissolution rate at 300 °C and near-neutral pH as a function of silica concentration: (a) dionized water; (b) 10^{-3} M $Pb(NO_3)_2$; (c) 0.05 M $NaNO_3$. The symbols correspond to experimental data but the dashed and solid lines were generated with Equation (20) (from Berger et al. 1994).

Similar to the formation of aqueous Al complexes, the concentration of $>AlO^-$ can be computed from the law of mass action for the reaction $>AlOH^\circ + OH^- = >AlO^- + H_2O$ which is given by $K_{21} = \{>AlO^-\}/(\{>AlOH^\circ\}\cdot a_{OH^-})$. Expressing the conservation of surface sites, $S = \{>AlO^-\} + \{>AlOH^\circ\}$ with $\{>AlO^-\} << \{>AlOH^\circ\}$, and the principle of detailed balancing yields the following expression for the overall rate of boehmite (or gibbsite) dissolution/precipitation

$$r = S \cdot k_{OH} \cdot a_{OH^-}(1-\exp(-A/RT)) = k'_{OH} \cdot a_{OH^-}(1-\exp(-A/RT)) = k'_{OH} \cdot a_{OH^-}(1-\Omega) \quad (22)$$

The dissolution and precipitation rates of boehmite (α-AlOOH) at 100 °C and the precipitation rates of gibbsite ($Al(OH)_3$) at 50 °C have been recently measured at near to equilibrium conditions in neutral to basic solutions by Bénézeth et al. (2008). These measurements were performed using a pH-relaxation technique in a hydrogen-electrode concentration cell. This elegant technique allows very precise measurement of the functional dependence of reactions rates on chemical affinity. Figure 21 shows the rates of boehmite dissolution and precipitation as a function of reaction affinity. It can be seen in this figures that measured rates are consistent with Equation (22). Significantly, as numerous measurements were obtained for very low A values ($A < 1$ kJ/mol), it can be verified that in accord with

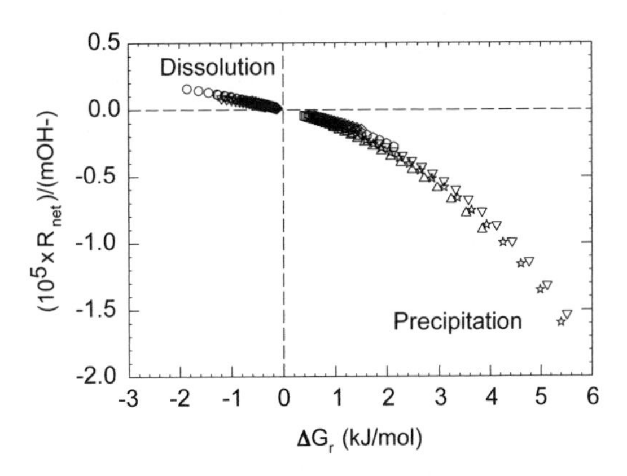

Figure 21. Boehmite dissolution and precipitation rates as function of chemical affinity at 100 °C and 6 < pH < 9.6 (from Bénézeth et al. 2008).

Equation (22), dissolution and precipitation rates very near to equilibrium are proportional to A. The same linear behavior has been observed by Bénézeth et al. (2008) for gibbsite precipitation at 50 °C and $8.2 < pH < 9.7$. These results for both boehmite and gibbsite contrast with those of Nagy and Lasaga (1992) who reported for gibbsite at 80 °C and pH = 3 a highly non linear dependence of dissolution rate on chemical affinity for $0.85 < A < 2.0$ kJ/mol. This apparent non-linear behavior may be an artifact of a poor choice for the gibbsite solubility product which strongly impacts the calculation of reaction chemical affinity.

THE DISSOLUTION MECHANISMS AND RATES OF MULTI-OXIDE MINERALS

Exchange reactions and the formation of leached layers at multi-oxide silicate mineral surfaces

We have seen that single oxide dissolution, a process which requires the breaking of only one type of bond, is controlled by the sorption of reactants at their surface and that it can be described within the framework of TST-SCM by assuming that precursor complexes are formed by the simple sorption of aqueous protons, water and/or various ligands which enter surface metals coordination spheres. The dissolution of many mixed oxide minerals, including many major rock forming silicates, requires the breaking of more than one type of metal-oxygen bond. ***The dissolution mechanism of multi-oxide minerals and glasses can involve sorption reactions, as well as exchange reactions between protons and constituting metals that are not essential to maintaining the mineral structure***. This is demonstrated, for example in the case of feldspars, by the formation of metal depleted leached layers during their dissolution and by surface titrations following their immersion in solution. The secondary ion mass spectrometry (SIMS) analysis of the surface of labradorite reacted for 1000 hours at pH 2 performed by Schweda et al. (1997) and shown in Figure 22 demonstrated the formation of a leached layer, several hundred nanometers thick, depleted in aluminum and formed by exchange with aqueous protons. The stoichiometry of this reaction, determined by surface acid-base titrations of albite powders (Oelkers et al. 2009), corresponds to exchange of 1 Al for three protons (Fig. 23) with the formation of a silica-rich surface.

These exchange reactions arise in multioxides because, as illustrated by the enormous variation of single oxides dissolution rates (see Fig. 10), metal-oxygen bonds break at very different rates depending on their relative strength. This relative bond strength is also reflected in the rate of exchange of water molecules in the coordination sphere of the corresponding aqueous ions. Thus multioxide mineral structure and composition has a strong impact on the extent of the non-stoichiometric release of elements and the thickness of formed leached layers. Whereas albite and enstatite dissolution readily become stoichiometric leading to the formation of thin leached layers of constant thickness, wollastonite dissolution at acidic pH

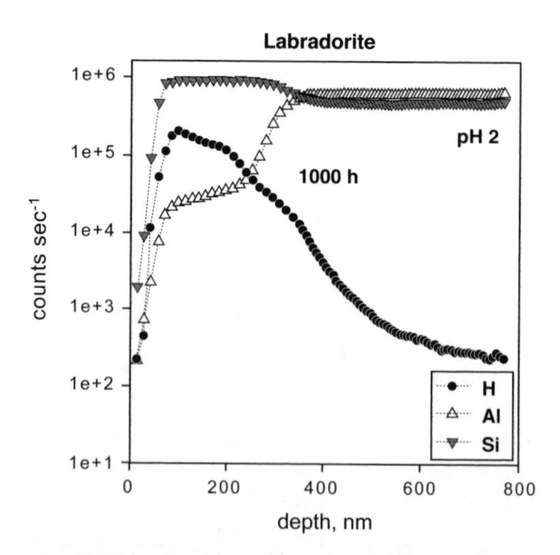

Figure 22. SIMS profile of H, Al and Si as a function of depth at the surface of a labradorite crystal reacted in 0.01 M LiCl at pH = 2 for 1000 h (from Schweda et al. 1997).

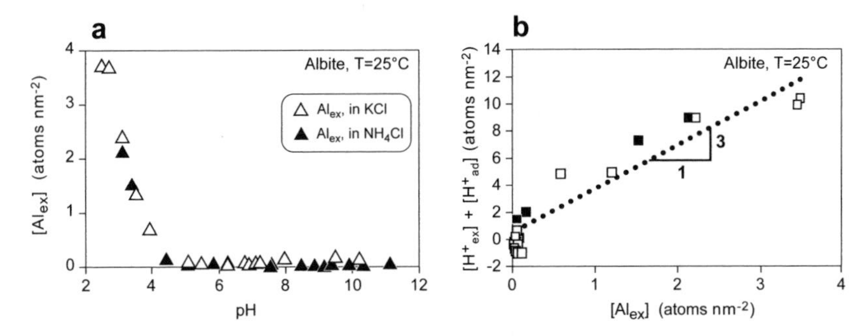

Figure 23. Surface titration of albite powders at 25 °C: (a) mass of Al exchanged during surface titrations as a function of pH, (b) number of protons incorporated onto/into the surface ($[H^+_{surf}]$) plotted as a function of the number of Al atoms exchanged during the 20 minute immersion of the albite powder (from Oelkers et al. 2009).

never reaches stoichiometry, leading to a total leaching of calcium from the mineral lattice (see Fig. 24) and the formation of thick leached layers which undergo important polymerization of silica tetrahedra as evidenced via ^{29}Si MAS NMR spectroscopy by the formation of Q^3 and Q^4 structural units at the expand of Q^2 units, and by intense crazing and spallation which affect the leached layers as shown in Figure 25a (Schott et al. 2002). It should be noted that it is unlikely that such leached layers are formed by stoichiometric mineral dissolution coupled to amorphous silica precipitation at the surface because i) the aqueous solution in contact with these leached layers is strongly undersaturated with respect to both quartz and amorphous silica ($\Omega \sim 0.1$ and 7×10^{-3}, respectively, see Fig. 24) and ii) the leached layers cover homogeneously the surface whereas preferential sites of precipitation would be normally expected (i.e., see Figs. 25a and 25b relative to wollastonite and bytownite, respectively, and Fig. 7 in Schott et al. 1981).

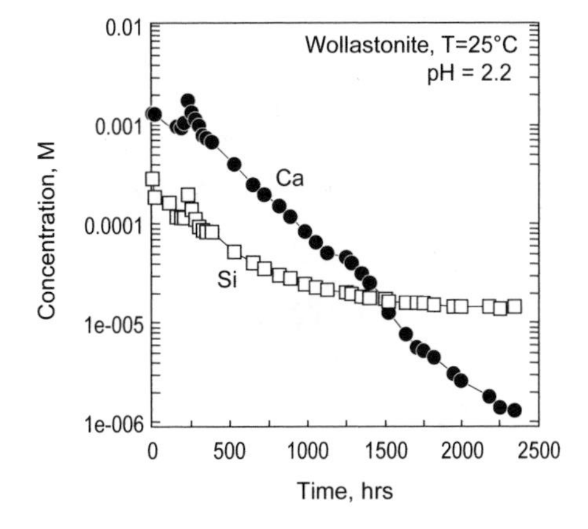

Figure 24. Oulet calcium and silica concentrations as a function of time during wollastonite dissolution at 25 °C and pH = 2.2 in a mixed flow reactor (after Schott et al. 2002).

Insight on the leached layer formation from solution chemistry

Insight on the metal-proton exchange reactions and the stability of leached layers that form during silicates minerals dissolution can be obtained from the analysis of analogous aqueous reactions, for example the formation of aqueous aluminum-silicate complexes. Figure 26 shows the geometry deduced from XAFS, NMR and titrations measurements of aqueous aluminum-silicate complexes with Al in octahedral or tetrahedral coordination as a function of solution pH and silicic acid concentration (Pokrovski et al. 1998, 2002). Pokrovski et al. (1996, 1998) and Salvi et al. (1998) determined the stability constant and enthalpy of the formation of several of these complexes including $Al(OH)_3H_3SiO_4^-$ (see Fig. 27) and $Al(OH)_2(H_3SiO_4)_2^-$ (see their optimized geometries (HF/6-31G*) in Fig. 27). The thermodynamic parameters of the aluminum-silicate trimer,

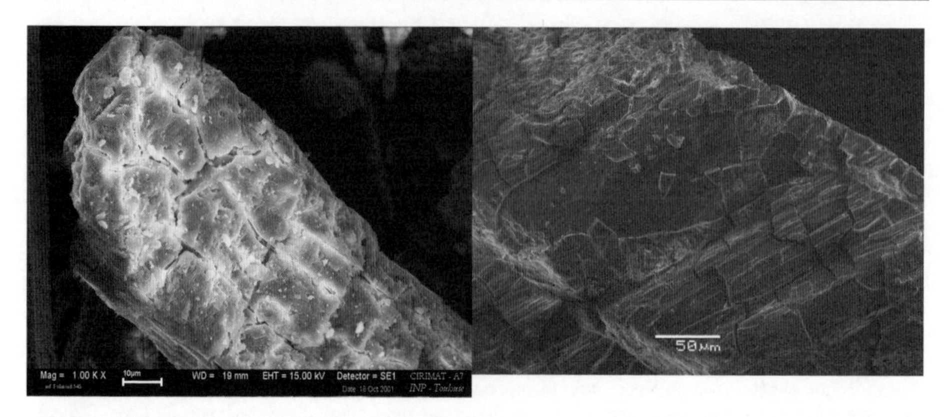

Figure 25. SEM images of leached layers formed at the surface of wollastonite (a) and bytownite (b) reacted at 25 °C and pH = 2 for few days. Note the important crazing and spalliation that affect the leached layers formed on wollastonite and bytownite. Partial removal of the leached layer allows seeing bytownite dissolving surface.

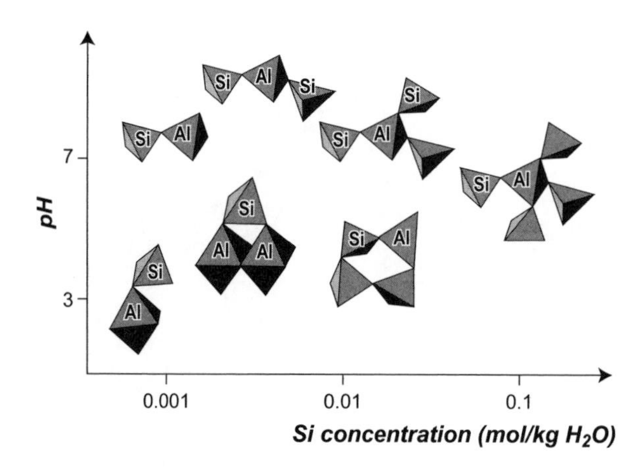

Figure 26. Schematic structure of Al(Ga)-silica complexes derived from EXAFS and NMR measurements (adapted from Pokrovski et al. 1998, 2002).

$$Al(OH)_2(H_3SiO_4)_2^- + 4\,H^+ = Al^{3+} + H_6Si_2O_7^0 + 3\,H_2O \tag{23}$$

which can be viewed as a proxy of feldspar surface, provide direct information on the thermodynamic properties of the Al-H^+ exchange reaction and the stability of the leached layer that control feldspar dissolution. The enthalpy of reaction (23) is negative ($\Delta_r H_{23} \sim -100$ kJ/mol) implying that the thickness and stability of the Al-depleted leached layer formed during feldspar dissolution decreases with increasing temperature. This result is in agreement with experimental observations reported by Hellmann (1994) and Chen et al. (2000) on albite leached layers.

Dissolution mechanisms of multioxide silicate minerals

During the dissolution of a multi-oxide mineral whose constituting single oxides have metal-oxygen bonds breaking at distinct velocities, the rate-limiting step is the destruction of the slowest breaking metal oxygen bond essential for maintaining the mineral structure. Taking account the relative dissolution rates of the single metal oxides, and the relative water exchange rates from the coordination sphere of aqueous metals (e.g., Figs. 9a, and 10b), Oelkers (2001) proposed that the relative rates of breaking each type of metal-oxygen bond was identical in all multi-oxide mineral. By comparing the relative rates for breaking metal-oxygen bonds and the structure of multi-oxide silicates, Oelkers (2001) proposed the dissolution mechanism for the

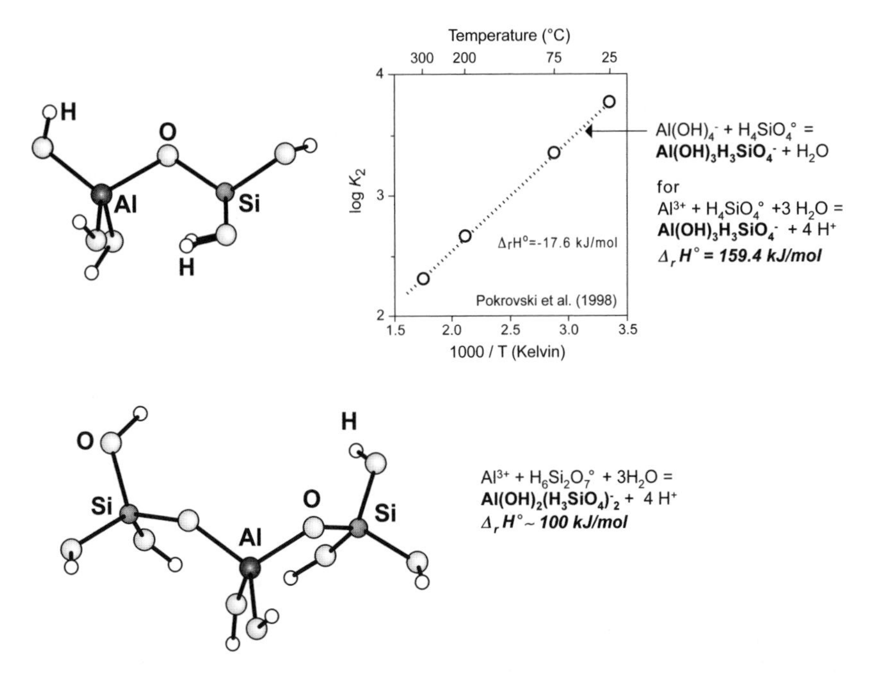

Figure 27. Optimized structures of aluminate-silicate dimer and trimer, and logarithm of aluminate-silicate dimer formation constant as a function of reciprocal absolute temperature (Pokrovski et al. 1998).

major multi-oxide silicates including aluminosilicates, basic silicates, and natural glasses. An example is found in Figure 28. In general, at acidic conditions, monovalent metal-oxygen bonds break more rapidly than divalent metal-oxygen bonds, which break faster than trivalent metal-oxygen bonds, which break faster than Si-O bonds. The dissolution of a multi-oxide silicate will proceed by the sequential breaking of metal-oxygen bonds resulting in the liberation of various metals from the mineral surface. Dissolution is initiated by the removal of the metal having the fastest breaking metal-oxygen bonds. The removal of metals from the structure is coupled to the addition of protons into the mineral structure (Oelkers et al. 2009) so the metal removing reactions are effectively metal-proton exchange reactions. Dissolution continues by the successive removal of metals in the order of the relative rates for breaking their corresponding metal-oxygen bonds, until the mineral structure is destroyed. The removal of the slowest metal essential to maintain the mineral structure is the slowest step of the process, so the breaking of these bonds are rate limiting (or sometimes referred to as rate controlling). Depending on silicate composition and structure, one or several H^+-cation exchange reactions may precede the rate limiting hydrolysis of Si-O bonds. An example of the dissolution mechanism for the alkali feldspars is shown in Figure 29. Alkali-feldspar dissolution proceeds by the initial removal of the alkali metals via their exchange with protons. The second step in the dissolution mechanism is the removal of Al by proton for Al exchange reactions. This exchange forms partially detached Si atoms at the alkali-feldspar surface. The breaking of Si-O bonds attached to this partially liberated Si atom is the slowest and rate liming step for alkali feldspar dissolution. Consideration of this rate mechanism leads directly to equations describing feldspar dissolution rates as a function of aqueous solution composition and degree of disequilibrium as will be shown below.

The mechanisms shown in Figure 28 suggests that for many major rock-forming silicates the slowest breaking bond at acidic pH is the Si-O bond and thus partially liberated Si could be considered the precursor complex. The relative order of breaking metal–oxygen bonds appears

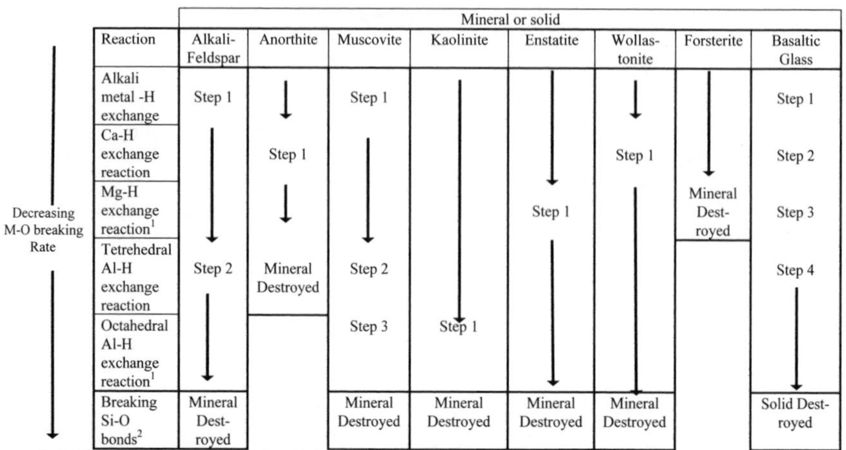

1) Evidence suggests that Mg-O and octahedral Al-H exchange is slower than Si-O bond breaking at basic conditions (see text).
2) The breaking of Si-O bonds most likely involves H_2O absorption rather than an Si for H exchange reaction (see Dove and Crerar, 1990).

Figure 28. Summary of dissolution mechanisms of some minerals and basaltic glass at acidic conditions (after Oelkers 2001).

to depend of pH to some extent. In accord with the relative dissolution rates of quartz versus gibbsite and corundum, the rates of breaking octahedral Al-O bonds appears to be slower than that of Si-O at alkaline pH. As such the mechanism for mineral containing octahedral Al will change with pH and the slowest step for the dissolution of such minerals at alkaline conditions will be the liberation of octahedral Al from the mineral structure. Similarly other evidence suggests that the breaking of Mg-O bonds will be slower than that of Si-O bonds at very basic conditions (pH > 9). These two cases will be explored in detail below.

Due to the role of metal-proton exchange reactions, the dissolution mechanism of multiple oxides will be different from that of single oxide minerals. Recall that for single oxides and minerals whose dissolution requires breaking of only one type of bonds, the rate controlling precursor complex could be formed by proton sorption reactions in accord with

$$n_H\, H^+ + {>M\text{-}OH} = P^* \qquad (24)$$

where n_H is the number of protonation steps necessary to form the precursor complex. The concentration of the precursor complex can be obtained from the law of

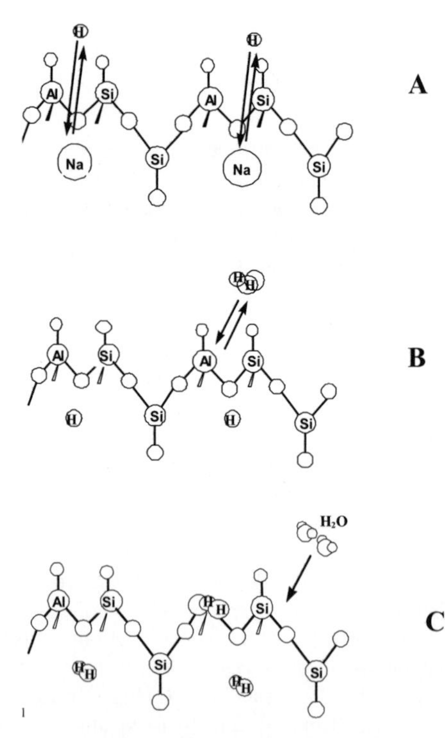

Figure 29 Mechanism of alkali feldspar dissolution (after Oelkers et al 1994).

mass action of reaction (24) and the assumption that the number of sites at the mineral surface is constant ([>M-OH] + [P*] = S = Constant)

$$[P^*] = S\frac{K_{24}^* \cdot a_{H^+}^n}{1 + K_{24}^* \cdot a_{H^+}^n} \qquad (25)$$

where K^*_{24} stands for the equilibrium constant for reaction (24). As far from equilibrium dissolution rates are proportional to the concentration of the precursor complex

$$r_+ = k_{P*}[P^*] \qquad (26)$$

where k_{P*} refers to a rate constant consistent with the precursor complex, it follows that

$$r_+ = k_{P^*} \cdot S\frac{K_{24}^* \cdot a_{H^+}^n}{1 + K_{24}^* \cdot a_{H^+}^n} \qquad (27)$$

which reduces to

$$r_+ = k_+ K_{24}^* . a_{H^+}^n \qquad (27a)$$

when [>M-O] >> [P*].

In the case of multi-oxide minerals (e.g., aluminosilicates, basic silicates, and natural glasses) whose dissolution requires breaking of more than one type of bond, the precursor complex is formed by exchange reactions with aqueous protons. In the case where only one type of metal is exchanged for precursor complex formation, the precursor forming reaction can be expressed as

$$z\, n\, H^+ + >M\text{-}O = P^* + n\, M^{z+} \qquad (28)$$

where z refers to the charge on the aqueous cation M, and n denotes the number of cations that need to be removed to create P*. Derivations similar to those performed for single oxides yield the following expression for the far from equilibrium dissolution rate of a multi-oxide mineral

$$r_+ = k_{p^*}\left[P^*\right] = k_{P^*}S\left(\frac{K_{28}^*\left(\frac{a_{H^+}^z}{a_{M^{z+}}}\right)^n}{1 + K_{28}^*\left(\frac{a_{H^+}^z}{a_{M^{z+}}}\right)^n}\right) \qquad (29)$$

which reduces to

$$r_+ = k_+ K_{28}^*\left(\frac{a_{H^+}^z}{a_{M^{z+}}}\right)^n \qquad (29a)$$

when [>M-OH] >> [P*].

Comparison of Equations (27a) and (29a) demonstrates that these two different mechanisms have a major impact on the dependence of dissolution rates on solution composition. Multi-oxide mineral dissolution rates depend on both H^+ concentration and the concentration of exchanged constituting metal which inhibits dissolution. The impact of H^+-metal exchange reactions on the rate of dissolution of silicates minerals can be further examined through the examples of both aluminosilicates and basic silicates.

Aluminosilicate minerals and glass dissolution rates

Far-from equilibrium dissolution rates. In accord with reaction (28), the formation of the precursor complex that controls the dissolution of numerous aluminosilicate minerals involves the exchange of Al and H on the mineral surface according to (Oelkers 1996, 2001; Schott and Oelkers 1995)

$$M - nAl + 3nH^+ + \sum_i \upsilon_i C_i \rightleftharpoons P^* + nAl^{3+} \tag{30}$$

where $M - nAl$ designates an Al-filled mineral surface site, n refers to a stoichiometric coefficient equal to the number of aluminum atoms in each potential precursor site, C_i stands for the ith any number of other aqueous species that might be involved in the formation of the precursor complex, and υ_i is a stoichiometric coefficient. The law of mass action for reaction (30) combined with the equation expressing the conservation of sites at mineral surface allows calculation of the precursor complex concentration as a function of aqueous solution composition. When combined with Equation (26), one obtains an expression describing far-from equilibrium dissolution rates as a function of solution composition (Oelkers et al. 1994)

$$r_+ = k_{p^*} S \frac{K^*_{30}\left(\dfrac{a^{3n}_{\mathrm{H}^+}}{a^n_{\mathrm{Al}^{3+}}}\right)\prod_i a^{\nu_i}_{C_i}}{1 + K^*_{30}\left(\dfrac{a^{3n}_{\mathrm{H}^+}}{a^n_{\mathrm{Al}^{3+}}}\right)\prod_i a^{\nu_i}_{C_i}} \tag{31}$$

Note that when the surface has relatively few precursor complexes, which seems to be the case in most experiments, Equation (31) reduces to

$$r_+ = k_+ K^*_{30}\left(\frac{a^3_{\mathrm{H}^+}}{a_{\mathrm{Al}^{3+}}}\right)^n \prod_i a^{\nu_i}_{C_i} \tag{31a}$$

The term $\Pi_i a^{\nu_i}_{C_i}$ includes the potential effects of the involvement of metals other than Al in the formation of the precursor complex. For example, data reported by Stillings and Brantley (1995) suggests that alkali metals could be involved in the formation of the alkali-feldspar precursor complexes. Similarly at basic conditions data suggest that some Si needs to be removed from the structure of minerals containing octahedral Al to form their rate controlling precursor complexes (see below).

Equations (31) and (31a) have been found to accurately describe albite (Chou and Wollast 1985; Knauss and Wolery 1986; Oelkers et al. 1994; Hellmann 1994; Brantley and Stillings 1996; Chen and Brantley 1997), K-feldspar (Gautier et al. 1994), kaolinite (Devidal et al. 1997), labradorite (Carroll and Knauss 2005), muscovite (Oelkers et al. 2008), and basaltic (Oelkers and Gislason 2001; Gislason and Oelkers 2003) and other natural glass (Wolff-Boenisch et al. 2004a,b) dissolution rates as a function of aluminum concentration. For example, it can be seen in Figure 30 that albite and K-feldspar dissolution rates at 150 °C and pH= 9 exhibit a close agreement to Equation (31a) for aqueous Al concentrations ranging from 5×10^{-7} to 1×10^{-3} mol/kg. Kaolinite dissolution rates at 150 °C and pH = 2 shown in Figure 31 are proportional to the reciprocal of Al molality which is consistent with Equation (28) and precursor complex formation according to

$$AlSiO_{5/2}(OH)_2 + 3\,H^+ \leftrightarrow Al^{3+} + SiO_2{\cdot}nH_2O^* + (5/2 - m)\,H_2O \tag{32}$$

It is worth noting that whereas feldspar dissolution rates continuously increases with decreasing aqueous Al concentration to at least to Al ~ 5×10^{-7} mol/kg, kaolinite dissolution rates become independent of Al concentration at Al $\leq 10^{-5}$ mol/kg. Because of a relatively high

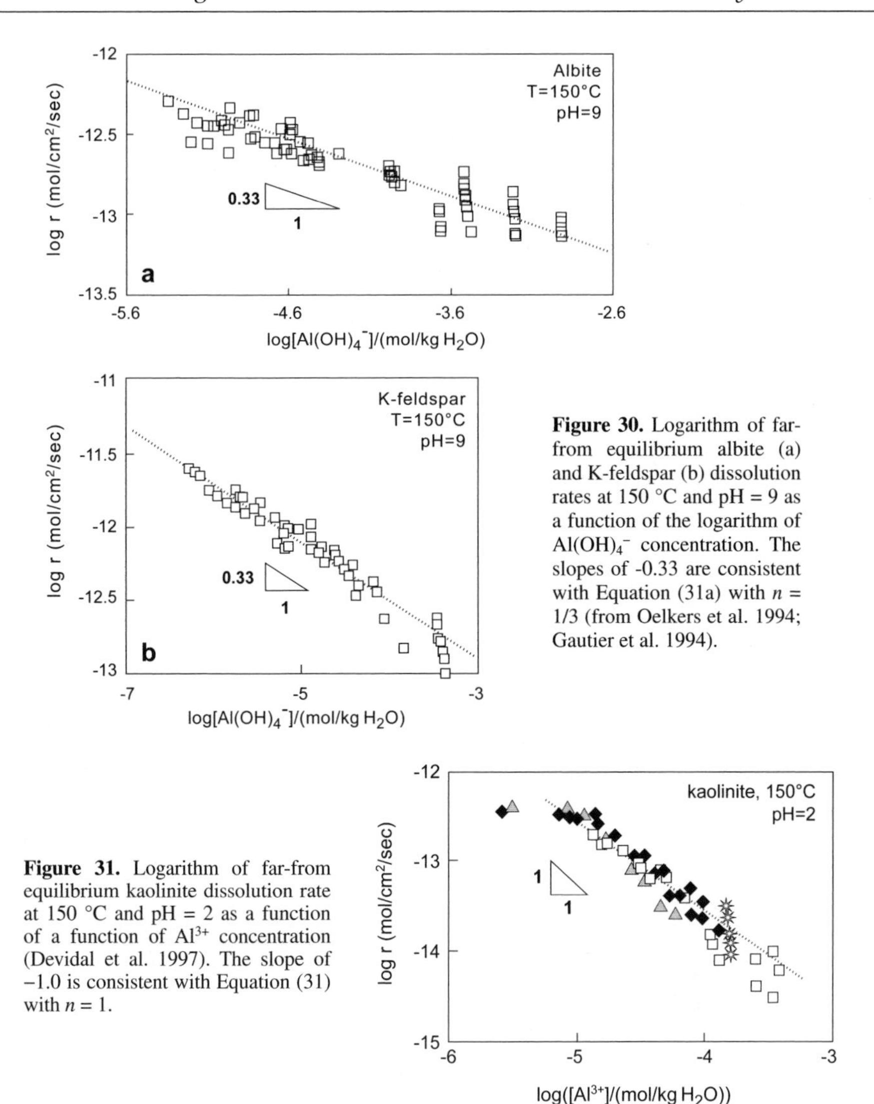

Figure 30. Logarithm of far-from equilibrium albite (a) and K-feldspar (b) dissolution rates at 150 °C and pH = 9 as a function of the logarithm of $Al(OH)_4^-$ concentration. The slopes of -0.33 are consistent with Equation (31a) with n = 1/3 (from Oelkers et al. 1994; Gautier et al. 1994).

Figure 31. Logarithm of far-from equilibrium kaolinite dissolution rate at 150 °C and pH = 2 as a function of a function of Al^{3+} concentration (Devidal et al. 1997). The slope of −1.0 is consistent with Equation (31) with $n = 1$.

K_{32}^* value for kaolinite; the product $K_{32}^*(a_{H^+}{}^3/a_{Al^{3+}})^n$ for kaolinite in Equation (31) quickly becomes substantially greater than 1 with decreasing Al concentration. At such conditions a true dissolution plateau is reached where rate is independent of both aqueous H^+ and Al^{3+} activities ($r_+ = k_+$).

As shown by Chou and Wollast (1985), Oelkers et al. (1994) and Oelkers (2001), the dependence of aluminosilicates dissolution rates on pH can be deduced from Equation (31) by expressing $a_{Al^{3+}}$ as a function of total dissolved Al, H^+ activity and Al hydrolysis constants (and eventually Al-L complexation constants if ligands L are present in solution). This yields

$$r_+ = k_+ \frac{\prod_i a_{C_i}^{\nu_i}}{[\mathrm{Al}_{tot}]^n} \left(\frac{1}{\gamma_{\mathrm{Al}^{3+}}} a_{\mathrm{H}^+}^3 + \frac{K_{\mathrm{Al(OH)}^{2+}}}{\gamma_{\mathrm{Al(OH)}^{2+}}} a_{\mathrm{H}^+}^2 + \frac{K_{\mathrm{Al(OH)}_2^+}}{\gamma_{\mathrm{Al(OH)}_2^+}} a_{H^+} + \frac{K_{\mathrm{Al(OH)}_3^0}}{\gamma_{\mathrm{Al(OH)}_3^0}} + \frac{K_{\mathrm{Al(OH)}_4^-}}{\gamma_{\mathrm{Al(OH)}_4^-}} a_{\mathrm{H}^+}^{-1} \right)^n \quad (33)$$

The close correspondence observed in Figure 32 between experimental data and the curves derived from Equation (33) supports the capacity of this equation to describe albite and muscovite dissolution as a function of pH. On Figure 33, that shows the variation of muscovite dissolution rate at 150 °C as a function of solution pH but also at constant pH where the concentration of Al and Si have been changed in the reactive solution, it can been seen that the impact on dissolution rate of the changes in Al and Si concentrations can be more important than pH changes.

Again it should be emphasized that the strong analogy between the dependence on pH of the kinetics of dissolution of aluminosilicates and the solubility of Al oxides/hydroxides is not fortuitous. This analogy stems directly from the law of mass action for the reactions forming the Al-depleted precursor complexes at the aluminosilicate surfaces (e.g., reaction 32). This connection can be formally derived by combining rate equations, such as Equation (33) with solute speciation calculations.

The relative order of breaking metal-oxygen bonds shown in Figure 28 was found to change somewhat depending on solution pH. As shown in Figure 34, the dissolution rates of octahedral Al-oxide minerals become lower than that for quartz at basic pH. This can be explained, within

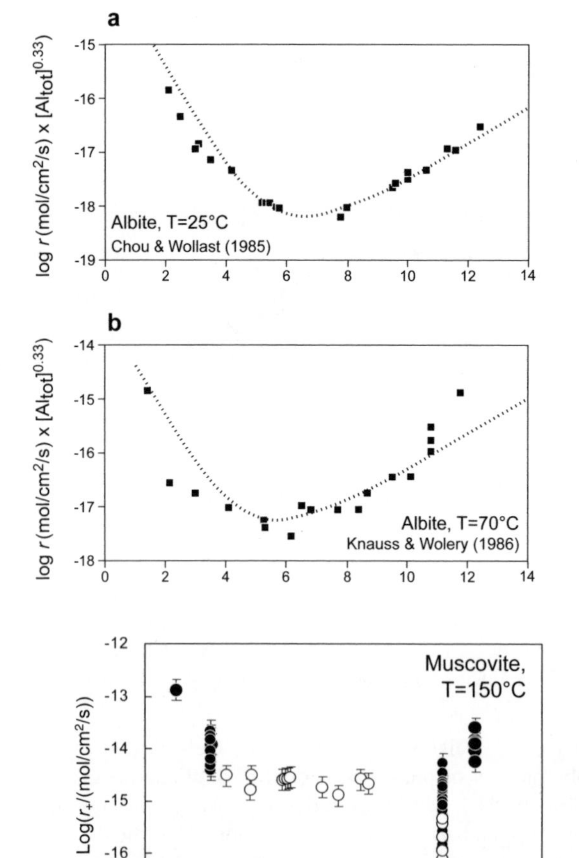

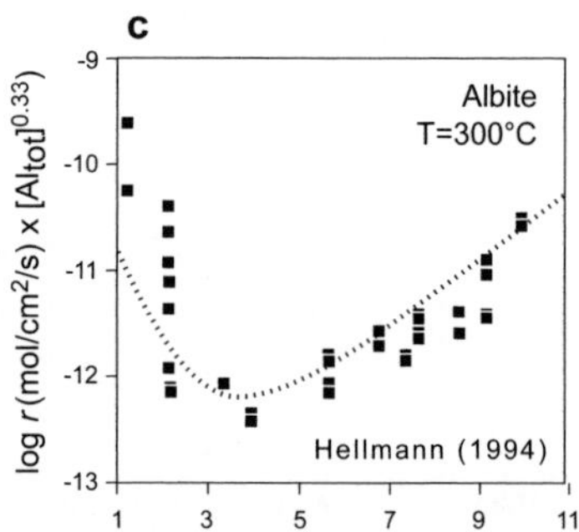

Figure 32. Far-from equilibrium albite dissolution rates as a function of pH illustrate adherence of experimental data to Equation (33). Symbols represent experimental data but the curves were generated using Equation (33) (from Oelkers et al. 1994 and Oelkers 1996).

Figure 33. Far-from equilibrium muscovite dissolution rates as a function of pH at 150 °C. The filled and open symbols correspond to rates that were undersaturated and supersaturated with respect to diaspore, respectively (from Oelkers et al. 2008).

the framework of surface coordination concepts, by a much lower pH_{zpc} of quartz ($pH_{zpc} \sim 2$) than Al oxides/hydroxides ($pH_{zpc} \sim$ 8-9) and thus by a polarization and weakening/breaking of SiO bonds by OH^- species developing at much lower pH than that of Al-O bonds. As a consequence, the breaking of octahedral-Al bonds becomes rate limiting in a number of Al-silicate minerals at basic pH. This mechanism change is manifested in variation of the dissolution rates of a number of silicates containing octahedral Al with aqueous Si and Al concentration at basic conditions. For example, far from equilibrium muscovite dissolution rates, which are proportional to the reciprocal of the square root of Al activity at acidic pH were found to depends on both reactive fluid Al and SiO_2 activity, which is consistent with (Oelkers et al. 2008)

$$r_{OH^-} = k_{OH^-}\left(\frac{a^3_{H^+}}{a_{Al^{3+}}}\right)^{0.5}\left(a_{SiO_2}\right)^{-1} \tag{34}$$

as it can be seen on Figure 35. It follows that the dissolution of aluminosilicates that contain octahedral Al in their structure like kaolinite (Devidal et al. 1997) and smectite (Cama et al. 2000) is controlled by precursor complexes formed by Al for proton exchange reaction only at acid to neutral pHs. At basic conditions the far from equilibrium dissolution rates of muscovite, kaolinite, illite, and smectite depend both on aqueous Al and SiO_2. In contrast, the feldspars

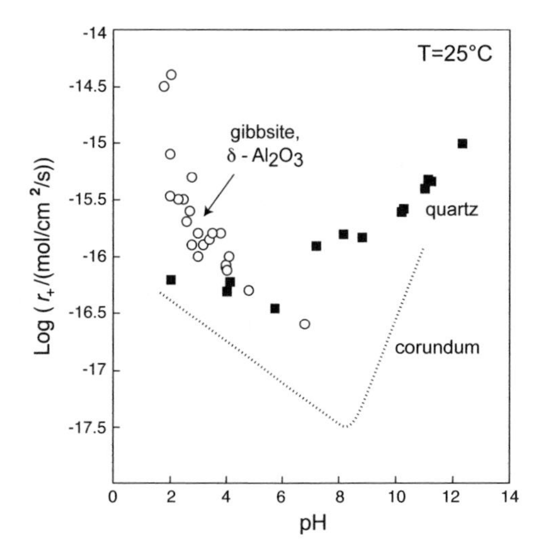

Figure 34. Comparison of quartz, gibbsite, δ-Al_2O_3 and corundum dissolution rates at 25 °C as a function of pH. Data from Brady and Walther 1990 (quartz); Bloom and Erich 1987; Dietzel and Bohme 2004 (gibbsite); Furrer and Stumm 1986 (δ-Al_2O_3); Carroll-Webb and Walther 1988 (corundum).

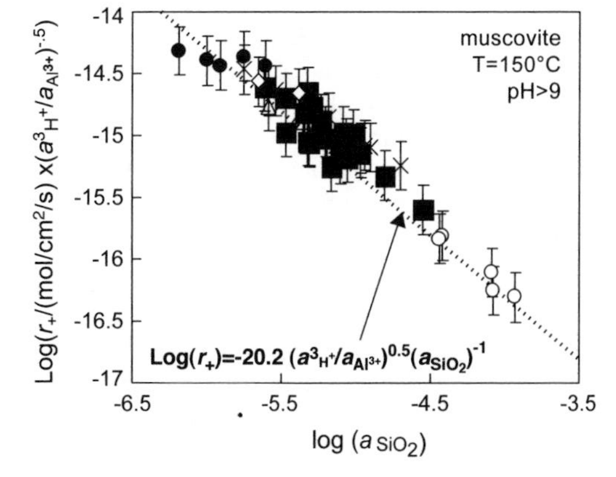

Figure 35. Muscovite dissolution rate at 150 °C and pH > 9: logarithm of the product $r_+ \times (a^3_{H+}/a_{Al3+})^{0.5}$ against aqueous silica concentration (from Oelkers et al. 2008).

structure contains only tetrahedral Al (which forms much weaker bonds). As such feldspar dissolution is controlled by precursor complexes formed by Al-H exchange both at acid and alkaline pH. Consequently, far from equilibrium dissolution rates of aluminosilicate minerals that contain Al only in tetrahedral coordination, like albite and K-feldspar, are functions of aqueous aluminum but independent of aqueous SiO_2 at all pH (Oelkers et al. 2008).

Dissolution rates at close-to-equilibrium conditions. Again invoking the law of detailed balancing (e.g., assuming the precursor complex is the same for the dissolution and precipitation reactions) and that each of the Si atoms in the mineral structure can create one rate controlling surface precursor complex (which is the case for feldspars at both acid and alkaline conditions, and for most other aluminosilicates at acid conditions), an equation describing the overall dissolution rates of aluminosilicates, both near and far from equilibrium, can be generated by combining Equations (8) and (31)

$$r = r_+ = k_{p^*} S \frac{K_{30}^* \left(\frac{a_{H^+}^{3n}}{a_{Al^{3+}}^n} \right) \prod_i a_{C_i}^{\nu_i}}{1 + K_{30}^* \left(\frac{a_{H^+}^{3n}}{a_{Al^{3+}}^n} \right) \prod_i a_{C_i}^{\nu_i}} \times \left(1 - \exp(-A / n_{Si} RT\right) \tag{35}$$

where n_{Si} represents the stoichiometric number of Si in one mole of the dissolving aluminosilicate. As the chemical formula of kaolinite contains only 2 Si, an expression describing kaolinite dissolution/precipitation at acid conditions (see Eqn. 32), is given by

$$r = k_+ \left(\frac{\frac{K_{32}^* a_{H^+}^3}{a_{Al^{3+}}}}{1 + \frac{K_{32}^* a_{H^+}^3}{a_{Al^{3+}}}} \right) \left(1 - \exp(-A / 2RT\right) \tag{36}$$

Kaolinite and K-feldspar dissolution/precipitation rates computed using Equations (35) and (36) as a function of chemical affinity in solution of different Si and Al initial concentrations are compared with their experimental counterparts in Figures 36-37. A close agreement is observed between the experimental and calculated rates which support the validity of Equations (35-36). Moreover, several important features of the rates dependence on chemical affinity should be highlighted. The most remarkable feature is the strong sigmoidal dependence of constant-pH dissolution rates on chemical affinity for each aqueous solution Al/Si ratio. For kaolinite (Fig. 36), rates are independent of A at far from equilibrium ($A > 40$ kJ/mol) owing to the fact that all surface sites contain precursor complexes, and relatively slow rates are observed for $A < 20$ kJ/mol (5 kcal/mol). This behavior, which dramatically contrasts with that of simple oxides like quartz (for which the rate becomes independent of chemical affinity for $A > 12$ kJ/mol), results from the form of Equation (36). The apparent variation of rates with chemical affinity depends on two terms: the rate dependence on reciprocal Al concentration and the slowing of rates near to equilibrium in accord with the Temkin's stoichiometric number ($n_{Si} = 2$ which implies $\sigma = 2$). It should be also noted that very close to equilibrium ($A < 8$ kJ/mol) dissolution and precipitation rates are proportional to chemical affinity in accord with the principle of detailed balancing.

K-feldspar and albite overall dissolution rates at 150 °C and pH = 9 measured by Oelkers et al. (1994) and Gautier et al. (1994) have been described as a function of chemical affinity using Equation (35) with $n = 1/3$, $n_{Si} = 3$ and $\nu_i = 0$. For example, the close correspondence observed in Figure 37 relative to K-feldspar between experimental data and the curves computed with Equation (35) demonstrates the validity of this equation for describing the dependence of K-feldspar dissolution rate as a function of chemical affinity and supports the hypothesis that

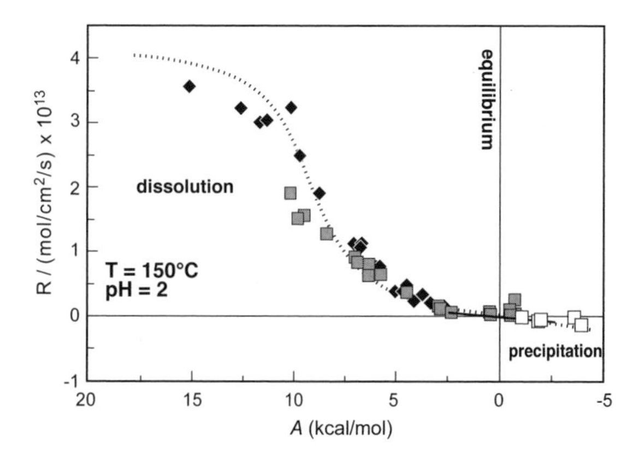

Figure 36. Kaolinite dissolution/precipitation rate at 150 °C and pH = 2 as a function of chemical affinity. Diamonds and squares stand for Al/Si free and [Si] = [Al] input solutions, respectively, whereas the curve was generated using Equation (36) (from Devidal et al. 1997).

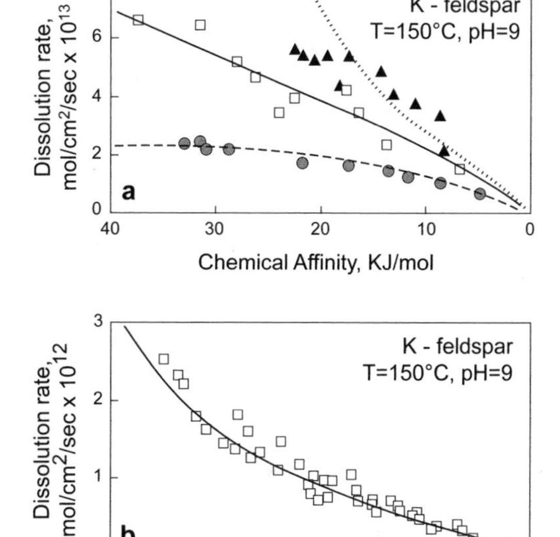

Figure 37. K-feldspar dissolution rate at 150 °C and pH = 9 as a function of chemical affinity: a) experiments performed in Si-rich (triangles), Al-rich (circles), and Si/Al-free (squares) input solutions; b) experiments performed in Si/Al-free solutions. The curves represent a fit of these data to Equation (35) (from Gautier et al. 1994).

these rates are controlled by Si-rich, Al-depleted surface precursor complexes. Like kaolinite dissolution rates, these rates exhibit a sigmoidal dependence on chemical affinity with very low values in Al-rich solutions close to equilibrium; however, in contrast to kaolinite, dissolution rates continue to increase with chemical affinity to at least 90 kJ/mol in Al- and Si-free input solutions. The non-attainment of a dissolution plateau, even at Al concentrations as low as $5{\times}10^{-7}$ M, suggests that all surface sites are not constituted of precursor complexes and, therefore, that the equilibrium constant K_i^* for reaction (32) is small, which probably reflects a strong tendency of the aluminate ion to reattach itself to the silica-rich surface precursor.

Among feldspars, albite, K-feldspar, and labradorite ($Ca_{0.6}Na_{0.4}Al_{1.6}Si_{2.4}O_8$; Carroll and Knauss 2008) dissolution rates exhibit the same inverse dependence on the concentration of dissolved aluminum. Such is not the case for anorthite (Oelkers and Schott 1995). It can be seen on Figure 38 that anorthite dissolution rate at 60 °C, pH = 2 and far from equilibrium is independent

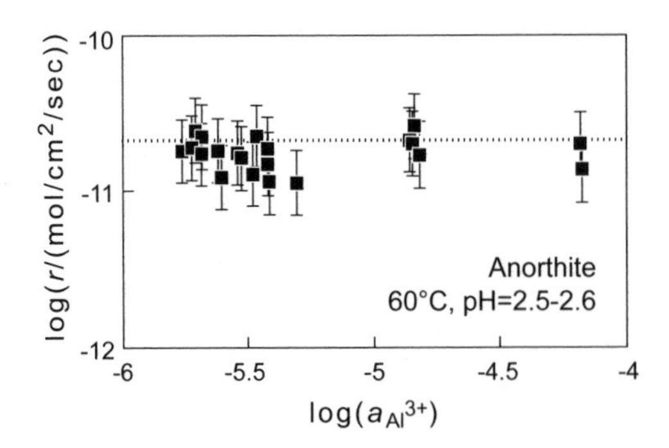

Figure 38. Anorthite dissolution rate at 60 °C and pH = 2.5-2.6 as a function of Al^{3+} activity (from Oelkers 2001).

on both aqueous aluminum concentration in the reacting fluid and chemical affinity. Because anorthite has a Si/Al ratio of 1, the removal of Al from its surface via an exchange reaction with H^+ leaves completely detached Si tetrahedra. As a result, the destruction of its Al-Si framework does not require the breaking of Si-O bonds; the precursor complex is directly formed by protons sorption on anorthite surface sites (like for a simple oxide).

The sigmoidal variation of aluminosilicate dissolution rates at constant aqueous Al/Si ratio when plotted as a function of chemical affinity has led several investigators to propose that the mechanism of dissolution for these minerals change with affinity (e.g., Burch et al. 1993, Taylor et al. 2000; Luttge et al. 2003, Hellmann and Tisserand 2006). These studies suggested that alkali-feldspar dissolution rates as a function of saturation state could be described using

$$r = k'\left\{1 - e^{-n\left(\frac{|\Delta G_r|}{RT}\right)}\right\}^{m_1} + k''\left\{1 - e^{-\left(\frac{|\Delta G_r|}{RT}\right)}\right\}^{m_2} \quad (37)$$

where k' and k'' denote the rate constants, and n, m_1, and m_2 are empirical parameters fitted from experimental data. It was argued that the alleged mechanism change was due to the opening of etch pits at dislocations for a critical chemical affinity resulting in accelerated dissolution at the defect sites. Conflicting analyses, however, have been presented in the literature regarding the effect of dislocations and etch pits on silicates dissolution rates. Several authors suggested that silicate mineral dissolution occurs preferentially at dislocations due to enhanced nucleation of dissolution pits at these sites (Brantley et al. 1986; Blum and Lasaga 1987). It was argued that the dissolution of quartz and aluminosilicates should be controlled by the nucleation of etch pits at dislocations at far from equilibrium conditions whereas at close to equilibrium, where etch pits nucleation ceases, dissolution should be controlled by step dissolution on etch pit–free surface. These arguments, however, are inconsistent with a large number of observations. For example, it was found that the far from equilibrium dissolution rate of rutile, bytownite, oligoclase, sanidine and calcite strained such that their dislocation densities increased by 3 to 4 orders of magnitude were essentially identical to that of the corresponding unstrained minerals (Boslough and Cygan 1988; Casey et al. 1988; Murphy 1989; Schott et al. 1989). Furthermore, a sigmoidal dependence of rate on chemical affinity was never observed for quartz dissolution (see above). Gautier et al (2001) found that etch pit formation, was not accompanied by any measurable increase of quartz dissolution rate. It was concluded by these last authors that etch pit walls, which constituted the bulk of the surface area increase in their experiments, are relatively unreactive negative crystal facets. Finally the empirical Equation (37) is not capable to describe the observed variation of rates as a function of aqueous Al concentration at constant saturation state.

Aluminosilicate near-equilibrium dissolution and precipitation rates computed from Equation (35) are much lower than those obtained using the general rate equation proposed for silicates minerals by Aagaard and Helgeson (1982) and Murphy and Helgeson (1989)

$$r = k \cdot a_{H^+}^{n_{H^+}} (1 - \exp(-A / RT)) \quad (38)$$

which is based on the assumption that, as for simple oxides, aluminosilicates precursor complex is formed by simple sorption of H^+ or other reactants at mineral surface. Knowledge of the exact dependence of reaction rates on chemical affinity is crucial because, in natural systems, reactions often occur close to equilibrium and are strongly influenced by chemical affinity. In this regard, it should be emphasized that the apparent discrepancy often noted between laboratory and field weathering rates (White and Brantley 2003; Zhu 2009) vanishes when feldspars weathering rates are calculated with kinetic rate laws derived from Equation (32). This has been recently shown by Maher et al. (2009) and Schott et al. (2008) in their modeling studies of soil profile development and mineral dissolution and precipitation rates documented by soils that developed from uniform parent materials.

Basic silicates

Leached layers, formed by H^+ exchange for Ca and/or Mg atoms at the mineral surface, develop during the dissolution of pyroxenes, pyroxenoids, talc and olivine (to a lesser extent for the later mineral) at acidic and neutral pH. This leaching is controlled by the hydrolysis of Si-O bonds, and the rate controlling precursor complexes for these minerals are formed in accord with Equation (28). It follows that at constant pH, like for aluminosilicates, the logarithm of these minerals far-from-equilibrium dissolution rates generally decrease linearly with the logarithm of the aqueous activity of constituting divalent metal; this is illustrated for enstatite on Figure 39. Similar to the case for many Al-silicates, the dissolution rates of enstatite, and several other Mg-silicates are controlled by the detachment of Si after adjacent Mg atoms have been removed via proton exchange reactions. This leads to far from equilibrium rate expressions of the form:

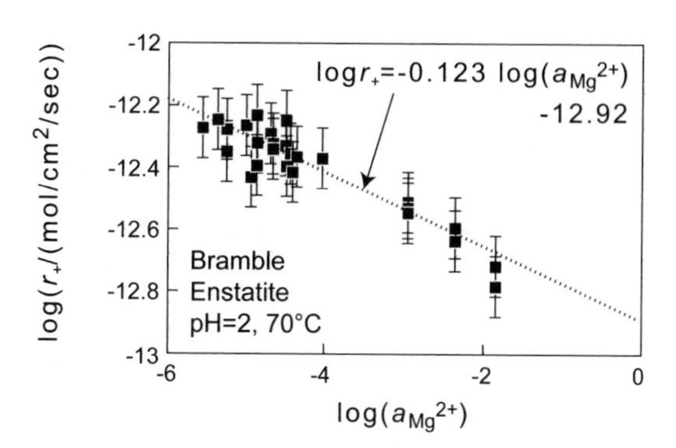

Figure 39. Enstatite dissolution rate at 70 °C and pH = 2 as a function of Mg^{2+} activity (from Oelkers and Schott 2001).

$$r_+ = k_{P*} S \frac{K_{28}^* \left(\dfrac{a_{H^+}^{2n}}{a_{Mg^{2+}}^{n}} \right) \prod_i a_{C_i}^{\nu_i}}{1 + K_{28}^* \left(\dfrac{a_{H^+}^{2n}}{a_{Mg^{2+}}^{n}} \right) \prod_i a_{C_i}^{\nu_i}} \tag{39}$$

which again reduces to

$$r = k_+ K_{28}^* \left(\frac{a_{H^+}^{2}}{a_{Mg^{2+}}} \right)^n \tag{39a}$$

when there are relatively few precursor complexes present at the surface. Note that in these expressions n refers to a stoichiometric coefficient equal to the number of precursor complexes created from the removal of each Mg atom from the surface. An exception to this behavior is found for olivine whose dissolution rates are independent of aqueous magnesium activity. Nevertheless, a thin Mg-leached layer evidenced both by surface titration and XPS analyses (Pokrovsky and Schott 2000a; Oelkers et al. 2009) forms in acid solution. Pokrovsky and Schott (2000b) proposed that this behavior is due to a control of olivine dissolution by the decomposition of a silica-rich, Mg-free, protonated surface precursor formed by the fast exchange of $2H^+$ for one Mg^{2+} ion ($\log K_{ex} > 22$ in reaction 28) followed by sorption of one proton on two polymerized silica tetrahedra. This mechanism is consistent with the development of a hydrogen permeated leached layer where residual hydrated silicate tetrahedra are polymerized and probably linked to Mg ions deeper in the mineral structure.

In alkaline solutions (pH > 9), preferential Si release is observed at the initial stage of olivine dissolution which results in the formation of a Mg-rich layer (Mg octahedral branched together via Mg-O-Mg bonds) as demonstrated by results of XPS analysis and column filtration experiments (Pokrovsky and Schott 2000a). As is also the case for crysotyle (Bales and Morgan 1985), brucite, and magnesite (Pokrovsky and Schott 1999), the hydrolysis of Mg-O bonds controls forsterite dissolution in alkaline solutions. This dissolution control, which results from slower breaking rates of Mg-O bonds compared to Si-O bonds at these pH, is consistent with quartz and amorphous silica dissolution rates being slower than that of brucite at pH > 10 (see Fig. 40). Note that brucite has a much higher pH_{zpc} than quartz (11 vs. 2). At 10 < pH <12 the brucite surface is dominated by unreactive neutral >MgOH° sites whereas deprotonated $>SiO^-$ sites which present a much greater affinity for hydrolysis than neutral >SiOH° (Nangia and Garrison 2008) predominate on the silica surface. The control of olivine dissolution by Si-free, MgOH surface complexes is consistent with its inhibition by silicic acid and carbonate ions. The inhibiting effect of CO_3^{2-} was attributed to the adsorption of carbonate ions on >MgOH/ $MgOH_2^+$ groups by Wogelius and Walther (1991) and Pokrovsky and Schott (2000). Similar inhibition of wollastonite dissolution by silicic acid (see Fig. 41; Pokrovsky et al. 2009) and CO_3^{2-} (Golubev et al. 2005) is observed in alkaline solutions, and it is likely that most Ca/ Mg silicates whose surface speciation is dominated by Mg(Ca)OH° surface sites in alkaline solutions exhibit the same behavior.

This example again illustrates the impact that metal oxide surface acidity and complexation have on the mechanisms and rates of dissolution of both single oxide and multioxide minerals. As the dissolution mechanism of basic silicates generally involve the same exchange reaction of divalent metals for protons, these minerals all exhibit a similar dependence on pH (1 < pH <12) including forsterite (Pokrovsky and Schott 2000), Ca-olivine (Westrich et al. 1993), wollastonite (Xie 1994; Schott et al. 2002; Pokrovsky et al. 2009), diopside (Schott et al. 1981; Knauss et al. 1993; Chen and Brantley 1998; Golubev et al. 2005), enstatite (Oelkers and Schott 2001), anthophylite (Mast and Drever 1987; Chen and Brantley 1998), crysotyle (Bales

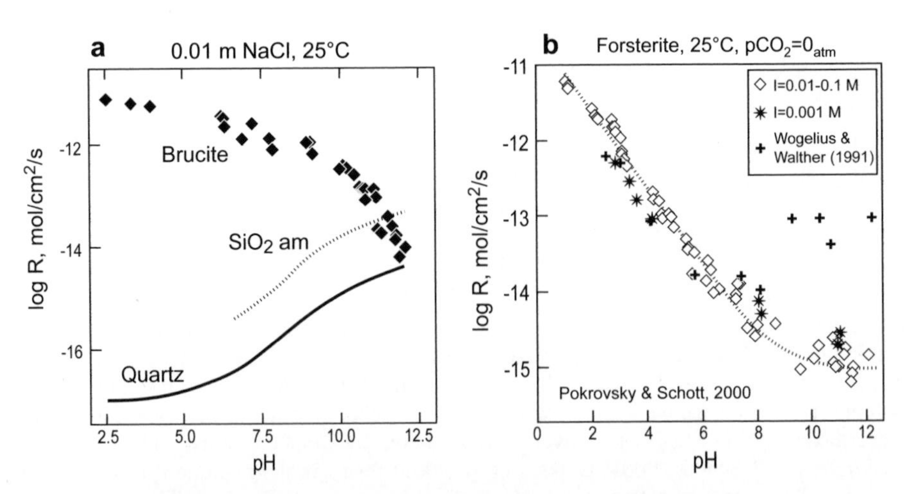

Figure 40. Comparison of quartz, amorphous silica, brucite and forsterite dissolution rates at 25 °C as a function of solution pH.

and Morgan 1985), and talc (Saldi et al. 2007). This pH variation is characterized by a continuous decrease of rates with increasing pH to at least mildly alkaline conditions. At high pH, these rates often become pH independent, consistent with the behavior of the aqueous complexation of these divalent metals (see Fig. 42). Brantley and Chen (1995) and Brantley (2008) concluded that dissolution rates decrease with increase in connectedness, the average number of bridging oxygens around each Si atoms, which increases from 0 for orthosilicates to 3 for phyllosilicates, and with the decrease of the ratio of non-tetrahedrally to tetrahedrally coordinated cations. This appears to be the case for the basic silicates as the dissolution rates of olivine far exceed those of talc. It is also evident that for a given connectedness, the rate of dissolution of Ca-silicates is much higher than that of Mg-silicates, in accord with much weaker Ca-O bonds than Mg-O bonds and relative rates of exchange of water molecules in calcium and magnesium coordination sphere (see also Schott et al. 1981 and Westrich et al. 1993). This is particularly evident at alkaline conditions where dissolution is controlled by the hydrolysis of Ca/Mg-O bonds.

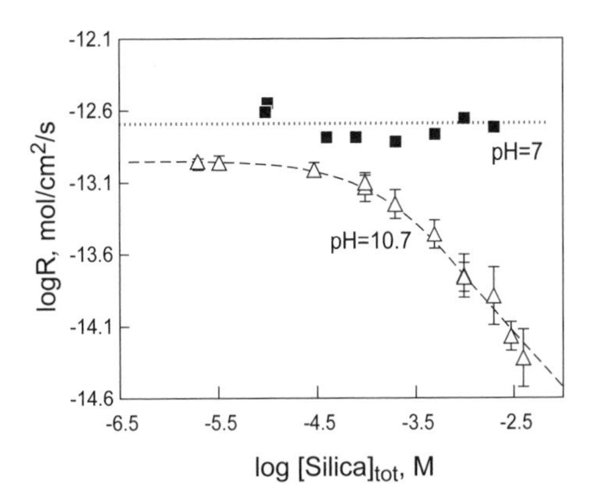

Figure 41. Wollastonite dissolution rate at 25 °C as a function of aqueous silica concentration.

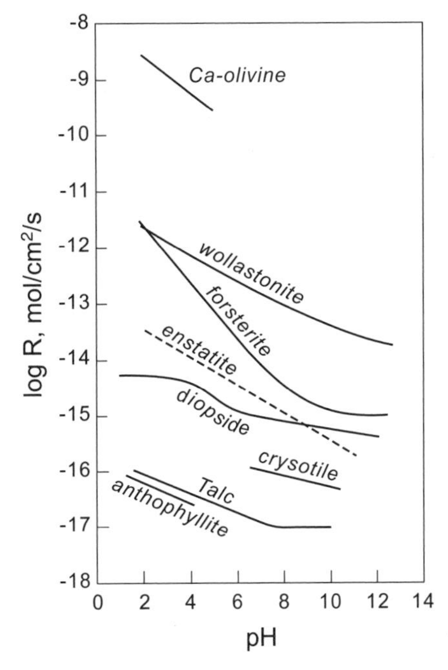

Figure 42. Dissolution rates of Ca/Mg silicates at 25 °C as a function of solution pH (enstatite dissolution rates derived from 70 °C rates measured by Oelkers and Schott (2001) using $r = A_A \exp(-E_A / RT)(a^2_{H^+} / a_{Mg^{2+}})^{1/8}$ with $A_A = 2.4\times10^{-4}$ (mol/cm²/s) and $E_A = 48.5$ kJ/mol).

The pH dependence of Ca-Mg silicates contrasts with that of aluminosilicates and quartz. Basic silicate dissolution rates decrease or remain constant with increasing pH at basic conditions, whereas aluminosilicate and quartz dissolution rates visibly increase with pH at basic conditions. This difference stems from the relative hydrolysis of divalent metal cations versus, aluminum and silicon in solution. Similarly the much stronger stability of aqueous aluminum - organic acid complexes compared to equivalent Ca^{2+} and Mg^{2+} complexes explains the much greater enhancement by organic acids of feldspar dissolution rates (see Oelkers and Schott 1995 and references therein) versus that of olivine (Wogelius and Walther 1991, 1992), diopside (Golubev and Pokrovsky 2006) and wollastonite (Pokrovsky et al. 2009) dissolution rates.

THE SCM/TST DESCRIPTION OF THE KINETICS OF CARBONATE MINERAL DISSOLUTION/PRECIPITATION

Analysis of the chemistry of the carbonate mineral-solution interface within the framework of surface coordination chemistry and transition state theory provides an improved description of the dissolution and precipitation rates of these minerals.

The carbonate-solution interface

In situ spectoscopic investigations of the carbonate-solution interface using XPS (Stipp and Hochella (1991), high-resolution X-ray reflectivity (Fenter et al. 2000), diffuse reflectance DRIT spectroscopy (Pokrovsky et al. 2000), electrokinetic measurements, and surface titration (Charlet et al. 1990; Pokrovsky et al. 1999a,b) indicate the presence of two primary hydration sites, $>CO_3H^\circ$ and $>MeOH^\circ$, at the surface of divalent carbonates where Me refers to a divalent metal (e.g., Ca^{2+}, Mg^{2+}). Subsequent hydration and adsorption of constituent ions from solution lead to the formation of the following surface species, $>MeOH_2^+$, $>MeOH^\circ$, $>MeO^-$, $>MeHCO_3^\circ$, $>MeCO_3^-$, $>CO_3Me^+$, $>CO_3H^\circ$, and $>CO_3^-$ (Van Cappellen et al. 1993; Pokrovsky et al. 1999a,b; Pokrovsky and Schott 2002). A schematic representation of the hydrated calcite-solution interface with the formation of various surface complexes is shown in Figure 43. The formation constants of these complexes have been derived for the main divalent metal carbonates from their pH dependent surface charges and zeta potentials together with correlations between formation constants for surface and aqueous complexes (Pokrovsky and Schott 2002). It is interesting to note that values of formation constants for surface reactions found for carbonates are similar to corresponding constants for homogeneous aqueous reactions. In accord with *in situ* X-ray reflectivity measurements of the calcite-water interface (see Fig. 43a), this similarity suggests the carbonate-water interface is highly hydrated. An example of speciation at the magnesite-solution interface for a 0.1 M NaCl, solution containing 0.01 M total CO_2 and 10^{-5} M of $MgCl_2$, calculated using the constant capacitance double layer model is shown in Figure 44. It can be seen that the dominant species on the metal sites are $>MgOH_2^+$ and $>MgCO_3^-$, the later prevailing at pH > 9. $>MgOH^\circ$ and $>MgO^-$ dominate only in strongly alkaline solutions and very low P_{CO_2}. In neutral to alkaline pH, the carbonate sites are dominated by $>CO_3^-$ whereas $>CO_3H^\circ$ becomes dominant at pH < 5.

Two new models for carbonate – solution interfaces have been recently published: 1) a

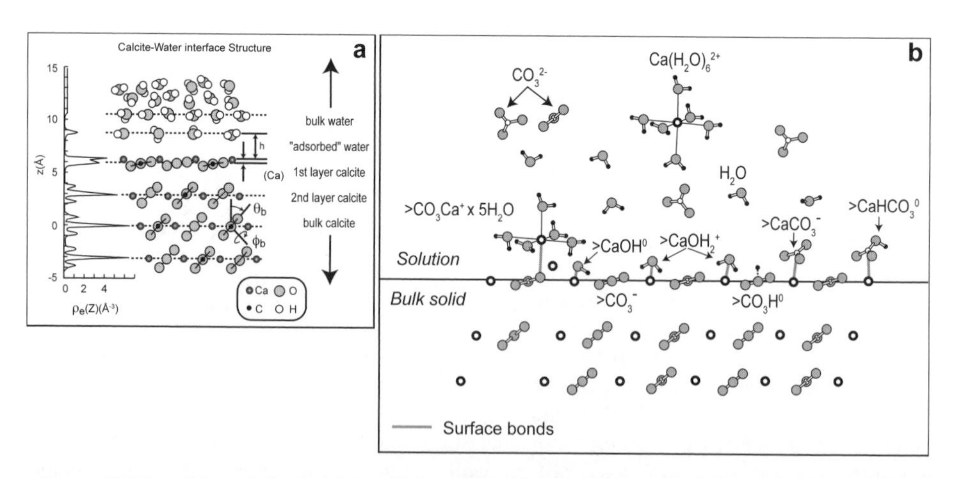

Figure 43. The calcite-solution interface: a) schematic cross section of the calcite-water interface structure deduced from X-ray reflectivity data at pH 8.3 (Fenter et al. 2000), b) species present at the calcite-solution interface (cross section of {1014} plane) from Pokrovsky et al. 2002.

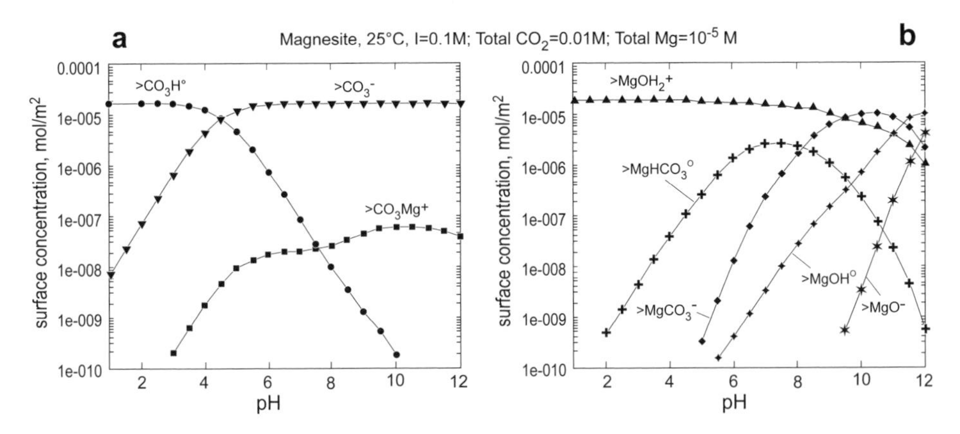

Figure 44. Calculated chemical speciation at magnesite-solution interface at 25 °C as a function of solution pH.

CD-MUSIC model that takes into account bond valence concepts to derive fractional charge values for surface sites at terraces, edges and corners of hydrated carbonate surfaces (Wolthers et al. 2008), and 2) a one-site model that is based on single, generic charge-neutral surface site ($\equiv MeCO_3 \cdot H_2O^0$, Villegaz-Jimenez et al. 2009). These models exhibit general agreement with the "classic" two-site model and provide similar surface speciation estimates. However, as the predictive capacities of the new models have not been yet tested for describing dissolution and precipitation rates, the next section focuses on the classic two-site approach.

Carbonate minerals dissolution and precipitation kinetics

Dissolution. According to the classical model of Plummer et al. (1978) and Wollast (1990), the dissolution kinetics of carbonate minerals is controlled by the following two parallel reactions

$$MeCO_3(cr) + H^+ = >HMeCO_3^+ \rightarrow Me^{2+} + HCO_3^- \quad (40)$$

$$MeCO_3(cr) + nH_2O = >MeCO_3 \cdot nH_2O \rightarrow Me^{2+} + CO_3^{2-} \quad (41)$$

which correspond to a protonation and hydration mechanism, respectively. The middle term of each reaction, which was not specifically mentioned, but rather implied by the original authors, corresponds to the rate controlling precursor complex of each mechanism. Reactions (40) and (41) prevail under acidic and alkaline conditions, respectively. Far from equilibrium, dissolution rate can thus be expressed as

$$r_+ = k_{H^+} \cdot a_{H^+} + k_{H_2O} \quad (42)$$

It should be noted that a third parallel reaction, surface carbonation by H_2CO_3, was envisaged by Plummer et al. (1978), but it was recently dismissed by Pokrovsky et al. (2005) who demonstrated that, at constant ionic strength, H_2CO_3 has no specific effect on carbonates dissolution rates and that measured increased rates as a function carbonic acid concentration can be accounted for by solution acidification resulting from CO_2 dissolution.

Examination of the pH dependent chemistry of the carbonate mineral-aqueous solution interface allows identification of the surface complexes that control dissolution under acidic and alkaline solutions. Because the metal surface sites are all protonated at pH < 7-8, enhancement of dissolution observed in acid solution results from the protonation of CO_3^- surface sites via the formation of $>CO_3H^\circ$. At circum neutral to basic pH, the dissolution of carbonate minerals, like that of oxides, is controlled by the hydration of surface metal sites, $>MeOH_2^+$. Thus within the framework of TST and surface coordination chemistry, carbonates far-from equilibrium dissolution rate can be expressed as

$$r_+ = r_{+,H^+} + r_{+,H_2O} = k_H\{>CO_3H^\circ\}^{n_H} + k_{H_2O}\{>MeOH_2^+\}^{n_{H_2O}} \quad (43)$$

where n_H and n_{H_2O} stand for the reaction order with respect to protonated carbonate and metal sites, respectively. The success of Equation (43) in describing carbonates dissolution kinetics can be assessed for magnesite on Figure 45 where are reported almost 200 magnesite steady-state dissolution rates measured at 25 °C as a function of pH and ionic strength. In accord with Equation (43), there is a linear correlation between the logarithm of dissolution rate and the logarithms of $>CO_3H^\circ$ and $>MeOH_2^+$ surface concentrations in acidic and circum-neutral to alkaline solutions, respectively, for all investigated pHs and ionic strengths. The reaction orders ($n_H = n_{H_2O} = 4$) suggest that magnesite proton-promoted dissolution requires the protonation of the four surface carbonates surrounding a hydrated surface Mg to detach this metal from the surface whereas, in neutral to alkaline condition, dissolution requires the hydration of the four surface Mg surrounding a surface carbonate site. The formation of the precursor surface activated complex, therefore, requires full protonation/hydration of all four surface sites around the site where the breaking of Mg-O bond occurs, consistent with roughly chess-like surface organization of magnesium and carbonate centers.

Equation (43) also describes calcite and dolomite far from equilibrium dissolution rates but with $n_H = n_{H_2O} = 2$ and $n_H = 2$, $n_{H_2O} = 1$ for dolomite and calcite, respectively (Van Cappellen et al. 1993; Pokrovsky and Schott 2001, 2002; Gautelier et al. 2007). The relative weakness of the Ca-O bond (2.36 Å in length) compared to the Mg-O bond (2.10 Å) likely explains the different reaction orders. For example, calcite and dolomite dissolution at acid pH requires protonation of two carbonate groups adjacent to a hydrated surface Ca and Mg, respectively. The relative strength of these bonds is illustrated by a rate of exchange of water molecules from aqueous calcium sphere that is almost 5 orders of magnitude higher that from magnesium sphere.

A major input of surface coordination compared to the classical Plummer et al. (1978) model for describing carbonate minerals reactivity is illustrated by dolomite dissolution rates shown as a function of pH in Figure 46. It can be seen that the Plummer et al. (1978) model (dashed line), which implies that rates are pH-independent at pH > 6 (see Eqn. 42), cannot

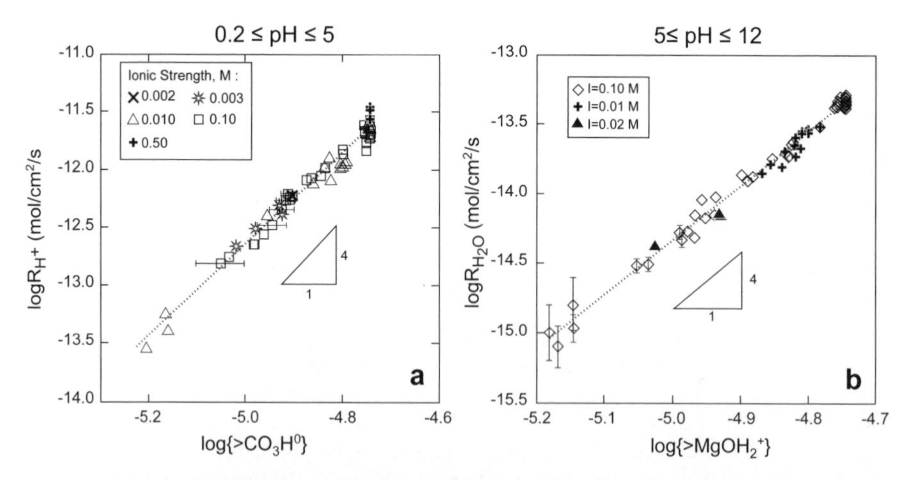

Figure 45. Magnesite dissolution rate at 25 °C as a function of $>CO_3H^\circ$ (a) and $>MgOH_2^+$ (b) concentrations (from Pokrovsky and Schott 1999).

describe experimental data obtained as a function of ΣCO_2 concentrations in neutral to alkaline solution. In contrast, all rate data are found to be proportional to the square of $MgOH_2^+$ surface concentration in accord with Equation (43) and $n_{H_2O} = 2$ (see Fig. 47a). The inhibiting effect of carbonate ions on far-from equilibrium dolomite (or magnesite) dissolution rates, which is illustrated in Figure 47b, is explained within the framework of the SCM/TST approach by the formation of $>MgCO_3^-$ at the expense of rate controlling $>MgOH_2^+$ according to

$$>MeOH_2^+ + CO_3^{2-} = >MgCO_3^- + H_2O \tag{44}$$

for which the equilibrium constant is $K^*_{CO_3}$. Note that the formation of the negatively charged $>MeO^-$ species at the expense of neutral $>MeOH^\circ$ allows description of the increase of calcite and dolomite dissolution rates with increasing pH observed in CO_2-free, strongly alkaline solutions (pH > 12) as illustrated in Figure 48 (Schott et al. 2004).

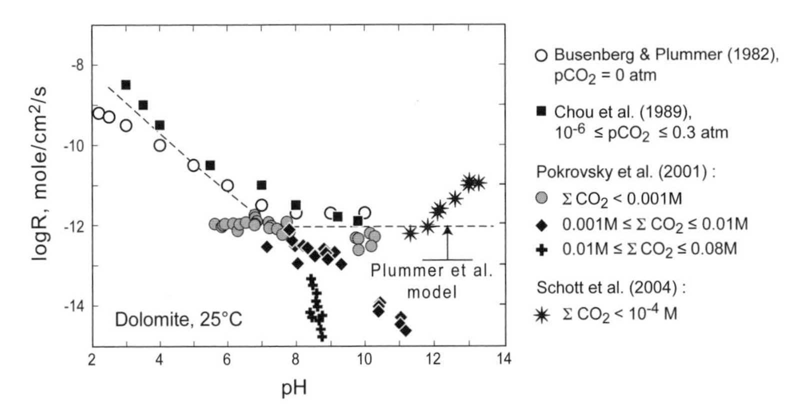

Figure 46. Dolomite dissolution rate at 25 °C as a function of solution pH.

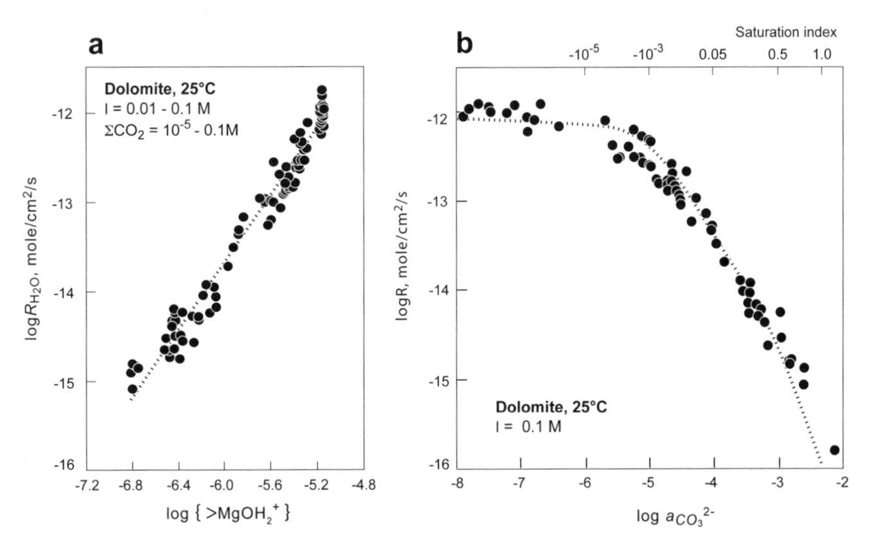

Figure 47. Dolomite dissolution rate at 25 °C as a function of $>MgOH_2^+$ concentration (a) and carbonate ions activity (b) (from Pokrovsky and Schott 2001). The dashed curve in Figure 47b has been genenerated using the surface coordination model for dolomite dissolution (Eqn. 15 in Pokrovsky and Schott 2001).

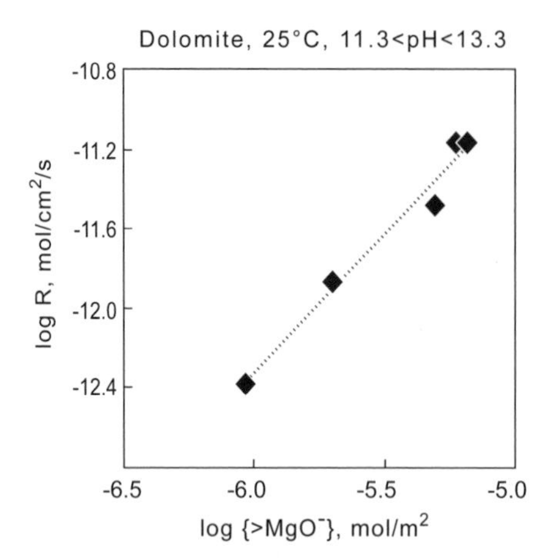

Figure 48. Dolomite dissolution rate at 25 °C as a function of $>MgO^-$ concentration.

Like for alkaline earth oxides whose dissolution rate in neutral to alkaline dissolution is controlled by $>MeOH_2^+$ concentration, the ligand promoted dissolution of carbonate minerals can be described by the formation of >MeL° according to the reaction

$$>MeOH_2^+ + L^- = >MeL^\circ + H_2O \quad (45)$$

for which the equilibrium constant is K^*_{MeL}. In the presence of ligands, the forward dissolution rate is the sum of H_2O- and ligand-promoted dissolution mechanisms in accord with

$$r_+ = k_{H_2O}\{>MeOH_2^+\}^{n_{H_2O}} + k_L\{>MeL^\circ\} \quad (46)$$

where n_{H_2O} again stands for a reaction order. Combining Equations (45) and (46) with metal site conservation yields (Pokrovsky and Schott 2001)

$$r_+ = k'_{H_2O}\left(1 - \frac{K^*_{MeL}\cdot[L^-]}{1+K^*_{MeL}\cdot[L^-]}\right)^{n_{H_2O}} + k'_L \frac{K^*_{MeL}\cdot[L^-]}{1+K^*_{MeL}} \quad (47)$$

which permits description of the ligand-promoted dissolution of carbonates minerals as a function of aqueous ligand activity. The capacity of Equation (47) to describe dolomite and magnesite dissolution rates in the presence of several organic ligands is demonstrated in Figure 49. A close correspondence is found between experimental data and the computed curves. Like for Ca/Mg oxides and silicates, and in contrast with aluminosilicates, the impact of organic ligands is generally weak except for certain ligands, like citrate and EDTA, that can form very stable five-membered chelate rings with surface metal which labilize critical metal oxygen bonds.

Surface coordination chemistry also provides an improved description of near-to equilibrium carbonate dissolution/precipitation rates compared to the Plummer et al. (1978) model.

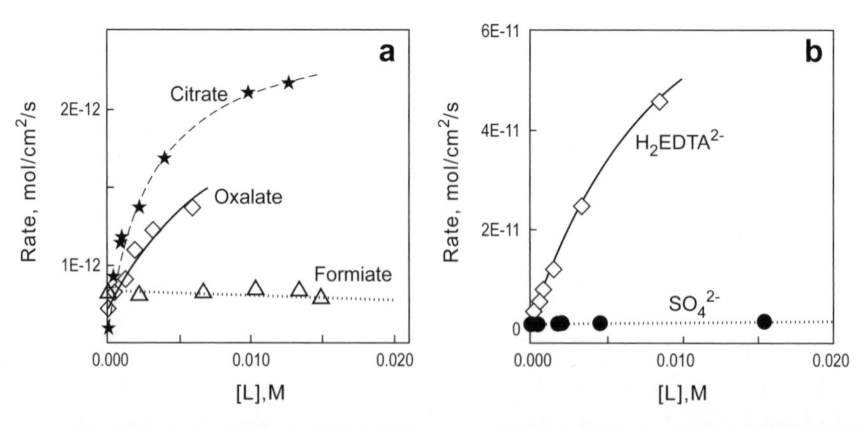

Figure 49. Dolomite dissolution rate at 25 °C in presence of various organic and inorganic ligands. The symbols represent the experimental data but the curves were generated using Equation (47) (from Pokrovsky and Schott 2001).

Assuming in accord with the law of detailed balancing that the precursor complex is the same for dissolution and precipitation, the following expression for carbonate minerals overall reaction rate can be derived from Equation (43) (Pokrovsky and Schott 1999)

$$r = \left[k_{\mathrm{H}} \{ >\mathrm{CO_3H^\circ} \}^{n_{\mathrm{H}}} + k_{\mathrm{H_2O}} \{ >\mathrm{MeOH_2^+} \}^{n_{\mathrm{H_2O}}} \right] \left(1 - \Omega^{n}\right) \tag{48}$$

where Ω again denotes the aqueous solution saturation index with respect to considered carbonate mineral. In neutral to alkaline solutions where close to equilibrium conditions are often achieved, Equation (48) simplifies to

$$r = k_{\mathrm{H_2O}} \{ >\mathrm{MeOH_2^+} \}^{n_{\mathrm{H_2O}}} \left(1 - \Omega^{n_{\mathrm{H_2O}}}\right) \tag{49}$$

In contrast, at these conditions, a corresponding derivation performed starting with the Plummer et al. (1978) model (Eqn. 42) yields

$$r = k_{\mathrm{H_2O}} \, (1 - \Omega) \tag{50}$$

Whereas the Plummer et al. (1978) model predicts that carbonates dissolution/precipitation rates in neutral to alkaline conditions should depend only on reaction chemical affinity (departure from equilibrium), the SCM model predicts a dependence of dissolution and precipitation rates on both chemical affinity and the activity of the aqueous species which control the surface concentration of metal hydrated sites ($>\mathrm{MeOH_2^+}$). For example, in ΣCO_2-bearing solutions Equation (49) can be combined with Equation (44) to generate the following expression describing carbonate dissolution and precipitation rate as a function of carbonate ion activity

$$r = k'_{\mathrm{H_2O}} \left(\frac{K^*_{\mathrm{CO_3}}}{K^*_{\mathrm{CO_3}} + a_{\mathrm{CO_3^{2-}}}} \right)^{n_{\mathrm{H_2O}}} \left(1 - \Omega^{n_{\mathrm{H_2O}}}\right) \tag{51}$$

where $K^*_{\mathrm{CO_3}}$ stands for $>\mathrm{MgCO_3^-}$ formation constant (see reaction 44).

Precipitation. Pokrovsky and Schott (1999, 2001, in prep) and Saldi et al. (2009) demonstrated that it was possible to accurately describe near-to equilibrium calcite, dolomite, and magnesite dissolution/precipitation rates within the SCM/TST framework using Equations (49) or (51). The validity of the equations presented above is illustrated in Figure 50. This figure compares experimentally measured calcite and magnesite precipitation rates reported

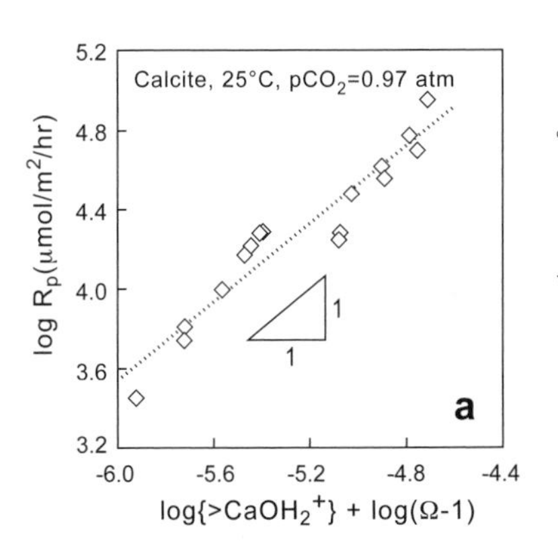

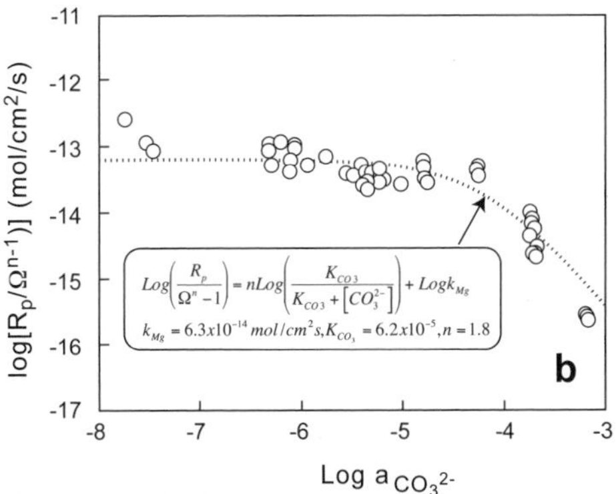

Figure 50. Description of calcite (a) and magnesite (b) precipitation rates at 25 and 200 °C, respectively within the framework of surface coordination concepts (from Pokrovsky and Schott 2009 and Saldi et al. 2009). The curve in Figure 50b was generated using Equation (51).

by Dromgoole and Walter (1990) and Saldi et al. (2009), respectively, with corresponding values calculated with Equation (49) and (51). For calcite, the slope of one observed in Fig 50a is consistent with calcite precipitation rate at 25 °C being proportional to the product $\{>CaOH_2^+\}(1 - \Omega)$. Alternatively this product can be rewritten in terms of aqueous species as $(K^*_{CO_3}/(K^*_{CO_3}+a_{CO_3^{2-}}))(1-\Omega)$. The rate of magnesite precipitation at 200 °C as a function of CO_3^{2-} activity is plotted in Figure 50b. It can be seen that magnesite precipitation rates are accurately described by Equation (51) with $n_{H_2O} = 2$ and that it decreases by about 2.5 orders of magnitude when aqueous carbonate ion activity increases from 10^{-7} to $10^{-3.2}$. This confirms that near-equilibrium dissolution and precipitation rates of divalent metals carbonates, like those of divalent metal oxides, depend both on the chemical affinity of the overall reaction and the activity of the aqueous species that control the concentration of $>MeOH_2^+$ precursor species (e.g., pH, concentrations of carbonate species, organic ligands...). An important outcome of this result is that carbonate reaction rates in most aquatic systems—oceans, diagenesis, CO_2 geological sequestration...—cannot be accurately quantified using kinetics models that only consider the effect of chemical affinity on reaction rate.

Carbonate mineral reactivity and the rate of exchange of water molecules in metal coordination sphere

In neutral to alkaline solutions, where the dissolution of divalent metal carbonate minerals is controlled by surface metal hydrolysis, it is expected that there should be a correlation between the carbonate dissolution rates and the rate of exchange of water molecules in metal coordination sphere. This appears to be the case as can be seen in Figure 51. This figure shows the far-from equilibrium dissolution rates at 25 °C and 6 < pH < 8 of eleven divalent carbonates. It can be seen that these rates are strictly proportional to the exchange rate of water molecules in the coordination sphere of corresponding metal constituting carbonate. Analysis of water molecule substitution rates may provide useful information on the mechanisms of dissolution and precipitation of the different carbonates. Within TST, the exchange of water molecules can be described by (Lincoln and Merbach 1995)

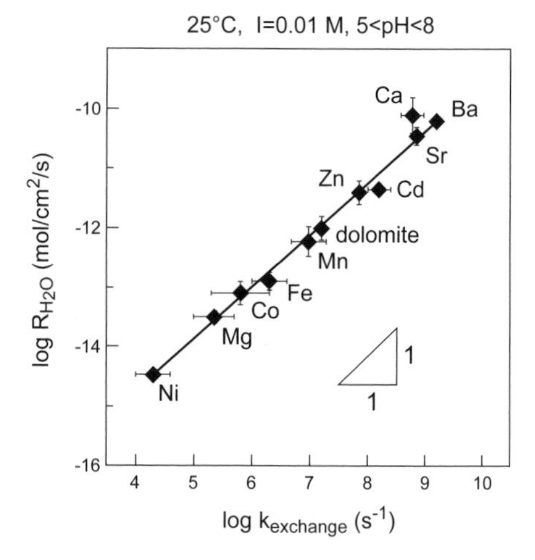

Figure 51. Rate of dissolution of divalent carbonate minerals at 25 °C and circumneutral pH as a function of water exchange rate constant from corresponding metal coordination sphere (from Pokrovsky and Schott 2002).

$$\mathrm{Me(H_2O)}_n^{m+} + \mathrm{H_2O} \rightleftharpoons \mathrm{Me(H_2O)}_{n\pm1}^{*m+}(+\mathrm{H_2O}) \rightleftharpoons \mathrm{Me(H_2O)}_n^{m+}(+\mathrm{H_2O}) \qquad (52)$$

where $\mathrm{Me(H_2O)}_n^{m+}$ can either lose one H_2O to form an activated complex of decreased coordination number or incorporate one H_2O to form an activated complex of increased coordination. Consequently, the rates of water exchange for different metals provides direct access to the energetics of $\mathrm{Me(H_2O)}_{n\pm1}^{*m+}$ activated complexes and thus to the water-promoted dissolution rates of carbonate minerals. Similarly, comparison of water molecule exchange rates in the $\mathrm{Me(H_2O)}_n^{m+}$ aquo ion and in the $\mathrm{MeL(H_2O)}_{n-1}^{(m-x)+}$ metal-ligand complex (see Fig. 52, Phillips et al. 1997b) should give insight to the rate constants of ligand L-promoted dissolution as proposed by Casey and Ludwig (1995).

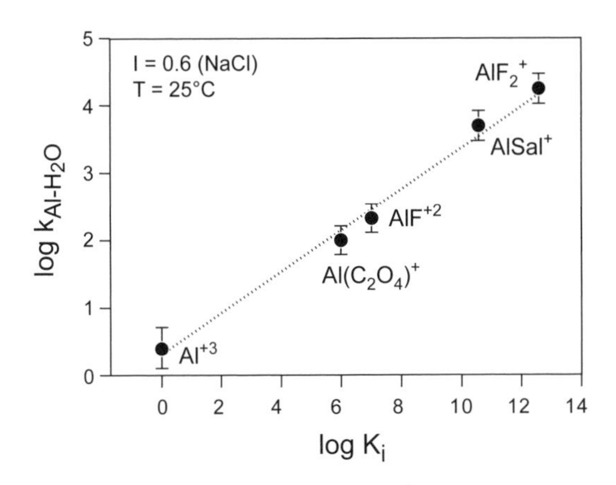

Figure 52. Water exchange rate constant in Al-organic and inorganic complexes against the stability constant of corresponding complex (from Phillips et al. 1997b and Sullivan et al. 1999).

The strong connection between carbonate dissolution rates and the rate of exchange of water molecules allows exploration of the pressure dependence of the former. The pressure dependence of the rate of exchange of water molecules is related to the volume of activation ΔV^* of reaction (52) according to

$$(\delta \ln k/\delta P)_T = -\Delta V^*/RT \qquad (53)$$

Equation (53) provides additional insights on the reaction mechanism of water exchange in the metal coordination sphere and thus on metal carbonate hydrolysis. Two extreme possibilities arise for making a complex between a ligand L^{x-} and a metal aquo ion (see Lincoln and Merbach 1996, and references therein). The first involves the removal of one water molecule from $Me(H_2O)_n^{m+}$ to form a reactive intermediate of decreased coordination number, $Me(H_2O)_{n-1}^{*m+}$. This complex can survive several molecular collision prior to reacting with L^{x-} to form $MeL(H_2O)_{n-1}^{(m-x)+}$; the rate-determining step for the formation of this complex is the dissociation of an H_2O molecule (i.e., the breaking of a Me-O bond), and the mechanism is said to be dissociatively activated and is classified as **D**. At the other extreme is the associatively activated mechanism (**A**), in which incorporation of L^{x-} proceeds through a reactive intermediate of increased coordination number $MeL(H_2O)_n^{*(m-x)+}$. This complex survives several molecular collisions before losing one H_2O molecule to form $MeL(H_2O)_{n-1}^{(m-x)+}$; the rate-controlling step is the creation of a Me-L bond. In between these two extremes, there exist a continuum of mechanisms, classified as interchange mechanisms (**I**), in which bond breaking and making may have similar or different relative importance. The volume of activation ΔV^* is a useful tool for the identification of the rate-controlling exchange mechanism because the **D** mechanism is characterized by a large increase of the volume of the transition complex resulting from the dissociation of the leaving ligand (ΔV^* is large and positive) whereas, at the other extreme, the **A** mechanism is characterized by a decrease in volume of the transition complex (ΔV^* is large and negative). The **I** mechanism is intermediate having a ΔV^* close to zero. Consequently, increasing pressure produces a decrease of the exchange rate of water molecules in a **D** mechanism and an increase in an **A** mechanism. This is illustrated in Figure 53 where the ratio (k_p/k_0) of the rate constant for water exchange of several metals coordination sphere measured at pressure P and ambient atmosphere is plotted as a function of pressure. The variation of the slopes of $\ln(k_p/k_0)$ for water exchange on $Me(H_2O)_6^{2+}$ reported in Figure 53 and derived values of ΔV^* listed in Table 1 imply that for Ca^{2+}, Mn^{2+} and also Ba^{2+}, Cd^{2+}, Zn^{2+} (see Fig. 51) water exchange in metal coordination sphere is associatively activated (**A**), but it is dissociatively activated (**D**) for Fe^{2+}, Co^{2+}, Mg^{2+} and Ni^{2+}. From the correlation shown in Figure 51, it can be concluded that the same two mechanisms control metal hydrolysis and

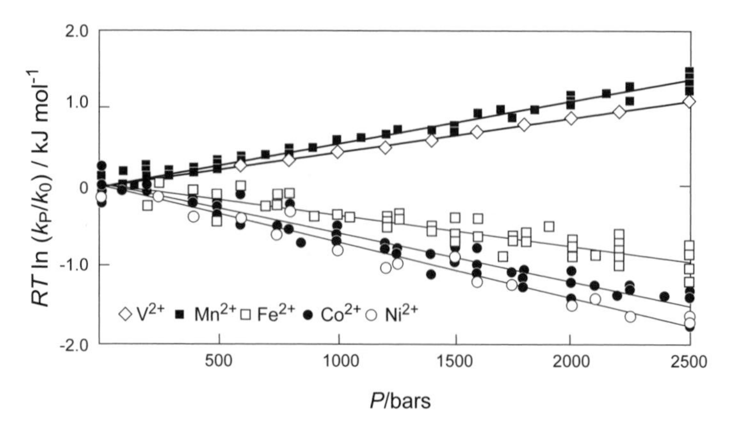

Figure 53. Influence of pressure on water exchange rate constants in the coordination sphere of divalent metals (from Lincoln and Merbach 1995).

Table 1. Values of the activated volume and mechanism which controls the rate of exchange of water molecules in the inner coordination sphere of some divalent metal ions.

	ΔV^* (cm³/mol)	Controlling mechanism
Ca^{2+}	−6.0	Associative
Mn^{2+}	−5.4	Associative
Fe^{2+}	+ 3.8	Dissociative
Co^{2+}	+ 6.1	Dissociative
Mg^{2+}	+ 6.5	Dissociative
Ni^{2+}	+ 7.2	Dissociative

ΔV^* values are from Ducommen et al. (1980, 1982)

dissolution at carbonate mineral surfaces in near neutral to alkaline solutions. In accord with these observations, siderite, magnesite, and gaspeite dissolution rates should decrease with increasing pressure whereas calcite dissolution rates should increase with increasing pressure which may impact the lysocline position in the oceans (Morse and Berner 1974). Interestingly, the dissociative mechanism **D** likely accounts for the much lower dissolution and precipitation rates of magnesite, siderite, and dolomite compared to calcite and the reluctance of Mg, Fe, Ni, Co, to form the anhydrous carbonates (magnesite, dolomite, siderite, gaspeite…) at ambient temperature in aquatic environments (magnesite and dolomite crystallization rate constants at 25 °C are respectively about 6 and 4 orders of magnitude lower than that of calcite according to Saldi et al. 2009). Moreover, it has been observed that the substitution of inorganic or organic ligands for one or two coordination sphere waters of the Mg aquo ion increases the rate of exchange of water molecules. It remains to be seen if a similar substitution of inorganic or organic ligands for one or two coordination sphere waters of the Mg at the magnesite and dolomite surface could enhance the crystallization rates of these minerals by decreasing the bond-breaking contribution to the volume of activation for water exchange, thus facilitating the entry of carbonate (bicarbonate) ions in magnesium inner coordination sphere.

TEMPERATURE DEPENDENCE OF MINERAL DISSOLUTION RATES

Activation energy

The temperature dependence of mineral dissolution rate constants are generally characterized using the Arrhenius equation

$$k = A_0 e^{-E_a/RT} \tag{54}$$

where A_0 refers to the temperature independent pre-exponential factor, E_a stands for the activation energy, and R again represents the molar gas constant. Experimentally, E_a is obtained from the slope of a plot of ln k vs. $1/T$ from the equation

$$\frac{d \ln k_{\text{exp}}}{dT} = \frac{E_a}{RT^2} \tag{55}$$

Within the framework of TST, the rate constant of an elementary reaction can be expressed as (see Fig. 3):

$$k = \frac{k_B T \gamma_A \gamma_B}{h} e^{\frac{\Delta S^{*0}}{RT}} \cdot e^{\frac{-\Delta H^{*0}}{RT}} \tag{56}$$

where ΔS^{*0} and ΔH^{*0} represent the activation entropy and enthalpy, respectively, k_B stands for Boltzman's constant, h refers to Plank's constant and γ_i refers to the activity coefficient of the subscripted species. Comparison of Equations 55 and 56, assuming that ΔS^{*0} and ΔH^{*0} do not vary with T, yields (Lasaga 1981)

$$E_a = RT + \Delta H^{*0} \tag{57a}$$

$$A_0 = e \cdot \frac{k_B T \gamma_A \gamma_B}{h} e^{\frac{\Delta S^{*0}}{RT}} \tag{57b}$$

For most experimental runs, RT value is less than 1 kcal/mol and so E_a and ΔH^{*0} values are almost the same. Therefore, activation energies retrieved from the temperature variation of rate constants can provide insight on the energetics of the precursor complex formation reactions and subsequently on the rate controlling dissolution/precipitation steps (i.e., surface reaction vs. aqueous diffusion of reactants). This is however not an easy task for the dissolution of most single and multi-oxide minerals which have rates controlled by activated complexes formed by adsorption and/or exchange reactions occurring at discrete surface sites. Characterization of E_a therefore requires separating the overall rate dependence on temperature into two parts: 1) the contribution of the enthalpy of formation of the rate controlling surface sites (for example protonated sites) and 2) the activation energy of the activated complex. For example, the far-from equilibrium proton-promoted dissolution of a metal oxide can be expressed as (see Eqn. 10)

$$r_+ = k_{\mathrm{H}} \left\{ >\mathrm{MeOH}_2^+ \right\}^{n_{\mathrm{H}}} \tag{58}$$

Combining Equation (58) with the formation reaction of $>MeOH_2^+$ from >MeOH and the conservation of sites at the solid surface and assuming for sake of simplicity {>M-OH}>> {$>MeOH_2^+$} yields

$$r_+ = k_{\mathrm{H}} (S \cdot K_{\mathrm{H}}\, a_{\mathrm{H+}})^{n_H} \tag{59}$$

Combining Equation (59) with Equation (56) results in the following equation describing the variation of dissolution rates with temperature:

$$r_+^T = k_+^0 e^{-E_a/RT} \left(S.K_{\mathrm{H}}^0 e^{-\Delta_r H_{\mathrm{H}}^0/RT} a_{\mathrm{H}^+} \right)^{n_{\mathrm{H}}} \tag{60}$$

Examination of Equation (60) clearly shows that the dependence on temperature of r_+ is a function of both E_a and $\Delta_r H_H^0$, the enthalpy of the formation reaction of protonated sites. As such, attempts to estimate the activation energy from plots of rates versus $1/T$ yield a combination of the activation energy and the enthalpy of the formation reaction of protonated sites. Activation energies generated in this way are therefore commonly referred to as apparent activation energies. Casey and Sposito (1993) discussed in detail the impact of surface proton or hydroxyl adsorption reaction enthalpy on apparent activation energies retrieved from oxide minerals dissolution rates and suggested that negative apparent activation energies may occur. It is interesting to note, however, that Arrhenius plots of dissolution rates measured at pH near the pH_{zpc} of oxide or carbonate minerals, whose dissolution at this pH is controlled by simple surface metal hydrolysis, are likely to generate true E_a. For example, the value of the apparent activation energy determined for magnesite dissolution at near neutral conditions reported by Pokrovsky et al. 2009, E_{exp} = 46 kJ/mol, is consistent with the enthalpy of activation of the rate of exchange of water molecules in Mg^{2+} coordination sphere (ΔH^* = 49 kJ/mol; Lincoln and Merbach 1995). The enthalpies, $\Delta_r H_{H\text{-}Me}$, of the H^+-Me^{n+} exchange reactions that control silicate dissolution minerals are not known. However, from the analogy between the mechanisms controlling the formation of metal leached layers at silicate minerals surface and in aqueous metal-silicate complex, negative values of $\Delta_r H_{H\text{-}Me}$ are possible. Consequently, the measured activation energy for dissolution is likely to be reduced by the proton for metal exchange reaction occurring at the solid surface.

Impact of solution chemistry on the temperature dependence of carbonate mineral dissolution at acid conditions

A recent study of the constant-ionic strength rate of magnesite, dolomite, and calcite dissolution rates as a function of temperature to 150 °C and pCO_2 to 55 atm provides interesting new insights on the temperature dependence of carbonate minerals dissolution (Pokrovsky et al. 2009). Rates reported by Pokrovsky et al. (2009) are shown in Figure 54. It can be seen in this figure that magnesite, dolomite, and calcite dissolution rates increase with temperature from 25 to 100 °C, but rates measured at 150 °C are the same or lower than those measured at 100 °C. For example, magnesite, dolomite, and calcite dissolution rates measured in acidic ($3 < pH < 4.2$), CO_2-rich solutions (Fig. 54b,c) are essentially the same at 150 °C and 60 °C, and about 3 to 10 times lower than that at 100 °C. This implies *negative values* for apparent activation energies of calcite, magnesite, and dolomite dissolution at $T > 100$ °C. This is the first time to our knowledge that negative values of apparent activation energies have been reported in the literature for mineral dissolution. These negative values stem from the carbonate minerals dissolution mechanism. Within the context of surface coordination and constant capacitance model of the electric double layer, the proton promoted dissolution of carbonate minerals at constant ionic strength can be described far-from equilibrium using Equation (43)

$$r_{+,H^+} = k_H \{>CO_3H^\circ\}^{n_H} \tag{61}$$

where the protonated complex is formed by $>CO_3^- + H^+ = >CO_3H^\circ$, which has the law of mass action

$$K_{>CO_3H} = \frac{\{>CO_3H^\circ\}}{\{>CO_3^-\} \cdot a_{H^+}} \tag{62}$$

where k_H and n_H are the rate constants and the reaction order with respect to protonated carbonate surface sites, $K_{>CO_3H}$ represents the equilibrium constant for the reaction of protonation of surface carbonate sites neglecting the electrostatic term, and a_{H^+} stands for aqueous proton activity. Expressing the total number of surface carbonate sites $S = \{>CO_3^-\} + \{>CO_3H^\circ\}$, and assuming at the weakly acid conditions of these experiments $\{>CO_3H^\circ\} << \{>CO_3^-\}$ yield the following expression for the far from equilibrium rate of dissolution at reference temperature T_r

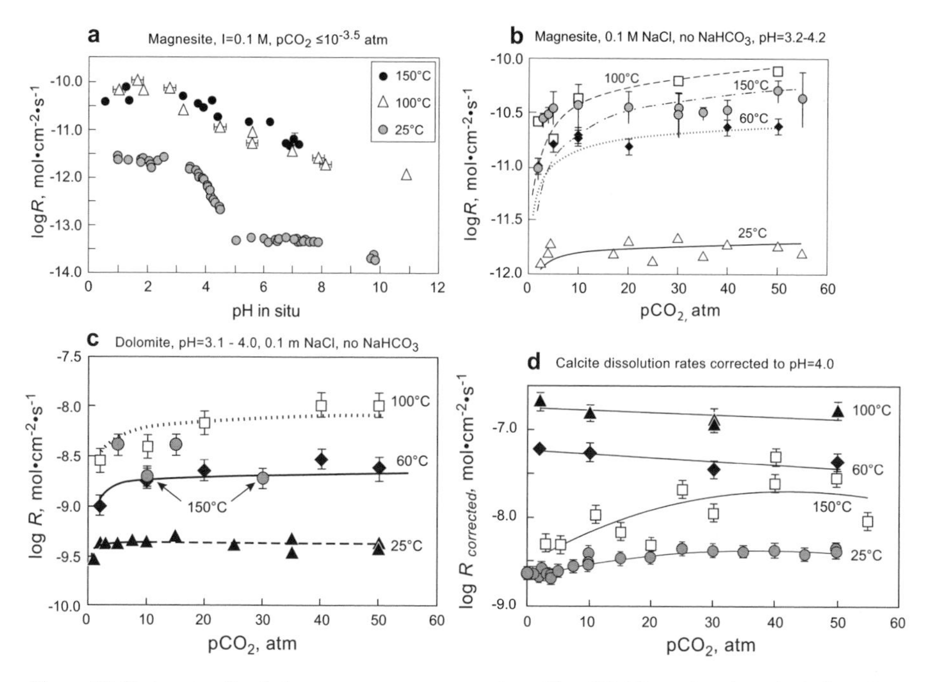

Figure 54. Carbonates dissolution rates at temperatures from 25 to 150 °C as a function of solution pH or pCO_2: a) magnesite as a function of pH, b) magnesite as a function of pCO_2, c) dolomite as a function of pCO_2 and d) calcite as a function of pCO_2 (from Pokrovsky et al. 2009).

$$r^{T_r}_{+,H^+} = k^0_H \left(K^0_{>CO_3H} \cdot S \cdot a_{H^+} \right)^{n_H} \tag{63}$$

whereas the rate of dissolution at temperature T is given by equating Equation (63) with Equation (56) yielding

$$r^{T}_{+,H^+} = k^0_H e^{-E_a/RT} \left(S.K^0_{>CO_3H} e^{-\Delta_r H^0_{>CO3H}/RT} a_{H^+} \right)^{n_H} \tag{64}$$

where E_a represents the activation energy for the dissolution rate constant and $\Delta_r H^0_{>CO_3H}$ stands for the standard enthalpy of surface carbonate protonation reaction and is assumed to be constant in the temperature interval considered. A straightforward derivation shows that dissolution rates become temperature independent and the apparent activation energy is equal to zero if

$$\Delta_r H^0_{>CO_3H} = -\frac{E_a}{n_H} \tag{65}$$

For the case of magnesite, for which $E_a = 46$ kJ/mol and $n_H = 4$, the critical value of $\Delta_r H^0_{>CO_3H}$ is −11.5 kJ/mol. For more negative values, magnesite dissolution rates will decrease with increasing temperature (see illustration in Fig. 55). For dolomite $E_a = 34$ kJ/mol, and $n_H = 2$, and for calcite $E_a = 21$ kJ/mol, and $n_H = 2$. Corresponding critical values of $\Delta_r H^0_{>CO_3H}$ are, therefore, −17 kJ/mol and −10.5 kJ/mol, respectively. These values are very close to the values reported by Patterson et al. (1984) for the apparent enthalpy of the first protonation reaction of aqueous carbonate ions at $75 < T < 150$ °C in 0.5-3 M NaCl solutions. Although the enthalpies of proton adsorption on carbonates surfaces have not been yet measured, it is likely, from

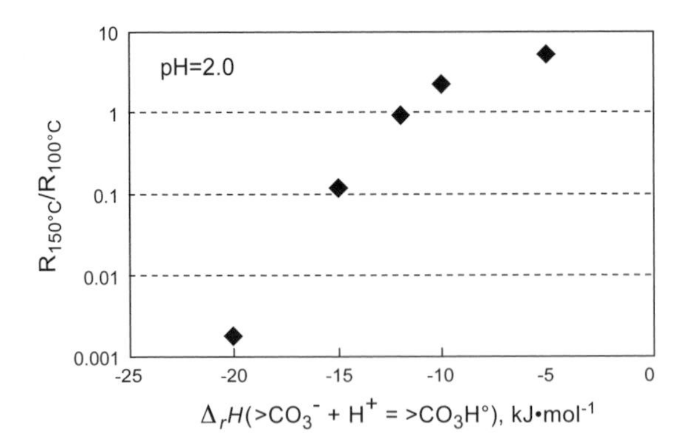

Figure 55. Effect of the value of the enthalpy of the surface carbonate sites protonation reaction on the ratio of magnesite dissolution rate at 150 °C to that at 100 °C, with E_a = 46 kJ/mol and n_H = 4. A decrease of $\Delta_r H^0_{>CO_3H}$ from −5 to −12 kJ/mol is sufficient to provide equal dissolution rates at 100 °C and 150 °C in acid solutions (from Pokrovsky et al. 2009)

the close correspondence between the surface and aqueous speciation of metal carbonates, that the decrease of carbonate dissolution rates in acidic 0.1 M NaCl solution above 100 °C results from the decrease of $\Delta_r H^0_{>CO_3H}$ above 50 °C. This decrease may reflect an increasing competition, as temperature increases, between H^+ and Na^+ for sorption on surface carbonate sites. Thus, contrary to general belief, increasing temperature does not necessarily lead to increasing dissolution rates.

CONCLUDING REMARKS

This review shows that the dissolution and precipitation rates of the major rock forming minerals can all be accurately described as a function of solution composition and distance from equilibrium by one simple unified theory based on Transition State Theory coupled to Surface Coordination Chemistry. Within this model, rates are found to be controlled by activated or precursor complexes that are formed on mineral surfaces by either adsorption or exchange reactions. Essential to this formalism is the accurate characterization of the species present at the mineral-aqueous fluid interface.

This review also showed a variety of evidence linking the chemistry of the mineral-aqueous fluid interface to that of corresponding aqueous complexes. The stabilities of the complexes that form at the solid-solution interface either via sorption or exchange reactions, which can be viewed as the precursors of the TST activated complex, are generally closely linked to the stabilities of corresponding metal aqueous complexes, including polynuclear species. This analogy facilitates the interpretation and modeling of single and multi-oxides dissolution/precipitation rates as a function of solution pH and chemical composition (including organic ligands). Knowledge of the stability constants of aqueous metal hydroxide complexes formed by corresponding aqueous metal ions not only allows prediction of the pH dependence of various metal (hydr)oxides but also the pH dependence of the dissolution rates of many multi-oxides as illustrated in Figure 56. Furthermore, information on the enthalpy of aqueous metal hydrolysis or complexation with various ligands may help us understand and model the dependence of mineral dissolution rates on temperature and aqueous solution composition.

Another outcome of this review is the demonstration that the dissolution/precipitation rates of numerous rock-forming minerals, including oxides, Ca-Mg silicates, carbonates and phosphates, is controlled by the hydrolysis of metal-oxygen bonds at the solid surface. Thereby, knowledge of the thermodynamics of solvent exchange and ligand substitution on corresponding metal aquo ions can provide quantitative information on the dissolution and precipitation rates and mechanisms of these minerals in solutions of various compositions. It is expected that these concepts can be coupled with new spectroscopic data on the solid-solution

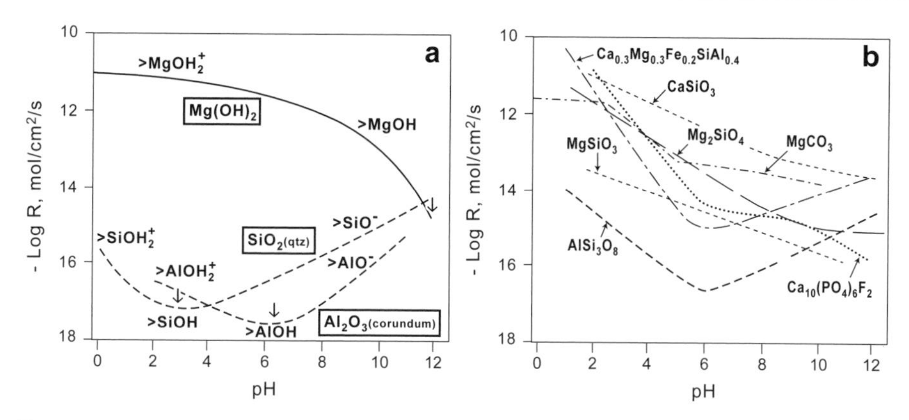

Figure 56. Illustration of the control exerted by metal hydrolysis affinity on the dissolution rates of single oxide (a) and multioxide (b) minerals as a function of solution pH. The pH_{zpc} of oxide/hydroxide minerals and the dominant species present at their surface are shown in Figure 56a.

interfaces and ab initio calculations of the energetic of relevant aqueous/surface reactions to create a comprehensive kinetics data base enabling real time modeling of reactive transport in natural and industrial systems.

ACKNOWLEDGMENTS

We would like to thank Pascale Bénézeth, Gleb Pokrovski, Jean-Louis Dandurand, Robert Gout, Sigurdur Gislason, and Stacey Callahan for helpful discussions during the course of this study. We are also grateful to all the PhD students and postdoctoral fellows of the Toulouse "Experimental Geochemistry" group who performed much of the experimental work analyzed in this study. Alain Castillo, Jean-Claude Harrichoury, and Carole Causserand provided technical assistance for much of the experimental work cited in this review. Christiane Cavaré-Hester is warmly thanked for her assistance in the realization of the numerous illustrations presented in this chapter. Support from Centre National de la Recherche Scientifique (CNRS), the Comité d'Etudes Pétrolières et Marines (CEP&M) through the project "Stockage de gaz acides dans les sites pétroliers," the ANR "Captage et Stockage du CO_2" through the project "GEOCARBONE-CARBONATATION" and the European Community through the MIR Early Stage Training Network (MEST-CT-2005-021120), MIN-GRO Research and training Network (MRTN-CT-2006-031482) and GRASP Marie Curie Research Training Network (MRTN-CT-2006-035868) is gratefully acknowledged.

REFERENCES

Aagaard P, Helgeson HC (1982) Thermodynamic and kinetic constraints on reaction rates among minerals and aqueous solutions: I. Theoretical considerations. Am J Sci 282:237-285

Abendroth RP (1970) Behavior of pyrogenic silica in simple electrolytes. J Colloid Interface Sci 34:591-596

Alkattan M, Oelkers EH, Dandurand JL, Schott J (1997) Experimental studies of halite dissolution kinetics. I. The effect of saturation state and the presence of trace metals. Chem Geol 137:201-221

Alkattan M, Oelkers EH, Dandurand JL, Schott J (1998) An experimental study of calcite and limestone dissolution rates as a function of pH from –1 to 3 and temperature from 25 to 80 °C. Chem Geol 151:199-214

Bales R, Morgan JJ (1985) Dissolution kinetics of crysotile at pH 7 to 10. Geochim Cosmochim Acta 49:2281-2290

Bénézeth P, Palmer DA, Wesolowsky DJ (2008) Dissolution/precipitation kinetics of boehmite and gibbsite: Application of a pH-relaxation technique to study near-equilibrium rates. Geochim Cosmochim Acta 72:2429-2453

Bennett PC, Melcer ME, Siegel DI, Hassett JP (1988) The dissolution of quartz in dilute aqueous solutions of organic acids at 25 °C. Geochim Cosmochim Acta 52:1521-1530

Berger G, Cadoré E, Schott J, Dove P (1994) Dissolution rate of quartz in Pb and Na electrolyte solutions. Effect of the nature of surface complexes and reaction affinity. Geochim Cosmochim Acta 58:541-551

Bloom PR, Erich MC (1987) Effect of solution composition on the rate and mechanism of gibbsite dissolution in acid solutions. Soil Sci Soc Am J 51:1131-1136

Blum A, Lasaga AC (1987) Monte Carlo simulations of surface reaction rate laws. *In:* Aquatic Surface Chemistry. Stumm W (ed) John Wiley, New York, p 255-292

Boslough MB, Cygan RT (1988) Shock-enhanced dissolution of silicate minerals and chemical weathering on planetary surfaces. Proc Eighteenth Lunar and Planet Sci Conf 443-453

Boudart A (1976) Consistency between kinetics and thermodynamics. J Phys Chem 80:2869-2870

Brantley SL, Crane SR, Crerar DA, Hellmann R, Stallard R (1986) Dissolution at dislocation etch pits in quartz. Geochim Cosmochim Acta 50:2349-2361

Brantley SL, Chen Y (1995) Chemical weathering rates of pyroxenes and amphiboles. Rev Mineral 31:119-172

Brantley SL, Stilling LL (1996) Feldspar dissolution at 25 °C and low pH. Am J Sci 296:101-127

Brantley SL (2008) Kinetics of mineral dissolution. *In:* Kinetics of Water-Rock Interaction. Brantley SL, Kubicki JD, White AF (eds) Springer, p 151-210

Brady PV, Walther JV (1990) Kinetics of quartz dissolution at low temperatures. Chem Geol 82:253-264

Burch TE, Nagy KL, Lasaga AC (1993) Free energy dependence of albite dissolution kinetics at 80 °C, pH 8.8. Chem Geol 105:137-162

Burgess J (1988) Ions in Solutions: Basic Principles of Chemical Interactions. Ellis Horwood

Busenberg E, Plummer LN (1982) The kinetics of dissolution of dolomite in CO_2-H_2O systems at 1.5 to 65 °C and 0 to 1 atm pCO_2. Am J Sci 282:45-78

Cadoré E (1995) Mécanismes de dissolution du quartz dans les solutions naturelles. Etude expérimentale et modélisation. PhD dissertation, Université Paul Sabatier, Toulouse, France, 141 p

Cama J, Ganor J, Ayora C, Lasaga AC (2000) Smectite dissolution kinetics at 80 °C and pH 8.8. Geochim Cosmochim Acta 64:2701-2717

Carroll SA, Knauss KG (2005) Dependence of labradorite dissolution kinetics on CO_2(aq), Al(aq) and temperature. Chem Geol 217:213-225

Carrol-Webb SA, Walther JV (1988) A surface complexation model for the pH dependence of corundum and kaolinite dissolution rates. Geochim Cosmochim Acta 52:2609-2623

Charlet L, Wersin P, Stumm W (1990) Surface charge of $MnCO_3$ and $FeCO_3$. Geochim Cosmochim Acta 54:2329-2336

Casey W, Carr MJ, Graham RA (1988) Crystal defects and the dissolution kinetics of rutile. Geochim Cosmochim Acta 52:1545-1556

Casey WH, Sposito G (1992) On the temperature dependence of mineral dissolution rates. Geochim Cosmochim Acta 56:3825-3830

Casey WH, Ludwig C (1995) Silicate mineral dissolution as a ligand-exchanged reaction. In: White AF, Brantley SL (eds) Chemical Weathering Rates of Silicates Minerals. Rev Mineral 31:87-117

Casey WH, Swaddle TW (2003) Why small? The use of small inorganic clusters to understand mineral surface and dissolution reactions in geochemistry. Rev Geophys 41:1008

Chen Y, Brantley SL (1997) Temperature and pH-dependence of albite dissolution rate at acid pH. Chem Geol 135:275-291

Chen Y, Brantley SL (1998) Diopside and anthophyllite dissolution at 25 °C and 90 °C and acid pH. Chem Geol 147:233-248

Chen Y, Brantley SL, Ilton ES (2000) X-ray photoelectron spectroscopic measurements of the temperature dependence of leaching of cations from the albite surface. Chem Geol 163:115-128

Chou L, Wollast R (1985) Steady state kinetics and dissolution mechanisms of albite. Am J Sci 285:963-993

Chou L, Garrels RM, Wollast R (1989) Comparative study of the dissolution kinetics and mechanisms of carbonates in aqueous solutions. Chem Geol 78:269-282

Devidal JL, Schott J, Dandurand JL (1997) An experimental study of kaolinite dissolution and precipitation kinetics as a function of chemical affinity and solution composition at 150 °C, 40 bars, and pH 2, 6.8 and 7.8. Geochim Cosmochim Acta 61:5165-5186

Dietzel M, Bohme G (2004) The dissolution rates of gibbsite in the presence of chloride, nitrate, silica, sulphate, and citrate in open and closed systems at 20 °C. Geochim Cosmochim Acta 69:1199-1211

Dove PM (1994) The dissolution kinetics of quartz in sodium chloride solutions at 25 to 300 °C. Am J Sci 294:665-712

Dove PM, Crerar DA (1990) Kinetics of quartz dissolution in electrolyte solutions using a hydrothermal mixed flow reactor. Geochim Cosmochim Acta 54:955-970

Dove PM, Elston SF (1992) Dissolution kinetics of quartz in sodium chloride solutions: Analysis of existing data and a rate model for 25 °C. Geochim Cosmochim Acta 56:4147-4156

Dromgoole EL, Walter LM (1990) Inhibition of calcite growth rates by Mn^{2+} in $CaCl_2$ solutions at 10, 25, and 50 °C. Geochim Cosmochim Acta 54:2991-3000
Ducommum Y, Newman KE, Merbach AE (1980) High-pressure NMR-kinetics. 11. High-pressure ^{17}O-labelled NMR evidence for a gradual mechanistic changeover from IA to ID for water exchange on divalent octahedral metal-ions going from manganese(II) to nickel(II). Inorg Chem 19:3696-3703
Ducommum Y, Zbinden D, Merbach AE (1982) High-pressure NMR-kinetics. 15. High-pressure ^{17}O-labelled NMR-study of vanadium(II) in water: a second example of an associative interchange mechanism (IA) for solvent exchange on an octahedral divalent transition-metal ion. Helv Chim Acta 65:1385-1390
Duval Y, Mielczarski JA, Pokrovsky OS, Mielczarski E, Ehrhardt JJ (2002) Evidence of the existence of three types of species at the quartz-aqueous solution interface at pH 0-10: XPS surface group quantification and surface complexation modeling. J Phys Chem B 106:2937-2945
Eyring H (1935) The activated complex in chemical reactions. J Chem Phys 3:107-115
Fenter P, Geissbühler P, DiMasi E, Srager G, Sorensen LB, Sturchio NC (2000) Surface speciation of calcite observed *in situ* by high-resolution X-ray reflectivity. Geochim Cosmochim Acta 64:1221-1228
Flukiger F, Bernard D (2009) A new numerical model for pore scale dissolution of calcite due to CO_2 saturated water flow in 3D realistic geometry: Principles and first results. Chem Geol (in press)
Furrer G, Stumm W (1986) The coordination chemistry of weathering: I Dissolution kinetics of δ-Al_2O_3 and BeO. Geochim Cosmochim Acta 50:1847-1860
Gautelier M, Oelkers EH, Schott J (2007) An experimental study of dolomite dissolution rates at 80° C as a function of chemical affinity and solution composition. Chem Geol 242:509-517
Gautier JM (1999) Etude expérimentale et modélisation de la cinétique de dissolution et de cristallisation des silicates en milieu hydrothermal: cas du quartz et du feldspath potassique. PhD dissertation, Université Paul Sabatier, Toulouse, France, 180 p
Gautier JM, Oelkers EH, Schott J (1994) Experimental study of K-feldspar dissolution rates as a function of chemical affinity at 150 °C and pH 9. Geochim Cosmochim Acta 58:4549-4560
Gautier JM, Oelkers EH, Schott J (2001) Are quartz dissolution rates proportional to B.E.T. surface areas? Geochim Cosmochim Acta 65:1059-1070
Gislason SR, Oelkers EH (2003) The mechanism, rates, and consequences of basaltic glass dissolution. II. An experimental study of the dissolution rates of basaltic glass as a function of pH at temperatures from 6 °C to 150 °C. Geochim Cosmochim Acta 67:3817-3832
Golubev SV, Pokrovsky OS, Schott J (2005) Experimental determination of the effect of dissolved CO_2 on the dissolution kinetics of Mg and Ca silicates at 25 °C. Chem Geol 217:227-238
Golubev SV, Pokrovsky OS (2006) Experimental study of the effect of organic ligands on diopside dissolution kinetics. Chem Geol 235:377-389
Gregory DP, Riddiford AC (1956) Transport to the surface of a rotating disc. J Chem Soc 3:3756-3764
Guy C, Schott J (1989) Multisite surface reaction versus transport control during the hydrolysis of a complex oxide. Chem Geol 78:181-204
Hellmann R (1994) The albite-water system: Part I. The kinetics of dissolution as a function of pH at 100, 200 and 300 °C. Geochim Cosmochim Acta 58:595-611
Hellmann R, Tisserand D (2006) Dissolution kinetics as a function of the Gibbs free energy of reaction: An experimental study based on albite feldspar. Geochim Cosmochim Acta 70:364-383
Hiemstra T, Van Riemsdijk WH (1990) Multiple activated complex dissolution of metal(hydr)oxides: A thermodynamic approach applied to quartz. J Colloid Interf Sci 136:132-149
Knauss KG, Wolery TJ (1986) Dependence of albite dissolution kinetics on pH and time at 70 °C. Geochim Cosmochim Acta 50:2481-2497
Kulik DA (2009) Thermodynamic concepts in modeling sorption at the mineral-water interface. Rev Mineral Geochem 70:125–180
Lasaga AC (1981) Transition State Theory. Rev Mineral 8:135-169
Lasaga AC (1995) Fundamental approaches in describing mineral dissolution and precipitation rates. Rev Mineral 31:23-86
Lincoln SF, Merbach AE (1995) Substitution reactions of solvated ions. *In:* Advances in Inorganic Chemistry. Sykes AG (ed) Academic Press San Diego, 42:1-88
Luttge A, Winkler U, Lasaga AC (2003) Variations in carbonate mineral dissolution rates: A new conceptual model for mineral dissolution. Geochim Cosmochim Acta 67:1099-1116
Maher K, Steelel CI, White AF, Stonestrom DA (2009) The role of reaction affinity and secondary minerals in regulating chemical weathering rates at the Santa Cruz Soil Chronosequence, California. Geochim Cosmochim Acta 73:2804-2831
Margerun DW, Cayley GR, Weatherburn DC, Pagenkopf GK (1978) Kinetics and mechanisms of complex formation and ligand exchange. *In:* Coordination Chemistry, vol 2., Martell A (ed) Am Chem Soc, Washington, DC. ACS Monograph 174:1-220
Mast MA, Drever JI (1987) The effect of oxalate on the dissolution rates of oligoclase and tremolite. Geochim Cosmochim Acta 51:2559-2568

Morse JW, Berner RA (1974) Dissolution kinetics of calcium carbonate in sea water: II. A kinetic origin for the lysocline. Am J Sci 272:840-851

Murphy WM (1989) Dislocations and feldspar dissolution. Eur J Mineral 1:315-328

Murphy WM, Helgeson HC (1989) Thermodynamic and kinetic constraints on reaction rates among minerals and aqueous solutions. IV. Retrieval of rate constants and activation parameters for the hydrolysis of pyroxene, wollastonite, olivine, andalusite, and quartz. Am J Sci 289:17-101

Nagy KL, Blum AE, Lasaga AC (1991) Dissolution and precipitation kinetics of kaolinite at 80 °C and pH 3: The dependence on solution saturation state. Am J Sci 291:649-686

Nagy KL, Lasaga AC (1992) Dissolution and precipitation kinetics of gibbsite at 80 °C and pH 3: The dependence on solution saturation state. Geochim Cosmochim Acta 56:3093-3111

Nangia S, Garrison BJ (2008) Reaction rates and dissolution mechanisms of quartz as a function of pH. J Phys Chem A 112:2027-2033

Oelkers EH (1996) Summary and review of the physical and chemical properties of rocks and fluids. Rev Mineral 34:131-191

Oelkers EH (2001) General kinetic description of multioxide silicate mineral and glass dissolution. Geochim Cosmochim Acta 65:3703-3719

Oelkers EH, Schott J, Devidal JL (1994) The effect of aluminum, pH, and chemical affinity on the rates of aluminosilicate dissolution reactions. Geochim Cosmochim Acta 58:2011-2024

Oelkers EH, Schott J (1995) Experimental study of anorthite dissolution and the relative mechanism of feldspar hydrolysis. Geochim Cosmochim Acta 59:5039-5053

Oelkers EH, Schott J (1998) Does organic adsorption affect alkali-feldspar dissolution rates? Chem Geol 151:235-245

Oelkers EH, Schott J (2001) An experimental study of enstatite dissolution rates as a function of pH, temperature, and aqueous Mg and Si concentration, and the mechanism of pyroxene/pyroxenoid dissolution. Geochim Cosmochim Acta 65:1219-1231

Oelkers EH, Gislason SR (2001) The mechanism, rates, and consequences of basaltic glass dissolution: I. An experimental study of the dissolution rates of basaltic glass as a function of aqueous Al, Si, and oxalic acid concentration at 25° C and pH = 3 and 11. Geochim Cosmochim Acta 65:3671-3681

Oelkers EH, Schott J, Gautier JM, Herrero-Roncal T (2008) An experimental study of the dissolution mechanism and rates of muscovite. Geochim Cosmochim Acta 72:4948-4961

Oelkers EH, Golubev SV, Chaïrat C, Pokrovsky OS, Schott J (2009) The surface chemistry of multi-oxide silicates. Geochim Cosmochim Acta (in press)

Patterson CS, Busey RH, Mesmer RE (1984) Second ionization of carbonic acid in NaCl media to 250 °C. J Solution Chem 13:647-661

Phillips BL, Casey WH, Neugebauer-Crawford S (1997a) Solvent exchange in $AlF_x(H_2O)_{6-x}{}^{3-x}$(aq) complexes: Ligand-directed labilization of water as an analogue for ligand-induced dissolution of oxide minerals. Geochim Cosmochim Acta 61:3041-3049

Phillips BL, Neugebauer-Crawford S, Casey WH (1997b) Rate of water exchange between $Al(C_2O_{4)}(H_2O)_4{}^+$(aq) complexes and aqueous solutions determined by ^{17}O-NMR spectroscopy. Geochim Cosmochim Acta 61:4965-4973

Plummer LN, Wigley TML, Parkhurst DL (1978) The kinetics of calcite dissolution in CO_2-water system at 5° to 60 °C and 0.0 to 1.0 atm CO_2. Am J Sci 278:179-216

Pokrovski GS, Schott J, Harrichoury JC, Sergeyev AS (1996) The stability of aluminum silicate complexes in acidic solutions from 25 to 150 °C. Geochim Cosmochim Acta 60:2495-2501

Pokrovski GS, Schott J, Salvi S, Gout R, Kubicki JD (1998) Structure and stability of aluminum-silica complexes in neutral to basic solutions. Experimental study and molecular orbital calculations. Mineral Mag 62A:1194-1195

Pokrovski GS, Schott J, Hazemann JL, Farges F, Pokrovsky OS (2002) An X-ray absorption fine structure and nuclear magnetic resonance spectroscopy study of gallium-silica complexes in aqueous solution. Geochim Cosmochim Acta 66:4203-4422

Pokrovsky OS, Schott J (1999) Processes at the magnesium-bearing-bearing carbonates/solution interface. II. Kinetics and mechanism of magnesite dissolution. Geochim Cosmochim Acta 63:881-897

Pokrovsky OS, Schott J, Thomas F (1999a) Processes at the magnesium-bearing interface. I. A surface speciation model for magnesite. Geochim Cosmochim Acta 63:863-880

Pokrovsky OS, Schott J, Thomas F (1999b) Dolomite surface speciation and reactivity in aquatic systems. Geochim Cosmochim Acta 63:3133-3143

Pokrovsky OS, Schott J (2000a) Forsterite surface composition in aqueous solutions: A combined potentiometric, electrokinetic, and spectroscopic approach. Geochim Cosmochim Acta 64:3299-3312

Pokrovsky OS, Schott J (2000b) Kinetics and mechanism of forsterite dissolution at 25 °C and pH from 1 to 12. Geochim Cosmochim Acta 64:3313-3325

Pokrovsky OS, Mielczarski JA, Barres O, Schott J (2000) Surface speciation models of calcite and dolomite/aqueous solution interfaces and their spectroscopic evaluation. Langmuir 16:2677-2688

Pokrovsky OS, Schott J (2001) Kinetics and mechanism of dolomite dissolution in neutral to alkaline solutions revisited. Am J Sci 301:597-626

Pokrovsky OS, Schott J (2002) Surface chemistry and dissolution kinetics of divalent metal carbonates. Environ Sci Technol 36:426-432

Pokrovky OS, Schott J, Mielczarski JA (2002) Surface speciation of dolomite and calcite in aqueous solution. In: Encyclopedia of Surface and Colloidal Science, Marcel Dekker, New York, p 5081-5095

Pokrovsky OS, Schott J (2004) Experimental study of brucite dissolution and precipitation in aqueous solution: Surface speciation and chemical affinity control. Geochim Cosmochim Acta 68:31-45

Pokrovsky OS, Schott J, Castillo A (2005) Kinetics of brucite dissolution in the presence of organic and inorganic ligands and divalent metals. Geochim Cosmochim Acta 69:905-918

Pokrovsky OS, Golubev SV, Schott J, Castillo A (2009) Calcite, dolomite and magnesite dissolution kinetics in aqueous solutions at acid to circumneutral pH, 25 to 150 °C and 1 to 55 atm pCO_2: New constraints on CO_2 sequestration in sedimentary basins. Chem Geol 260, doi:10.1016/j.chemgeo.2009.01.013

Pokrovsky OS, Shirokova LS, Bénézeth P, Schott J, Golubev SV (2009) Effect of organic ligands and herotropphic bacteria on wollastonite dissolution kinetics. Submitted to Am J Sci

Saldi GD, Köhler SJ, Marty N, Oelkers EH (2007) Dissolution rates of talc as a function of solution composition, pH and temperature. Geochim Cosmochim Acta 71:3446-3457

Saldi GD, Schott J, Pokrovsky OS, Gautier Q, Oelkers EH (2009) Experimental study of magnesite precipitation kinetics as a function of CO_2 pressure and solution chemistry. Submitted to Geochim Cosmochim Acta

Salvi S, Pokrovski GS, Schott J (1998) Experimental investigation of aluminum-silica complexing at 300 °C. Chem Geol 151:51-67

Samson SD, Eggleston CM (1998) Active sites at the non-steady-state dissolution of hematite. Environ Sci Technol 32:2871-2875

Samson SD, Stillings LL, Eggleston CM (2000) The depletion and regeneration of dissolution-active sites at the mineral-water interface: I. Fe, Al, and In sesquioxides. Geochim Cosmochim Acta 64:3471-3484

Schindler PW, Kamber HR (1968) Die Acidität von Silanolgruppen. Helv Chim Acta 51:1781-1786

Schindler PW, Stumm W (1987) The surface chemistry of oxides, hydroxides, and oxide minerals. *In:* Aquatic Surface Chemistry. Stumm W (ed) John Wiley, New York, p 83-110

Schott J, Berner RA, Sjöberg EL (1981) Mechanism of pyroxene and amphibole weathering: I. Experimental studies of iron-free minerals. Geochim Cosmochim Acta 45:2123-2135

Schott J, Brantley S, Crerar D, Guy C, Borcsik M, Willaime C (1989) Dissolution kinetics of strained calcite. Geochim Cosmochim Acta 53:373-382

Schott J, Oelkers EH (1995) Dissolution and crystallization rates of silicate minerals as a function of chemical affinity. Pure Appl Chem 67:903-910

Schott J, Pokrovsky OS, Golubev SV, Rochelle CA (2004) Surface coordination theory and the dissolution/ precipitation rates of carbonate minerals in a wide range of T, pCO2 and solution composition. Geochim Cosmochim Acta 68:A142

Schott J, Pokrovsky OS, Spalla O, Devreux F, Mielczarski JA (2002) The mechanism of altered layer formation on wollastonite revisited: a combined spectroscopic/kinetic study. Geochim Cosmochim Acta 66:A686

Schott J, Goddéris Y, Roelandt C, Williams J, Brantley S, Pollard D, François L (2008) Are we getting close to a mechanistic description of weathering in the field? Geochim Cosmochim Acta 72:A839

Schweda P, Sjöberg, Södervall U (1997) Near-surface composition of acid-leached labradorite investigated by SIMS. Geochim Cosmochim Acta 61:1985-1994

Stillings LL, Brantley SL (1995) Feldspar dissolution at 25° C and pH 3 – Reaction stoichiometry and the effect of cations. Geochim Cosmochim Acta 59:1483-1496

Stipp SL, Hochella MF Jr (1991) Structure and bonding environments at the calcite surface observed with X-ray photoelectron spectroscopy (XPS) and low energy electron diffraction (LEED). Geochim Cosmochim Acta 55:1723-1736

Stumm W (1992) Chemistry of the Solid-Water Interface. John Wiley & Sons, New York, 428 p

Stumm W, Wieland E (1990) Dissolution of oxide and silicate minerals: rates depend on surface speciation. *In:* Aquatic Chemical Kinetics. Stumm W (ed) John Wiley, New York, p 367-400

Stumm W, Furrer G, Kunz B (1983) The role of surface coordination in precipitation and dissolution of mineral phase. Croat Chem Acta 58:593-611

Stumm W, Furrer G, Wieland E, Zinder B (1985) The effects of complex-forming ligands on the dissolution of oxides and aluminosilicates. *In:* The Chemistry of Weathering. NATO ASI Series vol C149. Drever JI (ed) Kluwer Academic Publishers, Dordrecht, p 55-74

Sullivan DJ, Nordin JP, Phillipps BL, Casey WH (1999) The rates of water exchange in Al(III)-salicylate and Al(III)-sulfosalicylate complexes. Geochim Cosmochim Acta 63:1471-1480

Taylor AS, Blum JD, Lasaga AC (2000) The dependence of labradorite dissolution and Sr isotope release rates on solution saturation state. Geochim Cosmochim Acta 64:2389-2400

Van Cappellen P, Charlet L, Stumm W, Wersin P (1993) A surface complexation model of the carbonate mineral-aqueous interface. Geochim Cosmochim Acta 57:3505-3518

Villegas-Jiménez A, Mucci A, Pokrovsky OS, Schott J (2009) Defining reactive sites on hydrated mineral surfaces: rhombohedral carbonate minerals. Geochim Cosmochim Acta (in press)

Wehrhli B (1990) Redox reactions of metal ions at mineral surfaces. *In:* Aquatic Chemical Kinetics. Stumm W (ed) John Wiley, New York, p 311-336

Westrich HR, Cygan RT, Casey WH, Zemitis C, Arnold GW (1993) The dissolution of mixed-cation orthosilicate minerals. Am J Sci 293:869-893

White AF, Brantley SL (2003) The effect of time on the weathering of silicate minerals: why do weathering rates differ in the laboratory and field? Chem Geol 202:479-506

Wieland E, Werhli B, Stumm W (1988) The coordination chemistry of weathering: III. A generalization on the dissolution rates of minerals. Geochim Cosmochim Acta 52:1969-1981

Wogelius RA, Walther JV (1991) Forsterite dissolution kinetics at 25 °C: Effects of pH, CO_2 and organic acids. Geochim Cosmochim Acta 55:943-954

Wogelius RA, Walther JV (1992) Forsterite dissolution kinetics at near-surface conditions. Chem Geol 97:101-112

Wolff-Boenisch D, Gislason SR, Oelkers EH, Putnis CV (2004a) The dissolution rates of natural glasses as a function of their composition at pH 4 and 10.6, and temperatures from 25 to 74 °C. Geochim Cosmochim Acta 68:4843-4858

Wolff-Boenisch D, Gislason SR, Oelkers EH (2004b) The effect of fluoride on the dissolution rates of natural glasses at pH 4 and 25 °C. Geochim Cosmochim Acta 68:4571-4582.

Wollast R, Chou L (1988) Rate control of weathering of silicate minerals at room temperature and pressure. *In:* Physical and Chemical Weathering in Geochemical Cycles. NATO ASI Series vol C251. Lerman A, Meybeck M (eds) Kluwer Academic Publishers, Dordrecht, p 11-32

Wollast R (1990) Rate and mechanism of dissolution of carbonates in the system $CaCO_3$-$MgCO_3$. *In:* Aquatic Chemical Kinetics. Stumm W (ed) John Wiley, New York, p 431-445

Wolthers M, Charlet L, Van Cappellen PV (2008) The surface chemistry of divalent metal carbonate minerals; A critical assessment of surface charge and potential data using the charge distribution multi-site ion complexation model. Am J Sci 308:905-941

Xiao Y, Lasaga AC (1996) *Ab initio* quantum mechanical studies of the kinetics and mechanisms of quartz dissolution: OH^- catalysis. Geochim Cosmochim Acta 60:2283-2295

Zhu C (2009) Geochemical modeling of reaction paths and geochemical reaction networks. Rev Mineral Geochem 70:533–569

Zinder B, Furrer G, Stumm W (1986) The coordination of weathering: II. Dissolution of Fe(III) oxides. Geochim Cosmochim Acta 50:1861-1869

Reviews in Mineralogy & Geochemistry
Vol. 70 pp. 259-369, 2009

7

Organics in Water-Rock Interactions

Jiwchar Ganor, Itay J. Reznik and Yoav O. Rosenberg

Department of Geological and Environmental Sciences
Ben-Gurion University of the Negev
P. O. Box 653, Beer-Sheva 84105, Israel

contact e-mail: ganor@bgu.ac.il

INTRODUCTION

As reactive constituents, organic matter (OM) molecules interact with their surroundings. These interactions determine the fate of OM (e.g., degradation, mobilization), and may also alter the course and/or progress of different non organic reactions (e.g., mineral dissolution and precipitation). The goal of this chapter is to give an overview of these mutual effects between OM and water-rock interactions. The various interactions between organic molecules and minerals are of interest in a variety of scientific and engineering disciplines. Some of them are included in the classical framework of water-rock interactions (e.g., weathering processes, soil sciences, environmental geochemistry and petroleum geology). Other areas of interest include prevention of scale formation in oil production, geothermal energy conversion, desalination, and industrial water treatment (Amjad and Hooley 1986), pharmacy, rubber production, paper coatings, inks, and ceramics (Mortland 1986). Therefore, it is not surprising that the literature on the interactions between OM and minerals includes thousands of publications. The present chapter gives an overview of the interactions between OM and minerals and the mutual effects between OM and water-rock interactions. This overview reflects the perspective and understanding of the authors, and it does not represent all the important aspects of the field. As a result, a number of important studies are not included in this review.

The study of the origin of life is an example of a field of study that is not included in the present review, even though OM-mineral surface interactions have served as the cornerstone in this field. Following the hypothesis posed by Goldschmidt (1952 as cited in Schoonen et al. 2004), scientists have pursued the idea that mineral surfaces served as a catalyst for early prebiotic molecules. A recent review by Schoonen et al. (2004) discusses the mechanisms and possibilities with which different minerals might have promoted the synthesis of simple bio-molecules. Sorption, ion exchange, electron transfer, and photochemical reactions may all have served as mechanisms for mineral surface promoted reactions in the early synthesis of prebiotic molecules. In the present review these mechanisms are discussed from the standpoint of environmental and industrial processes.

As this chapter deals with OM-water-rock interactions it will begin with an introduction about organic compounds followed by their effects on solution chemistry and a short introduction on the characterization of the mineral-water interface. The main part of the review is divided into three subchapters dealing with the three main processes of water-rock interactions: sorption, dissolution and precipitation.

General classification, distribution and origin of organic compounds

OM comprises of chemical compounds whose molecules contain carbon. For historical reasons, a few types of compounds such as carbonates are considered inorganic. Typical organic molecules consist of a carbon bonded framework that contains different functional

1529-6466/09/0070-0007$15.00 DOI: 10.2138/rmg.2009.70.7

groups. Organic compounds are extremely diverse in their structure and functional groups. The classification and nomenclature of OM is usually based on the characteristics and size of the framework and on the number and type of the functional groups. A full classification and nomenclature can be found for example in Schwarzenbach et al.(2003). A list of the organic compounds that are included in the present review and their abbreviations is given in Table 1. In addition, Table 1 includes a summary of the main chemical characteristics of each OM and its role in water-rock interactions. Most organic compounds have more than one name. In the present review, we used the names as they appeared in the original paper that we cited or the abbreviations. Table 1 demonstrates the general importance of OM with carboxyl (-COOH) and hydroxyl (-OH) functional groups in water-rock interactions. Therefore, it is not surprising that organic acids are the most important group in this sense.

Natural organic matter (NOM) is abundant on the Earth surface and in near Earth surface environments and is mainly derived from natural biological activity. Soils, sediments, surface, and ground water contain a large variety of NOM. Although dissolved OM is present in all natural waters, their concentrations vary greatly in time and space and depend on the geological environment, seasonal variation, and biological and human activity. NOM contains identified compounds such as amino acids, fatty acids and nucleic acids, but also a mixture of macromolecules residues which have undergone *diagenesis* (partial degradation, rearrangement and recombination of the original molecules). This mixture is often divided into *humic substances* which are extractable in aqueous base and *humin* or *kerogen* which are not extractable in aqueous base.

Humic substances are the major organic constituents of soils and sediments and make up 60-70% of the total soil carbon content. They are derived from chemical and biological degradation of plant and animal residues and from synthetic activities of microorganisms. Humic substances are mixtures of many different organic molecules and therefore exhibit no specific physical and chemical characteristics (e.g., a sharp melting point, or an exact elementary composition). They are dark colored, acidic, predominantly aromatic, hydrophilic, chemically heterogeneous and complex, structurally complicated, polyelectrolyte-like substances that range in molecular weight from a few hundred to several thousand Dalton (Da) (Matthess 1984). Humic substances can be divided according to their behavior into *fulvic acids* which are soluble in both acidic and basic solutions and *humic acids* if they are not soluble at pH 2. The solubility in water, acidity and reducing properties decrease from fulvic acids to humin substances, and while molecular weight increases. Fulvic acids are comprised from numerous heterogeneous compounds that differ from humic acids and humins by having a lower molecular weight and a higher content of oxygen-containing functional groups such as COOH, OH and C = O.

The non-humic and humin substances in soils and sediments include carbohydrates, proteins, peptides, amino acids, fatty acids, waxes, and low molecular weight organic acids (e.g., oxalate, malonate). Carbohydrates account for 5-20% of soil organic matter. They are added to the soil in the form of plant tissues and are important constituents of soil microorganisms (Jackson et al. 1978).

The natural abundance of low molecular weight organic substances were summarized by Ullman and Welch (2002) and are given in Table 2. Detailed classification of organic compounds in a given environment is a hard task. Therefore, an analysis of total organic substances, referred as organic C content, is more common in the literature. The highest organic C contents are found in peat, lignite, coal and in oil and gas bearing deposits. A relatively high organic C content is found in top soils (usually in the range of 0.05-4.3 wt%). Typical ranges for organic C content in rocks are: shale 0.4-1.7 wt%, carbonate 0.15-0.5 wt%, and sandstone 0.04-0.4 wt%. OM usually increases with decreasing rock and soil grain size. In contrast to soils and sediments, the dominant OM in rivers and sea water is fulvic acid (Matthess 1984).

Table 1. A key table for the organic compounds names, abbreviation and their role in water-rock interaction.

Common name[a]/ abbreviation as in text	Other name/full name	Chemical properties	Role in water-rock interactions ([b]source)
2,4 dihydroxy-benzoic acid		aromatic compound, hydroxyl and carboxyl groups	catalytic effect on silicate dissolution (52)
3,4 dihydroxy-benzoic acid	See protocatechuic acid		
α-ketoglutaric acid (α-ketoglutarate)		dicarboxylic acid	catalytic effect on silicate dissolution (47)
acetic acid (acetate)		carboxylic acid	adsorption on silicate surfaces (9); OM degradation on silicate surfaces (1); catalytic effect on silicate dissolution (2,52,35,36,47); inhibition effect on silicate dissolution (51)
alanine		amino acid, amine and carboxyl functional groups	adsorption on mineral silicate (9); inhibition effect on carbonate dissolution (64)
albumin		protein, amine and carboxyl functional groups	inhibition effect on carbonate dissolution (63)
alginate		unbranched polysaccharides, hydroxyl and carboxyl groups	inhibition effect on silicate dissolution (51,56,57)
aliphatic amines		amines	adsorption on mineral silicate (9,21)
AMP	amino methylene phosphonic acid	phosphonate groups	inhibition effect on sulfate precipitation (96)
ascorbic acid		sugar acid, hydroxyl and carboxyl groups	catalytic effect on silicate dissolution (54)
aspartic acid		amino acid, amine and carboxyl functional groups	adsorption on silicate surfaces (18,19); catalytic effect on silicate dissolution (35,36); co-precipitation in carbonate minerals (8)
atrazine	2-chloro-4-(ethylamino)-6-(isopropylamino)-*s*-triazine	chlorinated triazine ring	photocatalysis of OM by oxide minerals(25)
benzoic acid (benzoate)		aromatic compound, carboxyl groups	catalytic effect on oxide dissolution (58)
BHTP	bishexamethylene-triaminetetra (methylene) phosphonic acid	phosphonate groups	inhibition effect on sulfate precipitation (80)
bisphenol A		raw materials of epoxy and polycarbonate resins	photocatalysis of OM by oxide minerals (24)
caffeic acid		aromatic moiety, hydroxyl and carboxyl groups	catalytic effect on silicate dissolution (37)
catechol		aromatic compound, hydroxyl groups	adsorption on oxide surfaces (13)
CE	cellulose ethers	polysaccharide, hydroxyl groups	inhibition effect on sulfate dissolution (105)
chondroitin sulfate		Sulfate groups	catalytic effect on hydroxyl-apatite precipitation (82)

Table 1. A key table for the organic compounds names, abbreviation and their role in water-rock interaction.

Common name[a]/ abbreviation as in text	Other name/full name	Chemical properties	Role in water-rock interactions ([b]source)
citric acid (citrate)		polycarboxylic acid	catalytic effect on silicate dissolution (36-39,42,43,47,49,51-53); catalytic effect on oxide dissolution (58); catalytic effect on silicate dissolution (2); co-precipitation in carbonate minerals (8)
CTAB	cetyltrimethyl-ammonium-bromide	cationic surfactant	inhibition effect on sulfate precipitation (81); effect on morphology and mineralogy of sulfates (81)
DDS	sodium dodecylsulfonate	sulfonate group	effect on morphology and mineralogy of carbonates (101)
DETPMP	diethylene-triamine-pentakis (methylene-phosphonic) acid	phosphonate groups	inhibition effect on carbonate precipitation (87)
DTPA	diethylene triamine pentaacetic acid	polycarboxyl acid	catalytic effect on sulfate dissolution (73-76)
EDTA	ethylenediamine-tetraacetic acid	poly amino carboxylic acid	catalytic effect on dissolution of silicates (51,53); carbonates (7) and sulfates (73,104); inhibition effect on mineral precipitation (94)
ENTMP	N,N,N',N'-ethylene-diaminetetra (methylene phosphonic acid)	phosphonate groups	inhibition effect on sulfate precipitation (93-94); effect on morphology and mineralogy of sulfates (94)
fatty acids		aliphatic monocarboxylic acids	adsorption on carbonate surfaces (12,17)
Fulvic acids			adsorption on oxide surfaces (14,16); ternary adsorption on surfaces of silicates (26) and oxides (29,30); catalytic effect on silicate dissolution (49,60); inhibition effect on precipitation of carbonates (98,103) and sulfates (99); effect on morphology and mineralogy of sulfates (99) and carbonates (103)
fumaric acid		dicarboxylic acid	inhibition effect on carbonate dissolution (66)
gallic acid		aromatic compound, hydroxyl and carboxyl groups	catalytic effect on silicate dissolution (37)
gluconic acid		aliphatic compound, hydroxyl and carboxyl groups	catalytic effect on silicate dissolution (46,52); inhibition effect on silicate dissolution (51,61)
glucosamine		amino sugar, amine and hydroxyl groups	inhibition effect on silicate dissolution (51)
glucuronic acid		carboxylic acid	inhibition effect on silicate dissolution (51)
glutamic acid		amino acid, amine and carboxyl functional groups	adsorption on silicate surfaces (9); inhibition effect on calcium oxalate precipitation (85), catalytic effect on calcium oxalate precipitation (85); co-precipitation in minerals (8)
glycine		amino acid, amine and carboxyl functional groups	inhibition effect on carbonate dissolution (64)

Table 1. A key table for the organic compounds names, abbreviation and their role in water-rock interaction.

Common name[a]/ abbreviation as in text	Other name/full name	Chemical properties	Role in water-rock interactions ([b]source)
HCBP	2,4,5,2',4',5'-hexachlorobiphenyl	polychlorinated biphenyl, an isomer of PCB	adsorption on silicate surfaces (31)
HDTMP	hexamethylenediamine-tetra (methylene phosphonic) acid	phosphonate groups	inhibition effect on sulfate precipitation (79)
HEDP	1- hydroxyethylidene-1,1-diphosphonic acid	phosphonate groups	inhibition effect on precipitation of sulfates (79,80,92,95,96) and carbonates (87)
Humic acids			adsorption on surfaces of oxides (14,23,33,34) and silicate (23); ternary adsorption on silicate surfaces (15,27,28); inhibition effect on carbonate dissolution (66); inhibition effect on carbonate precipitation (98); effect on morphology and mineralogy of sulfates (99)
L- arginine		amino acid, amine and carboxyl functional groups	inhibition effect on carbonate dissolution (64)
lysine		amino acid, amine and carboxyl functional groups	adsorption on silicate surfaces (9)
maleic acid		dicarboxylic acid	inhibition effect on carbonate dissolution (66,67)
malonic acid (malonate)		dicarboxylic acid	coordonative structure of adsorbed species on oxides (3); catalytic effect on dissolution of silicates (6,7)and oxides (58)
MDP	methylene diphosphonic acid	phosphonate groups	inhibition effect on sulfate precipitation (96)
myristic acid (myristate)		fatty acid	adsorption on carbonate surfaces (12)
NaPMA	polymethacrylate		inhibition effect on carbonate precipitation (83)
NTMP	nitro trimethyl phosphonic acid	phosphonate groups	inhibition effect on sulfate precipitation (96)
oxalic acid (oxalate)		dicarboxylic acid	adsorption on surfaces of silicates (11,19,20,31) and oxides (10); coordonative structure of adsorbed species on oxides (3-5); photocatalysis of OM by oxide minerals (24); catalytic effect on dissolution of silicates (6,11,20,37-41,44,45,47-53,55,59,61,62 and table 3) and oxides (58); inhibition effect on mineral dissolution (65)
PAA	polyacrylic acid (polyacrylates)	carboxylic groups, Various Mw	inhibition effect on precipitation of sulfates (77,84,88) and oxides (89); and carbonates(91); effect on morphology and mineralogy of carbonates (99) and sulfates (84)
PASP	polyaspartic acid	carboxyl groups	catalytic effect on carbonate dissolution (71,72); inhibition effect on carbonate dissolution (71)

Table 1. A key table for the organic compounds names, abbreviation and their role in water-rock interaction.

Common name[a]/ abbreviation as in text	Other name/full name	Chemical properties	Role in water-rock interactions ([b]source)
PBTC	sodium phosphonobutane-tricarboxylic acid	phosphonate groups	inhibition effect on sulfate precipitation (96)
PCAP	phosphino-polycarboxylic- acid polymer	polycarboxylate	inhibition effect on sulfate precipitation (79)
PGA	polyglutamic acid	carboxyl groups	inhibition effect on carbonate precipitation (78); catalytic effect on carbonate precipitation (78)
phenyl alanine		amino acid, amine and carboxyl functional groups	inhibition effect on carbonate dissolution (64)
phosphate ester		phosphonate groups	inhibition effect on sulfate precipitation (97)
phthalic acid (phthalate)		aromatic dicarboxylic acid	adsorption on oxide surfaces (13); catalytic effect on dissolution of silicates (6,53,54) and oxides (58); inhibition effect on carbonate dissolution (67)
p-hydroxybenzoic acid	4-hydroxybenzoic acid	aromatic compound, hydroxyl and carboxyl groups	adsorption on oxide surfaces (13); catalytic effect on silicate dissolution (37)
PMA-PVC		copolymer of maleic acid and vinyl sulphonic acid	inhibition effect on mineral precipitation (86); catalytic effect on sulfate precipitation (86)
polyacrylamide		amine groups	adsorption on silicate surfaces (22)
polygalacturonic acid		hydoxyl groups	catalytic effect on silicate dissolution (49); inhibition effect on silicate dissolution (56,57)
polymaleic acid		polycarboxylic acid	inhibition effect on carbonate dissolution (65)
polystyrene-maleic acid		aromatic moieties, carboxyl groups	inhibition effect on sulfate precipitation (90)
polystyrene-sulfonic acid		aromatic moieties, sulfonate functional groups	inhibition effect on sulfate precipitation (84)
PPPC	phosphinopolycarboxylic acid	polycarboxylate	inhibition effect on sulfate precipitation (79,80)
propionic acid (propionate)		carboxylic acid	catalytic effect on silicate dissolution (47)
protocatechuic acid	3,4 dihydroxybenzoic acid	aromatic compound, hydroxyl and carboxyl groups	catalytic effect on silicate dissolution (37)
PSA	co-polymer of PAA with polystyrene sulfonic acid	carboxyl groups, sulfonate groups	inhibition effect on sulfate precipitation (84); morphology and mineralogy of sulfates (84)
pyrophosphate		not an organic compound	inhibition effect on sulfate precipitation (77)
pyruvic acid (pyruvate)		carboxylic acid, keton group	catalytic effect on silicate dissolution (47)
resorcinol	Benzene-1,3-diol	aromatic compound, hydroxyl groups	adsorption on oxide surfaces (13)

Table 1. A key table for the organic compounds names, abbreviation and their role in water-rock interaction.

Common name[a]/ abbreviation as in text	Other name/full name	Chemical properties	Role in water-rock interactions ([b]source)
salicylic acid (salicylate)	2-hydroxybenzoic acid	aromatic compound, hydroxyl and carboxyl groups	adsorption on oxide surfaces (13); catalytic effect on dissolution of silicates (6,35-39,44) and oxides (58)
SDBS	sodium dodecylbenzensulfate	aromatic moiety, sulfonate group	inhibition effect on mineral precipitation (101); effect on morphology and mineralogy of carbonates (101)
SDS	sodium dodecyl sulfate	anionic surfactant, sulfate group	catalytic effect on precipitation of sulfates (81,102); inhibition effect on precipitation of carbonates (101) and sulfates (102); effect on morphology and mineralogy of carbonates (101) and sulfates (102)
sodium tetrametaphosphate		not an organic compound, cyclic polyphosphate	inhibition effect on carbonate precipitation (90)
sodium tetrapolyphosphate		not an organic compound, linear polyphosphate	inhibition effect on carbonate precipitation (90)
stearic acid	steric acid	fatty acid, carboxyl group	inhibition effect on carbonate dissolution (63); inhibition effect on carbonate precipitation (100)
succinic acid (succinate)		dicarboxylic acid	catalytic effect on dissolution of silicates (47,53) and oxides (58); inhibition effect on carbonate dissolution (67)
tannic acid		aromatic moieties, hydroxyl groups	catalytic effect on silicate dissolution (49,53)
tartaric acid		dicarboxylic acid, hydroxyl groups	catalytic effect on silicate dissolution (35); inhibition effect on carbonate dissolution (68)
TENTMP	N,N,N',N'-triethylene-diamine-tetra (methylene phosphonic acid)	phosphonate groups	inhibition effect on sulfate precipitation (93-94)
vanillic acid		aromatic compound, hydroxyl and carboxyl groups	catalytic effect on silicate dissolution (37)

[a]. the name of the acidic species is given together with the deprotonated species in parentheses.

[b]**Sources:** 1. Bell et al. (1994); 2. Hajash et al. (1998); 3. Persson and Axe (2005); 4. Yoon et al. (2004); 5. Axe and Persson (2001); 6. Chin and Mills (1991); 7. Knauss and Copenhaver (1995); 8. Phillips et al. (2005); 9. Wang and Lee (1993); 10. Fein and Brady (1995); 11. Cama and Ganor (2006); 12. Zullig and Morse (1988); 13. Gu et al. (1995); 14. Chi and Amy (2004); 15. Jada et al. (2006). 16. Kaiser et al. (1997); 17. Barcelona and Atwood (1979); 18. Carter (1978); 19. Poulson et al. (1997); 20. Stillings et al. (1998); 21. Baham and Sposito (1994); 22. Graveling et al. (1997); 23. Tombácz et al. (2004); 24. Li et al. (2007); 25. Lackhoff and Niessner (2002); 26. Theng (1976); 27. Theng and Scharpenseel (1975); 28. Liu and Gonzalez (1999); 29. Weng et al. (2005); 30. Weng et al. (2008); 31. Di Toro and Horzempa (1982); 32. Ganor et al. (2001); 33. Avena and Koopal (1998); 34. Avena and Koopal (1999); 35. Huang and Keller (1970); 36. Huang and Kiang (1972); 37. Manley and Evans (1986) ; 38. Bennett et al. (1988); 39. Blake and Walter (1999); 40. Bevan and Savage (1989); 41. Franklin et al. (1994); 42. Lundstrom and Ohman (1990); 43. Schweda (1989); 44. Wieland and Stumm (1992); 45. Stillings et al. (1996); 46. Vandevivere et al. (1994); 47. Welch and Ullman (1993); 48. Ganor and Lasaga (1994); 49. Zhang and Bloom (1999); 50. Oelkers and Gislason (2001); 51. Golubev et al. (2006); 52. Golubev and Pokrovsky (2006); 53. Grandstaff (1986); 54. Wogelius and Walther (1991); 55. Olsen and Rimstidt (2008); 56. Welch et al. (1999); 57. Welch and Vandevivere (1994); 58. Furrer and Stumm (1986); 59. Drever and Stillings (1997); 60. Schnitzer and Kodama (1976); 61. Welch and Ullman (2000); 62. Welch and Ullman (1996); 63. Suess (1970); 64. Hamdona et al. (1995); 65. Wilkins et al. (2001); 66. Compton et al. (1989); 67. Compton and Brown (1995); 68. Barwise et al. (1990); 69. Compton and Sanders (1993); 70. Fredd and Fogler (1998a); 71. Fredd and Fogler (1998b); 72. Burns et al.(2003); 73. Dunn and Yen, (1999); 74. Putnis et al., (1995b); 75. Putnis et al., (1995a); 76. Putnis et al., (2008); 77. Amjad and Hooley (1986); 78. Sarig and Kahana (1976); 79. He et al. (1994); 80. He et al. (1995); 81. Mahmoud et al. (2004); 82. Jiang (2005); 83. Williams and Ruehrwein (1957); 84. Lioliou (2006); 85. Azoury et al. (1983); 86. Van der Leeden (1993); 87. Jonasson, (1996); 88. Solomon and Rolfe (1966); 89. Baumgartner and Mijalchik (1991); 90. Smith and Alexander (1970); 91. Jada et al. (2007); 92. Weijnen and Van Rosmalen (1986); 93. Liu and Nancollas (1975); 94. Liu and Nancollas (1973); 95. Weijnen et al. (1983); 96. Becker et al. (2005); 97. El Dahan and Hegazy (2000); 98. Lebron and Suarez (1996); 99. Smith et al. (2004); 100. Mann et al. (1988); 101. Wei et al. (2005); 102. Rashad et al. (2003); 103. Hoch et al.(2000); 104. Bosbach et al. (1998); 105. Brandt and Bosbach (2001).

Table 2. Low molecular weights organic ligands in natural settings and some of their thermodynamic properties. All thermodynamic data from Smith and Martell (1998). [Used by permission of The Geochemical Society, From Ullman and Welch (2002), *Water-Rock Interactions, Ore Deposits, and Environmental Geochemistry: A Tribute to David A. Crearar*, Vol. 7, Table 1, pp. 8.].

Ligand (Acid Formula)	Concentrations	pK_a*	Al Complexes**
	Aliphatic Monocarboxylic Acids		
Formate (HCOOH)	Few hundred nmol/g in stone [(a)]; 5 to 174 μM in soil [(b)]; 10 to 20 μM in beech forest soil [(d)]; Few hundred nmol/g in soil extract [(e)]; 10 to 50 μM in groundwater [(f)]; 0.2 to 1.4 μM in forest floor solution [(g)]; 50 to 300 μmol/kg in soil leachate [(k)]	3.75	$\log \beta_1 = 1.36$ $\log \beta_2 = 2.02$ other complexes I = 1
Acetate (CH_3COOH)	10 to 20 μM in beech forest soil [(d)]; 140-1000 nmol/g in soil extract [(e)]; 10 to 50 μM in groundwater [(f)]; 0 to 50 μM in forest floor solution [(g)]; 50 μM to 250 mM in formation water [(h,j)]; 5 to 800 μmol/kg in soil leachate [(k)]	4.76	$\log \beta_1 = 1.51$ other complexes I = 1
Propionate (CH_3CH_2COOH)	Few μM in beech forest soil [(d)]; 30 μM to 24 mM in formation water [(h,j)]; 5 to 800 μmol/kg in soil leachate [(k)]	4.87	$\log \beta_1 = 1.69$ other complexes I = 1
Butyrate ($CH_3(CH_2)_2COOH$)	Up to 169 μM in formation water [(h,j)]; up to 10 μmol/kg in soil leachate [(k)]	4.82	$\log \beta_1 = 1.58$ (I = 1)
Valarate ($CH_3(CH_2)_3COOH$)	Up to 7 μM in formation waters [(j)]	4.83	No data
	Aliphatic Dicarboxylic Acids		
Oxalate (HOOCCOOH)	Few tens nmol/g in stone [(a)]; 25 to 1000 μM in soil [(b)]; 20 to 100 μmol/g dry sediment in reducing marine sediment [(c)]; Few μM in beech forest soil [(d)]; 40-100 nmol/g in soil extract [(e)]; 1-5 μM in forest floor solution [(g)]; up to 5 mM in formation water [(j)]; up to 200 μmol/kg in soil leachate [(k)]	1.25, 4.27	$\log \beta_1 = 5.97$ $\log \beta_2 = 10.93$ $\log \beta_3 = 14.88$ (I = 0.5)
Malonate ($HOOCCH_2COOH$)	Up to 25 mM in formation water [(j)]; up to 200 μmol/kg in soil leachate [(k)]	2.85, 5.70	$\log \beta_1 = 6.26$ $\log \beta_2 = 11.11$ $\log \beta_3 = 13.3$ (I = 0.15)
Succinate ($HOOC(CH_2)_2COOH$)	Few tens nmol/g in stone [(a)]; Few tens μmol/g dry sediment in reducing marine sediment [(c)]; 2 to 500 μM in formation water [(h)]; up to 200 μmol/kg in soil leachate [(k)]	4.21, 5.64	$\log \beta_1 = 3.20$ other complexes (I = 0.5)
Fumarate (HOOC-CH =CH-COOH)	Few tens nmol/g in stone [(a)]; up to 30 μmol/kg in soil leachate [(k)]	3.02, 4.48	No data
Malate ($HOOCCH_2CHOHCOOH$)	Few μM in beech forest soil [(d)]; Few nmol/g in soil extract [(e)]; 8 to 30 nmol/ml in soils and root leachate [(l)]; up to 30 μmol/kg in soil leachate [(k)]	3.46, 5.10	$\log \beta_1 = 4.60$ $\log \beta_2 = 7.62$ other complexes (I = 0.15)
	Aliphatic Tricarboxylic Acids		
Citrate ($COOH-CH_2-C(OH)COOH-CH_2-COOH$)	Few nmol/g in stone [(a)]; trace in soils [(b)]; Few tens (5 to 25) nmol/g in soil extract [(e)]; 0-0.13 μM in forest floor solution [(g)]; 70 to 600 nmol/ml in soils and root leachate [(l)]; up to 30 μmol/kg in soil leachate [(k)]	3.13, 4.76, 6.40	$\log \beta_1 = 7.14$ $\log \beta_2 = 12.9$ other complexes (I = 0.1)
Isocitrate ($COOH-CH_2-CH(COOH)-CHOH-COOH$)	Few μM in beech forest soil [(d)]	3.05, 4.30, 5.74 (I = 0.1)	No data

Aconitic (HOOC-CH=C-(COOH)CH_2-COOH)	Few μM in beech forest soil [d]; Few nmol/g in soil extract [e]; 1 to 20 nmol/ml in soils and root leachate [l]	2.8 4.5	No data
	α-Hydroxy- and α-Keto-carboxylic Acids		
Glycolate (CH_2OH-COOH)	<1 μmol/g dry sediment in reducing marine sediment [c]; up to 250 μmol/kg in leachate [k]	3.83	No data
Pyruvate ($CH_3CHOCOOH$)	Few tens nmol/g in stone [a]; Few μM in beech forest soil [d]; up to 40 μM/kg in soil leachate [k]	2.48	No data
Lactate ($CH_3CHOHCOOH$)	Few tens nmol/g in stone [a]; <1 μmol/g dry sediment in reducing marine sediment [c]; to 400 μmol/kg in soil leachate [k]	3.86	$\log \beta_1 = 2.36$ $\log \beta_2 = 4.42$ $\log \beta_3 = 5.8$ other complexes (I = 0.5)
Gluconate ($HOCH_2(CHOH)_4$ COOH)	Very reactive. Not reported in nature.	3.46, I = 0.1	$\log \beta_1 = 1.9$ (I = 0.5)
α-Ketoglutarate (HOOC-$(CH_2)_2$-(C = O)-COOH)	Very reactive. Not reported in nature.	1.9, 4.44, I = 0.5	No data
	Aromatic Carboxylic Acids		
Salicylic Acid (2-Hydroxybenzoic Acid)	2 - 5 nmol/g in rhizosphere [e]; 0 to 0.5 mM in formation waters [j]	13.7	$\log \beta_1 = 13.7$ ($I \leq 0.1$) $\log \beta_2 = 24.8$ ($I \leq 0.5$)
p-Hydroxybenzoic Acid	5 - 60 nmol/g in rhizosphere [e]	4.58, 9.46	No data
Protocatechuic Acid (3,4-Dihydroxybenzoic Acid)	1 – 5 nmol/g in rhizosphere [e]	4.49, 8.75, 13.0	$\log \beta_1 = 16.8$ $\log \beta_2 = 29.8$ $\log \beta_3 = 38.5$ other complexes (I = 0.1)
Caffeic Acid (3-(3,4-Dihydroxyphenyl) Propanoic Acid)	No concentration data found.	4.62, 9.07, 12.5, I = 0.1	No data
Gallic Acid (3,4,5-Trihydroxybenzoic Acid) (Note: Dissociation and stability constants refer to the singly protonated ligand. Full deprotonation is not observed)	No concentration data found.	4.4 9.11 11.4 ($I \leq 0.1$)	$\log \beta^*_1 = 6.4$ $\log \beta^*_2 = 9.0$ $\log \beta^*_3 = 7.4$ ($I \leq 0.1$)
Vanillic Acid (4-Hydroxy, 3-Methoxy-benzoic Acid)	1-14 nmol/g in rhizosphere [e]	No data	No data
	Other Aromatic Compounds		
Catechol (1,2-Dihydroxybenzene) (Note: Also forms strong complexes with Si; Smith and Martell, 1998)	No concentration data found.	9.45, 13.3	$\log \beta_1 = 16.7$ $\log \beta_2 = 30.35$ $\log \beta_3 = 39.4$ other complexes ($I \leq 0.1$)

• Acid dissociation constants at 25 °C and I = ionic strength = 0 unless noted

** Association constants for complexes of the form AlL_n where $\beta_n = [AlL_n]/[Al^{3+}][L]^n$. The stabilities of complexes of differing form (other complexes) are described in Smith and Martell (1998).

Concentrations reported by:

a) Palmer et al., 1991
b) Fox and Comerford, 1990
c) Peltzer and Bada, 1981
d) Shen et al., 1996
e) Baziramakenga et al., 1995
f) McMahon and Chapelle, 1991
g) Krzyszowska et al., 1996
h) Kharaka et al., 1986
i) Grierson, 1992
j) MacGowan and Surdam, 1988, 1990
k) Cordt and Kussmaul, 1992

In addition to natural occurring organic matter, organic chemicals of anthropogenic origin are introduced by man into the geochemical cycle. OM is present in sewage, industrial liquid and solid wastes. Synthetic polymers used to prevent scale formation (antiscalants) are increasingly used for industrial applications (e.g., desalinization, oil drillings) since the first successful scale suppression method (Rosenstein 1936). Pesticides and herbicides are commonly used in agriculture and forestry for the control of detrimental organisms.

The effect of OM on solution chemistry

OM may strongly affect solution chemical properties such as pH and Eh. In addition, dissolved species may adsorb onto OM and may form metal-organic complexes. Metal-organic complexes are formed when a metal cation is attached by direct bonding to electron-donor atoms other than carbon, e.g., oxygen, nitrogen and sulfur (Giordano 1994). These complexes between the metal and the organic ligand may be outer-sphere or inner-sphere complexes. In outer-sphere complexes water molecule(s) separate between the organic ligand and the metal, while in inner-sphere complexes the ligand is attached directly to the metal. As a result, the latter complexes are stronger. Chelation is a special case in which the metal ion is bound to two or more atoms of the ligand.

The complexation between organic ligand (L^{n-}) and a metal cation (M^{m+}) may be described by the equation:

$$L^{n-} + M^{m+} \Leftrightarrow LM^{(m-n)} \tag{1}$$

Sometimes the metal may be complexed by more than one ligand ion:

$$p \cdot L^{n-} + M^{m+} \Leftrightarrow L_p M^{(m-p \cdot n)} \tag{2}$$

where p is the number of ligand ions in the complex. The strength of the complex may be evaluated by the association constant of the complex:

$$\beta_i = \frac{a_{L_p M^{(m-p \cdot n)}}}{a^p_{L^{n-}} \cdot a_{M^{m+}}} \tag{3}$$

where β_i is the association constant for a complex with i ligands and a_j is the activity of species j in equilibrium. As an example, the log of the association constants of Al^{3+} with naturally occurring low molecular weight organic ligands are shown in Table 2. In general, complexes with ligands that form multidentate complexes (i.e., chelators) are more stable than those with ligands that form monodentate complexes. Also, multiprotic acids such as oxalic, citric and phthalic acids are usually much stronger complexing agents than monoprotic acids such as acetic and benzoic acids (Bennett and Casey 1994).

Both adsorption and chelation reduce the concentration of the free metal in the solution, and therefore affect water-rock interactions. For example, Figure 1 shows the change in the percentage of free Al^{3+} as a function of the total amount of oxalate in solution. The percentage of free Al^{3+} depends both on pH and total concentrations of aluminum and oxalate. In general, when the total amount of oxalate equals that of aluminum, less than 50% of the aluminum is free, and when the total amount of oxalate is an order of magnitude higher than that of aluminum, less than 1% of the aluminum is free.

The interactions between OM and dissolved metals and their effects on the mobility of cations were discussed in detail in the literature (e.g., Jackson et al. 1978). In the present review we will only discuss the effects of OM-metal complexation that influence water-rock interactions.

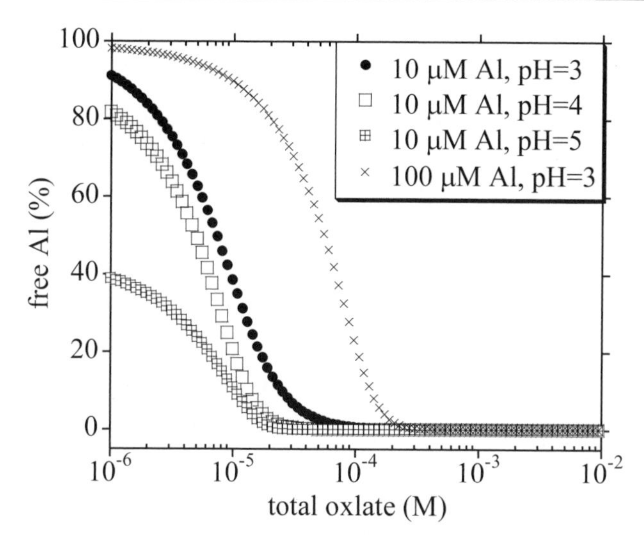

Figure 1. The effect of dissolved oxalate on the percentage of free aluminum in solution. The distribution of the aqueous species was calculated using the thermodynamic data in Table 4.

Characterization of the mineral-water interface

In order to understand the influence of OM on water-rock interactions it is essential to characterize the mineral-water interface. Mineral surface characterization should include the types of the mineral reactive sites, measurements or theoretical calculations of bulk surface area, surface charge, reactive site density and speciation.

Surface reactive sites form as a consequence of the termination of the three dimensional crystal lattice. This termination results in different properties of the surface compared to the bulk mineral (Koretsky 2000). Cations and anions forming the interface are coordinatively unsaturated (i.e., they are not fully bonded), therefore having an excess electrical charge. In contact with air and water these unsaturated cations and anions can either hydroxylate or protonate, to form a hydroxyl group (Fig. 2). Such hydroxyl groups are thought to be amphoteric, meaning that they may either protonate or deprotonate depending on the pH. This behavior can thus change the mineral surface charge. Hydroxyls form reactive sites on all mineral surfaces; these sites can then participate in the sorption process of inorganic and organic compounds and in mineral dissolution and precipitation reactions.

The reactivity of a given surface site depends on its chemical structure including the following factors: (1) the particular ion which is hydroxylated or protonated (e.g., Ca and Mg in dolomite or Al and Si in aluminosilicates); (2) the numbers of bonds with the mineral lattice (i.e., how much the site is coordinatively unsaturated). This property is reflected in the site geometry; and (3) the length of the cation-anion bonds in the lattice. Even equal coordinatively unsaturated sites may differ in the bond length (i.e., the bond *strength*) and therefore may exhibit different reactivity (Koretsky 2000).

It has long been recognized that mineral surfaces, even if they cut through the crystal cleavage, cannot be regarded as perfect planes. Any mineral interface has some topography and roughness and may also include impurities. For a given mineral, it is important to bear in mind that the type and density of surface reactive sites depend not only on the exposed crystallographic plane but also on its morphology (Graveling et al. 1997; Koretsky 2000, and references therein). At the microscopic level, mineral surfaces can be described as having steps

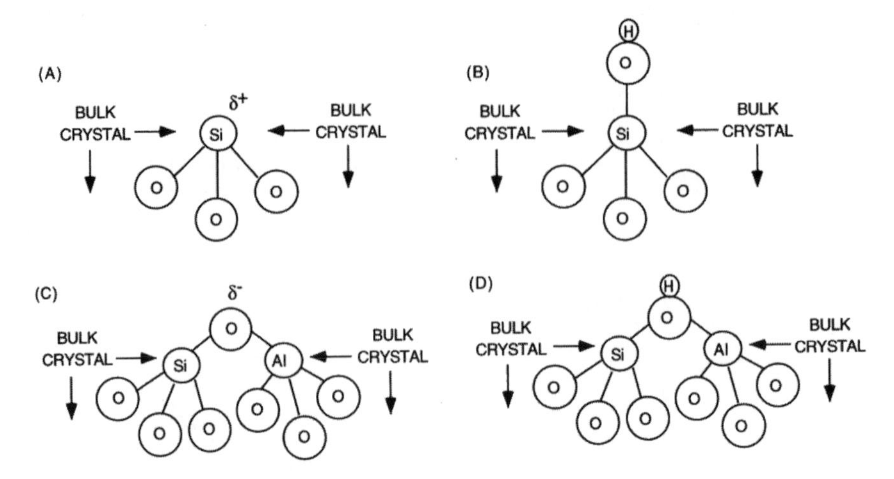

Figure 2. (A) Schematic representation of a three-fold coordinatively unsaturated silicon atom with a partial positive charge at the surface of a mineral such as quartz. (B) An isolated surface hydroxyl group is formed by the addition of a hydroxyl anion to the coordinatively unsaturated silicon atom depicted in (A). (C) Schematic representation of a two-fold coordinatively unsaturated oxygen atom with a partial negative charge at the surface of a mineral such as feldspar. (D) A bridging surface hydroxyl group is formed by protonation of the coordinatively unsaturated oxygen atom depicted in (C). [Used by permission of Elsevier, from Koretsky (2000), *Journal of Hydrology*, Vol. 230, Fig. 2, p. 132.]

and kinks of monomolecular height as presented in Figure 3 (Ives 1965). Apart from adatom sites which are the most thermodynamically unstable sites and therefore considered to be less abundant, kinks would be the most reactive surface sites. Mineral dissolution and precipitation progress mainly through attachment and detachment of ions at kink sites (Berner and Morse 1974; Nielsen 1981). Since kinks are also preferred sites for adsorption (Ives 1965) of both inorganic and organic compounds, adsorption to kinks may greatly affect mineral precipitation and dissolution reactions.

The overall *intrinsic surface charge* of a particle is comprised of two components of surface charge (Sposito 1989): (1) *permanent structural charge* is created by isomorphic substitutions, and especially produces significant surface charge in the 2:1 layer silicates, and (2) *net proton charge*, which is the difference between moles of protons and moles of hydroxide ions that are *complexed* by surface functional groups and is, therefore, pH dependent. The most important functional groups in soils that can protonate are located on the surfaces of hydrous oxides and 1:1 aluminosilicates. Aside of the *intrinsic surface charge* components, inner- and outer-sphere complexes of adsorbed ions other than protons may also contribute to the overall surface charge. In this context it is important to differentiate between the pH of the point of net zero proton charge (pH_{PNZPC}), which is when the *net proton charge* vanishes (i.e., when the sum of complexed protons and hydroxides equals zero) and the pH of net zero charge (pH_{PZC}). The latter is the pH at which the net *total* particle charge vanishes. pH_{PZC} includes *intrinsic surface charge* as well as adsorbed ions charges other than *proton charge*. The terms isoelectric point (IEP) and point of zero charge (PZC) are identical by definition (Parks 1965). A third term that is used in this context is the pH of the point of zero net charge (pH_{PZNC}), which is the point at which adsorbed ions charge other than *proton charge* vanishes (Sposito 1989).

Speciation of mineral surface sites is essential in determining which functional group may interact with which mineral surface group. Examples for mineral surface site speciation and modeling as a function of pH can be found in Wieland and Stumm (1992) for kaolinite, Fein and Brady (1995) for corundum, and Graveling et al. (1997) for quartz, kaolinite and

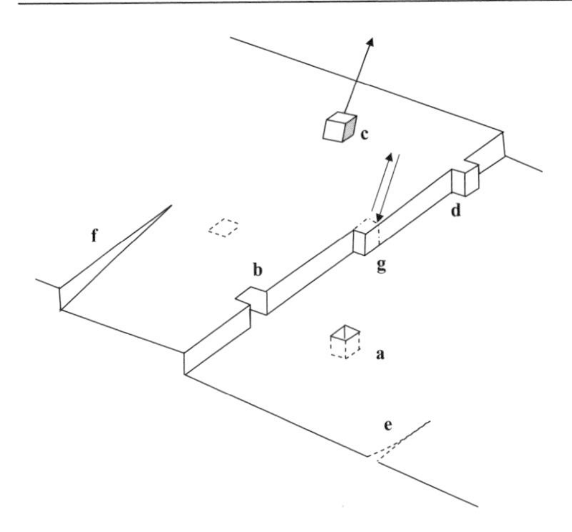

Figure 3. Diagram illustrating sites and defects on a mineral surface: (a) a vacancy on a terrace, (b) a vacancy or a kink-pairs at a step, (c) an adatom on a terrace which will tend to dissolve (upward arrow), (d) an adatom at a ledge or step, (e) an edge dislocation-surface intersection, (f) a screw dislocation-surface intersection, and (g) a kink site. Mineral dissolution and precipitation progress through detachment (upward arrow) or attachment (downward arrow) of an atom at the kink site, respectively. Thus, a new kink site is regenerated at either case. Modified after Lasaga (1998).

feldspar. For the latter mineral, Graveling et al. (1997) only semi-quantitatively predicted surface charge. The prediction of mineral surface charge and OM interaction (e.g., adsorption) can be improved by understanding the mineral site distribution, but the competition between different adsorption mechanisms remains an open question (Graveling et al. 1997). Some specific examples for surface reactive sites and their pH dependence are discussed later in the section on the effects of mineral surface properties on OM adsorption.

Experimental and measurement methods

The aim of the following section is to describe some of the methods used by investigators to study mineral surfaces, OM and the various interactions between them. It is important to bear in mind that NOM contains a large variety of organic compounds, having different sizes and functional groups. In order to separate variables, some investigators fractionate the NOM prior to their experiments. Such fractionations are made on the basis of molecular size and/or hydrophobic/hydrophilic character (e.g., Gu et al. 1995). It is important to emphasize that: (1) these fractionations are operational definitions, that is, the desired fraction is achieved and dependent on experimental techniques, (2) each fraction will still include, inevitably, a variety of organic compounds, and (3) fractionation of NOM is a natural process resulting from its adsorption (as will be discussed below). Therefore, artificial fractionation is aimed to discriminate between the different factors that affect this process in the environment.

Sterilization and OM degradation. Organic matter, especially aliphatic amines, amino acids, and fatty acids can be rapidly decomposed by bacteria and therefore compete with adsorption as a removal process. In order to ensure that the measurement of OM adsorption is not affected by biodegradation Wang and Lee (1993) and Zullig and Morse (1988) poisoned their solutions with $HgCl_2$ (100ppm and 5ppm, respectively), and Ganor et al. (2001) added an antibiotic mixture of penicillin-G, streptomycin, ampicillin and chloramphenicol. Wang and Lee (1993) also tested the possibility that $HgCl_2$ affects the adsorption of amino acids to clay minerals (kaolinite and montmorillonite) in sea water; at the maximum $HgCl_2$ concentration that was tested (2000 ppm); they concluded that it had an insignificant effect. Antibiotic was also used in order to differentiate between the catalytic effect of mineral surfaces on the decomposition of OM and biological degradation (Ganor et al. 2001). It is important to note that even by adding antibiotics to the experiment; one cannot prove that an observed effect is not influenced by microorganism activity as the microorganism may be antibiotic-resistant. In contrast, if an observed process is halted as antibiotic is added to the system, it is very probable that the process is affected by the presence of microorganisms (Ganor et al. 2001).

As will be shown below, the adsorption of OM affects the minerals surface area, and therefore it is essential to avoid degradation of OM when measuring the surface area of OM bearing samples. For example, Kaiser and Guggenberger (2000) degassed their samples (natural soils and synthetic amorphous $Al(OH)_3$ gel) at 25 °C under a continuous stream of helium to avoid both thermal degradation of adsorbed OM and conversion of amorphous mineral phases into more crystalline phases.

Pretreatment of minerals and their surfaces. Grinding and sieving mineral samples is usually the first procedure adopted in order to homogenize their surface area. However, the grinding also affects the surface properties of the minerals. Holdren and Berner (1979) showed that following grinding a large number of submicron particles are adhered to the surface of large grains. These fine particles may significantly affect the adsorption capability and kinetics of dissolution and precipitation. Dissolution of these ultrafine particles will result in the non-linear rates of reaction or parabolic kinetics (Holdren and Adams 1982). Therefore, Zullig and Morse (1988) sonicated their carbonate minerals samples in double distilled water (DDW) repeatedly (12 times) in order to dissolve and remove physically adhered submicron particles. Nagy et al. (1991) found that pretreatment of the mineral surfaces is also necessary to obtain stoichiometric dissolution of kaolinite. To avoid such effects, Cama and Ganor (2006) pretreated their samples with 1mM inorganic acid for several months at 70-80 °C, before measuring the effect of oxalate on kaolinite dissolution.

When investigating natural samples, such as river or sea sediments, a significant fraction of OM may already be adsorbed to the sediments at hand. The complete or partial removal of this fraction allows the investigator to study its role in the adsorption of additional OM. For example, Wang and Lee (1993) treated pond sediments with either peroxide or NaOH and compared the adsorption isotherms of the treated to untreated samples.

Measurements of surface charge. Acid-Base titration of mineral surfaces is used to determine the relative surface concentration of protons (i.e., *net proton charge*) via mass balance between added and measured protons (or hydroxides) according to:

$$\Delta C_S = (C_A - C_B - [H^+] + [OH^-])\frac{V}{A} \tag{4}$$

where ΔC_S is the change in surface proton concentration (mol m^{2-}), C_A and C_B are concentrations of acid and base added (mol L^-), respectively, $[H^+]$ and $[OH^-]$ are H^+ and OH^- solution concentrations at equilibrium (mol L^-), V is the solution volume (L) and A is the total mineral surface area (m^2). Surface titration curves are usually interpreted in terms of protonation and deprotonation of surface sites, although under low and high pH conditions other reaction (e.g., mineral dissolution, ion exchange and speciation in solution) may be responsible to significant proton consumption. Therefore, the proton mass balance should be corrected accordingly (e.g., Huertas et al. 1998). The interpretation of surface titration curves requires the choice of different models and assumptions which contain many degrees of freedom (e.g., Ganor et al. 2003). As a result, parameters estimated through surface titration curves may lack a true physical meaning. Ganor et al. (2003) discussed and demonstrated that differences in surface titration curves of kaolinite are the result of different ways in which the pH_{PZNPC} is determined (Fig. 4). It is important to bear in mind that until a value of zero charge is established, the value of the proton charge obtained by surface titration is arbitrary (Davis and Kent 1990). Different kaolinite titration curves published with different reference point (i.e., different pH_{PZNPC}, Fig 4a) were recalculated by Ganor et al. (2003) such that all curves were brought to pH_{PZNPC} of 5. This choice of pH_{PZNPC} represents an *absolute* value obtained by Schroth and Sposito (1997) and is also similar to a theoretical value estimated by Sverjensky and Sahai (1996). Figure 4b shows a good agreement between most of the recalculated curves demonstrating that pH_{PZNPC} in many studies is a relative definition.

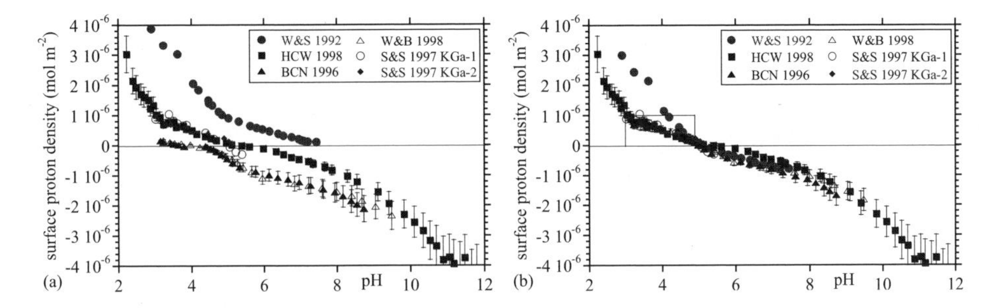

Figure 4. Surface titration curves of kaolinite from different studies: (a) comparison of the original curves at 25 °C; (b) recalculation of the data sets so that pH_{PZNPC} = 5. Source data: W&S 1992—(Wieland and Stumm 1992); BCN 1996—(Brady et al. 1996); HCW 1998—(Huertas et al. 1998); W&B 1998—(Ward and Brady 1998); S&S 1997—(Schroth and Sposito 1997). [Used by permission of Elsevier, from Ganor et al. (2003), *Journal of Colloid and Interface Science*, Vol. 264, Fig. 1, p. 68.]

Surface charge and pH_{PZC} can be determined by means of microelectrophoresis. In this method the particles move in an aqueous medium to which an electric voltage is applied producing a uniform electric field. The electrophoretic movement is directly dependent on the particle surface charge (Jada and Ait Chaou 2003).

Organic matter properties. OM molecular weight, functional groups, light absorptivity and charge are key factors affecting the molecule fate and adsorption affinity. For a mix of molecules (e.g., NOM) two molecular weights are commonly defined: *number-average molecular weight* (M_n) and *weight-average molecular weight* (M_w) (Cabaniss et al. 2000). These values can be calculated by supposing that the molecules are divided into N fractions according to molecular weight, the ith fraction having a characteristic molecular weight, M_i, and a characteristic molar frequency, f_i.:

$$M_n = \frac{\sum_{i=1}^{N} f_i M_i}{\sum_{i=1}^{N} f_i} \tag{5}$$

$$M_W = \frac{\sum_{i=1}^{N} W_i M_i}{\sum_{i=1}^{N} W_i} = \frac{\sum_{i=1}^{N} f_i M_i^2}{\sum_{i=1}^{N} f_i M_i} \tag{6}$$

where $W_i = f_i \cdot M_i$ is the weight of fraction i. While for a pure substance $M_w = M_n$, for a mix of molecules the ratio M_w/M_n is referred as the *polydispersivity* of the mix. For example, Chin et al. (1994) used the height of a high-pressure-size-exclusion chromatography peak to calculate these variables.

Another important organic matter property is the density of various functional groups. This property is usually measured by titration. For example, acid-base titration was used to quantify the density of carboxyl group and phenol of NOM. The solution was brought to pH 3 and thereafter titrated; base consumption in the range of pH 3 to pH 8 was attributed to carboxyl groups, while twice the base consumption from pH 8 to pH 11 was attributed to phenols (Chi and Amy 2004). Jada et al. (2006) measured the surface charge of negatively charged humic acid by titrating it with a cationic polyelectrolyte (see also Au et al. 1998, 1999). They measured the induced potential and determined the point of zero charge (pH_{PZC})

of the titration curve from which they calculated the surface charge. As in the case of pH_{PNPZC}, this measurement also allows the determination of relative potential values. The interested reader is referred to studies aimed exclusively at characterizing NOM physical and chemical properties such as molecular weight, light absorptivity and aromaticity (e.g., Chin et al. 1994; Cabaniss et al. 2000).

OM identification. Radioactive labeling, of either ^{14}C or ^{3}H in the organic compounds, was used by various groups of researchers (e.g., Theng 1976; Zullig and Morse 1988; Wang and Lee 1993; Gu et al. 1996). The benefits of such an approach are: 1. production of $^{14}CO_2$ during the adsorption experiment can serve as an assay for possible microbial consumption of OM; and 2. competitive studies between different OM can be performed by isotopically labeling one of the components. Measurement of ultraviolet absorbance (UVA) of dissolved organic matter (DOM) is also a useful method since it is characterized by the aromatic functional groups of the studied OM. Changes in the specific UVA (i.e., the ratio of UVA to DOM concentration) can therefore be used to deduce the role of functional groups and moieties of the OM in the adsorption reaction (Gu et al. 1996; Chi and Amy 2004).

Operationally dividing NOM is another methodology for studying its adsorption properties. For example, Arnarson and Keil (2000) have extracted two fractions of OM from sea sediments: pore water NOM and easily extracted OM extracted by centrifugation of the pore water-free sample with OM-free sea water. Additionally, each fraction was divided based on molecular weight into two size fractions with a 1000Da filter. Thereafter they studied the adsorption of this fraction onto clays.

Spectroscopic methods. Spectroscopic methods have been widely used in the field of OM-mineral interaction (e.g., Gu et al. 1995; Kaiser et al. 1997; Phillips et al. 2005). These powerful methods have added considerable knowledge of OM chemical and structure properties. For detailed description of different methods, including Ultraviolet Visible Spectroscopy, Fluorescence Spectroscopy, Infra Red (IR) and Nuclear Magnetic Resonance (NMR), the reader is referred elsewhere (Senesi and Loffredo 1999). IR (FTIR and DRIFT) and NMR techniques can provide qualitative and sometimes semiquantitative analysis of functional groups of humic substances which are involved in the adsorption as will be discussed below.

Desorption experiments. A methodological approach to study desorption is through consecutive desorption experiments (e.g., Di Toro and Horzempa 1982; Wang and Lee 1993; Gu et al. 1995). In this method some or all of the solution is replaced with OM-free solution after adsorption equilibrium was established. Thereafter, the desorption process follows until a new equilibrium is reached. OM concentrations are then measured and the procedure is repeated. This method allows investigators to study and model desorption and to elucidate the reversible and irreversible (or resistant) fractions of the reaction.

SORPTION OF OM ONTO MINERAL SURFACES

Sorption of OM on mineral surfaces influences the fate of both OM (Schwarzenbach et al. 2003) and of metals (e.g., Stevenson and Fitch 1986) in the environment. Adsorption affects the transport and reactivity of both natural organic matter (NOM) and pollutants (Schwarzenbach et al. 2003). Adsorbed OM can mask the properties of the underlying solid resulting in a surface with very different physicochemical properties (Hunter 1980). In addition, the adsorption affects the reactivity of the mineral surfaces and therefore the dissolution and precipitation rates of the minerals.

Adsorption isotherms

Adsorption isotherms describe the change of the surface coverage of an adsorbed molecule as a function of its concentration in the liquid phase at a fixed temperature. The amount of OM adsorbed to the mineral surface ($X_{i,ads}$, mol m^{-2}) may be described by a simple Langmuir adsorption isotherm (Stumm 1992):

$$X_{i,ads} = F_i \frac{b_i \cdot C_i}{1 + b_i \cdot C_i} \tag{7}$$

where F_i is the maximum surface coverage of i (mol m^{-2}), b_i is a constant related to the energy of adsorption and C_i is the concentration of i in the solution. Heterogeneous solid consists of various surface sites with different reactivity. Therefore, the energy released during adsorption is usually a nonlinear-decreasing function of the surface coverage ($X_{i,ads}$ / F_i). Therefore, b_i varies with adsorption energy of sites. By substituting a Gaussian shape adsorption energy distribution function into the Langmuir model one obtains an integrated adsorption isotherm of the nonlinear form (Adamson 1990):

$$X_{i,ads} = F_i \frac{b_i \cdot C_i^{n_i}}{1 + b_i \cdot C_i^{n_i}} \tag{8}$$

where n_i is a coefficient. This nonlinear adsorption isotherm reduces to the Freundlich empirical adsorption isotherm at low concentrations of i

$$X_{i,ads} = K_i \cdot C_i^{n_i} \tag{9}$$

where K_i is an adsorption coefficient. Both the Freundlich and the Langmuir isotherms are special cases of the general model of Equation (8).

Giles et al. (1960) divided adsorption isotherms into four main classes according to the initial slope of the isotherm, which they named the S-type, L-type (Langmuir type), H-type (high affinity), and C-type (constant partition) isotherms. They further subdivided each class to sub groups based on the shapes of the upper parts of the isotherm (see Fig. 5). This classical classification of Giles et al. (1960) is commonly used in describing adsorption data and is widely cited (almost 500 citations, 100 of them during the last two years). An example of an S-type adsorption of myristate (a fatty acid) on calcite is presented in Figure 6. The initial direction of curvature shows that adsorption becomes easier as concentration rises, whereas the slower increase above inflection point represent a "first degree saturation" of the surface, i.e., the condition in which all possible sites in the original surface are filled and further adsorption can take place only on new surfaces (Giles et al. 1960). The interpretation of the three regions in such adsorption isotherms will be discussed later on.

Mechanisms of organic matter sorption

While many insights can be derived from the shapes of the adsorption isotherms, information regarding the adsorption mechanism cannot be attained from such relationships (Sposito 1989). Many studies have tried to postulate the molecular mechanism involved in adsorption of the organic matter. OM may be a complex mixture containing various compounds differing in molecular size, functional groups and stereochemical structure. Similarly, mineral surfaces may be heterogeneous, containing variable surface sites. As a result, the sorption mechanisms of NOM on mineral surfaces are not completely understood (Gu et al. 1995; Chi and Amy 2004). Arnarson and Keil (2000) summarized six molecular based adsorption mechanisms: ligand exchange, anion exchange, cation exchange, cation bridge, Van der Waals interaction and hydrophobic effects. The adsorption of OM can be attributed to one dominant mechanism or too few of these mechanisms which may act in parallel or sequentially, depending on both the nature of the organic compound and mineral surface specific properties (Mortland 1986). Often, the OM is initially sorbed to the surface via one mechanism (e.g., Van der Waals interaction), and thereafter due to its proximity to the surface, the OM is readsorbed via another mechanism (e.g., ligand exchange). The first four mechanisms may involve the formation of

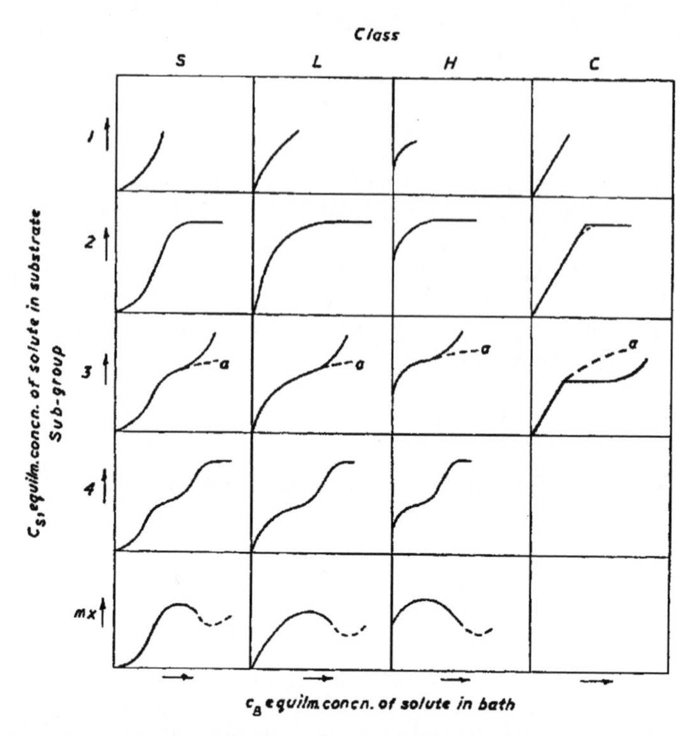

Figure 5. The isotherm classification of Giles et al. (1960). [Used by permission of The Royal Society of Chemistry, from Giles et al. (1960), *Journal of the Chemical Society*, Fig. 1, p. 3974, DOI: *http://dx.doi.org/10.1177/004051756103100207*.]

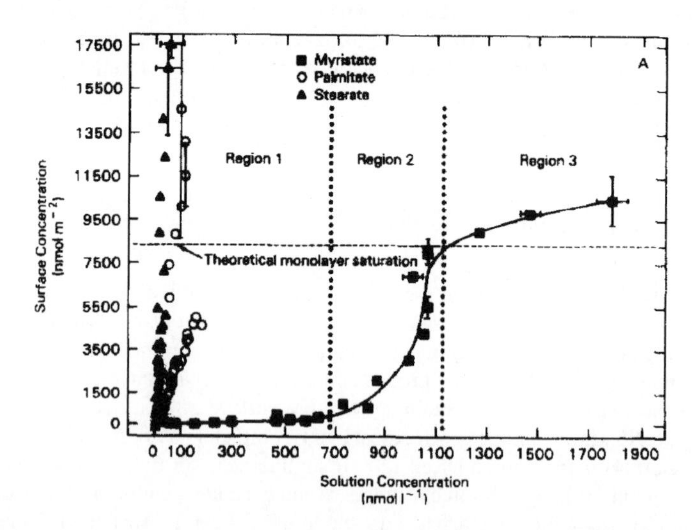

Figure 6. Adsorption isotherms for C_{14} (myristate), C_{16} (palmitate) and C_{18} (stearate) fatty acids on calcite. The adsorption isotherm of the myristate has an 'S' shape. The theoretical monolayer saturation (horizontal dashed line) was calculated by Zullig and Morse (1988) from perpendicular close packing of fatty acid molecules. An interpretation of the 3 regions in the adsorption isotherm is presented in Figure 14. [Used by permission of Elsevier, from Zullig and Morse (1988), *Geochimica et Cosmochimica Acta*, Vol. 52, Fig. 2a, p. 1670.]

a chemical bond and therefore are often termed as *chemisorption* reactions. Ligand exchange involves the replacement of hydroxyl group on the mineral surface by hydroxyl group from the adsorbed molecule. Anion and cation exchange involve the replacement of an anion or cation from the mineral surface with an organic anion or cation, respectively. For example, protonated amines and quaternary ammonium cations may be exchanged with metal cations in exchange sites on clay surfaces (Mortland 1986). Cation bridge is formed when a cation bridges between two anionic or polar groups, one on the mineral surface and the other on the adsorbed molecule. Cation bridge is analogous to the formation of a positively charged organic-metal complex or to the adsorption of a negatively-charged compound to a mineral surface bearing a positive site from the adsorption of cations (Baham and Sposito 1994). The simultaneous adsorption of OM and metal ion on mineral surfaces are commonly termed ternary adsorption, and they will be discussed separately below. Van der Waals interactions are intermolecular attractions (i.e., attractions between two neighboring molecules) and are short range (weak) additive associations. Van der Waals interactions can be significant in two scenarios: (a) in high ionic strength solutions, since then the solution masks the electrostatic repulsion between equal sign molecules, and therefore, they can closely approach each other; and (b) when Van der Waals interaction between two adjacent adsorbed molecules is energetically favored, leading to a further adsorption to the mineral surface (Arnarson and Keil 2000). Hydrophobic effects are the result of unfavorable interactions between water molecules and more hydrophobic moieties (i.e., the OM). The hydrophobic moieties are expelled from the water phase onto more hydrophobic phases (i.e., the mineral surface) and may be termed *physical adsorption*. Mechanistically speaking, physically adsorbed molecules will be more readily desorbed compared to chemically adsorbed molecules, since the later involve specific interactions between adsorbent and adsorbate.

Regardless the sorption mechanism, the OM must be close to the mineral surface in order that sorption will take place. Therefore, electrostatic attraction/repulsion between the OM and the mineral surface are expected to affect sorption. As a result, sorption is influenced by the charges of the mineral surface and OM functional groups as well as the solution chemistry that affects these charges.

Complexation and site coordination structure between OM and mineral surface site

Considerable effort is being made to recognize the type of adsorption bond and coordination structure between adsorbate and adsorbent. At first, it is essential to distinguish between inner-sphere complexes in which there is a direct bond between OM functional group and surface site, and outer-sphere complexes in which the bond is interposed by water molecule(s) and is held by hydrogen bonding and electrostatic interactions (Sposito 1989). The organic ligand may be connected to one (mononuclear) or more (binuclear, multinuclear) atoms on the mineral surface. With each of these atoms, they may form one (monodentate ligand) or more (bidentate, multidentate) bonds (Fig. 7). The coordination structure of inner-sphere complexes may exhibit various geometries, even for adsorption of relatively low molecular weight OM such as oxalate, malonate and citrate. Figure 7 is an example of possible inner-sphere complexation of oxalate and Al-(hydro)oxide (Yoon et al. 2004). The different coordination structures may induce different surface charges and therefore affect the reactivity of the mineral/water interface, and consequently affect the further adsorption of OM or other ionic species (Yoon et al. 2004; Persson and Axe 2005). Qualitative understanding of the nature of these interactions can be gained from adsorption-desorption experiments by changing experimental conditions (e.g., pH, solution ionic strength, competition with inorganic ions) as is discussed in following sections. A more comprehensive understanding of the coordination structure was obtained using IR spectroscopy combined with molecular orbital calculations (Axe and Persson 2001; Yoon et al. 2004; Persson and Axe 2005). Theoretical calculations, based on the density functional theory (DFT), are compared to experimental IR spectra as presented in Figure 8. These studies show that even for low molecular weight molecules more than one

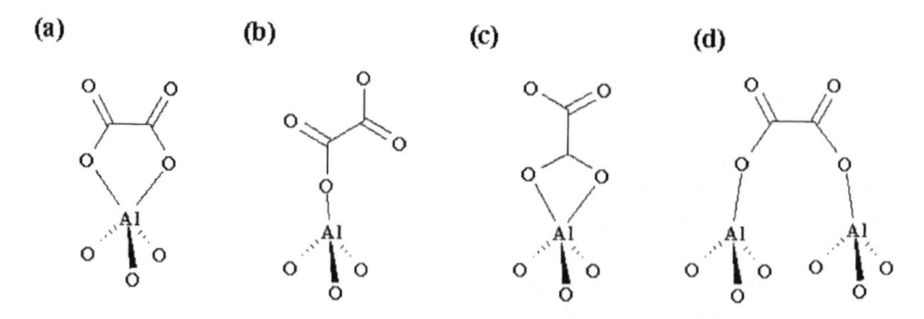

Figure 7. Schematic drawing of four possible inner-sphere coordination structures for oxalate adsorbed at the Al-(oxyhydr)oxide/water interface: (a) mononuclear bidentate side-on coordination with 5-membered ring; (b) mononuclear monodentate end-on coordination; (c) mononuclear bidentate end-on coordination with 4-membered ring; and (d) binuclear bidentate coordination. [Used by permission of Elsevier, from Yoon et al. (2004), *Geochimica et Cosmochimica Acta*, Vol. 68, Fig. 1, p. 4506.]

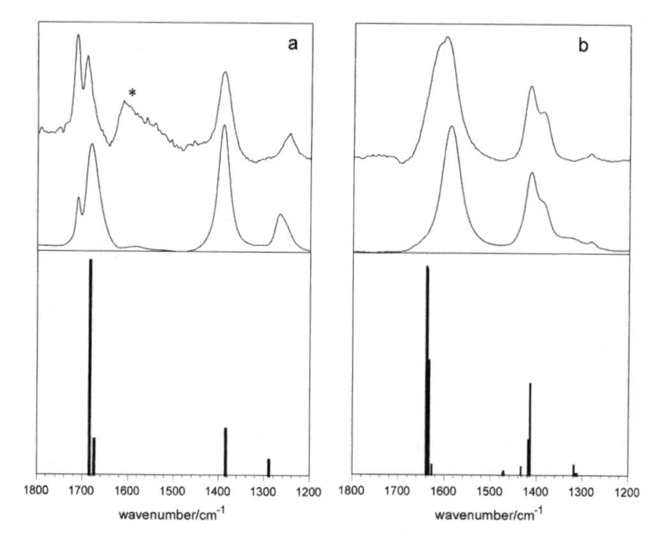

Figure 8. ATR-FTIR spectra of FeL^+ (upper spectrum) and FeL_3^{3-} (lower spectrum) where (a) L = oxalate, and (b) L = malonate. Solid bars are the theoretical IR spectra of FeL_3^{3-} from the DFT calculations. *This band is assigned to the bending mode of the four coordinated water molecules needed to complete the octahedral structure of FeL^+ (aq). [Used by permission of Elsevier, from Persson and Axe (2005), *Geochimica et Cosmochimica Acta*, Vol. 69, Fig. 3, p. 546.]

type of complex may coexist. For example, oxalate and malonate may form both outer-sphere and inner-sphere complexes on goethite, with the latter being more favored at low pH (Persson and Axe 2005). Based on theoretical considerations, inner sphere complexes of oxalate and malonate are assigned mononuclear five- and six- member ring chelate structures, respectively. Similar conclusions for oxalate adsorption on boehmite (γ-AlOOH) and corundum (α-Al_2O_3) were presented by Yoon et al. (2004).

The coordination structure(s) of adsorbed OM may change with time. Axe and Persson (2001) examined the IR spectra at different times following reaction completion. They observed that the outer-sphere complex of oxalate and boehmite was a transient species at low oxalate concentration and pH 5. During the first 10 hours, the concentration of the outer-sphere complexes decreased while that of the inner-sphere complexes increased by the same amount.

Therefore the authors suggested that a conversion between the two complexes took place. This observation of Axe and Persson (2001) may be an example of two adsorption mechanisms that act sequentially. Some more aspects of the coordination structures are related to the steric structure of the OM functional groups, and are further discussed below in the *OM functional groups, charge and steric structure* section.

The effects of environmental conditions on OM adsorption

Temperature. The mean isosteric enthalpy of adsorption of OM at constant surface concentration, $\Delta\bar{H}$, may be calculated using the equation (Baham and Sposito 1994):

$$\Delta\bar{H} = -\left(\frac{\partial \ln C_L}{\partial T}\right)_{C_S} RT^2 \tag{10}$$

where R is the molar gas constant (J mol^{-1} K^{-1}), T is temperature (K), and C_S and C_L are the adsorbed and equilibrium dissolved concentrations respectively. Depending on the mechanism, sorption reactions may be exothermic (i.e., leading to negative values of enthalpy) or endothermic (positive values). Ligand exchange reactions are characterized by large exothermic enthalpy values (Jardine et al. 1989 and references therein). A temperature independent adsorption, resulting in $\Delta\bar{H} \approx 0$, would suggest that the reaction is driven by entropy effects (Jardine et al. 1989; Baham and Sposito 1994). That is, OM adsorption in such cases is dominated by physical adsorption (hydrophobic effects) rather than by chemical adsorption.

Arnarson and Keil (2000) examined the adsorption of NOM onto montmorillonite at 3.5 °C and 21.5 °C. They found that adsorption decreased with increasing temperature. Although using only two points, they calculated a mean isosteric enthalpy of −7 kJ mol^{-1}, which is indicative of exothermic adsorption reaction. Arnarson and Keil (2000) concluded that hydrophobic effects are not of major importance in their study since the reaction was exothermic. In the adsorption of sewage sludge on clay surfaces (Baham and Sposito 1994) and soil organic matter on soil surfaces (Jardine et al. 1989) the adsorption was temperature independent, suggesting that the reaction is mainly driven by hydrophobic effects. An example of endothermic adsorption is given by Jada et al. (2006) for the adsorption of humic acid on quartz (T = 20, 40 and 60 °C). They did not estimate the mean isosteric enthalpy of adsorption, but speculated that the observed trend may be the result of increased mobility or transport of humic acid molecules from the bulk solution towards the mineral surface.

Nonetheless, thermodynamic calculations of temperature dependence cannot by themselves provide a thorough explanation for the mechanism involved (Baham and Sposito 1994). For example, Fein and Brady (1995) considered ligand-exchange mechanism for the adsorption of oxalate on γ-Al_2O_3 (supported also by Yoon et al. 2004) but found a small change in oxalate adsorption at 60 °C compared to 25 °C. Their interpretation was that a slight change in the thermodynamic stability of the complex occurred over the investigated range of temperatures. Similarly, Cama and Ganor (2006) found that the dependence of oxalate adsorption onto kaolinite on the sum of oxalate and bioxalate activities in solution is temperature independent (i.e., it is the same at 25, 50 and 70 °C; see Fig. 9). They further showed that kaolinite dissolution depends on the surface concentration of oxalate (see section below on *the case study of oxalate catalysis*). The observation that adsorbed oxalate affects kaolinite dissolution seems to indicate that oxalate is chemisorbed to kaolinite surface, even though it is temperature independent. A possible explanation of this apparent contradiction is that the sorption experiments of Cama and Ganor (2006) were of short duration (2 h) while the dissolution experiments were of longer duration (hundreds to thousands h). If oxalate initially forms outer-sphere complexes which are later converted to inner-sphere complexes, the initial physical adsorption may be temperature independent, while the chemisorption that formed later affects the kaolinite dissolution.

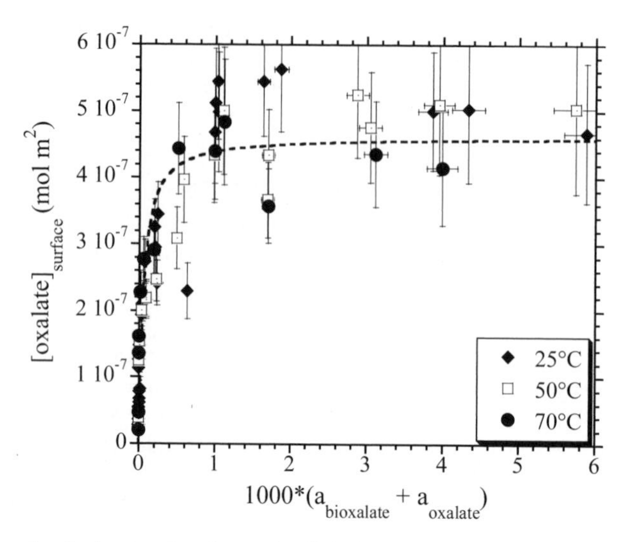

Figure 9. Adsorption isotherms of oxalate on kaolinite in experiments at pH 3, at 25, 50 and 70 °C. The data is from Cama and Ganor (2006). The observations are fitted to the simple Langmuir adsorption isotherm of Equation (47).

Dissolved ions. Dissolved anions and cations affect the adsorption of OM both directly and due to their effect on the ionic strength. It was proposed that dissolved bivalent cations will act as a cation bridge between two negatively charged groups, whereas dissolved bivalent anions will compete with OM adsorption over positive reactive sites (e.g., Arnarson and Keil 2000; Chi and Amy 2004). The existence of a cation bridge mechanism may be observed by comparing the adsorption of OM in the presence of divalent cations to that in the presence of monovalent cations that have a small tendency to form cation bridges. Competition of anions is usually studied by performing experiments using salts having a common cation. As will be shown below, the existence of anion competition is commonly interpreted as an indicative of a ligand exchange mechanism in the adsorption (e.g., Gu et al. 1995; Arnarson and Keil 2000; Chi and Amy 2004). These two opposite effects of anions and cations are demonstrated below mainly using the results of two experimental studies: (1) The dynamic column experiments of Chi and Amy (2004), in which they simulated groundwater flow in saturated medium and studied adsorption of fulvic and humic acids onto iron-oxide-coated quartz (Fe-quartz) and onto different natural sediments under variables conditions. Chi and Amy (2004) compared breakthrough curves (BTC) of OM in solution with ionic strength of 0.015 M containing sodium phosphate, sodium sulfate, calcium chloride or sodium nitrate (see Fig. 10). (2) The experiments of Arnarson and Keil (2000) on the adsorption of NOM extracted from sea sediments to montmorillonite in solution containing $CaCl_2$, NaCl, Na_2SO_4 or diluted sea water with variable ionic strength. In these experiments, the adsorption increased with ionic strength; the relative increase varied with the identity of the salt according to $CaCl_2$ > Seawater > NaCl >> Na_2SO_4 (see Fig. 11).

Phosphate ions strongly compete with DOM for adsorption sites (e.g., Chi and Amy 2004). Moreover, phosphate was shown to replace adsorbed OM, therefore, prompting its desorption (Kaiser and Guggenberger 2000). Figure 10 shows that the breakthrough of the fulvic acid with sodium phosphate precedes that of fulvic acid with sodium sulfate, indicating that sulfate competes with fulvic acid over the adsorption sites to a lesser extent than phosphate. Another example illustrating phosphate competition is given by Zullig and Morse (1988), who

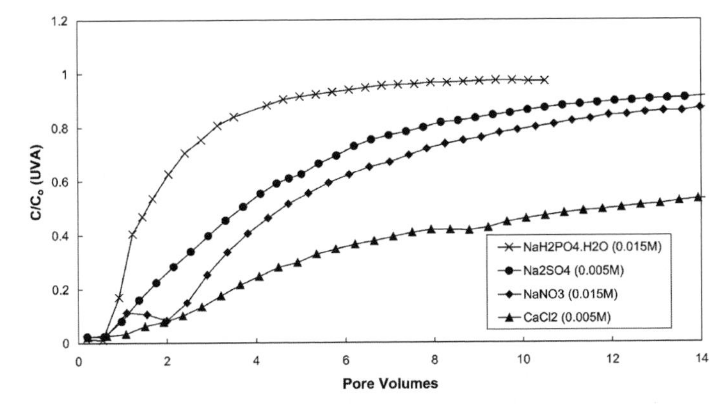

Figure 10. Breakthrough curves of fluvic acid through a column filled with Ringold sediment with different solution composition. The effluent ionic strength was 0.015M, the pH was 6.0, and the molar concentration of each solution appears in the legend in parentheses. [Used by permission of Elsevier, from Chi and Ami (2004), *Journal of Colloid and Interface Science*, Vol. 274, Fig. 2, p. 385.]

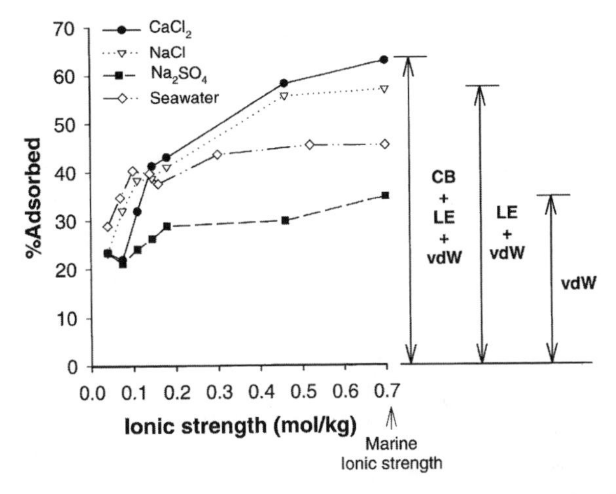

Figure 11. The effect of different solution composition at varying ionic strength on the adsorption of NOM extracted from sea sediment to montmorillonite. Arrows illustrate the contribution of each adsorption mechanism, see text for discussion. vdW = Van der Waals; LE = ligand exchange; CB = cation bridge. [Used by permission of Elsevier, from Arnarson and Keil (2000), *Marine Chemistry*, Vol. 71, Fig. 6, p. 314.]

showed that phosphate competes with fatty acids over adsorption sites of carbonate minerals. Based on their observations that phosphate competes with fulvic acid and the observation of Sigg and Stumm (1981) that phosphate forms inner-sphere complexes on goethite (by ligand exchange), Chi and Amy (2004) concluded that ligand exchange is an important mechanism for the adsorption of fulvic acid (as well as other NOM) onto soil minerals. This conclusion is in agreement with the conclusion of Gu et al. (1995) that the adsorption of both hydrophilic and hydrophobic NOM fraction onto iron oxide is via ligand exchange mechanism, since even a small phosphate concentration (1 mM) substantially decreased their adsorption. Similarly, Arnarson and Keil (2000) showed that at a given ionic strength, NOM adsorption is less significant in the presence of SO_4 ions than in the presence of Cl^- due to the stronger competition of the bivalent SO_4^{-2} for adsorption sites (see Fig. 11).

Different studies demonstrated that adsorption of OM is enhanced in the presence of Ca^{2+} relative to the adsorption in the presence of monovalent cations: for example, Figure 10 shows that $CaCl_2$ increased the adsorption of NOM by 3 fold compared to the $NaNO_3$ solution, and Figure 11 shows an increase in adsorption in the presence of $CaCl_2$ relatively to NaCl. These observations are in agreement with the above mentioned hypothesis that bivalent cations form cation bridges (e.g., Chi and Amy 2004). The favored adsorption of OM in the presence of Ca^{2+} was easily reversed when the column in the experiments of Chi and Amy (2004) was washed with deionized water. Following the wash, half of the NOM was desorbed, leaving only 1.5 fold more adsorbed NOM compared to the $NaNO_3$ case. Chi and Amy (2004) speculated that in addition to the formation of cation bridges (further discussed in *ternary adsorption* section), Ca^{2+} associated with NOM increased its adsorption thorough increasing NOM hydrophobicity and decreasing electrostatic repulsion between NOM and solid surface.

Arnarson and Keil (2000) suggested that the dominant adsorption mechanisms are Van der Waals interactions, ligand exchange, and cation bridging. The contributions of the mechanisms involved in NOM adsorption are illustratively presented in Figure 11. According to this interpretation adsorption via a cation bridge mechanism occurred only in the presence of Ca^{2+} (which is the only bivalent cation used in their study), a ligand exchange mechanism occurred in the presence of Cl^- and is inhibited in the presence of SO_4^{2-}, and Van der Waals interactions occurred in all the experiments, regardless of the specific composition of the solutions.

An increase in ionic strength lowers the electric repulsion between equal sign groups and, therefore, increases Van der Waals interactions. Arnarson and Keil (2000) observed that adsorption increased with increasing ionic strength for all solutions examined (Fig. 11). Similarly, Au et al. (1999) observed that NOM adsorption density on hematite increased with increasing ionic strength. Jada et al. (2006), also noticed an increase in the adsorption of humic acid onto clean quartz as ionic strength increased (NaCl 0.001-0.1 M). They attributed it to a decrease in the electrical double layer of both humic acid and quartz particles.

pH. pH affects the speciation and charge distribution on the surfaces of minerals and of DOM, and therefore affects the electrostatic attraction/repulsion between them. As pH decreases, both NOM acidic functional groups and surface charge of minerals become less negative due to an increase in protonation of functional groups. Thus, hydrophobic effects can be more significant in driving organic molecules toward the mineral surface, which in turn, can increase the possibility for adsorption via other mechanisms such as ligand exchange (Arnarson and Keil 2000). An increase in adsorption of NOM with increasing acidity was observed in several studies. For example, Arnarson and Keil (2000) showed that by decreasing pH from 8 to 4, adsorption of NOM on montmorillonite increased from 35% to 65%. Similarly, Gu et al. (1994, 1995) and Au et al. (1999) showed that adsorption of either bulk NOM or fractionated NOM onto iron oxide (hematite) increased as pH decreased, and Jada et al. (2006) observed that humic acid extensively adsorbed to quartz at pH 3 compared to pH 6. They attributed this behavior to an increase in the electrostatic repulsion between negatively charged humic acid and quartz surface at elevated pH. As previously mentioned, pH may also alter the mineral surface charge, which in turn, will affect OM-mineral surface interaction. The effect of surface charge on OM adsorption is further discussed below. It is important to notice that the effect of pH on OM adsorption is mostly attributed to physical adsorption. It does not necessarily follow that dissolution and precipitation of minerals will be accordingly affected since these processes are affected by chemisorption as discussed in the relevant sections.

The effects of OM properties on OM adsorption

OM functional groups, charge and steric structure. The adsorption of OM frequently depends on the presence of carboxyl groups, hydroxyl groups and aromatic moieties (e.g., Gu et al. 1995; Kaiser et al. 1997; Meier et al. 1999). The charge of these functional groups, and therefore their reactivity, depend on the solution chemistry (e.g., pH and ionic strength). For example, Tombácz et al. (2004) showed that humic acids become more negatively charged as either pH or ionic strength increases. The adsorption and desorption of amines and amino acids to clays (kaolinite and montmorillonite) and marsh sediment in seawater was studied by Wang and Lee (1993). They observed that in seawater positively charged compounds (aliphatic amines and lysine) adsorb greatly to clays compared to negatively charged (glutamic and acetic acids) compounds. As clay surfaces are negatively charged, the adsorption of positively-charged organic compounds is essentially a cation exchange process which involves electrostatic and Van der Waals forces (Wang and Lee 1993). Adsorption of lysine was extensive compared to the aliphatic amines, especially to kaolinite. This observation was explained by zwitter ion (i.e., a molecule carrying both a positive and a negative charges) association of two amino acids molecules. That is, the positively charged amine of one molecule is adsorbed to the mineral surface, while the free carboxyl group can interact with the free amine end of another molecule.

Low molecular weights (<200 Da) OM do not adsorb on oxide surfaces if they do not bear functional groups such as carboxylic, phenolic-OH or amino groups (Kummert 1979 cited in Kummert and Stumm 1980). These functional groups substitute with mineral surface hydroxyl groups. Nonetheless, the adsorption is intensified by either Van der Waals interaction or hydrophobic effects between OM and the mineral surface and among organic molecule themselves as discussed above. Such interactions may also lead to the adsorption of a second layer (Zullig and Morse 1988; Wang and Lee 1993).

Kaiser et al. (1997) used ^{13}C-NMR and DRIFT spectroscopy techniques to study the adsorption of soil-derived total DOM and acidic humic substances (mainly fulvic acid) to soil and minerals (goethite, ferrihydrite and amorphous $Al(OH)_3$). Liquid-state ^{13}C-NMR was divided to 4 regions: (a) the alkyl region (0-50 ppm), mainly representing C atoms bonded to other C atoms (methyl, methylene and methine groups); (b) the O-alkyl region (50-110 ppm), mainly representing C bonded to O (carbohydrates, alcohols, and ethers); (c) the aromatic region (110-160 ppm), representing C in aromatic systems and olefins, and (d) the carbonyl region (160-210 ppm), including carboxyl C (160-190 ppm). Resonance areas were calculated by electronic integration. They observed depletion in carbonyl and aromatic C of the solution examined before and after the adsorption experiment, and concluded that carboxyl groups and aromatic structures are preferentially adsorbed.

Gu et al. (1995) studied the pH dependency of adsorption onto iron oxide using 5 specific organic compounds, three of which were substituted benzoic acids (phthalic, salicylic and *p*-hydroxybenzoic acids) and the other two were phenols (catechol and resorcinol, see Fig. 12). This choice allowed studying both functional group type and steric arrangement of the functional groups. For all three benzoic acids adsorption increased as pH decreased, showing a similar trend to the NOM adsorption tested in their study. Moreover, the amount of phthalate carboxyl groups adsorbed at pH 4 agreed well with the amount of adsorbed NOM carboxyl groups. This observation led them to conclude that carboxyl groups are significantly involved in NOM adsorption onto iron oxide. The smaller adsorption of *p*-hydroxybenzoic acid (4-hydroxybenzoic acid) compared to salicylic acid (2-hydroxybenzoic acid) was explained in terms of their steric structure differences. These two compounds are isomers, i.e., they have exactly the same functional groups and chemical formula ($C_7H_6O_3$) but they differ in their structural arrangement. Both have a benzene ring with a carboxyl functional group and hydroxyl group. The difference between these two OM is the relative positioning of the hydroxyl and the carboxyl groups on the benzene ring. The salicylic acid is the ortho isomer (the hydroxyl and the carboxyl are attached to adjacent carbon atoms) and the *p*-hydroxybenzoic acid is the para isomer (the hydroxyl and the carboxyl are attached to carbon atoms which are separated by two unsubstituted carbon atoms). As a result, the adsorption of salicylic acid is more effective than that of *p*-hydroxybenzoic acid (see Fig. 12A). In contrast to benzoic acids adsorption, adsorption of catechol on iron oxide increased with increasing pH with a maximum at pH~9.5. This pH is close to the compound

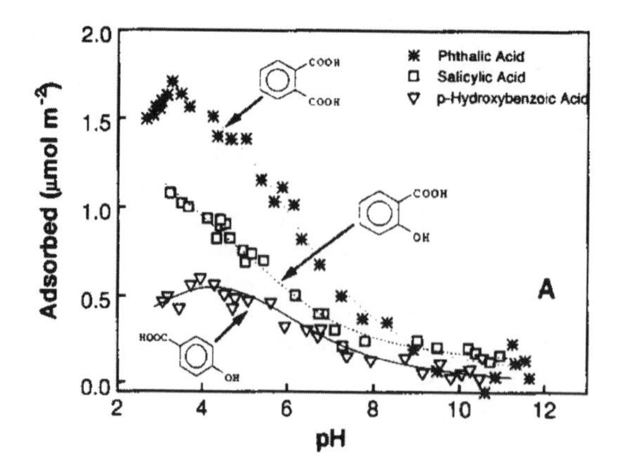

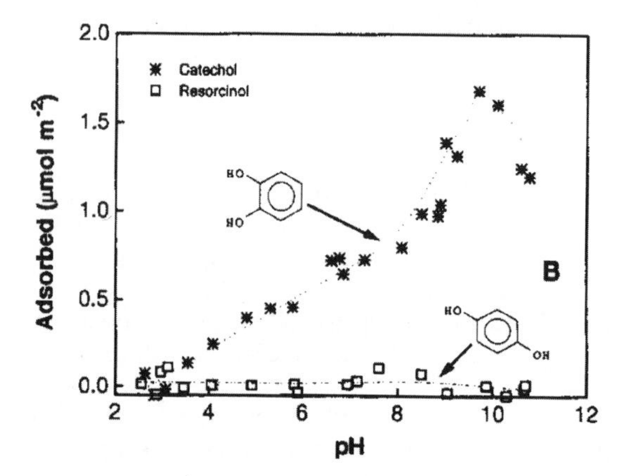

Figure 12. pH dependence adsorption of (A) three substituted benzoic acids and (B) two phenols on iron oxide (I = 0.01 M NaCl). Due to a typo in the original paper, the structure of the para isomer of resorcinol is drawn in Fig. B instead of the actual meta isomer. [Used by permission of Elsevier, from Gu et al. (1995), *Geochimica et Cosmochimica Acta*, Vol. 59, Fig. 7, p. 225.]

pK_a (9.45), suggesting that the adsorption of uncharged catechol is favored, in contrast to the adsorption of both the benzoic compounds and NOM. Steric effects are probably responsible for the insignificant adsorption of resorcinol in comparison to catechol (see Fig. 12B). Catechol and resorcinol are isomers of $C_6H_4(OH)_2$. Both have a benzene ring with two hydroxyl functional groups. The catechol is the ortho isomer and the resorcinol is the meta isomer (the hydroxyls are attached to carbon atoms which are separated by one unsubstituted carbon atom). This difference resulted in a substantially different adsorption behavior of these substituted benzoic phenols. Such steric effects are important, but poorly studied (Gu et al. 1995).

Gu et al. (1995) postulated that total carboxylic functional groups may not be critical in determining NOM adsorption, but rather the stereochemical arrangement of the carboxyl and hydroxyl group will determine which compound will be preferentially adsorbed. They suggested that only ortho-positioned carboxyls and/or hydroxyls may form complexes with iron oxide surfaces through the ligand exchange mechanism. This was tested by examining FTIR spectra of both hydrophilic and hydrophobic NOM (untreated and adsorbed), which were compared to FTIR spectra of the specific organic compounds. Gu et al. (1995) noticed a shift in the carboxyl band before and after adsorption for both NOM fractions. The shift was comparable to that of phthalate, which was different from that of *p*-hydroxybenzoic. Similarly, NOM hydroxyl band showed a similar pattern to that of catechol, indicating that –OH functional groups, especially ortho-positioned -OH, interacted with iron oxide surface.

Other studies that emphasized the effect of functional group steric structure on the adsorption affinity of OM include that of Kummert and Stumm (1980) on the adsorption of aromatic compounds with two ortho-positioned functional groups to γ-Al_2O_3 and that of Davis (1982) on NOM adsorption to γ-Al_2O_3 and quartz. Kummert and Stumm (1980) considered different surface coordinations between carboxyl group and surface sites. They concluded that only 1:1 complexes form via a ligand exchange mechanism, i.e., the carboxyl group of each aromatic acid interacts with one Al site. Davis (1982) observed that fractionation of OM to adsorbed and non-adsorbed fractions was not based on average acidity (i.e., functional groups density) and postulated that the stereochemical structure of the organic compounds determines which molecules adsorb and which do not.

Molecular weight. The solubility of OM decreases as a function of the size of the molecules. This trend is known as Traube's rule, which states that the concentration of a member of a homologous series required for equal lowering of the solution surface tension decreases threefold for each additional methylene group (Aranow and Witten 1958). As a result of the reduction of solubility with increasing size, adsorption of OM is expected to be greater for larger molecules. Indeed, preferential adsorption of large organic molecules over small molecules was observed by many investigators (e.g., Barcelona and Atwood 1979; Zullig and Morse 1988; Wang and Lee 1993; Gu et al. 1995; Meier et al. 1999; Arnarson and Keil 2000; Chi and Amy 2004). For example, Barcelona and Atwood (1979) observed that adsorption of fatty acids onto gypsum surface increases with molecule length, while their solubility decreases with increasing molecule length (see Fig. 13). Similarly, and more thoroughly discussed, are the observations of Zullig and Morse (1988) who studied the adsorption of fatty acids onto carbonate minerals (calcite, aragonite, dolomite and magnesite). They observed that short-chain molecules (C_4 to C_{12}) did not adsorb under their experimental conditions while chains longer than C_{12} adsorbed irreversibly. This trend was explained by both the effect of solubility and by the increase in Van der Waals interaction with the length of adsorbed fatty acids. Zullig and Morse (1988) summarized this relationship from a thermodynamic point of view using the Equation:

$$\Delta G^{\circ}_{ads} = \Delta G^{\circ}_{chem} + \Delta G^{\circ}_{hyd} - \Delta G^{\circ}_{solv} \quad (11)$$

where ΔG°_{ads} is the total Gibbs free energy of adsorption, ΔG°_{chem} is the energy of chemisorption (i.e., the energy for bonding carboxyl tail of a fatty acid to the mineral reactive site), ΔG°_{hyd}

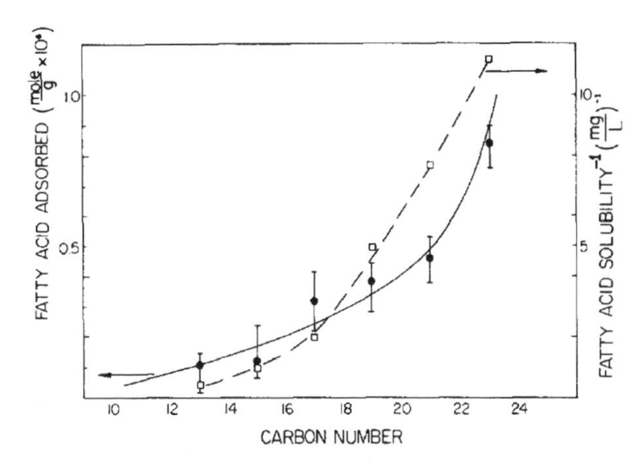

Figure 13. Adsorption of fatty acids onto gypsum (closed circles) and the reciprocal of their solubility in water (open squares) as a function of the length of the carbon chain. [Used by permission of Elsevier, from Barcelona and Atwood (1979), *Geochimica et Cosmochimica Acta*, Vol. 43, Fig. 4, p. 51.]

is the hydrophobic bonding energy (i.e., Van der Waals bonding energy) and ΔG°_{solv} is the solvation energy. For shorter fatty acids ($<C_{12}$) the solvation energy is larger than the adsorption energy, therefore adsorption is not favorable. The model for fatty acids adsorption suggested by Zullig and Morse (1988) is illustrated in Figure 14. According to their interpretation, at low surface coverage chemisorption of the anionic organic to surficial cation sites is the major mechanism for fatty acid adherence to the surface, although adsorption is primarily driven by decreasing solubility. As the surface coverage increases, adsorption is enhanced probably as a result of Van der Waals bonding between the alkyl chains of adjacent fatty acids. The transition between regions 1 and 2 is the point on the adsorption isotherm where Van der Waals bonding between alkyl chains results in the formation of two dimensional aggregates of chemisorbed fatty acids. As is shown in Figure 14, clusters of fatty acid molecules begin forming with their alkyl chains perpendicular to the surface of the mineral. Since Van der Waals forces are cumulative, it is expected that the total bonding energy between molecules will increase with increasing alkyl chain length.

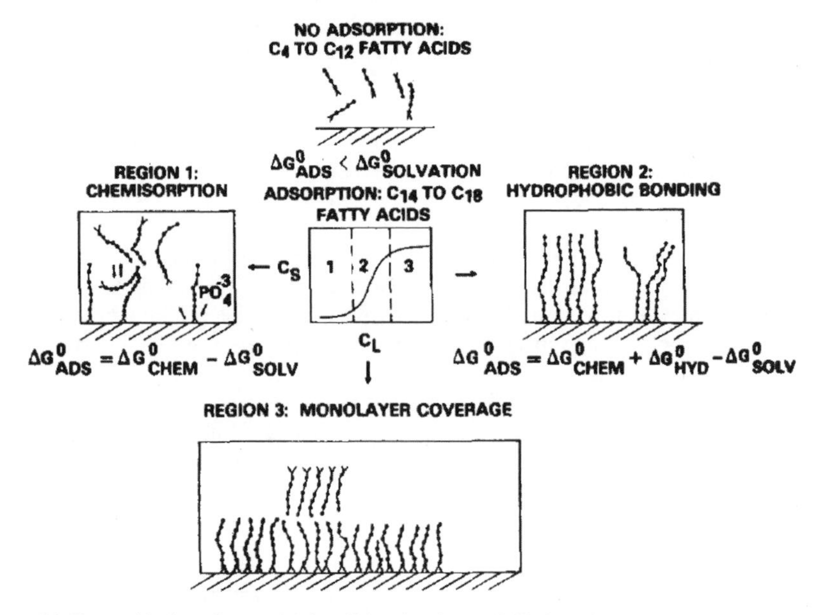

Figure 14. Fatty acid adsorption model describing the observed 'S' shape isotherm, shown in Figure 6, in terms of Gibbs free energy. [Used by permission of Elsevier, from Zullig and Morse (1988), *Geochimica et Cosmochimica Acta*, Vol. 52, Fig. 13, p. 1675.]

More examples of the effect of OM size are given by Gu et al. (1995) who fractionated NOM into hydrophilic and hydrophobic fractions and into two molecular size fractions (with a cutoff at 3000 Da). They examined NOM adsorption to iron oxide (hematite) at two pH values (4.3 and 6.3) and observed that the high M_w fraction was preferentially adsorbed compared to the low M_w. Wang and Lee (1993) also observed that high molecular size methylamines preferentially adsorbed to montmorillonite compared to low molecular size methylamines.

The effects of mineral surface properties on OM adsorption

Due to differences in the surface properties of different minerals, there is a significant variability in their tendencies to adsorb OM. Among the naturally abundant minerals, the most effective adsorbents are clays and oxides. Therefore, it is not surprising that most studies on adsorption of OM on mineral surfaces were conducted with oxides and clays. Other silicates, carbonates and sulfates were studied to a lesser degree.

Quartz. Surface charge of pure quartz was measured by Jada et al. (2006) and is mainly attributed to silanol (SiOH) dissociation at the mineral surface. The pH of isoelectric point (i.e., pH_{PZC} or zeta potential = 0) measured by Jada et al. (2006) was 2.44. Therefore, at lower or higher pH values, quartz surface is positively and negatively charged, respectively. Adsorption on quartz surfaces is reported for aspartic acid (Carter 1978), humic acid (Jada et al. 2006) and polyacrylamide (Graveling et al. 1997). Poulson et al. (1997) studied the adsorption of oxalate onto quartz surfaces using short-term (4-5 h, pH 4.3-7.1, oxalate = 5-20 μM) and long-term (1 week, autoclaved, pH 6.3-7.0, oxalate = 10-40 μM) experiments. Oxalate adsorption under these conditions was negligible, with a maximum adsorption of $3 \cdot 10^{-9}$ mol m^{-2}. They estimated that this value corresponds to an average adsorption of 1 oxalate ion per 2500 surface sites.

Feldspar. Aluminol (AlOH) and silanol sites are the reactive sites to which the adsorption is pH-dependent. At low pH ($2 < pH < 4$) a positive surface charge exists due to the formation of $>Al\text{-}O\text{-}H_2^+$ groups. As pH increases, deprotonation of surface sites lowers the overall surface charge where at $pH > 9$ the trend is reversed. Deprotonation of >Si-O-H sites causes an overall negative charge at high pH (Graveling et al. 1997). Although many studies suggested that adsorbed OM affects feldspar dissolution (see discussion below), only a few studies measured the adsorption of OM on feldspar surfaces. Among these studies are those of oxalate adsorption on andesine (Stillings et al. 1998) and hydrolyzed polyacrylamide adsorption on perthitic feldspar (Graveling et al. 1997).

Metal-oxides and clay minerals. Due to the differences in isomorphic substitutions and in mineralogical structure, the cation exchange capacity of montmorillonite (2:1 layer type) compared to kaolinite (1:1 layer type) is about 40 times higher. Indeed, Wang and Lee (1993) found that the adsorption of three positively-charged aliphatic amines was stronger to montmorillonite compared to kaolinite. However, the adsorption of a positively charged amino acid (lysine) was similar in both clays. This observation was explained in terms of the respective adsorbate properties: lysine-lysine associates through the interaction between amino and carboxyl functional groups enhanced the overall adsorption to both clays.

Tombácz et al. (2004) studied the role of reactive surface sites of two clay minerals (kaolinite and montmorillonite) and two metal-oxide minerals (hematite and magnetite). The pH-dependent charge of all 4 minerals was measured by acid-base titration as the net proton surface excess amount and is shown in Figure 15. For the metal-oxide minerals, surface charge is highly pH dependent, and the net proton charge should be proportional to the *intrinsic surface charge* (Tombácz et al. 2004). The curves of both clays reflect the difference between them: the 2:1 montmorillonite has large net proton consumption but also high permanent layer charge (Sposito 1989). Kaolinite, on the other hand, shows an amphoteric character, resembling the pH dependent surface charge of the metal-oxides. Protonation and deprotonation of Al-OH sites contribute to the surface charge of kaolinite, but the overall net proton consumption is low compared to the metal-oxide minerals.

The coordination of the Al and Si affects the affinity of the site for protons (Graveling et al. 1997). Whereas Si is always in a tetrahedral coordination, Al can be found in either tetrahedral (e.g., feldspar) or octahedral coordination (e.g., kaolinite). In their study, Graveling et al. (1997) observed that quartz and feldspar adsorbed polyacrylamide to the same degree whereas kaolinite adsorbed 3 times more polyacrylamide. They ascribed these differences to the greater number of reactive sites on kaolinite.

Carbonates. Suess (1968, cited in Zullig and Morse 1988) and Suess (1970) concluded that OM adsorption is specific to Ca^{2+} since dolomite showed half the sorption of calcite (in ordered dolomite 50% of the cation sites are occupied by calcium while the remaining 50% is occupied

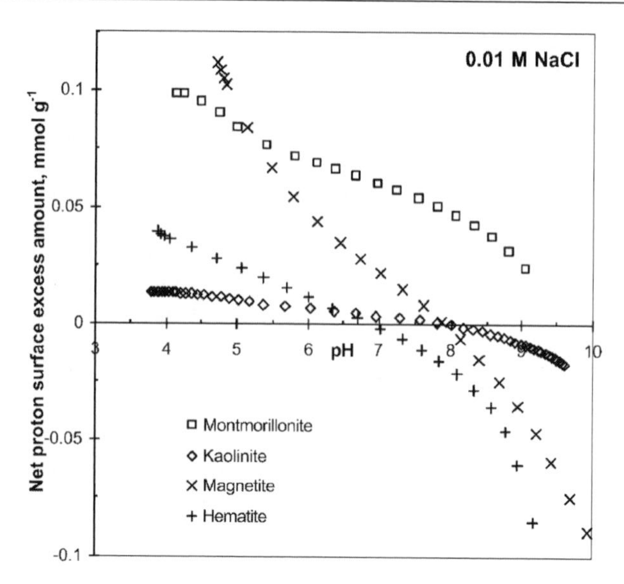

Figure 15. pH-dependent mineral surface charge expressed as the net proton surface excess amount for the studied minerals. [Used by permission of Elsevier, from Tombácz et al. (2004), *Organic Geochemistry*, Vol. 35, Fig. 1, p. 261.]

by magnesium; Suess 1970). In the systematic work of Zullig and Morse (1988) this result was not reproduced; dolomite adsorbed less OM but only by 10%. They argued that in the former study of Suess the solution was sometimes oversaturated with respect to the OM salt and that no attempt was made to prevent bacterial degradation. Thus, either the OM precipitated as a salt or undergone degradation. Moreover, Zullig and Morse (1988) studied the adsorption of fatty acids onto magnesite ($MgCO_3$) which was approximately 80% that of calcite. They hypothesized that the differences between the three minerals are the results of a smaller cation site associated with Mg^{2+} leading to less efficient ion packing on the mineral surface. To support their hypothesis they calculated the theoretical monolayer coverage and concluded that dolomite and magnesite coverage will be 91% and 82%, respectively, that of calcite. Several studies examined the effect of OM adsorption on the kinetics of dissolution and precipitation of carbonate minerals. These studies are described in the relevant sections below.

Gypsum. In their work on fatty acid adsorption to gypsum, Barcelona and Atwood (1979) relied upon early studies which concluded that adsorption of carboxyl groups is specific to Ca^{2+} sites both on carbonate surfaces (Suess 1970) and gypsum surfaces (Barcellona and Atwood 1978 cited in Barcelona and Atwood 1979). Assuming that Ca^{2+} sites are the reactive sites, they calculated that coverage of 11.5% of the total surface area by *n*-heptadecanoate will account for coverage of more than 72% of the Ca^{2+} sites.

The shape of the adsorption isotherm

Interaction of OM and mineral surface sites is affected by both adsorbent and adsorbate properties. These properties may be affected by the environmental conditions. The integration of these effects is reflected in the shape of the adsorption isotherm, of which some quality understanding of the interaction mechanism may be deduced. For example, the adsorption of fatty acids to carbonates resulted in an 'S' type isotherm which can be divided into three stages (see Figs. 6 and 14). According to the interpretation of Zullig and Morse (1988) discussed above, the relatively low initial adsorption reflects the chemisorption of the carboxyl tail to the reactive site (see Fig. 14). The increase in adsorption partition coefficient at the second stage is the result of Van der Waals interaction between adjacent fatty acids. A decrease in the partition coefficient finally occurs when the mineral surface coverage is close to saturation.

The adsorption of NOM to iron oxide surfaces resulted in 'L' shape curve isotherm (Gu et al. 1995; Chi and Amy 2004). This isotherm type is characterized by an initial steep slope (i.e., large partition coefficient) which reflects the high affinity of the adsorbate at low surface coverage of the adsorbent. As emphasized above, the shape of the adsorption isotherms cannot be interpreted in terms of any particular adsorption mechanism. For instance, it can be shown mathematically that any L-curve can be fitted with two Langmuir equations, having four adjustable parameters (Sposito 1989).

In the above sections we discussed how OM adsorption is affected by the properties of the solution, the OM and the mineral surface. As will be shown in the following sections, some of these properties are affected by the adsorption itself.

The effects of OM adsorption on solution and mineral properties

pH. Davis (1982) proposed that proton consumption (i.e., pH increase) following the adsorption of NOM onto γ-Al_2O_3 may be the result of at least three processes: 1.protonation of surface hydroxyl to form an OM-mineral surface complex; 2. protonation of ionized groups on the adsorbed organic molecule that are not directly involved in the complexation. This protonation can lower the electrostatic repulsion between adsorbed molecules; and 3. protonation of deprotonated reactive sites since the concentration of protonated sites is decreased by OM adsorption (i.e., re-equilibration between protonated and deprotonated sites). Davis (1982) calculated that the proton consumption was higher than the acidity of the NOM, and therefore concluded that processes 1 and 2 above cannot explain all the proton consumption by themselves. Furthermore, preliminary calculations showed that re-equilibrium of surface charge will tend to deprotonate surface hydroxyls (i.e., pH decrease). Therefore, Davis (1982) postulated that another mechanism should be responsible for the observed total proton consumption. He suggested that partial dissolution of γ-Al_2O_3 by OM complexation may release 3 hydroxyl for each Al^{3+} dissolved, resulting in an increase in proton consumption.

In their column experiments on the adsorption of NOM onto Fe-quartz, Chi and Amy (2004) measured an increase in solution pH. The pH increased from 6.0 to 7.2 and corresponded well with the NOM breakthrough curve (see Fig. 16). They also noticed that the pH increase

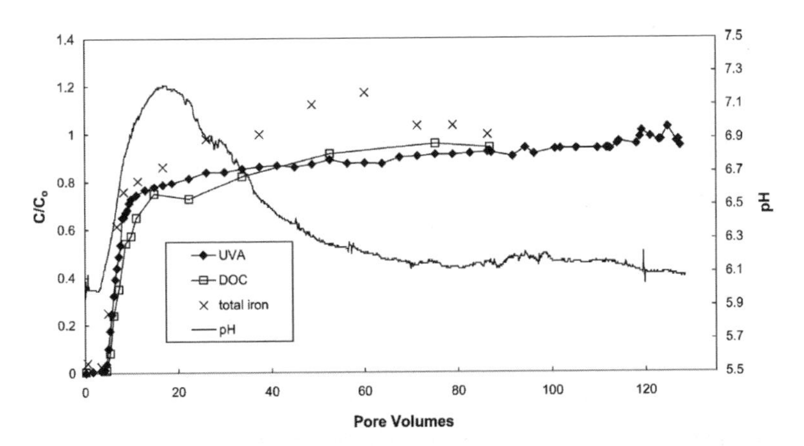

Figure 16. Breakthrough curves of humic acid through a column filled with Fe-quartz and changes in the effluent pH. [Used by permission of Elsevier, from Chi and Ami (2004), *Journal of Colloid and Interface Science*, Vol. 274, Fig. 7, p. 387.]

was delayed compared to the OM adsorption and suggested that this mismatch may be the result of a kinetically slower release of hydroxyl through a ligand exchange mechanism. According to their interpretation, OM initially forms outer-sphere complexes due to physical adsorption. Thereafter, hydroxyl is released due to ligand exchange with the outer-sphere OM causing an increase in pH. In their experiments, the pH BTC appears only after ~5 pore volumes had passed although a considerable amount of NOM was adsorbed already (Fig. 16). In another set of column experiments, performed in a close-loop arrangement, Chi and Amy (2004) reported that the mismatch between NOM adsorption and pH increase was about 15 pore volumes. Considering the flow rate, column volume, and matrix porosity reported by Chi and Amy (2004), it can be calculated that the conversion between outer- and inner-sphere complexes took between 1.5-2.5 hours. This conversion is of the same timescale (0.25-10 hours) reported by Axe and Persson (2001) for the conversion of outer- and inner-sphere complexes of oxalate on boehmite.

Mineral surface properties. Coverage of mineral surfaces by adsorbed OM can alter surface properties, especially surface area and surface charge. For example, Gu et al. (1994) calculated that the adsorbed NOM onto hematite contained more O moieties (i.e., hydroxyl and carboxyl groups) than the reported hydroxyl density of hematite. The resultant free O moieties can, thus, alter the mineral surface properties.

In the study of topsoils and subsoils, Kaiser and Guggenberger (2000) measured a decrease in surface area as a function of the concentration of hydrophobic OM (Fig. 17). This effect can be the result of either a decrease in surface roughness or the formation of microaggregates linked by the organic molecules. The interior of such microaggregates may not be accessible for N_2 which is used in the measurement of the surface area by the Brunauer-Emmett-Teller (BET) method (Brunauer et al. 1938).

Using electrophoretic mobility, Tombácz et al. (2004) studied the effect of humic acid adsorption on the surface charge of clays and metal-oxides. Figures 18 and 19 show the pH-dependent electric charge of metal-oxides and clays with and without the presence of humic acid, respectively. It is evident that at humic acid concentration of 0.005 mmol dm^{-3} the differences in surface charge between iron oxide and clay minerals disappear almost completely. The *intrinsic surface charge* is dominated by the adsorption of humic acids.

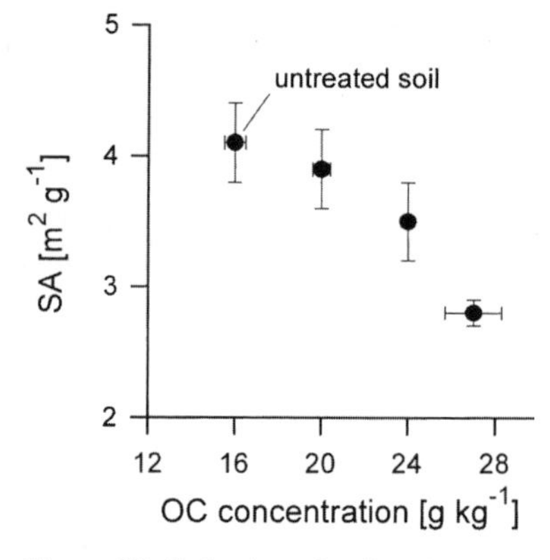

Figure 17. Reduction of soil surface area due to the adsorption of hydrophobic OM. [Used by permission of Elsevier, from Kaiser and Guggenberger (2000), *Organic Geochemistry*, Vol. 31, Fig. 3, p. 716.]

Particles aggregation/colloid stability. The pH-dependent aggregation of clays and metal-oxides particles was studied by Tombácz et al. (2004) and is shown in Figure 20. The presence of humic acid decreases the coagulation of all mineral particles. The adsorption of humic acid to Al-OH sites entirely hindered the edge-to-face coagulation of kaolinite particles. The study of particle flow, including a mix of iron oxide and clay, revealed that humic acid causes a liquefying effect on the suspension. Humic acid adsorption to Fe-OH sites and Al-OH sites significantly decreased the attraction between opposite charge particles of iron oxide and clay, therefore, aided breaking the heterocoagulated network formed between the particles.

The effects of OM adsorption on the fate of the OM itself

OM fractionation. As discussed above, the preferential adsorption of one OM compound over other compounds fractionates the OM between the aqueous and the solid interface

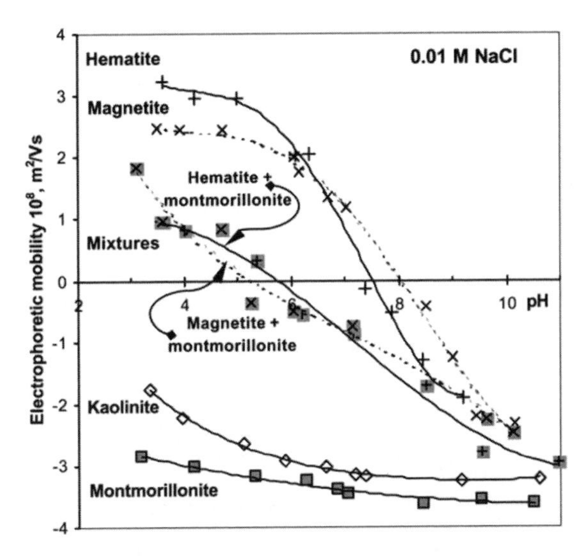

Figure 18. pH-dependent charge of OM-free iron-oxide and clay particles and their 1:1 mixture. [Used by permission of Elsevier, from Tombácz et al. (2004), *Organic Geochemistry*, Vol. 35, Fig. 2, p. 261.]

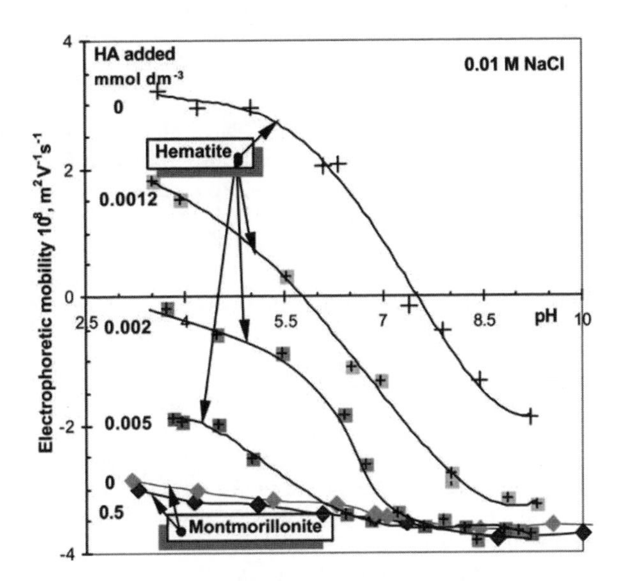

Figure 19. pH-dependent charge of hematite and montmorillonite. Black symbols indicate the presence of humic acid. [Used by permission of Elsevier, from Tombácz et al. (2004), *Organic Geochemistry*, Vol. 35, Fig. 5, p. 263.]

phases. Most of the studies show that molecules with high molecular weight, exhibiting more aromatic moieties tend to adsorb easily compared to those with small molecular weight and less aromatic moieties. An interesting observation for the well studied Suwannee River fulvic and humic acids was made in the study of Chi and Amy (2004). They observed that the fraction of high aromatic content and high molecular weight fulvic acid adsorbed preferentially while the opposite was observed for the humic acid, where the fraction of less aromatic content and

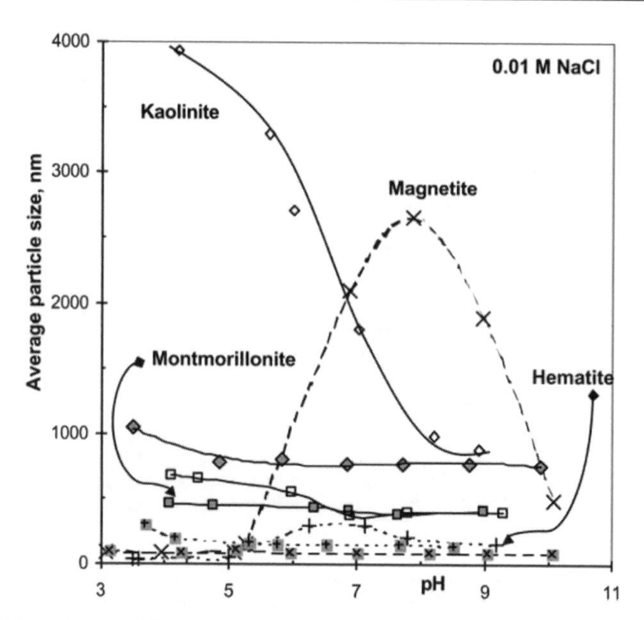

Figure 20. pH-dependent particle aggregation of iron oxide and clay particles for dilute systems in the absence (open symbols) and the presence (black symbols) of humic acid. [Used by permission of Elsevier, from Tombácz et al. (2004), *Organic Geochemistry*, Vol. 35, Fig. 6, p. 264.]

low molecular weight was preferentially adsorbed. This observation was explained by the differences in the carboxylic functional group density of the two acids. The corresponding low molecular weight fraction of both acids exhibit more carboxyl groups. The higher the carboxyl group density, the higher the molecular solubility on one hand, but its ligand exchange capability is also higher. Interplay between OM solubility and ligand exchange capability was, thus, suggested for the adsorption tendency of a given molecule. The high molecular weight fulvic acid is less soluble, thus, increasing its affinity to be adsorbed. The high molecular weight humic acid is also less soluble but its relatively low carboxyl group density reduces its adsorption affinity.

Another example of opposite trends, this time on the basis of the mineral surface is given by Carter (1978). He studied the adsorption of aspartic rich fulvic acid to calcite and quartz and found that calcite surfaces selectively adsorbed aspartic acid compared to quartz. Moreover, the fractionation increased as the ratio of total OM to calcite increased. Quartz surfaces adsorbed total amino acid to a similar degree as calcite but with no distinctive trend for any of the amino acids. Carter (1978) speculated that the selective adsorption of calcite is a consequence of the interaction between organic molecules and calcium sites on the mineral surface.

The fractionation process may greatly determine the fate of a given organic compound since processes such as transport and degradation may be enhanced or inhibited for the adsorbed fraction relative to the desorbed fraction.

Transport. Understanding transport processes of OM is essential to evaluate the fate of the OM itself and of contaminants and nutrient which may adsorb to it. Dunnivant et al. (1992) have studied the transport of naturally occurring DOM in laboratory columns containing aquifer sediments. Compared to a conservative tracer, the BTC of OM should be retarded if adsorption processes govern its fate. On the other hand, size exclusion process of OM can accelerate OM mobilization relative to a conservative tracer. Dunnivant et al. (1992) observed that the initial breakthrough of both OM and conservative tracer (Br^-) commenced almost simultaneously, but the OM BTCs were followed by extended tailing. They concluded that OM retardation was promoted by adsorption processes and that size exclusion did not play a significant role. To verify that size exclusion is not masked by adsorption, Dunnivant et al. (1992) carried out another column experiment in which the soil matrix was pre-equilibrated with DOM. They observed that OM and Br^- BTCs overlapped each other this time, suggesting again no significant size exclusion. Moreover, characterization of the DOM showed a preferential adsorption of hydrophobic components over hydrophilic components. Natural hydrophobic components were also found to be retained stronger compared to acidic hydrophobic components.

OM degradation and catalysis. Heterogeneous catalysis of OM on mineral surfaces is of paramount importance in the chemical industry and in environmental issues concerning the degradation of toxic and hazardous compounds. A review of heterogeneous catalysis by mineral surfaces, including photocatalysis, is discussed by Schoonen et al. (1998) while Liotta et al. (2009) reviewed advanced oxidative processes of phenols mediated by different mineral surfaces in the wastewater treatment industry. Two key characteristics that define the catalyst (i.e., the mineral surface) are *activity* and *selectivity* (Schoonen et al. 1998). *Activity* refers to what extent the catalyst could accelerate the reaction, which can be expressed as the decrease in activation energy relative to the uncatalyzed reaction. When more than one reaction pathway exists, then *selectivity* refers to the catalyst's ability to facilitate a specific reaction pathway.

An extensive study on the catalytic effect of mineral surfaces on acetate and acetic acid hydrothermal decarboxylation (i.e., breakage of the C-C bond between the carboxyl group and the root molecule) was presented by Bell et al. (1994). They investigated the catalytic potential of quartz, calcite, montmorillonite, pyrite, hematite and magnetite. As mentioned above, the most effective adsorbents are clays and oxides. Therefore, it is not surprising that that Ca-montmorillonite, hematite and magnetite were found to be the most effective catalysts. The destabilization of the C-C bond is promoted by chemisorption of the adsorbate to the mineral surface; therefore, the effectiveness of the catalyst depends on its ability to chemisorb the OM.

Photocatalysis may proceed via a number of mechanisms, of which the most important are: (1) photolysis of adsorbates; (2) photoelectrochemical reaction; and (3) charge injection. These were briefly reviewed by Schoonen et al. (1998) who included detailed references. Two recent examples for indirect and direct degradation by photocatalysis are those for bisphenol A (Li et al. 2007) and atrazine (Lackhoff and Niessner 2002). Bisphenol A degradation was examined in the absence and presence of iron oxide and oxalic acid. In the combination of both, oxalate is first adsorbed to the iron oxide surface, and under UV illumination the complex is excited to form a series of radicals. Under these conditions, Li et al. (2007) found that bisphenol A degradation was significantly accelerated compared to the sole presence of iron oxide. For the four different temperatures studied, Li et al. (2007) noted that the rate of bisphenol A degradation followed the order of oxalate rate degradation. That is, bisphenol A degradation is probably promoted by oxalate photocatalysis on iron oxide in an interactive pathway. The exact mechanism still needs to be explored.

The photolysis of atrazine by different synthetic and natural particles was studied by Lackhoff and Niessner (2002). When Fe containing minerals are used as catalysts, the degradation rate coefficients were significantly higher than direct photolysis (i.e., no mineral surface was present), but orders of magnitudes lower than in the presence of Ti and Zn containing materials. In contrast, degradation rate coefficients of atrazine in the presence of volcanic ash or fly ash were lower compared to direct photolysis, although the former had a high titanium content. Further experiments suggested that adsorption equilibrium was lowered both by the irradiation and the phosphate buffer that was used in the experiments, thus, masking atrazine degradation by photocatalysis.

Ternary adsorption

Ternary adsorption is the coupled adsorption of inorganic cations and anionic organic molecules. There are two types of ternary complexes: type A, in which the metal ion bridges between the OM and the mineral surface and type B in which the OM bridges between the metal ion and the mineral surface. A type A ternary complex is formed via a cation-bridge mechanism. Often, a water molecule remains between the cation and the functional group so the polymer is bonded by hydrogen bonds. In that case the bridges are termed water bridges (Arnarson and Keil 2000). Cation-bridge mechanism is analogous to the formation of a positively charged organic-metal complex that is then adsorbed to a negatively charged surface site, or alternatively, the adsorption of organic molecule to a mineral site on which a metal cation is already adsorbed (Baham and Sposito 1994). The formation of a 'type A' ternary complex with a bivalent cation may be written as (Stumm 1992):

$$\equiv S-OH+Me^{2+}+L^{-}\leftrightarrow\equiv S-OMe-L+H^{+} \tag{12}$$

where $\equiv S$, Me^{2+} and L^{-} are the mineral sorption site, cation and OM functional group (i.e., the ligand), respectively. A multidentate ligand is needed for the formation of a 'type B' ternary complex:

$$\equiv S-OH+Me^{2+}+L^{-}\leftrightarrow\equiv S-L-Me^{2+}+OH^{-} \tag{13}$$

The adsorption of fulvic acid to treated montmorillonite was studied by Theng (1976). Prior to the adsorption experiments the clay was treated with different cations to achieve the desired cationic form of the clay. At any given concentration the amount of fulvic acid adsorbed increased in the order of $Ba^{2+} < Ca^{2+} < Zn^{2+} < La^{3+} < Al^{3+} < Cu^{2+} < Fe^{3+}$ and showed linear isotherms (i.e., constant partition coefficients). Figure 21 presents the ionic potential of each ion against the respective logarithm of isotherm slope (which is proportional to the logarithm of the partition coefficient). The ionic potential is defined as the ratio between the valence (z) and the ionic radius (r). It combines the opposing effects of ionic charge and size, and it is therefore a measure of the polarizing power of the cation, i.e., the extent to which a cation exerts its influence on a neighboring molecule or anion. In addition to their data on adsorption of fulvic acid, Theng (1976) presented data of Theng and Scharpenseel (1975) on the adsorption of humic acid. Except for Al^{3+} and Cu^{2+}, the linear relationship between ionic potential and the logarithm of the partition coefficient indicates that ionic potential of the exchangeable cation significantly affects the adsorption of both fulvic and humic acid to montmorillonite. Similarly, Liu and Gonzalez (1999) have shown that heavy metals (Pb^{2+}, Cd^{2+} and Cu^{2+}) increase the adsorption of humic acid to montmorillonite.

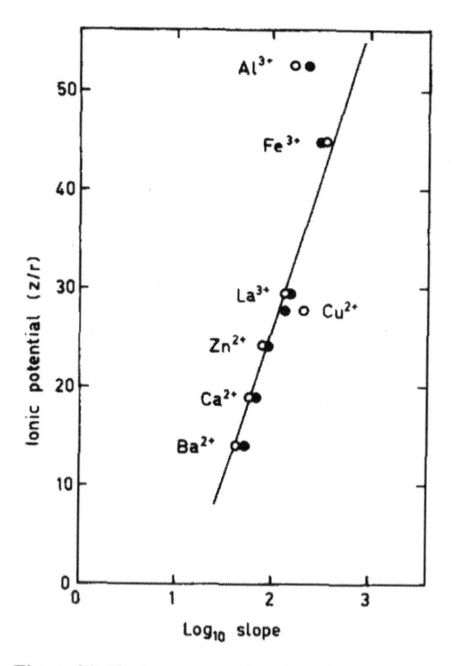

Figure 21. The ionic potential (z/r) of the exchangeable cation against the logarithm of the isotherm slope. The ionic radii (r) are expressed in nanometer units. Open circles are fulvic acid and close circles are humic acid. [Used by permission of Elsevier, from Theng (1976), *Geoderma* 15, Fig. 2, p. 248.]

Theng (1976) argued that polar organic compounds are generally held by clay mineral surfaces through a "water bridge," and that a direct association between cation and adsorbed organic species is possible only in situations where the cation has a low solvation energy or where it has previously been dehydrated. Support for the water bridge mechanism came from the observation that up to ¾ of the adsorbed fulvic acid was easily desorbed with one wash of deionized water. Similar observations of Chi and Amy (2004) for the adsorption-desorption experiments of NOM in the presence of Ca^{2+} were discussed above. Moreover, a linear relation exists between the logarithm of the isotherm slope and 'the energy by which water in the primary hydration shell is held to the cation' (Theng and Scharpenseel 1975 in Theng 1976).

Jada et al. (2006) examined the adsorption of humic acid onto quartz in solutions bearing Ca, Zn, Ba or Cu chlorine salts (Fig. 22). At the pH studied (6), the surfaces of both quartz and humic acid are negatively charged. Therefore, complexation of cations with either of them would lower the electrostatic repulsion and permit adsorption. Jada et al. (2006) found that Cu^{2+} and Zn^{2+} are more efficient than Ba^{2+} and Ca^{2+} in reducing the humic acid charge. Therefore, it is expected that the sorption of humic acid would be enhanced in the presence of these cations. In agreement with this expectation, the adsorption in the presence of Cu^{2+} was very effective. However, the adsorption in the presence of Zn^{2+} was significantly smaller than in the presence of Ba^{2+}. Jada et al. (2006) suggested that the enhanced effect of Ba^{2+}, compared to Zn^{2+}, is likely due to its preferential adsorption to the quartz surface as a result of its large ionic radius.

Weng et al. (2005) and Weng et al. (2008) thoroughly studied and modeled the adsorption of metal cations (Ca^{2+} and Cu^{2+}) to goethite in the absence (binary system) and in the presence of OM (fulvic acid, ternary system). The effect of fulvic acid is pH dependent for the adsorption of Cu^{2+}, increasing the cation adsorption at pH <5 but decreasing it at higher pH. Ca^{2+} adsorption increased in the presence of fulvic acid more pronouncedly. Weng et al. (2008) concluded that the OM-metal interactions at the mineral surface are stronger in the case of Ca^{2+} than in the case

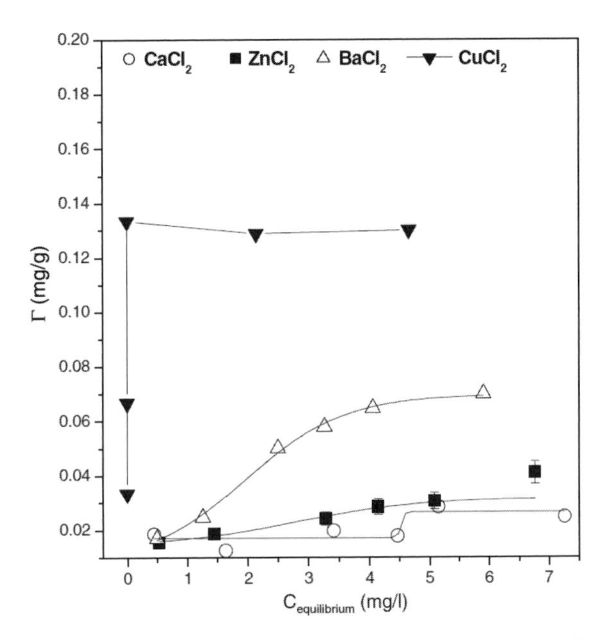

Figure 22. Adsorption isotherms of humic acids to quartz in the presence of various divalent cation (chlorine salt concentration = 3×10^{-4} M) at pH 6. [Used by permission of Elsevier, from Jada et al (2006), *Chemosphere*, Vol. 64, Fig. 6, p. 1292.]

of Cu^{2+}; this is in contrast to the fact that Cu^{2+} has a stronger affinity both to oxides and NOM. By modeling both the binary and ternary results, Weng et al (2008) concluded that charge distribution leads to stronger electrostatic interaction between Ca^{2+} and fulvic acid.

Desorption of OM

Essentially irreversible or slow desorption of OM is considered indicative of chemisorption (i.e., adsorption is site specific, Zullig and Morse 1988; Chi and Amy 2004) and the interaction may be referred as inner-sphere surface complexation (Sposito 1989). On the other hand, reversible desorption may reflect physical adsorption of OM (Chi and Amy 2004) or non-specific interaction, which may be referred as outer-sphere surface complexation (Sposito 1989; Stumm 1992).

When desorption occurs, it does not necessarily follow the reverse adsorption path, a phenomenon termed hysteresis. Adsorption-desorption hysteresis was reported by some investigators, for example for the adsorption of NOM onto iron oxides (Gu et al. 1994, 1995) and polychlorinated biphenyl (HCBP, an isomer of PCB) on clays (Di Toro and Horzempa 1982). Such behavior may be the result of NOM heterogeneity and fractionation (Gu et al. 1995): high affinity compounds adsorb to the mineral, thus, changing the solution composition. When part of the solution is replaced with NOM free solution to allow desorption, the solution composition and equilibrium are not identical between adsorption and desorption experiments. Therefore, an adsorption-desorption hysteresis is, to some extent, inevitable. Di Toro and Horzempa (1982) concluded that their observed hysteresis cannot be an experimental artifact. They described the adsorption reaction in terms of reversible and resistant components, the latter may or may not desorb. Another explanation is that the hysteresis phenomena may be the result of slow desorption kinetics compared to adsorption (Gu et al. 1995; Chi and Amy 2004), although Gu et al. (1994) and DiToro and Horzempa (1982) showed that even extended desorption time had little effect on the total amount desorbed. Chin and Amy (2004) measured the specific UVA and OM molecular weight during desorption experiments of adsorbed NOM induced by dissolved phosphate. They concluded that high molecular weight OM with higher aromaticity is more easily desorbed compared to lower molecular weight OM with less aromaticity.

Kaiser and Zech (1999) examined desorption from goethite of hydrophobic and hydrophilic DOM fractions derived from subsoil by different solutions (see Fig. 23). Strong

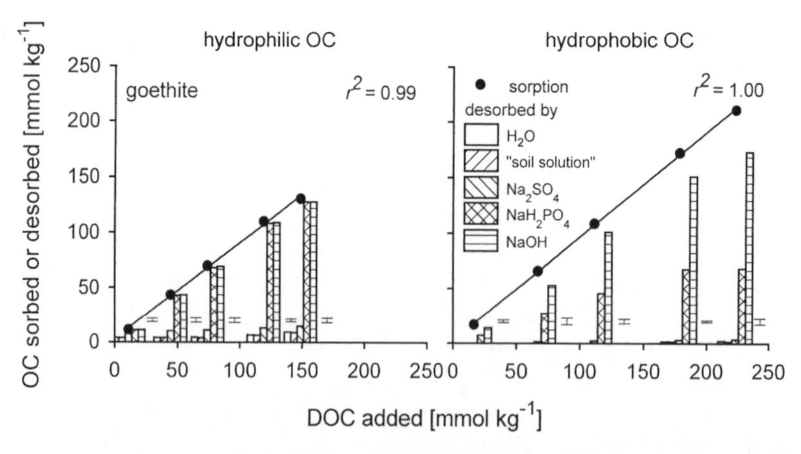

Figure 23. Desorption experiment of organic carbon from goethite by different solution. "Soil solution" is OM-free solution bearing the same inorganic composition used for the adsorption experiment. [Used by permission of Elsevier, from Kaiser and Guggenberger (2000), *Organic Geochemistry*, Vol. 31, Fig. 6, p. 720.]

desorption followed in the presence of inorganic oxyanions (e.g., SO_4^{2-}, $H_2PO_4^-$) especially with phosphate. The complete desorption of the hydrophilic fraction suggests that this fraction was initially adsorbed by weak non-specific interactions. Following the consecutive desorption methodology of DiToro and Horzempa (1982), Wang and Lee (1993) calculated that desorption of amines from clays is almost completely reversible. As discussed earlier, they observed that adsorption of methylamines increased with increasing molecular size. However, the smaller methylamines exhibit a larger irreversible fraction. Palmer and Bauer (1961) inferred that *gaseous* trimethylamine cannot adsorb into inter layer sites of montmorillonite since it is too big. Based on this assumption, Wang and Lee (1993) tried to explain that smaller methylamines had a larger irreversible fraction, since inter layer complex is more stable than the complex formed on the mineral surface.

Adsorption kinetics

So far adsorption and desorption reactions were treated as equilibrium reactions. Many investigators report that adsorption equilibrium is achieved within a few hours or even faster. For example, Ganor et al. (2001) reported that oxalate adsorption to kaolinite reached equilibrium within 30 minutes. By vigorously shaking the sample for 30 seconds, Ganor et al. (2001) were able to reduce the equilibration time to less than 1 min. From a kinetic point of view the adsorption rate can be depended on three sub-processes: 1. transport of the molecule towards the surface by diffusion and/or convection, 2. attachment to the surface and 3. re-conformation of the adsorbed molecules (Dijt et al. 1990; Stumm 1992). The latter stage deserves attention for the adsorption of large flexible molecules, especially since both conformation and adsorption reaction were found to be of similar time scales (Dijt et al. 1990 and references therein).

Dijt et al. (1990) used a reflectometry technique in a *stagnation point flow* experimental setup to continuously determine the adsorbed mass per unit area. A detailed description of the experimental setup and more theoretical background can be found elsewhere (Dijt et al. 1990; Avena and Koopal 1998, 1999). Briefly, in stagnation point flow a solution is injected perpendicular to the adsorbing surface, the intersection between the two is the stagnation point. This setup allows calculating the flux of adsorbing molecules (J) according to (Dijt et al. 1990; Avena and Koopal 1999):

$$J = KD^{2/3}(c_b - c_s) \tag{14}$$

where K is a constant for dilute solutions and is a function of the hydrodynamic properties of the experimental setup, D is the adsorbate diffusion coefficient and c_b and c_s are adsorbate concentrations in the bulk solution and in the subsurface adjacent to the adsorbent surface, respectively. For spherical molecules, D should increase with decreasing molecule size. For humic acid molecules, D should increase as the hydrodynamic volume of the molecule decreases. The latter is dependent on external properties such as pH and the electrolyte concentration. A decrease in pH and an increase in electrolyte concentration will decrease the hydrodynamic volume, therefore, increasing D.

The amount adsorbed is determined by means of a polarized laser beam reflected at the stagnation point. The technique of reflectometry makes use of the fact that the reflectivity of the adsorbing surface changes through the adsorption reaction. The amount of adsorbate adsorbed (Γ) is determined according to (Dijt 1993 in Avena and Koopal 1998):

$$\Gamma = \frac{1}{A_s} \cdot \frac{\Delta S}{S_0} \tag{15}$$

where A_s is a sensitivity factor that depends on geometrical and refractive indexes of the experimental setup, S_0 is the reflective signal of the bare adsorbent surface and ΔS is the measured change in surface reflectivity. Since mass transfer can be adequately calculated, this method provides the means to distinguish between transport and surface control processes; the latter are henceforth referred to as attachment processes (Dijt et al. 1990).

The above method was used by Avena and Koopal (1998) to study the kinetics of humic acid adsorption-desorption to iron oxides and by Avena and Koopal (1999) to study the kinetics of humic acid adsorption to hydrophobic (Fe_2O_3 and Al_2O_3) and hydrophilic (polystyrene and silanized SiO_2) surfaces. Figure 24 presents a typical adsorption experiment showing that adsorption initiates rapidly for 20-80s and then followed by a slower stage. The slower stage may last for hours and is probably related to long-range electrostatic repulsion, rearrangement in the adsorbed layer by conformational changes and exchange of small molecules adsorbed by large molecules. The initial adsorption rate, $(d\Gamma/dt)_i$, is plotted against the adsorbate concentrations in the bulk solution (c_b) in the inset of Figure 24 for iron oxide and against electrolyte concentration and pH for all adsorbent surfaces in Figures 25 and 26, respectively. At pH 3.9 for the case of iron and aluminum oxides, the electrostatic attraction should be considered due to the opposing charges of the adsorbent and adsorbate. Increasing ionic strength should, therefore, decrease the initial adsorption rate if attachment of adsorbate to adsorbent is the limiting process. Since up to an electrolyte concentration of ~0.03M, the opposite trend is observed (Fig. 25), Avena and Koopal (1999) considered the reaction to be transport-limited. Moreover, up to 0.03M an increase in the diffusion coefficient was calculated which supports an increase in $(d\Gamma/dt)_i$ for the case of transport-limited processes. Above an electrolyte concentration of ~0.03M a decrease in $(d\Gamma/dt)_i$ is probably due to screening of the electrolyte attraction (i.e., attachment processes become more limiting) while above 0.1M aggregation of adsorbent molecules also contributes to the decrease in $(d\Gamma/dt)_i$. For the pH case (see Fig. 26) the plots for the hydrophilic surfaces preserve a shape distinctive for electrostatic effect. When pH<IEP the opposing charges of adsorbent and adsorbate promote adsorption (i.e., adsorption is transport-limited), whereas for pH>IEP repulsion between adsorbent and adsorbate increases the reaction activation energy and, thus, lowers $(d\Gamma/dt)_i$ and adsorption becomes attachment-limited. Notice that up to pH≈5 the initial adsorption rate is not pH dependent.

Theories and models describing OM adsorption and desorption reactions

This section describes examples of different models that have been proposed to describe OM adsorption. As demonstrated above, OM adsorption depends on many variables related to all three phases (i.e., solution, OM, and mineral surface), whereas the adsorption of NOM is further complicated because it is a mixture of many compounds. Therefore, a theoretical and consistent description of NOM adsorption is difficult and may be impossible (Gu et al. 1995).

Surface complexation models. The most common model for describing adsorption of both inorganic and OM on mineral surface is the surface complexation (or coordination) model (SCM) which in its modern form was mainly developed in the works of Paul Schindler, Werner Stumm and their co-workers (Dzombak and Morel 1990). In this classical model, adsorption is described as chemical reactions between surface sites and dissolved species. Several variation of the SCM have been developed, all of them are based on four basic principles (Dzombak and Morel 1990):

(1) Sorption takes place at specific surface sites.

(2) Sorption reactions can be described quantitatively via mass law equations.

(3) Surface charge results from the sorption reactions themselves.

(4) The effect of surface charge on sorption can be taken into account by applying a correction factor derived from the electric double layer theory to mass law constant for surface reactions.

The product of a SCM is a set of simultaneous equations that can be solved numerically. These equations include (Dzombak and Morel 1990):

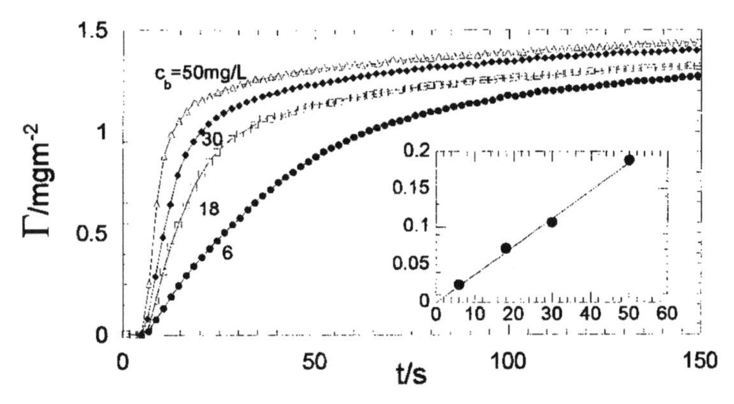

Figure 24. Effect of the concentration of humic acid (c_b) on its adsorption to iron oxide. Inset presents initial adsorption rate (mg m^{-2} s^{-1}) as a function of c_b (mg L^{-1}). [Used by permission of American Chemical Society, from Avena and Koopal (1999), *Environmental Science & Technology*, Vol. 33, Fig. 1, p. 2742.]

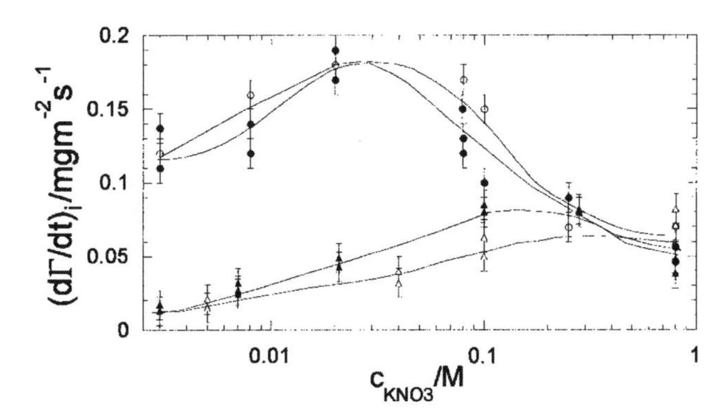

Figure 25. The effect of electrolyte concentration on the initial adsorption rate of humic acid for the different adsorbent surfaces. (●) Fe_2O_3; (○) Al_2O_3; (▲) silanized SiO_2; and (Δ) polystyrene. pH = 3.9±0.1, c_b = 30 mg L^{-1}. [Used by permission of American Chemical Society, from Avena and Koopal (1999), *Environmental Science & Technology*, Vol. 33, Fig. 2, p. 2742.]

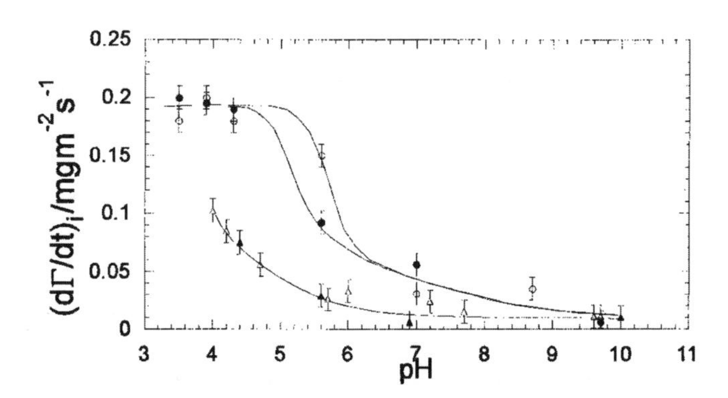

Figure 26. The effect of pH on the initial adsorption rate of humic acid for the different adsorbent surfaces. (●) Fe_2O_3; (○) Al_2O_3; (▲) silanized SiO_2; and (Δ) polystyrene. C_{KNO_3} = 0.1 M, c_b = 30 mg L^{-1}. [Used by permission of American Chemical Society, from Avena and Koopal (1999), *Environmental Science & Technology*, Vol. 33, Fig. 3, p. 2743.]

(1) Mass law equations for all possible surface reactions.

(2) A mass balance equation for each type of surface sites.

(3) An equation for the computation of the surface charge.

(4) A set of equations representing the constraints imposed by the model chosen for the structure of the electric double layer.

A typical SCM for the adsorption of an organic ligand takes into account both the surface protonation reactions and the adsorption reactions. For example, Kummert and Stumm (1980) presented a surface complexation model for the adsorption of organic acids on γ-Al_2O_3 using five equilibrium reactions:

$$\equiv AlOH_2^+ \leftrightarrow \equiv AlOH + H^+ \qquad K_{(16)} = \frac{\{\equiv AlOH\}\cdot[H^+]}{\{\equiv AlOH_2^+\}} \tag{16}$$

$$\equiv AlOH \leftrightarrow \equiv AlO^- + H^+ \qquad K_{(17)} = \frac{\{\equiv AlO^-\}\cdot[H^+]}{\{\equiv AlOH\}} \tag{17}$$

$$\equiv AlOH + H_2X \leftrightarrow \equiv AlXH + H_2O \qquad K_{(18)} = \frac{\{\equiv AlXH\}}{\{\equiv AlOH\}\cdot[H_2X]} \tag{18}$$

$$\equiv AlOH + H_2X \leftrightarrow \equiv AlX^- + H^+ + H_2O \qquad K_{(19)} = \frac{\{\equiv AlX^-\}[H^+]}{\{\equiv AlOH\}\cdot[H_2X]} \tag{19}$$

$$2 \equiv AlOH + H_2X \leftrightarrow \equiv Al_2X + 2H_2O \qquad K_{(20)} = \frac{\{\equiv Al_2X\}}{\{\equiv AlOH\}^2\cdot[H_2X]} \tag{20}$$

where X is the organic ligand, the symbol $\equiv$ denotes a surface site, values in { } are surface concentrations (mol/kg), values in [ ] are concentrations (mol/L) and $K_{(i)}$ is the equilibrium constant of the reaction in Equation (i). In this model, the surface charge (Q) was calculated from the equation:

$$Q = \frac{[\equiv AlOH_2^+] - [\equiv AlO^-] + [\equiv AlX^-]}{a} \tag{21}$$

where a is the amount of added γ-Al_2O_3 per liter solution (kg/liter).

SCM were used successfully for describing sorption of ions on sorbents having relatively well-defined surface groups such as metal oxides (Dzombak and Morel 1990). However, it was often found that the adsorption of weak polyelectrolytes on oxide surfaces over broad ranges of solution conditions cannot be described accurately with SCM (Au et al. 1998, 1999). SCM does not account for the macromolecular nature of many NOM. In addition, the SCM does not provide information about the interfacial conformation of adsorbed polyelectrolytes.

SC/SF hybrid model for the adsorption of polyelectrolytes. Au et al. (1998) and Au et al. (1999) combined two adsorption theories to elucidate the adsorption of NOM at oxide/water interface. The combined model was based on the self-consistent-field theory (SF) of Scheutjens and Fleer (1979) and on SCM.

The SF model is a statistical thermodynamic model which accounts for the adsorption of polyelectrolyte. In a nutshell, the SF model describes the adsorbed molecule as a mildly acidic, flexible polyelectrolyte which is comprised of an assemblage of segments. Each surface site can be occupied by one segment (not all segments have adsorption affinity). Adsorption equilibrium is driven by enthalpy effects and conformational entropy effects (i.e., the way polyelectrolytes are distributed within the solid/liquid interface: polyelectrolytes lose conformational entropy when they are adsorbed to any surface). A drawback of the SF model is that the specific chemical interactions between functional groups and the surface sites are not addressed mechanistically (Au et al. 1999). The non-electrostatic (specific) interactions are represented by an averaged constant parameter, denoted as χ_s, which is independent of the solution chemistry (e.g., pH, ionic strength). This approach is modified in the combined SC/SF model of Au et al. (1999) which accounts for both the polyelectrolyte and the specific interaction (ligand exchange reaction) of macromolecules with oxide surfaces. The non-electrostatic complexation parameter is assigned as χ_s(ads) for each specific-specific ligand and surface site species. This parameter is independent of the solution condition and is, thus, constant for the specific reaction. In the study of NOM adsorption (Au et al. 1999) the following simple reaction was found to be sufficient to describe the adsorption:

$$\equiv Me\text{OH} + L^- = \ \equiv Me\text{OH}L^- \tag{22}$$

where $\equiv Me$OH is the proton activated surface site and L^- is a deprotonated NOM group. The fraction of $\equiv Me$OH ($f_{Me\text{OH}}$) and L^- (f_L) are functions of the given solution conditions (pH, ionic strength). Au et al. (1998) assigned an average complexation parameter to all the surface sites species which was termed $\chi_s(hyb)$ and can be determined by:

$$\chi_s(hyb) = f_{Me\text{OH}} \cdot f_{L^-} \cdot \chi_s(ads) \tag{23}$$

According to Equation (23), $\chi_s(hyb)$ depends linearly on an intrinsic complexation parameter $\chi_s(ads)$, which is independent of the solution condition, and on the fractions of $MeOH$ (f_{MeOH}) and L^- (f_L). Since f_{MeOH} and $f_{L\text{-}}$ depend on the solution conditions, it follows that $\chi_s(hyb)$ also depends on these conditions. This dependency is the main difference between the SC/SF model and the SF model. As $\chi_s(ads)$ does not depend on the solution conditions, $\chi_s(hyb)$ must be determined only for a baseline set of solution conditions (*base*). Then $\chi_s(hyb)$ for any other set of solution conditions (m) can be related by the following expression:

$$\left[\frac{\chi_s(hyb)}{f_{Me\text{OH}} \cdot f_{L^-}}\right]_{base} = \left[\frac{\chi_s(hyb)}{f_{Me\text{OH}} \cdot f_{L^-}}\right]_m \tag{24}$$

Applying the SC/CF hybrid model to NOM adsorption allowed assessment of features such as adsorption density and hydrodynamic layer thickness (δ_h), the later reflects the conformational state of the adsorbed polyelectrolytes. Measurements showed that δ_h increased with increasing ionic strength indicating that the molecules protruded further into the solution (Au et al. 1999). The SC/CF hybrid model provided satisfactory agreement with experimental measurements in the case of oppositely charged NOM and mineral surfaces, thus, providing an insight into both specific interaction and the polyelectrolytic nature of the adsorbed molecule.

Sequential adsorption model. Arnarson and Keil (2000) studied the adsorption of NOM onto montmorillonite in sequential adsorption experiments. In these experiments the NOM solution was equilibrated with montmorillonite for 2 h, thereafter the montmorillonite was replaced with clean montmorillonite and the solution was re-equilibrated. The procedure was repeated 5 times and the results were fitted with the following equation:

$$C_{L,eq} = \Sigma f_i e^{-K_{d,i}X} \tag{25}$$

where $C_{L,eq}$ is the equilibrium DOM concentration, f_i is the fraction of OM component i in the solution, $K_{d,i}$ is the partition coefficient of component i and X is the cumulative concentration of solid added to the solution.

In their experiments Arnarson and Keil (2000) studied two fractions of NOM derived from sea sediments: 1. pore water NOM (pNOM) and easily extracted NOM (eNOM) extracted from the bulk sediments with organic-free sea water. The two fractions were further subdivided on the basis of molecular weight using 1000 Da filter. The results were fitted to the model assuming one surface reactive NOM component and one non-surface reactive component (see Fig. 27). Although the use of only two components is not realistic (i.e., NOM consist of a multiple components with a continuum of surface affinities), the authors found that increasing the number of components did not improve the fit of the model. Therefore, they concluded that the NOM heterogeneity can be averaged by these two components. By this method they

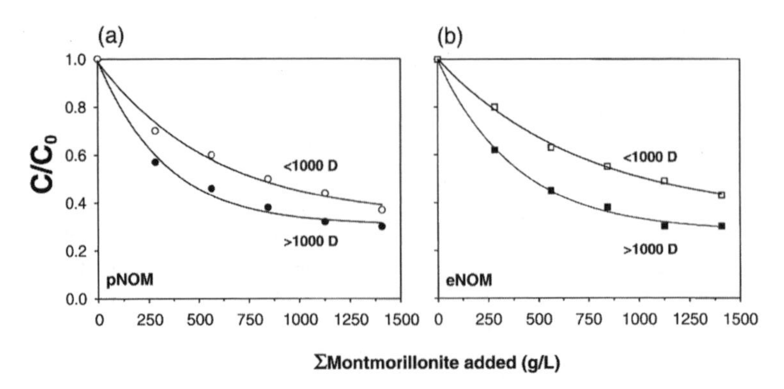

Figure 27. Sequential adsorption isotherms for (a) pNOM and (b) eNOM. Lines represent model fits using two components (surface reactive and non-surface reactive fractions). C/C_0 is the equilibrium DOC concentration after each addition of solid normalized to the initial concentration. [Used by permission of Elsevier, from Arnarson and Keil (2000), *Marine Chemistry*, Vol. 71, Fig. 3, p. 313.]

found that all NOM examined contained about ~30% non-surface reactive component. In regular equilibrium batch adsorption experiments this component tends to lower the partition coefficient. When correcting it according to the non-surface reactive component estimated by the sequential experiments, the authors reported an excellent match between partition coefficients calculated in the two experimental methods.

Modified Langmuir model. A decrease in adsorption energy is sometimes evident when the adsorption density on the mineral surface increases (Gu et al. 1994 and references therein). Gu et al. (1994) and Gu et al. (1995) proposed and tested a *modified Langmuir model* to account for this change and to account for adsorption-desorption hysteresis effect:

$$X_{i,ads} = F_i \frac{Ke^{-2b \cdot X_{i,ads}} \cdot C_i}{1 + Ke^{-2b \cdot X_{i,ads}} \cdot C_i} \tag{26}$$

where $X_{i,ads}$ is the amount of NOM adsorbed per surface area, F_i is the maximum surface coverage of the NOM, which is also equal to the maximum surface sites available for adsorption, C_i is the solution NOM concentration, K is an equilibrium constant of the reaction and b is a parameter which accounts for the changing adsorption energy with a changing surface coverage. Notice that if $b = 0$ (i.e., a constant energy of adsorption) Equation (26) is identical to the Langmuir isotherm (Eqn. 7). A detailed description of the model can be found in Gu et al (1994). Mathematically, the *modified Langmuir model* can be fitted to a variety of adsorption isotherms including L-type, S-type, H-type and C-type by modifying the two parameters, K and b, and holding F_i (Gu et al. 1994).

When desorption is completely reversible, Equation (26) describes both adsorption and desorption isotherms. However, as previously noted, adsorption-desorption hysteresis phenomena were reported in which the amount desorbed is smaller than the amount adsorbed. For these reasons Gu et al. (1994) introduced a correction factor for the desorption isotherm to account for hysteresis:

$$X_{i,ads} = F_i \frac{Ke^{-2b \cdot X_{i,ads}} \cdot C_i}{Ke^{-2b \cdot X_{i,ads}} \cdot C_i + \left(C_i / C_{i,a}\right)^h} \tag{27}$$

where $C_{i,a}$ is the equilibrium adsorbate concentration after adsorption but before commencing the desorption experiment and h is defined as a hysteresis coefficient. Desorption will commence when a fresh solution in which $C_i<C_{i,a}$ would be introduced. Theoretically speaking, C_i will equal 0 only after an infinite number of desorption cycles, and therefore, the condition $0 < C_i \leq C_{i,a}$ is required for Equation (27). When Equation (27) satisfies the condition that $h = 0$, it becomes identical to Equation (26) (i.e., the reaction is completely reversible sorption-desorption process). Also, at $h = 1$:

$$X_{i,ads} = F_i \frac{Ke^{-2b \cdot X_{i,ads}} \cdot C_{i,a}}{Ke^{-2b \cdot X_{i,ads}} \cdot C_{i,a} + 1} \tag{28}$$

and the reaction becomes independent of C_i but dependent on $C_{i,a}$ indicating a complete irreversible reaction. Therefore, the hysteresis coefficient, h, was considered by Gu et al. (1994) as a measure of sorption-desorption reversibility.

Gu et al. (1995) examined the adsorption of hydrophilic and hydrophobic fractions of NOM onto iron oxides (hematite), and showed that the adsorption affinity decreased when surface coverage increased (plotted as q, Fig. 28). These observations support their hypothesis that the adsorption affinity of NOM components decreases with surface coverage because of the heterogeneity of both NOM and mineral surface. That is, in the first adsorption stages the most reactive (high affinity) NOM components reacted with the mineral surface, followed by less reactive NOM components adsorbing in the following stages. Moreover, they speculated that not all anionic functional groups are directly involved in chemical bonding to surface sites. Free anionic moieties will repel each other, and therefore will lower the adsorption.

Figure 29 shows both adsorption and desorption experiments (symbols) and the fitted model (lines) for the hydrophilic and hydrophobic fractions at two pH values. Model parameters F_i, K, and b are the same for both adsorption and desorption, and the only parameter fitted for the desorption isotherm is h. Both hydrophilic and hydrophobic fractions exhibit a large

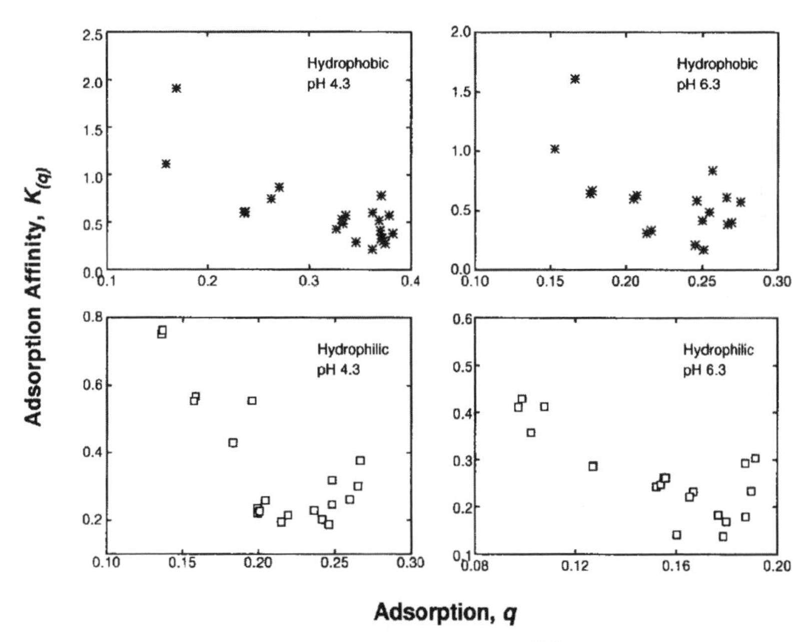

Figure 28. Relationships between the adsorption affinity ($K_{(q)} = Ke^{-2b \cdot X_{i,ads}}$ in Eqns. 26 and 27) and surface coverage ($q = X_{i,ads}$) for the hydrophilic and hydrophobic NOM fractions at two pH values. [Used by permission of Elsevier, from Gu et al. (1995), *Geochimica et Cosmochimica Acta*, Vol. 59, Fig. 3, p. 223.]

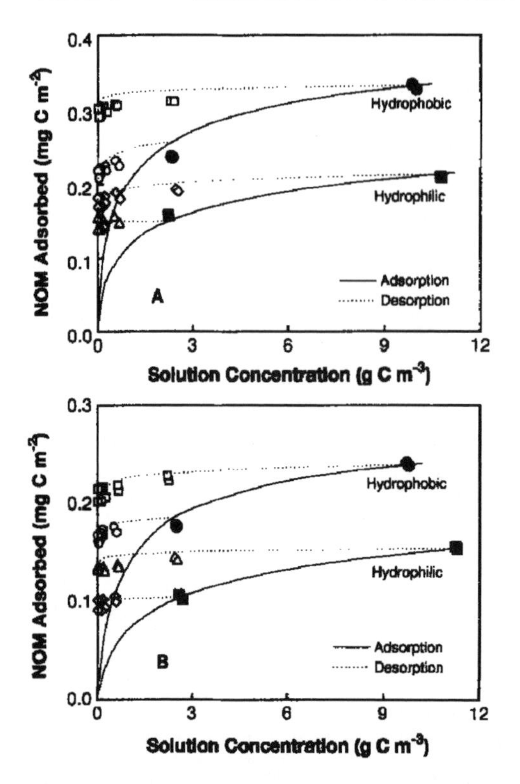

Figure 29. Adsorption (solid symbols) and desorption (open symbols) isotherms of hydrophobic and hydrophilic NOM fractions on iron oxide at pH 4.3 (A) and pH 6.5 (B). Solid and dashed lines were adsorption and desorption model simulations according to Equations (26) and (27), respectively. [Used by permission of Elsevier, from Gu et al. (1995), *Geochimica et Cosmochimica Acta*, Vol. 59, Fig. 5, p. 224.]

hysteresis coefficient ($0.81 < h < 0.98$) indicating their strong adsorption.

Summary

Sorption of OM to mineral surfaces has been extensively discussed in the literature. The above section does not intend to cover all the literature but merely to give representative examples. Since sorption affects greatly the fate of OM, as well as inorganic dissolved species, it is vital to define, qualitatively and quantitatively, the factors affecting it. It is evident that sorption is dependent on the mineral and OM, and solution properties but that these properties are interrelated. Any mechanistic description of OM sorption should be based on a separate analysis of both the OM and mineral surface at different solution conditions. Only then should adsorption or desorption experiments be investigated experimentally. It is important to bear in mind that sorption may be the synergic result of two or more mechanisms (e.g., ligand exchange and Van der Waals). Adsorption-desorption processes may be modeled with either a more theoretical approach (e.g., SC/SF model) or a more empirical approach (e.g., isotherms equations). OM diversity makes it complicated to define one consistent model which will generalize all these aspects. Some examples of how the mineral surface is affected by adsorption were also discussed, including the mineral surface area and charge and coagulation of particles. Water-rock interactions are further affected when dissolution and precipitation of minerals are subjected to the presence of OM. These issues are the topics of the next sections, where OM adsorption is many times the initial stage of these processes.

THE EFFECT OF OM ON MINERAL DISSOLUTION

General rate laws of dissolution reactions

In order to understand the effect of OM on mineral dissolution, it is necessary to understand the effects of other parameters on dissolution kinetics. These effects often depend on both the environmental conditions and the reaction mechanism(s). For example, the dissolution rates of many minerals as a function of pH exhibit a U shape (Fig. 30), i.e., dissolution rates decreases as a function of pH under acidic conditions, becomes independent of pH under circum neutral pH and increases as a function of pH under basic conditions. This U shape plot may be described using an overall empirical rate law such as:

$$Rate = k'_{\mathrm{H}} \cdot a_{\mathrm{H^+}}^{n_{\mathrm{H^+}}} + k'_{water} + k'_{\mathrm{OH}} \cdot a_{\mathrm{OH^-}}^{n_{\mathrm{OH^-}}} \tag{29}$$

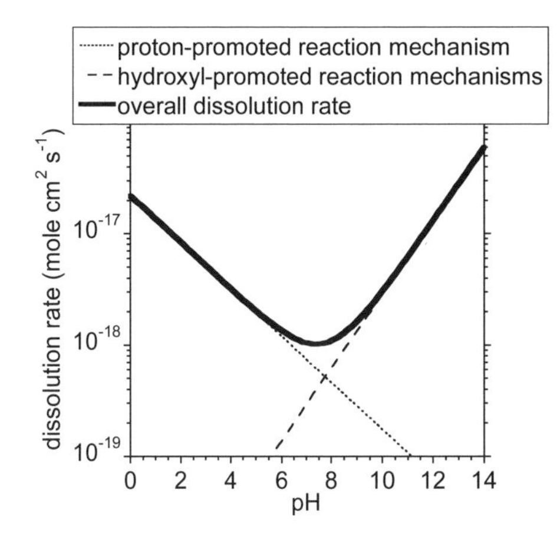

Figure 30. The pH dependence of the overall dissolution rate of smectite (solid line) demonstrates the typical U shape; the rate decreases as a function of pH under acidic conditions is independent of pH around neutral pH and increases as a function of pH under basic conditions. The curve was plotted using the empirical rate law of Golubev et al. (2006). The dotted and the dashed lines represent the rates of the proton-promoted and the hydroxyl-promoted reaction mechanisms, respectively (Eqn. 42). The rate of the hydrolytic-dissolution reaction mechanism (10^{-20}) is not shown as it is lower than the low limit of the plot and does not significantly influence the overall rate.

where k'_i (mol m^{-2} s^{-1}) is the rate coefficient of reaction mechanism i, and a_i is the activity of aqueous species i. The common interpretation is that the U shape dependence represents three parallel and independent reaction mechanisms proton-promoted, hydroxyl-promoted, and hydrolytic-dissolution reaction mechanisms (henceforth, PPRM, HPRM and HD, respectively). The PPRM dominates the overall rate under acidic conditions, HD at the intermediate pH range and HPRM at basic pH. The typical U shape pH dependence plot (see Fig. 30) indicates that under most pH conditions, only one of these reaction mechanisms dominates the overall rate.

In many studies the pH dependence of dissolution rate is modeled using a surface speciation model (e.g., Stumm and Wollast 1990; Blum and Lasaga 1991; Stumm 1992; Cama et al. 2005) in which the overall dissolution rate is described by:

$$Rate = k_{\mathrm{H}} \cdot C_{s,\mathrm{H}} + k_{water} + k_{\mathrm{OH}} \cdot C_{s,\mathrm{OH}} \tag{30}$$

where k_i (s^{-1}) is the rate coefficient of reaction mechanism i, and $C_{s,i}$ (mol·m^{-2}) is the mineral surface concentration of adsorbed species i.

The rate coefficients, k'_i, in Equation (29) are not constant and depend on the properties of the mineral, its surface area and the environmental conditions (e.g., temperature). A general form of a rate law for one mechanism of heterogeneous mineral surface reactions can be written as (Lasaga 1998):

$$Rate = k_0 \cdot S_a \cdot e^{-Ea/RT} \cdot a_H^{n_H} \cdot \prod_i a_i^{n_i} \cdot g(I) \cdot f(\Delta G) \tag{31}$$

where k_0 is a constant, S_a is the reactive surface area of the mineral, E_a is the apparent activation energy of the overall reaction, R is the gas constant, T is the temperature (K), a_i and a_H are the aqueous activities of species i and H^+, respectively, n_i and n_H are the orders of the reaction with respect to these species, $g(I)$ is a function of the ionic strength and $f(\Delta G)$ is a function of the Gibbs free energy. The much-studied pH dependence of dissolution reactions is represented explicitly by Equation (31). Terms involving activities of species in solution other than H^+ (i.e., a_i) incorporate other possible catalytic effects on the overall rate. The $g(I)$ term indicates a possible dependence of the rate on the ionic strength (I) in addition to that entering through the activities of a specific ion. The last term, $f(\Delta G)$, accounts for the important variation of the rate with deviation from equilibrium. As will be discussed below, one way by which OM affects the dissolution and precipitation rates of a mineral is by affecting one of the parameters that affect the rate in Equation (31). In particular, OM may affect the degree of saturation, and therefore the rate dependence on deviation from equilibrium must be discussed in detail.

The dependence of the dissolution rate on deviation from equilibrium ($f(\Delta G)$ term in Eqn. 31) may take a variety of forms (Lasaga et al. 1994). For example, an experimental $f(\Delta G)$ function of gibbsite dissolution at 80 °C and pH = 3 that was derived by Nagy and Lasaga (1992) is:

$$f(\Delta G) = -(1 - e^{(-8.1\pm1.0\cdot|\Delta G|/RT)^{3.01\pm0.05}}) \qquad (32)$$

where ΔG is the Gibbs free energy of the overall reaction (kcal/mol), R is the gas constant and T is the temperature (K). According to this function (Fig. 31), near equilibrium ($0 \geq \Delta G \geq -0.2$ kcal/mol) the rate increases gradually and approximately linearly with increasing undersaturation. Over the range $-0.2 \geq \Delta G \geq -0.5$ kcal/mol, the rates increase sharply. Far from equilibrium, at $\Delta G \leq -0.5$ kcal/mol, the dissolution rate is independent of the degree of saturation. This last region, where the function $f(\Delta G)$ is flat (Fig. 31), is termed the *dissolution plateau* (Lasaga et al. 1994).

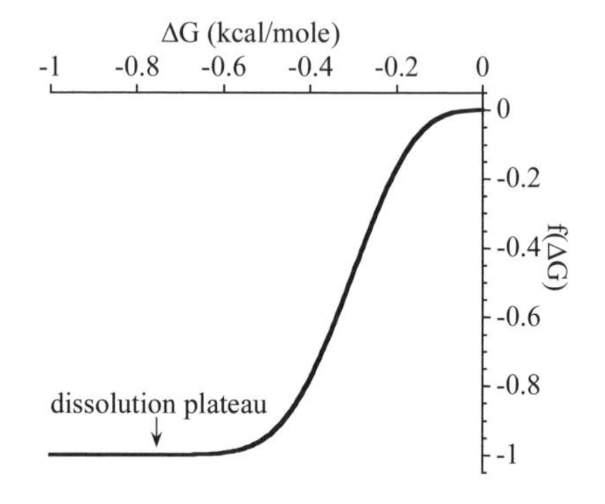

Figure 31. The effect of deviation from equilibrium on dissolution rate is described by the highly non linear $f(\Delta G)$ function observed originally by Nagy and Lasaga (1992) for synthetic gibbsite at pH = 3 and 80 °C. The far from equilibrium region where the function $f(\Delta G)$ is flat is termed the dissolution plateau.

OM may catalyze or inhibit dissolution rate. Whereas the a_i term in Equation (31) describes the possible effect of a catalyst, the role of inhibition is not described in Equation (31). Depending on the exact inhibition mechanism, the term a_i may also represent an inhibitor. However, commonly this is not the case, and other terms should be introduced into the equation (see for example Eqn. 33 of Brantley and Stillings 1996; Eqns. 14 and 43 of Ganor and Lasaga 1998; Eqns. 7.78 and 7.79 of Lasaga 1998; and Eqn. 44 of Ganor et al. 1999). In some of these inhibition mechanisms, competition between a catalyst and an inhibitor takes place and the general rate law cannot be described by a product of independent terms.

The formulation of Equation (31) is very useful because it relates the reaction rate to activities of ions in solution that may be obtained directly from the solution chemistry. However, mineral dissolution is a surface process and therefore it is more appropriate to express the dependence of the rate on the concentrations (or activities) of ions adsorbed on the mineral's surface rather than on the bulk activities in solution. Therefore, Equation (31) can be rewritten in terms of surface concentrations ($c_{s,i}$):

$$Rate = k_0 \cdot S_a \cdot e^{-E_a/RT} \cdot C_{s,\mathrm{H}}^{n_{\mathrm{H},ads}} \cdot \prod_i C_{s,i}^{n_{i,ads}} \cdot g(I) \cdot f(\Delta G) \qquad (33)$$

where the coefficients $n_{H,ads}$ and $n_{i,ads}$ are the reaction orders with respect to the adsorbed surface species. Equations (31) and (33) describe a general rate law for **one mechanism**. The overall rate of a reaction is the sum of the rates of all the **parallel mechanisms** involved.

The overall rate of dissolution in the presence of OM

OM may affect the overall rate of mineral dissolution either by catalyzing or inhibiting the "normal" reaction mechanism that dominates the overall rate in the absence of OM, or by promoting an alternative reaction mechanism. Such an alternative reaction mechanism, in which an OM ligand attacks the mineral surface, is usually referred to as ligand-promoted reaction mechanism (henceforth, LPRM). In the following section we will first discuss the effects of OM on the normal dominant reaction mechanism. This discussion will be followed by elaboration of the parameters that may be affected by OM as shown in Equation (31).

As was mentioned above, OM may affect the solution composition. The most important effect is due to the formation of metal-organic complexes, which reduces the solution concentration of the free metal. Of special importance is the ability of OM to chelate divalent and trivalent cations (Jackson et al. 1978). The degree of saturation of the solution with respect to dissolution of a mineral depends on the concentrations of the ions that are released by the reaction. If the concentration of any of these ions is reduced due to interactions with OM, the solution becomes more undersaturated and the dissolution rate may increase (Fig. 31). Such effects were proposed for oxides, silicates, sulfates and to leaser extent for carbonates.

According to this line of reasoning, this effect of OM on dissolution rate should be limited to near-equilibrium conditions since under far-from-equilibrium conditions the rate is independent of the degree of saturation (dissolution plateau region, Fig. 31). However, it has been shown in several studies that dissolution rate of kaolinite and feldspar is retarded in the presence of Al^{3+} even under far from equilibrium conditions (Gautier et al. 1994; Oelkers et al. 1994; Devidal et al. 1997; Cama et al. 2002). Chelation of such an inhibitor will reduce the concentration of its free species and as a result dissolution will be enhanced. Therefore, the chelation effect may be effective both under close-to-equilibrium and far-from-equilibrium conditions.

Organic acids may decrease soil and water pH and as a result may enhance the proton-promoted reaction mechanism and slow down the hydroxyl-promoted mechanism. As the dissolution rate of many silicates is pH independent under circum neutral pH conditions (i.e., at the pH range of many natural soil solutions), it is reasonable to assume that a slight lowering of the pH by organic acids will have a relatively small effect on weathering (Mast and Drever 1987). In few laboratory studies where organic acids enhanced dissolution rate, it was suggested that the enhancement was only due to pH effects (e.g., Manley and Evans 1986).

The work of Furrer and Stumm (1986) on the catalytic effect of dicarboxylic acids and aromatic acids on the dissolution of oxides, which has been cited more than 200 times since its publication, is an important benchmark in the study of OM effect on mineral dissolution. They proposed a detailed surface-controlled reaction mechanism, in which the overall dissolution rate in the presence of organic ligands under acidic conditions is the sum of the rates of two independent surface-reaction mechanisms: PPRM and LPRM. Both mechanisms involve the adsorption of either a proton or a ligand on the crystal surface, which results in a weakening of the bonds of the mineral, thereby prompting dissolution. Furrer and Stumm (1986) suggested that the rate of the LPRM is linearly proportional to the surface concentration of the respective adsorbed ligand. Therefore, the overall rate, in the presence of a ligand (L), can be described by:

$$Rate = k_{\mathrm{H}} \cdot C_{s,\mathrm{H}} + k_L \cdot C_{s,L} \tag{34}$$

Although a considerable volume of work assessing the catalytic effect of organic acids on mineral dissolution has been carried out since the study of Furrer and Stumm (1986), their basic concept may be used to describe many of the findings on the catalytic effect of OM on the dissolution of most groups of minerals.

Whereas some OM catalyze dissolution of minerals, others may inhibit dissolution by adsorbing to reactive sites on the mineral surface and blocking them. This inhibiting effect on mineral dissolution was mainly proposed for carbonate (e.g., Morse 1974; Compton et al. 1989; Compton and Brown 1995) although a few studies showed such inhibitory effects on silicates as well (Welch and Vandevivere 1994; Welch et al. 1999; Golubev et al. 2006). In contrast, catalysis by OM was mainly reported for silicates and oxides, while only few studies discussed catalysis of carbonate dissolution (e.g., Fredd and Fogler 1998a; Burns et al. 2003; Pokrovsky et al. 2009).

Silicate dissolution

The effects of OM on weathering of silicate minerals have been the subject of scientific discussions for several decades. The focus of these discussions has been on three major questions: First, is the dissolution rate faster in the presence of a specific OM than in its absence (under the same environmental conditions, such as pH)? Second, what are the mechanisms by which the dissolution rate is affected by the presence of OM? Third, to what extent does OM affect mineral dissolution rates in nature? The results of some of these studies were summarized by Bennett and Casey (1994), Drever and Vance (1994), Hajash (1994), Drever (1994), Stillings et al. (1996) , Drever and Stillings (1997), Stillings et al. (1998), Oelkers and Schott (1998), Blake and Walter (1999), van Hees et al. (2002), Ullman and Welch (2002) and Cama and Ganor (2006).

The debate over the possible catalytic effect of OM started during the 19th century when Sprengel (1826, in Drever and Stillings 1997) and later Julien (1879, in Bennett and Casey 1994) suggested that organic acids have an important role in mineral dissolution. These early suggestions were later examined in many experimental studies that measured enhancement in dissolution rates of many silicate minerals in the presence of various OM. The first experimental studies that reported catalysis of silicates dissolution rate by OM are probably those of Huang and Keller (1970, 1971, 1973) and Huang and Kiang (1972). In contrast, some later studies did not find a catalytic effect. For example, Mast and Drever (1987) showed that 0.5 and 1 mM solutions of oxalic acid had no effect on the steady-state dissolution rate of oligoclase at the pH range 4-9. In contrast, Amrhein and Suarez (1988) and Welch and Ullman (1993) reported that oxalic acid solutions increase dissolution rates of anorthite, bytownite and labradorite by factors of 2-7. Manley and Evans (1986) observed that microcline, albite, and labradorite dissolution rates increased in the presence of 0.1 mM oxalic acid, yet they concluded that the increase was due to the strength of the acid (a pH effect), rather than a metal-organic complexation effect. Bennett and Casey (1994) summarized the history of the debate, and concluded that the history of opposing conclusions suggests that the interaction of OM with silicates is complex, multifaceted, and somewhat difficult to characterize. Drever (1994) suggested that the contradictory observations regarding the catalytic effect of OM on silicate dissolution are mainly due to two reasons: (1) in some studies pH was not buffered so the effect of OM cannot be separated from the pH effect (e.g., Huang and Keller 1970) and (2) in some studies it is not possible to distinguish between catalytic effects on the dissolution of the primary minerals and inhibitory effects on the precipitation of the secondary minerals.

Today, after almost half a century of experiments there is a long list of studies that verified the catalytic effect of many OM on different silicate minerals. A partial list of the minerals that their dissolution was enhanced by OM includes: (1) olivine, augite, muscovite, labradorite, and microcline in the presence of acetic, aspartic, salicylic and tartaric acids (Huang and Keller 1970), (2) albite, oligoclase, labradorite, bytownite, anorthite and a high K plagioclase with acetic, aspartic, salicylic and citric acids (Huang and Kiang 1972), (3) calcic plagioclase, albite and microcline with citric, oxalic, salicylic, protocatechuic, gallic, *p*-hydroxybenzoic, vanillic, and caffeic acids (Manley and Evans 1986), (4) quartz in the presence of citrate, oxalate and salicylate (Bennett et al. 1988; Blake and Walter 1999), (5) perthitic K-feldspar with oxalate (Bevan and Savage 1989), (6) kaolinite with oxalate, malonate, salicylate and o-phthalate (Chin and Mills 1991), (7) albite with oxalate (Franklin et al. 1994), (8) albite with acetate and citrate (Hajash et al. 1998), (9) Orthoclase, albite and labradorite with oxalate and citrate (Blake and Walter 1999), (10) albite, quartz, and microcline with malonate (Knauss and Copenhaver 1995), (11) Oligoclase and K-feldspar with citrate (Lundstrom and Ohman 1990), (12) sanidine and microcline with citrate (Schweda 1989), (13) kaolinite with oxalate and salicylate (Wieland and Stumm 1992), albite, oligoclase, Labradorite, bytownite and andesine with oxalate (Stillings et al. 1996), (14) quartz, kaolinite, albite and bytownite with gluconate (Vandevivere et al. 1994), (15) bytownite with various polysaccharides (Welch et al. 1999), (16) bytownite and labradorite with acetate, propionate, oxalate, citrate, pyruvate, succinate, α-ketoglutarate (Welch and Ullman 1993), (17) kaolinite with oxalate (Ganor and Lasaga

1994; Cama and Ganor 2006), (18) hornblende with oxalic, citric, tannic, polygalacturonic and fulvic acids (Zhang and Bloom 1999), (19) basaltic glass with oxalate (Oelkers and Gislason 2001), (20) smectite with ethylenediaminetetraacetic acid (EDTA), 3,4 dihydroxybenzoic acid (protocatechuic acid), citrate and oxalate (Golubev et al. 2006), (21) diopside with EDTA, 2,4 dihydroxybenzoic acid, citrate, oxalate acetate and gluconate (Golubev and Pokrovsky 2006), (22) forsterite with citrate, EDTA, oxalate, tannic acid, succinate and phthalate (Grandstaff 1986), (23) forsterite with potassium hydrogen phthalate and ascorbic acid (Wogelius and Walther 1991) and (24) forsterite with oxalate (Olsen and Rimstidt 2008). The catalytic effect of these organic compounds on mineral dissolution is summarized in table 1 along with other possible roles they may exhibit in water-rock interactions.

In some of these studies the rate was only enhanced under particular experimental conditions. For example, Schweda (1989) reported that at pH 4, the dissolution rate of microcline is not affected by the presence of citrate and whereas at pH 5.7 the rate is approximately doubled compared to dissolution with inorganic acid. Similarly, Golubev and Pokrovsky (2006) showed that at basic pH conditions, citrate and acetate do not affect the dissolution rate of diopside, while at neutral pH they enhance the rate (see Fig. 32).

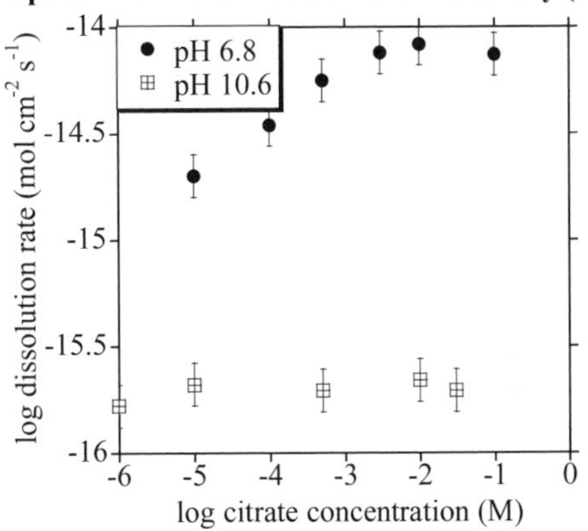

Figure 32. Log-log plot of the effect of citrate on the dissolution rate of diopside. At pH 10.6 the rate is independent of citrate concentration. At pH 6.8 the rate has non-linear dependence on citrate concentration. Data after Golubev and Pokrovsky (2006).

Some studies found that OM may also inhibit dissolution of silicates: bytownite with alginate and poly-galacturonic acid (Welch and Vandevivere 1994; Welch et al. 1999) and smectite with acetate gluconate, alginate, glucosamine and glucuronate (Golubev et al. 2006). Based on the extensive literature cited above, it is clear that although both inhibition and enhancement of dissolution rates were observed, the latter effect is much more common, which may lead to the conclusion that OM in most cases serves as a catalyst for dissolution of silicates.

Empirical rate laws. The effect of OM on silicate dissolution was described in many studies using an empirical rate law that relates the overall dissolution rate to the OM solution concentration. Some of the studies used a linear dependence:

$$Rate = k_{OM} \cdot C_{OM} \tag{35}$$

where k_{OM} is a rate coefficient and C_{OM} is the OM concentration. For example, Vandevivere et al. (1994) found that the rate of bytownite dissolution increases linearly with gluconate concentration. In other studies, the rate may be described using a linear rate law at low concentration but it becomes non-linear for higher concentrations (e.g., andesine and kaolinite with oxalate, Wieland and Stumm 1992; Stillings et al. 1996; Cama and Ganor 2006). By plotting the log of the overall rate versus log of total concentration of OM the data may be used to fit an empirical rate law of the form:

$$Rate = k_{OM} C_{OM}^{\ n} \tag{36}$$

where n is the reaction order with respect to OM concentration. For example, Cama and Ganor (2006) described the effect of oxalate on kaolinite dissolution using Equation (36) with n of 0.45, 0.24 and 0.55 at 25, 50 and 70 °C, respectively (see Fig. 33). In some cases (e.g., the effect of citrate on diopside at near neutral pH, Golubev and Pokrovsky 2006) the rate may be described using the non linear rate law of Equation (36) at low concentrations of OM, but it becomes independent of OM concentration at higher concentrations (e.g., rate at pH 6.8 in Fig. 32). In yet other cases (e.g., the effect of gluconate on diopside, Golubev and Pokrovsky 2006) the rate is independent of OM concentration at low concentrations and it increases non-linearly at higher concentrations (see Fig. 34). Empirical rate laws may be useful in reactive transport modeling in which experimentally determined rate laws can be utilized (Cama and Ganor 2006). However, in order to understand the observed catalytic effect and to extrapolate the rate to other environmental conditions, the experimental results should be examined using a mechanistic approach.

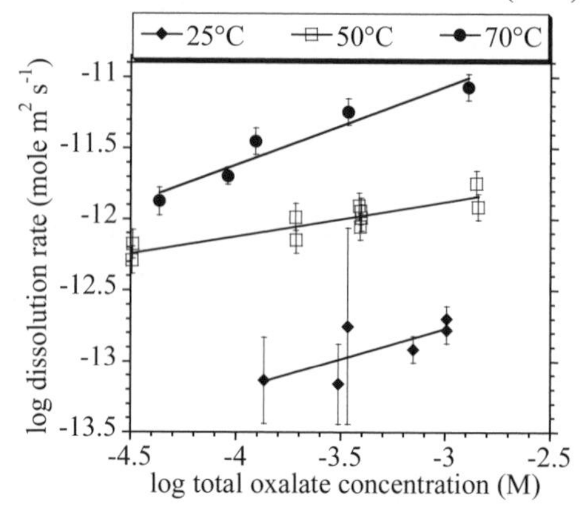

Figure 33. Log-log plot of the effect of oxalate concentration on the overall dissolution rate of kaolinite. The orders of the reaction with respect to oxalate concentration in solution (n in Eqn. 36) are 0.45, 0.24 and 0.55 at 25, 50 and 70 °C, respectively. Data after Cama and Ganor (2006).

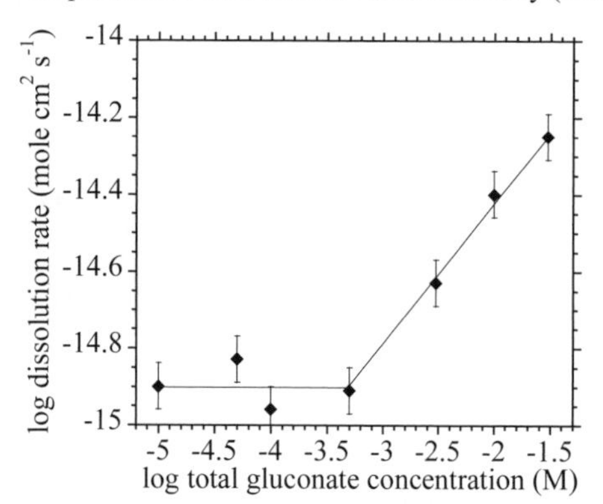

Figure 34. Log-log plot of the effect of gluconate on the dissolution rate of diopside. Dissolution rate is independent of gluconate concentration at low concentrations and it increases non-linearly at higher concentrations. Note that the trend looks linear due to the log-log scale. Data after Golubev and Pokrovsky (2006).

The case study of oxalate catalysis. The effect of oxalate is discussed below in order to demonstrate the complexity of the catalytic effect of OM on silicate dissolution. We chose oxalate as its concentrations in soil and formation water are relatively high, up to 0.001m and 0.005m, respectively (see Table 2), and its effect on dissolution rate has been widely studied (e.g., Grandstaff 1986; Mast and Drever 1987; Bennett et al. 1988; Bevan and Savage 1989; Chin and Mills 1991; Hajash et al. 1992; Welch and Ullman 1992; Wieland and Stumm 1992; Welch and Ullman 1993; Franklin et al. 1994; Ganor and Lasaga 1994; Stillings et al. 1996; Blake and Walter 1999; Oelkers and Gislason 2001; Cama and Ganor 2006; Golubev et al. 2006; Golubev and Pokrovsky 2006; Olsen and Rimstidt 2008).

Table 3 summarizes the enhancement ratio of oxalate on the dissolution kinetics of several silicates under variable conditions. The enhancement ratio is defined as the ratio between dissolution rate with OM and dissolution rate under the same temperature and pH but without organic acids. Comparison of results obtained in different laboratories is a major concern

Table 3. Calculated enhancement ratios of oxalate for different silicate minerals.

Mineral	Temperature (°C)	Oxalate concentration	pH	Enhancement ratio	Source	Comments
albite	?	2.5×10^{-4}	6	1.9	7	
albite	?	5.0×10^{-4}	6	2.1	7	
albite	?	1.0×10^{-3}	6	3.3	7	
albite	?	2.5×10^{-4}	4.5	1.2	7	
albite	?	5.0×10^{-4}	4.5	1.2	7	
albite	?	1.0×10^{-3}	4.5	1.4	7	
albite	?	2.0×10^{-3}	4.5	1.7	7	
albite	22	1.0×10^{-3}	6	1.2	16	
albite	22	1.0×10^{-3}	3	2.0	16	
albite	100	5.0×10^{-4}	4.8	0.73	8	a
albite	100	1.0×10^{-3}	4.6	1.4	8	a
albite	100	5.0×10^{-3}	5.1	24	8	a
albite	100	5.0×10^{-3}	4.6	5.5	8	a
albite	100	5.0×10^{-3}	3.3	2.1	8	a
albite	80	5.0×10^{-4}	6	0.85	4	c
albite	80	3.0×10^{-3}	6	1.4	4	c
albite	80	1.0×10^{-2}	6	2.3	4	c
andesine	?	2.5×10^{-4}	4.5	2.0	7	
andesine	?	5.0×10^{-4}	4.5	2.5	7	
andesine	?	1.0×10^{-3}	4.5	2.9	7	
andesine	?	2.0×10^{-3}	4.5	4.9	7	
andesine	25	1.0×10^{-3}	5.7	16	14	
andesine	25	1.0×10^{-3}	3	1.2	14	
andesine	25	2.0×10^{-3}	3	1.6	14	
andesine	25	4.0×10^{-3}	3	1.7	14	
andesine	25	8.0×10^{-3}	3	2.1	14	
andesine	25	1.0×10^{-3}	4	2.1	14	
andesine	25	2.0×10^{-3}	5	2.7	14	
andesine	25	4.0×10^{-3}	4	4.7	14	
andesine	25	8.0×10^{-3}	4	6.8	14	
andesine	25	1.0×10^{-3}	5	2.1	14	
andesine	25	2.0×10^{-3}	6.8	0.3	14	
andesine	25	4.0×10^{-3}	6.6	0.5	14	
andesine	25	8.0×10^{-3}	5.5	4.1	14	
andesine	25	1.0×10^{-3}	7.2	1.6	14	
andesine	25	2.0×10^{-3}	7.7	0.3	14	
andesine	25	4.0×10^{-3}	7.6	0.6	14	
andesine	25	8.0×10^{-3}	7.5	1	14	
anorthite	25	1.0×10^{-4}	4	2.2	1	
bytownite	?	5.0×10^{-5}	4.5	1.7	7	
bytownite	?	1.0×10^{-4}	4.5	2.0	7	
bytownite	?	5.0×10^{-4}	4.5	3.4	7	
bytownite	?	1.0×10^{-3}	4.5	7.6	7	
bytownite	25	1.0×10^{-3}	4.4	12	14	
bytownite	5	1.0×10^{-3}	6	2.5	17	
bytownite	20	1.0×10^{-3}	6	1.6	17	
bytownite	35	1.0×10^{-3}	6	1.3	17	
bytownite	20-22	1.0×10^{-4}	4	2.7	15	d
bytownite	20-22	1.0×10^{-3}	3	3.6	15	d
bytownite	20-22	1.0×10^{-4}	5.3	27	15	g

Table 3. Calculated enhancement ratios of oxalate for different silicate minerals.

Mineral	Temperature (°C)	Oxalate concentration	pH	Enhancement ratio	Source	Comments
bytownite	20-22	1.0×10^{-3}	3	2.2	15	g
bytownite	22	1.0×10^{-3}	2.95	2.2	16	
bytownite	22	1.0×10^{-3}	6	9.3	16	
diopside	25	1.0×10^{-5}	6±0.2	1.2	9	
diopside	25	1.0×10^{-4}	6±0.2	0.9	9	
diopside	25	5.0×10^{-4}	6±0.2	1.3	9	
diopside	25	3.0×10^{-4}	6±0.2	2.2	9	
diopside	25	1.0×10^{-3}	6±0.2	4	9	
diopside	25	1.0×10^{-2}	6±0.2	9	9	
forsterite	25	1.0×10^{-3}	4.5	≅30	11	
forsterite	25	1.0×10^{-3}	4.5	≅10	11	
forsterite	25	1.0×10^{-2}	4.0	4.4	13	b
forsterite	25	1.7×10^{-2}	4.0	5.6	13	b
forsterite	25	2.3×10^{-2}	4.0	5.9	13	b
forsterite	25	2.7×10^{-2}	4.0	5.6	13	b
forsterite	25	3.1×10^{-2}	4.0	5.8	13	b
forsterite	25	3.3×10^{-2}	0.6	1.2	13	b
forsterite	25	3.4×10^{-2}	4.1	7.5	13	b
forsterite	25	3.5×10^{-2}	3.6	5.5	13	b
forsterite	25	3.5×10^{-2}	4.1	8.3	13	b
forsterite	25	3.7×10^{-2}	4.0	7.2	13	b
forsterite	25	3.9×10^{-2}	4.0	7.2	13	b
forsterite	25	4.0×10^{-2}	3.2	6.0	13	b
forsterite	25	4.0×10^{-2}	3.6	9.0	13	b
forsterite	25	4.1×10^{-2}	4.0	7.5	13	b
forsterite	25	4.3×10^{-2}	3.1	5.8	13	b
forsterite	25	4.4×10^{-2}	2.8	4.4	13	b
forsterite	25	4.7×10^{-2}	2.3	3.7	13	b
forsterite	25	4.7×10^{-2}	4.0	8.6	13	b
forsterite	25	4.8×10^{-2}	2.6	3.9	13	b
forsterite	25	5.0×10^{-2}	3.8	9.0	13	b
forsterite	25	5.1×10^{-2}	3.6	8.0	13	b
forsterite	25	5.2×10^{-2}	4.0	9.1	13	b
forsterite	25	5.3×10^{-2}	0.6	1.2	13	b
forsterite	25	5.3×10^{-2}	1.9	3.1	13	b
forsterite	25	5.4×10^{-2}	2.1	3.1	13	b
forsterite	25	5.5×10^{-2}	3.7	8.1	13	b
forsterite	25	6.3×10^{-2}	2.2	3.4	13	b
forsterite	25	6.8×10^{-2}	1.8	2.7	13	b
forsterite	25	6.9×10^{-2}	0.7	1.2	13	b
forsterite	25	7.0×10^{-2}	2.3	3.5	13	b
forsterite	25	7.3×10^{-2}	1.9	2.5	13	b
forsterite	25	7.4×10^{-2}	1.7	3.1	13	b
forsterite	25	7.8×10^{-2}	1.5	2.3	13	b
forsterite	25	8.2×10^{-2}	1.5	2.4	13	b
forsterite	25	8.3×10^{-2}	1.0	2.3	13	b
forsterite	25	9.6×10^{-2}	0.5	1.3	13	b
forsterite	25	9.6×10^{-2}	1.2	2.0	13	b
forsterite	25	9.7×10^{-2}	0.5	1.3	13	b
forsterite	25	9.9×10^{-2}	0.8	1.4	13	b

Table 3. Calculated enhancement ratios of oxalate for different silicate minerals.

Mineral	Temperature (°C)	Oxalate concentration	pH	Enhancement ratio	Source	Comments
forsterite	25	1.0×10^{-1}	1.1	1.8	13	b
forsterite	25	1.4×10^{-1}	1.8	3.0	13	b
forsterite	25	2.1×10^{-1}	1.7	2.6	13	b
forsterite	25	3.5×10^{-1}	1.4	2.2	13	b
kaolinite	25	3.4×10^{-4}	3.29	3.8	5	
kaolinite	25	3.1×10^{-4}	3.33	2.5	5	
kaolinite	25	1.4×10^{-4}	3.15	2.8	5	
kaolinite	25	1.0×10^{-3}	2.88	6.2	5	
kaolinite	25	1.0×10^{-3}	2.92	5.8	5	
kaolinite	50	3.9×10^{-4}	3.26	2.3	5	
kaolinite	50	1.9×10^{-4}	3.13	1.9	5	
kaolinite	50	3.2×10^{-5}	2.98	1.9	5	
kaolinite	50	4.0×10^{-4}	3.24	1.9	5	
kaolinite	50	1.9×10^{-4}	3.14	1.5	5	
kaolinite	50	3.2×10^{-5}	3.08	1.0	5	
kaolinite	50	3.9×10^{-4}	3.28	1.6	5	
kaolinite	50	1.4×10^{-3}	2.77	3.3	5	
kaolinite	50	1.4×10^{-3}	3.52	2.3	5	
kaolinite	70	1.2×10^{-4}	3.19	4.0	5	
kaolinite	70	1.3×10^{-3}	3.45	9.5	5	
kaolinite	70	3.4×10^{-4}	3.46	6.3	5	
kaolinite	70	4.3×10^{-5}	3.10	1.4	5	
kaolinite	70	9.2×10^{-5}	3.13	2.3	5	
kaolinite	20	2.0×10^{-4}	3.7	1.0	6	
kaolinite	20	2.0×10^{-4}	4.2	1.1	6	
kaolinite	20	2.0×10^{-4}	4.9	1.0	6	
kaolinite	20	1.0×10^{-3}	3.7	2.3	6	
kaolinite	20	1.0×10^{-3}	4.2	1.7	6	
kaolinite	20	1.0×10^{-3}	4.9	1.7	6	
kaolinite	20	5.0×10^{-3}	3.7	6.3	6	
kaolinite	20	5.0×10^{-3}	4.2	2.7	6	
kaolinite	25	1.0×10^{-3}	2	1.5	18	
kaolinite	25	1.0×10^{-3}	3	1.6	18	
kaolinite	25	1.0×10^{-3}	4	1.9	18	
kaolinite	25	1.0×10^{-3}	5	2.2	18	
kaolinite	25	1.0×10^{-3}	6	2.6	18	
kaolinite	25	1.5×10^{-4}	4	1.1	18	
kaolinite	25	5.0×10^{-4}	4	1.5	18	
kaolinite	25	7.5×10^{-4}	4	2.1	18	
kaolinite	25	2.5×10^{-3}	4	2.7	18	
kaolinite	25	5.0×10^{-3}	4	3.0	18	
kaolinite	25	7.5×10^{-3}	4	3.3	18	
kaolinite	25	1.0×10^{-2}	4	3.8	18	
kaolinite	25	1.5×10^{-2}	4	3.9	18	
kaolinite	25	2.5×10^{-2}	4	4.4	18	
kaolinite	25	5.0×10^{-2}	4	5.9	18	
K-feldspar	70	2.0×10^{-2}	1	0.8	3	
K-feldspar	95	2.0×10^{-2}	1	0.7	3	
K-feldspar	70	2.0×10^{-2}	3.6	1.6	3	
K-feldspar	95	2.0×10^{-2}	3.6	3.9	3	

Table 3. Calculated enhancement ratios of oxalate for different silicate minerals.

Mineral	Temperature (°C)	Oxalate concentration	pH	Enhancement ratio	Source	Comments
K-feldspar	70	2.0×10^{-2}	9	1.6	3	
labradorite	22	1.0×10^{-3}	3	3.7	16	
labradorite	22	1.0×10^{-3}	4	2.2	16	
labradorite	22	1.0×10^{-3}	6	3.2	16	
labradorite	80	5.0×10^{-4}	6	1.4	4	c
labradorite	80	3.0×10^{-3}	6	1.7	4	c
labradorite	80	1.0×10^{-2}	6	3	4	c
labradorite	20-22	1.0×10^{-4}	3.7	2.4	15	g
labradorite	20-22	1.0×10^{-3}	3.1	3.9	15	g
labradorite	20-22	1.0×10^{-3}	5.9	3.1	15	g
oligoclase	25	1.0×10^{-3}	5.7	9.6	14	
oligoclase	22	1.0×10^{-3}	4	1	12	f
oligoclase	22	1.0×10^{-3}	5	1	12	f
oligoclase	22	1.0×10^{-3}	7	1	12	f
oligoclase	22	1.0×10^{-3}	9	1	12	f
oligoclase	22	5.0×10^{-4}	4	1	12	f
oligoclase	22	5.0×10^{-4}	5	1	12	f
orthoclase	80	5.0×10^{-4}	6	0.8	4	c
orthoclase	80	3.0×10^{-3}	6	1	4	c
orthoclase	80	1.0×10^{-2}	6	1.7	4	c
quartz	25	1.0×10^{-3}	7	1	2	
quartz	25	2.0×10^{-3}	7	1.4	2	
quartz	25	2.0×10^{-2}	7	3.5	2	
quartz	25	2.0×10^{-3}	3	1.3	2	
quartz	70	1.0×10^{-2}	6	2.1	4	
quartz	70	2.0×10^{-2}	6	2.3	4	
quartz	22	1.0×10^{-3}	3	4.5	16	
quartz	22	1.0×10^{-3}	6	1.1	16	
quartz	70	1.0×10^{-2}	6	1.2	4	e
quartz	70	2.0×10^{-2}	6	1.2	4	e
smectite	25	1.0×10^{-3}	7	1	10	
smectite	25	1.0×10^{-2}	7	1.7	10	
tremolite	22	5.0×10^{-4}	4	1	12	f
tremolite	22	5.0×10^{-4}	4	1	12	f
tremolite	22	1.0×10^{-3}	4	1	12	f
tremolite	22	1.0×10^{-3}	4	1	12	f
tremolite	22	1.0×10^{-3}	7	1	12	f
tremolite	22	1.0×10^{-3}	7	1	12	f

Sources: 1= Amrhein and Suarez (1988); 2= Bennett et al. (1988); 3= Bevan and Savage (1989); 4= Blake and Walter (1999); 5= Cama and Ganor (2006); 6= Chin and Mills (1991); 7= Fig. 4 of Ullman and Welch (2002); 8= Franklin et al. (1994); 9= Golubev and Pokrovsky (2006); 10= Golubev et al. (2006); 11= Grandstaff (1986); 12= Mast and Drever (1987); 13= Olsen and Rimstidt (2008); 14= Stillings et al. (1996); 15= Welch and Ullman (1993); 16= Welch and Ullman (1996); 17= Welch and Ullman (2000); 18= Wieland and Stumm (1992).

Comments: a= oxalate + 0.07 M acetate, enhancement is relative to 0.07M acetate with no oxalate; b=calculated with respect to fitting of pH effect on data with no oxalate; c=enhancement relative to acetate buffer (0.7 M lithium acetate/0.04 M acetic acid); d=fast flow rate (0.32 ml/min); e=in 1M NaCl solution; f=no effect on steady-state rate; g=slow flow rate (0.1 ml/min).

in interpreting kinetics data, as the reproducibility of dissolution rates is usually very poor. For example, the reproducibility of feldspar rate determinations by the same laboratory are almost always within a factor of two, while the agreement between rates obtained under similar conditions by different laboratories is generally within a factor of 5 (Blum and Stillings 1995). Ullman and Welch (2002) proposed that this variability may reflect the choice of experimental minerals, the preparation of the minerals for experiments, the presence of trace contaminants and compositional intergrowths in the minerals, the assignment of specific surface areas to the reacting phases, differences in the types of reactors used in the experiments and the length of the experiment. In order to minimize these effects, the enhancement ratios in Table 3 are given either as reported by the authors of the paper, or calculated using rate data that was reported within the same paper. Figure 35 shows the enhancement ratio distribution of 77 dissolution experiments of various feldspars which were published in ten different publications (Mast and Drever 1987; Amrhein and Suarez 1988; Bevan and Savage 1989; Welch and Ullman 1993; Franklin et al. 1994; Stillings et al. 1996; Welch and Ullman 1996; Blake and Walter 1999; Welch and Ullman 2000; Ullman and Welch 2002). The enhancement ratio was above 10 in only 4 experiments (5%), and above 5 in 5 more experiments. In 73% of the experiments the enhancement ratio was less than 3 and the median was only 2. This simple analysis indicates that the effect of oxalate is generally relatively small, and is much less than the between-laboratories agreement factor of 5. Therefore, the following discussion is based on comparisons between different sets of observations where each set includes only experiments that were conducted within the same laboratory.

Figure 36 shows the effect of oxalate concentration on the enhancement ratios of various silicates dissolution rates. When other factors are held constant, the dissolution rates in this figure increases monotonically with oxalate concentration. Note that Figure 36 does not include data from studies in which oxalate did not promote dissolution (e.g., Mast and Drever 1987). Figure 36 demonstrates that the enhancement ratio depends on pH, temperature and mineral composition. A thorough understanding of the reaction mechanism is required in order to explain these dependences.

Among the most natural abundant silicate minerals in the Earth's crust quartz is the only non-aluminosilicate mineral. As aluminosilicates weather substantially faster than quartz (Lasaga et al. 1994), it is not surprising that the major debate about the effects of OM on weathering revolves around the catalytic effect of OM on the dissolution of aluminosilicates in general and

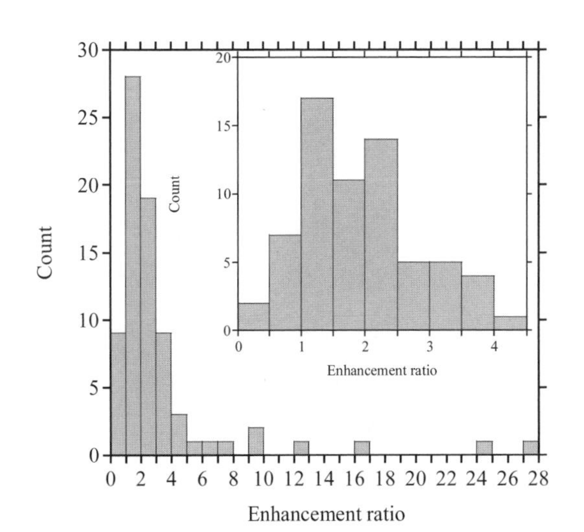

Figure 35. Distribution of the enhancement ratio of oxalate on the dissolution kinetics of various feldspars. The figure includes the results of 77 dissolution experiments. The insert shows in better resolution the data of the 59 experiments in which the enhancement ratios was below 5. The data is collected from Mast and Drever (1987), Amrhein and Suarez (1988), Bevan and Savage (1989), Welch and Ullman (1993), Franklin et al.(1994), Stillings et al. (1996), Welch and Ullman (1996), Blake and Walter (1999), Welch and Ullman (2000) and Ullman and Welch (2002).

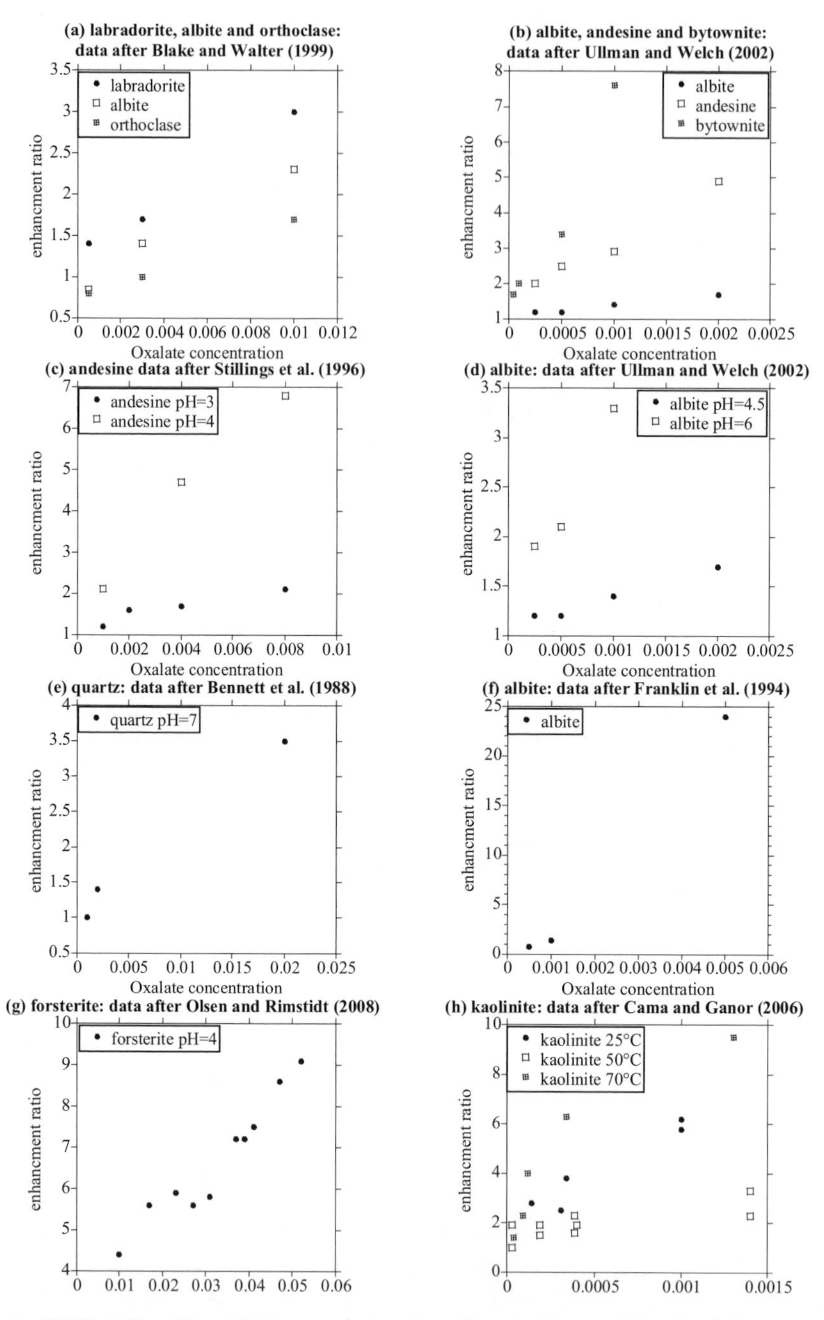

Figure 36. The effect of oxalate concentration on the enhancement ratios of the dissolution of various silicates: (a) labradorite, albite and orthoclase at pH 6 and 80 °C: data after Blake and Walter (1999); (b) albite, andesine and bytownite at pH 4.5: data after Ullman and Welch (2002); (c) andesine at pH 3 and 4 and 25 °C: data after Stillings et al. (1996); (d) albite at pH 4.5 and 6: data after Ullman and Welch (2002); (e) quartz at pH 7 and 25 °C: data after Bennett et al. (1988); (f) albite at pH 4.6 and 100 °C: data after Franklin et al. (1994); (g) forsterite at pH 4 and 25 °C: data after Olsen and Rimstidt (2008); (h) kaolinite at pH 2.5-3.5 and 25, 50 and 70 °C: data after Cama and Ganor (2006).

feldspars in particular. The debate over the catalytic mechanism of aluminosilicates dissolution revolved mainly around two questions: (1) does the observed enhancement of the overall dissolution rate result from an enhancement of the rate of a PPRM (through an effect of chelation of dissolved Al^{3+} by oxalate, which both reduces Al inhibition and alters the overall saturation state) or (2) does an inner-sphere oxalate- surface site complex facilitate the detachment of metal ions, thereby enhancing the dissolution rate (i.e., LPRM, Stumm 1992)? A simple LPRM consists of fast adsorption of the organic ion on the mineral surface followed by a slow catalyst-mediated hydrolysis step, which is the rate-determining step. The hydrolysis rate depends on the surface concentration of the adsorbed species. Therefore, the rate law must include the adsorption isotherms of the organic ion. However, only a few studies on the effect of organic acids on silicate dissolution have measured the adsorption isotherm (e.g., Stillings et al. 1998; Cama and Ganor 2006). The studies of Stillings et al. (1998) and Cama and Ganor (2006) will be used below to examine the two possible effects of oxalate on the overall dissolution rate of silicates.

Stillings et al. (1996) used oxalic acid to quantify the dissolution enhancement of six feldspars (microcline, albite, oligoclase, labradorite, bytownite and andesine) at acidic pHs. Cama and Ganor (2006) examined the effect of oxalate on dissolution of kaolinite at pH 2.5-3.5. As both studies were conducted under acidic conditions, it was suggested that oxalate may chelate free Al^{3+}, thereby reduce its concentration and inhibitory effect and consequently increase the overall rate by enhancing the PPRM (Gautier et al. 1994; Oelkers et al. 1994; Cama et al. 2002). To examine this possibility, the case studies of the effect of Al on kaolinite and feldspar dissolution will be discussed first. In the following we shall review several reaction mechanisms which were proposed in order to explain the Al inhibition. We will not repeat the theoretical background of the different mechanisms, which may be found in Oelkers et al. (1994), Oelkers and Schott (1998) and Ganor and Lasaga (1998), but merely present the proposed rate laws.

At the absence of OM, Oelkers et al. (1994) proposed that Al-silicate dissolution is controlled by a Si-rich precursor complex formed by the exchange of Al for protons at the surface. According to this mechanism, dissolution rate is expressed by:

$$Rate = k_1 \cdot \left(\frac{\left(\frac{a_{H^+}^3}{a_{Al^{3+}}} \right)^n}{1 + K_1 \cdot \left(\frac{a_{H^+}^3}{a_{Al^{3+}}} \right)^n} \right) \qquad (37)$$

where k_1 and K_1 are rate coefficient and equilibrium constant, respectively, n is a stoichiometric coefficient, and a_{H^+} and $a_{Al^{3+}}$ are the solution activities of H^+ and Al^{3+}, respectively. Ganor and Lasaga (1998) presented a mechanistic model describing the effects of an inhibitor (e.g., Al^{3+}) on mineral dissolution rate in the presence of a catalyst (e.g., proton). They proposed two end-member models: the competition model and the independent adsorption model. In the competition model, the catalyst and the inhibitor compete with each other over the same surface sites. The rate law for this mechanism is:

$$Rate = k_2 \frac{b_{H^+} \cdot a_{H^+}}{1 + b_{H^+} \cdot a_{H^+} + b_{Al^{3+}} \cdot a_{Al^{3+}}} \qquad (38)$$

where k_2 is a rate coefficient and b_{H^+} and $b_{Al^{3+}}$ are constants related to the energy of adsorption. This rate law is very similar to the rate law for competition which was proposed by Brantley and Stillings (1996):

$$Rate = k_3 \left[\frac{b_{H^+} \cdot a_{H^+}}{1 + b_{H^+} \cdot a_{H^+} + b_{Al^{3+}} \cdot a_{Al^{3+}}} \right]^{1/2} \qquad (39)$$

Due to the similarity between these two rate laws, only the rate law of the competition model of Ganor and Lasaga (1998) will be discussed. In the independent adsorption model proposed by Ganor and Lasaga (1998), the adsorption of the catalyst and the inhibitor are absolutely independent of each other. The rate law for this mechanism is:

$$Rate = k_4(1 - k_4 \cdot \frac{b_{Al^{3+}} \cdot a_{Al^{3+}}}{1 + b_{Al^{3+}} \cdot a_{Al^{3+}}}) \cdot \frac{b_{H^+} \cdot a_{H^+}}{1 + b_{H^+} \cdot a_{H^+}} \tag{40}$$

Figure 37a, b and c plot log dissolution rate versus log activity of Al at constant pHs calculated from Equations (37), (38) and (40), respectively. Figure 37 shows that there are several common characteristics for the three rate laws. In all of them, and regardless the pH, the dissolution rate is independent of Al^{3+} concentration in the very low concentration range of Al^{3+}. The maximum predicted dissolution rate for very low concentrations of Al^{3+} is independent of pH and equals k_1/K_1 in the Oelkers et al. (1994) model (Eqn. 37 and Fig. 37a). For higher pH, this rate will be approached at lower concentrations of Al^{3+}. For the independent adsorption model, the maximum predicted dissolution rate varies depending on the pH but it is approached at the same concentration of Al^{3+} (Eqn. 40 and Fig. 37c). In this sense, the competition model is similar to the Oelkers et al. (1994) model at low pH and to the independent adsorption model at high pH (Fig. 37). All the models predict that as the concentration of Al^{3+} increases, the dissolution rate decreases. In the case of the competition model, the rate has a negative linear dependence on the inhibitor concentration. This negative linear dependence is shown by the slope of −1 in the log rate vs. log Al^{3+} (Fig. 37b). According to the Oelkers et al. (1994) model the slope should be the stoichiometric coefficient $-n$, which is equals 1 for kaolinite and 0.33 for albite (Fig. 37a). Both the Oelkers et al. (1994) model and the competition model predict that the rate would continue decreasing with increasing Al^{3+} until the rate approaches zero (or until the dissolution would be dominated by an alternative reaction mechanism). Figure 37c shows that according to the independent adsorption model, the dissolution rate is independent of Al^{3+} for both very low and very high concentrations of Al^{3+}. Between these two regions of Al^{3+} where the rate is independent on Al^{3+} concentration, there is an intermediate region in which the rate decreases as a function of Al^{3+} concentration. The slope of log rate vs. log Al^{3+} changes with Al^{3+} concentration, and it

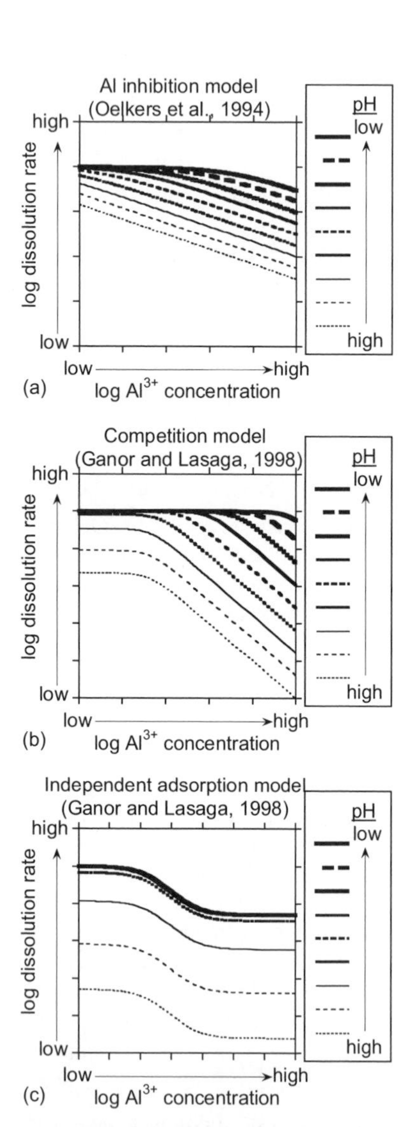

Figure 37. Theoretical plots showing the effect of pH and Al^{3+} on dissolution rate as predicted by (a) the model of Oelkers et al. (1994), (b) the competition model and (c) the independent adsorption model of Ganor and Lasaga (1998).

approaches a maximum at the inflection point. Another difference between the independent adsorption model and the two previous models is that in the former, the transitions between the different regions of Al^{3+} rate dependence occur at a constant concentration of Al^{3+}, which is independent of pH. This is illustrated by the parallel lines in Figure 37c. In contrast, the Al concentration in which the transition from Al^{3+} independent to Al^{3+} inhibition region occurs increases as the pH increases, both in the Oelkers et al. (1994) model and in the competition model. The common feature of the three models is that in the range in which log rate decreases as a function of log Al^{3+}, the slope of the log-log plot is pH independent. As mentioned above, it was proposed that the catalytic effect of oxalate reflects retardation of aluminum inhibition due to the formation of aluminum-oxalate complexes thereby enhancing the PPRM. In the following paragraphs, experimental data will be presented in attempt to examine the validity of this proposal. We will assume that the only catalytic effect of oxalate is related to the formation of Al-oxalate complexes and the inhibition effect of free Al^{3+}. If: (1) this assumption is true; (2) the concentration of free Al^{3+} is properly calculated; and (3) one of the Al^{3+} inhibition models discussed above appropriately describes Al inhibition, then log-log plots of rate vs. Al^{3+} should follow the predictions of the appropriate model. We will show that this is not the case, and therefore an alternative (LPRM) should be considered.

The distributions of Al, Si and oxalate aqueous species in the experiments of Cama and Ganor (2006), Stillings et al. (1996) and Stillings et al. (1998) were calculated using appropriate thermodynamic data (Table 4). Activity coefficients were calculated using the extended Debye-Huckel equation with parameters from Wolery (1979). The change in the

Table 4. Equilibrium constants (log K_{eq}) used in thermodynamic calculations.

Reaction	25 °C	50 °C	80 °C	Notes
$Al^{3+} + OH^- = Al(OH)^{+2}$	9.0	9.1	9.1	*1*
$Al^{3+} + 2OH^- = Al(OH)_2^+$	17.4	17.3	17.4	*1*
$Al^{3+} + 3OH^- = Al(OH)_3$	24.6	24.8	25.0	*1*
$Al^{3+} + 4OH^- = Al(OH)_4^-$	33.0	32.6	32.5	*1*
$Al^{3+} + C_2O_4^{-2} = AlC_2O_4^+$	7.1	8.7	10.1	*2*
$Al^{3+} + 2C_2O_4^{-2} = Al(C_2O_4)_2^-$	13.0	14.6	16.0	*3*
$Al^{3+} + 3C_2O_4^{-2} = Al(C_2O_4)_3^{-3}$	16.3	17.9	19.3	*3*
$H_2C_2O_4 = 2H^+ + C_2O_4^{-2}$	−4.3	−4.4	−4.5	*4*
$HC_2O_4^- = H^+ + C_2O_4^{-2}$	−5.5	−5.8	−6.0	*4*
$H^+ + OH^- = H_2O$	−14.0	−13.3	−12.8	*5*
$H_4SiO_4 = H_3SiO_4^- + H^+$	−9.7	−9.3	−9.1	*6*

Notes:

1 Wesolowski and Palmer (1994).

2 Calculated using a linear fit of the data of Harrison and Thyne (1992) at 0, 25, 60, 100,150, 200 °C.

3 25°C constant is from Graustein (1981). The constant at 50 and 70°C were calculated assuming that the slope is the same as that of AlC2O4+. The intercept was calculated using the 25 °C constant of Graustein (1981).

4 Kettler et al. (1991).

5 Busey and Mesmer (1978).

6 Equilibrium constants were obtained from standard state thermodynamic data of Naumov et al. (1974).

dissolution rates of andesine and kaolinite in the presence of different concentrations of oxalate as a function of Al^{3+} activity is shown in Figures 38 and 39, respectively. As predicted by the different models, log dissolution rate of andesine decreases with log Al^{3+}. However, a close examination of Figure 38 reveals that whereas all the models predict that the dissolution rate is independent of Al^{3+} for very low activity of Al^{3+}, the dissolution rate of andesine in the presence of oxalate continues to increase as the activity of Al^{3+} decreases below 10^{-12}. Moreover, in contrast to the predictions of all the models, the slope is not constant and it equals −0.08, −0.85 and −0.2 at pH 3, 4 and 5, respectively. In addition, the dissolution rate in the absence of oxalate (open symbols in Fig. 38) is much faster than predicted, based on the extrapolation of the trends of the experiments with oxalate at the same pH (closed symbols in Fig. 38). As noted above, such a pattern is in contrast with the prediction of the Oelkers et al. (1994) model and the competition model which predict a constant slope with increasing Al activity (Fig. 37a and b). According to the independent adsorption model, when the concentration of Al^{3+} is very high and the surface is saturated with Al, the dissolution rate may become independent of Al. Therefore, one could argue that the observations in Fig. 38 are predicted by the independent adsorption model. However, it is not reasonable to assume that the mineral surfaces are saturated with respect to Al^{3+} when its concentration is in the order of few μM or less, as in the oxalate free experiments in pH 4 and 5 (Fig. 38). Similar observations were made by Cama and Ganor (2006) for kaolinite dissolution: At 25 °C (Fig. 39a), Al inhibition is neither observed in the presence of oxalate nor in the absence of oxalate over a range of Al^{3+} concentrations from 2 to 80 μM. A possible Al inhibition effect is observed at 50 and 70 °C both with and without oxalate (Figs. 39b and c). However, the dissolution rates in experiments without oxalate and with Al^{3+} concentrations of few μM are within error the same as in the experiments conducted with relatively low concentrations of oxalate (few tens of μM). This is despite the fact that in the latter Al^{3+} concentration was two to three orders of magnitude lower. For example, increasing the concentration of free Al^{3+} by 3 orders of magnitude, from 10^{-8} M to 10^{-5} M, did not affect the dissolution rate (Fig. 39b). Devidal et al. (1997) and Cama and Ganor (2006) suggested that a minimum concentration of a few tens μM of free Al^{3+} is required to inhibit the dissolution rate of kaolinite. Therefore, reducing Al^{3+} concentrations further below 10μM did not enhance kaolinite dissolution rate. Because the Al^{3+} concentrations in the experiments with oxalate are far below this apparent minimum inhibition concentration, it is likely that the observed effect of oxalate does not reflect on retardation of the direct aluminum inhibition due to the formation of Al-oxalate complexes.

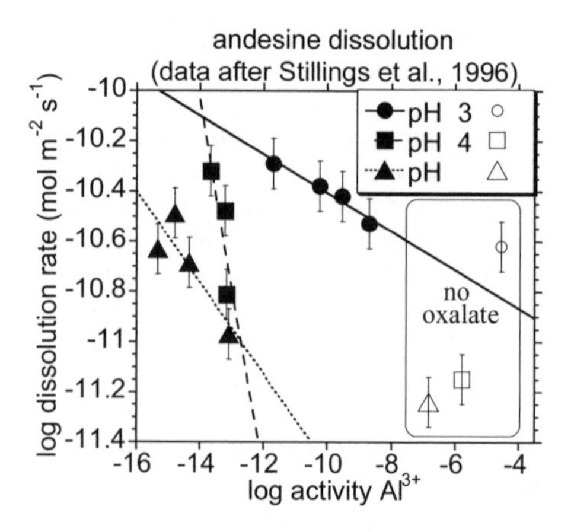

Figure 38. The possible effect of aluminum on andesine dissolution rate at pH = 3, 4 and 5. Experiments with oxalate are denoted by closed symbols and experiments without oxalate by open symbols. Data taken from Stillings et al. (1996).

Formation of Al-oxalate complexes and the resulting reduction in concentration of free Al may also influence the rate of the PPRM due to the decrease in the degree of undersaturation (Eqn. 31). In such a case, oxalate dependence on dissolution should occur in ΔG values in which the reaction is dependent on the degree of saturation. Burch et al. (1993) showed that the rate of albite dissolution at 80 °C and pH 8.8 depends on the saturation degree and exhibits a highly non linear increase at $\Delta G = -6.5$ kcal/mol. When $\Delta G_r < -9$ kcal/mol the *dissolution plateau* was approached and the dissolution reaction was found to be independent of the degree of undersaturation. Stillings et al. (1998) calculated that the values of ΔG with respect to andesine dissolution in the presence of oxalate, range between −17 and −36 kcal/mol in the experiments of Stillings et al. (1996). This range of ΔG is well within the *dissolution plateau* and is much lower than ΔG in which plagioclase is expected to be directly affected by the degree of undersaturation (> −9 kcal/mol). It follows that there is no reason to relate the observed effect of oxalate on andesine dissolution to the decrease in the degree of saturation due to the formation of Al-oxalate complexes.

In the absence of an adequate theory, the function $f(\Delta G_r)$ for overall reactions is difficult to predict *a priori*. Some attribute the non linear change in dissolution rate with the degree of undersaturation to the opening of etch pits (e.g., Lasaga and Blum 1986; Lasaga and Luttge 2003). For albite, the critical degree of undersaturation (ΔG_r^{crit}), beyond which a microscopic hole at a dislocation outcrop would expand to produce a macroscopic etch pit, lies between −2.4 and −7.6 kcal/mol depending on the burger vector and the type of dislocation (Burch et al. 1993). Burch et al. (1993) suggested that in this range the dissolution reaction is expected to be dependent on the saturation degree. The huge gap between the range of ΔG in the experiments of Stillings et al. (1996) and ΔG_r^{crit} observed and calculated by Burch et al. (1993) strengthens the conclusion that the effect of oxalate on andesine dissolution is not related to the decrease in the degree of saturation due to the formation of Al-oxalate complexes, even though the effect of deviation from equilibrium on andesine dissolution rate was not studied experimentally, and the experiments on albite dissolution were conducted under different temperature and pH than those used in the study of Stillings et al. (1996).

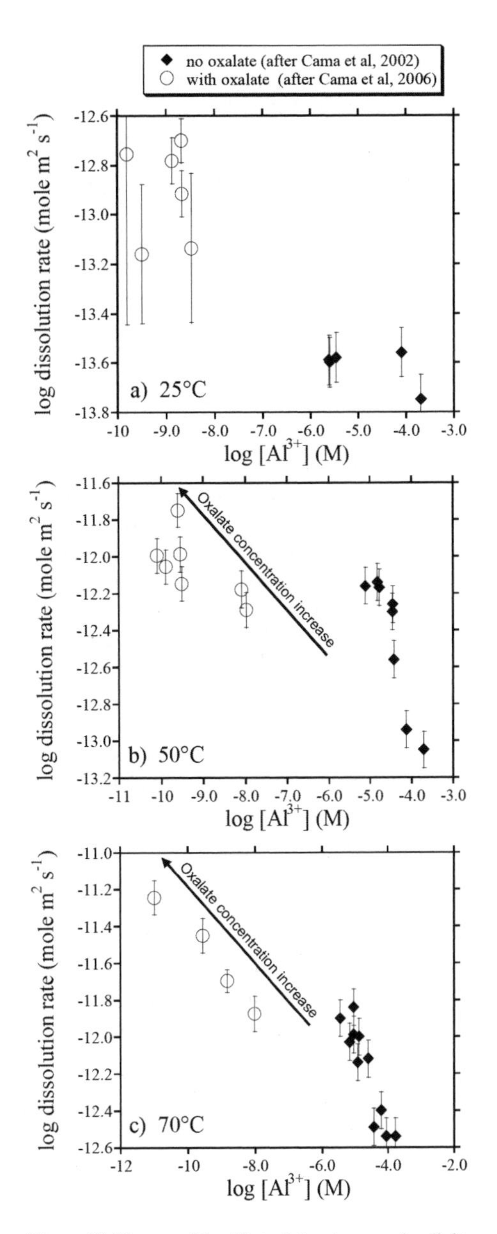

Figure 39. The possible effect of aluminum on kaolinite dissolution rate at pH≈3 and (a) 25 °C, (b) 50 °C and (c) 70 °C. Experiments with oxalate are denoted by open symbols and experiments without oxalate by closed symbols. Data taken from Cama and Ganor (2006).

For kaolinite, Nagy et al. (1991) showed that the *dissolution plateau* at 80 °C is reached at $\Delta G < -2$ kcal/mol. For gibbsite, the results of Mogollon et al. (1996) show that the dissolution plateau at 25 °C is in very good agreement with that of Nagy and Lasaga (1992) at 80 °C.

Assuming that the *dissolution plateau* for kaolinite is similarly independent of temperature, the experimental range of ΔG (−18 to −29 kcal/mol) for kaolinite dissolution in Cama and Ganor (2006) is well within the *dissolution plateau*. Therefore, the observed oxalate effect on kaolinite (Fig. 33) and albite dissolution rate did not result from the effect of the degree of saturation on the rate of the PPRM.

The total concentration of Al in a typical experiment of aluminosilicate dissolution ranges between a few μM and few tens of μM. In a few experiments Al concentrations are of the order of few hundreds of μM. When the effect of oxalate is examined in dissolution experiments, its concentration is usually in the range of 100 μM to few tens of mM. Therefore, in most of these experiments the concentration of oxalate is at least an order of magnitude higher than that of Al. As was shown in the introduction of this review (Fig. 1), the concentration of free Al is already very low when oxalate concentration is one order of magnitude higher than that of aluminum, and therefore further increase in the total concentration of oxalate cannot dramatically reduce the concentration of free Al. Therefore, the observation that the dissolution rates increase monotonically with oxalate concentration, cannot be explained by the effect of oxalate on the PPRM. We therefore conclude that the observed enhancement of aluminosilicates dissolution rate is the result of an oxalate-promoted reaction mechanism (LPRM henceforth named OXPRM), in which an inner-sphere oxalate adsorption complex facilitates the detachment of metal ions and thus enhances the dissolution rate (Stumm 1992). This conclusion is in accordance with the conclusions of many studies (e.g., Furrer and Stumm 1986; Wieland and Stumm 1992; Stillings et al. 1998; Ullman and Welch 2002; Cama and Ganor 2006). This conclusion challenges the suggestion of Oelkers and Schott (1998) that the affect of the presence of aqueous organic anions on alkali-feldspar dissolution rates stems solely from their aqueous complexing with aqueous Al. It is important to note that in most of the experimental work on mineral dissolution in general, and with OM in particular, experiments were conducted under far from equilibrium conditions and with Al-free initial solutions. Therefore, we do not argue that the PPRM is not influenced by the presence of oxalate under different experimental conditions, or under the generally near-equilibrium conditions in natural environments.

The OXPRM affects the minerals in addition to the PPRM. Assuming that the two reaction mechanisms occur independently (Furrer and Stumm 1986), the overall dissolution rate is the sum of at least two terms:

$$Rate_{overall} = Rate_{PPRM} + Rate_{OXPRM} \tag{41}$$

Stillings et al. (1998) noted that the assumption that OXPRM and PPRM occur independently is problematic, as the PPRM is proportional to the concentration of adsorbed H^+, and there is a possibility of competition between the adsorption of protons and oxalate on the surface. In a few studies (e.g., Golubev et al. 2006; Golubev and Pokrovsky 2006) the rate of the PPRM was calculated by assuming that the adsorption of the ligand reduces the number of sites available for proton attack. For example, Golubev et al. (2006) described the dissolution rate of smectite in the absence of OM by regressing their experimental observation to the empirical rate law of the form of Equation (29):

$$Rate_{no\ OM} = 2.2\times10^{-17} \cdot a_{\mathrm{H^+}}^{0.21} + 6\times10^{-17} \cdot a_{\mathrm{OH^-}}^{0.33} + 1.0\times10^{-20} \tag{42}$$

Their experiments with EDTA were conducted under an acidic pH range in which the rate is dominated by PPRM. Taking into account the presumed effect of the EDTA on the available surface sites for proton attack, they calculated the rate of PPRM using the equation:

$$Rate = Rate_{no\ OM} \cdot \left(1 - \frac{b_{EDTA} \cdot C_{EDTA}}{1 + b_{EDTA} \cdot C_{EDTA}}\right) \tag{43}$$

where C_{EDTA} is the concentration of EDTA in solution, b_{EDTA} (= 1000) is the adsorption constant of the Langmuir equation (Eqn. 7) and the Langmuirian term represents the fraction of sites that are occupied by EDTA and therefore are not available for proton attack. In order to estimate the effect of this correction on the determination of the LPRM rate, we plotted in Figure 40a the rate of PPRM of smectite dissolution for the experiments with EDTA of Golubev et al. (2006) using both Eqns. (42) and (43). The correction is relatively small for low concentration of EDTA, but it reaches 50% for an EDTA concentration of 1 mM (Fig. 40a). The rates of the EDTA-promoted reaction mechanism were calculated twice by subtracting separately these PPRM rates from the overall observed rates of Golubev et al. (2006) (Fig. 40b). The differences between the results were found to be insignificant. We suggest that the effect on PPRM driven by ligand competition over adsorption sites is expected to be small for any pair of OM and mineral. Therefore, LPRM rates calculations will not result in great differences. At low OM concentrations the effect of OM on the PPRM is small, and at high OM concentrations the rate of the PPRM is a relatively small fraction of the overall rate. In the absence of information regarding this possible effect of ligands on the PPRM, most studies did not consider such effects in the calculation of the LPRM (e.g., Franklin et al. 1994; Stillings et al. 1996; Cama and Ganor 2006).

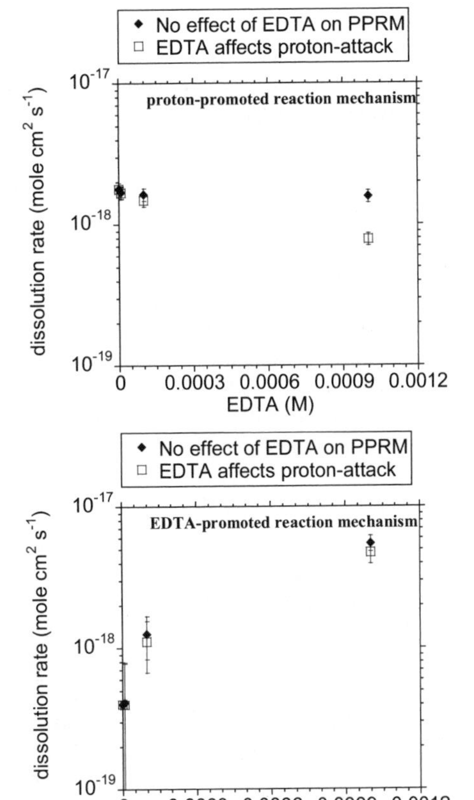

Figure 40. Rates calculation of (a) the proton-promoted and (b) the EDTA-promoted reaction mechanisms of smectite dissolution for the experiments with EDTA of Golubev et al. (2006). The calculations were conducted using the empirical pH dependence of the rate in the absence of organic matter (full symbols, Eqn. 42) and the correction used by Golubev et al. (2006) which assumes that the adsorption of EDTA reduces the amount of sites available for proton attack (open symbols, Eqn. 43).

Stillings et al. (1996), and Cama and Ganor (2006) calculated the rate of the OXPRM by subtracting the rate of the PPRM from the measured overall rate. According to the proposal of Furrer and Stumm (1986) the rate of the ligand promoted reaction mechanism depends linearly on the surface concentration of the ligand (Eqn. 34). For OXPRM:

$$Rate_{oxp} = k_{ox} \cdot C_{s,ox} \tag{44}$$

where k_{ox} (s^{-1}) is the rate coefficient of the OXPRM and $C_{s,ox}$ is the oxalate surface concentration. As we noted above, only a few studies on the effect of organic acids on silicate dissolution have measured the adsorption isotherm. Other studies fitted the experimental observations using an assumed adsorption isotherm (e.g., Franklin et al. 1994; Golubev et al. 2006; Golubev and Pokrovsky 2006). For example, Golubev et al. (2006) described the catalytic effects of EDTA, 3,4 dihydroxybenzoic acid (protocatechuic acid), citrate and oxalate on smectite dissolution by fitting dissolution rate to a rate law that include an assumed Langmuir adsorption isotherm:

$$Rate_{Lp} = k_L \cdot C_{s,L} = k'_L \cdot \frac{b_L \cdot C_L}{1 + b_L \cdot C_L} \tag{45}$$

where k'_L is the product of k_L and the maximum possible surface coverage of the ligand, b_L is an adsorption coefficient related to the energy of adsorption, and C_L is the ligand concentration in solution (Fig. 41).

Stillings et al. (1998) and Cama and Ganor (2006) measured the adsorption of oxalate on andesine and kaolinite surfaces, respectively. Before introducing the adsorption isotherm into a rate law, the speciation of oxalate in the solution and its effect on oxalate adsorption should be examined. Stillings et al. (1998) found that the total adsorbed oxalate varies with pH and with the total activity of oxalate in solution (Fig. 42a). The pH-dependent behavior of the total adsorbed oxalate disappeared when they plotted the total oxalate adsorbed as a function of the sum of bioxalate and oxalate activities (Fig. 42b). They suggested that both bioxalate and oxalate adsorb on the andesine surface. Stillings et al. (1998) suggested that oxalate and bioxalate may compete for the same surface sites, and used a Langmuir adsorption isotherm in which the same sites may be occupied by either oxalate or bioxalate:

$$C_{s,OxTot} = C_{s,Ox} + C_{s,HOx} = F\left[\frac{b_{Ox} \cdot C_{Ox}}{1 + b_{Ox} \cdot C_{Ox} + b_{HOx} \cdot C_{HOx}} + \frac{b_{HOx} \cdot C_{HOx}}{1 + b_{Ox} \cdot C_{Ox} + b_{HOx} \cdot C_{HOx}}\right] \tag{46}$$

where F is the maximum surface coverage of adsorbed oxalate species, $C_{s,OxTot}$ is the total concentration of all adsorbed oxalate species, and the subscripts Ox and HOx refer to oxalate and bioxalate, respectively. The oxalate/bioxalate ratios vary between 0.07 at pH 3 to 7 at pH 5, and therefore the sum of oxalate + bioxalate is about equal to the concentration of bioxalate at pH 3 and to the concentration of oxalate at pH 5. Therefore, the similarity of the adsorption isotherm at pH 3, 4 and 5 indicates that the two oxalate species have similar adsorption coefficients. As already discussed in the section on sorption, Axe and Persson (2001), Yoon et al. (2004) and Persson and Axe (2005) used Attenuated Total Reflectance Fourier Transform Infrared (ATR-FTIR) spectroscopy to study the types and structures of adsorption complexes of oxalate at mineral/water interfaces of boehmite, corundum and goethite. They identified both inner-sphere and outer-sphere complexes. Yoon et al. (2004) compared the adsorption of oxalate on boehmite at pH 2.5, where bioxalate is the dominant species in aqueous solution, to that at pH 5.1 where oxalate is the dominant species. Regardless of the differences in oxalate speciation in solution, similar spectral features of adsorbed oxalate were measured both at pH 2.5 and 5.1. This observation suggests that the coordination geometries of adsorbed oxalate and adsorbed bioxalate species are either the same or very similar. This suggestion is supported by the similarity of the adsorption isotherm at pH 3, 4 and 5 of Stillings et al. (1998) shown in Figure (42b). Therefore, it is also possible to fit the data of Stillings et al. (1998) using a simple Langmuir adsorption isotherm:

$$C_{ox,s} = F\frac{b'_{ox} \cdot (C_{HOx} + C_{Ox})}{1 + b'_{ox} \cdot (C_{HOx} + C_{Ox})} \tag{47}$$

The coefficients obtained by fitting the results to Equation (46) are very similar to those obtained by fitting them to Equation (47): $F_{(46)} = 2.48 \pm 0.1\times10^{-5}$ mol m^{-2} (15 sites nm^{-2}), $F_{(47)} = 2.3 \pm 0.1\times10^{-5}$ mol m^{-2} (14 sites nm^{-2}), and $b_{ox(46)} = 1966 \pm 400$ M^{-1}, $b_{Hox(46)} = 1120 \pm 400$ M^{-1}, $b'_{ox(47)} = 1674 \pm 400$ M^{-1}, and the regression coefficients $R^2_{(46)} = 0.960$ and $R^2_{(47)} = 0.955$. Figure 43 shows a good agreement between the results and the best fit curve of Equation (47). Following similar argumentation, Cama and Ganor (2006) modeled the total surface concentration of oxalate on kaolinite as a function of the sum of the activities of bioxalate and oxalate (Fig. 9) using an equation similar to Equation (47) (i.e., they used activities rather than concentrations):

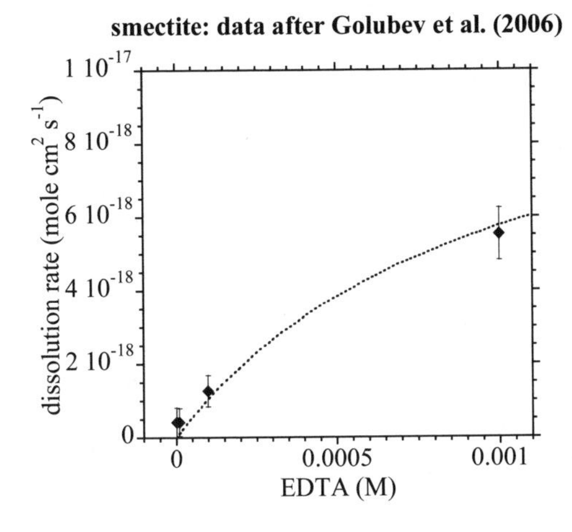

Figure 41. The effect of EDTA on the oxalate-promoted dissolution rate of smectite at 25 °C. The data are from Golubev et al. (2006). The dashed line a fit of the experimental data to a rate law that includes an assumed Langmuir adsorption isotherm (Eqn. 45).

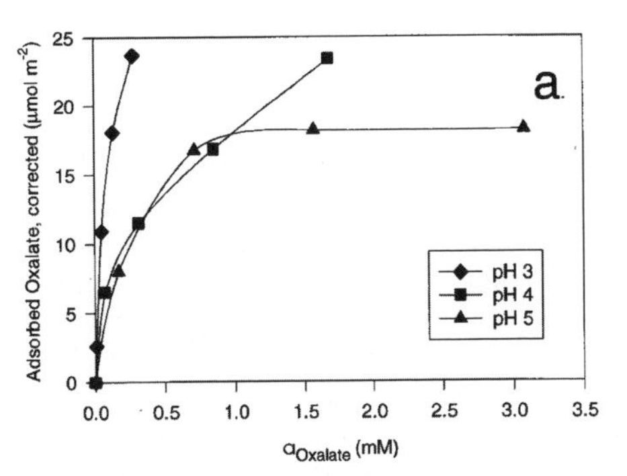

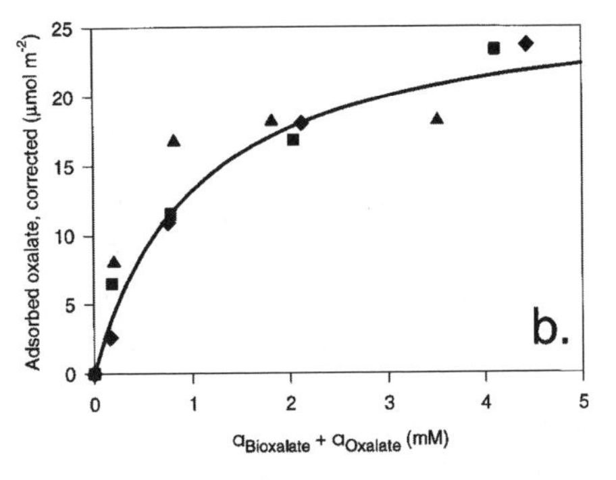

Figure 42. Total oxalate adsorption plotted against (a) oxalate activity and (b) the sum of the activities of oxalate and bioxalate at three pH values. The apparent pH dependence of the adsorption isotherm disappears when the adsorption isotherm is plotted as a function of the sum of the bioxalate and oxalate activities. [Used by permission of American Chemical Society, from Stillings et al. (1998), *Environmental Science and Technology* Vol. 32, Fig. 2, p. 2859.]

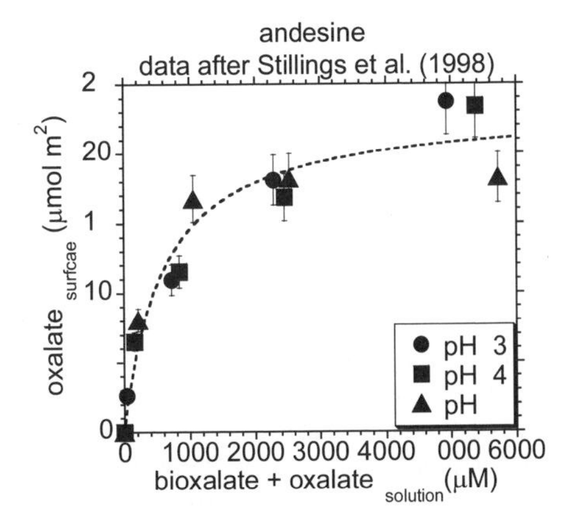

Figure 43. Adsorption isotherms of oxalate on andesine at pH = 3, 4 and 5 and 25 °C. The data is from Stillings et al. (1998). The observations are fitted to the simple Langmuir adsorption isotherm of Equation (47).

$$C_{ox,s} = F \frac{b''_{ox} \cdot (a_{HOx} + a_{Ox})}{1 + b''_{ox} \cdot (a_{HOx} + a_{Ox})} \tag{48}$$

The so obtained coefficients for oxalate adsorption on kaolinite are $F_{(48)} = 4.6 \pm 0.2\times10^{-7}$ mol m^{-2} (0.28 sites nm^{-2}), $b''_{ox(48)} = 18300\pm4000$, and the regression coefficients $R^2_{(48)} = 0.85$.

Stillings et al. (1998) additionally fitted the adsorption of oxalate and bioxalate to the empirical Freundlich adsorption isotherm (Eqn. 9) so that the overall oxalate concentration on the andesine surface was described by:

$$C_{s,OxTot} = C_{s,Ox} + C_{s,HOx} = 73.1 \cdot C_{Ox}^{0.244} + 1410 \cdot C_{HOx}^{0.841} \tag{49}$$

In this adsorption isotherm the number of surface sites is not limited and therefore there is no competition between oxalate and bioxalate. Cama and Ganor (2006) used the more general nonlinear adsorption isotherm of Equation (8) so that the overall oxalate concentration on the kaolinite surface was described by:

$$C_{s,ox} = 6.1\times10^{-7} \cdot \frac{64 \cdot (a_{HOx} + a_{Ox})^{0.48}}{1 + 64 \cdot (a_{HOx} + a_{Ox})^{0.48}} \tag{50}$$

Taking into account the strong complexation between Al and oxalate in solution, it seems reasonable that the adsorption of oxalate would occur on aluminol sites rather than on silanol sites. Using crystallographic data, Stillings et al. (1998) calculated that there are only 7.8 Al sites per nm^{-2} of andesine surface, a factor of 2 less than the number of sites that are predicted using a Langmuir adsorption isotherm (i.e., the maximum surface coverage, F, obtained from fitting the results to Eqn. 46). They therefore concluded that two oxalates can adsorb on each Al site. Cama and Ganor (2006) concluded that the total amount of adsorbed oxalate on kaolinite (0.28 and 0.37 sites nm^{-2} according to the Langmuir and the general adsorption isotherms, respectively) are significantly larger than the estimated number of available edge aluminol sites ($\approx$0.2 sites nm^{-2}), which was calculated following the crystallographic considerations of Sposito (1984). Based on Scanning Force Microscopy measurements, Brady et al. (1996) showed that the percentage of kaolinite edge surface area from the total surface area may be significantly larger than the value of 7.9% estimated by Sposito (1984). Therefore, the observation that the amount of adsorbed oxalate is somewhat larger than the estimated number of available edge aluminol does not prove that oxalate adsorbed on other sites, nor that more than one oxalate can adsorbed on the same aluminol site, as was suggested for andesine.

Both Stillings et al. (1998) and Cama and Ganor (2006) modeled their data using the OXPRM of Furrer and Stumm (1986). Stillings et al. (1998) described the overall dissolution rate ($Rate_T$) of andesine using the equation:

$$Rate_T = k_H \cdot C_{s,H} + k_{Ox} \cdot C_{s,Ox} + k_{HOx} \cdot C_{s,HOx} \tag{51}$$

They fitted their data to Equation (51) while calculating the surface concentration using either Langmuir or Freundlich adsorption isotherms. The fitting with Langmuir adsorption isotherm yields values of 8.13×10^{-13}, 2.18×10^{-12} and 9.30×10^{-13} (mol of feldspar/μmol of adsorbed ion/s) for the rate constants K_H, k_{HOx} and k_{Ox}, respectively. The fitting with Freundlich adsorption isotherms yields values of 9.61×10^{-13}, 1.16×10^{-12} and 1.05×10^{-13} (mol of feldspar/μmol of adsorbed ion/s) for K_H, k_{HOx} and k_{Ox}, respectively. Stillings et al. (1998) noted that the similarity between the rate coefficients indicates that the magnitude of the rates of each of the reaction mechanism is determined by the surface concentration of the various adsorbed species and not necessarily by the value of the rate coefficients. Or saying it in other words, adsorbed protons, adsorbed oxalate and adsorbed bioxalate have similar effects on the bond strength (i.e., similar k_i).

Cama and Ganor (2006) assumed that the different adsorbed oxalate species formed similar surface complexes (Yoon et al. 2004), and therefore have the same effect on the bond strength and hence on the dissolution rate. They calculated the OXPRM rate by further assuming that the OXPRM is independent of the PPRM. Initially they calculated the total surface concentration of oxalate using the general adsorption isotherm (Eqn. 50) and fitted the calculated rate of OXPRM to the simple rate law of Equation (44), where k_{ox} (s^{-1}) is the rate coefficient of the OXPRM of both the adsorbed oxalate and the adsorbed bioxalate. Figure 44 plots the dependence of the oxalate-promoted dissolution rate on the calculated oxalate surface concentration. As predicted by Equation (44), the observation may be adequately described using linear regression (solid lines in Fig. 44). The obtained slopes (k_{ox} in Eqn. 44) are $8 \pm 3{\times}10^{-7}$, $3.3 \pm 0.6{\times}10^{-6}$ and $2.5 \pm 0.3{\times}10^{-5}$ s^{-1} at 25, 50 and 70 °C, respectively. In contrast to the prediction of Equation (44), the regression lines in Figure 44 do not cross through the origin, and the oxalate-promoted rate approaches zero when oxalate surface concentrations are in the range of 0.1-0.2 μmol m^{-2} (i.e., when the sum of the activities of oxalate and bioxalate in solution is in the range of $5\text{-}40{\times}10^{-6}$). A similar dependence of the rate of OXPRM on oxalate adsorption to andesine is shown in Figure 45. The rate of the OXPRM in Figure 45 was calculated using the assumption that the OXPRM is independent of the PPRM, and the total surface concentration of oxalate was calculated using the simple Langmuir adsorption isotherm (Eqn. 47). The regression line in Figure 45 does not cross through the origin, and the rate of the oxalate-promoted mechanism approaches zero when the oxalate surface concentration is 13 μmol m^{-2}, i.e. when the sum of the concentrations of oxalate and bioxalate in solution is about 0.8 mM. Forcing the regression line to cross through the origin (dashed line in Fig. 45), as predicted by Equation (44), yields a line that does not fit the observations (R^2 = 0.37). As was

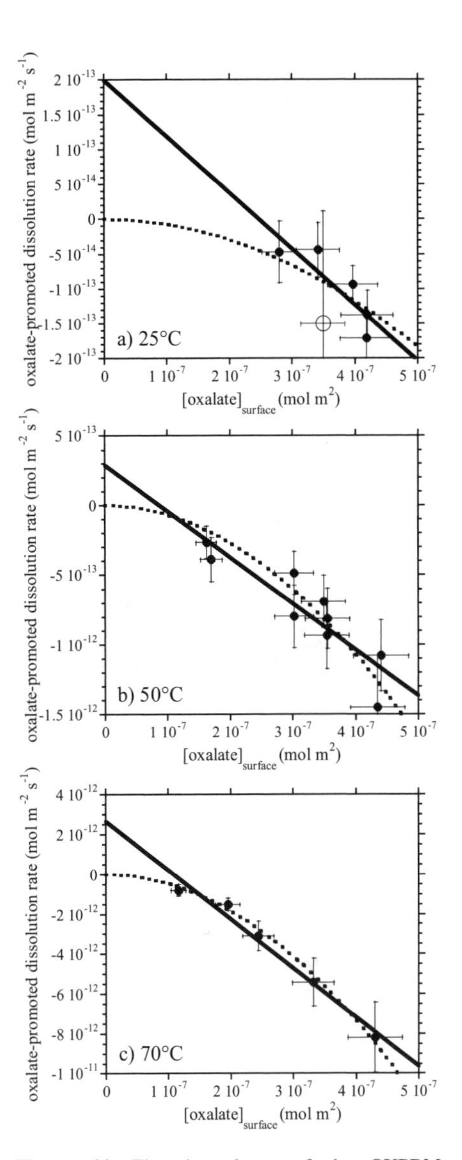

Figure 44. The dependence of the OXPRM rate of kaolinite dissolution on the total surface concentration of oxalate at (a) 25 °C, (b) 50 °C, and (c) 70 °C. The solid lines are fitting of the data to the traditional oxalate-promoted reaction mechanism of Equation (44). The dashed lines are the fitting of the experimental data to the quadratic rate law of Equation (52). The data point in (a) which is marked by an open symbol was not included in the regression calculations, due to its high uncertainty. [Used by permission of Elsevier, from Cama and Ganor (2006), *Geochimica et Cosmochimica Acta*, Vol. 70, Fig. 7, p. 2205.]

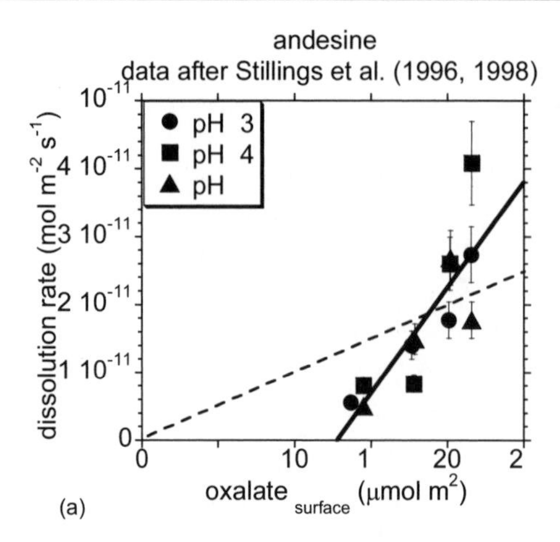

Figure 45. The dependence of OXPRM rate of andesine dissolution on oxalate surface concentration. The solid line is the fitting of the experimental data to a rate law which predict a linear dependence of the oxalate-promoted rate on the surface concentration of oxalate. The dashed line in is the best fit line that crosses through the origin, as predicted by Equation (44).

suggested by Cama and Ganor (2006) for kaolinite, there seems to be a threshold in oxalate concentration that is required to catalyze the dissolution reaction. Below this threshold, the OXPRM is not active, although a significant amount of oxalate is adsorbed to the mineral surface. The suggestion that there is a threshold of oxalate concentration that is required to catalyze the plagioclase dissolution is in agreement with the general observation that the effect of oxalate on silicate dissolution becomes significant at total oxalate concentration of around 1 mM (Drever and Stillings 1997).

Cama and Ganor (2006) proposed an alternative oxalate-promoted reaction mechanism in which the rate is not linearly proportional to the oxalate concentration on the surface. Indeed, the observations in Figure 44 are better described using a parabolic rate law of the form:

$$Rate_{oxp} = k_{2Ox} \cdot C_{s,ox}^2 \tag{52}$$

where k_{2Ox} (m^2 mol^{-1} s^{-1}) is the rate coefficient for the quadratic rate law. The parabolic rate law of Equation (52) was fitted to the calculated oxalate-promoted rates at each temperature using non linear regression (dashed lines in Fig. 44). The obtained k_{2Ox} values are 0.7 ± 0.1, 6.7 ± 0.4 and 46 ± 1 m^2 mol^{-1} s^{-1} at 25, 50 and 70 °C, respectively. Theoretically, a parabolic rate law may be justified if the reaction is catalyzed by the simultaneous adsorption of two ligands on or near the same surface site. As Cama and Ganor (2006) did not have any direct evidence that validate their proposed theoretical explanation for the OXPRM, one can only view the parabolic rate law of Equation (52) as an empirical rate law.

An alternative explanation for the observation that the rate of the OXPRM is insignificant for low surface concentration of oxalate may be related to our inability to accurately evaluate the rate of the PPRM in the presence of oxalate. If the rate of the PPRM in the presence of low concentrations of oxalate is significantly lower than assumed, the rate of the OXPRM is significantly higher and the presumed threshold of oxalate concentration is only an artifact related to the lack of understanding of the inhibitory effect of oxalate on the PPRM.

The two case studies of the catalytic effects of oxalate on kaolinite and andesine dissolution rate demonstrate that, regardless of the extensive amount of study of the effect of OM on silicate dissolution, the understanding of the exact mechanism of the LPRM is still imperfect. In particular, there is a lack of studies that combine dissolution and adsorption experiments and of studies on the mutual effects of the PPRM and the LPRM.

The relative effects of different OM on silicate dissolution. For oxides, Furrer and Stumm (1986) showed that oxalate, which forms a five-membered chelate ring, enhanced the dissolution rate more than malonate, which forms a six-membered ring, and that succinate, which forms a seven-membered ring, is even less efficient than malonate. Similarly for the aromatic ligands, Furrer and Stumm (1986) showed that the catalytic effect on oxides of the six-membered chelate ring formed with salicylate is approximately four times that of phthalate which forms a seven-membered ring. Similar trends were observed with silicate minerals such as quartz (Bennett 1991; Bennett and Casey 1994), feldspar (Ullman and Welch 2002) and kaolinite (Chin and Mills 1991). In general, OM that forms strong complexes with Al^{3+} in solution tend to form complexes with Al sites on silicate and oxides surfaces and therefore have a strong catalytic effects (Furrer and Stumm 1986; Welch and Ullman 1993; Bennett and Casey 1994; Ullman and Welch 2002). Correspondingly, polyfunctional acids such as oxalate, citrate succinate, α-ketoglutarate, and pyruvate enhance dissolution to a greater degree than monofunctional acids such as acetate and propionate (e.g., Huang and Kiang 1972; Welch and Ullman 1993; Franklin et al. 1994; Ullman and Welch 2002). This order of effectiveness of dissolution enhancement approximately follows the order of the association constants of Al^{+3}-ligand complexes (see β_1, in Table 2) (Ullman and Welch 2002). In contrast, Manley and Evans (1986) found that the amount of Al released from feldspar by organic acids appeared to be related more to their acidic strength than to their ability to form complexes.

Although most of the experiments with silicates were conducted with simple OM, mainly organic acids, several studies show that the dissolution rate of silicate is enhanced by complex natural mixtures of OM such as fulvic acid extracts from soil (Schnitzer and Kodama 1976), stream water, soil pore water, water extracts of peat and mor (Lundstrom and Ohman 1990), organic-rich groundwater (Hiebert and Bennett 1992), organic-rich peat bog (Bennett et al. 1991) and the extract from the O horizon of a forest soil (Van Hees et al. 2002). Ochs (1996) showed that exudates from ectomycorrhizal roots (a symbiosis between a fungus and plant roots) effectively enhance the dissolution rate of aluminum oxide (γ-Al_2O_3). In contrast, humic substances and exudates from non-mycorrhizal roots did not significantly affect weathering rates, and at pH $\geq$ 4 they may even inhibit dissolution rate. Ochs (1996) challenged the widespread belief that medium- to high-molecular-weight natural organic ligands, such as humic substances, significantly enhance mineral weathering similarly to simple organic acids such as salicylic or oxalic acids. They suggested that humified compounds may protect minerals from dissolution, rather than enhance weathering.

The relative effect of a specific OM on dissolution of different silicate minerals. Manley and Evans (1986) found that minerals dissolution rates in the presence of various OM (citric, oxalic, salicylic, protocatechuic, gallic, p-hydroxybenzoic, vanillic, and caffeic acids) at 13 °C followed the order: calcic plagioclase > microcline > albite. Similarly, Welch and Ullman (1993; 1996) showed that at 22 °C the effect of OM on dissolution rate followed the order: bytownite > labradorite > albite. At 80 °C, Blake and Walter (1999) also found that the effect of oxalate and citrate on dissolution rate of labradorite was stronger than that of albite. In contrast to the observations of Manley and Evans (1986), Blake and Walter (1999) found that microcline dissolved slower than albite in the presence of oxalate and citrate. Welch and Ullman (1996) and Stillings et al. (1996) generalized that dissolution rates tend to increase with increasing Al content of the mineral both in the presence and in the absence of OM (see Fig. 46). In the absence of OM, this effect was previously shown in several studies (e.g., Blum and Stillings 1995 and references therein). The increase in rate with Al content of the mineral is another indication that ligand attack (as well as proton attack) occurs predominantly on Al surface sites of aluminosilicates.

Schnitzer and Kodama (1976) examined the effect of fulvic acid on the dissolution of three micas and found that dissolution rate decreased in the following order: biotite > phlogopite > muscovite. They noted that the mineral that was most susceptible to an attack by fulvic acid contained the highest amount of Fe. This observation may indicate that, for iron rich minerals, ligands preferentially attack the Fe surface sites.

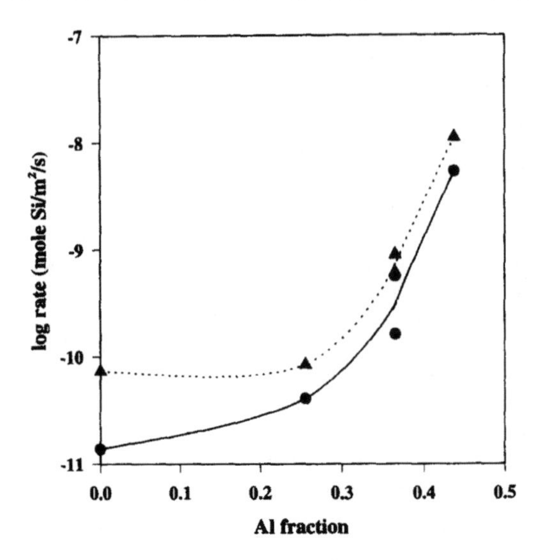

Figure 46. Steady-state dissolution rates for quartz, albite, bytownite and two labradorite. Black triangles and circles denote experiments that were conducted with 1 mM oxalate and 1 mM nitrate, respectively. Aluminum fraction equals Al/(Si+Al) of the initial solid material. [Used by permission of Elsevier, from Welch and Ullman (1996), *Geochimica et Cosmochimica Acta,* Vol. 60, Fig. 1, p. 2942.]

The effect of temperature on catalysis of silicate dissolution by OM. The effect of OM on silicate dissolution was examined at both near-room temperature conditions, which are relevant for studies of weathering processes, and at higher temperatures relevant to diagenetic conditions. The enhancement ratios in the two temperature regimes are similar (e.g., Table 3). As was mentioned above, comparison of results obtained in different laboratories is problematic because the reproducibility of dissolution rates is usually very poor. Unfortunately, only a few studies obtained rates of silicate dissolution in the presence of OM at different temperatures (e.g., Bevan and Savage 1989; Welch and Ullman 2000; Cama and Ganor 2006).

The temperature dependence of a chemical reaction is usually described using the Arrhenius Equation:

$$Rate = A \cdot e^{-E_a/RT} \tag{53}$$

where A is the pre-exponential factor, E_a is the activation energy, R is the gas constant and T is the temperature (K). Temperature variations are clearly very important in affecting the rates as well as in providing unique insights into the reaction mechanisms (Lasaga and Gibbs 1990). Activation energies are commonly calculated using Equation (53) and an Arrhenius Plot (a plot of the neutral logarithm of dissolution rate as a function of the temperature reciprocal) by a least squares estimate of the slope of ln(dissolution rate) vs. $1/T$. It is common to refer to such empirically derived activation energy as *apparent activation energy*. Calculated apparent activation energies of the overall rate in the presence of OM does not have a clear mechanistic meaning as the overall rate reflects the sum of at least two reaction mechanisms (Eqn. 34).

Welch and Ullman (2000) determined the temperature dependence of silica release from bytownite in solutions containing inorganic and organic ligands at neutral pH from 5 °C to 35 °C. The enhancement ratio decreased in their experiments with 1 mM oxalate from 2.5 at 5 °C to 1.3 at 35 °C. The apparent activation energy of the overall dissolution rate of the bytownite was 10 kcal mol^{-1} in the absence of OM and 6.6±1.0 and 7.1±0.5 kcal mol^{-1} in the presence of oxalate and gluconate, respectively (Fig. 47a). In order to estimate the apparent activation energy of the ligand attack, we calculated the rate of the OXPRM by subtracting the rate without ligand from the measured overall rate at each temperature. The calculated OXPRM rates were the same within error, and the calculated apparent activation energy was 0.8±1.7 kcal mol^{-1} (see Fig. 47b). Such low apparent activation energy is not expected, since the OXPRM is considered to be a surface-controlled mechanism. Usually, surface-controlled reactions have activation energies in the range of 10-20 kcal/mol, while low activation energies are typical to diffusion-controlled reactions in solution (E_a < 5 kcal/mol), (Lasaga 1984; Lasaga 1995). Nevertheless, we calculated the expected temperature dependence of the enhancement ratio using this low apparent activation energy (Fig. 48). According to this calculation, the enhancement ratio is expected to become insignificant (lower than 1.1) at 60 °C. Similar results were obtained when we assumed that the value of the activation energy was at the high limit of the uncertainty of the calculations (i.e., 2.5 = 0.8+1.7). As previously noted by Ullman and Welch (2002), the observation of Welch and Ullman (2000) that the relative importance of LPRM compared to PPRM decreases with increasing temperatures contradicts the observations of many studies that showed strong enhancement of silicate dissolution rates in the presence of OM at temperatures range of 70-150 °C (e.g. Bevan and Savage 1989; Stoessell and Pittman 1990; Gestsdottir and Manning 1992; Hajash et al. 1992; Huang and Longo 1992; Franklin et al. 1994; Blake and Walter 1999; Cama and Ganor 2006).

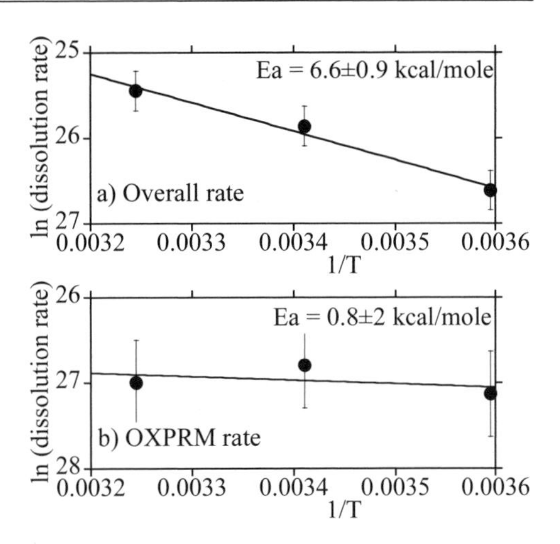

Figure 47. Arrhenius plots of the dissolution rate of bytownite in the presence of oxalate. (a) A plot of the overall dissolution rate and (b) a plot of the OXPRM which was calculated by subtracting the rate without oxalate from the measured overall rate at each temperature. Data after Welch and Ullman (2000).

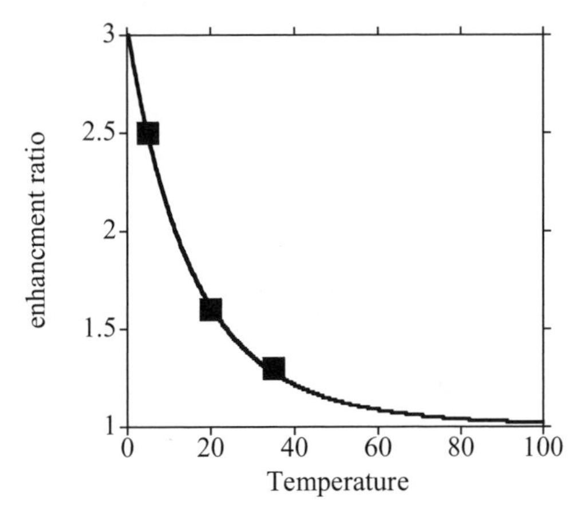

Figure 48. The expected temperature dependence (solid line) of the enhancement ratio of the dissolution rate of bytownite in the presence of oxalate which was calculated using the low apparent activation energy of the OXPRM that was obtained from the Arrhenius plot (Fig. 47). The enhancement ratio is expected to become insignificant (<1.1) at 60 °C.

Cama and Ganor (2006) determined the temperature dependence of kaolinite dissolution rates in the presence of oxalate at pH 3 from 25 °C to 70 °C. Using the rate constants obtained from their parabolic model of the OXPRM (Eqn. (52) and Fig. 44), Cama and Ganor (2006) obtained an apparent activation energy of 19 ± 1 kcal mol^{-1}. This value is very close to the apparent activation energy of the PPRM of kaolinite dissolution (22 kcal mol^{-1}), which was determined in the same laboratory (Cama et al. 2002). The similarity in apparent activation energy between the PPRM and OXPRM indicates that the weakening effect of adsorbed oxalate on the bond strength is similar to that of adsorbed H^+, which is consistent with the above mentioned finding of Stillings et al. (1998) that the rate constants of PPRM and OXPRM are quite similar.

The effect of pH on catalysis of silicate dissolution by OM. Welch and Ullman (1993) found that the enhancement ratio in the presence of various OM decreases as acidity increases, indicating that the LPRM becomes relatively more important as the rate of PPRM decreases. Welch and Ullman (1996) showed that the effect of pH on dissolution rate of tectosilicates in the presence of OM can be described using the same empirical pH dependence that is commonly used in the absence of OM (Schott 2009):

$$Rate = k \cdot a_H^{n_H} \tag{54}$$

where k and n_H are coefficients and a_H is the activity of H^+ in solution. Welch and Ullman (1996) found that the reaction order, n_H in Equation (54), was mostly similar in albite and labradorite dissolution experiments with 1mM oxalate and without OM (Fig. 49). For bytownite, n_H with oxalate was smaller than without oxalate. Ullman and Welch (2002) suggested that the similarities in reaction orders indicate that the effect of ligands on dissolution rates in highly acidic solution is to increase the rates of the PPRM reaction. This is perhaps achieved by buffering the reactive solution at acidic pH rather than to control a completely independent OXPRM.

Stillings et al. (1996) conducted systematic experiments that enabled quantification of the combined effects of oxalate and pH on andesine dissolution rate (Fig. 50). Between pH 3 and 5 the effect of pH on dissolution rate decreases with increasing oxalate concentration. Using the data of Stillings et al. (1996), we have calculated the reaction order of the overall dissolution rate with respect to the activity of H^+ (n_H in Eqn. 54) between pH 3 and 4. Figure 51 shows that n_H decreases as a function of the concentration of oxalate from 0.53 in the absence of oxalate to 0.03 in the presence of oxalate. We suggest that the reaction order with respect to the activity of H^+ reflects the combined effect of the PPRM rate, which decreases with pH, and the OXPRM rate, which is pH independent (or only has a slight pH dependency).

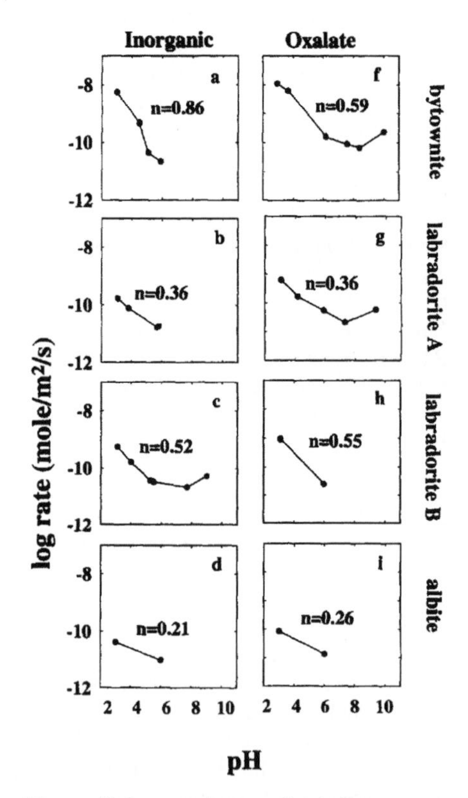

Figure 49. Log steady-state dissolution rates vs. pH for four tectosilicate minerals dissolved in inorganic (a through d) or 1 mM oxalate solutions (f through i). The exponent n in each figure is the slope of the lines at the acidic range (pH 3 to 7) of the curve. [Used by permission of Elsevier, from Welch and Ullman (1996), *Geochimica et Cosmochimica Acta*, Vol. 60, Fig. 3, p. 2942.]

As noted above, pH affects both the mineral surface site speciation and the ligand speciation in solution. If the charge of the dominant ligand species in solution is of the same sign as the surface, adsorption may be inhibited due to electrostatic repulsion. Therefore, it is expected that the rate of the LPRM should be greatest in those pH regions where surface and ligand charges are predominantly of opposite sign and where there are few ions that form stable complexes with the ligand in solution (Ullman and Welch 2002). This is clearly demonstrated in Figure 52 where the speciation of oxalate and the pH_{PZNPC} range of feldspar are shown together. In accordance with the prediction of Figure 52, Stillings et al. (1996) showed that

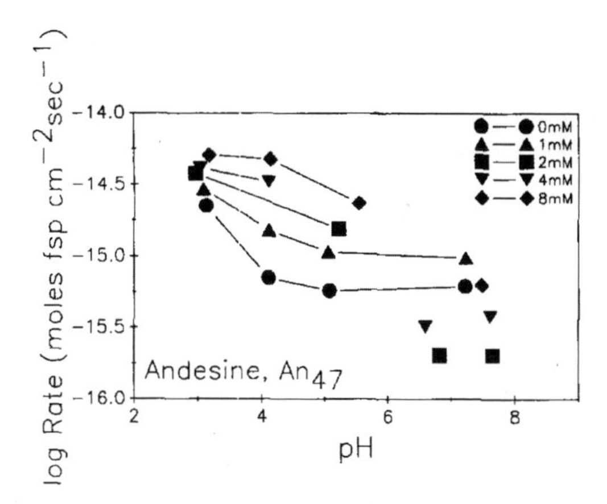

Figure 50. Andesine dissolution rate as a function of steady-state pH and oxalic acid concentration. In the title of the Y axis, the abbreviation fsp = feldspar phase. [Used by permission of Elsevier, from Stillings et al. (1996), *Chemical Geology,* Vol. 132, Fig. 2, p. 85.]

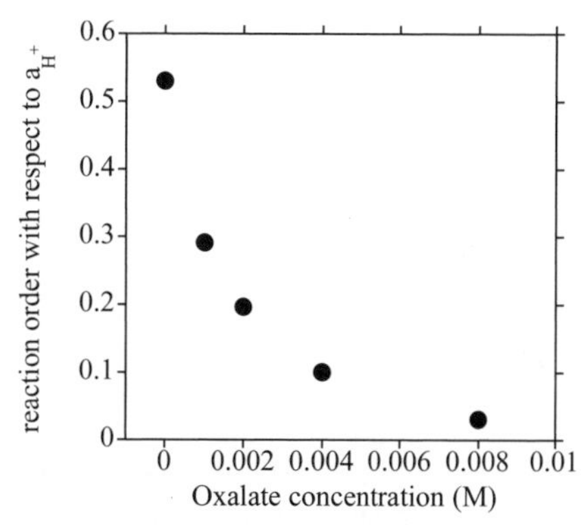

Figure 51. The change in reaction order with respect to $a_{H}+$ (n_H in Eqn. 54) as a function of oxalate concentration, calculated from data of andesine dissolution at 25 °C and pH 3 to 4 after Stillings et al. (1996).

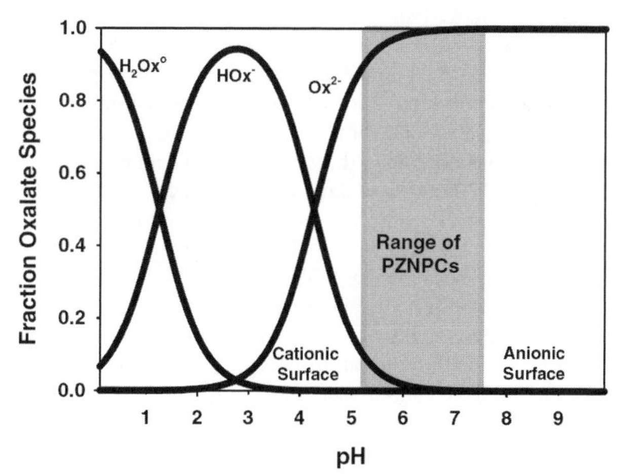

Figure 52. The importance of ligand and surface charges on dissolution rates. In the pH region below the point of zero net proton charge (PZNPC) the average charge on the feldspar surface is positive. In this range of pH, the adsorption of anionic organic ligands (oxalate is given as an example), which is the first step in the OXPRM, is favored. Above the PZNPC, the anionic surface will electrostatically inhibit ligand adsorption, and therefore LPRM as well. The pH_{PZNPC} is given as a range of values (stippled) in this diagram. [Used by permission of the Geochemical Society, from Ullman and Welch (2002), *Geochemical Society Special Publication.* No. 7. Fig. 2, p. 12.]

oxalate does not catalyze andesine dissolution at circum neutral pH (Fig. 50). At high oxalate concentrations (≥ 2 mM) oxalate even apparently inhibits the dissolution rate. It is important to note that at circum neutral pH, where the surfaces of silicates become negative and the expected catalytic effect of negatively charged ligands is expected to be relatively small, the solubility of silicates is low. Therefore, ligands may enhance dissolution rates by their effect on the degree of saturation as a result of complex formation in solution (e.g., Oelkers and Schott 1998; Ullman and Welch 2002).

Carbonate dissolution

While many investigators studied the effect of OM on the precipitation of carbonates, there have been surprisingly few studies on the influence of organic compounds on the dissolution of carbonates (see reviews in Morse and Arvidson 2002; and Morse et al. 2007). In contrast to silicates, the dissolution of carbonates is usually inhibited by the presence of OM. The suggestion that OM protects the calcite from dissolution in seawater was already proposed by the end of the 19th century by Murray and Renard (1891, as cited by Morse 1974). This suggestion was later confirmed by several studies (e.g., Suess 1970). Morse (1974) claimed that coccoliths (individual plates of calcium carbonate formed by coccolithophores) are very resistant to dissolution by acidic solutions prior to a treatment by strong oxidizing agent. Morse (1974) explained the preferential resistance of coccoliths, in comparison to foraminifera tests, by suggesting that a coating of OM is very important in protecting coccoliths, but not in protecting foraminifera.

Inhibition of calcite dissolution by OM was also observed in fresh water (e.g., Hamdona et al. 1995; Wilkins et al. 2001). Many types of OM were shown to inhibit calcite dissolution (e.g., stearic acid and albumin (Suess 1970), alanine, phenyl alanine and glycine, L- arginine (Hamdona et al. 1995), polymaleic acid and oxalic acid (Wilkins et al. 2001), maleic and fumaric acids (Compton et al. 1989), succinic acid, phthalic acid, and maleic acid (Compton and Brown 1995).

Several general mechanisms were suggested in order to explain the inhibiting effect of OM on the dissolution of calcite and other carbonates. In two of the mechanisms, OM blocks the active sites on the mineral surface: (a) the active sites may be blocked due to physical coatings such as bacterial or algal slime, or adhering dead OM (e.g., Morse 1974); and (b) the OM may be chemically adsorbed on the surface (e.g., Compton et al. 1989; Compton and Brown 1995). Such chemisorption may block the reactive sites. For example, Compton et al. (1989) studied the effect of maleic and fumaric acids on the dissolution rate of calcite. They determined adsorption isotherms showing that the di-anion of maleic acid (henceforth, Mal^{2-}) and fumaric acid (henceforth, Fum^{2-}) adsorbed on the calcite surface (Fig. 53). Compton et al. (1989) also showed that the morphologies of etch pits that are formed on a cleaved calcite crystal are affected by the presence of these carboxylic acids. The etch pits that are formed due to exposure to an inorganic acid (HCl) reflect the rhombohedral geometry of the (100) cleavage plane. These etch pits, which nucleate at sites of emergent dislocations, grow by recession of steps away from the dislocation. The recess may require the nucleation of a (double) kink site in a terrace within the pit, followed by the stripping away of the remaining ions in that particular ledge. Due to the rhombohedral cleavage of the calcite, there are two kinds of kink sites associated with the cleavage plane: (1) an obtuse kink that comprises a $[0\bar{1}0]$ ledge intersecting with a $[\bar{1}00]$ ledge, and (2) an acute kink comprising a $[0\bar{1}0]$ ledge intersecting a $[\bar{1}00]$ ledge (Fig. 54). In the absence of an inhibitor, the growth of the etch pit starts from both types of kinks, and as a result the shape of the edge pits is rhombohedral (Fig. 55). The nearest Ca^{2+}-Ca^{2+} distance between intersecting ledges is 4.04 Å for the acute type of kink and 4.99 Å for the obtuse kink. Since dissolution proceeds through the propagation of kinks, a given inhibitor will, if it is assumed to operate by adsorption at these sites, affect the dissolution in the [110] and $[\bar{1}\bar{1}0]$ directions differently to the $[\bar{1}10]$ and $[1\bar{1}0]$ directions, depending on its structure relative to that of the kinks.

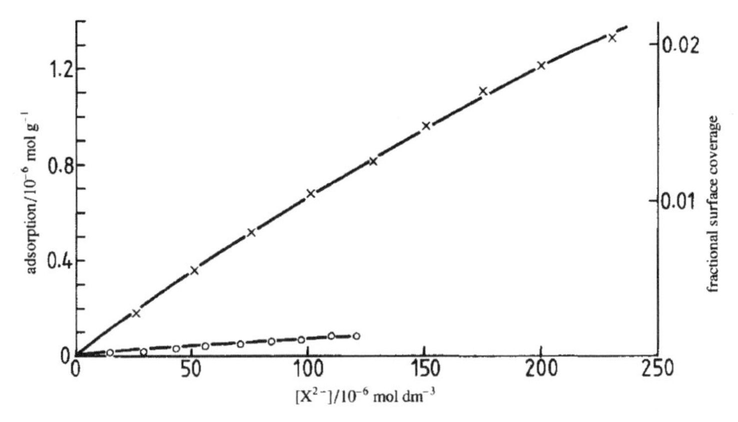

Figure 53. The adsorption isotherms of Mal^{2-} (x) and Fum^{2-} (O) on Calopake-F calcite powder. [Used by permission of Royal Society of Chemistry, from Compton et al. (1989), *Journal of the Chemical Society, Faraday Transactions 1: Physical Chemistry in Condensed Phases*, Vol. 85, Fig. 20, p 4359. DOI: *http://dx.doi.org/10.1039/F19898504335*]

In the presence of Mal^{2-} elongated rounded lens-shaped pits are formed (see Fig. 56). Compton et al. (1989) suggested that the Mal^{2-} anion acts as a very effective 'molecular clamp' between calcium ions in a kink formed between intersecting [$\bar{1}$10] and [010] ledges (Ca^{2+} separation 4.04 Å). In contrast, examination of the [010]/[100] kink reveals a much weaker interaction between the adsorbing anion and the calcium ions. The preferential adsorption of Mal^{2-} at the first (acute) type of kink site may therefore be taken to explain the marked reduction in the rate of dissolution in the [1$\bar{1}$0] direction, whilst dissolution in the [110] direction appears to be unchanged from that seen in the reaction of HCl with calcite. The observations of Compton et al. (1989) indicate that the di-anion of maleic acid inhibit calcite dissolution by chemical adsorption on a reactive site. Similar results were reported by Barwise et al. (1990) for the inhibition of calcite dissolution by tartaric acid.

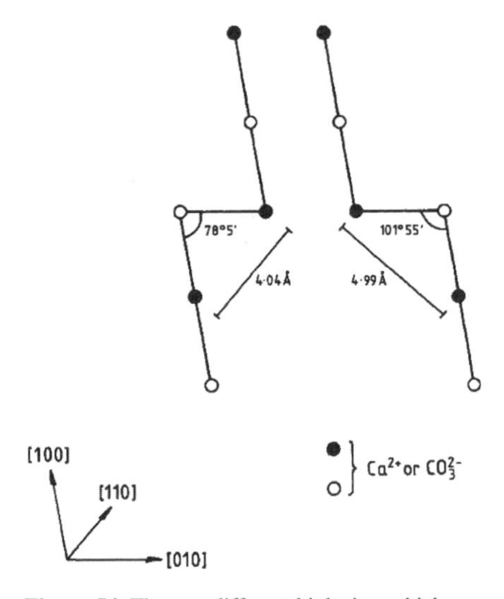

Figure 54. The two different kink sites which can be formed in the calcite (100) plane. [Used by permission of Royal Society of Chemistry, from Compton et al. (1989), *Journal of the Chemical Society, Faraday Transactions 1: Physical Chemistry in Condensed Phases*, Vol. 85, Fig. 11, p 4349. DOI: *http://dx.doi.org/10.1039/F19898504335*]

Whereas the di-anion of maleic acid adsorb on a surface site and block it, other organic anions such as succinate and phtalate compete with the carbonate anion on adsorption sites (Compton and Brown 1995). This mechanism inhibits calcite growth as well, as will be discussed later.

Compton and Sanders (1993) made the interesting observation that there is no inhibition by humic material in equilibrated acid solutions (pH < 4) and that dissolution proceeds at a rate simply determined by the solution pH. In contrast, the sodium salts of humic acids were found to have a significant inhibitory effect on the acid catalyzed dissolution.

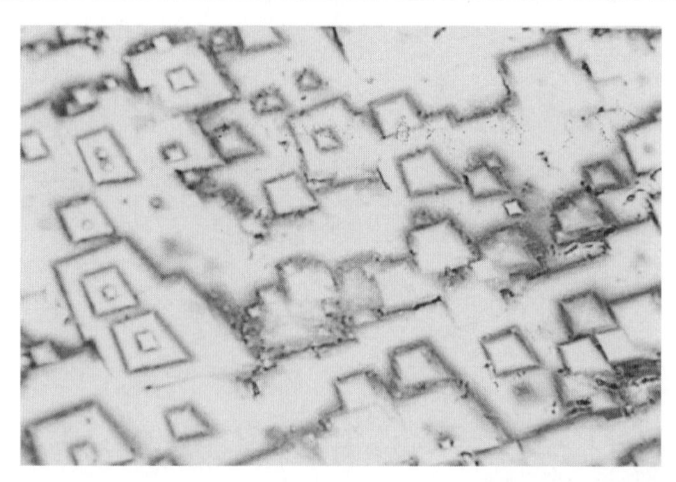

Figure 55. Interferogram (×220) of the (100) plane of cleaved calcite which has been subjected to dissolution in 2×10^{-3} mol dm^{-3} HCl for 50 min, in a channel flow cell. [Used by permission of Royal Society of Chemistry, from Compton et al. (1989), *Journal of the Chemical Society, Faraday Transactions 1: Physical Chemistry in Condensed Phases*, Vol. 85, Plate 2, facing page 4344. DOI: *http://dx.doi.org/10.1039/F19898504335*]

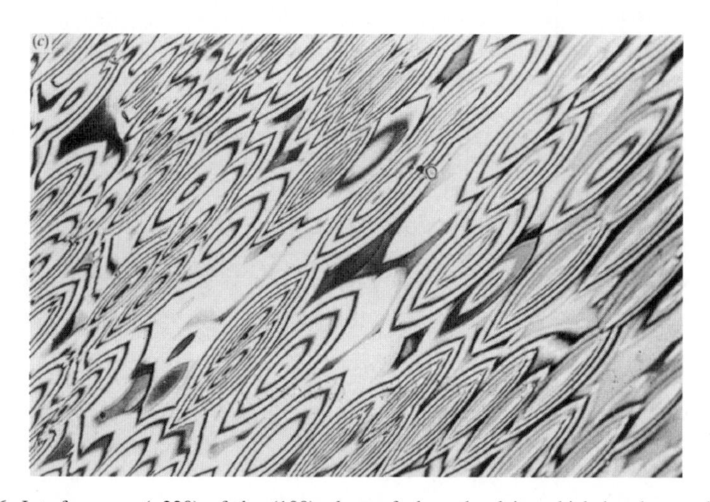

Figure 56. Interferogram (×220) of the (100) plane of cleaved calcite which has been subjected to dissolution in 5×10^{-3} mol dm^{-3} maleic acid solution for 30 min, under zero flow conditions. [Used by permission of Royal Society of Chemistry, from Compton et al. (1989), *Journal of the Chemical Society, Faraday Transactions 1: Physical Chemistry in Condensed Phases*, Vol. 85, Plate 7c, facing page 4344. DOI: *http://dx.doi.org/10.1039/F19898504335*]

Several studies showed that soaking calcite in solution containing OM prior to dissolution inhibits its dissolution rate (Morse 1974; Wilkins et al. 2001). Figure 57 compares the surface of freshly cleaved calcite before and after 5 h of exposure to 0.1 M oxalic acid. It can be seen that the calcite surface is well covered by Ca-oxalate that serves as a protective crust against dissolution. Indeed, following the exposure to oxalic acid, the rate of calcite dissolution by HCl (0.01 M) was slower by an order of magnitude compared to that of untreated calcite (Wilkins et al. 2001). A similar protective layer was formed by exposure to polymaleic acid (Wilkins et al. 2001) and polyacrylonitrile (Thompson et al. 2003b). Polyacrylic acid also forms a coverage of

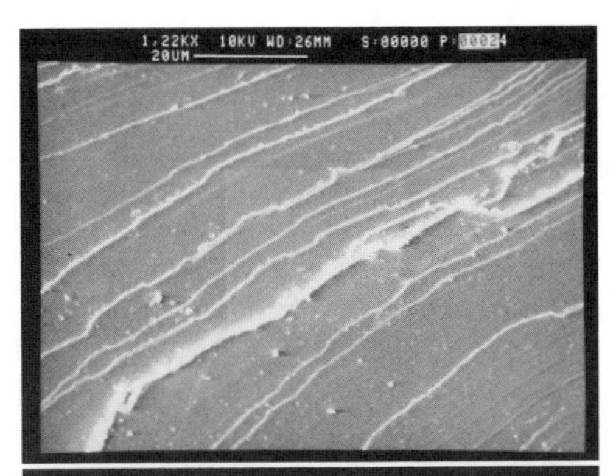

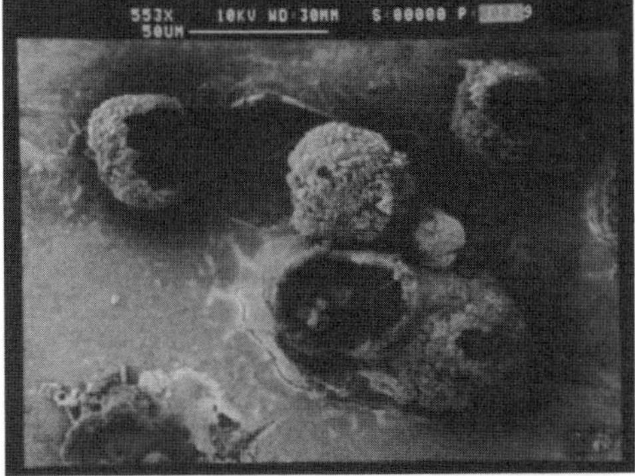

Figure 57. SEM images showing the surface of (a) untreated freshly cleaved calcite and (b) calcium oxalate on calcite after 5 h of exposure of calcite to 0.1 M oxalic acid. [Used by permission of Elsevier, from Wilkins et al. (2001), *Journal of Colloid and Interface Science*, Vol. 242, Fig. 2, p. 380.]

the calcite by the polymer but it erodes from the surface when exposed to an aqueous acid solution (Thompson et al. 2003b). The formation of a stable (organic or inorganic) layer on stone surface may be used to protect buildings and monuments from damaging effects of weathering (Price 1996). For such usage, the long term stability of the coatings should also be taken into account. For example, Thompson et al (2003a) showed that after exposure of calcite pretreated with polymaleic acid to environmental weathering for 6 months, the coating suffered from intensive weathering, and as a result calcite dissolution rate was only marginally slower than that obtained for the untreated exposed samples.

While most OM inhibits calcite dissolution, some may catalyze it (e.g., Fredd and Fogler 1998a; Burns et al. 2003; Pokrovsky et al. 2009). Calcium-chelating OM such as EDTA has been used for the removal of Ca-carbonate and Ca-sulfate in various industrial applications (Fredd and Fogler 1998a). As in the case of oxides and silicates discussed above, the main catalytic mechanism of aminopolycarboxylic ions (such as EDTA) is LPRM, which includes adsorption of the chelating agent onto the calcite. The adsorbed chelating agent weakens the bonds between calcium and carbonate components of the lattice by forming a calcium complex, which subsequently desorbs from the surface along with the carbonate products (Fredd and Fogler 1998a).

Some OM may both catalyze and inhibit the dissolution of carbonates. An interesting example is that of polyaspartic acid (PASP, $(C_4H_4NO_3)_x$). PASP is nontoxic, biodegradable, and water-soluble, and therefore may serve as a "green" alternative for chelating agents like EDTA. There are environmental concerns about heavy metal mobilization with EDTA in groundwater sources since it is non-biodegradable (Burns et al. 2003). At pH 10, calcite dissolution is inhibited by the presence of small amounts (0.001-0.01 M) of PASP. However, as PASP concentration increases to 0.1 M, the dissolution rate exceeds that without PASP. Fredd and Fogler (1998b) proposed that the mechanism of calcite dissolution at pH 10 is via water attack. In the presence of PASP, a parallel ligand attack may act, and therefore the overall dissolution rate would be the sum of the rates of the two parallel mechanisms (Burns et al. 2003). Therefore, PASP may affect the rate of dissolution, both as a catalytic surface ligand and by adsorbing onto the calcite surface and blocking the adsorption sites for the water. It seems that at low PASP concentration, its surface concentration is too low for significant ligand attack but high enough to partly inhibit water attack. At higher concentration of PASP the dominant mechanism is the ligand attack, which is significantly faster than the water attack.

Sulfate dissolution

In various types of industrial applications barite and gypsum tend to form mineral scales (see section below on the effect of OM on precipitation of minerals). Therefore, several studies have focused on the catalytic effect of OM on sulfate dissolution. Various studies have been made on the chemical removal of barium sulfate scale. Most of these processes utilized chelating or complexing agents such as DTPA (e.g., Putnis et al. 1995a,b, 2008; Dunn and Yen 1999) and EDTA (e.g., Bosbach et al. 1998; Dunn and Yen 1999). As previously discussed, chelating agents form strong complexes with metal ions in solution and are known to promote the dissolution of inorganic compounds by bonding with metal ions on the crystal surface and weakening the cation–anion bonds in the structure (i.e., LPRM). Therefore, it is expected that chelating agents such as EDTA will enhance barite dissolution by the formation of Ba-EDTA surface complexes resulting in an increased etch pit formation rate. Bosbach et al. (1998) distinguished between the etch pits forming on two different planes of barite. They showed that on different planes the etch pits exhibited different depths in the presence of EDTA. It was argued that the faster step velocities of retreating monolayer steps result in a shallower elongated etch pit morphology. When EDTA concentrations were increased by 3 orders of magnitude (from 0.1 mM to 100 mM), the velocity of steps was reduced by one order of magnitude (Fig. 58). This is in contrast to the expected increase of dissolution in the presence of a metal chelator. A similar observation was made for barite dissolution kinetics in the presence of DTPA where at high chelator concentrations (50 mM) the dissolution rate decreased with further increase of chelator concentration (Putnis et al. 1995a). These observations suggest that the LPRM is limited by another factor. The apparent activation energy for barite dissolution in the presence of DTPA is ~50 kJ mol^{-1} which suggests that desorption is the rate-controlling step. Due to the similarities between DTPA and EDTA they assumed that the same goes for EDTA. Therefore, the main conclusion from their studies was that lower concentrations of EDTA and DTPA are more effective at accessing the barite surface and therefore complexing with Ba^{2+} ions and removing them into the solution. The further consequence from their studies is that even a small amount of chelator in solution can effectively mobilize metal ions, which are then relocated in the environment (Putnis et al. 2008).

Another example pertaining the complexity and therefore, the inability to generalize the effect of OM on dissolution of minerals is the effect of cellulose ethers (CE) on bassanite ($CaSO_4 \cdot 0.5H_2O$) (Brandt and Bosbach 2001). It was showed that instead of the expected promotion effect, the dissolution kinetics were retarded by 50%. There was no indication for an inhibiting interaction between CE and the surface of the bassanite, which would imply that the reaction was surface-controlled. However, the apparent activation energy measured in the presence of CE was higher in comparison to pure water. It was therefore suggested that higher CE concentration changes the diffusion properties of the aqueous solution by increasing the viscosity of the solution, which, in turn, retards the dissolution kinetics.

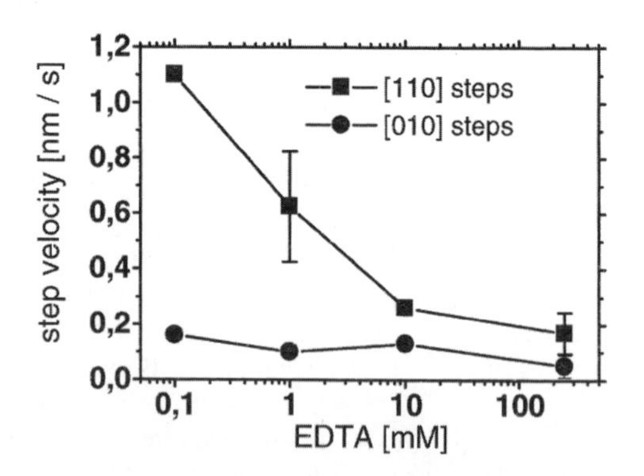

Figure 58. Step displacement velocity of retreating monolayer steps defining shallow etch pits on a dissolving barite surface in EDTA solution at pH 12. The step velocity of steps parallel to [110] is significantly influenced by the EDTA concentration. Consequently the lower the EDTA concentration the more elongated the etch pits are due to faster dissolution kinetics. [Used by permission of Elsevier, from Bosbach et al. (1998), *Chemical Geology*, Vol. 151, Fig. 6, p. 152.]

THE EFFECT OF OM ON MINERAL PRECIPITATION

Nucleation and crystal growth

Before discussing the effect of OM on precipitation, it is necessary to start with a short presentation of the two basic processes of precipitation of minerals: (1) *nucleation*, by which a new phase is formed and (2) *crystal growth* by which the crystal continues to grow. A thorough review of these processes may be found in Fritz and Noguera (2009).

Nucleation. Nucleation is an atomic process in which atoms of the reactant phase rearrange into a cluster of the product phase that is large enough to be thermodynamically stable. It is common to differentiate between homogeneous nucleation in which a three-dimensional nucleus nucleates from a homogeneous phase and heterogeneous nucleation in which the nucleus forms on a specific substrate or a nonspecific impurity particle. Direct observation of nuclei as they form is complicated due to their small size and the immediate occurrence of crystal growth once a stable nucleus is formed. Thus, indirect observations were developed in which the induction time is measured. The induction time is defined as the period of time between the establishment of a supersaturated solution and the first detectable appearance of a new phase. The appearance of a new phase may be detected by observing changes in solution properties and/or observing the forming crystals; as such, it is an operational definition. The induction time is affected by both nucleation and crystal growth rates. In order to differentiate between the two processes and derive the nucleation rate separately from the crystal growth rate it is possible to use carefully planned experimental designs with the aid of theoretical approaches (e.g., Verdoes et al. 1992; Barbier et al. 2009). Generally, it has been shown that the induction time of various minerals increases in the presence of a certain OM. For industrial purposes in which it is important to inhibit mineral precipitation, it is sufficient for the induction time to be much longer than the residence time of the solution in a particular system of interest. If this condition is met then the term 'total inhibition' is often used, although given longer time of observation, crystal precipitation may take place.

According to the classic nucleation theory (Nielsen 1964; Walton 1965; Furedi-Milhofer 1981), the nucleation rate (J_s), which is inversely related to the induction time, and the critical nuclei radius depend on the solution-mineral interfacial tension (σ, interfacial energy), the molecular volume of the mineral (V_m), supersaturation and temperature (T) according to He et al. (1995b):

$$J_s = k\exp\left\{-\frac{\beta V_m^2\sigma^3 N_A f(\theta)}{(\mathrm{R}T)^2\ln^2\Omega}\right\} \tag{55}$$

where k is a rate constant, β is a geometric shape factor, N_A is the Avogadro number, $f(\theta)$ is a correction factor for heterogeneous nucleation, R is the gas constant and Ω is the degree of saturation.

The interfacial tension is usually obtained by experimentally determining the induction time (t_{ind}) at various degrees of saturation (e.g., Furedi-Milhofer 1981; Verdoes et al. 1992; Van Der Leeden et al. 1993; He et al. 1994, 1995b; Saito et al. 1997; Mahmoud et al. 2004; Cundy and Cox 2005; Jiang et al. 2005). The experimental results are fitted to the equation (He et al. 1995a):

$$\log t_{ind} = B(\log\Omega)^{-2} - C \tag{56}$$

where C is a constant and B is a coefficient that may be determined from the slope of a plot of log (t_{ind}) vs. $(\log\Omega)^{-2}$ and is equal to:

$$B = -\frac{\beta V_m^2\sigma^3 N_A f(\theta)}{(2.3\mathrm{R}T)^3} \tag{57}$$

Equation (56) was verified by many studies which showed that a stronger thermodynamic driving force (higher supersaturation) results in a shorter induction time (all other parameters being constant), both in the absence and in the presence of OM.

Crystal growth. Further precipitation of a crystal on existing surfaces of the same mineral or on stable nuclei is referred to as crystal growth. Crystal growth is a process which is considered to occur in a succession of five stages (Furedi-Milhofer 1981): (1) transport of ions through the solution; (2) adsorption on the crystal surface; (3) surface diffusion; (4) reactions at the interface (nucleation, dehydration); and (5) incorporation of the reaction products into the crystal lattice. The rate of crystal growth is controlled by the slowest of these processes.

The rate of crystal growth depends both on the available surface area and the degree of deviation from equilibrium:

$$Rate = k \cdot S_s \cdot m \cdot f(\Delta G_r) \tag{58}$$

where *Rate* is the crystal growth rate (mol s^{-1}), k is a rate constant (mol s^{-1} m^{-2}), S_s is the specific surface area (m^2 g^{-1}), m is the mass of the crystal (g), and $f(\Delta G_r)$ is a function describing the dependency of the growth rate on deviation from equilibrium. When crystal growth involves surface defects, the function $f(\Delta G_r)$ takes the typical form (Lasaga 1998):

$$f(\Delta G_r) = \left(\exp\left(\frac{\Delta G_r}{RT} \right) - 1 \right)^n = (\Omega - 1)^n \tag{59}$$

where R is the gas constant, T is the temperature (K), n is a coefficient and Ω is the degree of saturation with respect to the precipitating mineral. Substituting Equation (59) into Equation (58) gives:

$$Rate = k \cdot S_s \cdot m \cdot (\Omega - 1)^n \tag{60}$$

According to the prediction of the BCF (Burton, Cabrera and Frank) crystal growth theory (Burton et al. 1951), $n = 2$ for spiral growth of screw dislocation. Equation (60) applies for solutions with stoichiometric activities of the molecules that compose the mineral. A more rigorous rate equation for spiral growth, which applies for both stoichiometric and non-stoichiometric solutions, is (Nielsen 1981):

$$Rate = k \cdot S_s \cdot m \cdot \left(\Omega^{1/\upsilon} - 1\right)^2 \tag{61}$$

where ν is the number of molecules in a formula unit of the crystal (e.g., two in the case of calcium carbonate, Nielsen and Toft 1984).

According to Equations (59)-(61), the reaction order with respect to the degree of saturation (n) is independent of the degree of saturation. In contrast to this prediction, several studies have observed that the reaction order is higher at a higher degrees of saturation than when the system is closer to equilibrium (e.g., Brandse et al. 1977; Witkamp et al. 1990).

Determination of the kinetics of crystal growth is usually achieved by introducing suitable crystallization seeds into a supersaturated solution and measuring the change in solution concentrations as a function of time. By repeating experiments with and without OM it is possible to measure its effect on the precipitation kinetics (e.g., inhibitor effectiveness). In contrast to the nucleation process, where it is not possible to physically examine the formation of nuclei, it is possible to visually track the crystal growth process if the crystals grow to a sufficient size.

Classification of OM known to affect precipitation

Due to the complex nature of OM-mineral interactions it is difficult to categorize them according to a specific role in the crystallization process (i.e., catalysts vs. inhibitors). Therefore, rather than classifying the OM according to its role in the precipitation process, a general classification is presented. This classification is based on the origin of the OM and its chemical properties. According to this classification there are three major groups of OM effectively influencing the crystallization processes of sparingly soluble salts: (1) synthetic soluble polymers or polyelectrolytes such as polycarboxylates (Weijnen and Van Rosmalen 1986), polyacrylates (Rolfe 1966; Solomon and Rolfe 1966), polyphosphonates (Liu and Nancollas 1975b; Hasson et al. 2003) and polyphosphates (Amjad 1985; Amjad and Hooley 1986); (2) natural molecules such as proteins, polyamino acids, polypeptides, proteoglycans, mucopolysaccharides (Furedi-Milhofer and Sarig 1996), humic substances (Lebron and Suarez 1996; Hoch et al. 2000; Alvarez et al. 2004) and low molecular weight organic acids (Cao et al. 2007); and (3) surfactants (Mahmoud et al. 2004; El-Shall 2005; Sikiric and Furedi-Milhofer 2006).

OM-mineral precipitation interaction– occurrences and processes

As was discussed in the introduction, the presence of OM is limited to the Earth's surface and near earth surface environments, and therefore, its presence in environments where igneous and metamorphic rocks crystallize is rare. Thus, effects of OM on precipitation are important only for minerals that precipitate in low temperature natural environments and industrial processes. Natural processes involving precipitation of minerals in the presence of OM include precipitation of secondary minerals (mainly clays) during weathering, formation of oxides, carbonates and sulfates from fresh water (Hoch et al. 2000), high saline lakes, seawater (Chave 1970) and evaporated sea water. These processes usually lead to precipitation of a wide range of minerals such as smectite, kaolinite, iron oxides, calcite, dolomite, gypsum, anhydrite, barite, and potassium and magnesium salts.

Industrial processes including oil recovery, desalinization, heating or evaporating subsurface water, sea water and brines are often associated with scale deposits. These deposits accumulate in pipes, membranes, and equipment where the degree of saturation changes due to variation in chemical composition, temperature and/or pressure. Scale may accumulate also as a result of mixing with other solutions or acids that affect the solution chemistry. Scale mostly contains carbonates (calcite ($CaCO_3$), aragonite ($CaCO_3$), vaterite ($CaCO_3$), magnesite ($MgCO_3$), strontianite ($SrCO_3$), witherite ($BaCO_3$) and cerussite ($PbCO_3$)), sulfates (gypsum ($CaSO_4 \cdot 2H_2O$), bassanite ($CaSO_4 \cdot 0.5H_2O$), anhydrite ($CaSO_4$), celestite ($SrSO_4$) and barite ($BaSO_4$)) and silica (SiO_2). Once the scale is formed, they are difficult to remove physically or chemically. Therefore, the most economical and efficient way to control scale is by inhibiting its nucleation and crystal growth using organic chemical inhibitors.

The effect of OM on mineral precipitation is relevant also to the field of bio-mineralization, mainly focusing on various polymorphs of calcium carbonate (Sarig and Kahana 1976), calcium phosphates (Nancollas 1992) and calcium oxalates (Khan 1997). Other minerals such as gypsum, barite, celestite, silica, and various iron oxides have been found to mineralize in biological environments, although to a lesser degree. The interested reader is referred to Mann (1988) where a comprehensive list of over 40 minerals and their polymorphs in various biological systems can be found according to their function. Common examples of bio-mineralization consist of repair of defective bones and teeth (Saito et al. 1997) or prevention of formation of urinary stones (Pak and Holt 1976) and unwanted pathological mineralization.

Nucleation is undoubtedly a crucial step in many geological and industrial processes. However, both theory and experimental methods are still in development. Part of the difficulty in exploring this field is that nucleation results in a very small decrease of the reactant phase concentration, typically involving 10-1000 atoms. Such small scale processes are difficult to observe directly. Furthermore, the microscopic process of nucleation is very hard to characterize theoretically using macroscopic theories. These processes are discussed in detail by Fritz and Noguera (2009).

As in the case of dissolution, the effects of OM on both solution and crystal surface properties may influence the rates and mechanisms of crystallization and aging processes. Specifically, OM may affect nucleation, crystal growth, aggregation, dispersion, and phase transformation processes. This may lead to changes in the number of crystals, their size distribution, morphology and type of precipitated polymorph.

The same organic compound may have different and sometime even opposite effects on different minerals. For example, pyrophosphate exhibits strong inhibitory effect on the crystal growth of calcium phosphates, calcium oxalates, barium sulfate and calcium carbonate minerals, but only a slight inhibitory affect on gypsum crystal growth (Amjad and Hooley 1986). Similarly, polyglutamic acid (PGA) is a retardant for precipitation of various carbonates (magnesite, calcite, strontianite and witherite), while for cerussite the reaction is accelerated (Sarig and Kahana 1976). Therefore, one should examine separately each specific mineral – specific OM interaction. The complexity is further enhanced since even in the presence of the same mineral the role of OM changes. As was shown above for dissolution and will be further discussed below, the role of OM in the precipitation process is influenced by the environmental conditions as well as by the OM/mineral ratio (Furedi-Milhofer and Sarig 1996). A comprehensive review of the literature indicates that the common effect of OM on a wide range of minerals is to inhibit precipitation rather than to catalyze it.

At relatively high supersaturation, a spontaneous homogenous nucleation may occur from solution. Heterogeneous nucleation is usually induced at much lower supersaturation due to the presence of impurities, which act as heteronuclei (Furedi-Milhofer and Sarig 1996). The possible effects of additives on crystallization processes are summarized in Figure 59. Depending on the

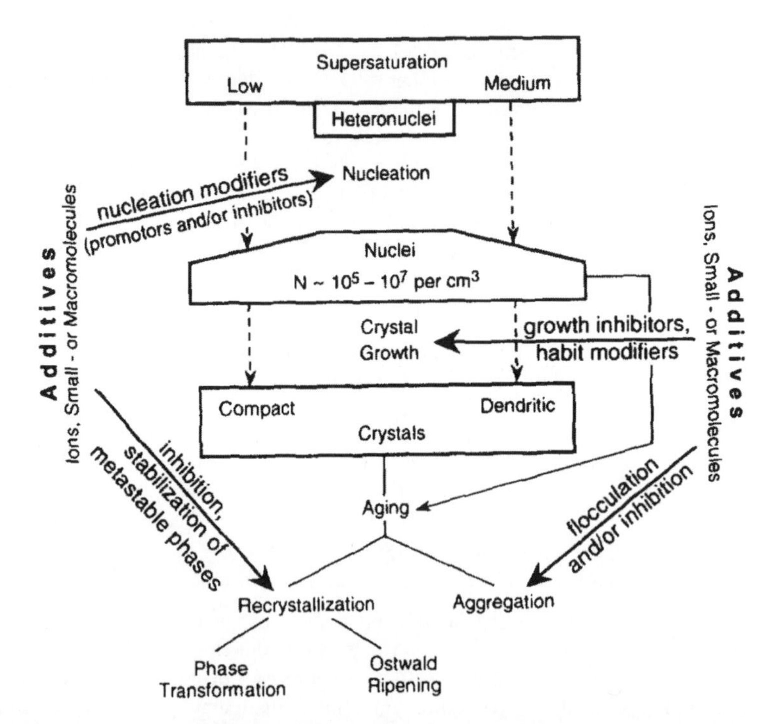

Figure 59. Schematic presentation showing the possible role of additives in the precipitation of slightly soluble salts initiated by heterogeneous nucleation. After Furedi-Milhofer et al. (1988). [Used by permission of Elsevier, from Furedi-Milhofer and Sarig (1996), *Progress in Crystal Growth and Characterization of Materials*, Vol. 32, Fig. 1, p. 46.]

nature and concentration of the additive, heterogeneous nucleation and crystal growth may be retarded and/or accelerated with significant consequences to the properties of the nascent solid phase(s). Aggregation, Ostwald ripening and/or phase transformation may commence at a very early stage of the crystallization process and may also be significantly influenced by additives (Fig. 59).

Mechanism of inhibition and promotion of precipitation

As in the case of catalyzing dissolution, two mechanisms may explain the inhibitory effects of OM on the rate of crystallization: (a) *Solution complexation* - complexation of anionic polyelectrolytes and cationic reactants which alters the activities of these ions and reduce the solution supersaturation with respect to a given mineral (Veintemillas-Verdaguer 1996); and (b) *Surface adsorption* - adsorption of OM to the crystal surface, particularly at dislocation points (Fig. 3). Such points provide sites for rapid crystal growth along steps and screws (Rolfe 1966; Mahmoud et al. 2004; Gloede and Melin 2006). The ability of many OM to retard nucleation was ascribed to preferential adsorption onto developing nuclei, thus prohibiting their outgrowth beyond the critical size required for further growth (Weijnen and Van Rosmalen 1985). In addition to the tendency of the OM to adsorb on the mineral surface, the rate of the adsorption is very important in determining the efficiency of an inhibitor. The rate at which bonding with the surface can be accomplished, will primarily depend on electrostatic attraction between the inhibitor ions and the crystal surface, whereas the type and thus the strength of the chemical bonds formed with the surface will determine the final attachment of the inhibitor ions to the surface (Weijnen and Van Rosmalen 1985). Therefore, excess of adsorbed OM, driven by hydrophobic effects (as discussed above in the sorption section), will not necessarily affect mineral precipitation.

Since nucleation is a nanoscale process, even trace amounts of additives (organic and inorganic) may affect it. This is not necessarily the case for crystal growth, because this process usually involves quantities of reactants of a larger order of magnitude. In the nucleation process, it is easy to grasp that trace amounts of OM may dramatically alter the nucleation rates, since this process involves mere clusters of atoms. When examining the crystal growth process, which occurs at a microscopic to macroscopic scale, one would think that larger amounts of OM are necessary in order to affect the crystal growth process. However, crystal growth mainly occurs on the surface reactive sites that are provided by kinks (Fig. 3), and a full blockage of these sites can still be attained with relatively small amounts of an inhibitor. For example, it has been reported that only a few percent of the crystal surface needs to be covered with an inhibitor to achieve total blockage of crystal growth process (Liu and Nancollas 1975b; Weijnen and Van Rosmalen 1986; Hoch et al. 2000). Weijnen and Van Rosmalen (1986) showed that the surface coverage needed to block gypsum growth by HEDP was only 4-5% of its total surface area.

As noted above, the achievement of 'total inhibition' depends on the duration of the observation. Therefore, even when a full blockage of surface reactive sites is attained, a complete inhibition of crystal growth depends on the time scale examined. After a certain time period, termed induction period (similarly to the induction period defined for nucleation – the time needed in order to detect changes in the solution properties), the seeded crystals may begin to grow. Similarly to nucleation, the duration of the crystal growth induction period depends on the additive concentration and molecular weight, temperature and amount of crystallization seeds added (Amjad and Hooley 1986). For example, Liu and Nancollas, (1973, 1975b) observed that complete inhibition of gypsum growth was primarily attained by ENTMP and TENTMP. After the induction period (up to several hours), crystal growth was initiated with a growth rate that followed a second order dependency on saturation state ($n = 2$ in Eqn. 59), in accordance with the prediction of the BCF crystal growth theory (Burton et al. 1951). The initiation of rapid crystal growth followed by induction period was explained by the consequential incorporation of the OM into the growing crystals (Liu and Nancollas 1975a).

During the induction period, most active growth sites may be blocked by the adsorbed additive. However, some of the growth sites may still be free to grow, and the reaction process proceeds in a very slow and immeasurable rate. The slow growth covers the blocked growth sites and the adsorbed OM is incorporated into the crystal lattice. Thereafter, the crystal resumes its growth at a rate comparable to that in the absence of OM. The co-precipitation of organic compounds in inorganic crystals will be discussed further in a section devoted to this process.

When OM shows inhibitory effects despite the fact that its concentrations are too low for solution complexation to be significant (i.e., inhibitor concentrations are 2-5 orders of magnitude lower than reactant concentrations), then surface adsorption may be considered as the dominant mechanism (Van Rosmalen et al. 1981; Baumgartner and Mijalchik 1991). On the other hand, when precipitation of minerals which are composed of less abundant elements is inhibited in the presence of OM, then solution complexation should be considered as well. Appreciable habit modifications in crystals grown in the presence of inhibitor are also indicative of surface adsorption mechanism. Such modifications result in profound changes in crystal morphology.

Not all the sorption mechanisms that were discussed in the section on OM sorption are effective in inhibiting precipitation. For adsorption to be effective in this sense, a dimensional stereochemical fit between the polar group of the inhibitor and the intercationic or interanionic distances is needed. Therefore in many cases, the crystal interface-OM interaction is highly specific. As the OM and a certain crystal face exhibit a higher resemblance in terms of structure, the effect becomes more habit specific (Furedi-Milhofer and Sarig 1996). Selective electrostatic interactions may occur between highly charged OM and a specific crystal faces, due to different ionic structures and charges of the crystal planes. Therefore, the developing crystal faces do not equally attract approaching OM molecules (Sikiric and Furedi-Milhofer 2006).

In cases where organic polymers exhibit a similar structure to that of the crystallizing molecules, the polymers can act as templates or microsubstrates and promote nucleation rather than inhibiting it (Davis et al. 2003). Addadi and Weiner (1985) have demonstrated that calcite crystals nucleated on acidic proteins that were previously adsorbed onto a rigid substrate. The presence of organic surfaces at the nucleation site may also lower the interfacial energy between the nucleating phase and the solution resulting in faster nucleation kinetics (Mann 1988).

Many of the studies described below are descriptive, and do not include mechanistic explanations for the observed effects. From an industrial and economical standpoint this may be sufficient. Often, some of the data needed for concise mechanistic description is missing. For example, similarly to studies on dissolution, most of the studies on precipitation kinetics did not investigate adsorption. A comprehensive understanding of the adsorption process under different environmental conditions would contribute to mechanistic understanding of the effect of OM on mineral precipitation. Similarly, better understanding of the thermodynamics of OM-cation complex equilibrium in aqueous solutions under various environmental conditions is of paramount importance to improve the understanding of the role of the *solution complexation* in precipitation mechanism.

Co-precipitation of OM in minerals

Evidence for co-precipitation of organic compounds in inorganic crystals was reported by Phillips et al. (2005). They synthesized calcite in a seeded constant-addition method in the absence (control) and presence of citrate, aspartic acid, glutamic acid or phthalate. In particular, aspartic acid is found to be associated with natural non-biogenic carbonates and it was speculated that it affects the growth of non-biogenic carbonates (Carter 1978). In the study of Phillips et al (2005), a portion of the synthesized calcite was digested and OM content was determined by ion chromatography. These analyses showed that overgrowth calcite (i.e., excluding the seeds) contained 1.1 wt% citrate. This amount was considered to be too high to solely be explained by adsorption. Phillips et al. (2005) have used $^{13}C[^{1}H]$NMR spectrometry techniques to examine their samples and argued that the co-precipitation occurs over a wide range of structural configurations. ^{1}H signals suggested that the OM was incorporated to the crystal together with water molecules. To a lesser degree the two amino acids were also incorporated into the crystal lattice, whereas no evidence for phthalate co-precipitation was found. Calculation of citrate speciation indicated that over 96% of it was present as 1:1 Ca-citrate complex, indicating that this species is probably involved in the co-precipitation. It is important to emphasize that this complex had insignificant influence on calcite saturation since Ca^{2+} concentration was much higher than citrate concentration.

Further evidence for co-precipitation was provided by Compton and Brown (1995) who applied a langmuirian adsorption term (Eqn. 7) to the adsorption of maleic acid onto calcite. It was suggested that the inhibitor specifically adsorbs at a growing sites, therefore preventing the incorporation of $CaCO_3$ units into the crystal lattice. It was further suggested that general adsorption of the inhibitor on the crystal surface might occur, so that the adsorbed inhibitor competes with the adsorption of Ca^{2+} or CO_3^{2-}. But sorption by itself cannot explain the partition of OM between a mineral and solution. Solution conditions on one hand play a significant role in this reaction and require further study, while, on the other hand, structural accommodation of OM, its conformation and charge balance in the mineral are also significant factors. Implications of such calcite-OM co-precipitation can be important in determining the fate of metals associated with OM and/or OM contaminates. As was noted above, incorporation of inhibitor into calcite was proposed to be the explanation for re-initiation of crystal growth of minerals that were blocked by adsorbed organic inhibitor.

The effects of environmental conditions on precipitation in the presence of OM

Supersaturation. The degree by which a change in the supersaturation affects the induction time depends on the mineral-solution interfacial tension. The mineral-solution interfacial tension is a fundamental parameter in understanding and modeling the rate of both nucleation and crystal growth. OM may alter mineral-solution interfacial tension by serving as surface active agents (surfactants), which consequently affect nucleation rates. Surfactants are organic compounds consisting of two parts: a hydrophobic portion usually including a long hydrocarbon chain and a hydrophilic portion that makes the compound sufficiently soluble in water and other polar solvents. For example HDTMP, PCAP, HEDP and PPPC increase the interfacial tension between gypsum and solution and barite and solution, and as a result, induction time of both gypsum and barite increases in the presence of these additives (He et al. 1994). Similarly, the induction time of celestine increases in the presence of HEDP, BHTP and PPPC (He et al. 1995b).

Mahmoud et al. (2004) have studied the affect of adding cationic (CTAB) and anionic (SDS) surfactants to gypsum supersaturated solutions. Figure 60 presents the log of the induction time, measured using a turbidimeter versus $(\log\Omega)^{-2}$. They derived the interfacial tension using Equations (56) and (57) from the slopes of the lines in Figure 60. Relative to experiments without any surfactant (baseline), the interfacial tension increased in the presence of the cationic surfactant (CTAB) and decreased in the presence of the anionic surfactant (SDS). Therefore, the addition of the cationic surfactant resulted in an inhibition of the reaction, while the addition of the anionic surfactant resulted in promotion of the reaction. Another example of reaction promotion due to the presence of OM was given by Jiang (2005), who showed that chondroitin sulfate promotes the formation of hydroxyapatite due to the suppression of the interfacial tension between the hydroxyapatite crystal and the solution.

Weijnen et al. (1983) compared the rate of gypsum crystal growth in experiments without inhibitors with those conducted in the presence of HEDP. The ratio between the two was termed effectiveness or strength of an inhibitor. Comparison of the effectiveness at two different degrees of saturation reveals that at a given HEDP concentration, effectiveness was lower at the higher supersaturation value.

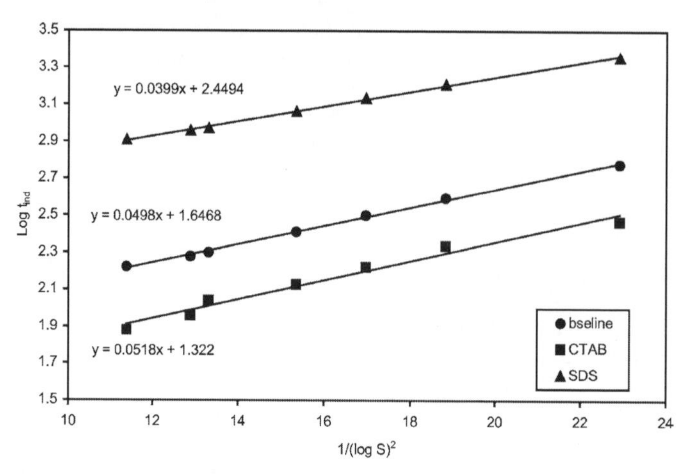

Figure 60. Relation between the log of induction time and $(\log\Omega)^{-2}$ with and without 100 ppm of CTAB and SDS surfactants. [Used by permission of Elsevier, from Mahmoud et al. (2004), *Journal of Colloid and Interface Science,* Vol. 270, Fig. 6, p. 102.]

OM concentration. As mentioned above, the nucleation retarding effect of OM is explained by preferential adsorption on the developing nuclei. Therefore, it is expected that increasing the concentration of such OM would decrease the percentage of nuclei which would grow beyond the critical size required for further growth, resulting in a decrease in nucleation rate. Indeed, increasing concentration of an organic inhibitor usually intensifies its inhibition effect on mineral nucleation, resulting in an increase in induction time. A partial list of studies that show such an increase in induction time include: calcium carbonate as a function of the concentrations of sodium polymethacrylate (NaPMA) (Williams and Ruehrwein 1957), gypsum with PAA and PSA (Lioliou et al. 2006), calcium carbonate with a wide range of sodium polyacrylates (NaPA) (Jada et al. 2007) and calcium oxalate with glutamic acid (Azoury et al. 1983). Some organic compounds may accelerate mineral precipitation when present in low concentrations, while at higher concentrations they may become an effective inhibitor. For example, Azaury et al. (1983) have observed that when increasing glutamic acid concentration in a solution supersaturated with respect to calcium oxalate, the reaction is first accelerated at low additive concentrations, and then retarded at higher additive concentrations. They suggested that at low concentrations of glutamic acid it serves as a microsubstrate which enhances the nucleation rate. A similar example was presented by van der Leeden et al., (1993) who found that PMA-PVC accelerates barite nucleation at concentration lower than 0.005 ppm but inhibits it at higher concentrations.

As for nucleation, the inhibition effectiveness of crystal growth commonly increases when increasing inhibitor concentration. Interestingly, a complete inhibition of crystal growth is not always attained, and the crystal growth rate may reach a plateau where it no longer depends on the inhibitor concentration. For example, Becker et al. (2005) have shown that the growth rate of barite is retarded in the presence of HEDP, NTMP, MDP, AMP and PBTC. A rapid decrease in growth rate (measured as step advancement rate) as a function of inhibitor concentrations was found for inhibitor concentrations lower than 10 μM. For the range higher than 10 μM, the retardation of the step advancement reached a plateau with little dependence between growth rate and inhibitor concentration. The extension and slope of the plateaus are different for each phosphonic acid. Only for PBTC, a complete inhibition at concentrations higher than 10 μM was observed. Similarly, El Dahan and Hegazy (2000) found that a higher inhibition of gypsum scaling is attained with higher concentrations of phosphate ester. At a concentration of several tens of ppm (depending on the temperature) nearly total inhibition was achieved whereas higher additive concentrations had insignificant effect.

Lebron and Suarez (1996) have examined the effect of varying concentrations of dissolved NOM (mainly fulvic and humic acid) on calcite precipitation. A strong reduction in precipitation rate is observed when DOC concentrations are increased. At concentrations higher than 0.01 mM DOC, total inhibition was achieved (Fig. 61).

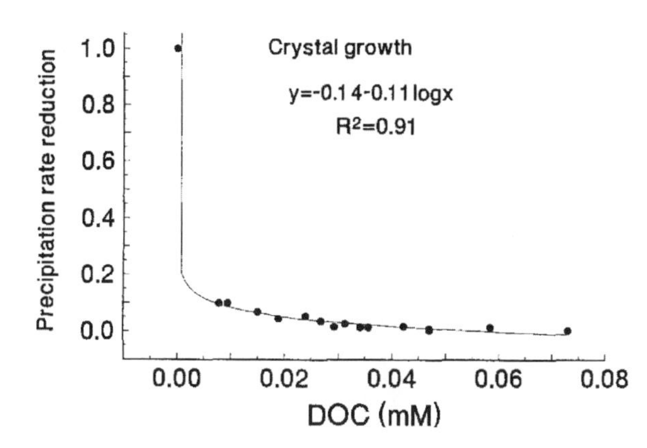

Figure 61. Reduction in the rate of calcite precipitation by crystal growth as a function of DOC. Experiments were conducted using the same degree of supersaturation and crystallization seeds. [Used by permission of Elsevier, from Lebron and Suarez (1996), *Geochimica et Cosmochimica Acta,* Vol. 60, Fig. 9, p. 2774.]

Temperature. Temperature may affect nucleation kinetics and crystal growth both directly as predicted by the Arrhenius equation (Eqn. 53) and indirectly by changing the mineral solubility, which leads to changes in the degree of saturation. Increasing temperature may result in either a decrease or an increase of solubility, depending on the mineral, which consequently may either increase or decrease the degree of saturation, respectively. These temperature effects on precipitation exist both in the presence and in the absence of OM. Temperature may affect the adsorption isotherm of an organic inhibitor on mineral surface as well as the concentration of the inhibitor due to thermal decomposition or solubility decrease of the OM-salt. Due to such decrease in concentration the inhibitor becomes less effective.

Liu and Nancollas (1975b) studied the effect of ENTMP and TENTMP on gypsum precipitation. They calculated the effect of degree of saturation at each temperature by fitting their experimental data to Equation (60) and found that the obtained rate constant (k in Eqn. 60) increased by approximately by a factor of 2 for each 10° rise in temperature. The corresponding apparent activation energy, 15 kcal mole^{-1}, was quite similar to that observed in the absence of OM. Activation energies in the range of 10 to 20 kcal mol^{-1} are typical to surface-controlled reactions (Lasaga 1984, 1998).

He et al. (1995b) studied the inhibition of celestite in 1 M NaCl solutions by HEDP, BHTP and PPPC at varying temperatures and concentrations (Fig. 62). In order to differentiate between the effect of temperature on the degree of saturation and a direct change on the activation energy term, the effect of temperature on the induction time should be studied in experiments that are conducted at a constant degree of supersaturation at varying temperatures. In the absence of OM, the nucleation experiments of He et al. (1995b) were conducted at constant degree of saturation (Ω = 1.4). They found an empirical linear relationship between the log of the induction time reciprocal ($1/t_{ind}$) and the reciprocal of the absolute temperature. Using an Arrhenius-like equation they calculated apparent activation energy of 12.8 kcal mol^{-1}, which suggest a surface-controlled mechanism.

The temperatures in the experiments in the presence of organic inhibitor by He et al. (1995b) were conducted at different saturation degrees, and therefore the comparison between

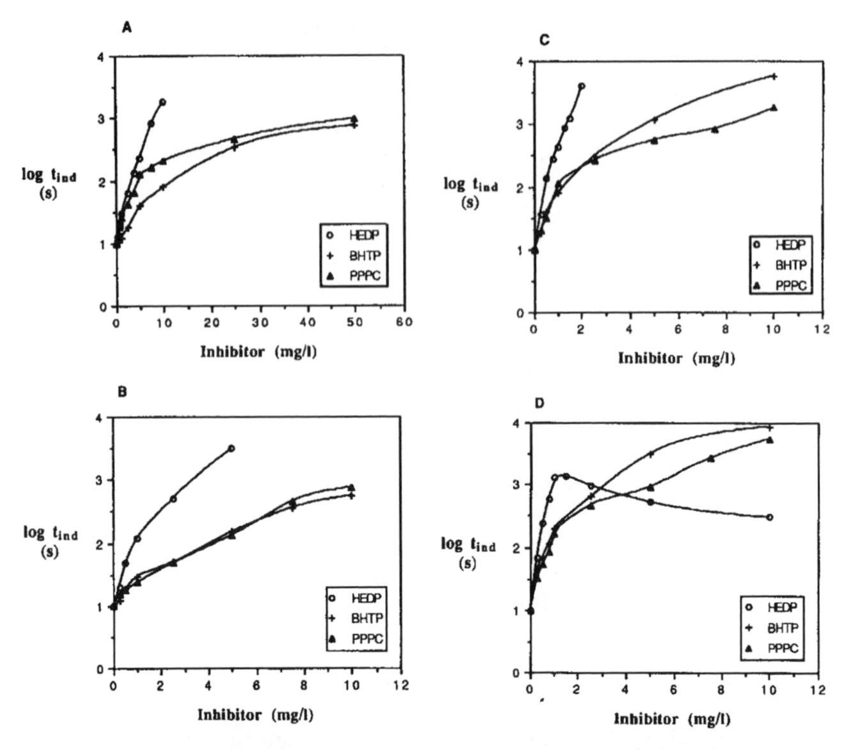

Figure 62. The inhibition of celestite ($SrSO_4$) nucleation in 1 M NaCl solutions by HEDP, BHTP, and PPPC at pH 6.5 and 25 °C (A), 50 °C (B), 70 °C (C), 90 °C (D). [Used by permission of Elsevier, from He et al. (1995b), *Journal of Colloid and Interface Science,* Vol. 74 Fig. 6, p. 333.]

them is not straightforward. Nevertheless they showed that up to 70 °C, induction time increases with increasing concentration of HEDP, whereas at 90 °C, for HEDP concentrations above 1 mg L^{-1}, a reduction occurred in the induction time (Fig. 62D). This reduction was attributed to the precipitation of strontium-HEDP salt which led to a decrease in inhibitor concentration, therefore making it less efficient. This effect was not observed with BHTP and PPPC. A similar phenomenon was shown by Jonasson (1996) who studied the effect of phosphonate inhibitors on calcite nucleation kinetics as a function of temperature. At temperatures between 20-100 °C phosphonates were either adsorbed and/or precipitated onto calcite nuclei and consequently retarded calcite nucleation kinetics. The study was carried out using X-ray photoelectron spectroscopy (XPS), which enabled the detection of surface monolayer deposits. At elevated temperatures (200 °C), calcium phosphonates (either HEDP or DETPMP) formed as separate crystalline phases, as opposed to an adsorbed monolayer at lower temperatures. None of the phosphonates used in the study of Jonasson (1996) affected calcite nucleation above 100 °C.

El Dahan and Hegazy (2000) have demonstrated that the dose of phosphate ester required for inhibition of gypsum scale is small at low temperatures and increases with a rise in temperature (Fig. 63). They noted that the extent of adsorption in terms of surface coverage of the phosphate ester was lowered as the temperature increased (Fig. 64).

pH. The adsorption of an organic inhibitor on the mineral surface is primarily due to interactions between anionic functional groups of the inhibitor and cation on the crystal (or nucleus) surface (Austin et al. 1975; Weijnen and Van Rosmalen 1985). The effect of pH on precipitation kinetics in the presence of such organic inhibitor is mainly a result of pH

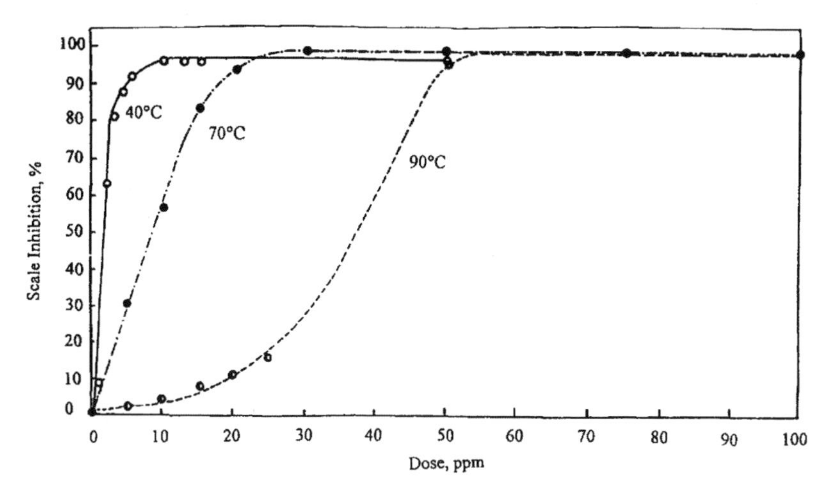

Figure 63. Influence of phosphate ester concentrations on its inhibition efficiency at different temperatures. [Used by permission of Elsevier, from El Dahan (2000), *Desalination,* Vol. 127, Fig. 2, p. 113.]

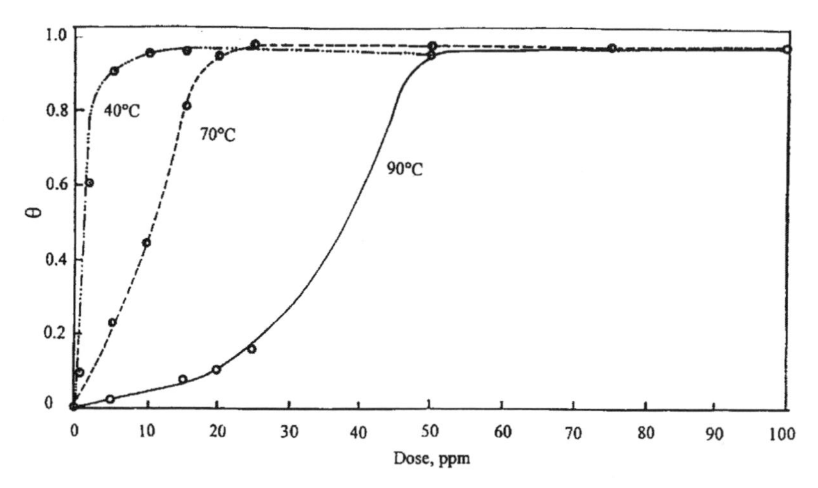

Figure 64. The effect of the concentration of phosphate ester antiscalant on its surface coverage (θ) at different temperatures. [Used by permission of Elsevier, from El Dahan (2000), *Desalination*, Vol. 127, Fig. 5, p. 116.]

influences on the adsorption, which result from pH effects on the degree of protonation of the functional groups and on the conformation of the organic additive (Benton 1993). As discussed above, at high pH values the degree of dissociation of acidic OM is greater, which enables the OM to remain ionized. The rate of adsorption of OM on the forming nuclei relative to the formation rate of stable nuclei is an important factor in inhibition of nucleation. As adsorption rate depends on the strength of the electrostatic interactions between the anionic groups and the surface cation, increase in the number of anionic surface groups would increase the inhibition. Weijnen and Van Rosmalen (1985) argued that at least several anionic functional groups are needed for an effective inhibitor performance over a wide range of pH-values.

As we showed in the section on adsorption, the attraction between OM and mineral surface depends not only on the charge of the OM but also on the surface charge of the mineral. The studies on antiscalants tend to concentrate on the charge of the target cation (e.g., Ca^{2+}) and not on the overall surface charge of the mineral. For calcite these common practice seems to be justified as Eriksson et al. (2007) showed that at pH range of 7.5-11 calcite exhibit a positive surface charge which they attributed to Ca^{2+}.

The effect of solution pH in the range of 1.5-12 on the inhibition of gypsum nucleation by HDTMP and PCAP and barite nucleation by HEDP and PPPC at a constant degree of saturation was studied by He et al. (1994) and is presented in Figure 65. Generally, the inhibitors were more effective at higher pH values than in low pH values. Above a certain pH, the induction time is no longer affected by pH, probably due to the degree of ionization or dissociation of the inhibitor at a certain pH. In the case of the phosphinopolycarboxylic acids (PPPC and PCAP), the maximum inhibition was reached when the solution pH was about 3 and 4, respectively.

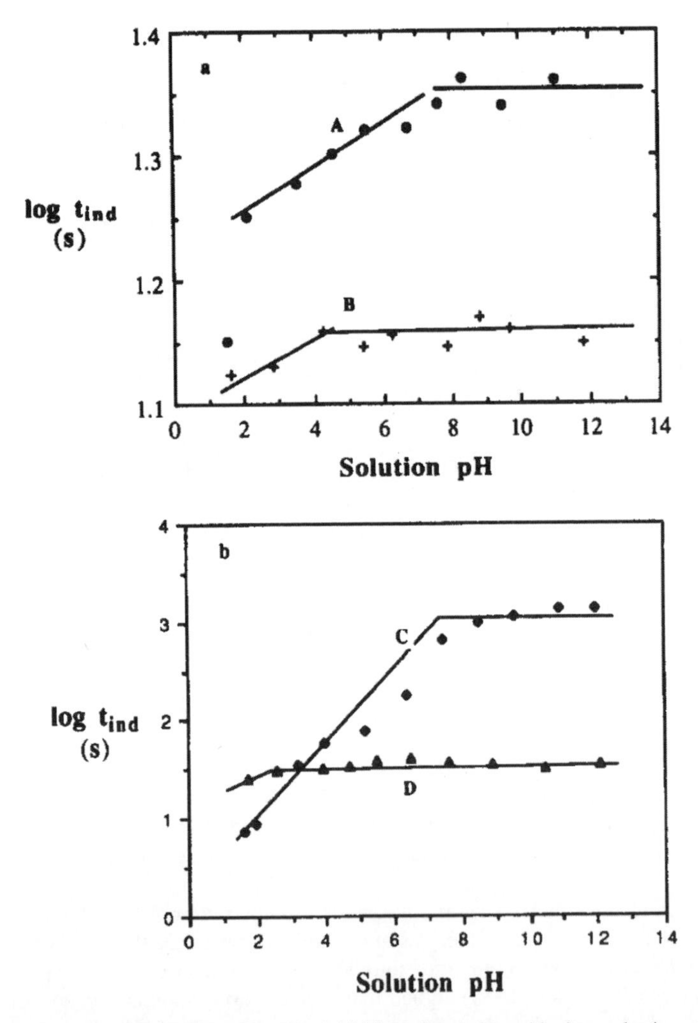

Figure 65. The effect of the solution pH on the inhibition of gypsum and barite nucleation at 25 °C. (a) Inhibition of gypsum by HDTMP (circles, A) and PCAP (crosses, B). (b) Inhibition of barite by HEDP (diamonds, C) and PPPC (triangles, D). [Used by permission of Elsevier, from He at al.(1994), *Applied Geochemistry,* Vol. 9, Fig. 5, p. 566.]

The maximum inhibition was reached at a higher pH (about 7) for the phosphonates HDTMP and HEDP.

From a comparative study of the growth retarding influence of various phosphonate, sulfonate and carboxylate inhibitors on calciumsulfate hemihydrate and anhydrite at high temperatures, Austin et al. (1975) concluded that an effective inhibitor for calcium sulfate needs to contain at least one acid group per molecule which still has an undissociated hydroxyl group. They proposed that the adsorption may be via OH^- bonding to Ca^{2+} on the crystallite growth face. In the case of phosphonate, Austin et al. (1975) found that OM with a singly ionized phosphonate group (PO_3H^-) are very effective in inhibiting the precipitation of $CaSO_4 \cdot XH_2O$, whereas the fully deprotonated phosphonate group (PO_3^{2-}) was ineffective. Weijnen and Van Rosmalen (1986) argued that both PO_3^{2-} and PO_3H^- groups are needed. The PO_3^{2-} groups provide the strong electrostatic interaction needed to induce the inhibitor ions to settle at the crystal surface. Once contact with the crystal surface has been established, the protonated acid groups, PO_3H^-, will be able to form stronger complexes with the calcium ions than the fully deprotonated acid groups. Therefore, at very high pH values, where even the weakly acidic PO_3H^- groups start to deprotonate, the phosphonate inhibitor effectiveness consequently starts to decrease.

Like the phosphonic acid groups in phosphonate inhibitors, the carboxylic acid groups in polyelectrolyte inhibitors are able to coordinate with the crystal cations on the surface. Again a certain degree of deprotonation is required for effective growth retardation, since the primary interaction of the polyelectrolyte inhibitor molecules with the crystal surface is due to electrostatic interaction. At high pH values the polyelectrolyte molecules exhibit a strong affinity towards the crystal surface as a result of their high anionic charge density (Weijnen and Van Rosmalen 1985). In analogy with the role of the fully and singly deprotonated phosphonic acid groups in growth inhibition by phosphonates, Weijnen and Van Rosmalen (1985) proposed that the ionized carboxylic acid groups are responsible for the electrostatic interaction of polyelectrolyte inhibitor molecules with the crystal surface, while the protonated groups establish the actual bonds with the cations in the mineral surface.

In essence, polyelectrolyte adsorption is supposed to be similar to that of NOM. As was demonstrated in the section on sorption, NOM sorption increases as pH decreases, probably due to the increase of hydrophobic effects (e.g., Gu et al. 1994, 1995; Au et al. 1999; Arnarson and Keil 2000; Jada et al. 2006). However, it does not necessarily follows that inhibition effectiveness should follow the same trend as adsorption since the additional OM adsorbed by hydrophobic effects may not be chemisorbed. In accordance, Weijnen and Van Rosmalen (1985) argued that low pH values and high ionic strengths will reduce the inhibitor performance and that hydrophobic groups have been showed to decrease the effectiveness of phosphonate inhibitors, are also detrimental to the performance of polyelectrolyte inhibitors. Such hydrophobic groups can be considered as "dummy" substituents not offering any additional bonding possibilities, but merely increasing the molecular weight and lowering the anionic charge density of the polyelectrolyte molecules (Weijnen and Van Rosmalen 1985).

Stoichiometry. The effect of the lattice ion molar ratio in the solution on nucleation inhibition can be explained in terms of nucleation theory. When the lattice cation/anion ratio in solution is >1, the nuclei surface tends to contain more cations and therefore bare a positive charge. Due to this positive charge, anionic inhibitors are attracted and their inhibiting effect increases (He et al. 1994). On the other hand, when the lattice anion in solution is dominant over the lattice cation, negatively charged nucleation surfaces tend to repulse anionic inhibitors, which results weakening of the retarding effect of anionic inhibitors. The effect of the reactant stoichiometry on the inhibition effectiveness of anionic inhibitors was studied on gypsum nucleation (HDTMP and PCAP inhibitors) and barite nucleation (HEDP and PPPC inhibitors) by He et al. (1994). The experiments were conducted at a constant degree of saturation, pH and ionic strength and their results are presented in Figure 66. As expected, inhibition increases with

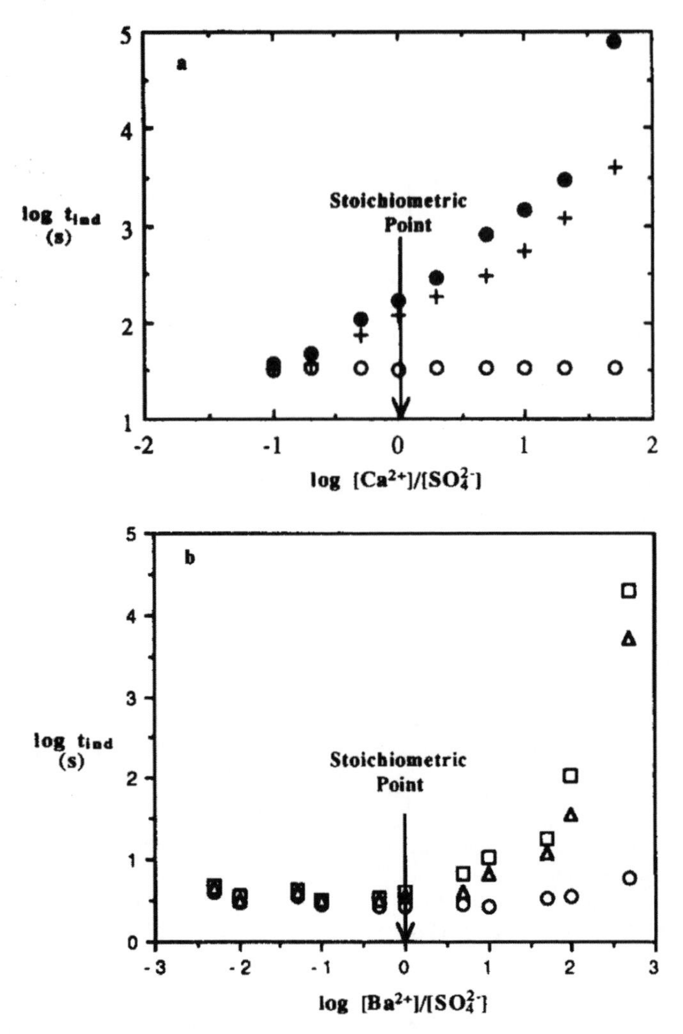

Figure 66. The effect of lattice cation/anion molar ratios on the inhibition of (a) gypsum nucleation at a pH of 5.5 by HDTMP (solid circles) and PCAP (crosses); and (b) barite nucleation at a pH of 6.5 by HEDP (squares) and PPPC (triangles) at 25 °C. The saturation index is 0.85 and 3.56 and the ionic strength is 5.0 and 1.0 M for calcium sulfate solutions and barium sulfate solutions, respectively. Open circles represent no inhibitors added. [Used by permission of Elsevier, from He et al. (1994), *Applied Geochemistry*, Vol. 9, Fig. 6, p. 566.]

increasing lattice cation to anion ratios. Where there are no inhibitors present in the solution, the nucleation induction period seems to be independent of the ratio in the case of calcium sulfate or slightly dependent on the ratio in the case of barium sulfate. Other studies (e.g., Alimi and Gadri 2004) showed that the stoichiometry ratio affects nucleation of gypsum even in the absence of OM, however discussion of these results are beyond the scope of the present review.

Crystallization seed concentration. Since crystal growth is a surface reaction, the precipitation rate depends on the surface area and therefore on the amount of crystallization seeds (Eqn. 58). Therefore, it is common to present growth rate data that are normalized to the total surface area (or total mass) of the seeds. In blank experiments (i.e., in the absence of an inhibitor), these normalized growth rates should be independent of seed concentration or mass. As discussed above, the mineral surface area may be reduced due to adsorption of OM (Fig. 17). Therefore, if inhibition is adsorption controlled, then in the presence of OM the normalized growth rate may depend on seeds concentration. For example, in the presence of a constant inhibitor concentration (HEDP) the inhibitor effectiveness decreased as the seed concentration of gypsum increased (Weijnen et al. 1983).

The effects of OM properties on precipitation in the presence of OM

OM molecular weight. As discussed above, the M_w of a polyelectrolyte greatly affects its adsorption. Therefore, inhibition efficiency may depend on the OM molecular weight. Two effects may be attributed to the polyelectrolyte M_w: (1) polyelectrolytes with high molecular weight would be expected to have lower adsorption rates if adsorption is transport limited, since their diffusion rate from the bulk of the solution to the crystal surface would be slower compared to low molecular weight polyelectrolytes (Eqn. 14). Therefore, in the presence of high M_W polymer a shorter induction time is expected compared to low M_w polymer. (2) polymers with high molecular weights usually exhibit a higher extent of adsorption compared to low molecular weight polymers, and inhibition may increase with increasing M_w. Thus, the balance between the contradictory kinetic and thermodynamic effects should result in an optimum molecular weight for retardation (Williams and Ruehrwein 1957). Solomon and Rolfe (1966) observed that inhibition of calcium sulfate scale formation with polyacrylic acids becomes less effective as the polyelectrolyte M_w exceeds the range of 15,000-30,000 g mol^{-1}. This was probably the result of the dominance of decreased solution mobility over increased adsorption onto surfaces and increased bridging effects associated with the higher molecular weights. A similar observation was made by Lioliou (2006), where polyacrylic acids of increasing molecular weights became progressively less effective in inhibiting calcium sulfate scale formation. Baumgartner and Mijalchik (1991) showed that polyacrylic acid ($4.4\text{-}8.8\times10^{-5}$ N) with a M_w of 4,000 g mol^{-1} inhibits hematite nucleation whereas those with higher M_W (up to 250,000 g mol^{-1}) had no inhibition effect. Smith and Alexander (1970) found an optimal molecular weight for inhibiting precipitation of gypsum using polystyrene-maleic acid with a molecular weight of 1,600 g mol^{-1}. Jada et al. (2007) have shown that for calcium carbonate an optimum inhibition occurs at a molecular weight of 5,530 g mol^{-1} of sodium polyacrylate (Fig. 67). Aside of the two factors discussed above, Jada et al. (2007)

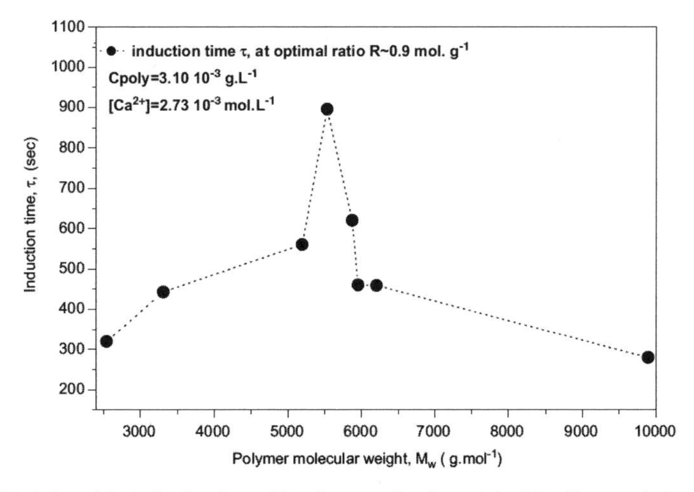

Figure 67. Variation of the induction time with polymer molecular weight. [Used by permission of Elsevier, from Jada (2007), *Journal of Crystal Growth,* Fig. 4, p. 378.]

suggested that polyelectrolyte with lower M_w would complex faster with calcium ions than high M_w polyelectrolyte. The complexation results in the neutralization of the polyelectrolyte, which then leads to the collapse of the polymer network and allowing precipitation to occur. Therefore, in the presence of low M_w polyelectrolytes the induction time should decrease. These effects combined, where lower molecular weight polymers are adsorbed faster but also neutralized faster while high molecular weight polymers has a higher degree of adsorption but exhibit slower adsorption rates, lead to an optimum.

Amjad and Hooly (1986) studied the effect of polyelectrolytes on the crystallization of gypsum. They found that crystallization in the presence of polyelectrolytes is preceded by an induction period following which crystal growth proceeds with a rate close to that in pure supersaturated solution. The induction period increased with decreasing molecular weight of polyacrylates ranging from 5,100-50,000 g mol^{-1} (Fig. 68). In contrast, the rate of crystal growth, following the induction period, has very little correlation with the molecular weight of the polyacrylate.

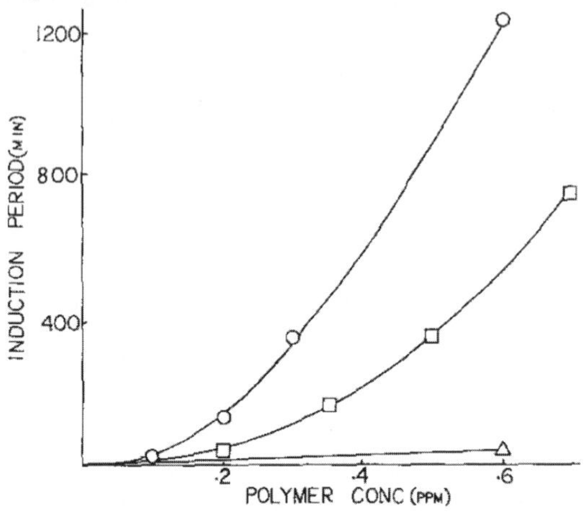

Figure 68. Relationships of induction periods with concentrations of polyacrylates of varying molecular weight at 35 °C: (○) 5100, (□) 10000 (△) 50000 g mol^{-1}. [Used by permission of Elsevier, from Amjad and Hooly (1986), *Journal of Colloid and Interface Science,* Vol. 111, Fig. 7, p. 501.]

OM structure, functional groups and charge. Once effects related to the molecular weight of OM are eliminated, it is possible to relate the differences in effectiveness to the OM structure and functional group. Changes in the valance of the OM may alter intermolecular repulsions between the various charged groups resulting in changes in the molecular dimensions. Both the charge and structure of the molecules are important variables that affect the adsorption mechanism and retarding effects. As discussed above, OM structure and functional groups may influence the steric effects and flexibility of the molecules. Similarly to the steric effect between mineral surface and OM (e.g., Gu et al. 1995), variation in OM properties may change it's interaction with dissolved ions. For example, it was argued by Smith and Alexander (1970) that the formation constant between calcium ions and styrene-maleic acid copolymer is significantly less than that between calcium ions and ethyl vinyl ether-maleic acid copolymer due to a variation in their structure. If a certain inhibiting mechanism is dependent on chelate formation (i.e., *solution complexation* mechanism), a difference in nucleation rates should be noticed. Smith and Alexander (1970) also showed that linear polyphosphates are more active inhibitors than cyclic polyphosphates due to a higher flexibility of the linear molecules. It was shown that sodium tetrametaphosphate with a cyclic structure had no inhibiting effect on calcium carbonate. However, when this compound was hydrolyzed to form a linear compound, sodium tetrapolyphosphate, a retarding effect was noticed.

Van der Leeden and Van Rosmalen (1990) tried to assemble a copolymer which would be an effective inhibitor for crystal growth of barite and gypsum. They used polyelectrolytes consisting of vinyl-based segments and two types of functional groups. All of them contained a carboxyl functional group, while the other type of functional group was a hydroxyl, a sulfonic acid or a phosphonic acid group. The functional groups were statistically distributed over the polymer chains. The functional groups replacing the carboxylic groups have to be more effective than the carboxylic group in order to overcompensate for the loss of the former group. The various inhibitors tested showed different results for barite and gypsum. For example, a sulphonic acid group that was incorporated on the backbone of the polyacrylate was one of the most effective inhibitors for barite, but one of the less effective inhibitors for gypsum.

The effect of OM on crystal size and morphology

Since each type of crystal face has a different structure and thus a different distribution of adsorption sites, a different degree of retardation for each type of crystal face is usually exhibited, which leads to the alteration of the crystal shape. This can result in one or more crystal faces either to appear or disappear with advancing outgrowth of the crystals, and eventually alter the crystal morphology (Liu and Nancollas 1973). The face that will have relatively slower growth rate in the direction perpendicular to it will be larger than faces with faster growth rates. Thus, observing changes in morphology may serve as another tool to assess organic-inorganic interactions.

For example, Figure 69 shows that calcite grown in seeded experiments in the presence of DOM did not exhibit smooth planes as would occur in the absence of DOM, but rather planes in which growth had been interrupted, showing broken or discontinuous growth (Hoch et al. 2000). Surface roughening in the presence of polyelectrolytes was reported for gypsum as well (Weijnen and Van Rosmalen 1985). Phosphonates used as barium sulfate scale inhibitors were shown to produce oblate spheroids and distorted star-like crystals which were 15-20 times smaller than crystals precipitated in the absence of additives (Benton 1993). A similar decrease in barium sulfate crystal size (up to 15 times in diameter) was obtained in the presence of humic and fulvic acids (Smith et al. 2004). Gypsum grown in the presence of polyacrlic acid and polystyrene sulfonic acid showed more plate like crystals compared to the thin elongated crystals obtained in the absence of organic acids (Fig. 70). The difference in the morphology was mainly attributed to the retardation of the [111] face in the presence of organic acids, whereas without organic acids the same plane grew rapidly. The presence of these acids also

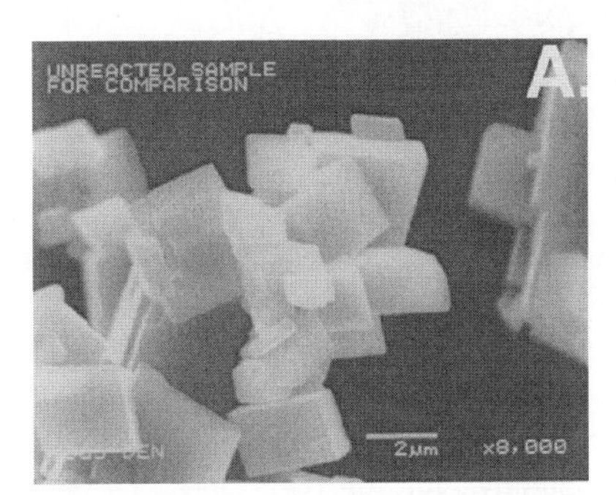

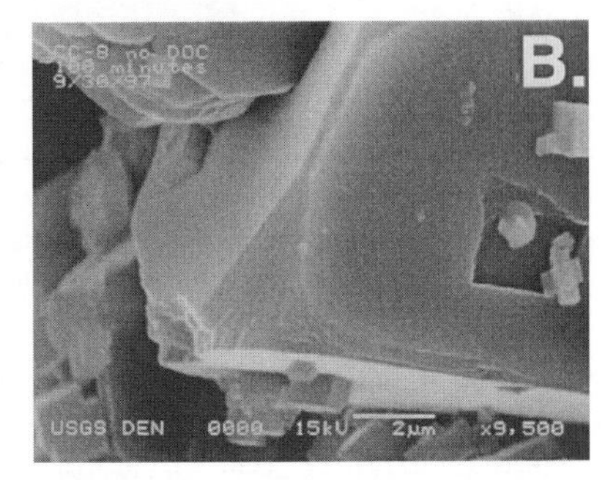

Figure 69. SEM photomicrographs of calcite seed material before and after crystal growth experiments: (A) unreacted calcite seed with straight edges and smooth surfaces; (B) calcite seed after 100 min growth in an experiment with no added organic materials. Note laterally continuous planes of growth; (C) calcite seed after growing for 100 min in a solution containing 0.5 mg/L hydrophobic acid. [Used by permission of Elsevier, from Hoch et al. (2000), *Geochimica et Cosmochimica Acta,* Vol. 64, Fig. 3, p. 67.]

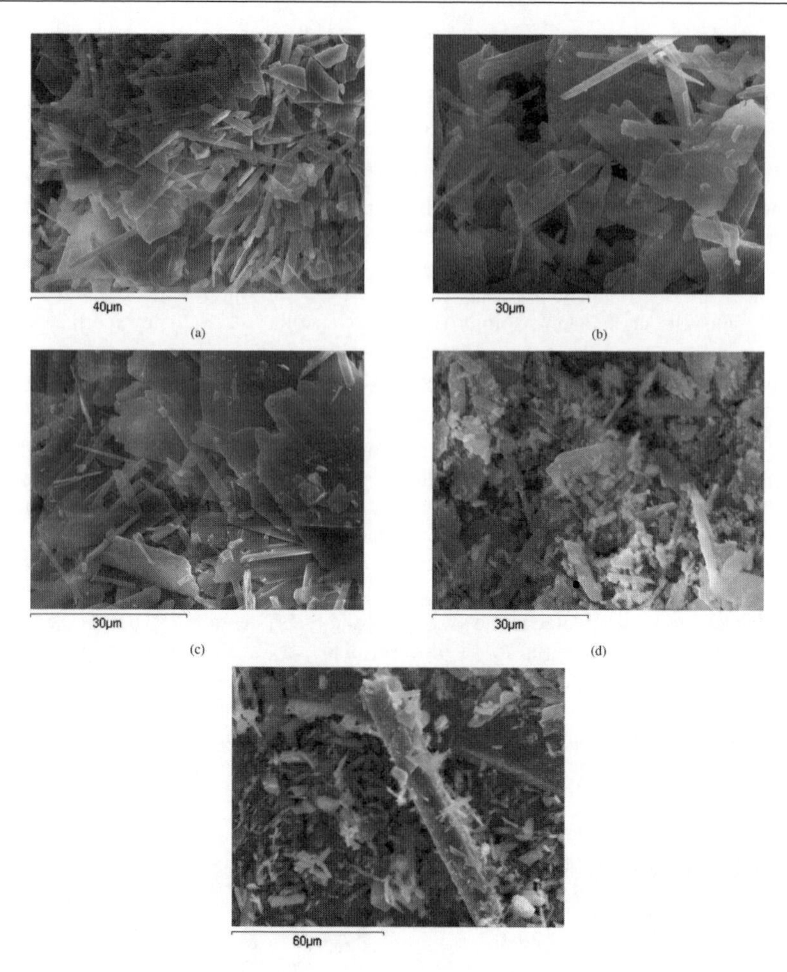

Figure 70. Scanning electron micrographs of gypsum crystals precipitated spontaneously at 25 °C; (a) no additive (b) 6 ppm PAA1 (c) 6 ppm PAA2 (d) 6 ppm PAA3 (e) 6 ppm PSA. [Used by permission of Elsevier, from Lioliou (2006), *Journal of Colloid and Interface Science*, Vol. 303, Fig. 5, p. 168.]

resulted in the enhancement of agglomeration and decrease of the crystal size (Smith and Alexander 1970; Lioliou et al. 2006). Similar results were obtained for gypsum in the presence of CTAB (Mahmoud et al. 2004). In the absence of additives, needle like crystals were obtained, whereas in the presence of CTAB the crystals were tabular. The crystals in the presence of SDS remained needle-like, similar to the crystals obtained in the control experiment.

The effect of OM on the mineralogy of the precipitated phase

In cases where the solution is supersaturated with respect to two or more polymorphs, the OM may preferentially adsorb to one polymorph. Such behavior may result in a change in the type of precipitated polymorph regardless of the thermodynamic state. For example, Rashad et al. (2003) found that the presence of SDS inhibits the crystallization of gypsum and leads to the formation of anhydrite and calcium sulfate hemihydrate. Several studies (e.g., Mann et al. 1988; Wei et al. 2005; Jada et al. 2007) showed that various OM affect the precipitation of calcium carbonate polymorphs. A wide range of calcium carbonate polymorphs exists including, in the order of decreasing solubility, calcium carbonate hexahydrate, calcium carbonate monohydrate, vaterite, aragonite, and calcite. Mainly the last three polymorphs are encountered since the first two structures are unstable and are converted into the thermodynamically more stable polymorphs. Mann et al. (1988) have showed that in presence of steric acid, vaterite precipitates while in the absence of steric acid calcite precipitates. Jada et al. (2007) have showed that the proportion of vaterite-calcite crystalline species is a function of sodium polyacrylates concentration and molecular weight. Various organic substances such as SDS, sodium dodecylsulfonate (DDS) and sodium dodecylbenzensulfate (SDBS) may also alter the precipitating polymorph of calcium carbonate varying from calcite to vaterite (Wei et al. 2005).

SUMMARY AND CONCLUSIONS

We have shown in this paper numerous examples of the role of OM in water-rock interactions. These interactions not only determine the fate of dissolved metals and OM, but also may alter the course and/or progress of different non organic reactions (e.g., mineral dissolution and precipitation), and are of interest in a variety of scientific and engineering disciplines. The literature on the interactions between a large variety of OM and many minerals includes thousands of publications. Naturally, the literature neither gives an equal representation to all minerals, nor a representation that reflects their natural abundances. Moreover, some minerals are overrepresented in the literature covering one type of interactions and underrepresented in other. For example, we found more papers that deal with adsorption on surfaces of clay minerals than on those of any other group of minerals, while we did not find papers on the effect of OM on clay precipitation. The scarcity of the latter studies is worth noting, as clay precipitation often occurs in organic-rich environment. This scarcity reflects the general deficiency in studies of the role of precipitation in natural weathering. As mentioned before, the majority of the papers that discuss adsorption deal with clays and oxides. A significant literature on adsorption of OM on carbonates exists as well. Studies of OM adsorption on sulfate and feldspars are relatively rare. Much was published on the effect of OM on dissolution of feldspars and quartz, while fewer publications were devoted to studies of the effect of OM on dissolution of clay and carbonates, and even less of sulfates. Publications studying the effect of OM on precipitation were almost entirely devoted to sulfate and carbonates.

Sorption of OM on mineral surfaces was widely studied, as it influences the fate of both OM and of metals in the environment. Adsorption affects the transport and reactivity of NOM as well as pollutants. Adsorbed OM can mask the properties of the underlying solid and present a surface with very different physicochemical properties. In addition, adsorption affects the reactivity of the mineral surfaces and therefore the dissolution and precipitation rates of the minerals. OM may be sorbed to the mineral via *chemisorption* (e.g., ligand exchange, anion exchange, cation exchange, and cation bridge) or via *physical adsorption* (e.g., Van der Waals interaction and hydrophobic effects). Often, the OM is initially sorbed onto the surface via one mechanism (e.g., Van der Waals interaction), and thereafter due to its proximity to the surface, the OM is chemisorbed via another mechanism (e.g., ligand exchange). It was shown that this initial adsorption is of time scales of seconds and minutes and depends on the environment conditions (e.g., pH). On the other hand, the chemisorption of the compound may be of longer time scale (hours) and therefore the rate of the adsorption may become significant in processes that are affected by it (e.g., dissolution and precipitation of minerals).

It is important to characterize the type of adsorption bond and coordinative structure between adsorbate and adsorbent. The organic ligand may be connected to one or more atoms on the mineral surface (i.e., mono/multinuclear). With each of these atoms, they may form one or more bonds (i.e., mono/multidentate). More than one type of complex may coexist and as we noted above, the coordination structure(s) of the adsorbed OM may change with time. Unfortunately, information on the characteristic of the bonds between a specific OM and a specific mineral is rare.

The relative charges of the functional groups of the OM and the mineral surface play an important role in adsorption. Regardless the sorption mechanism, the OM must be close to the mineral surface in order that sorption will take place. Therefore, electrostatic attraction/repulsion between the OM and the mineral surface affect sorption. Hence, sorption is influenced by the charges of the mineral surface and OM functional groups as well as the solution chemistry that affects these charges. It is important to note that although charges play an important role in sorption, often sorption may be more significant when the surfaces are less charged. For example, it was shown that as pH decreases, and both NOM acidic functional groups and surface charge of minerals become more protonated and therefore less negative, hydrophobic effects become more significant in driving organic molecules toward the mineral surface, which in turn, can increase the possibility for adsorption via other mechanisms.

Preferential adsorption of large organic molecules over small molecules was observed in many studies. This preferential adsorption was explained by the reduction of solubility with increasing size (i.e., Traube's Rule). Adsorption is also strongly influenced by the three dimensional structure of the OM. As a result, the adsorption of different isomers having the same chemical formula may be completely different. It was shown that ortho-positioned carboxyls and/or hydroxyls adsorbed more than the para- and meta- isomers. Such steric effects are important, but were poorly studied.

Both dissolution and precipitation of minerals are affected by the presence of OM. These effects can be due to complexation of OM with dissolved metals in solution or OM adsorption on the mineral surface. Interestingly, the different studies show that the common effects of OM are to catalyze mineral dissolution (mainly silicates) and to inhibit mineral precipitation (mainly carbonates and sulfates). The degree of saturation of the solution with respect to a mineral depends on the concentrations in solution of the lattice ions. If the concentration of any of these ions is reduced due to interactions with OM, the solution becomes more undersaturated or less supersaturated, and as a result dissolution may be faster while precipitation will be slower, respectively. One difference between dissolution and precipitation is the existence of a *dissolution plateau* (i.e., when dissolution occurs far enough from equilibrium, the degree of saturation does not affect the dissolution rate). As a result, this effect of OM on dissolution is effective only under close-to-equilibrium conditions. In contrast, there is no *precipitation plateau*, and precipitation rate always depends on saturation state and therefore complexation in solution of the lattice ions retards precipitation both under close-to-equilibrium and under far-from-equilibrium conditions. It is important to note that due to the inhibition effect of Al^{3+} on the dissolution of some aluminosilicates, OM complexes with Al^{3+} may enhance the dissolution rate of these minerals even at the *dissolution plateau* conditions.

Another difference between dissolution and precipitation reaction is the nucleation process. Nucleation is strongly dependent on the mineral-solution interfacial tension which is influenced by OM serving as surfactants, as well as the OM valance. These dependencies of the interfacial tension were shown to result in both promotion and retardation of nucleation kinetics.

Detailed examination of the literature data indicates that most of the experimental observations of OM effects on mineral dissolution and precipitation should be related to adsorption of OM on the surface of the minerals rather than to changes in solution chemistry. Therefore, one should examine separately each specific mineral – specific OM interaction. Whereas in some cases OM adsorption catalyzed dissolution of minerals, in others it may inhibit dissolution by adsorbing to reactive sites on the mineral surface and blocking them. This inhibiting effect on mineral dissolution was mainly reported for carbonate although few studies showed such inhibitory effects on silicates as well. Oppositely, catalysis by OM was mainly reported for silicates and oxides, while only few studies discussed catalysis of carbonate dissolution.

A simple ligand-promoted reaction mechanism consists of fast adsorption of the organic ion on the mineral surface followed by a slow catalyst-mediated hydrolysis step, which is the rate-determining step. It is generally agreed that organic ligands preferentially adsorb to aluminol sites of aluminosilicates and Al-oxides, Ca sites in carbonates and sulfates and probably Fe surface sites in iron rich minerals. The hydrolysis rate depends on the surface concentration of the adsorbed species, which depends both on the concentration of the ligand in solution and the surface concentration of the active site. Therefore, the increase in dissolution rate as a function of Al content of the mineral is another indication that ligand attack (as well as proton attack) occurs predominantly on Al surface sites of aluminosilicates. Correspondingly, it was found that OM that forms strong complexes with Al^{3+} in solution tends to form complexes with Al sites on silicates and oxides surfaces and therefore has a stronger catalytic effect. From the existing literature it is not possible to determine whether this stronger catalytic effect is because the stronger ligand forms more surface complexes than the weaker ligand or because the surface complexes of the former hydrolyzed faster than those of the latter. The similarity between the rate constants of oxalate attack, bioxalate attack, and proton attack on feldspar as well as the similarity between activation energies of proton-promoted and oxalate-promoted dissolution of kaolinite indicates that the magnitude of the rates of each of these reaction mechanisms is determined by the surface concentration of the various adsorbed species and not necessarily by the rate of the hydrolysis.

In order for an adsorbed OM to catalyze or inhibit dissolution it should form a strong complex with a metal ion on the surface. In the case of catalysis, the complexation weakens the strength of the bond between the metal and the bulk mineral, resulting in the release of the OM and the ion to solution. In the case of inhibition, this bond should remain strong, so the OM would remain on the surface and block the ion from being removed. Commonly, the same surface sites and faces on which crystal tends to grow under oversaturation conditions are the sites in which dissolution occurs at undersaturation conditions. It is therefore reasonable to assume that OM which effectively inhibits dissolution would inhibit precipitation as well. For catalysis, this is not the case, since adsorbed OM that weakens the bond strength is not expected to catalyze precipitation.

In order for an OM inhibitor to effectively inhibit the precipitation of a mineral, a dimensional stereochemical fit between the polar group of the inhibitor and the intercationic or interanionic distances is needed. When such a structural fit exists, the OM can also act as a template or microsubstrate and promote nucleation rather than inhibit it. Indeed, it was found that some organic compounds that accelerate mineral precipitation when present at low concentrations become an effective inhibitor at higher concentrations.

A comprehensive review of the literature indicates that the common effect of OM on a wide range of minerals is to inhibit precipitation rather than to catalyze it. It is possible that this observation represent a bias of the literature and not a real effect. The Earth science literature paid little attention to low temperature precipitation of non-carbonates in general and the effect of OM on precipitation in particular. The engineering literature is motivated mainly by scale deposits and therefore tends to study inhibition of its precipitation and catalysis of its dissolution.

Bearing in mind that dissolution and precipitation are affected by OM adsorption, it is surprising that only few studies on the effect of OM on dissolution and precipitation have measured the adsorption isotherm of the OM. Taking into account the numerous published papers on OM adsorption, this seems to be a niggling point. However, the effect of sorption on dissolution and precipitation depends on the sorption mechanism (e.g., *physical adsorption* does not enhance dissolution rate). Even if the adsorption of a given compound is studied, it is not trivial to characterize the mechanisms and, more important, the relative contribution of each of them to the overall adsorption. Moreover, sorption, dissolution and precipitation depend on both the characteristics of the mineral surface and the environmental conditions, and therefore it is complicated to combine adsorption experiments from one laboratory to dissolution or precipitation experiments that were conducted at another laboratory.

Even when the adsorption isotherms are conducted at the same laboratory as the dissolution/precipitation experiments, it is not straight forward to incorporate their results. One important limitation is related to the typical duration of the experiments. As sorption is fast, adsorption experiments are usually much shorter than dissolution experiments. However, few studies showed that the coordination structures of the adsorbed OM may change with time. Therefore, it is possible that OM adsorption during dissolution experiments is different than that during the adsorption experiments. Unfortunately, monitoring sorption during dissolution experiments is not an easy task. One problem is that changes in OM concentration due to decomposition (which may be enhanced in the presence of a mineral) may be more pronounced than that due to adsorption.

The kinetics of adsorption is generally considered to be fast in comparison with dissolution reactions. Therefore, the adsorption reactions are usually regarded as being in equilibrium. For nucleation inhibition, adsorption kinetics is important, as adsorption must take place before stable nuclei are formed.

The importance of OM-water-rock interactions varied in different processes and environments. Adsorption of OM on mineral surfaces, ternary adsorption of OM and metals and complexation between OM and dissolved metals all have a very important role in controlling the fate of both OM and of metals in the environment. The effect of OM on dissolution of minerals in nature seems to be of less significant. The analysis of literature data indicates that the effect of oxalate on silicate dissolution is generally small (median enhancement factor of 2), and is much less than the between-laboratories agreement factor of 5. The possible effects of OM on precipitation of minerals in nature haven't been properly studied, and therefore, we cannot estimate their importance. For industrial applications, the effect of organic antiscalants is most significant. Effective antiscalants can decrease the nucleation rates by orders of magnitude and thereby achieving a "total inhibition." OM has important catalytic effects on dissolution of scales as well. However, much less scientific effort was devoted to such application, as it is more economical and efficient to control scale by inhibiting its nucleation rather than by dissolving it. OM may also be important in formation of surface coating that protect buildings and monuments from damaging effects of weathering. However, the stability of such coatings over time is still in question.

ACKNOWLEDGMENT

We are grateful to Eric Oelkers and Jacques Schott for organizing the workshop on Thermodynamics and Kinetics of Fluid-Rock Interactions and for inviting us to write this review. It was a wonderful and unique learning experience. The corrections and suggestions of Eric Oelkers, Alan Mathews and Shimon Feinstein are greatly appreciated.

REFERENCES

Addadi L, Weiner S (1985) Interactions between acidic proteins and crystals: stereochemical requirements in biomineralization. Proc Nat Acad Sci USA 82:4110-4114

Alimi F, Gadri A (2004) Kinetics and morphology of formed gypsum. Desalination 166:427-434

Alvarez R, Evans LA, Milham PJ, Wilson MA (2004) Effects of humic material on the precipitation of calcium phosphate. Geoderma 118:245-260

Amjad Z (1985) Applications of antiscalants to control calcium sulfate scaling in reverse osmosis systems. Desalination 54:263-276

Amjad Z, Hooley J (1986) Influence of polyelectrolytes on the crystal growth of calcium sulfate dihydrate. J Colloid Interface Sci 111:496-503

Amrhein C, Suarez DL (1988) The use of a surface complexation model to describe the kinetics of ligand-promoted dissolution of anorthite. Geochim Cosmochim Acta 52:2785-2793

Aranow RH, Witten L (1958) Theoretical derivation of Traube's rule. J Chem Phys 28:405-409

Arnarson TS, Keil RG (2000) Mechanisms of pore water organic matter adsorption to montmorillonite. Mar Chem 71:309-320

Au K-K, Penisson AC, Yang S, O'Melia CR (1999) Natural organic matter at oxide/water interfaces: complexation and conformation. Geochim Cosmochim Acta 63:2903-2917

Au K-K, Yang S, O'Melia CR (1998) Adsorption of weak polyelectrolytes on metal oxide surfaces: A hybrid SC/SF approach. Environ Sci Technol 32:2900-2908

Austin AE, Miller JF, Vaughan DA, Kircher JF (1975) Chemical additives for calcium sulfate scale control. Desalination 16:345-357

Avena MJ, Koopal LK (1998) Desorption of humic acids from an iron oxide surface. Environ Sci Technol 32:2572-2577

Avena MJ, Koopal LK (1999) Kinetics of humic acid adsorption at solid-water interfaces. Environ Sci Technol 33:2739-2744

Axe K, Persson P (2001) Time-dependent surface speciation of oxalate at the water-boehmite (γ-AlOOH) interface: Implications for dissolution. Geochim Cosmochim Acta 65:4481–4492

Azoury R, Randolph AD, Drach GW, Perlberg S, Garti N, Sarig S (1983) Inhibition of calcium oxalate crystallization by glutamic acid: Different effects at low and high concentrations. J Cryst Growth 64:389-392

Baham J, Sposito G (1994) Adsorption of dissolved organic carbon extracted from sewage sludge on montmorillonite and kaolinite in the presence of metal ions. J Environ Qual 23:147-153

Barbier E, Coste M, Genin A, Jung D, Lemoine C, Logette S, Muhr H (2009) Simultaneous determination of nucleation and crystal growth kinetics of gypsum. Chem Eng Sci 64:363-369

Barcelona MJ, Atwood DK (1979) Gypsum-organic interactions in the marine environment: sorption of fatty acids and hydrocarbons. Geochim Cosmochim Acta 43:47-53

Barwise AJ, Compton RG, Unwin PR (1990) The effect of carboxylic acids on the dissolution of calcite in aqueous solution. Part 2.-d-, I- and meso-Tartaric acids. J Chem Soc Faraday Trans 86:137-144

Baumgartner E, Mijalchik M (1991) Polyelectrolytes as inhibitors of Fe(III) oxide precipitation. J Colloid Interface Sci 145:274-278

Baziramakenga R, Simard, RR, Leroux GD (1995) Determination of organic acids in soil extracts by ion chromatography. Soil Biol Biochem 27:349-356

Becker U, Biswas S, Kendall, T, Risthaus P, Putnis CV, Pina CM (2005) Interactions between mineral surfaces and dissolved species: From monovalent ions to complex organic molecules. Am J Sci 305:791-825

Bell JLS, Palmer DA, Barnes HL, Drummond SE (1994) Thermal decarboxylation of acetate: III. Catalysis by mineral surfaces. Geochim Cosmochim Acta 58:4155-4177

Bennett PC (1991) Quartz dissolution in organic-rich aqueous systems. Geochim Cosmochim Acta 55:1781-1797

Bennett PC, Casey W (1994) Chemistry and mechanisms of low-temperature dissolution of silicates by organic acids. *In:* Organic Acids in Geological Processes. Pittman ED, Lewan MD (eds), Springer-Verlag, Berlin, p 162-200

Bennett PC, Melcer ME, Siegel DI, Hassett JP (1988) The dissolution of quartz in dilute aqueous solutions of organic acids at 25 °C. Geochim Cosmochim Acta 52:1521-1530

Bennett PC, Siegel DI, Hill BM, Glaser PH (1991) Fate of silicate minerals in a peat bog. Geology 19:328-331

Benton JW, Collins RI, Grimsey MI, Parkinson MG, Rodger AS (1993) Nucleation, growth and inhibition of barium sulfate-controlled modification with organic and inorganic additives. Faraday Discuss 95:281-297

Berner RA, Morse JW (1974) Dissolution kinetics of calcium carbonate in sea water. IV, Theory of calcite dissolution. Am J Sci 274:108-134

Bevan J, Savage D (1989) Effect of organic acids on the dissolution of K-feldspar under conditions relevant to burial diagenesis. Mineral Mag 53:415-425

Blake RE, Walter LM (1999) Kinetics of feldspar and quartz dissolution at 70-80 °C and near-neutral pH: effects of organic acids and NaCl . Geochim Cosmochim Acta 63 2043-2059

Blum A, Stillings LL (1995) Feldspar dissolution kinetics. Rev Mineral 31:291-351

Blum AE, Lasaga AC (1991) The role of surface speciation in the dissolution of albite. Geochim Cosmochim Acta 55:2193-2201

Bosbach D, Hall C, Putnis A (1998) Mineral precipitation and dissolution in aqueous solution: In situ microscopic observations on barite (001) with Atomic Force Microscopy. Chem Geol 151:143-160

Brady PV, Cygan RT, Nagy KL (1996) Molecular control on kaolinite surface charge. J Colloid Interface Sci 183:356-364

Brandse WP, Vanrosmalen GM, Brouwer G (1977) The influence of sodium-chloride on crystallization rate of gypsum. J Inorg Nucl Chem 39 2007-2010

Brandt F, Bosbach D (2001) Bassanite ($CaSO_4{\cdot}0.5H_2O$) dissolution and gypsum ($CaSO_4{\cdot}2H_2O$) precipitation in the presence of cellulose ethers. J Cryst Growth 233:837-845

Brantley SL, Stillings L (1996) Feldspar dissolution at 25 °C and low pH. Am J Sci 296:101-127

Brunauer S, Emmett PH, Teller E (1938) Adsorption of gases in multimolecular layers. J Am Chem Soc 60:309-319

Burch TE, Nagy KL, Lasaga A C (1993) Free energy dependence of albite dissolution kinetics at 80 °C, pH 8.8. Chem Geol 105:137-162

Burns K, Wu Y-T, Grant CS (2003) Mechanisms of calcite dissolution using environmentally benign polyaspartic acid: A rotating disk study. Langmuir 19:5669–5679

Burton WK, Cabrera N, Frank FC (1951) The growth of crystals and the equilibrium structure of their surfaces. Philos Trans R Soc London 243:299-358

Busey RH, Mesmer RE (1978) Thermodynamic quantities for the ionization of water in sodium chloride media to 300 °C. J Chem Eng Data 23:175-176

Cabaniss SE, Zhou Q, Maurice PA, Chin Y-P, Aiken GR (2000) A Log-normal distribution model for the molecular weight of aquatic fulvic acids. Environ Sci Technol 34:1103-1109

Cama J, Ayora C, Querol X, Ganor J (2005) Dissolution kinetics of synthetic zeolite NaP1 and its implication to zeolite treatment of contaminated waters. Environ Sci Technol 39:4871-4877

Cama J, Ganor J (2006) The effects of organic acids on the dissolution of silicate minerals: A case study of oxalate catalysis of kaolinite dissolution. Geochim Cosmochim Acta 70:2191-2209

Cama J, Metz V, Ganor J (2002) The effect of pH and temperature on kaolinite dissolution rate under acidic conditions. Geochim Cosmochim Acta 66:3913-3926

Cao X, Harris WG, Josan MS, Nair VD (2007) Inhibition of calcium phosphate precipitation under environmentally-relevant conditions. Sci Total Environ 383 205-215

Carter PW (1978) Adsorption of amino acid-containing organic matter by calcite and quartz. Geochim Cosmochim Acta 42:1239-1242

Chave EK, Suess E (1970) Calcium carbonate saturation in seawater: Effects of dissolved organic matter. Limnol Oceanography 15:633-637

Chi F-H, Amy GL (2004) Kinetic study on the sorption of dissolved natural organic matter onto different aquifer materials: the effects of hydrophobicity and functional groups. J Colloid Interface Sci 274:380-391

Chin P-KF, Mills GL (1991) Kinetics and mechanisms of kaolinite dissolution: effects of organic ligands. Chem Geol 90:307-317

Chin Y-P, Aiken G, O'Loughlin E (1994) Molecular weight, polydispersity, and spectroscopic properties of aquatic humic substances. Environ Sci Technol 28:1853-1858

Compton RG, Brown CA (1995) The inhibition of calcite dissolution/precipitation: 1,2-dicarboxylic acids. J Colloid Interface Sci 170:586-590

Compton RG, Pritchard KL, Unwin PR, Grigg G, Silvester P, Lees M, House WA (1989) The effect of carboxylic acids on the dissolution of calcite in aqueous solution. Part 1. - Maleic and fumaric acids. J Chem Soc Faraday Trans 1: Phys Chem Cond Phases 85:4335-4366

Compton RG, Sanders GHW (1993) The dissolution of calcite in aqueous acid: the influence of humic species. J Colloid Interface Sci 158:439-445

Cordt T, Kussmaul H (1992) Characterization of some organic acids in the subsurface of the Sandhausen Ecosystem. *In:* Progress in Hydrogeochemistry; Organics; Carbonate Systems; Silicate Systems; Microbiology; Models. Matthess G, Frimmel FH, Schulz HD, Uschowski E (eds), Springer Verlag, Berlin, p 93-100

Cundy CS, Cox PA (2005) The hydrothermal synthesis of zeolites: Precursors, intermediates and reaction mechanism. Microporous Mesoporous Mat 82:1-78

Davis JA (1982) Adsorption of natural dissolved organic matter at the oxide/water interface. Geochim Cosmochim Acta 46:2381-2393

Davis JA, Kent DB (1990) Surface complexation modeling in aqueous geochemistry. Rev Mineral 23:177-260

Davis SA, Dujardin E, Mann S (2003) Biomolecular inorganic materials chemistry. Curr Opin Solid State Mater Sci 7:273-281

Devidal J-L, Schott J, Dandurand J-L (1997) An experimental study of kaolinite dissolution and precipitation kinetics as a function of chemical affinity and solution composition at 150 °C, 40 bars, and pH 2:6.8:and 7.8. Geochim Cosmochim Acta 61:5165-5186

Di Toro DM, Horzempa LM (1982) Reversible and resistant components of PCB adsorption-desorption: isotherms. Environ Sci Technol 16:594-602

Dijt JC, Stuart MAC, Hofman JE, Fleer GJ (1990) Kinetics of polymer adsorption in stagnation point flow. Colloids Surf 51:141-158

Drever JI (1994) The effect of land plants on weathering rates of silicate minerals. Geochim Cosmochim Acta 58:2325-2332

Drever JI, Stillings LL (1997) The role of organic acids in mineral weathering. Colloids Surf A 120:167-181

Drever JI, Vance GF (1994) Role of soil organic acids in mineral weathering processes. *In:* Organic Acids in Geological Processes. Pittman ED, Lewan MD (eds) Springer-Verlag, Berlin, p 138-161

Dunn K, Yen TF (1999) Dissolution of barium sulfate scale deposits by chelating agents. Environ Sci Technol 33:2821-2824 p

Dunnivant FM, Jardine PM, Taylor DL, McCarthy JF (1992) Transport of naturally occurring dissolved organic carbon in laboratory columns containing aquifer material. Soil Sci Soc Am J 56:437-444

Dzombak DA, Morel FMM (1990) Surface Complexation Modelling: Hydrous Ferric Oxides. John Wiley and Sons, New York

El-Shall H, Rashad MM, Abdel-Aal EA (2005) Effect of cetyl pyridinium chloride additive on crystallization of gypsum in phosphoric and sulfuric acids medium. Crystal Res Technol 40:860-866

El Dahan HA, Hegazy HS (2000) Gypsum scale control by phosphate ester. Desalination 127:111-118

Eriksson R, Merta J, Rosenholm JB (2007) The calcite/water interface: I. Surface charge in indifferent electrolyte media and the influence of low-molecular-weight polyelectrolyte. J Colloid Interface Sci 313:184-193

Fein JB, Brady PV (1995) Mineral surface controls on the diagenetic transport of oxalate and aluminum. Chem Geol 121:11-18

Fox TR, Comerford NB (1990) Low-molecular-weight organic acids in selected soils of the southeastern USA. Soil Sci Soc Am J 54:1139-1144

Franklin SP, Hajash AJ, Dewers TA, Tieh TT (1994) The role of carboxylic acids in albite and quartz dissolution: An experimental study under diagenetic conditions. Geochim Cosmochim Acta 58:4259-4279

Fredd CN, Fogler HS (1998a) The influence of chelating agents on the kinetics of calcite dissolution. J Colloid Interface Sci 204:187-197

Fredd CN, Fogler HS (1998b) The kinetics of calcite dissolution in acetic acid solutions. Chem Eng Sci 53:3863-3874

Fritz B, Noguera C (2009) Mineral precipitation kinetics. Rev Mineral Geochem 70:371–410

Furedi-Milhofer H (1981) Spontaneous precipitation from electrolytic solutions. Pure Appl Chem 53 2041-2055

Furedi-Milhofer H, Sarig S (1996) Interactions between polyelectrolytes and sparingly soluble salts. Prog Cryst Growth Charact Mater 32:45-74

Furrer G, Stumm W (1986) The coordination chemistry of weathering: I. Dissolution kinetics of δ-Al_2O_3 and BeO. Geochim Cosmochim Acta 50:1847-1860

Ganor J, Cama J, Metz V (2003) Surface protonation data of kaolinite - reevaluation based on dissolution experiments. J Colloid Interface Sci 264:67-75

Ganor J, Lasaga AC (1994) The effects of oxalic acid on kaolinite dissolution rate. Mineral Mag 58A:315-316

Ganor J, Lasaga AC (1998) Simple mechanistic models for inhibition of a dissolution reaction. Geochim Cosmochim Acta 62:1295-1306

Ganor J, Mogollon JL, Lasaga AC (1999) Kinetics of gibbsite dissolution under low ionic strength conditions. Geochim Cosmochim Acta 63:1635-1651

Ganor J, Nir S, Cama J (2001) The effect of kaolinite on oxalate (bio)degradation at 25 °C, and possible implications for adsorption isotherm measurements. Chem Geol 177:431-442

Gautier JM, Oelkers EH, Schott J (1994) Experimental study of K-feldspar dissolution rates as a function of chemical affinity at 150 °C and pH 9. Geochim Cosmochim Acta 58:4549-4560

Gestsdottir K, Manning DAC (1992) An experimental study of the dissolution of albite in the presence of organic acids. *In:* 7th international symposium on Water-Rock interaction. Kharaka YK, Maest AS (eds) Balkema, Park City, Utah, 315-318 p

Giles CH, MacEwan TH, Nakhwa SN, Smith D (1960) Studies in adsorption. Part XI. A system of classification of solution adsorption isotherms, and its use in diagnosis of adsorption mechanisms and in measurement of specific surface areas of solids. J Chem Soc 3973–3993

Giordano TH (1994) Metal transport in ore fluids by organic ligand complexation. *In:* Organic Acids in Geological Processes. Lewan MD, Prittman ED (eds), Springer-Verlag, Berlin Heidelberg, p 319-354

Gloede M, Melin T (2006) Potentials and limitations of molecular modelling approaches for scaling and scale inhibiting mechanisms. Desalination 199:26-28

Goldschmidt VM (1952) Geochemical aspects of the origin of complex organic molecules on the Earth, as precursors to organic life. New Biology 12:97-105

Golubev SV, Bauer A, Pokrovsky OS (2006) Effect of pH and organic ligands on the kinetics of smectite dissolution at 25 °C. Geochim Cosmochim Acta 70:4436-4451

Golubev SV, Pokrovsky OS (2006) Experimental study of the effect of organic ligands on diopside dissolution kinetics. Chem Geol 235:377-389

Grandstaff DE (1986) The dissolution rate of forsteritic olivine from Hawaiian beach sand. *In:* Rates of Chemical Weathering of Rocks and Minerals. Colman SM, Dethier DP (eds), Academic Press, Orlando, FL

Graustein WC (1981) The Effect of Forest Vegetation on Solute Acquisition and Chemical Weathering: A Study of the Tesuque Watersheds Near Santa Fe, New Mexico. PhD Dissertation, Yale University

Graveling GJ, Vala Ragnarsdottir K, Allen GC, Eastman J, Brady PV, Balsley SD, Skuse DR (1997) Controls on polyacrylamide adsorption to quartz, kaolinite, and feldspar. Geochim Cosmochim Acta 61:3515-3523

Grierson PF (1992) Organic acids in the rhizosphere of *Banksia integrifolia L.f.* Plant and Soil 144:259-265

Gu B, Mehlhorn TL, Liang L, McCarthy JF (1996) Competitive adsorption, displacement, and transport of organic matter on iron oxide: I. Competitive adsorption. Geochim Cosmochim Acta 60:1943-1950

Gu B, Schmitt J, Chen Z, Liang L, McCarthy JF (1994) Adsorption and desorption of natural organic matter on iron oxide: mechanisms and models. Environ Sci Technol 28:38-46

Gu B, Schmitt JR, Chen Z, Liang L, McCarthy JF (1995) Adsorption and desorption of different organic matter fractions on iron oxide. Geochim Cosmochim Acta 59:219-229

Hajash A, Carpenter TD, Dewers TA (1998) Dissolution and time-dependent compaction of albite sand: experiments at 100 °C and 160 °C in pH-buffered organic acids and distilled water. Tectonophysics 295:93-115

Hajash A, Franklin SP, Reed CL (1992) Experimental feldspar dissolution in acetic and oxalic acids at 100 °C, 345 bars. *In:* 7th international symposium on Water-Rock interaction. Kharaka YK, Maest AS (eds) Balkema, Park City, Utah, 325-328 p

Hajash AJ (1994) Comparison and evaluation of experimental studies on dissolution of minerals by organic acids. *In:* Organic Acids in Geological Processes. Pittman ED, Lewan MD (eds), Springer-Verlag, Berlin, p 201-225

Hamdona SK, Hamza SM, Mangood AH (1995) Effect of some amino acids on the rate of dissolution of calcite crystals in aqueous systems. Desalination 101:263-267

Harrison WJ, Thyne GD (1992) Predictions of diagenetic reactions in the presence of organic acids. Geochim Cosmochim Acta 56:565-586

Hasson D, Drak A, Semiat R (2003) Induction times induced in an RO system by antiscalants delaying $CaSO_4$ precipitation. Desalination 157:193-207

He S, Oddo JE, Tomson MB (1994) The inhibition of gypsum and barite nucleation in NaCl brines at temperatures from 25 to 90 °C. Appl Geochem 9:561-567

He S, Oddo JE, Tomson MB (1995a) The nucleation kinetics of barium sulfate in NaCl solutions up to 6 m and 90 °C. J Colloid Interface Sci 174:319-326

He S, Oddo JE, Tomson MB (1995b) The Nucleation kinetics of strontium sulfate in NaCl solutions up to 6 m and 90 °C with or without inhibitors. J Colloid Interface Sci 174:327-335

Hiebert FK, Bennett PC (1992) Microbial control of silicate weathering in organic-rich ground water. Science 258:278-281

Hoch AR, Reddy MM, Aiken GR (2000) Calcite crystal growth inhibition by humic substances with emphasis on hydrophobic acids from the Florida Everglades. Geochim Cosmochim Acta 64:61-72

Holdren GR, Adams JE (1982) Parabolic dissolution kinetics of silicate minerals: An artifact of non equilibrium precipitation processes. Geology 10:186-190

Holdren GR, Berner R (1979) Mechanism of feldspar weathering. I. Experimental studies. Geochim Cosmochim Acta 43:1161-1171

Huang W-L, Longo JM (1992) The effect of organics on feldspar dissolution and the development of secondary porosity. Chem Geol 98:271-292

Huang WH, Keller WD (1970) Dissolution of rock-forming silicate minerals in organic acids: simulated first-stage weathering of fresh mineral surfaces. Am Mineral 55:2076–2094

Huang WH, Keller WD (1971) Dissolution of clay minerals in dilute organic acids at room temperature. Am Mineral 56:1082-1095

Huang WH, Keller WD (1973) Kinetics and mechanisms of dissolution of Fithian illite in two complexing organic acid. *In:* Proceedings of the International Clay Conference, Madrid, Spain, 1972. Serratosa JM (ed), Div. Ciencias CSIC, Madrid, Spain, p 321-331

Huang WH, Kiang WC (1972) Laboratory dissolution of plagioclase feldspars in water and organic acids at room temperature. Am Mineral 57:1849-1859

Huertas JF, Chou L, Wollast R (1998) Mechanism of kaolinite dissolution at room temperature and pressure: Part 1. Surface speciation. Geochim Cosmochim Acta 62:417-431

Hunter KA (1980) Microelectrophoretic properties of natural surface-active organic matter in coastal seawater. Limnology Oceanography 25:807-822

Ives MB (1965) The action of adsorptive inhibitors in some corrosive systems. Ind Eng Chem 57:34-40

Jackson KS, Jonasson IR, Skippen GB (1978) The nature of metals--sediment--water interactions in freshwater bodies, with emphasis on the role of organic matter. Earth Sci Rev 14:97-146

Jada A, Ait Akbour R, Douch J (2006) Surface charge and adsorption from water onto quartz sand of humic acid. Chemosphere 64:1287-1295

Jada A, Ait Akbour R, Jacquemet C, Suau JM, Guerret O (2007) Effect of sodium polyacrylate molecular weight on the crystallogenesis of calcium carbonate. J Cryst Growth 306:373-382

Jada A, Ait Chaou A (2003) Surface properties of petroleum oil polar fraction as investigated by Zetametry and DRIFT spectroscopy. J Pet Sci Eng 39:287-296

Jardine PM, McCarthy JF, Weber NL (1989) Mechanisms of dissolved organic carbon adsorption on soil. Soil Sci Soc Am J 53:1378-1385

Jiang H, Liu X-Y, Zhang G, Li Y (2005) Kinetics and template nucleation of self-assembled hydroxyapatite nanocrystallites by chondroitin sulfate. J Biol Chem 280:42061-42066
Jonasson RG, Rispler K, Wiwchar B, Gunter WD (1996) Effect of phosphonate inhibitors on calcite nucleation kinetics as a function of temperature using light scattering in an autoclave. Chem Geol 132:215-225
Julien AA (1879) On the geological action of the humus acids. Proc Am Assoc Adv Sci 28:311-410
Kaiser K, Guggenberger G (2000) The role of DOM sorption to mineral surfaces in the preservation of organic matter in soils. Org Geochem 31:711-725
Kaiser K, Guggenberger G, Haumaier L, Zech W (1997) Dissolved organic matter sorption on sub soils and minerals studied by ^{13}C-NMR and DRIFT spectroscopy. Eur J Soil Sci 48:301-310
Kaiser K, Zech W (1999) Release of natural organic matter sorbed to oxides and a subsoil. Soil Sci Soc Am J 63:1157-1166
Kettler RM, Palmer DA, Wesolowski DJ (1991) Dissociation quotients of oxalic acid in aqueous sodium chloride media to 175 °C. J Solution Chem 20:905-927
Khan SR (1997) Calcium phosphate/calcium oxalate crystal association in urinary stones: implications for heterogeneous nucleation of calcium oxalate. J Urology 157:376-383
Kharaka YK, Law LL, Carothers WW, Goerlitz DF (1986) Role of organic species dissolved in formation waters from sedimentary basins in mineral diagenesis. *In:* Roles of Organic Matter in Sediment Diagenesis. Gautier DL (ed), Society of Economic Paleontologists and Mineralogists, 38:111-122
Knauss KG, Copenhaver SA (1995) The effect of malonate on the dissolution kinetics of albite, quartz, and microcline as a function of pH at 70 °C. Appl Geochem 10:17-33
Koretsky C (2000) The significance of surface complexation reactions in hydrologic systems: a geochemist's perspective. J Hydrology 230:127-171
Krzyszowska AJ, Vance GF, Blaylock MJ, David MB (1996) Ion-chromatographic analysis of low molecular weight organic acids in spodosol forest floor solutions. Soil Sci Soc Am J 60:1565-1571
Kummert R, Stumm W (1980) The surface complexation of organic acids on hydrous γ-Al_2O_3. J Colloid Interface Sci 75:373-385
Lackhoff M, Niessner R (2002) Photocatalytic atrazine degradation by synthetic minerals, atmospheric aerosols, and soil particles. Environ Sci Technol 36:5342-5347
Lasaga AC (1984) Chemical kinetics of water-rock interactions. J Geophys Res 89:4009-4025
Lasaga AC (1995) Fundamental approaches in describing mineral dissolution and precipitation rate. Rev Mineral 31: 23-86
Lasaga AC (1998) Kinetic Theory in the Earth Sciences. Princeton University Press, Princeton, NJ
Lasaga AC, Blum AE (1986) Surface chemistry, etch pits and mineral-water reactions. Geochim Cosmochim Acta 50:2363-2379
Lasaga AC, Gibbs GV (1990) Ab-initio quantum mechanical calculations of water-rock interaction: adsorption and hydrolysis reactions. Am J Sci 290:263-295
Lasaga AC, Luttge A (2003) A model for crystal dissolution. Eur J Mineral 15:603-615
Lasaga AC, Soler JM, Ganor J, Burch TE, Nagy KL (1994) Chemical weathering rate laws and global geochemical cycles. Geochim Cosmochim Acta 58:2361-2386
Lebron I, Suarez DL (1996) Calcite nucleation and precipitation kinetics as affected by dissolved organic matter at 25 °C and pH > 7.5. Geochim Cosmochim Acta 60:2765-2776
Li FB, Li XZ, Li XM, Liu TX, Dong J (2007) Heterogeneous photodegradation of bisphenol A with iron oxides and oxalate in aqueous solution. J Colloid Interface Sci 311:481-490
Lioliou MG, Paraskeva CA, Koutsoukos PG, Payatakes AC (2006) Calcium sulfate precipitation in the presence of water-soluble polymers. J Colloid Interface Sci 303:164-170
Liotta LF, Gruttadauria M, Di Carlo G, Perrini G, Librando V (2009) Heterogeneous catalytic degradation of phenolic substrates: Catalysts activity. J Hazard Mater 162:588-606
Liu A, Gonzalez RD (1999) Adsorption/desorption in a system consisting of humic acid, heavy metals, and clay minerals. J Colloid Interface Sci 218:225-232
Liu S-T, Nancollas GH (1973) The crystal growth of calcium sulfate dihydrate in the presence of additives. J Colloid Interface Sci 44:422-429
Liu S-T, Nancollas GH (1975a) The crystal growth and dissolution of barium sulfate in the presence of additives. J Colloid Interface Sci 52:582-592
Liu ST, Nancollas GH (1975b) A kinetic and morphological study of the seeded growth of calcium sulfate dihydrate in the presence of additives. J Colloid Interface Sci 52:593-601
Lundstrom U, Ohman L-O (1990) Dissolution of feldspars in the presence of natural organic solutes. J Soil Sci 41:359-369
MacGowan DB, Surdam RS (1988) Difunctional carboxylic acid anions in oil field waters. Org Geochem 12:245-259
MacGowan DB, Surdam RS (1990) Carboxylic acid anions in formation waters, San Joaquin Basin and Louisiana Gulf Coast, USA - Implications for clastic diagenesis. Appl Geochem 5:687-701

Mahmoud MHH, Rashad MM, Ibrahim IA, Abdel-Aal EA (2004) Crystal modification of calcium sulfate dihydrate in the presence of some surface-active agents. J Colloid Interface Sci 270:99-105
Manley EP, Evans LJ (1986) Dissolution of feldspars by low-molecular-weight aliphatic and aromatic acids. Soil Sci 141:106-112
Mann S (1988) Molecular recognition in biomineralization. Nature 332:119-124
Mann S, Heywood BR, Rajam S, Birchall JD (1988) Controlled crystallization of $CaCO_3$ under stearic acid monolayers. Nature 334:692-695
Mast AM, Drever JI (1987) The effect of oxalate on the dissolution rates of oligoclase and tremolite. Geochim Cosmochim Acta 51:2559-2568
Matthess G (1984) The role of natural organics on water interaction with soil and rock. *In:* Hydrochemical Balances of Freshwater Systems. Eriksson E (ed) Uppsala, p 11-22
McMahon PB, Chapelle FH (1991) Microbial production of organic acids in aquitard sediments and its role in aquifer geochemistry. Nature 349:233-235
Meier M, Namjesnik-Dejanovic K, Maurice PA, Chin Y-P, Aiken GR (1999) Fractionation of aquatic natural organic matter upon sorption to goethite and kaolinite. Chem Geol 157:275-284
Mogollon JL, Ganor J, Soler JM, Lasaga AC (1996) Column experiments and the full dissolution rate law of gibbsite. Am J Sci 296:729-765
Morse JW (1974) Dissolution kinetics of calcium carbonate in sea water. V. Effects of natural inhibitors and the position of the chemical lysocline. Am J Sci 274:638-647
Morse JW, Arvidson RS (2002) The dissolution kinetics of major sedimentary carbonate minerals. Earth-Sci Rev 58:51-84
Morse JW, Arvidson RS, Luttge A (2007) Calcium carbonate formation and dissolution. Chem Rev 107:342-381
Mortland MM (1986) Mechanism of adsorption of nonhumic organic species by clays. *In:* Interaction of Soil Minerals with Natural Organics and Microbes, SSSA Special Publ. No. 17. Huang PM, Schnitzer M (eds) Soil Science Society of America, Madison, WI
Murray J, Renard AF (1891) Report of the Deep-Sea Deposits in Report From the on the Scientific Results of the Voyage H.M.S. Challenger, Great Britain, Challenger Office
Nagy KL, Blum AE, Lasaga AC (1991) Dissolution and precipitation kinetics of kaolinite at 80 °C and pH 3: The dependence on solution saturation state. Am J Sci 291:649-686
Nagy KL, Lasaga AC (1992) Dissolution and precipitation kinetics of gibbsite at 80 °C and pH 3: The dependence on solution saturation state. Geochim Cosmochim Acta 56:3093-3111
Nancollas GH (1992) The involvement of calcium phosphates in biological mineralization and demineralization processes. Pure Appl Chem 64:1673-1678
Naumov GB, Ryzhenko BN, Khodakovsky IL (1974) Handbook of Thermodynamic Data (translation of Russian report) United States Geological Survey, 328 p
Nielsen AE (1964) Kinetics of Precipitation. Pergamon, Oxford
Nielsen AE (1981) Theory of electrolyte crystal growth - the parabolic rate law. Pure Appl Chem 53:2025-2039
Nielsen AE, Toft JM (1984) Electrolyte crystal growth kinetics. J Cryst Growth 67:278-288
Ochs M (1996) Influence of humified and non-humified natural organic compounds on mineral dissolution. Chem Geol 132:119-124
Oelkers EH, Gislason SR (2001) The mechanism, rates and consequences of basaltic glass dissolution: I. An experimental study of the dissolution rates of basaltic glass as a function of aqueous Al, Si and oxalic acid concentration at 25 °C and pH = 3 and 11. Geochim Cosmochim Acta 65:3671-3681
Oelkers EH, Schott J (1998) Does organic acid adsorption affect alkali-feldspar dissolution rates? Chem Geol 151:235-245
Oelkers EH, Schott J, Devidal J-L (1994) The effect of aluminum, pH, and chemical affinity on the rates of aluminosilicate dissolution reactions. Geochim Cosmochim Acta 58 2011-2024
Olsen AA, Rimstidt DJ (2008) Oxalate-promoted forsterite dissolution at low pH. Geochim Cosmochim Acta 72:1758-1766
Pak CYC, Holt K (1976) Nucleation and growth of brushite and calcium oxalate in urine of stone-formers. Metabolism 25:665-673
Palmer J, Bauer N (1961) Sorption of amines by montmorillonite. J Phys Chem 65:894-895
Palmer RJ Jr, Siebert J, Hirsch P (1991) Biomass and organic acids in sandstone of a weathered building: Production by bacterial and fungal isolates. Microb Ecol 21:253-266
Parks GA (1965) The isoelectric points of solid oxides, solid hydroxides, and aqueous hydroxo complex systems. Chem Rev 65:177-198
Peltzer ET, Bada JL (1981) Low molecular weight α-hydroxy carboxylic and dicarboxylic acids in reducing marine sediments. Geochim Cosmochim Acta 45:1847-1854
Persson P, Axe K (2005) Adsorption of oxalate and malonate at the water-goethite interface: molecular surface speciation from IR spectroscopy. Geochim Cosmochim Acta 69:541-552

Phillips BL, Lee YJ, Reeder RJ (2005) Organic co-precipitates with calcite: NMR spectroscopic evidence. Environ Sci Technol 39:4533-4539

Pokrovsky OS, Golubev SV, Jordan G (2009) Effect of organic and inorganic ligands on calcite and magnesite dissolution rates at 60 °C and 30 atm pCO_2. Chem Geol (in press)

Poulson SR, Drever JI, Stillings LL (1997) Aqueous S-oxalate complexing, oxalate adsorption onto quartz, and the effect of oxalate upon quartz dissolution rates. Chem Geol 140:1-7

Price CA (1996) Stone Conservation: An Overview of Current Research. The Getty Conservation Institute, Los Angeles

Putnis A, Junta J, Hochella MFJ (1995a) Dissolution of barite by a chelating ligand: an atomic force microscopy study. Geochim Cosmochim Acta 59:4623-4632

Putnis A, Putnis CV, Paul JM (1995b) The efficiency of a DTPA-based solvent in the dissolution of barium sulphate scale deposits. Proceedings of the 1995 SPE International Symposium on Oilfield Chemistry, San Antonio, Texas, 773-785 p

Putnis CV, Kowacz M, Putnis A (2008) The mechanism and kinetics of DTPA-promoted dissolution of barite. Appl Geochem 23:2778-2788

Rashad MM, Baioumy HM, Abdel-Aal EA(2003) Structural and spectral studies on gypsum crystals under simulated conditions of phosphoric acid production with and without organic and inorganic additives. Cryst Res Technol 38:433-429

Rolfe PF (1966) Polymers that inhibit the deposition of calcium sulphate: Some interesting conductance observations. Desalination 1:359-366

Rosenstein L (1936) Process of Treating Water. US Patent No. 2,038,316

Saito T, Arsenault AL, Yamauchi M, Kuboki Y, Crenshaw MA (1997) Mineral induction by immobilized phosphoproteins. Bone 21:305-311

Sarig S, Kahana, F (1976) On the association between sparingly soluble carbonates and polyelectrolytes. J Cryst Growth 35:145-152

Scheutjens JMHM, Fleer GJ (1979) Statistical theory of the adsorption of interacting chain molecules. 1. Partition function, segment density distribution, and adsorption isotherms. J Phys Chem 83:1619-1635

Schnitzer M, Kodama H (1976) The dissolution of micas by fulvic acid. Geoderma 15:381-391

Schoonen M, Smirnov A, Cohn C (2004) A perspective on the role of minerals in prebiotic synthesis. AMBIO J Human Eniron 33(8):539-551

Schoonen MAA, Xu Y, Strongin DR (1998) An introduction to geocatalysis. J Geochem Explor 62 201-215

Schott J, Pokrovsky OS, Oelkers EH (2009) The link between mineral dissolution/precipitation kinetics and solution chemistry. Rev Mineral Geochem 70:207–258

Schroth BL, Sposito G (1997) Surface charge properties of kaolinite. Clays Clay Mineral 45:85-91

Schwarzenbach RP, Gschwend PM, Imboden DM (2003) Environmental Organic Chemistry: Illustrative Examples, Problems, and Case Studies. Wiley, Hoboken, New Jersey

Schweda PS (1989) Kinetics of alkali feldspar dissolution at low temperature. *In:* 6th international symposium on Water-Rock interaction. Miles DL (ed), Balkema, 609-612 p

Senesi N, Loffredo E (1999) The Chemistry of Soil Organic Matter. *In:* Soil Physical Chemistry. Sparks DL (ed), CRC Press, p 232-346

Shen Y, Ström L, Jönsson J-Å, Tyler G (1996) Low-molecular organic acids in the rhizosphere soil solution of beech forest (Fagus sylvatica L.) cambisols determined by ion chromatography using supported liquid membrane enrichment technique. Soil Biol Biochem 28:1163-1169

Sigg L, Stumm W (1981) The interaction of anions and weak acids with the hydrous goethite (α-FeOOH) surface. Colloids Surf 2:101-117

Sikiric MD, Furedi-Milhofer H (2006) The influence of surface active molecules on the crystallization of biominerals in solution. Adv Colloid Interface Sci 128-130:135-158

Smith BR, Alexander AE (1970) The effect of additives on the process of crystallization II. Further studies on calcium sulphate (1). J Colloid Interface Sci 34:81-90

Smith E, Hamilton-Taylor J, Davison W, Fullwood NJ, McGrath M (2004) The effect of humic substances on barite precipitation-dissolution behaviour in natural and synthetic lake waters. Chem Geol 207:81-89

Smith RM, Martell AE (1998) NIST Critically Selected Stability Constants of Metal Complexes Database (Version 5.0) NIST Standard Reference Database 46. U.S. Department of Commerce

Solomon DH, Rolfe PF (1966) Polymers that inhibit the deposition of calcium sulphate. Desalination 1:260-266

Sposito G (1984) The Surface Chemistry of Soils. Oxford University Press, New York

Sposito G (1989) The Chemistry of Soils. Oxford University Press, New York

Sprengel C (1826) Über Pflanzenhumus, Huminsäure, und Humussaure Salze. Kastner's Arch. Ges. Nat. 8:145–220

Stevenson FJ, Fitch A (1986) Chemistry of complexation of metal ions with soil solution organics. *In:* Interaction of Soil Minerals with Natural Organics and Microbes, SSSA Special Publ. No. 17. Huang PM, Schnitzer M (eds), Soil Science Society of America, Madison, WI , p 29-58

Stillings LL, Drever JI, Poulson SR (1998) Oxalate adsorption at a plagioclase (An47) surface and models for ligand-promoted dissolution. Environ Sci Technol 32:2856-2864
Stillings LL, Drever JI, Poulson SR, Brantley SL, Yanting S, Oxburgh R (1996) Rates of feldspar dissolution at pH 3-7 with 0-8 m M oxalic acid. Chem Geol 132:79-89
Stoessell RK, Pittman ED (1990) Secondary porosity revisited: the chemistry of feldspar dissolution by carboxylic acids and anions. Am Assoc Petrol Geol Bull 74:1795-1805
Stumm W (1992) Chemistry of the Solid-Water Interface: Processes at the Mineral-Water and Particle-Water Interface in Natural Systems. John Wiley & Sons, Inc, New York
Stumm W, Wollast R (1990) Coordination chemistry of weathering : Kinetics of surface-controlled dissolution of oxide minerals. Rev Geophys 28:53-69
Suess E (1968) Calcium carbonate interaction with organic compounds. Ph.D. dissertation, 153 p, Lehigh University
Suess E (1970) Interaction of organic compounds with calcium carbonate--I. Association phenomena and geochemical implications. Geochim Cosmochim Acta 34:157-168
Sverjensky DA, Sahai N (1996) Theoretical prediction of single-site surface-protonation equilibrium constants for oxides and silicates in water. Geochim Cosmochim Acta 60:3773-3797
Theng BKG (1976) Interactions between montmorillonite and fulvic acid. Geoderma 15:243-251
Theng BKG, Scharpenseel HW (1975) The adsorption of ,4 C-labelled humic acid by montmorillonite. Proceedings International Clay Conference, Mexico City, p 643-653
Thompson M, Wilkins SJ, Compton RG, Viles HA (2003a) Channel flow cell studies on the evaluation of surface pretreatments using phosphoric acid or polymaleic acid for calcite stone protection. J Colloid Interface Sci 259:338-345
Thompson M, Wilkins SJ, Compton RG, Viles HA (2003b) Polymer coatings to passivate calcite from acid attack: polyacrylic acid and polyacrylonitrile. J Colloid Interface Sci 260 204-210
Tombácz E, Libor Z, Illés E, Majzik A, Klumpp E (2004) The role of reactive surface sites and complexation by humic acids in the interaction of clay mineral and iron oxide particles. Org Geochem 35:257-267
Ullman WJ, Welch SA (2002) Organic ligands and feldspar dissolution. *In:* Water-Rock Interactions, Ore Deposits, and Environmental Geochemistry: A Tribute to David A. Crearar. Hellmann R, Wood SA (eds), Geochemical Society Special Publication. No. 7, p 3-35
van der Leeden MC, Kashchiev D, van Rosmalen GM (1993) Effect of additives on nucleation rate, crystal growth rate and induction time in precipitation. J Cryst Growth 130:221-232
van der Leeden MC, van Rosmalen GM (1990) Inhibition of barium sulfate deposition by polycarboxylates of various molecular structures. Prod Eng 5:70-76
van Hees PAW, Lundstrom US, Morth C-M (2002) Dissolution of microcline and labradorite in a forest O horizon extract: the effect of naturally occurring organic acids. Chem Geol 189:199-211
Van Rosmalen GM, Daudey PJ, Marchee WGJ (1981) An analysis of growth experiments of gypsum crystals in suspension. J Cryst Growth 52:801-811
Vandevivere P, Welch SA, Ullman WJ, Kirchman DL (1994) Enhanced dissolution of silicate minerals by bacteria at near-neutral pH. Microb Ecol 27:241-251
Veintemillas-Verdaguer S (1996) Chemical aspects of the effect of impurities in crystal growth. Prog Cryst Growth Charact Mater 32:75-109
Verdoes D, Kashchiev D, van Rosmalen GM (1992) Determination of nucleation and growth rates from induction times in seeded and unseeded precipitation of calcium carbonate. J Cryst Growth 118:401-413
Walton AG (1965) Nucleation of crystals from solution: Mechanisms of precipitation are fundamental to analytical and physiological processes. Science 148:601-607
Wang X-C, Lee C (1993) Adsorption and desorption of aliphatic amines, amino acids and acetate by clay minerals and marine sediments. Mar Chem 44:1-23
Ward DB, Brady PV (1998) Effect of Al and organic acids on the surface chemistry of kaolinite. Clays Clay Mineral 46:453-465
Wei H, Shen Q, Zhao Y, Zhou Y, Wang D, Xu D (2005) On the crystallization of calcium carbonate modulated by anionic surfactants. J Cryst Growth 279:439-446
Weijnen MPC, Marchee WGJ, van Rosmalen GM (1983) A quantification of the effectiveness of an inhibitor on the growth process of a scalant. Desalination 47:81-92
Weijnen MPC, van Rosmalen GM (1985) The influence of various polyelectrolytes on the precipitation of gypsum. Desalination 54:239-261
Weijnen MPC, Van Rosmalen GM (1986) Adsorption of phosphonates on gypsum crystals. J Cryst Growth 79:157-168
Welch SA, Barker WW, Banfield JF (1999) Microbial extracellular polysaccharides and plagioclase dissolution. Geochim Cosmochim Acta 63:1405-1419
Welch SA, Ullman WJ (1992) Dissolution of feldspars in oxalic acid solutions. *In:* 7th international symposium on Water-Rock interaction. Kharaka YK, Maest AS (eds) Balkema, Park City, Utah, p 127-130

Welch SA, Ullman WJ (1993) The effect of organic acids on feldspar dissolution rates and stoichiometry. Geochim Cosmochim Acta 57:2725-2736

Welch SA, Ullman WJ (1996) Feldspar dissolution in acidic and organic acid solutions: Compositional and pH dependence of dissolution rate. Geochim Cosmochim Acta 60:2939-2948

Welch SA, Ullman WJ (2000) The temperature dependence of bytownite feldspar dissolution in neutral aqueous solutions of inorganic and organic ligands at low temperature (5-35 °C) Chem Geol 167:337-354

Welch SA, Vandevivere P (1994) Effect of microbial and other naturally occurring polymers on mineral dissolution. Geomicrobiology J 12:227-238

Weng L, Van Riemsdijk WH, Hiemstra T (2008) Cu^{2+} and Ca^{2+} adsorption to goethite in the presence of fulvic acids. Geochim Cosmochim Acta 72:5857-5870

Weng LP, Koopal LK, Hiemstra T, Meeussen JCL, Van Riemsdijk WH (2005) Interactions of calcium and fulvic acid at the goethite-water interface. Geochim Cosmochim Acta 69:325-339

Wesolowski DJ, Palmer DA (1994) Aluminium speciation and equilibria in aqueous solution: V. gibbsite solubility at 50 °C and pH 3 to 9 in 0.1 molal NaCl solutions, a general model for Al speciation, and analytical methods. Geochim Cosmochim Acta 58:2947-2969

Wieland E, Stumm W (1992) Dissolution kinetics of kaolinite in acidic aqueous solutions at 25 °C. Geochim Cosmochim Acta 56:3339-3355

Wilkins SJ, Compton RG, Viles HA (2001) The effect of surface pretreatment with polymaleic acid, phosphoric acid, or oxalic acid on the dissolution kinetics of calcium carbonate in aqueous acid. J Colloid Interface Sci 242:378-385

Williams FV, Ruehrwein RA (1957) Effect of polyelectrolytes on the precipitation of calcium carbonate. J Am Chem Soc 79:4898-4900

Witkamp GJ, Van der Eerden JP, Van Rosmalen GM (1990) Growth of gypsum : I. Kinetics. J Cryst Growth 102:281-289

Wogelius RA, Walther JV (1991) Olivine dissolution at 25 °C: Effects of pH, CO_2 and organic acids. Geochim Cosmochim Acta 55:943-954

Wolery TJ (1979) Calculation of chemical equilibrium between aqueous solution and minerals: the eq3/6 software package. Lawrence Livermore Laboratory, 41 p

Yoon TH, Johnson SB, Musgrave CB, Brown GEJ (2004) Adsorption of organic matter at mineral/water interfaces: I. ATR-FTIR spectroscopic and quantum chemical study of oxalate adsorbed at boehmite/water and corundum/water interfaces. Geochim Cosmochim Acta 68:4505-4518

Zhang H, Bloom PR (1999) Dissolution kinetics of hornblende in organic acid solutions. Soil Sci Soc Am 63:815-822

Zullig JJ, Morse JW (1988) Interaction of organic acids with carbonate mineral surfaces in seawater and related solutions: I. Fatty acid adsorption. Geochim Cosmochim Acta 52:1667-1678

Reviews in Mineralogy & Geochemistry
Vol. 70 pp. 371-410, 2009

Mineral Precipitation Kinetics

Bertrand Fritz

Laboratoire d'Hydrologie et de Géochimie de Strasbourg
Université de Strasbourg/EOST, CNRS
1 rue Blessig, F-67084 Strasbourg Cedex, France

bertrand.fritz@illite.u-strasbg.fr

Claudine Noguera

CNRS, Institut des Nanosciences de Paris, UMR 7588
140 rue de Lourmel, 75015 Paris, France

INTRODUCTION

Precipitation of solid phases in aqueous solutions which deviate from thermodynamic equilibrium is an ubiquitous kinetic phenomena. It may occur in soft chemistry experiments where the concentration of a solute in the aqueous solution, is prepared in excess with respect to the solubility of a desired compound (Adamson 1960; Markov 1995). As first recognized by Gibbs, the solid phase first appears as small nuclei which subsequently grow. The understanding and control of these nucleation and growth processes is of prime importance in modern technology which aims at producing artificial nano-size objects, such as ultra-thin films, nano-dots, nano-clusters, whether by molecular beam epitaxy methods (Pimpinelli and Villain 1998), chemistry in aqueous solution (Jolivet et al. 2004), electro-deposition (Schmickler 1996), etc. For example, it has recently been possible (Jolivet et al. 2006) to synthesize oxide nano-particles with a high surface to volume ratio displaying interesting properties: catalytic activity, optical, electronic and magnetic properties, liquid crystal properties, etc. The control of many of their size and shape dependent characteristics can be obtained by playing with external parameters, such as the acidity of the solution, the presence of specific anions, the temperature, etc.

In the geochemical context, nucleation and growth processes are also particularly relevant for complex oxides, such as alumino-silicates or clays, which are ubiquitous in our environment, being often produced as secondary nano- to micro-phases in all alteration processes of rock forming minerals: weathering processes near the Earth's surface, diagenetic or hydrothermal processes in the Earth's crust, etc.

The geochemical modeling of water-rock interactions has been intensively developed since the pioneering work of Helgeson and co-workers (1970). The aim of geochemists was to be able to better understand the evolution of the aqueous solution (AS) composition in the different water cycles as summarized by Drever (1997): in weathering systems involving dilute rain waters, spring or well waters (alteration processes, formation and/or evolution of soils); in underground systems at moderate to important depths, with increasing temperatures, commonly up to 100 to 200 °C in hydrothermal systems or in geological reservoirs concerned by diagenetic processes (Berner 1980). In all these systems, rock-forming minerals produced under higher P and T conditions are generally thermodynamically unstable and may dissolve partially or totally (Lasaga et al. 1994). The induced evolution of the fluid composition may then allow the systems to produce secondary mineral phases among which clay minerals are the most common in near surface weathering conditions (Oelkers et al. 1994), or even at greater depth in geological basins or in hydrothermal systems at higher P and T conditions.

1529-6466/09/0070-0008$05.00 DOI: 10.2138/rmg.2009.70.8

The description of out-of-equilibrium processes in the natural environment is much more complex than in the context of soft chemistry. There are, at least, three levels of additional complexity. First, the supersaturation is not fixed at the beginning of an experiment, but rather results from the dissolution of primary minerals. The aqueous solution composition thus not only evolves due to the precipitation of solid particles, but also gets continuously enriched by the ions resulting from primary rock dissolution. Second, the minerals that form very rarely have a fixed chemical composition. The precipitation of solid-solutions is the rule rather than the exception, as clearly evidenced by the multiple substitutions displayed by phyllo-silicate minerals. Finally, most often, water-rock interactions are not closed system processes, but are associated with transport of the aqueous solutions through rock porosities.

Most of the first experimental and modeling studies concerned the irreversible dissolution of minerals (Busenberg 1978; Stumm and Wieland 1990; Huertas et al. 1999). A recent evolution of the geochemical modeling also concerns the so called "coupled models" which combine the transport and the reactions processes in water-rock systems where the fluids are circulating. This is the aim of various works (Lichtner 1985) and codes like PHREEQC (Parkhurst and Appelo 1999), or CHESS (van der Lee 1998; van der Lee and de Windt 2000) and CrunchFlow (Steefel and Lasaga 1994; Steefel et al. 2005; Steefel 2008). This is also the aim of the code KIRMAT (Gérard et al. 1998a,b) with recent applications to long term behaviour of clay barriers in nuclear waste storages (Montes et al. 2005, Marty et al. 2008).

Much less information exist on the precipitation of secondary phases. They are more difficult to follow in the aqueous reacting phase due to the small amounts of minerals generally precipitated among dominant preexisting and reacting phases. Field work results integrate long term complex sequences of precipitation, from which kinetic laws are difficult to extract. For this reason, geochemical models first considered equilibrium conditions for the precipitation of secondary phases (Helgeson et al. 1970 in the code PATH1; Fritz 1981) or highly simplified kinetic models for precipitation without simulating the nucleation step (Madé et al. 1994 in the code KINDIS). The same is true as regards solid solution precipitation (Glynn and Reardon 1990, Blanc et al. 1997, Nourtier-Mazauric et al. 2005). Very rarely have the first stages of nucleation been partially (Steefel and van Cappellen 1990) or totally (Noguera et al. 2006a,b, 2008, 2009; Fritz et al. 2008, 2009) accounted for.

It is the aim of this course to give some theoretical background on nucleation and growth processes, for minerals of both fixed and variable chemical composition, and to describe how they are accounted for in the recently conceived NANOKIN geochemical code. Then, several examples of applications will be given.

THEORETICAL APPROACH

A typical system under consideration involves a set of minerals M_i which interact with an aqueous solution (AS), in a closed medium. Let M be a given mineral with a fixed composition, obeying the following equilibrium equation:

$$M \leftrightarrow \sum_i \alpha_i E_i \tag{1}$$

with α_i the well-defined stoichiometric coefficients of the reaction. Let $[E_i]$ and $[M]$ be respectively the activities of the aqueous species E_i and of M in the solid phase. If equilibrium is achieved for M, its equilibrium constant K reads:

$$K = \frac{\prod_i [E_i]_{eq}^{\alpha_i}}{[M]_{eq}} \tag{2}$$

If M is out of equilibrium with the solution, its ion activity product Q is different from K and the saturation state of the solution I with respect to M is different from 1:

$$Q = \frac{\prod_i [E_i]^{\alpha_i}}{[M]} \qquad I = \frac{Q}{K} \tag{3}$$

Depending upon the value of I, precipitation ($I > 1$) or dissolution ($I < 1$) are thermodynamically allowed.

For minerals with variable composition, a model of solid-solution (SS) can be applied, each time that the substitutions do not induce a change of the crystallographic structure. A mineral M which involves a single type of substitution of iso-valency ions, such as in $Ba_{1-x}Sr_xCO_3$ or $Ba_{1-x}Sr_xSO_4$, may be written $A_{1-x}B_xC$ ($0 < x < 1$), with the valencies of A, B and C species equal to $+q$, $+q$, and $-q$, respectively. AC and BC are the end-members of the SS. The products of dissolution of one formula unit (f.u.) of $A_{1-x}B_xC$ are:

$$A_{1-x}B_xC \leftrightarrow (1-x)A^{q+} + xB^{q+} + C^{q-} \tag{4}$$

They are equivalent to those resulting from the dissolution of $(1-x)$ f.u. of AC and x f.u. of BC:

$$AC \leftrightarrow A^{q+} + C^{q-} \quad ; \quad BC \leftrightarrow B^{q+} + C^{q-} \tag{5}$$

The solubility products of AC and BC, denoted K_{AC} and K_{BC} are associated to changes in Gibbs free energy during dissolution, ΔG_{AC} and ΔG_{BC}, in the following way:

$$\begin{aligned} K_{AC} &= \exp\left(-\Delta G_{AC} / k_B T\right) = [A^{q+}]_{eq}[C^{q-}]_{eq} \\ K_{BC} &= \exp\left(-\Delta G_{BC} / k_B T\right) = [B^{q+}]_{eq}[C^{q-}]_{eq} \end{aligned} \tag{6}$$

in which $[X]_{eq}$ refers to the activity of the aqueous species X at equilibrium, k_B is the Boltzmann constant and T the temperature. The change of Gibbs free energy $\Delta G(x)$ for the SS dissolution reads:

$$\Delta G(x) = \Delta H(x) - k_B T\left(x \ln x + (1-x)\ln(1-x)\right) \tag{7}$$

The last term is the mixing entropy of A and B, upon the assumption of a total disorder in the $A \rightarrow B$ substitution. $\Delta H(x)$ is the enthalpy change, whose most general dependence upon x is given by the Margules development (Wagner 1952):

$$\Delta H(x) = (1-x)\Delta G_{AC} + x\Delta G_{BC} + Ax(1-x) + \ldots. \tag{8}$$

Keeping only the two first terms is relevant for *ideal* solid solutions, in which $\Delta H(x)$ is the weighted average of the AC and BC enthalpies of dissolution. The additional terms represent the enthalpy of mixing, which originates from A-B interactions. In *regular* solid solutions, these interactions are pairwise and only involve the mean probability to find an AB pair. The mixing enthalpy is then proportional to $x(1-x)$.

The following will be concerned with the behavior of ideal solid solutions, whose solubility product $K(x)$ thus reads:

$$K(x) = \exp\left(-\frac{\Delta G(x)}{k_B T}\right) = K_{AC}^{1-x} K_{BC}^{x} x^x (1-x)^{(1-x)} \tag{9}$$

The saturation state $I(x)$ of the solution, with respect to $A_{1-x}B_xC$, is then equal to:

$$I(x) = \frac{I_{AC}^{(1-x)} I_{BC}^{x}}{x^x (1-x)^{(1-x)}} = \left(\frac{I_{AC}}{1-x}\right)^{1-x} \left(\frac{I_{BC}}{x}\right)^{x} \tag{10}$$

I_{AC} and I_{BC} are the saturation states of the solution with respect to the AC and BC end-members of the SS, respectively. It should be noted that, while I_{AC} and I_{BC} are instantaneous functions of time, I depends upon both time and composition. Usually, $I(x)$ presents a maximum for a value $0 < x_0 < 1$. The SS is in equilibrium with the solution if $I(x_0) = 1$, in addition to the condition $dI(x_0)/dx = 0$. Using Equation (10), these two conditions are equivalent to:

$$I_{BC} = x_0 \quad ; \quad I_{AC} = 1 - x_0 \quad ; \quad I_{AC} + I_{BC} = 1 \tag{11}$$

According to Equation (11), at equilibrium, each end-member is represented in the SS with a molar fraction equal to its saturation state.

Nucleation

When the saturation state of the AS exceeds 1, precipitation is thermodynamically allowed. This means that a first order phase transformation takes place. However, as first recognized by Gibbs, the solid phase does not immediately gather all the excess of solute. Small nuclei first appear, which contain a few *growth units* (molecular constituents of same composition as the solid phase) in coexistence with the AS. These nuclei, being small, have a very large surface to volume ratio (typically equal to $4\pi\rho^2/(4\pi\rho^3/3) = 3/\rho$ for spherical particles of radius ρ), which means that there is a very large contribution of the surface energy to their free energy of formation. It is the competition between surface and volume effects which drives the kinetics of nucleation.

One distinguishes two families of nucleation theories: the atomistic theory, appropriate when large super-saturations are considered, because nuclei then consist of a small number of growth units, and the classical (or capillary) nucleation theory (CNT), which involves a continuous description of the solid phase and is well suited when the nuclei are larger, associated to a small or moderate degree of supersaturation (Adamson 1960; Markov 1995). The atomistic approach is very material dependent and will not be described here. The CNT is more generic, and, through the use of a few parameters, can be applied to a larger class of situations. It is described in the following for three-dimensional (3D) homogeneous nucleation of spherical particles. The extension to heterogeneous nucleation (i.e., nucleation of particles on foreign solids) and non-spherical particle shapes is given in Appendix 1, while two-dimensional (2D) nucleation is developed in Appendix 2. Nucleation of particles of fixed composition is first considered, then the formalism is generalized to account for ideal solid solution nucleation.

Particles of fixed composition. When the precipitation of a single mineral of fixed composition and solubility product K is considered, the change in Gibbs free energy in the nucleation process may be written:

$$\Delta G = -nk_BT\ln I + 4\pi\rho^2\sigma \tag{12}$$

The first term ($-nk_BT\ln I$) is equal to the difference $\delta\mu$ between the chemical potentials of n growth units in solution and in the solid (hereafter n will be referred to as the particle "size"). It is negative when $I > 1$, which means that the *infinite* solid phase is thermodynamically stable. However, for finite size particles, one has to take into account surface effects, included in the second term, which is positive and proportional to the particle surface area and surface energy σ (assumed isotropic). This term can be expressed as a function of n, using the volume v of a growth unit in the solid, such that $nv = 4\pi\rho^3/3$.

After establishment of the super-saturation state, during some time interval, in the AS, growth units constantly clusterize, forming embryos which can subsequently dissociate. These fluctuations are associated to energy fluctuations driven by the dependence of ΔG upon size. When $I > 1$, ΔG presents a maximum ΔG^* as a function of n, which is the maximum activation barrier that nuclei have to overcome during their size fluctuations (see Fig. 1a). This maximum point defines the characteristics of the *critical nuclei*, whose size n^* and radius ρ^* are thus

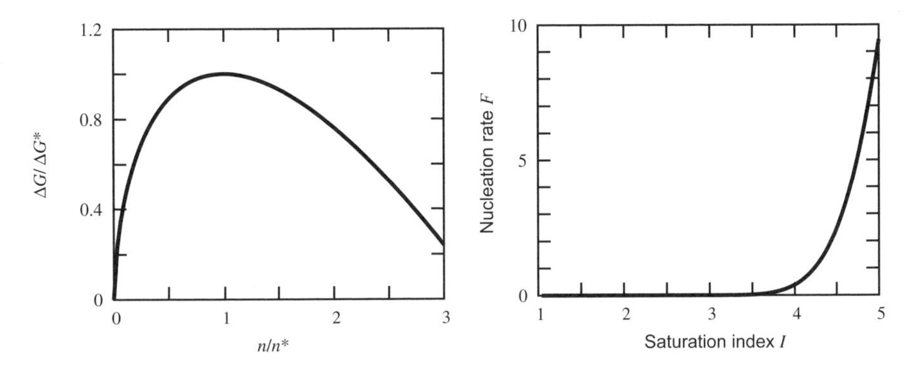

Figure 1. (*left*) Plot of the change in Gibbs free energy for the nucleation of a spherical nucleus of size n (n = number of growth units). (*right*) Variation of the nucleation rate F with the AS saturation index I.

equal to:

$$n^* = \frac{2u}{\ln^3 I} \quad \rho^* = \frac{2\sigma v}{k_B T \ln I} \quad \text{with } u = \frac{16\pi\sigma^3 v^2}{3(k_B T)^3} \tag{13}$$

The maximum point of ΔG represents an *unstable* equilibrium with the surrounding phase. The addition or substraction of one growth unit to the critical nuclei irreversibly drive them towards increasing sizes or resorption, respectively.

Equation (13) is the equation of Gibbs-Thomson. By inverting it, it is found that the environment with which a particle of size n or radius ρ is in equilibrium, is characterized by a value of the saturation state I_{eff} equal to:

$$I_{\text{eff}} = \exp\left(\frac{2\sigma v}{k_B T \rho}\right) = \exp\left(\frac{2u}{n}\right)^{1/3} \tag{14}$$

I_{eff} may be written $I_{eff} = K(\rho)/K$, which defines the solubility $K(\rho)$ of a particle of finite radius ρ, and shows that it is always larger than the solubility K of an infinite crystal. Such a relationship between K and ρ was for example confirmed in a recent study of calcite precipitation (Kile et al. 2000).

More generally, an equivalent relationship applies to the Laplace excess pressure $P(\rho) = P_0\exp(2\sigma v/k_B T\rho)$ inside a droplet of radius ρ which condensates in the vapor phase, and to the melting temperature $T_m(\rho)$ of finite size particles (Noguera et al. 2006a). $K(\rho)$, $P(\rho)$ and $T_m(\rho)$ converge to their macroscopic values K_0, P_0 and T_m in the limit of infinite sizes.

The nucleation barrier ΔG^* reads:

$$\Delta G^* = k_B T \frac{u}{\ln^2 I} = \frac{1}{3} 4\pi\rho^{*2}\sigma \tag{15}$$

It is equal to one third of the surface energy and this result also applies to more complex particle shapes. Nucleation occurs more easily (smaller ΔG^*) if v is small (smaller growth units nucleate more easily), σ is small (softer materials), or I is high (larger departure from equilibrium favors nucleation).

As mentioned above, the nuclei of the solid phase appear and resorb randomly, with large size fluctuations. After some transient time (Frenkel 1946), a steady state regime of nucleation is established. In order to estimate the stationary nucleation rate F, one has to calculate the average (over a statistical ensemble) nuclei number per unit time and volume (or per kg H_2O), taking into account the elementary probability ω of attachment of growth units on the nuclei. The attachment involves a desolvation of the growth units, by which they break their bond with the solvent molecules. The stationary nucleation rate can be written:

$$F = \omega c Z \exp\left(-\frac{\Delta G^*}{k_B T}\right) \tag{16}$$

with c the concentration of growth units and Z the so-called Zeldovich factor. The latter is inversely proportional to the curvature of ΔG at its maximum:

$$Z \propto \left(\frac{\Delta G^*}{3k_B T n^{*2}}\right)^{\frac{1}{2}} \tag{17}$$

A similar factor appears in the transition state theory for chemical reactions.

The pre-factor $F_0 = \omega c Z$ of the exponential in Equation (16) mostly contains slowly varying quantities. There have been attempts to theoretically estimate it for specific systems. However, in most cases, it has resulted in huge (several orders of magnitude) discrepancies with measured values. For this reason, it is often assumed to be a constant, whose value has to be empirically determined.

For not too large a supersaturation, F monotonically increases with I (see Fig. 1b). There exists a metastable zone (sometimes called latency zone) for $1 < I < I_c$, in which F is exponentially small; I_c is arbitrarily defined by the condition that $F(I_c) = 1$ particle par second in the reference volume:

$$\ln I_c = \sqrt{\frac{u}{\ln F_0}} \tag{18}$$

In the range $1 < I < I_c$, nucleation is thermodynamically allowed but it is unobservable in reasonable time scales. When I exceeds I_c, the nucleation rate suddenly increases, which corresponds to the observed bloom of particle nucleation.

Particles of variable composition. The treatment of nucleation for particles of variable composition follows similar lines. However, studies of nucleation of binary droplets (Reiss 1950; Doyle 1961; Heady and Cahn 1972; Reiss and Shugard 1976; Wilemski 1984; Laaksonen et al. 1993, 1999; Noppel et al. 2002; Gaman et al. 2005) and alloy particles (Legrand et al. 1994; Soisson and Martin 2000; Varadarajan et al. 2001; Senthil-Kumar et al. 2004; Los and Matovic 2005; Moreno et al. 2006; Nagano and Enomoto 2006; Shcheritsa et al. 2006) have shown that, as soon as the surface energy σ of the nucleus varies with the composition x, there are inhomogeneities of composition throughout the nuclei, and some (algebraic) surface enrichment takes place. This case is treated in Appendix 3. Here, we present the simplest approach which assumes that σ is independent of x.

The change in Gibbs free energy $\Delta G(\rho,x)$ in the formation of a spherical nucleus then reads:

$$\Delta G(\rho,x) = -(1-x)n(x)k_B T \ln\left(\frac{I_{AC}}{1-x}\right) - xn(x)k_B T \ln\left(\frac{I_{BC}}{x}\right) + 4\pi\rho^2\sigma(x) \tag{19}$$

For later use, one has explicitly written the changes in AC and BC chemical potentials, (for example, $\delta\mu_{BC} = -k_B T\ln(I_{BC}/x)$, the equivalent of $-k_B T\ln I$ in Eqn. 12), weighted by the respective amounts of the end-members in the nucleus, rather than the equivalent expression $-n(x)k_{BT}\ln I(x)$. The size and volume of the particles depend upon composition: $n(x)v(x) = 4\pi\rho^3/3$. Consistently with the assumption of ideality for the SS, the following derivation will take $v(x)$ as a linear function of x, equal to $v(x) = (1-x)v_{AC} + xv_{BC}$.

When $I(x) > 1$, the size and composition of the critical nuclei are determined by the condition that $\Delta G(\rho,x)$ is maximum with respect to ρ, as for the nucleation of minerals with fixed composition, and by the additional condition that the nucleation rate is maximum with respect to x (assuming that the flow of embryos through size and composition space is confined to a path through this point only; Reiss and Shugard 1976). The maximum of $\Delta G(\rho,x)$ with respect to ρ is obtained for $\rho = \rho_m(x)$ and $n(x) = n_m(x)$ such that:

$$\rho_m(x) = \frac{2\sigma v(x)}{k_B T \ln I(x)}; \; n_m(x) = \frac{2u(x)}{\ln^3 I(x)} \quad \text{with} \quad u(x) = \frac{16\pi\sigma^3 v(x)^2}{3(k_B T)^3} \tag{20}$$

It is equal to:

$$\Delta G_m(x) = \frac{1}{3} 4\pi\rho_m^2(x)\sigma \tag{21}$$

an expression similar to that found for nuclei of fixed composition (Eqn. 15).

The composition $x = x^*$ of the critical nuclei is obtained from the maximum of the nucleation rate $F(x) = F_0(x)\exp\{-\Delta G_m(x)/k_BT\}$. If the pre-factor $F_0(x)$ is composition independent, this condition is equivalent to minimizing the nucleation barrier $\Delta G_m(x)$, in which case the critical nuclei correspond to a saddle point of $\Delta G(\rho,x)$. However, more generally, x obeys the implicit equation:

$$\frac{d \ln F_0(x)}{dx} = \frac{1}{k_B T} \frac{d\Delta G_m(x)}{dx} \tag{22}$$

which also reads:

$$\left(\frac{I_C}{1-x^*}\right)^{v_{BC}} = \left(\frac{I_M}{x^*}\right)^{v_{AC}} \exp\left(\frac{v(x^*)}{n_m(x^*)} \frac{d \ln F_0(x^*)}{dx}\right) \tag{23}$$

The exponential term is equal to 1 if $F_0(x)$ is a constant. Even in this case, x^* generally differs from the equilibrium composition x_0 (Eqn. 11) unless the end-members AC and BC of the SS have identical molar volumes v_{AC} and v_{BC}. For example, in a study of $(Ba,Sr)SO_4$ there is a quantitative example in which the composition x^* for the nucleation maximum does not coincide with the maximum x_0 of $I(x)$ (Pina et al. 2000). In the following, ρ^* and n^* will denote $\rho_m(x^*)$ and $n_m(x^*)$, respectively.

Finally, the nucleation barrier and the nucleation rate can be estimated:

$$\Delta G^* = \frac{4\pi}{3}\rho^{*2}\sigma \quad ; \quad F = F_0(x^*)\exp\left(\frac{-\Delta G^*}{k_B T}\right) \tag{24}$$

All quantities related to a given critical nucleus: ρ^*, x^*, n^*, ΔG^* and F, depend upon the time t_1 at which nucleation has occurred. This time dependence comes from the instantaneous values of the saturation state $I(t_1)$ in the case of particles with fixed composition, or from the instantaneous values of the saturation states $I_{AC}(t_1)$ and $I_{BC}(t_1)$ entering the equation which determines $x^*(t_1)$, for particles of variable composition.

Growth

Since critical nuclei are in unstable equilibrium with the solution (ΔG is *maximum*), they either grow, if they become super-critical ($\rho > \rho^*$) or resorb, if they become sub-critical ($\rho < \rho^*$), depending upon the instantaneous value of the saturation state I. A growth law which correctly describes such processes has to be size-dependent, but its expression depends upon the rate limiting process: diffusion in the liquid or the gaseous phase, continuous interfacial effects, two-dimensional nucleation on flat faces, spiral growth (Burton et al. 1951; Baronnet 1982; Parbhakar et al. 1995). In the following, only the two first processes will be considered.

Particles of fixed composition. In a large range of supersaturation values and when the particle surface is rough at the atomic scale, so that all its surface sites are available for incorporation of growth units, growth is limited by interfacial reactions. In that case, the growth rate can be assumed to be proportional to $I - I_{eff}$ in first approximation, i.e. with κ the proportionality constant:

$$\frac{d\rho}{dt} = \kappa\left[I - \exp\left(\frac{2\sigma v}{k_B T \rho}\right)\right] \tag{25}$$

A similar expression would apply to diffusion limited growth with κ replaced by κ/ρ. This algebraic law in Equation (25) allows particle to increase in size (if $I > I_{eff}$, i.e., $\rho > \rho^*$) as well as to resorb (if $I < I_{eff}$, i.e., $\rho < \rho^*$), and thus, it naturally encompasses Ostwald ripening (Myhr and Grong 2000; Ratke and Voorhees 2002). Indeed, Ostwald ripening, also named coarsening, involves the disappearance of sub-critical particles and the growth of super-critical ones. It is recognized to be one of the most efficient mechanisms for the decrease of the particle number in the ageing phase. However, the examples of applications given in the second part of this course will demonstrate that resorption is also at work since the first stages of the dynamics of precipitation.

For particles much larger than the critical nuclei, Equation (25) reduces to:

$$\frac{d\rho}{dt} = \kappa(I - 1) \tag{26}$$

which can be used for the dissolution of pre-existing (macroscopic) minerals in the system (if $I < 1$). Written as in Equation (26), and assuming the principle of micro-reversibility to apply, κ represents a macroscopic *dissolution* constant. Such an equation has also been used to account for the growth of small particles in most previous geochemical codes (when $I > 1$). Consequences of such a drastic approximation for the precipitation characteristic times will be discussed in the second part of the course.

The knowledge of the global time evolution of the system requires keeping track of the time dependence of the saturation indices $I_j(t)$ of the AS with respect to all minerals present, and of the particle populations at all times. At a given time t, for a given mineral M, the latter consists in particles which have nucleated at all anterior times $t_1 < t$ and have grown in the time interval between t_1 and t. It is thus necessary to introduce the two-variable functions $n(t_1,t)$ and $\rho(t_1,t)$ giving the size and the radius at time t of particles nucleated at t_1. By using Equations (13) and (25) and the dimensionless parameter $w = 3\kappa(4\pi/3v)^{1/3}$, their time evolution is given by (t' the time integration variable):

$$n(t_1,t) = \frac{2u}{\ln^3 I(t_1)} + \int_{t_1}^{t} w n^{2/3}(t_1,t')\left(I(t') - \exp\left[\frac{2u}{n(t_1,t')}\right]^{1/3}\right)dt' \tag{27}$$

If growth is limited by diffusion in the solution, Equation (27) transforms into:

$$n(t_1,t) = \frac{2u}{\ln^3 I(t_1)} + \int_{t_1}^{t} w' n^{1/3}(t_1,t')\left(I(t') - \exp\left[\frac{2u}{n(t_1,t')}\right]^{1/3}\right)dt' \tag{28}$$

with $w' = 3\kappa\left(4\pi/3v\right)^{2/3}$.

In a closed system, the nucleation and (algebraic) growth of M exert a feedback effect on the AS, which modifies not only the saturation state with respect to M but also with respect to other minerals. At time t, the total number $N(t)$ of growth units of M which have changed phase, is equal to:

$$N(t) = \int_0^t F(t_1)\big(n(t_1,t) - 1\big)\, dt_1 \tag{29}$$

with the instantaneous nucleation rate $F(t_1)$ given by Equation (16), and $n(t_1,t) - 1$ is written rather than $n(t_1,t)$ to signal that more than one growth unit is necessary to tell if a new phase is formed.

From the knowledge of $N_j(t)$ for all minerals M_j, it is possible to estimate the activity of all ions in solution, using matter and charge conservation equations and an aqueous speciation model, and thus deduce the ion activity products Q_j and saturation states I_j for all minerals M_j.

Particles of variable composition. The additional difficulty in treating the case of SS precipitation lies in the determination of the composition of the deposited layers as the particle radius grows. For this, one has to write the variation of Gibbs free energy for the deposition of growth units at the surface of the particles. It is formally similar to that written for the critical nucleus, but interfacial ionic activities have to be used in the expressions of I, I_{AC} and I_{BC}. In the following, they will be denoted I_i, I_{ACi} and I_{BCi}, respectively. Thus:

$$\Delta G(\rho,x) = -(1-x)n(x)k_BT\ln(\frac{I_{ACi}}{1-x}) - xn(x)k_BT\ln(\frac{I_{BCi}}{x}) + 4\pi\rho^2\sigma \tag{30}$$

Assuming local equilibrium at the particle-solution interface amounts to canceling the derivatives of $\Delta G(\rho,x)$ with respect to ρ and x, which yields the equalities:

$$\ln I_i(x) = \frac{2\sigma v(x)}{k_BT\rho} \quad ; \quad \left(\frac{I_{BCi}}{x}\right)^{\nu_{AC}} = \left(\frac{I_{ACi}}{1-x}\right)^{\nu_{BC}} \tag{31}$$

In order to fully determine the interfacial activities and the composition of the deposited layer, the condition of mass balance at the surface of the particle must be added to Equation (31) (Maugis and Gouné 2005):

$$\begin{aligned} D_A\big([A^{+q}]_i - [A^{+q}]\big) &= (1-x)D_C\big([C^{-q}]_i - [C^{-q}]\big) \\ D_B\big([B^{+q}]_i - [B^{+q}]\big) &= xD_C\big([C^{-q}]_i - [C^{-q}]\big) \end{aligned} \tag{32}$$

in which D_A, D_B and D_C are the diffusion coefficients of the various species in solution. Equations (31) and (32) allow a full characterization of the composition of the particles during growth.

They take a simplified form if cation diffusion in the AS is much faster than anion diffusion (D_A, $D_B >> D_C$). Then, Equation (32) reduces to:

$$[A^{+q}]_i = [A^{+q}] \quad ; \quad [B^{+q}]_i = [B^{+q}] \tag{33}$$

and Equation (31) yields the interfacial activity of C anions:

$$\ln\left(\frac{[C^{-q}]_i}{[C^{-q}]}\right) = \frac{2\sigma v(x)}{k_BT}\left(\frac{1}{\rho} - \frac{1}{\rho_m(x)}\right) \tag{34}$$

The composition x of the growing layer can then be expressed as a function of x^* and the function $\rho_m(x)$ (Eqn. 20):

$$\nu_{AC}\ln\frac{x^*}{x} - \nu_{BC}\ln\frac{1-x^*}{1-x} = \frac{2\sigma v(x)}{k_BT}\left(\frac{1}{\rho} - \frac{1}{\rho_m(x)}\right)(\nu_{BC} - \nu_{AC}) + \frac{v(x^*)}{n_m(x^*)}\cdot\frac{d\ln F_0(x^*)}{dx} \tag{35}$$

Usually, and especially at low temperatures, solid state diffusion is very slow compared to all other characteristic times and can be neglected. The composition of a deposited layer thus remains fixed once formed. The solid particles display composition profiles (x as a function of ρ) given by Equation (35).

In the regime of increasing particle size ($d\rho(t_1,t)/dt > 0$)), the rate equation used for the growth of particles of fixed composition can be straightforwardly generalized. If growth is limited by interfacial reactions, it reads:

$$\frac{d\rho(t_1,t)}{dt} = \kappa\left(I\left(t,x(t_1,t)\right) - \frac{\exp 2\sigma v\left(x(t_1,t)\right)}{k_B T \rho(t_1,t)} \right) \tag{36}$$

while, if it is limited by diffusion in the AS, κ should be replaced by $\kappa/\rho(t_1,t)$. In Equation (36), it is the saturation state relative to the mineral of composition $x(t_1,t)$, given by Equation (10), which has to be used.

During resorption ($d\rho(t_1,t)/dt < 0$), layers formed at anterior times are progressively dissolved. A layer of radius $\rho(t_1,t)$ which reaches the particle/solution interface at a given time t, had been deposited at t_2 such that

$$\rho(t_1,t_2) = \rho(t_1,t) \tag{37}$$

t_2 is specific to the particle and to the time t and should thus be written $t_2(t_1,t)$. At time t_2, the layer composition was equal to $x\{t_1,t_2(t_1,t)\}$. Consequently, in the resorption regime, using the short-hand notation $\tau = \{t_1,t_2(t_1,t)\}$, the growth rate reads:

$$\frac{d\rho(t_1,t)}{dt} = \kappa\left(I\left(t,x(\tau)\right) - \frac{\exp 2\sigma v\left(x(\tau)\right)}{k_B T \rho(t_1,t)} \right) \tag{38}$$

At a given time t, the particle population consists in all the particles which have nucleated at times $t_1 < t$, and whose nucleation rates, size and compositions are equal to $F(t_1)$, $n(t_1,t)$ and $x(t_1,t)$, respectively. From them, the amounts $N_A(t)$, $N_B(t)$ and $N_C(t)$ of A^{+q}, B^{+q} and C^{-q} species which have been withdrawn from the AS at time t can be calculated. For example, $N_A(t)$ reads:

$$N_A(t) = \int_0^t F(t_1)(n^*(t_1)-1)(1-x(t_1,t))dt_1 + \int_0^t F(t_1)dt_1 \int_{t_1}^t dt_2 \frac{4\pi\rho(t_1,t_2)^2}{v(t_1,t_2)} \frac{d\rho(t_1,t_2)}{dt_2}(1-x(t_1,t_2)) \tag{39}$$

A similar expression holds for $N_B(t)$ with $\{1 - x(t_1,t)\}$ replaced by $x(t_1,t)$ and $N_C(t) = N_A(t) + N_B(t)$. From these quantities, one can calculate all ion activities in the AS and the values of the bulk and local saturation states $I_{AC}(t)$, $I_{BC}(t)$, $I_{ACi}(t_1,t)$, $I_{BCi}(t_1,t)$, $I(x,t)$ and $I_i(t_1,t)$.

The equations giving $N_A(t)$, $N_B(t)$, $N_C(t)$, together with those which fix $x^*(t_1)$, $n^*(t_1)$, $\Delta G^*(t_1)$, $F(t_1)$, $x(t_1,t)$, and $\rho(t_1,t)$ form a complete set which, together with the speciation equations, allows the full determination of the precipitation dynamics.

The NANOKIN code

The above equations have been embedded into a computer code called NANOKIN, resulting from a transformation of the KINDIS code (Madé et al. 1990, 1994). The set of nucleation and growth equations are solved by dividing the time axis in finite intervals and using an explicit method to approximate the solution of this integro-differential system. All particles nucleated during the same time interval (same sub-class) are assumed to have the same size and the same time evolution. At each time step t_n, the change of size of the particles formed at all previous time steps $t_i < t_n$ is calculated, and the new sub-class of particles formed at t_n is added.

In order to speed-up the execution and limit the necessary memory size, it is possible to group a new sub-class of particles to an older one, provided that the particle sizes differ by less than a given quantity. In order to make the algorithm more efficient, flexible time steps are adopted: at each step, the next time interval $\Delta t_n = t_{n+1} - t_n$ is evaluated from the slopes of ion activities in solution and particle size, so that they do not exceed a given percentage initially chosen. It is also possible to fix an upper limit to Δt_n in order to increase the precision of the calculation, during some periods of the simulation. A third optimization concerns the limit of very low nucleation frequencies, namely when the saturation index gets lower than the critical supersaturation. Below a given infinitesimal value, the nucleation rate is set to zero, an no further new particle class is considered.

Compared to KINDIS, NANOKIN represents an advance on various aspects: 1) nucleation step: KINDIS assumed the sudden appearance of a given quantity of secondary minerals as soon as I exceeds the critical value I_c, while NANOKIN takes into account the correct relationship between critical nucleus size and I; 2) growth: KINDIS used a size independent growth law. In addition, NANOKIN produces new information: instead of predicting an amount of a given mineral phase versus time, the mineral phase is thoroughly described as a population of particles, whose first steps of nucleation are accounted for. At present, NANOKIN can account for the dissolution of primary rocks, the precipitation of minerals of fixed or variable composition (limited to two end-member ideal solid solutions), and the exchange with a gas phase (partial pressures of CO_2, O_2, H_2). A number of functionalities exist: choice of the particle shapes (not limited to spheres), heterogeneous nucleation driven by an adhesion energy term (rather than systematically yielding a semi-hemispherical shape), 2D nucleation and growth.

NANOKIN allows, either to directly compute the time evolution of the crystal size distribution function (CSD) as a function of time or to fully characterize the particle population (if all classes of particles are followed). In the latter case, the code keeps track of the number of particles nucleated at each time t_1 and their subsequent evolution of size, via the function $n(t_1,t)$. It is thus possible, at any time t, not only to know how many particles have a given size, but also at which time(s) t_1 they had been produced. This additional information is of prime importance, for example, when one wishes to understand the genetic conditions prevailing during the crystallization of magmatic and metamorphic rocks (Baronnet 1984; Cabane et al. 2005).

In addition to the thermodynamic data bases, NANOKIN needs a kinetic one, including surface energies, dissolution constants and pre-factors of nucleation rates. Such a kinetic data base presently does not exist. Applications to mineral precipitation in the following section will exemplify the difficulty associated with this point, together with the sensitivity of the results to the choice of the parameters. No doubt that kinetic data bases will progressively be enriched, in a way similar to what happened to thermodynamic data bases in the 80s and 90s.

APPLICATIONS

In the following, we develop several applications pertaining to the field of soft chemistry and geochemistry. Presented in order of increasing complexity, they aim at introducing the characteristics of the precipitation dynamics, in as a pedagogical manner as possible: competition between nucleation, growth and resorption to fix the evolution of the AS composition; shapes of CSD depending upon growth laws; competition between the precipitation of two or more minerals; composition profile of SS particles as a function of end-member solubilities, etc.

Precipitation of a single mineral phase of fixed composition

The simplest possible scenario concerns the precipitation of a single mineral, in a solution which is prepared in an initial super-saturation state and when this mineral is associated to a single type of aqueous species (e.g., SiO_2 associated to H_4SiO_4). Such a situation can be met in lab experiments, for example as in the recent work by Tobler et al. (2008) who studied the initial steps of nucleation and growth of silica nanoparticles by *in situ* Small angle X-Ray scattering.

General dynamics characteristics. Figure 2 displays a typical time evolution of the saturation index of the AS. In this example (Noguera et al. 2006b), I was assigned an initial value $I(t = 0)$ about three times larger than I_c. The initial nucleation rate is thus high and induces a bloom

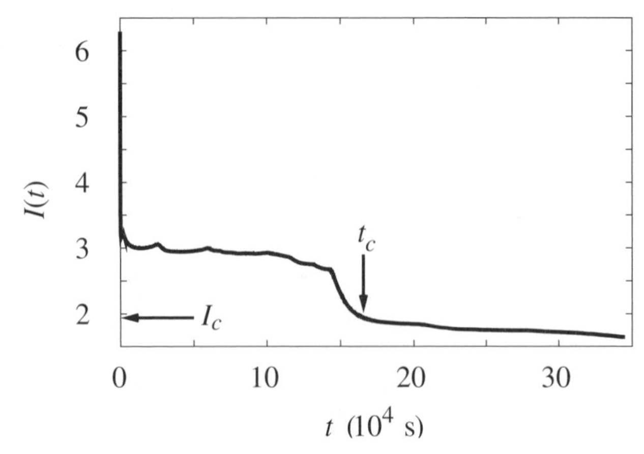

Figure 2. Time dependence of the AS saturation index $I(t)$ (t in seconds) (obtained with $u = 24$, $w = 2 \times 10^{-4}$, $J = 1.26 = 1.26 \times 10^3$ and $S_0 = 6.3$). [Used with permission of Elsevier, from Noguera et al. (2006), *J. Cryst. Growth*, Vol. 297, Fig. 1, p. 189]

of solid particles. This first step of the dynamics is revealed by a rapid decrease of I, on a short time interval, during which growth is not yet active. As I becomes smaller, the initial size of the critical nuclei grow $\{n(t_1,t_1) = 2u/\ln^3 I(t_1)\}$, and by and by growth starts being more efficient.

When this happens, a second regime appears which corresponds to a quasi-plateau in the $I(t)$ curve. Two processes are at work simultaneously at this stage: the formation of critical nuclei and the size evolution of the particles which have been previously formed. Figures 3a-d illustrate the particle population size $n(t_1,t)$ (represented as a function of t_1, time of formation of the particles) at four successive times t: $t_2 < t_3 < t_4 < t_5$. Also represented is the initial size $n(t_1,t_1)$ of the particles at their nucleation time, which allows to visualize size variations between t_1 and t.

What Figure 3 clearly shows is that most of the particles resorb, in particular those formed at the beginning of the dynamics, while only some classes (denoted #1 and #2), formed during finite time intervals in this plateau regime, are able to grow. Such opposite behaviors between particles formed at different times originates from the size-dependent growth law, which states that only super-critical particles $\{n(t_1,t) > n(t,t)\}$ grow in size. In Figure 3, one can check whether this condition is fulfilled, by comparing the values of $n(t_1,t)$ with the point located at the extreme right ($t_1 = t$) of each $n(t_1,t_1)$ diagram. At times t_2 and t_3, all particles are sub-critical; however, at t_3, some particles (class #1) have a size larger than at their time of nucleation; this means that they have been supercritical at some time, but, at t_3, they have started resorbing. This process goes on (times t_4 and t_5): particles in class #1 eventually completely resorb, while a second class of particles (class #2) becomes super-critical and its size will not stop growing until the end of the ageing process.

Another presentation of the particle population is given in Figure 4a in a grey level (t_1,t) plot (t_1 on the horizontal axis, $t > t_1$ on the vertical axis). Moving vertically allows one to follow the size evolution of a particle since its nucleation time ($t = t_1$ on the diagonal line). Moving horizontally shows the size, at a given time t, of all particles (nucleation times in the range $[0,t]$). One can recognize classes #1 and #2, formed at times $t_1 \approx 15000$ s and $t_1 \approx 40000$ s, the first of which eventually completely resorbs, while the second one keeps growing with time.

Figure 4b shows a similar plot for the sign of $dn(t_1,t)/dt$, which highlights the ranges of t_1 and t values during which particles grow. From a comparison with Figure 2, one can remark that

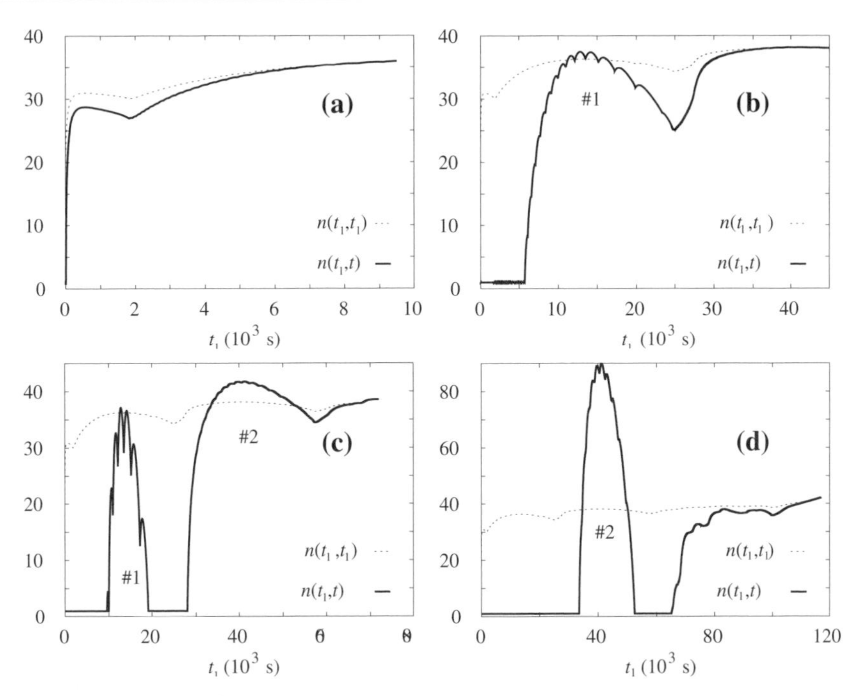

Figure 3. Dependence of the particle size $n(t_1,t)$ (solid line) and critical nucleus size $n(t_1,t_1)$ (dashed line) (both expressed in number of growth units) upon t_1 ($t_1 < t$) at four successive times t : $t_2 < t_3 < t_4 < t_5$ (from left to right and top to bottom). Behavior obtained for $u = 24$, $w = 2 \times 10^{-4}$, $J = 1.26 \times 10^3$ and $S_0 = 6.3$; times in seconds. [Used with permission of Elsevier, from Noguera et al. (2006), *J. Cryst. Growth*, Vol. 297, Fig. 4, p.190]

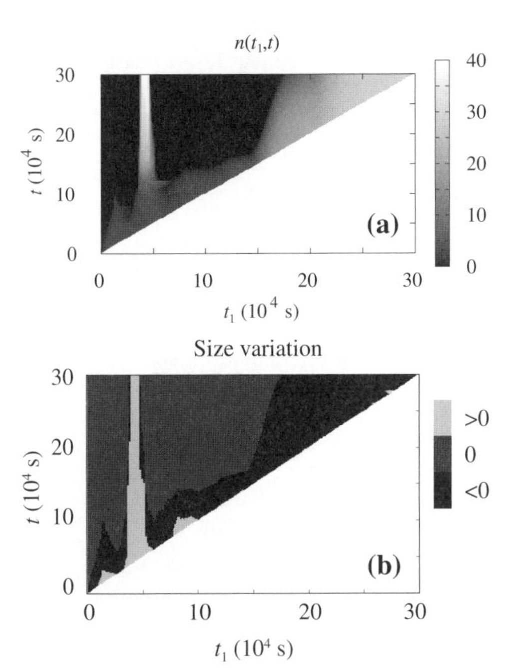

Figure 4. (a) Particle size $n(t_1,t)$ (expressed in number of growth units), in a (t_1,t) plot ($t > t_1$, t_1 on the horizontal axis, t on the vertical axis) with grey levels. Note lighter and lighter contrast along the diagonal line as t increases, which indicates that the critical nucleus size increases as $I(t)$ globally decreases. (b) Similar plot for the sign of $dn(t_1,t)/dt$. The lower right of these figures are blank due to causality reasons—particle sizes can only by considered at times t greater than their creation time t_1. Same numerical simulation parameters as in Figure 2. [Used with permission of Elsevier, from Noguera et al. (2006), *J. Cryst. Growth*, Vol. 297, Fig. 5, p.191]

they correlate exactly with the time intervals during which $dI/dt > 0$. This may be shown analytically, calculating the first and second derivatives of $n(t_1,t)$ (starting from Eqn. 27) at $t = t_1$:

$$\frac{dn(t_1,t)}{dt} = wn^{2/3}(t_1,t)\left(I(t) - \exp\left[\frac{2u}{n(t_1,t)}\right]^{1/3}\right) \tag{40}$$

and:

$$\frac{d^2n(t_1,t)}{dt^2} = wn^{2/3}(t_1,t)\frac{dI(t)}{dt} + \frac{2w}{3n^{1/3}(t_1,t)}\left(I(t) - \exp\left[\frac{2u}{n(t_1,t)}\right]^{1/3}\right)\frac{dn(t_1,t)}{dt} \tag{41}$$

In the limit $t \to t_1$, the factor in parentheses in Equation (40) vanishes, and so does the second term on the right hand side of Equation (41). The curve $n(t_1,t)$ thus starts from its initial value $n(t_1,t_1)$ with a zero slope and a curvature whose sign is the same as that of $dI(t)/dt$. Newly formed nuclei can thus increase in size when $dI(t)/dt > 0$. The existence of oscillations of $I(t)$ on the plateau is thus the necessary condition to a particle (positive) growth. However, since the time intervals in which the slope of $I(t)$ is positive represent a very small part of the total dynamics time, one comes to the striking conclusion that most nuclei formed since the start of the process rapidly disappear and that particles which will be observable in the ageing regime are the very few created during very specific time intervals. In a water-rock interaction process, if the long term surviving particles are not those formed at the start of the nucleation process, this has important consequences on the estimation of any dating of geochemical process.

Actually, the plateau period corresponds to a very complex stage of the dynamics. It is characterized by a nearly constant value of the supersaturation, which results from a quasi balance between two opposing forces: nucleation and growth on the one hand and resorption on the other hand. The small departures from complete equilibrium between these forces give rise to oscillations of I. These oscillations are the necessary conditions for particles to grow. All particles nucleated out of the small time intervals during which $dI/dt > 0$ eventually resorb and so will be absent in the ageing regime.

In the very last stage of the plateau, the particle size in class #2 reaches very large values: $n(t_1,t) >> n(t,t)$. Due to the non-linearity in Equation (27), their growth rate thus also strongly increases. Resorption cannot counterbalance such large growth which induces a sudden drop of $I(t)$. The quasi-balance between nucleation, growth and resorption of particles is broken; it is the end of the plateau.

The system then enters the metastable zone in which nucleation becomes negligible ($I < I_c$ and thus $F < 1$). This initiates the third stage of the dynamics: the ageing regime. The critical time t_c at which $I = I_c$ is inversely proportional to the growth constant κ and slowly varying with u and $I(t = 0)$. It is usually in this regime that one speaks of Ostwald ripening (Bigelow and Trimble 1927; Ratke and Voorhees 2002), assuming that the total fraction of newly formed phase remains virtually constant. The growth of some classes of particles are then totally dependant on the dissolution of other classes of particles. The total number of particles decreases and so does the total particle surface area. It ultimately leads to the survival of a single particle that gathers all convertible matter in the system. This is the state of thermodynamic equilibrium, theoretically reached at infinite time. However, it is practically rarely reached, since time limitation in the lab or in the field, mass balance in the system, steric constraints in the system, etc, may constrain the dynamics.

Crystal Size Distribution function (CSD). Statistical properties of the particle population are given by the normalized crystal size distribution (CSD) $P(n,t)$, whose expression as a function of $n(t_1,t)$ and $I(t)$ is the following (δ the Kronecker symbol):

$$P(n,t) = \frac{\int_0^t \exp\left(-\frac{u}{\ln^2 I(t_1)}\right)\delta(n - n(t_1,t))dt_1}{\int_0^t \exp\left(-\frac{u}{\ln^2 I(t_1)}\right)dt_1} \quad (42)$$

The information on the time at which the particles nucleated is lost when performing the integration. Indeed, a CSD may be calculated within the so-called population balance models (Langer and Schwartz 1980; Myhr and Grong 2000; Ramkrishna 2000; McCoy 2001), which do not give a full description of the particle population dynamics. It is also a by-product of NANOKIN simulations. A typical CSD, in the ageing regime, is presented in Figure 5a. It displays a cut-off for large n values and a negative asymmetry tail at small n. The time evolution of the mean particle size $\langle n(t)\rangle$, obtained as the first moment of $P(n,t)$, and that of the total number of particles N in the system are displayed in Figures 5b and 5c, respectively. N grows nearly exponentially during the nucleation stage, passes through a broad maximum during the plateau regime, and decreases during coarsening. The mean particle size, on the other hand, is a reasonably linear function of time for $t > t_c$, an indication that a law of the type $\langle \rho(t)\rangle^m - \langle \rho(t')\rangle^m \propto t - t'$ for the particle radius is obeyed with $m \approx 3$. In this regime, it is found that $\langle n(t)\rangle \approx n^*(t)$, that the cut-off value $n_{max}(t)$ of the CSD is also a linear function of time, and that the CSD is stationary, when plotting $P(n)/P_{max}$ as a function of $n/\langle n\rangle$ or n/n_{max}.

In the past, properties of the CSD in the ageing regime have been scrutinized by numerous mineralogists or geochemists, in order to try and reconstitute the conditions under which rocks or secondary minerals had been formed in the past, in nature (e.g., Eberl et al. 1998). The authors usually made the hypothesis that observed particles had been formed since $t = 0$, and postulated some initial shape of the CSD at the start of the ageing regime. However, the full account of the dynamics presented above contradicts this hypothesis, and requires no hypothesis on the CSD at $t = t_c$. The characteristics of the CSD found by NANOKIN are reasonably consistent with the Lifschitz-Slyozov-Wagner model of diffusion controlled

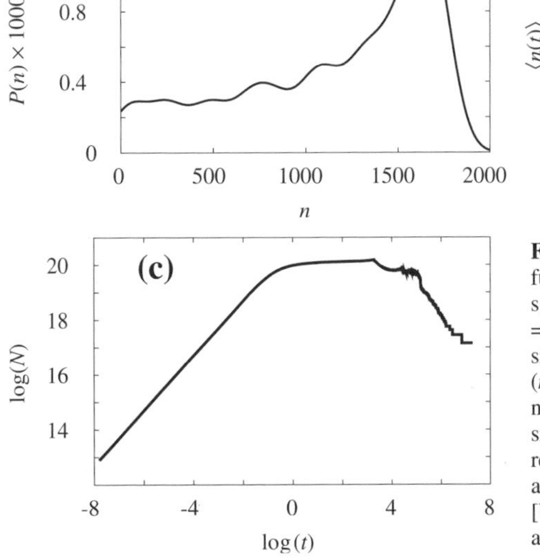

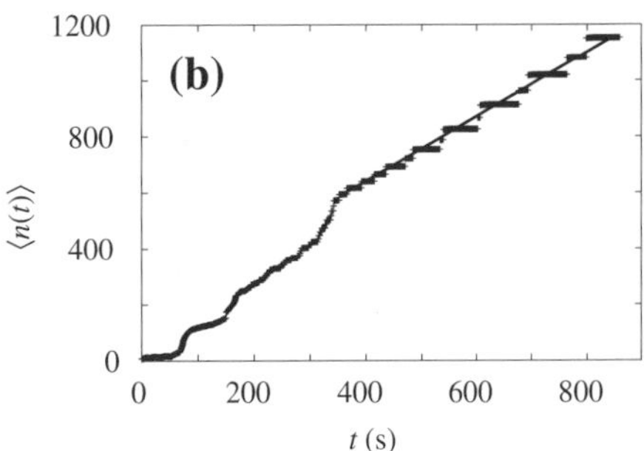

Figure 5. Normalized crystal size distribution function $P(n,t)$ for t in the ageing regime ($t = 860$ s; convolution by a Gaussian function of width $\sigma = 100$) (*left*). Time evolution of the mean particle size $\langle n(t)\rangle$ (expressed in number of growth units) (*middle*). Total number of particles (*right*). Same numerical simulation parameters as in Figure 2. The small wiggles in the middle panel correspond to resorption of individual classes of particles and are an artefact of the time discretization method used. [Used with permission of Elsevier, from Noguera et al. (2006), *J. Cryst. Growth*, Vol. 297, Fig. 9, p.194]

growth (Lifschitz and Slyozov 1961; Wagner 1961; Chai 1975) but less consistent with the De Hoff model (De Hoff 1984), although the latter is designed for describing a size evolution controlled by interfacial growth-dissolution mechanisms as in the simulation presented here. The absence of consideration of the pre-ageing stages and of explicit treatment of the particle population in the De Hoff model may be responsible for this discrepancy.

Relative strength of nucleation and growth processes. As illustrated above, the precipitation dynamics is an intricate result of nucleation, growth and resorption processes. The plateau-regime is a consequence of this competition, which may thus strongly depend upon the relative values of the parameters: u and $J = F_0/KN_{Av}$ (N_{Av} is the Avogadro number) for nucleation and w for (algebraic) growth. Indeed, at short times, the time dependence of $I(t)$ is not always that presented in Figure 2. Exploring the configuration space, it is found that $I(t)$ may only display two different behaviors, which are summarized in Figure 6. Scenario #1 (left panel) has already been described. In Scenario #2 (right panel), the initial rapid decrease of I does not exist. The plateau regime starts at $t = 0$ and, for some time interval, I remains close to its initial value $I(t = 0)$. The range of parameters for which one or the other of the two scenarios takes place is shown in Figure 7. At constant u values, scenario #1 takes place above a critical value of the ratio J/w. Scenario #2 may be considered as mainly driven by growth (large w values), while scenario #1 is favored by nucleation (large F_0, i.e., J values). It may be surprising that, despite the huge range of variations of w and J (note the logarithm scale in Fig. 7), these two scenarios remain unchanged and no other one could be found. This demonstrates their robustness.

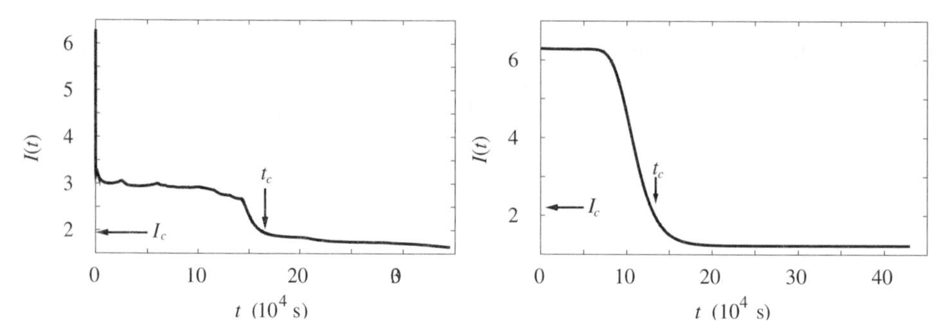

Figure 6. (*left*) Time dependence of the AS saturation index $I(t)$ (t in seconds). Scenario #1 obtained with same numerical simulation parameters as in Figure 2. (*right*) Scenario #2 obtained by decreasing J by seven orders of magnitude. [Used with permission of Elsevier, from Noguera et al. (2006), *J. Cryst. Growth*, Vol. 297, Fig. 1, p.189]

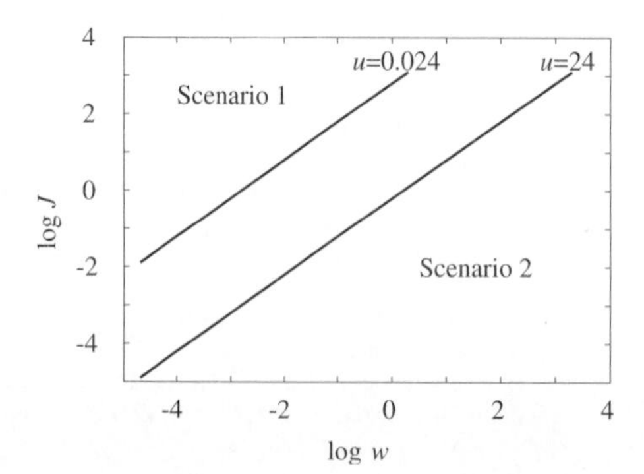

Figure 7. Range of (J,w) values for which the scenarios #1 and #2 are found. The line of separation between the two scenarios is drawn for two values of u: u = 24 and u = 0.024. [Used with permission of Elsevier, from Noguera et al. (2006), *J. Cryst. Growth*, Vol. 297, Fig. 2, p.189]

Dynamics with other growth laws. It is instructive to compare the previous results with those obtained with different growth laws. Two alternative models can be considered: the first one includes only nucleation (growth constant κ = 0):

$$n(t_1,t) = \frac{2u}{\ln^3 I(t_1)} \tag{43}$$

The second treats growth as a size-independent process. It is prototypical of the growth law used in most geochemical simulations (Steefel and Van Cappellen 1990; Madé et al. 1994):

$$n(t_1,t) = \frac{2u}{\ln^3 I(t_1)} + \int_{t_1}^{t} wn^{2/3}(t_1,t')\left(I(t')-1\right)dt' \tag{44}$$

Figure 8 compares the time evolution of I and the CSDs in the ageing regime for these two models together with those obtained with the size dependent law included in NANOKIN.

In the absence of growth (upper panels), the supersaturation I displays a first initial decrease due to nucleation, but, as expected, there is no plateau nor ageing regime. I rapidly

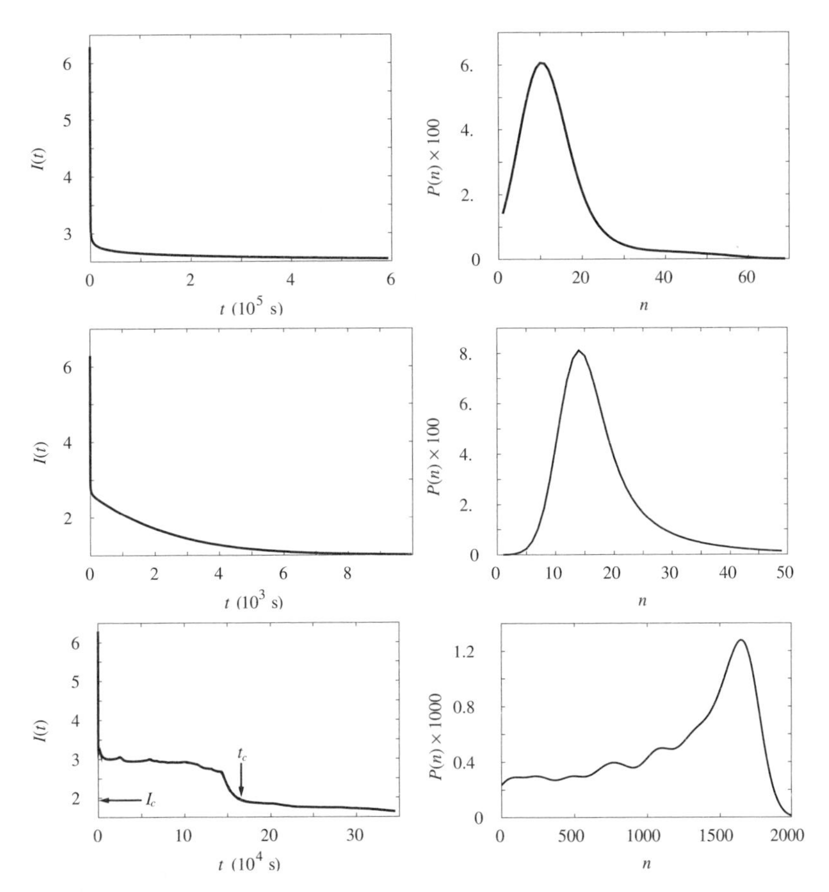

Figure 8. Time dependence of the AS saturation index $I(t)$ (t in seconds) (*left*) and CSD in the ageing regime (*right*). From top to bottom: in the absence of growth (model #1, Eqn. 43), with a size independent growth law (model #2, Eqn. 44) and with the size dependent law (Eqn. 25). Same numerical simulation parameters as in Figure 2. [Used with permission of Elsevier, from Noguera et al. (2006), *J. Cryst. Growth*, Vol. 297, Fig. 1, 9, 11 and 12]

converges towards an asymptotic value slightly larger than I_c. Since, in this model, nucleation is the only process which can lower I, the system remains stuck in a metastable state reflecting (slightly in anticipation) the nucleation metastable zone. The CSD has a positive asymmetry since the most numerous particles are those which were first formed (close to $t = 0$ when I was the largest), with the highest nucleation rate and the smallest size.

With the size independent growth law (middle panels in Fig. 8), I also displays a first initial decrease due to nucleation, and then a second decrease, assigned to growth. The plateau does not exist. Contrary to the preceding case, $I(t)$ rapidly drops below I_c and tends towards 1. The critical time t_c, at which $I = I_c$ is roughly two orders of magnitude shorter than when a size-dependent law is used (Fig. 8, lower panels). The absence of plateau and the rapidity of the dynamics come from the fact that no resorption can counterbalance the effect of nucleation and (positive) growth on I. For this same reason, all particles formed remain in the AS and can be observed at the end of the process. Their sizes keep increasing with time, with a rate which slows down as I approaches 1. The size distribution function displays a right skewed asymmetry.

Competition between two mineral phases of same composition in an initially supersaturated AS

As a second application, we now describe the simultaneous precipitation of two minerals of same composition, produced in an initially supersaturated AS. Halloysite and kaolinite are chosen as model minerals of relatively simple chemical composition $Al_2Si_2O_5(OH)_4$, differing only by their water content (halloysite: $Al_2Si_2O_5(OH)_4{\cdot}2H_2O$), whose competition has often been considered by authors in natural or experimental systems. La Iglesia and Galan (1975), in a batch-type experiment, studied the halloysite-kaolinite transformation at room temperature during short period of time (up to 90 days). Busenberg (1978) experimented the feldspars-AS interaction with production of halloysite as unique product for aluminium and silica. Kretzschmar et al. (1997) focused on the weathering of biotite with secondary halloysite dominant in the saprolite but diminishing in favor of kaolinite in the soil profiles. Kleber at al. (2007) studied the competition between halloysite and kaolinite in tropical rain forest soils. de Oliveira et al. (2007) showed the coexistence of halloysite and kaolinite in kaolin clays from alteration of volcanic rocks and proposed that some kaolinite may be inherited from halloysite precursors. From a theoretical point of view, Trolard et al. (1990) considered the possible kaolinite-halloysite competition on a thermodynamical basis and Steefel and Van Cappellen (1990) simulated their formation as a result of granite dissolution, in a reaction-flow system.

Simulation set-up. Kaolinite generally crystallizes as thin platelets with an hexagonal basal unit cell. Despite its importance in the natural environment, little quantitative information has been gathered on its surface properties. The basal face, which is the most stable surface, has a measured surface energy equal to 250 mJ/m^2 in dry atmosphere (Helmy et al. 2004) and between 34 mJ/m^2 and 95 mJ/m^2 when it is in contact with water, with a strong sensitivity to adsorbed ions on the platelet surfaces (Janczuk et al. 1989; Helmy at al. 2004). Estimation of the kaolinite/water interfacial energy by classical pair potential simulation methods yields a value equal to 48 mJ/m^2 (Warne et al. 2000). One should note that previous approximate simulations of kaolinite precipitation have used values as high as 200 mJ/m^2 (Steefel and Van Cappellen 1990). For other surface orientations, it seems that no measurements nor calculations exist. It has been argued (Fritz et al. 2009) that kaolinite nucleation and growth are quasi-2D processes, in which the platelet thickness remains constant and growth only proceeds laterally. The growth along the c direction then requires 2D nucleation on already formed particles, in agreement with the widely accepted statement that kaolinite generally nucleates on preexisting mineral surfaces as proposed by Lanson et al. (2002) and Meunier (2006). Then, the surface energy of the basal face is no longer relevant and only the value of σ_{lat} is needed, which is tentatively taken equal to $\sigma_{lat} = 0.1$ J/m^2 for kaolinite and half this value for halloysite (Table 1).

Table 1. Parameters used in NANOKIN for the precipitation of secondary minerals at 25 °C: rhombohedral particles of 2D or 3D form, K their solubility product (Helgeson et al. 1978), v their formula unit volume, κ their dissolution constant (Bloom (1983) for gibbsite and Stumm and Wieland (1990) for all others), $\bar{\sigma}$ their mean surface energy in contact with water, e the sheet thickness for 2D crystals, F_0 the pre-factor of their nucleation frequency is taken equal to 10^{20} nuclei/s for all minerals. The last line gives the critical supersaturation for each mineral. [Used with permission of Elsevier, from Fritz et al. (2009), *Geochim Cosmochim Acta*, doi: 10.1016/j.gca.2008.11.043, Table 1]

Mineral	**kaolinite**	**halloysite**	**gibbsite**	**Ca-mont.**
Form	2D	2D	3D	2D
Log K	−39.14	−37.98	−15.05	−48.10
v(Å^3)	164	167	53	349
κ(10^{-17}m/s)	2.786	2.838	0.0474	2.786
$\bar{\sigma}$ (J/m^2)	0.1	0.055	0.22	0.05
e(Å)	7	10		15
I_c	162	9.4	30.8	333

Mineral	**Reaction**
kaolinite	$Si_2Al_2O_5(OH)_4 + 7H_2O \leftrightarrow 2H_4SiO_4 + 2Al(OH)_4^- + 2H^+$
halloysite	$Si_2Al_2O_5(OH)_4 + 7H_2O \leftrightarrow 2H_4SiO_4 + 2Al(OH)_4^- + 2H^+$
gibbsite	$Al(OH)_3 + H_2O \leftrightarrow Al(OH)_4^- + H^+$
Ca-mont.	$Si_{3.67}Al_{2.33}O_{10}(OH)_2Ca_{0.17} + 12H_2O \leftrightarrow 3.67H_4SiO_4 + 2.33\ Al(OH)_4^- + 0.165Ca^{2+} + 2H^+$

Regarding the pre-factor F_0 of the nucleation rate, a huge uncertainty exists in the literature, within a range of 10^{10}-10^{25} nuclei per second and per kg of H_2O (Van Cappellen 1990; Markov 1995). However, changing F_0 by several orders of magnitude in most cases does not qualitatively modify the dynamics of precipitation. A discussion of the sensibility of the results to changes in F_0 and σ is given in Noguera et al. (2006b) and Fritz et al. (2009). Table 1 summarizes the values of the mineral parameters used in the present and following sections.

The precipitation of halloysite (H) and kaolinite (K) from an initially oversaturated AS is simulated at 25 °C in a closed system. The initial composition of the solution (AS1) is reported in Table 2. It is initially supersaturated with respect to both phases: $I_K = 488$ ($I_c = 162$), $I_H = 33.7$ ($I_c = 9.4$) and since the saturation states are both greater than their critical value for nucleation, both minerals start precipitating at $t = 0$.

Precipitation dynamics. The time evolution of I_K and I_H is reported in Figure 9. Since the two minerals have the same composition, the ratio between I_K and I_H remains constant at all times, inversely proportional to the solubility ratio. Both saturation indices decrease initially, as an effect of the rapid nucleation. In the long term, they stabilize on a kind of plateau ($I_K \approx 16$ and $I_H \approx 1.14$ at $t \approx 10^9$ s) during which no nucleation can take place since I_K and I_H are both smaller than their critical values. During this time interval, kaolinite particles continue to grow while the size of halloysite particles decreases. This is a kind of Ostwald ripening effect between two phases of the same composition. Eventually, the increase of kaolinite particle sizes disrupts the plateau, halloysite reaches an under-saturation state ($I_H < 1$) and it resorbs in favor of kaolinite.

Figure 10 displays the time evolution of the total amount of precipitates for the two phases on this longer time scale. Halloysite is largely dominant in the first stages of the dynamics. It

Table 2. Composition of the reacting AS. AS1: Solution presenting an initial supersaturation with respect to kaolinite and halloysite. AS2: Initial state of the AS for granite dissolution. [Used with permission of Elsevier, from Fritz et al. (2009), *Geochim Cosmochim Acta*, doi : 10.1016/j.gca.2008.11.043, Table 2]

Element	Concentration (mol/kg of H_2O) (AS1)	Concentration (mol/kg of H_2O) (AS2)
Al	1×10^{-7}	1×10^{-10}
K		2.9×10^{-6}
Na	1×10^{-6}	12.5×10^{-6}
Ca		8.4×10^{-6}
Mg		2.5×10^{-6}
Si	1.5×10^{-4}	1×10^{-10}
Cl	1×10^{-6}	43.1×10^{-6}

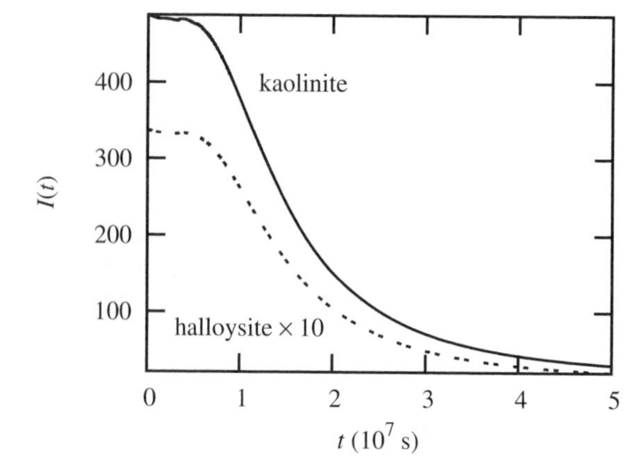

Figure 9. Time evolution of the AS saturation states $I(t)$ with respect to kaolinite (*solid line*) and halloysite (*dashed line*) precipitation. The critical values I_c for kaolinite and halloysite are equal to 162 and 9.4, respectively. [Used with permission of Elsevier, from Fritz et al. (2009), *Geochim Cosmochim Acta*, doi : 10.1016/j.gca.2008.11.043, Fig. 5, p.9]

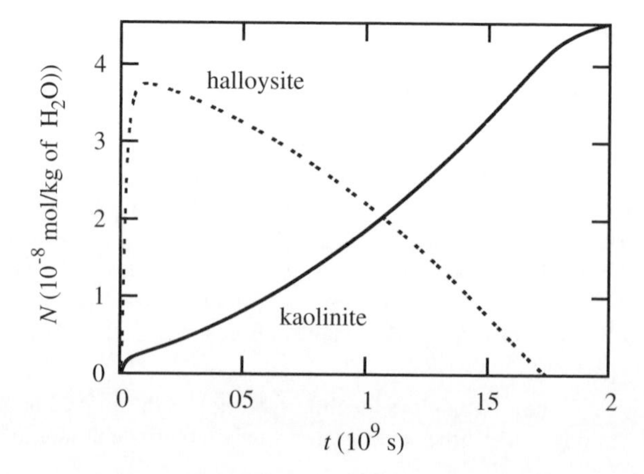

Figure 10. Time evolution of the total amount of kaolinite (*solid line*) and halloysite (*dashed line*) precipitates from an initially over-saturated AS. [Used with permission of Elsevier, from Fritz et al. (2009), *Geochim Cosmochim Acta*, doi : 10.1016/j.gca.2008.11.043, Fig. 6, p.9]

reaches a maximum value at $t \approx 10^8$ s and then steadily decreases while the amount of kaolinite becomes more and more significant. Both curves cross at $t \approx 10^9$ s (about 30 years). This is a scenario in which the more soluble phase (halloysite, which is the less stable with respect to the solution, in terms of thermodynamics) is initially the most abundant one. But, in the long term, kaolinite, which is the more stable but the less rapidly produced, eventually dominates.

The mean sizes of the halloysite and kaolinite particles remain in the expected range for these clay minerals. The maximum value reached by halloysite particles before the start of the resorption is 0.014 microns, while it amounts to 1.1 microns for kaolinite at $t = 3 \times 10^9$ s (100 years).

The simulation presented here represents a batch-type experiment of halloysite-kaolinite precipitation as described by authors mentioned above in this section and by Tsuzuki and Kawabe (1983) who showed the occurrence of halloysite on short term reactions of the order of a few months, an order of magnitude in reasonable agreement with NANOKIN findings. The long term transformation of halloysite into kaolinite is of course very difficult to approach experimentally because of time limitations in the experiments: the model results show clearly that the tendency exists to transform progressively early produced halloysite into kaolinite but this process requires more than tens of years: this timing is of course out of the frame of experiments even if it is still very short if one considers geological systems.

Precursors of precipitation. Halloysite, although being more soluble than kaolinite, thus plays the role of precursor to kaolinite precipitation. In order to more generally assess the potential outcome of the competition between two minerals of same composition, tests can be performed by successively modifying the key parameters of the dynamics: σ, κ, and F_0.

Changing the halloysite parameters and keeping constant the kaolinite ones induce systematic variations of the crossing point t_1 of the curves which display the time evolution of the total precipitated amount of the two minerals (Fig. 10):

- increasing σ_H pushes t_1 towards lower values. A mere increase of σ_H by 2 mJ/m^2 decreases t_1 by a factor of 2. This shift is due to the increase of the nucleation barrier which makes halloysite disappear more quickly.
- decreasing κ_H also pushes t_1 towards lower values, since it hinders the growth of halloysite particles.
- decreasing F_{0H} has the same effect, since it reduces the number of halloysite particles.

This shows that, in large domains of variation of the key-parameters for nucleation and growth, the scenario is very robust: kaolinite, because it is the more stable phase from a thermodynamic point of view (smaller solubility product K), remains the long term dominant phase, which eventually destabilizes the halloysite precipitate, the first phase to be produced abundantly in the same process. This result has to be considered when one refers to alteration processes occurring at low temperatures (25 °C in the calculations) like in weathering conditions (Kretzschmar et al. 1997).

The only possibility for having an opposite situation, where kaolinite would not be the long term survivor, is to have a much higher nucleation rate for halloysite ($F_{0H} > 10^{28}$ keeping the other parameters identical to those in the above calculation, but the critical value is a function of the initial supersaturation state of the solution). Halloysite nucleation would then induce a very rapid decrease of the saturation index for both phases. The initial kaolinite nuclei would resorb during the phase of halloysite dominance, and no new nuclei would be formed because the saturation state I_K would decrease too quickly below its critical value. The time at which kaolinite would fully resorb is a decreasing function of F_{0H}.

A symmetric situation is found if kaolinite nucleation rate F_{0K} is largely increased (e.g., $F_{0K} = 10^{25}$), in which case kaolinite, although less soluble, is dominant at all times.

So, it would be incorrect to systematically conclude that the more soluble mineral is always the precursor phase to the less soluble. For example, it may not apply at higher temperature, e.g., in diagenetic conditions, up to 150-200 °C (Srodon et al. 2000) where kaolinite may play a more important role initially, without halloysite precursor as shown by Ruiz Cruz and Reyes (1998).

Generalizing these findings, three scenarios may thus be found for different minerals of same composition competing for precipitation from the same AS:

(1) As a general rule the less soluble one is the one which survives on very long time scales, but the more soluble one plays the role of precursor, in a way which depends upon the ratio of solubilities

(2) It may happen that the more soluble phase is not initially dominant if the nucleation frequency and/or the growth rate of the less soluble mineral are very high.

(3) It may also happen that the less soluble phase is totally resorbed in the first stages of the dynamics and does not succeed in re-nucleating : this situation takes place if the F_0 parameter for the more soluble phase is high enough: its surviving time t_1 becomes infinite and the other mineral is totally resorbed during the rapid decrease of the saturation state.

Precipitation in response to granite dissolution

To go one step further towards a more realistic geochemical situation, we now consider water-rock interactions in which the evolution of the AS is due to the alteration of a primary rock. At all stages of the dynamics, the source of elements in the solution not only comes from the resorption of previously precipitated particles, but also from the dissolution of the primary rock-forming minerals as long as they remain thermodynamically unstable in the aqueous phase. As primary rock, we consider a granitic rock, source of the elements (Al, Si).

The AS chemical composition (Table 2, AS2) aims at representing rain water in a weathering process in industrial countries (Probst et al. 1999; Aubert et al. 2002). It is thus slightly acidic (pH = 4.5), with small amounts of all dissolved cations, and with CO_2 fugacity equal to ten times that of atmosphere, as expected in most of the weathering zones in soils ($f_{CO_2} = 10^{-2.5}$).

The granite is composed of quartz, microcline, low albite and anorthite for representing a plagioclase, and muscovite of macroscopic size, as recently described in the literature (Sausse et al. 2001) for water-rock interactions in a fractured rock (Table 3). In contact with the AS, all these primary minerals are under-saturated ($I_j = Q_j / K_j << 1$), and thus suffer dissolution. The alteration of this granite produces a solution successively supersaturated with respect to gibbsite, kaolinite (K) and halloysite (H), and finally a Ca-montmorillonite (M), representative of a smectite phase. Nucleation and growth parameters for these minerals are given in Table 1.

Precipitation dynamics. In the first stages of the process, the chemical composition of the AS evolves, with an increase of the pH from the initial 4.5 value to about 7.5, an increase of silica and sodium concentrations and a decrease of aluminium concentration. Its saturation states I_i with respect to the secondary minerals present a time evolution completely different from that obtained when precipitation is induced by an initially super-saturated AS. As shown in Figure 11, the $I_j(t)$ plots present two well defined regimes. In a first time interval (at the extreme left of the time axis), the solution, first under-saturated with respect to all secondary minerals, gets enriched due to the dissolution of the primary minerals forming the granite; a rapid increase of all I_j takes place together with nucleation of the secondary minerals. A competition between the enrichment of the solution by granite dissolution and an impoverishment by nucleation and growth of solid particles occurs, which eventually turns in favor of the second process. The

Table 3. Mineralogical composition of the granitic rocks and kinetic parameters for the dissolution of their mineral constituents; κ_j calculated from (a) Murphy and Helgeson (1989), (b) Siegel and Pfannkuch (1984), (c) Nickel (1973), (d) Lin and Clemency (1981), (e) Helgeson et al. (1984); solubility product (log *K*) from Helgeson et al. (1978); *S* the reaction area per kg of H_2O, derived from Sausse et al. (2001). [Modified from Fritz et al. (2009), *Geochim Cosmochim Acta*, doi : 10.1016/j.gca.2008.11.043, Table 3]

Mineral	**κ (10^{-12} mol/m²/s)**	**Log *K***	**Volume (%)**	***S* (m²)**
quartz	0.0159[a]	–4.00	36	84.24
microcline	1.86[b]	–23.19	22	51.48
low albite	3.16[c]	–20.18	23.6	55.3
muscovite	0.01[d]	–55.27	9	21
anorthite	10.[e]	–19.49	9.4	22

Mineral	**Reaction**
quartz	$SiO_2 + 2H_2O \leftrightarrow H_4SiO_4$
microcline	$KAlSi_3O_8 + 8H_2O \leftrightarrow 3H_4SiO_4 + Al(OH)_4^- + K^+$
low albite	$NaAlSi_3O_8 + 8H_2O \leftrightarrow 3H_4SiO_4 + Al(OH)_4^- + Na^+$
muscovite	$Si_3Al_3O_{10}(OH)_2K + 12H_2O \leftrightarrow 3H_4SiO_4 + 3Al(OH)_4^- + K^+ + 2H^+$
anorthite	$CaAl_2Si_2O_8 + 8H_2O \leftrightarrow 2H_4SiO_4 + 2Al(OH)_4^- + Ca^{2+}$

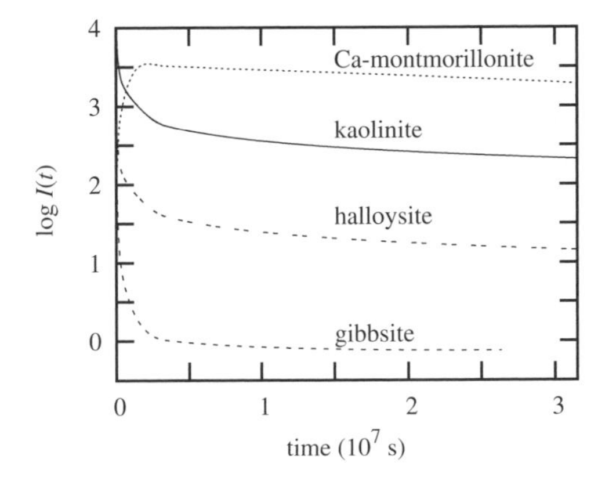

Figure 11. Time evolution of the saturation states *I*(*t*) of halloysite, kaolinite, gibbsite and Ca-montmorillonite during alteration of granite. [Used with permission of Elsevier, from Fritz et al. (2009), *Geochim Cosmochim Acta*, doi : 10.1016/j.gca.2008.11.043, Fig. 11, p.11]

$I_j(t)$ curves then pass through a maximum and decrease until nucleation stops when $I_j < I_{cj}$. All critical nuclei which are produced during this second phase resorb, because $dI_j/dt < 0$.

The amounts of precipitated secondary phases are shown in Figure 12—halloysite is the most abundant one and gibbsite is a minor phase which rapidly disappears as its saturation state becomes less than one. The competition between halloysite and kaolinite is clearly in favor of halloysite for the time period considered in the simulation ($t < 10^8$ s). However, as the supersaturation states of the minerals decrease, the halloysite/kaolinite ratio steadily decreases. This means that, as a result of the dissolution of aluminium rich silicates during weathering, halloysite will slowly, but progressively, be transformed into kaolinite. Here again, halloysite

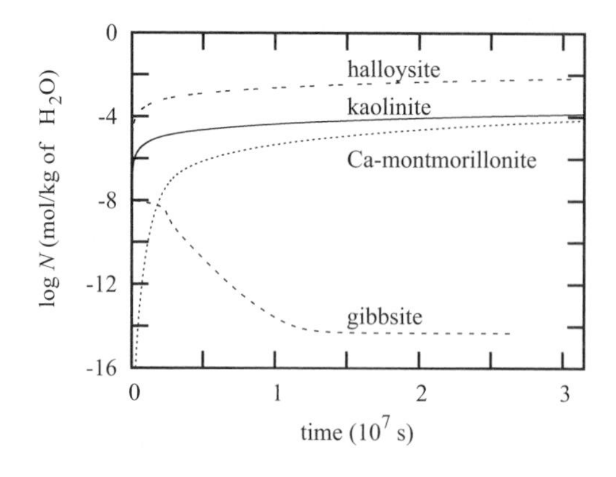

Figure 12. Time evolution of the amount of precipitated halloysite, kaolinite, gibbsite and Ca-montmorillonite during alteration of granite. [Used with permission of Elsevier, from Fritz et al. (2009), *Geochim Cosmochim Acta*, doi : 10.1016/j.gca.2008.11.043, Fig. 12, p.11]

plays the role of a precursor for kaolinite. At later times, Ca-montmorillonite appears as a significant precipitate: the respective amounts of halloysite, kaolinite and Ca-montmorillonite are as follows $N_H > N_K > N_M$ but the long term evolution shows a systematic decrease of the N_H/N_K, N_H/N_M and N_K/N_M ratios.

Chemical pathway in a Khorzhinskii-type diagram. It is interesting to locate the time evolution of the AS composition in a Korzhinskii diagram (Khorzhinskii 1965), in which the thermodynamic stability domains of the minerals under consideration are represented : in a purely thermodynamic approach, in a given domain, only one phase can be present. Notice that halloysite, having the same ion activity product as kaolinite (if water activity remains near 1 in dilute solutions), but a larger solubility, does not appear in such a diagram, because it is thermodynamically less stable than kaolinite.

The pathway during precipitation is shown in Figure 13. It goes from the gibbsite field to that of kaolinite and finally Ca-montmorillonite. Contrary to thermodynamic predictions, kinetic effects allow the solution to be simultaneously supersaturated with respect to several mineral phases. Moreover, the most abundant phase does not necessarily correspond to the largest super-saturation in solution: for example, halloysite is more abundant than kaolinite although their saturation states obey the reverse inequality $I_K > I_H$. It is important to have in mind this type of result when one compares themodynamic and kinetic simulations.

CSD and maximum particle sizes. The mean particle sizes of halloysite, kaolinite and Ca-Montmorillonite are recorded all along the precipitation dynamics. The largest sizes reached during the simulation are of the order of 0.35 microns for kaolinite, 0.025 microns for halloysite and 0.5 to 2 microns for Ca-montmorillonite. Although the kaolinite particle sizes are about ten times those of halloysite, their total precipitated amount is about 50 times less (8.2 vs. 380 micro-moles per kg of H_2O). The CSD for Ca-Montmorillonite (Fig. 14) clearly shows that, when the simulation is stopped ($t \approx 3 \times 10^8$ s), different classes of particles are still in evolution while particles in the dominant class have about 2 microns size. All these mineral phases remain in the size range of clay phases. However, the montmorillonite particles would continue to grow at later times and their sizes would become unrealistic.

This is a puzzle that remains for future investigations. In the first example of application, we have seen that when precipitation is driven by the establishment of an initial supersaturation in the solution, the dynamics ends with a single class of particle of (theoretically) infinite size. On the other hand, when dissolution of primary minerals is responsible for precipitation, it could happen that all primary minerals reach saturation and no longer dissolve. In a closed system, the dynamics would thus end like in the case of initial supersaturation, by ageing processes leading to a single class of particles of infinite size.

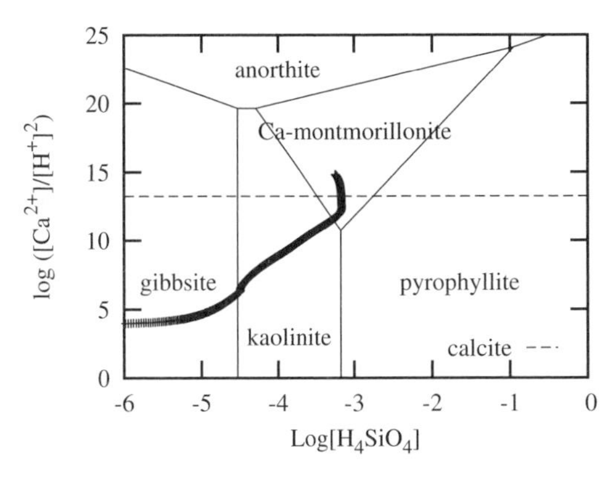

Figure 13. Reaction pathway of the AS in the Ca-Si Korzhinskii diagram for the alteration of granite at 25 °C. [Used with permission of Elsevier, from Fritz et al. (2009), *Geochim Cosmochim Acta*, doi : 10.1016/j.gca.2008.11.043, Fig. 13, p.12]

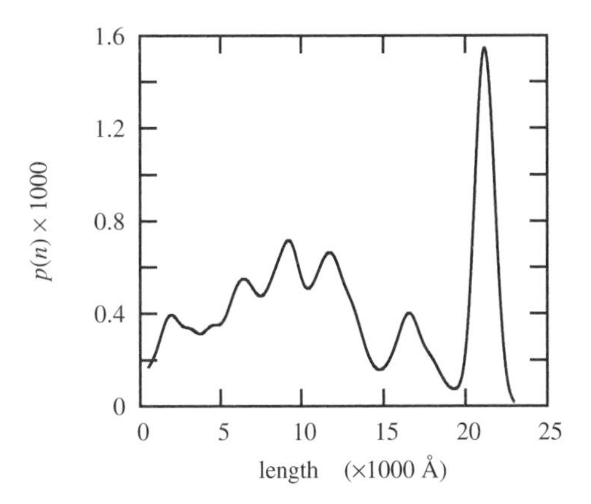

Figure 14. CSD $P(l,t)$ for Ca-montmorillonite after ≈ 1 year; l is the lateral size of the particles. [Used with permission of Elsevier, from Fritz et al. (2009), *Geochim Cosmochim Acta*, doi: 10.1016/j.gca.2008.11.043, Fig. 14, p.12]

However, this is a theoretical end-point (infinite time, infinite size) never reached in natural rock-solution systems which are at least partially open and for which the water-rock interaction is limited in time and space. The transfers in the fluid and the finite sizes of reactive elementary volumes are taken into account in coupled models, and have been the object of recent works (Steefel et al. 2005; Lichtner and Carey 2006; Marty et al. 2008). They will be treated in other courses in this Workshop.

Additionally, in order to correctly describe granite weathering on the very long term, it appears mandatory to extend the approach to include the formation of more complex clay phases involving one or several substitutions in their layers. Indeed, the chemical pathway in the Khorzhinskii diagram (Fig. 13) leads to stability regions of such complex clays. The Ca-Montmorillonite used here as representative of smectites clearly remains too simplistic, lacking elements like Fe and Mg. The chemical variability within classes of particles has thus to be combined with the size variability. This is a development presently under progress but not yet applied to clay minerals (Amal 2008; Noguera et al. 2008, 2009).

It is, however, not obvious that coupling of nucleation, growth and chemical variability will be sufficient to explain the small sizes of clay phases. As discussed by Meunier (2006), other processes may be at work, like disorders at the surface of the growing particles, incomplete filling of inter-layer sites, heterogeneity of the hydration layer or turbostratic organization of the layer stacking in clay minerals. The abundance of defects in the structure could explain that growth requires too much energy at low temperature, compared to the continuation of nucleation. Following this author, the variability in the maximum sizes of particles could be linked to the different defect frequencies (for example, higher for montmorillonite than for illites or kaolinites).

Precipitation of mineral phases of variable composition

As a last example of application, and as an illustration of the latest developments in NANOKIN (Amal 2008; Noguera et al. 2009), we discuss here the dynamics of precipitation of binary ideal solid solutions, induced by an initial super-saturation of an AS. The $(Ba,Sr)CO_3$

is considered first, as prototypical of such processes. It intends to illustrate how it is possible to coherently account for the size *and* composition of precipitating particles, with correct inclusion of the nucleation step, growth and feed-back on the AS. It raises questions on the relative strength of thermodynamic and kinetic effects. To answer them, another solid-solution is considered, namely $(Ba,Sr)SO_4$, whose end-members have a solubility ratio several orders of magnitude larger than $(Ba,Sr)CO_3$.

Precipitation of $(Ba,Sr)CO_3$. Experimental data suggest that $(Ba,Sr)CO_3$ forms a complete SS between its two end-members witherite ($BaCO_3$) and strontianite ($SrCO_3$) (Chang 1971), but analyses of natural witherites show a mean content of 3.3 mol% Sr, and also show that Ba is present in only minor amounts in strontianite (Baldasari and Speer 1979; Speer and Henseley-Dunn 1976). In the laboratory, Prieto et al. (1997) have carried out $(Ba,Sr)CO_3$ crystallization experiments by counter-diffusion of reactants through a column of porous silica hydrogel. They have observed the particles after one month of ageing and recorded their composition profile.

The simulation set-up considers precipitation in a closed system at ambient temperature, as a result of the application of an initial supersaturation. Initial concentrations of Ba, Sr and carbon are taken equal to $c_{0,Ba} = 10^{-3}$ and $c_{0,Sr} = 10^{-3}$ and $c_{0,C} = 6 \times 10^{-3}$, respectively, and solute speciation in the AS is determined, using thermodynamic data taken from Shock and Helgeson (1988) and Kharaka and Barnes (1973). Initial pH is fixed at a rather high value of 9, a range in which the contribution of CO_3^{2-} to the total carbon concentration is not negligible compared to that of HCO_3^- and $H_2CO_3^0$. At the beginning of the calculation, the CO_3^{2-} activity corresponds to an equivalent CO_2 partial pressure of the order of 3×10^{-4}, typical of ambient atmosphere. The witherite and strontianite minerals have elemental volumes equal to 7.605×10^{-29} m^3 and 6.477×10^{-29} m^3, respectively, (WebMineral 2008), solubility products equal to $10^{-8.56}$ (Busenberg and Plummer 1986) and $10^{-9.27}$ (Busenberg et al. 1984), respectively and surface energies close to 110 mJ/m^2 (Nielsen and Söhnel 1971; Söhnel 1982). The growth constant κ of the solid-solution, assumed independent on composition, is taken equal to 5×10^{-12} m/s, a mean value between those of strontianite (Sonderegger et al. 1976) and witherite (Chou et al. 1989). As regards the prefactor F_0 for the nucleation rate, in the absence of values in the literature, we have assumed an average value of 10^{25} nuclei per second and kg of H_2O.

Figure 15 (left panel) shows the time evolution of the saturation state of the aqueous solution $I(x^*)$, together with those of the partial saturation states I_{BaCO_3} and I_{SrCO_3} with respect to the two end-members. Despite initial conditions with an equal amount of Sr and Ba, I_{SrCO_3} is nearly five times larger than I_{BaCO_3}, in inverse ratio to the solubility products of the two carbonates. The three saturation states decrease as nucleation and growth of SS particles take place and as ions are incorporated into the solid phase. A quasi-thermodynamic equilibrium is reached rather quickly: I approaches 1 and becomes equal to the sum of I_{BaCO_3} and I_{SrCO_3} in approximately 100 hours. Simultaneously, the global witherite and strontianite contents in the solid phase N_{BaCO_3} and N_{SrCO_3} reach a saturation plateau, as represented in the right panel of Figure 15. At that stage, much of the initial Sr and Ba remain in solution—with the present set-up, the amount of precipitate is limited by carbonate availability.

Figure 16 shows the time dependence of the critical nuclei composition $x^* = x(t,t)$, together with the mean composition $\bar{x}$ of the solid phase: $\bar{x} = N_{SrCO_3}/(N_{SrCO_3} + N_{BaCO_3})$, on the same time scale as Figure 15. Actually, only particles nucleated in the very first seconds of the process can grow, since, in the rest of the dynamics $dI/dt < 0$. Due to the higher value of I_{SrCO_3} compared to I_{BaCO_3}, and according to Equation (23), the first nuclei to form have a much higher content in Sr than in Ba ($x^* >> 0.5$). As nucleation proceeds, all saturation states decrease. However, since the consumption of Sr is larger than that of Ba, x^* decreases, as well as the composition of the deposited layers on preformed particles and as well as $\bar{x}$. When $I < I_c$, nucleation stops and some further time is needed to resorb all particles except those which

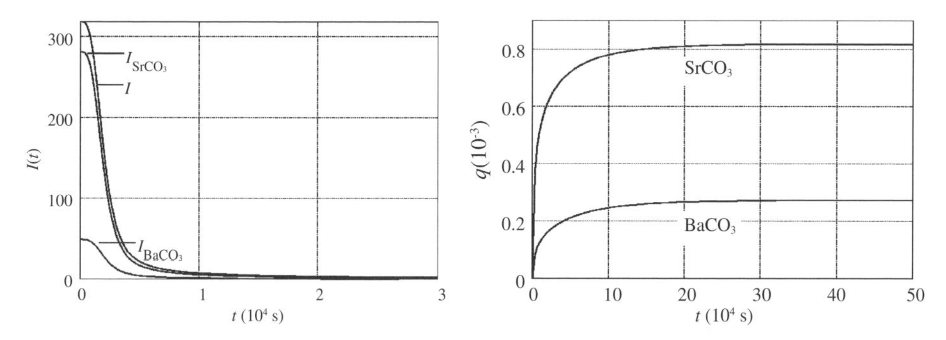

Figure 15. (*left*) Time dependence of the total (I) and partial (I_{BaCO_3} and I_{SrCO_3}) saturation states of the AS during the precipitation of (Ba,Sr)CO_3. I_c = 131 with the present set-up. (*right*) Time dependence of the content of witherite and strontianite in the solid phase during the precipitation of (Ba,Sr)CO_3 (see text).

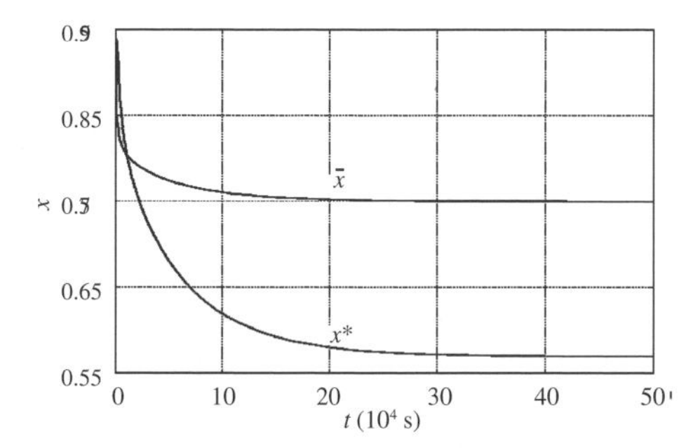

Figure 16. Time evolution of the (Ba,Sr) CO_3 critical nuclei composition $x^* = x(t,t)$ and mean particle composition $\overline{x} = N_{SrCO_3}/(N_{SrCO_3} + N_{BaCO_3})$.

had been nucleated during the first moments of the dynamics. At that stage, a thermodynamic equilibrium is nearly reached, with particle sizes of the order of 5 μm. Beyond this point, on a much longer time scale (10^9-10^{10} s, thus unobservable in a lab experiment), ageing takes place, that we will not discuss here.

Figure 17 shows the time evolution of the particle radius (left panel), for the last surviving particles, together with their composition profile ($x(t_1,t)$ as a function of $\rho(t_1,t)$ at fixed value of t_1) (right panel). It is seen that the composition of the deposited layers qualitatively follows the behavior of x^*: according to Equation (35), these two quantities indeed become equal in the limit of large radii $\rho \rightarrow \infty$ and when $I \rightarrow 1$ (which implies $\rho_m \rightarrow \infty$). Figure 17 shows that most of the matter has precipitated in the first 100 hours of the process; beyond this time, the particle radii remain nearly constant. The radius and composition characteristics obtained in these modelings are reminiscent of those found experimentally (Prieto et al. 1997): radii of the order of several micrometers on a month time scale, and smooth composition profile, although the precipitation conditions are different.

Precipitation of (Ba,Sr)SO₄. As evidenced above, the complete kinetic modeling of precipitation in $A_{1-x}B_xC$ mixed systems is quite complex, and this is likely the reason why, in the past, most geochemical approaches have assumed that SS formed at equilibrium. Indeed, at stoichiometric saturation, Equation (11) yields a simple relationship between the ion partitioning in the AS ($X_{aq} = [B^{+q}]/[B^{+q}]+[A^{+q}]$) and in the SS ($x_0$):

$$X_{aq} = \frac{x_0}{x_0 + (1-x_0)\frac{K_{AC}}{K_{BC}}} \tag{45}$$

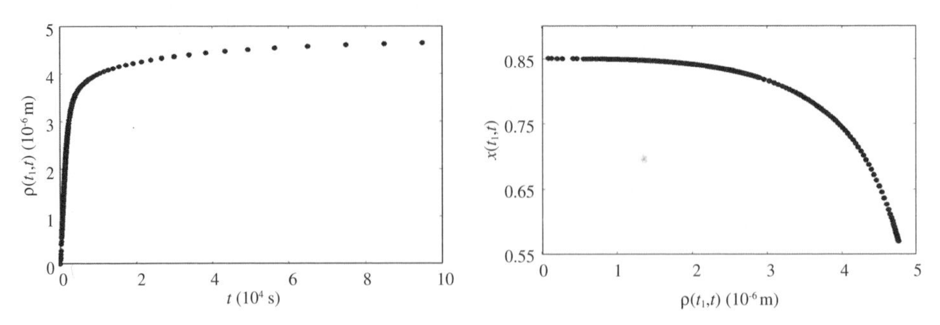

Figure 17. Time evolution of the $(Ba,Sr)CO_3$ particle radius, for the last surviving particles (*left*) and composition profile along the radius of these particles (*right*).

The diagram which represents x_0 as a function of X_{aq} is known as the Roozeboom plot (Mullin 1993). Each point on this diagram depicts a stoichiometric saturation.

The representation of Equation (45) is plotted for the $(Ba,Sr)CO_3$ SS in Figure 18 (left panel), together with the values of x^* and X_{aq} found in the course of the model calculation. The discrepancy between x^* and x_0 for a given value of X_{aq} is a measure of how far the system is from equilibrium. The out-of-equilibrium situation is evidenced in Figure 18 for times less than ≈ 100 h before x^* stabilizes on the lower plateau. x^* and x_0 thereafter remain nearly indistinguishable from each other. The equilibrium approach is thus found to be inaccurate as far as the composition of the SS is concerned, especially in the first steps of nucleation processes. In addition, it is unable to provide composition profiles such as those shown on Figure 17.

However, Roozeboom diagrams give useful hints on the relationship between x^* and X_{aq}. In the case where the ratio of the end member solubility products K_{AC}/K_{BC} is of order unity, x_0 smoothly follows the evolution of X_{aq}. The same is true for x^*, as seen in the case of the (Ba,Sr) CO_3 SS, for which $K_{AC}/K_{BC} \approx 5$. However, when K_{AC}/K_{BC} largely exceeds (respectively if largely below) 1, x_0 remains close to 0 (respectively 1) for nearly all values of X_{aq} and suddenly changes when X_{aq} reaches 1 (respectively 0).

The simulation of the precipitation of $(Ba,Sr)SO_4$ illustrates this situation. This SS, whose end-members are barite $BaSO_4$ and celestite $SrSO_4$, has been the object of various experimental laboratory studies (Malinin and Urusov 1983; Prieto et al. 1993; Pina et al. 2000; Putnis et al. 2003; Sanchez-Pastor et al. 2006). Becker et al. (2000) have simulated its thermodynamic stability, using classical inter-ionic potentials, and shown that, although it is not an ideal SS, non-ideality effects diminish as temperature is raised, and turn out to be weak even at ambient temperature. The solubility products of the two end-members differ by more than three orders of magnitude: $10^{-9.98}$ (Blount 1977) and $10^{-6.63}$ (Reardon and Armstrong 1987), for barite and celestite, respectively and are associated with large differences in their surface energies: 130 mJ/m^2 and 90 mJ/m^2 respectively (Nielsen and Söhnel 1971; Söhnel 1982). Prieto et al. (1993), Pina et al. (2000) and Putnis et al. (2003) have achieved growth of $(Ba,Sr)SO_4$ by counter-diffusion in aqueous solutions. After one month of growth, the morphology of the resulting crystals was studied by scanning electron microscopy and showed various zoning textures. In cases where the concentrations of the reservoir solutions were large, oscillatory zoning was observed, which has given impulse to the development of various theoretical models (L'Heureux and Jamtveit 2002; L'Heureux and Katsev 2006).

Figure 18 (right panel) and Figure 19 show the result of a NANOKIN simulation for the precipitation of $(Ba,Sr)SO_4$, with a set-up comparable to that used for $(Ba,Sr)CO_3$. Initial concentrations of Ba and Sr are taken equal to $c_{0,Ba} = c_{0,Sr} = 10^{-3}$ and $c_{0,S} = 3 \times 10^{-3}$ and pH is fixed at pH = 9. The barite and celestite minerals have elemental volumes equal to 8.651×10^{-29}

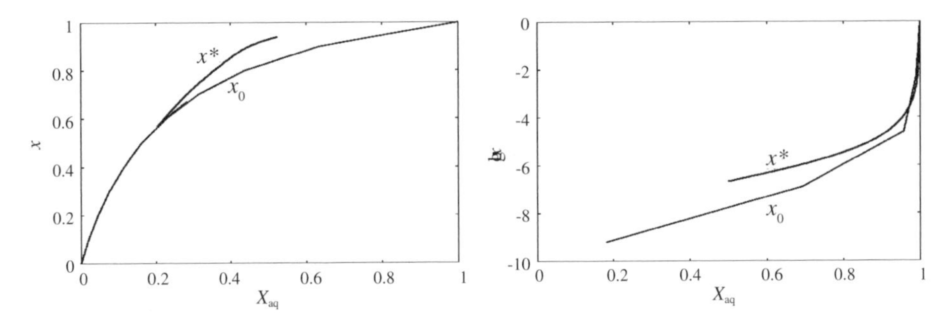

Figure 18. Roozeboom plots for the (Ba,Sr)CO_3 (*left*) and (Ba,Sr)SO_4 (*right*) solid solutions: dark line: equilibrium composition of the solid solution x_0 as a function of aqueous solution partition X_{aq} = $[Sr^{++}]$/ $[Sr^{++}]+[Ba^{++}]$. Light grey line: evolution of x^* as a function of X_{aq} during a precipitation process. Note the vertical logarithm scale in the right panel.

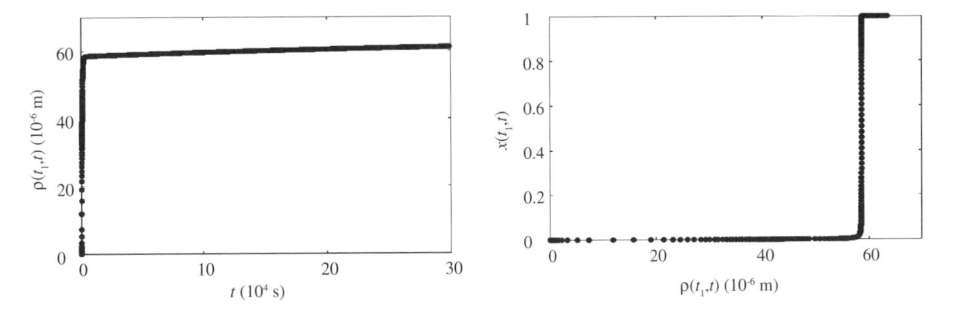

Figure 19. Time evolution of the radius (*left*) and composition profile (*right*) of the last surviving (Ba,Sr) SO_4 particles, produced from AS with initial amounts of Ba, Sr and S, equal to 10^{-3}, 10^{-3} and 3×10^{-3}, respectively, and pH = 9.

m^3 and 7.68×10^{-29} m^3, respectively, (WebMineral 2008). The growth constant κ of the SS, assumed independent of composition, is taken equal to 5×10^{-12} m/s, an intermediate value between those of barite and celestite (Dove and Czank 1995). The pre-factor F_0 for the nucleation rate, also assumed independent of composition, is taken equal to 10^{25} nuclei per second and kg of H_2O, intermediate between those quoted by Walton (1969) for barite and celestite.

In the Roozeboom diagram, Figure 18 (right panel), a logarithm scale is necessary in order to make the relationship between X_{aq} and x_0 visible. The figure clearly demonstrates that, both at stoichiometric saturation and in the dynamic run, x^* suffers a sudden jump as X_{aq} approaches 1. In contrast with the case of (Ba,Sr)CO_3, there is a brutal transition from 0 to 1 in the time dependence of the critical nuclei composition x^* (Fig. 19). The same effect is found for the composition of the growing layers: the last surviving particles have a core which is nearly pure $BaSO_4$, while their outermost layer is nearly pure $SrSO_4$. This is in reasonable agreement with experimental findings. However, at the present stage of the study (Noguera et al. 2008, 2009), the necessary ingredients to find oscillatory zoning are probably not yet introduced in the code NANOKIN. This example still demonstrates how core-shell particles may be produced, whose technological interest is now well recognized (e.g., Portehault et al. 2008 and references therein).

FUTURE EVOLUTIONS: NEEDS AND DREAMS

The various applications which have been described have shown that kinetic effects associated to precipitation in aqueous solutions can now be accounted for, especially those related to the first stages of nucleation and growth, as embedded in NANOKIN. This opens the way to simulations of processes taking place in soft chemistry experiments, relevant in modern technology, or to processes taking place in the natural environment, with various applications to societal questions.

The NANOKIN code is now available within direct scientific cooperation, upon request to the first author. However, as discussed in this course, there is an actual need to measure the parameters entering the nucleation and growth model: surface energies, parameters for the nucleation rate, etc, for many minerals of interest in the field. Also interesting would be experimental data on the evolution of sizes of particles created in over-saturated solutions, necessary to progressively validate these models. An important research effort has thus to be promoted in this field, as it has been done for dissolution processes, although it is, of course, a more difficult challenge when one focuses on the complex phases as those occurring in natural environments.

The latest development of NANOKIN for binary solid solutions should be considered as a first step but the final aim is to account for more complex phases with n-end-members SS like clay minerals. Such a generalization remains to be done on the conceptual level, and will not be easily validated without new experimental data on precipitations of these phases. This is a real challenge for future studies of nucleation and growth of these phases which are as complex as ubiquist in the natural environment.

Such a development could be interesting for predicting the long term evolution of populations of particles which may vary simultaneously in size and compositions. Will this be a key for understanding the size limitation of clay phases in the nano- to micro-meter range, which is so important for their reactivity in the environment?

ACKNOWLEDGMENT

A large part of the work presented in this course was performed under contract of the French ANR-PNANO 2006 (project “SIMINOX” 0039).

REFERENCES

Adamson AW (1960) Physical Chemistry of Surfaces. Interscience Publishers

Amal Y (2008) Modèles de nucléation et croissance de particules en solution: propriétés mathématiques et simulation numérique. Ph. D. Dissertation, University Louis Pasteur, Strasbourg, France.

Aubert D, Probst A, Stille P, Viville D (2002) Evidence of hydrological control of Sr isotopic ratio in surface waters draining a granitic catchment. Implication for weathering. Appl Geochem 17:285-300

Baldasari A, Speer JA (1979) Witherite composition, physical properties, and genesis. Am Mineral 64:742-747

Baronnet A (1982) Ostwald ripening in solution. The case of calcite and mica. Estudios Geol 38:185-198

Baronnet A (1984) Growth kinetics of the silicates - a review of basic concepts. Fortschr Mineral 62:187-232

Becker U, Fernandez-Gonzalez A, Prieto M, Harrison R, Putnis A (2000) Direct calculation of thermodynamic properties of the bariter-celestite solid solution from molecular principles. Phys Chem Mineral 27:291-300

Berner RA (1980) Early Diagenesis: A Theoretical Approach. Princeton University Press, Princeton, NJ.

Bigelow SL, Trimble HM (1927) The relation of vapor pressure to particle size. J Phys Chem 31:1798

Blanc Ph, Bieber A, Fritz B, Duplay J (1997) A short range interaction model applied to illite smectite mixed layers minerals. Phys Chem Mineral 24:574-581

Bloom PR (1983) The kinetic of gibbsite dissolution in nitric acid. Soil Sci Soc Am J 47:164-168

Blount CW (1977) Barite solubilities and thermodynamic quantities up to 300 °C and 1400 bars. Am Mineral 62:942-957

Burton W K, Cabrera N, Frank F (1951) The growth of crystals and the equilibrium structure of their surfaces. Philos Trans R Soc 243:299-358

Busenberg E (1978) The products of the interaction of feldspars with aqueous solutions at 25 °C. Geochim Cosmochim Acta 42:1679-1683

Busenberg E, Plummer LN, Parker VB (1984). The solubility of strontianite ($SrCO_3$) in CO_2-H_2O solutions between 2 and 91 °C, the association constants of $SrHCO_3^+$ and $SrCO_3^0$ (aq) between 5 and 80 °C, and an evaluation of the thermodynamic properties of Sr^{2+} (aq) and $SrCO_3$ (cr) at 25 °C and 1 atm total pressure. Geochim Cosmochim Acta 48:2021-2035

Busenberg E, Plummer LN (1986) The solubility of $BaCO_3$ (cr) (witherite) in CO_2-H_2O solutions between 0 and 90 °C, evaluation of the association constants of $BaHCO_3^+$ (aq) and $BaCO_3^0$ (aq) between 5 and 80 °C, and a preliminary evaluation of the thermodynamic properties of Ba^{2+} (aq). Geochim Cosmochim Acta 50:2225-2233

Cabane H, Laporte D, Provost A (2005) An experimental study of Ostwald ripening of olivine and plagioclase in silicate melts: implications for the growth and size of crystals in magmas. Contrib Mineral Petrol 150:37-53

Chai BHT (1975) The kinetics and mass transfer of calcite during hydrothermal recrystallization process. PhD Dissertation, Yale University

Chang LLY (1971) Sub-solidus phase relations in the aragonite type carbonates I. The system $CaCO_3$-$SrCO_3$-$BaCO_3$. Am Mineral 56:1660-1673

Chou L, Garrels RM, Wollast R (1989) Comparative study of the kinetics and mechanisms of dissolution of carbonate minerals. Chem Geol 78:269-282

De Hoff RT (1984) Generalized microstructural evolution by interface controlled coarsening. Acta Metall Mater 32:43-47

de Oliveira MTG, Furtado SMA, Formoso MLL, Eggleton RA, Dani N (2007) Coexistence of halloysite and kaolinite - a study on the genesis of kaolin clays of Campo Alegre Basin, Santa Catarina State, Brazil. An. Acad. Bras. Ciencias 79 (4):665-681

Dove PM, Czank CA (1995) Crystal chemical controls on the dissolution kinetics of the isostructural sulfates: Celestite, anglesite and barite. Geochim Cosmochim Acta 59:1907-1915

Doyle GJ (1961) Self nucleation in the sulfuric acid/water system. J Chem Phys 35:795-799

Drever JI (1997) The Geochemistry of Natural Systems. Surface and Groundwaters Environments. Prentice Hall, Inc. Third Edition

Eberl DD, Kile DE, Dritz VA (1998) Deducing growth mechanisms for minerals from the shapes of crystal size distributions. Am J Sci 298:499-533

Frenkel J (1946) Kinetic Theory of Liquids. Oxford.

Fritz B (1981) Etude thermodynamique et modélisation des réactions hydrothermales et diagénétiques. Mem Sci Géol 65:1-197, Strasbourg

Fritz B, Clément A, Amal Y, Noguera C (2008) Modelling nucleation and growth of nano- to micro-size secondary clay particles in weathering processes. Geochim Cosmochim Acta 72(12S):A285-A285

Fritz B, Clément A, Amal Y, Noguera C (2009) Simulation of the nucleation and growth of simple clay minerals in weathering processes: the NANOKIN code. Geochim Cosmochim Acta 73:1340-1358, doi: 10.1016/j.gca.2008.11.043

Gaman AI, Napari I, Winkler PM, Vehkamäki H, Wagner PE, Strey R, Viisanen Y, Kulmala M (2005) Homogeneous nucleation of n-nonane and n-propanol mixtures: A comparison of classical nucleation theory and experiments. J Chem Phys 123:244502-244512

Gérard F, Clément A, Fritz B (1998a) Numerical validation of an Eulerian hydrochemical code using a 1-D multisolute mass transport system involving heterogeneous kinetically-controled reactions. J Contam Hydrol 30(3-4):199-214

Gérard F, Fritz B, Clément A, Crovisier JL (1998b) General implications of aluminum speciation-dependent kinetic dissolution rate law in water-rock modelling. Chemical Geology 151:247-258

Glynn PD, Reardon EJ (1990) Solid solution aqueous-solution equilibria : thermodynamic theory and representation. Am J Sci 290:164-201

Heady RB, Cahn JW (1972) Experimental test of classical nucleation theory in a liquid-liquid miscibility gap system. J Chem Phys 58:896-910

Helgeson HC, Brown TH, Nigrini A, Jones TA (1970) Calculation of mass transfer in geochemical processes involving aqueous solutions. Geochim Cosmochim Acta 34:455-481

Helgeson HC, Delany JM, Nesbitt HW, Bird DK (1978) Summary and critic of the thermodynamic properties of rock-forming minerals. Am J Sci 278A:1-229

Helgeson HC, Murphy WM, Aagard P (1984) Thermodynamic and kinetic constraints on reaction rates among minerals and aqueous solutions. Geochim Cosmochim Acta 48:2405-2432

Helmy AK, Ferreiro EA, De Bussetti SG (2004) The surface energy of kaolinite. Colloid Polym Sci 283:225-228

Huertas FJ, Chou L, Wollast R (1999) Mechanisms of kaolinite dissolution at room temperature and pressure. Part II: kinetic study. Geochim Cosmochim Acta 63(19-20):3261-3275

Israelachvili JN (1985) Intermolecular and Surface Forces, with Applications to Colloidal and Biological Systems. Academic Press

Janczuk B, Chibowki E, Hajnos M, Biakopiotrowicz T, Stawinski J (1989) Influence of exchangeable cations on the surface free energy of kaolinite as determined from contact angles. Clays Clay Mineral 37:269-272

Jolivet JP, Froidefond C, Pottier A, Chaneac C, Cassaignon S, Tronc E, Euzin P (2004) Size tailoring of oxide nanoparticles by precipitation in aqueous medium. A semi-quantitative modelling. J Mater Chem 14:3281-3288

Jolivet JP, Tronc E, Chaneac C (2006) Iron oxides: From molecular clusters to solid. A nice example of chemical versatility. Comptes Rendus Geosci 338:488-497

Kharaka YK, Barnes I (1973) SOLMINEQ - Solution-mineral equilibrium computation. U.S. Dept. of the Interior. Geological Survey Computer Contribution, Report # USGS-WRD-73-002

Kile DE, Eberl DD, Hoch AR, Reddy MM (2000) An assessment of calcite crystal growth mechanisms based on crystal size distributions. Geochim Cosmochim Acta 64:2937-2950

Kleber M, Schwendenmann L, Vedkamp E, Rössner J, Jahn R (2007) Halloysite versus gibbsite: Silicon cycling as a pedogenetic process in two lowland neotropical rain forest soils of La Selva, Costa Rica. Geoderma 138(1-2):1-11

Korzhinskii DS (1965) The theory of systems with perfectly mobile components and processes of mineral formation. Am J Sci 263:193-205

Kretzschmar R, Robarge WP, Amoozegar A, Vepraskas MJ (1997) Biotite alteration to halloysite and kaolinite in soil-profiles developed from mica schist and granite gneiss. Geoderma 75:155-170

Laaksonen A, Kumala M, Wagner PE (1993) On the cluster composition in the classical binary nucleation theory. J Chem Phys 99:6832-6835

Laaksonen A, McGraw R, Vehkamäki H (1999) Liquid-drop formalism and free-energy surfaces in binary homogeneous nucleation theory. J Chem Phys 111:2019-2027

La Iglesia A, Galan E (1975) Halloysite-kaolinite transformation at room temperature. Clays Clay Mineral 23:109-113

Langer JS, Schwartz AJ (1980) Kinetics of nucleation in near-critical fluids. Phys Rev A 21:948-958

Lanson B, Beaufort D, Berger G, Bauer A, Cassagnabère A, Meunier A (2002) Authigenic kaolin and illitic minerals during burial diagenesis of sandstones : a review. Clay Minerals 37:1-22

Lasaga AC, Soler JM, Ganor J, Burch TE, Nagy KL (1994) Chemical weathering rate laws and global geochemical cycles. Geochim Cosmochim Acta 58:2361-2386

Legrand B, Saul A, Treglia G (1994) Chemical and topological structure of alloy surfaces: from equilibrium to kinetics. Mater Sci Forum 155-156:165-188

L'Heureux I, Jamtveit B (2002) A model of oscillatory zoning in solid solutions grown from aqueous solutions: applications to the $(Ba,Sr)SO_4$ system. Geochim Cosmochim Acta 66:417-429

L'Heureux I, Katsev S (2006) Oscillatory zoning in a $(Ba,Sr)SO_4$ solid solution: macroscopic and cellular automata models. Chem Geol 225:230-243

Lichtner PC (1985) Continuum model for simultaneous chemical reactions and mass transport in hydrothermal systems. Geochim Cosmochim Acta 49:779-800

Lichtner PC, Carey JW (2006) Incorporating solid solutions in reactive transport equations using a kinetic discrete-composition approach. Geochim Cosmochim Acta 70(6):1356-1378

Lifschitz IM, Slyozov VV (1961) The kinetics of precipitation from supersaturated solid solutions. J Phys Chem Solids 19:35-50

Lin FC, Clemency CV (1981) The kinetics of dissolution of muscovites at 25 °C and 1 atm CO_2 partial pressure. Geochim Cosmochim Acta 45:571-576

Los JH, Matovic M (2005) Effective Kinetic Phase Diagrams. J Phys Chem B109:14632-14641

Madé B, Clément A, Fritz B (1990) Modélisation cinétique et thermodynamique de l'altération: le modèle géochimique KINDIS. CR Acad Sci Paris 310 (II):31-36

Madé B, Clément A, Fritz B (1994) Modeling mineral/solution interactions : The Thermodynamic and Kinetic Code KINDISP. Comput Geosci 20(9):1347-1363

Malinin SD, Urusov VS (1983) The experimental and theoretical data on isomorphism in the $(Ba,Sr)SO_4$ system in relation to barite formation. Geokhimiya 9:1324-1334

Markov IV (1995) Crystal growth for Beginners: fundamentals of nucleation, crystal growth and epitaxy, World Scientific, Singapore, New Jersey, London, Hong Kong.

Marty N, Fritz B, Clément A, Michau N (2008) Modelling the long term alteration of the engineered bentonite barrier in an underground radioactive waste repository. Appl Clay Sci DOI : 10.1016/j.clay.2008.10.002

Maugis P, Gouné M (2005) Kinetics of vanadium carbonitride precipitation in steel: A computer model. Acta Materialia 53:3359-3367

McCoy BJ (2001) A new population balance model for crystal size distributions: Reversible size-dependent growth and dissolution. J Colloid Interf Sci 240:139-149

Meunier A (2006) Why are clay mineral small? Clay Minerals 41:551-566

Montes-Ha G, Fritz B, Clément A, Michau N (2005) Modeling of transport and reaction in an engineered barrier for radioactive waste confinement. Appl Clay Sci 29:155-171
Moreno V, Creuze J Berthier F, Mottet C, Treglia G, Legrand B (2006) Site segregation in size-mismatched nanoalloys: Application to Cu-Ag. Surf Sci 600:5011-5020
Müller P, Kern R (2000) Equilibrium nanoshape changes induced by epitaxial stress (generalized Wulff Kaishef theorem) Surf Sci 457:229-253
Mullin JW (1993) Crystallization. Butterworth-Heinemann
Murphy WM, Helgeson HC (1989) Thermodynamic and kinetic constraints on reaction rates among minerals and aqueous solutions. Am J Sci 289:17-101
Myhr OR, Grong O (2000) Modelling of non-isothermal transformations in alloys containing a particle distribution. Acta Mater 48:1605-1615
Nagano T, Enomoto M (2006) Simulation of the growth of copper critical nucleus in dilute bcc Fe-Cu alloys. Scripta Materialia 55:223-226
Nickel E (1973) Experimental dissolution of light and heavy minerals in comparison with weathering and intrastratal solutions. Contrib Sediment 1:1-68
Nielsen AE, Söhnel O (1971) Interfacial tensions electrolyte crystal-aqueous solution, from nucleation data. J Cryst Growth 11:233-242
Noguera C, Fritz B, Clément A, Baronnet A (2006a) Nucleation, growth and ageing in closed systems I : a unified model for precipitation in solution, condensation in vapor phase and crystallization in the melt. J Cryst Growth 297:180-186
Noguera C, Fritz B, Clément A, Baronnet A (2006b) Nucleation, growth and ageing in closed systems II: simulated dynamics of a new phase formation. J Cryst Growth 297:187-198
Noguera C, Fritz B, Amal A, Clément A (2008) Simulation of the nucleation and growth of solid solutions in aqueous solutions. Geochim Cosmochim Acta 72(12):A687-A687
Noguera C, Fritz B, Amal A, Clément A (2009) Simulation of the nucleation and growth of solid solutions in aqueous solutions. Chem Geol (accepted)
Noppel M, Vehkamäkia H, Kulmala M (2002) An improved model for hydrate formation in sulfuric acid-water nucleation. J Chem Phys 116:218-228
Nourtier-Mazauric E, Guy B, Fritz B, Brosse E, Garcia D, Clément A (2005) Modelling the dissolution/ precipitation of ideal solid solutions. Oil Gas Sci Technol - Rev IFP 60 (2):401-415
Oelkers EH, Schott J, Devidal JL (1994) The effect of aluminum, pH and chemical affinity on the rates of aluminosilicate dissolution reactions. Geochim Cosmochim Acta 58:661-669
Parbhakar K, Lewandowski J, Dao LH (1995) Simulation model for Ostwald ripening in liquids. J Colloid Interf Sci. 174: 142-147
Parkhurst DL, Appello CAJ (1999) User's guide to PHREEQC (Version 2). A computer program for speciation, batch-reaction, one-dimensional transport, and inverse geochemical calculations. U.S. Department of the Interior, U.S.G.S.
Pimpinelli A, Villain J (1998) Physics of Crystal Growth. Cambridge University Press, Cambridge
Pina CM, Enders M, Putnis A (2000) The composition of solid solutions crystallising from aqueous solutions: The influence of supersaturation and growth mechanisms. Chem Geol 168:195-210
Portehault D, Cassaignon S, Baudrin E, Jolivet JP (2008) Design of hierarchical core-corona architectures of layered manganese oxides by aqueous precipitation. Chem Mater 20:6140-6147
Prieto M, Putnis A, Fernández-Díaz L (1993) Crystallization of solid solutions from aqueous solutions in a porous medium: Zoning in (Ba-Sr) SO_4. Geol Mag 130:289-299
Prieto M, Fernandez-Gonzalez A, Putnis A, Fernandez-Diaz L (1997) Nucleation, growth and zoning phenomena in crystallizing $(Ba,Sr)CO_3$, $Ba(SO_4,CrO_4)$, $(Ba,Sr)SO_4$ and $(Cd,Ca)CO_3$ solid solutions from aqueous solutions. Geochim Cosmochim Acta 61:3383-3397
Probst A, Party JP, Février C, Dambrine E, Thomas AL, Stussi JM (1999) Evidence of springwater acidification in the Vosges mountains (NorthEast of France): influence of bedrock buffering capacity. Water Air Soil Pollut 114:395-411
Putnis A, Pina CM, Astilleros JM, Fernández-Díaz L, Prieto M (2003) Nucleation of solid solutions crystallizing from aqueous solutions. Phil Trans R Soc London A 361:615-632
Ramkrishna D (2000) Population balances, Academic Press. San Diego
Ratke L, Voorhees PWW (2002). Growth and Coarsening: Ostwald Ripening in Material Processing. Springer, Germany
Reardon EJ, Armstrong DK (1987) Celestite ($SrSO_4$ (s)) solubility in water, seawater and NaCl solution. Geochim Cosmochim Acta 51:63-72
Reiss H (1950) The kinetics of phase transitions in binary systems. J Chem Phys 18:840-848
Reiss H, Shugard M (1976) On the composition of nuclei in binary systems. J Chem Phys 65:5280-5293
Ruiz Cruz MD, Reyes E (1998) Kaolinite and dickite formation during shale diagenesis: isotopic data. Appl Geochem 13:95-104

Sanchez-Pastor N, Pina CM, Fernandez-Diaz L (2006) Relationships between crystal morphology and composition in the (Ba,Sr)SO_4-H_2O solid solution-aqueous solution system. Chem Geol 225:266-277
Sausse J, Jacquot E, Fritz B, Leroy J, Lespinasse M (2001) Evolution of crack permeability during fluid-rock interaction. Example of the Brezouard granite (Vosges, France). Tectonophysics 336:199-214
Schmickler W (1996) Interfacial Electrochemistry. Oxford University Press, New York
Senthil Kumar O, Soundeswaran S, Dhanasekaran R (2004) Nucleation kinetics and growth of ZnS_xSe_{1-x} single crystals from vapour phase. Mater Chem Phys 87:75-80
Shcheritsa OV, Mazhorovaa OS, Denisov IA, Popova YuP, Elyutin AV (2006) Numerical study for diffusion processes in dissolution and growth of Cd_xHg_{1-x}Te/CdTe heterostructures formed by LPE. Part I. Isothermal conditions. J. Crystal Growth 290:350-356
Shock EL, Helgeson HC (1988) Calculation of the thermodynamic and transport properties of aqueous species at high pressures and temperatures: Correlation algorithms for ionic species and equation of state predictions to 5 kb and 1000 °C. Geochim Cosmochim Acta 52:2009-2036
Siegel DI, Pfannkuch HO (1984) Silicate mineral dissolution at pH 4 and near standard temperature and pressure. Geochim Cosmochim Acta 48:197-201
Söhnel O (1982) Electrolyte crystal aqueous solution interfacial tensions from crystallization data. J Cryst Growth 57:101-108
Soisson F, Martin G (2000) Monte Carlo simulations of the decomposition of metastable solid solutions: Transient and steady-state nucleation kinetics. Phys Rev B 61:203-214
Sonderegger JL, Brower KR, Lefebre VG (1976) A preliminary investigation of strontianite dissolution kinetics. Am J Sci 276:997-1022
Speer J. A., Henseley-Dunn M. L. (1976) Strontianite composition and physical properties. Amer Mineral. 61:10001-10004
Srodon J, Eberl DD, Drits VA (2000) Evolution of fundamental-particle size during illitization of smectite and implications for reaction mechanism. Clays Clay Mineral 48:446-458
Steefel CI, Van Cappellen P (1990) A new kinetic approach to modeling water-rock interactions: the role of nucleation, precursors and Ostwald ripening. Geochim Cosmochim Acta 54:2657-2677
Steefel CI, Lasaga AC (1994) A coupled model for transport of multiple chemical species and kinetic precipitation/dissolution reactions with application to reactive flow in single phase hydrothermal systems. Am J Sci 294:529-592
Steefel CI, DePaolo D, Lichtner PC (2005) Reactive transport modeling: An essential tool and a new research approach for the Earth sciences. Earth Planet Sci Lett 240:539-558
Steefel CL (2008) CrunchFlow. Software for Modeling Multicomponent Reactive Flow and Transport. User's Manual. http://www.csteefel.com/CrunchPublic/WebCrunch.html
Stumm W, Wieland E (1990) Dissolution of oxide and silicate minerals. Aquatic Surface Chemistry. J. Wiley and Sons, New York.
Tobler D, Benning LG, Shaw S (2008) Kinetics and mechanisms of silica nanoparticle formation. Geochim Cosmochim Acta 72:A949-A949
Trolard F, Bilong P, Guillet B, Herbillon AJ (1990) Halloysite - kaolinite - gibbsite - boehmite: A thermodynamical modelisation of equilibria as function of water and dissolved silica activities. Chem Geol 84(1-4):294-297
Tsuzuki Y, Kawabe I (1983) Polymorphic transformations of kaolin minerals in aqueous solutions. Geochim Cosmochim Acta 47:59-66
Van Cappellen P (1990) The formation of marine apatite: a kinetic study. PhD dissertation, Yale Unversity
van der Lee J (1998) Thermodynamic and mathematical concepts of CHESS. Technical Report # LHM/RD/98/39, Ecole des Mines de Paris, Fontainebleau, France.
van der Lee J, De Windt L (2000) CHESS Tutorial and Cookbook. User's Guide Nr. LHM/RD/99/05, Ecole des Mines de Paris, Fontainebleau, France.
Varadarajan E, Dhanasekaran R, Ramasamy P (2001) Modeling of nucleation kinetics of ternary nitrides from vapour phase. J Crystal Growth 230:258-262
Wagner C (1952) Thermodynamics of Alloys. Addison-Wesley Publishing Company Inc.
Wagner C (1961) Theorie der Alterung von Niederschlagen durch Umlosen (Ostwald-Reifung). Z Elektrochem 65:650
Walton AG (1969) Nucleation in liquids and solutions. *In*: Nucleation. Zettlemoyer AC (ed) Marcel Dekker, NY, p 225-307
Warne MR, Allan NL, Cosgrove T (2000) Computer simulation of water molecules at kaolinite and silica surfaces. Phys Chem Chem Phys 2:3663-3668
WebMineral (2008) Atlas Minéralogique: http://webmineral.brgm.fr:8003/mineraux/Main.html
Wilemski G (1984) Composition of the critical nucleus in multicomponent vapor nucleation. J Chem Phys 80:1370-1372
Zangwill A (1990) Physics at Surfaces. Cambridge University Press

APPENDIX 1

The formalism associated to the 3D homogeneous nucleation and growth of spherical particles of fixed composition has been presented in the main text. However, most solids, except amorphous ones, are non-isotropic and their external shape, which departs from the sphere, reflects the relative energies of their low index faces, as recognized by Wulff (Müller and Kern 2000 and references therein). Indeed, Wulff theorem states that, at equilibrium, the distance from the center of a particle to its external facets is proportional to the surface energy of these surfaces. For example, according to Wulff theorem, the aspect ratio of tetragonal particles (basal dimensions $l \times l$ and thickness e), is given by the ratio between the surface energies of the basal and lateral faces:

$$\frac{e}{l} = \frac{\sigma_{bas}}{\sigma_{lat}} \tag{46}$$

This result can be extended to the case of particles in equilibrium with a substrate on which they lie on their basal face. In this case, σ_{bas} is relevant for the face in contact with the AS, and $\sigma_{bas} - W_{adh}$ for the one in contact with the substrate (W_{adh} the adhesion energy). Their aspect ratio is then given by the Wulff-Kaishev theorem:

$$\frac{e}{l} = \frac{\sigma_{bas} - \frac{W_{adh}}{2}}{\sigma_{lat}} \tag{47}$$

If, instead of the basal face, one of their lateral face is in contact with the substrate, their equilibrium shape involves three inequivalent dimensions l, l' and e. The ratio between these lengths is then equal to:

$$\frac{e}{\sigma_{bas}} = \frac{l}{\sigma_{lat}} = \frac{l'}{\sigma_{lat} - \frac{W_{adh}}{2}} \tag{48}$$

With these elements in mind, one can consider heterogeneous nucleation of particles of various shapes. It will become clear that the barrier to nucleation is lowered when particles germinate on a foreign solid, so that the probability of heterogeneous nucleation is usually much higher than the probability of homogeneous nucleation.

Spherical particles on a substrate

Considering spherical particles which nucleate on a substrate, the change in Gibbs free energy in the nucleation process reads:

$$\Delta G = -nk_BT\ln I + A\sigma + A_i(\sigma - W_{adh}) \tag{49}$$

The second and third terms are surface and interface energy terms, respectively, for the particle which has a contact area A with the solution and A_i with the substrate. The equilibrium shape is a spherical cap of curvature radius ρ and wetting angle θ with the substrate (Fig. 20). θ is given by the Young-Dupré equation $-\sigma\cos\theta = \sigma - W_{adh}$ (Israelachvili 1985). For a strong adhesion to the substrate, the wetting angle is equal to 0°, while it is equal to 180° when no wetting occurs (which is also the case for homogeneous nucleation). Writing:

$$A = 2\pi\rho^2(1-\cos\theta) \quad ; \quad A_i = \pi\rho^2\sin^2\theta \tag{50}$$
$$n = 4\pi\rho^3\Phi(\theta)/3v \quad ; \quad \Phi(\theta) = (1-\cos\theta)^2(2+\cos\theta)/4$$

ΔG reads:

$$\Delta G = -nk_BT\ln I + \sigma n^{2/3}v^{2/3}(36\pi\Phi(\theta))^{1/3} \tag{51}$$

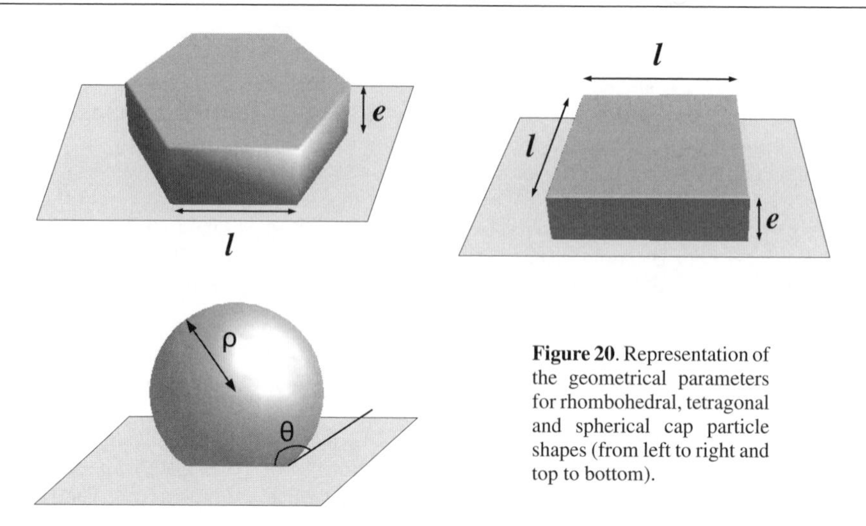

Figure 20. Representation of the geometrical parameters for rhombohedral, tetragonal and spherical cap particle shapes (from left to right and top to bottom).

Tetragonal particles

Tetragonal particles nucleating with their basal face in contact with a substrate have a volume $V = nv = l^2e$ and an aspect ratio $e/l = (\sigma_{bas} - W_{adh}/2)/\sigma_{lat}$. The surface contribution to the change in Gibbs free energy during nucleation is equal to $E_s = l^2(2\sigma_{bas} - W_{adh}) + 4le\sigma_{lat}$, so that:

$$\Delta G = -nk_BT\ln I + 6l^2\left(\sigma_{bas} - \frac{W_{adh}}{2}\right) \tag{52}$$

i.e.:

$$\Delta G = -nk_BT\ln I + 6n^{2/3}v^{2/3}\left(\sigma_{lat}^2\left(\sigma_{bas} - \frac{W_{adh}}{2}\right)\right)^{1/3} \tag{53}$$

If tetragonal particles nucleate on one of their lateral face, their volume is equal to $V = nv = ll'e$ and the surface contribution to ΔG is equal to $E_s = 2ll'\sigma_{bas} + 2l'e\sigma_{lat} + le(2\sigma_{lat} - W_{adh})$. Considering the equilibrium ratio between l, l' and e given by the Wulff-Kaishev theorem, ΔG is equal to:

$$\Delta G = -nk_BT\ln I + 6l^2\left(\sigma_{lat} - \frac{W_{adh}}{2}\right)\frac{\sigma_{bas}}{\sigma_{lat}} \tag{54}$$

i.e.:

$$\Delta G = -nk_BT\ln I + 6n^{2/3}v^{2/3}\left(\sigma_{lat}\sigma_{bas}\left(\sigma_{lat} - \frac{W_{adh}}{2}\right)\right)^{1/3} \tag{55}$$

Rhombohedral particles

A similar derivation can be made for rhombohedral particles nucleating on their hexagonal basal face (side length l, thickness e). The volume of the particles is equal to $V = nv = 3el^2\sqrt{3}/2$ The equilibrium ratio between l and e is given by the Wulff-Kaishev equation:

$$\frac{e}{l} = \sqrt{3}\,\frac{\sigma_{bas} - (W_{adh}/2)}{\sigma_{lat}} \tag{56}$$

The surface contribution to ΔG is equal to $E_s = 3\sqrt{3}l^2(\sigma_{bas} - W_{adh}/2) + 6le\sigma_{lat}$. ΔG is equal to:

$$\Delta G = -nk_BT\ln I + 9l^2\sqrt{3}\left(\sigma_{bas} - \frac{W_{adh}}{2}\right) \tag{57}$$

i.e.:

$$\Delta G = -nk_BT\ln I + 36^{1/3}\sqrt{3}n^{2/3}v^{2/3}\left(\sigma_{lat}^2\left(\sigma_{bas} - \frac{W_{adh}}{2}\right)\right)^{1/3} \tag{58}$$

Nucleation

To summarize, the change in Gibbs free energy during nucleation can be written with the general form:

$$\Delta G = -nk_BT\ln I + n^{2/3}v^{2/3}X\bar{\sigma} \tag{59}$$

with ($\Phi(\theta)$given by Eqn. 50):

$$X = \left(36\pi\Phi(\theta)\right)^{1/3}\ ;\ \bar{\sigma} = \sigma \quad \text{spherical particles} \tag{60}$$

$$X = 6\ ;\ \bar{\sigma} = \left(\sigma_{lat}^2\left(\sigma_{bas} - \frac{W_{adh}}{2}\right)\right)^{1/3} \quad \text{tetragonal particles on basal face}$$

$$X = 6\ ;\ \bar{\sigma} = \left(\sigma_{lat}\sigma_{bas}\left(\sigma_{lat} - \frac{W_{adh}}{2}\right)\right)^{1/3} \quad \text{tetragonal particles on lateral face}$$

$$X = 36^{1/3}\sqrt{3}\ ;\ \bar{\sigma} = \left(\sigma_{lat}^2\left(\sigma_{bas} - \frac{W_{adh}}{2}\right)\right)^{1/3} \quad \text{hexagonal particles on basal face}$$

As a consequence, the critical nucleus size and nucleation barrier have the same expression as in Equation (13), with a mere generalization of the u parameter:

$$n^* = \frac{2u}{\ln^3 I};\quad \frac{\Delta G^*}{k_BT} = \frac{u}{\ln^2 I};\quad u = \frac{4X^3\bar{\sigma}^3v^2}{27(k_BT)^3} \tag{61}$$

Since $\Phi(\theta)$ monotonically increases from zero to one when the wetting angle varies from zero (total wetting) to π (no wetting; homogeneous nucleation), and since $\bar{\sigma}^3$ is always smaller than $\sigma_{bas}\sigma_{lat}^2$, the barrier ΔG^* is always higher for homogeneous than for heterogeneous nucleation.

Growth

The equation driving the time evolution of tetragonal and rhombohedral particle dimensions is similar to Equation (25):

$$\frac{dl}{dt} = \kappa\left[I - \exp\left(\frac{2u}{n}\right)^{1/3}\right] \tag{62}$$

Variation of the particle thickness e is deduced from the Wulff-Kaishev equation, assuming that the particles retain their equilibrium shape during growth. From this, the time dependence of the size n of the particle is obtained, which can be written under the general form:

$$n(t_1,t) = \frac{2u}{\ln^3 I(t_1)} + \int_{t_1}^{t} wn^{2/3}(t_1,t')\left(I(t') - \exp\left[\frac{2u}{n(t_1,t')}\right]^{1/3}\right)dt' \tag{63}$$

with:

$$w = \frac{2\pi\kappa(1-\cos\theta)}{v}\left(\frac{3v}{4\pi\Phi(\theta)}\right)^{2/3} \quad \text{spherical particles} \tag{64}$$

$$w = \frac{3\kappa}{v^{1/3}}\left(\frac{\sigma_{bas}-(W_{adh}/2)}{\sigma_{lat}}\right)^{1/3} \quad \text{tetragonal particles on basal face}$$

$$w = \frac{3\kappa}{v^{1/3}}\left(\frac{(\sigma_{lat}-(W_{adh}/2))\sigma_{bas}}{\sigma_{lat}^2}\right)^{1/3} \quad \text{tetragonal particles on lateral face}$$

$$w = \frac{3\kappa}{v^{1/3}}\left(\frac{9(\sigma_{bas}-(W_{adh}/2))}{2\sigma_{lat}}\right)^{1/3} \quad \text{hexagonal particles on basal face}$$

APPENDIX 2

Some lamellar minerals, like clay minerals (e.g., kaolinite), have a very small aspect ratio, which varies during growth in such a way as to nearly conserve the layer thickness. For them, the assumption that all dimensions of the particle vary with the same growth law, made in the 3D formalism, is non-valid. For example, in the Wulff-Kaishev equation, which gives the equilibrium shape of a rhombohedral particle lying on its hexagonal basal face (see Appendix 1), the two-dimensional (2D) limit is obtained by simultaneously assuming that $\sigma_{bas} - W_{adh}/2 = 0$ and $e \to 0$ (Müller and Kern 2000). The former condition means that the interfacial energy between the particle and the substrate is equal to zero, which is fulfilled when the particle nucleates and grows on a substrate of same nature (homo-epitaxy). It is also necessary that the layer thickness be negligible with respect to other dimensions. In that case, the change in Gibbs free energy during nucleation becomes:

$$\Delta G = -nk_BT\ln I + 4le\sigma_{lat} \tag{65}$$

for tetragonal particles ($nv = el^2$), or:

$$\Delta G = -nk_BT\ln I + 6le\sigma_{lat} \tag{66}$$

for hexagonal particles ($nv = 3el^2\sqrt{3}/2$), which gives the characteristics of the critical nuclei:

$$n^* = \frac{u}{\ln^2 I} \qquad \frac{\Delta G^*}{k_BT} = \frac{u}{\ln I} \quad \text{with} \quad u = \frac{Xev\sigma_{lat}^2}{(k_BT)^2} \tag{67}$$

$X = 4$ and $X = 2\sqrt{3}$ for tetragonal and rhombohedral particles respectively. The overall size variation of the particles reads:

$$n(t_1,t) = \frac{u}{\ln^2 I(t_1)} + \int_{t_1}^{t} wn^{1/2}(t_1,t')\left(I(t') - \exp\left[\frac{u}{n(t_1,t')}\right]^{1/2}\right)dt' \tag{68}$$

with $w = 2\kappa\sqrt{e/v}$ for tetragonal particles and $w = (6\sqrt{3})^{1/2}\kappa\sqrt{e/v}$ for rhombohedral ones.

When growth is limited by diffusion in the solution, Equation (68) transforms into:

$$n(t_1,t) = \frac{u}{\ln^2 I(t_1)} + \int_{t_1}^{t} w' \left(I(t') - \exp\left[\frac{u}{n(t_1,t')}\right]^{1/2} \right) dt' \tag{69}$$

with $w' = 2\kappa e/v$ for tetragonal particles and $w' = 3\kappa\sqrt{3}e / v$ for rhombohedral ones.

It is important to note that in such a situation, the surface energy of the basal face no longer enters the nucleation and growth equations.

APPENDIX 3

This appendix develops the modifications to introduce in the nucleation and growth equations for SS precipitation, when the surface energy depends on composition. It follows treatments made in the context of binary droplet condensation (Reiss 1950; Doyle 1961; Heady and Cahn 1972; Reiss and Shugard 1976; Wilemski 1984; Laaksonen et al. 1993, 1999; Noppel et al. 2002; Gaman et al. 2005), formation of precipitates in metallurgy or semi-conducting devices (Varadarajan et al. 2001; Senthil-Kumar et al. 2004; Los and Matovic 2005; Nagano and Enomoto 2006; Shcherista et al. 2006) and segregation at alloy surfaces (Legrand et al. 1994) or in alloy nanoparticles (Soisson and Martin 2000; Moreno et al. 2006).

The main difference with the simplified treatment, presented in the first part of this course, comes from the existence of surface enrichment effect, associated to (algebraic) excess quantities n_{ACs} and n_{BCs} of AC and BC growth units at the surface of the nucleus. The change in Gibbs free energy $\Delta G(\rho,x)$ is now written :

$$\Delta G(\rho,x) = -[(1-x)n(x) + n_{ACs}]k_B T \ln(\frac{I_{AC}}{1-x}) - [xn(x) + n_{BCs}]k_B T \ln(\frac{I_{BC}}{x}) + 4\pi\rho^2\sigma(x) \tag{70}$$

and its maximum value (with respect to its variation with ρ) amounts to:

$$\Delta G_m(x) = 4\pi 3\rho_m^2(x)\sigma(x) - n_{BCs}k_B T \ln(\frac{I_{BC}}{x}) - n_{ACs}k_B T \ln(\frac{I_{AC}}{1-x}) \tag{71}$$

with $\rho_m(x)$ given by Equation (20).

In order to obtain the critical nucleus composition, via the maximum of the nucleation rate, the derivative of $\Delta G_m(x)$ with respect to x has to be performed. The part which depends upon $d\sigma/dx$ and the excess quantities n_{ACs} and n_{BCs} vanishes, because it represents the Gibbs adsorption isotherm equation (Zangwill 1988) $4\pi\rho^2 d\sigma = -n_{ACs}d\mu_{AC} - n_{BCs}d\mu_{BC}$ (equivalent of Gibbs-Duhem equation for adsorption processes), which reads:

$$4\pi\rho_m^2(x)\frac{d\sigma(x)}{dx} = k_B T\left(\frac{n_{ACs}}{1-x} - \frac{n_{BCs}}{x}\right) \tag{72}$$

The implicit equation which determines x^* is unchanged (Eqn. 23), but the excesses n^*_{ACs} and n^*_{BCs} can now be determined, with the additional assumption that the surface energy does not depend upon the curvature of the surface ($n_{ACs}v_{AC} + n_{BCs}v_{BC} = 0$):

$$n^*_{ACs} = \frac{4\pi\rho^{*2}}{k_B T}x^*(1-x^*)\frac{d\sigma(x^*)}{dx}.\frac{v_{BC}}{v(x^*)} \quad ; \quad n^*_{BCs} = -\frac{4\pi\rho^{*2}}{k_B T}x^*(1-x^*)\frac{d\sigma(x^*)}{dx}.\frac{v_{AC}}{v(x^*)} \tag{73}$$

They are proportional to the critical nucleus area, while n^* is proportional to the nucleus volume. They vanish if the surface energy is composition independent. They contribute to the nucleation barrier:

$$\Delta G^* = \frac{4\pi}{3}\rho^{*2}\sigma(x^*) - \frac{4\pi\rho^{*2}x^*(1-x^*)}{n_m(x^*)}\frac{d\sigma(x^*)}{dx}\frac{d\ln F_0(x^*)}{dx} \tag{74}$$

being responsible for the last term in Equation (74). The latter vanishes if F_0 or σ are independent on x. In that case, surface excesses have no influence on the barrier. In Gibbs dividing surface model, this is consistent with the meaning of excess quantities which do not modify extensive thermodynamic quantities such as free energy and volume.

Excess quantities also contribute to the variation of Gibbs free energy during growth. Using I_i, I_{ACi} and I_{BCi} notations for interfacial saturation states:

$$\Delta G(\rho,x) = -[(1-x)n(x)+n_{ACs}]k_BT\ln(\frac{I_{ACi}}{1-x}) - [xn(x)+n_{BCs}]k_BT\ln(\frac{I_{BCi}}{x}) + 4\pi\rho^2\sigma(x) \tag{75}$$

The cancelation of the derivatives of $\Delta G(\rho,x)$ with respect to ρ and x, in order to write local equilibrium at the particle surface, yields the set of equations:

$$\ln I_i(x) = \frac{2\sigma(x)v(x)}{k_BT\rho} \quad ; \quad \left(\frac{I_{BCi}}{x}\right)^{v_{AC}} = \left(\frac{I_{ACi}}{1-x}\right)^{v_{BC}} \tag{76}$$

and:

$$n_{BCs} = -\frac{4\pi\rho^2}{k_BT}x(1-x)\frac{d\sigma(x)}{dx}.\frac{v_{AC}}{v(x)} \quad ; \quad n_{ACs} = \frac{4\pi\rho^2}{k_BT}x(1-x)\frac{d\sigma(x)}{dx}.\frac{v_{BC}}{v(x)} \tag{77}$$

The composition x of the deposited layers is unchanged with respect to the simplified derivation, and can be introduced in Equation (77) to determine n_{ACs} and n_{BCs}. The time evolution of the particle radius is unchanged and only the amounts $N_A(t)$, $N_B(t)$ and $N_C(t)$ of A^{+q}, B^{+q} and C^{-q} species have to be modified:

$$N_A(t) = \int_0^t F(t_1)n_{ACs}(t_1,t)dt_1 + \int_0^t F(t_1)(n^*(t_1)-1)(1-x(t_1,t))dt_1 \tag{78}$$
$$+\int_0^t F(t_1)dt_1\int_{t_1}^t dt_2\frac{4\pi\rho(t_1,t_2)^2}{v(t_1,t_2)}\frac{d\rho(t_1,t_2)}{dt_2}(1-x(t_1,t_2))$$

A similar expression holds for $N_B(t)$ with $(1 - x(t_1,t))$ replaced by $x(t_1,t)$ and $n_{ACs}(t_1,t)$ replaced by $n_{BCs}(t_1,t)$. $N_C(t)$ is the sum of $N_A(t)$ and $N_B(t)$.

Reviews in Mineralogy & Geochemistry
Vol. 70 pp. 411-434, 2009

Towards an Integrated Model of Weathering, Climate, and Biospheric Processes

Yves Goddéris, Caroline Roelandt, Jacques Schott

Laboratoire d'étude des Mécanismes et Transferts en Géologie
Observatoire Midi-Pyrénées
CNRS-Université de Toulouse
Toulouse, France

Marie-Claire Pierret

Centre de Géochimie de la Surface
Université de Strasbourg
Strasbourg, France

Louis M. François

Laboratoire de Physique Atmosphérique et Planétaire
Université de Liège
Liège, Belgium

Contact e-mail: godderis@lmtg.obs-mip.fr

INTRODUCTION

Subaerial weathering of the continental rocks is an important component of global biogeochemical cycles. During the dissolution of continental rocks, atmospheric CO_2 is consumed resulting in alkalinity production and its transfer to the ocean via river transport. Atmospheric carbon is also consumed during the dissolution reaction itself, as illustrated by the dissolution equation of plagioclase:

$$4Na_{0.5}Ca_{0.5}Al_{1.5}Si_{2.5}O_8 + 17H_2O + 6CO_2 \rightarrow 3Al_2Si_2O_5(OH)_4 + 2Na^+ + 2Ca^{2+} + 6HCO_3^- + 4H_4SiO_4 \quad (1)$$

Plagioclase dissolution produces dissolved species (basic cations, bicarbonate ions, and silica) and clay minerals which can precipitate locally. Dissolution of carbonate minerals also consumes atmospheric carbon, for example:

$$CaCO_3 + H_2O + CO_2 \rightarrow Ca^{2+} + 2HCO_3^- \quad (2)$$

Bicarbonate ions generated from either reaction can be transported to the ocean where it is mixed by the oceanic thermohaline circulation. Both reactions (1) and (2) thus remove carbon from the atmosphere and store it into the oceanic dissolved inorganic carbon reservoir at a timescale of the oceanic mixing (around 5000 yrs today). The total amount of atmospheric carbon being removed through this process has been estimated from the inventory of the amount of bicarbonate ions carried by the world major rivers. It reaches today 0.288 gigatons of carbon per year (GtC/yr) (Gaillardet et al. 1999), 0.14 GtC/yr being consumed by silicate weathering (a number updated to 0.163 GtC/yr by Dessert et al. 2003), and 0.148 GtC/yr by carbonate minerals dissolution. These fluxes are comparable to the net exchange fluxes between the ocean and the atmosphere

1529-6466/09/0070-0009$05.00 DOI: 10.2138/rmg.2009.70.9

(0.6 GtC/yr, pre-industrial state) and between the atmosphere and the land biosphere (0.4 GtC/yr, pre-industrial state), making continental weathering a potentially important component of the short term anthropogenic global carbon cycle. This importance has been recently stressed in two studies showing that continental weathering is strongly affected by human activities. Indeed, the HCO_3^- discharge of the Mississippi river has increased by about 25% over the last 40 years probably in response to anthropogenic land use change and agricultural practices (Raymond et al. 2008). CO_2 consumption by basalt weathering has increased by about 4 to 14% since 1960 in Iceland, an increase probably driven by glacier melting and rising temperatures forced by anthropogenic climatic change (Gislason et al. 2008).

FACTORS CONTROLLING WEATHERING AT THE WATERSHED SCALE

Considerable efforts are currently focused on modeling the response of the atmospheric, oceanic, and biospheric carbon cycle components to the ongoing climate change, including geophysical transport and biogeochemical processes. But very little attention has paid to weathering. Weathering processes are not easy to model because they heavily depend on numerous processes which are listed below (Brantley et al. 2008).

(1) The presence and the type of vegetation above the bedrock are among the most important driving factor. Indeed, land plants enhance weathering rates through root respiration (increasing subsurface CO_2 concentrations, and hence the acidity of the soil solutions), through the production of organic acids, and through the mechanical effect of roots (Drever 1995; Moulton et al. 2000; Berner 2004). Land plants also heavily impact the silica cycle in soils (Gérard et al. 2008), especially in tropical environments, via the accumulation of phytolites in soils (Lucas 2001).

(2) The soil hydrological behavior is another major factor, itself heavily affected by the climatic regime and the type of the standing vegetation. Water content in the weathering profile defines the wetness state of the mineral surface, and water fluxes through the weathering profile (either vertical or horizontal) remove the weathering products. Both define the residence time of the soil solution.

(3) The weathering rates are also correlated to mechanical erosion (Millot et al. 2002), which removes the soil cover, thus exposing fresh mineral surfaces to weathering agents.

(4) The mineral composition of the bedrock is also a first order controlling factor of weathering intensity. For instance, basaltic rocks weather about 8 times faster than granitic rocks, and are thus more efficient CO_2 consumers (Dessert et al. 2001, 2003). Carbonate rocks dissolve up to 8 orders of magnitude faster than silicate minerals, explaining why their contribution to CO_2 consumption is equivalent to that of silicates despite their small outcrop area (10.4% of the continental area (Dürr et al. 2005)).

(5) The thermodynamic properties of clay minerals appear to be a key factor. Clay minerals are the products of silicate rock weathering and they build up in the weathering profile with time. As such, they control the concentration of silica, cations, and H^+ in soil interstitial solutions, and hence the solution saturation index with respect to both primary and secondary silicate minerals. As weathering rates of primary minerals near equilibrium depends on departure from equilibrium (i.e., saturation index), CO_2 consumption linked to silicate mineral dissolution depends on the thermodynamic parameters of the precipitating clay minerals. Those parameters, especially the solubility product, are well constrained only for few "model" well crystallized clay minerals. But in general poorly crystallized solid-solution whose thermodynamic parameters are poorly known form in the soil. For instance, values of the solubility product of a smectite mineral (namely the Ca-montmorillonite)

may span 15 order of magnitude in field modeling studies (Goddéris et al. 2006; Maher et al. 2009) depending on its crystallinity and chemical composition (i.e., Fe^{3+} content).

All those critical parameters are intimately intertwined. For instance, climate controls vegetation, which strongly affects the subsurface hydrology, which in turn defines the residence time of water and the saturation state of the solution with respect to silicate minerals. A complex chain of environmental and mineralogical parameters thus control CO_2 consumption by weathering reactions at different scales from the soil profile to the watershed scale.

PHENOMENOLOGICAL MODELING OF WEATHERING PROCESSES AT THE WATERSHED SCALE

Modeling this complex chain of interactions is a challenge. The easiest way to work is to let nature to integrate all the factors by itself. The perfect integrators of weathering processes are the rivers, and hence the watershed. Numerous studies have focused on measurements of dissolved products at the outlet of a large number of catchments, relating those fluxes to weathering rates. The calculated weathering rates can then be related to environmental parameters through phenomenological laws (Meybeck 1987; Brady 1991; Bluth and Kump 1994; Amiotte-Suchet and Probst 1995; White et al. 1999; Dessert et al. 2003; Oliva et al. 2003). One of the most extensive study has been performed by Oliva and co-authors (Oliva et al. 2003) for 99 small granitic watersheds. They correlate the CO_2 consumption rate (F_{CO2}) to the annual runoff (*Run*) and mean annual temperature (T).

$$F_{CO_2} = k \cdot Run \cdot \exp\left(\frac{-E_a}{R} \cdot \left(\frac{1}{T} - \frac{1}{T_0}\right)\right) \tag{3}$$

R is the gas constant, E_a is an apparent activation energy calculated at 48.7 kJ/mol, T_0 is a reference temperature and k is a constant. Although there is a rather good linear correlation between runoff and weathering rates (Fig. 1), the dependence on temperature is much less obvious, and is only suggested at runoffs above 1000 mm/yr. The scatter of the data points is not surprising given the complex chain of parameters listed above. Mathematically speaking, this scatter means that k is not a true constant. Indeed, the impact of the various vegetation types from one catchment to another, the influence of clay minerals, the role of mechanical erosion,

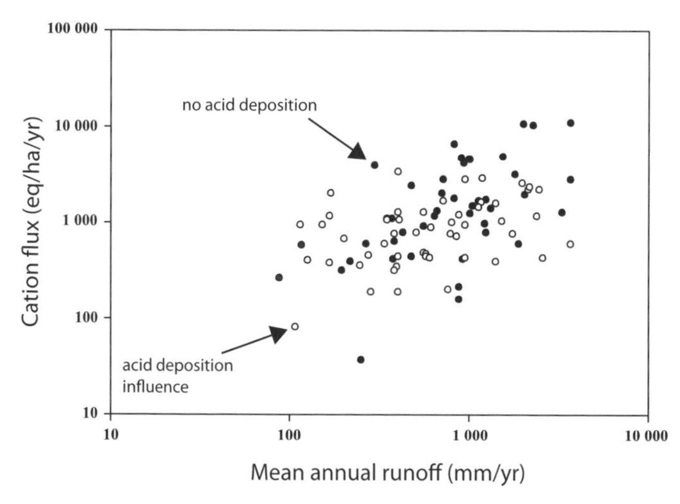

Figure 1. Specific cation export as a function of mean annual runoff for 99 small granitic watersheds. The atmospheric deposition has been removed from the export fluxes. Adapted from (Oliva et al. 2003).

the variability in the mineralogical composition of the granitic bedrock, are all hidden within the *k* constant.

An analogous phenomenological law has been derived for basaltic lithologies from the compilation of discharge data from 12 watersheds (Dessert et al. 2003). The apparent activation energy is calculated at 42.3 kJ/mol and the scatter of the data points is less important, possibly reflecting the lower mineralogical variability of the basaltic outcrops.

Those phenomenological rate laws have been extensively used for quantifying the CO_2 weathering sink in the geological past, and hence for reconstructing the past climatic evolution of the Earth (Wallmann 2001; Berner 2004, 2006; Donnadieu et al. 2004, 2006; Goddéris and Joachimski 2004; Goddéris et al. 2008). However, such relationships suffer from several shortcomings:

(1) They only give an instantaneous snapshot of the relationship between large scale climatic parameters (e.g., temperature and runoff). Indeed they reflect a geographical diversity of the weathering dependence on climate, and ignore the weathering system dynamics.

(2) They predict that the most efficient weathering and CO_2 consumption occurs in humid equatorial environments, where temperature and runoff are maximal. However, Millot and co-authors (Millot et al. 2002) have shown that high weathering rates can only be sustained if it is combined with an intense mechanical erosion. This is not the case for the major equatorial watersheds, including most of the Amazon basin, the Congo river, and the Orinoco (c.f. Braun et al. 2005 for a description of weathering processes in low erosion tropical environments). Since humid equatorial areas cover about 20% of the continents, the contribution of weathering processes to global geochemical cycles may be overestimated. The discrepancy between these phenomenological model predictions and the observation of weathering efficiency in tropical areas is related to the low number of tropical watershed considered in the climate/weathering correlations. Most of the monitored catchments in the world are located in the mid-latitudes of the Northern Hemisphere, where temperature and/or runoff increase will indeed enhance weathering rates. Phenomenological models are thus biased against tropical environments.

(3) As mention above, the rate constant *k* is implemented to account for a variety of parameters (lithology, land plants, physical erosion, clay mineral thermodynamic). However as a consequence of being integrated within the proportionality constant, these parameters are not explicitly modeled. As anthropogenic perturbations may cause wide fluctuations in those parameters (biome distribution, enhanced mechanical erosion), their impact is complex and cannot be explored with the simple phenomenological models.

The weathering system is highly dynamic. As mentioned in the introduction, anthropogenic perturbations are suspected to cause wide and rapid fluctuations in weathering rates at the decedal scale in response to land use and climatic changes, and the mechanisms at play are complex. Such a reactive behavior cannot be explored and understood with simple phenomenological models because these are too integrative. Several key questions cannot be assessed with integrative phenomenological models. For instance, what is the role of trace minerals reactivity on the CO_2 consumption budget of the watersheds? What is the response of the CO_2 consumption through weathering to biospheric changes, either induced by climatic changes or human activities? What is the importance of clay minerals in the weathering budget and how will their stability evolves in the near future for the various environments encountered on Earth? Therefore, we focus our efforts on the building up and use of mechanistic process-based models describing weathering in the natural environments. Such models are based on laboratory derived kinetic laws that describe the dissolution/precipitation rates of minerals

(Sverdrup and Warfinge 1995; Goddéris et al. 2006; Boyle 2007; Maher et al. 2009). To further account for the role of the continental biosphere, we have coupled a mechanistic model to a regional vegetation model describing the soil water fluxes, CO_2 pressure in soil solutions, and cation uptake and release by the dead and living biomass (Rasse et al. 2001), all parameters that have a major impact on weathering processes. This coupled model has been successfully used to simulate weathering fluxes in a small granitic watershed (80 ha) located in a temperate forested environment (Goddéris et al. 2006).

In the present study, we apply the same coupling strategy to estimate silicate weathering over large spatial scales. The mechanistic weathering model WITCH is coupled to a large scale dynamic vegetation model (LPJ, Sitch et al. 2003). The coupled model is applied to a tropical continental watershed (the Orinoco) to assess its capabilities to estimate the major cation export of a continental scale drainage basin.

THE WEATHERING MODEL

The weathering model calculates the time evolution of the chemistry of the soil solutions as they percolate along the weathering profile. The model has the structure of a vertical box model. The time evolution of the soil solution chemistry is calculated at each time step solving the following equation:

$$\frac{d\left(z\cdot\theta\cdot C_j\right)}{dt} = F_{top} - F_{bot} + \sum_{i=1}^{n_m} F^i_{weath,j} - \sum_{i=1}^{n_m} F^i_{prec,j} \pm R_j \tag{4}$$

where C_j is the concentration of the species j. WITCH includes a budget equation for Ca^{2+}, Mg^{2+}, K^+, Na^+, SO_4^{2-}, total alkalinity, total aluminum, total silica, and total phosphorus. z is the thickness of the considered layer, and θ the water volume fraction. F_{top} is the input of species from the layer above the considered layer in moles per unit of time through drainage, and F_{bot} is the removal through downward drainage. Both the drainage and the water volume fraction are taken from the biospheric model explained below, and can potentially fluctuate through time. R_j stands for the cation flux between the soil solution and the clay-humic complex. This cation exchange flux is set to zero when a pluri-annual steady-state is calculated to save computation time. However, it is fully accounted for in a dynamic mode for seasonal simulations through the computation of a Fick diffusion law.

$F^i_{weath,j}$ is the release of the species j through dissolution of the mineral i in the considered box, and $F^i_{prec,j}$ is the removal through precipitation when saturation is reached (see below). n_m is the total number of minerals considered in the simulation. Those terms are calculated for silicate minerals through laboratory kinetic laws derived from transition state theory (TST) (Eyring 1935) and rates are normalized to the BET surface. The overall dissolution rate F_g of mineral g inside a given layer is the sum of four parallel elementary reactions promoted by H^+, OH^-, water and organic ligands, respectively:

$$F_g = A_g' \cdot \left[\sum_l k_{l,g} \cdot \exp\left(\frac{-E^l_{a,g}}{RT}\right) \cdot a_l^{n_{l,g}} \cdot f_{inh}\right] \cdot \left(1 - \Omega_g^s\right) \tag{5}$$

In this relation a_l and $n_{l,g}$ stands for the activity of the lth aqueous species and the reaction order with respect to the lth species, respectively. For organic ligands, a_l equals the activity of a generic organic species $RCOO^-$ (conjugate of RCOOH). $k_{l,g}$ is the elementary reaction dissolution rate constant of mineral g with respect to the lth species, and $E^l_{a,g}$ is the activation energy for this reaction. f_{inh} stands for the inhibition factors, and the last factor described the effect on rates of the departure from equilibrium, where Ω_g is the solution saturation index with respect to mineral g, and s is a stoichiometric number. Mineral dissolution or precipitation occurs for $\Omega_g < 1$ and $\Omega_g > 1$, respectively. A' is the BET reactive surface area calculated through a phenomenological law for each layer of the weathering profile:

$$A' = \left(8.0x_{clay} + 2.2x_{silt} + 0.3x_{sand} + 0x_{coarse}\right) \cdot \rho \quad (6)$$

where x_{clay}, x_{silt}, x_{sand} and x_{coarse} are the textural fractions of the soils clay (size < 2 μm), silt (2 μm < size < 60 μm), sand (60 μm < size < 250 μm), and coarse (size > 250 μm), materials (Sverdrup and Warfinge 1995). The sum $x_{clay}+x_{silt}+x_{sand}+x_{coarse}$ always equal 1. ρ is the soil density in g/m^3 for each layer. This total reactive surface calculated for each layer is then distributed among the various minerals according to their volume abundance.

Carbonate mineral dissolution has been added to WITCH and modeled within the framework of TST and surface coordination chemistry concepts. In this study we only consider calcite whose overall reaction rate F_{cal} is given by:

$$F_{cal} = \left(k_{\mathrm{H}}^{cal} \cdot a_{\mathrm{H}^+} + \frac{k_o}{1 \cdot 10^{-5} + a_{\mathrm{CO_2}}}\right) \cdot \left(1 - \Omega_{cal}^{1.0}\right) \quad (7)$$

where k_H^{cal} equals $10^{-0.659}$ mol/m^2/s and k_o 10^{-11} mol/m^2/s at 25 °C (Chou et al. 1989). Activation energies for k_H and k_o are set to 8.5 and 30 kJ/mol, respectively. Additional details about the weathering model can be found in Goddéris et al. (2006).

THE COUPLED WEATHERING/BIOSPHERIC MODEL

Generic features

The water volume fraction and the vertical drainage required by the weathering model are calculated by the hydrological module of the biospheric models (Rasse et al. 2001; Sitch et al. 2003; Gerten et al. 2004). The uptake of dissolved species from the soil solutions by the living vegetation are proportional to the primary productivity calculated by the biospheric model and to a constant species/carbon ratio (taken as the Redfield ratio of the continental vegetation). Uptake only occurs in the root zone, and the distribution of the uptake all along the root depth parallels the calculated uptake of water by plants. The release of species through the decay of organic matter in the soil horizons is calculated proportional to the organic carbon decay estimated by the biospheric model, and to the Redfield ratio. On an interannual basis, the vegetation is assumed to be at steady-state, meaning that the mean annual uptake of species equals the mean annual release. Once the water enters the soil, only vertical drainage is supposed to occur (see Fig. 2).

Specific features

Depending on the investigated spatial scale (from small watershed to the continental scale), the chemical weathering model WITCH is coupled to various vegetation productivity models. The investigated sites are also briefly described below.

Small watershed scale

The site. The study site is the Strengbach catchment (80 ha) located in the Vosges mountains, 58 km SW from Strasbourg. This watershed is an equipped environmental observatory with permanent sampling and measuring stations. It is a base-poor granitic monolithological catchment. The elevation ranges from 883 m at the outlet to 1146 m at the catchment divide. Monthly mean temperature ranges from –2 to +14 °C (Probst et al. 1990). Mean annual rainfall is about 1400 mm/yr and the total runoff averaged 853 mm/yr over 1986-1995. The vegetation consists of 130-year-old beech trees covering 20% of the catchment surface located on the northern flank. The remaining 80% are covered with 80-year-old spruces. The average thickness

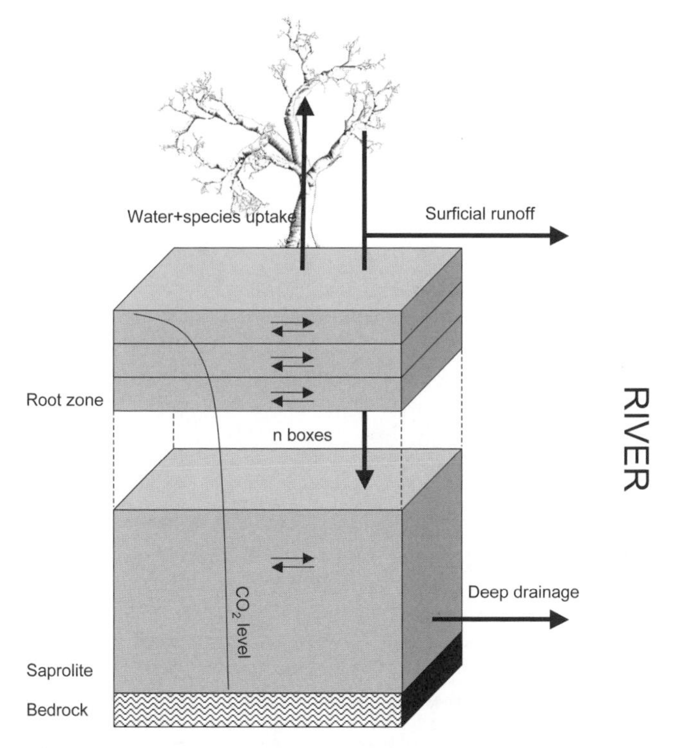

Figure 2. Vertical structure of the coupled biospheric-weathering model. The double arrows in the boxes stand for the weathering/precipitation and cation exchange with the soil exchange complex.

of the soil is 80-100 cm and the saprolite can reach 10 m at some place. The catchment has been monitored for 20 years and a time series of meteorological, hydrological, geochemical data are available online (*http://ohge.u-strasbg.fr/index.html*). Water composition has been analyzed both in the main stream and at different depth in various soil profiles.

The coupled biospheric-weathering model. The biospheric productivity model used for regional scale simulation is the ASPECTS model (Rasse et al. 2001). ASPECTS is a fully coupled scheme of the water and carbon cycles in the vegetation and soils of temperate forests. Water/carbon stocks and fluxes are calculated every 2 minutes. The ASPECTS model is forced by hourly meteorological data (air temperature, precipitation, global radiation, air relative humidity and wind speed) measured at the site of study (see Table 1). Soil water content is computed by solving the Richards' equation (Viterbo and Beljaars 1995), setting the vertical water flux to the infiltration rate at the soil surface and imposing free drainage at the bottom. Soil temperature is calculated by solving the heat diffusion equation assuming, as boundary conditions, that temperature is equal to the measured surface air temperature at the soil surface and to its measured annual mean surface temperature at the bottom of the lowest soil layer. Soil CO_2 production and pressure is calculated from the CO_2 production rate (soil microbial respiration + root respiration) in each layer and by solving a diffusion equation (Fang and Moncrieff 1999). The coupled ASPECTS-WITCH model is run over 7 vertical layers for two locations in the catchment, from the surface down to 10 m depth: one below beech trees, and one below spruce trees (Table 2). As explained in (Goddéris et al. 2006), the integration of the weathering fluxes over the whole catchment assumes that 20% of the catchment behave as the simulated weathering profile below beech trees, and the remaining 80% as the profile below

Table 1. Forcing data

	Regional simulation ASPECTS + WITCH Strengbach watershed	**Continental simulations LPJ + WITCH Orinoco watershed**
Climate data (rainfall, temperature, global radiation)	On site 1994-1995 hourly measurements	Climate research unit monthly data, spatialized on a 05°×0.5° grid (New et al. 2000; Mitchell and Jones 2005)
Rain chemistry	On site weekly measurements over the 1994-1995 hydrological year	Constant rain chemistry over the catchment (central Amazon values; Tardy et al. 2005), except for Na^+ and Cl^- (concentrations decreasing with the nearest coastline along the main wind flow)
Soil mineralogy	On site measurements (Fichter 1997)	Inferred from the ISRIC world soil database (Batjes 2005)
Bedrock and saprolite mineralogy	On site measurements (Fichter 1997)	Inferred from Amiotte-Suchet et al.'s (2003) lithological map (1°×1° spatial resolution)
DOC soil content	On site measurements (constant value) (Dambrine et al. 1995)	Fixed values from 12 mg/l in the top 50 cm to 0 mg/l in the saprolitic horizon
Textural fraction	On site measurements (Aubert 2001)	ISRIC world soil database (Batjes 2005)

spruce trees. The spatial scale of the simulations is thus metric, and then upscaled to 80 ha. This procedure is supported by the uniformity of the vegetation cover and of the underlying bedrock. In this regional scale modeling exercise, no surficial runoff has been considered. So the river is fully fed by the deep drainage through the complete weathering profile. This is justified by the fact that the Strengbach soils are sandy and stony, and thus characterized by an efficient drainage. However, peak runoff events cannot be accurately modeled since surficial runoff is potentially an important contributor to the total runoff during these short time events. Again a full description of the simulation setting can be found in (Goddéris et al. 2006).

Simulation scheme. Weathering processes are simulated over the hydrological year 1994-1995. ASPECTS is first integrated over several decades to simulate the growth of trees (Rasse et al. 2001). Then the WITCH model is run for 20 years assuming constant temperature, vertical drainage, and water volumetric contents so that initial conditions are relaxed and the soil solution chemical composition reaches a steady-state. The constant forcing is set to their averaged simulated 1994-1995 values by the ASPECTS model. Then the WITCH model restarts from this steady-state now driven by the seasonal output from ASPECTS for the hydrological year 1994-1995. The model is run cyclically for 5 years with the 1994-1995 forcing, and the results for the last year are investigated.

Continental scale

The site. The chosen site for investigation is the Orinoco watershed. The Orinoco River originates in the Sierra Parima at an elevation of 1074 m, and receives more than 2000 tributaries during its 1970 km course to the Atlantic Ocean. With a surface of 836,000 km^2, it is the third largest drainage basin in South America. With a mean annual discharge of 32,980 m^3/s, the Orinoco is the third highest water discharge in the world. The climate of the watershed is

Table 2. Mineralogical composition of the Strengbach watershed weathering profiles in volume percent.

	HP site (under beech trees)	PP site (under spruce trees)
Albite An6	0-20 cm: 16% 20-70 cm: 15.3% below 70 cm: 11%	0-30 cm: 5% 30-90 cm: 5.6% below 90 cm: 10%
K-feldspar	0-20 cm: 27.8% 20-70 cm: 22.1% below 70 cm: 15.8%	0-30 cm: 16.2% 30-90 cm: 12.2% below 90 cm: 9.9%
Quartz	0-20 cm: 8.5% 20-70 cm: 7.6% below 70 cm: 6%	0-30 cm: 7.9% 30-90 cm: 6.6% below 90 cm: 5.4%
Muscovite	0-20 cm: 12.1% 20-70 cm: 18.8% below 70 cm:24.6%	0-30 cm: 19.8% 30-90 cm: 27.8% below 90 cm: 42.6%
Biotite	0-20 cm: 4.1% 20-70 cm: 8.1% below 70 cm: 10.6%	0-30 cm: 6.1% 30-90 cm: 5.5% below 90 cm: 8.5%
Kaolinite	0-20 cm: 1.1% 20-70 cm: 1.8% below 70 cm: 0.7%	0-30 cm: 2.6% 30-90 cm: 2.4% below 90 cm: 0.9%
Smectite (Montmorill.)	0-20 cm: 15.2% 20-70 cm: 14.4% below 70 cm: 20.8%	0-30 cm: 11.6% 30-90 cm: 13.2% below 90 cm:10.4%
Illite	0-20 cm: 15.1% 20-70 cm: 11.5% below 70 cm: 10.3%	0-30 cm: 31% 30-90 cm: 26.4% below 90 cm: 11.6%
Apatite	0-150 cm: 0% below 150 cm: 0.3%	0-180 cm: 0% below 180 cm: 0.8%

driven by the oscillations of the Intertropical Convergence Zone (ITCZ). The period from June through November is characterized by heavy rainfall while a dry season extends from December to April. The strong seasonality in rainfall induces strong differences in discharge volume between low and high stage. Mean monthly discharge measured at Ciudad Bolivar (ORE HYBAM data from *http://www.ore-hybam.org* using "IRD-LMTG/UCV" convention) for the period 1926 to 2004 varied between 2135 m^3/s and 87530 m^3/s. The mean monthly temperature ranges between 20 °C and 32 °C. The Andes and the Caribbean coastal ranges constitute the Western and Northern border of the basin. The Northern part of the watershed is dry and covered by grasslands. In the South, precipitation is higher than 2000 mm/year; the vegetation cover is composed of tropical rain forests. Agriculture impact is very small and limited to a few percent of the watershed surface.

The Orinoco basin's lithology is described by (Edmond et al. 1996) as "the result of the complex inter-play between sea level changes, the emergence of the Guyana shield weathering and tectonics." The Guyana Shield on the right riverbank is a very old, deeply eroded silicate complex relatively uniform and characterized by the nearly complete lack of limestone and evaporites (Edmond et al. 1995). On the left riverbank, the plains are composed of very young sedimentary terrains.

The coupled biospheric-weathering model. We apply the same modeling strategy than described above except that the biospheric model used at the continental scale is the LPJ global dynamic vegetation model (Sitch et al. 2003) (see Fig. 3). Indeed, the modeling of weathering

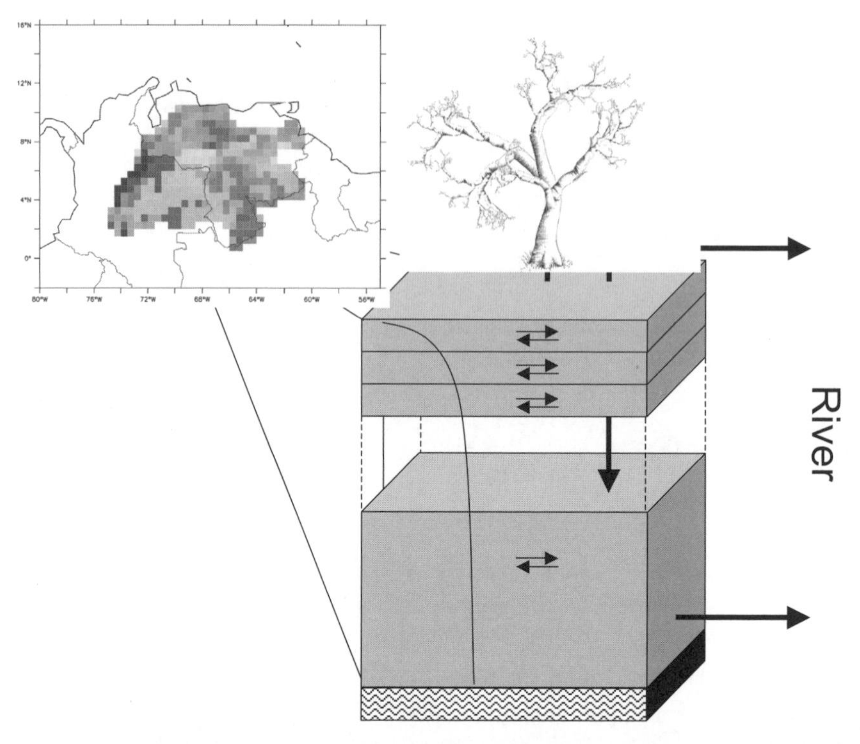

Figure 3. Simulation scheme at the continental scale. The vertical profile is applied to each continental grid cell. No transfer of water from one grid cell to the neighboring is accounted for owing to the large size of the grid elements (0.5° lat × 0.5° long).

processes at the continental scale raises the important question concerning upscaling. The biospheric-weathering model is a process-based model. Hence our aim is to apply the same description of the physical processes from the soil to the continental scales. But two difficulties arise: (1) the high spatial variability of the vegetation, and (2) the high variability of the underlying bedrock and overlying soil mineralogical compositions. Whereas both parameters were rather uniform for the Strengbach catchment, such is not the case here. Global biospheric models solve the first problem through the introduction of plant functional types (PFTs), instead of modeling the true vegetation itself. The physical parameters (hydrological behavior, carbon allocation process, etc.) are known for each PFT. The complexity of the continental vegetation is reduced to a number of key plant types. The continental surface is thus represented as a patchwork of PFTs, allowing the use of a process based model to describe the continental vegetation dynamic: each PFT is defined by a variety of optical, morphological, and physiological parameters (see Fig. 4). Regarding the bedrock mineralogy, (Amiotte-Suchet et al. 2003) define world lithological maps defining the main bedrock type with a spatial resolution of 1° lat × 1° long (see Fig. 5). Each continental grid element can thus be assigned a lithology map, and a standard mineralogy for each lithological type can be defined. This method is not as efficient as the PFT representation of the continental vegetation. Indeed, a single lithological type is defined for each grid element owing to the spatial resolution of the available data, while various PFTs may share the same area. As a result, the local variability of the lithology is not described. This is not a major limitation when simulating weathering above relatively uniform geological features (such as shields, or igneous provinces), but becomes critical when the lithology is more complex (namely with the so-called shale lithology). The

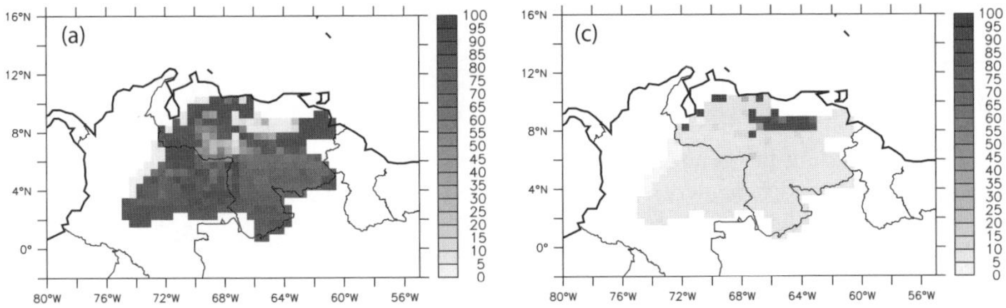

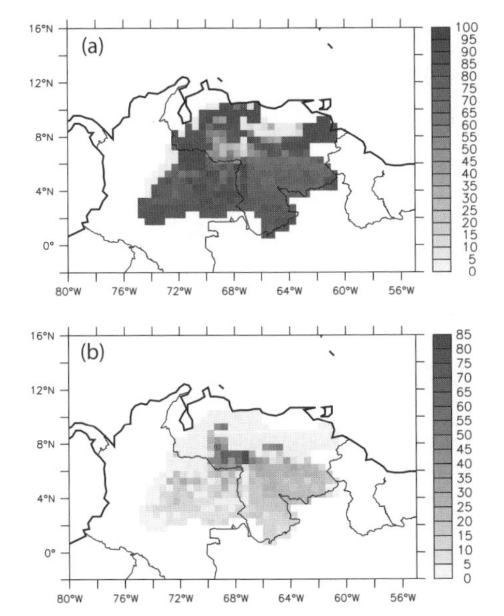

Figure 4. Potential vegetation of the Orinoco watershed simulated by the LPJ model. The color indicates the in percentage of surface covered with each type vegetation. (a) tropical broad-leaved evergreen trees, (b) tropical broad-leaved raingreen trees, (c) tropical herbaceous.

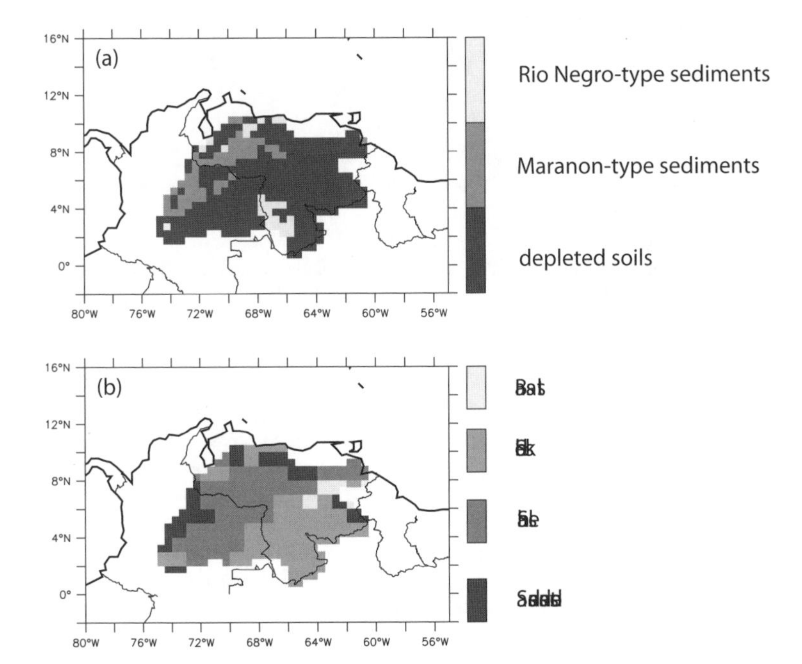

Figure 5. (a) Soil type of the Orinoco watershed deduced from the ISRIC database. (b) lithological map of the Orinoco watershed.

(Amiotte-Suchet et al. 2003) map accounts for 6 lithological types (sands and sandstones, shales, shield rocks, basalts, acid volcanic rocks, and carbonates). Soil definition is also a critical boundary condition when modeling weathering at the continental scale. Data about the soil (or weathering profile) thicknesses from one grid element to the other, and about the mineralogical composition of those layers are generally not available at the continental scale. We thus define the following procedure.

First, regarding the geometry of the weathering profile, we stick to the geometry of the LPJ biospheric model, namely a two layer model (L1: 0 to 0.5 m depth; L2: 0.5 to 1.5 m depth). We add a third layer accounting for the saprolitic horizon L3 with a prescribed depth ranging from 1.5 to 6.5 m, assuming that below this depth the water is no more percolating. The water volume fractions and the vertical water fluxes (including water uptake through evapotranspiration) are taken from the LPJ model. The water volume fraction calculated for L2 is applied to L3. Given the size of the grid element (about 50 km × 50 km) we only account for the vertical drainage, assuming that all the water flowing through the grid element reaches a river flowing above the element. No horizontal transfer is thus modeled in this study.

Then each grid cell is divided into fractions of carbonated and non-carbonated lithology. The carbonate fraction of each grid cell is determined from (Lopez et al. 2003). The weathering calculations are performed independently for each subgrid fraction. The mineralogical composition of carbonated rocks is assumed to be 100% calcite for all the layers (from layer 1 down to layer 3). Dissolution calculation of those outcrops is assumed to drive instantaneously the draining waters to equilibrium with respect to calcite, an assumption valid on a mean annual basis.

For non-carbonated rocks, the mineral composition of the soil profile was defined by a two step procedure consisting of defining a mineralogical composition for the L1 and L2 layers independently of the deep soil layer L3 (Fig. 5). Soil layers 1 and 2 properties are the result of the in-situ soil evolution (generally characterized by highly depleted soils in tropical humid environment) and the possible addition of materials produced by physical weathering elsewhere (which can be a mixing of lightly weathered materials).

The surficial clay CEC data (Batjes 2005) were therefore used as a proxy to define the age of the soil profile, assuming that a low CEC value is characteristic of highly cation-depleted soils. Threshold in the CEC values were defined so that the soil age distribution corresponded to the soil types described in the Soil Map of the World (FAO-UNESCO) and the best agreement between the model output and the data is reached. The value of the threshold thus requires calibration, rather than based on a true physical boundary, especially given the large uncertainties inherent to the ISRIC global database. Below 36 cmol/kg it is hypothesized that the soils are mainly composed of kaolinite (90%) and quartz (10%) (see Table 3). For values equal or higher than 36 cmol/kg the soils are supposed to be young and not well developed, their mineralogy is defined based on the sediment composition of the Maranon river for rivers draining the Andes (Guyot et al. 2007) (see Table 3). We use the observed Rio Negro sediment composition when young soils occur on the Guyana Shield (Guyot et al. 2007). As a result of this procedure, young undifferentiated soils are located along the western boundary of the watershed, thus mainly in the uplands where physical erosion should produce such sediments. The Guyana shield is mostly covered with highly weathered soils (Fig. 5). Finally, 0.5% calcite are added to the young undifferentiated sediments of the western boundary because calcite weathering has been identified as a source of Ca^{2+} in the Andes (Stallard 1985). Such calcite mineral sources are hidden by the low spatial resolution of the lithological database (Amiotte-Suchet et al. 2003). Furthermore, the mineralogical analysis of the Maranon river sediments was made only on the decarbonated fraction. We fix the amount of calcite in these soils so that the calculated Ca^{2+} dissolved load of the Guaviare and Inirida rivers are comparable to the data.

The mineralogical composition of layer 3 (assumed to be the saprolite zone for shield rocks) is summarized in Table 4 as a function of the lithological type. In case of shale lithology, it is

Table 3. Mineral abundances in volume percent in soil layers above the saprolite as a function of the cation exchange capacity and spatial location (Guyot et al. 2007).

Mineral	CEC < 36 cmol/kg	CEC > 36 cmol/kg	CEC > 36 cmol/kg Guyana shield
Quartz	10%	25.3%	
Kaolinite	90%	14.3%	81.1%
Albite		5.2%	
K-feldspar		5.1%	
Chlorite		10.2%	
Montmorillonite		12.5%	3%
Illite		19.4%	7.7%
Calcite		0.5%	

Table 4. Bedrock mineralogical composition of the Orinoco watershed in volume percent.

	Sands and sandstones	Shales	Shield rock	Basalts
Albite	10.7	—	—	—
K-feldspar	8.25	5	16	—
Quartz	63.3	47	22	—
Kaolinite	7.2	31	—	—
Chlorite	2.9	1	—	—
Biotite	—	—	6.5	—
Andesine	—	—	49	—
Hornblend	—	—	6.5	—
Ca-montmorill.	—	1	—	—
Mg-montmorill.	—	0.1	—	—
Illite	—	15	—	—
Apatite	—	—	—	4.3
Basaltic glass	—	—	—	9.2
Forsterite	—	—	—	1.3
Diopside	—	—	—	31.2
Labradorite	—	—	—	54

assumed that the depth of the weathering profile is so important that the bedrock is unreachable by the draining waters. The composition of layer 3 is thus assumed to be mainly clays together with mineralogical remnants of the underlying bedrock such as quartz and K-feldspar, both of which are poorly weatherable minerals.

Soil CO_2 concentrations are calculated based on the root respiration calculated by the LPJ model for each continental grid element. The diffusion coefficient $D^s_{CO_2}$ (cm^2/s) of gaseous CO_2 in the weathering profile is assumed to be linearly dependent on soil porosity (ϕ given by the ISRIC database) and soil tortuosity (Hillel 1998). τ , the tortuosity is the ratio of straight flow paths to the average roundabout path (Hillel 1998) and is fixed here to 0.45, assumed to be a standard value in the absence of constraint (Gwiazda and Broecker 1994). $D^s_{CO_2}$ is calculated according to

$$\frac{D^s_{CO_2}}{D^a_{CO_2} \cdot [T/273.15]^2} = \phi \cdot \tau \tag{8}$$

where $D^a_{CO_2}$ is the diffusion coefficient of CO_2 in air at 273.15 K and is set to 0.139 cm^2/s (Mattson 1995). The maximum CO_2 pressure reached below the root zone $P^{max}_{CO_2}$ in ppmv (assumed to be the only CO_2 production zone) is calculated as (Van Bavel 1951):

$$P^{max}_{CO_2} = P^{atm}_{CO_2} + \frac{5.7 \cdot 10^{-7} \cdot CO_2^{pr} \cdot \left(root_{depth}\right)^2}{2 \cdot \phi \cdot D^s_{CO_2}} \tag{9}$$

where $P^{atm}_{CO_2}$ is the atmospheric pressure in ppmv, $root_{depth}$ is the root depth calculated by LPJ at each time step and for each continental grid element in cm, and CO_2^{pr} is the production of CO_2 in the root zone in gC/m^2. To save computation time, an exponential function of the depth is used to define the CO_2 profile from the surface boundary condition (atmospheric value) down to the root depth (where $P^{max}_{CO_2}$ is reached, see Eqn. 3) instead of solving the complete diffusion equation. Owing to the lack of data, DOC levels are set to constant values across the weathering profile (from 12 mg/l in layer L1 to 0 mg/l in the saprolite zone L3). Mineral reactive surfaces are calculated from the grain size distribution given by the ISRIC global database.

Simulation scheme. LPJ is first run from 1901 to 2002 using monthly climatic data from the Climate Research Unit to establish and stabilize the potential vegetation cover. Then the WITCH model is run for 20 years using the mean annual soil balance calculated by LPJ for the 1983-2002 period, as well as the mean climatic conditions over the same period. The result of the WITCH model after a 20 years run are then investigated. Only mean annual weathering processes are model, and the model is run until a steady-state has been reached (i.e., until a constant chemical composition of the waters draining the Orinoco watershed is reached at the end of the 20-year run).

RESULTS

The goal of this contribution is to emphasize the key role played by the continental vegetation on the weathering flux estimates. We will thus focus on a set of sensitivity tests rather than on fully describing the reference simulations. In both cases (continental and regional scale), the coupled biospheric-weathering model is able to simulate the observed export of major species at the outlet of the catchments on a mean annual basis (see Figs. 6 and 7). Therefore, we will only investigate the sensitivity tests. A full description of the reference results can be found in (Goddéris et al. 2006) and in (Roelandt et al. 2009). Table 4 summarized the performed sensitivity tests.

We performed a series of tests where the impacts of the land plants on the weathering profile chemistry and hydrology are progressively removed. First, the soil CO_2 is set to the atmospheric value. Then all organic acids are removed from the profile. Finally the uptake of water and chemical species is removed in the last test.

Regional scale: the Strengbach catchment

Soil CO_2 sensitivity (W+A atm simulation). When below ground CO_2 levels are set to the 1994-1995 atmospheric values (320 ppmv at the Strengbach altitude; *W+A atm* test, Table 5), contrasted results are observed (see Fig. 8). The mean annual Mg^{2+} concentration in the main stream (and thus the catchment export flux) decrease by a factor of 1.45. This is expected since less acidic conditions should result in enhanced Mg^{2+} removal through smectite precipitation. Indeed, while smectites dissolve all along the PP weathering profile in the reference run (PP is the main hydrological contributor to the stream, contributing to 80% of the water flow), it is reduced by up to a factor of 4 above 1 m depth in the W+A atm run, and becomes negative (e.g., precipitation) in the saprolite (Fig. 9) This explains the strong depletion of Mg^{2+} at the outlet. Indeed, the primary minerals contributing to the Mg^{2+} flux (here biotite) are only

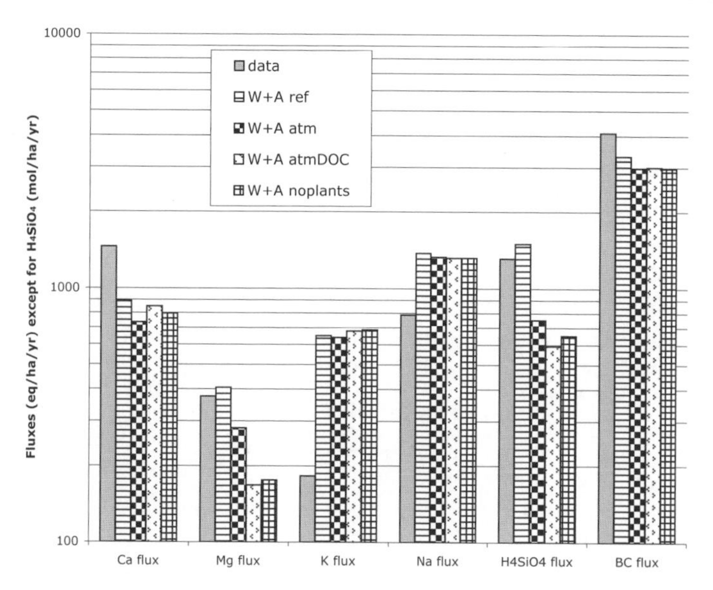

Figure 6. Mean annual flux of the cations and silica exported by the main stream of the Strengbach watershed as simulated by the WITCH + ASPECTS model, and compared to the available data (*http://ohge.u-strasbg.fr/index.html*). The year simulated is the hydrological 1994-1995 year.

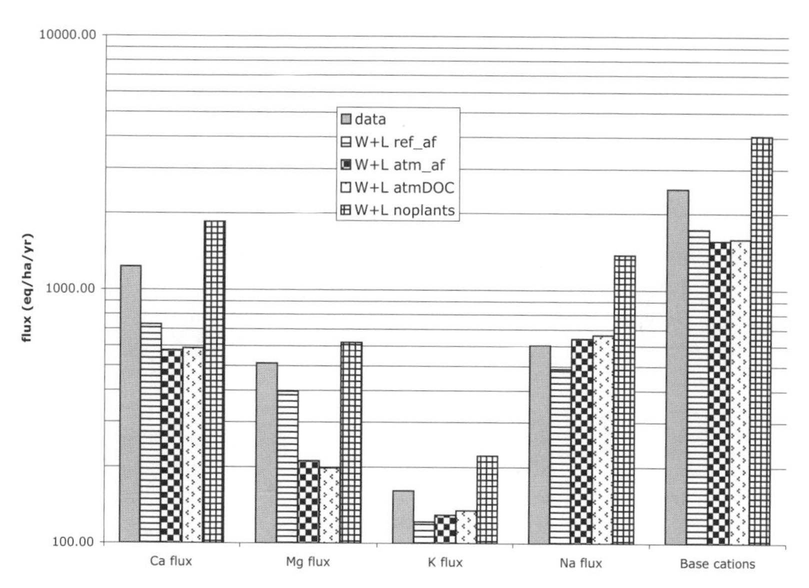

Figure 7. Mean annual flux of the cations exported by the main stream of the Orinoco watershed as simulated by the WITCH + LPJ model, and compared to the available data (Gaillardet et al. 1999). The mean annual flux is the steady-state flux calculated using the 1982-2002 mean annual meteorological forcings.

marginally affected by the CO_2 decrease (not shown). The W+A atm test demonstrates that smectite dissolution/precipitation is the dominant controlling factor. However, the specific behavior of each smectite mineral (namely Ca-, Mg-, K- and Na-montmorillonite) is rather complex. The Ca^{2+} export flux also decreases by a factor of 1.2. Decreasing dissolution rates of

Table 5. Summary of the sensitivity tests performed in this study. W+A stands for Witch+ASPECTS, while W+L stands for Witch+LPJ.

	Simulation objectives	Soil pCO_2	Soil DOC	Soil hydrology
Regional scale				
W+A ref	reference	CO_2 from ASPECTS	DOC decreasing from 20 mg/l at the surface to 0 mg/l below the root depth	Land plant control
W+A atm	No increase in soil CO_2	CO_2 fixed at the atmospheric level	DOC decreasing from 20 mg/l at the surface to 0 mg/l below the root depth	Land plant control
W+A atmDOC	No geochemical effect of land plants	CO_2 fixed at the atmospheric level	DOC set to 0 mg/l	Land plant control
W+A noplants	No acidification neither hydrological effect of land plants	CO_2 fixed at the atmospheric level	DOC set to 0 mg/l	Drainage set at effective rain, water volumetric contents held at their mean annual values
Continental scale				
W+L ref	reference	CO_2 from LPJ	DOC decreasing from 12 mg/l at the surface to 0 mg/l below the root depth	Land plant control
W+L atm	No increase in soil CO_2	CO_2 fixed at the atmospheric level	DOC decreasing from 12 mg/l at the surface to 0 mg/l below the root depth	Land plant control
W+L atmDOC	No acidification effect of land plants	CO_2 fixed at the atmospheric level	DOC set to 0 mg/l	Land plant control
W+L noplants	No acidification neither hydrological effect of land plants	CO_2 fixed at the atmospheric level	DOC set to 0 mg/l	No land plants

apatite by a factor of 1.2 (only present in the saprolite) is the main culprit. So Ca^{2+} is controlled by the response of trace mineral dissolution to deacidification of the soil solution. The less influenced cations are the K^+ and Na^+, with only a slight decrease in their concentration in the main stream. For K^+, the decrease in the dissolution rate of K-montmorillonite is compensated for by a decrease in illite precipitation rate and an increase of K-feldspar dissolution rate owing to a decrease in their respective saturation index. This saturation decrease is triggered by the removal of cations and silica through enhanced smectite precipitation. The dissolution/precipitation rates of Na-bearing minerals (albite and Na-montmorillonite) are almost unchanged. The overall CO_2 consumption rate decreases by 10% when below ground CO_2 is set to the atmospheric value of 320 ppmv.

This first test illustrates the complex behavior of the cation release as a function of soil CO_2 for the studied site: Mg^{2+} is controlled by smectites, Ca^{2+} by the behavior of apatite, a trace mineral, K^+ by a mix between smectite/illite and primary mineral (K-feldspar) responses,

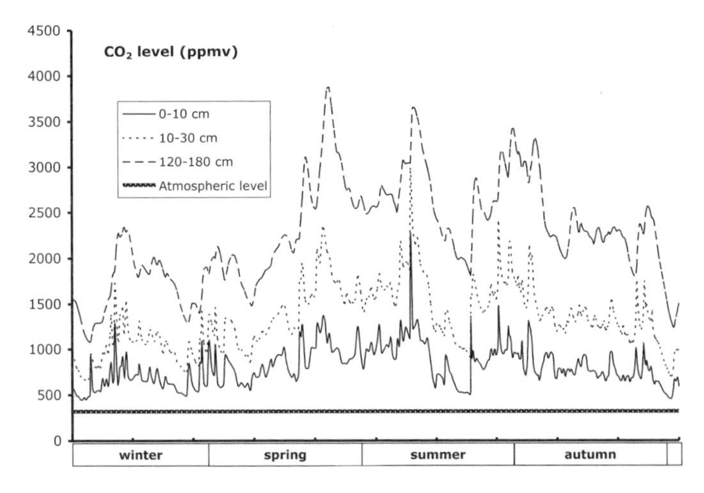

Figure 8. Calculated seasonal fluctuations of the below ground atmospheric CO_2 pressure in the Strengbach watershed below spruce trees. The different curves correspond to the depth in the soil of the calculation.

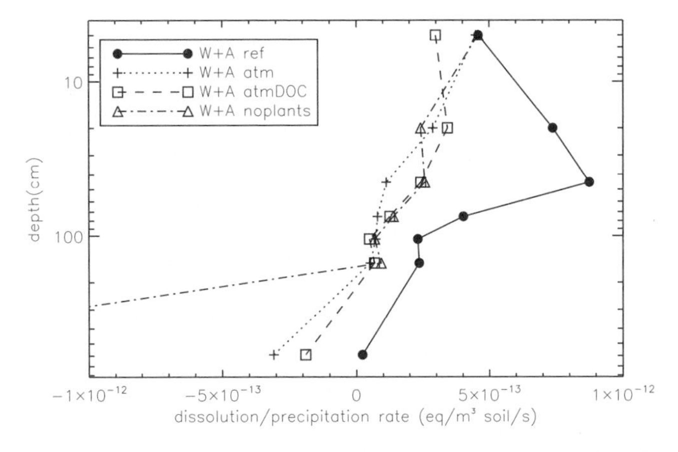

Figure 9. Calculated dissolution/precipitation rates of the smectite minerals in the soil profile below spruce trees in eq/m^3 of soil/s. When the rate is negative, it stands for a precipitation.

and Na^+ is not affected.

Sensitivity to the soil CO_2 and the removal of organic acids (simulations W+A atmDOC). In addition to setting the soil CO_2 levels to the atmospheric pressure, we remove all organic acids from the weathering profile. The most striking effect of removing the organic acid is a strong decrease in the Mg^{2+} concentration in the main stream (see Fig. 6). It decreases by a factor of 1.7 compared to the W+A atm run, while other cation fluxes are not significantly modified. Although this behavior cannot be fully explained by the response of a single mineral phase, the main culprit is a 60% decrease of the smectite dissolution rates above 1m depth, below the beech trees (HP site) (see Fig. 10), while illite saturation coevally decreases in the same depth range. On the other hand, the smectite dissolution rate below the spruce trees is not significantly different than that found in the W+A atm run. The main difference between the upper soil layers of the two sites is the prescribed abundance of illite, which is two times more abundant below spruce trees than below beech trees (Table 2). This difference in reactive surface introduces a

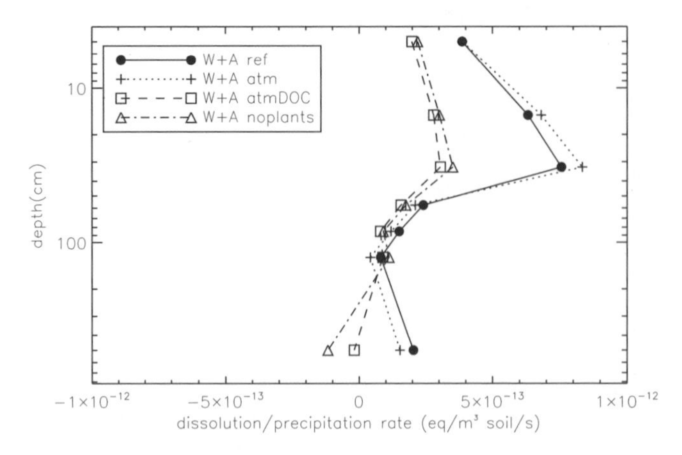

Figure 10. Calculated dissolution/precipitation rates of the smectite minerals in the soil profile below beech trees in eq/m^{3}soil/s. When the rate is negative, it stands for a precipitation.

non linearity in the sensitivity of the system to the CO_2 and organic acid level in soil solution. Below the beech trees, the weathering system is more sensitive to changes in the acid load at the top of the weathering profile (organic acid concentration decreasing as depth increase), while below spruce trees, the acid load at the bottom of the weathering profile is more critical (CO_2 levels rising as depth increase). The overall decrease in CO_2 consumption (including the CO_2 set at the atmospheric level and organic acid removal) reaches 9.5% compared to the reference run, which is lower than the decrease only triggered by the CO_2 decrease.

No land plants (simulation W+A noplants). Strictly speaking, a simulation with absolutely no land plants cannot be performed with the ASPECTS biospheric model for numerical stability reasons. Here, in addition to setting CO_2 levels to the atmospheric value and to the removal of organic acid from the weathering profile, we further remove any uptake or release of species by the land plants and we prescribe the hydrological behavior by assuming that no water uptake occurs in the root zone. All the water percolating at the top of the pile is assumed to drain down to the bedrock and then to the main stream. This hydrological behavior probably mimics the one in the absence of land plants. The catchment overall simulated runoff rises from about 919 mm/yr up to 1120 mm/yr.

The most obvious consequence of plant removal is a further decrease in all dissolved cation concentration in the upper part of the weathering profile, triggered by the dilution of atmospheric SO_4^{2-} deposition by a higher downward drainage flux (see Fig. 6). Indeed, the sulfate concentration in the throughfall water reaches 100 μeq/l for the year 1994-1995. Since we do not account for sulfate uptake by land plants, the sulfate concentration rises due to water uptake by the roots. Here, this water uptake is set to zero, and the sulfate concentration decreases accordingly. pH rises above 100 cm depth thus decreasing the dissolution rates and cation release. Cation concentrations in the main stream decrease accordingly, but the amount of water circulating through the system has increased. As a result, the total CO_2 consumption decreases only by 10.4% compared to the reference run whereas total base cation concentration in the main stream falls by 27%.

Continental scale: the Orinoco watershed

In addition to the spatial scale, several parameters are significantly modified when moving from the Strengbach to the Orinoco. First, the Orinoco watershed display a patchwork of soil and bedrock mineralogies, from the Guyana shield to the sandstones, and including some basaltic outcrops close to the outlet. Calcite is also present in many weathering profiles. Second, the climate is now a humid and tropical, implying an intense evapotranspiration. The vertical water transfer is much more dependent on the presence of vegetation than in a temperate environment. Despite these large differences between the two study sites, the sensitivity tests to the land vegetation display at first order similar results in terms of exported cations, except when the continental vegetation has been removed.

Soil CO_2 sensitivity (simulation W+L Atm). The soil CO_2 level calculated in the reference run ranges from 20 to 80 times the atmospheric CO_2 pressure (see Fig. 11). This is generally much higher than the maximum CO_2 level calculated for the Strengbach forested temperate ecosystem (8 times the atmospheric level). When CO_2 is forced to the atmospheric level through the complete weathering profile, the base cation concentration and hence the CO_2 consumption calculated at the outlet of the watershed decreases by 21% (see Fig. 7). The major decrease is calculated for Mg^{2+}; its flux is decreased by a factor of 2 from 400 to about 200 mol/ha/yr. Ca^{2+} release is also reduced by about 20%. K^+ is almost unchanged while the Na^+ flux increases by 30% increase. These contrasted behaviors are explained as follows (see Fig. 12):

(1) For Mg^{2+}, the deacidification caused by the CO_2 decrease results in enhanced smectite precipitation for the shale lithology. But an important contributor to the calculated decrease in Mg^{2+} is also the drastic reduction of its release by the dissolution of the basaltic lithology.

(2) For Ca^{2+}, things are more complicated. First, Ca^{2+} release by calcite dissolution in the undifferentiated sediments along the Andes decreases due to increasing pH of the percolating waters. Ca^{2+} production, however, increases over the Guyana shield. Indeed, the pH of the water entering the saprolite rises above 7 when CO_2 is set to its atmospheric value. As a result, the dissolution rate of andesine is enhanced. This increase is only partly compensated for by an increase in the saturation state of the soil solutions with respect to andesine so net Ca^{2+} flux increases.

(3) For Na^+, its release by the Guyana shield weathering increases for the same reason: andesine dissolution in neutral to slightly basic environments increases.

(4) We observe an increase in K^+ release by the shale lithologies, corresponding to a decreasing illite precipitation rate triggered by the decreasing saturation state of the

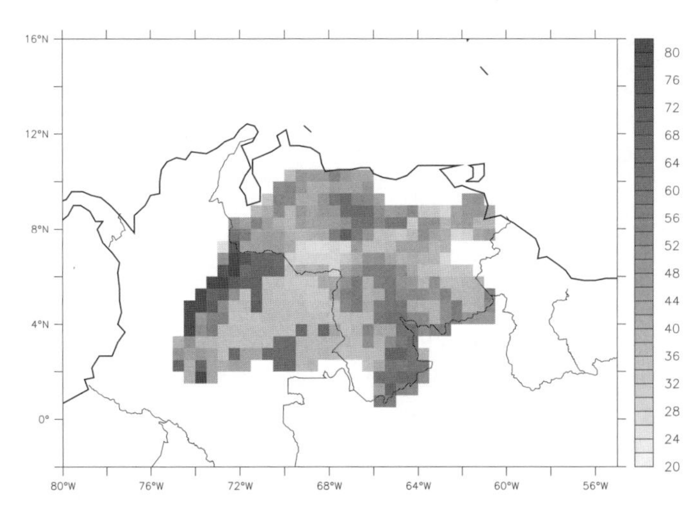

Figure 11. Calculated CO_2 pressure in the saprolite zone over the Orinoco watershed in PAL (Present Atmospheric Level: 1 PAL= 280 ppmv).

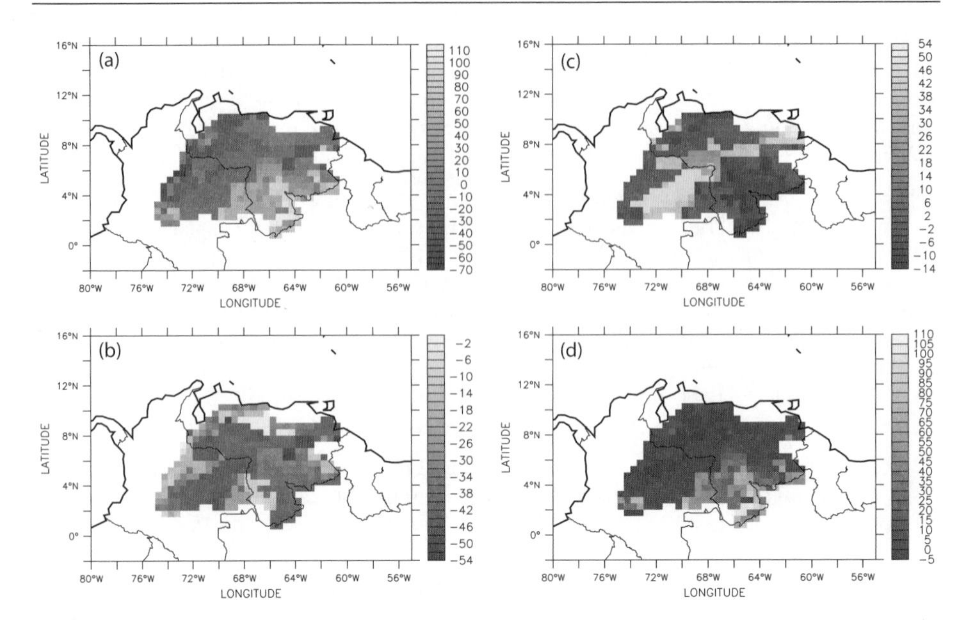

Figure 12. Change in the riverine flux of the 4 main cations from the reference run (W+L ref) to the constant soil CO_2 run (W+L atm) (see Table 4). The percent change in the fluxes is given by the intensity of the grey scale.

percolating waters with respect to illite owing to smectites precipitation. But a large area of the watershed also experiences a decrease in the K^+ production rate (triggered by a decrease in K-feldspar dissolution in the saprolite of the Guyana shield), and the net flux is almost unchanged (Figs. 7 and 12).

In conclusion, decreasing soil CO_2 does not produce a unique response in CO_2 consumption through weathering. This results from a complex interplay between pCO_2, solution pH, chemical affinity, soil mineralogy, on the control of cations and silica release to soil solutions. Decreasing the CO_2 level in soils does not lead to a general decrease of element release when modeled with a mechanistic tool.

Soil DOC sensitivity (simulation W+L AtmDOC). Removing the organic acid content of the soil solutions in addition to setting the CO_2 level to its atmospheric value does not change further the cation budget of the Orinoco watershed (see Fig. 7). The only area affected by this change is the south-western boundary of the Guyana shield, were the bedrock is overlain by the young sediments carried by the Rio Negro river. While the rest of the Guyana shield is assumed to be covered by kaolinite, the soil layers in this area (layer 1 and 2) contain 3% smectite and 8% illite. By removing the DOC from the soil layers, smectites precipitation is further facilitated and Mg^{2+} release is decreased by an additional 50% compared to the CO_2 test simulation (W+L Atm). At the same time, soil solutions become undersaturated with respect to illite and K^+ release increases by up to 80% in this area. Once again, the model simulations emphasize a strong coupling between illite and smectite precipitation/dissolution rates affecting the chemical composition of the main stream.

No plants run (simulation W+L no plants). In this simulation, the climate is held constant (including rainfall and temperature) compared to the reference run. PFT is not allowed to settle over the watershed, the soil CO_2 level is set its atmospheric value, and organic acids are removed from the profile. The impact of the complete removal of land plants is spectacular (see Fig. 7). The flux of base cations exported by the watershed increases by 130%. This increase is driven by a profound modification of the water cycle in the watershed. Removing plants from the watershed results in an increase of the effective rain from a mean value of 1806 mm/yr in the reference simulation to 2348 mm/yr, because of the absence of interception and subsequent evaporation. As a result, mean surficial runoff rises from 680 to 1274 mm/yr, and mean drainage into the saprolite zone from 153 to 572 mm/yr. Water contents of the soil boxes also increase. This augmentation of the water cycle results in an enhancement of cation released by weathering sufficient to compensate for the dilution effect so that concentrations remain roughly unchanged. This result thus suggests that, for the Orinoco type of environment, the first order impact of land plants on weathering occurs through water cycling above and below ground, owing to the intense evapotranspiration in this warm tropical environment. Indeed, the evapotranspiration is so high that vertical drainage fluxes are small. Removing the vegetation cover decreases significantly evapotranspiration, and vertical drainage is strongly enhanced. This effect also largely compensates for the de-acidification of the soil solutions, by removal of organic acids and drawdown of CO_2 levels, which decreases the weathering rates.

LAND PLANTS, SOIL HYDROLOGY, AND WEATHERING

The measured cation export for the Strengbach catchment is above 4000 eq/ha/yr, while it only slightly exceeds 2500 eq/ha/yr for the Orinoco. The modeling also reproduces these contrasted behaviors: the calculated specific export by the Orinoco is 1.9 times lower than the calculated specific export of the Strengbach. This difference arises despite that the mean annual temperature of the Orinoco is around 24 °C while it is only 5.9 °C for the Strengbach. One might expect that this temperature difference would inhibit the dissolution rates of silicate minerals in the Strengbach compared to the tropical environment. But this temperature effect is largely compensated by two causes. First, the soil mineralogy of the Orinoco is much more depleted in cations than the Strengbach soils. The dominant clay mineral for the Orinoco is the kaolinite, while the Strengbach soils are dominated by smectites and illite. But the effects of hydrology are also significant. Indeed, a contrasted behavior is observed for both watersheds when the land plants are totally removed: while the calculated cation export in the absence of plants rises by a factor of 2.3 for the Orinoco watershed, it decreases by a factor of 1.1 for the Strengbach. This is directly related to the modifications in the hydrological behavior of the weathering profiles when land plants are removed. While the drainage of the saprolite zone increases by almost a factor of 4 for the Orinoco when the vegetation is removed, it only rises by a factor of 1.2 in the Strengbach. The soil hydrological control exerted by land plants is much larger in the humid tropical environment than in the temperate forest. This is because the high temperature in the tropical environment sustains a high evapotranspiration compared to a temperate forest. These sensitivity tests suggest that the intense evapotranspiration is one of the main inhibitors of weathering processes in the tropical environment.

CONCLUSIONS

Continental weathering is a complex process. Many feedback loops are intertwined and an accurate model of CO_2 consumption by continental weathering and of its sensitivity to environmental parameters (climate, vegetation, etc.) requires the use of a mechanistic description of the coupled processes. In this study, we coupled a transport-reactive model describing the dissolution/precipitation of minerals in the soil through laboratory kinetic laws to a model of the continental biospheric productivity. Land plants heavily impact weathering, particularly through the hydrological behavior of the soil profiles, and the concentrations of acid species in the soil including CO_2 and organic acids. We explored the sensitivity of the weathering profile, from the soil surface down to the bedrock, to the continental vegetation at the regional and continental scale, and for two largely different climatic regimes (cold temperate and tropical).

We show through the use of a process-based modeling method that there is strong coupling between the hydrological and geochemical consequence of the land plants, the thermodynamic behavior of clay minerals, and the dissolution rates of primary minerals. For both case studies, we demonstrate a strong coupling between illite and smectite precipitation/dissolution rates itself dependent on the acid species supplied by the land plants productivity.

We also show that the response of weathering systems to the removal of land plants is largely dependent on the climatic regime. In both cases, drainage increases when land plants are removed due to decreasing evapotranspiration. But for the temperate environment, an overall decrease of the base cation export by 10% is predicted. The dilution effect seems to dominate. For the tropical environment, the base cation export increases by 130% when land plants are removed. While a dilution of the cationic species is predicted in the main stream, the dramatic increase in the soil profile drainages largely overwhelm this dilution effect. This is the result of a much higher evapotranspiration in tropical environments compared to the temperate environments when plants are present.

The demonstration presented above, that coupled models can accurately reproduce field observations on a variety of scales demonstrates that these models can now be used to explore the response of the continental weathering to anthropogenic land use and climatic changes.

REFERENCES

Amiotte-Suchet P, Probst JL (1995) A global model for present-day atmospheric/soil CO_2 consumption by chemical erosion of continental rocks (GEM-CO2). Tellus 47B: 273-280

Amiotte-Suchet P, Probst JL, Ludwig W (2003) World wide distribution of continental rock lithology: implications for atmospheric/soil CO_2 uptake by continental weathering and alkalinity river transport to the oceans. Global Biogeochemical Cycles 17: doi:10.1029/2002GB001891

Aubert D (2001) Contribution de l'altération et des apports atmosphériques aux transferts de matières en milieu silicaté: traçage par le strontium et le terres rares. Cas du bassin versant du Strengbach (Vosges, France). PhD Thesis, Université Louis Pasteur, Strasbourg, 225 pp.

Batjes NH (2005) ISRIC-WISE global data set of derived soil properties on a 0.5 by 0.5 degree grid (version 3.0). 2005/08, ISRIC-World Soil Information.

Berner RA (2004) The Phanerozoic Carbon Cycle. Oxford University Press, New York

Berner RA (2006) GEOCARBSULF: A combined model for Phanerozoic O_2 and CO_2. Geochim Cosmochim Acta 70:5653-5664

Bluth GJS, Kump LR (1994) Lithologic and climatic control of river chemistry. Geochim Cosmochim Acta 58:2341-2359

Boyle JF (2007) Simulating loss of primary silicate minerals from soil due to long-term weathering using Allogen: comparison with the soil chronosequence, lake sediment and river solute flux data. Geomorph 83:121-135

Brady PV (1991) The effect of silicate weathering on global temperature and atmospheric CO_2. J Geophys Res 96:18101-18106

Brantley SL, Goldhaber MB, Vala Ragnarsdottir K (2008) Crossing disciplines and scales to understand the Critical Zone. Elements 3:307-314

Braun J-J, Ndam Ngoupayou JR, Viers J, Dupré B, Bedimo Bedimo J-P, Boeglin J-L, Robain H, Nyeck B, Freydier R, Sigha Nkamjou L, Rouiller J, Muller J-P (2005) Present weathering rates in a humid tropical watershed: Nsimi, South Cameroon. Geochim Cosmochim Acta 69:357-387

Chou L, Garrels RM, Wollast R (1989) Comparative study of the kinetics and mechanisms of dissolution of carbonate minerals. Cheml Geol 78:269-282

Dambrine E, Sverdrup H, Warfinge P (1995). Atmospheric deposition, forest management and soil nutrient availability: a modeling exercise. *In:* Forest Decline and Atmospheric Deposition Effects in the French Mountains. Landmann G, Bonneau M (Eds), Springer-Verlag, Berlin

Dessert C, Dupré B, François LM, Schott J, Gaillardet J, Chakrapani GJ, Bajpai S (2001) Erosion of Deccan Traps determined by river geochemistry: impact on the global climate and the $^{87}Sr/^{86}Sr$ ratio of seawater. Earth Planet Sci Lett 188:459-474

Dessert C, Dupré B, Gaillardet J, François LM, Allègre CJ (2003) Basalt weathering laws and the impact of basalt weathering on the global carbon cycle. Chem Geol 202:257-273

Donnadieu Y, Goddéris Y, Pierrehumbert RT, Dromart G, Fluteau F, Jacob R (2006) A GEOCLIM simulation of climatic and biogeochemical consequences of Pangea breakup. Geochemy Geophys Geosys 7: doi: 10.1029/2006GC001278

Donnadieu Y, Goddéris Y, Ramstein G, Nédelec A, Meert JG (2004) Snowball Earth triggered by continental break-up through changes in runoff. Nature 428:303-306

Drever JI (1995) The effect of land plants on weathering rates of silicate minerals. Geochim Cosmochim Acta 58: 2325-2332

Dürr HH, Meybeck M, Dürr SH (2005) Lithologic composition of the Earth's continental surfaces derived from a new digital map emphasizing riverine material transfer. Global Biogeochem Cycles 19: doi:10.1029/2005GB002515

Edmond JM, Palmer MR, Measures CI, Brown ET, Huh Y (1996) Fluvial geochemistry of the eastern slope of the northeastern Andes and its foredeep in the drainage of the Oricono in Columbia and Venezuela. Geochim Cosmochim Acta 60:2949-2976

Edmond JM, Palmer MR, Measures CI, Grant B, Stallard RF (1995) The fluvial geochemistry and denudation rate of the Guyana Shield in Venezuela, Colombia, and Brazil. Geochim Cosmochim Acta 59:3301-3325

Eyring H (1935) The activated complex in chemical reactions. J Chem Phys 3:107-115

Fang C, Moncrieff JB (1999) A model for soil CO_2 production and transport 1: Model development. Agric For Meteorol 95:225-236

FAO-UNESCO Soil map of the world, digitized by ESRI. Soil climate map, USDA-NRCS, Soil Survey Division, World Soil Resources, Washington D.C.

Fichter J (1997) Minéralogie quantitative et flux d'éléments minéraux libéré par altération des minéraux des sols dans deux écosystèmes sur granite (bassin versant du Strengbach, Vosges). PhD Thesis, Université Henri Poincaré, INRA, Strasbourg, 284 pp.

Gaillardet J, Dupré B, Louvat P, Allègre CJ (1999) Global silicate weathering and CO_2 consumption rates deduced from the chemistry of the large rivers. Chem Geol 159:3-30

Gérard F, Mayer KU, Hodson MJ, Ranger J (2008) Modeling the biogeochemical cycle of silicon in soils: application to a temperate forest ecosystem. Geochim Cosmochim Acta 72:741-758

Gerten D, Schaphoff S, Haberlandt U, Lucht W, Sitch S (2004) Terrestrial vegetation and water balance - hydrological evaluation of a dynamic global vegetation model. J Hydrol 286:249-270

Gislason SR, Oelkers EH, Eiriksdottir ES, Kardijlov MI, Gisladottir G, Sigfusson B, Snorrason A, Elefsen S, Hardardottir J, Torssander P, Oskarsson N (2008) Direct evidence of the feedback between climate and weathering. Earth Planet Sci Lett 277:213-222

Goddéris Y, Donnadieu Y, De Vargas C, Pierrehumbert RT, Dromart G, van de Schootbrugge B (2008) Causal of casual link between the rise of nannoplankton calcification and a tectonically-driven massive decrease in the Late Triassic atmospheric CO_2 ? Earth Planet Sci Lett 267:247-255

Goddéris Y, François LM, Probst A, Schott J, Moncoulon D, Labat D, Viville D (2006) Modeling weathering processes at the catchment scale with the WITCH numerical model. Geochim Cosmochim Acta 70:1128-1147

Goddéris Y, Joachimski MM (2004) Global change in the late Devonian: modeling the Frasnian-Famennian short-term carbon isotope isotope excursions. Palaeogeogr Palaeoclimatol Palaeoecol 202: 309-329

Guyot J-L, Jouanneau J-M, Soares L, Boaventura GR, Maillet N, Lagane C (2007) Clay mineral composition of river sediments in the Amazon Basin. Catena 71:340-356

Gwiazda RH, Broecker WS (1994) The separate and combined effect of temperature, soil pCO_2, and organic acidity on silicate weathering in the soil environment: formulation of a model and results. Global Biogeochem Cycles 8:141-155

Hillel D (1998) Environmental Soil Physics. Academic Press

Lopez VM, Guerrero G, Bertorelli J (2003) Rocas Industriales de Venezuela. Fundacite Aragua

Lucas Y (2001) The role of plants in controlling rates and products of weathering. Annu Rev Earth Planet Sci 29:135-163

Maher K, Steefel CI, White AF, Stonestrom DA (2009) The role of reaction affinity and secondary minerals in regulating chemical weathering rates at the Santa Cruz Soil Chronosequence, California. Geochim Cosmochim Acta 73: 2804-2831

Mattson KG (1995) CO_2 efflux from coniferous forest soils: comparison of measurements methods and effects of added nitrogen. *In:* Soils and Global Change. Kumble JM, Stuart B, Levine E (Eds), CRC Press, p 329-342

Meybeck M (1987) Global chemical weathering of surficial rocks estimated from river dissolved loads. Am J Sci 287:401-428

Millot R, Gaillardet J, Dupré B, Allègre CJ (2002) The global control of silicate weathering rates and the coupling with physical erosion: new insights from rivers of the Canadian Shield. Earth Planet Sci Lett 196:83-98

Mitchell TD, Jones PD (2005) An improved method of constructing a database of monthly climate observations and associated high-resolution grids. Int J Climatol 25:693-712

Moulton KL, West J, Berner RA (2000) Solute flux and mineral mass balance approaches to the quantification of plant effects on silicate weathering. Am J Sci 300:539-570

New M, Hulme M, Jones PD (2000) Representing twentieth century space-time climate variability. Part 2: development of 1901-96 monthly grids of terrestrial surface climate. J Climate 13:2217-2238

Oliva P, Viers J, Dupré B (2003) Chemical weathering in granitic crystalline environments. Chem Geol 202:225-256

Probst A, Dambrine D, Viville D, Fritz B (1990) Influence of acid atmospheric inputs on surface water chemistry and mineral fluxes in a declining spruce stand within a small granitic catchment (Vosges massif, France). J Hydrol 116:101-124

Rasse DP, François LM, Aubinet M, Kowalski AS, Vande Walle I, Laitat E, Gérard JC (2001) Modeling short-term CO_2 fluxes and long-term tree growth in temperate forests with ASPECTS. Ecol Modeling 141:35-52

Raymond PA, Neung-Hwan O, Turner RE, Broussard W (2008) Anthropogenically enhanced fluxes of water and carbon from the Mississippi River. Nature 451:449-452

Roelandt C, Goddéris Y, Bonnet M-P, Sondag F (2009) Coupled modeling of biospheric and chemical weathering processes at the continental scale. Global Biogeochemical Cycles (in press)

Sitch S, Smith B, Prentice IC, Arneth A, Bondeau A, Cramer W, Kaplan J, Levis S, Lucht W, Sykes MT, Thonicke K, Venevsky S (2003) Evaluation of ecosystem dynamics, plant, geography, and terrestrial carbon cycling in the LPJ dynamic global vegetation model. Global Change Biology 9:161-185

Stallard RF (1985) River chemistry, geology, geomorphology, and soils in the Amazon and Orinoco basins. *In:* The Chemistry of Weathering. Drever JJ (Ed) Reidel, Dordrecht, p 293-316

Sverdrup H, Warfinge P (1995) Estimating field weathering rates using laboratory kinetics. Rev Mineral 31:485-541

Tardy Y, Bustillo V, Roquin C, Mortatti J, Victoria R (2005) The Amazon. Biogeochemistry applied to river basin management. Part I: Hydro-climatology, hydrograph separation, mass transfer balances, stable isotopes and modeling. Appl Geochem 20:1746-1829

Van Bavel CHM (1951) A soil aeration theory based on diffusion. Soil Science 72:33-46

Viterbo P, Beljaars CM (1995) An improved land surface parametrization scheme in the ECHWF model and its validation. J Climate 8:2716-2748

Wallmann K (2001) Controls on the Cretaceous and Cenozoic evolution of seawater composition, atmospheric CO_2 and climate. Geochim Cosmochim Acta 65:3005-3025

White AF, Blum AE, Bullen TD, Vivit DV, Schulz MS, Fitzpatrick J (1999) The effect of temperature on experimental and natural chemical weathering rates of granitoid rocks. Geochim Cosmochim Acta 63:3277-3291

Reviews in Mineralogy & Geochemistry
Vol. 70 pp. 435-484, 2009

Approaches to Modeling Weathered Regolith

Susan L. Brantley

Earth and Environmental Systems Institute, Department of Geological Sciences
Pennsylvania State University
University Park, Pennsylvania 16802, U.S.A.

brantley@essc.psu.edu

Art F. White

U. S. Geological Survey
Menlo Park, California 94025, U.S.A.

INTRODUCTION

Sustainable soils are a requirement for maintaining human civilizations (Carter and Dale 1974; Lal 1989). However, as the "most complicated biomaterial on the planet" (Young and Crawford 2004), soils represent one of the most difficult systems to understand and model with respect to chemical, physical, and biological coupling over time (Fig. 1).

Despite the complexity of these interactions, certain patterns in soil properties and development are universally observed and have been used in soil science as a means for classification. Elemental, mineralogical, or isotopic concentrations in soils plotted versus depth beneath

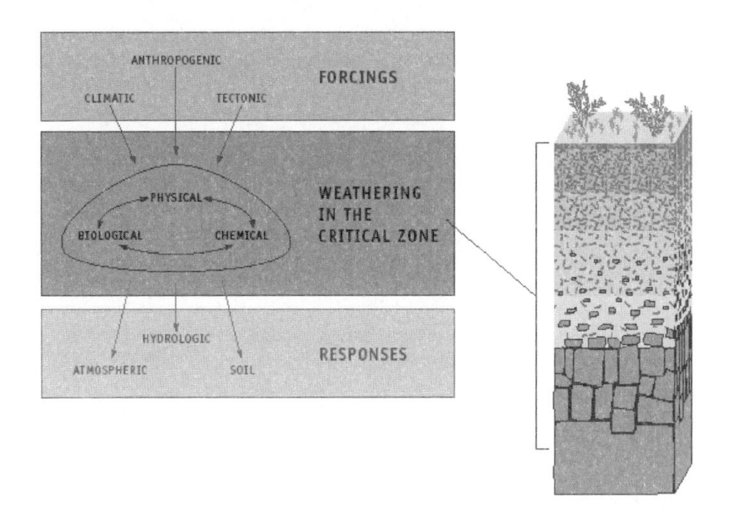

Figure 1. A schematic picture of the "weathering engine" at the Earth's surface. This weathering engine is part of the Critical Zone that extends from the vegetation canopy down through the saturated zone. The regolith-bedrock interface lowers at the weathering advance rate, ω. The rate of removal of material at the surface is the erosion rate, E. Some regolith profiles grow with time in a transient mode while others may attain steady state where $\omega = E$. As shown, many physical, chemical, and biological processes combine to control regolith in the Critical Zone. Climatic, anthropogenic, and tectonic forcings affect these processes; the sum total of weathering processes can then be read in changes in the atmosphere, hydrosphere, and pedosphere. [Used with permission of the American Geophysical Union from Anderson et al. 2004.]

1529-6466/09/0070-0010$10.00 DOI: 10.2138/rmg.2009.70.10

the land surface comprise such patterns. Soil depth profiles are often reported for solid soil materials, and, less frequently, for solutes in soil pore waters. These profiles cross a large range in spatial scales that traditionally have been studied by different disciplines. For example, shallow, biologically active horizons are commonly defined as the soil zone in agronomic studies whereas the mobile layer of the regolith is referred to as soil in geomorphological studies. In contrast, many geochemical studies target chemical weathering to tens or even hundreds of meters in depth, sometimes extending the definition of "soils" to include the entire regolith down to parent bedrock or alluvium.

Soil profiles also exhibit a large range in temporal scales (Amundson 2004; Brantley 2008b). Solid-state profiles document chemical and mineralogical changes integrated over the time scales of evolution of regolith from protolith. This "geologic time" can vary from tens to hundreds of years for weathered material developed on moraines deposited by active glaciers (Anderson et al. 1997), to millions or possibly hundreds of millions of years of regolith evolution as documented in laterites and bauxites on stable cratons (Nahon 1986). In contrast, solute profiles reflect much shorter time scales corresponding to the residence time of the soil water which commonly ranges from days to decades (Stonestrom et al. 1998). Factors impacting soil minerals can therefore be related to geologically old processes while those impacting pore waters are related to contemporary processes.

We first discuss a geochemical frame work for modeling soil profiles, including a simple scheme that depends on the extent of enrichment or depletion. Such profiles are comprised of reaction fronts affected by chemical, hydrologic, geologic and biologic processes that control soil evolution. We then present a hierarchy of models that have been used to interpret both solid state and solute compositions in regolith. The more simple approaches to model depletion in soils, using analytical models, are first described. The most elementary of these is a linear model that calculates rate constants from the slopes of either solid or solute weathering gradients: these rate constants represent lumped parameters that describe weathering in terms of an integrated reaction rate. Two other analytical models are then presented that have been used to fit solid state elemental profiles with exponential and sigmoidal functions. All of these analytical approaches are derived for models of soils as containing a limited number of components, phases, and species.

At a more complex level, numerical models are then presented to elucidate how kinetic and transport parameters as well as chemical, hydrologic, and physical soil data can be incorporated. We consider two forms of these models, first relatively simple spreadsheet calculators and then more sophisticated multi-component, multi-phase reactive-transport numerical codes. Our treatment of reactive transport modeling is relatively cursory, in recognition of the treatment in the chapter by Steefel and Maher (2009, this volume). Because these models incorporate more phases, components, and species than the other approaches and explicitly model the more fundamental reaction mechanisms involved, they generally have a greater need for parameterization. In our conclusion section, we discuss how this hierarchy of approaches can yield generalizations about soils that are often complementary.

A FRAMEWORK FOR SOIL PROFILE MODELING

This section will discuss several of the basic parameters used to quantify regolith-forming processes. Brief descriptions of the specific sites targeted in this discussion are contained in an Appendix at the end of this chapter. Note that throughout this chapter, we compare equations and models derived by different authors in different publications. Because symbols are used differently in papers by different authors, the reader is advised to note that symbol definitions differ throughout this review.

Deformation and mass transfer

While soil profiles are most commonly described in terms of chemical concentration (commonly, in units of mg g^{-1}, kg m^{-3} and mol l^{-1}), such concentrations do not define the amount of a specific component mobilized during weathering. The determination of *changes* in mass requires a comparison of the concentration of a weatherable component *j* in a soil or regolith, $C_{j,w}$, against its corresponding concentration in the parent rock or protolith, $C_{j,p}$. Such concentrations can refer to elements, components, isotopes, or minerals. Changes in the value of $C_{j,w}$ reflect not only mass changes in *j* but also relative changes due to gains and losses of other components in the regolith, as well as factors such as compaction or dilation that may change the original parent volume V_p to a new regolith volume V_w. The most common method for overcoming these latter effects is to compare the ratio of the mobile component *j* to an additional component *i* which is chemically inert during weathering (Merrill 1906; Barth 1961).

The concentration ratios of weatherable components in the regolith and protolith have been formally defined based on the relationship (Brimhall and Dietrich 1987):

$$\frac{C_{j,w}}{C_{j,p}} = \frac{\rho_p}{\rho_w} \frac{1}{(\varepsilon_j + 1)} (1 + \tau_j) \tag{1}$$

Here, the change in the mobile element concentration ratio $C_{j,w}/C_{j,p}$, is dependent on 3 factors: (i) the ratio of the bulk densities of the parent and weathered material, ρ_p/ρ_w; (ii) the deformation or volume change resulting from soil compaction or extension described by the strain factor $1/(\varepsilon_j + 1)$ where $\varepsilon_j = V_w/V_p - 1$; and (iii) the mass transport coefficient τ_j which describes the loss, gain, or conservative nature of component *j* within a unit volume of regolith.

The volumetric strain or volume change, ε_j, is calculated from the ratios of densities and concentrations of the inert element *i* in regolith and protolith:

$$\varepsilon_j = \frac{\rho_p}{\rho_w} \frac{C_{i,p}}{C_{i,w}} - 1 \tag{2}$$

Positive values of ε_j indicate regolith expansion and negative values indicate collapse. A value of $\varepsilon_j \approx 0$ indicates isovolumetric weathering.

The mass transfer coefficient τ_j in Equation (1) can be simplified (Brimhall and Dietrich 1987; Anderson et al. 2002):

$$\tau_j = \frac{C_{j,w}}{C_{j,p}} \frac{C_{i,p}}{C_{i,w}} - 1 \tag{3}$$

When $\tau_j = 0$, no mobilization of the weatherable component *j* has occurred in comparison to the immobile component *i*. When $\tau_j = -1$, complete depletion has occurred. When $\tau_j > 0$, addition of component *j* has occurred. Note that in this chapter and in the literature, both ε and τ are sometimes shown without subscripts, sometimes with one subscript (either *i* or *j*), and sometimes with both subscripts *i* and *j*, depending upon whether the immobile (*i*) and mobile (*j*) elements are denoted explicitly. Regardless of whether the subscripts are included, the values of these terms can vary with both the choice of *i* and *j*.

In the absence of erosion, the total mass of a component lost from or added to the weathered profile, ΔM_j, is obtained by integrating the mass transfer coefficient τ_j over the regolith thickness *z* for isovolumetric soils (Chadwick et al. 1990; White et al. 1996):

$$\Delta M_j = \left(\rho_p \frac{C_{j,p}}{m_j^{\circ}} 10^4 \right) \int_0^z -(\tau_j) dz \tag{4}$$

Here, $m°_j$ is the atomic weight of element j. The integration of the soil profile in Equation (4) provides a quantitative determination of the total amount of weathering or the intensity of weathering in a non-eroding soil. The average long term weathering flux Q_i (mol m^{-2} yr^{-1}) is then defined as

$$Q_j = \frac{\Delta M_j}{\Delta t} \tag{5}$$

where Δt (yrs) is the duration of weathering. Mass fluxes are a useful approach for comparing weathering in different environments (White 2008).

The mass balance approach has been extended (Riebe et al. 2003) to characterize concurrent chemical and physical denudation under steady state conditions such that

$$D = E\left(\frac{C_{i,w}}{C_{i,p}}\right) \tag{6}$$

Here, D is the total denudation rate and E is the physical denudation rate (or erosion rate in mass, or mol, per unit area per unit time). The chemical flux for individual weatherable species is defined by substitution to derive

$$W_j = D\left(C_{j,p} - C_{j,w}\frac{C_{i,p}}{C_{iw}}\right) \tag{7}$$

If the weathering flux W (mass, or mol, per unit area per unit time) is expressed in a non-dimensional form by normalizing to the total denudation rate for component j, the following relationship is obtained:

$$\frac{W_j}{D \cdot C_{j,p}} = \left(1 - \frac{C_{j,w}}{C_{j,p}}\frac{C_{i,p}}{C_{i,w}}\right)_j \tag{8}$$

Riebe et al. (2003) defined the terms in parentheses in this equation as the *chemical depletion factor* or CDF. The CDF is the negative of the value of τ_j as defined in Equation (3) (Brimhall and Dietrich 1987; Anderson et al. 2002).

Importantly, calculations of strain ε_j (Eqn. 2), mass transfer τ_j (Eqn. 3), mass change ΔM_j (Eqn. 4), and chemical depletion (Eqn. 8) require that at least one component in the regolith (i) remains inert or immobile during chemical weathering. Commonly assumed conservative elements include Zr (Harden 1987), Ti (Johnsson et al. 1993) and rare earth elements such as Nb (Brimhall and Dietrich 1987). However, for sediments, these elements are often present in high-density grains and are thus susceptible to density segregation; therefore, an alternate approach is to use SiO_2 as the inert component (i then refers to quartz). Quartz has long been considered useful as an indicator of weathering intensity (Ruhe 1956) and has been used as a conservative component in mass balance calculations (Sverdrup and Warfvinge 1995; White et al. 1996). Due to similar grain sizes and densities, quartz, unlike heavy minerals, behaves similarly to more reactive aluminosilicates such as feldspars in depositional environments. Although quartz does weather, its dissolution rate in most weathering environments is slow compared to aluminosilicates.

An assessment of the composition of the parent material or protolith, $C_{j,p}$, is also required in calculating the weathering parameters. This composition is generally easy to determine for regoliths developed *in situ* on relatively homogenous bedrock. The determinations of the parent concentrations are more difficult for heterogeneous protoliths such as sediments. One approach is to equate the parent material to similar recent unweathered sediments, assuming that the depositional environment has not changed significantly. An example is the weathering of beach sands in marine soil chronosequences (Chadwick et al. 1990). The second approach utilizes the common observation that weathering intensity decreases with depth in a regolith. Samples acquired from deep in the regolith may therefore reflect the unweathered protolith composition (White et al. 2008).

The preceding characterization of deformation and mass transfer provides a quantitative basis for describing weathering patterns in soils. One simple graphical approach (Brimhall et al. 1991) involves plotting soil compositions over four quadrants defined by values of ε_j (Eqn. 2) and τ_j (Eqn. 3). Values of $\varepsilon_j > 1$ define volume increases or dilation. Values of $\varepsilon_j < 1$ define soil collapse. Concurrently, $\tau_j > 1$ defines addition of mass to the profile while $\tau_j < 1$ denotes mass loss. The utility of this approach is illustrated in Figure 2 which contains selected elemental compositions (Si, Ca, Al and Fe) from 3 soil profiles of varying age from the Santa Cruz chronosequence in California (White et al. 2008, 2009) (see Appendix I). In calculations of deformation and mass transfer, quartz is employed as the conservative component, i.e., $i = SiO_2$ in quartz for defining $C_{i,p}$ in Equations (1)-(4), and the protolith is assumed to be equivalent to the composition of the deepest sample in each profile.

Symbols corresponding to positive values of ε_j correspond generally to shallow soils that reflect dilation due to bioturbation, addition of organics, and inputs of eolian material. In contrast, soils at greater depths reflect collapse as a result of chemical weathering (values

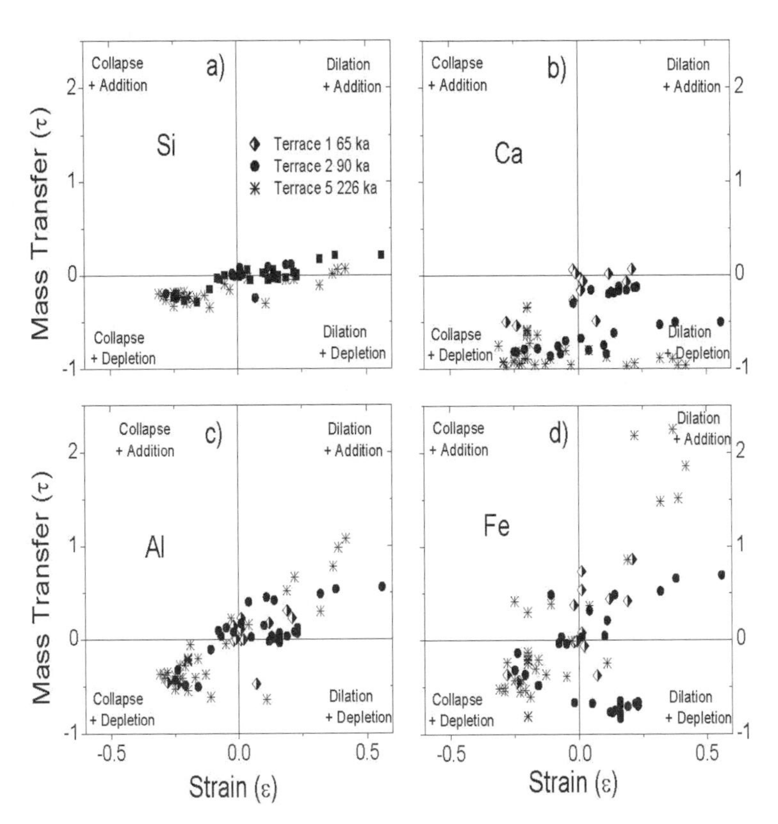

Figure 2. Relationships between volumetric strain (Eqn. 2) and the mass transfer coefficient (Eqn. 3) for selected elements in the the Santa Cruz terraces (see Appendix). Strain values ε of < 0 and > 0 define zones of soil compaction and dilation respectively while values of the mass transfer coefficient τ < 0 and > 0 define zones of mass depletion and enrichment (data from White et al. 2008). See text for further explanation.

of ε_j are negative). The relative element mobility occurring during this weathering is reflected in the mass transfer coefficient. Si in Figure 2a exhibits insignificant mobility ($\tau_{Si} \approx 0$) in both the collapsed and dilated soil horizons because it is principally contained in quartz which has been assumed to be inert during weathering (White et al. 2008). Divergence from zero must be related to loss of Si from minerals other than quartz. In contrast, Ca in Figure 2b is significantly depleted as a result of plagioclase weathering in soils that experienced both compaction and dilation, and is also strongly correlated with the age of the soil in the chronosequence. Values of $\tau_{Ca} \approx -1$ denote near-complete depletion throughout much of the oldest terrace (226 ka) while Si remains unweathered, i.e., $\tau_{Si} \approx 0$, in much of the deeper parts of the younger profile (90 k.y.).

Aluminum and Fe are mobilized in the Santa Cruz soils during weathering of primary silicates but are subsequently incorporated into secondary clays. In Figure 2c and d, these elements exhibit depletion ($\tau_j < 1$) in zones of collapse associated with chemical weathering at depth in the older terrace. In contrast, Al and Fe are added in zones of dilation which correspond to relatively shallow soils in which a clay and Fe oxide argillic zone progressively forms in terraces of increasing age. The patterns for Al and Fe denote external inputs at shallow depths from dust or other sources or else the mobilization by weathering at depth and re-precipitation in shallower soils as a result of biological uptake (bio-lifting).

Profile categories

The same mass transfer approach can be extended to individual soil profiles such as those derived from soils in Pennsylvania as shown in Figure 3. Such soil profiles can be most easily interpreted when they are developed on ridgetops, flat-lying landscapes such as lavaflows, or terraces. For such simple examples, the regolith can be conceptualized as columns of homogenous protolith with minimum amounts of lateral sediment input and with chemical redistributions reflecting net one-dimensional downward fluid flow.

For such systems, three basic categories of profile are apparent: i) *immobile profiles* exhibiting parent concentrations of a given component ($\tau = 0$); ii) at all depths *depletion profiles* exhibiting depletion of the element of interest with respect to the immobile element at the surface ($\tau < 0$); iii) *addition profiles* exhibiting enrichment of the mobile element with respect to the immobile element ($\tau > 1$).

For example, in Figure 3A, τ values for both Nb and Zr equal 0 throughout the profile. Those elements are therefore neither added nor removed from the soil, as compared to Ti, the immobile element used in Equation (3) for that calculated figure. By inference then, Ti, Nb, or Zr could have been used as the conservative component in Equations (3) and (4) for that soil. Importantly, the relative mobility of such refractory elements varies under differing weathering conditions and as a function of the mineralogy of the parent material (Gardner 1980; Kurtz et al. 2000; Hodson 2002; Neaman et al. 2006). In Figure 3A, however, the general agreement in the calculations for Ti, Zr, and Nb (i.e., $\tau \approx 0$) supports the conclusion of conservative behavior for these elements during weathering of the Rose Hill shale PA (USA).

In contrast to the immobility of Zr, Ti, and Nb, Figure 3B shows depletion of Mg during weathering of the Rose Hill shale. Depletion profiles can be subdivided into *partially developed profiles* (*i.e.*, showing *partial depletion,* $-1 < \tau < 0$ at the surface) or *fully developed profiles* (i.e., showing *complete depletion* of the component of interest, $\tau = -1$, at the surface). Such a classification is useful in understanding the soil development, as discussed later in this chapter. The thickness of the zone over which the element or mineral of interest becomes depleted by reaction in the profile is often referred to as the *reaction front*. For example, mobilization of Mg out of the Rose Hill shale has been attributed to loss of clay minerals in the upper soil zone.

External inputs to soils create *addition profiles* that are characterized by net addition: $\Delta M_j > 0$ in Equation (4). Figure 3C shows the addition profiles of Mn and C developed on weathered Rose Hill shale. Such profiles commonly document inputs of wet or dry deposition

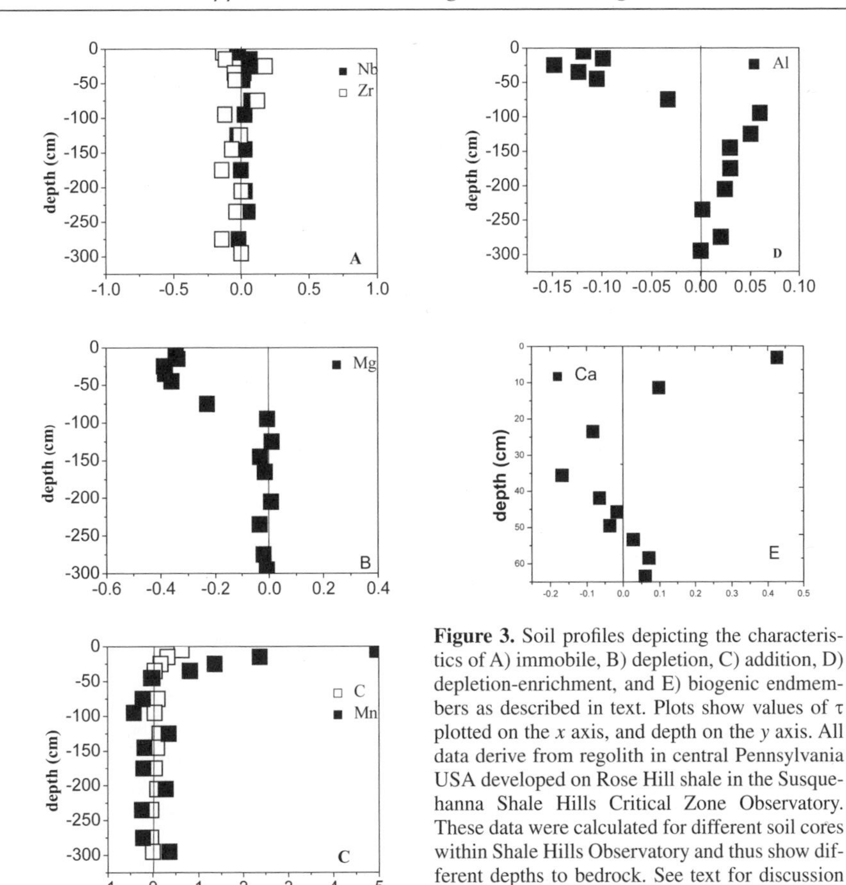

Figure 3. Soil profiles depicting the characteristics of A) immobile, B) depletion, C) addition, D) depletion-enrichment, and E) biogenic endmembers as described in text. Plots show values of τ plotted on the x axis, and depth on the y axis. All data derive from regolith in central Pennsylvania USA developed on Rose Hill shale in the Susquehanna Shale Hills Critical Zone Observatory. These data were calculated for different soil cores within Shale Hills Observatory and thus show different depths to bedrock. See text for discussion of endmember profiles.

(e.g., Mn in Fig. 3C), or inputs of elements due to biological fixation (carbon, C, in Fig. 3C). Addition profiles generally are characterized by both addition at the surface and redistribution downward by leaching or mixing due to bioturbation. For example, in Figure 3C, Mn is inferred to have been added to the soil through atmospheric inputs, but Mn also moved downward into the soil due to remobilization as well as bioturbation. These profiles are therefore characterized by concentrations that are highest at the surface, but that grade back to parent concentration at depth.

When an element that is leached at the surface reprecipitates at depth in the regolith, the profile is described as a *depletion – enrichment profile* (Fig. 3D). An example of a depletion-enrichment profile is the E-Bs, E-Bt sequence commonly observed in forest soils where dissolution of Al occurs near the top due to organic ligand concentrations. Organic ligands lower in the profile are present at lower concentrations, causing precipitation of Al at depth. Thus, Al and Fe often exhibit depletion-enrichment profiles, as do other elements such as P that have affinity for Fe or Al. Depletion-enrichment profiles may be such that the element is conservative, $\Delta M_j = 0$, or may demonstrate net loss, $\Delta M_j < 0$. Where $\Delta M_j > 0$, additional inputs to the soil from aeolian sources or from lateral mixing must have occurred.

Biogenic profiles commonly exhibit trends that are opposite to depletion-enrichment profiles. These profiles often characterize nutrients such as K or Ca that are released from parent material but are redistributed by bio-lifting (Fig. 3E). Profiles characterized as biogenic therefore generally show depletion at depth but enrichment at the surface. Such redistributions result from the secretion of organic acids by plants or fungi through roots that dissolve primary minerals at depth and recycle them in the upper layers of the soil. Compilations of global soil data (Jobbagy and Jackson 2001) document that biogenic profiles are common for nutrient elements.

Some profiles do not fall easily into a category. For example, P is often ultimately derived from the primary mineral apatite as it is dissolved by organic acids secreted by organisms—if net loss of P occurs, it can define a depletion profile. However, because P is taken up into biomass, it often demonstrates biogenic characteristics. Furthermore, when P is released upon decomposition of that biomass, it is often sorbed onto Fe oxyhydroxides in the soil. These processes imprint a depletion-enrichment character to the P profile, overprinting the biogenic character and yielding very complex P profiles in some regoliths.

Of the modeling approaches discussed in this paper, the analytical models have generally been used to interpret depletion or addition profiles. In contrast, reactive transport models are needed when interpreting depletion-enrichment profiles. Few modeling efforts have addressed biogenic profiles. We focus largely on depletion profiles in this chapter.

Time evolution of profiles

The previous section did not consider how the various types of concentration-depth profiles evolved with time. Where erosion is insignificant—for example in a stable alluvial terrace—a depletion profile such as that shown in Figure 3B moves downward with time at the *regolith advance* or *weathering advance rate*. As shown in Figure 4, as the element of interest is dissolved and removed, the geometry of the profile changes with time and the profile transforms from a *partially depleted* (*incompletely developed*) profile, to a *complete depletion* (*fully developed*) profile. Such partially depleted profiles that change in time are often referred to as depicting the *transient phase* of profile development.

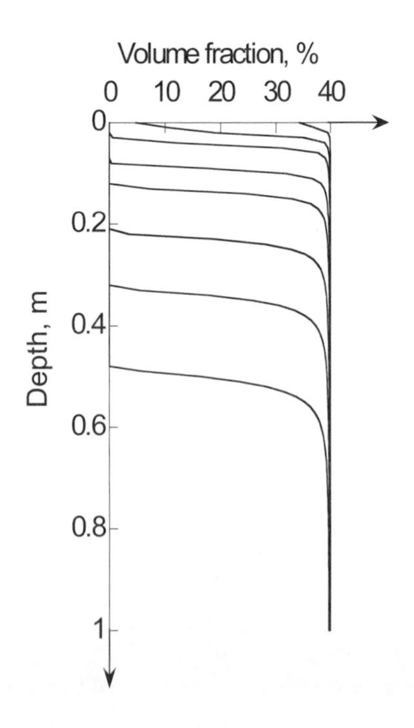

Figure 4. The volume fraction of albite plotted versus depth as calculated for an isovolumetric quartz + albite soil undergoing insignificant erosion (Lebedeva et al. 2007). Profiles are plotted as time snapshots, where elapsed time increases for each deeper curve. The uppermost two profiles show reaction fronts that are *partially developed* (partially depleted at land surface) while deeper profiles demonstrate fronts that are *fully developed* (completely depleted at land surface). As shown for this theoretical model, the partially developed profiles characterize the *transient period* because they change in geometry and position while the completely developed profiles characterize the *quasi-stationary period* because they show constant geometry while moving downward with time (Lichtner 1988). See text and Figure 5 for more model information.

If such downward movement and increased depletion occurs for the profile, the normalized concentrations of component j at the top of the profile may eventually become invariant with time once complete depletion is attained, i.e., $\tau_j = -1$—this is the point when the profile can be described as *fully developed* and the profile itself is *quasi-stationary* (Lichtner 1988). Once a profile is quasi-stationary, it is predicted that the profile retains its geometry but moves downward at the weathering advance rate.

Given these considerations, the easiest type of profile to interpret (other than an immobile profile) is a depletion profile where the dissolving element does not reprecipitate and is not taken up significantly by biota. Often, geochemists interpret Na profiles, often reflecting plagioclase feldspar weathering, as one of the best example of such profiles, since Na is not commonly reprecipitated in high concentrations in clays or oxides nor is it an important nutrient (White and Brantley 2003). Several interpretations of Na depletion profiles are included later in this paper. For depletion profiles, two important observations must be fit by the model: i) the *depth to the reaction front*, and ii) the *thickness of the reaction front*. In turn, the observables are related to i) the weathering (regolith) advance rate, and ii) the mineral reaction rate, respectively. Figure 4 demonstrates that for a non-eroding profile, the depth of the weathering front increases with time. In contrast, the reaction front thickness—the depth interval over which a given component is depleted within a profile—can remain relatively constant in time once the quasi-stationary profile develops. Importantly, as discussed in the next section, the depth to the front and the reaction front thickness both theoretically remain constant with time for an eroding profile that attains steady state.

Analytical models utilizing parameters for flow and rate constants for dissolution have been derived to treat depletion profiles for quasi-stationary systems assuming constant surface area, porosity, diffusivity, and fluid flow velocity when only a single component reacts within a single phase system (Lichtner 1988). In such models, the thickness of the reaction front is related to the ratio of the rates of transport to dissolution. For example, for diffusion-limited weathering rind formation in low porosity rock where transport is dominantly by diffusion, the thickness of the reaction front (h) represents a balance between the product of the porosity (ϕ, %) and effective diffusivity (D, m^2 s^{-1}) divided by the product of the linear rate constant for solubilization of a given element (k, m s^{-1}) and the mineral-water interfacial area (S_v, m^2 m^{-3}) (Lichtner 1988):

$$h = \left(\frac{\phi D}{kS_v}\right)^{1/2} \tag{9}$$

In contrast, for a reaction front developed on a higher permeability matrix such that advection (v: advection velocity, m s^{-1}) dominates the reaction front thickness h can be expressed (Lichtner 1988):

$$h = \frac{v}{kS_v} \tag{10}$$

Multiphase multicomponent chemical systems incorporate species that are coupled between reacting phases and thus cannot generally be described analytically without simplifying assumptions (Lebedeva et al. 2007). For example, many analytical models for soils either explicitly or implicitly rely on the assumption that the reacting mineral, e.g., albite, is dissolving within a largely nonreactive matrix. For such a treatment, the rate constant fit to the data is an operational rate constant that may or may not be comparable to laboratory measurements where dissolution was measured without the presence of other phases. This is one of the reasons that rate constants from field observations do not match those from laboratory observations.

Such analytical models can effectively be used to interpret a "depletion" profile even if an element is reprecipitated at some depth below the dissolution front as long as the distance between net depletion and net addition (precipitation) is large, i.e., the reactions are spatially separated. For example, dissolution of calcic plagioclase may result in loss of Ca at the top of a profile but precipitation of smectite lower in the profile. If there is no overlap in the zone of dissolution and precipitation then such a depletion profile could be treated by any of the analytical models discussed herein.

However, once the distance between these fronts decreases so that plagioclase dissolves simultaneously with reprecipitation of clay, then this depletion-enrichment profile can only be modeled with an analytical model if the precipitation reactions are described with a judicious choice of the reacting component, and the rate constant must be corrected for other reactions that are occurring in the real system (Lebedeva et al. 2007). We have also treated such a system by deriving an equation for one chemical system reacting in a matrix that can be described with a buffering capacity to model the effect of other reactants (Brantley et al. 2008). Alternately, such profiles can instead be modeled using a reactive transport code that incorporates a number of the competing reactions. Given the large number of reacting phases and species in soils, even the use of a reactive transport code generally requires a simplification of the system by choice of reacting phases.

Coupling of weathering and permeability

Theoretical calculations of profile development such as shown in Figure 4 are based upon the assumption of a hydrologically homogenous matrix that is experiencing weathering under the condition of constant permeability. In reality, the permeability may vary significantly with depth across a weathering profile. For example, a common but extreme example is the weathering of low permeability bedrock to a highly porous saprolite. Under such conditions, mineral weathering and depletion profile development (Fig. 3B) represents a coupling between the physical and chemical processes that affect both hydraulic conductivity and weathering intensity (Fig. 1).

The schematic in Figure 5 shows the hypothetical weathering of bedrock to saprolite as described from granite weathering at Panola Georgia (White et al. 2001). In the schematic, the arrow represents increasing weathering with decreasing depth. A zone of "pristine rock" is present at depth (Fig. 5, region I) which is isolated from surficial processes because the primary permeability is so low (hydraulic conductivity, q_p, in Fig. 5) and/or the weathering time interval is so short that meteoric waters have not flushed through the bedrock and chemically equilibrated with the minerals. At some point, meteoric fluids are introduced into the bedrock and aqueous species react with one or more mineral phases. The products of this reaction begin

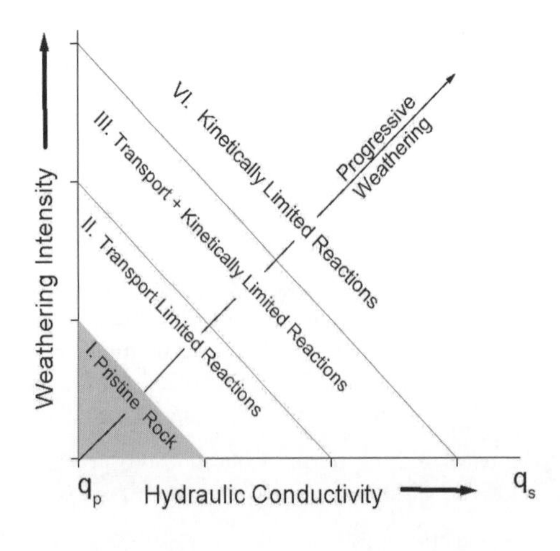

Figure 5. Schematic showing weathering regimes as coupled functions of increasing weathering intensity and permeability as denoted by increases in hydraulic conductivities (q_p = primary permeability of bedrock and q_s = secondary permeability, or permeability of saprolite, reflecting mass depletion due to weathering). In this schematic, the arrow, signifying progressive increases in weathering, can be considered either to represent duration of weathering, or decreasing soil depth (after White et al. 2001).

to be transported out of the granite, creating mass depletion and a secondary permeability q_h which is greater than that of the pristine bedrock q_p (Fig. 5 region II). Under these conditions, however, the limited volume of pore water rapidly saturates thermodynamically with respect to the primary minerals. Weathering rates at this stage of regolith development proceed very close to equilibrium and are thus limited by rates of transport through the weathered granite rather than by the kinetic rates of reaction of the primary minerals. These transport rates may be in turn limited by physical factors such as ongoing fracturing.

With increasing time, an increase in weathering intensity and mass depletion increases the secondary hydraulic conductivity such that the fluid pore volume becomes sufficiently large and one or more of the most soluble reactive minerals becomes thermodynamically undersaturated (Fig. 5 region III). Weathering of these minerals is now dependent on the kinetic rate of reaction and not on the fluid transport rate defined by q_s. However, mineral phases with lower solubilities will continue to remain thermodynamically equilibrated. Finally, sufficient mass transfer occurs such that $q_s >>> q_p$, which increases the pore water volume to the extent that more primary minerals become unsaturated. As shown in the diagram, weathering is thus expected to progress from a regime that is limited by fluid transport (transport-limited) to a regime that is described by mixed limitation (both transport- and kinetic-limited) to a regime that is described by kinetic limitation (Fig. 5).

The analysis in Figure 5 indicates that weathering is initiated under low permeability conditions within the bedrock environment and that such profile development can be limited by the initial formation of porosity and permeability due to dissolution of early reacting minerals. For argument sake, we assume here that the reaction of only one mineral characterizes this incipient weathering zone, and we call this a *profile-controlling mineral* because its reactivity affects access of fluids and rate at which regolith/protolith boundary advances downward. We will now discuss two examples of mineral weathering which could be described as profile-controlling. For both cases, the initial concentration of this mineral is very small, and yet its reaction significantly impacts the weathering development.

Oxidation of Fe (II) in biotite is observed to be the first reaction occurring in the otherwise pristine Rio Blanco quartz diorite in the Luquillo Experimental Forest in Puerto Rico (Buss et al. 2008) (Fig. 6). This reaction is hypothesized to produce a form of altered biotite that has a larger volume than the initial biotite (Fletcher et al. 2006; Buss et al. 2008). Resulting cracking of the rock creates cm-thick rindlets surrounding the pristine corestone in a process known as spheroidal weathering (Turner et al. 2003). This fracturing enhances the influx of reactive solution, causing the initiation of dissolution of plagioclase feldspar which creates greater secondary porosity (Fig. 6). Modeling is consistent with the conclusion that the very high rates of chemical weathering of the quartz diorite are at least partly due to this spheroidal fracturing (Fletcher et al. 2006).

Rindlets comprise onionskin-like layers of intact rock around the corestone. The intensity of weathering increases in rindlets further from the corestone (Buss et al. 2008). As weathering intensity increases over the duration of exposure to water, porosity increases until outer rindlets begin to show dissolution of hornblende (Fig. 6). Eventually, the loss of plagioclase and hornblende in rindlets is so extensive that an outermost rindlet transforms to saprolite. At this outermost edge of the rindlet zone, the Fe(II) content of the biotite has been greatly diminished (measured at ~7% of total Fe in biotite) and remains at these low levels through the saprolite (Murphy et al. 1998). The sequence of mineral reactions is consistent with the effects of progressive permeability increases proposed in Fig. 5.

The reaction front for the FeO component in the biotite comprises a completely depleted or fully developed profile across the rindlet zone, i.e., the Fe(II) content is reduced to near zero across the rindlets. In contrast, the resulting oxidized ferric biotite continues to react to kaolinite in the overlying saprolite until approximately 50 cm depth where the rest of the

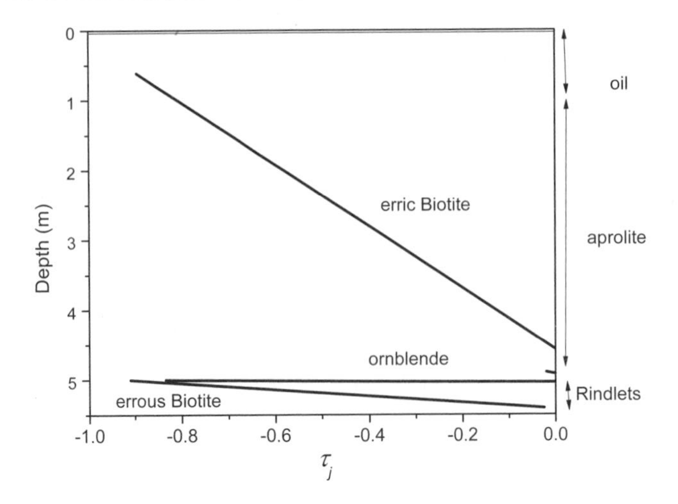

Figure 6. Plot showing reaction fronts for three minerals, j, denoted as τ_j versus depth, as observed in regolith cores sampled from the Rio Blanco quartz diorite in the Luquillo Experimental Forest, Puerto Rico (after Buss et al. 2008). Oxidation of ferrous biotite occurs deepest in the profile (reaction front labeled as *ferrous biotite*), followed by plagioclase dissolution (reaction front not shown here for clarity), hornblende dissolution (reaction front labeled as *hornblende*), and ferric biotite conversion to kaolinite (labeled *ferric biotite*). The plagioclase τ-depth data is not shown here because it overlies the ferrous biotite data at the scale of this plot, although it is shifted slightly upward. The completely developed reaction fronts for plagioclase, hornblende and ferrous biotite all lie largely within the rindlet zone. The completely developed reaction front for ferric biotite lies largely in the saprolite. In contrast, quartz, not plotted here, is characterized as a partially developed reaction profile because quartz is observed to dissolve in the upper regolith but remains in the profile even to the land surface. This plot is a schematic version based on data from several cores taken from several locations as described previously (Murphy et al. 1998; White et al. 1998; Turner et al. 2003; Buss et al. 2008). As such, the actual depth values may vary for different cores. The ferric biotite may be a profile-controlling mineral in this system (see text).

biotite is totally consumed (Murphy et al. 1998). Thus, two fully developed profiles are defined for biotite: one across the rindlets for the FeO component, and one across the depth range from rindlets to 50 cm depth for the ferric biotite phase (Fig. 6). The loss of the FeO component in the biotite is hypothesized here as the profile-controlling reaction.

A second example of a profile controlling reaction is that of calcite, which occurs as a minor but ubiquitous mineral in granitic rocks (White et al. 1999, 2001). At near-neutral pH, the solubility of calcite is about 4 orders of magnitude greater than those of silicate minerals such as feldspars (Bandstra and Brantley 2008). Therefore calcite is expected to weather preferentially in low-permeability bedrock environments. This is substantiated by the presence of a calcite weathering front (White et al. 2001) that precedes the plagioclase-weathering front by about a meter into the Panola Granite in Georgia (Fig. 7). One unique feature of the Panola granite is the relatively high content of disseminated calcite (0.33 wt%) relative to most other granites (White et al. 1999). The close spatial correlation between weathering of calcite and plagioclase, minerals characterized by very different reactivities, is consistent with initial bedrock weathering being controlled by transport rather than reaction kinetics, i.e., zones I and II in Figure 5.

The bedrock/saprolite interface is not the only type of weathering profile which develops across a strong permeability gradient. For example, depletion-addition profiles described in Figure 3C are often characterized by the production of secondary products of weathering at depth, including clays, iron oxides and amorphous silica. These products can form low permeability horizons such as fragipans, duripans and ferricretes across which the permeability

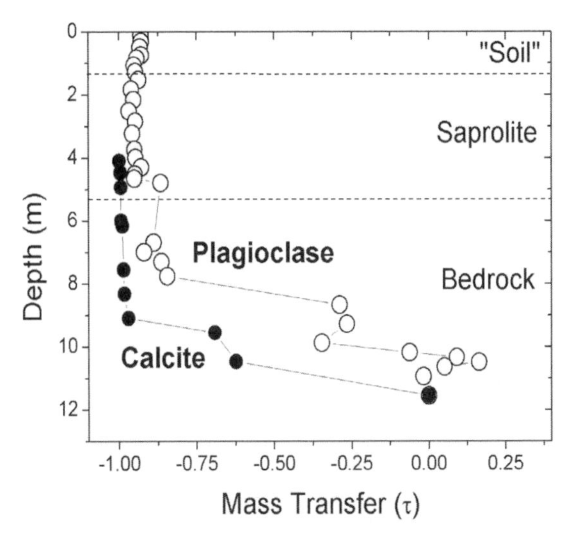

Figure 7. Plot of mass transfer coefficient (τ) versus depth for weathering profiles for plagioclase and trace amounts of disseminated calcite developed on the Panola Granite, Georgia (after White et al. 2001). The close correspondence between these two completely developed reaction fronts is consistent with both reactions being limited by transport rather than kinetics, i.e., regions I and II in Figure 5. The calcite may be a profile-controlling mineral in this system (see text).

decreases significantly in progressively older soils (White et al. 2008). The development of such features and the coupled feedback between weathering and fluid transport is a fruitful area for future modeling efforts.

Effects of physical erosion

The preceding sections define soil profile development strictly in terms of chemical weathering, In reality, most regoliths are impacted by physical erosion which can produce profile surfaces that are either aggrading or degrading. Eventually, for some systems, a situation of true steady state (Fig. 8), as opposed to a quasi-steady state (Fig. 4), can develop in which the concentration-depth geometry remains constant in position with respect to the soil surface and the rate of erosion, E, equals the rate of weathering advance, ω. Solid material in a profile that experiences net erosion can therefore be conceptualized as moving upward through the weathering window as porefluids move downward. The thickness of a layer in a steady-state weathering system—the zone of incipiently weathered bedrock, saprolite, or soil—and the rate of transformation of each layer into the overlying layer must therefore be constant in time (Fig. 8).

In the absence of physical erosion, incompletely-developed profiles (τ at the land surface lies between −1 and 0) represent transient phenomena, which given time, will evolve to quasi-steady state conditions (Fig. 4). However in the presence of constant physical erosion, incompletely-developed profiles can represent true steady state. Such profiles might look like the model calculations shown in Figure 8 for relatively fast erosion rates (E > 0.0075 m/k.y.). Alternately, for slower erosion rates, a steady-state profile could develop into a completely developed profile (E < 0.0075 m/k.y., Fig. 8).

The models plotted in Figure 8 were calculated (Lebedeva et al. 2009) for model "rocks" containing albite + quartz weathering to kaolinite + quartz. These models, calculated assuming that the dominant transport mechanism is diffusion, should be most applicable for weathering at the low-porosity bedrock-saprolite interface. For steady-state weathering systems as depicted in Figure 8, the linear rate of erosion equals the rate of weathering advance. The

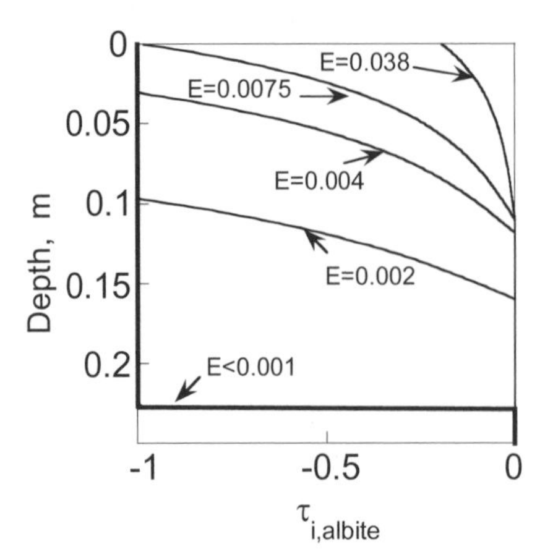

Figure 8. Albite depletion profiles plotted versus depth for albite weathering to kaolinite as calculated within diffusion-limited isovolumetric regolith initially containing only albite + quartz (Lebedeva et al. 2009). The steady-state curves document depletion of albite as a function of depth, contoured for various rates of erosion, E (expressed as length time^{-1}), as indicated. When E is small compared to the albite weathering rate constant, k, a very sharp reaction front is observed; in contrast, when E is fast compared to k, a steep and thicker reaction front is observed. A sharp reaction front within a completely depleted profile is characteristic of the local equilibrium regime, $E < 0.001$, while a partially depleted profile is characteristic of the kinetic regime, $E > 0.0075$ (Lebedeva et al. 2009). Profiles that show complete depletion at the land surface but also show relatively wide reaction fronts result in the transition regime. ($0.001 < E < 0.0075$). These regimes observed for simulated profiles for model "rocks" containing albite + quartz + kaolinite are compared in the text to the transport-limited (similar to local equilibrium) and the weathering-limited regimes (similar to kinetic regime) hypothesized for catchments. This figure derives from an analytical model derived assuming that the specific surface area for reacting albite is equal to geometric surface area, e.g., all curves represent "rock" with the same grain size albite. Inclusion of advection into these models results in thicker reaction zones than the diffusion-limited case shown here.

models show that, as erosion (E) increases, an endmember regime is approached where the chemical weathering fluxes of elements such as Na and Si become a linear function of the albite dissolution rate constant. In this endmember, the model albite profiles are observed to be *incompletely developed* (e.g., E > 0.0075 m/k.y. in Fig. 8). In contrast, this chemical weathering flux is *not* a direct linear function of the dissolution rate constant for endmember systems with fully developed reaction fronts that are thin (e.g., E < 0.001 m/k.y. in Fig. 8). Respectively, these cases were described as the *kinetically limited* and *local equilibrium regimes* (Lebedeva et al. 2007). In the *transition regime*, chemical weathering fluxes vary with *both* dissolution and erosion rate (0.001 < E < 0.0075 m/k.y. in Fig. 8).

Based on these models, chemical fluxes do not vary strongly with erosion rate in the kinetic regime and vary more strongly on erosion rate in the transition and local equilibrium regimes. This derives from the observation that the chemical flux in these diffusion-limited models is a function of the concentration gradient in the porefluid at the surface, and this curve varies insignificantly for models in the kinetic regime as erosion rate is varied. In contrast, this porefluid gradient varies as a function of E in the transition and local equilibrium regimes.

Figure 8 can be related to Figures 6 and 7. Importantly, while the profiles plotted in Figure 8 were calculated assuming transport by diffusion, Lebedeva et al. (2009) have shown that for similar models derived in the presence of advection, the thicknesses of reaction zones increase. Therefore, although the fronts in Figure 8 are thinner than observed in many natural systems (e.g., Fig. 6, 7), this may be explained by the assumption of diffusion only. For both the FeO and calcite examples in these figures, the reaction goes to completion, defining completely developed profiles. By analogy with the models in Figure 8, these profiles may therefore lie in transition or local equilibrium regimes. The relatively thick reaction front for calcite (Fig. 7)—meters—is most consistent with a transition regime. Although the reaction front for the FeO component is only tens of centimeters, modeling by Fletcher et al. (2006) is consistent with chemical fluxes that depend nonlinearly upon the rate of reaction of the FeO—i.e., transition regime. Therefore, for both of these cases, the chemical weathering fluxes are hypothesized to depend on both the rate of erosion and dissolution.

This discussion highlights the important hypothesis that the type of profile that characterizes a given mineral in a steady state regime—partially or completely developed—may indicate whether the chemical flux due to reaction of that mineral is strongly dependent upon dissolution kinetics or transport or both (kinetic-controlled vs. transport-controlled vs. mixed control). Furthermore, for a steady state thickness of regolith, these profiles may elucidate whether chemical weathering fluxes vary strongly with dissolution rate constants (kinetic regime) or with the erosion rate (local equilibrium regime) or both (transition regime). As discussed for Figure 6, the reaction fronts for the FeO component of biotite and the ferric biotite are both fully developed and relatively thick, and are therefore probably occurring within a transition regime. Apparently, the FeO component is closer to the local equilibrium regime endmember. In contrast to these two reactions, quartz dissolution occurs throughout the entire regolith (Schulz and White 1999) and is still present at the land surface. Quartz is therefore hypothesized to be reacting within the kinetic regime where the chemical flux from dissolving quartz should vary directly with the quartz dissolution rate constant. Consistent with these observations, researchers have documented that the rate of weathering of biotite in this system is significantly slower than rates in the laboratory (Murphy et al. 1998), as expected for the transition regime, while the rate of weathering of quartz is relatively close to the values measured in the laboratory (Schulz and White 1999), as expected for the kinetic regime.

Intriguingly, solute outfluxes from both small and large watersheds worldwide increase with sediment outfluxes (Gaillardet et al. 1999; Millot et al. 2002; Dupre et al. 2003). If such watersheds are characterized by steady state regolith thickness, then such an observation may be consistent with reaction within local equilibrium or transition regimes and coupling between chemical and physical weathering. Indeed, coupling between physical and chemical erosion is necessary if regolith thickness is maintained at steady state regionally. In other words, if no such coupling occurred, then ultimately the surface would be largely devoid of regolith (i.e., $\omega < E$) or covered by extremely thick regolith ($E < \omega$). The common presence of relatively thin regolith across much of Earth's surface may be a strong argument that many systems are maintained close to steady state due to coupling of physical and chemical processes.

A model has been advanced to explore how coupling between physical and chemical weathering might happen as described earlier in the Puerto Rico example of Figure 6 (Fletcher et al. 2006; Lebedeva et al. 2007). In that model, the weathering advance rate into the quartz diorite is related to the oxidative reaction of biotite (Buss et al. 2008). Fletcher et al. (2006) therefore argued that the rate of weathering advance is a function of the porefluid concentration of oxygen, and that, if erosion increased, this would decrease regolith thickness and increase the porefluid oxygen concentration at the bedrock interface. Thus, the porefluid concentration of reactant is hypothesized to couple the rate of erosion to the rate of weathering advance at depth, maintaining steady state regolith thickness.

By analogy, the reactant concentration in porefluid could also couple erosion to weathering advance for systems where the profile-determining mineral is calcite (Fig. 7). For calcite, the "coupling" reactant is most likely the proton concentration. Thus, both the concentrations of oxygen and protons could couple the rate of erosion to the weathering advance rate for profile-controlling minerals in the transition or local equilibrium regimes. For regolith where the profile-controlling mineral demonstrates strict kinetic regime behavior, the rate of weathering advance at depth may instead be coupled to the erosion rate at the surface by the grain size distribution (Fletcher and Brantley 2009).

The importance of porefluid concentrations or grain size distributions toward coupling physical and chemical weathering points to difficulties in using single mineral profiles such as those in Figure 8 to interpret natural weathering or erosion. Specifically, the porefluid oxygen or proton concentrations in regolith do not depend only on one reacting mineral. In the Puerto Rico example, while the rate of oxidation of the FeO component could control the rate of weathering advance, the porefluid oxygen concentration is not strictly controlled by only that component. Therefore, oxidation reactions with other phases affect the rate of weathering advance rate. In addition, the impacts of biota throughout the regolith on porefluid concentrations of oxygen (and protons) must also be important as discussed next.

Role of biology in profile development

These purely physico-chemical models for weathering in Puerto Rico or Panola incorporate no biology. Importantly, however, the concentrations of both oxygen and protons (as well as other reactants and products) in regolith porewaters are strongly affected by the presence of biota. These effects open up the possibility—indeed probability—that biota contribute toward maintaining steady state thicknesses of regolith. Indeed, for the quartz diorite in Puerto Rico, it has been observed that a microbial ecosystem lives at the bedrock-saprolite interface that may be supported by or even driving weathering reactions (Buss et al. 2005). In general, modeling approaches that incorporate biological processes into the fully coupled physical-chemical-biological system (Fig. 1) are still in their infancy despite the fact that many lines of evidence show that ecosystems affect weathering and vice versa.

Many researchers have shown that the presence of biota can influence the rate of chemical (Drever 1994; Drever and Stillings 1997; Van Scholl et al. 2008) and physical weathering (Wischmeier and Smith 1978; Vanacker et al. 2007). These authors cite the fact that biota produce organic acids, change the concentration of oxygen in pore fluids, take up and release nutrients, affect cycling of water, bind fine particles, cause formation of organic-mineral aggregates, and cause fracturing and bioturbation of regolith. At present, our ability to model these processes is limited at best and therefore these processes are not treated in depth in this chapter.

The nature of the element-depth profile is also useful in understanding the effect of biota on weathering. An incompletely developed steady state regolith profile such as those depicted in Figure 8 can support ecosystems with constant nutrient delivery at a rate dictated by the weathering advance rate (Porder et al. 2005). In contrast, in a quasi-stationary no-erosion profile that moves downward with time (Fig. 4) or in a completely developed steady-state profile (Fig. 8), regolith at the surface can become nutrient-poor. In those regimes, ecosystems may become limited by nutrients derived from the parent material (Wardle et al. 2004). In such nutrient-limited regimes, biota may ultimately be supported only by atmospheric additions of nutrients (Chadwick et al. 1999).

INTRODUCTION TO CASE STUDIES

As described previously, modeling of weathering profiles are most likely to be successful for systems developed on a monolithologic protolith located at ridgetop or terrace-like locations such that weathering is "one-dimensional". In this regard, net fluid flow can be conceptualized as largely 1-D (downward through the system, e.g., rainfall minus evapotranspiration), and no sedimentary inputs to the system need be considered other than atmospheric input. In contrast, to model profiles developed on regolith on hillslopes entails incorporation of downslope transport of sediments (Yoo and Mudd 2008) and this is not considered here.

Once such a regolith site has been chosen, the solid-state or solute profiles can be analyzed and plotted, often in normalized form (e.g., Fig. 3). In this section, we discuss weathering profiles at six localities: Santa Cruz CA, Davis Run VA, Panola GA, Puerto Rico, Costa Rica and Svalbard, Norway. In spite of a wide range of parent material, age and spatial scale, these sites are all amenable to the type of profile characterization described in the preceding sections (sites are further described in the Appendix). Two of the sites are represented by chronosequences (Fig. 9) that show weathering on the same lithology versus time (basaltic material for Costa Rica and granitic sands for Santa Cruz). For the Costa Rica profiles, the three time snapshots all com-

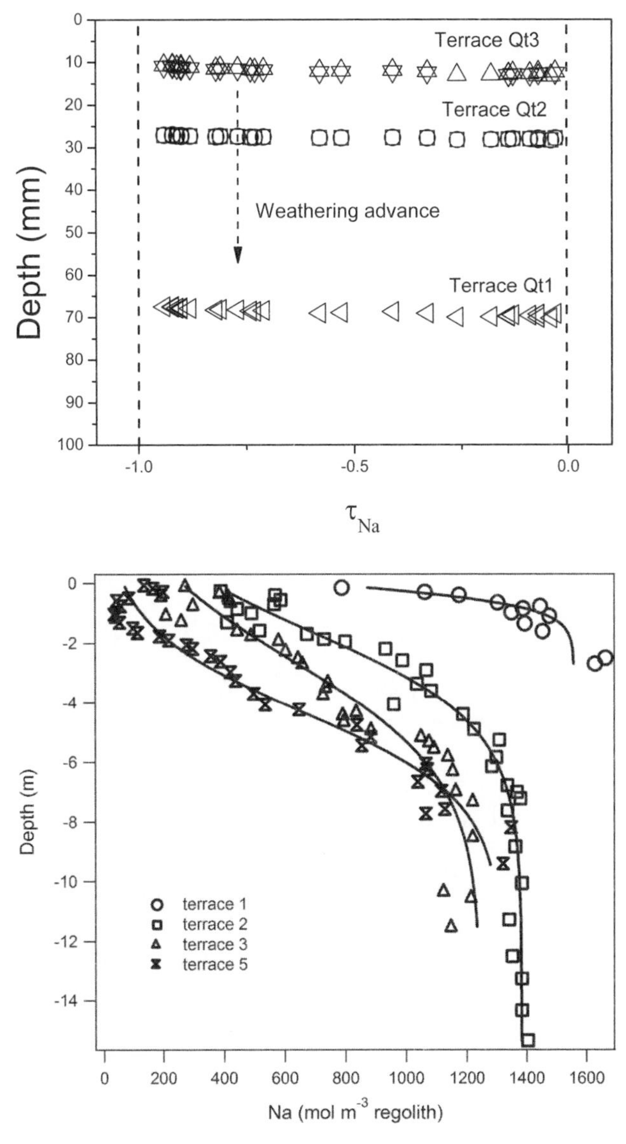

Figure 9. Reaction fronts for Na plotted for the two chronosequences represented in our case studies (see text and Appendix). Included are the data measured on basalt clasts weathering in three alluvial terraces in Costa Rica (upper figure), and for unconsolidated sediments in four marine terraces at Santa Cruz (lower figure). The symbols for Costa Rica system are based on unpublished electron microprobe analyses measured across the core-rind boundary of three basalt clasts weathered for increasing durations from terrace Qt3 to Qt2 to Qt1, i.e., roughly 35000, 120000, and 240000 yr respectively (Navarre-Sitchler et al. 2009). Dashed lines represent parent ($\tau_{Na} = 0$) and fully depleted compositions ($\tau_{Na} = -1$). The Santa Cruz profiles (White et al. 2008) show an evolution from partial to complete depletion at the land surface (i.e., from partially to completely developed profiles). The youngest terraces thus demonstrate *transient, incompletely developed profiles*. In contrast, of the three terraces in Costa Rica, no clast demonstrated an incompletely developed depletion front. Instead, profiles are all interpreted as *quasi-stationary and completely developed*. Curves plotted overlying the data for Santa Cruz terraces represent previously published fits of Equation (30) to the data (Brantley et al. 2008).

prise completely developed profiles that show quasi-stationarity (compare Fig. 4). In contrast, for the profiles identified in Santa Cruz, the Na profiles developed on the younger terraces are incompletely developed and only the oldest soil demonstrates a completely developed profile. Therefore, the younger terraces are transient, i.e., the geometry of the profile is evolving with time (e.g., Fig. 4, upper profile) while the oldest terrace is interpreted to be in a quasi-stationary state (Fig. 4, lower profile).

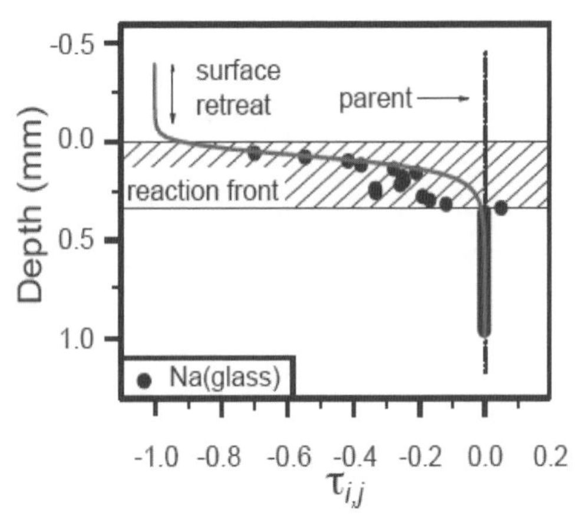

Figure 10. $\tau_{Ti,Na}$ plotted versus depth for a piece of weathered basalt regolith in Svalbard (see text and Appendix). Na depletion in the sample is due to dissolution of glass. The curve shows a model simulation based on the reactive transport code CrunchFlow where diffusion was the dominant mode of transport (Hausrath et al. 2008a). As described by Hausrath et al. parameters for the model such as glass surface area and duration of weathering were varied until the modeled output for the reaction front geometry fit the observed data. For that model-data fit, there was a mis-match such that the depth of the reaction front in the model was greater than that on the weathered sample (labeled in figure as "surface retreat"). Given that many similar samples show surface-parallel cracks, it was therefore assumed that more than half of the depleted rind had spalled off. Data points represent averages of 9 profiles measured by electron microprobe. Symbols were calculated with five-point running averages to smooth variations related to the fact that the sizes of the probe spots were smaller than grain size. Symbols for parent values in the unweathered core are not measured values but just average parent composition. Weathering of the Svalbard basalt is interpreted to occur by cyclic Na leaching and surface-parallel cracking and spalling which creates a new Na-replete surface. [Used with permission of The Geological Society of America from Hausrath et al. 2008a.]

Like the profile in Figure 8 for E = 0.0075 m/k.y., the Na depletion from glass at Svalbard appears to be a completely developed profile, but without significant weathering advance downward (Fig. 10). As discussed in the Appendix, this surface spalls off rapidly and intermittently, and thus samples from Svalbard show variability in surface depletion. The remaining case studies at Puerto Rico, Panola, and Davis Run comprise completely developed Na profiles that have formed from weathering of granitic rock to saprolite (Fig. 11).

It is important to note that we are mixing two different scales of observation: four of the profiles developed across regolith zones of meters to tens of meters depth (Santa Cruz CA, Davis Run VA, Panola GA, Puerto Rico) and two represent profiles developed upon hand samples from within a regolith profile (Svalbard, Costa Rica). Such weathering rinds, cm-thick permeable crusts enriched in the oxides of Fe, Ti, and Al, often envelop unweathered cores of low-porosity rock. Where rinds have not been removed, they represent regolith developed *in situ* without erosion, and we argue here that comparison of these examples to regolith profiles elucidates controls on weathering. For example, weathering rinds developed on clasts from alluvial terraces deposited and exposed to weathering at different times demonstrate the time evolution of weathering (Colman and Pierce 1981; Oguchi and Matsukura 1999; Sak et al. 2004). Such clast systems are largely free of many of the complexities related to fracturing, heterogeneity in parent material, and differences in microclimate that complicate studies of soil profiles or watersheds.

It is also important to note that for the weathering rinds and for one of the regolith studies, Panola GA, the reaction fronts for Na cross into the bedrock. In such cases, modeling generally must incorporate the development of secondary porosity to characterize the reaction front thickness accurately. For example, where the reaction front lacks fracturing it is thinner than

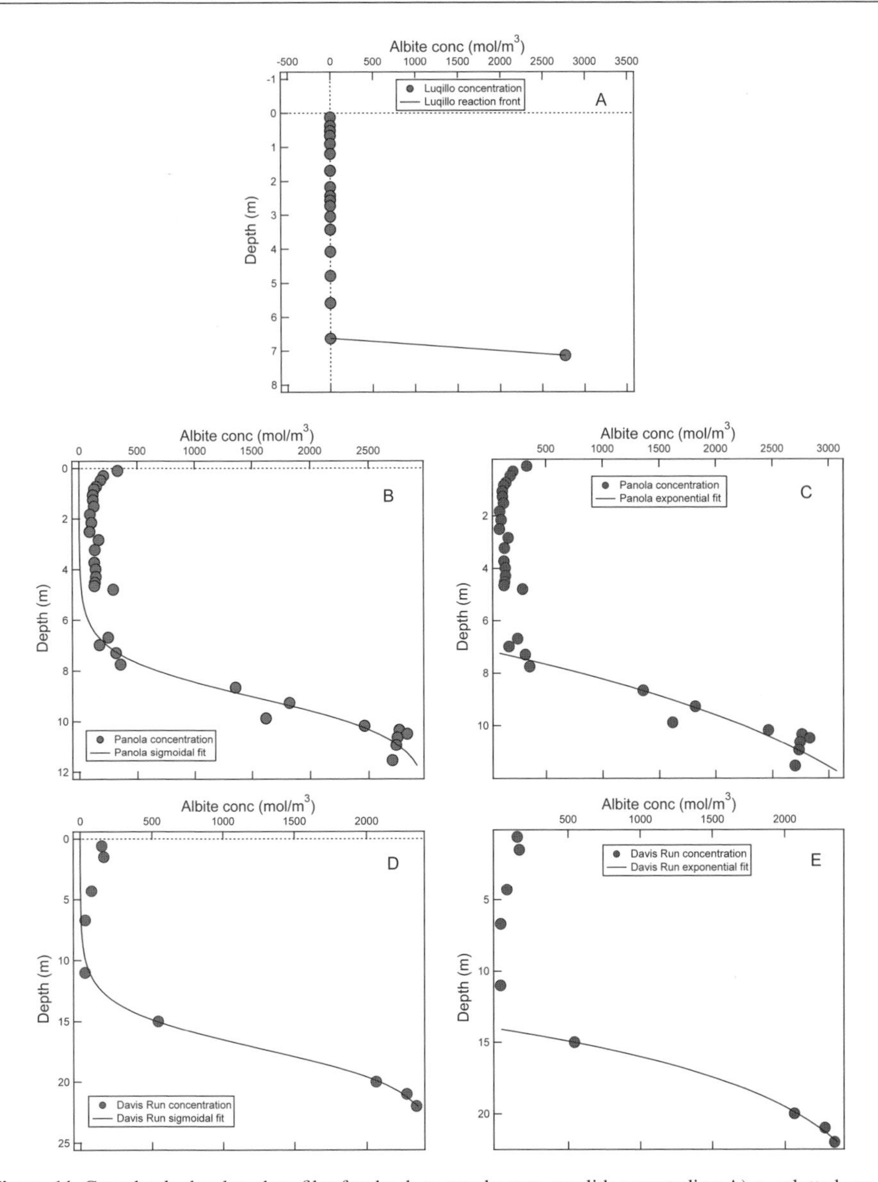

Figure 11. Completely developed profiles for the three steady-state regolith case studies: A) τ_{Na} plotted versus depth for one core into the Rio Blanco quartz diorite, Luquillo Experimental Forest, Puerto Rico (Buss et al. 2008); B) Na (albite) concentration versus depth for Panola GA (White et al. 2001); C) Na (albite) concentration versus depth for Davis Run, VA (White et al. 2001). The reaction front thickness for Na decreases from Davis Run granite (~10 m) to Panola granite (~6 m) to Puerto Rico quartz diorite (~40 cm) as the ratio of transport to dissolution rate decreases. A first-order control on this ratio is the physical state of the weathering material: at Davis Run, weathering of Na-feldspar is occurring in saprolite where advective transport dominates, while Na release from Panola occurs within the reacting low-porosity bedrock where diffusion dominates. In Puerto Rico, fracturing due to spheroidal weathering has been observed to create onionskin-like rindlets around weathering bedrock, accelerating the transport of reactants into the weathering bedrock (Fletcher et al. 2006). Curve fits to the data are also shown for the sigmoidal (B, D, Eqn. 30) and exponential models (C, E, Eqn. 35). The loss of Na in the Rio Blanco quartz diorite occurs in the 30-40 cm thick rindlet zone that defines the reaction front for Na plagioclase at the bedrock-saprolite transition. In Panola, the bedrock-saprolite transition lies just deeper than 5 m. For Davis Run, the bedrock-saprolite transition is more than a meter wide and lies below 20 m.

where the front is characterized by fracturing. Even where fracturing is not observed such as in Costa Rica, the opening of porosity at the reaction front must be incorporated into modeling in order to predict reaction front thickness (Navarre-Sitchler et al. 2009).

Given all these considerations, the thickness of the reaction front for the Na-containing phase generally decreases in the following order for the case examples: Santa Cruz CA, Davis Run VA, Panola GA, Puerto Rico, Costa Rica, Svalbard. As shown in Equations (9) and (10), the thickness of a reaction front for a single-component single-phase system is controlled by the ratio of transport of reactant into (or product out of) the protolith divided by the product of the rate constant for dissolution multiplied by the reactive surface area of the reacting phase. By analogy to the theoretical, single-phase, single-component system, we therefore hypothesize that the ratio of transport to reaction decreases in the order Santa Cruz CA > Davis Run VA > Panola GA > Puerto Rico > Costa Rica > Svalbard. In other words, this ordering is expected to progress from transport largely controlled by advection, where the reaction front lies within unconsolidated material, i.e., Santa Cruz CA, Davis Run VA, to control by diffusion where the reaction front lies across consolidated material—i.e., the bedrock-saprolite or rindlet-saprolite interface in Panola GA, Puerto Rico, Costa Rica, and Svalbard. A first order control on the thickness of the reaction front is apparently the dominant rate of transport of reactants and products, a property which often correlates with the extent of consolidation or disaggregation of the rock material comprising the reaction front.

Petrographic evidence suggests that the initial stages of weathering of low-porosity bedrock may be dominated by the rate of diffusion into mineral and rock matrix along grain boundaries, internal mineral porosity, or microfractures. One of the major difficulties in modeling such weathering is the fact that chemical weathering results in physical changes such as increases in porosity and permeability due to pore growth or fracturing driven by reaction (Navarre-Sitchler et al. 2009). In general, such coupling between chemical dissolution and surface area or pore space during weathering has only been included in a small number of weathering models.

MODELING APPROACHES

Linear gradient approaches

Linear gradient approaches toward modeling weathering profiles for solutes and solids are based upon the assumption of mass balance. At the catchment scale, solute mass balances have been related to rates of weathering of minerals since at least the 1960s (Johnson et al. 1968). In contrast, estimates of chemical weathering based upon solid phase concentrations versus depth (solid state gradients) were not developed extensively in the literature until the 1980s (April et al. 1986). Modeling of solid and solute gradients in weathering profiles is now well developed in the literature.

Idealized fully developed weathering profiles, defined by either solute or solid compositions, consist of three segments (Fig. 12a,b). The vertical segments at shallow depth corresponds to the zone in which the dissolving mineral has been completely depleted. The corresponding deep vertical segment corresponds to the protolith where weathering has not yet commenced. The reaction gradients for the solute, b_{solute}, and for the solid, b_{solid} (White et al. 2002), define the reaction front between these end-member conditions. The solid gradient reflects decreasing weathering intensities with depth. The corresponding solute profile reflects the cumulative increase in aqueous weathering products produced across the solid state gradient during the downward advection of water. The solute and solid gradients shown as solid lines in Figure 12 are only linear approximations to the actual weathering fronts which are suggested by the sigmoidal dashed curves. These weathering gradients develop over very different time spans. Solute gradients are produced over the decadal-scale residence times of pore waters while the

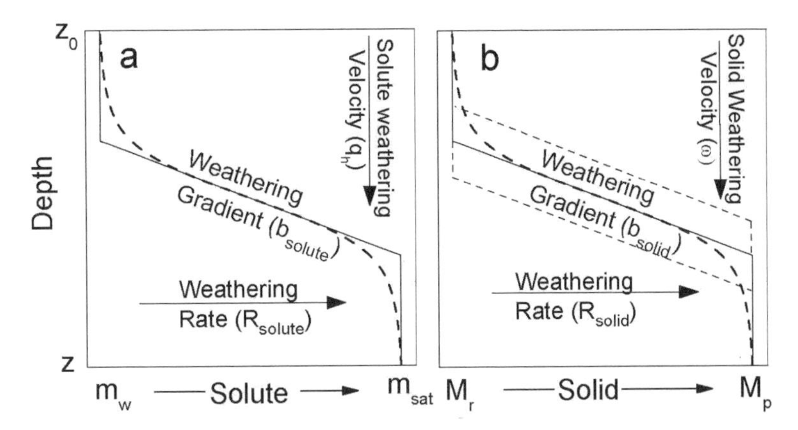

Figure 12. Schematic showing the relationships between (a) solute and (b) solid-state weathering rates, gradients and velocities (after White 2002). The soil surface is defined as z_0, m_w is the net solute composition derived from weathering and m_{sat} is the solute composition in thermodynamic equilibrium with the protolith. M_r and M_p are the respective solid regolith and protolith compositions. Dashed lines denote progressive advance of the solid-state profile with time. Sigmoid-shape curves are representations of the non-linear aspects of weathering profiles described in more inclusive models described below (White et al. 2008; Brantley et al. 2008).

solid-state gradients reflect the cumulative history of weathering, commonly on the order of tens of thousands to millions of years. In addition, the absolute mass change with depth per volume of regolith is much greater in the solid profile than in the solute profile or, conversely, $b_{solute} >> b_{solid}$.

In Figure 12, the solute and solid weathering gradients can be viewed as a balance between weathering rates R_{solute} and R_{solid} (mol m^{-2} s^{-1}) which define mass fluxes normalized to the volumetric surface area of a mineral phase and the solute and solid weathering velocities v and ω (m s^{-1}). The solute velocity (m s^{-1}) is equivalent to the hydrologic flux q_h (Eqn. 11) while the solid-state weathering velocity ω (m s^{-1}) in Equation (12) defines the vertical propagation rate of a weathering profile into the protolith as shown schematically in Figures 4 and 12b.

Mineral weathering rates can be determined from the weathering velocities and gradients using the following equations (White 2002):

$$R_{solute} = \frac{q_h}{b_{solute} \beta S_v} \tag{11}$$

and

$$R_{solid} = \frac{\omega}{b_{solid} \beta S_v} \tag{12}$$

Here β is the stoichiometric coefficient relating the elemental concentrations in minerals and S_v is the volumetric surface area of the mineral phase per m^3 of regolith (m^2 m^{-3}).

Solute gradients. Weathering rates based on solutes require the determination of the fluid flux q_h which is the volume of pore water moving through a cross section of soil. This flux is commonly defined by Darcy's law as the product of the hydraulic conductivity K_m (m s^{-1}) and the hydraulic gradient ∇H (m m^{-1}) (Hillel 1982):

$$q_h = -K_m \nabla H \tag{13}$$

Even for the case of simple vertical infiltration, as assumed for the models in the present paper, determining fluid fluxes in unsaturated soils, based on Equation (13), is difficult due to the non-linearities involved. The change in head with depth, z, is described such that (Hillel 1982):

$$\frac{dH}{dz} = \frac{dh_g}{dz} + \frac{dh_p}{dz} \tag{14}$$

where h_g is the gravitational head at any point and h_p is the head related to the matric potential of the soil mineral surfaces. In addition, for unsaturated flow, common in soils, K_m can vary by orders magnitude in response to relatively small changes in moisture saturation. For some special cases for Equations (13) and (14) (Stonestrom et al. 1998), constant matric potentials or tensions exist within a vertical section of a weathering regolith (i.e., $dh_p/dz = 0$).

An alternate, commonly used approach for estimating hydrologic fluxes q_h is based on the balance between precipitation and evapotranspiration fluxes q_p and q_{ET} such that

$$q_h = q_p - q_{ET} = q_p \left(\frac{c_{Cl,precip.}}{c_{Cl,pw}} \right) \tag{15}$$

Using the common assumption that the only source of pore water Cl is from precipitation and that Cl subsequently behaves conservatively in the soil pore water, the hydraulic flux becomes equal to the precipitation flux times the ratio of the respective Cl concentrations (Eqn. 15). The other term commonly used to characterize fluid flow in soil profiles is the advective flux or infiltration rate v ($m\ s^{-1}$) which is equal to the hydraulic flux density divided by the soil porosity φ ($m^3\ m^{-3}$) and degree of moisture saturation Γ ($m^3\ m^{-3}$):

$$\nu = \frac{q_h}{\phi \Gamma} \tag{16}$$

Dividing the infiltration rate by the depth of the weathering profile yields the total amount of time that a given volume of pore water resides in the weathering profile. Such residence times, commonly on the order of years, are short. Thus, these fluxes cannot integrate the effects of longer-term processes such as physical erosion.

The linear gradient approach (Eqn. 11) was first applied to solute distributions for Mg (Murphy et al. 1998) and for K and Mg (White 2002) to quantify contemporary biotite weathering rates in saprolite in the Rio Icacos regolith (Fig. 13). Due to significant variability in the overlying soil horizon, linear regression fits are confined to pore water Mg and K increases from the top (c_0) to the base (c_w) of the underlying saprolite. The slope for the Mg regression in Figure 13 is $b_{solute} = 1.13 \times 10^5$ m L mol^{-1} ($r^2 = 0.69$). Based on experimental unsaturated conductivities in the Rio Icacos saprolite, a flux of $q_h = 3.4 \times 10^{-8}$ m s^{-1} was derived (Stonestrom et al. 1998; White et al. 1998). This is in relatively close agreement with a flux of $q_h = 2.1 \times 10^{-8}$ m s^{-1} for averaging base discharge in the Rio Icacos watershed. The specific surface area ranged between 8.1 to 8.3 m^2 g^{-1} for biotite (Murphy et al. 1998).The biotite weathering rates in the Rio Icacos are therefore estimated as $R = 1.8$ to 3.6×10^{-16} mol m^{-2} s^{-1} based on solute Mg weathering gradients (Fig. 13).

Solid state gradients in the presence of steady state denudation. While the need to consider the effects of concurrent physical denudation on solute weathering profiles is minimal due to short pore water residence times, the impact of physical denudation must be considered when characterizing solid state profiles which are developed over much longer times. Two conditions have been considered in the linear characterizations of soil profiles, that of minimum physical erosion and that of steady state physical erosion.

Weathering of the Panola granite, as previously discussed (Fig. 11), has a ^{36}Cl-based erosion rate of E = 6 to 10 mm per thousand years (Bierman et al. 1995). Regional geomorphic

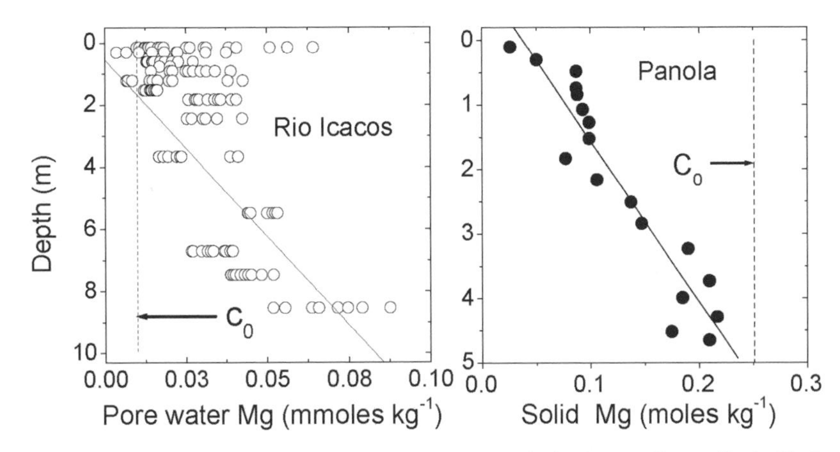

Figure 13. Pore water Mg concentrations reflecting biotite weathering in saprolite profiles in Rio Icacos and solid state Mg concentration reflecting biotite weathering in saprolite profiles at Panola, GA. The solute concentration c_0 refers to average rainfall at Rio Icacos and the solid Mg concentrations C_0 to the unweathered Panola granite. The diagonal lines represent linear regressions through the data with the slopes corresponding to the solute and solid weathering gradients (see Eqns. 11 and 12).

evidence indicates steady state conditions suggesting that soil production rates are comparable to denudation rates. In this case, the weathering profile remains spatially fixed relative to the land surface (Fig. 8). The solid state velocity of saprolite production is equated to the denudation rates such that $\omega = E$. As discussed by White (2002), decreases in solid-state Mg at progressively shallower depths in the Panola saprolite are attributed to increasing intensity of biotite weathering (Fig. 13B). Note that this is analogous to increases in solute Mg with depth in the Rio Icacos saprolite (Fig. 13A). Biotite BET surface areas in the Panola saprolite ranged between 1.5 and 7 $m^2\ g^{-1}$ (White et al. 2001). The resulting weathering rates for biotite were $R = 3.3$ to 6.8×10^{-17} mol $m^{-2}\ s^{-1}$ (Eqn. 12). These values are somewhat lower than those calculated for solute gradients in the Rio Icacos saprolite (Eqn. 11), a difference ascribed by White (2002) to less precipitation at the Panola site.

Solid state gradients in the absence of denudation. An alternate scenario is for solid state weathering velocities without significant physical denudation (Fig. 4). An example of such a situation is the paleo-beach terraces at Santa Cruz (Fig. 9b) in which ^{10}Be profiles indicate a minimum of soil removal from the topographically flat terrace surfaces that are up to 226 k.y. old (Perg et al. 2001). Normalized Na distributions (Eqn. 3), indicative of plagioclase weathering, are shown for terraces of two ages, SCT 2 = 90 k.y. (partially developed) and SCT 5 = 226 k.y. (fully developed) in Figure 14. As indicated, the weathering fronts are translocated downward with time from the surface at rates proportional to the solid-state weathering velocity (ω). In the absence of physical erosion, the weathering velocity ω (Eqn. 12) is estimated by simply dividing the average depth of the profile by the age of the profile surface. The resulting weathering velocities are estimated to be 2.9 and 4.0×10^{-5} m yr^{-1} respectively for the two terraces.

Solid state weathering velocities can also be estimated for quasi-steady state profiles under fluid transport-limited conditions by considering that the vertical displacement of the profiles with time is primarily controlled by the fluid flux. The weathering velocity can be defined as

$$\omega = q_h \cdot \left[\frac{m_{sol}}{M_{total}} \right] \tag{17}$$

where ω is proportional to q_h and the ratio of m_{sol}/M_{total} (both units are mol m^{-3}). This latter term describes the mass of a mineral, e.g., albite that can be dissolved in a thermodynamically-saturated volume of pore water divided by the initial mass of albite present in the protolith (White et al. 2008). For sparingly soluble phases such as silicates, the amount of mineral that can be dissolved in a single volume of pore water is very small compared to the amount initially present in the regolith, i.e., $m_{sol} << M_{total}$. Therefore, many pore volumes of water are required to completely solubilize M_{total}, the condition of a completed depleted or fully developed profile. The consequence is that the solid-state weathering velocity is much slower than the fluid flux, i.e., $\omega << q_h$ (m yr^{-1}).

The respective fluid fluxes through the SCT 2 and SCT 5 terraces, based on Cl balances (Eqn. 15) are 0.12 and 0.095 m yr^{-1} respectively (White et al. 2009). The mass of plagioclase per unit volume of each protolith, based on quantitative X-ray diffraction analyses, is 2300 and 2360 mol m^{-3} respectively. Based on the average pH and the solubility quotients the amount of plagioclase that can be dissolved in pore water in both terraces is 0.77 mol m^{-3}. The resulting weathering velocities for SCT 2 and SCT 4 are ω = 0.040 and 0.031 m k.y.$^{-1}$ values which is similar to the velocities estimated from the profile geometries.

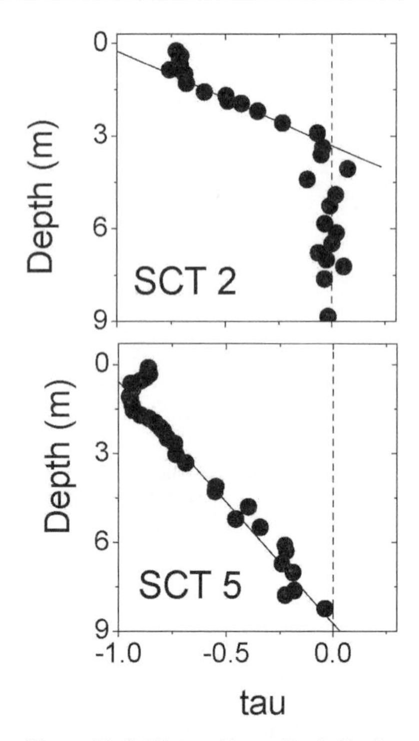

Figure 14. Solid state Na profiles indicative of plagioclase weathering in the Santa Cruz soil chronosequence. Linear fits correspond to solid state weathering gradients, b_{solid} (Fig. 12b). Note that in the younger SCT 2, the weathering front has not penetrated into the deeper part of the profile, corresponding to the vertical protolith gradient shown in Figure 12b (the reaction front is not fully developed).

Based on the geometric interpretations summarized in Figure 12, the solute and solid-state weathering profiles should be directly related under steady state weathering conditions. Assuming constant weathering rates and surface areas, Equations (11) and (12) can be equated such that

$$\left[\frac{q_h}{\omega}\right]\cdot\left[\frac{b_{solid}}{b_{solute}}\right]=1 \quad (18)$$

where the product of the ratio of the weathering velocities and inverse ratio of the weathering gradients is equal to one. The extent to which this condition is met, i.e., the extent to which the product of the measured parameters on the left side of Equation (18) approach unity, is a gauge of the correctness of the steady-state weathering assumption (Brantley et al. 2008; White et al. 2009). Combining Equations (12) and (17) produces an expression for the solid state weathering rate R:

$$R=\left[q_h\cdot\frac{m_{sol}}{M_{total}}\right]\cdot\left[\frac{1}{S_v\cdot b_s}\right] \quad (19)$$

The product of the parameters contained in the first set of brackets is constant across a weathering profile and equal to the weathering velocity ω (Eqn. 17) while the parameters within the second set of brackets contain the profile gradient and surface area. The product of these two terms characterize the reaction rate. The surface area S_v is equal to the product of the mass of residual mineral present in a volume of regolith and the specific surface area defined per unit of mineral mass S_{min} ($m^2\ kg^{-1}$).

$$S_v = M_r \cdot S_{min} \tag{20}$$

The residual mass decreases with decreasing depth in a profile while specific mineral surface area may be expected to increase, reflecting decreases in residual grain sizes and increases in surface pitting and roughness (White et al. 1996). These opposing trends may result in relatively small variations in volumetric surface areas across significant portions of a weathering profile and may be responsible for the commonly observed pseudo-linear nature of many weathering gradients (Lichtner 1988; Brantley et al. 2008). In estimating both the solute and solid-state weathering rates of the Santa Cruz terraces, the specific mineral surface area S_{min} is calculated from the geometric surface area based on average particle diameters and normalized to a BET equivalent value using age-dependent surface roughness factors estimated (White and Brantley 2003).

Analytical models

The "geometric model" just presented can be conceptualized simply as a linear fit to the reaction fronts as shown in Figure 12. Other functionalities that can be used to describe reaction fronts include exponential and sigmoidal functions (e.g., Torrent and Nettleton 1979; Ortoleva et al. 1986; Lichtner 1988; Mathieu et al. 1995; Brantley et al. 2008). We present a generalized derivation of these equations for either quasi-stationary or steady-state fronts based on our earlier treatment (Brantley et al. 2008). Lichtner (1988) earlier derived analytical expressions such as those developed here and pointed out that generally they can only be derived for one component, single phase systems. For multicomponent multiphase systems such as soils, simplifications must be made in description of components and phases. As a result of these simplifications, the derived values for rate constants differ from the measured rate constants reported in the literature (Bandstra et al. 2008). In the example developed below (Brantley et al. 2008), the analytical expression is derived for the H^+ concentration in porefluids. In contrast, in the work used to derive Figures 4 and 8 (Lebedeva et al. 2007; Lebedeva et al. 2009), modeling is completed with respect to a thermodynamic component defined as Na_2O-½H_2O-$2SiO_2$. In those latter papers, the rate constant for albite dissolution used in the model is not explicitly equivalent to measured rate constants but rather differs by a term noted as Ψ.

In deriving an equation to fit a quasi-stationary soil profile where transport is by advection, we assume a "front-fixed" reference frame defined such that 0 is the top of the reaction front. Thus, for a non-eroding quasi-stationary profile, the reference frame moves downward with respect to the land surface at the rate of weathering advance. In contrast, for a steady-state eroding profile, this same reference frame lies at a constant depth with respect to the land surface. Assuming the front-fixed reference frame in both cases, mineral grains move upward through the origin at the same time that pore fluid moves downward at a rate v (m/s). Note that Equation (16) relates the advective flux (υ) used in derivations in this section to the hydraulic flux (q_h) used in the preceding section. To maintain consistency with the published derivation (Brantley et al. 2008), here, $\nu > 0$ and $\omega < 0$ and the weathering advance rate (the rate that the weathering front propagates downward for the non-eroding case) equals $-\omega$. In general, ν (or q_h) >> $-\omega$ as discussed in the previous section.

Mass balance on the total concentration (dissolved + solid) Na (mol m^{-3}), here noted as C^{Na}_{total}, in an infinitely thin layer of regolith at the top of the reaction front at the origin requires that the rate of change of Na in the layer must equal the rate of delivery of aqueous Na, C^{Na}_{aq}, and solid Na, C^{Na}_{solid}:

$$\frac{\partial C^{Na}_{total}}{\partial t} = -\upsilon \frac{\partial C^{Na}_{aq}}{\partial x} - \omega \frac{\partial C^{Na}_{solid}}{\partial x} \tag{21}$$

In this equation, t represents time and x represents depth below the origin. In effect, because we have chosen a front-fixed reference frame, the variable x is related to the depth z as used in the previous section (see Fig. 12) by a correction factor related to the depth to the reaction front. Equation (21) reduces at steady state to:

$$\omega \frac{dC^{Na}_{solid}}{dx} = -\upsilon \frac{dC^{Na}_{aq}}{dx} \tag{22}$$

According to Equation (22), the Na concentration gradient across the steady state profile is related to the gradient in dissolved Na in the porefluid at steady state by the ratio $-\nu/\omega$. Note that this equation is identical to the previously derived Equation (18) where differential expressions are used for the values of the two slopes (b_{solid} and b_{solute}). Note that the sign difference between the two equations is because the sign convention chosen for ω for this derivation is the opposite of the sign convention in deriving Equation (18). Importantly, if ν is known for a steady state profile, and if both the solid and porefluid gradients in Na concentration are known, then ω can be estimated. Alternately, where both ν and ω are known, Equation (22) can be checked to determine whether the soil is weathering at steady state (analogous to Eqn. 18).

Equation (22) can be integrated under the boundary conditions such that at the origin $x = 0$ the dissolved and solid Na concentrations can be defined to equal C^{Na}_{input} and $C_{x=0}$ respectively. Furthermore, if Na is only present in the profile as albite, then the concentration of Na can be related to dissolution of albite and precipitation of a phase such as kaolinite by choosing a reaction stoichiometry:

$$NaAlSi_3O_{8(solid)} + 0.5H_2O + H^+_{(aq)} = Na^+_{(aq)} + 0.5Al_2Si_2O_5(OH)_{4(solid)} + 2SiO_{2(aq)} \tag{23}$$

The rate of this reaction, R, can be defined as the rate of change in the reaction extent, ξ, of Equation (23) as a function of time:

$$R = \frac{d\xi}{dt} = \frac{-dC^{Na}_{solid}}{dt} = \frac{dC^{Na}_{aq}}{dt} = \frac{-dC^{H}_{react}}{dt} \tag{24}$$

As protons are consumed in the reaction, the proton concentration will decrease below the starting proton concentration at the top of the reaction front, C^H_{input}. However, the observed change in proton concentration will differ from that predicted based on mass balance on Equation (23) alone because of other proton-consuming or proton-producing reactions in the soil matrix + solution system. We therefore define a buffer capacity, β, to describe how dC^H_{react}, the differential change in concentration of protons added to the soil-water system by reaction (23), translates to dC^H_{aq}, the differential change in the actual proton concentration. This latter change will generally be small due to buffering by other phases, biota, and organic matter:

$$\beta = \frac{dC^H_{react}}{dC^H_{aq}} \tag{25}$$

The stoichiometry of reaction (23) also dictates that $dC^{Na}_{aq} = -\beta dC^H_{aq}$. After insertion of this equation into Equation (22), the equation can be integrated over the reaction front from $C^{Na}_{solid} = C^{Na}_{x=0}$ to C^{Na}_{solid} at some depth x, and from $C^H_{aq} = C^H_{input}$ to C^H_{aq}. At the bottom of the profile, the Na concentration in the solid must be identical to the parent Na concentration in the solid, C_o. At that point, $C^H_{aq} = C^H_{eq}$, the concentration of protons in porefluid equilibrated with the parent rock. All of these considerations lead to a clarification of the meaning of the *buffer capacity*, β, and demonstrate the utility of defining a new parameter, the *proton driving force for reaction*, m:

$$m = C^H_{input} - C^H_{eq} = -\frac{\omega}{\upsilon}\frac{(C_o - C_{x=0})}{\beta} \tag{26}$$

$$\beta = \frac{\omega}{v} \frac{C_0 - C_{x=0}}{C_{eq}^H - C_{input}^H} \tag{27}$$

The proton driving force quantifies how far the concentration of protons in the input water, C_{input}^H, is from the value at equilibrium with respect to the parent material, C_{eq}^H. The buffer capacity quantifies how much albite will dissolve before the proton concentration reaches equilibrium.

Following standard geochemical treatments of kinetics (Lasaga 1981; Helgeson et al. 1984; Brantley 2008a), the net rate of the reaction can be written as a function of k, the rate constant for dissolution of albite (mol Na m^{-2} s^{-1}) at pH 0 and A_{vol}^{BET}, the albite-water interfacial area (measured by BET adsorption isotherm, normalized per unit volume of regolith (m^2 m^{-3})). Note that A_{vol}^{BET} is equivalent to volumetric surface area term S_v used in the preceding section, i.e., Equations (11) and (12). Defining a_{H+} as the activity of H^+ and n as the reaction order that relates the rate of plagioclase dissolution to pH (Brantley 2008a), then mass balance on dissolved Na in the porefluid in the infinitely thin layer of regolith at the origin requires the following:

$$v \frac{dC_{aq}^{Na}}{dx} = kA_{vol}^{BET} (\gamma C_{aq}^H)^n f(\Delta G) \tag{28}$$

Here $f(\Delta G)$ is a function of the chemical affinity and γ is the activity coefficient. The activity of protons, a unitless term, is equal to γC_{aq}^H.

Again, if steady state is obtained, then the dissolved concentration of Na^+ or H^+ can be assumed to stay constant with time and the derivative in time can be replaced by a derivative in space. By rearrangement, and by choosing a simple form for the affinity function (Brantley et al. 2008) and setting n to unity, we derive

$$\frac{dC_{solid}^{Na}}{dx} = -\frac{kA_{vol}^{BET}\gamma}{v\beta}\left(C_{solid}^{Na} - C_0\right) \tag{29}$$

Note that A_{vol}^{BET} in Equation (29) varies with depth, i.e., with C_{solid}^{Na}. Therefore, this is a rate equation that is not strictly first-order with respect to the Na-containing reactant. The reaction order depends instead on how the surface area term varies. To integrate this equation depends upon choosing a function for A_{vol}^{BET} as a function of depth. Importantly, these surface area changes are responsible for the upward concavity at the top of the front while the downward concavity at the bottom of the front is due to approach to equilibrium. These two effects compensate at the center of the front to produce a linear profile (Brantley et al. 2008; White et al. 2008).

Importantly, integration of this equation under different assumptions for the affinity or the surface area term will yield sigmoidal or exponential model equations. For example, under one simple assumption for how surface area varies with depth, the sigmoidal equation is derived (Brantley 2008a):

$$C = \frac{C_0}{\frac{C_0 - C_{x=0}}{C_{x=0}} \exp\left(\Gamma_{ini} \cdot \hat{k} \cdot x\right) + 1} \tag{30}$$

Here, the notation has been simplified so that C is the concentration of Na in regolith solid phase and C_o is the Na concentration in the protolith. $C_{x=0}$ is the concentration of solid-phase Na at the top of the front. Γ_{ini} describes the roughness of the albite in the protolith, and is equivalent to the BET surface area of the albite divided by the geometric surface area. Geometric surface area is generally calculated assuming a grain size and a spherical geometry for the particles.

In this equation, the kinetic parameter, $\hat{k}$, can be shown to be related to other parameters:

$$\hat{k} = \frac{kA_{ini}\gamma C_0}{v\beta(C_0 - C_{x=0})} \approx -\frac{kA_{ini}\gamma}{\upsilon\beta} \tag{31}$$

Here A_{ini} is the initial geometric surface area of the mineral in the parent (m^2 m^{-3}), v is the advective flux of porefluid through the front, β is the buffer capacity of the rock-water system, ω is the velocity of the weathering front, γ is an activity coefficient (m^3 mol^{-1}), and m (Eqn. 26) is the difference in concentration of protons in the incoming fluid at the top of the front and the equilibrium value at depth in the protolith. Model Equation (30) is a function that is sigmoidal with depth—fits of this equation to Santa Cruz terraces, to Panola, and to Davis Run are shown in Figures 9 and 11.

One interpretation of the sigmoidal model equation is that it is consistent with a fractal description of the area of mineral surfaces as a function of depth (Brantley et al. 2008). In particular, the surface areas of the dissolving mineral grains are not well described using an Euclidean geometry; instead the surface area of the mineral grains are more space-filling (more rough). The roughness of the particles reflects changes in grain size distribution as well as the roughening of the internal and external surface of individual particles during reaction. Brantley et al. pointed out that the equation is formally equivalent to the steady-state solution for a simple autocatalytic reaction in a plug-flow reactor where the "product" that autocatalyzes the reaction is surface area (Moore and Pearson 1981).

This fractal description may be one reason why the sigmoidal equation describes the curvature of reaction fronts—concave-upward at the top, sublinear throughout the middle, and concave-downward at the bottom of the profile. Although this equation has been derived assuming a system with only plagioclase transforming to kaolinite, it should have utility in exploring soil profiles for other elements given that the sigmoidal curvature of reaction fronts is shared by many systems for depletion profiles (Figs. 9-10). Brantley et al. (2008) showed that the parameter $\hat{k}$ for a front can be written as a simple function of the weathering advance rate:

$$\hat{k} \approx -\frac{kA_{ini}\gamma m}{\omega C_0} \tag{32}$$

This formulation shows that this parameter is the inverse of the thickness of the reaction front under certain conditions as derived by Lichtner (1988) and discussed earlier (Eqn. 10). When Equation (31) was used to calculate values of the dissolution rate constant k from the fits in Figures 9 and 11 (γ = 1) for comparison to literature values at 25 °C, the calculated values for k were observed to be within a factor of 60 of the published dissolution rate constants for albite at 25 °C (Brantley et al. 2008).

The value of $\hat{k}$ is related to the slope of the sublinear section of the elemental or mineralogical profile. As described in the previous section on geometrical treatments of profiles, this slope is often used to determine the apparent rate constant of dissolution. The value of dC/dx or $1/b_{solid}$, using the notation of geometrical treatment, evaluated at the inflection point can be derived:

$$b_{solid} = \frac{4\omega}{kA_{ini}\gamma\Gamma_{ini}m} \approx \frac{4\omega}{\phi kA_{ini}\gamma\Gamma_{ini}C_{input}^{H,meas}} \tag{33}$$

Here, $C_{input}^{H,meas}$ describes the measured concentration (mol m^{-3} water) of protons in porefluids at the top of the reaction profile. The denominator of the righthand expression in (33) is the rate of reaction of feldspar, R_{solid} (where the reaction order with respect to protons, $n = 1$), times the porosity. This expression for the slope of the linear part of the reaction front is similar to that derived based upon geometrical considerations as described in the previous section in Equation (12) (Murphy et al. 1998; White 2002). Thus, the geometrical approach described in a previous section can be thought of as an approximation of the full sigmoidal model. The geometrical approach also results from the implicit assumption that the rate of albite dissolution is zeroth order with respect to albite (surface area and affinity are both constant with depth).

The second parameter of interest that describes reaction fronts is the weathering advance rate, $|\omega|$. This rate is a function of the porefluid velocity (Ortoleva et al. 1986; Lichtner 1988). However, as pointed out by Ortoleva et al. (1986) and discussed previously, $|\omega|$ is generally much smaller than v:

$$|\omega| = \upsilon \frac{\beta m}{C_0 - C_{x=0}} \approx \upsilon \frac{\beta m}{C_0} \tag{34}$$

This equation is equivalent to the equation derived previously as Equation (17) although in this case, the reaction is parameterized in terms of proton consumption in a buffered solution. Consistent with arguments by Ortoleva and coworkers, the lag between ω and v is dependent upon the ratio m/C_0, which is the ratio of the change in fluid reactant (here, protons) to the change in mineral reactant (here, albite) across the front. Based upon this model where we have calculated reactions on the basis of consumption of protons, the lag is also affected by the buffer capacity of the system. The buffer capacity of the Santa Cruz system was estimated from fits (Fig. 9) to equal ~10^5. Similar calculations based on fits for Panola (Fig. 11) yielded a value of ~10^7, and for Davis Run (Fig. 11) a value of ~10^6 (Brantley et al. 2008).

In contrast to allowing the surface area term to vary with depth as discussed for the sigmoidal model (Brantley et al. 2008), if we assume surface area is constant across the reaction front, then integration of Equation (29) (n = 1) yields an exponential equation:

$$\frac{C_0 - C}{C_0} = \exp\left(-\frac{kA_{ini}\gamma x}{v\beta}\right) \tag{35}$$

Such an exponential equation has been used previously in the literature to fit soil profiles successfully and is consistent with the assumption of a rate equation for dissolution of albite that is first order with respect to the reacting albite (Torrent and Nettleton 1979). A version of such an exponential equation is commonly used in treating U reaction kinetics in soils (Mathieu et al. 1995). Exponential fits to Panola and Davis Run are shown in Figure 11.

In general, of the three analytical models presented here—linear, exponential, sigmoidal—the linear model has been most fully developed in the literature (White 2002). However, often the sigmoidal equation best fits the data: the linear model only fits the sublinear part of the curve whereas the exponential model incorporates curvature at the bottom of the front that is related to affinity and the sigmoidal model incorporates both the affinity-related bottom curvature and the surface area-related curvature at the top of the front. This is exemplified, for example, in the fits of two of the models to chemical data for Davis Run and Panola as shown in Figure 11. Use of the sigmoidal model, derived for completely developed profiles, has even been shown to have utility toward fitting transient, incompletely developed profiles such as shown in Figure 4 (Brantley et al. 2008).

Spreadsheet models

An approach for interpreting soil profile data using EXCEL spreadsheets has also been developed which focuses on field-based parameters and using several of the parameters described in the preceding section. Two of these *weathering profile calculators* have been developed, one based on constant permeability in unconsolidated sediments, i.e., the Santa Cruz chronosequence (White et al. 2008, 2009) and the other describing weathering across a permeability gradient at a bedrock/saprolite interface, i.e., Panola granite (White et al. 2005). The calculators use very limited solute inputs and, as such, represent significant simplifications of the more complex aspects of weathering described by modeling simulations that couple solute compositions, fluid transport and, increasingly, the impacts of biology in reactive transport codes (see the next section and Steefel and Maher 2009 in this volume). However, the profile calculators address a common situation in which weathered regoliths are characterized in terms of solid element and mineral distributions with depth but for which solute data are lacking. In addition, the spreadsheet models have more versatility than the analytical models presented in the last section. Results demonstrate that the calculators reproduce the major profile features and produce consistent estimates of chemical weathering rates as described below.

Only the weathering reactions of plagioclase (albite) and K-feldspar to kaolinite are considered in Santa Cruz and Panola calculations (White et al. 2005, 2009). The spreadsheets employed forward calculations based on matrix arrays were the mass of mineral reacted at depth z (m) and time increment t (k.y.) is defined by the expression

$$\Delta M_{z,t} = S_{z,t} \cdot t \cdot R \tag{36}$$

where $S_{z,t}$ is the volumetric surface area of this increment. The weathering rate R is defined in terms of a simplified transition state (TST) rate expression such that

$$R = k_r(1-\Omega)^n \tag{37}$$

where k_r is the rate constant, Ω is the saturation index and n is an affinity exponent. Aqueous concentrations are used only in determining the values of the saturation index Ω. The effects of specific aqueous species, such as H^+ and Al ions, on the weathering kinetics are not considered (see other chapters in this volume). Likewise, the simple dependence of the weathering rate on the saturation index does not incorporate more complex interpretations of the role of reaction affinities on kinetic mechanisms (see a brief discussion of this topic in the next section on reactive transport codes).

The resulting aqueous concentration of a given aqueous species is then calculated

$$c_{z,t} = \frac{\beta\left(\sum_{i=1}^{z-1} M_{z,t} - \sum_{i=1}^{z-1} M_{z,t-1}\right)}{q_h \cdot (t_t - t_{t-1})} \tag{38}$$

where β is the stoichiometric coefficient for the feldspar components and q_h is the hydraulic flux (m s^{-1}) which describes the rate of pore water movement through the profile. Equation (38) defines the solute concentration at a given depth as the amount of solute produced from feldspar reaction in the overlying regolith $M_{z,t}$ during the time increment between t and $t-1$ divided by the total volume of water that moved through the system. This volume is the product of the hydraulic flux and time. The Si concentration is equal to the sum of parallel calculations for both plagioclase and K-feldspar. The saturation index Ω is then calculated using the appropriated values of $c_{z,t}$ for each species and a fixed pH corresponding to average pore water and the rate calculations repeated in the next time step.

The initial geometric surface area of a mineral in the protolith, S°_{geo}, defined on a volumetric basis, (m^2/m^3), is

$$S^{\circ}_{geo} = \frac{6}{\rho d} \cdot 10^4 \cdot m_w \cdot M_p \tag{39}$$

where ρ is the mineral density (g cm^{-3}), m_w is molar weight (mol g^{-1}), d is the average feldspar grain diameter (cm), and M_p is the initial molar concentration of the reacting mineral. The volumetric feldspar surface $S_{z,t}$, in a given reaction increment, is related to the mass of residual feldspar present, i.e., $(M_i - M_{t\text{-}1,d})/M_p$, by the relationship

$$S_{z,t} = \lambda S^{\circ}_{geo} \cdot \left(\frac{M_p - M_{z,t-1}}{M_p} \right)^{\alpha} \tag{40}$$

Here λ is the surface roughness factor described on a time-dependent basis by White and Brantley (2003). The exponent α relates the mineral surface area to volume and can be related to the fractal analysis described in the previous section. For a model of surface area that follows Euclidean geometry, α must equal 2/3, whereas in the fractal treatment mentioned with respect to the sigmoidal model in the previous section, the value of this term increases above 2/3 as surfaces become more space-filling.

Fitting profiles with constant permeability. Only four variables were considered in the profile calculator used to fit the SCT 5 weathering profile of the Santa Cruz chronosequence (White et al. 2008). These were the fluid flux q_h (Eqn. 38) the rate constant k_r (Eqn. 37), the saturation index exponent n (Eqn. 37) and the surface area exponent α (Eqn. 40). A simple least-square-minimization was applied between the fitted and field-based feldspar profiles using variations in only q_h and k_r, the two parameters found to have the largest impact upon the shape and position of the calculated weathering profiles. For the SCT 5 plagioclase profile, a least-squares minimization produced a unique pair of fitting parameters (q_h = 0.058 m yr^{-1} and k_{plag} = 4.8 × 10^{-16} mol m^{-2} s^{-1}) shown in Figure 15A and B.

Fluxes slower than the best fit value (q_h/2 and q_h/4) produced shallower weathering profiles

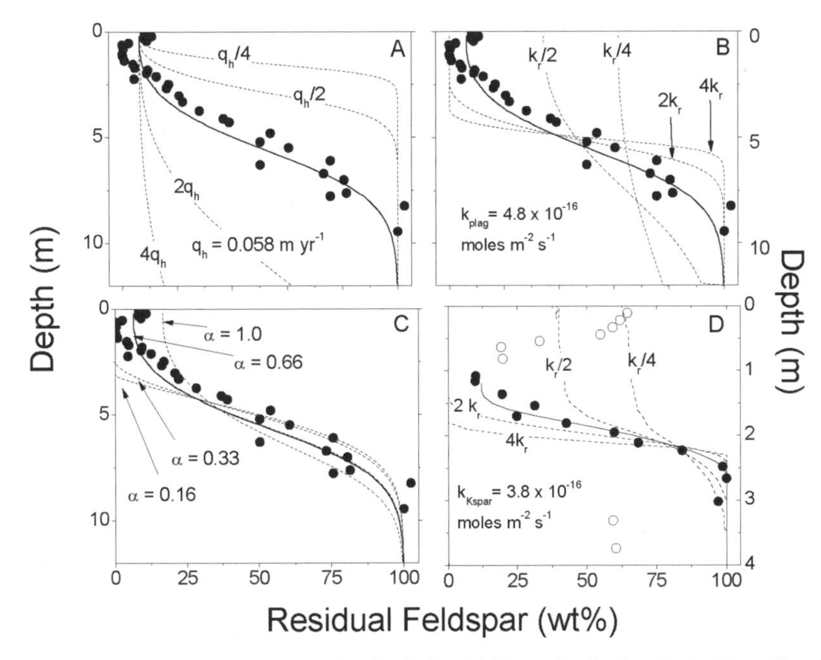

Figure 15. Comparison between measured and calculated feldspar distributions in the Santa Cruz SCT 5 weathering profile. In plots A, B and C, the solid lines are the best least squares fits to the measured plagioclase profiles (•) using a fluid flux of q_h = 0.058 m yr^{-1} and a weathering rate constant of k_{plag} = 4.8 × 10^{-16} mol m^{-2} s^{-1}. Plot C shows the effects of variation in the surface area exponent α (Eqn. 40). In plot D, the best fit to the K-feldspar profile is shown using q_h=0.058 m yr^{-1} and a weathering rate constant of k_{plag} = 3.8 × 10^{-16} mol m^{-2} s^{-1}. Dashed lines in each plot show the effects on calculated profiles by increasing and decreasing q_h, k_{plag} and the surface area exponent α (Eqns. 36-40). Open circles indicate field data not fitted in calculation. [Used with permission of Elsevier from White et al. 2008.]

(lower values of b_s in Figure 12 because pore waters became thermodynamically saturated with plagioclase at progressively shallower depths (Fig. 16). Higher values of $2k_{plag}$ and $4k_{plag}$ produced progressively flatter, shallower gradients (lower values of b_s) with larger changes in residual plagioclase contents over smaller depth intervals These results are consistent with Equation (10) (Lichtner 1988). Unlike for pore water flow, increases in the rate constant k_{plag} did not significantly affect the total amount of plagioclase reacted in the profile ($\Delta M_z t$ in Eqn 38). Graphically, this is apparent in Figure 15B, where increases in k_{plag} rotate the weathering gradient b_s, increasing the weathering intensity at shallower depth but also decreasing it at greater depths. The net result is that the amounts of plagioclase reacted underneath the different curves in Figure 15B are about the same. The total mass loss, representing the weathering intensity, is dependent principally on the depth at which pore waters reach thermodynamic saturation (Fig. 16) which is ultimately controlled by the volume of water that moves through the regolith and not on kinetics which control how fast the plagioclase dissolves.

The persistence of K-feldspar compared to plagioclase is a significant weathering characteristic in the Santa Cruz Terraces (Fig. 15D) as well in many natural environments. This difference is not replicated experimentally, where K-feldspar and plagioclase rates are similar (Blum and Stillings 1995; White and Brantley 2003). This contrast is commonly attributed to the fact that experimental dissolution studies are generally conducted far from equilibrium while natural weathering generally occurs much closer to equilibrium. Such natural weathering commonly reflects differences in feldspar solubilities rather than differences in kinetic rates (White et al. 2001). Indeed, where K is recycled upward by plants to produce a biogenic K profile such as documented in Figure 3E, this higher concentration of solid and solute K in surface layers will suppress K-feldspar dissolution rates due to affinity effects.

Fitting profiles across a strong permeability gradient. The second application of the spreadsheet calculator is in describing granite/ saprolite weathering of the Panola Granite (White et al. 2001). As previously discussed, this type of weathering is characterized in the initial stages of bedrock weathering dominated by very slow rates of transport that is controlled by mineral and rock fabrics, grain boundaries and internal mineral porosity (see correlations between calcite and plagioclase weathering profiles in Fig. 7) followed by weathering-induced mass transfer ultimately producing a high permeability saprolite.

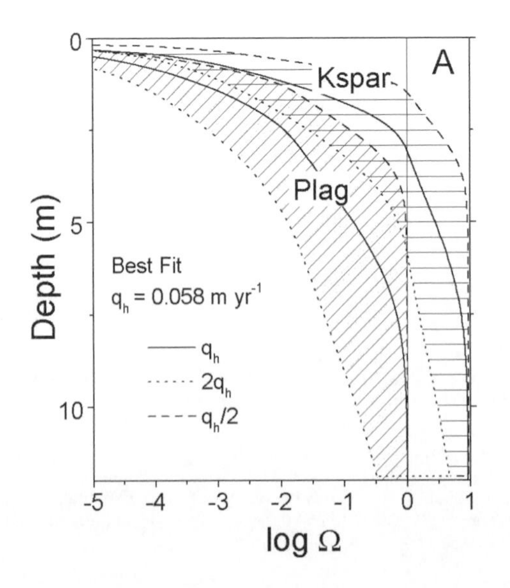

Figure 16. Predicted effects of thermodynamic saturation on feldspar weathering in the Santa Cruz SCT 5 terrace profile. Fluid saturation indexes for plagioclase and K-feldspar (Ω in Eqn. 37) plotted as functions of depth. Results are generated using both the best-fit fluid flux q_h (solid lines) and lower and higher q_h values (dashed lines). $\Omega < 0$ indicates undersaturation and $\Omega > 0$ indicates supersaturation.

A critical assumption that underlies the calculations is that the weathering front advances at the rate at which chemically unsaturated meteoric water penetrates into fresh granodiorite bedrock to initiate the weathering process (Bierman et al. 1995). A propagation rate of the weathering front of 7 m per 10^6 yr under steady state conditions at Panola translates into a primary conductivity of $q_p = 7 \times 10^{-6}$ m yr^{-1}. This value is more than an order of magnitude slower than the slowest hydraulic conductivity measured for fresh granites in laboratory and field studies (see Table 1), but may be more representative of the pervasive nature of plagioclase weathering requiring slower fluid transport on a microscopic scale into the granite matrix.

Table 1. Hydraluic conductivities in granite rocks and weathered regoliths (myr^{-1}) After White et al., 2001

	Rock	Method	Reference
Primary granite conductivities q_p			
$7x10^{-6}$	Fresh Panola Granite	Weathering rate	White et al. 2001
$6x10^{-6}$	Fresh Occoquan Granite	Weathering rate	White et al., 2001
$3x10^{-4}$ to $6x10^{-4}$	Stripa Granite, Sweden	In situ, tracer	Abelin et al, 1991
$3x10^{-4}$ to $3x10^{-1}$	Bukit Timah Granite, Singapore	Insitu, falling head	Zhao, 1998
$6x10^{-4}$ to $2x10^{-2}$	Beauvoir Granite, France	Core N_2 permeability	Galle, 1994
$1x10^{-2}$ to $3x10^{0}$	Granite, Illinois, USA	Core H_2O permeability	Morrow and Lockner, 1997
Secondary regolith Conductivities q_s			
$1x10^{-3}$	Weathered Panola Granite	Weathering rate	White et al., 2001
$>4.0 \times 10^{-3}$	Davis Run Saprolite/Soil	Weathering rate	White et al., 2001
$3x10^{-2}$ to $1x10^{1}$	Soil/Saprolite, Panola, GA USA	Exp. sat. conductivity	Stonestrom et al., 1998
$4x10^{-1}$ to $1x10^{2}$	Soil/Saprolite, NC, USA	Exp. sat. conductivity	Schoeneberger et al., 1995
$5x10^{-1}$ to $2x10^{0}$	Saprolite, NC, USA	Exp. sat. conductivity	Vepraskas and Williams, 1995
$3x10^{0}$	Saprolite, Rio Icacos PR	Exp. sat. conductivity	White et al. 1998

Variables fitted to feldspar weathering in the Panola bedrock are the kinetic rate constants for plagioclase and K-feldspar, k_{plag} and k_{Kspar}, the surface area exponent α in Equation (40), and the secondary hydraulic conductivity, q_s. A value of $k_{plag} = 3 \times 10^{-16}$ mol m^{-2} s^{-1} produces the best fit for plagioclase weathering data (Fig. 17A, solid line). Increasing k_{plag} by a factor of 5 results in a very flat weathering profile in which plagioclase is completely removed over a very small depth interval immediately above the unweathered granite. (Fig. 17A, dashed line). Decreasing k_{plag} by factors of 2 and 5 result in progressively steeper plagioclase profiles in which up to 50% of the plagioclase persists in the upper soil horizon.

The exponent α, which relates the surface area to the mass of residual feldspar (Eqn. 40) is the second fitting parameter describing the distribution of feldspar reacted in the Panola Granite. Increasing the value of α produces progressively stronger curvatures to the fit of the data in which mass of reacted plagioclase asymptotically decreases with decreasing depth (Fig. 17B). The secondary hydraulic conductivity q_s is the final fitting parameter in the spreadsheet calculations (Fig. 17C). The value for q_s fixes the pore water volume that passes through the weathered granite for each time increment. This volume, coupled with the mass of reacted mineral, determines the thermodynamic saturation state of the feldspars. The secondary hydraulic conductivity q_s can be expressed as a ratio relative to the primary conductivity, which is independently estimated based on assumed steady weathering and cosmogenic age dating of the regolith ($q_p = 7.2 \times 10^{-6}$ m yr^{-1}). The best fit to the plagioclase data in Figure 17C is produced when the secondary conductivity is significantly larger than of the primary conductivity, i.e., $q_s/q_p = 150$ ($q_s = 1.1 \times 10^{-3}$ m yr^{-1}). This ratio insures that plagioclase does not reach thermodynamic saturation in the bedrock but rather continues to react until its mass is exhausted at a depth of 6.5 m.

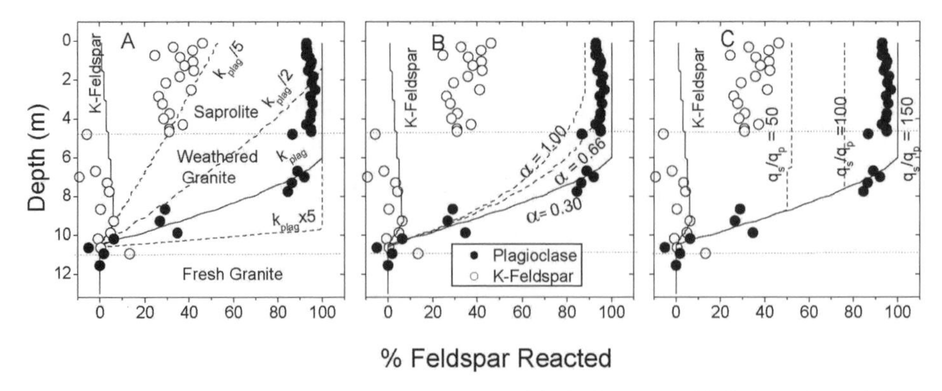

Figure 17. Percent of feldspars reacted plotted versus depth in the Panola regolith. Solid and open circles are the respective plagioclase and K-feldspar losses based on elemental mass balances. Solid lines are the best fits of the spreadsheet calculations for plagioclase loss with depth. Dashed lines show the sensitivity of the calculations to changes (= factor of 5) in (A) the plagioclase rate constant where $k_{plag} = 2.8 \times 10^{-16}$ mol m^{-2} s^{-1} (B) the exponential volumetric surface area term α and (C) the ratio of secondary to primary permeability q_s/q_p where $q_p = 7 \times 10^{-6}$ m yr^{-1}. [Used with permission of Elsevier from White et al. 2001.]

Reactive transport modeling

Perhaps the most complex approach to weathering systems are reactive transport (RT) models that incorporate chemical reaction and transport, usually set up to include reactions of either all or some of the major components in the system. RT models are continuum models which are unlike pore network models that incorporate explicit descriptions of the pore matrix. RT models are further described in this volume by Steefel and Maher (2009), and so this section is limited to their application to weathering profiles. Again, given the importance of high versus low porosity and permeability toward weathering behavior, we initially discuss weathering of alluvium (Santa Cruz) and then weathering of low porosity bedrock (Costa Rica and Svalbard). As discussed below, the model for the alluvium incorporated purely advective transport while the model for the weathering bedrock was largely based upon diffusive transport at the bedrock-rind interface.

A reactive transport model for the Santa Cruz terrace profiles. The multi-component reactive transport model CrunchFlow (see Steefel and Maher 2009, this volume) was used to interpret soil profile development and mineral precipitation and dissolution rates in the Santa Cruz chronosequence (Maher et al. 2009). The model incorporates mineral dissolution rates from the literature under the conditions of transport dominated by advective flow. The results of this more comprehensive approach can be compared to previously described results based on the linear, analytical and spreadsheet approaches described in the preceding sections.

For the Santa Cruz terraces, a model formulation was developed that described the progressive evolution of terrace weathering in order to reproduce the observed data for the oldest 226 ka terrace Santa Cruz terrace (White et al. 2008). In addition to individual rate laws describing mineral dissolution and precipitation, the model also includes mineral-specific cation exchange and the measured flow rate and water contents to describe unsaturated zone flow and transport (White et al. 2009).

For the model simulations, the measured precipitation was corrected for evaporation (ET) on the basis of the measured soil water chloride concentrations (Eqn. 15). The measured solute concentrations within the profile contained in White et al. (2009) were used to calculate the saturation state of the minerals for comparison with model results. In order to reflect the changing composition of the total cation exchange pool, a model for mineral-specific cation exchange was developed for use in CrunchFlow. The results of the coupled transport model-produced weathering profiles for SCT 5 terrace that were generally similar to those produced using the spreadsheet model (compare Fig. 18 with Fig. 15). Plagioclase weathering rates were also comparable to those predicted from both the spreadsheet and linear approximation approaches.

A significant advantage of the reactive transport model approach is the ability to investigate in detail a number of mechanistic parameters that impact chemical weathering rates. Two of these, the role of reaction affinity and of the consideration of secondary clay precipitation, i.e., the simultaneous formation of a depletion-addition profile (see Fig. 3), will now be briefly discussed.

Role of reaction affinity. The significant discrepancy commonly observed between field-based weathering rates and kinetic rate constants determined from experimental studies has often

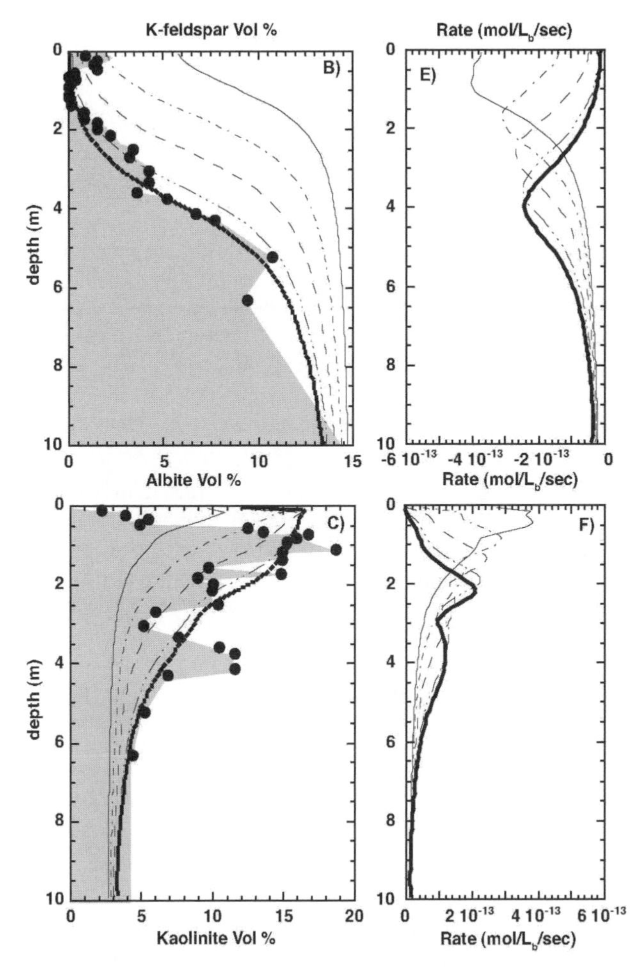

Figure 18. Model results (solid lines) corresponding to the SCT 5 (226 ka) profiles for primary plagioclase and secondary kaolinite determined using the linear TST formulation (see Eqn. 37, $n = 1$). Calculated profiles corresponding to the ages of the younger terraces are shown for reference (after Maher et al. 2009). The right hand set of figures indicates the zones of maximum plagioclase and kaolinite reactivities in the profile. [Used with permission of Elsevier from Maher et al. 2009.]

been explained by the fact that natural weathering commonly occurs close to thermodynamic equilibrium whereas experimental solutions are commonly far from equilibrium. Several proposed kinetic rate laws are available to describe aluminosilicate dissolution in terms of reaction affinity or the degree to which a solution is under- or super-saturated with respect to the reacting phase. These rate law variations for reaction affinity were evaluated by Crunchflow in terms of observed profile development at Santa Cruz (Maher et al. 2009). The variations that were evaluated included the transition state (TST) approach (Lasaga 1984), the Al-inhibition model (AIM) (Gautier et al. 1994; Oelkers et al. 1994; Carroll and Knauss 2005) and the parallel rate law approach (PRL) (Burch et al. 1993; Hellmann and Tisserand 2006).

As indicated in Figure 19, Santa Cruz pore waters were close to thermodynamic saturation and therefore were found to be strongly dependent on the fluid flux. Under these conditions, the three rate laws produced field weathering rates that were comparable because the reaction kinetics did not control the overall reaction as much as does fluid transport. This is consistent with the interpretation advanced earlier that SCT5 is best described as operating in the transition regime (compare Figs. 8 and 9). However, extrapolation to far-from-equilibrium conditions, showed that the TST rate law resulted in rate constants for the Santa Cruz terraces that are 2 orders of magnitude slower than the established laboratory rate constant of ca. $10^{-12.5}$ to $10^{-12.8}$ mol/m^2/sec based on an equivalent interpretation of experimental data obtained under far-from-equilibrium conditions. In contrast, the other two non-linear rate law formulations (PRL and AIM) resulted in rate constants that are comparable to laboratory values.

The role of secondary clay precipitation. Soil profile development in the Santa Cruz terraces is characterized by the formation of progressively more intensely developed argillic horizons in soils of increasing terrace age. The increase of the kaolinite in these zones is approximately the inverse of the plagioclase profile, with kaolinite accumulation occurring at increasingly greater depths over time as a result of depletion in primary minerals at the top and the downward transport of aqueous weathering products (Fig. 18). This indicates that Al is largely conserved in this pH range, as expected (a depletion-enrichment profile is shown in Fig. 3 where Al is largely conservative). The maximum accumulation of kaolinite corresponds to the point where the primary mineral reaction fronts overlapped.

Kaolinite and other secondary minerals are often treated as passive by-products of primary mineral dissolution with infinite rates of precipitation as in the case of the previously described spreadsheet approach (Figs. 15 and 17). However, the incorporation of kaolinite kinetics into CrunchFlow demonstrated that secondary clay formation served to draw down solute concentrations which in turn provided a substantial driving force for dissolution of both plagioclase and K-feldspar (Fig. 20). Changes in the rate of kaolinite precipitation do not affect the primary mineral kinetic rate constants or weathering gradients. However, an increase in the kaolinite rate constant removes weathering products from solution faster and therefore results in an increase the weathering velocity of the primary feldspars As shown in Figure 20, a factor of 10 decrease in the rate of kaolinite precipitation produces changes in the profiles comparable to the effect of a factor of 1.5 decrease in the flow rate q_h or about 100 ka in profile development (Maher et al. 2009).

Reactive transport modeling of low-porosity systems. In continuum modeling of low porosity rocks, transport is described using terms for effective diffusivity (Steefel and MacQuarrie 1996; Steefel 2008). Such effective transport parameters in porous rocks must account for tortuosity (τ), a parameter that is always less than 1 and that represents the square of the path length for solute in water, L, divided by the tortuous path length in rock, L_e:

$$\tau = \left(\frac{L}{L_e} \right)^2 \tag{41}$$

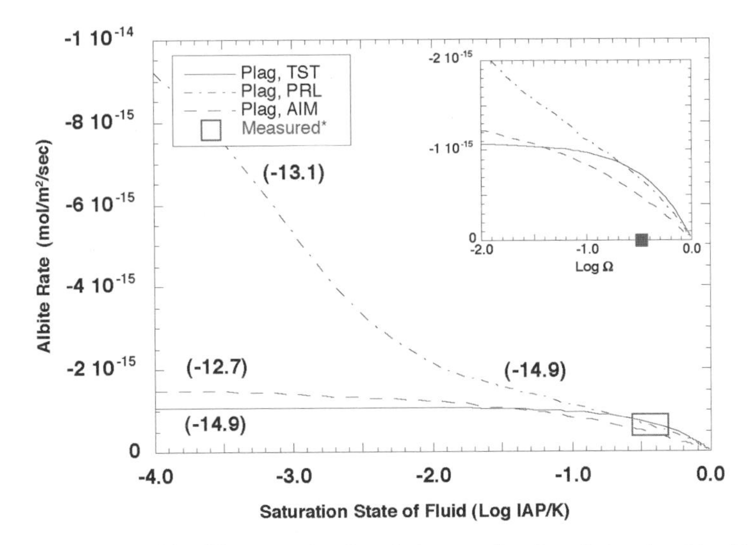

Figure 19. Comparison of the different rate law formulations as a function of saturation state of the Santa Cruz pore waters. Note that the model rates are all within a similar range for the field of observed near-saturation for Santa Cruz pore waters but deviate substantially as undersaturation is increased. See text for definition of acronyms for affinity functions. [Used with permission of Elsevier from Maher et al. 2009.]

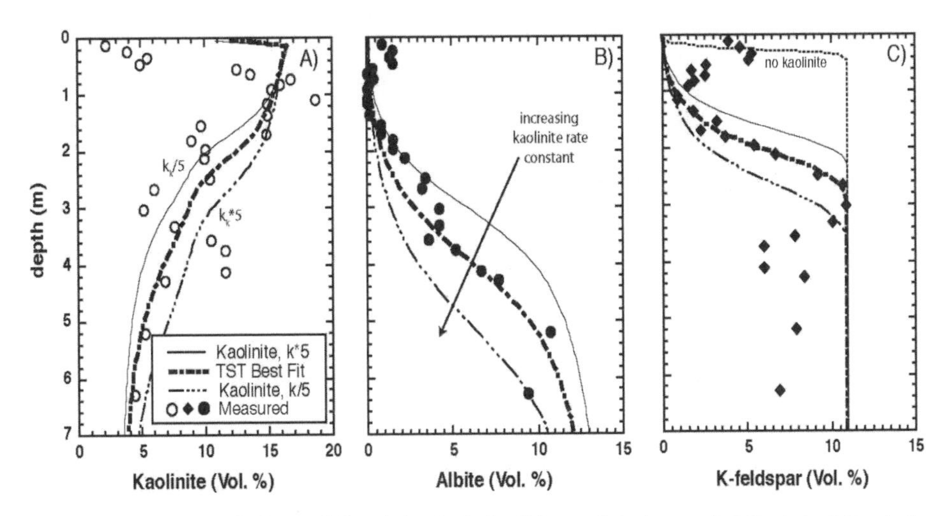

Figure 20. The effect of a factor of 10 variation in the kaolinite precipitation rate (relative to best-fit value) on kaolinite, plagioclase (albite) and K-feldsapr profiles in the SCT 5 Santa Cruz terrace. Solid points represent measured data (White et al. 2008). Lines correspond to the best fits using the TST affinity model with additional lines showing the effects of a factor of 5 increase and decrease in the kaolinite precipitation rates. Note that increasing the kaolinite precipitation constant has the effect of increasing the mineral profile depths. [Used with permission of Elsevier from White et al. 2008.]

The effective diffusion coefficient (D_e, cm^2 s^{-1}) in porous media is a function of the tortuosity:

$$D_e = \tau D_0 \qquad (42)$$

where D_0 is the diffusion coefficient in pure water (cm^2 s^{-1}). Tortuosity, however, changes with changing porosity and pore geometry as alteration progresses. Therefore, in a model of a diffusion-dominated system where porosity varies with reaction, the tortuosity and effective diffusivity vary as well.

An alternate way to model diffusivity—i.e., without knowledge of tortuosity—is to use Archie's Law (Archie 1942):

$$D_e = D_0 \varphi^m \qquad (43)$$

Here φ is the measured total porosity of the porous media and m is the experimentally determined cementation exponent which varies in geological samples from about 1 to 5 (Oelkers 1996). As discussed below, use of Equation (43) is successful in simulating *the weathering advance rate* for the basalt clasts weathering in Costa Rica (Fig. 9). We present the details of that modeling below.

Reactive transport modeling of Costa Rican clasts. We describe here how CrunchFlow (Steefel 2008) was used to forward-model the rind developed on one of the low-porosity basalts weathered in Costa Rica (Fig. 9A, Appendix). The clast was modeled as a one dimensional system using laboratory rate constants for mineral dissolution (White and Brantley 1995; Bandstra et al. 2008). It was assumed that, given the low porosity of the parent material, transport was dominated by diffusion (Fig. 21). This Costa Rican clast had been weathered for approximately 35,000 yr (Sak et al. 2004). In the rind on this clast, pyroxene and plagioclase were completely dissolved and the Ca, Mg, and Na profiles thus define fully developed profiles (Fig. 21). The rind consists predominantly of Fe oxide, minor kaolinite, and gibbsite. Ti, contained in primary mineral oxides, was observed to be largely immobile (Sak et al. 2004). Oxidation of Fe in augite is the alteration process observed microscopically to occur deepest in the core material, i.e., it may be the "profile-controlling mineral" as discussed earlier. The dissolution of plagioclase is the first alteration process to generate secondary porosity. In consequence, the reaction fronts of pyroxene and plagioclase are almost identical.

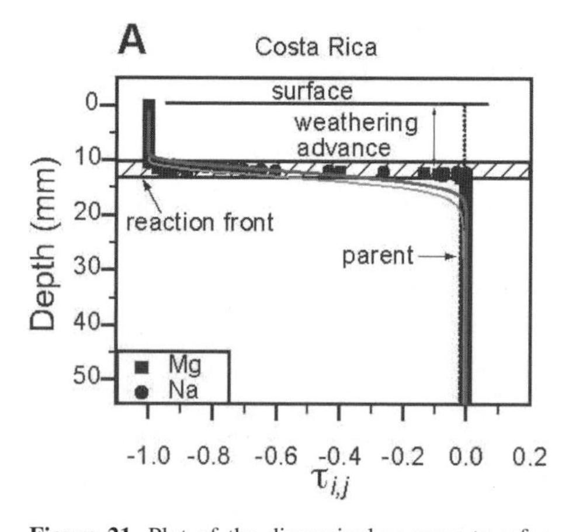

Figure 21. Plot of the dimensionless mass transfer coefficient $\tau_{Ti,j}$ calculated using Equation (3) based on electron microprobe data on a clast weathered in an alluvial terrace in Costa Rica for 35,000 yr. The double subscript on τ as plotted here indicates that Ti was used as the immobile element in Equation (3) while j = Mg or Na as indicated in the legend. Data points represent averages of 9 profiles. Data were calculated with three-point running averages to smooth variation due to the fact that the sizes of the probe spots were smaller than grain size for many minerals. Na and Mg are largely in plagioclase and pyroxene respectively in parent. Parent values are shown as constant with depth to deeper depths than measured under microprobe: symbols in the unweathered core are average parent values. Similarly, the rind values of $\tau_{Ti,j}$ are plotted further into the rind than measured. CrunchFlow model outputs for Na and Mg are shown as curves and the thickness of the reaction front is delineated by the hachured zone. The weathering advance distance for this clast—the distance from outer edge of rind to outer edge of the reaction front—was generally well modeled. [Used with permission of The Geological Society of America from Hausrath et al. 2008.]

In the model (Hausrath et al. 2008a), the parent rock was defined to contain 66% plagioclase + 26% pyroxene + 2.6% quartz + 1.8% alkali feldspar + 1% magnetite + 1.5% ilmenite + 1% porosity. Reaction rate constants for feldspars and augite were set equal to laboratory-determined reaction rate constants. Because no dissolution of quartz, ilmenite, or magnetite was observed, their reaction rate constants were set to essentially zero. Secondary minerals defined in the model included kaolinite, $Fe(OH)_3$, gibbsite, and siderite.

Grain sizes of the two dissolving phases were estimated based upon microscopic imaging. These values were used to calculate geometric surface area based on the assumption of spherical particles. Of course, the total surface area of grains did not initially interact with fluids during weathering because of the low porosity of the parent material. Therefore, the BET surface areas of plagioclase and pyroxene were determined by fitting the observed to the modeled reaction fronts (5 and 3.9 cm^2/g respectively) at the initiation of the model. These surface area values were within a factor of approximately 100 of the estimated geometric values based on microscopic measurement of grain size.

In the model, the clast was assumed to be continuously bathed in constant-chemistry pore-water based on concentrations of solutes estimated for soil porefluids (see Appendix). This was achieved by setting the boundary between the soil and the clasts to a fixed chemical composition representing the soil pore waters. For example, the pH was maintained equal to 5 at the outer surface. Temperature was maintained at 25 °C and the system was maintained at 100% hydrologic saturation. Effective diffusivity was calculated by using Archie's Law (Eqn. 43) for each grid cell at each time step as porosity increased in the cells. The cementation exponent was varied in an attempt to fit the thickness of the reaction front. The best fit cementation exponent for this system (Hausrath et al. 2008a) was 2.5 as shown in Figure 21. With these models, CrunchFlow was consistent with a linear advance rate for the rind-core interface with time, consistent with geological evidence and measured rind thickness as shown in Figure 9 (Sak et al. 2004).

Such a linear advance rate contradicts the general expectation of time-dependent, parabolic kinetics for a diffusion-limited system as described in many examples in the literature (Fletcher et al. 2006). It is interesting that, for both the weathering of the low-porosity basalt clasts in Costa Rica (Fig. 9) and the weathering of the low-porosity quartz diorite in Puerto Rico (Fig. 6), the weathering advance rate is inferred to be a linear function of time, even when the rate of reaction in the underlying bedrock is at least partially rate-limited by diffusion. Apparently, a linear weathering advance rate can be maintained when porosity increases significantly in the alteration layer. For example, the weathering process is accompanied by fracturing in Puerto Rico. However, fractures which were not observed in the basalt clasts (Sak et al. 2004), are apparently not necessary during the weathering of low-porosity rock to maintain linear weathering advance rates. Other processes of porosity opening may maintain linear advance rates.

Importantly, an advance rate that is linearly dependent on time is probably necessary if weathering systems are to achieve steady-state thickness. If the weathering advance rate on low-porosity bedrock is generally linearly dependent upon time due to time-dependent porosity and diffusivity increases, then the maintenance of steady-state thickness of regolith on altering bedrock may be possible for many other systems as well (Sak et al. 2004; Fletcher et al. 2006; Navarre-Sitchler et al. 2009).

Reactive transport modeling of Svalbard basalt. The approach to the modeling of the Costa Rica clasts was used with CrunchFlow to also model dissolution of Na-containing glass in Svalbard (see Appendix for description of this system). The weathering exposure period was assumed equal to 2500 years due to the Arctic location. This value was estimated as ¼ of the total exposure period for the basalt flow since during most of the year, weathering was assumed to cease while temperatures dropped below freezing. No secondary minerals were included in the model since few were observed petrographically (Hausrath et al. 2008a). Given that we inferred that the surface was spalling periodically (see discussion in Appendix), we chose a model fit to the glass dissolution that was consistent with significant surface retreat of the initial rock surface. The best fit of model output to the measured Na profile was consistent with surface retreat ≅ 0.4 mm (Fig. 10).

In this case, the modeled system was defined as a simplified system of 25% glass + 73% inert mineral + 2% porosity, consistent with observations that only glass was dissolving (Hausrath et al. 2008a). The outer surface of the reacting rock was assumed to be bathed in a pore fluid of constant concentration set equal to observed values (Hausrath et al. 2008a). Appropriate pH based on pore water observations was maintained at the outer surface (pH = 8.65) and the model rock was 100% hydrologically saturated (i.e., no air-filled porosity). Similar to the Costa Rica example, the surface areas of reacting phases were determined by fitting the modeled to the observed reaction fronts. Tortuosity was set equal to 1×10^{-6} following laboratory values for basalt (Sato et al. 1997).

In contrast to the modeling effort described above for the Costa Rican basalt, CrunchFlow simulations showed reaction fronts that were very similar to the observed Na profiles in Svalbard when diffusivities calculated from Equation (42) were used (Fig. 10). In other words, it was not necessary to use Archie's law (as shown for the Costa Rica sample in Fig. 21) within CrunchFlow to model the thickness of the reaction front on Svalbard samples accurately. Importantly however, CrunchFlow could not fit the reaction front unless it was inferred that significant surface retreat had occurred (as indicated in Fig. 10). The conclusion that the CrunchFlow-calculated distance of weathering advance was greater than the observed rind thickness on individual weathered specimens such as that shown in Figure 10 is consistent with the observation of cyclic spalling as described in the Appendix (Hausrath et al. 2008b).

The modeling effort for the Svalbard sample highlights that even without knowledge of the weathering advance distance or surface area term, the geometry of the reaction front can be used to constrain a CrunchFlow simulation to produce a model that is consistent with all observations. For example, for the sample shown in Figure 10, the minimum weathering advance distance based on petrographic observation was 20-40 μm, while the inferred weathering advance distance based on the modeling is larger (almost 400 μm). Cyclic spalling of more than half the Na-depleted rind is inferred to have occurred. It is tempting to conclude from this example that with increasing use of models such as CrunchFlow, it may become possible in some cases to estimate erosion rates of rinds as well as weathering rates.

CONCLUSIONS

The variations in concentration of chemical species versus depth in weathering regolith can often be considered to comprise combinations of the three simple endmember types shown in Figure 3: immobile, depletion, addition. Importantly, although measurement of profiles in regolith is common in both soil science and geology, only recently have such distributions been interpreted quantitatively in terms of rates and mechanisms of chemical weathering. Significant advance has been made in terms of modeling the weathering of such systems. One goal that has permeated many modeling efforts is the desire to predict rates of weathering and soil formation based on laboratory measurements. To do this will always require extrapolating models across large scales of space and time. In this paper, we have reviewed several of the types of quantitative models that are currently in use for this purpose.

Analytical and geometric models

The simplest models represent linear fits to the reaction fronts documented in solid-phase or solute concentration versus depth profiles. In this approach, the weathering rate is described as a linear function of the measured weathering gradient and the weathering velocity. The solid state velocity defines the rate of regolith advance into the protolith and the solute velocity is equivalent to the solute flux or advection. These gradients, while spatially similar, are temporally very different, with solid gradients representing geological scale weathering processes and solute gradients corresponding to the fluid residence time. A comparison of weathering rates derived from the two gradients provides a means of interpreting the consistency of weathering conditions over time. These geometric models require a minimum of input information and yield field weathering rates that are comparable to more detailed approaches. However, field rates consistent with this approach differ significantly from those measured in the laboratory (White and Brantley 2003). One way to rationalize this discrepancy is to compare these geometric models with the next level of model – the analytical models.

Analytical models for weathering can only be strictly derived for single-component single-phase systems (Lichtner 1988). Thus, when analytical models are developed, generally a very simplistic system such as quartz + water is modeled (Lichtner 1988). If a more complex system is modeled such as quartz + albite weathering to quartz + kaolinite, simplifying assumptions must be made to cast the model in terms of mass balance on simplified components (Lebedeva et al. 2007; Brantley et al. 2008). Even with such simplifications, a correction factor—the buffer capacity for the part of the system that was not explicitly included in the model (Brantley et al. 2008) or a correction factor for the rate constant (Lebedeva et al. 2007)—must be incorporated to reasonably compare model parameters to laboratory-measured or field-estimated parameters.

In this chapter, we presented analytical models based upon linear, exponential, and sigmoidal functions. The linear model was shown to be equivalent to the geometric model where the reaction rate is approximated as a zeroth order reaction. In contrast, the exponential function models the reaction rate as a first-order reaction. These models are both approximations to the full sigmoidal model that incorporates both a term related to changes in surface area and to affinity as a function of depth. The sigmoidal shapes of the solid-phase concentration versus depth reaction fronts are described by upward-concave curvature at the top of the front and downward-concave curvature at the bottom of the front. The upward and downward curvature is related largely to changes across the reaction front in surface area of the dissolving solid and chemical affinity of the solution, respectively. Furthermore, the sigmoidal function is consistent with a fractal description of surface area during weathering.

Comparisons of these models leads to the conclusion that the linear model is only a fit of the slope of the weathering gradient (reaction front), the exponential model is a fit of the slope and the concave-downward curvature at the bottom of the front, and the sigmoidal model is a fit of the slope, the concave-downward and concave upward curvature at the bottom and top of the front respectively. Our own experience suggests that the sigmoidal model provides the best fit to soil profiles where many data are available across the front.

After consideration of all these models, it becomes clear that discrepancies among model- and laboratory-derived rate constants are due to the fact that the model-derived parameters are not equivalent to the rate parameters derived from the laboratory. In particular, the model-derived parameters necessarily contain factors that correct the laboratory-derived rate constant from the simple laboratory system to the more complex natural system. Of particular interest in this regard are corrections related to divergence from equilibrium due to the chemical affinity driving force and changes in surface area due to roughening or grain size evolution. Both of these factors largely contribute to the curvature of the reaction front as discussed previously.

Numerical models

Given these issues, numerical models, either spreadsheet or reactive transport approaches, are attractive. In theory, these models contain some or all of the perceived important reactions and transport equations that are needed to model the natural system. Excel spreadsheets have been developed which use forward calculations in terms of depth and time to fit measured solid state weathering gradients with a limited number of variables such as a kinetic rate constant, a hydrologic flux term, an exponent or surface area parameter, and a simple reaction affinity term (White et al. 2001, 2008). Results have been applied to both weathering in sediments with assumed constant permeability as well as weathering across a strong permeability gradient such as occurs at the bedrock-saprolite interface. Results indicate that weathering in both environments become transport limited at deepest depths and therefore most strongly depend on the rate of pore water flow. Consistent with the discussion of analytical models, the non-linear aspects of the soil profiles are reproduced based on changes in the mineral surface to volume ratio during weathering and on the reaction affinity defining the approach to thermodynamic equilibrium.

Given the large number of parameters that need to be constrained while using spreadsheet or RT codes, it is helpful to have a systematic three-step approach toward the modeling. In modeling the Santa Cruz sediments and Costa Rican clasts using CrunchFlow, we first found it useful to determine the best fit value for q_h or the diffusivity D_o that allowed the calculated *weathering advance rate* to match that of the natural system. Next, using this transport parameter, we found it useful to constrain the reactivity term (reaction rate constant and surface area) to best match the slope of the reaction front. In effect, this slope or weathering gradient is a function of the ratio of the rate of transport to reaction. Third, modeling was completed by determining how to treat surface area, affinity terms, and other mineral reactions, to match the curvature of the front.

This general approach worked well for both the Santa Cruz and Costa Rica examples. However, the approach did not work well for weathering of the Svalbard sample because significant erosion of surface material was inferred to have occurred and the model did not explicitly include erosion. In the Svalbard case, therefore, the model was first fit to the geometry of the reaction front, and then the model – observation discrepancy for weathering advance was used to estimate the erosion rate (surface retreat). Modeling of this system was less well constrained than the Santa Cruz or Costa Rican examples. Much more work on eroding and weathering systems will be needed to evaluate all of these approaches.

The general subject of reactive transport (RT) modeling is described in detail by Steefel and Maher (2009, this volume) and is considered here only in terms of demonstrating utility in deciphering processes associated with weathering profiles. As with the spreadsheet calculations, examples of weathering are considered in both high and low permeability environments. Important ramifications resulting from application of these RT codes include the coupling of dissolution/precipitation processes in controlling overall rates of profile development and the ability to explain some of the apparent discrepancies in field and laboratory rates. Also important are results pertaining to the nature and rate of diffusive fluxes in low permeability environments that control the initiation of weathering processes in corestones and weathering rinds.

Reaction fronts and weathering advance

Six case studies were emphasized in the chapter that showed a large variation in the thickness of the reaction front for Na-containing phases. In general, this thickness increased from low-porosity diffusion-dominated systems to higher-porosity advection-dominated systems. In spite of the range of site studies considered and some of the complexities outlined above, we have emphasized throughout this chapter the interpretation of both this reaction front thickness as well as the weathering advance distance. The weathering advance distance is largely constrained by the important transport parameters related to advection or diffusion. In contrast, the reaction front thickness depends upon the ratio of transport to reaction rate. Therefore, once the transport parameters are constrained by weathering advance rate, the reaction front thickness can often be modeled through parameterization of the rate constant and reacting surface area.

We have explored the hypothesis that completely developed reaction fronts develop when minerals dissolve in the transition or local equilibrium regimes. In contrast, when reaction fronts are incompletely developed, this may imply that the system has not yet reached a steady state (i.e., the system represents a transient) or the system is a steady-state system operating in the kinetic regime. Such interpretations of the geometry of reaction fronts may be useful in explaining why some reaction rates inferred from field systems are close to laboratory rates, as would be inferred for the kinetic regime, while others are slower than laboratory rates, as would be inferred for transition or local equilibrium regimes.

Modeling frontiers

Each of the types of models that have been presented here have utility: in other words, just because reactive transport (RT) codes have the most power, this should not preclude the use of simpler analytical models or spreadsheet models that may point the way toward deeper understanding than the complex RT codes. Almost universally, our computation power greatly exceeds our ability to constrain sensitive field-based parameters over large spatial and temporal scales. The entire hierarchy of models have utility, and different complexity of models should be used depending upon the question under investigation and the insight that they can provide.

Observations from the review of modeling presented in this chapter point out the importance of the coupling between physical, hydrological and chemical processes. Examples include the contrasts in the weathering of bedrock versus unconsolidated materials. In low-porosity bedrock, fluid transport is often controlled by poorly understood matrix diffusion. We have hypothesized that there may often be a weathering precursor or "profile-controlling" mineral, although possibly minor in concentration, that controls the initial pore water penetration into otherwise pristine bedrock. In contrast, in high permeability environments, fluid flow occurs under unsaturated conditions and is highly variable, producing wetting and drying cycles whose effect on chemical weathering rates is uncertain. Some preliminary ideas about the coupling of physical erosion and chemical weathering are addressed in this review, and a hypothesis is advanced that the concentration of reactants in porefluids and the distribution of grain sizes in regolith may contribute toward this coupling. Such mechanisms controlling the feedback between soil thickness and chemical and erosion rates are not at all well understood and represent the frontier in modeling. Similarly, simulations of the role of biologic inputs both as instigators of chemical weathering and as sinks for weathering products are also currently only on the horizon in terms of quantitative modeling.

ACKNOWLEDGMENTS

This material is based upon work supported by the U.S. National Science Foundation under Grant Number CHE-0431328 and by the U.S. Dept. of Energy Grant number DE-FG02-05ER15675 to SLB. NSF funding for the Critical Zone Exploration Network, to SLB, is also acknowledged. We thank A. Navarre-Sitchler, J. Bandstra, E. Hausrath, H. Buss, J. Williams, and L. Jin for help with figures and for their insights. We thank E.Oelkers and J. Rosso for editorial patience.

REFERENCES

Abelin H, Birgerrson L, Moreno L, Widen H, Agen T, Neretnieks I (1991) A large-scale flow and tracer experiment in granite: Results and interpretation. Water Resour Res 27:3119-3136

Amundson R (2004) Soil Formation. *In:* Treatise in Geochemistry: Surface and Ground Water, Weathering, and Soils. Volume 5. Drever JI (ed) Elsevier Press, Amsterdam, p 1-35

Anderson SP, Blum J, Brantley SL, Chadwick O, Chorover J, Derry LA, Drever JI, Hering JG, Kirchner JW, Kump LR, Richter D, White AF (2004) Proposed initiative would study Earth's weathering engine. EOS Trans AGU 85:265-269

Anderson SP, Dietrich WE, Brimhall GH (2002) Weathering profiles, mass-balance analysis, and rates of solute loss: Linkages between weathering and erosion in a small, steep catchment. Geol Soc Am Bull 114:1143-1158

Anderson SP, Drever JI, Humphrey NF (1997) Chemical weathering in glacial environments. Geology 25:399-402

April R, Newton R, Coles LT (1986) Chemical weathering in two Adirondack watersheds: past and present-day rates. Geol Soc Am Bull 97:1232-1238

Archie GE (1942) The electrical resistivity log as an aid in determining some reservoir characteristics. Trans Am Inst Mining Metall Pet Eng 146:54-62

Bandstra JZ, Brantley SL (2008) Data fitting techniques with applications to mineral dissolution kinetics. *In:* Kinetics of Water-Rock Interaction. Brantley SL, Kubicki JD, White AF (eds) Springer, New York, p 211-257

Bandstra JZ, Buss HL, Campen RK, Liermann LJ, Moore J, Hausrath EM, Navarre-Sitchler A, Jan J-H, Brantley SL (2008) Compilation of Mineral Dissolution Rates. *In:* Kinetics of Water-Rock Interaction. Brantley SL, Kubicki JD, White AF (eds) Springer, New York, p 737-823

Barth TF (1961) Abundance of the elements, aereal averages and geochemical cycles. Geochim Cosmochim Acta 23:1-8

Bierman P, Gillespie A, Caffee M, Elmore D (1995) Estimating erosion rates and exposure ages with ^{36}Cl produced by neutron activation. Geochim Cosmochim Acta 59:3779-3798

Blum A, Stillings LL (1995) Feldspar dissolution kinetics. Rev Mineral 31:291-351

Brantley SL (2008a) Kinetics of mineral dissolution. *In:* Kinetics of Water-Rock Interaction. Brantley SL, Kubicki JD, White AF (ed) Springer-Kluwer, New York, p 151-196

Brantley SL (2008b) Understanding soil time. Science 321:1454-1455

Brantley SL, Bandstra J, Moore J, White AF (2008) Modeling chemical depletion profiles in regolith. Geoderma 145:494-504

Brimhall G, Dietrich WE (1987) Constitutive mass balance relations between chemical composition, volume, density, porosity, and strain in metasomatic hydrochemical systems: results on weathering and pedogenisis. Geochim Cosmochim Acta 51:567-587

Brimhall GH, Chadwick OA, Lewis CJ, Compston W, Williams IS, Danti KJ, Dietrich WE, Power ME, Hendricks D, Bratt J (1991) Deformational mass transport and invasive processes in soil evolution. Science 255:695-702

Burch TE, Nagy KL, Lasaga AC (1993) Free energy dependence of albite dissolution kinetics at 80°C and pH 8.8. Chem Geol 105:137-162

Buss HL, Bruns MA, Schultz MJ, Moore J, Mathur CF, Brantley SL (2005) The coupling of biological iron cycling and mineral weathering during saprolite formation, Luquillo Mountains, Puerto Rico. Geobiology 3:247-260

Buss HL, Sak PB, Webb SM, Brantley SL (2008) Weathering of the Rio Blanco quartz diorite, Luquillo Mountains, Puerto Rico: Coupling oxidation, dissolution and fracturing. Geochim Cosmochim Acta 72:4488-4507

Carroll S, Knauss KG (2005) Dependence of labradorite dissolution kinetics on $CO_{2(aq)}$, $Al_{(aq)}$, and temperature. Chem Geol 217:213-225

Carter VG, Dale T (1974) Topsoil and Civilization. University of Oklahoma Press, Norman

Chadwick OA, Brimhall GH, Hendricks DM (1990) From black box to a grey box: a mass balance interpretation of pedogenesis. Geomorphology 3:369-390

Chadwick OA, Derry LA, Vitousek PM, Huebert BJ, Hedin LO (1999) Changing sources of nutrients during four million years of ecosystem development. Nature 397:491-497

Drever JI (1994) The effect of land plants on weathering rates of silicate minerals. Geochim Cosmochim Acta 58:2325-2332

Drever JI, Stillings L (1997) The role of organic acids in mineral weathering. Colloids Surf 120:167-181

Dupre B, Dessert C, Oliva P, Godderis Y, Viers J, Francois L, Millot R, Gaillardet J (2003) Rivers, chemical weathering and Earth climate. C R Geosci 335:1141-1160

Fletcher RC, Brantley SL (2009) A model interpretation of weathering of bedrock blocks into corestones. Am J Sci (in review)

Fletcher RC, Buss HL, Brantley SL (2006) A spheroidal weathering model coupling porewater chemistry to soil thickenesses during steady-state denudation. Earth Planet Sci Lett 244:444-457

Gaillardet J, Dupre B, Louvat P, Allegre CJ (1999) Global silicate weathering and CO_2 consumption rates deduced from the chemistry of large rivers. Chem Geol 159:3-30

Galle C (1994) Neutron porosity logging and core porosity measurements in the Beauvoir Granite, Massif Central Range, France. J Appl Geophys 32:125-137

Gardner LR (1980) Mobilization of Al and Ti during weathering--isovolumetric geochemical evidence. Chem Geol 30:151-165

Gautier J-M, Oelkers EH, Schott J (1994) Experimental study of K-feldspar dissolution rates as a function of chemical affinity at 150 °C and pH 9. Geochim Cosmochim Acta 58:4549-4560

Harden JW (1987) Soils developed in granitic alluvium near Merced, California. US Geol Survey Bull # 1590-A

Hausrath EM, Navarre-Sitchler A, Sak P, Steefel CI, Brantley SL (2008a) Basalt weathering rates on Earth and the duration of liquid water on the plains of Gusev Crater, Mars. Geology 36:67–70; doi: 10.1130/G24238A

Hausrath EM, Treiman AH, Vicenzi E, Bish DL, Blake D, Sarrazin P, Hoehler T, Midtkandal I, Steele A, Brantley SL (2008b) Short- and long-term olivine weathering in Svalbard: implications for Mars. Astrobiology 8:1079-1092, DOI: 10.1089/ast.2007.0195

Helgeson HC, Murphy WM, Aagard P (1984) Thermodynamic and kinetic constraints on reaction rates among minerals and aqueous solutions, II. Rate constants, effective surface area, and the hydrolysis of feldspar. Geochim Cosmochim Acta 48:2405-2432

Hellmann R, Tisserand D (2006) Dissolution kinetics as a function of the Gibbs free energy of reaction: An experimental study based on albite feldspar. Geochim Cosmochim Acta 70:364-383

Higgins MW, Atkins RL, Crawford TJ, Crawford RF, III, Brooks R, Cook R (1988) The structure, stratigraphy, tectnostratigraphy and evolution of the southernmost part of the Appalachian orogen. USGS Prof Paper 1475:173

Hillel D (1982) Introduction to Soil Physics. Academic Press, Orlando, FLA

Hodson ME (2002) Experimental evidence for the mobility of Zr and other trace elements in soils. Geochim Cosmochim Acta 66:819-828

Jobbagy EG, Jackson RB (2001) The distribution of soil nutrients and depth: Global patterns and the imprint of plants. Biogeochem 53:51-77

Johnson NM, Likens GE, Bormann FH (1968) Rate of chemical weathering of silicate minerals in New Hampshire. Geochim Cosmochim Acta 32:531-545

Johnsson MJ, Ellen SD, McKittrick MA (1993) Intensity and duration of chemical weathering: An example from soil clays of the southeastern Koolau Mountains, Oahu, Hawaii. Geol Soc Am Spec Pub 284:147-170

Kurtz AC, Derry LA, Chadwick OA, Alfano MJ (2000) Refractory element mobility in volcanic soils. Geology 28:683-686

Lal R (1989) Land degradation and its impact on food and other resources. *In:* Food and Natural Resources. Pimentel D (ed) Academic Press, San Diego, p 85-140

Lasaga AC (1981) Transition state theory. Rev Mineral 8:135-169

Lasaga AC (1984) Chemical kinetics of water-rock interactions. J Geophys Res 89:4009-4025

Lebedeva MI, Balashov VN, Fletcher RC, Brantley SL (2009) Mathematical model of steady-state regolith production at a constant erosion rate. Earth Surf Process Landforms (in review)

Lebedeva MI, Fletcher RC, Balashov VN, Brantley SL (2007) A reactive diffusion model describing transformation of bedrock to saprolite. Chem Geol 244:624-645

Lichtner PC (1988) The quasi stationary state approximation to coupled mass transport and fluid-rock interaction in a porous medium. Geochim Cosmochim Acta 52:143-165

Maher K, Steefel CI, White AF, Stonestrom DA (2009) The role of reaction affinity and secondary minerals in regulating chemical weathering rates at the Santa Cruz Soil Chronosequence, California. Geochim Cosmochim Acta 73:2804-2831

Mathieu D, Bernat M, Nahon D (1995) Short-lived U and Th isotope distribution in a tropical laterite derived from granite (Pitinga river basin, Amazonia, Brazil): Application to assessment of weathering rate. Earth Planet Sci Lett 136:703-715

Merrill GP (1906) A Treatise on Rocks, Rock Weathering and Soils. MacMillian Inc., New York

Millot R, Gaillardet J, Dupre B, Allegre CJ (2002) The global control of silicate weathering rates and the coupling with physical erosion: new insights from rivers of the Canadian Shield. Earth Planet Sci Lett 196:83-93

Moore JW, Pearson RG (1981) Kinetics and Mechanism. Wiley & Sons, Inc., New York

Morrow CA, Lockner DA (1997) Permeability and porosity of the Illinois UPH 3 drillhole granite and a comparison with other deep drillhole rocks. J Geophys Res 102 B2:3067-3075

Murphy SF, Brantley SL, Blum AE, White AF, Dong H (1998) Chemical weathering in a tropical watershed, Luquillo Mountains, Puerto Rico: II. Rate and mechanism of biotite weathering. Geochim Cosmochim Acta 62:227-243

Nahon DB (1986) Evolution of iron crusts in tropical landscapes. *In:* Rates of Chemical Weathering of Rocks and Minerals. Colman SM, Dethier DP (ed) Academic Press, Orlando, FL, p 169-187

Navarre-Sitchler A, Steefel C, Yang L, Tomutsa L, Brantley SL (2009) Evolution of porosity and diffusivity during chemical weathering of a basalt clast. J Geophys Res Earth Sci (in press)

Neaman A, Chorover J, Brantley SL (2006) Effects of organic ligands on granite dissolution in batch experiments at pH 6. Am J Sci 306:451-473

Oelkers EH (1996) Physical and chemical properties of rocks and fluids for chemical mass transport calculations. Rev Mineral 34:335-371

Oelkers EH, Schott J, Devidal J-L (1994) The effect of aluminum, pH, and chemical affinity on the rates of aluminosilicate dissolution reactions. Geochim Cosmochim Acta 58:2011-2024

Oguchi CT, Matsukura Y (1999) Effect of porosity on the increase in weathering-rind thicknesses of basaltic andesite gravel. Eng Geol 55:77-89

Ortoleva P, Auchmuty G, Chadam J, Hettmer J, Merino E, Moore CH, Ripley E (1986) Redox front propagation and banding modalities. Physica D 19(3):334-354

Pavich M, Leo GW, Obermeier SF, Estabrook JR (1989) Investigations of the characteristics, origin, and residence time of the upland residual mantle of the Piedmont of Fairfax County, Virginia. USGS Professional Paper # 1352, U.S. Geological Survey

Pavich MJ (1986) Processes and rates of saprolite production and erosion on a foliated granitic rock of the Virginia Piedmont. *In:* Rates of Chemical Weathering of Rocks and Minerals. Colman SM, Dethier DP (ed) Academic Press, Orlando, FL, p 551-590

Pavich MJ, Brown L, Harden J, Klein J, Midden R (1986) ^{10}Be distribution in soils from Merced River terraces, California. Geochim Cosmochim Acta 50:1727-1735

Pavich MJ, Brown L, Valette-Silver JN, Klein J, Middleton R (1985) ^{10}Be analysis of a quaternary weathering profile in the Virginia Piedmont. Geology 13:39-41

Perg LA, Anderson RS, Finkel RC (2001) Use of a new ^{10}Be and ^{26}Al inventory method to date marine terraces, Santa Cruz, California, USA. Geology 29:879-882

Porder S, Paytan A, Vitousek PM (2005) Erosion and landscape development affect plant nutrient status in the Hawaiian islands. Oecologia 142:440-449

Riebe CS, Kirchner JW, Finkel RC (2003) Long-term rates of chemical weathering and physical erosion from cosmogenic nuclides and geochemical mass balance. Geochim Cosmochim Acta 67:4411-4427

Ruhe RV (1956) Geomorphic surfaces and the nature of soils. Soil Science 82:441-4445

Sak PB, Fisher DM, Gardner TW, Murphy KM, Brantley SL (2004) Rates of formation of weathering rinds on Costa Rican basalt Geochim Cosmochim Acta 68:1453-1472

Sato H, Shibutani T, Yui M (1997) Experimental and modeling studies on diffusion of Cs, Ni, and Sm in granodiorite, basalt, and mudstone. J Contam Hydrol 26:119-133

Schoenberger PJ, Amoozegar A, Buol SW (1995) Physical property of a soil and saprolite continuum at three geomorphic positions. Soil Sci Soc Am 59:1389-1397

Schulz MS, White AF (1999) Chemical weathering in a tropical watershed, Luquillo Mountains, Puerto Rico: III. Quartz dissolution rates. Geochim Cosmochim Acta 63:337-350

Seiders VM, Mixon RB, Stern TW, Newall MF, Thomas CB (1975) Age of plutonism and tectonism and a new minimum age limit on the Glenarm Series in the Northeast Piedmont near Occoquan. Am J Sci 275:481-511

Steefel CI (2008a) Geochemical kinetics and transport. *In:* Kinetics of Water-Rock Interaction. Brantley SL, Kubicki JD, White AF (eds) Springer, New York, NY, p 545-590

Steefel CI, MacQuarrie KTB (1996) Approaches to modeling reactive transport in porous media. Rev Mineral 34:83-125

Steefel CI, Maher K (2009) Fluid-rock interaction: a reactive transport approach. Rev Mineral Geochem 70:485–532

Stonestrom DA, White AF, Akstin KC (1998) Determining rates of chemical weathering in soils-solute transport versus profile evolution. J Hydrol 209:331-345

Sverdrup K, Warfvinge P (1995) Estimating field weathering rates using laboratory kinetics. Rev Mineral 31:485-542

Torrent J, Nettleton WD (1979) A simple textural index for assessing chemical weathering in soils. Soil Sci Soc Am J 43:373-377

Turner B, Stallard R, Brantley SL (2003) Investigation of in situ weathering of quartz diorite bedrock in the Rio Icascos basin, Luquillo Experimental Forest, Puerto Rico. Chem Geol 202:313-341

Van Scholl L, Kuyper TW, Smits MM, Landeweert R, Hoffland E, Van Breemen N (2008) Rock-eating mycorrhizas: their role in plant nutrition and biogeochemical cycles. Plant Soil 35:35-47

Vanacker V, Von Blanckenburg F, Govers G, Molina A, Poesen J, Deckers J (2007) Restoring dense vegetation can slow mountain erosion to near natural benchmark levels. Geology 35:303-306

Vepraska MJ, Williams JP (1995) Hydraulic conductivity of saprolite as a function of sample dimensions and measurement technique. Soil Sci Am J 59:975-981

Wardle DA, Walker LR, Bardgett RD (2004) Ecosystem properties and forest decline in contrasting long-term chronosequences. Science 305:509-513

White A, Bullen TD, Vivit DV, Schulz MS, Clow DW (1999) The role of disseminated calcite in the chemical weathering of granitoid rocks. Geochim. Cosmochim Acta 63:1939-1953

White AF (2002) Determining mineral weathering rates based on solid and solute weathering gradients and velocities: application to biotite weathering in saprolites. Chem Geol 190:69-89

White AF (2008) Quantitative approaches to characterizing natural chemical weathering rates. *In:* Kinetics of Water-Rock Interaction. Brantley SL, Kubicki JD, White AF (eds) Springer, New York, p 469-544

White AF, Blum AE, Schulz MS, Bullen TD, Harden JW, Peterson ML (1996) Chemical weathering rates of a soil chronosequence on granitic alluvium: I. Quantification of mineralogical and surface area changes and calculation of primary silicate reaction rates. Geochim Cosmochim Acta 60:2533-2550

White AF, Blum AE, Schulz MS, Vivit DV, Stonestrom DA, Larsen M, Murphy SF, Eberl D (1998) Chemical weathering in a tropical watershed, Luquillo Mountains, Puerto Rico: I. Long-term versus short-term weathering fluxes. Geochim Cosmochim Acta 62:209-226

White AF, Brantley SL (eds) (1995) Chemical Weathering Rates of Silicate Minerals. Rev Mineral Vol. 31, Mineralogical Society of America, Washington, D.C.

White AF, Brantley SL (2003) The effect of time on the weathering of silicate minerals: why do weathering rates differ in the laboratory and field? Chem Geol 202:479-506

White AF, Bullen TD, Schultz MS, Blum AE, Huntington TG, Peters NE (2001) Differential rates of feldspar weathering in granitic regoliths. Geochim Cosmochim Acta 65:847-869

White AF, Schulz MS, Stonestrom DA, Vivit DV, Fitzpatrick J, Bullen TD, Maher K, Blum AE (2009) Chemical weathering of a marine terrace chronosequence, Santa Cruz, California. Part II: Solute profiles, gradients and the comparison of contemporary and long-term weathering rates. Geochim Cosmochim Acta 73:2804-2831

White AF, Schulz MS, Vivit DV, Blum AE, Stonestrom DA, Anderson SP (2008) Chemical weathering of a marine terrace chronosequence, Santa Cruz, California I: Interpreting rates and controls based on soil concentration-depth profiles. Geochim Cosmochim Acta 72:36-68

White AF, Schulz MS, Vivit DV, Blum AE, Stonestrom DA, Harden JW (2005) Chemical weathering rates of a soil chronosequence on granitic alluvium: III. Hydrochemical evolution and contemporary solute fluxes and rates. Geochim Cosmochim Acta 69:1975-1996

Wischmeier WH, Smith DD (1978) Predicting rainfall erosion losses - a guide to conservation planning Agriculture Handbook no. 537. U. S. Department of Agriculture

Yoo K, Mudd SM (2008) Discrepancy between mineral residence time and soil age: implications for the interpretation of chemical weathering rates. Geology 36:35-38

Young IM, Crawford JW (2004) Interactions and self-organization in the soil-microbe complex. Science 304:1634-1637

Zhao J (1998) Rock mass hydraulic conductivity of the Bukit Timah granite, Singapore. Eng Geol 50:211-216

APPENDIX
CASE STUDIES

The field sites used to test the various approaches to weathering profile characterization have been described in detail in earlier papers. A brief summary is as follows.

Santa Cruz Marine Terrace Chronosequence – Santa Cruz, California, U.S.A.

Perhaps the best characterized set of soil profiles for a chronosequence are the variations in solid-state and solute elemental concentrations for marine terraces near Santa Cruz, California (Figs. 2, 9, 14, 15 ,16 18. 19 and 20 in this paper) (White et al. 2008, 2009; Maher et al. 2009). The entire terrace series was age-dated between 65 to 220 k.y. using ^{10}Be (Perg et al. 2001). Soil chronosequences are ideal environments to simultaneously compare weathering processes over both geologic and contemporary time scales and to determine if present day weathering fluxes and rates are comparable to those of the past.

Profiles in Na versus depth in the Santa Cruz terraces are generated by a single weathering reaction:

$$NaAlSi_3O_{8\ (plag)} + H^+ + 9/2\ H_2O \rightarrow Na^+ + 2H_4SiO_4 + 1/2\ Al_2Si_2O_{15}(OH)_{4\ (kaolinite)}$$

The plagioclase, found in marine sediments deposited on the terraces, originated from long shore transport of locally-derived granitic riverine sediments produced as erosion products from the adjacent Santa Cruz Mountains (Perg et al. 2003). The decrease in plagioclase with depth in the profile reflect the fact that it, like most primary silicate minerals, is inherently thermodynamically unstable in many surficial environments, and it progressively interacts with acid-containing reactants provided at or near the top of the profile to produce clay (kaolinite). Mineralogical analyses of the regolith indicate that significant Na is contained only in plagioclase (White et al. 2008) and therefore the increases in solid state Na with depth closely parallel the decrease in the weathering intensity of plagioclase.

In situ weathering of the terrace deposits and, where present, the underlying Santa Margarita sandstone, produces depth-dependent mineral weathering profiles not only for plagioclase but also for the slower weathering K-feldspar. Shallow horizons, particularly in the older terraces, exhibit significant mobility reversals reflecting eolian deposition as a result of the exposure of the continental shelf during Wisconsin glaciation. The average plagioclase compositions in the terrace sediments (An_{28} - An_{33}) are compositionally similar to the Ben Lomond Granite in the adjacent Santa Cruz Mountains. The terrace K-feldspars are dominantly microcline. Kaolinite abundances increase with terrace age, becoming more concentrated in argillic horizons at depths of 0.5 to 1.5 meters.

Davis Run, Virginia, U.S.A.

One particularly well-characterized granitic weathering sequence is in the Davis Run watershed in the Piedmont Province in northern Virginia, USA (Pavich 1986; Pavich et al. 1986, 1989). The regolith in the upland area of the watershed consists of residual soil (1.5 m) overlying a thick saprolite (19 m) resting on granitic bedrock. Average mineralogy of the Occoquan Granite, described as an adamellite (Seiders et al. 1975), is quartz = 37%; K-feldspar = 27%; plagioclase An_6 = 21%; muscovite = 13%; and biotite = 2.1%. With measured ^{10}Be concentrations, a minimum age of the Davis Run regolith surface was calculated to be 8×10^5 yrs (Pavich et al. 1985). A weathering advance rate for formation of saprolite from bedrock consistent with this age, 4 m myr^{-1}, is in agreement with weathering rates based on fluxes of dissolved solids from the Davis Run watershed (Pavich et al. 1985; Pavich 1986).

White et al. (2001) used the Ti concentration to calculate volumetric strain (Eqn. 2) and showed that saprolite weathering at Davis Run is essentially isovolumetric. The mobilities of Ca and Na, based on Equation (3), are very different in the Davis Run regolith compared to the Panola regolith (Fig. 11) in that the plagioclase weathering front, reflected in Ca and Na τ values, does not persists into bedrock in Davis Run. In addition, in the upper tens of meters, the behavior of K is also significantly different between Davis Run and Panola. At Davis Run, no K decrease occurs at the saprolite/bedrock boundary. Rather, K decreases through the entire thickness of saprolite and soil.

Panola Granite – Atlanta, Georgia, U.S.A.

Detailed weathering studies have been conducted within the Panola Mountain Research Watershed located in the Piedmont province of the eastern United States, 25 km southeast of Atlanta, Georgia. The relief in the catchment is 56 m with flat ridge tops and generally steep hill slopes. The regolith profile modeled in the present study is situated on a ridge top site in the southwestern quadrant of the watershed and is underlain by the 320 myr old Panola Granite described as a biotite-muscovite-oligioclase-quartz-microcline granodiorite (Higgins et al. 1988). Bierman et al. (1995) measured ^{36}Cl concentrations in surface outcrops of granodiorite within the Panola watershed and calculated exposure ages between 250-500 k.y..

The ridge top soils in the Panola watershed are highly weathered ultisols developed from bedrock residuum. Soils consist of a dense dark red unstructured clay-rich A-horizon (~0.5 m thick), underlain by a less dense orange-brown B-horizon (~1 m thick). The soils are underlain by a porous gray saprolite (3-4 m thick), which retains the original granodiorite texture. The saprolite/bedrock interface grades from friable saprock to competent bedrock over an interval of several cm (Fig. 11). Average mineralogy of the fresh Panola granodiorite indicates a preponderance of plagioclase (32%), quartz (28%), K feldspar (21%), biotite (13%) and muscovite (7%) with much lesser amounts of hornblende (<2%). Plagioclase compositions in fresh granodiorite, as determined by microprobe analysis, range between An_{16} and An_{27} with an average of An_{23}. K-feldspar compositions range between Ab_2 and Ab_4 with an average of Ab_{31}. A ridge-top weathering profile characterized by White (2001) indicates a saprolite containing quartz, bioitite and K-feldspar but completely depleted in plagioclase. Strain calculations indicated that weathering was essentially isovolumetric. Additional drilling into bedrock revealed a plagioclase weathering profile extending several meters into the granite (Fig. 11).

Puerto Rico quartz diorite

The reaction front for albite in regolith developed on the Rio Blanco quartz diorite in ridgetops within the Rio Icacos watershed in Puerto Rico's Luquillo Experimental Forest is approximately 30-40 cm thick (Fig. 6, 11). Depth to this reaction front varies from one location to another. The mean annual temperature in this watershed is 22.8 °C, and the mean annual precipitation is 346 cm. The quartz diorite is comprised of plagioclase, quartz, hornblende, partially chloritized biotite, minor iron oxides, and accessory minerals (Murphy et al. 1998; White et al. 1998; Buss et al. 2008). The quartz diorite weathers spheroidally, forming corestones surrounded by rindlets which eventually disaggregate to form saprolite. The corestones are unweathered except in their outermost surface.

The first reaction observed during chemical weathering of this bedrock is oxidation of ferrous Fe in biotite (Buss et al. 2008). The volume expansion during oxidation is thought to crack the rock, allowing water access (Fletcher et al. 2006). Relatively large macrocracks form first that are onionskin-like, wrapping around the corestone, and defining rindlets. Once a rindlet forms, micro-cracks also start forming within rindlets. Presumably, water in macro- and microcracks explains why plagioclase, hornblende dissolve completely in the rindlet zone (Buss et al. 2008). The thickness of the Na-feldspar reaction front (rindlet zone) is attributed to the balance between rindlet formation, driven by diffusion-limited Fe oxidation in biotite in corestones, and rindlet transformation into saprolite, driven by plagioclase weathering to kaolinite. Plagioclase is only found in the rindlets and not in the saprolite.

Costa Rican basalt clasts

Weathered basalt clasts were collected in Costa Rica from the B horizon of three regionally extensive alluvial fill terraces (Qt 1, Qt 2, Qt 3) exposed for approximately 35, 120, and 240 ka respectively along the central Pacific coast of Costa Rica. The mean annual temperature and precipitation for 1941 to 1982 for this location are 27.3 °C and 3085 mm, respectively (Sak et al. 2004). The basaltic parent material is primarily plagioclase (67%) and augite (27%) with trace glass (2%), ilmenite (2%) and magnetite (1%). Plagiocase and augite occur both as fine-grained matrix and as phenocrysts of dimension up to ~ 50 μm diameter (augite) or several hundred μm long (plagioclase). The porosity of the basalts is low (1-3%), consistent with insignificant advective flow through the cores. Diffusion is therefore inferred to control mass transfer of reactants and products to or from the weathering interface whereas advection presumably controls transport in the surrounding soil matrix.

The weathering products in the rind were observed by X-ray diffraction to include goethite (FeOOH), gibbsite ($Al(OH)_3$), and trace kaolinite ($Si_2Al_2O_5(OH)_4$). Both plagioclase and augite weather within the core material (Sak et al. 2004; Hausrath et al. 2008a) and the generation of porosity due to dissolution of plagioclase is an especially important control on the rate of weathering advance into the low porosity basalt (Navarre-Sitchler et al. 2009). Dissolution of primary minerals decreases the density from 2.8 g cm^{-3} in the core to ~1 g cm^{-3} in the rind . The porosity increases from about 3 to 50% across a narrow reaction front that is < 20 mm thick as analyzed by bulk sampling analysis (Sak et al. 2004) or less than several mm in thickness as analyzed under electron microprobe (Navarre-Sitchler et al. 2009). The advance of this weathering front into basalt is documented by analysis of clasts from all three of the analyzed terraces (Fig. 9). No lysimeters were installed in the Costa Rica soils, and thus porewater samples were unavailable. However, deionized water was mixed in the laboratory with soil samples and then analyzed for dissolved solute concentrations to constrain porefluid chemistry for CrunchFlow modeling.

Svalbard basalt

Of the reaction fronts that are discussed in this chapter, the thinnest is the reaction front for Na-containing glass dissolving on basalt samples from the Sverrefjell Volcano in the Bockfjord volcanic complex in northern Spitsbergen (Hausrath et al. 2008a). The front was imaged using back-scattered electron microscopy as hundreds of microns in thickness (Hausrath et al. 2008a). The thin front documented in Figure 10 is presumably related to the climate in Spitsbergen, which is cold (yearly average = −5 °C) and dry (<200 mm/yr), and the exposure age: the reaction front was developed on Quaternary basalt lava weathered *in situ* since glaciation last stripped the landscape ~10 ka ago. Even more importantly, as described below, the weathered surface was significantly eroded by spalling.

Samples of weathered basalt lava were picked from rubbly soil developed on the flat-lying flanks of the volcano. Porefluids in the rubbly soil showed pH ranging from ~7-9 (Hausrath et al. 2008b). Samples were thin-sectioned across their exposed surfaces and imaged under optical and electron microscopy. Four characteristic types of surfaces were observed (Hausrath et al. 2008b): i) Na-depleted (e.g., Fig. 10); ii) relatively pristine surfaces that exhibited little to no Na depletion; iii) surfaces covered by thin Si-rich layers; iv) lichen-covered surfaces. The Si-rich layers, when observed, were non-crystalline, consisting presumably of amorphous Si that precipitated from evaporating fluids in the very dry climate. No attempt was made to model these precipitates or the lichen-covered layers.

On many samples, surface-parallel microfractures were observed to be causing spalling of the alteration layers. The variety of surfaces that were observed, from relatively pristine to Na-depleted, is consistent with cycles of spalling and depletion.

The deepest Na depletion was observed to be >250 μm (e.g., Fig. 10). Based on indirect evidence from microscopic observations (Hausrath et al. 2008b), it was inferred that the sample shown in Figure 10 had lost a surface layer of at least ~20–40 μm due to spalling. Model calculations, described in the text and Figure 10, are consistent with loss of more than 250 μm of the surface during weathering. This modeled erosion advance is referred to as "surface retreat" on Figure 10.

Reviews in Mineralogy & Geochemistry
Vol. 70 pp. 485-532, 2009

Fluid-Rock Interaction: A Reactive Transport Approach

Carl I. Steefel

Earth Sciences Division
Lawrence Berkeley National Laboratory
Berkeley, California 94720, USA

CISteefel@lbl.gov

Kate Maher

Dept. of Geological & Environmental Sciences
School of Earth Sciences
Stanford University
Stanford, California 94305, USA

INTRODUCTION

Fluid-rock interaction (or water-rock interaction, as it was more commonly known) is a subject that has evolved considerably in its scope over the years. Initially its focus was primarily on interactions between subsurface fluids of various temperatures and mostly crystalline rocks, but the scope has broadened now to include fluid interaction with all forms of subsurface materials, whether they are unconsolidated or crystalline ("fluid-solid interaction" is perhaps less euphonious). Disciplines that previously carried their own distinct names, for example, basin diagenesis, early diagenesis, metamorphic petrology, reactive contaminant transport, chemical weathering, are now considered to fall under the broader rubric of fluid-rock interaction, although certainly some of the key research questions differ depending on the environment considered.

Beyond the broadening of the environments considered in the study of fluid-rock interaction, the discipline has evolved in perhaps an even more important way. The study of water-rock interaction began by focusing on geochemical interactions in the absence of transport processes, although a few notable exceptions exist (Thompson 1959; Weare et al. 1976). Moreover, these analyses began by adopting a primarily thermodynamic approach, with the implicit or explicit assumption of equilibrium between the fluid and rock. As a result, these early models were fundamentally static rather than dynamic in nature. This all changed with the seminal papers by Helgeson and his co-workers (Helgeson 1968; Helgeson et al. 1969) wherein the concept of an irreversible reaction path was formally introduced into the geochemical literature. In addition to treating the reaction network as a dynamically evolving system, the Helgeson studies introduced an approach that allowed for the consideration of a multicomponent geochemical system, with multiple minerals and species appearing as both reactants and products, at least one of which could be irreversible. Helgeson's pioneering approach was given a more formal kinetic basis (including the introduction of real time rather than reaction progress as the independent variable) in subsequent studies (Lasaga 1981, 1984; Aagaard and Helgeson 1982). The reaction path approach can be used to describe chemical processes in a batch or closed system (e.g., a laboratory beaker), but such systems are of limited interest in the Earth sciences where the driving force for most reactions is transport. Lichtner (1988) clarified the application of the reaction path models to water-rock interaction

1529-6466/09/0070-0011$05.00 DOI: 10.2138/rmg.2009.70.11

involving transport by demonstrating that they could be used to describe pure advective transport through porous media. By adopting a reference frame which followed the fluid packet as it moved through the medium, the reaction progress variable could be thought of as travel time instead.

Multi-component reactive transport models that could treat any combination of transport and biogeochemical processes date back to the early 1980s. Berner and his students applied continuum reactive transport models to describe processes taking place during the early diagenesis of marine sediments (Berner 1980). Lichtner (1985) outlined much of the basic theory for a continuum model for multicomponent reactive transport. Yeh and Tripathi (1989) also presented the theoretical and numerical basis for the treatment of reactive contaminant transport. Steefel and Lasaga (1994) presented a reactive flow and transport model for non-isothermal, kinetically-controlled water-rock interaction and fracture sealing in hydrothermal systems based on simultaneous numerical solution of both reaction and transport

This chapter begins with a review of the important transport processes that affect or even control fluid-rock interaction. This is followed by a general introduction to the governing equations for reactive transport, which are broadly applicable to both qualitative and quantitative interpretations of fluid-rock interactions. This framework is expanded through a discussion of specific topics that are the focus of current research or are either incompletely understood or not fully appreciated. At this point, the focus shifts to brief discussion of the three major approaches to modeling multi-scale porous media 1) continuum models, 2) pore scale and pore network models, and 3) hybrid or multi-continuum models. From here, the chapter proceeds to investigate some case studies which illuminate the power of modern numerical reactive transport modeling in deciphering fluid-rock interaction.

TRANSPORT PROCESSES

Transport is a fundamental part of the fluid-rock interaction process for two reasons: 1) it provides the driving force for many of the reactions that take place by continuously introducing fluid out of equilibrium with respect to the reactive solid phase, and 2) it provides a characteristic time scale to be compared with the rates of reaction. In a batch system, the time scale of interest is the time over which equilibrium is achieved (where net reaction rates are, by definition, = 0), or rates become so slow that the system effectively no longer changes. In an open system, a good laboratory analogue of which is the continuously stirred tank reactor (CSTR) or well-mixed flowthrough reactor, the continuous injection of solution that is out of equilibrium (either with itself, or with a solid phase inside the reactor) causes the solution composition coming out of the reactor to reach a steady-state. If the residence time in the CSTR is long enough, equilibrium between the solution and solid phase may be achieved—otherwise, the solution and solid phase may remain some distance from equilibrium at steady state. The steady state behavior is controlled by the ratio of the characteristic fluid residence time, τ_{res}, to the characteristic reaction time, τ_{react}

$$\frac{\tau_{res}}{\tau_{react}} = \frac{kV}{Q} \tag{1}$$

where k is the reaction rate constant (s^{-1}), Q is the volumetric flow rate ($cm^3\ s^{-1}$), and V is the volume of the reactor (cm^3).

From this simple analysis, it should be clear that the approach to equilibrium has nothing to do with geologic time, but is controlled instead by the ratio of the characteristic residence (or transport) time and reaction time in the reactor. Similar concepts apply in the more complicated case of flow and transport through porous media, although the detailed transport processes require additional discussion.

The continuously stirred tank reactor provides perhaps the simplest example where rates of transport can be compared to rates of reaction. But in porous and/or fractured media, reactions can be driven by a variety of transport processes and changes in porosity, mineral saturation states and mineral abundances can be distributed along a flow path. The most important of these transport processes are advection, molecular diffusion, and mechanical dispersion. Although widely neglected, electrochemical migration is a flux created by the diffusion of charged species at differing rates which may be important in some cases. Below, these transport processes are defined mathematically in terms of fluxes, that is, the amount of solute passing through a unit area per unit time.

Advection

Advection involves the translation in space of dissolved or suspended material at the rate of movement of the bulk fluid phase. No modification of the shape of a front and no dilution occurs when transport is purely via advection—a sharp front remains so when undergoing purely advective transport. The advective flux, J_{adv}, of a dissolved species in porous media can be described mathematically as

$$J_{adv} = \phi v C_i \tag{2}$$

where ϕ is the porosity, v is the average linear velocity in the media, and C_i is the concentration of the ith species. In most (but not all) examples of water-rock interaction, flow is through a porous medium and is described with Darcy's Law.

Darcy's Law

The fluid velocity in porous and fractured media is usually calculated from Darcy's Law, which states that the volumetric flux of water ($m^3_{fluid}/m^2_{medium}/s$), $\mathbf{q}$, is a vector proportional to the gradient in the hydraulic head, h

$$\mathbf{q} = \phi \mathbf{v} = -\mathbf{K}\nabla h \tag{3}$$

where $\mathbf{K}$ is the hydraulic conductivity (here a second-order tensor) in units of m/s (Darcy 1856). One can also write Darcy's Law in terms of fluid pressure by defining the hydraulic head as

$$h = z + \frac{P}{\rho g} \tag{4}$$

where z is the depth, P is the fluid pressure, ρ is the fluid density, and g is the acceleration due to gravity. The hydraulic conductivity is defined as

$$\mathbf{K} = \frac{\kappa \rho g}{\mu} \tag{5}$$

where κ is the permeability and μ is the dynamic viscosity. One can also write Darcy's Law explicitly in terms of the fluid pressure, permeability, and the viscosity

$$\mathbf{q} = \phi \mathbf{v} = -\frac{\kappa}{\mu}\left[\nabla P - \rho \mathbf{g}\right] \tag{6}$$

Darcy's Law is applied where the porous medium can be treated as a continuum in which a representative elementary volume (REV) is significantly larger than the average grain size. In the case of flow at the pore scale, the averaging does not apply and Darcy's Law cannot be used. Representative values of the hydraulic conductivity (or permeability) are given in Table 1 for various subsurface materials.

Table 1. Range of values of hydraulic conductivity and permeability. Modified from Bear (1972).

<table>
<tr><th>K (cm/s)</th><th>10^{2}</th><th>10^{1}</th><th>10^{0}</th><th>10^{-1}</th><th>10^{-2}</th><th>10^{-3}</th><th>10^{-4}</th><th>10^{-5}</th><th>10^{-6}</th><th>10^{-7}</th><th>10^{-8}</th><th>10^{-9}</th><th>10^{-10}</th></tr>
<tr><td>κ (cm²)</td><td>10^{-3}</td><td>10^{-4}</td><td>10^{-5}</td><td>10^{-6}</td><td>10^{-7}</td><td>10^{-8}</td><td>10^{-9}</td><td>10^{-10}</td><td>10^{-11}</td><td>10^{-12}</td><td>10^{-13}</td><td>10^{-14}</td><td>10^{-15}</td></tr>
<tr><td>Unconsolidated Sand & Gravel</td><td colspan="2">Clean Gravel</td><td colspan="3">Clean Sand or Sand & Gravel</td><td colspan="5">Very Fine Sand, Silt, Loess, Loam</td><td colspan="3"></td></tr>
<tr><td>Unconsolidated Clay & Organic</td><td colspan="4"></td><td colspan="2">Peat</td><td colspan="3">Stratified Clay</td><td colspan="4">Unweathered Clay</td></tr>
<tr><td>Consolidated Rocks</td><td colspan="4">Highly Fractured Rocks</td><td colspan="3">Oil Reservoir Rocks</td><td colspan="2">Sandstone</td><td colspan="2">Limestone</td><td colspan="2">Granite</td></tr>
</table>

Molecular diffusion

In water-rock interaction, we are often concerned with transport through low porosity and permeability material, in which case molecular diffusion needs to be included in addition to flow. The treatment of molecular diffusion has also suffered to some extent from the assumption that Fick's First Law (described below) is sufficient, but as shown below, this is true only in the case of diffusion of uncharged species in dilute solutions. A more general formulation relates the diffusive flux linearly to gradients in the chemical potential rather than concentration, but we begin with the simpler Fickian description of diffusion for the sake of simplicity.

Fick's First Law. Molecular diffusion is usually described in terms of Fick's First Law, which states that the diffusive flux (here shown for only a single coordinate direction x) is proportional to the concentration gradient

$$J_i = -D_i \frac{\partial C_i}{\partial x} \tag{7}$$

D_i is referred to as the diffusion coefficient and is specific to the chemical component considered as indicated by the subscript i. Fick's First Law is a phenomenological theory for diffusion that relates diffusion to the "driving force" provided by the concentration gradient, although it can also be derived atomistically (Lasaga 1998). In the case of diffusion in porous media, it is normally necessary to include a tortuosity correction as well (see discussion below). Values of diffusion coefficients for selected ion at infinite dilution in water at 25 °C are given in Table 2.

Fick's Second Law. Integrating the diffusive fluxes over a control volume leads to Fick's Second Law (Bear 1972; Lasaga 1998). By defining the fluxes at the faces of an elemental volume as in Figure 1 (here shown for only a single coordinate direction x) a general form of the continuity equation can be derived by considering the change in concentration inside an elemental volume for a given time increment:

$$\frac{dC}{\Delta t} = \frac{1}{\Delta x}\left[J_x - J_{x+\Delta x}\right] = -\frac{1}{\Delta x}\left[J_{x+\Delta x} - J_x\right] \tag{8}$$

Written as a partial derivative as Δx and $\Delta t \rightarrow 0$, and substituting Fick's First Law in Equation (7), Equation (8) becomes

$$\frac{\partial C_i}{\partial t} = -\frac{\partial}{\partial x}\left[J_i\right] = \frac{\partial}{\partial x}\left[D_i \frac{\partial C_i}{\partial x}\right] \tag{9}$$

which provides an expression for the change in concentration in terms of the divergence of the diffusive flux.

Table 2. Tracer diffusion coefficients of ions at infinite dilution in water at 25 °C. Modified from Lasaga (1998).

Cation	$D_i \times 10^5$ cm²/s	Anion	$D_i \times 10^5$ cm²/s
H^+	9.31	OH^-	5.27
Li^+	1.03	F^-	1.46
Na^+	1.33	Cl^-	2.03
K^+	1.96	Br^-	2.01
Rb^+	2.06	I^-	2.00
Cs^+	2.07	IO_3^-	1.06
NH_4^+	1.98	HS^-	1.73
Ag^+	1.66	HSO_4^-	1.33
Mg^{2+}	0.705	NO_2^-	1.91
Ca^{2+}	0.793	NO_3^-	1.90
Sr^{2+}	0.794	HCO_3^-	1.18
Ba^{2+}	0.848	$H_2PO_4^-$	0.846
Mn^{2+}	0.688	$H_2AsO_4^-$	0.905
Fe^{2+}	0.719	$H_2SbO_4^-$	0.825
Co^{2+}	0.699	SO_4^{2-}	1.07
Ni^{2+}	0.679	SeO_4^{2-}	0.946
Cu^{2+}	0.733	CO_3^{2-}	0.955
Zn^{2+}	0.715	HPO_4^{2-}	0.734
Cd^{2+}	0.717	CrO_4^{2-}	1.12
Pb^{2+}	0.945	MoO_4^{2-}	0.991
UO_2^{2+}	0.426	PO_4^{3-}	0.612
Cr^{3+}	0.594		
Fe^{3+}	0.607		
Al^{3+}	0.559		

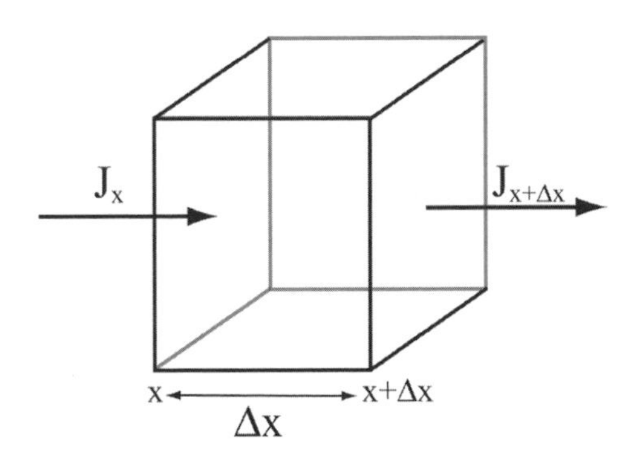

Figure 1. Derivation of a continuity equation (Fick's Second Law).

Integration of Equation (9) provides a solution for the time and space evolution of the concentration field. This equation is quite straightforward to integrate numerically, but analytical solutions are also available, with the general form

$$C(x,t) = A + B\,\text{erf}\left[\frac{x}{2\sqrt{Dt}}\right] \tag{10}$$

where A and B are constants that depend on the initial and boundary conditions. For the case where the initial mass is located between $-h$ and h, with concentrations equal to zero elsewhere, the analytical solution is given by

$$C(x,t) = \frac{1}{2} C_0 \left[\operatorname{erf}\left(\frac{x+h}{2\sqrt{Dt}} \right) - \operatorname{erf}\left(\frac{x-h}{2\sqrt{Dt}} \right) \right] \tag{11}$$

where erf is the error function and C_0 is the initial concentration between $-h$ and h (Lasaga 1998). Figure 2 shows the evolution of such an initially narrow and sharp zone with time as a result of molecular diffusion.

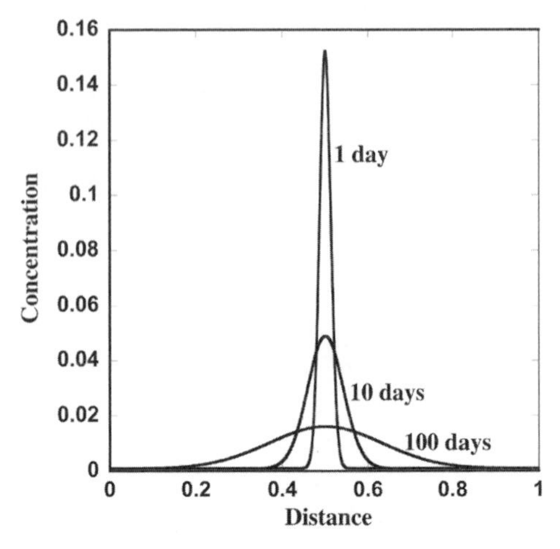

Figure 2. Spreading of an initially sharp solute plume as a result of molecular diffusion.

Electrochemical migration

Fick's First Law is strictly applicable only in the case of an infinitely dilute solution with uncharged chemical species. In electrochemical systems or systems containing charged species, it is also necessary to consider an electrochemical migration term. While this could be considered as a separate flux on the same level as advection and diffusion, this effect is most commonly related to diffusion of charged species at different rates. The full expression for the migration of a charged species in an electric field (written again in terms of a single coordinate direction, x, for the sake of simplicity) is given by (Newman 1991)

$$J_i^{migr} = -z_i u_i F C_i \frac{\partial \Phi}{\partial x} \tag{12}$$

where z_i is the charge of the species, u_i is its mobility, F is Faraday's constant (= 96,487 Coulombs/equivalent), and Φ is the electrical potential. The mobility refers to the average velocity of a species in solution acted upon by a unit force, independent of the origin of the force. The flux of charged species in an electric field gives rise to a current, which can be expressed as

$$i = F \sum_i z_i J_i \tag{13}$$

where i is the current density in units of ampere per m^2. Expanding Equation (13) in terms of the migration term (Eqn. 12) and the diffusive flux (Eqn, 7), the current density can be written as

$$i = -F^2 \frac{\partial \Phi}{\partial x} \sum_i z_i^2 u_i C_i - F \sum_i z_i D_i \frac{\partial C_i}{\partial x} \tag{14}$$

Where no concentration gradients are present, the current is given by the first term on the right hand side

$$i = -\kappa_e \frac{\partial \Phi}{\partial x} \tag{15}$$

where

$$\kappa_e = F^2 \sum_i z_i^2 u_i C_i \tag{16}$$

is the conductivity of the solution. Note that using Equation (16), it is possible to determine the mobility of an ion by measuring the conductivity of a solution. The mobility in turn can be used to determine the diffusion coefficient for an ion from the Nernst-Einstein equation (Newman 1991; Lasaga 1998)

$$D_i = RTu_i \tag{17}$$

where R is the gas constant and T is the temperature on the Kelvin scale. Rearranging Equation (14) to obtain an expression for the gradient in the electrical potential

$$\frac{\partial \Phi}{\partial x} = -\frac{i}{\kappa_e} - \frac{F}{\kappa_e}\sum_i z_i D_i \frac{\partial C_i}{\partial x} \tag{18}$$

it is apparent that even in the absence of an electrical current, i, it is possible to have a gradient in the electrical potential as a result of concentration gradients of charged species. The second term on the right hand side of Equation (18) is known as the diffusion potential, which vanishes when all of the diffusion coefficients for the charged species are the same. To proceed further, it is useful to define the fraction of the current carried by a species j (or the transference number)

$$t_j = \frac{z_j^2 u_j C_j}{\sum_i z_i^2 u_i C_i} \tag{19}$$

which can be used along with the substitution of Equation (18) and Equation (16) into Equation (12) to obtain an expression for the migration flux

$$J_j^{migr} = \frac{t_j i}{z_j F} + \frac{t_j}{z_j}\sum_i z_i D_i \frac{\partial C_i}{\partial x} \tag{20}$$

This equation (still only strictly applicable to relatively dilute solutions) can be used to describe the case where an electrical current is present (e.g., an electrochemical cell), although more commonly in fluid-rock interaction one encounters only the second term.

In the absence of a current and advection, the total flux (combining Fickian diffusion and electrochemical migration) is then given by

$$J_j = -D_j \frac{\partial C_j}{\partial x} + \frac{t_j}{z_j}\sum_i z_i D_i \frac{\partial C_i}{\partial x} \tag{21}$$

As noted above, the second term vanishes when all of the diffusion coefficients for the charged species are the same, but since this is in general not the case, this term should normally be retained along with the first term. Note also that the effect of electrochemical migration may be important even in dilute systems.

Electrochemical migration has several important consequences. Inclusion of electrochemical migration prevents the unrealistic build up of charge in calculations using differing diffusivities for charged species. The typical way to avoid this problem in transport calculations is to use the same diffusion coefficient for all of the charged species, in which case the diffusion potential driving electrochemical migration disappears. However, this approach does not provide the same results as would a full treatment of electrochemical migration. An example from Steefel (2007) illustrates some of the potential effects. A fixed boundary condition on the left side is separated by 20 mm of 10% porosity material from a reacting feldspar (albite) grain (Fig. 3). At 25 °C, the dissolution rate of feldspar is sufficiently slow that concentration gradients will not develop (Murphy et al. 1989). At higher temperatures where the rate of surface reaction is more rapid, however, it is possible to develop diffusion-limited dissolution (the rate of surface reactions is significantly faster than the rate at which

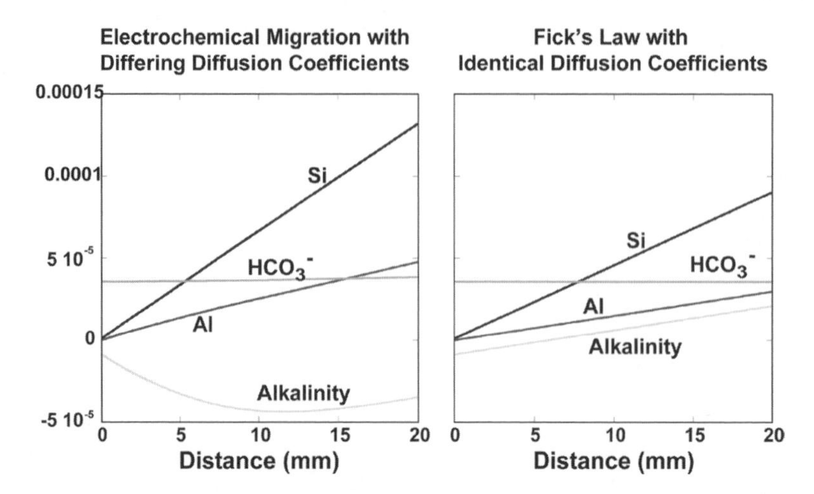

Figure 3. Comparison of diffusion profiles in the case of a dilute, mildly acidic solution (pH 4) on the left reacting with an albite grain located 2 cm to the right. The left panel shows a calculation using differing diffusion coefficients for the ions (see Table 2) and includes electrochemical migration. The right panel shows a calculation in which identical diffusion coefficients for all species are used—in this case, electrochemical migration vanishes and diffusion occurs according to Fick's Law. From Steefel (2007).

ions can diffuse across the 20 mm distance, and thus a gradient develops). Comparison of the full multicomponent diffusion and electrochemical migration calculation (essentially the Nernst-Planck equation) with one based on identical diffusion coefficients (no electrochemical migration) for all species indicates that conservative quantities like alkalinity need not have linear profiles where electrochemical migration is important. Comparing the two cases with identical albite dissolution rates, the concentrations of dissolved species next to the albite grain are not even the same. This is primarily due to the rapid diffusion of hydrogen ion, which results in a more nearly linear profile for pH in the multicomponent diffusion/electrochemical migration case as compared to the identical diffusion coefficient case.

Diffusion in concentrated solutions

Ficks's First Law runs into additional difficulties in concentrated aqueous solutions as discussed in Steefel (2007). A more rigorous and general expression is given by

$$J_j^{diff} = -L_{ji}\frac{\partial \mu_j}{\partial x} \tag{22}$$

where the L_{ji} are the phenomenological coefficients introduced in the theory of irreversible thermodynamics (Onsager 1931; Prigogine 1967; Lasaga 1998) and μ_j is the chemical potential of the jth species. Here, the fluxes are linearly related to gradients in the chemical potentials of the solutes rather than to their concentrations as in Fick's Law. The phenomenological coefficients, L_{ji}, can be linked back to measurable quantities by making use of the mobility again as the "velocity" of a particle acted upon by a force, with the force in this case provided by the chemical potential rather than the concentration

$$J_j^{diff} = -u_j C_j \frac{\partial \mu_j}{\partial x} \tag{23}$$

In the case where a gradient in chemical potential rather than concentration is used, the migration flux (compare to Eqn. 20 in the absence of a current) becomes

$$J_j^{migr} = \frac{t_j}{z_j} \sum_i z_i D_i \frac{\partial \mu_i}{\partial x} \tag{24}$$

By writing the chemical potential as (Denbigh 1981)

$$\mu_j = \mu_j^{\,0} + RT \ln(\gamma_j C_j) \tag{25}$$

where γ_j is the activity coefficient for the species, and differentiating with respect to x, we obtain the following combined expression for the pure diffusive flux and electrochemical migration by making use again of the definition of the ion mobility (Eqn. 17)

$$J_j = -D_j \frac{\partial C_j}{\partial x} - D_j C_j \frac{\partial \ln \gamma_j}{\partial x} + \frac{t_j}{z_j} \sum_i z_i D_i \frac{\partial C_i}{\partial x} + \frac{t_j}{z_j} \sum_i z_i D_i \frac{\partial \gamma_i}{\partial x} \tag{26}$$

This more general expression, which reduces to Equation (21) where gradients in activity coefficients are negligible (for example, for diffusion of a trace species in a strong electrolyte), makes clear that both the diffusive and electrochemical migration fluxes can depend on the activity coefficients for the species. If gradients in one or more activity coefficients are negative, it is possible for "uphill diffusion," in which a species diffuses up its own concentration gradient, to occur (Lasaga 1998).

Tortuosity

Since water-rock interaction commonly takes place in porous materials, it is important to account for the effect of tortuosity (Fig. 4), which is defined as the ratio of the path length the solute would follow in water alone, L, relative to the tortuous path length it would follow in porous media, L_e (Bear 1972)

$$T_L = \left(L / L_e \right)^2 \tag{27}$$

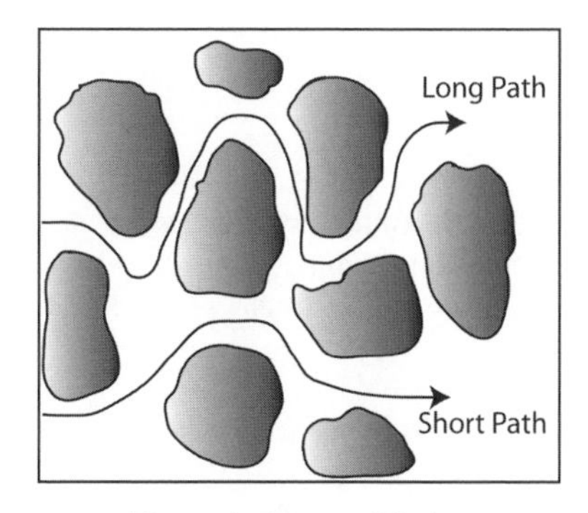

Figure 4. Tortuous diffusion paths in porous material.

In this definition of tortuosity (sometimes the inverse of Eqn. 27 is used), its value is always < 1 and the effective diffusion coefficient in porous media is obtained by multiplying the tortuosity by the diffusion coefficient for the solute in pure water. With this formulation, the diffusion coefficient in porous media is given by

$$D_i^* = T_L D_i \tag{28}$$

The diffusive flux, then, is given by

$$J_j^{diff} = -\phi D_j T_L \frac{\partial C_j}{\partial x} = -\phi D_j^* \frac{\partial C_j}{\partial x} \tag{29}$$

An alternative formulation for the coefficient for molecular diffusion in porous media is given by the formation factor, F, defined as (Bear 1972)

$$F = \frac{(L_e/L)^2}{\phi} = \frac{1}{\phi T_L} \tag{30}$$

in which case the diffusive flux in porous media becomes

$$J_j^{diff} = -\frac{D_j}{F}\frac{\partial C_j}{\partial x} \tag{31}$$

Various approaches for calculating formation factors (and thus, the diffusion coefficient in porous medium) are in use, with a formulation based on Archie's Law being the most common

$$F = \frac{1}{a\phi^m} \tag{32}$$

where a is a fitted constant and m is the cementation exponent. Values of $a = 0.71$ and $m = 1.57$ have been suggested for unfractured granite (Pharmamenko 1967), although values of $m = 2$ have been reported for many more porous and permeable subsurface materials (Oelkers 1996). Values for the formation factor in igneous intrusive rock in Sweden range from 10^3 to 10^7 (Skagius and Neretnieks 1986). In the case of low porosity materials, especially those with low pore connectivity, the formation factor may be quite large (and thus, the diffusion coefficient in porous media quite small).

Hydrodynamic dispersion

The phenomenon of hydrodynamic dispersion was noted as early as 1905 when Slichter reported that the concentration of an electrolyte monitored in an aquifer downstream of an injection point increased only gradually, with the plume shape becoming longer and wider as it advanced (Slichter 1905). The spreading of the solute mass as a result of dispersion is a diffusion-like process that has led to the use of Fick's First Law to describe the process in one dimension as

$$J_j^{disp} = -D_h\frac{\partial C_j}{\partial x}, \tag{33}$$

where D_h is the hydrodynamic dispersion coefficient. The coefficient of hydrodynamic dispersion is defined as the sum of molecular diffusion and mechanical dispersion, since these effects are not separable where flow is involved (Bear 1972)

$$D_h = D^* + D \tag{34}$$

The representation of dispersion as a Fickian process, however, has been questioned extensively (Dagan 1988). As applied to reactive transport problems, the main problem is that dispersion does not involve mixing in the same way that molecular diffusion does (Cirpka 2002). This is discussed further below in the section on macrodispersion.

Mechanical dispersion. Mechanical dispersion is a result of the fact that variations in the flow velocities exist, even where an average flow rate (as in Darcy's Law) can be defined for a particular representative elementary volume (REV). If all the detailed flow paths could be captured, then there would be no need for inclusion of a dispersion coefficient.

Mechanical dispersion was first clearly discussed for the case of laminar flow and transport within a cylindrical tube (Taylor 1953). In the case where a non-reactive tracer is instantaneously released at the inlet of the tube and complete mixing does not occur, the tracer pulse will arrive first where the flow is fastest, later where the flow is slower. At any one time, the bulk concentration of the fluid leaving the tube represents a flux-weighted average of the concentration in the fast and slow moving flow paths, so the breakthrough of the tracer will appear as disperse or gradual rather than sharp. This effect is referred to as Taylor dispersion (Taylor 1953; Aris 1956). Assuming Poiseuille's equation for flow in a cylinder holds, the steady-state parabolic velocity distribution as a function of radius r can be calculated from (Daugherty and Franzini 1965)

$$u(r) = \frac{2Q}{\pi R^2}\left[1-\left(\frac{r}{R}\right)^2\right] = 2U\left[1-\left(\frac{r}{R}\right)^2\right] \tag{35}$$

where $u(r)$ is the local fluid velocity within the cylinder, Q is the volumetric flow rate, R is the radius of the cylinder, and U is the average velocity (Fig. 5A). Using Equation (35), Taylor derived an analytical expression for the dispersion in the case of Poiseuille flow

$$D_h = D + \frac{UR^2}{D} \tag{36}$$

where D_h is the dispersion coefficient, and D is the molecular diffusion coefficient. Note that the dispersion coefficient depends on the molecular diffusion coefficient for its contribution to longitudinal spreading (the first term on the right hand side of Eqn. 36), but also inversely on the molecular diffusion coefficient in the second term on the right hand side because radial diffusion acts to eliminate the gradients in concentration developed by the non-uniform velocity profile (Fig. 5A). A comparison of the breakthrough curve for a non-reactive tracer calculated with the Taylor-Aris dispersion coefficient given in Equation (36) and a full two-dimensional axisymmetric cylindrical Poiseuille flow calculation in which dispersion is represented only through the variation in flow velocities and through molecular diffusion (i.e., no explicit dispersion coefficient is included) is given in Figure 5B (Steefel 2007).

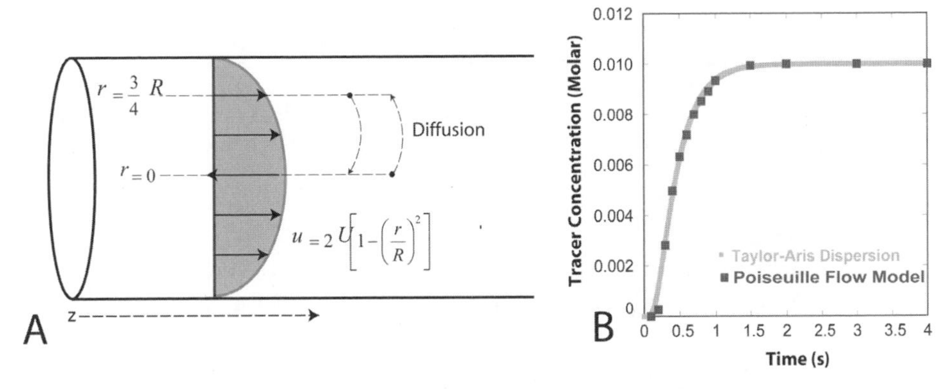

Figure 5. (A) Schematic representation of the parabolic velocity distribution that develops in the case of Poiseuille flow. (B) Comparison of non-reactive tracer breakthrough profile using a Taylor-Aris dispersion coefficient (Taylor 1953) and a full two-dimensional calculation of transport in a cylindrical tube using the analytical solution for Poiseuille flow (Eqn. 35). From Steefel (2007).

A similar effect occurs within porous media (which are often conceptualized as bundles of capillary tubes), since the velocity adjacent to a solid grain is slower than the velocities in the center of the pore. Even in the case of a perfectly homogeneous porous medium (e.g., glass beads all of the same size), therefore, spreading of a solute plume will occur. Additional effects in porous media are related to the tortuosity of flow paths (Fig. 4) and to differences in pore size, since velocities will be higher in those pores with wider apertures (Bear 1972).

Dispersion needs to be quantified at much larger scales in porous media, however, so some averaging of these dispersion mechanisms is usually required. Dispersion in porous media is typically defined as the product of the fluid velocity and dispersivity, α, with longitudinal and transverse components

$$D_L = \alpha_L V_i \quad (37)$$
$$D_T = \alpha_T V_i$$

where V_i refers to the average velocity in the principal direction of flow, and the subscripts L (longitudinal) and T (transverse) refer to the dispersion coefficient parallel and perpendicular to the principal direction of flow respectively (Bear 1972).

Macrodispersion. The treatment of hydrodynamic dispersion as a Fickian process (that is, one in which the flux is linearly proportional to the concentration gradient) has been widely criticized (Gelhar and Axness 1983; Gelhar 1986; Dagan 1988; Gelhar et al. 1992). As pointed out by a number of investigators, dispersion is scale-dependent, with larger dispersivities observed for larger observation scales. At the column scale, a typical dispersivity may be on the order of a cm, but at the field scale, apparent dispersivities of 10 to 100 m or more are common. Gelhar (1986) showed that the dispersivity increased with scale of observation, although even at a single observation scale, dispersivities could range over 2-3 orders of magnitude due to differing degrees of aquifer heterogeneity at the different sites (Gelhar 1986; Gelhar et al. 1992). The increase with dispersivity as a function of the scale of measurement is related to the hierarchical nature of natural porous medium and is often referred to as macrodispersion. At the pore-scale, variations in flow velocity and direction are related to pore-scale heterogeneities like the variation in grain size and pore apertures, but at the field scale, additional heterogeneity features (sedimentary features, fractures, faults) are encountered which add to the spreading of the solute plume. As in the case of Taylor dispersion, the spreading of a solute plume is fundamentally related to the variations in flow velocity, with fast-flowing pathways leading to earlier breakthrough than slow flow pathways. In the case of Poiseuille flow in a tube, the effect is predictable and rigorously quantifiable because of the well-defined physics of the system, but in natural porous media where multi-scale heterogeneities typically occur, the problem of determining dispersivities is much more difficult and usually requires a stochastic treatment.

The effect of macrodispersion can be illustrated with a simple example as considered in Steefel (2007). Consider a permeability field with approximately six orders of magnitude variability in permeability (Fig. 6A). The heterogeneous distribution of and strong contrast in permeability results in preferred flow paths that lead to fingering of a non-reactive tracer plume as it moves with the flow from left to right (Fig. 6B). The two-dimensional numerical simulation

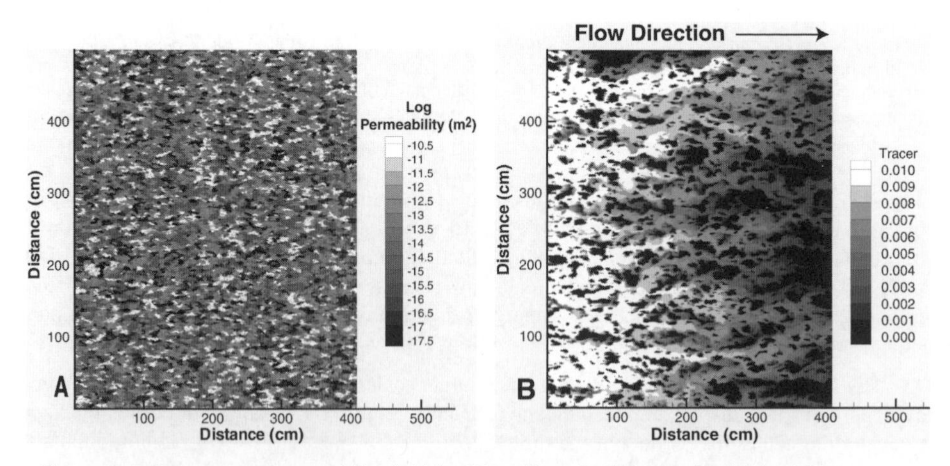

Figure 6. A. Heterogeneous permeability field for 2D problem. B. Concentration field at 0.25 years as a result of flow from left to right across the heterogeneous domain. After Steefel (2007).

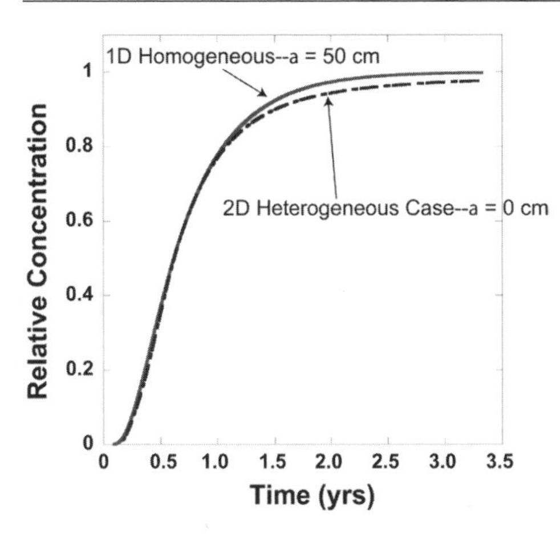

Figure 7. Comparison of solute breakthrough curves for the case of the heterogeneous permeability field shown in Figure 6A, calculated with a dispersivity (α) = 0, and a 1D homogeneous permeability field calculated with α = 50 cm. After Steefel (2007).

was conducted without including a Fickian treatment of hydrodynamic dispersion and using an accurate Total Variation Diminishing (TVD) scheme that minimizes numerical dispersion (Steefel and MacQuarrie 1996). When a flux-weighted average of the solute concentration is calculated over the entire length of the capture plane on the right hand side, perpendicular to the flow direction (similar conceptually to what occurs where flow converges into a pumping well), one observes a gradual, disperse breakthrough as a result of the mixing of the early and late arriving fluid packets (Fig. 7). To match the disperse breakthrough curve in the heterogeneous flow field with a one-dimensional simulation of flow in a homogeneous permeability field, it is necessary to use a dispersivity, α, of about 50 cm.

Stochastic descriptions of dispersion. There is much interest now in more rigorous descriptions of macrodispersion in particular. One approach has been to treat the parameters controlling transport (primarily the hydraulic conductivity or permeability, but also potentially such parameters as mineral volume fractions and/or reactive surface area) as stochastic rather than deterministic parameters, since the exact distribution of these parameters in the subsurface is typically poorly known (Gelhar 1986, 1993; Dagan 1989). This is a particularly useful approach when the primary data available are breakthrough curves for tracers at an observation point. The shape of the curve can be used to deconvolve a stochastic distribution of hydraulic conductivities in the aquifer, implying that while we do not know the exact path a particular solute particle followed, we do have a statistical representation of the ensemble of solute travel times that could have produced the observed tracer concentrations at the observation point. Tracer breakthrough data, therefore, is often used to generate probability density functions for hydraulic conductivities and travel times, although this concept need not be strictly stochastic.

Reaction-induced changes in transport properties

One of the most difficult problems in reactive transport today is the treatment of cases in which the physical properties of the porous medium evolve as a result of chemical reactions. The most commonly identified coupling involves the porosity and permeability as affected by chemical reactions (dissolution or precipitation), but other properties like the reactive surface area may evolve as well. A number of experimental studies have been conducted in which reaction-induced permeability change was identified (Dobson et al. 2003). Modeling studies have also been carried out in which flow and reaction have been explicitly coupled and allowed to evolve (Ortoleva et al. 1987; Steefel and Lasaga 1990, 1994; Steefel and Lichtner 1994; Steefel and Lichtner 1998b; Dobson et al. 2003; Steefel 2007; Cochepin et al. 2008). A smaller set of studies have combined experimental and/or characterization studies with reactive transport modeling (Dobson et al. 2003; Navarre-Sitchler et al. 2009).

Porosity is typically the first order parameter that is predicted from reactive transport simulations, since it is directly related to the sum of the mineral volumes precipitated or dissolved. It is possible to predict the evolution of the total porosity, therefore, with some degree of confidence. More difficult is to predict the transport parameters, especially permeability and diffusivity, since they depend on the detail pore geometry. To make progress in this area, it will be necessary to combine the modeling with microscopic characterization of the porous medium as mineral dissolution or precipitation occur. A first attempt at this has been carried out by Navarre-Sitchler et al. (2009) who studied the evolution of the porosity, pore connectivity, pore geometry, and diffusivity as a function of the extent of chemical weathering. This study showed that as the chemical weathering process advanced within the approximately 1 mm wide weathering interface developed within a basalt clast, porosity increased as the primary minerals dissolved (as one would expect), but that pore connectivity increased as well once a critical porosity of ~9% was achieved. The porosity and pore connectivity as a function of weathering were imaged with Xray microtomography at the Advanced Light Source (Fig. 8). The 3D data was also used to construct a pore network model that was used for tracer diffusion calculations (Navarre-Sitchler et al. 2009). A comparison of unweathered and weathered basalt using both μXray fluorescence (Panel A) and 3D tracer diffusion simulations using the pore network model constructed from the microtomography (Panel B) indicates that the weathering process substantially increased the diffusivity as a result of weathering (Fig. 9).

REACTIVE TRANSPORT BASICS

We are now ready to combine the transport processes outlined above with expressions for kinetically-controlled geochemical and biogeochemical reaction. A comprehensive and modern treatment requires the use of a sophisticated numerical simulator capable of handling variations in system properties (hydraulic conductivity, porosity, tortuosity, etc.) in both space and time, along with rigorous treatments of multicomponent diffusion and advection. In addition, the modern numerical simulators are capable of considering multicomponent reactions along with nonlinear rate laws, features that are well beyond what is possible with analytical solutions. Analytical solutions, while exact by definition for the particular equation and boundary and initial conditions considered, often force the would-be user into approximations (physical or chemical) that may or may not be warranted. Our interest here initially, however, is in exploring the first-order dynamics when reaction and transport are coupled, and for this purpose a simplified version of the advection-dispersion-reaction equation is adequate.

For a system with transport of a non-reactive tracer, an expression for the conservation of solute mass can be derived by accounting for the flux of solute across the faces of a volume element. For a one-dimensional system (fluxes in the Y and Z directions = 0), the net flux is obtained from

$$\frac{\partial J_i}{\partial x} = \lim_{\Delta x \to 0} \frac{J_i\big|_{x+\Delta x} - J_i\big|_x}{\Delta x} \tag{38}$$

In a multidimensional system involving porous media, the accumulation of solute mass is given by the difference (that is, the divergence) of the fluxes summed over all of the faces of the element

$$\frac{\partial(\phi C_i)}{\partial t} = -\nabla \cdot \mathbf{J}_i = -\left(\frac{\partial J_i}{\partial x} + \frac{\partial J_i}{\partial y} + \frac{\partial J_i}{\partial z}\right) \tag{39}$$

where J_i is the flux vector. Substitution of Equations (2) and (7) into Equation (39) (neglecting electrochemical migration for simplicity) yields

$$\frac{\partial(\phi C_i)}{\partial t} = -\nabla \cdot (\phi \mathbf{v} C_i) + \nabla \cdot (\phi D_i^* \nabla C_i) \tag{40}$$

To include reactions, it is more instructive to use the one-dimensional version of the advection-dispersion equation, particularly since non-dimensionalization of the equation is more

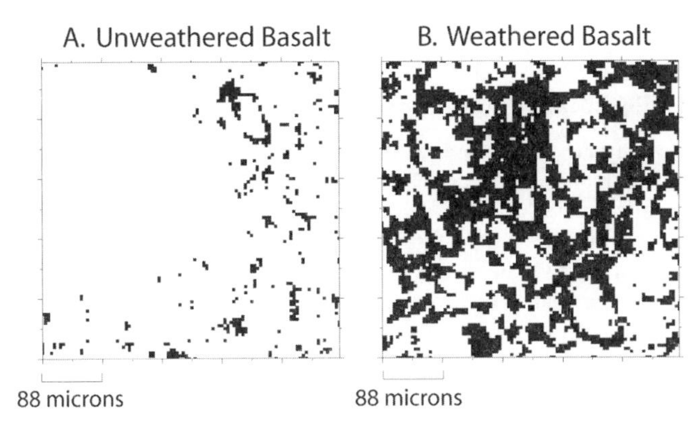

Figure 8. Microtomographic images of porosity (dark voxels) in unweathered (Panel A) and weathered (Panel B) basalt. Note the increase in pore connectivity in the weathered sample, which has the effect of increasing the diffusivity of the material. From Navarre-Sitchler et al. (2009).

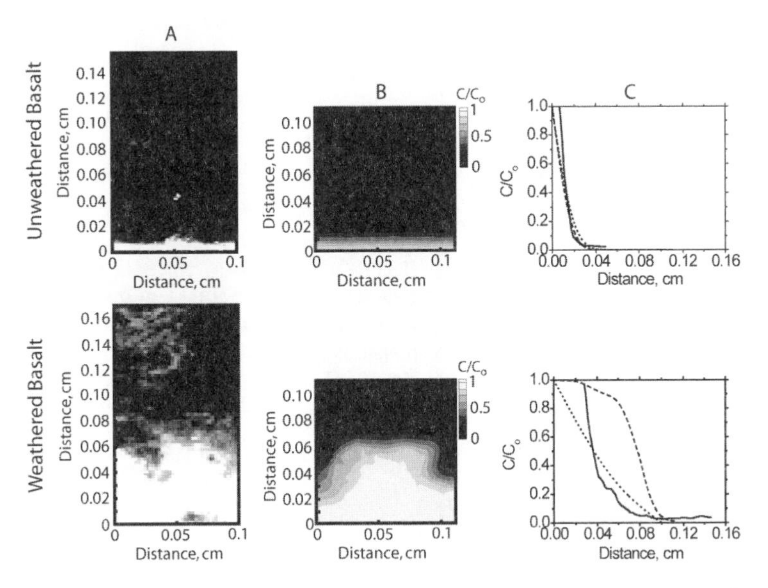

Figure 9. Comparison of results from diffusion experiments and models for unweathered (top) and weathered (bottom) basalt. A) μXRF image of Br concentrations in physical diffusion experiments after 7 days. Higher concentrations of Br appear whiter. B) Contour plot of Br concentrations from 3D pore network simulation at same scale as μXRF images for comparison. C) Concentration profiles of Br in physical diffusion experiments (solid line), 1D model results (dotted line) and 3D model results (dashed line). From Navarre-Sitchler et al. (2009).

straightforward in this case. For a constant porosity, tortuosity, and flow system characterized by a first-order precipitation and dissolution reaction that can be described in terms of a single chemical component (e.g., $SiO_{2[aq]}$ reacting with the single mineral quartz) , the advection-dispersion-reaction equation becomes

$$\phi \frac{\partial C}{\partial t} = -\phi v \frac{\partial C}{\partial x} + \phi D^* \frac{\partial^2 C}{\partial x^2} + Ak\left(1 - \frac{C}{C_{eq}}\right) \tag{41}$$

where k is the rate constant in units of moles $m^{-2}\ s^{-1}$, A is the reactive surface area of the mineral in units of $m^2\ m^{-3}$, and C_{eq} is the solubility of the mineral in moles m^{-3}.

Important non-dimensional parameters

By converting a partial differential equation to its non-dimensional form, it is possible to define a series of parameters that describe the relative importance of dynamic processes like chemical reaction, advection, and hydrodynamic dispersion. The first step is to define a length scale, l, which may be a characteristic length scale defined by the geology or hydrology (e.g., the flow length through an aquifer or weathering profile), or more commonly a length scale which reflects the spatial distribution of observations. Using this length scale, a non-dimensional or fractional distance can be defined as

$$x' = \frac{x}{l} \tag{42}$$

In addition, this allows us to define a characteristic time for dispersive transport

$$t_D = \frac{l^2}{D^*} \tag{43}$$

and advective transport

$$t_A = \frac{l}{v} \tag{44}$$

When combined with a characteristic time for reaction

$$t_R = \frac{\phi C_{eq}}{Ak} \tag{45}$$

it is possible to define a series of non-dimensional parameters that control the behavior of a water-rock interaction system involving transport: the Damköhler number for advective, Da_I, and diffusive (or dispersive), Da_{II}, systems respectively

$$Da_I = \frac{t_A}{t_R} = \frac{Akl}{\phi v C_{eq}} \tag{46}$$

$$Da_{II} = \frac{t_D}{t_R} = \frac{Akl^2}{\phi D^* C_{eq}} \tag{47}$$

In addition, the relative importance of advective versus diffusive (or dispersive) transport is compared in the Péclet number

$$Pe = \frac{t_D}{t_A} = \frac{vl}{D^*} \tag{48}$$

Note that each of these numbers depends on the characteristic length scale considered.

After introducing a non-dimensional form of the concentration

$$C' = \frac{C - C_{eq}}{C_0 - C_{eq}} \tag{49}$$

where C_0 is the concentration of the fluid injected into the system (Lichtner 1998) and a dimensionless time

$$t'_D = \frac{D^* t}{l^2} \tag{50}$$

Equation (41) becomes

$$\frac{\partial C'}{\partial t'_D} = \frac{\partial^2 C'}{\partial x'^2} - Pe\frac{\partial C'}{\partial x'} - Da_{II}C' \quad (51)$$

which makes clear how the Péclet and Damköhler numbers control the behavior of the advection-dispersion-reaction equation.

At high Péclet numbers ($Pe >> 1$), advective transport dominates and dispersive and diffusive transport are negligible, with the result that the concentration front is relatively sharp (Fig. 10). Similarly, for Damköhler numbers >> 1, reaction times are much faster than transport times for a given length scale. Where this is the case, local equilibrium is achieved for a reaction of the form given in Equation (41) over a length scale, l (Bahr and Rubin 1987). In addition, reaction fronts are relatively sharp where the Damköhler number is large, since the approach to equilibrium requires a small amount of time relative to the time needed for a solute to traverse a distance (Figure 11).

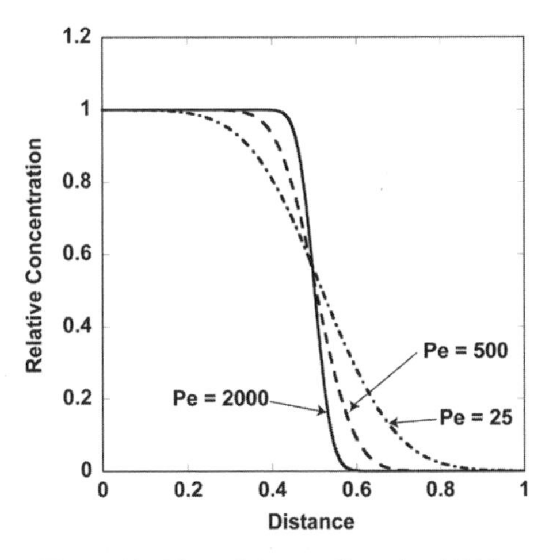

Figure 10. Effect of the non-dimensional Péclet number on the shape of a concentration profile. The length scale is given by the flow distance, with flow from left to right.

Transport versus surface reaction control

Under the steady-state conditions considered here, there is always some theoretical length scale over which local equilibrium could be attained (assuming such an equilibrium state exists), although this length scale may not be realized in a particular geological environment (Steefel 2007). Where the concentration in the aqueous phase increases or decreases to the point where equilibrium or near-equilibrium is achieved, the overall rate of reaction within the spatial domain defined by the length scale l becomes *transport-controlled*. This implies that the rate-limiting process in the overall reaction evaluated over the length scale l is transport, rather than the rate of attachment and detachment of ions from the mineral surface. Transport control has often been discussed in the geochemical literature, but usually with regard to experimental studies where transport involved diffusion across a boundary layer next to a reacting mineral (Berner 1980). For example, numerous studies have shown that the dissolution of calcite is diffusion rather than surface-reaction controlled at pH val-

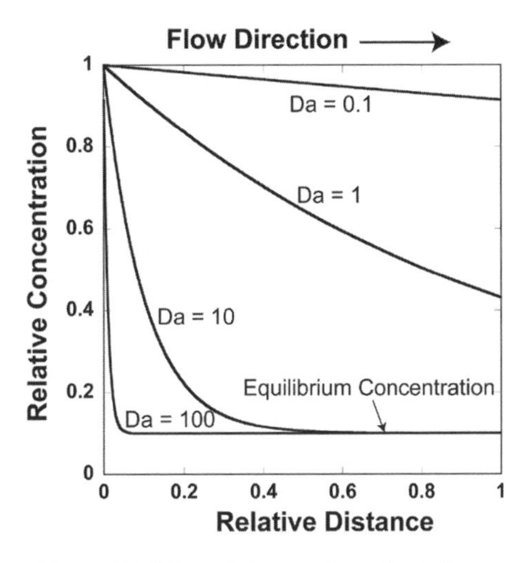

Figure 11. Effect of the non-dimensional Damköhler number on the shape of a concentration profile at steady-state. A large Damköhler number indicates that the reaction kinetics are rapid compared to transport rates over the length scale considered, or equivalently, that the characteristic time for transport is significantly less than the characteristic time for reaction. The local equilibrium approximation is justified in such cases.

ues < 4. This can be verified in rotating disk reactors by increasing the speed of rotation—in the case of a diffusion control on the rate of dissolution, the faster rate of spinning reduces the size of the diffusion boundary layer adjacent to the crystal, thus increasing the overall rate of dissolution (Pokrovsky et al. 2005).

While the rotating disk experiment involving calcite is a particularly good example of a transport-controlled system, the concept is much more general and may be applied at any scale. Consider a flow path through a granite undergoing weathering. If a fluid percolating through the granite approaches equilibrium over a flow distance of about 25 m, but we collect water samples issuing from the profile 50 m from the start of flow path, then we are collecting waters close to or at equilibrium that will express a transport control on the overall rate. Operationally, this means that the flow rate through the weathering profile (due, for example, to an increase in rainfall) controls the weathering flux, $J_{weather}$, in the advection-dominant case over the 50 m according to

$$J_{weather} = qC_{eq} \tag{52}$$

where q is the flow rate and C_{eq} is the equilibrium concentration. If the scale of observation, l, is >> than the equilibration length scale λ, then the weathering rate will not depend on the reaction kinetics. This is a general result that applies to all water-rock interaction systems (Lichtner 1993). A direct comparison of laboratory-determined rates of reaction, which are carried out in most cases under far from equilibrium conditions and thus under surface reaction control, and natural weathering rates evaluated at length scales greater than the equilibration length scale, will be meaningless in this case, since the rate control is not the same (Schnoor 1990). By definition, the field rate when transport-controlled will be slower than the surface reaction-controlled rate. But, as stressed above, the control on rates is scale-dependent, so data collected at a length scale << than the equilibration length scale may show evidence for a surface reaction control on rates and far from equilibrium conditions. Thus, observations of etch pitting in primary minerals, generally taken as good evidence of far from equilibrium conditions (White and Brantley 2003), may be fully compatible with an overall transport control on weathering rates, depending on how the scales of observations compare to the equilibration length scale.

Rates of fluid-rock interaction in heterogeneous systems

Most studies of the kinetics of fluid-rock interaction carried out to date have assumed flow and reaction through a homogeneous, well-mixed system, as would be the case with a single continuously stirred flowthrough reactor (Haggerty et al. 1998; Snodgrass and Kitanidis 1998; White and Brantley 2003; Zhu 2005). This is perhaps somewhat strange given the enormous number of studies on macrodispersion in the hydrologic literature indicating that travel times, and thus residence times, in the subsurface are not all the same. Fortunately, the roles of both physical and chemical heterogeneities and their role in water-rock interaction are now beginning to be discussed (Malmström et al. 2000; Li et al. 2006, 2007a,b, 2008; Meile and Tuncay 2006). The role of physical heterogeneities can be clarified by returning to the heterogeneous permeability field shown in Figure 6. If we now include reactions along with the flow and transport, in this case the dissolution of plagioclase according to the kinetic scheme proposed by Oelkers et al. (1994), we see that the numerical modeling of the reactive transport predicts a steady-state fingering of the pore water pH that follows the high permeability zones (Fig. 12A). Similarly, a contour plot of the logarithm of the mineral saturation states (the distance the pore water is from equilibrium with respect to plagioclase) shows a similar strong spatial variability—a transect in the Y direction at X = 10 cm would encounter up to 10 orders of magnitude variability, ranging from far from equilibrium to relatively close to equilibrium pore waters (Fig. 12B). Therefore, a single transect could encounter both surface-reaction and at least partly transport-controlled rates. While this is a model result, it should be broadly representative of what would be expected for any reactive system with seven orders of magnitude variation in the local permeability.

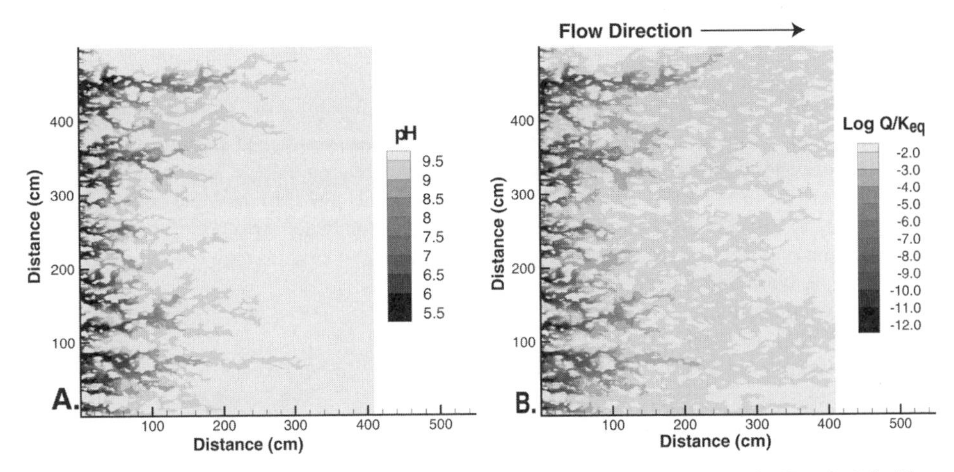

Figure 12. (A) pH distribution for the 2D heterogeneous flow field shown in Figure 6. pH at the left side of the flow domain is fixed at 5. The principal reaction is the dissolution of plagioclase with a rate law that includes dissolved aluminum inhibition (Oelkers et al. 1994). (B) The saturation state of the pore fluid with respect to plagioclase (log Q/K_{eq}) varies spatially, resulting in a heterogeneous distribution of surface reaction versus transport control of the rate. From Steefel (2007).

Residence time distributions. The concept of macrodispersion has important implications for the kinetics of water-rock interaction. One can also think of these variations in flow velocity and/or flow length as leading to a distribution of residence times for aqueous species migrating through reactive material (Steefel 2007). A sample collected from a stream draining a catchment, for example, typically includes waters that have moved through a myriad of flow paths and distances. Some water may have moved through relatively shallow portions of the soil profile—these flow paths would typically involve relatively short residence times, with less exposure to primary reactive phases forming the bedrock. In contrast, other water may have moved along deeper, slower flow paths, encountering a greater abundance of reactive primary minerals along the way. The residence time of reactive water in contact with rock, therefore, can vary substantially in a natural system. Locally, the system may be far from equilibrium, elsewhere close to equilibrium. Conceptually, one might think of the overall system as an ensemble of flow-through reactors, some with short residence times and/or small amounts of reactive material, others with much longer residence times. In some cases, the residence time will have been long enough that the fluid actually equilibrates within the reactor, leading to a rate that is transport rather than surface reaction-controlled. If the effluent of this ensemble of flow-through reactors were to converge into a single sampling tube (the experimental analogue of a stream gauging station at the catchment scale or a pumping well drawing water from a heterogeneous aquifer), then the total extent of reaction (i.e., the upscaled, volume-averaged reaction rate) would reflect the full distribution of residence times and reactive mineral mass.

Many examples of reactive flow through permeable rocks or sediment exist where strong contrasts in porosity and permeability result in regions characterized by very different transport regimes. This is often referred to as multi-region transport (Harvey and Gorelick 2000). The most common example is a dual permeability (or porosity) system, where a portion of the system is sufficiently permeable that macroscopic flow occurs within it, another portion is so impermeable that flow is negligible and transport is by molecular diffusion. Perhaps the best example of a dual permeability system is provided by fractured rock, where the permeability of the fractures is typically many orders of magnitude higher than the permeability of the rock matrix (Steefel and Lichtner 1998a).

Much of the evidence that multi-region transport systems exist in the subsurface comes from measurements of hydraulic conductivities on small laboratory-scale samples. Another source of evidence is provided by groundwater ages determined with various methods (Plummer et al. 2001). Although these groundwater ages are often interpreted as the result of mixing between two end members (old and young water), what we have learned about the distribution of hydraulic conductivities, and thus residence times, in heterogeneous 3D subsurface media argues against this (Varni and Carrera 1998; Tompson et al. 1999; Bethke and Johnson 2002; Weissmann et al. 2002). For example, Figure 13 shows groundwater ages determined using the chlorofluorocarbon (CFC) technique compared to CFC ages based on high resolution groundwater flow simulation of a heterogeneous alluvial fan aquifer in California (Weissmann et al. 2002). The high resolution flow modeling argues that the individual CFC ages are real, and not simply the result of mixing of old and young waters.

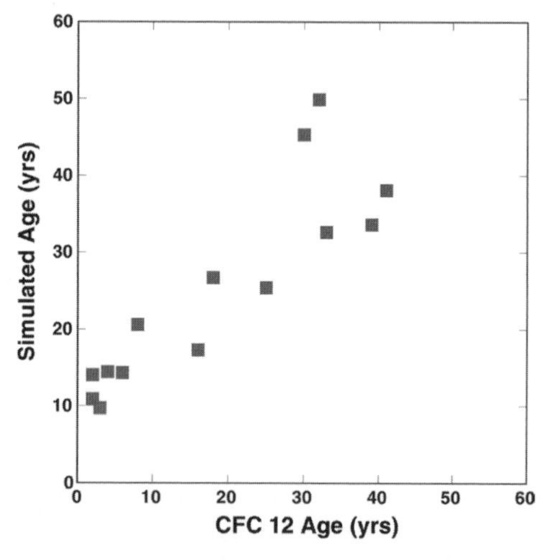

Figure 13. Chlorofluorocarbon (CFC) ages compared to simulated CFC ages based on high resolution groundwater flow modeling from a heterogeneous alluvial fan aquifer in California (Weissmann et al. 2002).

Evidence also comes from non-reactive tracer concentration profiles (or "breakthrough curves") determined in subsurface wells. Tracer concentration profiles at an observation well (or at the end of a laboratory column) characterized by multi-region effects typically show an early, usually sharp breakthrough, followed by a long gradual approach to a maximum value in the case of a constant injection. In the case of a transient pulse of a tracer followed by another fluid (a common case in contaminant hydrology), the tracer is eluted only very slowly from the column or aquifer. This behavior occurs because it requires some time for the tracer to be transported into the low permeability material, and once elution begins, then to be flushed from this zone.

Tracer breakthrough curves can be used to estimate a residence time distribution. With release of a non-reactive tracer pulse, fast pathways will deliver tracer first. Lower permeability regions within the flow domain will flush the tracer more slowly, providing information on the longer residence times. Using the first derivative of the concentration in the breakthrough curve in Figure 7, the formal probability density function for residence can be determined (Fig. 14).

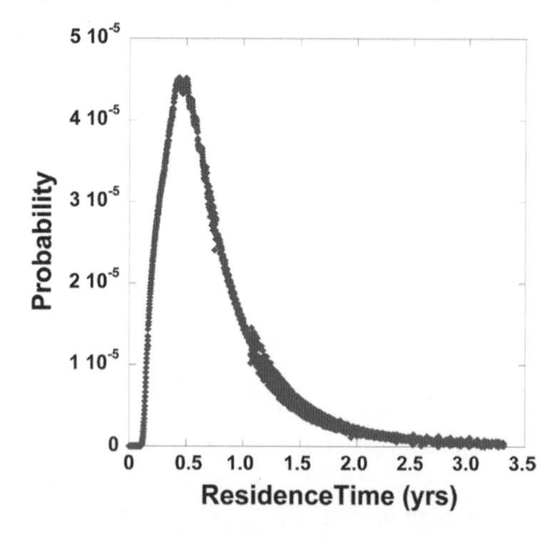

Figure 14. Residence time distribution for the two-dimensional flow case presented in Figure 9. Residence times are presented in terms of probabilities and are calculated recursively based on the breakthrough curve shown in Figure 7. From Steefel (2007).

Upscaling reaction rates in heterogeneous media. Mineral reaction rates are typically measured in well-stirred laboratory systems where the intent is to eliminate gradients in concentration within the solution. An individual crystal is expected to experience the same chemical environment as its neighbor. To the extent that a diffusion boundary layer can be eliminated via stirring, the rate of the mineral reaction is surface reaction-controlled—this is potentially a problem with a rapidly reacting mineral like calcite at low pH (Pokrovsky et al. 2005). In porous media where water-rock interaction takes place, however, physical, chemical, and microbiological heterogeneities can lead to the development of gradients in concentration, and thus to a spatially variable reaction rates.

Perhaps the clearest example of where rates are not the same at all scales is provided by the case where a reacting phase is distributed heterogeneously—obviously, in this case the reaction rate in the microenvironment next to the mineral is greater than what is determined over a larger volume average. If the system is well-mixed via either molecular diffusion or dispersion, however, then the rate should be scale-independent if the reactive surface area is calculated correctly over the larger volume. At fairly high flow rates in single pores, it may even be possible to develop diffusion boundary layers along the reacting mineral surfaces, which results in gradients in concentration and reaction rate at this scale (Li et al. 2008). However, the Li et al. study showed that given the typical flow rates encountered in the subsurface, the efficiency of molecular diffusion is high enough that complete mixing occurs at the scale of a single pore (Li et al. 2008). However, in the case where incomplete mixing at the larger pore network scale occurs, significant spatial variations in concentration, mineral saturation state, and reaction rates are likely to occur and thus result in rates that are scale dependent (Li et al. 2006).

A similar scaling effect may occur at larger scales characterized by slow transfer between low permeability zones (often referred to as "immobile") and higher permeability ("mobile") zones. In these cases, disequilibrium between these zones may prevail, creating gradients in concentration and thus a spatial variability in reaction rates (Harvey and Gorelick 2000).

From the preceding discussion, it should be clear that upscaling reaction rates is not as simple as normalizing the rates to the total reactive surface area in the water-rock system. Rates in natural heterogeneous porous media will always be scale-dependent where incomplete mixing occurs.

A REACTIVE TRANSPORT APPROACH TO MODELING FLUID-ROCK INTERACTION

The chapter began with an overview of the transport processes that are important in fluid-rock interaction and was followed with a very basic introduction to the essential equations that describe fluid-rock interaction concisely. We have relied so far primarily on analytical expressions, although some examples have been provided that make use of numerical models. The analytical solutions for reactive transport discussed in Section 5 are extremely useful for several reasons: 1) they have a pedagogical value in that fundamental system behavior can often be captured with relatively simple expressions or with non-dimensional parameters, 2) in most cases they are easier to solve than are the corresponding numerical representations, and 3) they are exact, and therefore offer a reliable comparison for other methods (chiefly numerical) where errors of various kinds can arise. There are some significant limitations, however, to relying strictly or even mostly to analytical solutions for the reactive transport problem. Perhaps most important is the fact that multicomponent, nonlinear reaction rate laws and reaction networks can rarely be handled with analytical approaches. Another potential danger in an over-reliance on analytical solutions is that they often lead to approximations that may or may not be justifiable. Examples abound, but to mention a few: 1) it is common to take the nonlinear reaction kinetics found in most experimental studies and treat these as linear for the purposes of including them in an analytical expression, significantly changing in some cases the system behavior, 2) the treatment of contaminant retardation is often carried out with the use of linear distribution coefficients, or K_ds values, despite the fact that nonlinear expressions for sorption (multicomponent ion exchange or surface complexation methods) typically provide far better descriptions of the processes involved, especially under chemically variable conditions (Steefel et al. 2005), and 3) assumptions of constant and/or uniform physical or chemical properties can lead to completely erroneous predictions of system behavior.

Modern numerical software for reactive transport simulations provides some powerful additional capabilities that make it the ideal tool to analyze and interpret fluid-rock interaction. The potential pitfalls of numerical reactive transport modeling are many and should not be discounted (see Steefel and MacQuarrie (1996) for a discussion of many of the errors that can arise when numerical methods are used). Other problems have to do with the extensive thermodynamic and kinetic databases that need to be used with this approach—each of these is subject to its own errors and these can propagate through to the dynamic system behavior in ways that may or may not be intuitively obvious. That said, the features to be found in these software packages are constantly evolving in response to the development of new conceptual models, but even now they can handle a variety of coupled biogeochemical and transport processes that are beyond the capabilities of an analytical model. Perhaps their key strength for the geochemist and mineralogist is their ability to handle multicomponent systems of arbitrary complexity. For the mulicomponent case, the governing advection-dispersion-reaction equation becomes

$$\frac{\partial(\phi C_i)}{\partial t} = \nabla \cdot (\phi D_i \nabla C_i) - \nabla \cdot (\phi \mathbf{u} C_i) - \sum_{r=1}^{Nr} \nu_{ir} R_r - \sum_{m=1}^{N_m} \nu_{im} R_m \tag{53}$$

where we now have a set of coupled partial differential equations for the chemical components, i, in the system. These solutes may be affected by both the rates of the N_r aqueous (homogeneous) reactions, R_r, and the rates of the N_m solid phase (mineral) reactions, R_m, each reacting according to the stoichiometry given in the coefficients ν_{ir} and ν_{ir}, respectively. The numerical software also has no problem handling both parallel and sequential kinetic pathways, which are common in microbially-mediated reaction networks.

Equally important is the ability to handle variable physical and chemical properties when a numerical approach is used. For some problems (e.g., the wormholing example shown in Fig. 15 below), this is essential for describing the dynamics of fluid-rock interaction.

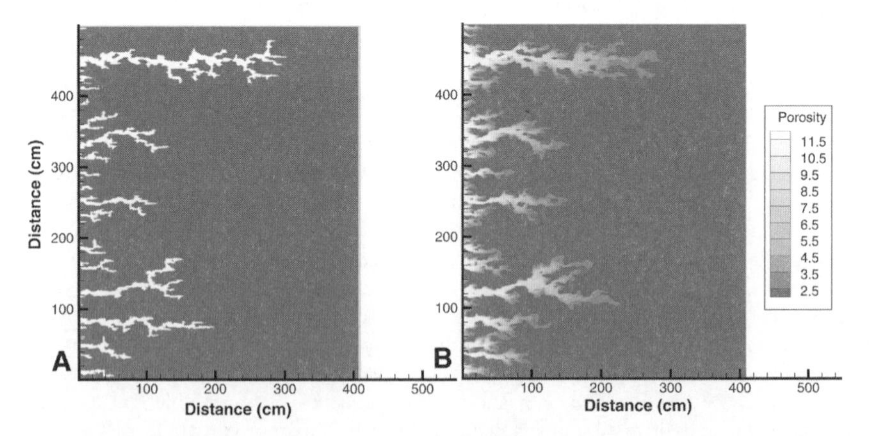

Figure 15. The reactive-infiltration instability results in "wormholing," as regions experiencing an increase in porosity and permeability as a result of mineral dissolution accelerate their growth at the expense of adjacent regions. The wormholing rate is at a maximum under transport-controlled conditions (Panel A)—at lower Damköhler numbers (Panel B), fingers are more diffuse and the difference between the growth rates of individual fingers is less. From Steefel (2007).

Approaches to modeling multi-scale porous media

Depending on the scale of interest, three different types of models are used to describe reactive transport in porous media today, 1) continuum models, 2) pore scale models, and 3) multiple continua or hybrid models involving a combination of scales (Steefel et al. 2005). Flow, transport, and reaction are conventionally described by macroscale models based on a continuum formulation of the underlying microscale equations (Lichtner 1985, 1996). The continuum approach is the most advanced in terms of its treatment of multicomponent chemical processes. The continuum equations are often derived intuitively by averaging over a representative elementary volume (REV) or control volume resulting in effective macroscale parameters. Pore scale models include pore network and lattice Boltzmann models, but few efforts in reactive transport have been attempted. Multiple continuum models have been in use for some time (Pruess and Narisimhan 1985)—hybrid models which combine pore scale and continuum scale behavior are only now beginning to be considered seriously (Lichtner and Kang 2007). These different approaches are considered in more detail in what follows.

Continuum models. Subsurface reactive flow and transport models are most commonly based on the continuum representation of porous media. This approach depends on averaging system properties over a macroscopic length (control volume or REV) containing many solid grains (Lichtner 1996; Steefel et al. 2005). Because of this averaging of properties, however, the continuum approach is not valid for length scales on the order of a single pore dimension. Since many of the physical, chemical, and biological processes actually take place at the pore scale, developing averaging approaches for these coupled processes at larger scales is essential.

Solid properties may be either fixed (frozen) or evolving. The latter is much more difficult to simulate and involves using various phenomenological approaches for describing changes in such macroscale parameters as porosity, permeability, capillarity, tortuosity, and reactive surface area. The formation of wormholes, which takes place in carbonate rocks due the infiltration of fluids undersaturated with respect to the carbonate minerals, is an extreme example of evolving heterogeneity that can still be modeled at the continuum scale (Fig. 15).

The wormholes form as a result of a reactive infiltration instability in which the enhancement of permeability resulting from mineral dissolution causes the wormholes to capture progressively increasing amounts of flow, thus accelerating their propagation (Ortoleva et al. 1987; Steefel and Lasaga 1990; Spiegelman et al. 2001; Spiegelman and Kelemen 2003; Steefel et al. 2005; Steefel 2007). Wormholes are observed in ideally homogeneous materials, such as plaster of Paris (gypsum), indicating that only slight variations in pore size are needed to form wormholes given the presence of high flow velocities and fast reaction kinetics.

Basic assumptions invoked in the continuum formulation are (Steefel et al. 2005) 1) solid, liquid, and gaseous phases all coexist at a single point (control volume) in space; 2) heterogeneous reactions involving two separate phases which interact across a common interface are treated as if they were homogeneous reactions uniformly distributed throughout a control volume. Interfacial surface area is represented as a uniform surface area concentration; 3) flow is described in terms of Darcy's law with velocity proportional to the pressure gradient. The proportionality coefficients are the permeability divided by the fluid viscosity; 4) the fluid, solid, and gas phases are well-mixed and therefore without concentration gradients, thus resulting in uniform reaction rates within the control volume.

The continuum description breaks down at the pore scale where the Navier-Stokes equations derived from fluid mechanics are required and where it may be necessary to capture microscopic-scale gradients in concentration resulting from transport and a non-uniform distribution of reactive material.

Pore-scale models. Pore scale models have been considered for some time in hydrology and petroleum engineering, but only very recently have such models included coupled reactive transport processes. Network models aim to capture pore scale behavior through a set of rules governing mass transport and chemical/biological reactions within and between individual pores. While they don't capture sub-pore scale gradients and processes that might be present, they offer the advantage of being able to treat processes like wormholing described above with only modest computational expense (Li et al. 2006). Li and co-workers used this approach to investigate various upscaling procedures for reactive transport processes controlled by a combination of transport and reaction kinetics at the pore scale. They demonstrated that continuum or volume averaging approaches often introduce significant errors, in some cases not even capturing the correct reaction direction. A more rigorous approach because of its ability to capture intra-pore gradients and geometries is offered by lattice Boltzmann methods (LBM) (Kang et al. 2006, 2007; Lichtner and Kang 2007), but these are limited with present day computational facilities to relatively small scale problems on the order of millimeters. In addition, there is some question how far up in scale the approach can be extended due to the computational difficulties associated with resolving enough pores to describe the equivalent of a representative elementary volume. The approach is potentially very powerful for developing formulations for use in hybrid models where pore scale processes such as biofilms, diffusion, and fracture or pore sealing can occur.

Multiple continuum or hybrid models. Fractured porous media, and more generally hierarchical media involving multiple length scales, play an important role in subsurface flow and transport processes. Fracture-dominated flow systems are involved in numerous subsurface geochemical processes, including contaminant migration, ore deposition, weathering, enhanced oil recovery, geothermal energy, degradation of concrete, and subsurface carbon sequestration. A description based on a single continuum formulation is unable to capture the unique features of a fractured system, and more generally, a hierarchical system involving multiple characteristic length scales, and can lead to serious error (Lichtner and Kang 2007).

If it were possible to model subsurface reactive flow problems at the micro scale over any desired macro length scale, there would be no need to consider other approaches. However, this is obviously not the case even with access to the world's fastest computers. Consequently, the problem must be simplified and approximations made in such a way that at least the first order effects of multi-scale processes are accommodated. One mathematical formulation of this problem is through a hierarchical set of continua, with each characteristic length scale represented by a different continuum that is coupled to the next continuum in the hierarchy (Lichtner and Kang 2007).

Approaches and constraints for determining rates

Rates of reaction can be determined in water-rock interaction systems affected by transport using several approaches. In general, however, intrinsic rates of reaction (e.g., moles/m^2/s) cannot be determined uniquely without independently determining transport rates. An exception is the case where one process or the other (surface reaction or transport) is clearly the slowest and therefore rate-limiting step in the overall process. Unfortunately, a surface reaction or transport control is often assumed to apply in advance of a rigorous analysis that demonstrates this control.

Using aqueous concentration profiles. From Equation (41), it is clear that rates of reaction can be determined from a concentration profile if the rates of transport are known. One can determine rates of water-rock interaction even where non-steady state conditions prevail, but the steady state case is much simpler to deal with. In the case of a steady state system, the rates of reaction in a pure diffusion system will be given by the curvature or second derivative of the concentration

$$R = \phi D^{*} \frac{\partial^2 C}{\partial x^2} \tag{54}$$

while in a pure advective system, the rate is given by the first derivative or slope of the concentration

$$R = -\phi v \frac{\partial C}{\partial x} \tag{55}$$

Equations (54) and (55) can be used with finite differencing of discrete data to determine local reaction rates, respectively

$$R_j = \left[\left(\phi D^*\right)_{j+\frac{1}{2}} \frac{C_{j+1} - C_j}{x_{j+1} - x_j} - \left(\phi D^*\right)_{j-\frac{1}{2}} \frac{C_j - C_{j-1}}{x_j - x_{j-1}} \right] \tag{56}$$

$$R_{j-\frac{1}{2}} = -\left(\phi v\right)_{j-\frac{1}{2}} \left[\frac{C_j - C_{j-1}}{x_j - x_{j-1}} \right] \tag{57}$$

where j refers to the discrete data points and x_j is the location of the data in space. In this form, the finite difference equations are written without assuming equally spaced data (implicitly assumed in Eqns. 54 and 55). Alternatively, the rates can be fit using the analytical or numerical solution to the governing reactive transport equation, taking into account the appropriate boundary and initial conditions as necessary. If a species is affected by more than one reaction, then all of the relevant reactions need to be factored into the analysis and this is generally best accomplished with modern numerical reactive transport software. An example in which rates were determined using pore water solute data from marine sediments and the numerical software CrunchFlow is shown in Figure 16 (Maher et al. 2006).

Using mineral profiles. Determining rates from mineral profiles is generally more difficult, both because the mineral profiles reflect integrated rates, but also because the movement of

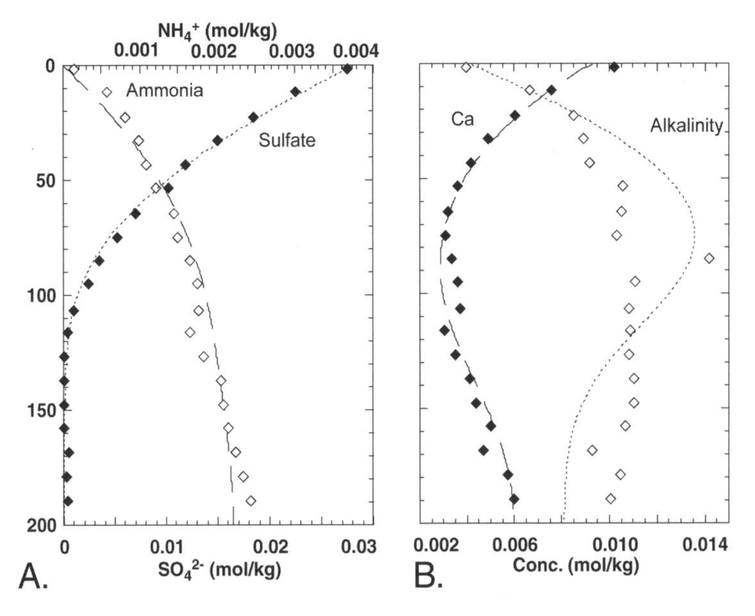

Figure 16. (A) Fit of the sulfate and ammonia profiles at Site 984 in the North Atlantic with the reactive transport code CrunchFlow allows the determination of the rate of microbially-mediated sulfate reduction. (B) Fit of the calcium and alkalinity profiles provides the rates of both plagioclase dissolution and calcite precipitation (Maher et al. 2006).

the reaction fronts can introduce additional difficulties of interpretation. Determining intrinsic rates (per unit surface area mineral) can be difficult because the spatial variation of reactive surface area across the reaction front needs to be factored in. An example involving the Santa Cruz Chronosequences is shown in Figure 17 and is discussed in more detail in the Case Studies section below (Maher et al. 2009).

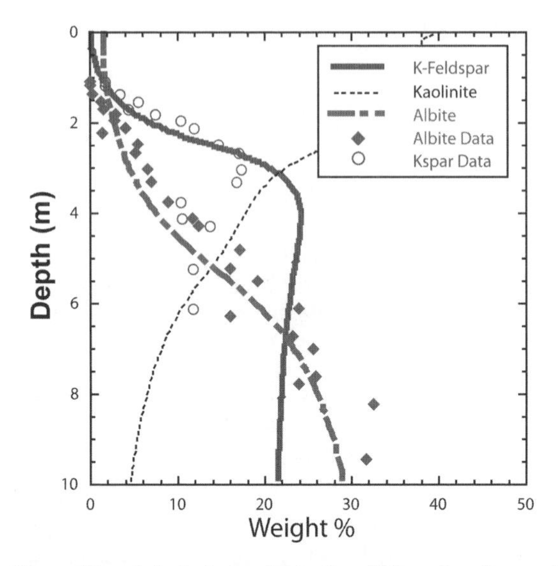

Figure 17. Fitting of the profiles of plagioclase and potassium feldspar from Terrace 5 (226,000 years) at the Santa Cruz Chronosequence (Maher et al. 2009) using the reactive transport code CrunchFlow.

CASE STUDIES

To make clear the power and flexibility of a reactive transport approach in fluid-rock interaction, it is perhaps most instructive at this point to consider some specific case studies. Several topics are considered, including 1) estimation of rates at the field scale, with examples coming from the field of biogeochemical cycling (Dale et al. 2008), chemical weathering (Maher et al. 2009), and contaminant remediation (Mayer et al. 2006). We conclude this section with a discussion of the ways in which isotopic studies can be integrated with multicomponent reactive transport modeling to improve our understanding and quantification of individual pathways within the overall fluid-rock interaction process.

Rates of anaerobic methane oxidation in marine sediments

The first example involves the use of reactive transport modeling to estimate the rate of anaerobic oxidation of methane (AOM) in marine sediments and is based on the study of Dale et al. (2008). The study was sufficiently detailed that, in addition to providing estimates of field-scale rates of reaction, it yielded important new insights in to the form of the rate law for AOM, with consideration of both thermodynamic and kinetic effects. Specifically, the dataset and the modeling allowed the authors to explore the bioenergetic limitations on AOM by comparing observed and simulated *in situ* catabolic energy yields.

The Skagerrak study area is a sequence of marine sediments located in the Norwegian Trench parallel to the southern coast of Norway, with a maximum depth of 700 m (Knab et al. 2008). The reactive transport modeling framework is validated with two gravity cores from the site that show a zone of anaerobic oxidation of methane 0.5 meter thick at 1 meter depth. Dale et al. (2008) developed a one-dimensional diffusion-reaction model for these marine sediments following the general approach outlined in Berner (1980) and Boudreau (1997), writing a set of coupled partial differential equations for the depth profile of solutes and solids. While tortuosity-corrected molecular diffusion is the dominant transport process considered, they also included a bioirrigation term to capture the non-local transport of solutes (especially sulfate) at the top of the profile.

The reaction network considered in the model includes a complex suite of organic carbon degradation pathways (Table 3 and Fig. 18). Labile and refractory organic carbon undergo extracellular hydrolysis to low molecular weight dissolved organic compounds (LMW-DOC), defined as glucose for the purposes of the mass balance and Gibbs energy calculations, described with reactions R_1 and R_2 respectively (Table 3). The LMW-DOC then undergoes fermentation to produce acetate and H_2 (R_3, ferm), which then provide the electrons and Gibbs energy for hydrogenotrophic (R_4, hySR) and acetotrophic (R_5, acSR) sulfate reduction and for hydrogenotrophic (R_6, hyME) and acetotrophic (R_7, acME) methanogenesis. Additionally, the H_2 and acetate drive acetogenesis (R_9, acet) and acetotrophy (R_{10}, actr) respectively. The remaining microbial reaction is AOM, which produces H_2 (R_8, AOM). These electron donor reactions represent catabolic pathways in which energy is generated. Anabolic pathways that synthesize biomass are presumably active as well (Rittmann and McCarty 2001; Dale et al. 2006), although it was shown in an earlier paper that the assumption of a steady-state biomass was justified for modeling microbial respiration in most marine sediments (Dale et al. 2006). This eliminates the need for an explicit calculation of the time and space evolution of the biomass for this case.

Table 3. Reaction network implemented in reactive transport modeling of Skagerrak sediments undergoing anaerobic oxidation of methane (Dale et al. 2008).

Rate	Type	Reaction Stoichiometry
R_1	hydr	$CH_2O_{LAB} \rightarrow \frac{1}{6}C_6H_{12}O_{6(aq)}$
R_2	hydr	$CH_2O_{REF} \rightarrow \frac{1}{6}C_6H_{12}O_{6(aq)}$
R_3	ferm	$\frac{1}{24}C_6H_{12}O_{6(aq)} + \frac{1}{6}H_2O \rightarrow \frac{1}{12}CH_3COO^- + \frac{1}{6}H_{2(aq)} + \frac{1}{6}H^+ + \frac{1}{12}HCO_3^-$
R_4	hySR	$\frac{1}{2}H_{2(aq)} + \frac{1}{8}SO_4^{2-} + \frac{1}{8}H^+ \rightarrow \frac{1}{8}HS^- + \frac{1}{2}H_2O$
R_5	acSR	$\frac{1}{8}CH_3COO^- + \frac{1}{8}SO_4^{2-} \rightarrow \frac{1}{8}HS^- + \frac{1}{4}HCO_3^-$
R_6	hyME	$\frac{1}{2}H_{2(aq)} + \frac{1}{8}HCO_3^- + \frac{1}{8}H^+ \rightarrow \frac{1}{8}CH_4 + \frac{3}{8}H_2O$
R_7	acME	$\frac{1}{8}CH_3COO^- + \frac{1}{8}H_2O \rightarrow \frac{1}{8}CH_4 + \frac{1}{8}HCO_3^-$
R_8	AOM	$\frac{1}{8}CH_4 + \frac{3}{8}H_2O \rightarrow \frac{1}{2}H_2 + \frac{1}{8}HCO_3^- + \frac{1}{8}H^+$
R_9	acet	$\frac{1}{2}H_{2(aq)} + \frac{1}{4}HCO_3^- + \frac{1}{8}H^+ \rightarrow \frac{1}{8}CH_3COO^- + \frac{1}{2}H_2O$
R_{10}	actr	$\frac{1}{8}CH_3COO^- + \frac{1}{2}H_2O \rightarrow \frac{1}{2}H_2 + \frac{1}{4}HCO_3^- + \frac{1}{8}H^+$

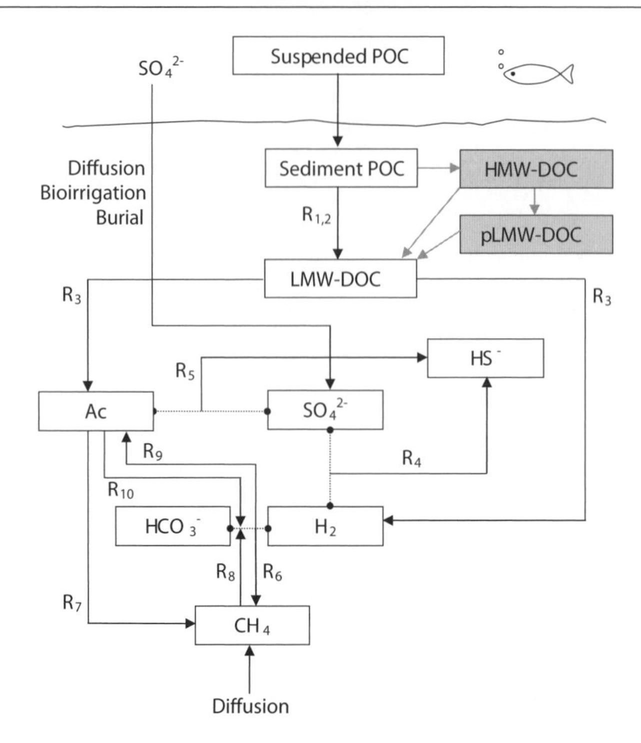

Figure 18. Schematic representation of the reaction network in which particulate organic carbon (POC) undergoes hydrolysis to LMW-DOC, which is then fermented to H_2 and acetate. From Dale et al. (2008).

The reactions in Table 3 represent the catabolism between electron donors and acceptors written in each case as the transfer of one electron. The rate of uptake of an electron donor by a particular catabolic pathway therefore follows the stoichiometries in Table 3 and is modulated by both a kinetic term, F_K, and a thermodynamic term, F_T, following the pioneering work of Jin and Bethke (2005). According to their approach as implemented in the Skagerrak study, the rate is given by

$$R_D = \upsilon_{\max} F_K F_T \tag{58}$$

where $\upsilon_{\max}$ is the maximum rate of electron donor (ED) utilization, with the biomass implicit in the rate constant. The kinetic factor is given by a standard dual Monod expression

$$F_K = \left(\frac{[E_D]}{K_{E_D} + [E_D]} \right) \left(\frac{[E_A]}{K_{E_A} + [E_A]} \right) \tag{59}$$

where [ED] and [EA] refer to the activities of the electron donor and acceptor respectively and K_{E_D} and K_{E_A} are the half-saturation constants for the electron donor and acceptor respectively. The thermodynamic driving force, which is assumed to drive the reaction in only one direction, is given by (Jin and Bethke 2005)

$$F_T = \max \left[0, \left(1 - \exp \left(\frac{\Delta G_{NET}}{\chi RT} \right) \right) \right] \tag{60}$$

where ΔG_{NET} is the fraction of the Gibbs energy of catabolism that provides a thermodynamic drive for the reaction (further defined below) and χ is the average stoichiometric coefficient, R is the gas constant, and T is the absolute temperature. The average stoichiometric coefficient, χ, is equivalent to the number of protons translocated across the cell membrane during catabolism and is assumed equal to 1 in anaerobic metabolism (Jin and Bethke 2005). As is apparent from Equation (59) and (60), both F_K and F_T are dimensionless and vary from 0 to 1. As pointed out, inclusion of such a thermodynamic term removes the need for kinetic inhibition terms as are often found in microbial rate models (Dale et al. 2008).

In the formulation of Dale et al. (2008) as adapted from Jin and Bethke (2005), the ΔG_{NET} term is the sum of two Gibbs energy terms of opposite sign

$$\Delta G_{NET} = \Delta G_{INSITU} + \Delta G_{BQ} \tag{61}$$

where ΔG_{INSITU} is the *in situ* Gibbs energy yield of the catabolic process (Table 3) and ΔG_{BQ} is the bioenergetic energy quantum (Hoehler 2004; Dale et al. 2006)

$$\Delta G_{INSITU} = \Delta G^{0'} + RT \ln[Q] \tag{62}$$

Here $\Delta G^{0'}$ is the standard Gibbs energy of catabolism at the *in situ* temperature and corrected for neutral pH (implying the same conditions for ΔG_{INSITU}). This was done to provide an ion activity product, Q, defined only in terms of electron donors and acceptors, but one could also define these more generally in terms of a standard Gibbs energy of reaction if the hydrogen ion was included in the expression. The minimum bioenergetic energy, ΔG_{BQ}, is the minimum that can be exploited by living cells to synthesize adenosine tri-phosphate (ATP) and is coupled to the transfer of 3-4 protons across the cellular membrane. Under actively growing conditions, this is usually considered to require ~60 kJ mol ATP^{-1} (Schink 1997; Jin and Bethke 2005). Hoehler (2004) argued instead that a lower limit of 9-12 kJ mol ATP^{-1} was more reasonable in energy-starved communities, leading to a suggested energy minimum on an electron equivalent basis of 1.125 to 1.5 kJ e-mol^{-1}.

The authors of the Skagerrak study proceeded to calibrate the model on the basis of data collected from the S13 station. The first step was to fit the sulfate data (Fig. 19a, g), which was carried out by including a non-local source of sulfate due to bioirrigation (Dale et al. 2008). The measured CH_4 profile (Fig. 19a) shows the characteristic concave-up shape of AOM. However, the expanded scale in Figure 19b also shows that the CH_4 exhibits tailing from depth into the sulfate reduction zone, where it persists at a concentration of 10-30 μM CH_4. To capture this behavior, some tuning of the kinetic and bioenergenetic parameters of the reaction network was required.

The modeling shows that at the depth of the maximum AOM rate, F_K and F_T for AOM are equal to 0.18 and 0.002, respectively (Fig. 20a,b). The maximum AOM rate, therefore, is strongly reduced by the low bioenergetic drive as a result of thermodynamic conditions very close to the bioenergetic energy minimum ΔG_{BQ}. In contrast, hydrogenotrophic sulfate reduction (hySR), which is considered to consume the H_2 and therefore provide the thermodynamic drive for spontaneous methane oxidation (Hoehler 2004), is limited by the kinetic drive, F_K, rather than by the thermodynamic drive, F_T (Fig. 20b, f). The modeling clearly demonstrates that the CH_4 tail in the sulfate reduction zone is very sensitive to the value of the bioenergetic minimum, ΔG_{BQ}, for AOM (Fig. 21). Based on the simulations, the field data suggest a value for $\Delta G_{BQ\text{-}AOM}$ of about 1.42 kJ e-mol^{-1} (compare simulated profiles using 2 kJ e-mol^{-1} and 0 kJ e-mol^{-1}).

The simulations demonstrate that the ΔG_{BQ} for AOM acts as a kind of thermodynamic switch which enables AOM only when H_2 levels are maintained at low levels by the sulfate reducing bacteria (Dale et al. 2006, 2008). Furthermore, it is clear that the pore water concentrations in a natural setting of this kind are functions of the kinetic and bioenergetic driving forces of potentially all of the processes affecting reactive intermediates, not just those of the reaction

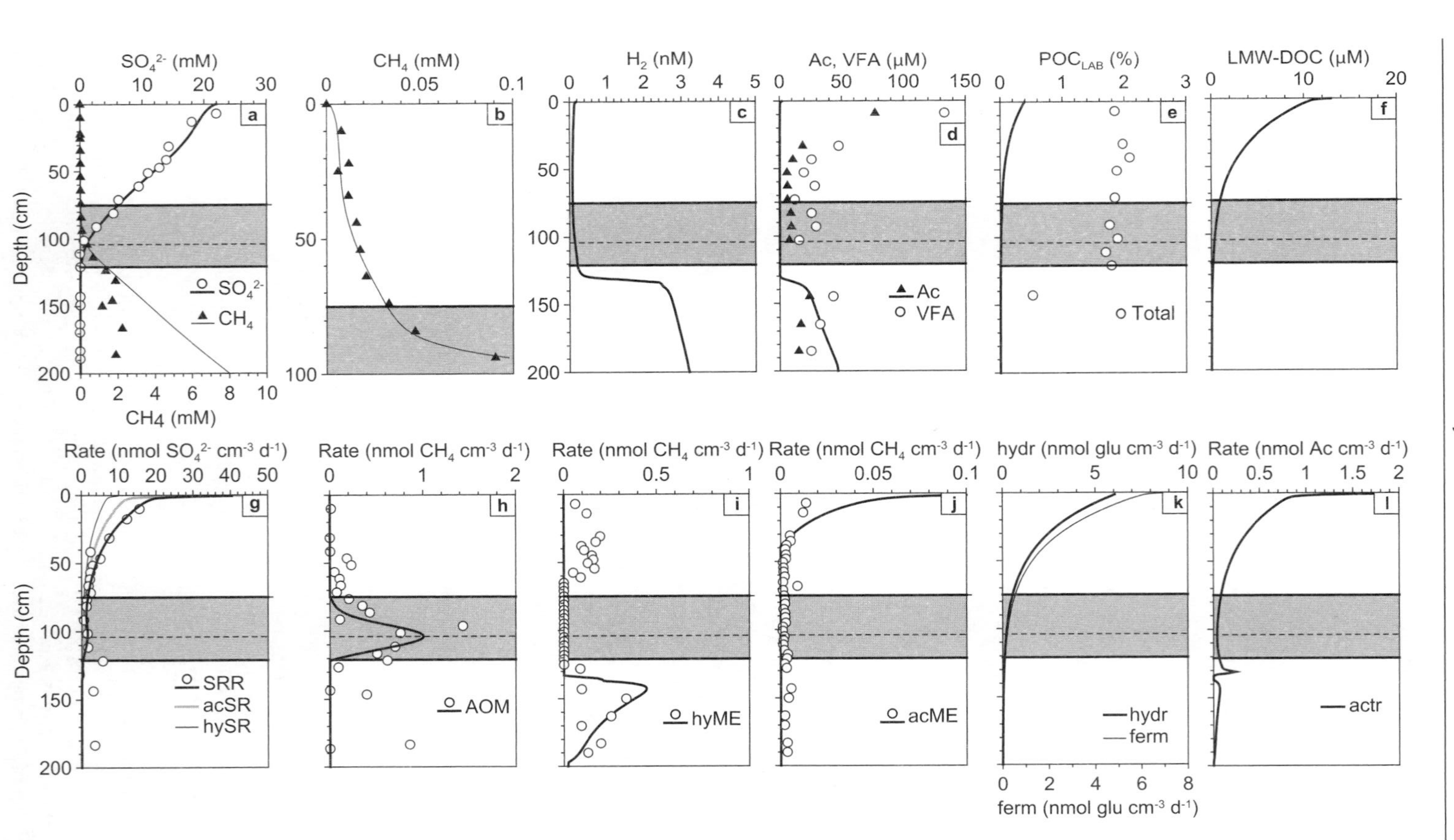

Figure 19. Measured (symbols) and modeled (lines) concentrations (top row) and rates (bottom row) for station S13 at Skagerrak. All rates are in production or consumption of solutes per unit volume of pore water. The shaded band indicates the sulfate-methane transition zone and the horizontal dashed line shows the depth of maximum AOM rate. See Table 3 for definition of individual reaction pathways. From Dale et al. (2008).

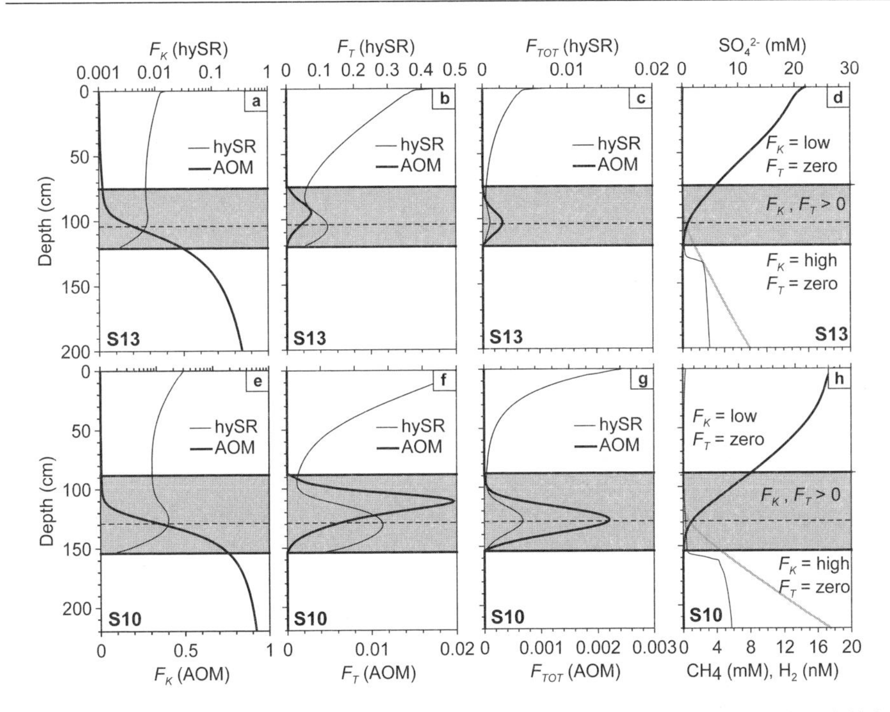

Figure 20. Model results for kinetic driving forces, F_K, and thermodynamic forces, F_T, and total driving forces, F_{TOT} for various reaction pathways for stations S13 and S10 at Skagerrak. From Dale et al. (2008).

pathway itself. Although kinetically possible, a particular reaction pathway may be limited bioenergetically by other reactions taking place within the microbial network (Dale et al. 2008).

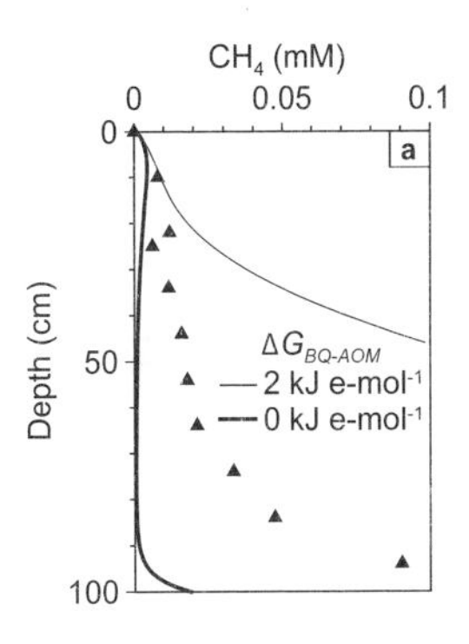

Figure 21. The effect of changes in $\Delta G_{BQ\text{-}AOM}$ on the CH_4 tail in the sulfate reduction zone. The measured data is shown as symbols. From Dale et al. (2008).

Rates of chemical weathering

The second example involves the use of reactive transport modeling to estimate the rates of chemical weathering at the Santa Cruz chronosequence (soils of different ages) and relies primarily on the study by Maher et al. (2009). The chronosequence is defined by a series of five marine terraces at different elevations within and northwest of the city of Santa Cruz, California. The local climate is Mediterranean, with cool wet winters and dry hot summers. Precipitation averages 727 mm annually. The ages of Terraces 1-5 are estimated to be 65 kyr, 90 kyr, 137 kyr, 199 kyr, and 226 kyr, respectively (White et al. 2009). The local bedrock is covered by 1-10 m of sediment, principally sands, derived from long shore marine transport of local fluvial sediment from the Santa Cruz Mountains. The spatial distribution of moisture varies seasonally in the upper terrace soils. However, beneath an argillic horizon at about 1 m depth moisture contents remain essentially constant throughout the course of the year. Pump tests suggested minimal lateral flow through the argillic horizon, so the deeper portions of the weathering profile behave essentially as a one-dimensional column characterized by steady unsaturated flow at 0.088 m/

yr as suggested by chloride mass balance (White et al. 2009). The sediments making up the soils are uncemented, so the estimate of a specific surface area of 1.94 m^2/g for the primary minerals based on grain size and a surface roughness factor is likely a maximum value (Maher et al. 2009). The volume fractions of the primary minerals K-feldspar and plagioclase along with the secondary mineral kaolinite are shown in Figure 22 for the youngest terrace (SCT 1) and the oldest terrace (SCT 5).

Present-day and long-term weathering rates were estimated using the code CrunchFlow. Unsaturated flow was assumed to be steady-state at 0.088 m/yr. Molecular diffusion was calculated with the diffusion coefficients given in Table 2. Electrochemical migration was also calculated using the approach outlined above and more thoroughly described in Giambalvo et al. (2002). K-feldspar and plagioclase dissolution rates were calculated with a variety of rate laws, but we will focus here on the two providing a good fit to the data while also minimizing the discrepancy between experimental and field weathering rates. The first rate law considered was proposed by Hellmann and Tisserand (2006) based on their extensive experimental studies of albite feldspar as a function of Gibbs energy. They proposed a composite rate law very similar in form to that suggested by Burch et al. (1993) that consists of two individual parallel rate laws: 1) a far from equilibrium rate law, R_1, with a strongly nonlinear dependence on the Gibbs energy, and 2) a close to equilibrium rate law, R_2, with a nearly linear dependence on the Gibbs energy

$$R_1(\text{mol/m}^2/\text{s}) = 1.02\times10^{-12}\left\{1-\exp\left[-0.0000798\left(\frac{|\Delta G|}{RT}\right)^{3.81}\right]\right\} \tag{63}$$

$$R_2(\text{mol/m}^2/\text{s}) = 1.81\times10^{-14}\left[1-\exp\left(-\frac{|\Delta G|}{RT}\right)\right]^{1.17} \tag{64}$$

A second rate law that is considered is referred to as the Al-inhibition model (AIM) and is based on the principle that aluminosilicate dissolution rates are controlled by the formation

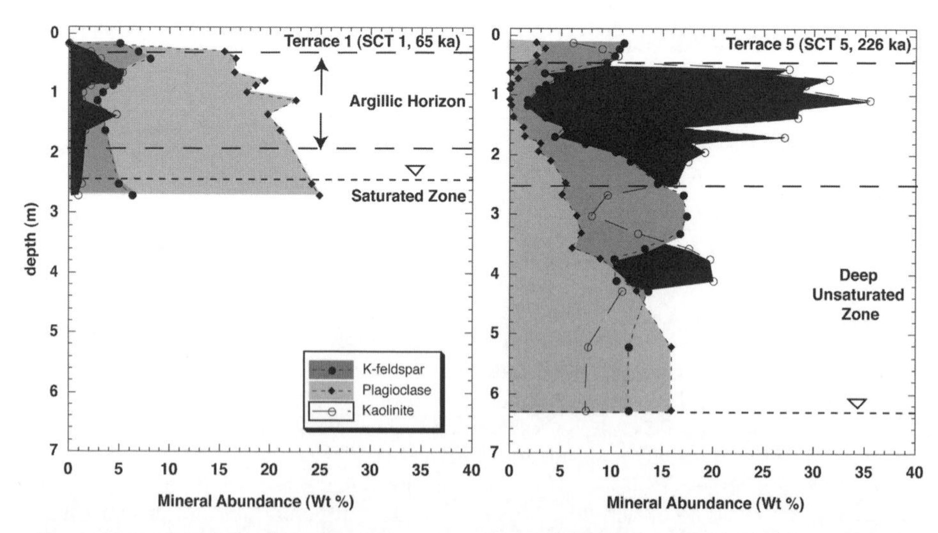

Figure 22. Profile evolution between SCT 1 (65 ka) and SCT 5 (226 ka). The zone of peak kaolinite accumulation referred to as the argillic horizon is shown as thick dashed lines, and the average water table depth is shown as a light stippled line. Data summarized from White et al. (2009).

and destruction of an Al-deficient and Si-rich precursor complex (Gautier et al. 1994; Oelkers et al. 1994; Carroll and Knauss 2005). The rate law considered has the form:

$$R = A \cdot k_1 \frac{K_f'}{K_f' + a_{Al(OH)_3}^{1/3}} \left[1 - \exp\left(-\frac{|\Delta G|}{3RT} \right) \right] \tag{65}$$

where K_f' is the equilibrium constant for the reaction describing the precursor complex and is calculated from the experimental data of Carroll and Knauss (2005) and described in Maher et al. (2006). The factor of 3 in the denominator of the exponential expression is the Temkin coefficient (Oelkers et al. 1994). For kaolinite precipitation, a linear TST model is used (Aagaard and Helgeson 1982).

Focusing on the mineral volume fractions in Terrace 5, Figure 23 shows that the reactive transport modeling integrated out to 226 kyr captures the geometry of the steep K-feldspar and relatively broad plagioclase fronts very accurately. It is noteworthy that these fits were obtained using virtually the same rate constant and rate law for K-feldspar and albite—the difference in the profiles, in fact, reflects the differences in the solubilities of the two minerals. The model also captures the kaolinite profile fairly well, which indicates that most of the clay developed in the argillic horizon is the result of the early coincidence of the plagioclase and K-feldspar dissolution fronts and does not involve significant quantities of translocated clays (Maher et al. 2009). It should be noted that experimentally-determined rate laws from either the parallel rate law of Hellmann and Tisserand (2006) or the aluminum inhibition model of (Oelkers et al. 1994) can used to match the mineral profiles with virtually no adjustment from experimentally determined values. The rate constants determined in this study are summarized in Table 4. Thus, the oft-cited discrepancy between laboratory and field rates (White and Brantley 2003) does not exist when the non-linear approach to equilibrium is accounted for, although the supersaturation of secondary phases is an important aspect and also needs to be considered, as discussed below.

To use either of the experimentally-determined rate laws described above and to have them succeed in matching the field weathering data, it is also necessary to allow for significant supersaturation with respect to kaolinite. The effect of the kaolinite supersaturation, either due to very slow kaolinite growth or to formation of a more soluble precursor phase (Steefel and Van Cappellen 1990), is to allow the pore water fluid to approach closer to equilibrium where the rates of dissolution are slower, particularly in the case where a nonlinear dependence

Table 4. Summary of rate constants for Santa Cruz, terrace 5 (SCT 5).

		Albite[a] log rate constant ($mol/m^2/sec$)	**K-feldspar**[b] log rate constant ($mol/m^2/sec$)	**Kaolinite**[c] log rate constant ($mol/m^2/sec$)
Maher et al. (2009)		*25 °C*	*25 °C*	*25 °C*
TST Rate Law		−14.3± 0.3	−14.7± 0.2	−19.6 ± 3
Al-Inhibition Rate Law		−12.1± 0.3	−12.4± 0.2	−19.4 ± 3
Close-to-Equilibrium Rate Law	k_1	−12.5± 0.3	−13.6± 0.2	−19.7 ± 3
	k_2	−14.3± 0.3	−14.8± 0.2	
White et al. (2009)				
Weathering Gradient Model[d]		−15.3	−15.3	n.d.

a: surface area of 1.94 m^2/g (White et al. 2009), activation energy of 69.8 kJ/mol (Hellmann and Tisserand 2006).
b: surface area of 1.94 m^2/g (White et al. 2009), activation energy of 38 kJ/mol (Palandri and Kharaka 2004).
c: surface area of 10 m^2/g (White et al. 2009), activation energy of 22.2 kJ/mol (Palandri and Kharaka 2004).
d: at a flow rate of 0.058 m/yr determined from fit to solid data using spreadsheet model.
n.d.: not determined

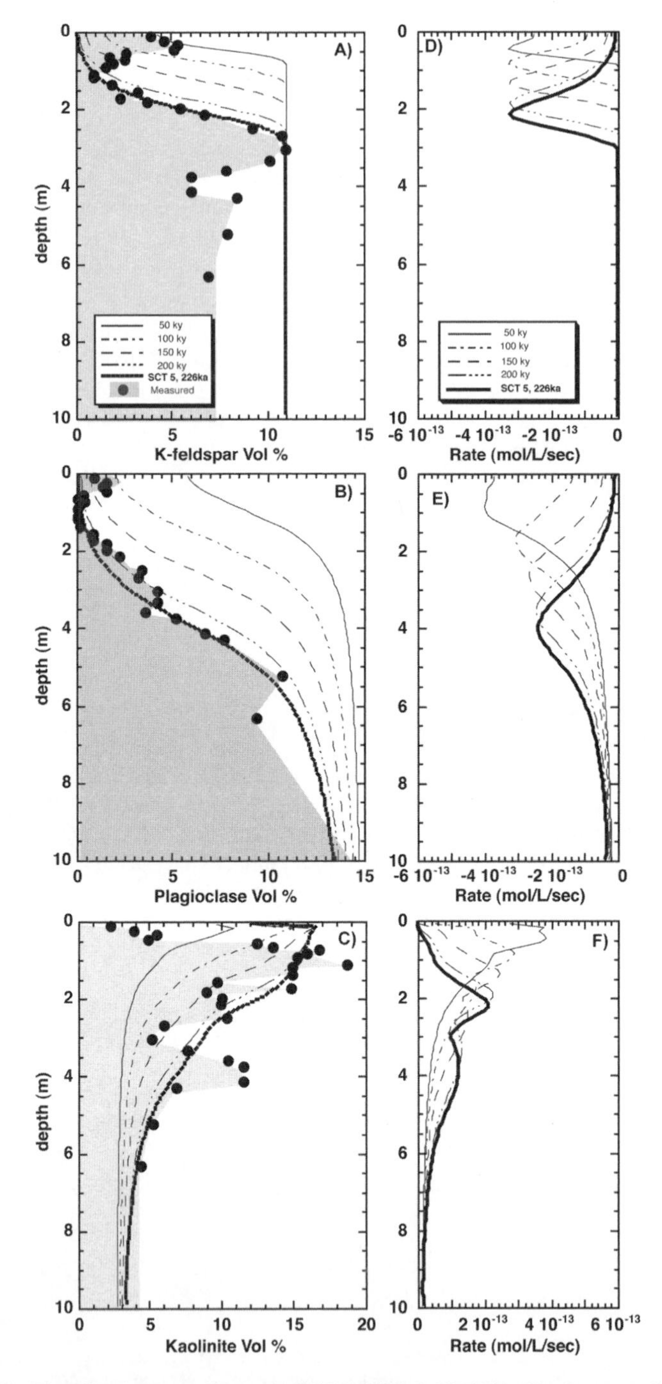

Figure 23. Model profiles corresponding to the SCT 5 (226 ka). Model Profiles corresponding to the ages of the younger terraces are shown for reference and the SCT 5 profile is shown in bold. (A)-(C) Profile evolution for K-feldspar and albite dissolution and kaolinite precipitation. (D)-(E) overall reaction rates (mol/L(porous media)/sec) as a function of depth for individual minerals. From Maher et al. (2006).

on Gibbs energy prevails. The role of slow kaolinite precipitation was hypothesized in an earlier study of mineral dissolution in marine sediments (Maher et al. 2006), but pore water data to constrain the saturation state of kaolinite was lacking. In the study of the Santa Cruz Chronosequences, however, pore water data provide an independent, present-day constraint on the kaolinite saturation state. The data indicate that if the kaolinite equilibrium constant determined by (Yang and Steefel 2007) at 22 °C and pH 4 is used, the pore waters are 3-4 orders of magnitude supersaturated with respect to this phase. Pore waters that were closer to equilibrium with respect to kaolinite would have the effect of making the system more undersaturated with respect to the feldspars, since clay precipitation modulates their saturation state by uptake of aluminum and silica (Steefel and Van Cappellen 1990; Maher et al. 2009). As is apparent from Figure 24, the same model used to closely reproduce mineral profiles in Figure 23 (a result that involves integrating over 226 kyr) succeeds in capturing the present-day pore water chemistry and the resulting mineral saturation states.

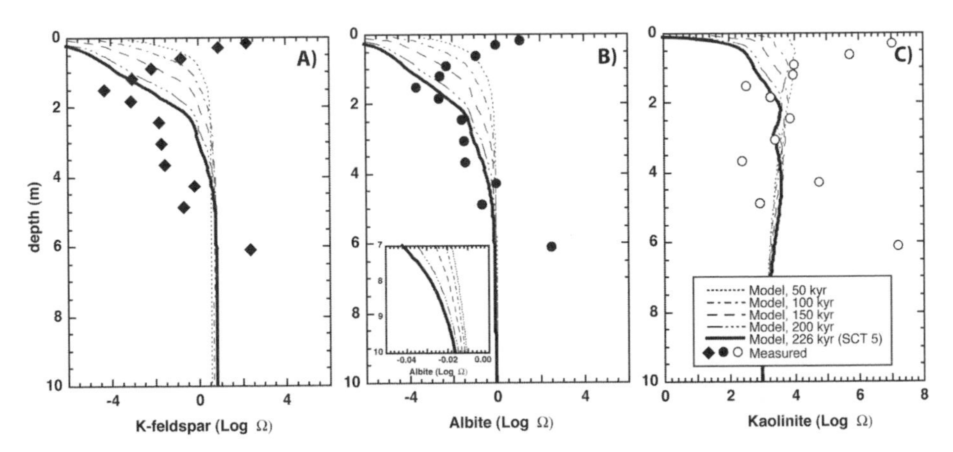

Figure 24. Mineral saturation state profiles from model simulations compared to values calculated from measured pore water concentrations. From Maher et al. (2009).

Relative mineral solubilities from reaction front separation

Reactive transport modeling of the separation of mineral fronts can be used to estimate their relative solubilities with a fair degree of accuracy. This was pointed out as a possibility indirectly by Lichtner (1993), but appears not to have been pursued in the context of a real field system until the study of chemical weathering at the Santa Cruz Chronosequences by Maher et al. (2009). In the study at Santa Cruz, the reactive transport modeling showed that the use of equilibrium constants for the specific compositions of feldspar reported, namely a plagioclase with 30% anorthite (An_{30}) and pure K-feldspar (Arnorsson and Stefansson 1999), could not match the observed separation of the K-feldspar and plagioclase weathering fronts. A significantly better match (see Fig. 23) was achieved by using end-member compositions of pure K-feldspar and sodic plagioclase (An_0). As it turns out, the front positions are quite sensitive to the equilibrium constants used, since in the local equilibrium limit the front separation should be a function only of the thermodynamics (Lichtner 1993). Maher et al. (2009) reached the same conclusion using numerical modeling with the reactive transport software CrunchFlow. These results, which are considered to be robust, suggested a model in which the An_{30} plagioclase was better described as an intergrowth of fine scale lamellae of albite and anorthite such that the anorthitic component dissolved first, leaving the albite as the surviving reactant controlling the front positions once weathering had proceeded to a sufficient extent (Maher et al. 2009).

Rates of biogeochemical reactions in heterogeneous media

A third example involves simulations of rates in the subsurface to predict the effectiveness and longevity of a remediation treatment for acid mine drainage. This is a topic of interest to researchers and engineers interested in contaminant transport and remediation, in addition to geochemists, and we now see these groups exploring the topic of fluid-rock interaction with much greater depth and focus than was the case previously. The study area is a permeable reactive barrier for treatment of mine drainage high in iron and sulfate and low in pH at the Nickel Rim mine site in Ontario, Canada (Mayer et al. 2006). The permeable barrier consists of an organic carbon mixture designed to promote microbial reduction of sulfate so as to precipitate sparingly soluble iron sulfide (FeS). The aquifer is bounded by bedrock and is well constrained hydrologically, with a known flux across the domain. The water chemistry monitoring network, the most important of the data used to constrain the rates of reaction and thus the performance of the barrier, consists of 12 well nests aligned parallel to the direction of groundwater flow (Fig. 25). Solid phase data (Herbert et al. 2000) was also used to constrain the reaction network.

Significant spatial variations are observed in the amount of sulfate and iron removed within the barrier (Fig. 26), with the differences attributed primarily to variations in groundwater flow velocities. The rate of sulfate reduction in the barrier is observed to decline over time as the organic carbon is depleted (Benner et al. 2002).

The modeling of the biogeochemical reaction network at the site is carried out with a simpler approach than was used in Dale et al. (2008). The principal reaction involves the oxidation of organic carbon by sulfate to produce dissolved sulfide according to

$$CH_2O_{(s)} + \frac{1}{2}SO_4^{2-} \rightarrow HCO_3^- + \frac{1}{2}H_2S_{(aq)} + H_2O \tag{66}$$

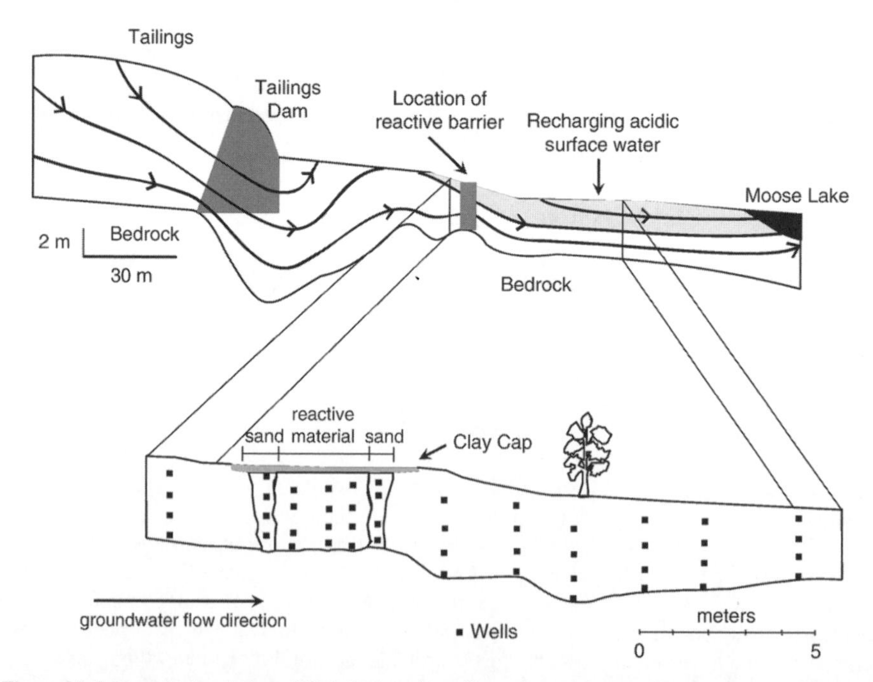

Figure 25. Schematic description of Nickel Rim mine tailings site and the investigation area for permeable reactive barrier (Mayer et al. 2006).

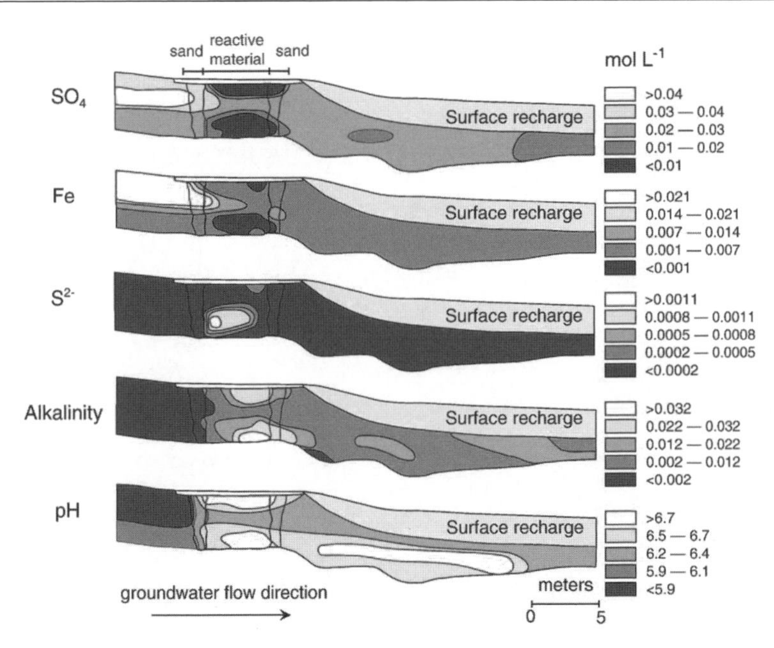

Figure 26. Concentration contours of dissolved sulfate, iron, sulfide, alkalinity and pH in the Nickel Rim permeable reactive barrier (Mayer et al. 2006).

This reaction is modeled using a standard dual Monod kinetic scheme, that is, essentially the formulation given in Equation (58) without the thermodynamic limitation term, F_T. The sulfide produced in this reaction combines with dissolved Fe(II) to precipitate mackinawite (FeS). Elemental sulfur is also detected and is modeled as another parallel sulfate reduction reaction that goes to an intermediate redox state.

Combining the groundwater flow field and reaction network developed for the site, Mayer et al. (2006) were able to match the observed solute concentration distribution remarkably well (compare Figs. 26 and 27).

In matching the concentration contours at 23 months, Mayer et al. (2006) demonstrated that such data, when interpreted with appropriately mechanistic reactive transport modeling, could provide a reasonable estimate of spatially and temporally variable reaction rates in the subsurface (Fig. 28). The analysis thus allows them to further develop estimates for the efficacy and longevity of the permeable reactive barrier. Such analyses, to the extent that they can be fully validated, suggest the possibility that reactive transport modeling can provide significant cost savings over expensive field tests, although there is probably no scenario in which the approach dispenses with field testing altogether.

Isotopic reactive transport approaches for deciphering physical and chemical processes

Early reactive transport modeling studies involving isotopes tend to use lumped parameter models describing a generic isotopic exchange reaction (Richter and Liang 1993; Johnson and DePaolo 1997; Maher et al. 2003). Other approaches use dimensional analysis to simplify systems in order to quantify key parameters (Fantle and DePaolo 2007). Both approaches generally consider the isotopic variations in isolation from the overall chemical evolution of the system, thus limiting the amount of information that can be extracted from the isotopic data. Most importantly, isotopic variations can provide an additional tracer of reaction pathways within ambiguous reaction networks. An important new development is the rigorous incorporation of

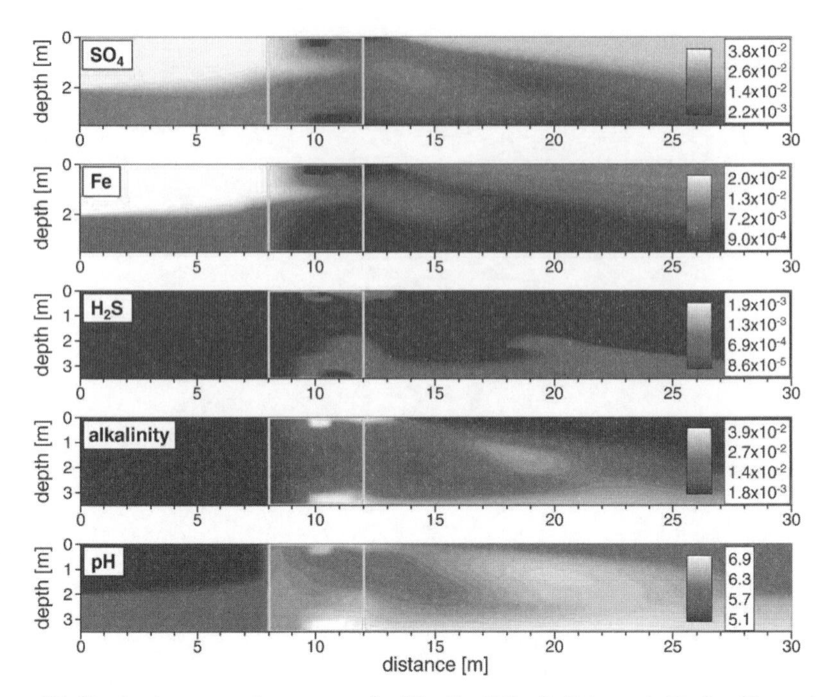

Figure 27. Simulated concentration contours for SO_4, Fe, H_2S, alkalinity, and pH after 23 months of operation of the PRB (Mayer et al. 2006).

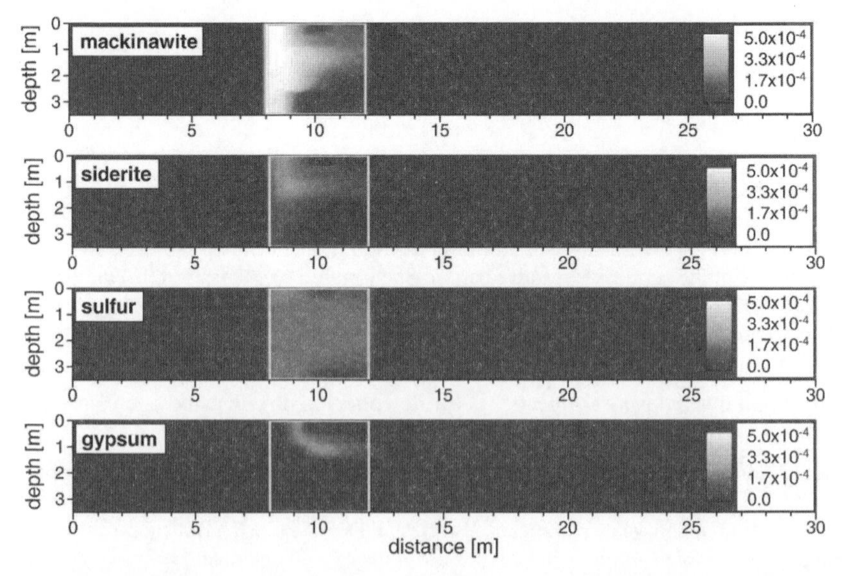

Figure 28. Simulated contours of mineral volume fractions for mackinawite, siderite, elemental sulfur, and gypsum after 23 months of operation of the PRB (Mayer et al. 2006).

isotopes into multicomponent models such that the isotopic variations are considered in conjunction with the major element chemistry and thus provide an additional framework for identifying and quantifying physical and chemical processes. As we view the development of isotopic reactive transport models as an important new frontier in fluid-rock interaction, we provide a few examples to illustrate the variety of questions that can be approached with the coupling.

With respect to physical processes such as transport and transport pathways, naturally occurring variations in the isotopic composition of waters can be used to determine preferential flow paths (Luo et al. 2000) or to determine net recharge rates throughout an aquifer (Maher et al. 2003; Singleton et al. 2006). Singleton et al. (2006) used Sr isotopic variations in an unconfined basaltic aquifer to calculate the long-term net infiltration flux along a flow path. The approach relied on the isotopic variations imparted to the waters from the lithology, with high $^{87}Sr/^{86}Sr$ values (0.712) occurring in the unsaturated zone due to presence of granitic sediments (Maher et al. 2003), and relatively low values (0.707) associated with underlying basalts. By accounting for the basalt-fluid interaction along the flow path, regions of high infiltration rates stemming from human activities could be quantitatively distinguished from natural recharge conditions. This example highlights the ability to use isotopically distinct sources to quantify additional flux terms in the reactive transport equation (e.g., infiltration fluxes), thus obtaining additional information that could not be extracted from a simple mixing model.

Isotopes can also be used to decipher individual reaction rates in systems where a key component (e.g., Ca, Na, Si, Al) is derived from multiple reactions, or cycled within a reaction network (Maher et al. 2006). For example, in deep-sea sediments silicate dissolution results in the release of Ca to the pore water. Some of this Ca may then be incorporated into newly formed calcite along with the initial seawater Ca. Concurrently, changes in the porewater chemistry and/or Ostwald ripening may result in calcite recrystallization, whereby the porewater Ca is exchanged with both the initial marine calcite and the newly formed calcite. To constrain the rates of mineral transformation in this system, Maher et al. (2006) incorporated the ($^{234}U/^{238}U$) isotopic system into a reactive transport model (CrunchFlow) in order to use the U isotopic variations in pore fluid, silicate minerals and bulk calcite as an additional mass transfer constraint. In this model, the processes of preferential release of ^{234}U from solids to the fluid due to α-recoil (during the decay of ^{238}U), dissolution of silicate minerals, and the co-precipitation of U into calcite were considered. Although the primary mineral dissolution rates could also be constrained by the aqueous Ca and Si profiles, the measured isotopic data in the fluid and the calcite could only be reconciled with the model a substantial fraction (ca. 30%) of the primary calcite was recrystallized (Fig. 29). This isochemical process would otherwise be undetectable and the sensitivity of isotopes to mineral-fluid exchange provided a unique means of detecting the substantial recrystallization that took place. Mineral-fluid exchange during early diagenesis has important implications for the integrity of a material such as calcite that is commonly used for paleoceanographic studies and studies of this type can be used to understand more clearly how the isotope systems may be modified.

The previous examples highlight isotopic reactive transport approaches involving relatively heavy isotope pairs where mass dependent fractionation is of minor importance relative to other signatures such as those imposed by variations in lithology and radioactive decay. However, for many isotopic systems of interest in biogeochemical cycles, isotopic variations derive from both kinetic and equilibrium effects. An advantage of the integration of isotopes into reactive transport models is that diffusion coefficients specific to each isotope can be implemented to account for kinetic effects resulting from aqueous or gaseous diffusion. For kinetic isotope effects that occur as a result of biological or enzymatic processes, the fractionation factor can be defined as the ratio of the rate constants for each isotope. Similarly, equilibrium constants can be defined for each isotope system to reflect the equilibrium fractionation factors. This ability to track the systematics of individual isotope pairs in conjunction with major element chemistry offers unique opportunities that are just beginning to be explored.

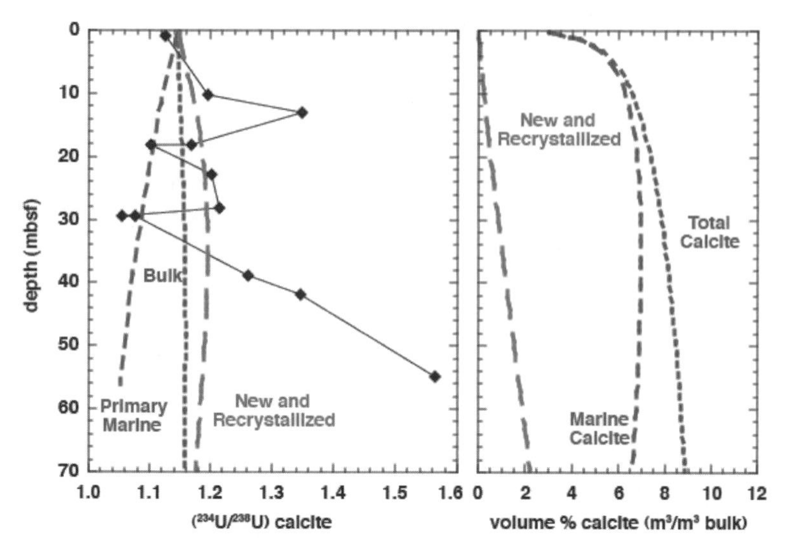

Figure 29. Summarized from Maher et al. (2006). (A) Measured bulk (symbols) and model ($^{234}U/^{238}U$) of the calcite fractions. The ($^{234}U/^{238}U$) of the new and recrystallized calcite is greater than the bulk calcite or primary marine due to the α-recoil enrichment of the pore fluid. The recrystallization process not only enriches the calcite ($^{234}U/^{238}U$) but also lowers the pore fluid ($^{234}U/^{238}U$) at depth (not shown) due to exchange of U. The primary marine calcite reflects the ($^{234}U/^{238}U$) for closed system decay of U in calcite. B) The resulting distribution of different calcite pools showing the increase in new and recrystallized calcite components.

A recent study by Dale et al. (2009) provides an interesting example where the coupling of the major element chemistry is used to interpret sulfur isotopic data from organic rich muds of the Namibian Shelf. At this site, the supply of hypoxic water coupled with high primary productivity results in anoxic conditions at the sediment-water interface. The anoxic conditions favor the accumulation of particulate organic carbon, resulting in extremely high rates of sulfate reduction (100 mM/year of SO_4^{2-}). The high rates of sulfate reduction coupled with the notable lack of oxidants results in H_2S concentrations of approximately 10mM. In this setting, *Thiomargarita namibiensis*, a non-phototrophic large sulfur bacteria produces SO_4^{2-} and NH_4^+ by catalyzing the reaction between H_2S and NO_3^- (Schulz et al. 1999; Bruchert et al. 2003). The SO_4^{2-} then becomes available to oxidize organic carbon or methane. In order to understand the coupling between the C, N, and S cycles, and the role of *Thiomargarita* in the oxidation-reduction cycle of S, a multicomponent reactive transport model was developed that includes an explicit representation of the individual S isotopes (^{34}S, ^{32}S) in addition to the major chemical components (Dale et al. 2009). The measurements used to constrain the 1-D model include pore water concentrations, solid phase analyses of C, N, S, reactive Fe and elemental S, and rates of sulfate reduction determined using ^{35}S-sulfate whole-core incubations. The porosity and sedimentation rate over the 5 meter deep core were also constrained to account for the effect of compaction during burial.

The reaction network and major geochemical pathways in relationship to expected isotopic fractionations (ε) are shown in Figure 30. The particulate organic carbon (POC) is characterized as three reactive fractions according to a multi-G model (Westrich and Berner 1984) and is degraded via sulfate reduction, methanogenesis and dissimilatory Fe reduction. CH_4 diffusing upwards was allowed to be consumed by anaerobic oxidation of methane (AOM). The NO_3^- is assumed to be confined within the *Thiomargarita* cell and thus denitrification is not considered. The maximum fractionation for dissimilatory sulfate reduction (ε_{SR}) derived

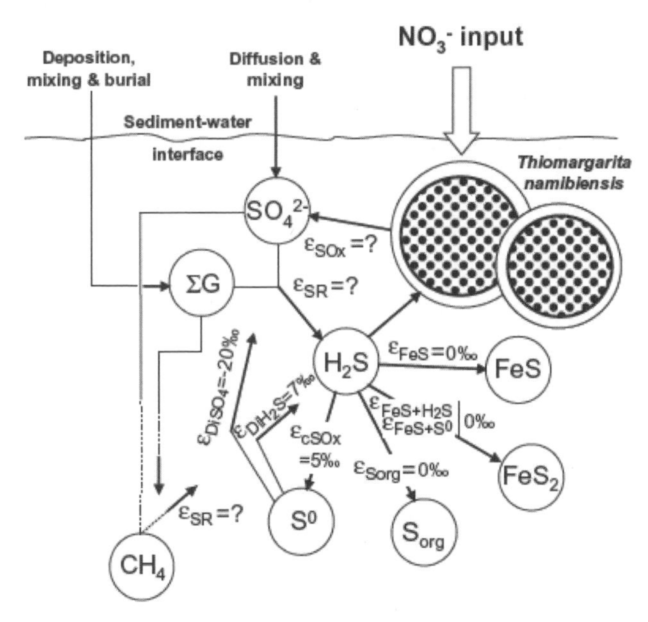

Figure 30. Conceptual model of the S isotope fractionations at the Namibia site from Dale et al. (2009). The ε_{SOx} and ε_{SR} fractionation factors that were constrained with the model are indicated by "?". The parameters used in the model accompanied with a detailed description of the reaction network can be found in Table 3 of Dale et al (2009).

from both reactive transport model approaches for marine sediments (Goldhaber and Kaplan 1980; Rudnicki et al. 2001; Wortman et al. 2001) and theoretical approaches suggest ε_{SR} values of 70 to 80‰ (Farquhar et al. 2003), whereas the maximum value measured under laboratory conditions is on the order of 47‰ (Bruchert 2004). Because sulfide oxidizing bacteria have not been grown in pure cultures, the fractionation of sulfide during sulfide oxidation (ε_{SOx}) is not well known, but is expected to be ~ 5 ‰. By using the constraints from detailed modeling of the C, N, S cycles to balance the S isotope system, the values for ε_{SR} and the value of ε_{SOx} due to the activity of *Thiomargarita* can also be constrained.

To formulate the isotopic system, rate constants are prescribed for each isotope for each reaction pathway so that the overall rate (R_i) is simply the sum of the two reaction rates:

$$R_i = {}^{32}R_i + {}^{34}R_i = {}^{32}k_i\,{}^{32}S + {}^{34}k_i\,{}^{34}S \tag{67}$$

where ${}^{32}k_i$ and ${}^{34}k_i$ are the corresponding rate constants. The individual reaction rates for the isotopic species can then be formulated as:

$${}^{32}R_i = R_i\left(\frac{\alpha_i\,{}^{32}S}{{}^{34}S + \alpha_i\,{}^{32}S}\right) \tag{68}$$

using the ratio of the two rate constants to define the fractionation factor α_i:

$$\alpha_i = \frac{{}^{32}k}{{}^{34}k} = 1 + \frac{\varepsilon_i}{1000} \tag{69}$$

Because the fractionation factor associated with sulfate reduction is observed to increase with a decrease in the rates of sulfate reduction, ε_{SR} was equated with the rate of sulfate reduction using a linear function:

$$\varepsilon_{SR} = (\varepsilon_{SR})_{max} - m_{SR} r_i \quad (70)$$

where m_{SR} describes the gradient of change of ε_{SR} as a function of rate (r_i) of sulfate reduction or anaerobic methane oxidation (AOM). Figure 31 shows the results of using the major element chemistry alone to constrain the values for ε_{SR} and ε_{SOx}. The $\delta^{34}S_{SO_4}$ increases with depth in the zone of sulfate reduction. The $\delta^{34}S_{H_2S}$ rapidly decreases to a minimum of −24% by 10 cm depth, followed by a progressive increase with depth. The values of ε_{Sox} and $\varepsilon_{SR\text{-}max}$ of 25‰ and 100‰ are the model values required to fit the isotope profiles in Figure 31c,d, and a corresponding sensitivity analysis revealed the uniqueness of these parameters. However, these model-derived fractionations are far greater than previous laboratory based estimates (as summarized in Bruchert (2004)) suggesting that the model for the SOx pathway may be missing an additional process. Dale et al. (2009) suggest that this process may be a stepped disproportionation of S in the cell from H_2S to S° to SO_4^{2-}. The disproportionation step (e.g., S° to SO_4^{2-}.) would occur with a large and opposite fractionation between H_2S and SO_4^{2-}. Including this additional step in the model resulted in model values for ε_{Sox} and $\varepsilon_{SR\text{-}max}$ of 5‰ and 78‰ that are consistent with the theoretical and laboratory values. While multi-stepped disproportionation by *Thiomargarita* has not been determined experimentally, this example highlights the utility of the reactive transport approach in using natural isotopic data to infer reaction pathways.

An additional important aspect of the data and model presented by Dale et al. (2009) is shown in Figure 32: the measured $\delta^{34}S$ values for the solid phases show greater depletion in FeS_2 suggesting fractionation during precipitation, although the fractionation factor that would be predicted is zero. In addition, both the H_2S and values of the authigenic phases continue to change with depth beneath the zone of sulfate reduction, suggestive of ongoing exchange between the aqueous H_2S pool and the reactive portions of the solids. To consider this possibility, Dale et al. (2009) considered a separate exchange process where discrete fractions of the solids (f_{ex}) are allowed to undergo isotopic exchange with the fluid. The model was found to reproduce all of the observed data if f_{ex} was on the order of 0.6 to 0.8 which means that 60 to 80% of the solid phase exchanges with the fluid. This result is similar to the isotopic exchange observed in the U isotopic data discussed above and seems to suggest that even during early diagenesis modification of the isotopic and possibly trace element features are likely. In addition, this ongoing exchange during diagenesis may be overlooked in settings where complete profiles are not available.

As the field of isotope geochemistry considers more complex biogeochemical and geochemical systems, the ability to constrain the isotope systematics using major element chemistry offers a number of advantages for determining isotopic fractionations associated with specific processes. Conversely, as in the case of the Maher et al. (2006) and the Dale et al. (2009) studies, inclusion of the isotope variations can reveal intricacies of the reaction network and mechanisms that were not detectable from the major element chemistry alone. The presence of multiple reaction pathways can make the interpretation of measured data difficult, but the incorporation of isotopic data into overall framework improves the chances of deciphering the system behavior.

CONCLUDING REMARKS

A good understanding of the coupling between transport and (bio)geochemical kinetics is essential in the study of fluid-rock interaction. In this chapter, we have shown that in open systems characterized by transport, the characteristic time scales of reaction need to be compared with the characteristic time scales for transport—geologic time is often irrelevant. The transport processes that are relevant for the study of water-rock interaction include advection, due in most cases to Darcy flow through porous media, molecular diffusion modified by electrochemical migration where charged species are present, and hydrodynamic dispersion. An idealized

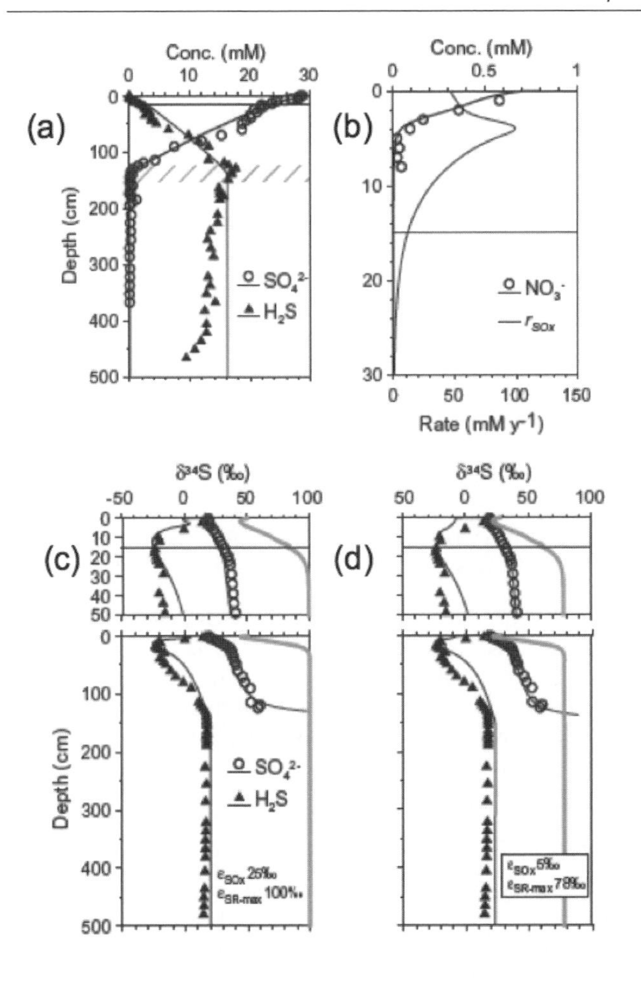

Figure 31. Summarized from Dale et al. (2009). (a) measured (symbols) and modeled lines (lines) for SO_4^{2-} and $H_2S(aq)$ profiles, (b) measured and model pore water NO_3^- profiles and the rate of microbial sulfide oxidation (r_{SOx}) for the top 30 cm; (c) baseline model simulations using rates determined from elemental chemistry and values ε_{Sox} and $\varepsilon_{SR\text{-}max}$ of 25 and 100‰ determined by best fit to the data; (d) case with sulfide oxidation occurring with an additional disproportionation step and ε_{Sox} and $\varepsilon_{SR\text{-}max}$ of 5 and 78‰. Thick grey line shows the rate-dependent fractionation for sulfate reduction.

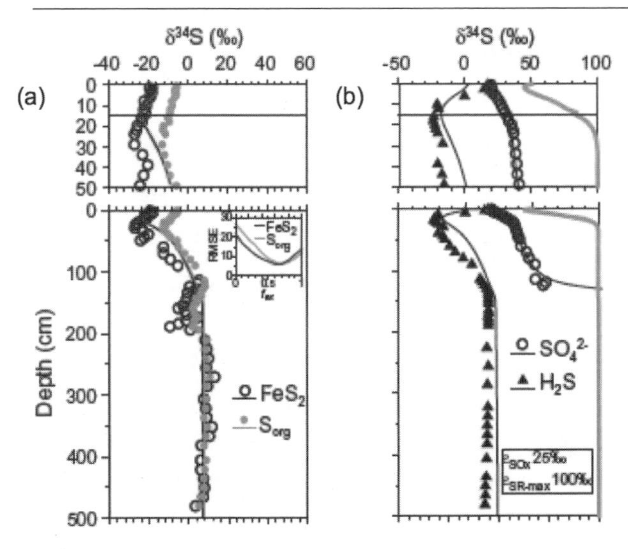

Figure 32. Summarized from Dale et al. (2009). (a) Model and measured $\delta^{34}S$ profiles for solid phases showing the model for the downhole decrease in $\delta^{34}S$ due to exchange between solids and the aqueous H_2S; (b) measured (symbols) and modeled lines (lines) for SO_4^{2-} and $H_2S(aq)$ profiles. Thick grey line shows the rate-dependent fractionation for sulfate reduction.

version of the reactive transport (or advection-dispersion-reaction) equation was presented and the important non-dimensional quantities were briefly derived and discussed.

The treatment of fluid-rock interaction systems as well-mixed reactors was challenged and it was shown how an analysis of macrodispersion, which has been widely discussed in the hydrologic literature, can be used to produce a residence time probability density function that predicts the ensemble of reaction times for flow paths through heterogeneous media. Real water-rock systems are typically characterized by physical, chemical, and microbiological heterogeneities that can give rise to concentration gradients, and thus to spatially variable reaction rates. It was shown that it is necessary to upscale reaction rates in fluid-rock systems that are not perfectly mixed—simple averaging that does not take into account the multi-scale character of most porous media will inevitably result in rates that are scale dependent.

The remainder of the chapter examined the potential of modern numerical reactive transport modeling for solving important problems in fluid-rock interaction. Three different types of models were proposed to describe reactive transport in porous media, with the choice depending to some extent on the scale of interest and the architecture of the porous medium: 1) continuum models, 2) pore scale models, and 3) multiple continua or hybrid models involving a combination of scales. It was shown that various data, ranging from pore water chemistry to mineral volume fractions, may be used to constrain reactive transport models for fluid-rock interaction. Case studies are presented that demonstrate that reactive transport modeling can be used to estimate rates in 1) marine sediments characterized by anaerobic oxidation of methane, 2) chemical weathering, and 3) a permeable reactive barrier used to remediate groundwater contamination. Finally, the power of integrating isotopic systematics into mechanistic reactive transport modeling was briefly explored and the case was made that incorporation of isotopic data into overall framework of multicomponent reactive transport models improves our chances of deciphering complex biogeochemical systems.

ACKNOWLEDGMENTS

We are pleased to be able to contribute this chapter to a volume honoring the career of Jacques Schott, who has championed a mechanistic and quantitative understanding of geochemistry now for over 30 years. Funding from the Office of Science at the U.S. Department of Energy through the Geoscience Program in the Office of Basic Energy Sciences (Contract No. DE-AC02-05CH11231) is gratefully acknowledged.

REFERENCES

Aagaard P, Helgeson HC (1982) Thermodynamic and kinetic constraints on reaction rates among minerals and aqueous solutions, I. Theoretical considerations. Am J Sci 282:237-285

Aris R (1956) On the dispersion of a solute in a fluid flowing through a tube. Proc R Soc London Ser A 235:67-77

Arnorsson S, Stefansson A (1999) Assessment of feldspar solublity constants in water in the range 0 degrees to 350 degrees C at vapor saturation pressures. Am J Sci 299:173-209

Bahr JM, Rubin J (1987) Direct comparison of kinetic and local equilibrium formulations for solute transport affected by surface reactions. Water Resour Res 23:438-452

Bear J (1972) Dynamics of Fluids in Porous Media. Dover, New York

Benner SG, Blowes DW, Ptacek CJ, Mayer KU (2002) Rates of sulfate reduction and metal sulfide precipitation in a permeable reactive barrier. Appl Geochem 17:301-320

Berner RA (1980) Early Diagenesis: A Theoretical Approach. Princeton University Press, Princeton

Bethke CM, Johnson TM (2002) Paradox of groundwater age. Geology 30:107-110

Boudreau BP (1997) Diagenetic Models and Their Implementation. Springer-Verlag,

Bruchert V (2004) Physiological and ecological aspects of sulfur isotope fractionation during bacterial sulfate reduction. *In*: Sulfur Biogeochemistry--Past and Present. AP Amend, KJ Edwards, TW Lyons (eds) Geological Society of America, p 1-16

Bruchert V, Joergensen BB, Neumann K, Riechmann D, Schloesser M, Schulz HN (2003) Regulation of bacterial sulfate reduction and hydrogen sulfide fluxes in the central Namibian coastal upwelling zone. Geochim Cosmochim Acta 67:4505-4518

Burch TE, Nagy KL, Lasaga AC (1993) Free-energy dependence of albite dissolution kinetics at 80 degrees C and pH 8.8. Chem Geol 105:137-162

Carroll SA, Knauss KG (2005) Dependence of labradorite dissolution kinetics on CO_2(aq), Al(aq), and temperature. Chem Geol 217:213-225

Cirpka OA (2002) Choice of dispersion coefficients in reactive transport calculations on smoothed fields. J Contam Hydrol 48:261-282

Cochepin B, Trotignon L, Bildstein O, Steefel CI, Lagneau V, Van der Lee J (2008) Intercomparison of predictions on a 2-D cementation experiment in porous medium. Adv Water Resour 31:1540-1551

Dagan G (1988) Time-dependent macrodispersion for solute transport in anisotropic heterogeneous aquifers. Water Resour Res 24:1491-1500

Dagan G (1989) Flow and Transport in Porous Formations. Springer-Verlag, Berlin; New York

Dale AW, Bruchert V, Alperin M, Regnier P (2009) An integrated sulfur isotope model for Namibian shelf sediments. Geochim Cosmochim Acta 73:1923-1944

Dale AW, Regnier P, Knab NJ, Joergensen BB, Van Cappellen P (2008) Anaerobic oxidation of methane (AOM) in marine sediments from Skagerrak (Denmark): II. Reaction-transport modeling. Geochim Cosmochim Acta 72:2880-2894

Dale AW, Regnier P, Van Cappellen P (2006) Bioenergetic controls on anaerobic oxidation of methane (AOM) in coastal marine sediments: A theoretical analysis. Am J Sci 306:246-294

Darcy H (1856) Les fontaines publiques de la ville de Dijon. Dalmont, Paris

Daugherty RL, Franzini JB (1965) Fluid Mechanics with Engineering Applications. McGraw-Hill, New York

Denbigh K (1981) The Principles of Chemical Equilibrium. Cambridge University Press, Cambridge

Dobson PF, Kneafsey TJ, Sonnenthal EL, Spycher NF, Apps JA (2003) Experimental and numerical simulation of dissolution and precipitation: implications for fracture sealing at Yucca Mountain. J Contam Hydrol 62-63:459-476

Fantle MS, DePaolo DJ (2007) Ca isotopes in carbonate sediment and pore fluid from ODP Site 807A: The Ca^{2+}(aq)-calcite equilibrium fractionation factor and calcite recrystallization rates in Pleistocene sediments. Geochim Cosmochim Acta 71:2524-2546

Gautier JM, Oelkers EH, Schott J (1994) Experimental study of K-feldspar dissolution rates as a function of chemical affinity at 150 °C and pH 9. Geochim Cosmochim Acta 58:4549-4560

Gelhar LW (1986) Stochastic subsurface hydrology from theory to applications. Water Resour Res 22:135S-145S

Gelhar LW (1993) Stochastic Subsurface Hydrology. Prentice-Hall, Englewood Cliffs, N.J.

Gelhar LW, Axness CL (1983) Three-dimensional stochastic analysis of macrodispersion in aquifers. Water Resour Res 19:161-180

Gelhar LW, Welty C, Rehfeldt KR (1992) A critical-review of data on field-scale dispersion in aquifers. Water Resour Res 28:1955-1974

Giambalvo ER, Steefel CI, Fisher AT, Rosenberg ND, Wheat CG (2002) Effect of fluid-sediment reaction on hydrothermal fluxes of major elements, eastern flank of the Juan de Fuca Ridge. Geochim Cosmochim Acta 66:1739-1757

Goldhaber MB, Kaplan IR (1980) Mechanisms of sulfur incorporation and isotope fractionation during early diagenesis in sediments of the Gulf of California. Mar Chem 9:95-143

Haggerty R, Schroth MH, Istok JD (1998) Simplified method of "Push-Pull" test data analysis for determining in situ reaction rate coefficients. Ground Water 36:314-324

Harvey C, Gorelick SM (2000) Rate-limited mass transfer or macrodispersion: Which dominates plume evolution at the Macrodispersion Experiment (MADE) site? Water Resour Res 36:637-650

Helgeson HC (1968) Evaluation of irreversible reactions in geochemical processes involving minerals and aqueous solutions—I. Thermodynamic relations. Geochim Cosmochim Acta 32:853-877

Helgeson HC, Garrels RM, MacKenzie FT (1969) Evaluation of irreversible reactions in geochemical processes involving minerals and aqueous solutions—II. Applications. Geochim Cosmochim Acta 33:455-482

Hellmann R, Tisserand D (2006) Dissolution kinetics as a function of Gibbs free energy of reaction: an experimental study based on albite feldspar. Geochim Cosmochim Acta 70:364-383

Herbert Jr. RB, Benner SG, Blowes DW (2000) Solid phase iron-sulfur geochemistry of a reactive barrier for treatment of mine drainage. Appl Geochem 15:1331-1343

Hoehler TM (2004) Biological energy requirements as quantitative boundary conditions for life in the subsurface. Geobiology 2:205-215

Jin QA, Bethke CM (2005) Predicting the rate of microbial respiration in geochemical environments. Geochim Cosmochim Acta 69:1133-1143

Johnson TM, DePaolo DJ (1997) Rapid exchange effects on isotope ratios in groundwater systems, 1. Development of a transport-dissolution-exchange model. Water Resour Res 33:187-195

Kang Q, Lichtner PC, Zhang D (2007) An improved lattice Boltzmann model for multicomponent reactive transport in porous media at the pore scale. Water Resour Res 43, W12S14, doi:10.1029/2006WR005551
Knab NJ, Cragg BA, Borowski C, Parkes JR, Pancost R (2008) Anaerobic oxidation of methane (AOM) in marine sediments from Skagerrak (Denmark): I. Geochemical and micrbiological analyses. Geochim Cosmochim Acta 72:2868-2879
Lasaga AC (1981) Rate laws in chemical reactions. Rim Mineral 8:135-169
Lasaga AC (1984) Chemical kinetics of water-rock interactions. J Geophys Res 89:4009-4025
Lasaga AC (1998) Kinetic Theory in the Earth Sciences. Princeton University Press, Princeton
Li L, Peters CA, Celia MA (2006) Upscaling geochemical reaction rates using pore-scale network modeling. Adv Water Resour 29:1351-1370
Li L, Peters CA, Celia MA (2007a) Applicability of averaged concentrations in determining geochemical reaction rates in heterogeneous porous media Am J Sci 307:1146-1166
Li L, Peters CA, Celia MA (2007b) Effects of mineral spatial distribution on reaction rates in porous media. Water Resourc Res 43, doi: 10.1029/2005WR004848
Li L, Steefel CI, Yang L (2008) Scale dependence of mineral dissolution rates within single pores and fractures. Geochim Cosmochim Acta 72:99-116
Lichtner PC (1985) Continuum model for simultaneous chemical reactions and mass transport in hydrothermal systems. Geochim Cosmochim Acta 49:779-800
Lichtner PC (1988) The quasi-stationary state approximation to coupled mass transport and fluid-rock interaction in a porous medium. Geochim Cosmochim Acta 52:143-165
Lichtner PC (1993) Scaling properties of time-space kinetic mass transport equations and the local equilibrium limit. Am J Sci 293:257-296
Lichtner PC (1996) Continuum formulation of multicomponent-multiphase reactive transport. Rev Mineral 34:1-81
Lichtner PC (1998) Modeling reactive flow and transport in natural systems. Proceedings of the Rome Seminar on Environmental Geochemistry, 5-72
Lichtner PC, Kang Q (2007) Upscaling pore-scale reactive transport equations using a multiscale continuum formulation. Water Resour Res 43, doi:10.1029/2006WR005664
Luo SD, Ku TL, Roback R, Murrell MT, McLing TL (2000) In-situ radionuclide transport and preferential groundwater flows at INEEL (Idaho): Decay-series disequilibrium studies. Geochim Cosmochim Acta 64:867-881
Maher K, DePaolo DJ, Conrad ME, Serne RJ (2003) Vadose zone infiltration rate at Hanford, Washington, inferred from Sr isotope measurements. Water Resour Res 39:1029-1043
Maher K, Steefel CI, DePaolo DJ, Viani BE (2006) The mineral dissolution rate conundrum: Insights from reactive transport modeling of U isotopes and pore fluid chemistry in marine sediments. Geochimica et Cosmochimica Acta 70:337-363
Maher K, Steefel CI, White AF, Stonestrom DA (2009) The role of reaction affinity and secondary minerals in regulating chemical weathering rates at the Santa Cruz Soil Chronosequence, California. Geochim Cosmochim Acta 73: 2804-2831
Malmström ME, Destouni G, Banwart SA, Strömberg BHE (2000) Resolving the scale-dependence of mineral weathering rates. Environ Sci Technol 34:1375-1378
Mayer KU, Benner SG, Blowes DW (2006) Process-based reactive transport modeling of a permeable reactive barrier for the treatment of mine drainage. J Contam Hydrol 85:195-211
Meile C, Tuncay K (2006) Scale dependence of reaction rates in porous media. Adv Water Resour 29:62-71
Murphy WM, Oelkers EH, Lichtner PC (1989) Surface reaction versus diffusion control of mineral dissolution and growth rates in geochemical processes. Chem Geol 78:357-380
Navarre-Sitchler A, Steefel CI, Yang L, Tomutsa L, Brantley SL (2009) Evolution of porosity and diffusivity associated with chemical weathering of a basalt clast. J Geophys Res 114, doi:10.1029/2008JF001060
Newman JS (1991) Electrochemical Systems. Prentice-Hall, Englewood Cliffs, New Jersey
Oelkers EA (1996) Physical and chemical properties of rocks and fluids for chemical mass transport calculations. Rev Mineral 34:131-191
Oelkers EH, Schott J, Devidal J-L (1994) The effect of aluminum, pH, and chemical affinity on the rates of aluminosilicate dissolution reactions. Geochim Cosmochim Acta 58:2011-2024
Onsager L (1931) Reciprocal relations in irreversible processes II. Phys Rev 38:2265-2279
Ortoleva P, Chadam J, Merino E, Sen A (1987) Geochemical self-organization II: The reactive-infiltration instability. Am J Sci 287:1008-1040
Palandri JL, Kharaka YK (2004) A compilation of rate parameters of water-mineral interaction kinetics for application to geochemical modeling. U.S. Geological Survey Water-Resources Investigations Report 04-1068
Pharmamenko EI (1967) Electrical Properties of Rock. Plenum Press

Plummer LN, Busenberg E, Bohlke JK, Nelms DL, Michel RL, Schlosser P (2001) Groundwater residence times in Shenandoah National Park, Blue Ridge Mountains, Virgina, USA: a multi-tracer approach. Chem Geol 179:93-111

Pokrovsky OS, Golubev SV, Schott J (2005) Dissolution kinetics of calcite, dolomite and magnesite at 25 degrees C and 0 to 50 atm pCO(2). Chem Geol 217:239-255

Prigogine I (1967) Introduction to the Thermodynamics of Irreversible Processes. Interscience, New York

Pruess K, Narisimhan TN (1985) A practical method for modeling fluid and heat flow in fractured porous media. Soc Pet Eng J 25:14-26

Richter FM, Liang Y (1993) The rate and consequences of Sr diagenesis in deep-sea carbonates. Earth Planet Sci Lett 117:553-565

Rudnicki MD, Elderfield H, Spiro B (2001) Fractionation of sulfur isotopes during bacterial sulfate reduction in deep ocean sediments at elevated temperatures. Geochim Cosmochim Acta 65:777-789

Schink B (1997) Energetics of syntrophic cooperation in methanogenic degradation. Microbiol Molec Biol Rev 61:262-280

Schnoor JL (1990) Kinetics of chemical weathering: a comparison of laboratory and field weathering rates. *In* Aquatic Chemical Kinetics: Reaction Rates of Processes in Natural Waters. W Stumm (ed), John Wiley and Sons, p 475-504

Schulz HN, Brinkhoff T, Ferdelman TG, Hernandez Marine M, Teske A, Joergensen BB (1999) Dense populations of a giant sulfur bacterium in Namibian shelf sediments. Science 284:493-495

Singleton MJ, Maher K, DePaolo DJ, Conrad ME, Dresel PE (2006) Dissolution rates and vadose zone drainage from strontium isotope measurements of groundwater in the Pasco Basin, WA unconfined aquifer. J Hydrol 321:39-58

Skagius K, Neretnieks I (1986) Diffusivity measurements and electrical-resistivity measurements in rock samples under mechanical stress. Water Resour Res 22:570-580.

Snodgrass MF, Kitanidis PK (1998) A method to infer in-situ reaction rates from push-pull experiments. Ground Water 36:645-650

Spiegelman M, Kelemen PB, Aharonov E (2001) Causes and consequences of flow organization during melt transport: The reaction infiltration instability in compactible media. J Geophys Res 106:2061-2077

Steefel CI (2007) Geochemical kinetics and transport. *In* Kinetics of Water-Rock Interaction. Brantley SL, Kubicki J, White AF (ed) Springer, New York, p 545-589

Steefel CI, DePaolo DJ, Lichtner PC (2005) Reactive transport modeling: An essential tool and a new research approach for the Earth Sciences. Earth Planet Sci Lett 240:539-558

Steefel CI, Lasaga AC (1990) The evolution of dissolution patterns: Permeability change due to coupled flow and reaction. *In* Chemical Modeling of Aqueous Systems II. Vol 416. Melchior D, Bassett RL (ed) American Chemical Society, Washington, p 212-225.

Steefel CI, Lasaga AC (1994) A coupled model for transport of multiple chemical species and kinetic precipitation/dissolution reactions with application to reactive flow in single phase hydrothermal systems. Am J Sci 294:529-592

Steefel CI, Lichtner PC (1994) Diffusion and reaction in rock matrix bordering a hyperalkaline fluid-filled fracture. Geochim Cosmochim Acta 58:3595-3612

Steefel CI, Lichtner PC (1998a) Multicomponent reactive transport in discrete fractures: II. Infiltration of hyperalkaline groundwater at Maqarin, Jordan, a natural analogue site. J Hydrol 209:200-224

Steefel CI, Lichtner PC (1998b) Multicomponent reactive transport in discrete fractures: I. Controls on reaction front geometry. J Hydrol 209:186-199

Steefel CI, MacQuarrie KTB (1996) Approaches to modeling of reactive transport in porous media. Rev Mineral 34:83-130

Steefel CI, Van Cappellen P (1990) A new kinetic approach to modeling water-rock interaction: the role of nucleation, precursors, and Ostwald ripening. Geochim Cosmochim Acta 54:2657-2677

Taylor GI (1953) The dispersion of soluble matter in a solvent flowing through a tube. Proc R Soc London Ser A 219:196-203

Thompson JB (1959) Local equilibrium in metasomatic processes. Res Geochem 1:427-457

Tompson AFB, Carle SF, Rosenberg ND, Maxwell RM (1999) Analysis of groundwater migration from artificial recharge in a large urban aquifer: A simulation perspective. Water Resour Res 35:2981-2998

Varni M, Carrera J (1998) Simulation of groundwater age distributions. Water Resour Res 34 :3271-3281

Weare JH, Stephens JR, Eugster HP (1976) Diffusion metasomatism and mineral reaction zones; general principles and application to feldspar alteration. Am J Sci 276:767-816

Weissmann GS, Zhang Y, LaBolle EM, Fogg GE (2002) Dispersion of groundwater age in an alluvial aquifer system. Water Resour Res 38:1198

Westrich JT, Berner RA (1984) The role of sedimentary organic matter in bacterial sulfate reduction--the G model tested. Limnol Oceanograph 29:236-249

White AF, Brantley SL (2003) The effect of time on the weathering of silicate minerals: why do weathering rates differ in the laboratory and field? Chem Geol 202:479-506

White AF, Schulz MS, Stonestrom DA, Vivit DV, Fitzpatrick V, Bullen TD, Maher K, Blum AE (2009) Chemical weathering of a marine terrace chronosequence, Santa Cruz, California: Controls on solute fluxes and comparisons of long-term and contemporary mineral weathering rates. Geochim Cosmochim Acta 73:2769-2803

Wortman UG, Bernasconi SM, Boettcher ME (2001) Hypersulfidic deep biosphere indicates extreme sulfur isotope fractionation during single-step microbial sulfate reduction. Geology 29:647-650

Yang L, Steefel CI (2007) Kaolinite dissolution and precipitation kinetics at 22 °C and pH 4. Geochim Cosmochim Acta 72:99-116

Yeh GT, Tripathi VS (1989) A critical evaluation of recent developments in hydrogeochemical transport models of reactive multichemical components. Water Resour Res 25:93-108

Zhu C (2005) In situ feldspar dissolution rates in an aquifer. Geochim Cosmochim Acta 69:1435-1453

Reviews in Mineralogy & Geochemistry
Vol. 70 pp. 533-569, 2009

12

Geochemical Modeling of Reaction Paths and Geochemical Reaction Networks

Chen Zhu

Department of Geological Sciences
Indiana University
Bloomington, Indiana 47405 U.S.A

chenzhu@indiana.edu

INTRODUCTION

From the discussions in previous chapters, we can see that in order to understand the kinetics of dissolution and precipitation reactions, we really need to understand geochemical reaction networks. The previous chapters have depicted a complex picture of the details of chemical reactions and kinetics. The complexity comes from the nature of chemical kinetics. Unlike thermodynamics, which describes one state versus another, independent of the reaction paths, the kinetics of chemical reactions is path dependent.

Lasaga (1998) cited Benson (1960) who described the time- and path-dependent nature of reaction kinetics "*a body of water on top of a hill may be described thermodynamically in terms of its composition, pressure, and temperature. At a later time, this same body of water may find its way to a lake below. The thermodynamic description of the body of water in the lake is again well defined. However, if we try to describe the transition—the water in process of flowing from the hilltop—we see that it may depend on almost innumerable factors: on the outlets, on the contour of the hillside, on the structural stability of the contour, and on the numerous subterranean channels through the hillside that may exist and permit seepage.*"

What this familiar quote highlights is the path dependent nature of chemical kinetics, which helps to weave together a bewildering variety of geochemical reaction networks in many geologic systems. How can we handle this complexity and apply chemical kinetics to the geological systems? It is certainly impossible to conduct laboratory experiments to determine every reaction path and explore every type of geochemical reaction network. It appears that geochemical modeling, if standing on solid building blocks of theory and experimental data, can play a critical role in exploring the ranges of behaviors of geochemical reaction networks and extrapolate laboratory experimental data to field applications.

Before we proceed, let's define some terms. The term *reaction path* will be used here in the traditional geochemical sense. It means tracing the sequences of states of aqueous solution composition and speciation and mineral paragenesis through time as a result of irreversible reactions or processes (Helgeson 1968; Steinmann et al. 1994). It does not refer to a mechanism or the elementary reaction steps as commonly used in chemistry.

The term *geochemical reaction network* means the finite array of reactions in a geochemical system, which can be defined by a set of ordinary differential equations for a well-mixed batch

1529-6466/09/0070-0012$05.00 DOI: 10.2138/rmg.2009.70.12

system and a set of transport equations for a reactive transport system[1]. The dynamic evolution over time and the coupling and feedback of reactions in this network is the subject of this review. Note that this definition is not strictly mathematical in nature (c.f. Feinberg 1979) and is different from those used in biochemistry or isotope geochemistry.

Currently, there are several pressing societal issues for which chemical reaction kinetics play a significant role; these include carbon sequestration, nuclear waste disposal, oil and gas exploration and recovery, and remediation of environmental contamination. Let's take carbon sequestration as an example. Geological carbon sequestration—the injection of carbon dioxide (CO_2) into deep geological formations—is presently the most promising method of sequestering CO_2 released from the burning of fossil fuels (IPCC 2005; Oelkers and Cole 2008).

A key challenge to geological carbon sequestration involves the accurate prediction of reaction kinetics among CO_2, brine, and minerals in the target geological formations, in which CO_2 is stored, and in the cap rocks (e.g., shale/mudstone), which prevent CO_2 seeping upward to the ground surface (IPCC 2005). These reactions may alter the porosity and permeability of rocks, compromise the integrity of the cap rocks, and precipitate carbon-containing minerals that permanently trap CO_2 underground (Gunter et al. 1997; Gunter et al. 2000; Perkins et al. 2002; Strazisar et al. 2006). Risk assessment of CO_2 sequestration projects and the ultimate public acceptance of them depend, in part, on accurate estimates of the rates of these reactions (White et al. 2003). Since these reactions occur kilometers below the ground surface, they are difficult to measure. Furthermore, there are gaps in our knowledge of the kinetics of geochemical reactions, which is one of the most challenging problems for modern geochemistry (Velbel 1990; Brantley 1992; Blum and Stillings 1995; Drever and Clow 1995; Nagy 1995; Drever 1997; Drever 2004). All of these factors make our understanding of the kinetics of geochemical reactions an urgent global problem.

As geologists, we care about how reactions proceed in geological systems, how these reactions explain the natural world that we observe, and, hopefully, how these reactions affect our societal concerns and solutions. Kinetics theories are fundamentally based at the molecular level, while field applications are fundamentally macroscopic. By way of macroscopic scale modeling (e.g., thermodynamics is a macroscopic scale science), geochemical modeling bridges the molecular details from laboratory kinetics studies and the field applications. In particular, as we will see, it can deal with the complexity of reaction networks.

In this chapter, we will take a slightly different perspective on the kinetics problem. First, we will look at field examples of geochemical reaction networks. Second, we will review the state of knowledge as how chemical kinetic concepts are applied or can be applied to model reaction networks. Third, we will consider geochemical modeling research needs to address societal issues. All the while, we balance the review of recent research results with development of some fundamental concepts.

---

[1] Mathematically, for a geochemical system that has n species, the following ordinary differential equations completely define the geochemical reaction network (Helgeson et al. 1970),

$$\frac{dC_i}{dt} = \sum_j \nu_j r_i, i \in n$$

where C_i denotes the concentrations of ith species, t the time, ν_j the stoichiometric coefficient for i^{th} species in the jth reaction, and r_i the production or consumption rate of the ith species in the jth reaction. For a reactive transport system, the geochemical reaction network is defined by the transport equations,

$$\frac{\partial C_i}{\partial t} + L(C_i) = \sum_j \nu_j r_i$$

where L is the advection, dispersion, diffusion operator (Fang et al. 2003).

FIELD EVIDENCE OF REACTION NETWORKS

It is beyond the scope of this chapter to comprehensively review the data or results concerning all field and laboratory studies. Instead, let us use a few illustrative examples. Kharaka et al. (2006) reported a field carbon sequestration test in the Frio Formation in Texas, USA. About 1600 metric tons of CO_2 were injected at 1500 m depth into a 24 m thick sandstone section of the Frio Formation. They collected water and gas samples before, during, and after the injection experiment in both injection and observation wells. The data show a decrease of formation water pH from 6.5 to 5.7 in the observation well in a matter of days (Fig. 1). This pH drop, however, occurred before the breakthrough of supercritical CO_2. Furthermore, the magnitude of the pH decrease was much less than what one would expect from dissolution of supercritical CO_2 into the brine alone. Kharaka et al. (2006) calculated that the pH, if determined by the CO_2 solubility alone, should be around 3 at the reservoir temperature, pressure and salinity. The discrepancy between the observed pH and CO_2 solubility-determined pH was attributed to extensive water-rock interactions including reactions involving calcite, feldspar, and iron oxide dissolution. They used a modified version of the geochemical modeling code SOLMINEQ and performed speciation–solubility geochemical modeling calculations.

As we have learned from previous chapters, we know that silicate reactions are slow. Yet, measurable changes of chemical concentrations were observed in the Frio Formation in a matter of days. Presently, standards of performance assessment for geological carbon sequestration have yet to be determined by regulatory agencies, but if the regulatory framework for high level nuclear waste disposal is of any guide, the performance period must be over thousands to a million years. The field data presented by Kharaka et al. (2006) demonstrated that reactions in the sedimentary formation where CO_2 is going to be stored are fast with respect to human time scales and significant enough that we must take them into consideration for safe CO_2 disposal.

Another field example of measurable water-rock interactions comes from Houston et al. (2007) in the North Sea oil fields. In an attempt for secondary recovery of oil and gas, sea water was injected to maintain pressure. A time series of produced water from a well was analyzed. Deviation of the water chemistry from simple linear mixing of seawater and formation water

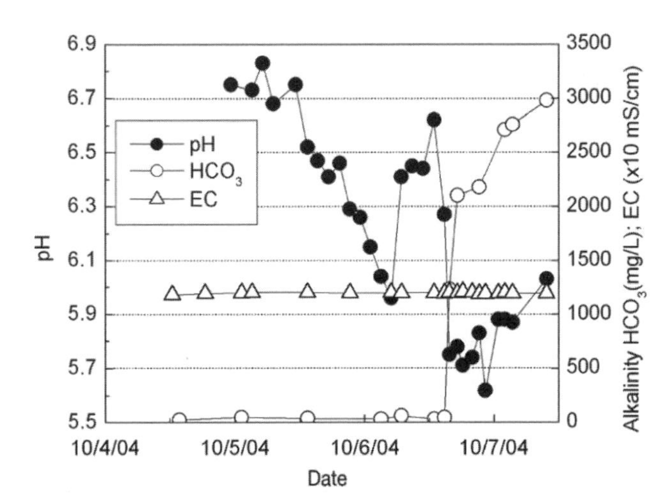

Figure 1. Electrical conductance (EC), pH, and alkalinity of Frio brine samples from "C" sandstone of observation well determined on-site during CO_2 injection on 4–7 October 2004. Note sharp drop of pH and alkalinity increase with breakthrough of CO_2 on 6 October. [Used by permission of The Geological Society of America, from Kharaka et al. (2006), *Geology* Vol. 34, Fig. 1, p. 578.]

was interpreted as resulting from water-rock interactions. In particular, silica concentrations were shown to have remained almost constant or exhibiting a slight increase, despite the fact that mixing with seawater would decrease the silica concentrations in the formation water because the silica concentration in the ambient sea water is much lower than that of the formation water at ~120 °C. Their favored interpretation of the silica data was K-feldspar dissolution and illite precipitation. Again, the reaction was observed in the time scale of months! Other chemical reactions were barite scale formation and calcite dissolution.

A third example is from the Weyburn CO_2-injection enhanced oil recovery (EOR) demonstration project in southern Saskatchewan, sponsored by the International Energy Agency (IEA). The oil field was discovered in 1954. Secondary recovery by way of water flooding began in 1964. Injection of CO_2 constitutes tertiary recovery to get more oil production from the system. However, the project also served as a demonstration for CO_2 geological storage. Emberley et al. (2005) reported baseline geochemical data before the injection and 44 monitoring samples after about a year of injection. The chemical data indicated CO_2 dissolution into the formation water, and dissolution of carbonate minerals in this carbonate reservoir. Emberley et al. (2005) also conducted speciation–solubility modeling to calculate the saturation indices of minerals. From these data and calculations, they deduced that silicate minerals were also dissolving to buffer the pH of fluids in which measurable bicarbonate (HCO_3^-) was detected.

In all three field examples of water–CO_2–rock interactions, we see that geochemical modeling was used to aid the interpretation of field data. However, the modeling work was limited to speciation–solubility models. In each case, there are some difficulties encountered in applying the modeling techniques to the field situations.

The geochemical problems at hand are a network of geochemical reactions—the dissolution of primary minerals (defined here as minerals already in the aquifers) and precipitation of secondary minerals and aqueous complex formation and dissociation. Each of the heterogeneous reactions can be controlled by its own peculiar kinetics. The solution chemistry undergoes high gradient variations in both space and time, causing feedback to the dissolution and precipitation kinetics (e.g., reaction free energy, Al concentrations). Moreover, reactions may result in de-coating of existing clay and iron mineral rinds, which may disperse, migrate, bridge, and therefore impair permeability, or result in formation of a secondary rind on dissolving primary minerals, diminishing the reactivity of the primary minerals (see below).

In addition to understanding the coupling among the reactions themselves, these chemical reactions are also coupled with other physical processes: advection, dispersion, diffusion, heat conduction, and mechanical deformation. In terms of modeling, coupled reactive mass transport models are the ultimate tools for this task (Steefel et al. 2005).

With applications to such field problems in mind, and through the lenses of reactions coupled with flow and transport processes in porous media, we will review what we know and what we don't know about geochemical modeling of these systems.

SPECIATION–SOLUBILITY MODELING

In the field examples described above, speciation–solubility modeling was the primary tool used to calculate the mineral saturation indices in order to decipher the chemical reactions that produced the observed temporal changes in chemical concentrations. Speciation modeling calculates the distribution of aqueous species based on Ion Association (IA) Theory (e.g., total dissolved inorganic carbon can be distributed among HCO_3^-, $H_2CO_3^0{}_{(aq)}$, CO_3^{2-}, $NaHCO_3^0$, and other species) according to mass balance and mass action equations.

The saturation state of the aqueous solution with respect to a specific mineral is evaluated according to the calculated Saturation Indices (*SI*). *SI* is defined as

$$SI = \log\left(\frac{IAP}{K}\right) \tag{1}$$

where K stands for the equilibrium constant of the solubility reaction (see below) and IAP stands for the Ion Activity Product. When $SI = 0$, the mineral is at equilibrium with the aqueous solution. When $SI < 0$, the aqueous solution is undersaturated with respect to the mineral of concern and the mineral will dissolve. When $SI > 0$, the aqueous solution is supersaturated with respect to the mineral and the mineral will precipitate.

As elaborated in Zhu and Anderson (2002) and in numerous books on thermodynamics, saturation indices show us the direction of the chemical reactions, as dictated by the second law of thermodynamics, but tell us nothing about the rate of reactions. For example, it is well known in geochemistry that natural waters are often supersaturated with respect to crystalline quartz, but the rate of quartz precipitation is too slow for these waters to reach equilibrium with quartz even after thousands of years at low temperatures. However, the field examples above show clearly that reaction direction is the first thing we must know before we can extract reaction rates (you need to know in which direction the reactions will go before you can estimate how fast or slow they will do it).

Speciation–solubility modeling has become a routine exercise since Garrels and Thompson (1962) calculated the aqueous speciation in seawater and saturation states with respect to mineral solubilities. The principles are well known, and the numerical modeling techniques and their tweaking are mature. There are hundreds of computer codes available for this kind of calculation. For further reading, refer to Anderson and Crerar (1993) and Wolery (1992). In geochemistry, the computer code EQ3/6 (Wolery 1992), SOLMINEQ.88 (Kharaka et al. 1988), PHREEQC (Parkhurst and Appello 1999), MINTEQA2 (Allison et al. 1991) are essentially free of charge and are widely used (see Oelkers et al. 2009).

In nearly all laboratory experiments of kinetic studies, speciation–solubility calculations have been routinely performed to calculate the *in situ* pH in the reactors (Reed and Spycher 1984), the SI or Gibbs free energy of reaction, ΔG_r, to derive the relationship between rate of reaction and ΔG_r (e.g., Gautier et al. 1994; Oelkers et al. 1994; Beig and Lüttge 2006; Hellmann and Tisserand 2006), and the dominant Al species (Carroll and Knauss 2005). Over the years, many studies have been conducted to evaluate mineral saturation states in surficial natural waters (Arnórsson and Stefánsson 1999; Stefansson and Arnórsson 2000; Arnórsson et al. 2002; Gudmundsson and Arnorsson 2005).

Speciation–solubility modeling provides a "snapshot" of a dynamic system, and represents the basic building block for more advanced process modeling. In general, it is assumed that aqueous species in the solution are in mutual equilibrium (homogeneous equilibria). The exception to this rule is redox species, which are well known to not be at equilibrium in surficial water bodies (Lindberg and Runnells 1984; Stumm and Morgan 1996). The calculated activities of the various ionic and molecular species give the IAP for the saturation state evaluation. However, despite the maturity of the modeling techniques, a number of challenges exist, which make the evaluation of the reaction directions (such as described in the field examples above) a non-trivial task. Below, we will elaborate the progress and research needs of these challenges.

Thermodynamic properties for minerals and solids

The following example demonstrates that *standard state thermodynamic properties* are needed to calculate equilibrium constants, which are in turn used for speciation–solubility modeling. For example, calcite solubility can be evaluated from the following reaction,

$$CaCO_3 = Ca^{2+} + CO_3^{2-} \tag{2}$$

The law of mass action states that,

$$K = \frac{a_{Ca^{2+}} a_{CO_3^{2-}}}{a_{CaCO_3}} \tag{3}$$

where K, as noted earlier, denotes the equilibrium constant, and a_i the activity of the ith aqueous or mineral species. K can be calculated from the relationship,

$$K = e^{-\Delta G_r^o / RT} \tag{4}$$

where R stands for the gas constant and T temperature in Kelvin. $\Delta G_r°$ is the standard state Gibbs free energy of the reaction, which in turn is calculated from

$$\Delta G_r^\circ = \Delta G_{f,CO_3^{2-}}^\circ + \Delta G_{f,Ca^{2+}}^\circ - \Delta G_{f,CaCO_3}^\circ \tag{5}$$

where $\Delta G_f°$ stands for the standard state Gibbs free energy of formation. An excellent tutorial of this important thermodynamic relationship (Eqn. 4) was provided by Anderson and Crerar (1993).

From the above derivation, it is clear that we need standard state thermodynamic properties for minerals and aqueous species in order to calculate equilibrium constants, which are used in the geochemical modeling codes mentioned above (the other modeling approach is Gibbs free energy minimization, see Anderson and Crerar (1993) and the chapter in this volume by Kulik 2009). But first, let's define standard states.

Commonly, for the geochemical modeling studies we describe in this chapter, the standard states for mineral species (solids) are defined as unit activity for pure end-member solids at the temperature and pressure of interest. The standard state for water is the unit activity of pure water. For aqueous species other than H_2O, the standard state is the unit activity of the species in a hypothetical one molal solution referenced to infinite dilution at the temperature and pressure of interest. Equilibrium constants compiled in the thermodynamic databases for the computer programs PHREEQC, EQ3/6, MINTEQA2, SOLMINEQ.88 adopt this set of standard states. For further reading on standard states, the readers are encouraged to read Anderson (1970) for an elegant narration on the meaning of standard states in thermodynamics.

There are several available compilations of standard state properties for minerals. Helgeson et al. (1978) provided the first comprehensive and internally consistent set of standard state thermodynamic properties for mineral end-members. Other mineral databases include those from Berman (1988, 1990), Holland and Powell (1998), Nordstrom et al. (1990), and Robie and Hemingway (1995). More specialized databases are available for uranium (Grenthe et al. 1992) and feldspars (Arnórsson and Stefánsson 1999).

Equilibrium constant databases accompanying the geochemical modeling computer codes use these standard thermodynamic property databases. For example, the databases for EQ3NR were compiled by Jim Johnson (Johnson et al. 1992; Johnson and Lundeen 1994). The core for the mineral data set is from Helgeson et al. (1978). The equilibrium databases for PHREEQC and MINTEQA2 programs are mostly those from Nordstrom et al. (1990).

We should note that the standard state thermodynamic properties in the mineral databases (Helgeson et al. 1978; Berman 1988, 1990; Nordstrom et al. 1990; Holland and Powell 1998) are *internally consistent*. According to Nordstrom and Munoz (1994), a set of thermodynamic properties is internally consistent if :

(1) Data are consistent with thermodynamic relationships (e.g., tabulated Gibbs free energy, enthalpy (ΔH), and entropy (S) values obey the relationship $\Delta G_f° = \Delta H_f° - TS$);

(2) Common scales are used for temperature, energy, atomic mass, and fundamental physical constants;

(3) Conflicts among measurements have been resolved;

(4) The same mathematical model is used to fit different data sets;

(5) The same chemical model is used to fit different data sets;

(6) Appropriate consideration has been given to the starting point in applying item 1;

(7) Appropriate choice of standard state has been made, and the same standard states have been used for all similar substances.

The above list is far from exhaustive. As will be shown below, if the users are not careful, new inconsistencies could arise in kinetic studies when the free energy of reaction, ΔG_r, used for deriving the rate law functions are different from those used in applications of the rate laws.

Because these internally consistent databases only contain a limited number of minerals while applications of geochemical modeling to a variety of geological and environmental topics require a more wide range of minerals and solids, additional minerals are added to these equilibrium constant databases (e.g., DATA0.COM for EQ3/6, PHREEQC.DAT for PHREEQC, MINTEQA2.DAT for MINTEQA2). For example, in addition to the "core" of internally consistent mineral database of Helgeson et al. (1978), numerous other minerals, particularly those relevant to the disposal of high level nuclear wastes have been added to various EQ3/6 databases. Not all these "extra" minerals are internally consistent with those in Helgeson et al. (1978) or among each other!

An example of this lack of internal consistency can be found in the arsenic mineral data. The WATEQ4F.DAT database contains a self-consistent set of equilibrium constants for arsenic species As_2O_3 (arsenolite), As_2O_3 (claudetite), As_2S_3 (orpiment), As_2S_3 (amorphous), AsS (realgar), AsS (β-realgar), FeAsS (arsenopyrite), which were evaluated and compiled carefully by Nordstrom and Archer (2002). However, the same WATEQ4F.DAT database distributed with the PHREEQC code also contains $FeAsO_4{\cdot}2H_2O$ (scorodite) and $Ba_3(AsO_4)_2$, which Nordstrom and Archer (2002) did not review. It turns out that the thermodynamic properties for scorodite are the subject of considerable study and debate (Dove and Rimstidt 1985; Robins 1987; Krause and Ettel 1988; Robins 1990; Zhu and Merkel 2001; Langmuir et al. 2006). Langmuir et al. (2006) recently evaluated the solubility products of scorodite by taking into account ion activity coefficients, aqueous ferric hydroxide, ferric sulfate, and ferric arsenate complexes. Their derived solubility product of $10^{-26.12}$ for crystalline scorodite is approximately six orders of magnitude lower than the previous estimate ($10^{-20.24}$) (Chukhlantsev 1956). A much larger stability field of scorodite will result in if Langmuir et al.'s (2006) solubility product is adopted.

Although the equilibrium constant for $Ba_3(AsO_4)_2$ is included in the same WATEQ4F.DAT database, it is evident that less confidence should be put on these values than on those minerals that Nordstrom and Archer (2002) have recently evaluated. Direct application of the $Ba_3(AsO_4)_2$ equilibrium constant together with the arsenic aqueous species proposed by Nordstrom and Archer (2002) would result in a large stability field for $Ba_3(AsO_4)_2$ on Eh-pH diagrams. One could easily draw the (though erroneous) conclusion that barium arsenate could exert control over dissolved arsenic concentrations in a wide range of aquifers.

In general, there is a lack of standard and mixing properties for clay minerals that have complex chemistry and structural variations. Although some progress has been already made on the experiments and theoretical predictions (Aja and Rosenberg 1992; Ransom and Helgeson 1994; Aja 1995, 2002; Ransom and Helgeson 1995; Aja and Rosenberg 1996; Aja and Small 1999; Aja and Dyar 2002; Vieillard 2002), more work is needed. For thermodynamics of solid solution -aqueous solution systems, see Prieto (2009, this volume).

Thermodynamic properties for aqueous species and speciation models

For aqueous species, an internally consistent database, which includes a large number of aqueous species that can be extrapolated to high temperatures and pressures, has been widely used in the field of geochemistry (Helgeson et al. 1981; Shock and Helgeson 1988; Shock et al. 1989; Sverjensky et al. 1997). In this database, the temperature and pressure dependences of thermodynamic properties for aqueous species were predicted using the parameters of the revised Helgeson-Kirkham-Flowers (HKF) equations of state for aqueous species (Helgeson et al. 1981; Tanger and Helgeson 1988; Shock et al. 1992). Activity coefficients for the charged aqueous species were calculated from the extended Debye-Hückel equation or B-dot equation fitted to mean salt NaCl activity coefficients (Oelkers and Helgeson 1990). The computer program SUPCRT92 can be used to generate equilibrium constants at elevated temperatures and pressures (Johnson et al. 1992). Equilibrium constants have been calculated using the standard state properties from this database and compiled into databases that accompany the program EQ3/6, which was then adopted to other programs, e.g., LLNL.DAT in PHREEQC and THERMO.DAT in GEOCHEMIST'S WORK BENCH (GWB)©.

However, controversies still exist regarding the aluminum speciation. Different competing Al hydrolysis constants and Al-metal complexes were proposed at various temperatures and pressures (among them: Apps and Neil 1990; Bourcier et al. 1993; Pokrovskii and Helgeson 1995; Shock et al. 1997; Tagirov and Schott 2001). Zhu and Lu (2009) carried out speciation–solubility modeling for calculating the *SI* of kaolinite in a batch experiment of feldspar dissolution and secondary mineral precipitation at 200 °C and 300 bars (Fu et al. 2009). Figure 2a shows the kaolinite *SI* calculated from four different sets of thermodynamic properties. The calculated *SI* can vary by up to two *SI* units, which result from the discrepancy of both mineral properties and aqueous speciation of Al.

Figure 2b shows calculated kaolinite saturation indices in groundwater from Black Mesa, Arizona. Georg et al. (2009) used the computer program PHREEQC and both databases PHREEQC.DAT and LLNL.DAT. As shown before, the former database has minerals mainly from Nordstrom et al. (1990) while the latter database contains minerals from Helgeson et al. (1978). Geochemical modeling results using the LLNL.DAT database show that groundwater is supersaturated with respect to kaolinite while results using the PHREEQC.DAT database show

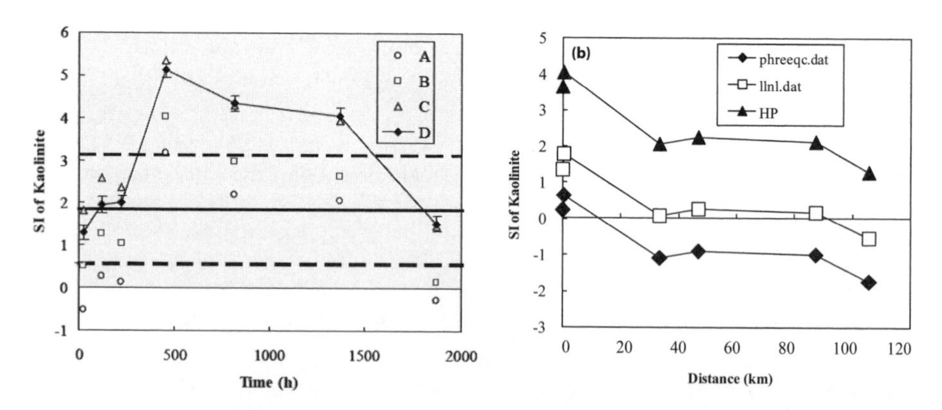

Figure 2. Calculated saturation indices (*SI*) of kaolinite for (a) experimental fluids (Zhu and Lu 2009) and for (b) groundwater at Black Mesa, Arizona (Georg et al. 2009). A, B, C, D represent *SI* values using different sets of thermodynamic datasets (see Zhu and Lu 2009), and *.dat in (b) represents databases distributed with computer program PHREEQC. [Used by permission of Elsevier, (a) from Zhu and Lu (2009), *Geochimica et Cosmochimica Acta*, doi:10.1016/j.gca.2009.03.015, Fig. 3, p. 9; and (b) from Georg et al. (2009), *Geochimica et Cosmochimica Acta* Vol. 73, Fig. A3, p. 2238].

groundwater are undersaturated with respect to kaolinite in areas away from the recharge zone. A third thermodynamic dataset, labeled as HP, using the kaolinite properties from Holland and Powell (1998) and Al speciation model of Tagirov and Schott (2001), result in supersaturation of kaolinite in all water samples.

Many of the basinal brines related to geological carbon sequestration have concentrations of dissolved solids up to 300,000 mg/L. In dealing with concentrated solutions, Pitzer's ion interaction approach to calculate ionic activities using virial specific interaction equations is generally preferred over the ion association theory mentioned earlier. The Pitzer's model, commonly the Harvie-Moller-Weare (HMW) formulation of it (Harvie et al. 1984), has been incorporated into geochemical modeling codes EQ3/6, PHRQPITZ (Plummer et al. 1988), and TOUGHREACT (Zhang et al. 2006). Although progress has been made in compiling the Pitzer interaction parameters (Wolery et al. 2004), the lack of Pitzer's activity coefficient parameters at elevated temperatures and for minor or trace elements remains a barrier to accurate calculation of solubility and saturation indices for highly saline fluids.

From the above discussion, it is clear that, while computer codes provide adequate modeling tools, the availability and accuracy of thermodynamic properties for minerals and aqueous species (not to say solid solutions for minerals with complex structure and chemistry) are still lacking for speciation and solubility modeling.

KINETIC REACTION PATH MODELING

Once we have the results of aqueous speciation and solubility calculations (i.e., calculated saturation indices) for a given temperature, pressure, and instance of time, we can model processes. The simplest next step is *reaction path modeling*, tracing the evolution of aqueous solution composition and speciation and mineral paragenesis through time or reaction progress as a result of irreversible reactions (e.g., feldspar dissolution) or processes (e.g., titration, mixing, or increase or decrease of temperature or pressure). The modeling is accomplished by applying the principle of mass balance, thermodynamics that governs the equilibrium between species, and kinetics that governs the rate of mass transfer among phases. The concept and mathematical foundation of reaction path modeling was introduced to geochemistry by Harold Helgeson (Helgeson 1968). Numerous articles and books have described this approach (Helgeson et al. 1970; Helgeson 1979; Wolery 1992; Anderson and Crerar 1993). Computer codes EQ3/6, PHREEQC, MINTEQA2, SOLMINEQ.88, and GWB© all can perform these kinds of calculations.

In the past decade, reaction path modeling has advanced on two major fronts. First, advances in computer code development now allow incorporation of various forms of rate laws and inclusion of an almost unlimited number of reactions into a single model. Second, conceptual developments in geochemical reaction networks now allow us to explore the intricacies of feedback mechanisms for complex geochemical systems. It is important to acknowledge, when we review the historical development of these concepts, the foundation established by Helgeson and co-workers forms a quantitative basis for interpretation of an otherwise chaotic geochemical system. Given the enduring value of their pioneering work, it is almost immaterial whether the earlier predictions for the specific systems are literally precise.

In contrast to speciation–solubility modeling, reaction path modeling introduces the time variable. However, it essentially simulates reactions in a well-mixed batch reactor. One may argue that, without addressing transport processes (advection, dispersion, diffusion) and providing spatial information, reaction path modeling has limited application to geological systems. This is very true. However, reaction path modeling in a closed system (a laboratory batch reactor) still serves as a staging ground for simulating real geological and environmental processes (see Table 1). We will see below that we still cannot accurately model feldspar dissolution in a batch reactor—the original example Helgeson explored in the 1960s.

Table 1. Progression of complexity in a simplified and non-rigorous sense.

Reactors	Measurements	Pro/con
Mixed flow reactors	k, $f(\Delta G_r)$, Al other aqueous solution influences	Intrinsic and isolated properties
Batch reactors	As reactions progress and coupling of reactions	Not resemble field conditions
Column reactors, field sites	Coupled processes	Additional assumptions and parameters for flow and transport, physical and chemical heterogeneities

To introduce spatial information and mass transport induced by fluid flow, a plethora of uncertainties in transport parameters and boundary conditions are introduced, and the number of non-unique solutions quickly multiplies. As we will show below, the coupling of advective, dispersive, and diffusive transport, thermal conduction, and mechanical deformation with chemical reactions can now be simulated with reactive mass transport models. The development of coupled reaction mass transport (CRMT) computer codes makes some time honored varieties of reaction path models obsolete.

Historic development and the partial equilibrium assumption

The first quantitative model for a network of geochemical reactions can be traced back to the classical reaction path model of K-feldspar hydrolysis in a batch system (Helgeson 1968, 1971, 1979; Helgeson et al. 1969) although one might argue that the feldspar work was preceded by the modeling work on the evaporation of spring waters in the Sierra Nevada mountains of California and Nevada and the resultant deposition of salt minerals by Garrels and Mackenzie (1967). However, the feldspar hydrolysis example is well known. In fact, a generation of geochemists have been taught by this example in widely used geochemistry textbooks (see Anderson and Crerar 1993; Krauskopf and Bird 1995; Drever 1997; Faure 1998).

A classic example of feldspar hydrolysis reaction path modeling was reported by Helgeson and Murphy (1983). In a batch system with a feldspar and pure water, because it is out of equilibrium, feldspar dissolves. The rate of mass transfer from feldspar to the aqueous solution is governed by the kinetic rate law for feldspar dissolution. A succession of secondary minerals (gibbsite, kaolinite, and muscovite) is precipitated. The precipitation is assumed to be so fast that the secondary minerals are always at equilibrium with the aqueous solutions. Because the overall system is out of equilibrium, but the aqueous species are at equilibrium with each other, and secondary minerals are at equilibrium with the aqueous solution, it is called a *partial equilibrium* system.

Here we see the coupling of dissolution and precipitation reactions. The precipitation of the secondary minerals removes Al, Si, K, preventing their build up in the solution and hence slowing the equilibration of feldspar with the aqueous solution. This constitutes a positive feedback loop. The more secondary minerals precipitate, the more feldspar dissolves. The more feldspar dissolves, the more secondary minerals precipitate. The factor limiting the speed of secondary mineral precipitation is the feldspar dissolution rate. The constraints for the feedback loop are the increasing saturation of feldspars as the solution chemistry evolves toward the field of feldspar stability. Eventually (in fact, quickly), feldspar reaches equilibrium with the aqueous solution and the secondary minerals. The whole system is then at overall equilibrium.

Here, dissolution and precipitation reactions are coupled, but the coupling is weak.

Deviations from partial equilibrium

When the secondary minerals are not at equilibrium with the aqueous phase or the secondary minerals do not precipitate instantaneously, the specific reaction path can be different from the classic reaction path model described above and a stronger, more complex coupling of dissolution and precipitation kinetics can result. For example, Lasaga (1981) showed the coupling of nepheline dissolution and gibbsite precipitation. A steady state of Al concentration was reached with the aqueous species as the intermediate.

A seminal paper on the possible deviation from the partial equilibrium model was made by Steefel and Van Cappellen (1990). They also used K-feldspar dissolution and kaolinite precipitation as an example. The overall reaction proceeds in two steps,

$$2KAlSi_3O_8 + 8H^+ \xrightarrow{k_1} 2K^+ + 2Al^{3+} + 6SiO_{2(aq)} + 4H_2O \qquad (6)$$

$$2Al^{3+} + 2SiO_{2(aq)} + 5H_2O \xrightarrow{k_2} Al_2Si_2O_5(OH)_4 + 6H^+ \qquad (7)$$

where k_1 and k_2 denote feldspar dissolution and kaolinite precipitation rate constants, respectively.

The interplay of dissolution and precipitation rates occurs via the common aqueous species appearing in the chemical affinity or Gibbs Free Energy ΔG_r term in the rate equations for both precipitation and dissolution reactions (Lasaga 1981; Steefel and Van Cappellen 1990). For example, using the simple Transition State Theory (TST) linear form of the rate law for illustration,

$$r_1 = k_1 S_1 (1 - e^{-\Delta G_{r1}/RT}) \qquad (8)$$

$$r_2 = k_2 S_2 (1 - e^{-\Delta G_{r2}/RT}) \qquad (9)$$

where k_j denotes the rate constant of the jth reaction, S_j the surface area, R the gas constant, and T temperature (Aagaard and Helgeson 1982; Lasaga 1998). The Gibbs free energy of reaction, ΔG_r, is a measure of disequilibrium or the driving force for a reaction (Prigogine and Defay 1965). Expanding the ΔG_r, we have

$$\Delta G_{r1} = -RT \ln\left[\frac{a^2_{K^+} a^6_{SiO_2} a^3_{Al^{3+}}}{a^8_{H^+}}\right] + RT \ln K_1 \qquad (10)$$

$$\Delta G_{r2} = -RT \ln\left[\frac{a^6_{H^+}}{a^2_{SiO_2} a^2_{Al^{3+}}}\right] + RT \ln K_2 \qquad (11)$$

where a denotes the activities of the subscripted aqueous components, and K represents equilibrium constants.

Thus, the feldspar dissolution and kaolinite precipitation reactions are linked by sharing the common species Al^{3+}, H^+, and SiO_2(aq) in the ΔG_r terms and the mass balance equations for Al, H^+, and Si. In the case of partial equilibrium, kaolinite precipitation proceeds very fast, – fast enough so that partial equilibrium can be assumed. ΔG_r is zero at all times for precipitation reactions. Precipitation of kaolinite is a positive feedback to K-feldspar dissolution as Al and Si are continuously removed from the solution, slowing the attainment of equilibrium between K-feldspar and the aqueous solution. However, the coupling is weak.

In the case that precipitation of kaolinite is also kinetically controlled, more complicated feedback effects result, leading to strong coupling between the two reactions. The rate of production of Al^{3+} (neglecting the Al speciation and solubility dependence on pH here for arguments sake) from feldspar dissolution is countered by the rate of Al^{3+} consumption from kaolinite precipitation by a factor of 2 due to the stoichiometry of the overall reaction. A steady state of aqueous components (i.e, Al^{3+}) is reached, with the concentrations determined by the relative rates of feldspar dissolution and kaolinite precipitation. In order for a steady state to be reached, the product (k_1^*) of the second reaction constant (k_2) and surface area (S_2) must be much slower than that of the first (Lasaga 1981). Steefel and Van Cappenllen (1990) used a ratio of $k_1^*/k_2^* = 1000$ in their example.

Another important and insightful result from Steefel and Van Cappellen (1990) is their observation of a quasi-steady state. They stated, "*This fundamental tendency of the coupled dissolution–precipitation reaction system to evolve towards a steady state in which the rates of the reactions are nearly in balance is a result of the feedback of the saturation state of the solution on the rates of the heterogeneous reactions.*" We shall see below that recent experimental data confirm their prediction.

Lasaga et al. (1994) and Lasaga (1998) went further, and used different ratios of rate constants k_i^*/k_{feld}^*, where i stands for the secondary minerals gibbsite, kaolinite, or muscovite, and showed that reaction paths deviate from the classic reaction path model if the secondary minerals are not at equilibrium with the aqueous fluid. The deviations are manifested in two forms. First, the fluid chemistry no longer evolves along the mineral stability boundaries, but enters into another mineral stability field across the boundary at an oblique angle (hence, the reaction path is different from the classic reaction path model in terms of fluid chemistry). Second, the secondary minerals precipitated earlier no longer dissolve completely as fluid moves into the stability field of another mineral that is lower in the paragenesis sequence, rather it persists for a time determined by the rate constant ratios. For example, some gibbsite persists when the solution chemistry is in the stability field of kaolinite. There may be a region of coexistence of gibbsite and kaolinite, and even a region of coexisting gibbsite, kaolinite, and muscovite. These regions become larger as the ratios of k_i^*/k_{feld}^* become smaller.

Although Lasaga and co-workers mapped out some expected consequences for deviation from the partial equilibrium assumption, they did not test their ideas with laboratory or field data. The examples given were illustrative, with many simplifying assumptions. Nevertheless, their work provided a background for further research as described below.

Advancements in geochemical modeling code developments

Modeling a complex network of reactions needs ever more computing power and sophisticated computer codes. In the last decade, we have seen tremendous advancements in code developments and exponential growth in computing powers. The earliest work on reaction path modeling was carried out by hand calculations (Garrels and Mackenzie 1967). Helgeson (1968) was the first to introduce computerized reaction path geochemical modeling.

The necessary tools for complex reaction path modeling have arrived in the last few years. New developments have allowed customized rate laws, and expanded the number of species and components allowed in a model. Version 2.0 of PHREEQC or later (*http://wwwbrr.cr.usgs.gov/projects/GWC_coupled/phreeqc/*) allows customized rate laws to be programmed into a program. For example, for the albite dissolution rate dependency on Al concentrations (Gautier et al. 1994; Oelkers et al. 1994), the rate law can be programmed in BASIC language as follows:

```
RATES
Albite
-start
 10 REM Al inhibition rate law
 20 IF EXISTS(1) = 0 THEN PUT(M, 1)
 41 n = 1/3
```

```
  42 X = 1/(7+(ACT("A(OH)4-")*ACT("H+"))^n)
  45 SR_Ab = SR("Albite") # define saturation ratio
  50 rate = 0.12*(3.2625e-6)*X*(1-(SR_Ab)^n)
  90 moles= rate * TIME
  100 PUT(rate,1)
  150 SAVE area
  200 SAVE moles
 -end
```

Such function for kinetics studies has also been later adapted by GWB© and other programs.

REACTION NETWORK CALCULATIONS OF FELDSPAR HYDROLYSIS

Linking to the field–lab rate discrepancy

Lasaga (1981, 1998) and Steefel and Van Cappellen (1990) did not explore the link between the coupling of dissolution and precipitation reaction rates and its potential significance to the well known field–laboratory dissolution rate discrepancy. That was later explored by Zhu et al. (2004a,b), Zhu (2005), and Ganor et al. (2007).

It has been consistently found that field feldspar dissolution rates are typically two to five orders of magnitude slower than those determined in laboratory settings under similar pH and temperature conditions (Paces 1983; Velbel 1985, 1990; Brantley 1992; Blum and Stillings 1995; Drever and Clow 1995; White et al. 2001; White and Brantley 2003; Zhu 2005). The widely reported discrepancy between laboratory and field-based weathering rates of feldspars always needs to be qualified to ensure that we are comparing dissolution rates under similar conditions (e.g., chemical affinity, Al concentrations, and see below). Nevertheless, this gap in our knowledge about laboratory measurements versus field applications is large enough to underscore the existence of general gaps in our understanding, which inhibits our efforts to reliably quantify a number of geological and environmental processes, including geological carbon sequestration mentioned earlier.

Zhu et al. (2004a) showed that when the k_i^*/k_{feld}^* (i stands for i^{th} secondary mineral) are in the 10^{-2} to 10^{-4} range, secondary mineral precipitation becomes rate limiting and a steady state persists at which feldspar continues to dissolve near equilibrium at a much reduced rate due to the free energy effect. In an attempt to explain the 10^2 to 10^5 times slower K-feldspar dissolution rates in the Navajo Sandstone aquifer at Black Mesa (Zhu 2005), they stated "*We conclude that the most important mechanism controlling the slow dissolution rates feldspars in aquifers is the slow precipitation kinetics of kaolinite and smectite. The precipitation of clays would remove solutes from solution, undersaturate the solution with respect to feldspars, and make additional feldspar dissolution possible. In our studies of the subsurface systems, in situ clay formation as a result of feldspar dissolution is ubiquitous. For reactions in series, the slowest reaction controls the overall rate. Thus, the overall rate of the feldspar dissolution along the flow path may be controlled by the precipitation rates of kaolinite and smectite if this is the slow step in the overall reaction.*" They further speculated "*A similar mechanism of clay precipitation controlling the overall feldspar weathering rates may also potentially operate in unsaturated soil systems.*"

The details of this hypothesis were demonstrated by kinetic reaction path calculations. For simplicity, they modeled the classic K-feldspar hydrolysis and sequential formation of kaolinite and muscovite (Helgeson et al. 1969). In this model, they used the surface area, feldspar content, and porosity data from Black Mesa (Zhu 2005), and let K-feldspar react with the solution chemistry of the groundwater at the starting point of the flow path (recharge area water). The laboratory derived k_1 value of $10^{-12.5}$ mol s^{-1} m^{-2} (Blum and Stillings 1995) was used together with the rate law of Equations (8) and (9). Kaolinite and muscovite (a proxy for smectite) precipitation was modeled with the same rate law but with a different set of values for k_i^*, to evaluate the influence of secondary mineral precipitation rates on feldspar dissolution.

The results show that if feldspar initially dissolves at the laboratory rate (measured at far from equilibrium conditions) and secondary minerals precipitate instantaneously (i.e., assuming partial equilibrium; Helgeson 1971; Helgeson and Murphy 1983), feldspar dissolution rates would decrease rapidly as ΔG_r drops, and the system would reach equilibrium ($\Delta G_r = 0$) in a few years (Fig. 3). However, when $k_i^* << k_{feld}^*$, the feldspar dissolution rate initially decreases rapidly due to the ΔG_r effect, as in the partial equilibrium case described above, but reaches a steady state when ground water is near (but not at) equilibrium with feldspar ($\Delta G_r \approx -0.3$ J/mol for Model 3 in Fig. 3b). A k_i^*/k_{feld}^* ratio of 10^{-4} for muscovite effectively reduces the feldspar dissolution rate by a factor of 10^3, and a k_i^*/k_{feld}^* ratio of 10^{-6} would reduce the dissolution rate by the observed factor of 10^5.

There are two basic parts to the Zhu, Blum, Veblen (ZBV) hypothesis: (1) the rates of feldspar dissolution are greatly retarded because slow secondary mineral precipitation holds the solution close to equilibrium with respect to feldspars; and (2) slow secondary mineral

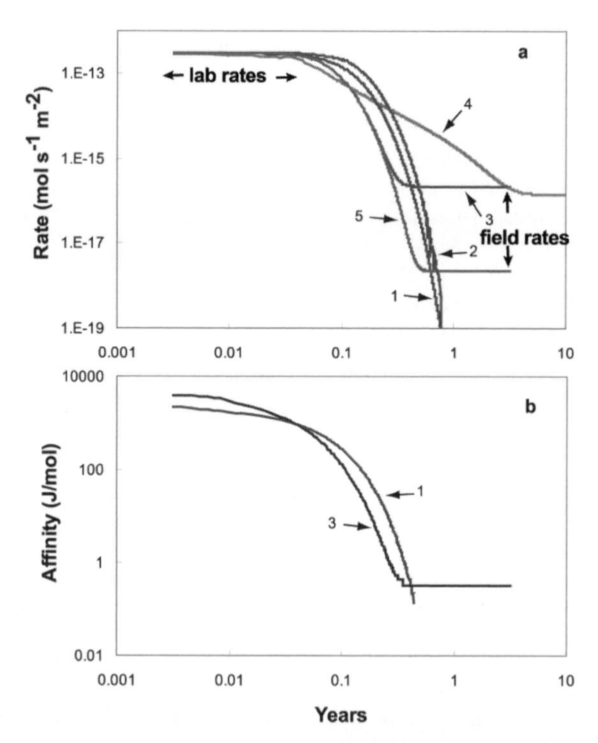

Figure 3. Results of numerical experiments to demonstrate the coupled slow secondary mineral precipitation and slow feldspar dissolution. The five reaction path models simulated with the aid of computer code PHREEQC with thermodynamic data for minerals and aqueous species compiled by Nordstrom et al. (1990) and the kinetic expression $r_j = k_j(1 - e^{-\Delta G/RT})$. (a) Temporal evolution of feldspar dissolution rates. Feldspar effective dissolution rate constant $k^*_1 = k_1 S$, where S stands for surface area. Model 1, $k^*_1 = 10^{-10.3}$ mol s^{-1} m^{-2}, $k^*_{clay} >> k^*_1$; 2. $k^*_{clay} = k^*_1$; 3. $k^*_1 = k^*_{kaol}$, and $k^*_{mus} = k^*_1 \times 10^{-4}$; 4. $k^*_{kaol} = k^*_1 \times 10^{-2}$, but $k^*_{mus} = k^*_1 \times 10^{-4}$; 5. $k^*_1 = k^*_{kaol}$, and $k^*_{mus} = k^*_1 \times 10^{-6}$. (b) affinity of feldspar dissolution reactions with different relative rates of feldspar dissolution and clay precipitation. [Used by permission of Elsevier, from Zhu et al. (2004b), *Geochimica et Cosmochimica Acta*, Vol. 68, Fig. 1, p. A148.]

precipitation provides a regulator for close to equilibrium reaction affinity to persist for a long time and over a long distance in a geological system (Zhu 2005). As predicted by Steefel and Van Cappellen (1990), a steady state of slow feldspar dissolution and secondary mineral precipitation is maintained.

Most laboratory based kinetics studies (e.g., using mixed flow reactors) strive to measure rate constants and their dependence on temperature, saturation states, and solution chemistry. The precipitation of secondary minerals is carefully avoided by controlling solution chemistry (e.g., undersaturation with respect to secondary minerals) and rate of fluid circulation. On the other hand, being close to equilibrium in a field system (as compared to laboratory experiments) has been cited as a possible reason for the observed laboratory-field discrepancy in numerous publications, but a *regulator* that holds the solution close to equilibrium persistently in natural systems has not previously been identified. The slow precipitation of secondary minerals provides this regulator.

Ganor et al. (2007) further explored the possibility of the ZBV hypothesis to "bridge the gap between laboratory measurements and field estimations of silicate weathering rates." While Zhu et al. (2004a,b) used the simple TST rate law, Ganor et al. (2007) used the empirical parallel rate law of Burch et al. (1993) in the form of,

$$r = k'\left\{1 - e^{-n\left(\frac{|\Delta G_r|}{RT}\right)^{m1}}\right\} + k''\left\{1 - e^{-\left(\frac{|\Delta G_r|}{RT}\right)}\right\}^{m2} \tag{12}$$

where k' and k'' denote the rate constants in units of mol s^{-1} m^{-2}, and n, m_1, and m_2 are empirical parameters fitted from experimental data.

They used the rate data from the column experiments on Panola Granite by White and Brantley (2003) as a proxy for the value of k' at ambient temperature and pH 6.5, and retained the exponent parameter values and the k'/k'' ratio from Burch et al. (1993). Their simulations of oligoclase dissolution and kaolinite precipitation in batch systems have shown that the oligoclase dissolution rates and kaolinite precipitation rates are coupled through the saturation state terms in Equation (12). For this particular case, the r_{Oli}/r_{Kln} = 1.626, where r stands for rate in unit of mol s^{-1} L^{-1}. This ratio is determined by the stoichiometry of the minerals. Ganor et al. (2007) analyzed the feedback between the two rates on saturation states. The ratios are fixed even when the effective rate constant of kaolinite varied orders of magnitude (Fig. 4).

These results, even from grossly simplified reaction path modeling, nevertheless demonstrated the importance of the rate laws in the coupled relationships between dissolution and precipitation reactions. Note that the linear TST rate law in Equations (8-9) requires conditions very close to equilibrium, e.g., ΔG_r = 0.3 J/mol to reduce rates by three orders of magnitude. This contradicts field data. At Black Mesa, Arizona, the *SI* for K-feldspar, depending on the databases used, varied from −0.5 to −3 or ΔG_r −2.8 to −17 kJ/mol (Georg et al. 2009). Similarly, even though ΔG_r for albite and K-feldspar in pore waters is not very close to equilibrium, White et al. (2001) found the plagioclase and K-feldspar dissolution rates in the granitic regoliths are three to four orders slower than the laboratory rates at comparable pH and temperature (Fig. 5). Ganor et al.'s (2007) simulations showed that with the Burch et al. (1993) type rate law, the retarded rates could easily be within the range of field rates.

Testing the ZBV hypothesis with laboratory data

From previous sections, we can see that if partial equilibrium is not maintained and the secondary mineral precipitation rate is slow, it is possible that the field–laboratory discrepancy regarding dissolution rate of the primary minerals could be partly reconciled (the Zhu-Blum-Veblen Hypothesis). Previously, inhibition of primary mineral dissolution by the slow precipi-

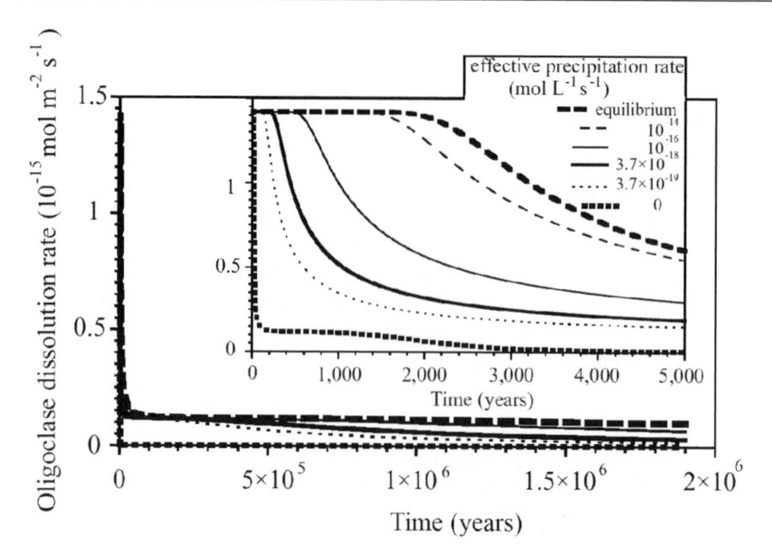

Figure 4. Comparison of the change with time of oligoclase dissolution rate in simulations with different effective rate coefficients of kaolinite precipitation (see text). Enlargement of the first 5000 hours of the simulations is shown in the insert. From Ganor et al. (2007).

tation of secondary minerals has also been proposed to explain the delayed smectite to illite conversion, metastable coexistence of kaolinite, K-feldspar, smectite, quartz in clastic sandstone diagenesis processes, and several metamorphic reactions (Abercrombie et al. 1994; Oelkers et al. 1996; Berger et al. 1997; Thyne et al. 2001). However, these are demonstrated by numerical modeling of kinetic reaction path in simplified and idealized systems. Whether these concepts can be verified in laboratory experiments and field systems has yet to be seen.

Recently, there are increasing amounts of laboratory experimental evidence that support this idea. Alekseyev et al. (1997) showed experimentally that the precipitation of secondary minerals slows down sanidine and albite dissolution at 300 °C and pH 9. Murakami et al. (1998) conducted anorthite dissolution experiments at 90, 150, and 210 °C and observed boehmite, modified boehmite, and kaolinite precipitation. Zhu and co-workers conducted a series of batch reactor experiments to assess alkali-feldspar dissolution and secondary mineral formation in initially acidic fluids (pH = 3-4) at 200 °C and 300 bars (Fu et al. 2009). Temporal evolution of fluid chemistry

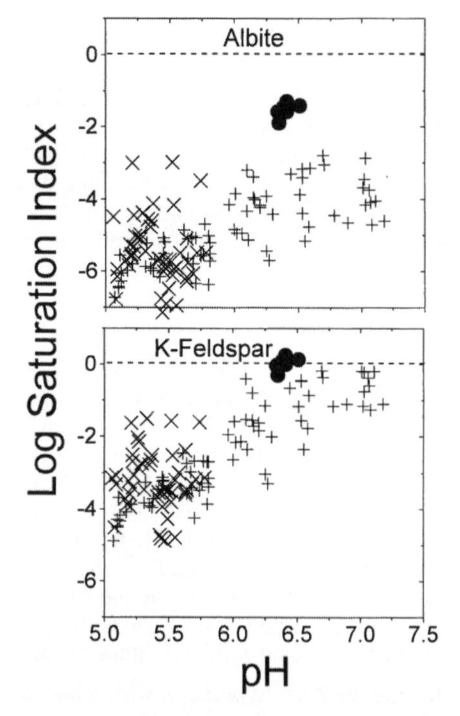

Figure 5. Saturation indexes for regolith pore waters (cross for ridge top pore water, plus for ridge slope pore water) and ground waters (solid circles) with respect to albite and K-feldspar. [Used by permission of Elsevier, from White et al. (2001), *Geochimica et Cosmochimica Acta*, Vol. 65, Fig. 7, p. 856.]

was monitored by major and trace element analysis of *in situ* fluid samples. Solid reaction products were analyzed with scanning electron microscopy, X-ray diffraction, transmission electron microscopy, and X-ray photoelectron spectroscopy. Their experimental results also demonstrated coupled effects.

To model the coupling between dissolution and precipitation reactions is a challenge, even in batch reactor systems. Zhu and Lu (2009) recognized the myriad of assumptions that are necessary to simulate the kinetic reaction path in these batch experiments, and conducted speciation and solubility geochemical modeling to compute the saturation indices (*SI*) for product minerals and to trace the reaction paths on activity-activity diagrams. The modeling results for the experimental data of Fu et al. (2009), Alekseyev et al. (1997), and Muraikai et al. (1998) demonstrated: (1) the experimental aqueous solutions were supersaturated with respect to secondary minerals for almost the entire duration of the experiments; (2) the aqueous solution chemistry did not evolve along the phase boundaries but crossed the phase boundaries at oblique angles; and (3) the earlier precipitated product minerals did not dissolve but continued to precipitate even after the solution chemistry had evolved into the stability fields of minerals lower in the paragenesis sequence. The experimental evidence is unambiguous that the partial equilibrium assumption does not hold in the feldspar-water system for the temperature range of 90-300 °C.

The next step is to figure out how the reactions of primary mineral dissolution and secondary mineral precipitation are coupled. This requires reaction path modeling, which would numerically solve a set of ordinary differential equations and algebraic equations that define the geochemical reaction network. The experimental dataset by Alekseyev et al. (1997) for albite dissolution and sanidine precipitation provide detailed observations of the evolution of a batch system at 300 °C and pressure of vapor saturation. Numerical reaction path modeling of their experiments shows that a quasi-steady state was reached with albite dissolution and sanidine precipitation. The rates are almost equal and Al and Si concentrations near constant. It requires a $k_{san}^*/k_{alb}^* < 10^{-1}$ ratio to fit the experimental data. What is most interesting is that neither of the dissolution nor the precipitation reaction was able to approach equilibrium, but instead, in the words of Oelkers and co-workers (Cubilas et al. 2005) the dissolution was "arrested" (according to the Oxford Dictionary, the word "arrest" can mean "bring to a stop of a process") at some fixed ΔG_r. Albite dissolution reaction proceeded at a near constant $\Delta G_r = \sim -16$ kJ/mol until all albite was converted to sanidine.

In the case of the experiments conducted by Cubilas et al. (2005), coupled dissolution of calcite and precipitation of otavite ($CdCO_3$) resulted in an otavite coating layer on the calcite surface, which reduced calcite dissolution rates by close to two orders of magnitude.

The numerical kinetic reaction path modeling interpretation of the batch experimental data requires the assumption of proper rate laws (see next section) and reactive surface areas. The subject of surface area in kinetics studies is controversial (White and Peterson 1990; Blum 1994; Gautier et al. 2001; Metz et al. 2005; Zhu et al. 2006). Although the concept of "reactive surface area" (Helgeson et al. 1984) is most consistent with theories of surface controlled reaction kinetics and the reactive surface area is ultimately related to the available reactive surface sites (Furrer and Stumm 1986; Wieland et al. 1986), this concept is difficult to implement at the present. The Brunauer-Emmett-Teller (BET) surface areas are most widely used in geochemistry as a proxy for reactive surface areas. However, how the surface areas change as reactions progress, even in batch reactors, is not known. Alekseyev et al. (1997) reported "horseshoe" shaped $r - \Delta G_r$ relationships in their experimental data on albite and sanidine dissolution, which at the first glance defies the TST theory. Without dramatic changes of "reactive surface areas," e.g., due to nucleation of secondary minerals on primary mineral surfaces, these experimental data are difficult to interpret. BET surface area-normalized dissolution rates can decrease with time as a result of the extinction of highly reactive fine particles (e.g., Helgeson et al. 1984) and decrease the ratio of reactive to un-reactive sites (e.g., Gautier et al. 2001).

Earlier simple models which illustrate the concepts either assumed a constant surface area (Helgeson and Murphy 1983, 1990; Zhu et al. 2004a,b) or were assumed to vary according to $S_{A,j} = S_{A,j}^{\circ} (N_j^{\circ}/N_j^{t})^{2/3}$, where $S_{A,j}^{\circ}$ is the initial surface area of the jth reactant, N_j° is the initial moles of reactant, and N_j^{t} is the moles of reactant at a given time (e.g. Parkhurst and Appello 1999). The latter formulation is only true when all grains have the geometry of perfect spheres with uniform sizes, and obviously ignore surface roughness. Apparently, another possible change of reactive surface area is the formation of a coating layer on the mineral surface (Cubilas et al. 2005).

It is even more difficult to estimate the surface areas for precipitating secondary phases. For modeling, this presents a dilemma: precipitation cannot proceed without a surface area first; and without precipitates, there are no surface areas for the secondary phases. The current practice commonly assumes a "seed" surface area to start the precipitation. Rate laws of nucleation have seldom been used in the modeling of experiments.

As modeling studies in the simplified systems have shown (Steefel and Van Cappellen 1990; Zhu et al. 2004a; Zhu et al. 2004b; Ganor et al. 2007), with a different effective rate constant k^*, the rates of dissolution and precipitation adjust in tandem, but the ratios of dissolution and precipitation rates remain nearly constant. Besides the surface area, the rate laws are also important regarding how the dissolution and precipitation reactions are exactly coupled.

Rate laws for dissolution and precipitation reactions

Earlier exploration of the coupling between dissolution and precipitation reactions used the linear TST rate law of Equations (8) and (9) (Steefel and Van Cappellen 1990; Lasaga 1998; Zhu et al. 2004a,b). One feature of the TST rate law is that there is a plateau of constant rate at a solution chemistry condition far from equilibrium pertinent to a specific k^* if there are no other effects, but the rates plunge exponentially near equilibrium. As a consequence, the range of ΔG_r where the rates are influenced by it is narrow (cf. Fig. 6c). For the case of Hellmann and Tisserand (2006), a linear TST expression only starts to affect the rate at ΔG_r >10 kJ/mol at 150 °C. It is well understood that the theoretical underpinning for Equations (8) and (9) holds only for elementary reactions (Lasaga 1981). The application of Equations (8) and (9) to the overall reaction, such as silicate dissolution reactions, implies that only one elementary reaction is the limiting step (Aagaard and Helgeson 1982).

However, a number of experiments near equilibrium have shown that the actual relationships between r_j and ΔG_r deviate from this so-called linear kinetics (Schramke et al. 1987; Nagy et al. 1991; Nagy and Lasaga 1992, 1993; Burch et al. 1993; Gautier et al. 1994; Oelkers et al. 1994; Alekseyev et al. 1997; Taylor et al. 2000; Beig and Lüttge 2006; Hellmann and Tisserand 2006). In particular, the $r - \Delta G_r$ relationship for feldspars exhibit a sigmoidal shape, for which the empirical rate law of Burch et al. (1993) is sufficient to fit the experimental data although it is by no means a mathematically unique expression (Fig. 6). Burch et al. (1993) assigned the physical meaning to Equation (12) as two parallel mechanisms operating simultaneously. At far from equilibrium conditions, $\Delta G_r < \Delta G_{r,critical}$, dissolution occurs mainly through the opening of etch pits. Near equilibrium, dissolution occurs on the planar surface sites, which have far less energy and hence slow rates. Alternatively, Oelkers and co-workers interpret this sigmoidal shape as stemming from the dissolution mechanism that involves removal of one or more cations to make the activated complex. By taking this into account, one can accurately fit not only the variation of rates with affinity but also their variation on aqueous solution composition (Oelkers 2001). Indeed, if one plots the dissolution rate as a function of Al concentrations, the relationships for sanidine and albite from dissolution experiments by Alekseyev et al. (1997) are linear (Fig. 7) even though the relationships $r - \Delta G_r$ show sigmoidal shapes (Fig. 6i, k)

Recently, Beit and Lüttge (2006) and Arvidson and Lüttge (2009) provided critical experimental data and more theoretical interpretations. They conducted paired experiments on albite dissolution. One experiment used pristine surfaces but another used etched surfaces. They observed dissolution rates that differed over 1 to 1.5 orders of magnitude, where the experiment with etch pits dissolved faster. They concluded that dissolution occurs mainly at the screw dislocations when $\Delta G_r < \Delta G_{r,critical}$, which can be described by their stepwave model (Lasaga and Luttge 2001, 2003), but at $\Delta G_r > \Delta G_{r,critical}$ dissolution primarily occurs at point defects and preexisting edges and corners. Furthermore, they believe that the experimental data points in the transition region merely reflect an unsteady state when one mechanism is switched to another.

The experimental evidence that the dissolution rate is also controlled by the history or heritage of the mineral surface is consistent with earlier suspicions when comparing laboratory and field rates (White and Brantley 2003). A detailed electron microscopic study of the feldspars in the Navajo sandstone shows that naturally weathered feldspars can have complex surface features and history (Zhu et al. 2006). In the ~200 m.y. old Jurassic Navajo sandstone at Black Mesa, Arizona, the K-feldspars are first incompletely covered with a layer of well oriented, tightly adhered kaolinite, and then enveloped with a complete layer of smectite. A 10 nanometer amorphous layer appears on the top of K-feldspar surfaces and beneath the clay coatings.

From the viewpoint of geochemical modeling, the $r - \Delta G_r$ experimental data by Hellmann, Lüttge, and their colleagues (Beig and Lüttge 2006; Hellmann and Tisserand 2006; Hellmann et al. 2007; Arvidson and Lüttge 2009) represent a significant advance in our understanding of how dissolution and precipitation are coupled. The sigmoidal shaped $r - \Delta G_r$ relationships make the ZBV hypothesis more applicable to bridging the gap between laboratory and field data (see reactive transport modeling below). As Ganor et al. (2007) have shown, dissolution of the primary mineral starting from a rate similar to the laboratory measured k^* at far from equilibrium can quickly be retarded by two orders of magnitude due to the $r - \Delta G_r$ relationship prescribed in the Burch et al. (1993) rate law. The reduction of Gibbs free energy of reaction does not need to be at a state very close to equilibrium as when the linear TST rate law was used (Zhu et al. 2004a, b), but can be achieved when the solution is still quite undersaturated with the primary mineral plagioclase. The dissolution rate of plagioclase and precipitation rate of kaolinite was locked in tandem through shared components in the $f(\Delta G)$ term, and that happened at a ΔG_r not very close to zero.

Applications of this sigmoidal relationship in geochemical modeling to laboratory and field data still face challenges. As shown in Figure 6, the available $r - \Delta G_r$ data for albite, plagioclase, and sanidine show a variety of slopes, k'/k'' ratios, and values of $\Delta G_{r,critical}$. According to Lüttge and co-workers, the "transition" regions not only represent an unsteady state but are also path dependent, e.g., whether dissolution is approached far from equilibrium moving toward equilibrium or approached from the near equilibrium region and then move away from it. That makes the fitting parameters n, m_1, m_2 also path dependent, and not likely to exhibit predictable correlations or universal applicability, if Lüttge and his colleagues are correct.

It is worthwhile to visit the mathematical formulae of Burch et al. (1993) and experimental $r - \Delta G_r$ relationships compiled in Figure 6. If k'/k'' are small, e.g., ~11 in Burch et al. (1993), the second term in Equation (12) also contributes to the dissolution rates at far from equilibrium. A second dissolution plateau is prominent and the two mechanisms are parallel. This is also the case for Beig and Lüttge (2006) and Taylor et al. (2000). When k'/k'' ratios are large, e.g., ~55 in Hellmann and Tisserand (2006), there is no contribution from the second term to the overall dissolution rate at far from equilibrium. The second plateau is missing.

The sigmoidal shape of the curve as commonly described is only controlled by the mathematical formulae and the parameters n, m_1 in the first term of Equation (12). Most experiments do not define a continuous non-linear curve due to absence of experimental data points in the transition region. However, the experiments by Alekseyev et al. (1997) provided perhaps the most well-defined continuous sigmoidal curves because their batch reactors and "initial rate method" caught the transition in the first few hours of experimental runs (Fig. 6i-l). Among

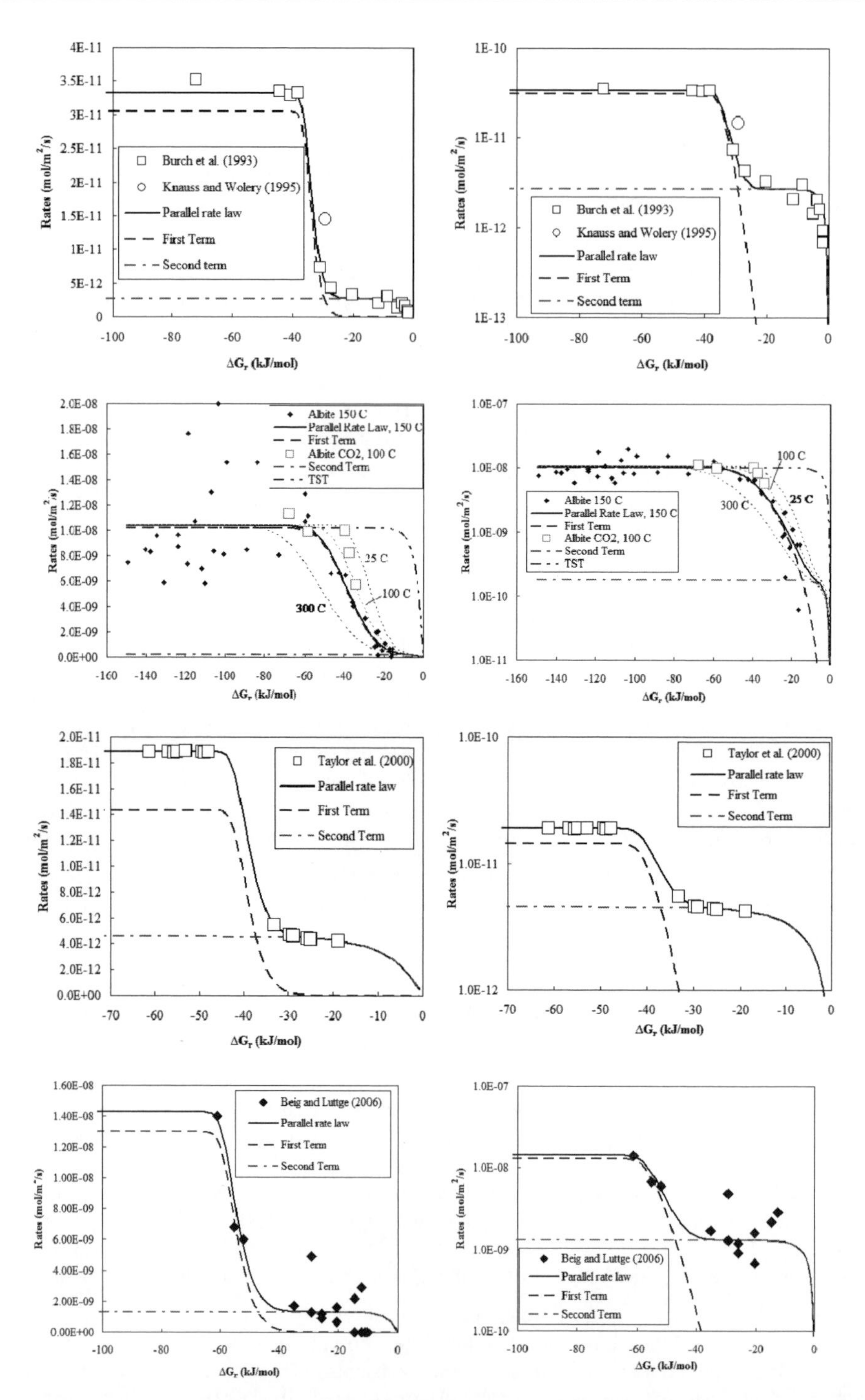

Figure 6. *Figure and caption are continued on facing page*

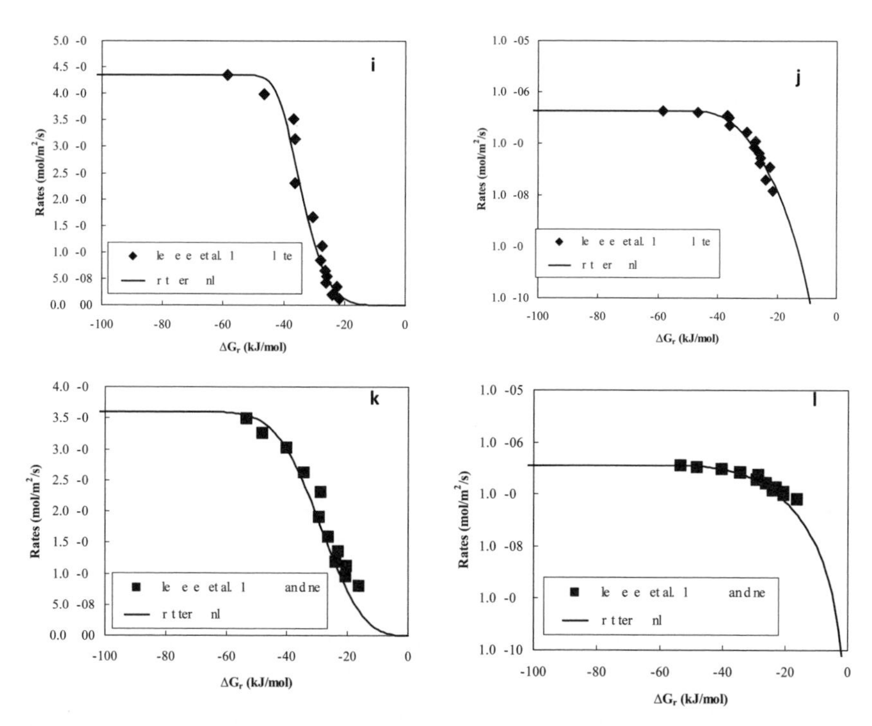

Figure 6 cont. Experimental dissolution rates as a function of ΔG_r. The right panel shows log scale on the vertical axis. The lines are fitted to Equation (12) either using both terms or the first term on the right (cf. Table 2 for experimental details)

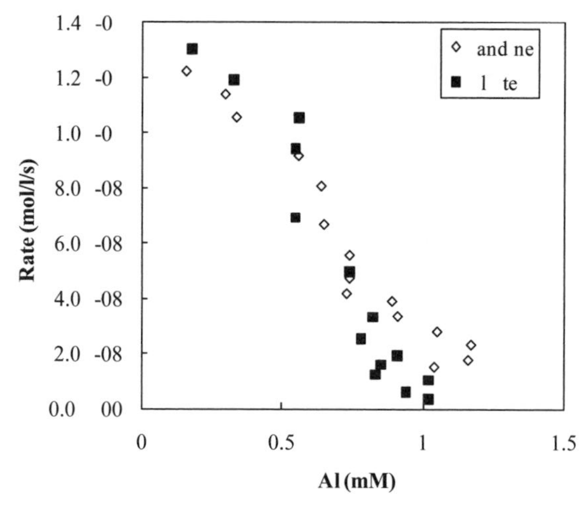

Figure 7. Sanidine and albite dissolution rates as a function of Al concentrations during the first a few hours of congruent dissolution of feldspars in batch reactors. Data are from Alekseyev et al. (1997).

the continuously mixed flow reactor experiments, Hellmann and Tisserand (2006) provided the most abundant data points that allow us to define a continuous sigmoidal curve.

We must note that the experimental data of Hellmann and Tisserand (2006) can be fitted with the first term only and the second term in Equation (12) is unnecessary (Fig. 6d). In fact, the values of k'' and m_2 in Equation (12) cannot be determined by their experimental data because their experiments only approached equilibrium at the largest ΔG_r of ~ −16 kJ/mol. Similarly, the second term cannot be determined by the experimental data of Alekseyev et al. (1997) for albite and sanidine (Fig. 6i-l).

Therefore, the comparable field derived k'' with Hellmann and Tisserand's (2006) laboratory k'' in the study of Maher et al. (2009) are not meaningful evidence for reconciliation of the discrepancy between field and lab rates. The Hellmann and Tisserand (2006) k'' and m_2 parameters were also used by Yang and Steefel (2008) to predict soil formation at 25 °C.

It should be noted that the regressed parameters m_1 and n in the first term of Equation (12) are not mathematically unique. An infinite number of m_1-n pairs can satisfy the experimental data within a given distribution of uncertainties and define a sigmoidal shaped curve (Fig. 8).

The mathematical formulae of the second term would produce a second sigmoidal curve if m_2 is large. However, most authors have used a small m_2 (Table 2), which produces a linear $r - \Delta G_r$ relationship close to equilibrium.

The difficulties involved in obtaining experimental data close to equilibrium pose serious challenges for geochemical modeling as different values of n, m_1, m_2 proposed by different authors result in orders of magnitude differences in the close to equilibrium region (Fig. 9b). This situation is further exacerbated by the potential for inconsistency among the thermodynamic data. The standard state thermodynamic properties for minerals and aqueous speciation models used to calculate ΔG_r (from speciation and solubility modeling) in deriving n, m_1 and m_2 in Equation (12) can be different from those actually used in reaction path and coupled reactive transport modeling. For example, if the thermodynamic properties from Holland and Powell (1998) and Al species from Tagirov and Schott (2001) are used to calculate *SI* for albite in the experimental fluids analyzed by Alekseyev et al. (1997), the calculated ΔG_r are about 4 kJ/mol smaller than those calculated by Alekseyev et al. (1997), who used different sets of standard thermodynamic properties. Both Hellmann and Lüttge used EQ3/6, which uses the Helgeson et al. (1978) mineral databases and the Shock et al. (1998) Al species. If one prefers a different database for minerals and Al species from Tagirov and Schott (2001), an inconsistency is then introduced, which can cause large discrepancies in the near equilibrium region in calcualted dissolution rates.

Experimental evidence for whether there are pH and temperature dependences on the parameters n, m_1, m_2 are conflicting. On one hand, Hellmann et al. (2007) show that the same set of parameters n, m_1, m_2 can be applied to their Amelia albite dissolution experiments at 150 °C and pH 9, and at 100 °C and pH 3.2. However, the albite dissolution data at the congruent stage in Alekseyev et al. (1997) at 300 °C and pH ~9 clearly showed a different set of values for n, m_1, m_2 are needed to represent the experimental $r - \Delta G_r$ relationship. (In fact, our preceding discussion shows only n, m_1 are legitimate parameters for both experiments.) The temperature dependence of n, m_1, m_2 is not clear from the experimental data (Fig. 9a), but in practice, Ganor et al. (2007) assumed the same n, m_1, m_2 parameters developed by Burch et al. (1983) at 80 °C are applicable to 25 °C, and Yang and Steefel (2008) and Maher et al. (2009) used the n, m_1, m_2, k'/k'' parameters from Hellmann and Tisserand (2006) developed at 150 °C for 25 °C numerical simulations.

Experimental data show that the transition regions are wider at low temperatures (Fig. 9c). This implies that: (1) r is more likely influenced by ΔG_r at low temperature; (2) the coupling of reactions at low temperature is strong; and (3) opening up etch pits require more negativeΔG_r at low temperatures. Furthermore, the slow rate of reactions at low temperatures means many

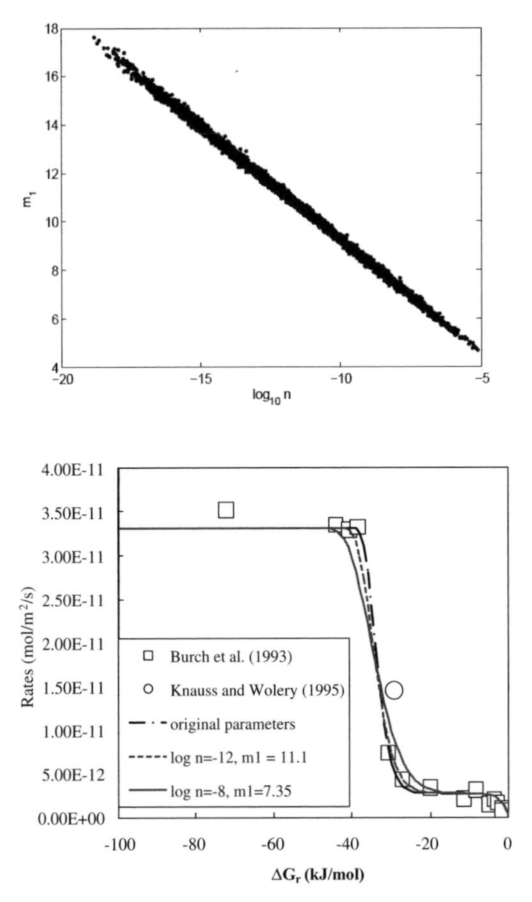

Figure 8. (a) Monte Carlo simulation demonstrates the linear covariance of parameters of m_1 and n in Equation (12). All m_1-n pairs in the diagram can fit the Burch et al. (1993) experimental $r - \Delta G_r$ data within the (assumed) normally distributed uncertainties. (b) examples of three different pairs of m_1-n that all can define a sigmoidal shape of the curve.

reactions in geological systems stay in the transition region for a long time (Arvidson and Lüttge 2009).

It is important to note that a significant amount of experimental data clearly show that the $r - \Delta G_r$ relationship is not always sigmoidal for feldspars (Gautier et al. 1994; Oelkers et al. 1994). Beig and Lüttge (2006) attempted to explain this category of experimental $r - \Delta G_r$ relationships from the perspective of reaction history. One must also note that in explaining the $r - \Delta G_r$ relationship, the solution chemistry effects should not be neglected. As shown in Figure 7, Alekseyev et al.'s (1997) sanidine and albite dissolution rates can be plotted as a nearly linear function of Al concentrations for the congruent stage even though a sigmoidal shaped curves are defined on the $r - \Delta G_r$ plots. Numerical modeling of reaction path and coupled reactive transport using Oelkers and co-workers' formulation is needed to explore the role of Al inhibition in coupling between dissolution and precipitation reactions.

Even after we have a good handle on the dissolution rate law, the influence of the rate law for the precipitation reactions is not well defined. We can easily postulate that the rate laws for the secondary mineral precipitation will also have a profound influence on how dissolution

Table 2. Laboratory experiments from which parallel rate law functions were investigated

Reference	Minerals	T (°C)	pH	k_1	k_1/k_2	n	m_1	m_2
Burch et al. (1993)	Amelia albite	80	8.8	3.05×10^{-11}	11.2	8.40×10^{-17}	15	1.45
Hellmann and Tisserand (2006)	Amelia albite	150	9.2	1.02×10^{-8}	56.65	7.98×10^{-5}	3.81	1.17
Hellmann et al. (2007)	Amelia Albite	100	3.25	1.02×10^{-8}	56.65	7.98×10^{-5}	3.81	1.17
Taylor et al. (2000) §	Labradorite ($Ab_{37}An_{61}Or_2$)	25	3.08 3.2	1.43×10^{-11}	3.16	1.3×10^{-17}	14	
Beig and Lüttge (2006)†	Amelia albite	185	9	1.3×10^{-8}	10	7×10^{-16}	13	1.17
Alekseyev et al. (1997)*	Albite $Na_{0.97}K_{0.02}Al_{1.00}Si_{3.01}O_8$	300	9	4.35×10^{-7}		5×10^{-6}	6	
Alekseyev et al. (1997)*	Sanidine $K_{1.00}Na_{0.03}Al_{0.99}Si_{3.00}O_8$	300	9	3.6×10^{-7}		3×10^{-3}	3	

Parameters corresponding to the following parallel rate law function:

$$r = k'\left\{1 - e^{-n\left(\frac{|\Delta G_r|}{RT}\right)^{m1}}\right\} + k''\left\{1 - e^{-\left(\frac{|\Delta G_r|}{RT}\right)}\right\}^{m2}$$

§for Taylor et al. (2000):

$$r = -k\left\{0.76\times\left[1-\exp\left((-1.3\times10^{-17})\times\left(\frac{|\Delta G_r|^{14}}{RT}\right)\right)\right] - 0.24\times\left[1-\exp\left(-0.35\times\frac{|\Delta G_r|}{RT}\right)\right]\right\}$$

† Parallel rate law function has not been provided, and thus being fitted in this study.
* Rate law functions fitted in this study. Only first term in the parallel rate law was used.

and precipitation reactions are coupled because the coupling is through the $f(\Delta G_r)$ term for fixed k_1^*/k_2^*. Most modeling studies have assumed a linear TST rate for secondary mineral precipitation (Steefel and Van Cappellen 1990; Lasaga et al. 1994; Lasaga 1998; Zhu et al. 2004a,b; Ganor et al. 2007; Yang and Steefel 2008).

While I am obligated to lay out the unknowns and deficiencies in our knowledge on rate laws in this review, I do not want to leave this section with a pessimistic tone. My point here on the rate laws is that, while the rate laws were fitted to a set of specific experimental data with controlled chemistry, significant implications of these rate laws are manifested in geochemical modeling of reaction networks. Applications of a sigmoidal $r - \Delta G_r$ relationship for feldspar dissolution, however empirical it may be and perhaps a better mathematic form should be found, brings us closer to bridging the apparent gap between field and laboratory dissolution rates (Ganor et al. 2007; Maher et al. 2009). As a consequence, the ZBV hypothesis of negative feedback of slow secondary mineral precipitation to the dissolution of feldspars becomes viable. The calculations are also consistent with field observations which show that groundwater is close to equilibrium, but not that close to equilibrium as the linear TST rate law requires in order to retard primary mineral dissolution rates by the orders of magnitude.

Implications for modeling geochemical reaction networks

From the above discussion, we can see that what geochemists have attempted with numerical

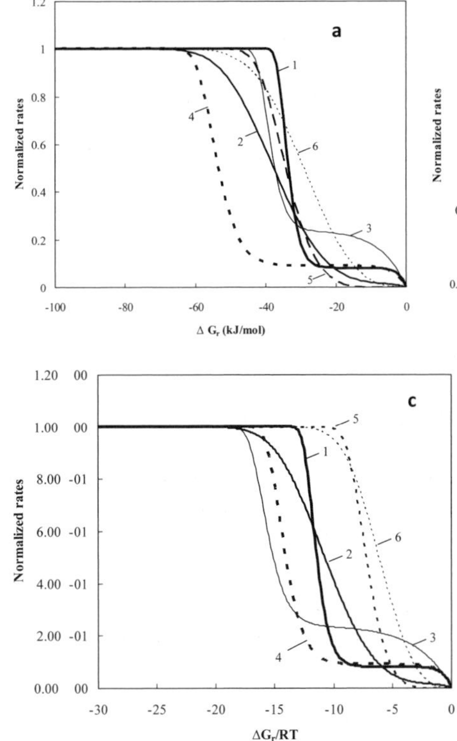

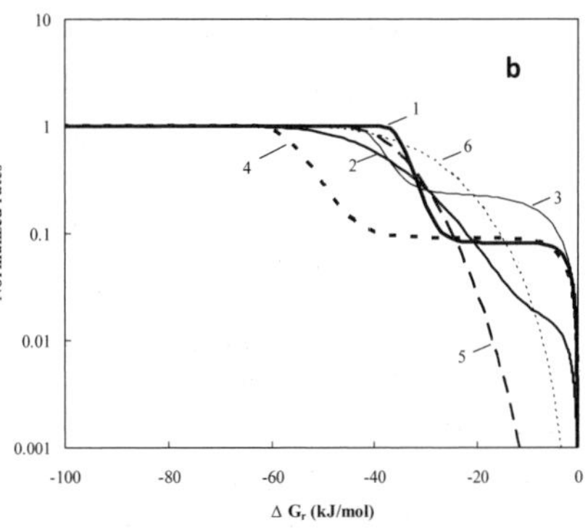

Figure 9. (a) Comparison of parallel rate law functions in normalized rates ($r' = r/r_{max}$) as a function of ΔG_r at the temperatures of experiments; (b) the same plot as in (a) but vertical axis is on logarithm scale; (c) Comparison of parallel rate law functions in normalized rates as a function of $\Delta G_r/RT$. Experimental data sources: 1 = (Burch et al. 1993), 2 = (Hellmann and Tisserand 2006), 3 = (Taylor et al. 2000), 4 = (Beig and Lüttge 2006), 5 = albite (Alekseyev et al. 1997), 6 = sanidine (Alekseyev et al. 1997)

modeling of kinetic reaction path in a batch system is no different from what mathematicians, chemical engineers, and biochemists do to model their reaction networks (Feinberg 1979). We numerically solve a set of ordinary differential equations and non-linear algebraic equations for mass action, mass balance, and rate functions (Fang et al. 2003). The behaviors of the geochemical reaction networks depend on the types of chemical reactions and the physical parameters that we assign in the models.

Geochemical systems with partial equilibrium between secondary mineral phases and the aqueous solution represent relatively simple geochemical reaction networks. When the precipitation of secondary minerals is also controlled by the kinetics, more complicated geochemical reaction networks, with strong coupling between some reactions, become manifest as a result. Different secondary minerals can behave differently, some fast, some slow with respect to the dissolution of primary minerals. Different values of the rate constants of primary and secondary minerals, different rate laws for dissolution and precipitation, and different paths of temporal evolution of the reactive surface areas could lead to widely different behaviors of the geochemical reaction networks. We can easily imagine that the networks can be extremely complex. Geochemical modeling will undoubtedly be an indispensible tool as we begin to understand the dynamics of geochemical reaction networks in geological systems.

At the present time, the tools are available, but both our theoretical understanding and experimental data are lacking. Clearly, more experimental data are needed for various minerals at various temperatures and pH. With its promise and attention, more experimental data will surely come in the near future.

COUPLED REACTIVE TRANSPORT MODELING

As we discussed in the INTRODUCTION, field applications of geochemical modeling to geological systems, such as geological carbon sequestration, almost always require the consideration of mass transport processes (advection, dispersion, diffusion). The use of coupled reactive mass transport (CRMT) models is more appropriate. This is because fluid flow brings material and heat fluxes and creates chemical potential gradients, which are the driver for chemical reactions. Volume 34 of the Reviews in Mineralogy and Geochemistry was devoted to this subject (Lichtner et al. 1996). The readers are referred to this volume and Yeh and Tripathi (1989) for the fundamentals of the subject. A recent review paper by Steefel and co-workers articulates the promise of CRMT as a modeling tool (Steefel et al. 2005).

Historically, chemical reactions and reaction kinetics in natural systems have mainly been studied using the *inverse mass balance modeling* approach pioneered by Garrels and Mackenzie (1967) and popularized by the development of the modeling code NETPATH and a series of seminal papers by Niel Plummer (Plummer et al. 1990, 1994). Numerous studies have utilized the mass balance of dissolved constituents and deduced silicate dissolution rates in small streams, watersheds, soil profiles, and groundwater aquifers (e.g., Paces 1973; Velbel 1985; Rowe and Brantley 1993; Drever and Clow 1995; Obrien et al. 1997; Brantley et al. 2001; Zhu 2005; Hereford et al. 2007). This modeling method and its applications have recently been reviewed by Bowser and Jones (2002) and Bricker et al. (2004).

The inverse mass balance modeling approach is intended to derive order of magnitude estimates of reaction rates in natural systems. The advantage of circumventing the use of thermodynamics and kinetic rate laws is countered by the disadvantage that the derived rates are only effective and apparent rates. For example, these rates do not automatically tell us the influences of saturation states or Al concentrations on the obtained rates. As shown in the previous sections, the coupling of reactions in a geochemical reaction network, in which slow secondary mineral precipitation can retard the dissolution of the primary minerals, would make the field rates as apparent rates. This makes it an inappropriate comparison between field derived rates and laboratory rates measured far from equilibrium and under particular solution chemistry conditions.

Furthermore, the hydrogeological conditions for the applications of the inverse mass balance modeling method have not always been met. Zhu and Anderson (2002) emphasized the following assumptions that we must bear in mind: (a) that the initial and final waters are along the same flow path or a stream tube; (b) dispersion and diffusion have a negligible effect on water chemistry; (c) the system has been in a chemical steady state. Additionally, hydraulic heterogeneity is the rule rather than the exception for nearly all geological systems. The inverse mass balance modeling approach may be biased in sampling the fast flow paths in an aquifer or soil profile, over-estimating the reactive surface areas (Drever and Clow 1995). Therefore, these calculated rates are integrated over flow paths, as shown by the integral equation given in Lichtner (1996).

On the other hand, the coupled reactive transport modeling approach is process-based and reaction mechanisms-oriented. In principle, it allows spatial distribution and temporal variations of physical and chemical parameters. Therefore, it overcomes the shortcomings and limitations of the inverse mass balance modeling approach and allows us to explore the effects of hydraulic and chemical heterogeneities, different mineralogy, the role of microbial activities, and the coupling of mass and heat transport processes with chemical reactions. Although these models have been around for a long time, recent development of computer codes, an increase in computing power, and advances in our knowledge of chemical kinetics have made more sophisticated applications to explore geochemical reaction networks possible. One must be cautioned, however, given the heavy demands of field data for these types of models (see Table 3). As experience of decades of groundwater flow modeling has shown, the most difficult challenge is often the characterization and representation of heterogeneity and boundary conditions in a geological system.

Geochemical modeling using the CRMT approach, by default, deals with a network of geochemical reactions. Earlier modeling work used the assumption of partial and local equilibrium, which are perhaps good approximations at high temperatures or when the minerals dissolved and precipitated are calcite, amorphous Fe and Al hydroxides, and sulfates (e.g., Zhu and Burden 2001; Zhu et al. 2001). The simulation of dolomitization has the interesting case that dolomite precipitation was modeled as a kinetic process while dissolution of aragonite, calcite, and precipitation of gypsum was modeled as an equilibrium process, apparently because of the well known slow dolomite precipitation (Jones and Xiao 2005, 2006). In general, kinetics models tend to use the linear TST rate law for dissolution reactions and secondary mineral precipitation, typically assuming k for precipitation is equal or one order of magnitude smaller than k for dissolution reactions. That may not set off strong coupling between reactions. In general, these models are better tools for exploring concepts rather than interpreting actual field data.

The applications of CRMT models have resolved some long-time concerns in the study of kinetics. It is well known that analyses of groundwater samples represent average concentrations drawing from a large number of pores. Whether the averaging nature of groundwater sample chemistry contribute to the well known field and laboratory rate discrepancy was recently examined by Li et al. (2007). They developed pore-scale network model and calculated "true reaction rates" that take into account variability in individual pore properties. The results show

Table 3. Assumptions and data needs in the inverse mass balance and coupled reactive mass transport models.

Inverse	**Forward**
Based on principle of mass balance only;	Also based on principle of mass balance;
Thermodynamics not considered;	Thermodynamic properties required for • Minerals • aqueous species • activity coefficient models • solid solutions
Kinetics is not explicitly in the equation	Must provide rate laws *a prior*
Additional Assumptions: 1. the two water analyses from the "initial" and "final" wells should represent packets of water that flows along the same path; 2. dispersion and diffusion do not significantly affect solution chemistry; 3. a chemical steady state prevailed during the time considered; 4. the mineral phases used in the calculation are or were present in the aquifer	Additional Assumptions: Groundwater flow: • Boundary conditions and recharge rates • Temporal evolution of recharge rates and boundary conditions • Spatial distribution of hydraulic conductivity Mass transport: • Effective porosity and spatial distribution • Dispersion coefficients and spatial distribution Reactive mass transport: • Mineral abundance and spatial distribution; • Reactive surface areas for minerals • Reactive surface area for precipitating minerals; • Temporal evolution of mineral surface areas • Rate laws for dissolution reactions • Rate laws for precipitation reactions
Results: Bulk, effective rates	Results: Mechanistic, process-based, spatial variable reaction rates; geological processes.

that the effects of incomplete mixing and heterogeneous distribution of anorthite in the pore network resulted in errors by only a factor of three in calculated dissolution rates using the average concentrations (Li et al. 2007).

The possible effects of scale-dependent mineral dissolution rates on the field–laboratory discrepancy was also studied by Li et al. (2008) using pore scale CRMT models. The modeling results show that concentration gradients in single pores have only negligible effects on the slow dissolution rates of plagioclase but can have large effects for fast calcite reaction rates under most hydrological conditions. Li et al. (2008) concluded that the discrepancy between laboratory and field rates must be attributed to other factors.

Recently, a state of the art field application of the CRMT model has been conducted by Maher at el. (2009) in an attempt to bridge the gap between field and laboratory rates. Their field site is a soil profile at a 226 ka marine chronosequence near Santa Cruz, CA. Abundant mineralogical and fluid chemistry data have been carefully obtained for this chronosequence (White et al. 2008). Their CRMT model used the format of the Burch et al. (1993) rate law with parameters from Hellmann and Tisserand (2006) for albite, assuming n, m_1, m_2 are temperature and pH independent while k' and k'' were adjusted for pH and temperature dependence. The same parameters for K-feldspar appear to have been derived from model fitting. The precipitation of kaolinite was simulated with a TST rate law. The parameters for groundwater flow and transport in this unsaturated and saturated hydrologic regime were only described sketchily, but they used a constant and average moisture and water flux for the last 226 k.y. based on the chloride mass balance method. Research results show that at other localities, the chloride deposition rates and groundwater recharge rates varied significantly from late Pleistocene to Holocene due to the well known glacial–inter-glacial climate changes (e.g., Yucca Mountain, Nevada and Black Mesa, Arizona; Zhu et al. 2003).

Maher et al. (2009) concluded that, when the Burch et al. (1993) rate law is used in simulations, they could reconcile the two orders of magnitude discrepancy between field derived rates and laboratory measured rates that were far from equilibrium while model application using the linear TST rate law would not. The Burch et al.'s (1993) parallel formulation allowed the reduction of dissolution rates two orders of magnitude lower as compared to the far from equilibrium rates due to the Gibbs free energy effect in groundwaters that are near equilibrium with albite. In order to match the field solute and mineralogical profiles, Maher et al. (2009) needed to include a rate constant for kaolinite that is five orders of magnitude smaller than that of albite ($\log k = -19.9$ mol s^{-1} m^{-2} for kaolinite versus k'' of -14.9 mol s^{-1} m^{-2} for albite), echoing the finding of Zhu et al. (2004a,b) and Ganor et al. (2007). The five orders of magnitude smaller k for kaolinite precipitation as comparing to k for albite dissolution in the near equilibrium region ensured strong coupling between dissolution and precipitation reactions.

Although Maher et al.'s (2009) work is not the final word for the vexing conundrum of laboratory and field rate discrepancy, as a plethora of assumptions of groundwater flow, mass transport, and reactions is unavoidable in this deterministic forward model, it is encouraging that the ZBV hypothesis has successfully explained the field rates, at least in a plausible if not unique model. The model sensitivity analyses in this paper especially show the power of the CRMT modeling tool, which allows one to explore the different factors controlling reaction rates and reaction fronts.

The message from this section is that CRMT models are useful to further understand geochemical reaction networks. Typically, CRMT modeling generates a great amount of numerical experimental data. Significant amounts of time and energy are necessary to dissect and distill the information on what reactions have happened and how reactions are coupled. With the advancement of our understanding of rates and rate laws, it is promising to use CRMT to explore geochemical reaction networks.

BIOGEOCHEMICAL REACTION NETWORKS

Paradoxically, the concept of a geochemical reaction network appears to have been more accepted in dealing with biogeochemical reactions. Perhaps this is because studies of biogeochemical reactions concern soils or shallow aquifers, where abundant data can be collected. One of the pioneering works for modeling biogeochemical reactions as a network was by Hunter et al. (1998). The coupling of hydrological, microbial, and geochemical processes has received increasing emphasis recently (Jardine 2008).

A good field example of a network of biogeochemical reactions is the biostimulation experiment near Oak Ridge National Laboratory in Tennessee (Reeder and Zhu 2005). A time series of groundwater samples and major, minor, and trace element analyses as well as C, N, and S stable isotopes demonstrated NO_3^-, Mn^{IV}, Fe^{III}, U^{VI}, Tc^{VII}, and SO_4^{-2} reduction (see Fig. 10). A zeroth order rate law generally describes the data well, but the derived *in situ* rates are effective

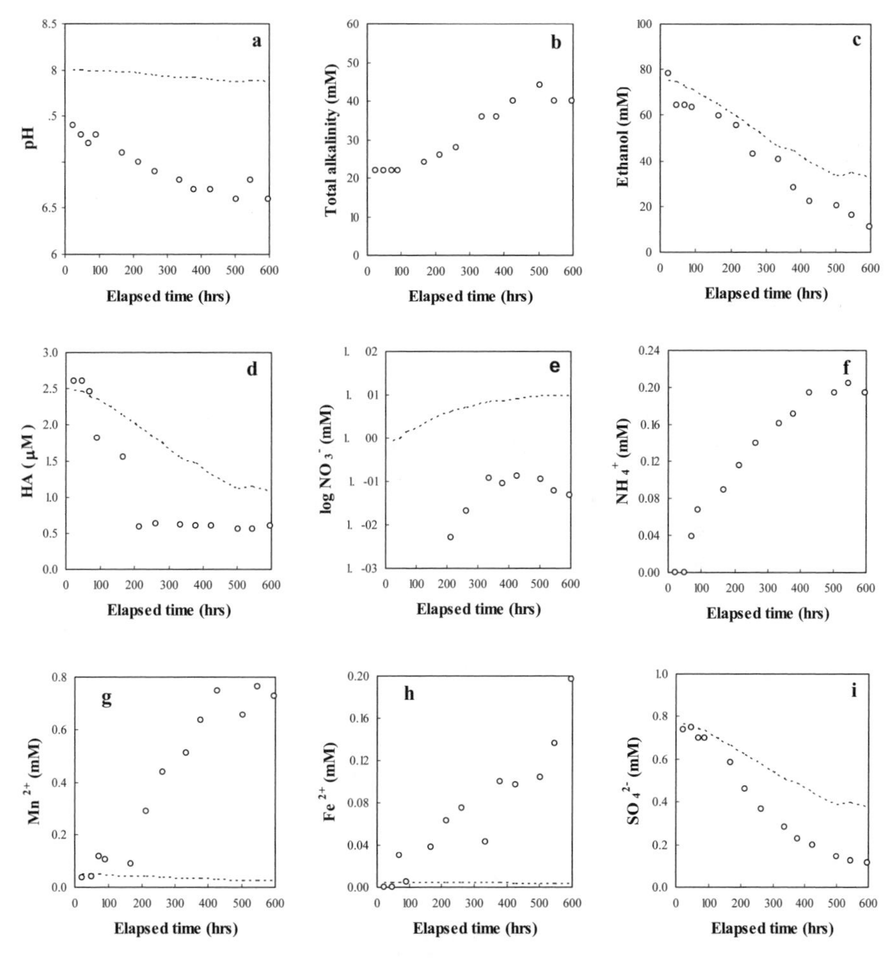

Figure 10. Unpublished field experimental data from a push-pull biostimulation experiment at Oak Ridge, Tennessee. The symbols are experimental measurements, the dashed lines denote mixing of native and injected water if there are no chemical reactions or other processes going on. The deviations from the mixing lines indicate possible chemical reactions. HA: Humic and fulvic acids.

rates that mask many simultaneous and competing reactions and processes. A case in point is uranium (see Fig. 11). Uranium concentrations in groundwater first increased, probably due to desorption from Fe and Mn oxides as Fe and Mn oxides were reductively dissolved, and then decreased due to reduction of U^{VI} to U^{IV}. Similar bioremediation experiments are conducted at the Old Rifle site in Colorado (*http://esd.lbl.gov/research/projects/ersp/field_research/UMTRA.html*) and numerous other sites.

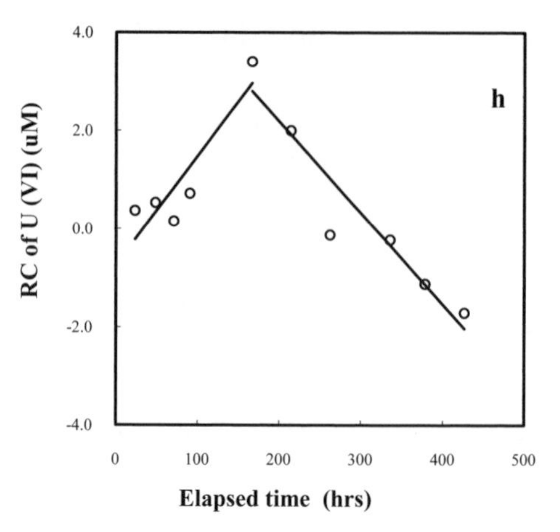

Figure 11. Temporal evolution of U(VI) concentrations (symbols) subtracted from the mixing effects from a push-pull biostimulation experiment at Oak Ridge, Tennessee in 2004 (from unpublished experimental data of Chen Zhu and Jack Istok). The lines represent pseudo-zeroth order rates of UO_2^{2+} production and consumption. However, a complex biogeochemical network regarding UO_2^{2+} probably operated during the field biostimulation experiment, including competing reactions of desorption of UO_2^{2+} from iron and manganese oxides and reduction of UO_2^{2+} to U(IV).

If sufficient chemical data are available, biogeochemical modeling can simulate a network of redox and non-redox reactions. Note that recent development of computer software, such as the freely available PHREEQC (one needs to modify the database in order to simulate redox disequilibrium) and commercially available GWB© allows modeling a biogeochemical reaction network that includes microbial metabolism, reaction kinetics, and partial equilibrium reactions. A variety of reaction types was included: abiotic redox reactions, mineral precipitation and dissolution, surface adsorption, and ion exchange (Jin and Zhu, in prep.). As stated by Steefel et al. (2008), historically microbial mediated reaction were modeled as semi-empirical kinetics, decoupled to a whole suit of inorganic geochemical reactions as a consequence of biostimulation or bioremediation. The functions of customizable rate laws in new versions of the modeling software allow incorporation of rigorous thermodynamic and kinetic treatment of the reactions (Liu et al. 2001; Jin and Bethke 2005). Clearly, this is an emerging field that now that coupling between microbially mediated reactions and inorganic geochemical reactions can be explored.

While some computer software is fully capable of performing such complex computations, the requirements of modeling parameters for the biogeochemical reactions are daunting (Roden 2008). As we have found, simulations for the Oak Ridge site can provide plausible explanations of field data (Jin and Zhu 2006), but numerous non-unique interpretations of field data may be possible. In other words, biogeochemical reaction models by themselves are more a fiction, than a representation of reality at the present time. However, geochemical modeling can help to integrate processes and reactions in a quantitative manner and provide a more comprehensive understanding of biogeochemical processes than simply providing apparent zeroth order and first order rates.

RESEARCH NEEDS AND CHALLENGES

Below is an incomplete list of research needs to model networks of geochemical reactions:

(1) Internally consistent standard state thermodynamic properties for minerals, particularly minerals with complex and variable chemical compositions and structures like

smectite. More accurate thermodynamic properties for common minerals like feldspars would help to resolve the controversy on the Al-bearing minerals.

(2) More experimental data and resolution of ambiguities surrounding the speciation of aqueous elements like that for Al species;

(3) Improvements in sampling and filtration of natural and laboratory samples for better saturation state assessments;

(4) Solid solution models for feldspar and clay minerals;

(5) More measurements of $r - \Delta G_r$ relations at a variety of temperature and pH conditions leading to accurate theoretical or empirical correlations;

(6) Rates and rate laws for precipitation reactions;

(7) Pitzer activity coefficient parameters for trace elements and for all elements at elevated temperatures;

(8) More rigorous treatment of experimental data with statistical analysis; and

(9) Assessment of error propagations.

CONCLUDING REMARKS

The above review demonstrates that geochemical modeling, as a macroscopic modeling tool, is a necessary step to upscale our knowledge from molecular level to field applications. The theoretical and mathematical foundation for geochemical modeling was developed by Helgeson decades ago, and the computational tools (namely, the computer software) are already at hand. However, for geochemical modeling to be effective, thermodynamic databases for minerals and aqueous speciation need to be improved. Application of geochemical modeling to study mineral dissolution and precipitation reaction kinetics at laboratory and field scale is in its infancy.

Recent advances show that dissolution and precipitation reactions are closely coupled, in fact, a network of closely coupled geochemical reactions appear to operate in some laboratory experiments and most field systems. Due to societal needs, e.g., to understand what happens when CO_2 is injected into geological formations, a flood of new laboratory and field geochemical data will flow into the literature. An understanding of geochemical reaction networks appears to be key.

ACKNOWLEDGMENTS

I want to thank Eric Oelkers and Jacques Schott for this opportunity to contribute a chapter. My work has been supported by the U.S. Department of Energy under Award No. DE-FG26-04NT42125 and by the National Science Foundation under Award No. EAR0423971 and EAR0509775. Any opinions, findings, and conclusions or recommendations expressed in this material, however, are those of the authors and do not necessarily reflect the views of the United States Government or any agency thereof. I am grateful to my students and collaborators Peng Lu, Bill Seyfried, Jack Istok, and Qusheng Jin; some of the materials presented here result from discussion and collaboration with them. Kaj Johnson helped with the Monte Carlo simulation. Review comments by Eric Oelkers, Jiwchar Ganor, Jim Brophy, and Kurt Swingle are greatly appreciated.

REFERENCES

Aagaard P, Helgeson HC (1982) Thermodynamic and kinetic constraints on reaction rates among minerals and aqueous solutions. I. Theoretical considerations. Am J Sci 282:237-285

Abercrombie HJ, Hutcheon IE, Bloch JD, de Caritat P (1994) Silica activity and the smectite-illite reaction. Geology 22:539-542

Aja SU (1995) Thermodynamic properties of some 2/1 layer clay-minerals from solution-equilibration data. Eur J Mineral 7:325-333

Aja SU (2002) The stability of Fe-Mg chlorites in hydrothermal solutions: II. Thermodynamic properties. Clays Clay Miner 50:591-600

Aja, SU, Dyar MD (2002) The stability of Fe-Mg chlorites in hydrothermal solutions - I. Results of experimental investigations. Appl Geochem 17, PII S0883-2927(01)00131-7

Aja SU, Rosenberg PE (1992) The thermodynamic status of compositionally-variable clay-minerals - A discussion. Clays Clay Miner 40:292-299

Aja SU, Rosenberg PE (1996) The thermodynamic status of compositionally-complex clay minerals: Discussion of ''Clay mineral thermometry - A critical perspective''. Clays Clay Miner 44:560-568

Aja SU, Small JS (1999) The solubility of a low-Fe clinochlore between 25 and 175 °C and P_{v,H_2O}. Eur J Mineral 11:829-842

Alekseyev VA, Medvedeva LS, Prisyagina NI, Meshalkin SS, Balabin AI (1997) Change in the dissolution rates of alkali feldspars as a result of secondary mineral precipitation and approach to equilibrium. Geochim Cosmochim Acta 61:1125-1142

Allison JD, Brown DS, Novo-Gradac KJ (1991) MINTEQA2/PRODEFA2, a geochemical assessment model for environmental systems, version 3.0 user's manual. U.S. Environmental Protection Agency Report EPA/600/3-91/021

Anderson GM (1970) Standard states at fixed and variable pressures. J Chem Educ 47:676-679

Anderson GM, Crerar DA (1993) Thermodynamics in Geochemistry: The equilibrium model. Oxford University Press, New York.

Apps JA, Neil JM (1990) Solubilities of aluminum hydroxides and oxyhydroxides in alkaline solutions. *In:* Chemical Modeling of Aqueous Solutions II. Melchior DA, Bassett RL (eds) ACS Symposium Series # 416, Washington DC

Arnórsson S, Stefánsson A (1999) Assessment of feldspar solubility constants in water in the range of 0 to 350 °C at vapor saturation pressures. Am J Sci 299:173-209

Arnórsson SN, Gunnarsson I, Stefansson A, Andresdottir A, Sveinbjornsdottir AE (2002) Major element chemistry of surface- and ground waters in basaltic terrain, N-Iceland. I. Primary mineral saturation. Geochim Cosmochim Acta 66:4015-4046

Arvidson RS, Lüttge A (2009) Mineral dissolution kinetics as a function of distance from equilirbium - experimental results. . Chemical Geology in press

Beig MS, Lüttge A (2006) Albite dissolution kinetics as a function of distance from equilibrium: Implications for natural feldspar weathering. Geochim Cosmochim Acta 70:1402-1420

Benson SM (1960) The Foundation of Chemical Kinetics. Mcgraw-Hill, New York

Berger G, Lacharpagne JC, Vedel B, Beaufort D, Lanson B (1997) Kinetic constraints on illitization reactions and the effects of oragnic diagenesis in sandstone/shale sequences. Appl Geochem 12:23-35

Berman RG (1988) Internally-consistent thermodynamic data for minerals in the system Na_2O-K_2O-CaO-MgO-FeO-Fe_2O_3-Al_2O_3-SiO_2-TiO_2-H_2O-CO_2. J Petrol 29:445-522

Berman RG (1990) Mixing properties of Ca-Mg-Fe-Mn garnets. Am Mineral 75:328-344

Blum A, Stillings L (1995) Feldspar dissolution kinetics. Rev Mineral Geochem 31:291-346

Blum AE (1994) Feldspars in weathering. *In:* Feldspars and Their Reactions. Proceedings of NATO Advanced Study Institute. Parsons I (ed) Kluwer Academic Publication 421:595-629

Bourcier WL, Knauss KG, Jackson KJ (1993) Aluminum hydrolysis constants to 250 °C from boehmite solubility measurements. Geochim Cosmochim Acta 57:747-762

Bowser CJ, Jones BF (2002) Mineralogic controls on the composition of natural waters dominated by silicate hydrolysis. Am J Sci 302:582-662

Brantley SL (1992) Kinetics of dissolution and precipitation-experimental and field results. *In:* Proceedings of the Seventh International Conference on Water-Rock Interactions, Park City, Utah. Kharaka Y, Maest A (eds) Rotterdam, Balkema, p 3-6

Brantley SL, Bau M, Yau S, Alexander B (2001) Interpreting kinetics of groundwater-mineral interaction using major element, trace element, and isotopic tracers. *In:* Proceedings of the Tenth International Conference on Water-Rock Interactions. Cidu R (ed), Balkema, p 13-18

Bricker OP, Jones B, Bowser CJ (2004) Mass-balance approach to interpreting weathering reactions in watershed systems. *In:* Surface and Ground Water. Drever JI (ed) Elsevier, Oxford, 5:119-132

Burch TE, Nagy KL, Lasaga AC (1993) Free energy dependence of albite dissolution kinetics at 80 °C and pH 8.8. Chem Geol 105:137-162

Carroll SA, Knauss KG (2005) Dependence of labradorite dissolution kinetics on $CO_{2(aq)}$, $Al_{(aq)}$, and temperature. Chem Geol 217:213-225

Chukhlantsev VG (1956) Solubility products of arsenates. J Inorg Chem (USSR) 1:1975-1982

Cubilas P, Kohler S, Prieto M, Causserand C, Oelkers EH (2005) How do mineral coating affect dissolution rates? An experimental study of coupled $CaCO_3$ dissolution -- $CaCO_3$ precipitation. . Geochim Cosmochim Acta 69:5459-5476

Dove PM, Rimstidt JD (1985) The solubility and stability of scorodite, $FeAsO_4{\cdot}2H_2O$. Am Mineral 70:838-844

Drever J (2004) Volume editor's introduction. *In:* Treatise on Geochemistry. Holland HD, Turekian KK (eds) Elsevier 5:xv-xvii

Drever JI (1997) The Geochemistry of Natural Waters: Surface and Groundwater Environment. Prentice-Hall, Englewood Cliffs, New Jersey. 2nd

Drever JI, Clow DW (1995) Weathering rates in catchments. Rev Mineral Geochem 31:463-481

Emberley S, Hutcheon I, Shevalier M, Durocher K, Mayer B, Gunter WD, Perkins EH (2005) Monitoring of fluid-rock interaction and CO_2 storage through produced fluid sampling at the Weyburn CO_2-injection enhanced oil recovery site, Saskatchewan, Canada. Appl Geochem 20:1131-1157

Fang YL, Yeh GT, Burgos WD (2003) A general paradigm to model reaction-based biogeochemical processes in batch systems. Water Resour Res 39: HWC2.1-HWC2.25

Faure G (1998) Principles and Applications of Inorganic Geochemistry Prentice Hall, New York

Feinberg M (1979) Lectures on Chemical Reaction Networks - written version of lectures given at the Mathematical Research Center, University of Wisconsin, Madison, WI (1979. Available online at *http://www.chbmeng.ohio-state.edu/~feinberg/research/.*

Fu Q, Lu P, Konishi H, Dilmore R, Xu H, Seyfried WE Jr., Zhu C (2009) Coupled alkali-feldspar dissolution and secondary mineral precipitation in batch systems: 1. New experiment data at 200 °C and 300 bars. Chem Geol 91:955-964

Furrer G, Stumm W (1986) The coordination chemistry of weathering I. Dissolution kinetics of δ-Al_2O_3 and BeO. Geochim Cosmochim Acta 48:2405-2432

Ganor J, Lu P, Zheng Z, Zhu C (2007) Bridging the gap between laboratory measurements and field estimations of weathering using simple calculations. Environ Geol 53:599-610

Garrels RM, Mackenzie FT (1967) Origin of the chemical composition of some springs and lakes. *In:* Equilibrium Concepts in Natural Water Systems. Gould RF (ed) American Chemical Society, Washington, D.C., 67:222-242

Garrels RM, Thompson ME (1962) A chemical model for sea water at 25 °C and one atmospheric pressure. Am J Sci 260:57-66

Gautier J-M, Oelkers EH, Schott J (1994) Experimental study of K-feldspar dissolution rates as a function of chemical affinity at 150 °C and pH 9. Geochim Cosmochim Acta 58:4549-4560

Gautier JM, Oelkers EH, Schott J (2001) Are quartz dissolution rates proportional to B.E.T. surface areas? Geochim Cosmochim Acta 65:1059-1070

Georg RB, Zhu C, Reynolds RC, Halliday AN (2009) Stable silicon isotopes of groundwater, feldspars, and clay coatings in the Navajo Sandstone aquifer, Black Mesa, Arizona, USA. Geochim Cosmochim Acta 73:2229-2241

Grenthe I, Fuger J, Konings R, Lemire RJ, Muller AB, Wanner J (1992) The Chemical Thermodynamics of Uranium. Elsevier: New York

Gudmundsson BT, Arnorsson S (2005) Secondary mineral-fluid equilibria in the Krafla and Namafjall geothermal systems, Iceland. Appl Geochem 20:1607-1625

Gunter WD, Perkins EH, Hutcheon I (2000) Aquifer disposal of acid gases: modeling of water-rock reactions for trapping of acid wastes. Appl Geochem 15:1085-1095

Gunter WD, Wiwchar B, Perkins EH (1997) Aquifer disposal of CO_2-rich greenhouse gases: extension of the time scale of experiment for CO2-sequestering reactions by geochemical modeling. Mineral Petrol 59:121-140

Harvie CE, Moller N, Weare JH (1984) The predication of mineral solubilities in natural waters: the Na-K-Mg-Ca-H-Cl-SO_4-OH-HCO_3-CO_3-CO_2-H_2O system to high ionic strength at 25 °C. Geochim Cosmochim Acta 48:723-751

Helgeson HC (1968) Evaluation of irreversible reactions in geochemical processes involving minerals and aqueous solutions-1. Thermodynamic relations. Geochim Cosmochim Acta 32:853-877

Helgeson HC (1971) Kinetics of mass transfer among silicates and aqueous solutions. Geochim Cosmochim Acta 35:421-469

Helgeson HC (1979) Mass transfer among minerals and hydrothermal solutions. *In:* Geochemistry of Hydrothermal Ore Deposits. Barnes HL (ed) John Wiley & Sons, New York, p 568-610

Helgeson HC, Brown TH, Nigrini A, Jones TA (1970) Calculation of mass transfer in geochemical processes involving aqueous solutions. Geochim Cosmochim Acta 34:569-592

Helgeson HC, Delany JM, Nesbitt HW, Bird DK (1978) Summary and critique of the thermodynamic properties of rock forming minerals. Am J Sci 278A:569-592

Helgeson HC, Garrels RM, Mackenzie FT (1969) Evaluation of irreversible reactions in geochemical processing involving minerals and aqueous solutions- II. Applications. Geochim Cosmochim Acta 33:455-481

Helgeson HC, Kirkham DH, Flowers GC (1981) Theoretical prediction of the thermodynamic behavior of aqueous electrolytes at high pressures and temperatures. IV. Calculation of activity coefficients, osmotic coefficients, and apparent molal and standard and relative partial molal properties to 600 °C and 5 kb. Am J Sci 281:1249-1516

Helgeson HC, Murphy WM (1983) Calculation of mass transfer among minerals and aqueous solutions as a function of times and surface area in geochemical processes, I. Computational approach. Math Geol 15:109-130

Helgeson HC, Murphy WM, Aagaard P (1984) Thermodynamic and kinetic constraints on reaction rates among minerals and aqueous solutions II. Rate constants, effective surface area, and the hydrolysis of feldspar. Geochim Cosmochim Acta 48:2405-2432

Hellmann R, Daval D, Tisserand D, Renard F (2007) Albite feldsapr dissolution kinetics as a function of the Gibbs free energy at high PCO2. *In:* Water-Rock Interaction 12. Wanty RB, Seal RRI (eds) Balkema, p 895-899

Hellmann R, Tisserand D (2006) Dissolution kinetics as a function of the Gibbs free energy of reaction: An experimental study based on albite feldspar. Geochim Cosmochim Acta 70:364-383

Hereford AG, Keating E, Guthrie G, Zhu C (2007) Reactions and reaction rates in the aquifer beneath Pajarito Plateau, north-central New Mexico. Environ Geol 52:965-977

Holland TJB, Powell R (1998) An internally consistent thermodynamic data set for phases of petrological interest. J Metamorph Geol 16:309-343

Houston SJ, Yardley BWD, Smalley PC, Collins I (2007) Rapid fluid-rock interaction in oilfield reservoirs. Geology 35:1143-1146

IPCC (2005) Special Report on carbon Dioxide Capture and Storage. http://www.unep.ch/ipcc/activity/srccs/.

Jardine PM (2008) Influence of coupled processes on contaminant fate and transport in subsurface environments. Adv Agronomy 99:1-99

Jin Q, Bethke CM (2005) Predicting the rate of microbial respiration in geochemical environments. Geochim Cosmochim Acta 69:1133-1143

Jin Q, Zhu C (2006) A bioenergetics-kinetics coupled modeling study on subsurface microbial metabolism in a field biostimulation experiment. AGU 2006 fall meeting Suppl. EOS Trans B53B-0343

Johnson JW, Lundeen SR (1994) GEMBOCHS thermodynamic data files for use with the EQ3/6 software package. Lawrence Livermore National Laboratory

Johnson JW, Oelkers EH, Helgeson HC (1992) SUPCRT92 - A software package for calculating the standard molal thermodynamic properties of minerals, gases, aqueous species, and reactions from 1-bar to 5000-bar and 0 °C to 1000 °C. Comput Geosci 18:899-947

Jones GD, Xiao YT (2005) Dolomitization, anhydrite cementation, and porosity evolution in a reflux system: Insights from reactive transport models. AAPG Bulletin 89:577-601

Jones GD, Xiao YT (2006) Geothermal convection in the Tengiz carbonate platform, Kazakhstan: Reactive transport models of diagenesis and reservoir quality. AAPG Bulletin 90:1251-1272

Kharaka YK, Cole DR, Hovorka SD, Gunter WD, Knauss KG, Freifeld BM (2006) Gas-water-rock interactions in Frio Formation following CO_2 injection: Implications for the storage of greenhouse gases in sedimentary basins. Geology 34:577-580

Kharaka YK, Gunter WD, Aggarwal PK, Perkins EH, DeBraal JD (1988) SOLMINEQ.88: A computer program for geochemical modeling of water-rock interactions. U.S. Geological Survey, Water-Resources Investigations Report 88-4227

Krause E, Ettel VA (1988) Solubility and stability of scorodite, $FeAsO_4 \cdot 2H_2O$. New data and further discussion. Am Mineral 73:850-854

Krauskopf KB, Bird DK (1995) Introduction to Geochemistry. WCB McGraw-Hill, Boston

Kulik DA (2009) Thermodynamic concepts in modeling sorption at the mineral-water interface. Rev Mineral Geochem 70:125–180

Langmuir D, Mahoney J, Rowson J (2006) Solubility products of amorphous ferric arsenate and crystalline scorodite ($FeAsO_4 \cdot 2H_2O$) and their application to arsenic behavior in buried mine tailings. Geochim Cosmochim Acta 70:2942-2956

Lasaga AC (1981) Transition State Theory. Rev Mineral 8:135-169

Lasaga AC (1998) Kinetic Theory in the Earth Sciences. Princeton University Press, New York.

Lasaga AC, Luttge A (2001) Variation of crystal dissolution rate based on a dissolution stepwave model. Science 291:2400-2404

Lasaga AC, Luttge A (2003) A model for crystal dissolution. Eur J Mineral 15:603-615

Lasaga AC, Soler JM, Ganor J, Burch TE, Nagy KL (1994) Chemical weathering rate laws and global geochemical cycles. Geochim Cosmochim Acta 58:2361-2386

Li L, Peters CA, Celia MA (2007) Applicability of averaged concentrations in determining geochemical reaction rates in heterogeneous porous media. Am J Sci 307:1146-1166

Li L, Steefel CI, Yang L (2008) Scale dependence of mineral dissolution rates within single pores and fractures. Geochim Cosmochim Acta 72:360-377

Lichtner PC (1996) Continuum formulation of multicomponent-multiphase reactive transport. Rev Mineral 34:1-82

Lichtner PC, Steefel CI, Oelkers EH (eds) (1996) Reactive Transport in Porous Media. Reviews in Mineralogy Volume 34. Mineralogical Society of America, Washington, D.C.,

Lindberg RD, Runnells DD (1984) Groundwater redox reactions - an analysis of equilibrium state applied to Eh measurements and geochemical modeling. Science 225:925-927

Liu C, Kota S, Zachara JM, Fredrickson JK, Brinkman CK (2001) Kinetic analysis of the bacterial reduction of goethite. Environ Sci Technol 35:2482-2490

Maher K, Steefel CI, White AF, Stonestrom DA (2009) The role of reaction affinity and seocndary mienrals in regulating chemical weathering rates at the Santa Cruz Soil Chronosequence, California. Geochim Cosmochim Acta, doi:10.1016/j.gca.2009.01.030

Metz V, Raanan H, Pieper H, Bosbach D, Ganor J (2005) Towards the establishment of a reliable proxy for the reactive surface area of smectite. Geochim Cosmochim Acta 69:2581-2591

Murakami T, Kogure T, Kadohara H, Ohnuki T (1998) Formation of secondary minerals and its effects on anorthite dissolution. Am Mineral 83:1209-1219

Nagy KL (1995) Dissolution and precipitation kinetics of sheet silicates. Rev Mineral 31:173-225

Nagy KL, Blum AE, Lasaga AC (1991) Dissolution and precipitation kinetics of kaolinite at 80 °C and pH 3: the effect of deviation from equilibrium. Am J Sci 291:649-686

Nagy KL, Lasaga AC (1992) Dissolution and precipitation kinetics of gibbsite at 80 °C and pH 3. The dependence on solution saturation state. Geochim Cosmochim Acta 56:3093-3111

Nagy KL, Lasaga AC (1993) Kinetics of simultaneous kaolinite and gibbsite precipitation. Geochim Cosmochim Acta 57:4329-4337

Nordstrom DK, Archer DG (2002) Arsenic thermodynamic data and environmental geochemistry. *In:* Arsenic in Ground Water: Geochemistry and Occurrence. Welch AH, Stollenwerk KG (eds) Springer, p 1-26

Nordstrom DK, Munoz JL (1994) Geochemical Thermodynamics. 2nd Edition. Blackwell Scientific Publications

Nordstrom DK, Plummer LN, Langmuir D, Busenberg E, May HM, Jones B, Parkhurst DL (1990) Revised chemical equilibrium data for major water-mineral reactions and their limitations. *In:* Chemical Modeling of Aqueous Systems II. Melchior DC, Bassett RL (eds) American Chemical Society, p 398-413

Obrien AK, Rice KC, Bricker OP, Kennedy MM, Anderson RT (1997) Use of geochemical mass balance modelling to evaluate the role of weathering in determining stream chemistry in five mid-atlantic watersheds on different lithologies. Hydrol Processes 11:719-744

Oelkers EH (2001) General kinetic description of multioxide silicate mineral and glass dissolution. Geochim Cosmochim Acta 65:3703-3719

Oelkers EH, Bjorkum PA, Murphy WM (1996) A petrographic and computational investigation of quartz cementation and porosity reduction in North sea sandstones. Am J Sci 296:420-452

Oelkers EH, Cole DR (2008) Carbon Dioxide Sequestration: A Solution to a Global Problem. Elements 4:305-310

Oelkers EH, Helgeson HC (1990) Triple-ion anions and polynuclear complexing in supercritical electrolyte-solutions. Geochim Cosmochim Acta 54:727-738

Oelkers EH, Schott J, Devidal JL (1994) The effect of aluminum, pH, and chemical affinity on the rates of aluminosilicate dissolution reactions. Geochim Cosmochim Acta 58:2011-2024

Oelkers EH, Bénézeth P, Pokrovski GS (2009) Thermodynamic databases for water-rock interaction. Rev Mineral Geochem 70:1–46

Paces T (1973) Steady-state kinetics and equilibrium between ground water and granitic rocks. Geochim Cosmochim Acta 37:2641-2663

Paces T (1983) Rate constants of dissolution derived from the measurements of mass balance in hydrological catchments. Geochim Cosmochim Acta 37:1855-1863

Parkhurst DL, Appello AAJ (1999) User's guide to PHREEQC (version 2)-a computer program for speciation, batch-reaction, one dimensional transport, and inverse geochemical modeling. Water-Resource Investigation Report. U.S. Geological Survey

Perkins EH, Gunter WD, Hutcheon I, Shevalier M, Durocher K, Emberley S (2002) Geochemical modelling and monitoring of CO_2 storage at the weyburn site, Saskatchewan, Canada. GSA annual meeting & exposition abstracts with programs, Denver, Colorado, 34, 174-11, Denver, Colorado.

Plummer LN, Busby JF, Lee RW, Hanshaw BB (1990) Geochemical modeling of the Madison-aquifer in parts of Montana, Wyoming and South Dakota. Water Resour Res 26:1981-2014

Plummer LN, Parkhurst DL, Fleming GW, Dunkle SA (1988) A computer program incorporating Pitzer's equations for calculation of geochemical reactions in brines. U.S. Geol. Surv. Water Resour. Invest. Rep. 88-4153, 310 pp

Plummer LN, Prestemon EC, Parkhurst DL (1994) An interactive code (NETPATH) for modeling NET Geochemical Reactions Along a flow PATH -- version 2.0. U.S. Geological Survey Water-resources investigations report 94-4109, 130p.

Pokrovskii VA, Helgeson HC (1995) Thermodynamic properties of aqueous species and the solubilities of minerals at high pressures and temperatures: The system Al_2O_3-H_2O-NaCl. Am J Sci 295:1255-1342

Prigogine I, Defay R (1965) Chemical Thermodynamics. Longmans Green, London.

Ransom B, Helgeson HC (1994) Estimation of the standard molal heat-capacities, entropies, and volumes of 2/1-clay-minerals. Geochim Cosmochim Acta 58:4537-4547

Ransom B, Helgeson HC (1995) A chemical and thermodynamic model of dioctahedral 2-1 layer clay-minerals in diagenetic processes - dehydration of dioctahedral aluminous smectite as a function of temperature and depth in sedimentary basins. Am J Sci 295:245-281

Reed MH, Spycher NF (1984) Calculation of pH and mineral equilibria in hydrothermal waters with application to geothermometry and studies of boiling and dilution. Geochim Cosmochim Acta 48:1479-1492

Reeder M, Zhu C (2005) Determination of in situ reaction rates as a result of biostimulation at the Field Research Center, Oak Ridge, TN, Geological Society of America Abstr. with Programs. Vol. 37, p 381

Robie RA, Hemingway BS (1995) Thermodynamic properties of minerals and related substances at 298.15 K and 1 bar (10^5 pascals) pressure and at higher temperatures. U.S. Geological Survey Bulletin 2131

Robins RG (1987) Solubility and stability of scorodite, $FeAsO_4 \cdot 2H_2O$: Discussion. Am Miner 72:842-844

Robins RG (1990) The stability and solubility of ferric arsenate-an update. *In:* EPD Congress ' 90, TMS Annual Meeting. Gaskell DR (ed) p 93-104

Roden EE (2008) Microbiological controls on geochemical kinetics 1: Fundamentals and case study on microbial Fe(III) oxide reduction. *In:* Kinetics of Water-Rock Interaction. Brantley SL, Kubicki J, White AF(eds) Springer, p 335-415

Rowe G, Brantley SL (1993) Estimation of the dissolution rates of andestic glass, plagioclase, and pyroxene in a flank aquifer of Poas Volcano, Costa Rica. Chem Geol 105:71-88

Schramke JA, Kerrick DM, Lasaga AC (1987) The reaction muscovite + quartz <=> andalusite + K-feldspar + water. Part I. Growth kinetics and mechanism. Am J Sci 287:517-559

Shock EL, Helgeson HC (1988) Calculation of the thermodynamic and transport properties of aqueous species at high pressures and temperatures: Correlation algorithms for ionic species and equation of state predictions to 5 kb and 1000 °C. Geochim Cosmochim Acta 52:2009-2036

Shock EL, Helgeson HC, Sverjensky DA (1989) Calculations of the thermodynamic and transport properties of aqueous species at high pressures and temperatures: Standard partial molal properties of inorganic neutral species. Geochim Cosmochim Acta 53:2157-2183

Shock EL, Oelkers EH, Sverjensky DA, Johnson JW, Helgeson HC (1992) Calculation of thermodynamic and transport properties of aqueous species at high pressures and temperatures. Effective electrostatic radii, dissociation constants and standard partial molal properties to 1000 °C and 5 kb. J Chem Soc London, Faraday Trans 88:803-826

Shock EL, Sassani DC, Willis M, Sverjensky DA (1997) Inorganic species in geologic fluids: Correlations among standard molal thermodynamic properties of aqueous ions and hydroxide complexes. Geochim Cosmochim Acta 61:907-950

Steefel CI, DePaolo DJ, Lichtner PC (2005) Reactive transport modeling: An essential tool and a new research approach for the Earth sciences. Earth Planet Sci Lett 240:539-558

Steefel CI, Van Cappellen P (1990) A new approach to modeling water-rock interaction: The role of precursors, nucleation, and Ostwald ripening. Geochim Cosmochim Acta 54:2657-2677

Stefansson A, Arnórsson S (2000) Feldspar saturation state in natural water. Geochim Cosmochim Acta 64:2567-2584

Steinmann P, Lichtner PC, Shotyk W (1994) Reaction-path approach to mineral weathering reactions. Clays Clay Miner 42:197-206

Strazisar BR, Zhu C, Hedges SW (2006) Preliminary modeling of the long-term fate of CO_2 following injection into deep geological formations. Environ Geosci 13:1-15

Stumm W, Morgan JJ (1996) Aquatic Chemistry, Chemical Equilibria and Rates in Natural Waters, 3rd ed. John Wiley & Sons, Inc., New York

Sverjensky DA, Shock EL, Helgeson HC (1997) Prediction of the thermodynamic properties of aqueous metal complexes to 1000 degrees C and 5 kb. Geochim Cosmochim Acta 61:1359-1412

Tagirov B, Schott J (2001) Aluminum speciation in crustal fluids revisited. Geochim Cosmochim Acta 65:3965-3992

Tanger JC, Helgeson HC (1988) Calculations of the thermodynamic and transport properties of aqueous species at high pressures and temperatures: Revised equation of state for the standard partial molal properties of ions and electrolytes. Am J Sci 288:19-98

Taylor AS, Blum JD, Lasaga AC (2000) The dependence of labradorite dissolution and Sr isotope release rates on solution saturation state. Geochim Cosmochim Acta 64:2389-2400

Thyne G, Boudreau BP, Ramm M, Midtbo RE (2001) Simulation of potassium feldspar dissolution and illitization in the Statford Formation, North Sea. AAPG Bulletin 85:621-635

Velbel MA (1985) Geochemical mass balances and weathering rates in forested watersheds of the southern Blue Ridge. Am J Sci 285:904-930

Velbel MA (1990) Influence of temperature and mineral surface characteristics on feldspar weathering rates in natural and artificial systems: A first approximation. Water Resour Res 26:3049-3053

Vieillard P (2002) New method for the prediction of Gibbs free energies of formation of phyllosilicates (10 angstrom and 14 angstrom) based on the electronegativity scale. Clays Clay Miner 50:352-363

White AF, Brantley SL (2003) The effect of time on the weathering of silicate minerals: why do weathering rates in the laboratory and field? Chem Geol 202:479-506

White AF, Bullen TD, Schulz MS, Blum AE, Huntington TG, Peters NE (2001) Differential rates of feldspar weathering in granitic regoliths. Geochim Cosmochim Acta 65:847-869

White AF, Peterson ML (1990) Role of reactive-surface-area characterization in geochemical kinetic models. *In:* Chemical Modeling in Aqueous Systems II. Am Chem Soc Symp series 416:461-475

White AF, Schulz MS, Vivit DV, Blum AE, Stonestrom DA, Anderson SP (2008) Chemical weathering of a marine terrace chronosequence, Santa Cruz, California I: Interpreting rates and controls based on soil concentration-depth profiles. Geochim Cosmochim Acta 72:36-68

White C, Strasizar BR, Granite EJ, Hoffman JS, Pennline HW (2003) Separation and capture of CO_2 from large stationary sources and sequestration in geological formations -- coalbeds and deep saline aquifers. J Air Waste Management Assoc 53:645-715

Wieland E, Wehrli B, Stumm W (1986) The coordination chemistry of weathering: III. A generalization on the dissolution rates of minerals. Geochim Cosmochim Acta 52:1969-1981

Wolery T, Jove-Colon C, Rard J, Wijesinghe A (2004) Pitzer Database Development: Description of the Pitzer Geochemical Thermodynamic Database data0.ypf. Appendix I in In-Drift Precipitates/Salts Model (P. Mariner) Report ANL-EBS-MD-000045 REV 02. Las Vegas, Nevada: Bechtel SAIC Company

Wolery TJ (1992) EQ3/6, A software package for geochemical modeling of aqueous systems: Package overview and installation guide (version 7.0): URCL-MA-110662-PT-I, Livermore, Calif., Univ. California, Lawrence Livermore Laboratory

Yang L, Steefel CI (2008) Kaolinite dissolution and precipitation kinetics at 22 °C and pH 4. Geochim Cosmochim Acta 72:99-116

Yeh GT, Tripathi VS (1989) A critical evaluation of recent development of hydrogeochemical transport models of reactive multi-components. Water Resource Res 25:93-108

Zhang G, Spycher NF, Xu T, Sonnenthal E, Steefel CI (2006) Reactive Geochemical Transport Modeling of Concentrated Aqueous Solutions: Supplement to TOUGHREACT User's Guide for the Pitzer Ion-Interaction Model. LBNL-62718

Zhu C (2005) In situ feldspar dissolution rates in an aquifer. Geochim Cosmochim Acta 69:1435-1453

Zhu C, Anderson GM (2002) Environmental Applications of Geochemical Modeling. Cambridge University Press, London

Zhu C, Blum AE, Veblen DR (2004a) Feldspar dissolution rates and clay precipitation in the Navajo aquifer at Black Mesa, Arizona, USA. *In:* Water-Rock Interaction, Proceedings of the 11th Water/Rock Interaction Conference. Wanty RB, Seal RRI (eds) AA Balkema, p 895-899

Zhu C, Blum AE, Veblen DRD (2004b) A new hypothesis for the slow feldspar dissolution in groundwater aquifers. Geochim Cosmochim Acta 68:A148

Zhu C, Burden DS (2001) Mineralogical compositions of aquifer matrix as necessary initial conditions in reactive contaminant transport models. J Contam Hydrol 51:145-161

Zhu C, Hu FQ, Burden DS (2001) Multi-component reactive transport modeling of natural attenuation of an acid ground water plume at a uranium mill tailings site. J Contam Hydrol 52:85-108

Zhu C, Lu P (2009) Coupled alkali feldspar dissolution and secondary mineral precipitation in batch systems: 3. Saturation indices of product minerals and reaction paths. Geochim Cosmochim Acta, doi:10.1016/j.gca.2009.03.015

Zhu C, Veblen DR, Blum AE, Chipera SJ (2006) Naturally weathered feldspar surfaces in the Navajo Sandstone aquifer, Black Mesa, Arizona: Electron microscopic characterization. Geochim Cosmochim Acta 70:4600-4616

Zhu C, Winterle JR, Love EI (2003) Estimate of Pleistocene and Holocene recharge rates from the chloride mass balance method and chloride-36 data. Water Resour Res 39:1982

Zhu Y, Merkel BJ (2001) The dissolution and solubility, $FeAsO_4{\cdot}2H_2O$ evaluation and simulation with PHREEQC2. Wiss. Mitt. Inst. Fur Geologie, TU Bergakedemie Freiberg, Germany 18:1-12